THE ELECTRIC CIRCUITS PROBLEM SOLVER®

REGISTERED TRADEMARK

Staff of Research and Education Association,

Dr. M. Fogiel, Director

REA

Research and Education Association
505 Eighth Avenue
New York, N. Y. 10018

THE ELECTRIC CIRCUITS PROBLEM SOLVER®

Printed in the United States of America

Library of Congess Catalog Card Number 79-92401

International Standard Book Number 0-87891-517-6

Revised Printing, 1984

WHAT THIS BOOK IS FOR

For as long as electric circuits has been taught in schools, students have found this subject difficult to understand and learn because of the unusually large number of principles and their complex interrelationships. Despite the publication of hundreds of textbooks in this field, each one intended to provide an improvement over previous textbooks, electric circuits remains particularly perplexing and the subject is often taken in class only to meet school/departmental requirements for a selected course of study.

In a study of the problem, REA found the following basic reasons underlying students' difficulties with electric circuits taught in schools:

(a) No systematic rules of analysis have been developed which students may follow in a step-by-step manner to solve the usual problems encountered. This results from the fact that the numerous different conditions and principles which may be involved in a problem, lead to many possible different methods of solution. To presribe a set of rules to be followed for each of the possible variations, would involve an enormous number of rules and steps to be searched through by students, and this task would perhaps be more burdensome than solving the problem directly with some accompanying trial and error to find the correct solution route.

(b) Textbooks currently available will usually explain a given principle in a few pages written by a professional who has an insight in the subject matter that is not shared by students. The explanations are often written in an abstract manner which leaves the students confused as to the application of the principle. The explanations given are not sufficiently detailed and extensive to make the student aware of the wide range of applications and different aspects of the principle being studied. The numerous possible variations of principles and their applications are usually not discussed, and it is left for the students to discover these for themselves while doing exercises. Accordingly, the average student is expected to rediscover that which has been long known and practiced, but not published or explained extensively.

(c) The examples usually following the explanation of a topic are too few in number and too simple to enable the student to obtain a thorough grasp of the principles involved. The explanations do not provide sufficient basis to enable a student to solve problems that may be subsequently assigned for homework or given on examinations.

The examples are presented in abbreviated form which leaves out much material between steps, and requires that students derive the omitted material themselves. As a result, students find the examples difficult to understand--contrary to the purpose of the examples.

Examples are, furthermore, often worded in a confusing manner. They do not state the problem and then present the solution. Instead, they pass through a general discussion, never revealing what is to be solved for.

Examples, also, do not always include diagrams/graphs, wherever appropriate, and students do not obtain the training to draw diagrams or graphs to simplify and organize their thinking.

(d) Students can learn the subject only by doing the exercises themselves and reviewing them in class, to obtain experience in applying the principles with their different ramifications.

In doing the exercises by themselves, students find that they are required to devote considerably more time to electric circuits than to other subjects of comparable credits, because they are uncertain with regard to the selection and application of the theorems and principles involved. It is also often necessary for students to discover those "tricks" not revealed in their texts (or review books), that make it possible to solve problems easily. Students must usually resort to methods of trial-and-error to discover these "tricks", and as a result they find that they may sometimes spend several hours to solve a single problem.

(e) When reviewing the exercises in classrooms, instructors

usually request students to take turns in writing solutions on the boards and explaining them to the class. Students often find it difficult to explain in a manner that holds the interest of the class, and enables the remaining students to follow the material written on the boards. The remaining students seated in the class are, furthermore, too occupied with copying the material from the boards, to listen to the oral explanations and concentrate on the methods of solution.

This book is intended to aid students in electric circuits to overcome the difficulties described, by supplying detailed illustrations of the solution methods which are usually not apparent to students. The solution methods are illustrated by problems selected from those that are most often assigned for class work and given on examinations. The problems are arranged in order of complexity to enable students to learn and understand a particular topic by reviewing the problems in sequence. The problems are illustrated with detailed step-by-step explanations, to save the students the large amount of time that is often needed to fill the gaps that are usually found between steps of illustrations in textbooks or review/outline books.

The staff of REA considers electric circuits a subject that is best learned by allowing students to view the methods of analysis and solution techniques themselves. This approach to learning the subject matter is similar to that practiced in various scienticfic laboratories, particularly in the medical fields.

In using this book, students may review and study the illustrated problems at their own pace; they are not limited to the time allowed for explaining problems on the board in class.

When students want to look up a particular type of problem and solution, they can readily locate it in the book by referring to the index which has been extensively prepared. It is also possible to locate a particular type of problem by glancing at just the material within the boxed portions. To facilitate rapid scanning of the problems, each problem has a heavy border around it. Furthermore, each problem is identified with a number immediately above the problem at the right-hand margin.

To obtain maximum benefit from the book, students should familiarize themselves with the section, "How To Use This Book," located in the front pages.

To meet the objectives of this book, staff members of REA have selected problems usually encountered in assignments and examinations, and have solved each problem meticulously to illustrate the steps which are usually difficult for students to comprehend. For outstanding effort and competence in this area, special gratitude is due to

Peter Halatyn and Timothy Waters

The following persons also have contributed a great deal of support and much patient work to achieve the objectives of the book:

Prof. John P. Basart
Iowa State University

Prof. T.K. Ishii
Marquette University

Prof. Artice M. Davis
San Jose State University

Prof. Wesley K. Kay
U.S. Naval Academy

Prof. F.J. Eberhardt
U.S. Naval Academy

Prof. William L. Kuriger
University of Oklahoma

Prof. Paul G. Gray
Michigan State University

Prof. Tim D. Slagh
Auburn University

Gratitude is also expressed to the many persons involved in the difficult task of typing the manuscript with its endless changes, and to the REA art staff who prepared the numerous detailed illustrations together with the layout and physical features of the book.

The difficult task of coordinating the efforts of all persons was carried out by Carl Fuchs. His conscientious work deserves much appreciation. He also trained and supervised art and production personnel in the preparation of the book for printing.

Finally, special thanks are due to Helen Kaufmann for her unique talents to render those difficult border-line decisions and constructive suggestions related to the design and organization of the book.

Max Fogiel, Ph.D.
Program Director

HOW TO USE THIS BOOK

This book can be an invaluable aid to students in electric circuits as a supplement to their textbooks. The book is subdivided into 19 chapters, each dealing with a separate topic. The subject matter is developed beginning with basic circuit analysis methods and extending through RLC circuits, Laplace transforms, and topological analysis. Sections on matrix methods, three-phase circuits, frequency domain analysis, and discrete systems have also been included, since these appear to be most troublesome to students.

TO LEARN AND UNDERSTAND A TOPIC THOROUGHLY

1. Refer to your class text and read there the section pertaining to the topic. You should become acquainted with the principles discussed there. These principles, however, may not be clear to you at that time.

2. Then locate the topic you are looking for by referring to the "Table of Contents" in the front of this book, "The Electric Circuits Problem Solver".

3. Turn to the page where the topic begins and review the problems under each topic, in the order given. For each topic, the problems are arranged in order of complexity, from the simplest to the more difficult. Some problems may appear similar to others, but each problem has been selected to illustrate a different point or solution method.

To learn and understand a topic thoroughly and retain its contents, it will be generally necessary for students to review the problems several times. Repeated review is essential in order to gain experience in recognizing the principles that should be applied, and select the best solution technique.

TO FIND A PARTICULAR PROBLEM

To locate one or more problems related to a particular subject matter, refer to the index. In using the index, be certain to note that the numbers given there refer to problem numbers, not to page numbers. This arrangement of the index is intended to facilitate finding a problem more rapidly, since two or more problems may appear on a page.

If a particular type of problem cannot be found readily, it is recommended that the student refer to the "Table of Contents" in the front pages, and then turn to the chapter which is applicable to the problem being sought. By scanning or glancing at the material that is boxed, it will generally be possible to find problems related to the one being sought, without consuming considerable time. After the problems have been located, the solutions can be reviewed and studied in detail. For this purpose of locating problems rapidly, students should acquaint themselves with the organization of the book as found in the "Table of Contents."

In preparing for an exam, it is useful to find the topics to be covered in the exam from the Table of Contents, and then review the problems under those topics several times. This should equip the student with what might be needed for the exam.

CONTENTS

Chapter No. **Page No.**

1 RESISTIVE CIRCUITS 1

Voltage, Current, and Power Relationships 1
Energy 6
Circuit Reduction and Voltage/Current Division 9
Kirchhoff's Voltage Law 18
Kirchhoff's Current Law 32

2 BASIC CIRCUIT ANALYSIS METHODS 45

Nodal Method 45
Mesh Method 64
Norton's Equivalent, Thevenin's Equivalent, and Superposition 84

3 MATRIX METHODS 107

Matrix Math 107
KCL with Matrices 123
KVL with Matrices 126

4 INDUCTORS AND CAPACITORS 147

Voltage - Current Relationships for Inductors 147
Voltage - Current Relationships for Capacitors 156
Energy, Charge, and Power 165

5 NATURAL RESPONSE OF RL AND RC CIRCUITS 180

RL Circuits 180
RC Circuits 187

6 FORCED RESPONSE OF RL AND RC CIRCUITS 199

The Unit Step Function 199
RL Circuits 204
RC Circuits 225
Differential Equations 243

7 RLC CIRCUITS 250

Series RLC 250
Parallel RCL 258
Combination 270

8 RMS VALUES, PHASORS, AND POWER 308

RMS Values 308
Phasors 335
Power 339

9 STEADY - STATE ANALYSIS 362

Impedance - Admittance Calculations 362
Equivalent Circuits 365
Response 370
Power 443

10 THREE - PHASE CIRCUITS 463

Y Connected 463
Δ Connected 475
Combination of Y and Δ 482

11 LAPLACE TRANSFORM TECHNIQUES 490

Simple Time Functions 490
Laplace Transform of Time Functions 496
Laplace Transform Properties 506
Convolution 514
Expansion by Partial Fractions 521
Inverse Laplace Transforms 525

12 LAPLACE TRANSFORM APPLICATIONS 546

Series Circuits 546
Parallel Circuits 581
Mixed Circuits 589
Inverse Laplace Transforms 607
Thevenin's and Norton's Equivalent Circuits 667
State Equations 680

13 FREQUENCY DOMAIN ANALYSIS 685

Poles and Zeros 685
Frequency Response 689
Resonance 717

14 FOURIER ANALYSIS 736

Fourier Techniques 736
Applications to Circuit Theory 774

15 DISCRETE SYSTEMS AND Z - TRANSFORMS 807

Discrete Elements and Equations 807
Steady State and Homogeneous Solutions 818
Digital Solution of Analog Systems 828
Z - Transform Definitions and Properties 831
Z - Transform Applications 840

16 TWO - PORT NETWORKS 847

Transformers and Mutual Inductance 847
Network Parameters 865
Reciprocity 913
π - T Conversion 922

17 STATE EQUATIONS 927

Definitions and Properties 927
Applications to RC Circuits 938
Applications to RL Circuits 939
Applications to RLC Circuits 943
Applications to Nonlinear and Time - Varying Circuits 977

18 TOPOLOGICAL ANALYSIS 984

Definitions 984
The Incidence Matrix 986
The Loop Matrix 991
Applications 996

19 NUMERICAL METHODS 1005

Trial and Error Procedure 1005
Newton's Method 1008
Simpson's Rule 1014
Runge - Kutta Method 1027
Other Methods 1030

INDEX 1039

CHAPTER 1

RESISTIVE CIRCUITS

VOLTAGE, CURRENT, & POWER RELATIONSHIPS

• PROBLEM 1-1

With reference to Fig. 1, find v if;

(a) $G = 10^{-2}$ ℧ and $i = -2.5A$

(b) $R = 40\Omega$ and the resistor absorbs 250W.

(c) $i = 2.5A$ and the resistor absorbs 500W.

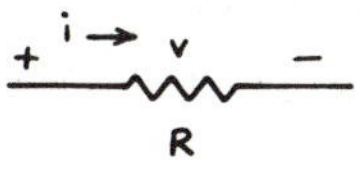

FIG. 1

Solution: (a) Apply Ohm's Law.

$$v = iR = \frac{i}{G}$$

$$v = \frac{-2.5}{10^{-2}} = -250V.$$

(b) We know that p = vi since v = iR;

$$p = i^2 R = \frac{v^2}{R} \qquad (1)$$

Solving for v

$$v = \sqrt{pR}$$

$v = \sqrt{250(40)} = \pm 100V$. Since R absorbs power, v = +100V.

(c) Solving for R in equation (1)

$$R = \frac{p}{i^2} = \frac{500}{(2.5)^2} = 80 \text{ ohms.}$$

Using Ohm's Law;

$$v = iR = 2.5(80) = 200V.$$

Determine the power being supplied by the sources shown in Fig. 1a, b, and c.

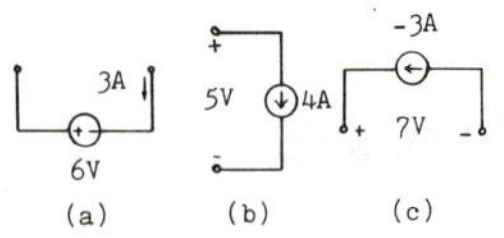

Figure 1

Solution: In order for a voltage source to deliver power, current must flow out of the positive terminal. For a current source to deliver power, there must be a positive potential at the terminal where current (positive current) flows out.

a) Current flows out of the positive terminal, therefore $p = \nu i$, $(3)(6) = 18$ W.

b) Current flows out the negative terminal, therefore the current source absorbs power or delivers negative power.

$$p = \nu i,\ (5)(-4) = -20 \text{ W}.$$

c) Here, a negative current flows out the positive terminal, therefore

$$p = \nu i,\ (7)(-3) = -21 \text{ W}.$$

• PROBLEM 1-3

Figure 1 shows the voltage waveform, $V_r(t)$, observed across the terminals of a resistor. Calculate the current, i(t), through a resistor of: (a) 1 ohm; (b) 2 ohms; (c) ∞ ohms.

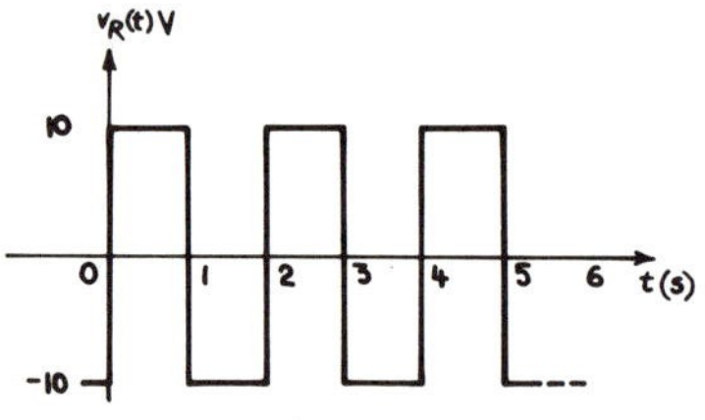

FIG. 1

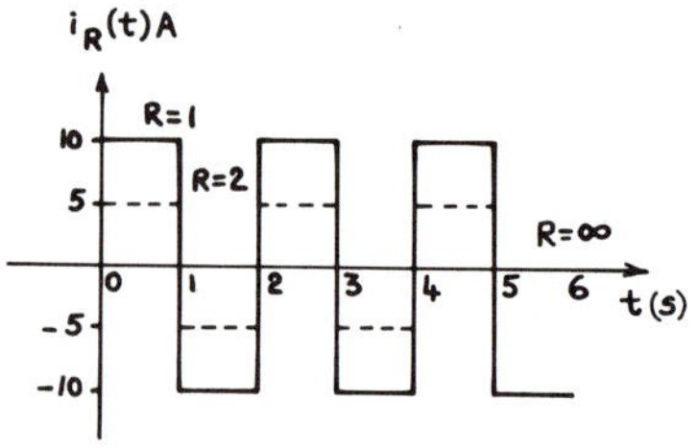

FIG. 2

Solution: We apply Ohm's Law to find the current by dividing $V_R(t)$ by the resistance values given.

We ignore the discontinuites at t = 1, 2, 3 etc. and observe that the voltage is constant between certain time values. We observe that there are only two values of interest, + 10V and -10 V.

Fig. 2. shows the results.

$$\text{(a)}\quad i_R = \frac{\pm 10}{1} = \pm 10 \text{ A}$$

(b) $i_R = \frac{\pm 10}{2} = \pm 5$ A

(c) $i_R = \frac{\pm 10}{\infty} = 0.$

• PROBLEM 1-4

The v-i relation of a nonlinear resistor is given by

$$v(t) = \frac{1}{2}i^2(t)$$

Is this resistor voltage-controlled? Current-controlled?

Solution: If we solve the above v-i relation for i(t) in terms of $v(t)$ we obtain

$$i(t) = \pm \sqrt{2\, v(t)}.$$

It is seen that a single value of $v(t)$ will not produce a single value of i(t). However, a single value of i(t) will produce a single value of $v(t)$. Thus the resistor is current controlled.

• PROBLEM 1-5

The output of a certain independent current source is a function of temperature T(°K), $i_s = 10^{-6}T^2 - 6\times 10^{-4}T + 0.1$ A. (a) At what temperature in the range $250 \le T \le 320$°K is i_s a maximum? (b) At a constant temperature of 320°K, how much charge can the source deliver in 10 minutes?

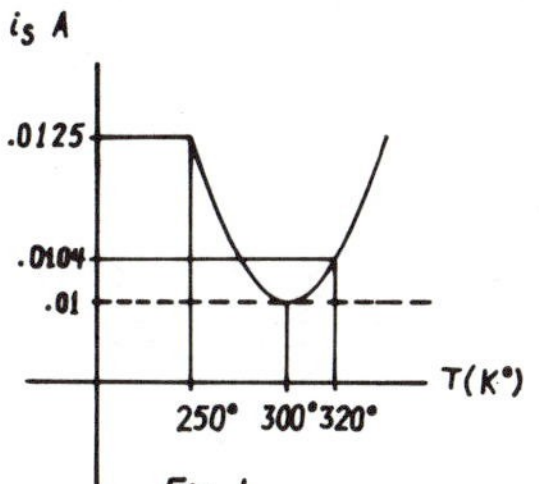

Fig. 1

Solution: (a) The function i_s is a parabolic function with a minimum at 300°K; see Fig. 1.

the maximum value for the range $250° \le T \le 320$°K would be 250°K.

(b) If i_s remains constant at 320°K, $i_s = .0104$ A; the charge delivered would be

$$q = i\cdot t = .0104\ \frac{c}{sec}\ (10\ min)\ 60\ \frac{sec}{min} = 6.24\ C.$$

since i_s is not a function of time and assuming

$$q(0-) = 0.$$

• PROBLEM 1-6

A non-linear resistor is connected in series with a 5-V source and a linear 2Ω resistor. Use the v-i characteristics of the non-linear resistor in Fig. 1 to find the current through the circuit.

Fig. 1

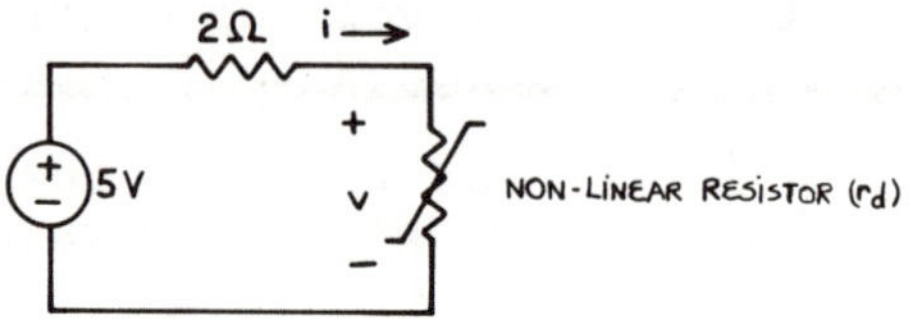

Fig. 2

Solution: Fig. 2 shows the circuit described in this example.

Examining the v-i characteristics for the non-linear resistor, notice that the slope of the curve at any point yields the resistor's conductance,

$$g_d = \frac{1}{r_d}.$$

This value can be either positive or negative. Since each value of voltage gives only a single resistance value, the resistor is voltage controlled. It is not current controlled since certain values of current give more than one value of resistance.

In order to find the current in the circuit we must find the v-i characteristics for the voltage source-linear resistor combination shown in Fig. 3.

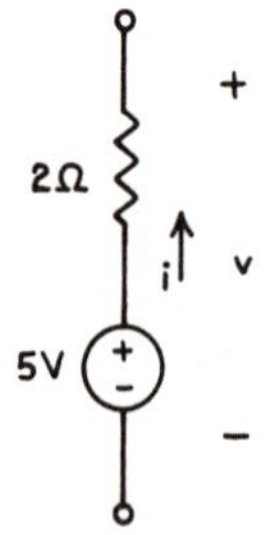

Fig. 3

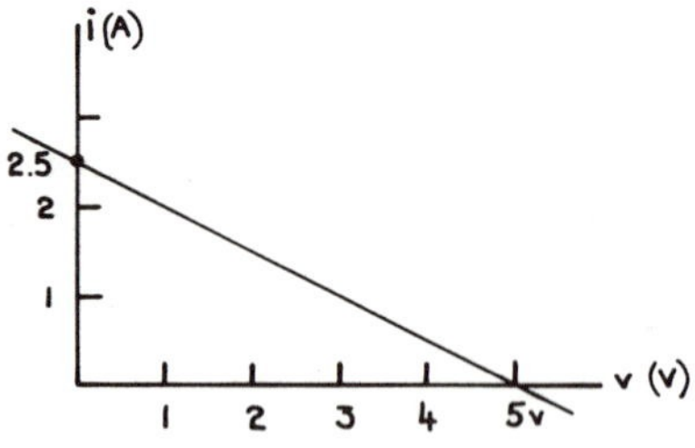

Fig. 4

To do this write the equation

$$v = 5 - 2i.$$

Note that in a v-i plot this is a straight line with v = 5V when i = 0 and i = 2.5A when v = 0. Fig. 4 shows the v-i characteristics of the voltage source-linear resistor combination.

Since the v and i for the voltage source-linear resistor combination is the same as that for the non-linear resistor, one can find the points of operation for the non-linear resistor by overlapping their v-i characteristics as shown in Fig. 5.

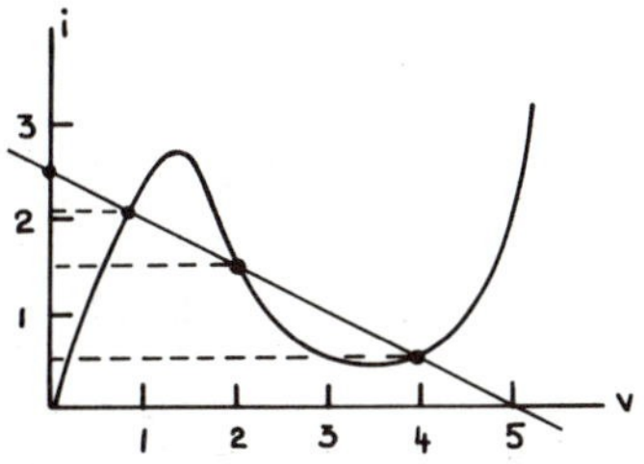

FIG. 5

Notice the 3- points of intersection from which we obtain the three possible currents 2.2A,1.9A, and 0.3A.

• PROBLEM 1-7

> The current through a resistor of 2Ω is given by
>
> $$i(t) = \cos^2 \pi t.$$
>
> Find the energy dissipated in the resistor from $t_0 = 0$ to $t_1 = 5$ sec.

Solution: Energy is defined as

$$E(t_0, t) = \int_{t_0}^{t} p(t)\, dt$$

where p(t) is the power dissipation. In this example, $p(t) = [i(t)]^2 R = 2 \cos^4 \pi t$. Hence,

$$E(0,5) = \int_0^5 2 \cos^4 \pi t$$

In order to evaluate the above integral, it is necessary to write $\cos^4 \pi t$ in terms of simpler and more easily integratable components.

Using the trigonometric identity,

$$\cos^2 x = \frac{1}{2}(1 + \cos 2x)$$

we find that we can express

$$\cos^4 \pi t = (\cos^2 \pi t)(\cos^2 \pi t) = \left[\frac{1}{2}(1 + 2\ t)\right]^2$$

as $$\cos^4 \pi t = \frac{1}{8}(3 + 4 \cos 2\pi t + \cos 4\pi t).$$

Evaluating the above integral

$$E(0,5) = \frac{1}{4}\int_0^5 (3 + 4 \cos 2\pi t + \cos 4\pi t)\ dt$$

$$E(0,5) = \frac{1}{4}\left[3t + \frac{2}{\pi} \sin 2\pi t + \frac{1}{4\pi} \sin 4\pi t\right]_0^5$$

$$E(0,5) = \frac{1}{4}\left[3(5) + \frac{2}{\pi} \sin (10\pi) + \frac{1}{4\pi} \sin (20\pi)\right]$$

$$- \frac{1}{4}\left[0 + \frac{2}{\pi} \sin 0 + \frac{1}{4\pi} \sin 0\right]$$

$$E(0,5) = 3.75\ \text{J}$$

ENERGY

• PROBLEM 1-8

A linear time-invariant resistor of $4\ \Omega$ has a current through it given by $i(t) = \sin \pi t$. If $E(0) = 0$, find the energy dissipated in the resistor at $t = 1, 2, 3, 4, 5, 6$ sec.

Solution: We are asked to find the energy, $E(t)$, dissipated in a $4\ \Omega$ resistor at $t = 1, 2, 3, 4, 5, 6$ s given $i(t) = \sin \pi t$. Since $E(t)$ is defined as

$$\int_0^t p(t)\ dt \qquad \text{where}$$

$$p(t) = [i(t)]^2\ R$$

Hence,

$$E(t) = 4 \int_0^t \sin^2 \pi t\ \ dt$$

since $\sin^2 \pi t$ can be written $\frac{1}{2}(1 - \cos 2\pi\ t)$

$$E(t) = 4 \int_0^t \left(\frac{1}{2} - \frac{1}{2} \cos 2\pi\ t\right)\ dt$$

$$E(t) = 4\left[\ \frac{1}{2}t - \frac{1}{4\pi}\sin 2\pi\ t\ \right]$$

$$E(1) = 4\ (\ \frac{1}{2} - \frac{1}{4\pi}\sin 2\pi\ ) = 2\ J$$

similarly

$E(2) = 4\ J$

$E(3) = 6\ J$

$E(4) = 8\ J$

$E(5) = 10\ J$

$E(6) = 12\ J$

• PROBLEM 1-9

Assume the waveform in Fig. 1 represents a voltage across a 10Ω resistor. Find the energy delivered to the resistor over the time intervals (a) (0, 1), (b) (0, 2); and (c) (0, 5) seconds.

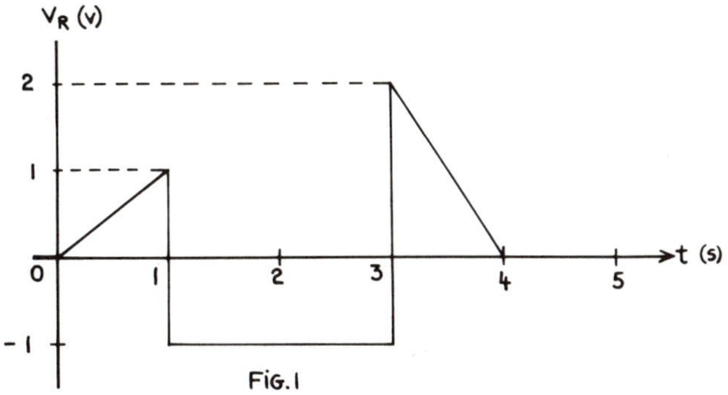

FIG. 1

Solution: Energy is defined as

$$E[t_0,\ t_1] = \int_{t_0}^{t_1} p(t)\ dt$$

where p(t) is the power dissipation, v(t) i(t). For a resistor,

$$p(t) = [v_R(t)]^2\ \frac{1}{R}\ .$$

Hence

$$E[t_0,\ t_1] = \frac{1}{R}\int_{t_0}^{t_1} [v_R(t)]^2\ dt$$

(a) For the time interval [0, 1];

$v_R(t) = t; \qquad 0 \le t < 1.$

Hence, $E[0, 1] = \frac{1}{10} \int_0^1 t^2 \, dt$

$$E[0, 1] = \frac{1}{10} \left[\frac{t^3}{3} \right]_0^1$$

$$E[0, 1] = \frac{1}{30} \text{ J.}$$

(b) For the interval [0, 2] seconds

$$E[0, 2] = E[0, 1] + E[1,2]$$

where $E[1, 2] = \frac{1}{10} \int_1^2 (-1)^2 \, dt$

$$= \frac{1}{10} [t]_1^2 = \frac{2}{10} - \frac{1}{10}$$

$$= \frac{1}{10} \text{ J.}$$

Hence, $E[0, 2] = \frac{1}{30} + \frac{1}{10} = \frac{4}{30}$ J.

(c) For the interval [0, 5] seconds

$v_R(t) = t \qquad 0 \le t < 1$

$v_R(t) = -1 \qquad 1 \le t < 3$

$v_R(t) = -2(t - 4) \qquad 3 \le t \le 4$

$v_R(t) = 0 \qquad 4 \le t < 5$

Hence, $E[0, 5] = E[0, 1] + E[1, 3] + E[3, 4] + E[4, 5]$

$$E[0, 5] = \frac{1}{30} + \frac{1}{10} \int_1^3 (-1)^2 \, dt + \frac{1}{10} \int_3^4 [-2(t - 4)]^2 \, dt + 0$$

$$E[0, 5] = \frac{1}{30} + \frac{1}{10} [t]_1^3 + \frac{1}{10} \int_3^4 (4t^2 + 32t + 64) \, dt$$

$$E[0, 5] = \frac{1}{30} + \frac{3}{10} - \frac{1}{10} + \frac{1}{10} \left[\frac{4t^3}{3} - \frac{32t^2}{2} + 64t \right]_3^4$$

$$E[0, 5] = \frac{1}{30} + \frac{3}{10} - \frac{1}{10} + \frac{1}{10} \left[\frac{256}{3} - \frac{512}{2} + 256 \right]$$

$$- \frac{1}{10} \left[\frac{108}{3} - \frac{288}{2} + 192 \right]$$

$$E[0, 5] = \frac{1}{30} + \frac{2}{10} + \frac{1}{10}\left[\frac{148}{3} - \frac{224}{2} + 64\right]$$

$$E[0, 5] = \frac{7}{30} + \frac{148}{30} - 48$$

$$E[0, 5] = \frac{7}{30} + \frac{148}{30} - \frac{144}{30}$$

$$E[0, 5] = \frac{7}{30} + \frac{4}{30} = \frac{11}{30} \text{ J.}$$

CIRCUIT REDUCTION AND VOLTAGE/ CURRENT DIVISION

• PROBLEM 1-10

Find R_{eq} for the network of Fig. 1; (a) as it is shown; (b) with the 5-Ω resistor replaced by a short circuit (0 Ω); (c) with the 5-Ω resistor replaced by an open circuit (∞ Ω).

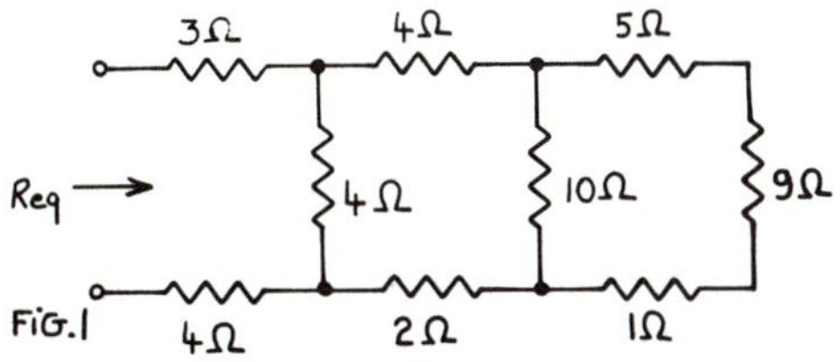

FIG.1

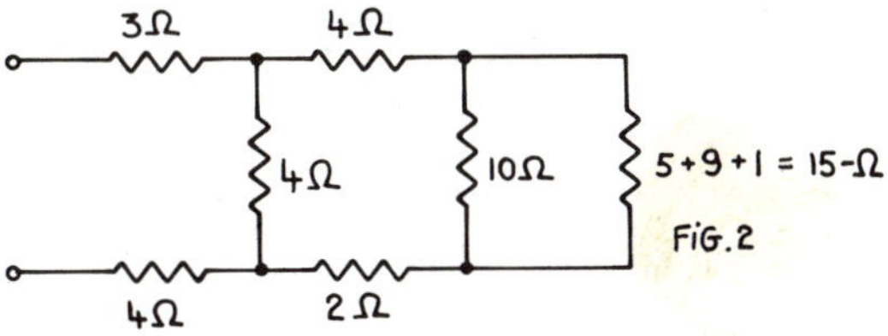

FIG.2

Solution: (a) Start to combine resistors at the right side of the circuit.

Add the 5, 9, and 1-Ω resistors, which now form a resistor parallel to the 10-Ω resistor.

Next, $\frac{15 \cdot 10}{15 + 10} = 6\Omega$

Fig. 3 shows the resulting circuit. This circuit can be further simplified similarly.

and $\frac{12 \cdot 4}{12 + 4} = 3\Omega$

$$R_{eq} = 3 + 3 + 4 = 10\Omega.$$

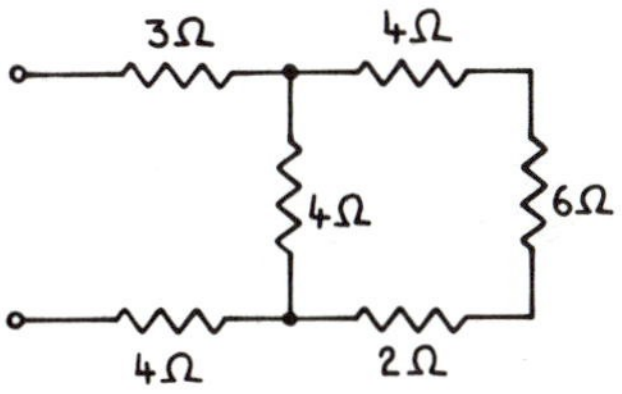

FIG.3

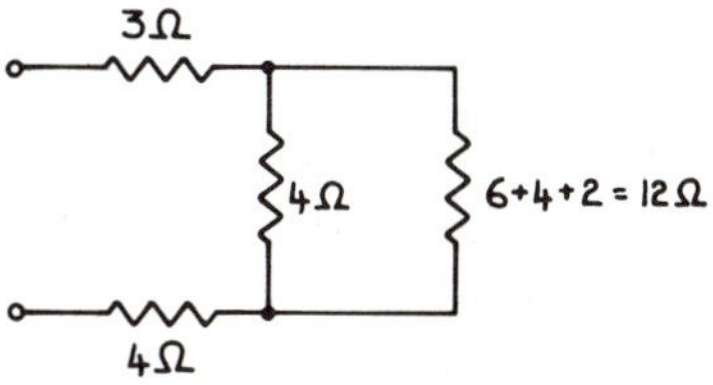

FIG.4

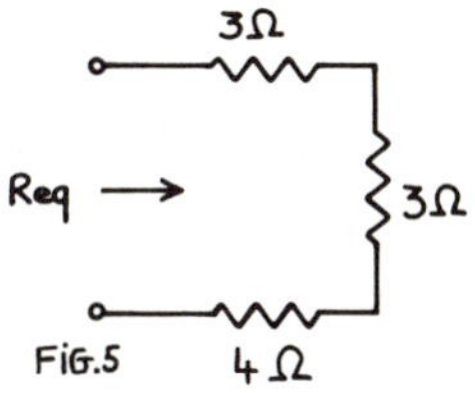

FIG.5

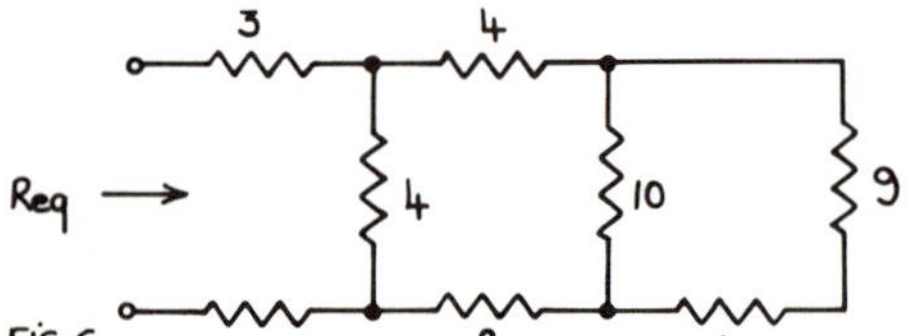

FIG.6

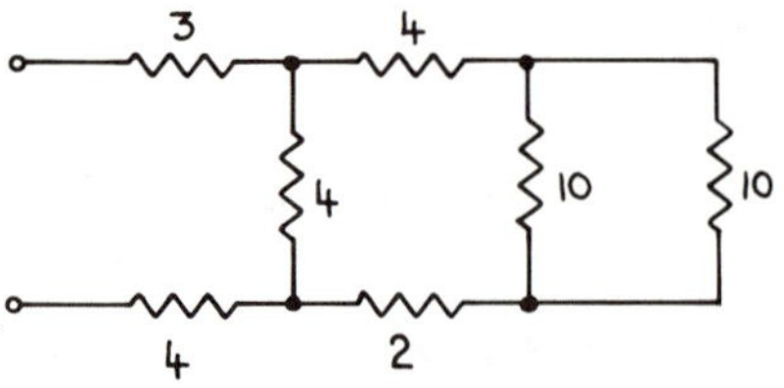

FIG.7

(b) Replacing the 5Ω resistor by a short circuit, proceed in a similar manner. Figs. 6 - 10 show the process.

$$R_{eq} = 3 + 4 + 2.933 = 9.933\ \Omega.$$

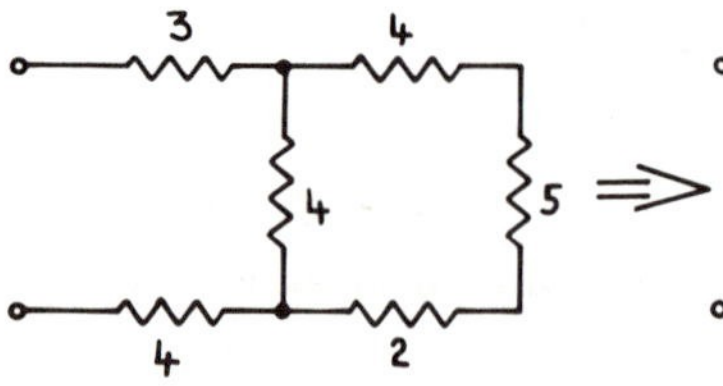

FIG.8

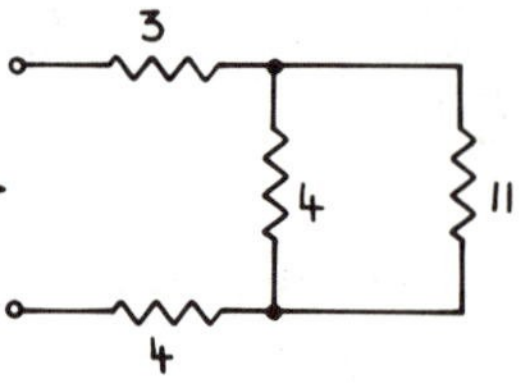

FIG.9

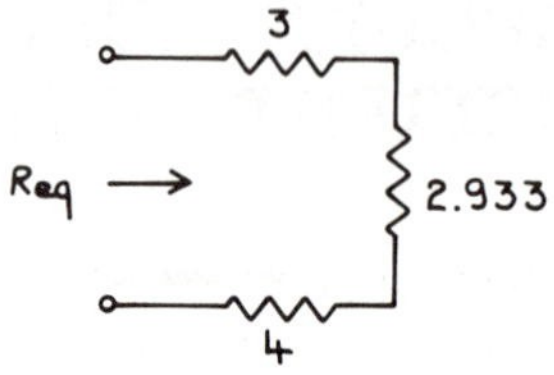

FIG.10

(c) Replace the 5Ω resistor with an open circuit. This isolates the 9 and 1-Ω resistors. Fig. 11 demonstrates the process.

$$R_{eq} = 3 + 4 + 3.2 = 10.2\ \Omega.$$

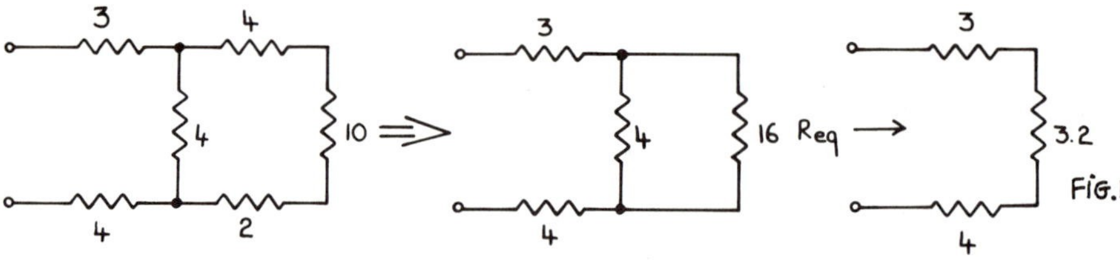

FIG.11

• PROBLEM 1-11

Find G_{eq} for the network of Fig. 1. (a) as it is shown; (b) if it is disconnected and the six elements reconnected in parallel; (c) if it is disconnected and the six elements reconnected in series.

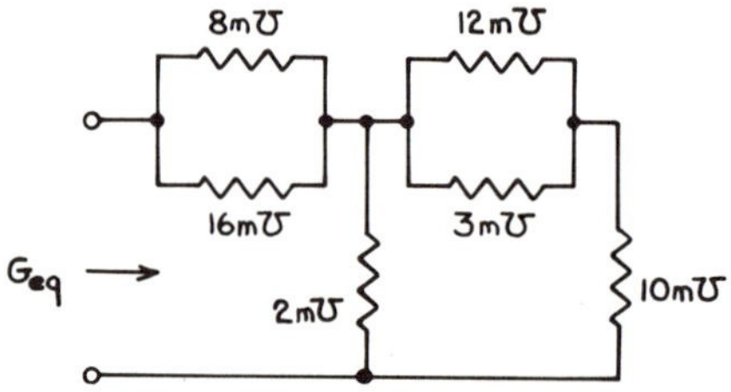

FIG.1

Solution: (a) Start combining conductances at the right of the circuit. Note that the conductances are in parallel. Simply add,

$$\frac{15 \cdot 10}{15 + 10} = 6\ \text{m℧}$$

$$G_{eq} = \frac{24 \cdot 8}{24 + 8} = 6 \text{ m℧}$$

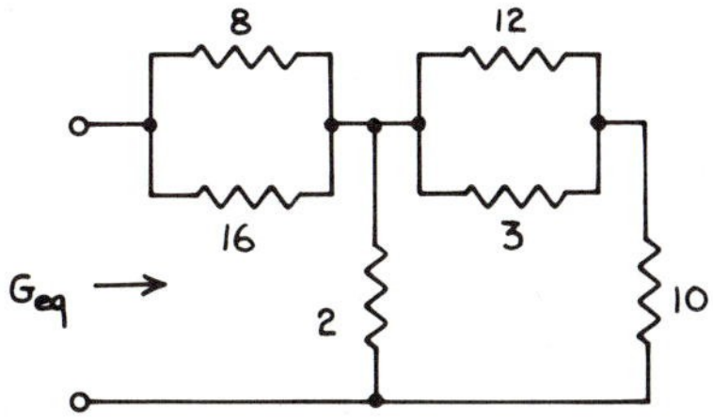

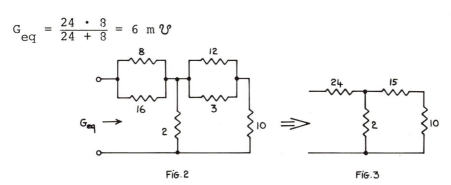

Fig. 2

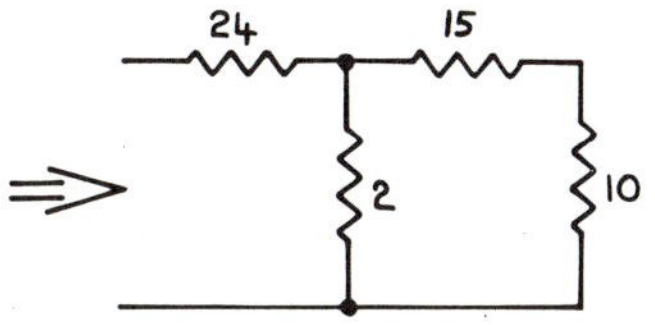

Fig. 3

(b) If all the elements are reconnected in parallel, the conductances would equal

$$G_{eq} = 8 + 16 + 2 + 12 + 3 + 10 = 51 \text{ m℧}$$

(c) If all the elements are connected in series, G_{eq} would be the reciprocal of the sum of the reciprocal of the conductances.

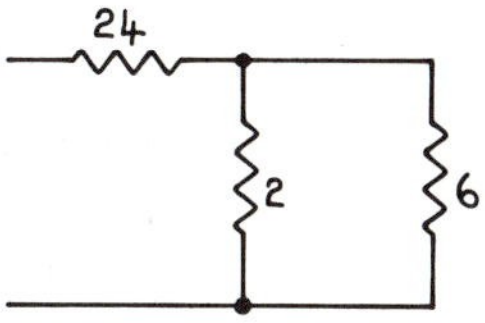

Fig. 4

$$\frac{1}{G_{eq}} = \frac{1}{8} + \frac{1}{16} + \frac{1}{2} + \frac{1}{12} + \frac{1}{3} + \frac{1}{10}$$

$$\frac{1}{G_{eq}} = \frac{30 + 15 + 120 + 20 + 80 + 24}{240}$$

$$\frac{1}{G_{eq}} = \frac{289}{240}$$

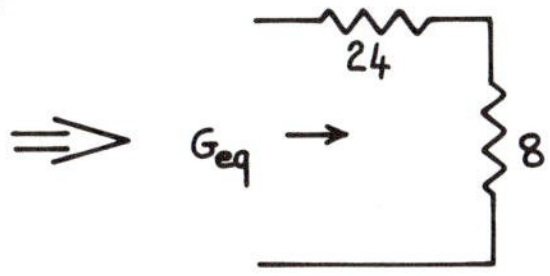

$$G_{eq} = \frac{240}{289} \text{ m℧} = 0.830 \text{ m℧}.$$

Fig. 5

• PROBLEM 1-12

> In the circuit shown in Fig. 1, it is known that $v_1 = 6$ V. Find i_s and v.

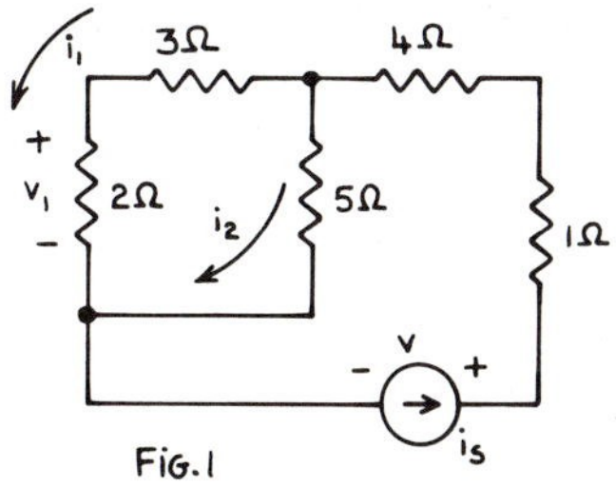

Fig. 1

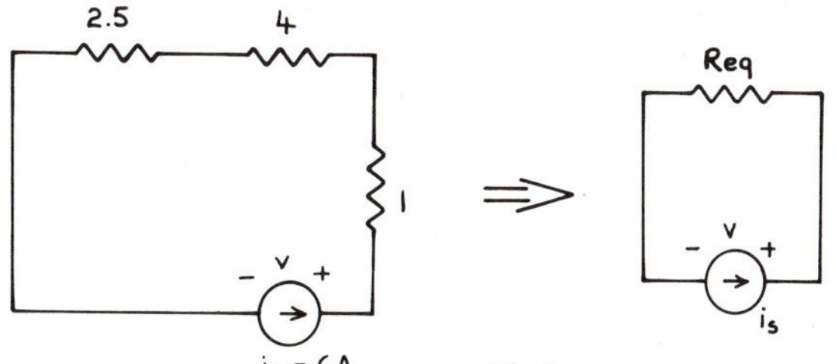

Fig. 2

Solution: Knowing the voltage across the 2-Ω resistor makes it possible to find the current through the branch containing the 2-Ω and the 3-Ω resistors.

$$i_1 = \frac{v}{R} = \frac{6}{2} = 3\text{A}$$

i_1 must be equal to i_2 because they both flow through 5-Ω resistances. The series resistor equivalent

$$R = 3 + 2 = 5\Omega.$$

Since $i_s = i_1 + i_2$ we have, $i_s = 6A$.

To find v, calculate the equivalent resistance across the terminals of the source.

$$R_{eq} = 7.5\text{-}\Omega$$

It follows that

$$v = R_{eq}\, i_s = 7.5(6) = 45V.$$

• PROBLEM 1-13

In the circuit shown in Fig. 1, find v_1 and the voltage v across both sources if i_s is given as 12A.

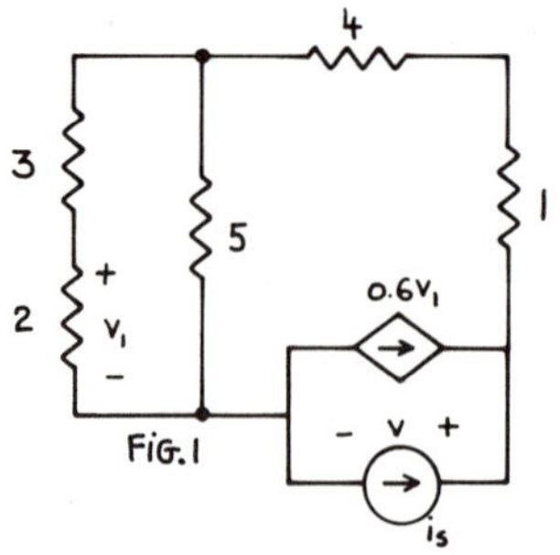

FIG. 1

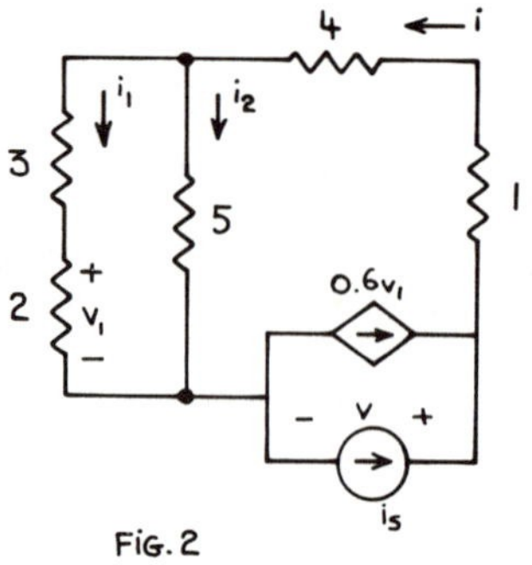

FIG. 2

Solution: In Fig. 2 observe that

$$i = i_s + 0.6v_1 = i_1 + i_2.$$

Since i_1 and i_2 both flow through a branch with the same equivalent resistances, they must be equal. Thus we have

$$i = i_s + 0.6v_1 = 2i_1 = 2i_2.$$

We can write $v_1 = i_1R$ where R = 2-Ω. Since $i = 2i_1$ we have

$$i_1 = \tfrac{1}{2}i = \tfrac{1}{2}(i_s + 0.6\, v_1)$$

$$v_1 = i_1R = \tfrac{1}{2}(i_s + 0.6v_1)R$$

Substituting 2-Ω of R and 12A for i_s, solve for v_1

$$v_1 = \tfrac{1}{2}(12 + 0.6v_1)\ 2$$

$$v_1 = 12 + 0.6v_1$$

$$0.4\,v = 12$$

giving $$v_1 = \frac{12}{0.4} = 30V.$$

v is found by calculating i and the equivalent resistance R_{eq} across the terminals of the two sources.

From Fig. 3 we find $v = R_{eq}\, i = 7.5(30) = 225V.$

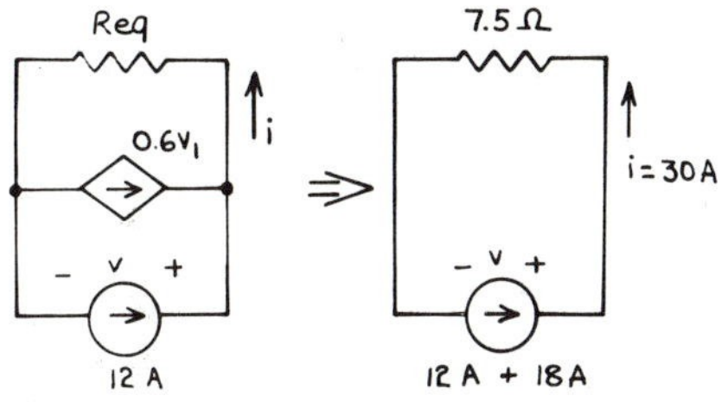

Fig. 3

• PROBLEM 1-14

The destruction (by open circuit) of which resistor in the circuit of Fig. 1 would cause the greatest increase in the power supplied by the current source? What is this power?

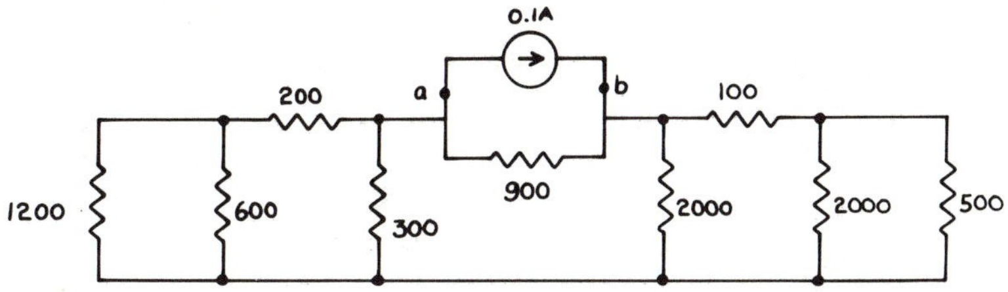

Fig. 1

Solution: In order for the current source to supply increased power, the voltage across its terminals v_{ba} must be at a maximum. This will happen only if the equivalent resistance across its terminals is at a maximum.

The equivalent resistance for the complete circuit is 360-Ω.

A larger equivalent resistance can be obtained by removing either the 900-Ω or 100-Ω resistor.

If the 900-Ω resistor is removed, the equivalent resistance is 600-Ω. However, if we remove the 100-Ω resistor the equivalent resistance becomes 639-Ω. The power supplied by the source becomes

$$p = iv = i^2R = (.1)^2(639) = 6.39W.$$

• PROBLEM 1-15

What must be the resistance of R in the circuit of Fig. 1.

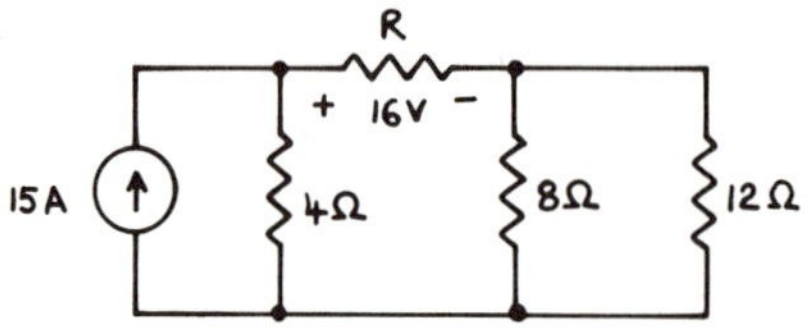

FIG. 1

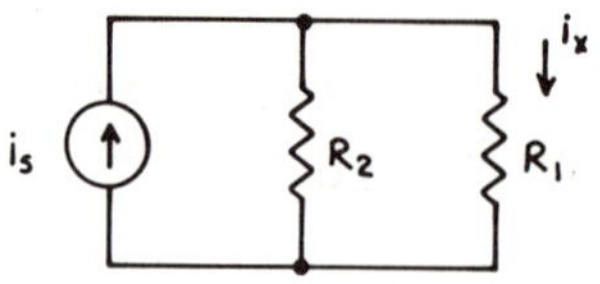

FIG. 2

Solution: By applying the concept of current division illustrated in Fig. 2, it is possible to solve for R directly.

$$i_x = \frac{i_s R_2}{R_1 + R_2} \tag{1}$$

First we combine the 8- and 12-Ω resistors and write an equation for current division in Fig. 3.

$$i = \frac{15(4)}{4 + R + 4.8}$$

Now substitute $\frac{16}{R}$ for i and thus solve for R.

$$\frac{16}{R} = \frac{60}{R + 8.8}$$

$$60R = 16R + 140.8$$

$$44R = 140.8$$

$$R = \frac{140.8}{44} = 3.2\Omega$$

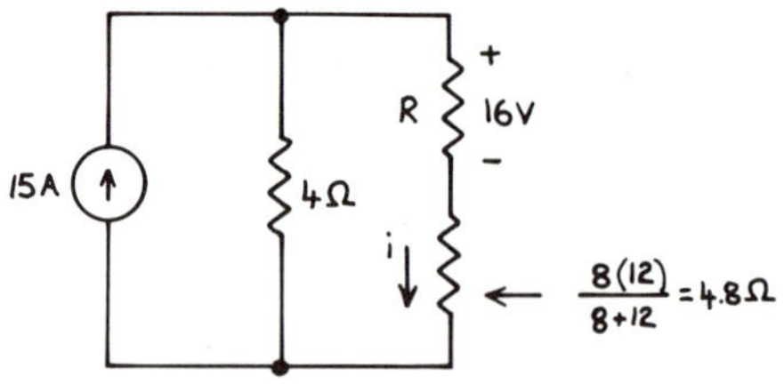

FIG. 3

• PROBLEM 1-16

Let v_s = 50 V in the circuit of Fig. 1. Using the concepts of voltage division, current division, and resistance combination, write a single expression that will yield the current i_x.

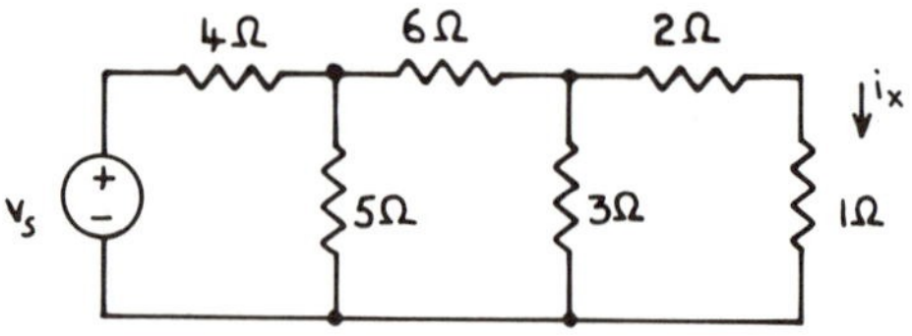

FIG. 1

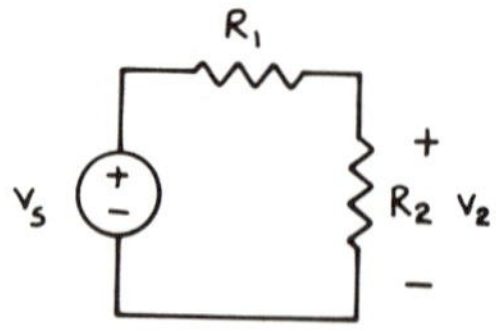

FIG. 2

Solution: The basic concept of voltage division is illustrated in Fig. 2.

$$v_1 = \frac{v_s R_2}{R_1 + R_2} \; ; \qquad (1)$$

In Fig. 1 the voltage division concept can be applied to consecutive stages to compute the voltage across the 1-Ω resistor and finally its current i_x.

For each stage, write an expression for R_{eq} that is substituted for R_2 in Eq. (1).

$$v_1 = \frac{v_s R_{eq_1}}{4 + R_{eq_1}} \qquad (2)$$

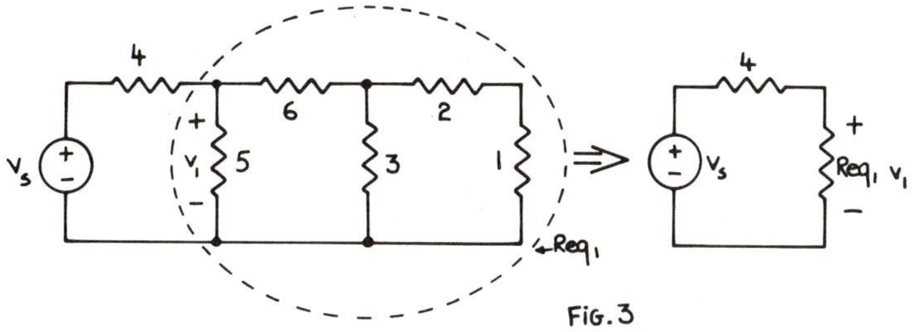

FiG. 3

Fig. 3 shows voltage division for the first stage.

$$v_2 = \frac{v_1 R_{eq_2}}{6 + R_{eq_2}}$$

Fig. 4 shows voltage division for the second stage.

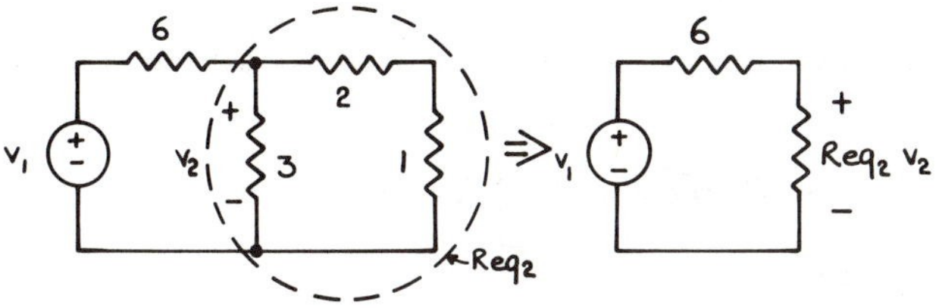

FiG. 4

Finally, the third stage is shown in Fig. 5.

$$v_x = \frac{v_2(1)}{2 + 1}$$

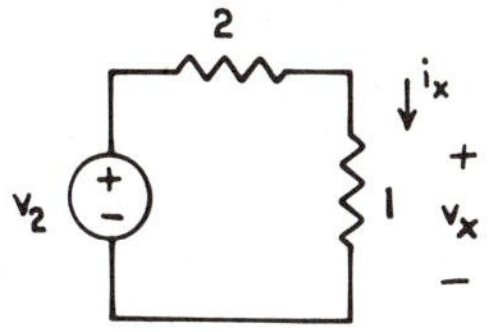

$$\frac{v_x}{1} = i_x$$

FiG. 5

The entire expression is written piece by piece:

$$R_{eq_2} = \frac{3(2 + 1)}{3 + (2 + 1)} \qquad (2)$$

$$R_{eq_1} = \frac{5(R_{eq_2} + 6)}{5 + (R_{eq_2} + 6)} \qquad (3)$$

$$v_1 = v_s \left(\frac{R_{eq_1}}{4 + R_{eq_1}}\right) \tag{4}$$

$$v_2 = v_1 \left(\frac{R_{eq_2}}{6 + R_{eq_2}}\right) \tag{5}$$

$$i_x = \frac{v_x}{1} = v_2 \left(\frac{1}{2 + 1}\right) \tag{6}$$

By combining Eqs. (2) - (6) and substituting 50V for v_s, we have

$$\frac{50}{4+\dfrac{5\left[6+\frac{3(2+1)}{3+(2+1)}\right]}{5+\left[6+\frac{3(2+1)}{3+(2+1)}\right]}} \cdot \frac{5\left[6+\frac{3(2+1)}{3+(2+1)}\right]}{5+\left[6+\frac{3(2+1)}{3+(2+1)}\right]} \cdot \frac{\frac{3(2+1)}{3+(2+1)}}{6+\frac{3(2+1)}{3+(2+1)}} \cdot \frac{1}{2+1}$$

$$= \frac{v_x}{1} = i_x = 1.429 \text{ A.}$$

• PROBLEM 1-17

For the circuit of Fig. 1 find (a) v_1; (b) v_2; (c) v_3. At $t = \infty$.

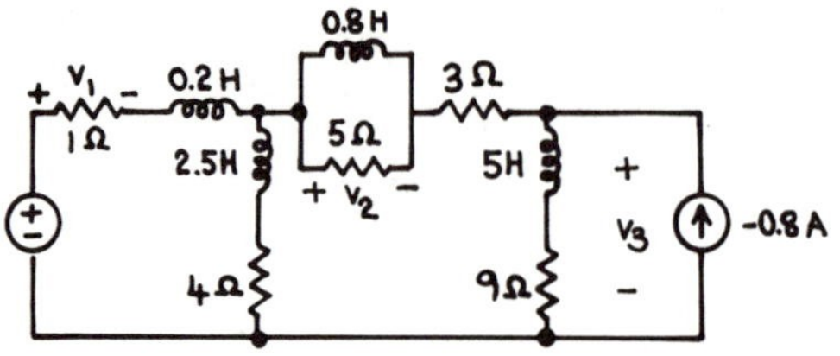

FIG. 1

Solution: Redraw the circuit in Fig. 1 and replace each inductor by a short circuit.

Immediately apparent is that $v_2 = 0$.

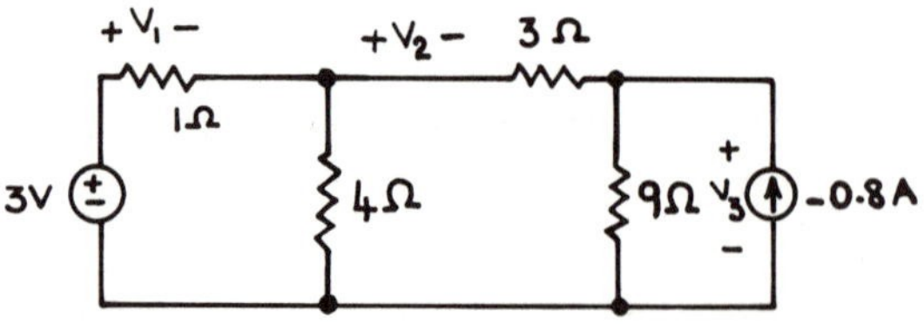

FIG. 2

By transforming the practical voltage source on the left and combining resistances, one obtains the circuit seen in Fig. 3.

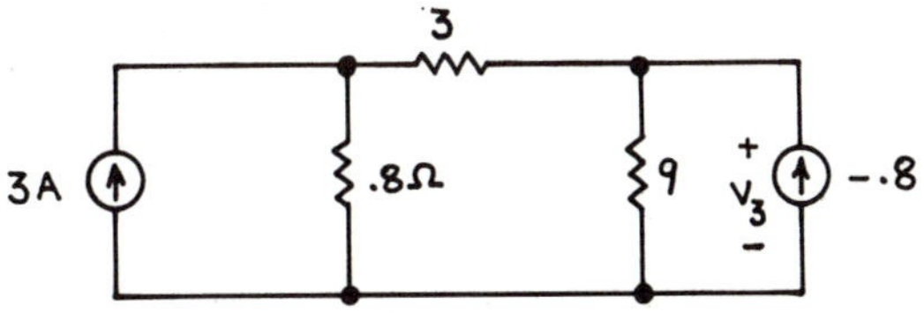

Fig. 3

Transform the practical current source on the left and combine resistances to obtain the circuit in Fig. 4.

Writing a KCL equation for node n_1 in Fig. 4 gives

$$-.8 = \frac{v_3 - 2.4}{3.8} + \frac{v_3}{9}$$

$$\left(\frac{1}{3.8} + \frac{1}{9}\right) v_3 = -.8 + \frac{2.4}{3.8}$$

$$v_3 = \frac{-.8 + \frac{2.4}{3.8}}{\left(\frac{1}{3.8} + \frac{1}{4}\right)} = -0.45 \text{ V}$$

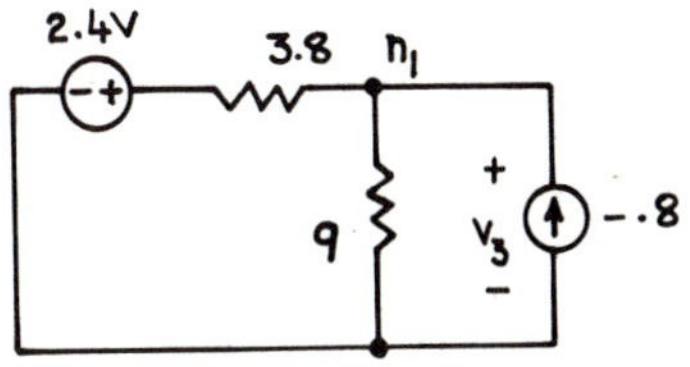

Fig. 4

By repeated source transformations of the practical current source on the right, the single loop circuit in Fig. 5 is obtained.

A KVL equation around loop ℓ_1 gives

$$i_1 = (1)i_1 = v_1.$$

$$3 - (-1.8) = i_1(1 + 3)$$

$$v_1 = (1)i_1 = \frac{3 - (-1.8)}{1 + 3} = 1.2 \text{ V}.$$

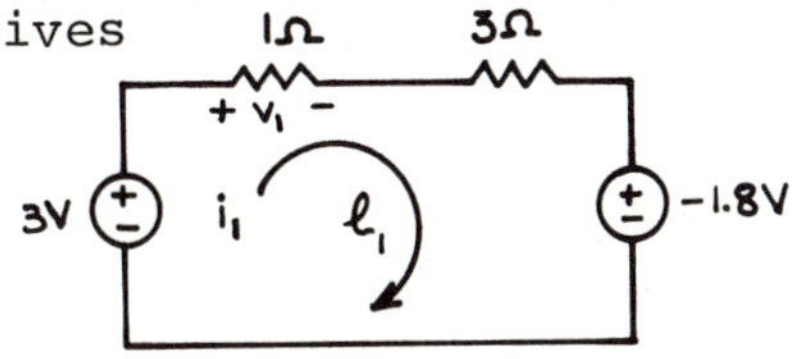

Fig. 5

KIRCHHOFF'S VOLTAGE LAW

• PROBLEM 1-18

Determine the number of branches and nodes present in each of the circuits of Fig. 1.

Fig. 1

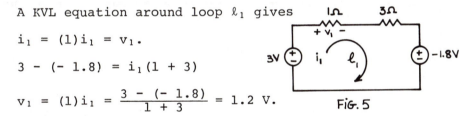

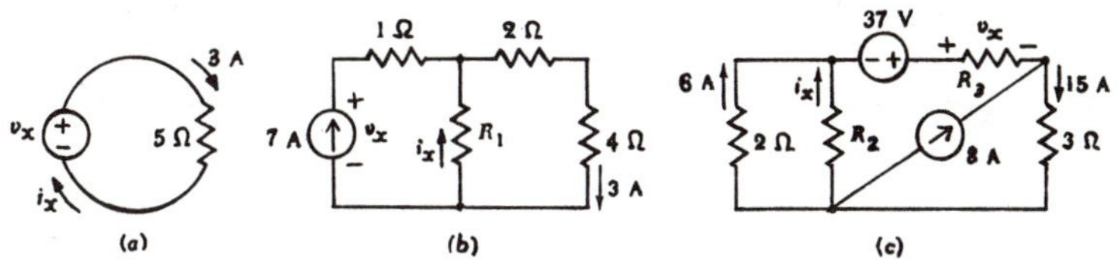

Solution: To determine the nodes and branches of the circuits, draw a circuit using a minimum number of nodes without the circuit elements.

a) 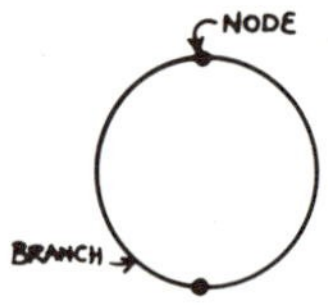

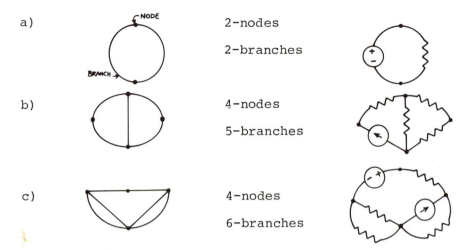

2-nodes

2-branches

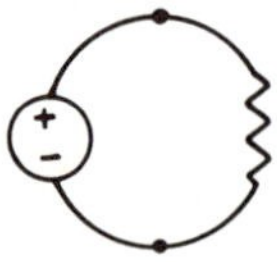

b) 

4-nodes

5-branches

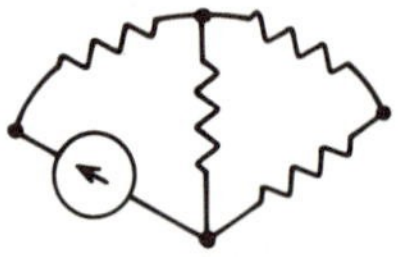

c) 

4-nodes

6-branches

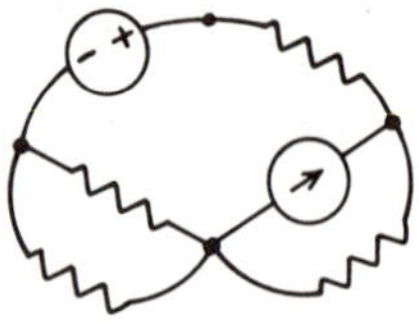

The lines that interconnect the nodes are branches.

• PROBLEM 1-19

A circuit contains six nodes lettered A, B, C, D, E, and F. Let v_{AB} be the voltage between nodes A and B with its positive reference at the first-named node, here A. Find v_{AC}, v_{AD}, v_{AE}, and v_{AF} if $v_{AB} = 6$ V, $v_{BD} = -3$ V, $v_{CF} = -8$ V, $v_{EC} = 4$ V, and: (a) $v_{DE} = 1$ V; (b) $v_{CD} = 1$ V; (c) $v_{FE} = 4$ V.

<u>Solution</u> (a) Position the nodes in any arbitrarily chosen order and indicate the voltages between the nodes.

In order to find the voltages in question, write a KVL equation around a loop which contains the unknown voltage.

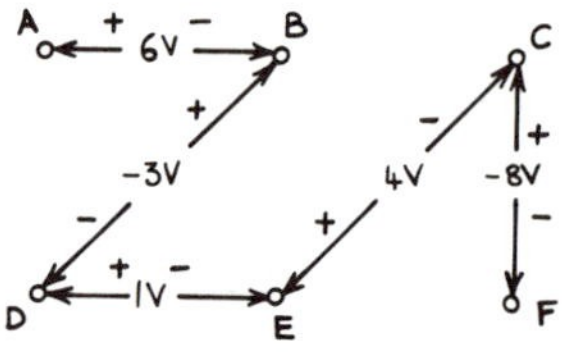

Fig. 1

To find v_{AC} write,

$$v_{AC} = v_{AB} + v_{BD} + v_{DE} + v_{EC}$$

$$v_{AC} = 6 + (-3) + 1 + 4$$

yielding $v_{AC} = 8V.$

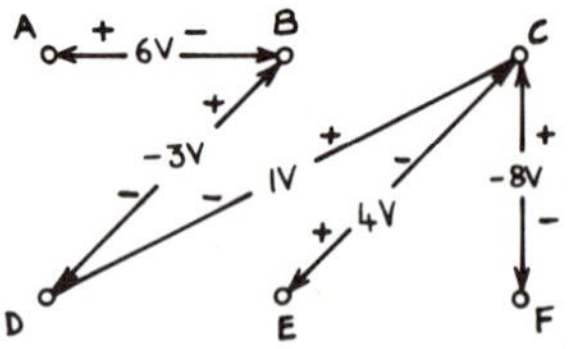

Fig. 2

To solve for v_{AD} write,

$$v_{AD} = v_{AB} + v_{BD}$$

$$v_{AD} = 6 + (-3)$$

yielding $v_{AD} = 3V.$

To find v_{AE} write,

$$v_{AE} = v_{AB} + v_{BD} + v_{DE}$$

$$v_{AE} = 6 + (-3) + 1$$

yielding $v_{AE} = 4V.$

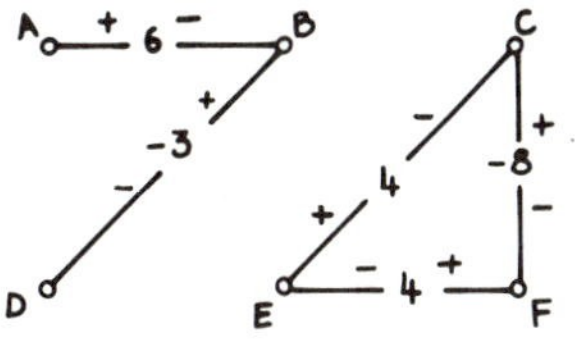

FIG.3

To find v_{AF}:

$$v_{AF} = v_{AB} + v_{BD} + v_{DE} + v_{EC} + v_{CF}$$

$$v_{AF} = 6 + (-3) + 1 + 4 + (-8)$$

yielding $v_{AF} = 0.$

(b) $v_{AC} = 6 + (-3) - 1 = 2V$

$$v_{AD} = 6 + (-3) = 3V$$

$$v_{AE} = 6 + (-3) - 1 - 4 = -2V$$

$$v_{AF} = 6 + (-3) - 1 - 8 = -6V.$$

(c)

Only one loop can be formed to include the unknown voltages and that loop is

$$v_{AD} = v_{AB} + v_{BD} = 6 + (-3) = 3V.$$

• PROBLEM 1-20

Determine v_x in each of the circuits of Fig. 1

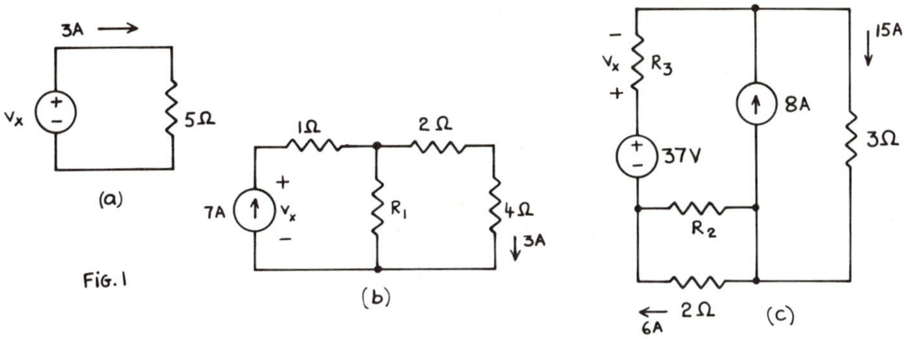

Solution: (a) Apply Kirchhoff's Voltage Law to establish the sum of the voltage drops around a loop. The sum of the voltage drops around a loop must be equal to the sum of the sources around that loop.

In Fig. 1(a) the voltage drop across the 5Ω resistor is $3 \cdot 5 = 15V$. Therefore, v_x must be 15V.

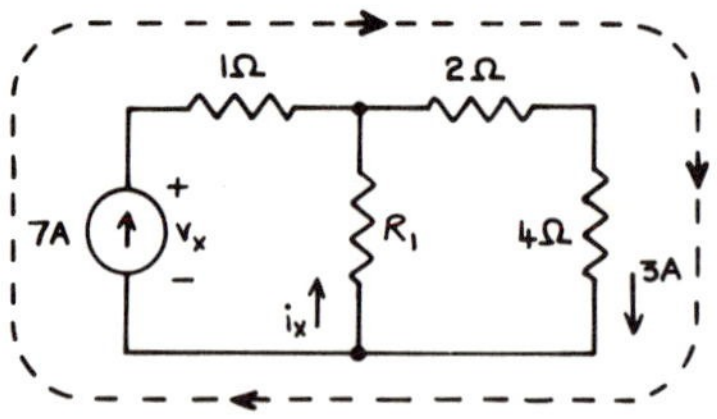

FiG.2

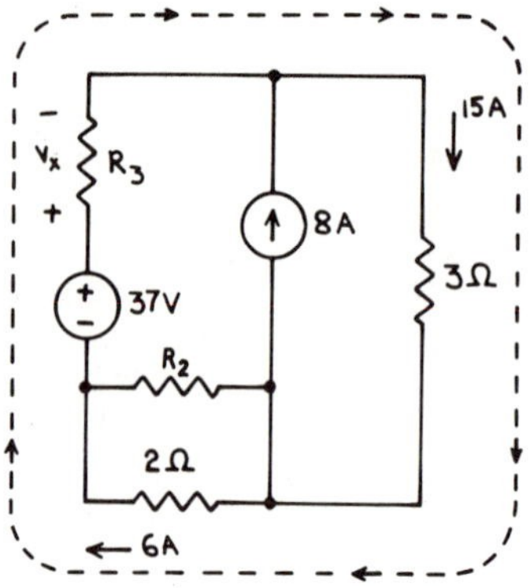

FiG.3

(b) Fig. 2 shows the loop path taken for this problem.

$$\text{The sum of the voltage drops} = 7 \cdot 1 + 3(2 + 4)$$

$$= 25.$$

$$\text{The source voltage } v_x = 25V.$$

(c) Fig. 3 illustrates the loop path taken to solve this problem.

$$\text{The sum of the voltage drops} = v_x + 15 \cdot 3 + 2 \cdot 6$$

$$= v_x + 57V$$

$$\text{The voltage source} = 37V.$$

Therefore $$37V = v_x + 57V$$

$$v_x = -20V.$$

Note: There are several methods one can use to apply Kirchhoff's laws. In all methods, one must pay careful attention to the signs of voltage sources and current directions.

• PROBLEM 1-21

A series loop contains the following circuit elements in order: a 6-V source, a 2-kΩ resistor, a 3-kΩ resistor, an 18-V source, and a 7-kΩ resistor. The voltage sources add.
(a) Find the magnitude of the voltage across each resistor.
(b) Determine the power absorbed by each element.

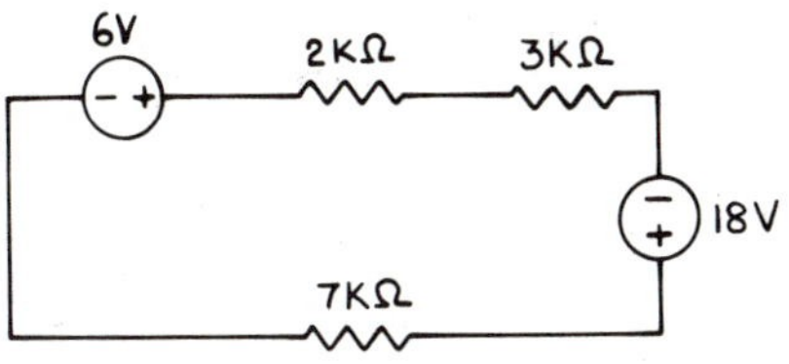

FiG. 1

Solution: Begin by drawing the circuit diagram for the problem. This is shown in Fig. 1.

Voltage sources add as you proceed around the loop.

(a) To find the voltage across each resistor, write a KVL equation around the loop which enables us to find the current in the loop. We assume a clockwise current i.

$$6V + 18V = i(2 + 3 + 7) \times 10^3$$

$$i = \frac{24}{12 \times 10^3} = 2 \times 10^{-3} \text{ A} = 2 \text{ mA}$$

Thus,

$$v_{2k\Omega} = iR$$

$$= (2 \times 10^{-3})(2 \times 10^3) = 4V$$

$$v_{3k\Omega} = (2 \times 10^{-3})(3 \times 10^3) = 6V$$

$$v_{7k\Omega} = (2 \times 10^{-3})(7 \times 10^3) = 14V.$$

(b) The power absorbed by each element would be p = vi.

Consider that the power absorbed by the two sources is negative.

Thus, for the power in each element p = vi

$$p_{6 \text{ volt source}} = -(6)(2 \text{ mA}) = -12 \text{ mW}$$

$$p_{2k\Omega} = (4)(2 \text{ mA}) = 8 \text{ mW}$$

$$p_{3k\Omega} = (6)(2 \text{ mA}) = 12 \text{ mW}$$

$$p_{18 \text{ volt source}} = -(18)(2 \text{ mA}) = -36 \text{ mW}$$

$$p_{7k\Omega} = (14)(2 \text{ mA}) = 28 \text{ mW}.$$

• PROBLEM 1-22

For the circuit shown in Fig. 1, find the power: (a) delivered to the 5-KΩ resistor; (b) supplied by the -25-V source; (c) supplied by the 10-V source.

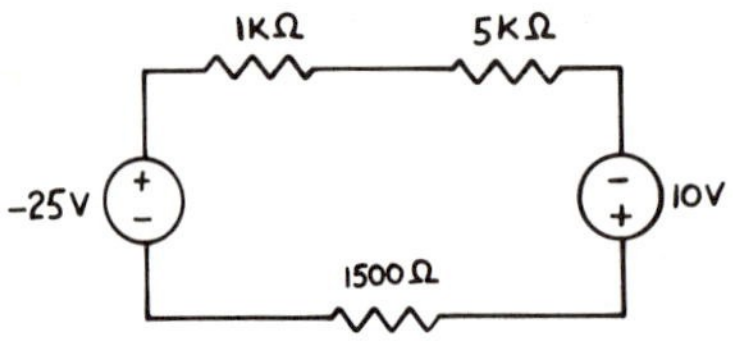

Fig. 1

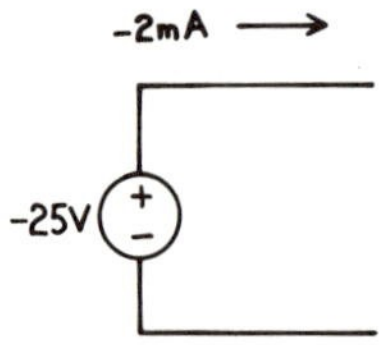

Fig. 2

Solution: Write a KVL equation for the single loop. Assume the current i flows in a clockwise direction.

$$-25V + 10V = i\,(1k\Omega + 5k\Omega + 1500\Omega).$$

Solve for i

$$i = \frac{-25 + 10}{(1 + 5 + 1.5) \times 10^3}$$

$$i = \frac{-15}{7.5 \times 10^3} = -2.0 \times 10^{-3}A.$$

$$i = -2\ mA.$$

(a) The power delivered to the 5kΩ resistor would be

$$p = i^2R$$

$$p = (-2 \times 10^{-3})^2\, 5000$$

$$p = (4 \times 10^{-6})5000 = 20\ mW.$$

(b) In order for a voltage to supply (positive) power, a (positive) current must flow from the positive terminal.

Fig. 2 shows the current and voltage for the - 25 V supply.

Fig. 3

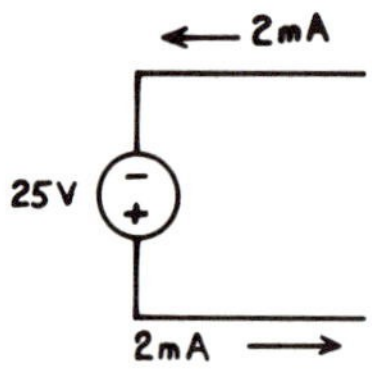

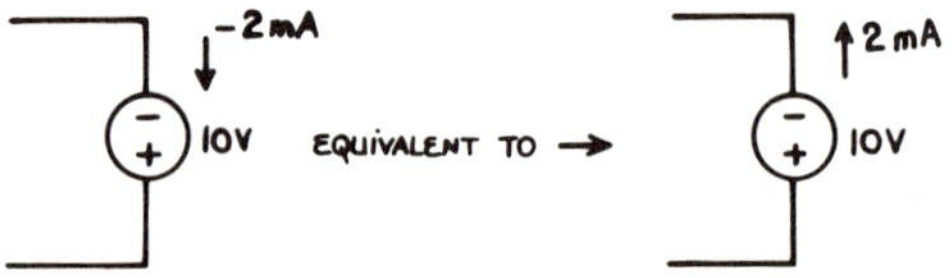

Fig. 4

Reverse the signs for the voltage source and change it to + 25V, reverse the arrow for the current, and change it to + 2 mA. The resulting diagram in Fig. 3 demonstrates that the power supplied is

$$(+2\ mA)(+25V) = 50\ mW.$$

(c) Use the same methods as in part (b).

Power supplied is negative = - (2 mA)(10V)

= - 20 mW.

With reference to the circuit of Fig. 1, find the power absorbed by: (a) the 100-Ω resistor; (b) the 300-V source; (c) the dependent source.

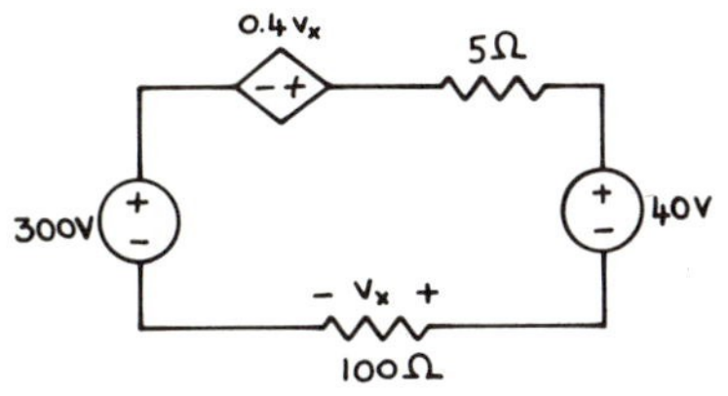

Fig. 1

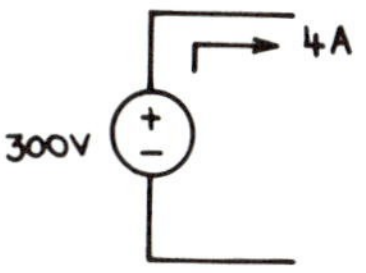

Fig. 2

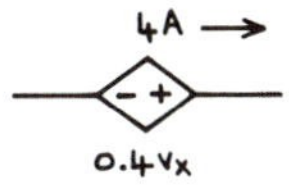

Fig. 3

Solution: Write a KVL equation for the single loop. Assume the current i flows in a clockwise direction.

$$\overbrace{300V + 0.4v_x - 40V}^{\text{voltage sources}} = \overbrace{i(5\Omega + 100\Omega)}^{\text{voltage drops}}$$

Notice that the voltage across the 100-Ω resistor is 100i. This can be substituted for v_x in the above KVL equation.

Thus, $300 + 0.4(100i) - 40 = i\ (5 + 100)$.

Now solve for the single unknown variable i:

$$300 - 40 = i(5 + 100) - (0.4)(100)i$$

$$i\ (5 + 100 - 40) = 300 - 40$$

$$i = \frac{260}{65} = 4A.$$

(a) Power absorbed by the 100-Ω resistor is

$$p = i^2R = (4)^2(100) = 1600\ W.$$

(b) For a source to absorb power (positive) current must flow out of its negative terminal.

For the 300V source, Fig. 2 shows that current flows from the positive terminal. Thus,

$$p = vi = -\ (300)(4) = -\ 1200\ W.$$

(c) Similarly, the dependent source absorbs negative power as shown in Fig. 3. The power is found by

$$p = vi = -\ [0.4\ v_x\ (4)]$$

since $v_x = 100i = 100(4)$

we have $p = -\ [(0.4)(100)(4)(4)] = -\ 640\ W.$

• PROBLEM 1-24

Specify i, v, and the power absorbed by the unknown circuit element in Fig. 1 if the 100-V source supplies (a) 100 W; (b) 300 W.

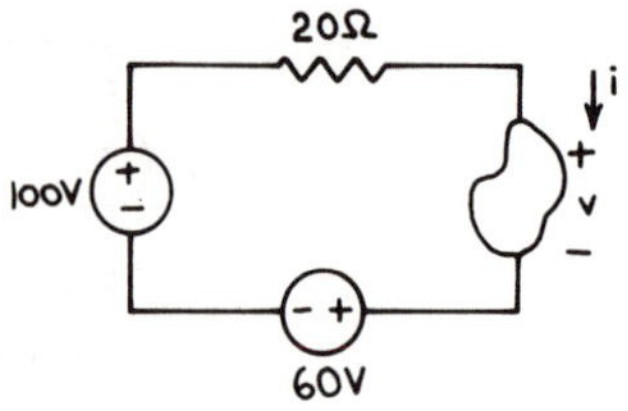

Fig. 1

Solution: **(a)** Consider that the 100-V source supplies 100W. Thus, the positive current through the loop flows clockwise and its value is

$$\frac{p}{v} = \frac{100W}{100V} = 1\ A.$$

Knowing this current makes it possible to write a KVL equation around the loop and solve for v.

$$100V - 60V = (1A)\ (20\Omega) + v$$

$$v = 20V.$$

The power absorbed is positive and is

$$20(1) = 20W.$$

(b) With the 100-V source supplying 300W, it can be seen that the positive current direction is still the same and the current is

$$i = \frac{p}{v} = \frac{300W}{100V} = 3\ A.$$

The KVL equation

$$100V - 60V = (3A)(20\Omega) + v \quad \text{yields}$$

$$v = 100 - 60 - 60 = -\ 20V.$$

The power absorbed now becomes negative

$$-\ (20)(3) = -\ 60W.$$

• PROBLEM 1-25

In the circuit of Fig. 1, what value should v be so that a charge of 600 C is delivered to the 100-V source in 1 min?

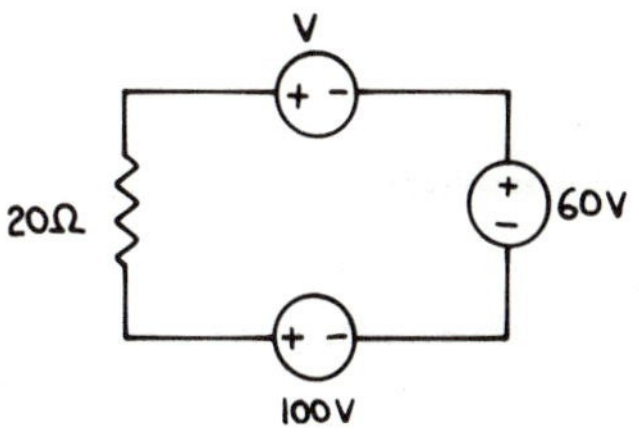

FIG. 1

Solution: In order for 600 C of charge to be delivered to the 100-V source, the current must flow counterclockwise.

To find the current, use the definition

$$i = \frac{dq}{dt} = \frac{600\ C}{1\ min \times 60\ s} = 10\ A.$$

Now apply a KVL equation for the loop to solve for v.

$$v + 60 - 100 = 10(20)$$

$$v = 200 + 100 - 60 = 240V.$$

• PROBLEM 1-26

(a) In the circuit shown in Fig. 1, find i_1. (b) Replace the dependent source by a 2-kΩ resistor and again find i_v.

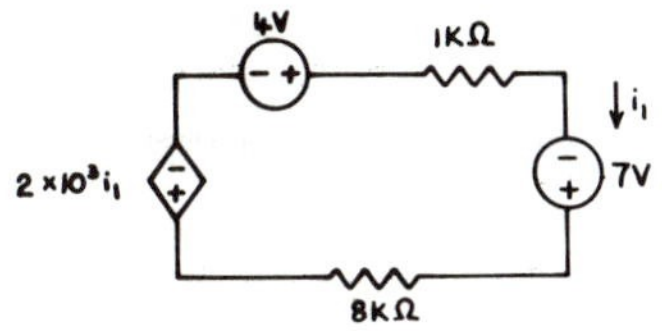

FIG. 1

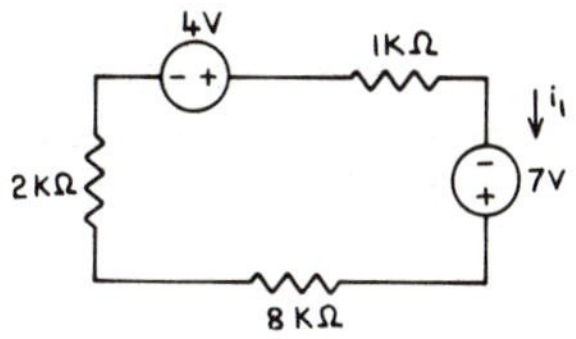

FIG. 2

Solution: (a) To accomplish this, write a KVL equation around the loop in Fig. 1.

$$4V + 7V - (2 \times 10^3\ i_1) = i_1(1 + 8) \times 10^3$$

$$11V = i(11 \times 10^3)$$

$$i = \frac{11}{11 \times 10^3} = 1\ mA.$$

(b)

Write a KVL equation around the loop in Fig. 2.

$$4V + 7V = i_1(1 + 8 + 2) \times 10^3$$

$$11V = i(11 \times 10^3)$$

$$i = \frac{11}{11 \times 10^3} = 1\ mA\ .$$

• PROBLEM 1-27

In the circuit shown in Fig. 1 find i_1.

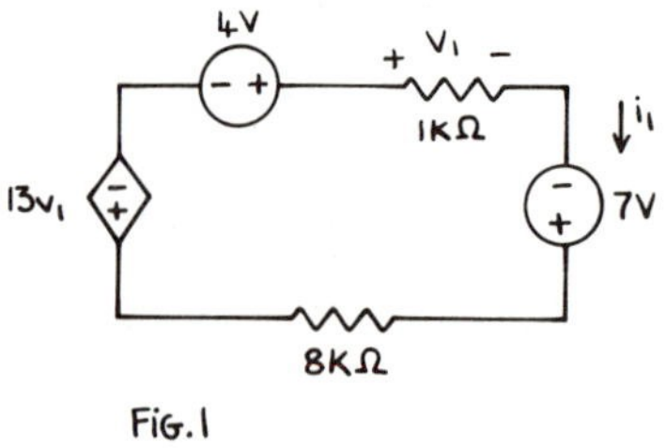

Fig.1

Solution: Writing a KVL equation for the loop, we obtain

$$4V + 7V - 13v_1 = i_1(1 + 8) \times 10^3$$

substitute $\qquad v_1 = 1 \times 10^3\, i_1$

$$4 + 7 - 13 \times 10^3\, i_1 = 9 \times 10^3\, i_1$$

$$11 = 22 \times 10^3\, i_1$$

$$i_1 = \frac{11}{22 \times 10^3} = 0.5 \times 10^{-3}\ A = 0.5\ mA.$$

• PROBLEM 1-28

Find the power delivered to each of the 10-Ω resistors in the circuit of Fig. 1.

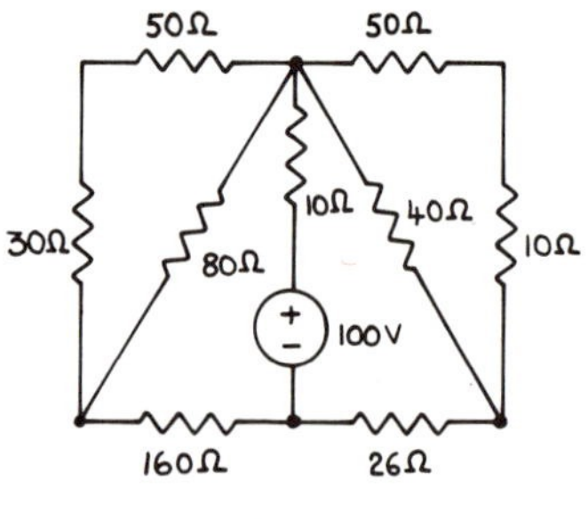

Fig.1

Solution By combining resistances, it is possible to reduce the circuit in Fig. 1 to a single loop circuit. Now the voltage drop for the center 10-Ω resistor can be found.

Combining on the left side yields

Series Resistor Equivalent,

$$R_1 = (50 + 30)\Omega = 80\Omega$$

$$R_2 = (50 + 10)\Omega = 60\Omega$$

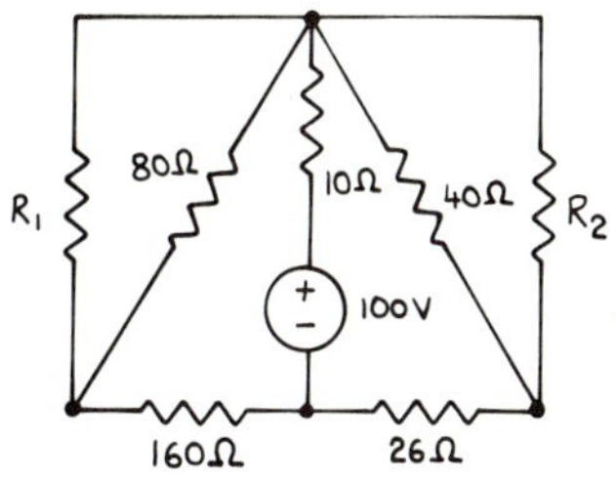

Fig. 2

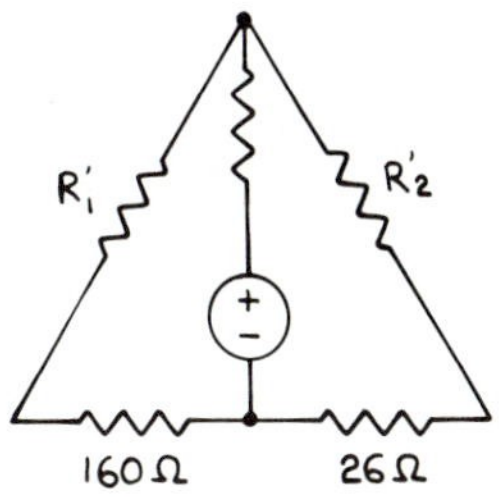

Fig. 3

Parallel Resistor Equivalent,

$$R_1' = \frac{(80)(80)}{80 + 80}\Omega = \frac{(80)^2}{2(80)} = 40\Omega$$

$$R_2' = \frac{40(60)}{40 + 60} = \frac{24 \times 10^2}{100}$$

$$= 24\Omega$$

Combining series resistors we get,

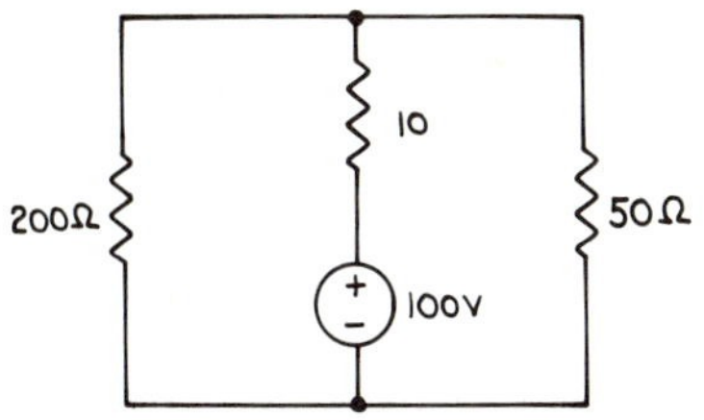

Fig. 4

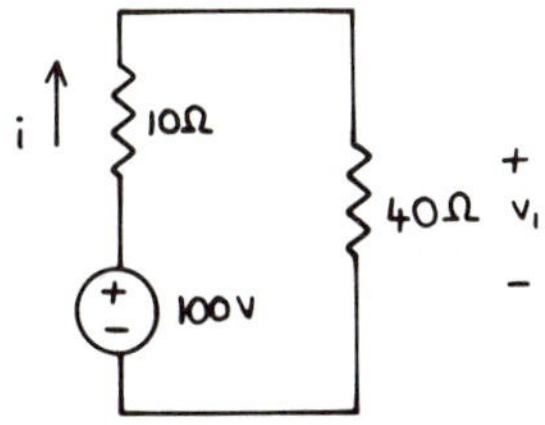

Fig. 5

The 200-Ω and the 50-Ω resistors are in parallel and are combined to obtain the single loop circuit of Fig. 5.

The KVL equation for this loop

$$100 = i(10 + 40)$$

gives $$i = \frac{100}{50} = 2A$$

Power absorbed by the center 10-Ω resistor is,

$p = i^2R = (2)^2\ 10 = 40W.$

The voltage drop v_1 in Fig. 5,

$v_1 = i\ R = (2)(40) = 80V,$

is the voltage across the 50-Ω resistor of Fig. 4, which is in turn the voltage across the left-half of the circuit in Fig. 4.

From Fig. 4

$i_1 = \frac{V}{R} = \frac{80}{50} = 1.6A,$

the current through the left side of the circuit is known (see Fig. 6). Now the voltage drop across the 50 and 10-Ω resistor combination in Fig. 7 can be found.

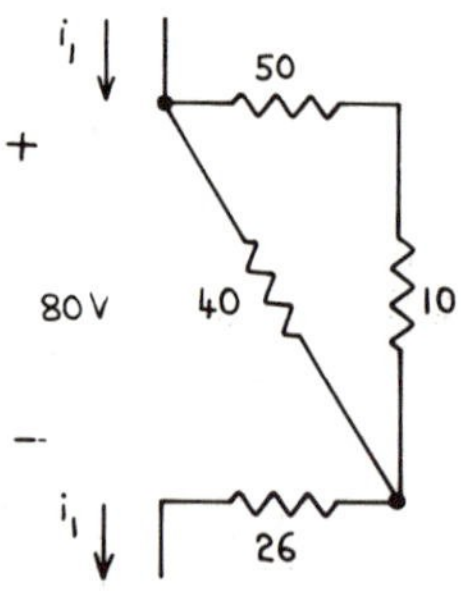

FiG. 6

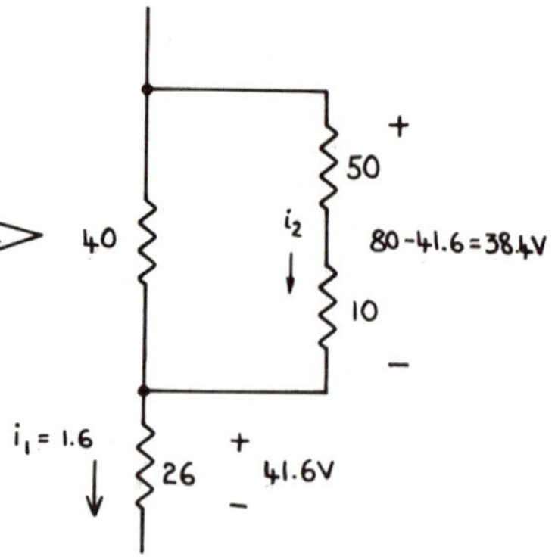

FiG. 7

It follows that the current i_2 is

$$i_2 = \frac{V}{R} = \frac{38.4}{50 + 10} = 0.64A$$

and the power absorbed by the 10-Ω resistor is

$$p = i^2 R = (0.64)^2 \, 10 = 4.096W.$$

• PROBLEM 1-29

By combining resistances in the circuit of Fig. 1, find the power supplied by the source and also the power absorbed by the 900-Ω resistor.

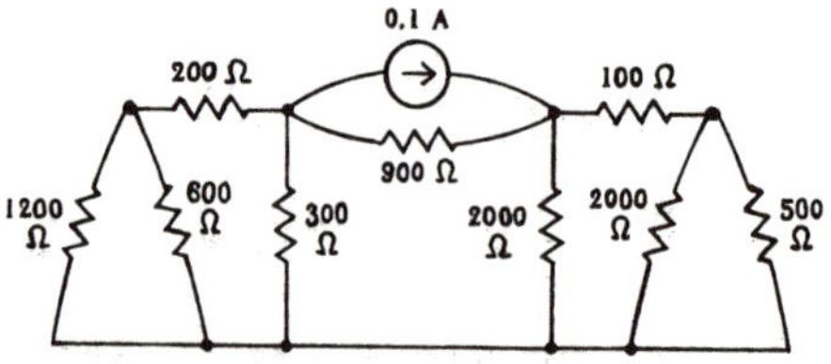

FiG. 1

Solution: Figs. 2 - 5 demonstrate the process of resistor combination.

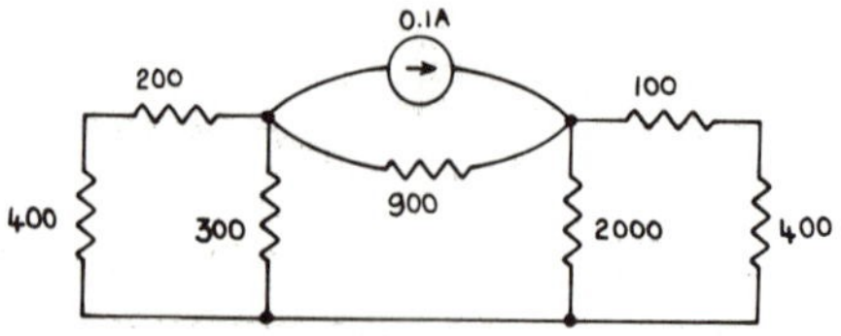

FiG. 2

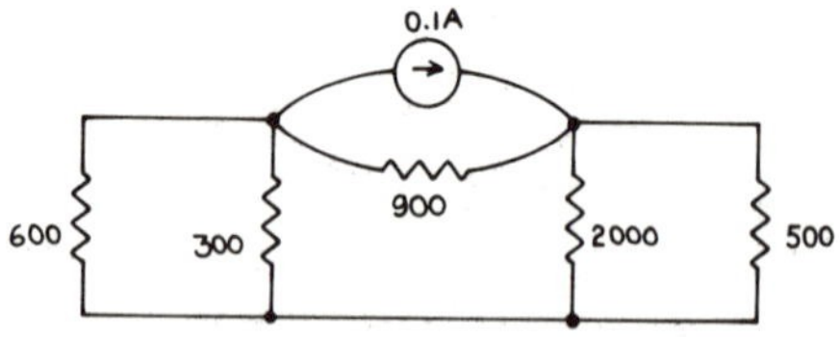

FiG. 3

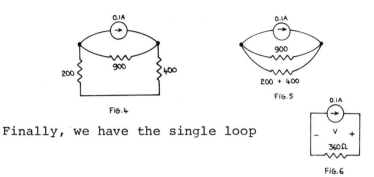

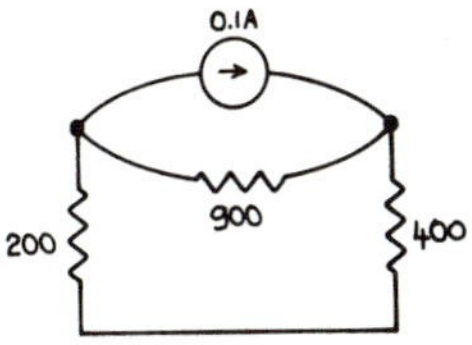

Fig.4

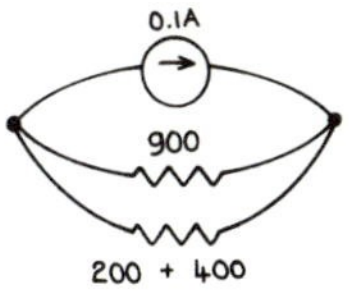

Fig.5

Finally, we have the single loop

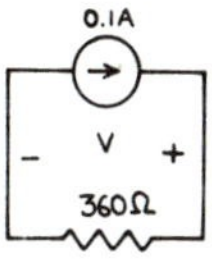

Fig.6

where $v = (0.1)(360) = 36V$. It follows that the power supplied by the current source is

$$p = vi = 36(0.1) = 3.6W.$$

The power absorbed by the 900-Ω resistor is

$$p = \frac{v^2}{R} = \frac{(36)^2}{400} = 1.44W.$$

• PROBLEM 1-30

> Find v_x in the circuits of Fig. 1a, b, and c.

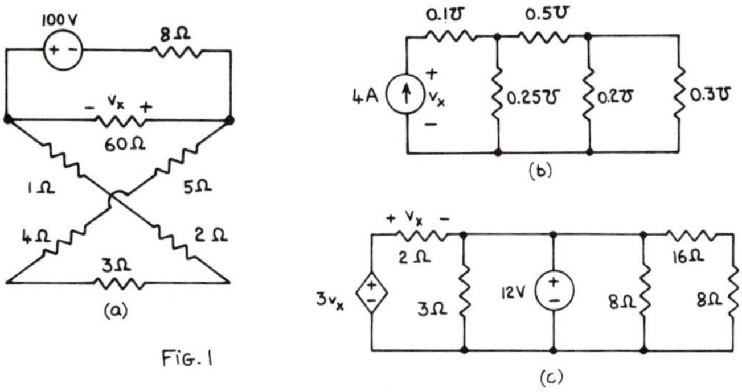

Fig. 1

Solution: (a) In order to find v_x, combine resistors to form a single loop circuit.

The circuit of Fig. 1a reduces to the circuit shown in Fig. 3.

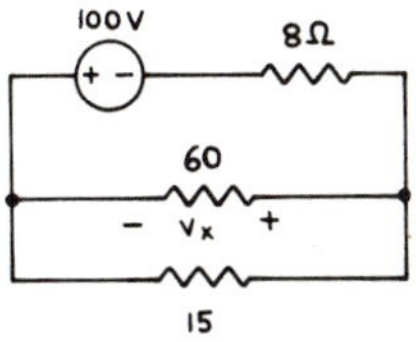

Fig.2

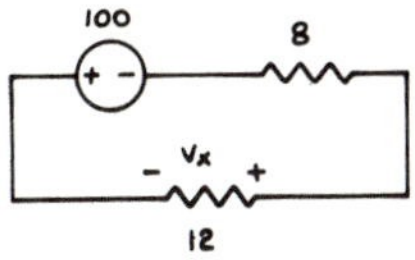

Fig.3

Write a KVL equation for the single loop circuit. Assume that i flows counterclockwise.

$$100 = i(8 + 12)$$

$$i = 5A.$$

v_x voltage sense is opposite that of the current. Thus $v_x = -12(5) = -60V$.

(b) To find v_x, combine conductances to form a single loop circuit.

The circuit of Fig. 1b reduces to

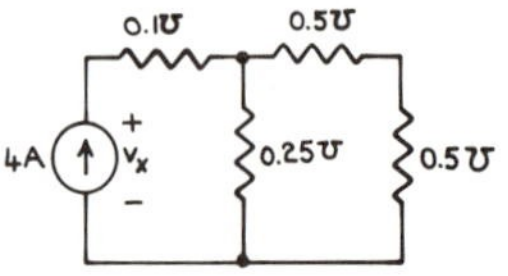

Fig. 4

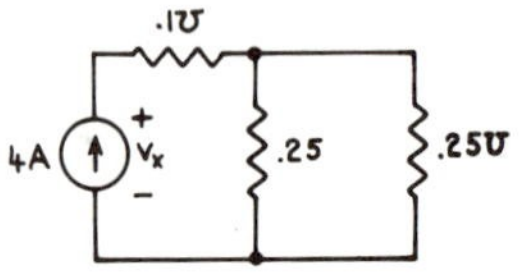

Fig. 5

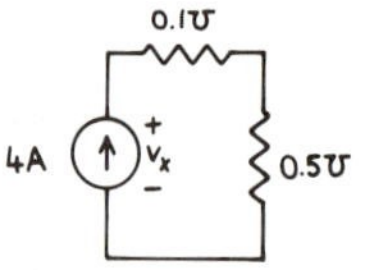

Fig. 6

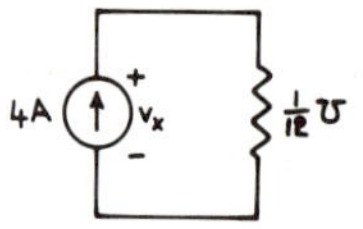

Fig. 7

Write a KVL equation for the single loop circuit in Fig. 7.

$$v_x = \frac{4A}{\frac{1}{12}\,℧} = 4 \cdot 12 = 48V$$

(c) Combine resistances in Fig. 1c to form a simple circuit.

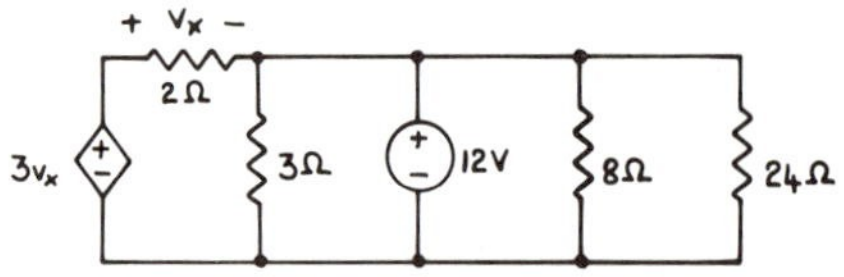

Fig. 8

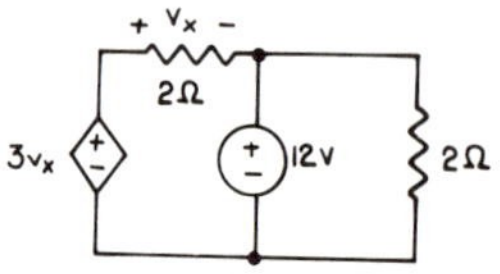

Fig. 9

In Fig. 8 the 3, 8 and 24Ω resistors are in parallel and can be combined to form the new circuit in Fig. 9.

Observing this simplified circuit, notice that the voltage across the rightmost 2-Ω resistor must be 12V, since a 12V voltage source is connected across it. Thus a single loop circuit for the purpose of finding v_x can be formed.

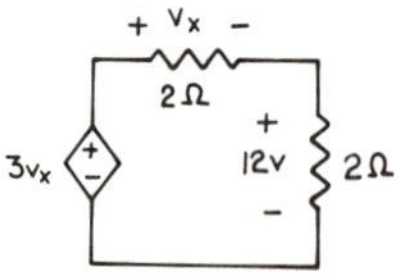

Fig. 10

A single loop KVL equation can be written for the circuit in Fig. 10, if we assume the current flows clockwise.

$$3v_x = i(2\Omega) + 12V$$

since $\quad v_x = 2i \quad$ we can substitute

$$3(2i) = 2i + 12$$

$$4i = 12$$

$$i = 3A; \qquad v_x = 2(3) = 6V.$$

• PROBLEM 1-31

Find i in the circuits of Fig. 1a, b, and c.

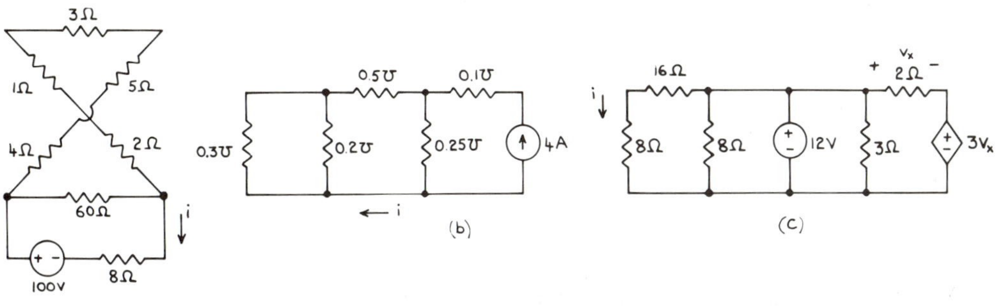

(b) (c)

Fig.1 (a)

Solution: (a) To find the current i, combine resistors to form a single loop circuit. Note that, despite its complicated appearance, the five resistors in the upper part of the circuit are all in series.

$$R_{eq} = \frac{60(15)}{75} = 12\Omega$$

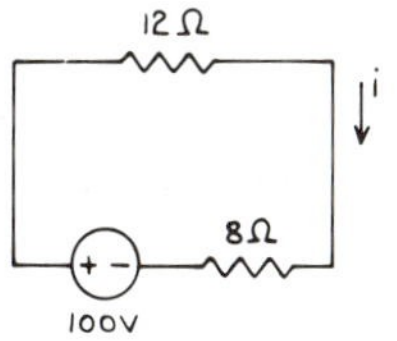

Fig.2

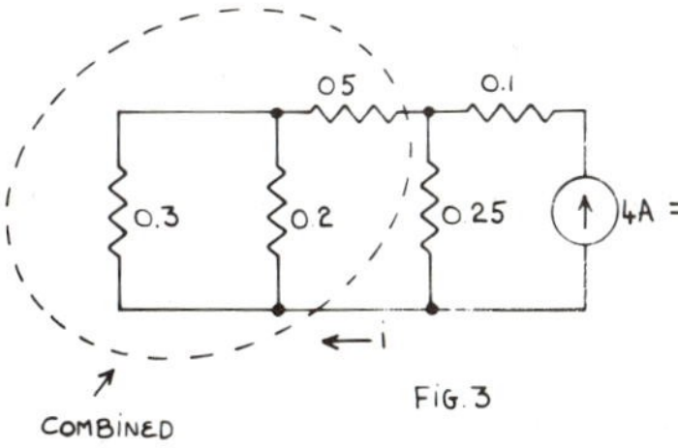

Fig.3

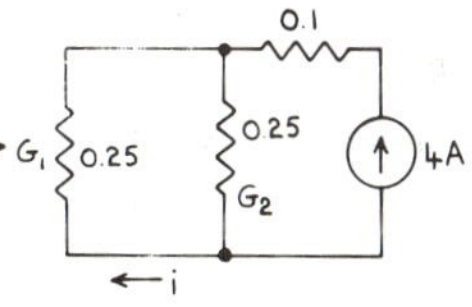

Fig 4

The sum of the resistors in the upper half of the network is in parallel with the 60Ω resistor. Fig. 2 demonstrates the resulting circuit. KVL around the loop

$$100 = i\ (12 + 8)$$

gives $\quad i = \dfrac{100}{20} = 5A.$

(b) Fig. 3 shows the conductances that are combined to

form the circuit in Fig. 4.

To find i, which is the current through the 0.25 conductance G_1 on the left, it is necessary to know the voltage across G_1. Since the voltage across G_1 and G_2 are the same, one can combine them. Remembering that 4A flows through the combined conductance of 0.5 , calculate the voltage.

$$v = \frac{1}{G} = \frac{4}{0.5} = 8V.$$

Fig.5

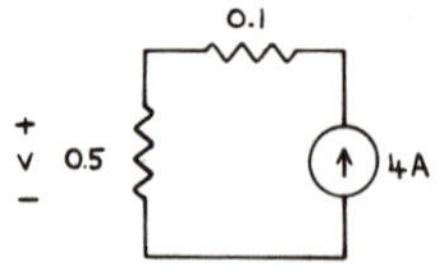

The current sense of i is opposite the convention adopted, thus

$$i = -\ 8(0.25)$$

$$i = -\ 2A.$$

Fig.6

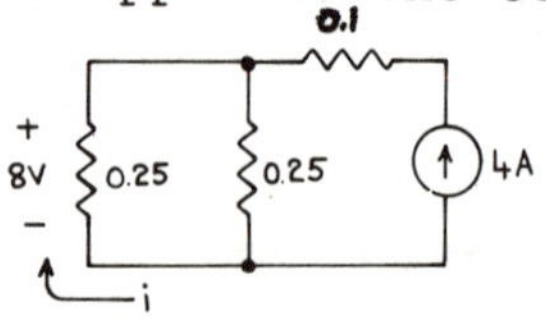

(c) Observe in Fig. 1c that the 12-V source is connected across the 8 and 16Ω series combination on the left.

Hence,

$$i = \frac{V}{R} = \frac{12}{8 + 16} = \frac{12}{24} = 0.5A.$$

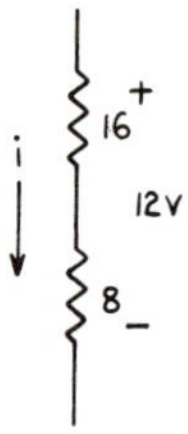

KIRCHHOFF'S CURRENT LAW

• PROBLEM 1-32

A circuit contains four nodes lettered A, B, C, and D. There are six branches, one between each pair of nodes. Let i_{AB} be the current in branch AB directed from node A to node B through the element. Then given i_{AB} = 16 mA, and i_{DA} = 39 mA, find i_{AC}, i_{BD} if i_{CD} = : (a) 23 mA; (b) - 23 mA.

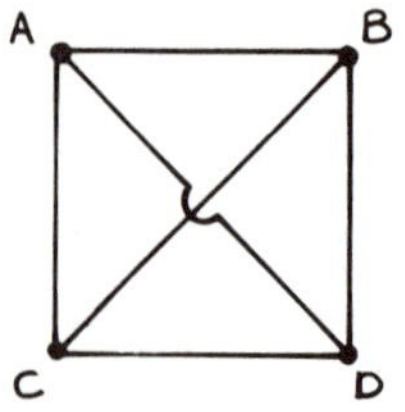

Fig. 1

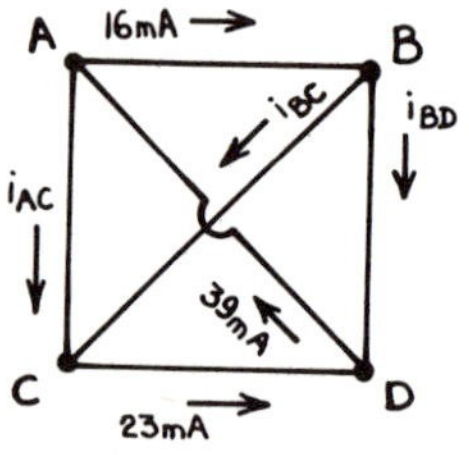

Fig. 2

Solution: Fig. 1 represents the circuit described in this problem.

All possible interconnections of 6 different branches

between the 4 nodes will give the same results.

(a) Fig. 2 shows the current values given and those to be found.

We can proceed to write a KCL equation for either node A or D.

Choosing node A,

$$39 \text{ mA} = i_{AC} + 16 \text{ mA}$$

$$i_{AC} = 23 \text{ mA}.$$

Now an equation for either node C or D can be written.

Choosing node C:

$$i_{AC} + i_{BC} = 23 \text{ mA}$$

$$23 \text{ mA} + i_{BC} = 23 \text{ mA}$$

$$i_{BC} = 0$$

At node B, since $i_{BC} = 0$,

$$i_{BD} = i_{AB} = 16 \text{ mA}$$

(b) If $i_{CD} = -23$ mA, then $i_{DC} = 23$ mA.

The current i_{AC} remains unchanged $i_{AC} = 23$ mA.

At node C we have,

$$i_{AC} + 23 \text{ mA} + i_{BC} = 0$$

$$i_{BC} = -23 - 23 = -46 \text{ mA}$$

At node D we have,

$$i_{BD} = 39 \text{ mA} + 23 \text{ mA}$$

$$i_{BD} = 62 \text{ mA}.$$

• PROBLEM 1-33

If $i_s = 4$ A in the circuit of Fig. 1, calculate in a single step (each) the current, voltage, and power associated with the 3-Ω resistor.

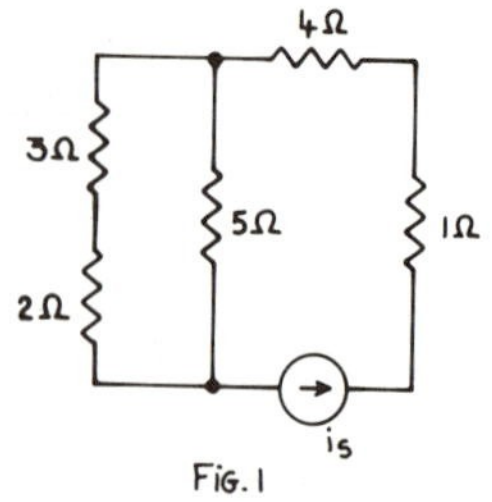

Fig. 1

Solution: When i_s flows through two branches of equal resistance, the current through each branch must be $i_s/2$. The current in the 3-Ω resistor is

$$\frac{i_s}{2} = \frac{4}{2} = 2A. \qquad \text{Voltage,} \quad v = \frac{i_s}{2} R = 2(3) = 6V.$$

Power, $p = vi = 2(6) = 12W.$

• PROBLEM 1-34

Determine i_x in each of the circuits of Fig. 1.

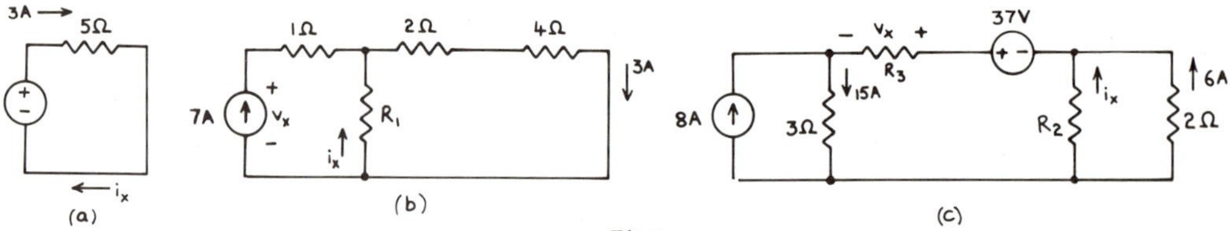

Fig. 1

Solution: (a) The circuit in Fig. 1(a) is a single loop circuit. Hence, the current flowing through the loop is the same everywhere on the loop. Thus we can conclude

$$i_x = 3A.$$

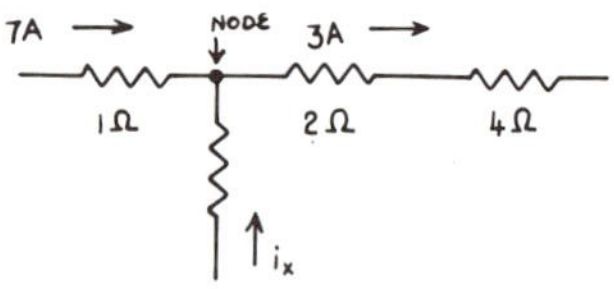

Fig. 2

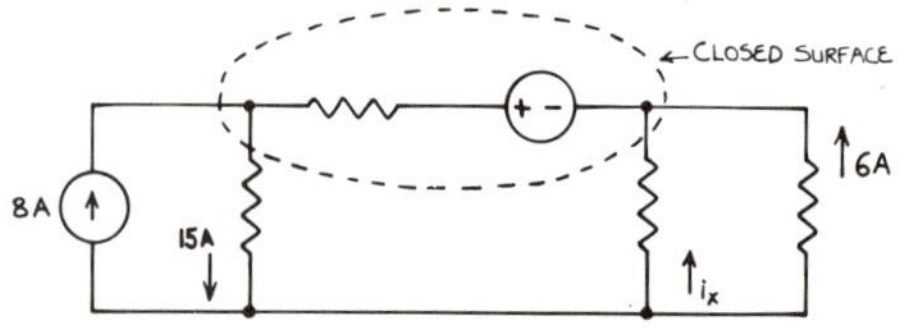

Fig. 3

(b) Use Kirchhoff's Current Law to sum the currents at the node in Fig. 2.

The sum of the currents entering the node must be equal to the sum of the current(s) leaving the node. Thus,

$$7A + i_x = 3A.$$

Solving for i_x we obtain

$i_x = 3A - 7A = -4A.$

(c) Kirchhoff's current law can be extended by summing the currents entering and leaving a closed surface. Fig. 3 shows the closed surface used in this problem.

It follows:

$8A + 6A + i_x = 15A$

solving for i_x,

$i_x = 15A - 8A - 6A = 1A.$

• PROBLEM 1-35

(a) Given $i_x = 1$ A in the circuit of Fig. 1, find i and v_s. (b) Use the results of (a) to determine i_x if $v_s = 70$ V; (c) if $v_s = 100$ V.

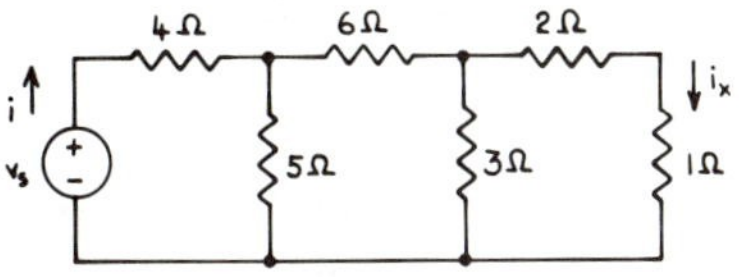

FIG. 1

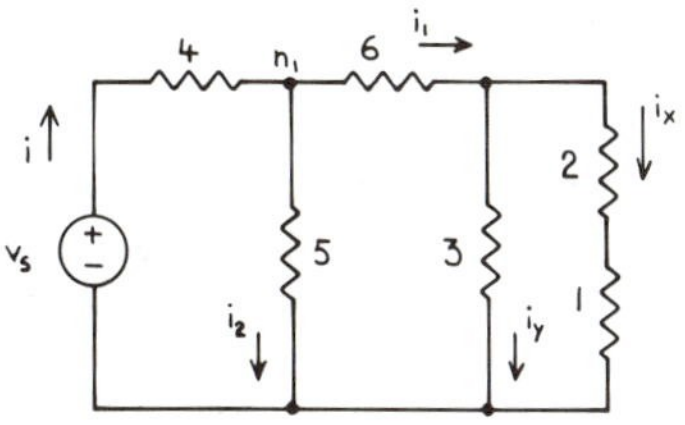

FIG. 2

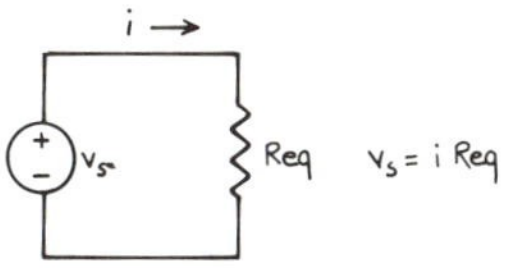

FIG. 3

Solution: Fig. 2 shows that i_x and i_y are equal and that $i_1 = 2i_x = 2i_y$.

$$i_y = i_x$$

$$i_1 = i_y + i_x$$

$$i_1 = 2i_x = 2i_y$$

This yields $i_1 = 2A$, $i_y = 1A$.

We find,

v(across 5-Ω) = v(across 6-Ω) + v(across 3-Ω) giving

$v_5 = i_1\ (6\Omega) + i_y\ (3\Omega) = 2(6) + 1(3) = 15V.$

The current through the 5-Ω resistor is,

$i_2 = \frac{v_5}{R} = \frac{15}{5} = 3A.$

i can now be found by adding the currents at node n_1.

$$i = i_1 + i_2 = 2 + 3 = 5A.$$

To find v_s, the equivalent resistance R_{eq} is calculated.

$$v_s = i\, R_{eq}$$

R_{eq} is calculated to be 7-Ω. From Fig. 3 we have

$$v_s = i\, R_{eq} = (5)(7) = 35V.$$

• PROBLEM 1-36

For the circuit of Fig. 1, find: (a) i_1; (b) i_2; (c) i_3.

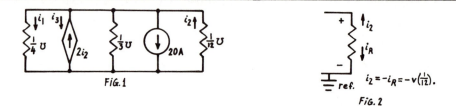

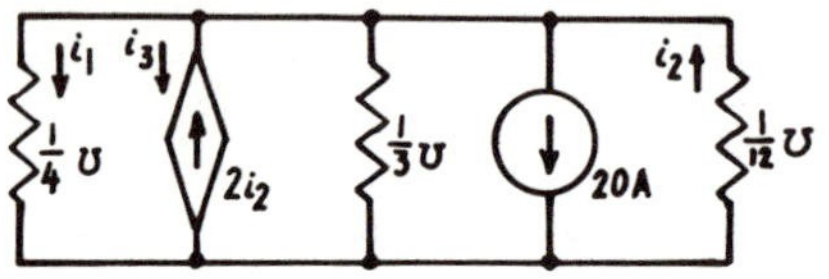

FiG.1

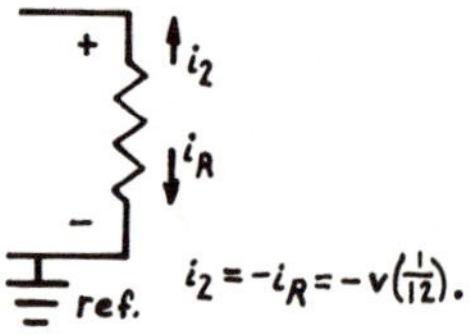

FiG.2

Solution: To find the voltage across all the branches, write a KCL equation for the circuit. Assign the lower node to be the reference node and the upper node to be at voltage v.

$$\overbrace{2i_2 - 20A}^{\text{current sources}} = \overbrace{v\left(\frac{1}{4} + \frac{1}{3} + \frac{1}{12}\right)}^{\text{branch currents}}$$

Notice that i_2 can be written as $-\frac{1}{12}V$. The reason for this is that the current direction for i_2 is opposite the current direction for the passive element as indicated in figure 2. Substituting this in the equation yields,

$$-\frac{2}{12}v - 20A = v\left(\frac{1}{4} + \frac{1}{3} + \frac{1}{12}\right)$$

solving for v

$$v\left(\frac{1}{4} + \frac{1}{3} + \frac{1}{12} + \frac{2}{12}\right) = -20$$

$$v = \frac{-20}{\frac{10}{12}} = \frac{-20(12)}{10} = -24V.$$

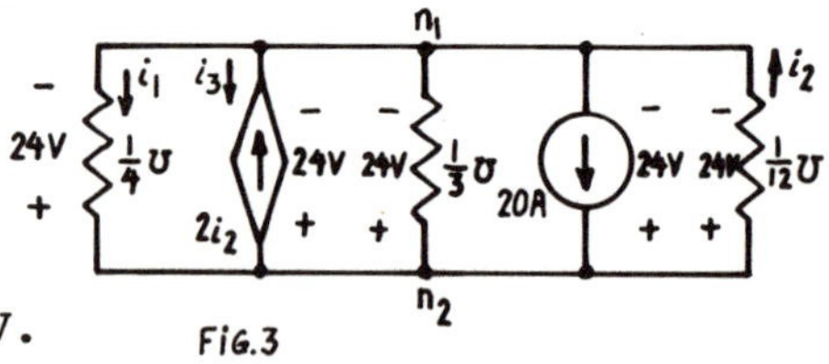

FiG.3

Fig. 3 shows the circuit with all the correct voltages and current directions.

According to the voltage indicated on the circuit, which

was obtained from the calculations above, the positive current direction on this circuit flows from n_2 to n_1. Therefore we will obtain a negative value for i_1 and i_3 and a positive value for i_2. These current values can now be calculated as follows

$$i_1 = vG = -24\left(\frac{1}{4}\right) = -6A$$

$$i_2 = vG = 24\left(\frac{1}{12}\right) = 2A$$

$$i_3 = -2i \;\; = -4A.$$

• PROBLEM 1-37

The following circuit elements are connected between a pair of nodes: a 2-m℧ conductance, a 6-mA source, a 3-m℧ conductance, an 18-mA source, and a 7-m℧ conductance. The current sources are both directed into the upper node.
(a) Find the magnitude of the current through each conductance. (b) Determine the power absorbed by each element.

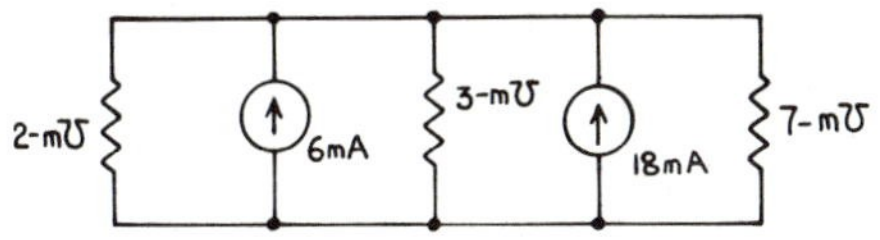

Fig. 1

Solution: Fig. 1 shows that the circuit diagram for the problem

(a) By writing a KCL equation using the lower node as the reference, the voltage between the two nodes can be found.

$$6 \text{ mA} + 18 \text{ mA} = v(2 + 3 + 7) \times 10^{-3}$$

$$24 \times 10^{-3} = 12 \times 10^{-3} \text{ v}$$

$$v = \frac{24 \times 10^{-3}}{12 \times 10^{-3}} = 2V$$

The current through each of the conductances is

$$i = vG.$$

For the 2-m℧ conductance,

$$i = (2)(2 - m℧) = 4 \times 10^{-3} \text{ A} = 4 \text{ mA}$$

3-m℧ conductance,

$$i = (2)(3 - m℧) = 6 \times 10^{-3} \text{ A} = 6 \text{ mA}$$

7-m℧ conductance

$$i = (2)(7 - m℧) = 14 \times 10^{-3} \text{ A} = 14 \text{ mA}.$$

(b) The power absorbed by each element is p = vi. Realizing that the sources absorb negative power, we have, from left to right,

$$p_{2m℧} = (2)\ (4\ \text{mA}) = 8\ \text{mW}$$

$$p_{6\ \text{mA source}} = -\ (2)\ (6\ \text{mA}) = -\ 12\ \text{mW}$$

$$p_{3m℧} = (2)\ (6\ \text{mA}) = 12\ \text{mW}$$

$$p_{18\ \text{mA source}} = -\ (2)(18\ \text{mA}) = -\ 36\ \text{mW}$$

$$p_{7m℧} = (2)(14\ \text{mA}) = 28\ \text{mW}.$$

• PROBLEM 1-38

Specify v, i, and the power absorbed by the unknown circuit element in Fig. 1 if the 0.1-A source supplies (a) 1 W; (b) 3 W.

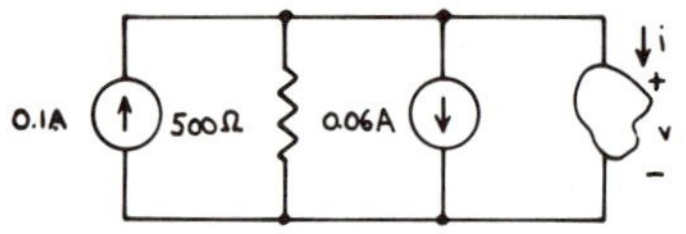

FIG. 1

Solution: (a) The voltage, v, can be found by

$$\frac{p}{i} = \frac{1W}{0.1\ A} = 10V.$$

A KCL equation can be applied to find i.

$$0.1\text{-A} - 0.06\text{-A} = 10\ v\left(\frac{1}{500}\right) + i$$

solving for i we have

$$i = 0.1 - 0.06 - 0.02 = 0.02\ A.$$

The power absorbed is

$$p = vi = (10)(.02) = .2W.$$

(b) We find v,

$$\frac{p}{i} = \frac{3W}{0.1} = 30V.$$

Now our KCL equation becomes

$$0.1\ A - 0.06\ A = 30V\left(\frac{1}{500}\right) + i$$

solving for i we have

$$i = 0.1 - 0.06 - 0.06 = -0.02 \text{ A}.$$

The power absorbed is negative.

$$p = vi = -(30)(.02) = -0.6\text{W}.$$

• PROBLEM 1-39

In the circuit of Fig. 1, find the energy absorbed in 1 hr. by the: (a) 1-mA source; (b) - 2-mA source; (c) 40-$\mu\mho$ conductance.

Solution: Write a KCL equation using the lower node as a reference node.

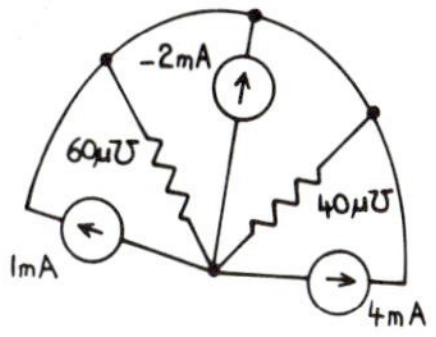

Fig. 1

$$\overbrace{(1 - 2 + 4) \times 10^{-3}}^{\text{current sources}} = \overbrace{v(60\mu\mho + 40\mu\mho)}^{\text{branch currents}}$$

solve for v

$$v = \frac{3 \times 10^{-3}}{100 \times 10^{-6}} = 30\text{V}.$$

(a) In order to find the energy absorbed, find the power and integrate with respect to time, remembering that when energy is absorbed, the current must flow out of the negative side of the source.

For the 1-mA source, note that the current flows out of the positive side. This means the source absorbs negative power and energy.

$$E = \int_{t_0}^{t} p\, dt = -\int_{0}^{3600} (1 \text{ mA})(30\text{v})\, dt = -108 \text{ J}.$$

(b) Fig. 2 illustrates the voltage and current direction for the - 4-mA source,

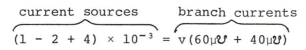

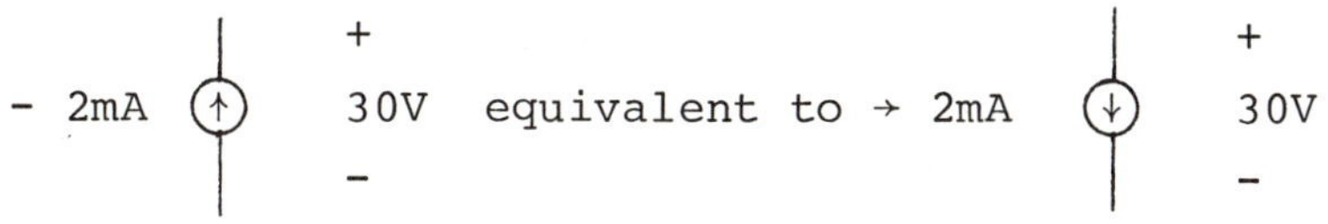

Fig. 2

In this case, power and energy are absorbed.

$$E = \int_{t_0}^{t} p\, dt = \int_{0}^{3600} (2\text{mA})(30)\, dt = 216 \text{ J}.$$

(c) Resistors are passive elements that always absorb power and energy.

$$E = \int_{t_0}^{t} p\,dt = \int_{t_0}^{t} v^2 G\,dt = \int_{0}^{3600} (30)^2 (40\text{-}\mu\mho)\,dt$$

$E = 129.6$ J.

• PROBLEM 1-40

For the circuit shown in Fig. 1, find v, i, and the power absorbed by element X.

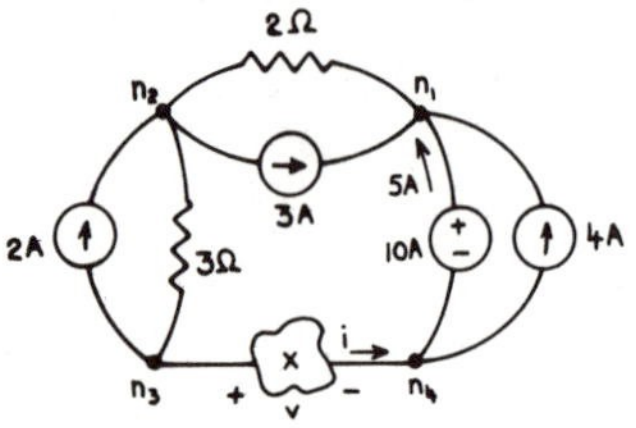

Fig. 1

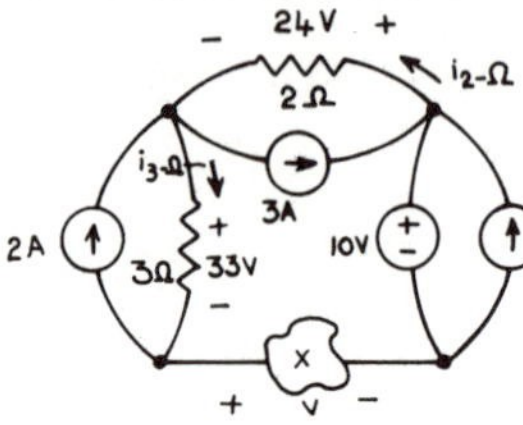

Fig. 2

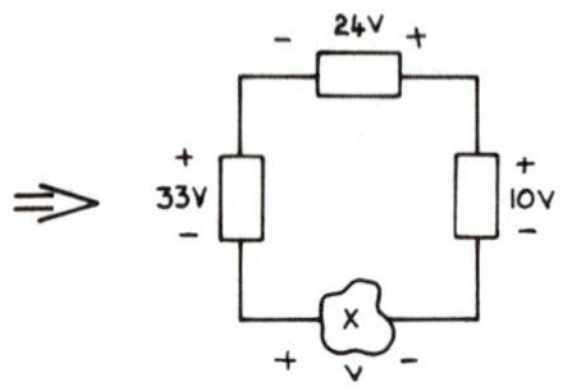

Fig. 3

Solution: To solve for i and v for the element X, the currents and voltages across the 2 and 3 Ω resistors must be found. This can be done by writing KCL equations for nodes n_1 and n_2, respectively.

For node n_1, KCL yields

$3A + 5A + 4A = i_{2-\Omega} = 12A.$

The voltage across the 2Ω resistor is

$12 \cdot 2 = 24V.$

For node n_2, KCL yields

$12A + 2A = 3A + i_{3\Omega}$

$i_{3\Omega} = 12 + 2 - 3 = 11A.$

The voltage across the 3Ω resistor is

$11 \cdot 3 = 33V.$

i can now be found by using a KCL equation for node n_3.

$11A = 2A + i$

$i = 9A$

To solve for v, write a KVL equation for the loop shown in Fig. 2 and 3.

KVL equation

$10 = 24 + 33 + V$

yields $V = 10 - 24 - 33 = -47V$

Observe that the element absorbs negative power or supplies power. Thus

$$p = -47V \cdot 9A = -423\ W.$$

• PROBLEM 1-41

In the circuit of Fig. 1 consider the unknown as a current source. What would be the value of i if an energy of 120 J is delivered to the 0.1-A source in 0.5 min?

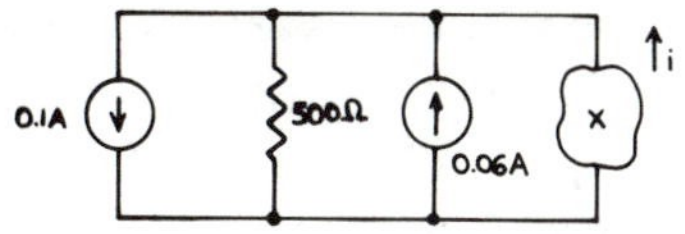

Fig. 1

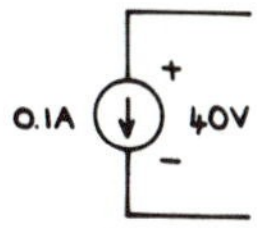

Fig. 2

Solution: To solve for i, it is essential to know the voltage across the branches. Since the energy delivered per time to the 0.1 A source is known, the power and then the voltage can be calculated. This voltage value makes it possible to calculate i by using a KCL equation.

Consider that energy is defined as

$$E = \int^{t} p\ dt = \int^{t} vi\ dt$$

Taking the derivative with respect to time for each side of the equation, yields

$$\frac{dE}{dt} = p = vi$$

Since power is a constant in this problem, we can substitute for dE/dt,

$$\frac{E}{t} = \frac{120\ J}{0.5\ \text{min} \cdot 60\ \text{sec}} = \frac{120}{30} = 4W.$$

The current source this power is delivered to is a 0.1 A current source.

The voltage across the branches is

$$v = \frac{p}{i} = \frac{4W}{0.1\ A} = 40V.$$

The next step in the problem is to determine which node is the positive reference node, the upper node or lower node?

It is known that the 0.1 A source absorbs power. This means the current through the source must enter the positive terminal and leave the negative.

Fig. 2 illustrates that the upper node is the positive reference node.

Write a KCL equation for the pair of nodes. The lower node is ground (reference datum).

$$0.06\text{A} - 0.1\text{ A} + i = 40\left(\frac{1}{500}\right)$$

solving for i

$$i = 0.08 - 0.06 + 0.1 = 0.12\text{ A.}$$

• PROBLEM 1-42

(a) In the circuit of Fig. 1, find v_1. (b) Replace the dependent source by a 500-Ω resistor and again find v_1.

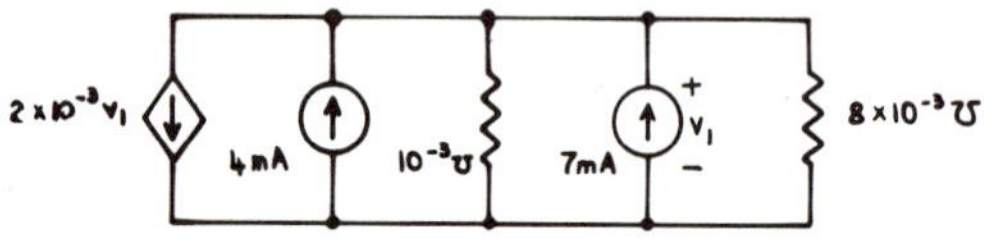

Fig. 1

Solution: (a) Write a KCL equation for the pair of nodes. The lower node is the reference.

$$(4 + 7 - 2v_1) \times 10^{-3}\text{ A} = v_1(1 + 8) \times 10^{-3}\,℧$$

$$11 \times 10^{-3}\, v_1 = 11 \times 10^{-3}$$

$$v_1 = 1\text{V.}$$

(b)

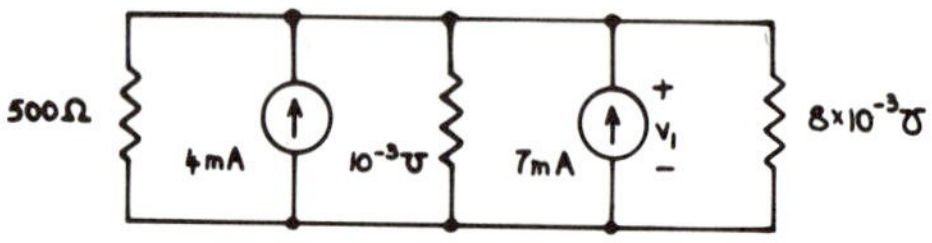

Fig. 2

The KCL equation for this circuit becomes

$$(4 + 7) \times 10^{-3}\text{ A} = v_1\left(\frac{1}{500\Omega} + 1 \times 10^{-3} + 8 \times 10^{-3}\,℧\right)$$

$$11 \times 10^{-3} = v_1(2 + 1 + 8) \times 10^{-3}$$

$$11 \times 10^{-3}\, v_1 = 11 \times 10^{-3}$$

$$v_1 = 1\text{V.}$$

• PROBLEM 1-43

In the circuit shown in Fig. 1 find v_1.

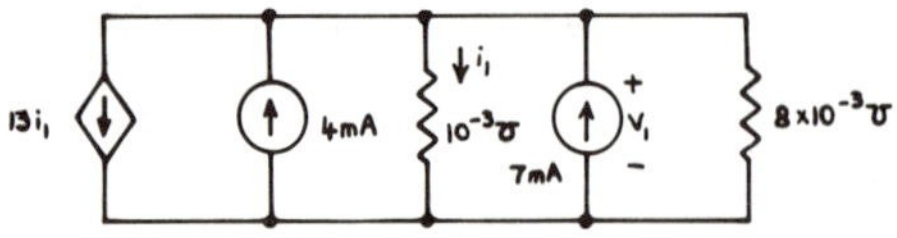

Fig. 1

Solution: Write the KCL equation for the pair of nodes.

$$(4 + 7) \times 10^{-3} - 13i_1 = v_1(1 + 8) \times 10^{-3}$$

Note that $i_1 = v_1 G = v_1(1 \times 10^{-3})\mho$

substituting into the equation and solving for v_1

$$11 \times 10^{-3} - (13 \times 10^{-3})v_1 = v_1(9 \times 10^{-3})$$

$$22 \times 10^{-3} v_1 = 11 \times 10^{-3}$$

$$v_1 = \frac{11 \times 10^{-3}}{22 \times 10^{-3}} = 0.5\text{V}.$$

• PROBLEM 1-44

Analyze the circuit in the figure below using Kirchoff's Rules. Find the magnitude and direction of I_2, the current in the upper 2Ω resistor.

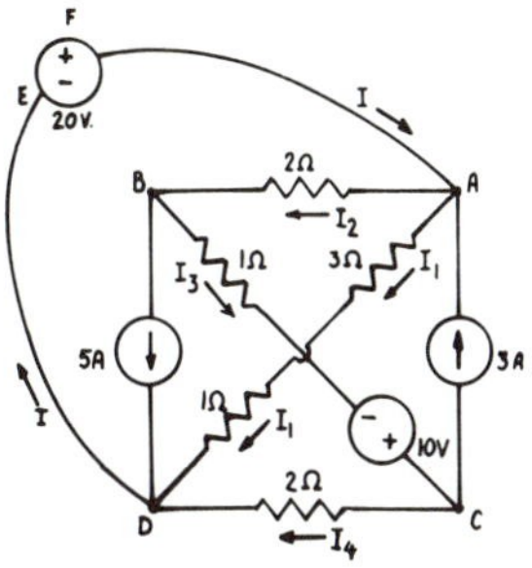

Solution: The circuit contains two constant voltage sources and two constant current sources. It has four junctions where three or more wires join, labelled A, B, C, and D, as shown. To begin, we label the unknown currents I, I_1, I_2, I_3, and I_4, as indicated in the figure.

This problem has five unknowns (the five unknown currents) and, therefore, requires five independent equations. Three independent equations are obtained by applying Kirchoff's Current Sum rule at any three of the four junctions. For convenience, we ignore units and choose junctions A, B, and C:

at A, $I + 3 = I_2 + 5;$ (1)

at B, $I_2 = I_3 + 5;$ (2)

at C, $I_3 = I_4 + 3.$ (3)

The two additional independent equations required for the solution are obtained by applying Kirchoff's Voltage Sum rule about closed loops in the circuit. As it is difficult to represent the potential drops across the constant current sources, we choose the loops ADEF and ABCDEF:

loop ADEF, $20 = 3I_1 + I_1$, (4)

loop ABCDEF, $20 = 2I_2 + I_3 - 10 + 2I_4$. (5)

From (4), by inspection, $I_1 = 5$ amperes. Multiplying (2) by minus two gives

$$-2I_2 + 2I_3 = -10,$$

and, rewriting (5),

$$2I_2 + I_3 + 2I_4 = 30.$$

The sum of these two equations is

$$3I_3 + 2I_4 = 20. \quad (6)$$

Multiplying (3) by two gives

$$2I_3 - 2I_4 = 6$$

which, when summed with (6), results in

$$5I_3 = 26$$

or, $I_3 = 26/5$ amperes. Substituting this result back into (6) gives

$$3(26/5) + 2I_4 - 20,$$

or $I_4 - 11/5$ amperes.

The remaining two currents are found by substituting these results, first into (2) and then into (1):

$$I_2 = (26/5) + 5,$$

$$I = I_2 + 2.$$

Therefore, the current in the upper 2Ω resistor has the magnitude $I_2 = 51/5$ amperes, and flows in the direction of the arrow shown in the figure. The total current flowing in the circuit is $I = 61/5$ ampere.

CHAPTER 2

BASIC CIRCUIT ANALYSIS METHODS

NODAL METHOD

• PROBLEM 2-1

Use Kirchhoff's current law to write an integrodifferential equation for v(t) for the circuit shown.

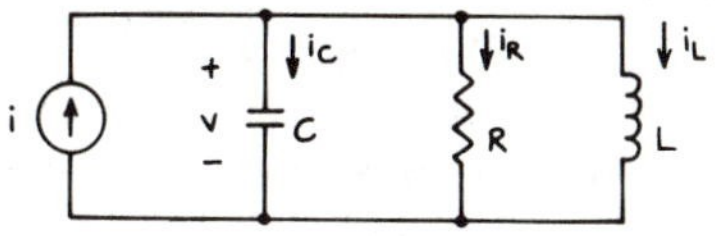

Solution: Kirchhoff's current law applied to the upper node of the circuit yields the equation

$$i = i_C + i_R + i_L$$

The currents for each element i_C, i_R and i_L can be expressed in terms of the same voltage v,

$$i_C = C\frac{dv}{dt}, \qquad i_R = \frac{1}{R}v, \qquad i_L = \frac{1}{L}\int_{-\infty}^{t} v\, d\tau$$

Substituting these terms into the KVL equation, yields the required integrodifferential equation.

$$i = C\frac{dv}{dt} + \frac{1}{R}v + \frac{1}{L}\int_{-\infty}^{t} v\, d\tau.$$

• PROBLEM 2-2

In the network of Fig. 1, write a set of linearly independent integrodifferential equations that can be solved for node voltages v_1 and v_2.

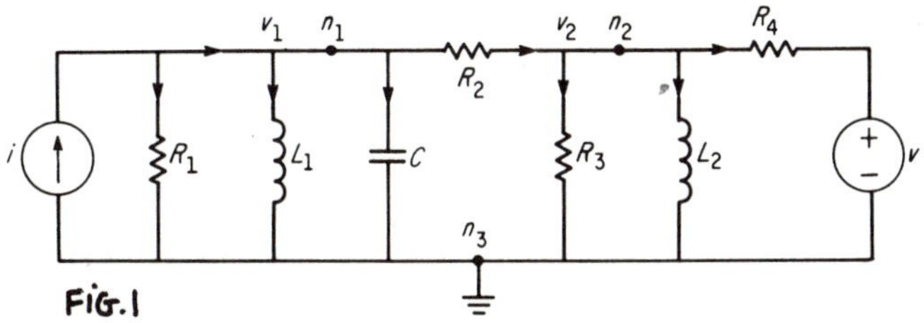

FIG. 1

Solution: Choosing n_3 as the reference node write a KCL equation for node n_1

$$n_1: \; i = \frac{v_1}{R_1} + \frac{1}{L_1}\int_{t_0}^{t} v_1 \; (\tau)d\tau + C\frac{dv_1}{dt} + \frac{v_1 - v_2}{R_2} \; .$$

For node n_2

$$n_2: \; 0 = \frac{v_2 - v_1}{R_2} + \frac{v}{R_3} + \frac{1}{L_2}\int_{t_0}^{t} v_2 \; (\tau)d\tau + \frac{v_2 - v}{R_4} \; .$$

These equations completely describe the network. All other information can be derived from these equations.

• PROBLEM 2-3

In Fig. 1, let circuit element X be a 2-℧ conductance. Find the power supplied by the 10-A source.

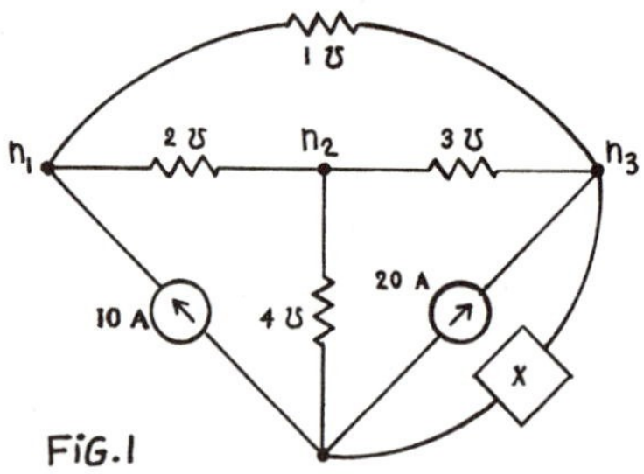

FiG.1

Solution: Labeling the bottom node the reference node, and writing KCL equations for each of the other three nodes produces 3 equations and 3 unknowns.

$$n_1: \; 10 = (v_1 - v_2)\,2 + (v_1 - v_3)\,1$$

$$n_2: \; 0 = (v_2 - v_3)\,3 + (v_2 - v_1)\,2 + v_2\,(4)$$

$$n_3: \; 20 = (v_3)\,2 + (v_3 - v_2)\,3 + (v_3 - v_1)\,1$$

$$10 = 3v_1 - 2v_2 - 1v_3$$

$$0 = -\,2v_1 + 9v_2 - 3v_3$$

$$20 = -\,1v_1 - 3v_2 + 6v_3 .$$

Write the determinate Δ

$$\Delta = \begin{vmatrix} 3 & -2 & -1 \\ -2 & 9 & -3 \\ -1 & -3 & 6 \end{vmatrix}$$

$$\Delta = 3\,(54 - 9) - (-\,2)(-\,12 - 3) + (-\,1)(6 + 9)$$

$\Delta = 3\ (45) - (-2)(-15) + (-1)(15)$

$\Delta = 135 - 30 - 15 = 90.$

Solving for v_1 yields

$$v_1 = \frac{\begin{vmatrix} 10 & -2 & -1 \\ 0 & 9 & -3 \\ 20 & -3 & 6 \end{vmatrix}}{\Delta} = \frac{10\ (45) - (0)(-15) + 20(15)}{90}$$

$$v_1 = \frac{450 + 300}{90} = 8.333\ \text{V}.$$

The power supplied by the 10-A source is

$p = vi = 8.333\ (10) = 83.33\ \text{W}.$

• PROBLEM 2-4

In Fig. 1, find the power supplied by the 10-A current source. Note conductance values are given.

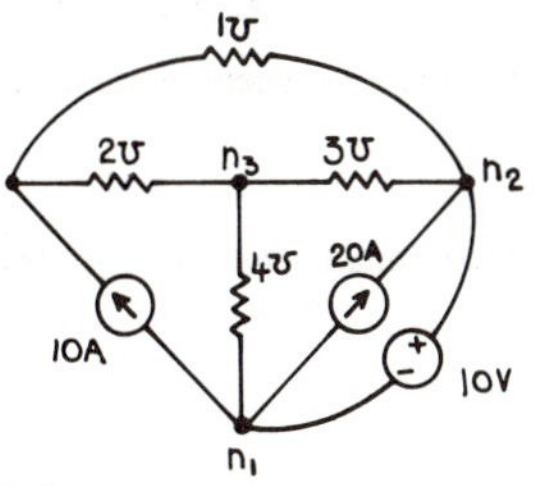

FiG. 1

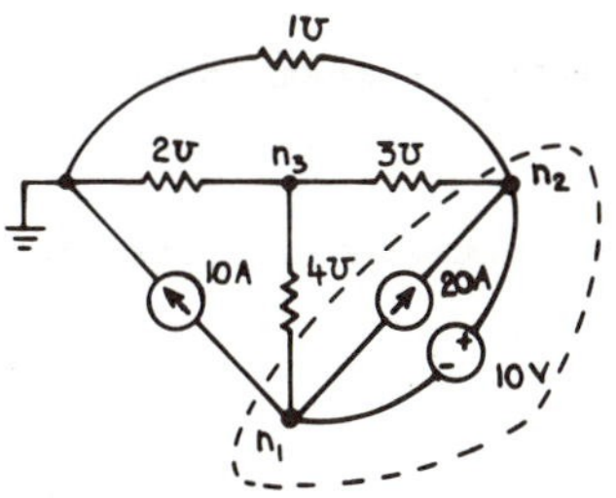

FiG. 2

Solution: Using the left-most node as the reference node and treating nodes n_2 and n_1 as one node, write two KCL equations.

The dotted line in Fig. 2 shows the combining of nodes n_1 and n_2.

The reason for combining nodes n_1 and n_2 into a single node is because

$v_2 - v_1 = 10\text{V},$

the voltage difference between the two nodes. This makes it possible to write only two KCL equations and substitute for v_2 or v_1 from the above ($v_2 - v_1 = 10\text{V}$), expression.

The KCL equations are

n_1: $\quad 0 = (v_3 - v_1)\ 4 + (v_3)\ 2 + (v_3 - v_2)\ 3$

$n_2 + n_1$: $-10 = (v_1 - v_3)\ 4 + (v_2 - v_3)\ 3 + (v_2)\ 1$

Simplifying:

$$0 = -4v_1 - 3v_2 + 9v_3$$

$$-10 = 4v_1 + 4v_2 - 7v_3 .$$

Substituting, $10 + v_1$ for v_2 in the 2 equations above

$$30 = -7v_1 + 9v_3$$

$$-50 = 8v_1 - 7v_3 .$$

Using determinates to solve for v_1

$$v_1 = \frac{\begin{vmatrix} 30 & 9 \\ -50 & -7 \end{vmatrix}}{\begin{vmatrix} -7 & 9 \\ 8 & -7 \end{vmatrix}} = \frac{-210 - (-450)}{49 - 72} = \frac{240}{-23} = -10.43 \text{ V.}$$

The power supplied by the 10-A source is

$$p = vi = (10.43)(10) = 104.3 \text{ W.}$$

• PROBLEM 2-5

In Fig. 1 find v_A.

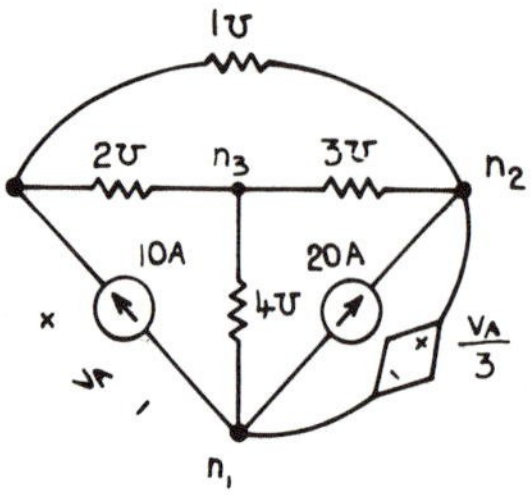

Fig. 1

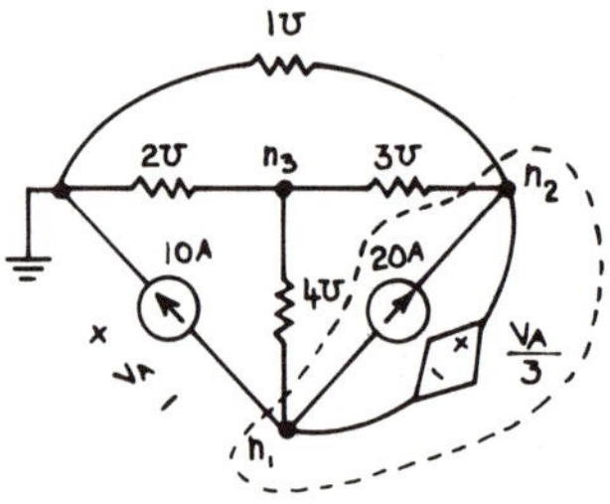

Fig. 2

Solution: Defining the left-most node as the reference node and treating nodes n_2 and n_1 as one node, write two KCL equations. The dotted line in Fig. 2 shows the combining of nodes n_1 and n_2.

First, establish the relationship between nodes n_2 and n_1.

$$v_2 - v_1 = \frac{v_A}{3}$$

Since $v_A = -v_1$ we have

$$v_2 - v_1 = -\frac{v_1}{3}$$

$$v_2 = \frac{2}{3} v_1. \qquad (1)$$

The KCL equations for the two nodes are

n_3: $\quad 0 = (v_3 - v_1) 4 + (v_3) 2 + (v_3 - v_2) 3$

$n_2 + n_1$: $- 10 = (v_1 - v_3) 4 + (v_2 - v_3) 3 + (v_2) 1.$

Simplifying, we obtain

$$0 = - 4v_1 - 3v_2 + 9v_3 \qquad (2)$$

$$- 10 = 4v_1 + 4v_2 - 7v_3 \qquad (3)$$

Substituting Eq. (1) into (2) and (3);

$$0 = - 6v_1 + 9v_3$$

$$- 10 = \frac{20}{3} v_1 - 7v_3.$$

Using determinates to solve for v_1 yields

$$v_1 = \frac{\begin{vmatrix} 0 & 9 \\ -10 & -7 \end{vmatrix}}{\begin{vmatrix} -6 & 9 \\ \frac{20}{3} & -7 \end{vmatrix}} = \frac{-(-90)}{42 - 60} = \frac{90}{-18} = -5V.$$

Since $\quad v_A = - v_1 \quad ; \quad v_A = 5V.$

• PROBLEM 2-6

In Fig. 1 find v_1.

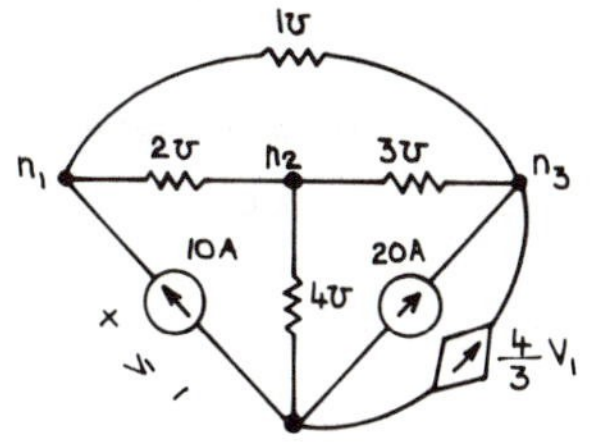

Fig. 1

Solution: Use the bottom node as the reference node and write KCL equations for each of the other nodes.

n_1: $10 = (v_1 - v_2) 2 + (v_1 - v_3) 1$

n_2: $\ 0 = (v_2 - v_3) 3 + (v_2 - v_1) 2 + (v_2) 4$

$$n_3:\ 20 + \frac{4}{3} v_1 = (v_3 - v_2)\,3 + (v_3 - v_1)\,1.$$

Simplifying the above equations yields

$$10 = 3v_1 - 2v_2 - v_3$$

$$0 = -2v_1 + 9v_2 - 3v_3$$

$$20 = -\frac{7}{3} v_1 - 3v_2 + 4v_3.$$

Solving for v_1

$$v_1 = \frac{\begin{vmatrix} 10 & -2 & -1 \\ 0 & 9 & -3 \\ 20 & -3 & 4 \end{vmatrix}}{\begin{vmatrix} 3 & -2 & -1 \\ -2 & 9 & -3 \\ -\frac{7}{3} & -3 & 4 \end{vmatrix}}$$

$$= \frac{10\,(36 - 9) + 20\,(6 - (-9))}{3\,(36 - 9) + 2\,(-8 - 3) - \frac{7}{3}\,(6 - (-9))}$$

$$v_1 = \frac{10\,(27) + 20\,(15)}{3\,(27) + 2\,(-11) - \frac{7}{3}\,(15)}$$

$$= \frac{270 + 300}{81 - 22 - 35} = \frac{570}{24} = 23.75\text{V}.$$

• PROBLEM 2-7

For the circuit shown in Fig. 1: (a) let R = 4 Ω and use nodal methods to find the power supplied by the 7-A source; (b) repeat if R is infinite.

Solution: (a) Taking the upper node as the reference, write three KCL equations

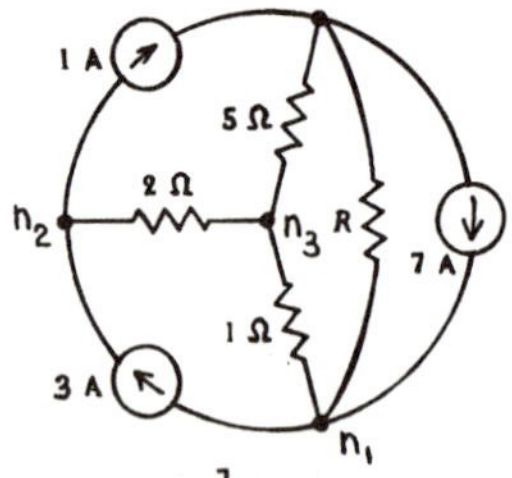

FIG.1

$$n_1:\ 7 - 3 = (v_1)\,\frac{1}{4} + (v_1 - v_3)\,\frac{1}{1}$$

$$n_2:\ 3 - 1 = (v_2 - v_3)\,\frac{1}{2}$$

$$n_3:\ 0 = (v_3)\,\frac{1}{5} + (v_3 - v_2)\,\frac{1}{2} + (v_3 - v_1)\,\frac{1}{1}.$$

Simplifying:

$$4 = \frac{5}{4} v_1 \qquad - v_3$$

$$2 = \frac{1}{2} v_2 - \frac{1}{2} v_3$$

$$0 = - v_1 - \frac{1}{2} v_2 + \frac{17}{10} v_3 .$$

Solving for v_1

$$v_1 = \frac{\begin{vmatrix} 4 & 0 & -1 \\ 2 & \frac{1}{2} & -\frac{1}{2} \\ 0 & -\frac{1}{2} & \frac{17}{10} \end{vmatrix}}{\begin{vmatrix} \frac{5}{4} & 0 & -1 \\ 0 & \frac{1}{2} & -\frac{1}{2} \\ -1 & -\frac{1}{2} & \frac{17}{10} \end{vmatrix}} = \frac{4\left(\frac{17}{20} - \frac{1}{4}\right) - 2\left(\frac{1}{2}\right)}{\frac{5}{4}\left(\frac{17}{20} - \frac{1}{4}\right) - 1\left(\frac{1}{2}\right)}$$

$$v_1 = \frac{4\ (.6) - 2\ (-\ .5)}{1.25\ (.6) - 1\ (.5)} = \frac{3.4}{0.25} = 13.6.$$

The power supplied by the 7-A source is

$p = vi = (13.6)(7) = 95.2W.$

(b) With $R = \infty$ there is no current through that branch, so that our equation for node n_1 changes to

n_1: $7 - 3 = (v_1 - v_3).$

All the other equations remain unchanged.

Solving for v_1

$$v_1 = \frac{\begin{vmatrix} 4 & 0 & -1 \\ 2 & \frac{1}{2} & -\frac{1}{2} \\ 0 & -\frac{1}{2} & \frac{17}{20} \end{vmatrix}}{\begin{vmatrix} 1 & 0 & -1 \\ 0 & \frac{1}{2} & -\frac{1}{2} \\ -1 & -\frac{1}{2} & \frac{17}{20} \end{vmatrix}} = \frac{3.4}{1\ (.6) - 1\ (.5)} = \frac{3.4}{.1} = 34V.$$

The power supplied by the 7-A source is

$p = vi = (34)(7) = 238W.$

If $v_y = 12$ V and $G = 1/6\,\mho$ in the circuit of Fig. 1, find the power absorbed by each of the five circuit elements.

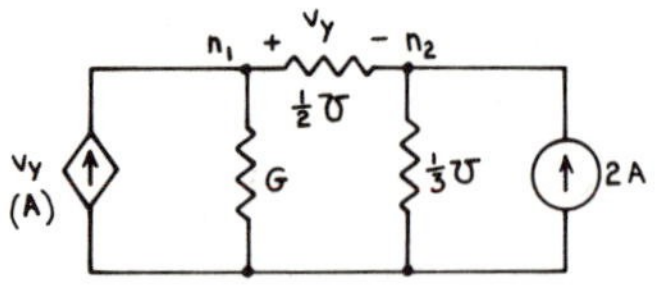

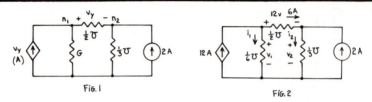

Fig. 1

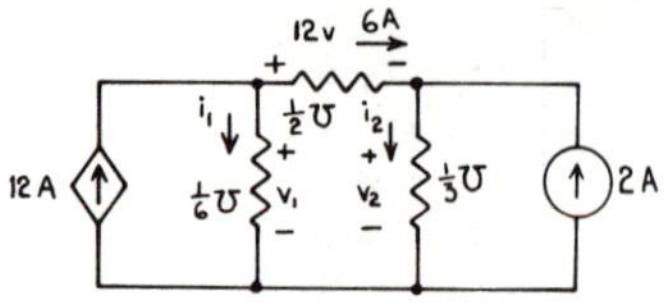

Fig. 2

Solution: To find the power absorbed for each element one must know the voltages and currents for each element. Start by finding the current in each branch. For the top branch with the $\frac{1}{2}\mho$ conductance we have

$$i = v_y G = 12(\tfrac{1}{2}) = 6\text{A}.$$

By writing a KCL equation for node n_1 find the current through the $1/6\,\mho$ conductance.

$$12\text{A} = i_1 + 6\text{A}; \qquad i_1 = 6\text{A}$$

KCL equation of node n_2

$6\text{A} + 2\text{A} = i_2$ yields $i_2 = 8\text{A}$. (See Fig. 2.)

We obtain the voltages

$$v_1 = \frac{i_1}{G} = \frac{6}{\frac{1}{6}} = 36\text{V}$$

$$v_2 = \frac{i_2}{G} = \frac{8}{\frac{1}{3}} = 24\text{V}.$$

For each element from left to right the power is

$$p = vi = -\,36\text{V} \cdot 12\text{A} = -\,432 \text{ W}$$
$$= 36\text{V} \cdot 6\text{A} = 216 \text{ W}$$
$$= 12\text{V} \cdot 6\text{A} = 72 \text{ W}$$
$$= 24\text{V} \cdot 8\text{A} = 192 \text{ W}$$
$$= 24\text{V} \cdot 2\text{A} = -\,48 \text{ W}$$

Check (sum) 0

With reference to Fig. 1, let the lower node be the reference, and write the single nodal equation required to find v_x. How much power is delivered by the dependent source?

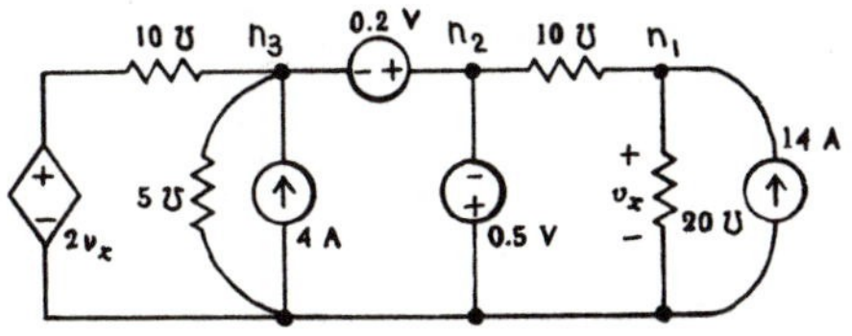

FiG.1

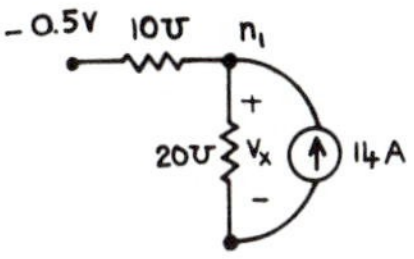

Fig. 2

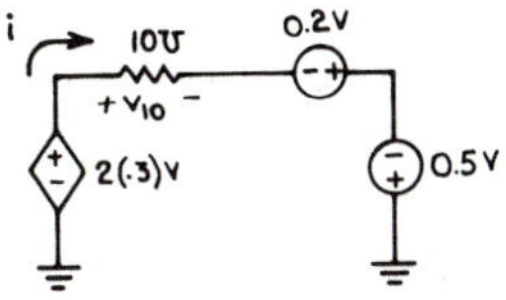

Fig. 3

Solution: One can find v_x by a single nodal equation at node n_1 since we know the voltage at node n_2. Fig. 2 shows the isolated node.

Write the KCL equation for node n_1

$$14 = [v_x - (-0.5)]\ 10 + (v_x)\ 20.$$

Solving for v_x:

$$14 = 10v_x + 5 + 20v_x$$

$$30v_x = 9$$

$$v_x = \frac{9}{30} = .3\text{V}.$$

To find the power delivered by the dependent voltage source, find the current flowing through it. We can find the current through the 10-$\mho$ conductance since the voltage across it is known. The current through the dependent current source and the 10-$\mho$ conductance is the same since they are in the same branch.

Fig. 3 shows the voltages across the 10-$\mho$ conductance.

$$v_{10} = 2\ (.3) + .2 + 0.5$$

$$v_{10} = 1.3\text{V}$$

$$i = v_{10}\ 10\text{-}\mho = 1.3\ (10) = 13\text{A}.$$

The power supplied by the dependent voltage source is

$$p = vi = (.6)(13) = 7.8\text{W}.$$

• PROBLEM 2-10

(a) Find the power absorbed by each element in the circuit of Fig. 1. (b) How would the results change if the 2-Ω resistor were reduced to 0?

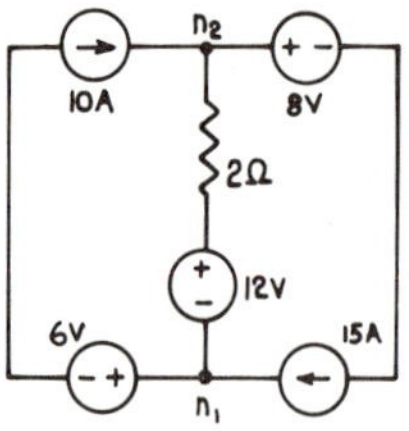

Fig. 1

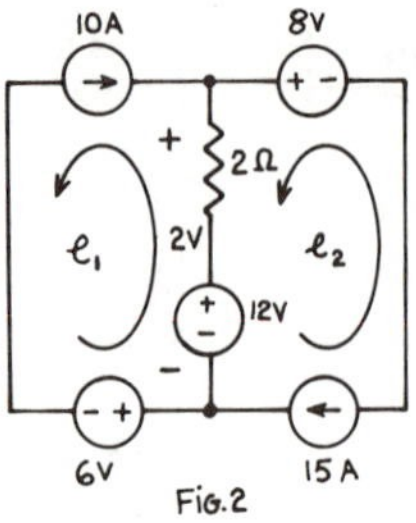

Fig. 2

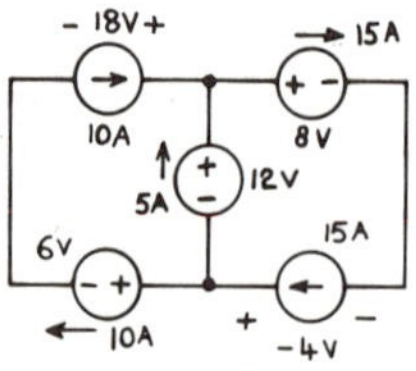

Fig. 3

Solution: (a) If a KCL equation for the node n_1 is written, the current in the center branch can be found.

$$15A - 10A = i \text{ (center branch)}$$

$$i = 5A$$

The voltage drop across the 2-Ω resistor is

$$iR = 5(2) = 10V.$$

The voltage from nodes n_2 to n_1 is

$$12 - 10 = 2V.$$

Thus, the voltages across each of the current sources can be found by writing a KVL equation for each loop.

$$\text{loop}_1 : \quad 6 + 2 = v \text{ (across the 10A source)}$$
$$= 8v$$

$$\text{Loop}_2 : \quad 8 - 2 = v \text{ (across the 15A source)}$$
$$= 6V$$

Knowing the voltage across each element and the current in each branch makes it possible to calculate the power absorbed.

$$p_{10A} = -(10)(8) = -80W$$

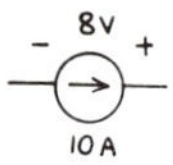

$$p_{8V} = (15)(8) = 120W$$

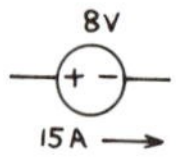

$$p_{2\Omega} = (5)(10) = 50W$$

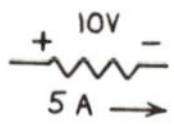

$$p_{12V} = -(5)(12) = -60W$$

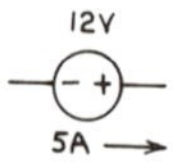

$p_{6V} = (10)\ (6) = 60W$

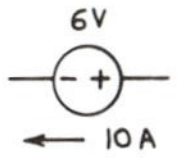

$p_{15A} = -\ (15)\ (6) = -\ 90W$

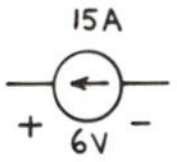

(b) Current through the center branch would remain at 5A. The voltage from node n_2 to n_1 would be equal to the 12V source.

Apply a KVL equation for $loop_1$ and $loop_2$ as in part (a)

$loop_1$: $6 + 12 = v$ (across the 10A source)

$= 18V$

$loop_2$: $8 - 12 = v$ (across the 15A source)

$= -\ 4V$

From Fig. 3 we find

$p_{10A} = -\ 180W,\quad p_{8V} = 120W,\quad p_{12V} = -\ 60W$

$p_{6V} = 60W$ and $p_{15A} = 60W.$

• PROBLEM 2-11

Find the current i_x in Fig. 1 by: (a) changing the two practical voltage sources to practical current sources and then using nodal analysis; (b) changing the practical current source to a practical voltage source and again using nodal analysis.

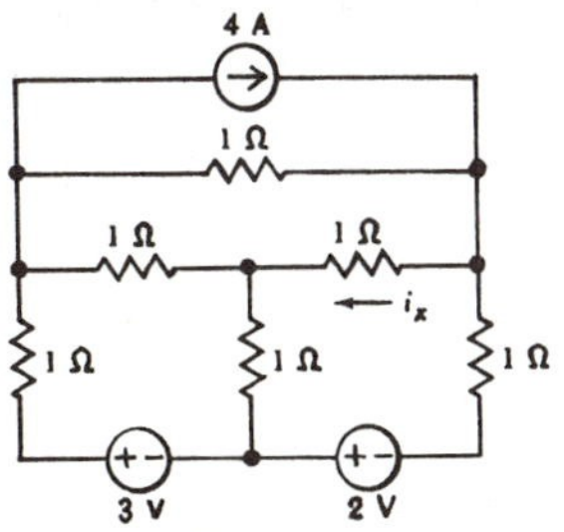

Fig.1

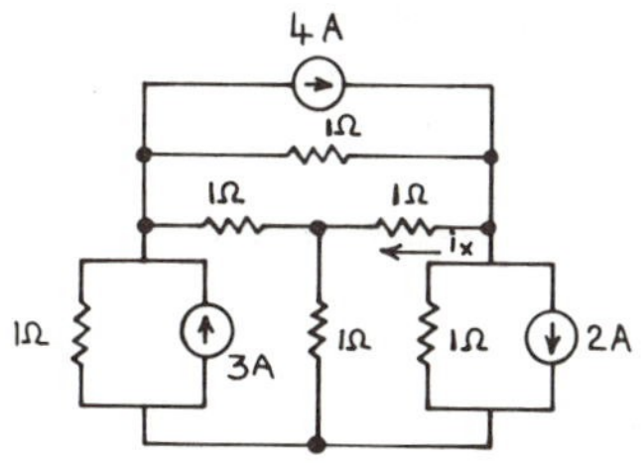

Fig. 2

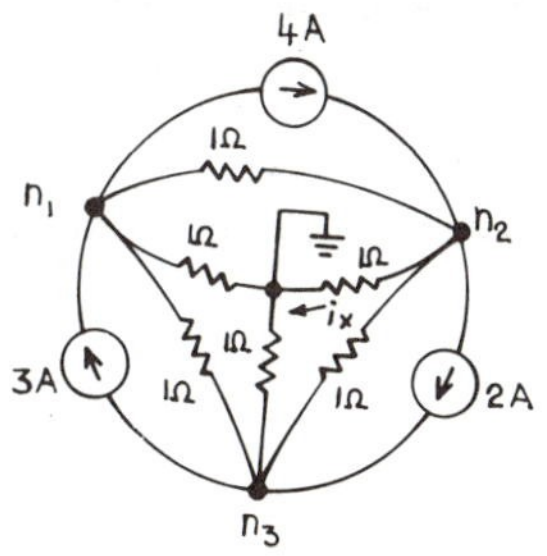

Fig. 3

Solution: (a) Fig. 2 shows the converted circuit with the two voltage sources changed to current sources.

Fig. 3 shows the circuit in Fig. 2 re-drawn to indicate its nodes.

Writing the KCL equations for the three nodes:

n_1: $3 - 4 = v_1 + (v_1 - v_3) + (v_1 - v_2)$

n_2: $4 - 2 = v_2 + (v_2 - v_3) + (v_2 - v_1)$

n_3: $2 - 3 = v_3 + (v_3 - v_2) + (v_3 - v_1)$

Simplifying, yields

$$-1 = +3v_1 - v_2 - v_3$$

$$2 = -v_1 + 3v_2 - v_3$$

$$-1 = -v_1 - v_2 + 3v_3.$$

Solving for v_2 and i_x

$$v_2 = \frac{\begin{vmatrix} 3 & -1 & -1 \\ -1 & 2 & -1 \\ -1 & -1 & 3 \end{vmatrix}}{\begin{vmatrix} 3 & -1 & -1 \\ -1 & 3 & -1 \\ -1 & -1 & 3 \end{vmatrix}}$$

$$= \frac{+1(-3-1) + 2(9-1) + 1(-3-1)}{+1(-3-1) + 3(9-1) + 1(-3-1)}$$

$$v_2 = \frac{1(-4) + 2(8) + 1(-4)}{1(-4) + 3(8) + 1(-4)} = \frac{8}{16} = 0.5\text{V}.$$

Thus $\quad i_x = \frac{v_2}{1} = 0.5$ A.

• PROBLEM 2-12

In the circuit of Fig. 1, use nodal analysis to find: (a) v_A; (b) v_B; (c) v_C.

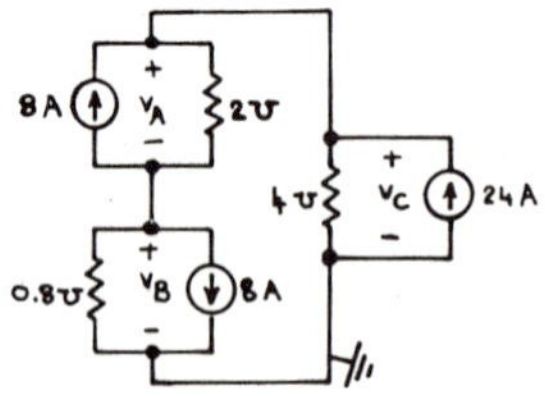

FiG. 1

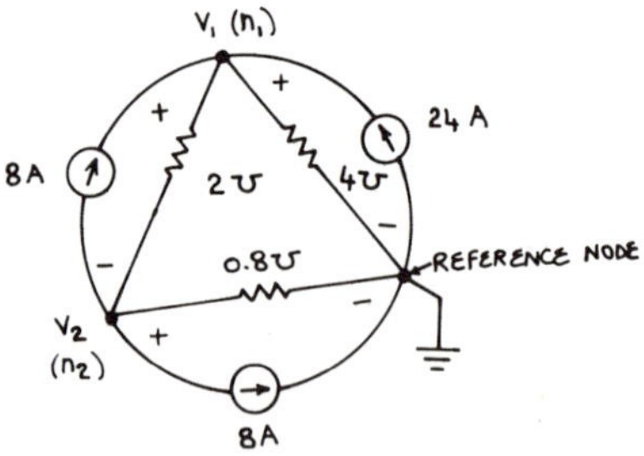

FiG. 2

Solution: Observe that this network has three nodes. If the bottom node becomes the reference node, there will be two linearly independent KCL equations needed. Also note that the values for the resistors are given in conductance, not resistance. This is indicated by the symbol ℧ instead of Ω. The conductance, measured in mhos, is $\frac{1}{R}$ where R is in ohms, thus the inverted omega. When conductance $\left(G = \frac{1}{R}\right)$ is used, Ohm's law takes the form $i = vG$.

In Fig. 2 the circuit in Fig. 1 is redrawn to show how the elements are interconnected between 3-nodes. Note that $v_A = v_1 - v_2$, $v_C = v_1$ and $v_B = v_2$. Writing the KVL equations

$$n_1: \quad 8 + 24 = (v_1 - v_2)\,2 + v_1\,4$$

$$n_2: \; -8 - 8 = v_2\,(.8) + (v_2 - v_1)\,2.$$

Simplifying each equation yields

$$6v_1 - 2v_2 = 32$$

$$-2v_1 + 2.8v_2 = -16.$$

It is possible to solve these two equations by algebraic methods or determinants as shown below.

$$v_1 = \frac{\begin{vmatrix} 32 & -2 \\ -16 & 2.8 \end{vmatrix}}{\begin{vmatrix} 6 & -2 \\ -2 & 2.8 \end{vmatrix}} = \frac{(89.6) - (32)}{(16.8) - (4)} = \frac{57.6}{12.8} = 4.5\text{V}$$

$$v_2 = \frac{\begin{vmatrix} 6 & 32 \\ -2 & -16 \end{vmatrix}}{\begin{vmatrix} 6 & -2 \\ -2 & 2.8 \end{vmatrix}} = \frac{(-96) - (-64)}{(16.8) - (4)} = \frac{-32}{12.8} = -2.5\text{V}.$$

Thus,
$$v_A = 4.5 - (-2.5) = 7\text{V}$$

$$v_B = v_2 = -2.5\text{V}$$

$$v_C = v_1 = 4.5\text{V}.$$

• PROBLEM 2-13

In the circuit in Fig. 1, use nodal analysis to find: (a) v_A; (b) v_B; (c) v_C. Note that the resistors are labeled with their respective conductance.

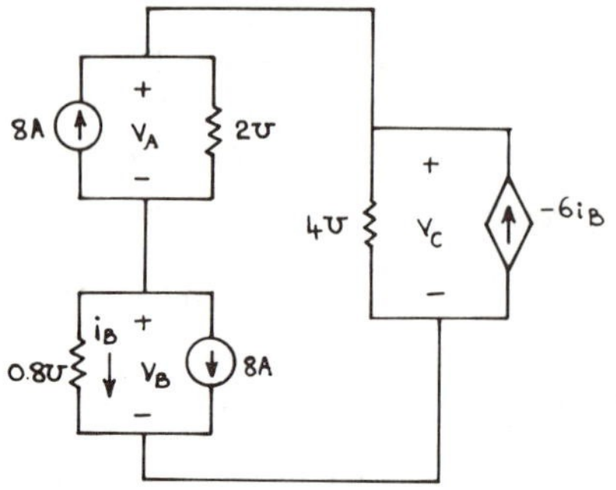

FIG. 1

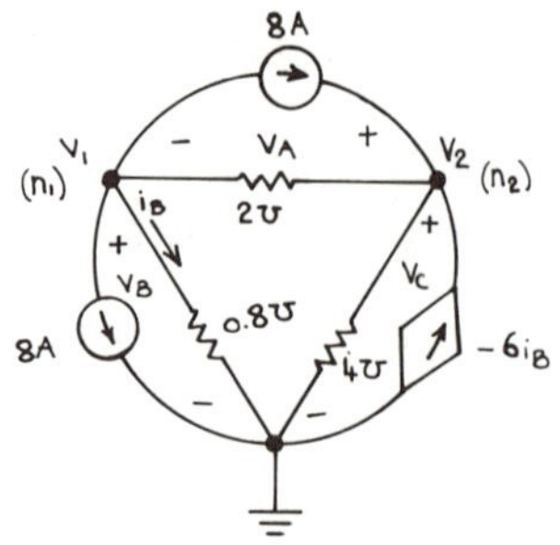

FIG. 2

Solution: Choosing the bottom node for the reference node, one can write two KCL equations for the remaining nodes. First, we draw the circuit.

The two KCL equations are:

$$n_1: \quad -8-8 = (v_1 - v_2)\,2 + v_1\,0.8$$

$$n_2: \quad 8 + (-6\,i_B) = (v_2 - v_1)\,2 + v_2\,4.$$

Observe that in the above equations there are 3 unknowns. In order to solve them an expression must be found for i_B in terms of the other variables. The current i_B is the current through the branch with 0.8℧ conductance which is written $v_1\,0.8$.

Thus, $i_B = v_1\,0.8$. Substitute this into the equation for node n_2. Simplify both equations to obtain

$$2.8\,v_1 - 2\,v_2 = -16$$

$$2.8\,v_1 + 6\,v_2 = 8.$$

Solve these two equations by algebraic methods or determinants. Using determinates yields

$$v_2 = \frac{\begin{vmatrix} 2.8 & 8 \\ 2.8 & -16 \end{vmatrix}}{\begin{vmatrix} 2.8 & 6 \\ 2.8 & -2 \end{vmatrix}} = \frac{(-44.8)-(22.4)}{(-5.6)-(+16.8)} = \frac{-67.2}{-22.4} = 3$$

$$v_1 = \frac{\begin{vmatrix} 8 & -2 \\ -16 & 6 \end{vmatrix}}{\begin{vmatrix} 2.8 & 6 \\ 2.8 & -2 \end{vmatrix}} = \frac{(-16)-(-96)}{(-5.6)-(16.8)} = \frac{80}{-22.4} = -3.57.$$

And $v_2 - v_1 = v_A = 3 - (-3.57) = 6.57$ V

$v_1 = v_B = -3.57$ V

$v_2 = v_C = 3$ V.

• PROBLEM 2-14

In the circuit in Fig. 1 use nodal analysis to find: (a) v_A; (b) v_B; (c) v_C. Note that the resistors are labeled with their respective conductances.

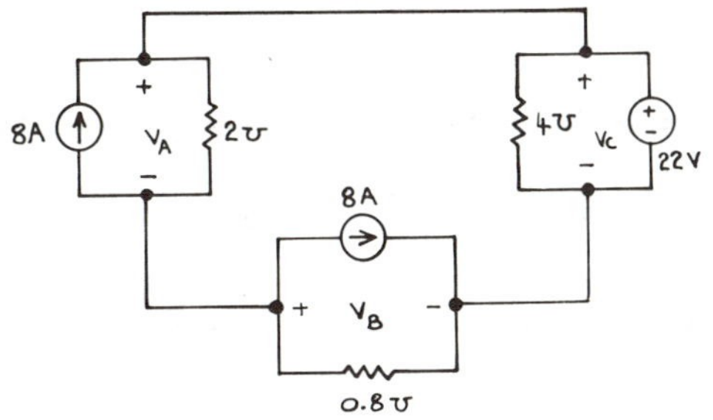

Fig. 1

Solution: By examining the circuit note that with the voltage source fixing the potential, $v_C = 22V$.

Re-drawing the circuit we have:

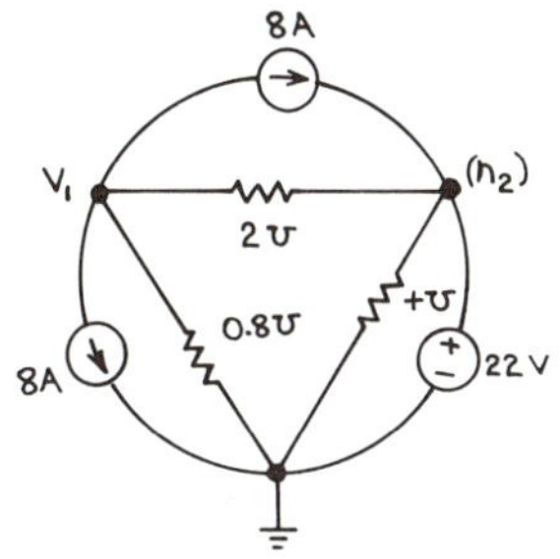

Fig. 2

The circuit in Fig. 2 can be simplified further:

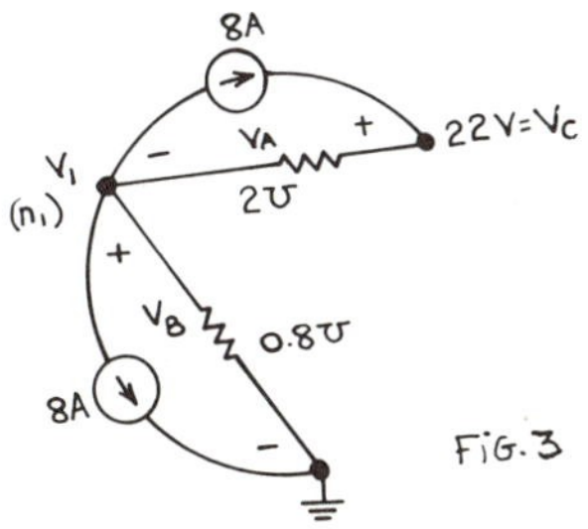

Fig. 3

In Fig. 3 it can be seen that one KCL equation for node n_1 is needed.

$$-8 - 8 = (v_1 - 22)\ 2 + v_1(0.8)$$

$$-16 = 2\ v_1 - 44 + 0.8\ v_1$$

$$28 = 2.8\ v_1$$

$$v_1 = 10V.$$

Yielding $\quad v_1 = v_B = 10V$

$$22 - v_1 = v_A = 12V$$

$$v_C = 22V.$$

• PROBLEM 2-15

In the circuit in Fig. 1 use nodal analysis to find: (a) v_A; (b) v_B; (c) v_C. Note that the resistors are labeled with their respective conductances.

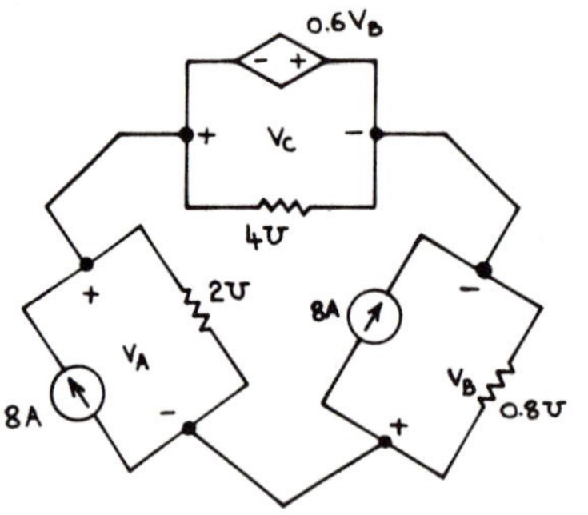

Fig. 1

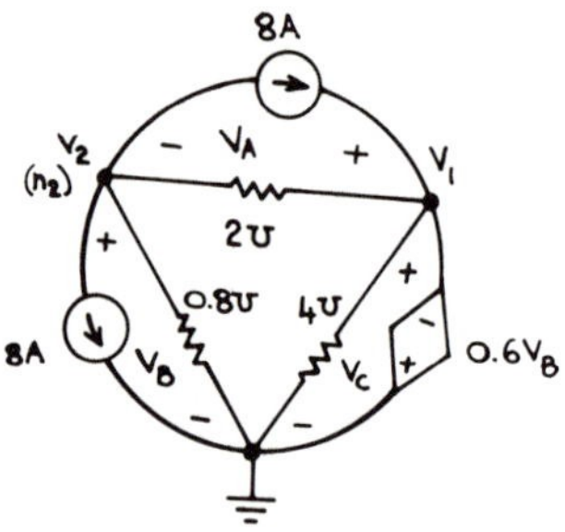

Fig. 2

Solution: The right node is made the reference node and voltages at the left and bottom are labeled node v_1 and v_2 respectively. Redraw the circuit in Fig. 1.

From Fig. 2, we can write

$$v_1 = v_C = -\ 0.6\ v_B \tag{1}$$

$$v_B = v_2 \tag{2}$$

$$v_1 - v_2 = v_A.$$

The KCL equation for node n_2

$$-\ 8 - 8 = (v_2 - v_1)\ 2 + v_2(0.8). \tag{3}$$

Substitute Eq. (2) into (1) and the resulting equation into (3).

This yields

$$-\ 8 - 8 = (v_2 - [-\ 0.6\ v_2])\ 2 + v_2(0.8)$$

$$-\ 16 = 3.2\ v_2 + 0.8\ v_2$$

$$-16 = 4v_2$$

$$v_2 = -4V = v_B.$$

We find v_A and v_C

$$v_1 = v_C = -0.6\, v_B = 2.4V$$

$$v_A = v_1 - v_2 = 2.4 - (-4) = 6.4V.$$

• PROBLEM 2-16

An equivalent network for an operational amplifier circuit is shown in Fig. 1. Perform a nodal analysis on this network.

Solution: Using the nodes shown, write

node n_1: $i = i_{C_1} + i_{R_1}$

node n_2: $i_{C_1} + i_{R_1} = i_{C_2} + i_{R_2}$

node n_2: $i_{C_2} + i_{R_2} = i_R$

where $$i = \frac{kv_1 - v}{R}$$

$$i_{R_1} = \frac{v - v_1}{R_1}$$

$$i_{C_1} = C_1 \frac{d}{dt}(v - v_1)$$

$$i_{R_2} = \frac{v_1 - kv_1}{R_2}$$

$$i_{C_2} = C_2 \frac{d}{dt}(v_1 - kv_1) \quad .$$

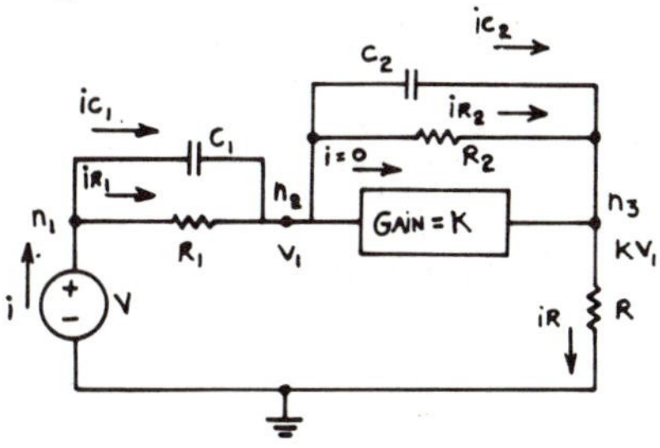

Fig. 1

• PROBLEM 2-17

In the resistive ladder network of Fig. 1, deduce the relationship between v_4 and v_0, between i_4 and v_0, and between i_0 and v_0.

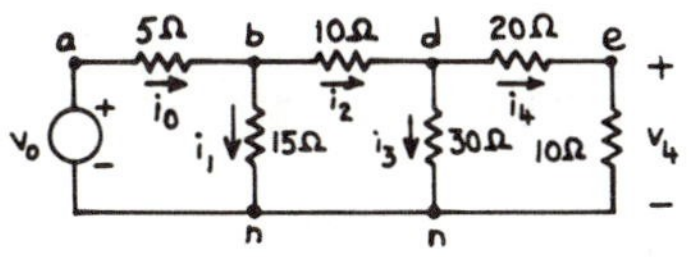

Fig. 1

Solution: By using the nodes and current directions indicated in Fig. 1,

$$i_4 = \frac{1}{10} v_4 \qquad \text{(from Ohm's Law.)}$$

Since, $v_d = v_{de} + v_4$ and $v_{de} = 20\ i_4 = 20\left(\frac{1}{10} v_4\right) = 2\ v_4$, then $v_d = 2\ v_4 + v_4 = 3\ v_4$.

The current i_3 through the 30 Ω resistance is

$$\frac{v_d}{30} = \frac{3\ v_4}{30} = \frac{v_4}{10}$$

Summing the current at node d yields

$$i_2 = i_3 + i_4 = \frac{v_4}{10} + \frac{1}{10} v_4 = \frac{v_4}{5}$$

Since, $v_b = v_{bd} + v_d$ and $v_{bd} = 10 i_2 = 2\ v_4$, then

$$v_b = 2\ v_4 + 3\ v_4 = 5\ v_4$$

The current i_1 through the 15Ω resistor is

$$\frac{v_b}{15} = \frac{5\ v_4}{15} = \frac{v_4}{3}$$

Summing the currents at node b gives

$$i_0 = i_1 + i_2 = \frac{v_4}{3} + \frac{v_4}{5} = \frac{8}{15} v_4$$

Since $v_0 = v_{ab} + v_b$ where $v_{ab} = 5 i_0 = 5\left(\frac{8}{15} v_4\right) = \frac{8}{3} v_4$, gives, $v_0 = \frac{8}{3} v_4 + 5\ v_4 = \frac{23}{3} v_4$ the relationship between v_4 and v_0.

Since we have already found

$$i_4 = \frac{1}{10} v_4\ ; \qquad v_4 = 10 i_4 \text{ gives}$$

$$v_0 = \frac{23}{3} v_4 = \frac{230}{3} i_4 \text{ the second required relationship.}$$

Finally, $i_0 = \frac{8}{15} v_4$ which gives

$$v_0 = \frac{24}{3} v_4 = \frac{23}{3}\frac{15}{8} i_0 = \frac{115}{8} i_0$$

which is also a required result.

• PROBLEM 2-18

In the capacitance network of Fig. 1, deduce the relationship between i_3 and i, v_4 and i, and v_0 and i.

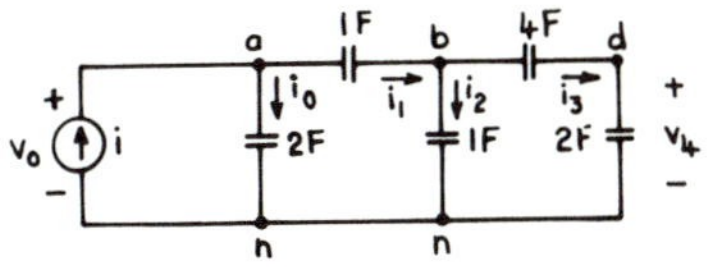

FIG. 1

Solution: Since the voltage across a capacitor is defined as

$$v_{c_1} = \frac{1}{C_1}\int_{-\infty}^{t} i_{c_1}(t)\,dt ,$$

We can replace the operation $\int_{-\infty}^{t} i_{c_1}(t)\,dt$, with the symbol W_1.

Hence v_4 can be written as $\frac{1}{2}W_3$.

The voltage across the 4-F capacitor v_{bd}, is $\frac{1}{4}W_3$.

Hence, $v_b = v_{bd} + v_4 = \frac{1}{2}W_3 + \frac{1}{4}W_3 = \frac{3}{4}W_3$.

Since $i_2 = 1\,\frac{dv_b}{dt} = \frac{d}{dt}\,\frac{3}{4}W_3 = \frac{3}{4}i_3$, we can write

$$i_1 = i_2 + i_3 = \frac{3}{4}i_3 + i_3 = \frac{7}{4}i_3.$$

The voltage v_{ab} across the 1-F capacitor is $\frac{7}{4}W_3$, and

$$v_a = v_{ab} + v_b = \frac{7}{4}W_3 + \frac{3}{4}W_3 = \frac{10}{4}W_3.$$

Hence, $i_0 = 2\,\frac{dv_a}{dt} = \frac{20}{4}i_3$

and $i = i_1 + i_0 = \frac{7}{4}i_3 + \frac{20}{4}i_3 = \frac{27}{4}i_3$

gives the relationship between i_3 and i. Since the current through the 2-F capacitor at the extreme right

$$2\,\frac{dv_4}{dt} = i_3,$$

we have $\frac{dv_4}{dt} = \frac{i_3}{2}$ where it was found that $i = \frac{27}{4}i_3$.

Hence, $\frac{dv_4}{dt} = \frac{2}{27} i$

$$i = \frac{27}{2} \frac{dv_4}{dt} .$$

Since $2 \frac{dv_0}{dt} = i_0 = \frac{20}{4} i_3$

$$\frac{dv_0}{dt} = \frac{20}{8} \frac{4}{27} i$$

$$i = 2.7 \frac{dv_0}{dt} .$$

MESH METHOD

• PROBLEM 2-19

Use Kirchhoff's voltage law to write an integrodifferential equation for i(t) for the circuit shown.

Solution: Kirchhoff's voltage law applied to the single loop circuit yields the equation

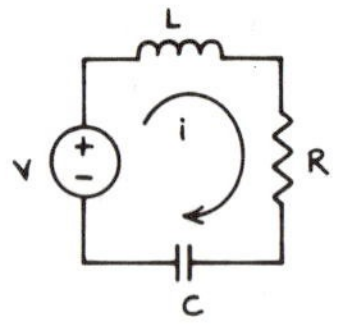

$$v = v_L + v_R + v_C.$$

The voltages for each element v_L, v_R and v_C can all be expressed in terms of current

$$v = L \frac{di}{dt} \quad ; \quad v_R = Ri \quad ; \quad v_C = \frac{1}{C} \int_{-\infty}^{t} i \, d\tau.$$

Substituting these terms into the KVL equation yields the required integrodifferential equation;

$$v = L \frac{di}{dt} + Ri + \frac{1}{C} \int_{-\infty}^{t} i \, d\tau .$$

• PROBLEM 2-20

How many independent loops are there in the network shown? Hence write down the corresponding loop equations.

Solution: We can draw the planar equivalent circuit shown in Fig. 2.

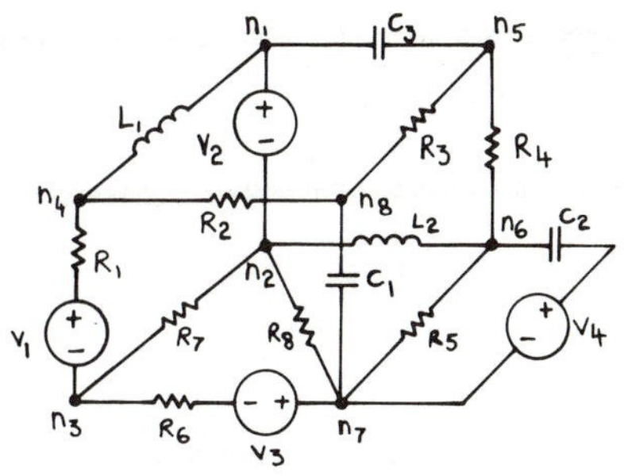

FiG. 1

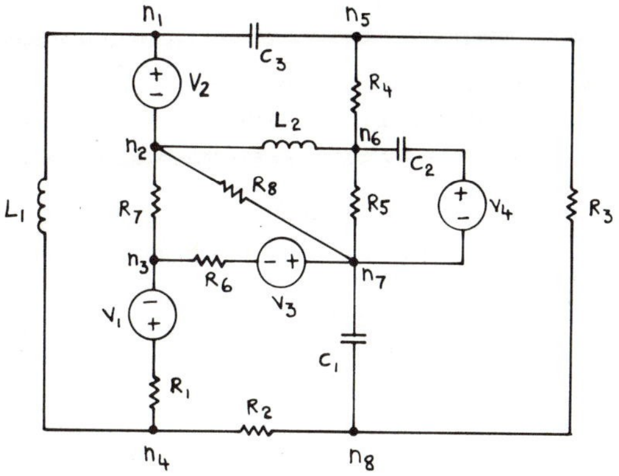

FiG. 2

Fig. 3 shows the loops chosen.

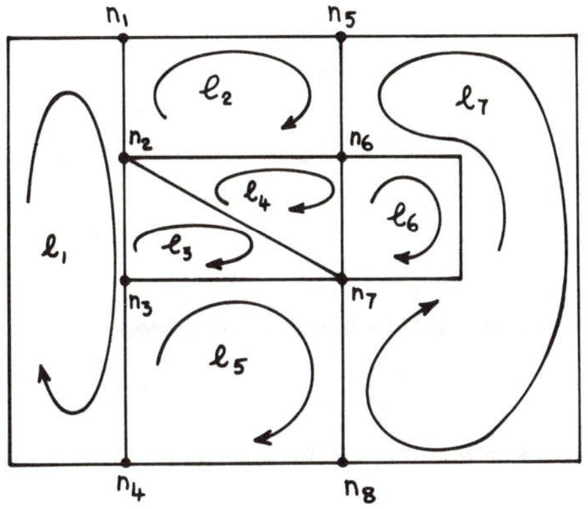

FiG. 3

Since there are 8 nodes and 14 branches, there are 14 - (8 + 1) = 7 independent KVL equations. The equations are:

ℓ_1: $\quad v_1 - v_2 = v_{R_1} + v_{R_7} + v_{L_1}$

ℓ_2: $\quad v_2 = v_{C_3} + v_{R_4} + v_{L_2}$

ℓ_3: $\quad -v_3 = v_{R_6} + v_{R_8} - v_{R_7}$

ℓ_4: $\quad 0 = v_{R_5} - v_{R_8} - v_{L_2}$

ℓ_5: $\quad -v_1 + v_3 = v_{C_1} - v_{R_2} - v_{R_1} - v_{R_6}$

ℓ_6: $\quad -v_4 = v_{C_2} - v_{R_5}$

ℓ_7: $\quad v_4 = v_{R_3} - v_{C_1} - v_{C_2} - v_{R_4}.$

• PROBLEM 2-21

Use the loop-analysis method to develop KVL equations for the network shown in Fig. 1.

Solution: In this network only one loop is necessary since the current through R_1 is i.

Consider that

$$i = i_1 + i_2$$

for the loop ℓ_2

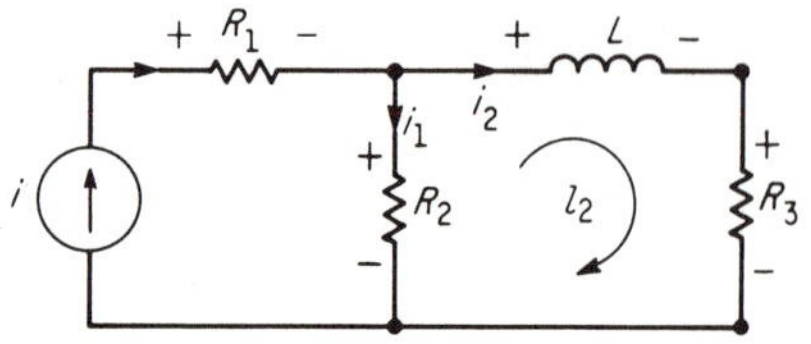

FiG. 1

$$L\frac{di_2}{dt} + i_2R_3 = i_1R_2$$

Substitute $i - i_2$ for i_1.

$$L\frac{di_2}{dt} + i_2R_3 = (i - i_2)R_2$$

This is a first order differential equation that can be solved for i_2.

• PROBLEM 2-22

In Fig. 1, let circuit element X be a 2-Ω resistance. Find the power supplied by the 6-V source.

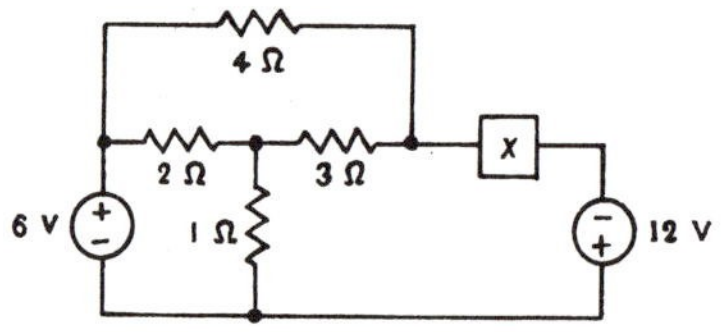

FiG. 1

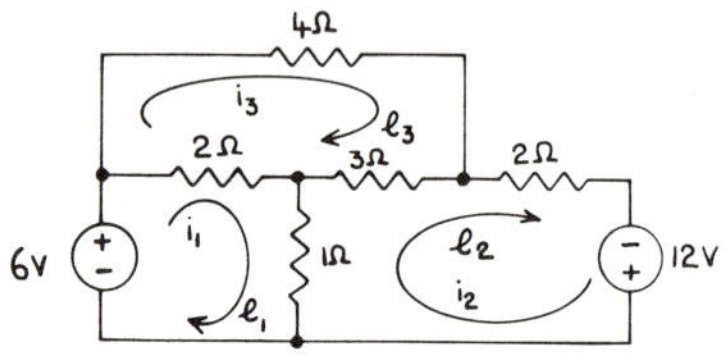

FiG. 2

Solution: Fig. 2 indicates the three loops and their assigned currents used to write the three KVL equations below.

ℓ_1: $6 = 2(i_1 - i_3) + (i_1 - i_2)$

ℓ_2: $12 = (i_2 - i_1) + 3(i_2 - i_3) + 2i_2$

ℓ_3: $0 = 2(i_3 - i_1) + 4i_3 + 3(i_3 - i_2)$.

Simplifying:

$$6 = 3i_1 - i_2 - 2i_3$$

$$12 = -i_1 + 6i_2 - 3i_3$$

$$0 = -2i_1 - 3i_2 + 9i_3 .$$

Solving for i_1

$$i_1 = \frac{\begin{vmatrix} 6 & -1 & -2 \\ 12 & 6 & -3 \\ 0 & -3 & 9 \end{vmatrix}}{\begin{vmatrix} 3 & -1 & -1 \\ -1 & 6 & -3 \\ -2 & -3 & 9 \end{vmatrix}}$$

$$= \frac{6(54 - 9) - 12(-9 - 6)}{3(54 - 9) + 1(-9 - 6) - 2(3 - (-12))}$$

$$i_1 = \frac{6(45) - 12(-15)}{3(45) + 1(-15) - 2(15)} = \frac{450}{90} = 5A.$$

The power supplied by the 6-V source is

$p = vi = (6)(5) = 30W.$

• PROBLEM 2-23

If $i_x = 4$ A and $R = 3\Omega$ in the circuit of Fig.1 find the power absorbed by each of the five circuit elements.

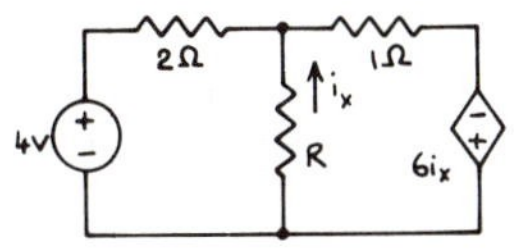

Fig. 1

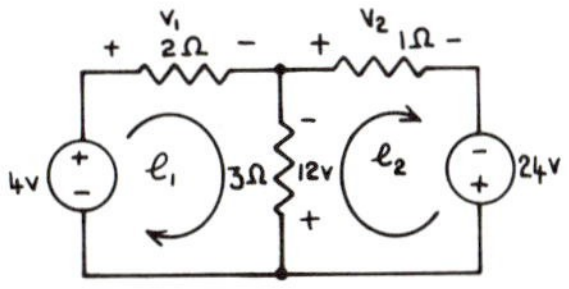

Fig. 2

Solution: In order to solve for the power for each element, the voltages and currents for each element must be found. Begin by finding the voltages across each of the three resistors. The voltage across R is

$$i_x R = 4 \cdot 3 = 12V.$$

The dependent voltage source can be replaced with an independent source

$$6i_x = 6 \cdot 4 = 24V.$$

Fig. 2 demonstrates how to find the voltages across the two remaining resistors by using Kirchhoff's Voltage Law

The KVL equation around loop ℓ_1,

$$V_1 - 12V = 4V$$

gives $$V_1 = 16V$$

The KVL equation around loop ℓ_2

$$V_2 + 12V = 24V$$

yields $$V_2 = 24 - 12 = 12V$$

The current through the 2-Ω resistor is:

$$\frac{V_1}{2} = \frac{16}{2} = 8A.$$

The current through the 1-Ω resistor is:

$$\frac{V_2}{1} = 12A .$$

It can be seen from the voltage signs and current directions that the power absorbed by the two voltage sources must be negative and the power absorbed by the resistors is positive. Calculate the power by using the voltage-current product, starting with the element farthest to the left, and proceeding to the right.

$$p = vi = -4V \cdot 8A = -32W$$
$$= 16V \cdot 8A = 128W$$
$$= 12V \cdot 4A = 48W$$
$$= 12V \cdot 12A = 144W$$
$$= -24V \cdot 12A = -288W$$

To check the results, consider that the total power for all elements must be zero, according to the law of conservation of energy.

Thus, $$-32 + 128 + 48 + 144 - 288 = 0.$$

• PROBLEM 2-24

For the network shown in Fig. 1, perform the loop analysis.

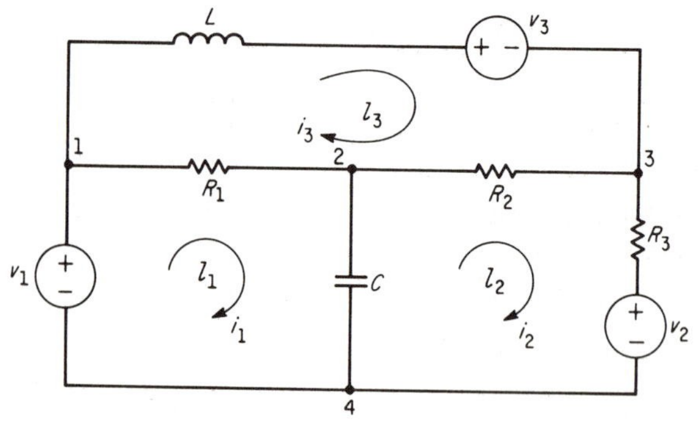

FIG.1

Solution: The network in Fig. 1 indicates that all the elements are covered by three loops. We denote the corresponding loop currents i_1, i_2 and i_3.

Consider loop ℓ_1, the voltage drop across R_1 which is $(i_1 - i_3)R_1$ and the voltage drop across the capacitor which is $\frac{1}{C}\int (i_1 - i_2)\,dt$. Application of KVL yields

$$\ell_1:\quad v_1 = (i_1 - i_2)R_1 + \frac{1}{C}\int (i_1 - i_2)dt$$

$$\ell_2:\quad v_2 + i_2R_3 + (i_2 - i_3)R_2 = \frac{1}{C}\int (i_1 - i_2)dt$$

$$\ell_3:\quad v_3 + C\frac{di_3}{dt} = (i_1 - i_3)R_1 + (i_2 - i_3)R_2 \quad .$$

• PROBLEM 2-25

In the circuit of Fig. 1, use mesh analysis to find: (a) i_A; (b) i_B; (c) i_C.

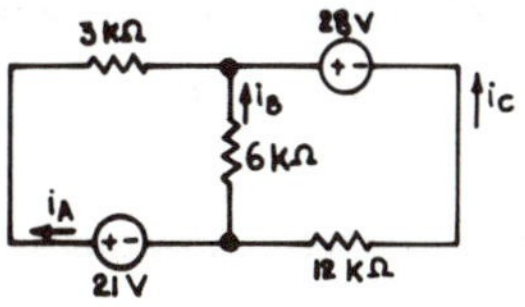

Fig. 1

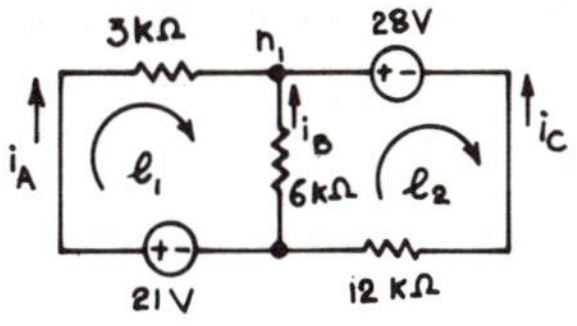

Fig. 2

Solution: Using KVL for the two loops in Fig. 2 yields

$$21 = i_A\, 3K - i_B\, 6K \tag{1}$$

$$-28 = -i_C\, 12K + i_B\, 6K \tag{2}$$

Apply Kirchhoff's current law at node n_1 and obtain the expression

$$i_A = -i_B - i_C$$

Substituting into Eq. (1) and simplifying both (1) and (2) produces

$$21 = (-9i_B - 3i_C) \times 10^3$$

$$-28 = (6i_B - 12i_C) \times 10^3.$$

Using determinants, solve for i_B and i_C

$$i_B = \frac{\begin{vmatrix} 21 & -3 \times 10^3 \\ -28 & -12 \times 10^3 \end{vmatrix}}{\begin{vmatrix} -9 & -3 \\ 6 & -12 \end{vmatrix} \times 10^3} = \frac{(-252) - (84)}{108 - (-18)} \times 10^3$$

$$= \frac{-336}{126} \times 10^{-3} = -2.667 \text{ mA}$$

$$i_C = \frac{\begin{vmatrix} -9 \times 10^3 & 21 \\ 6 \times 10^3 & -28 \end{vmatrix}}{\begin{vmatrix} -9 & -3 \\ 6 & -12 \end{vmatrix} \times 10^3} = \frac{(252) - (126)}{108 - (48)} \times 10^3$$

$$= \frac{126}{126} \times 10^{-3} = 1 \text{ mA}$$

$$i_A = -i_B - i_C = -(-2.667) - (1) = 1.667 \text{ mA.}$$

• PROBLEM 2-26

In the circuit of Fig. 1 use mesh analysis to find: (a) i_A ; (b) i_B; (c) i_C.

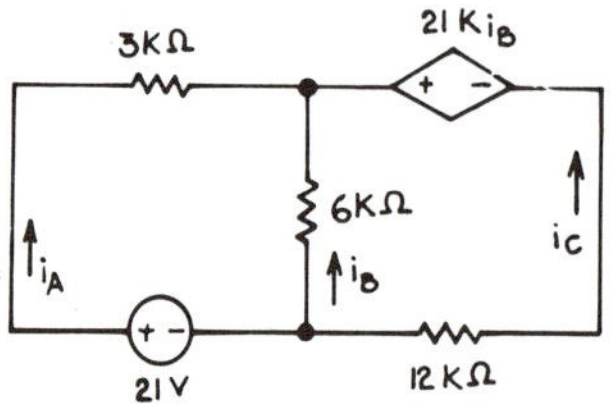

FiG. 1

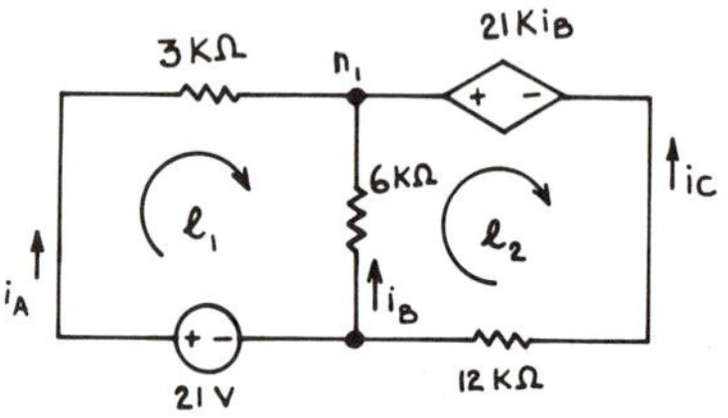

FiG. 2

Solution: Use KVL for the two loops in Fig. 2.

Now we have

$$21 = i_A\, 3K - i_B\, 6K \tag{1}$$

$$-21K\, i_B = -i_C\, 12K + i_B\, 6K. \tag{2}$$

KCL analysis at node n_1 yields

$$i_A = -i_B - i_C. \tag{3}$$

Substituting equation (3) into (1) and simplifying both (1) and (2) yields

$$21 = (-\ 9i_B - 3i_C) \times 10^3$$

$$0 = (27i_B - 12i_C) \times 10^3.$$

Using determinants, solve for i_B and i_C

$$i_B = \frac{\begin{vmatrix} 21 & -\ 3 \times 10^3 \\ 0 & -\ 12 \times 10^3 \end{vmatrix}}{\begin{vmatrix} -\ 9 & -\ 3 \\ 27 & -\ 12 \end{vmatrix} \times 10^3} = \frac{-\ 252 \times 10^{-3}}{108 - (-\ 81)}$$

$$= \frac{-\ 252}{(189)} \times 10^{-3} = -\ 1.333 \text{ mA}$$

$$i_C = \frac{\begin{vmatrix} -\ 9 \times 10^3 & 21 \\ 27 \times 10^3 & 0 \end{vmatrix}}{\begin{vmatrix} -\ 9 & -\ 3 \\ 27 & -\ 12 \end{vmatrix} \times 10^3} = \frac{-\ (567) \times 10^{-3}}{108 - (-\ 81)}$$

$$= \frac{-\ 567}{(189)} \times 10^{-3} = -\ 3 \text{ mA}.$$

Solve Eq. (3) for i_A

$$i_A = -\ i_B - i_C = -\ (-\ 1.333) - (-\ 3) = 4.333 \text{ mA}.$$

• PROBLEM 2-27

In the circuit of Fig. 1, use mesh analysis to find: (a) i_A; (b) i_B; (c) i_C.

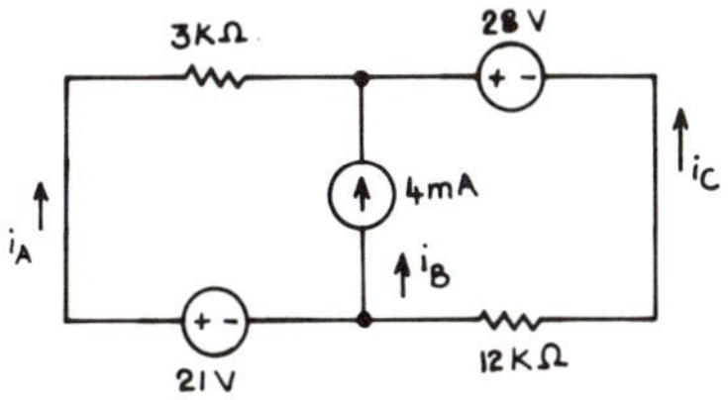

FiG. 1

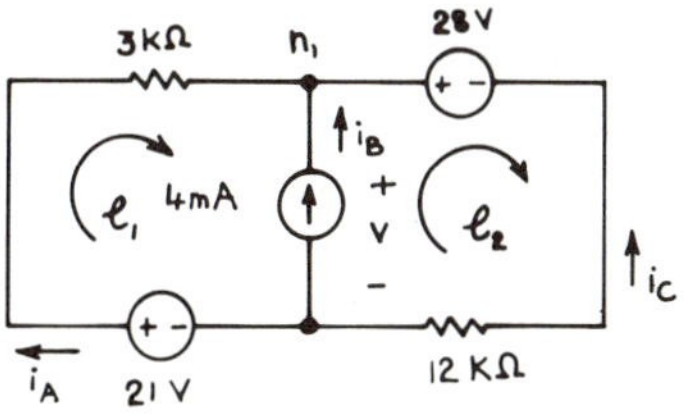

Fig. 2

Solution: It is immediately apparent that $i_B = 4$ mA. Write two KVL equations for the two loops in Fig. 2 if the voltage across the current source is v.

ℓ_1: $\quad 21 = 3K\ i_A + v \quad (1)$

ℓ_2: $- 28 = - 12K\ i_C - v$ (2)

A KCL equation at the node n_1 yields

$$i_A = - i_B - i_C$$

Since $i_B = 4$ mA, then

$$i_A = - 4m - i_C \qquad (3)$$

Substitute Eq. (3) into (1) and simplify both (1) and (2) giving

$$33 = - 3K\ i_C + v$$

$$- 28 = - 12K\ i_C - v$$

Solve for i_C

$$i_C = \frac{\begin{vmatrix} 33 & 1 \\ -28 & -1 \end{vmatrix}}{\begin{vmatrix} -3 \times 10^3 & 1 \\ -12 \times 10^3 & -1 \end{vmatrix}} = \frac{(-33) - (-28)}{(3) - (-12)} \times 10^{-3}$$

$$= \frac{-5}{15} \times 10^{-3} = - .333 \text{ mA.}$$

Using the value for i_C:

$$i_A = - i_B - i_C = - 4 \text{ mA} + .333 \text{ mA} = - 3.667 \text{ mA.}$$

• PROBLEM 2-28

In the circuit of Fig. 1, use mesh analysis to find: (a) i_A; (b) i_B; (c) i_C.

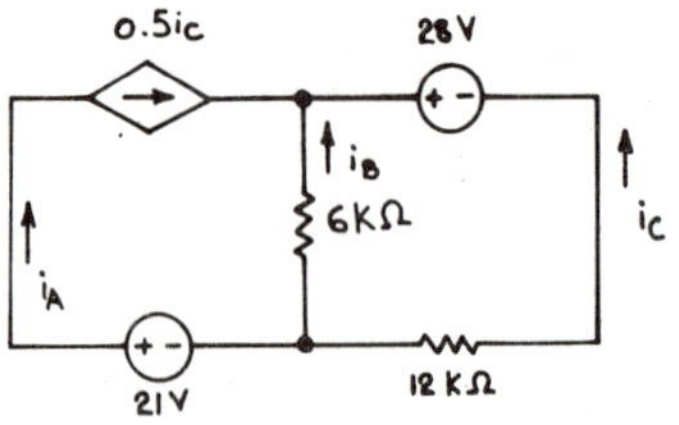

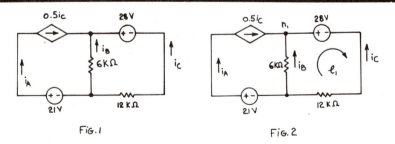

FiG. 1

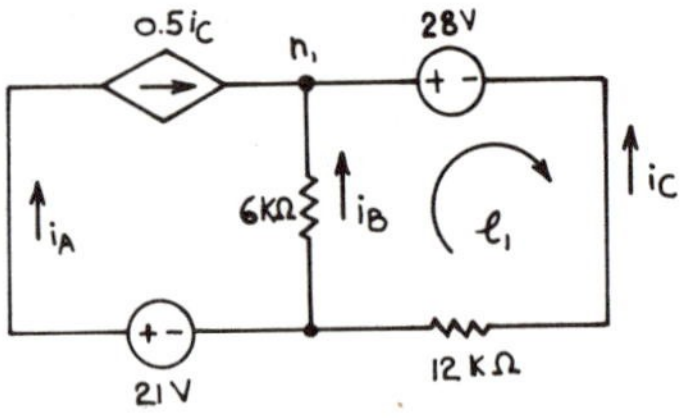

FiG. 2

Solution: Observe that $i_A = 0.5\ i_C$. This enables us to solve for all three variables using the loop shown in Fig. 2 and the nodal equation for node n_1.

ℓ_1: $- 28 = (- 12\, i_C + 6\, i_B) \times 10^3.$ (1)

Using KCL

$n_1 = i_A = - i_B - i_C.$ (2)

Since $i_A = 0.5\, i_C$, then

$0.5\, i_C = - i_B - i_C$

$15\, i_C = - i_B$

$$i_C = \frac{- i_B}{1.5} \quad . \tag{3}$$

Substituting Eq. (3) into (1) and solving for i_B yields

$- 28 = (8\, i_B + 6\, i_B) \times 10^3$

$- 28 = 14 \times 10^3\, i_B$

$i_B = - 2$ mA.

Substituting i_B into Eq. (1):

$(- 12\, i_C) \times 10^3 = - 28 - 6\,(- 2)$

$$i_C = \frac{- 16}{- 12 \times 10^3} = 1.333 \text{ mA.}$$

i_A becomes $0.5\, i_C = 0.5\,(1.333) = 0.667$ mA.

• **PROBLEM 2-29**

In the circuit shown in Fig. 1, $v_1 = 3e^{-2\times10^4 t}$ V. Find: (a) v_2; (b) v_3; (c) v_4.

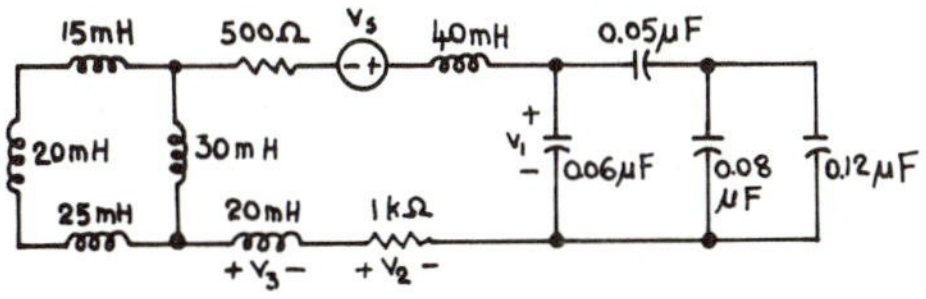

FiG. 1

Solution: Combining the inductors on the left and the capacitors on the right, forms a single loop circuit shown in Fig. 2.

The current can be found through the .1μF capacitor, hence the current through the loop by use of

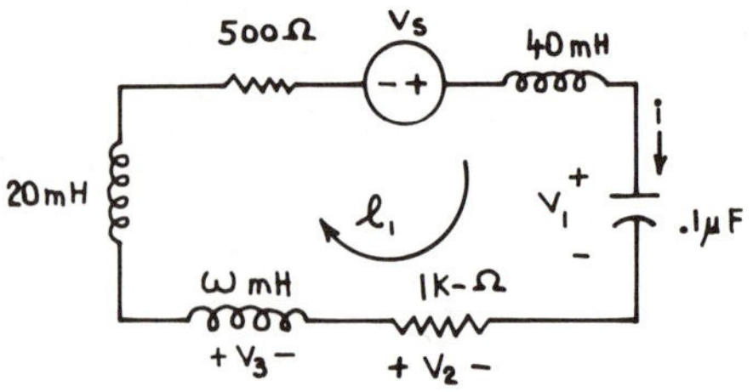

Fig. 2

$$i = C\,\frac{dv_1}{dt} \qquad \text{is}$$

$$i = C\,\frac{d\left[3e^{-2\times10^4 t}\right]}{dt}$$

$$i = 0.1 \times 10^{-6}\,(-\,2 \times 10^4)(3)\;e^{-\,2\,\times\,10^4 t}$$

$$i = -\,6\;e^{-\,2\,\times\,10^4 t}\ \text{mA.}$$

Find $v_2 = iR$

$$v_2 = iR = -\left[-\,6\,(1k)\;e^{-2\times10^4 t}\right] = 6e^{-2\times10^4 t}\ \text{V}$$

Also find $v_3 = L\,\frac{di}{dt}$

$$v_3 = -\left[(.02)\,\frac{d\left[-\,6e^{-2\times10^4 t}\ \text{mA}\right]}{dt}\right] = -\,2.4e^{-2\times10^4 t}\ \text{V}$$

v_s can be found by summing the voltages around the loop ℓ_1 in Fig. 2.

$$v_s = 40\ \text{mH}\,\frac{di}{dt} + v_1 - v_2 - v_3 + 20\ \text{mH}\,\frac{di}{dt} + 500\ i$$

Note that

$$v_1 = 3e^{-2\times10^4 t}$$

$$v_2 = 6e^{-2\times10^4 t}$$

$$v_3 = -\,2.4\ e^{-2\times10^4 t}\ ; \quad -\,v_3 = 20\ \text{mH}\,\frac{di}{dt}$$

$$-\,\frac{v_2}{2} = 500i = -\,3e^{-2\times10^4 t}$$

$$-\,(2\,v_3) = 40\ \text{mH}\,\frac{di}{dt} = 4.8\ e^{-2\times10^4 t}$$

thus, $v_s = [4.8 + 3 - 6 - (-\,2.4) + 2.4 - 3]e^{-2\times10^4 t}$

$$v_s = 3.6e^{-2\times10^4 t}\ .$$

Find v_x in the circuit shown in Fig. 1 by: (a) nodal analysis; (b) mesh analysis; (c) beginning on the right side of the circuit and alternating source transformations and source and resistance combinations until only a single loop circuit remains.

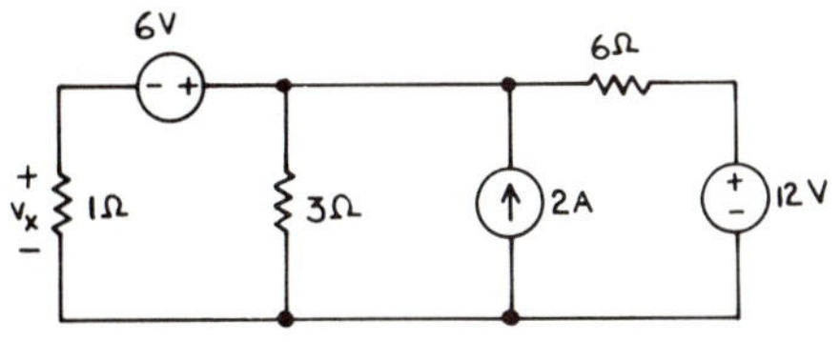

Fig. 1

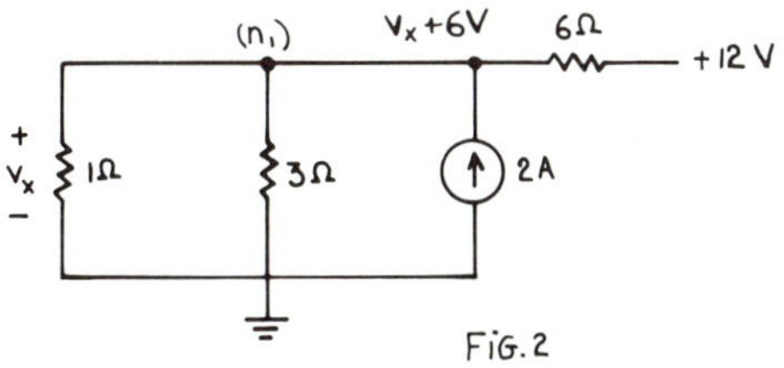

Fig. 2

Solution: (a) Remove the 6-V and the 12-V source from Fig. 1 and re-draw a new circuit as shown in Fig. 2.

By setting the voltage at node n: $v_x + 6$, a single KCL equation can be derived and solved for v_x directly.

Thus, for node n_1,

$$2 = (v_x + 6)\,\frac{1}{3} + ([v_x + 6]) - 12)\,\frac{1}{6} + \frac{v_x}{1}$$

$$2 = \frac{v_x}{3} + 2 + \frac{v_x}{6} + 1 - 2 + v_x$$

$$1 = \frac{9v_x}{6}$$

$$v_x = \frac{2}{3}\text{ V.}$$

(b) Write three KVL equations for the three loops shown in Fig. 3.

However, there is a current source in the circuit. Thus, the analysis must be modified. One standard procedure is to remove the current source leaving an open circuit in its place, writing down the reduced-number of loop equations and, by relating the current source to the currents, pick up the extra equations necessary. This yields for $\ell_1 - \ell_2$:

$$12 + 6i_2 - 3(i_1 + i_3) = 0$$

for ℓ_3

$$6 - 3\,(i_1 + i_3) - 1i_3 = 0$$

and $\quad i_1 + i_2 = 2$

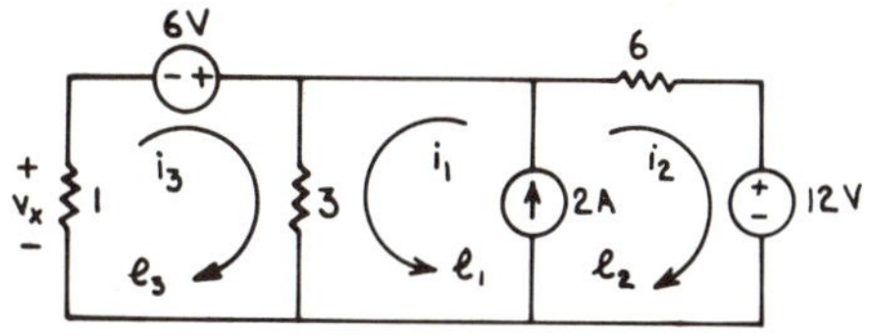

Fig. 3

Substituting i_1 from the current equation into the voltage equations gives us

$$6 + 9i_2 - 3i_3 = 0$$

$$0 + 3i_2 - 4i_3 = 0.$$

By direct substitution of $i_3 = 3/4\ i_2$ into the first equation results in $i_2 = -8/9$ which then gives $i_3 = -2/3$. This means $i_3 = 2/3$ A flowing down the 1Ω resistor giving us $v_x = 2/3$ V with the polarity as indicated in Figure 3.

(c) The 12-V source and the 6-Ω resistor in the circuit of Fig. 1 can be transformed into a current source and a parallel resistor as shown in Fig. 4.

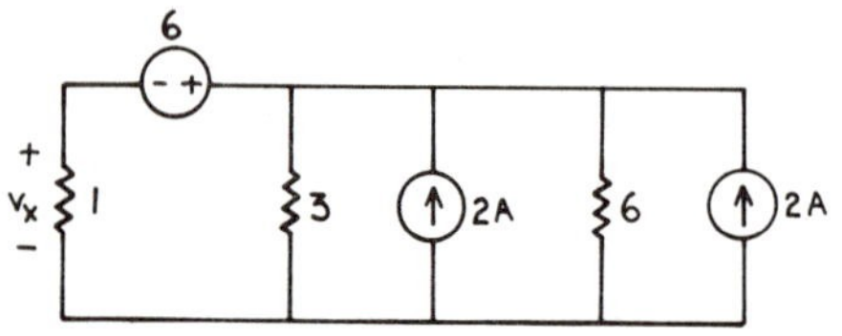

Fig. 4

The 3-Ω and 6-Ω resistors are combined. The expression

$$\frac{1}{R_T} = \frac{1}{R_1} + \frac{1}{R_2}$$

gives us that the resulting resistance is 2Ω. The two 2-A sources add to form a single 4-A source and the resulting practical current source can be transformed into a voltage source and series resistor. The circuit shown in Fig. 5 is the resulting single loop circuit.

Writing a KVL equation around the loop shown

$$8 - 6 = i(2 + 1)$$

we have $i = \frac{8 - 6}{2 + 1} = \frac{2}{3}$ A

giving $v_x = (1)\ \frac{2}{3} = \frac{2}{3}$ V.

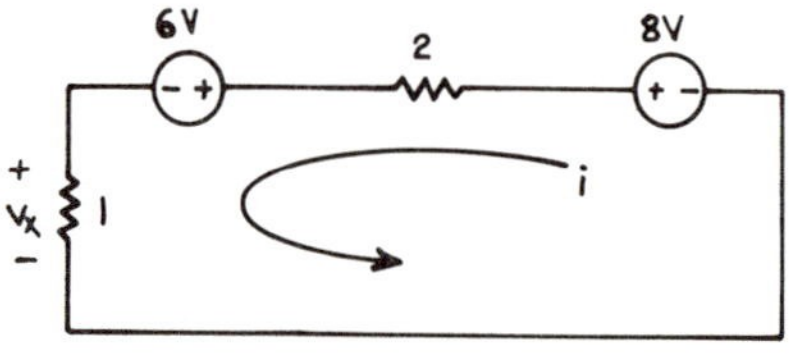

Fig. 5

• PROBLEM 2-31

For the circuit shown in Fig. 1; (a) use nodal analysis to find the power applied by the 3-A source; (b) transform the practical voltage source into a practical current source and use nodal analysis to find the power supplied by the new ideal current source; (c) transform the practical current source into a practical voltage source and use mesh analysis to find the power supplied by the new ideal voltage source.

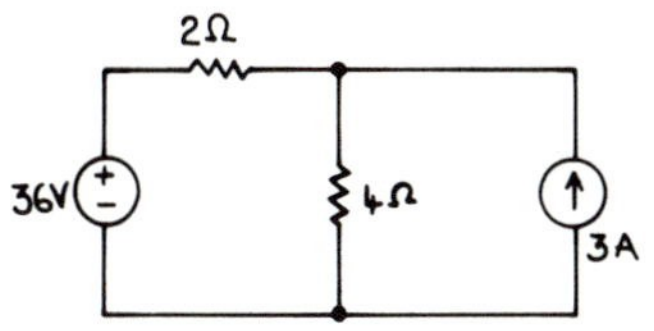

Fig. 1

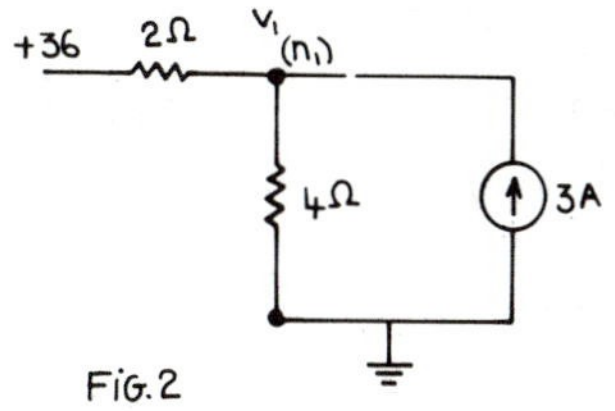

Fig. 2

Solution: (a) Writing a KCL equation for the node n_1 in Fig. 2 we have

$$3 = v_1 \frac{1}{4} + (v_1 - 36)\frac{1}{2}$$

$$3 = v_1 \frac{3}{4} - 18$$

$$21 = v_1 \frac{3}{4}$$

$$v_1 = 28V.$$

The power supplied by the current source is

$$p = vi = 28V\ (3A) = 84W.$$

(b) Transform the practical voltage source shown in Fig. 3 to the current source of Fig. 4.

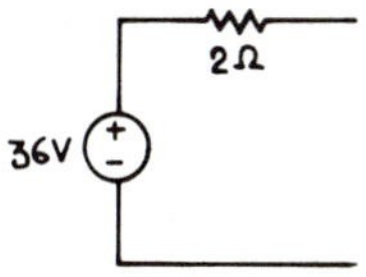

Fig. 3

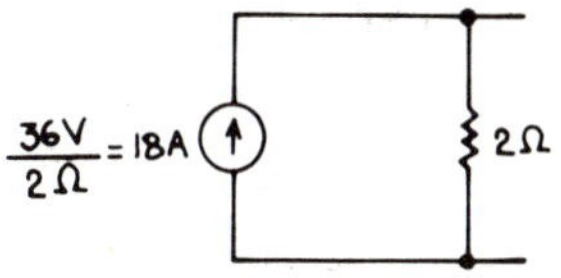

Fig. 4

This gives the new circuit:

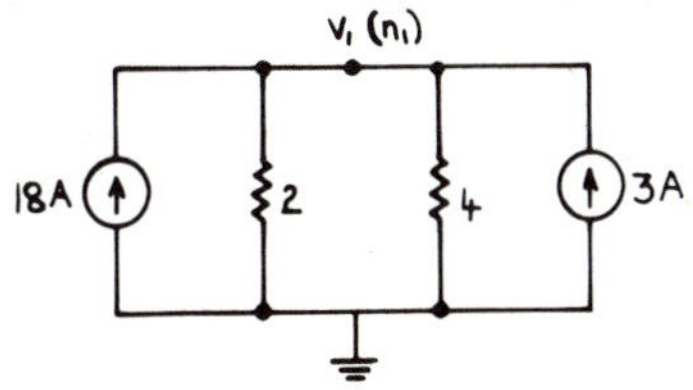

Fig. 5

The KCL equation for node n_1 is

$$18 + 3 = v_1\left(\frac{1}{2} + \frac{1}{4}\right)$$

$$21 = v_1 \frac{3}{4}$$

$$v_1 = 28V.$$

The power supplied by the 18A current source is

$p = vi = (28V)(18A) = 504W.$

(c) Transform the practical current source shown in Fig. 6 to the voltage source of Fig. 7.

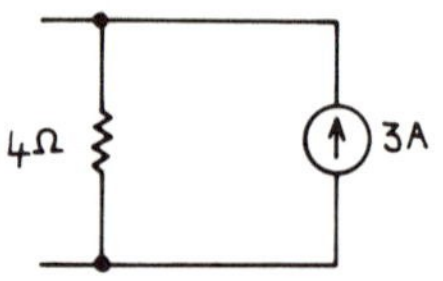

Fig.6

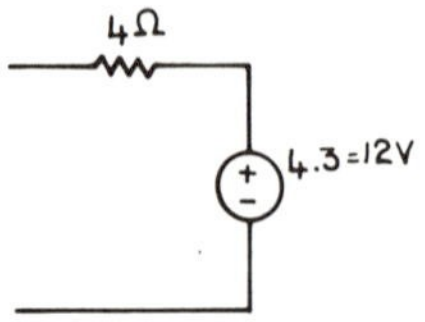

Fig.7

This yields the new circuit:

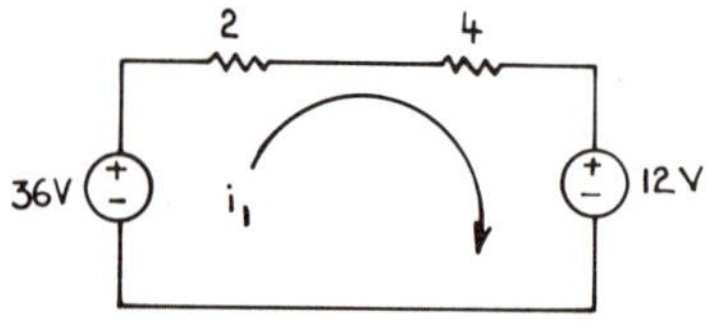

Fig.8

The KVL equation around the loop in Fig. 8 gives us

$$36 - 12 = i_1 (2 + 4)$$

$$i = \frac{24}{6} = 4A.$$

The power supplied by the 12-V source is negative since current flows opposite the conventional power applying source thus,

$$p = vi = -(12)(4) = -48 \text{ W}.$$

• PROBLEM 2-32

Consider the network shown in Fig. 1. Write all the KCL and KVL equations for this network.

Solution: Adopting the reference directions given in Fig. 1, the KCL equations can be written as:

node n_1: $i_a + i_b = i_c$

node n_2: $i_e = i_a + i_d$

node n_3: $i_d = i_b$

node n_4: $i_c = i_e$

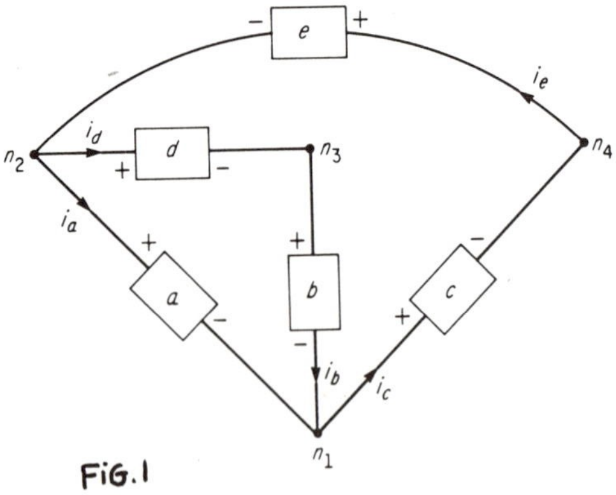

Fig.1

Write the KVL equations assuming all these loops are clockwise.

loop dba : $v_a = v_d + v_b$

loop ecbd: $v_e + v_c + v_b + v_d = 0$

loop eca : $v_e + v_c + v_a = 0$.

• PROBLEM 2-33

Write all the linearly independent KCL and KVL equations for the circuit shown in Fig. 1.

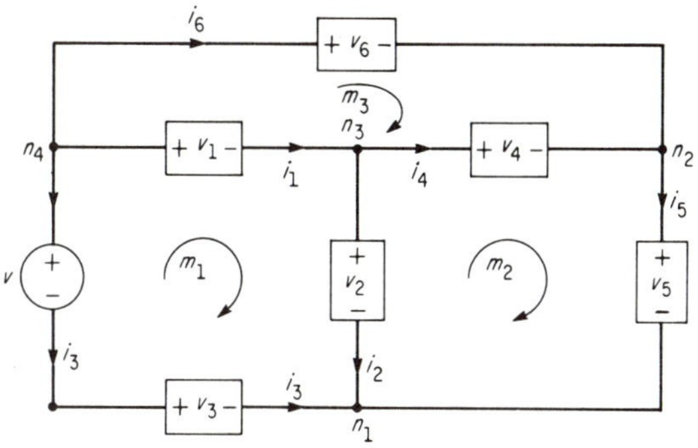

FIG. 1

Solution: Taking n_1 as the reference node, it is possible to write three linearly independent KCL equations for the remaining nodes:

n_2: $i_4 + i_6 = i_5$

n_3: $i_1 = i_2 + i_4$

n_4: $i_3 + i_1 + i_6 = 0$

Subtracting the number of KCL equations from the number of branches in the network, gives the number of loops that must be chosen, 6 - 3 = 3, and then gives the number of linearly independent KVL equations.

m_1: $v_1 + v_2 = v_3 + v$

m_2: $v_4 + v_5 = v_2$

m_3: $v_6 = v_4 + v_1$.

• PROBLEM 2-34

Write the independent KVL and KCL equations for the network shown in Fig. 1 using the loops and nodes labeled.

Solution: The network has 4 nodes and 6 branches, therefore, write, 6 - 4 + 1 = 3, independent KCL equations and three independent KCL equations. Taking n_1 as the reference node the independent KCL equations are

$n_2: \quad i_6 - i_1 = 0$

$n_3: \quad i_1 - i_2 - i_3 - i_4 = 0$

$n_4: \quad i_4 + i_3 - i_5 = 0$

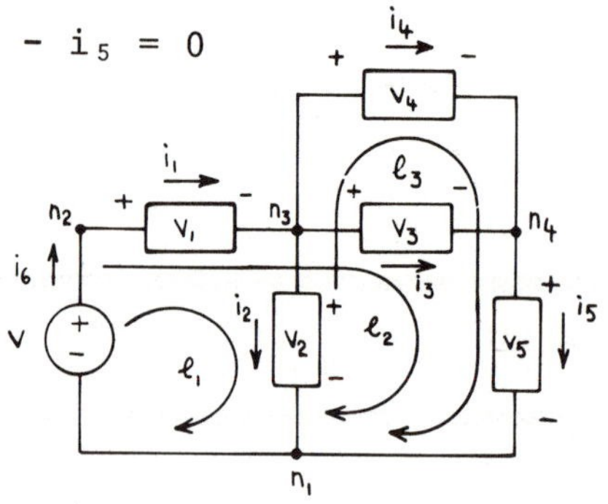

FIG. 1

Summing the voltages around the loops shown, write the independent KVL equations as follows:

$\ell_1: \quad v = v_1 + v_2$

$\ell_2: \quad v = v_1 + v_3 + v_5$

$\ell_3: \quad 0 = -v_2 + v_4 + v_5 \quad .$

• PROBLEM 2-35

Write all KCL and KVL equations for the networks shown, and choose node voltages and branch currents arbitrarily.

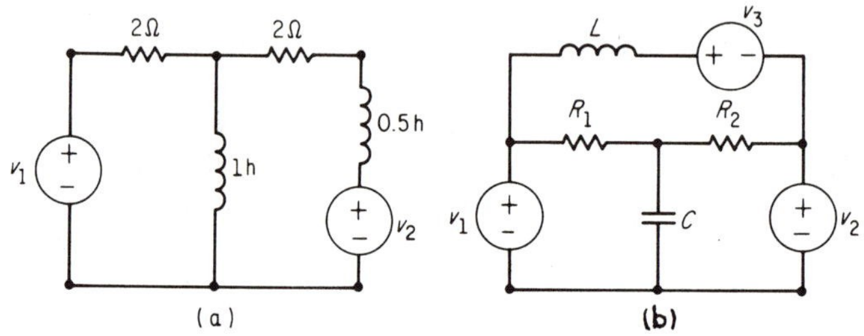

FIG. 1

Solution: Before writing the KVL equations for Fig. 1(a), label the loops used. Fig. 2 shows the three possible loops.

The directions are chosen arbitrarily. Hence, the KVL equations are

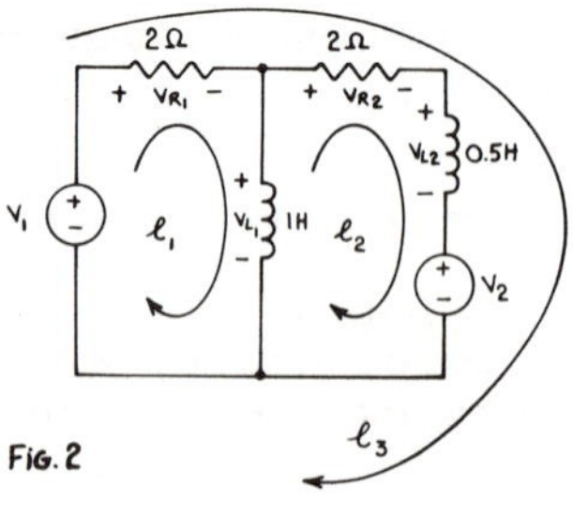

loop $\ell_1: \quad v_1 = v_{R_1} + v_{L_1}$

loop $\ell_2: -v_2 = -v_{L_1} + v_{R_2} + v_{L_2}$

loop $\ell_3: v_2 - v_1 = v_{R_1} + v_{R_2} + v_{L_2}$

FIG. 2

Writing the KCL equations for the nodes shown in Fig. 3 below gives,

node n_1: $i_1 - i_2 - i_3 = 0$

node n_2: $i_3 - i_4 = 0$

node n_3: $i_2 + i_4 - i_3 = 0$

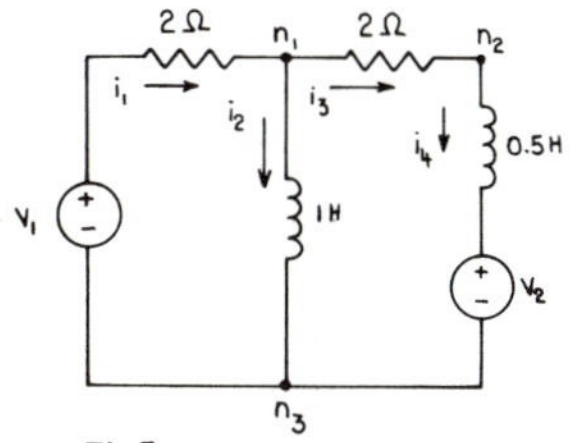

Fig.3

Labeling the loops in the circuit of Fig. 1(b) (shown in Fig. 4), gives the KVL equations,

loop ℓ_1: $v_1 = v_{R_1} + v_C$

loop ℓ_2: $-v_2 = v_{R_2} - v_C$

loop ℓ_3: $-v_3 = v_L - v_{R_1} - v_{R_2}$

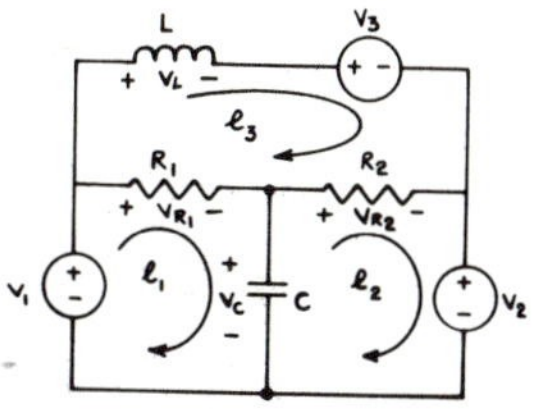

Fig.4

These however are not all the possible KVL equations, Fig. 5 below shows 4 more possible loops.

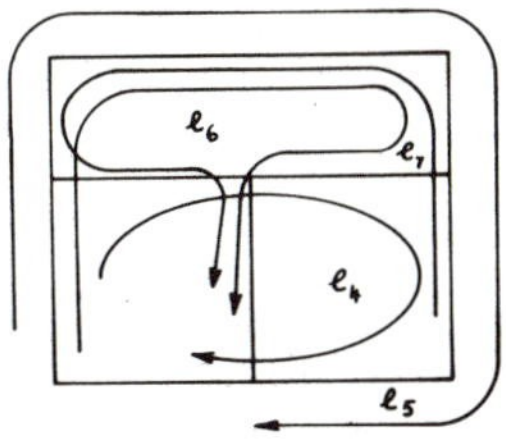

Fig.5

• PROBLEM 2-36

Fig. 1 shows a circuit containing a n-channel enhancement mode MOS-FET semiconductor which can be approximated by an equivalent network shown in Fig. 2. Write the independent KVL and KCL equations of the network.

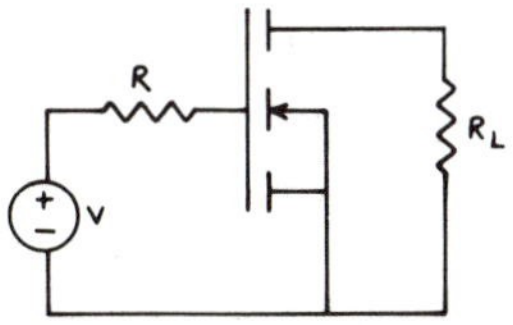

Fig.1

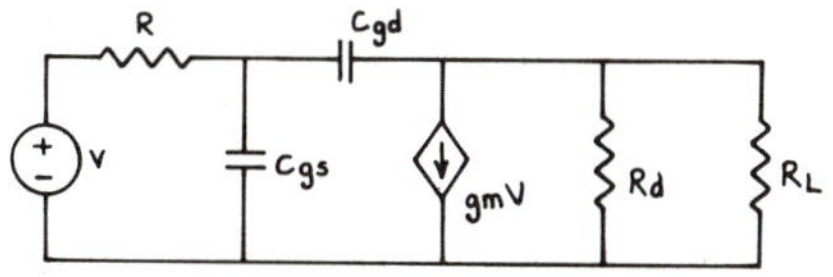

Fig. 2

Solution: Fig. 3 shows the loops selected in the network of Fig. 2.

Note that there are 7 branches and 4 nodes. Hence, N = 4 - 1 = 3, there are B - N = 7 - 3 = 4 independent KVL equations. Since the current in one branch is already defined as $g_m v$, in the branch containing the dependent current source, we have 3 independent KVL equations.

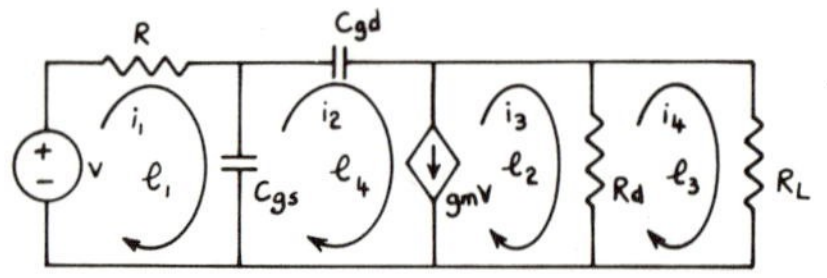

Fig. 3

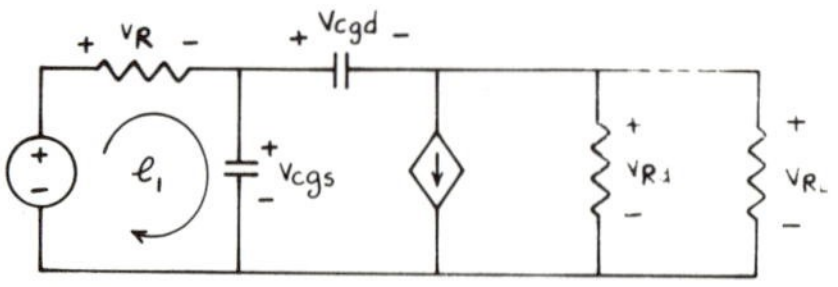

Fig. 4

To prove this, start by writing the KVL equations for the loops in Fig. 3. But first we define the voltages around each loop as shown in Fig. 4.

Hence $v = v_R + v_{C_{gs}}$

For loop ℓ_1, this yields,

$$v = i_1 R + \frac{1}{C_{gs}} \int_{-\infty}^{t} (i_1 - i_2)\, dt .$$

For loop ℓ_2, we run into problems. Since the resistance of an ideal current source is infinite, we cannot include the branch with the current source in our loop equation. Therefore, replace the current source with an open circuit and form a "super-loop" as shown in Fig. 5.

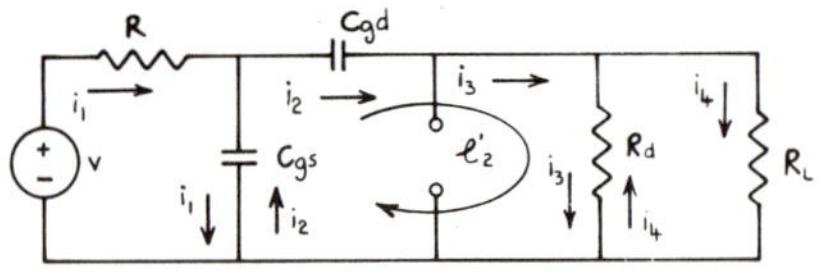

Fig. 5

For the "super-loop" ℓ'_2 we obtain the KVL equation

$$0 = v_{C_{gs}} + v_{C_{gd}} + v_{R_d}$$ where

$$-v_{C_{gs}} = \frac{1}{C_{gs}} \int_{-\infty}^{t} (i_2 - i_1)\, dt,$$

$$v_{C_{gd}} = \frac{1}{C_{gd}} \int_{-\infty}^{t} i_2\, dt$$

and $v_{R_d} = R_d\,(i_3 - i_4)$. Hence we have

$$0 = \frac{1}{C_{gs}} \int_{-\infty}^{t} (i_2 - i_1)\, dt + \frac{1}{C_{gd}} \int_{-\infty}^{t} i_2\, dt + R_d (i_3 - i_4).$$

For the loop ℓ_4

$$0 = -v_{rd} + v_{R_L} \text{ , thus}$$

$$0 = R_d\ (i_4 - i_3) + i_4\ R_i$$

Since there are now three equations and four unknowns, make the substitution $g_m v - i_3$ for i_2 or $i_2 - g_m r$ for i_3 in the above three KVL equations.

Substituting for i_3 gives:

$$v = i_1 R_1 + \frac{1}{C_{gs}} \int_{-\infty}^{t} (i_1 - i_2)\ dt$$

$$0 = \frac{1}{C_{gs}} \int_{-\infty}^{t} (i_2 - i_1) dt + \frac{1}{C_{gd}} \int_{-\infty}^{t} i_2\ dt$$

$$+ R_d\ (i_2 - g_m v - i_4)$$

$$- g_m v R_d = - i_2 R_d + i_4 (R_d + R_L)$$

Since there are 4 nodes in the network of Fig. 2 there are, at most, N = 4 - 1 = 3 linearly independent KCL equations.

Fig. 6 shows nodes selected including the reference node n_0.

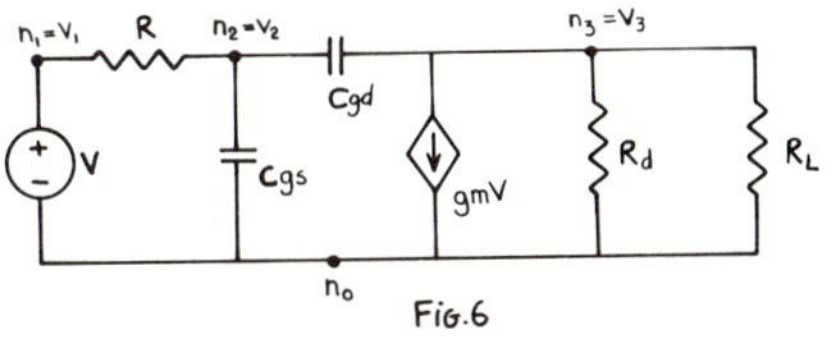

Fig.6

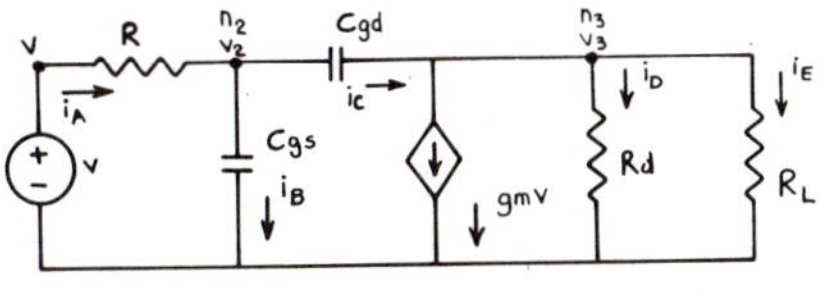

Fig.7

In the KVL equations, the unknowns were the loop currents i_1, i_2, i_3, i_4. In the KCL equations the unknowns are the node voltages labeled v_1, v_2, and v_3 in Fig. 6. Note that the voltage at node n_1 is already defined, $v_1 = v$.

Summing the currents at each node in Fig. 7, gives the two independent KCL equations.

node n_2 : $i_A = i_B + i_C$

node n_3 : $i_C = g_m v + i_D + i_E$

where $\quad i_A = \dfrac{v - v_2}{R}, \quad i_B = C_{gs} \dfrac{dv_2}{dt}, \quad i_C = C_{gd} \dfrac{d(v_2 - v_3)}{dt}$

$\quad i_D = \dfrac{v_3}{R_d} \quad$ and $\quad i_E = \dfrac{v_3}{R_L}$

Hence, $\quad \dfrac{v - v_2}{R} = C_{gs} \dfrac{dv_2}{dt} + C_{gd} \dfrac{d(v_2 - v_3)}{dt}$

$$C_{gd} \frac{d(v_2 - v_3)}{dt} = g_m v + \frac{v_3}{R_d} + \frac{v_3}{R_L}$$

are the KCL equations.

NORTON'S EQUIVALENT, THÉVENIN'S EQUIVALENT, & SUPERPOSITION

• PROBLEM 2-37

(a) By beginning with the practical current source at the right of Fig. 1, make repeated source transformation and resistance combinations in order to find the power supplied by the 24-V source. (b) What power does the 18-A source provide?

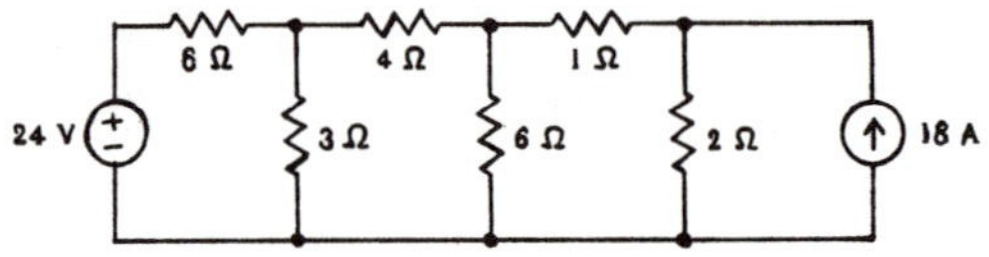

FIG. 1

Solution: (a) Figs. 2 - 6 show each transformation.

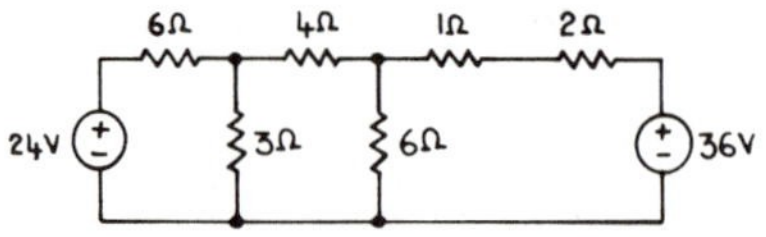

FIG. 2

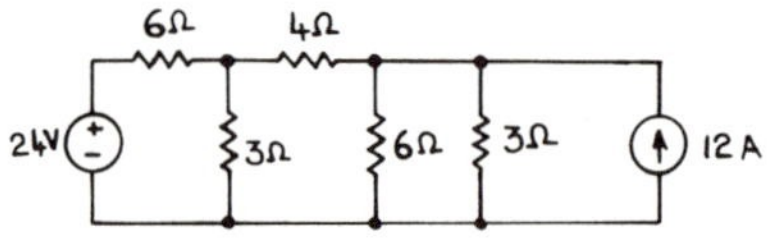

FIG. 3

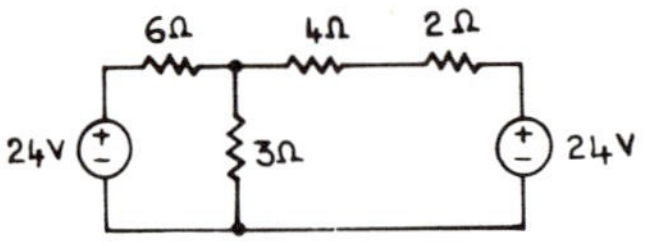

FIG. 4

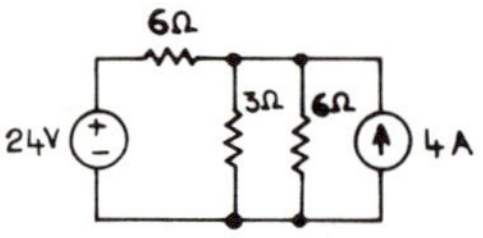

FIG. 5

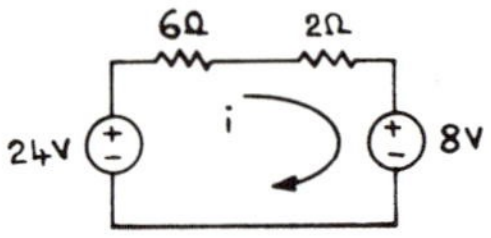

FIG. 6

In Fig. 6, find i by the KVL equation

$$24 - 8 = i\ (8)$$

$$i = \frac{16}{8} = 2A.$$

The power supplied by the 24-V source is

$$p = vi = (24)(2) = 48W.$$

(b) Starting at the left in the circuit of Fig. 1, and making repeated source transformations as shown in Figs. 7 - 11

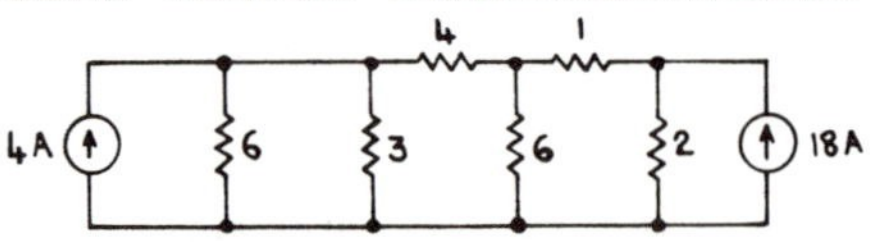

FIG. 7

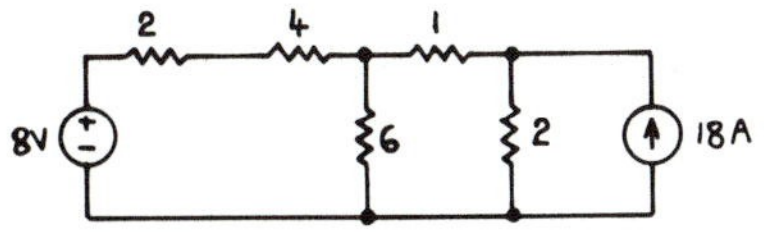

Fig. 8

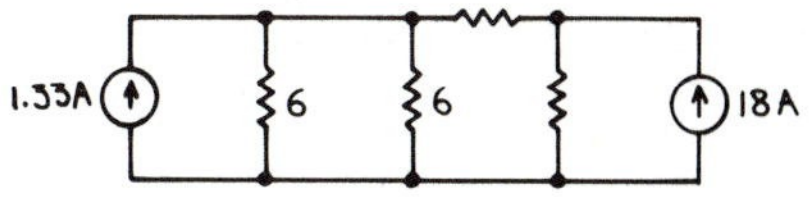

Fig. 9

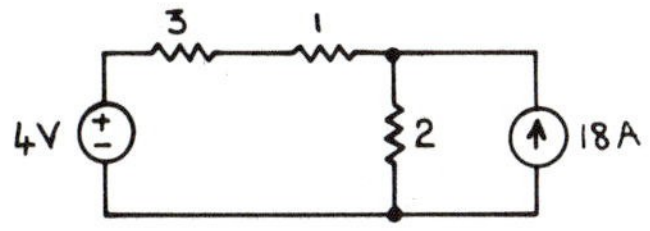

Fig. 10

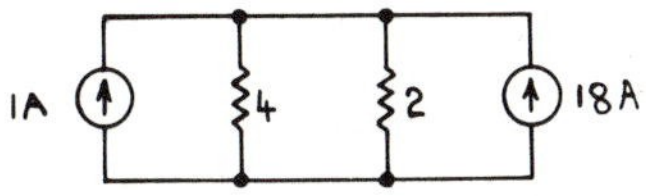

Fig. 11

gives the single node-pair circuit in Fig. 11. The KCL equation for this circuit is

$$1 + 18 = v\left(\frac{1}{4} + \frac{1}{2}\right)$$

$$v = \frac{19}{.75} = 25.33$$

The power supplied by the 18-A source is

$$p = vi = (25.33)(18) = 456 \text{ W}.$$

• PROBLEM 2-38

In Fig. 1 obtain Thevenin's equivalent circuit with respect to terminals a-b, and use the result to determine (a) the value of a resistance R to be connected between terminals a-b so that R receives maximum power from the active terminal pair; (b) the value of the maximum power delievered to R.

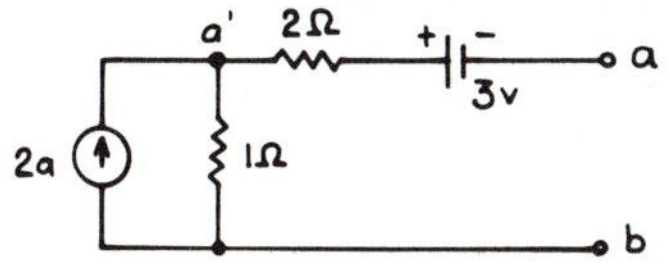

Fig. 1

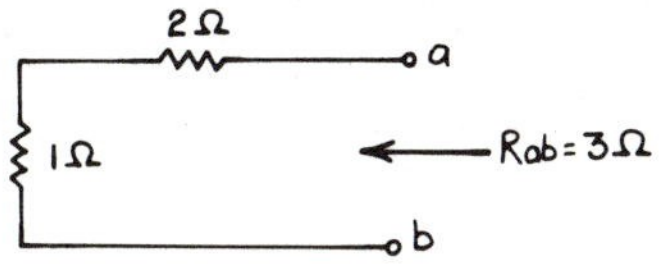

Fig. 2

Solution: (a) The Thevenin equivalent of any circuit can be found by solving for the open circuit voltage v_{oc} across the output terminals and placing this voltage v_{oc} in series with the Thevenin resistance which is the resistance across the terminals found by setting all independent voltage and current sources to zero (i.e. short circuits and open circuits, respectively).

The open circuit voltage v_{ab} can be found by noting that $v_{ab} = v_{a'b} + v_{aa'}$.

Since the terminals a-b are open circuited, all the source current must flow through the 1-Ω resistor. Therefore, $v_{a'b} = 2V$ and $v_{aa'} = -3V$ since no current flows through the 2-Ω resistor. Hence, the open circuit voltage is $v_{ab} = 2 - 3 = -1$ V.

The Thevenin resistance R_{ab} is found by open-circuiting the current source shown in Fig. 2.

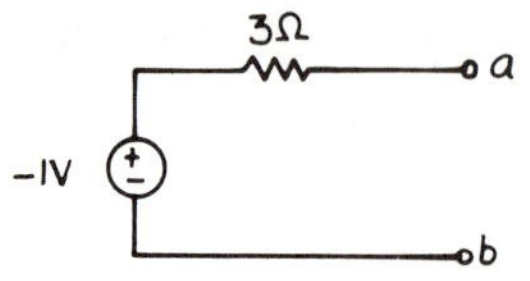

FiG.3

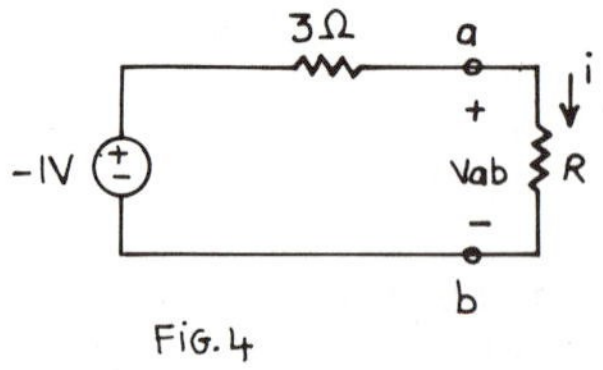

FiG.4

The Thevenin equivalent circuit is shown in Fig. 3.

(b) Maximum power transfer can be calculated by noting that

$$i = \frac{v_{ab}}{R} = \frac{-1}{3 + R}$$

$$v_{ab} = iR$$

$$p = i\,v_{ab} = \left(\frac{-1}{3 + R}\right)^2 R$$

from Fig. 4.

Hence, $$p = \frac{R}{R^2 + 6R + 9} \qquad (1)$$

for maximum power transfer $p = \infty$ when the denominator of Eq.(1) is 0. The value of R that makes $R^2 + 6R + 9$ zero is found by solving the quadratic equation

$$R = \frac{6 \pm \sqrt{6^2 + 4(9)}}{2} = \frac{6 \pm \sqrt{36 - 36}}{2}$$

giving $R = 3\ \Omega$.

• PROBLEM 2-39

Find the Thevenin and Norton equivalents of the circuits shown in: (a) Fig. 1a; (b) Fig. 1b with $R = \infty$; (c) Fig. 1b with $R = 1.6$ kΩ.

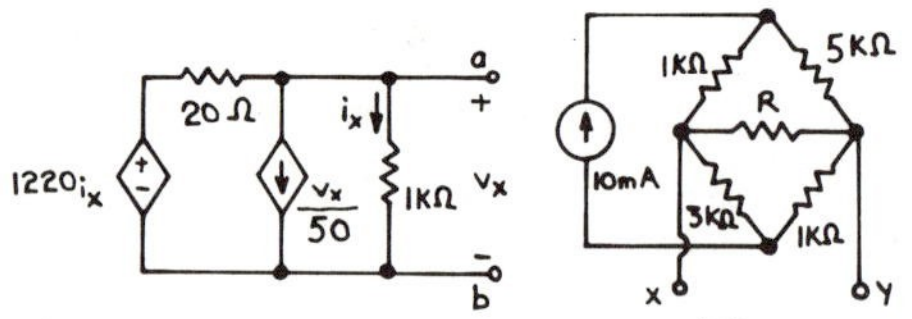

(a) (b)

Fig. 1

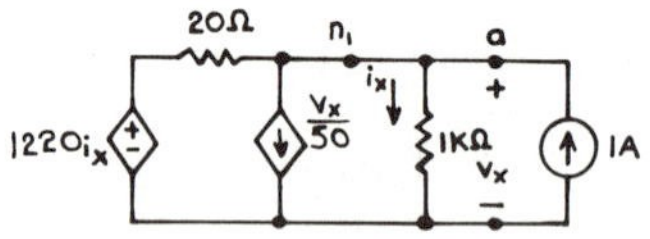

Fig. 2

Solution: (a) Since the circuit in Fig. 1(a) contains no independent sources, the Thevenin and Norton equivalents will have 0-A and 0-V sources respectively. To find R_{th}, connect a 1-A current source across terminals ab and find the voltage v_x. We find R_{th} directly from a nodal equation since $R_{th} = \frac{v_x}{1}$.

Fig. 2 shows the modified circuit.

Writing a nodal equation for node n_1 yields,

$$1 - \frac{v_x}{50} = \frac{v_x}{1000} + \frac{v_x - 1220\ i_x}{20}$$

Substitute $\frac{v_x}{1000}$ for i_x

$$1 - \frac{v_x}{50} = \frac{v_x}{1000} + \frac{v_x - 1.22\ v_x}{20}$$

$$1 -.02v_x = .001v_x - .011v_x$$

$$.01v_x = 1$$

$$v_x = 100v; \quad R_{th} = \frac{v_x}{1} = 100\ \Omega.$$

(b) The circuit shown in Figure 1(b) becomes as shown in Figure 3 when $R = \infty$, since an infinite resistance is equivalent to an open circuit. First, find the open circuit voltage v_{oc} across the terminals xy.

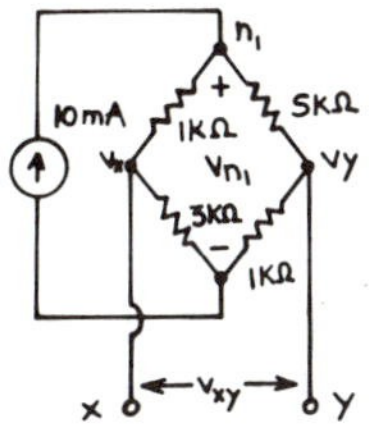

Fig.3

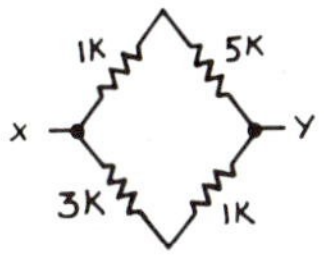

Fig.4

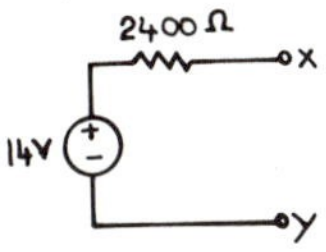

Fig.5

By writing a KCL equation for node n_1 in Fig. 3 the voltage v_{xy} can be found

$$10 \times 10^{-3} = \frac{v_{n_1}}{4 \times 10^3} + \frac{v_{n_1}}{6 \times 10^3}$$

$$v_{n_1} = \frac{10 \times 10^{-3}}{\left(\frac{1}{4} + \frac{1}{6}\right) \frac{1}{10^3}} = 24V.$$

Since $v_{xy} = v_x - v_y$,

and $$v_x = 24 - \left[\frac{24}{4 \times 10^3}(1 \times 10^3)\right] = 18V$$

$$v_y = 24 - \left[\frac{24}{(6 \times 10^3)}(5 \times 10^3)\right] = 4V,$$

we have $$v_{xy} = v_{oc} = 18 - 4 = 14V.$$

Find R_{th} by open circuiting the 10 mA source and by finding the equivalent resistance across terminals xy.

This is now a 6K resistance in parallel with a 4K resistance so

$$R_{th} = \frac{6 \cdot 4}{6 + 4} \times 10^3 = 2400\ \Omega.$$

The Norton equivalent current source is

$$\frac{14V}{2400\Omega} = 5.833\ mA.$$

(c) Find the open circuit voltage, v_{oc}, by applying KCL to the three nodes labeled in Fig. 6.

$$\left(\frac{1}{10^3}\right) \times 10 = \left[\frac{(v_1 - v_2)}{1} + \frac{v_1}{5}\right]\frac{1}{10^3}\ ; \quad \text{for node } n_1$$

$$0 = \left[\frac{(v_2 - v_1)}{1} + \frac{(v_2 - v_3)}{3} + \frac{(v_2)}{1.6}\right]\frac{1}{10^3}\ ; \quad n_2$$

$$-\left(\frac{1}{10^3}\right) \times 10 = \left[\frac{(v_3 - v_2)}{3} + \frac{v_3}{1}\right]\frac{1}{10^3}\ ; \quad n_3.$$

Simplifying,

$$10 = \left(1 + \frac{1}{5}\right)v_1 - v_2$$

$$0 = -v_1 + \left(1 + \frac{1}{3} + \frac{1}{1.6}\right)v_2 - \frac{1}{3}v_3$$

$$-10 = -\frac{1}{3}v_2 + \left(1 + \frac{1}{3}\right)v_3.$$

Solving for v_2 and v_{oc},

$$v_2 = v_{oc} = \frac{\begin{vmatrix} 1.2 & 10 & 0 \\ -1 & 0 & -.33 \\ 0 & -10 & 1.33 \end{vmatrix}}{\begin{vmatrix} 1.2 & -1 & 0 \\ -1 & 1.96 & -.33 \\ 0 & -.33 & 1.33 \end{vmatrix}}$$

$$= \frac{-10(-1.33) + 10(-.396)}{1(-1.33) + (1.96)(1.2)(1.33) + .33(-.396)}$$

$$v_2 = v_{oc} = \frac{9.34}{1.67} = 5.6\text{V}.$$

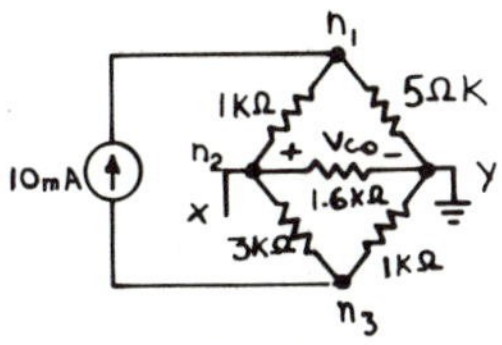

FiG.6

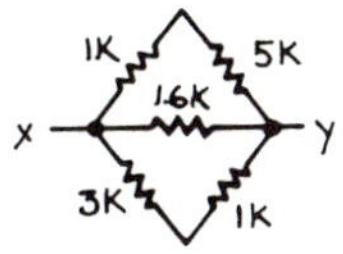

FiG.7

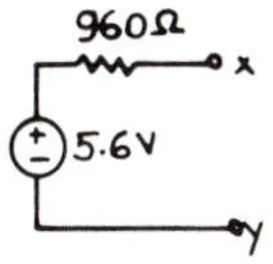

FiG.8

THÉVENIN EQUIVALENT CIRCUIT

Find R_{th} by open-circuiting the 10 mA source and by finding the equivalent resistance across terminals xy.

This is a 6K resistance in parallel with a 4K resistance which is parallel with a 1.6K resistance. Thus,

$$R_{th} = \frac{\left[\frac{6 \cdot 4}{6 + 4}\right] 1.6}{1.6 \left[\frac{6 \cdot 4}{6 + 4}\right]} \times 10^3 = 960\ \Omega.$$

the Norton equivalent current source is

$$\frac{5.6\text{V}}{900\Omega} = 5.833\ \text{mA}.$$

• PROBLEM 2-40

For the two-terminal network shown in Fig. 1, (a) obtain a single equation relating v and i; (b) plot a curve of i versus v.

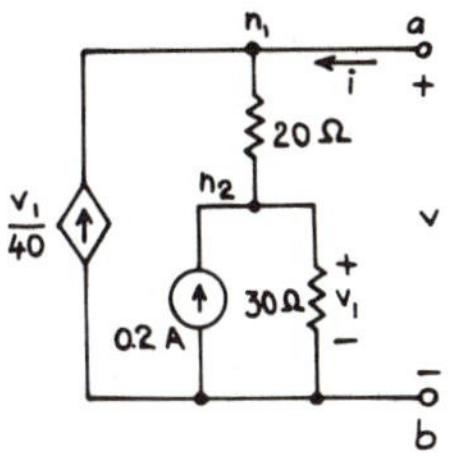

FiG.1

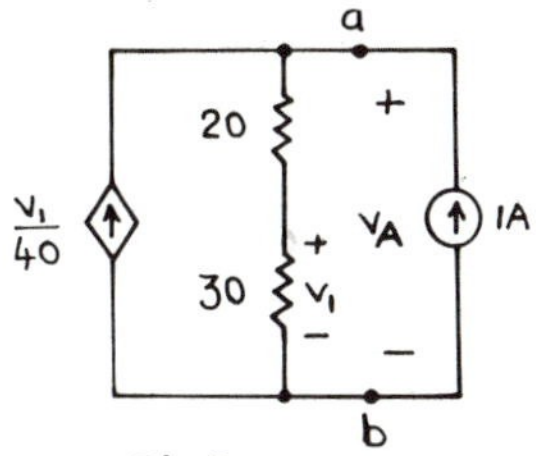

FiG.2

Solution: In order to write an equation relating i and v, first find the Thevenin-Norton equivalent of the circuit in Fig. 1. In doing so, the circuit shown is isolated from all interactions (through the terminals a-b) with any external circuitry. Therefore the current i shown in Figure 1 may be disregarded until the proper equivalents are obtained.

Write two KCL equations for nodes n_1 and n_2.

$$n_1: \quad \frac{v_1}{40} = \frac{v - v_1}{20} \tag{1}$$

$$n_2: \ 0.2 = \frac{v_1 - v}{20} + \frac{v_1}{30} \qquad (2)$$

Solving, for v_1 in Eq. (1) and substituting into Eq. (2) aids in solving for v.

$$\frac{v_1}{40} = \frac{v}{20} - \frac{v_1}{20}$$

$$v_1\left(\frac{1}{40} + \frac{1}{20}\right) = \frac{v}{20}$$

$$v_1 = \frac{v}{20\ (.075)} = \frac{2}{3}v.$$

Then,
$$0.2 = \frac{\frac{2}{3}v - v}{20} + \frac{\frac{2}{3}v}{30}$$

$$0.2 = -\frac{v}{60} + \frac{v}{45}$$

$$v = \frac{0.2}{\left(\frac{1}{45} - \frac{1}{60}\right)} = 36V.$$

In order to find R_{th}, open circuit the 0.2-A source and connect a 1-A source to terminals ab as shown in Fig. 2.

Writing a KCL equation for the top node yields

$$\frac{v_1}{40} + 1 = \frac{v_A}{50}\ . \qquad (3)$$

Write the voltage across the 20-Ω resistor as, $\frac{v_A}{50}(20)$. This allows us to write an expression for v_1 in terms of v.

$$v_1 = v_A - \frac{v_A}{50}(20)$$

$$v_1 = v_A - .4v_A$$

$$v_1 = .6v_A. \qquad (4)$$

Substituting Eq. (4) into (3):

$$1 = \frac{v_A}{50} - .015v_A$$

$$1 = (0.02 - .015)v_A$$

$$v_A = \frac{1}{.005}\ \text{volts}$$

$$R_{th} = \frac{v_A}{1} = 200\ \Omega.$$

The Norton equivalent current source is $\frac{36V}{200\Omega} = 0.18A$.

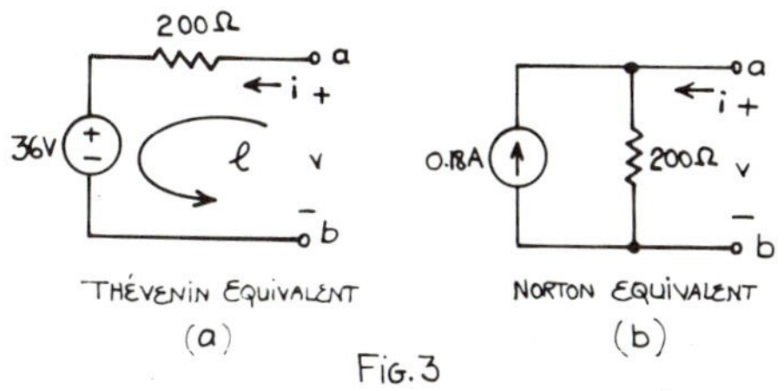

FIG. 3

Now, considering the external current input as well in Fig. 3(a), write a KVL equation around the loop shown.

$$v = i\ 200 + 36.$$

By solving the above equation for i, we have

$$i = \frac{v}{200} - 0.18. \qquad (5)$$

This equation is the equation of a straight line in the form,

$$y = x(m) + b.$$

Below is a graph of equation (5).

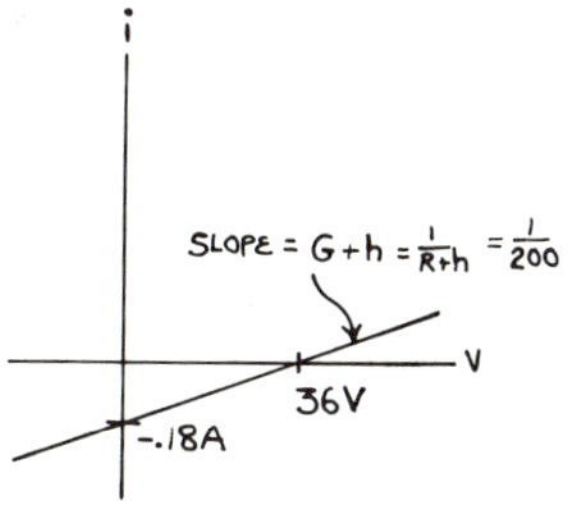

• **PROBLEM 2-41**

Find the Norton equivalent of the network to the left of terminals ab in Fig. 1 if element X is: (a) a 5-Ω resistor; (b) a 3-A independent current source, arrow directed upward; (c) a dependent voltage source, of magnitude $2i_1$, positive reference at top.

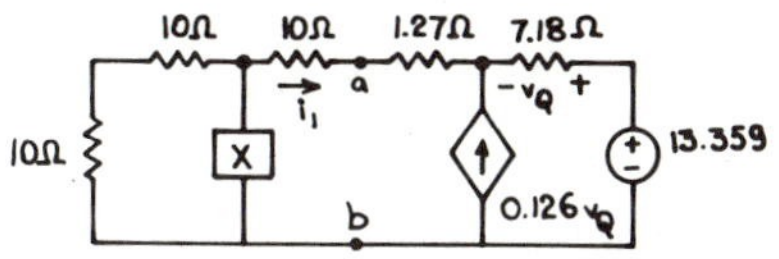

FIG. 1

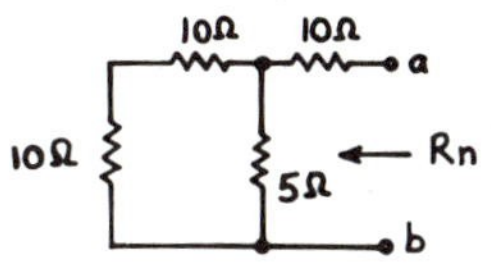

FIG. 2

Solution: Since the problem is to find the Norton equivalent of the network to the left of terminals ab, all the circuitry to the right is not important.

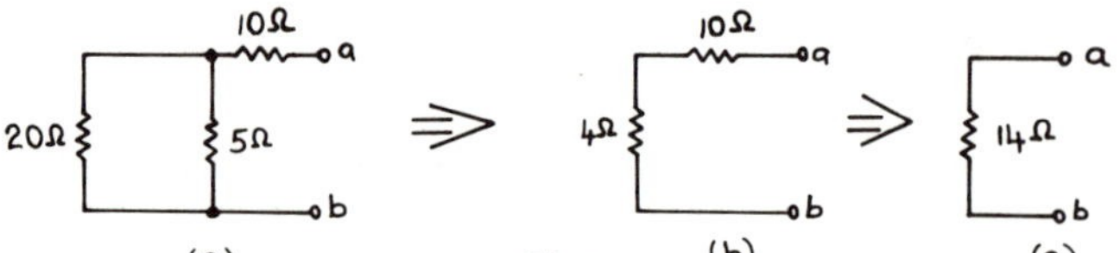

Fig. 3

With element X a 5-Ω resistor, $i_{sc} = 0$ and R_n is found

$$R_n = 14\Omega.$$

(b)   Fig. 4 shows the circuit for which the Norton equivalent must be determined.

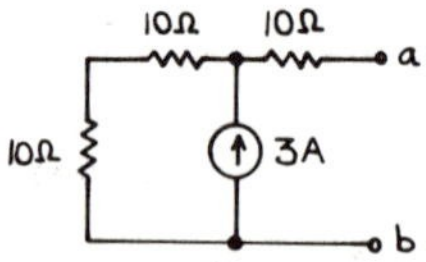

Fig. 4   Fig. 5

If we short terminals ab we can write a KCL equation and solve for i_{sc}. Fig. 5 shows the circuit.

$$3 = v\left(\frac{1}{20} + \frac{1}{10}\right) \quad v(G_1+G_2) \text{ at single node}$$

$$G_1 \qquad G_2$$

$$v = \frac{3}{0.15} = 20V$$

$$i_{sc} = \frac{20V}{10\Omega} = 2A.$$

Replacing the 3-A source with an open circuit gives R_n.

$$\leftarrow R_n = 10+10+10 = 30\Omega$$

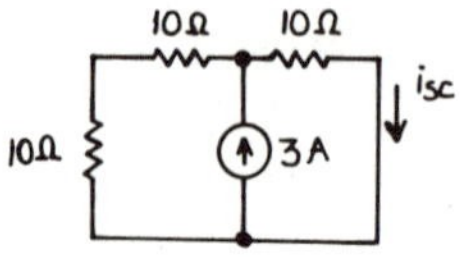

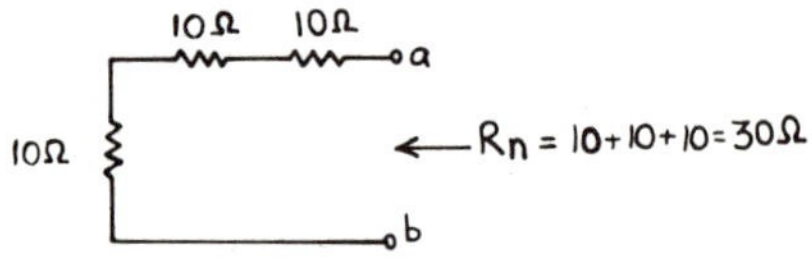

Fig. 6   Fig. 7

Fig. 7 shows the resulting Norton equivalent circuit.

(c)   Fig. 8 shows the new circuit.

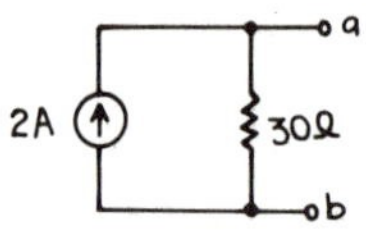

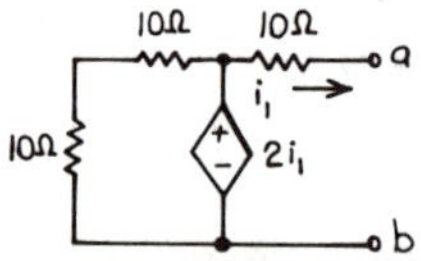

Fig. 8   Fig. 9

Connect a 1-A current source across the terminals ab as shown in Fig. 9.

By writing a KVL equation for the loop on the right $\frac{v}{I} = R_n$ is found.

$$v = v_R + v_{TERMINAL}$$

$$2i_1 = i_1(10) + v$$

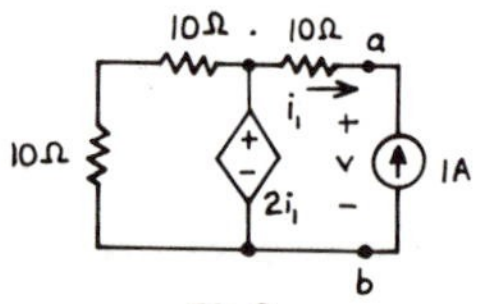

Since $i_1 = -1$

We have $2(-1) = -1(10) + v$

$$v = 8V$$

$$R_n = \frac{v}{1} = 8\Omega.$$

With no independent sources to the left of terminals ab $i_{sc} = 0$.

• PROBLEM 2-42

Find the Thevenin equivalent of the network shown in Fig. 1 if: (a) $C_1 = 0$, $C_2 = 12$ V; (b) $C_1 = 0.2$ A/V, $C_2 = 12$ V; (c) $C_1 = 0.2$ A/V, $C_2 = 0$.

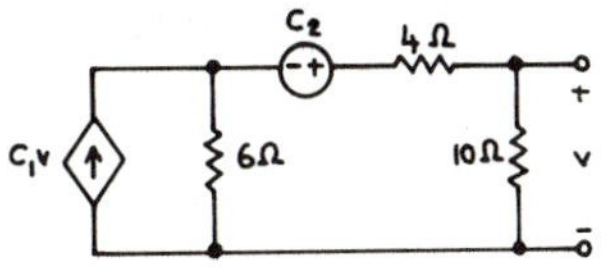

Fig. 1

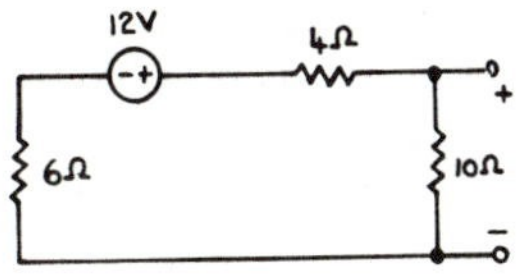

Fig. 2

Solution: Substituting $C_1 = 0$ and $C_2 = 12V$ into the circuit of Fig. 1 produces the new circuit of Fig. 2.

Add the 6-Ω and 4-Ω resistors and transform the 12-V source into a current source.

This results in Fig. 3.

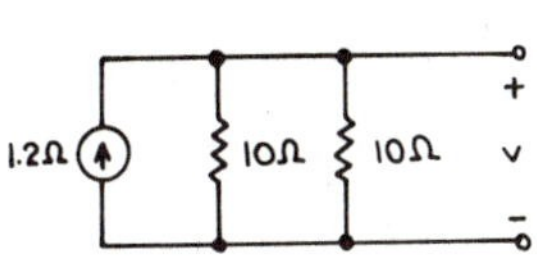

Fig. 3

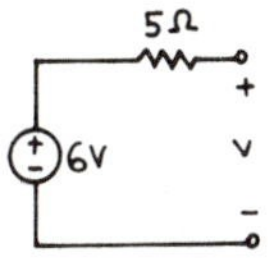

Fig. 4

Combining the two 10-Ω resistors and transforming the current source back to a voltage source gives the final Thevenin equivalent circuit shown in Fig. 4.

In the above method source transformations were used to find the Thevenin equivalent circuit.

In the following method, apply Thevenin's theorem to find the Thevenin equivalent circuit.

In Fig. 2 the voltage v can be found by use of a KVL equation around the single loop circuit. This yields

$$12 = i(20)$$

$$i = \frac{12}{20},$$

$$v = \frac{12}{20}(10) = 6V.$$

Thevenin's theorem states that the Thevenin equivalent circuit can be formed by a voltage source equal to v in series with the equivalent resistance R_{th} when all independent voltage sources and independent current sources are replaced by short circuits and open circuits, respectively.

Replacing the 12-V source by a short circuit yields

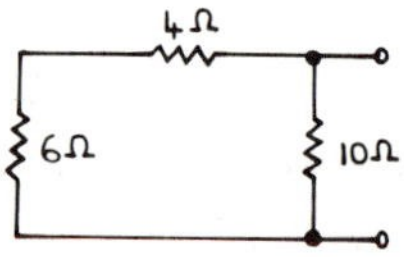

$$\leftarrow R_{th} = \frac{10\ (6 + 4)}{10 + (6 + 4)} = 5\Omega$$

Fig. 5

Finally, we have the same circuit as in Fig. 4.

(b) When there are dependent voltage sources, Thevenin's theorem must be applied.

First find the open circuit voltage v_{oc} which is equal to v.

Write a KCL equation for node n_1 in the circuit of Fig. 6.

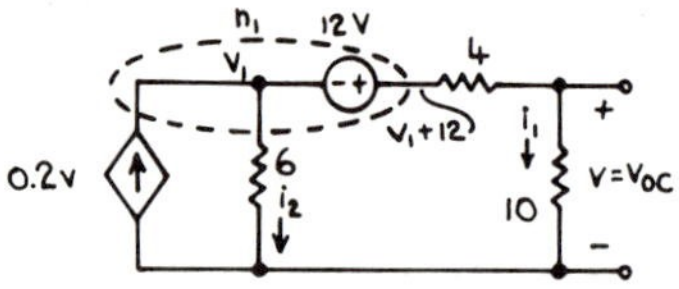

Fig. 6

$$0.2\ v = \underbrace{v_1\left[\frac{1}{6}\right]}_{i_2} + \overbrace{(v_1 + 12)\left(\frac{1}{4 + 10}\right)}^{i_1} . \qquad (1)$$

Knowing that $\qquad v = i_1\ (10)$ gives

$$v = (v_1 + 12)\left(\frac{1}{14}\right)(10).$$

Solving for v_1,

$$v_1 = \frac{14}{10}\ v - 12 .$$

Substitute this expression into Eq. (1) and solve for v.

$$0.2\ v = \left(\frac{14}{10}\right)\frac{1}{6}\ v - \frac{12}{6} + \left[\left(\frac{14}{10}\ v - 12\right) + 12\right]\left(\frac{1}{14}\right)$$

$$\frac{1}{5}\ v = \frac{7}{30}\ v - 2 + \frac{1}{10}\ v$$

$$v\left(\frac{1}{5} - \frac{7}{30} - \frac{1}{10}\right) = -\ 2$$

$$v\left(\frac{6-7-3}{30}\right) = -2$$

$$v\left(-\frac{4}{30}\right) = -2$$

$$v_{oc} = v = \left(-\frac{30}{4}\right)(-2) = \frac{60}{4} = 15V.$$

To find R_{th}, find i_{sc} (the short circuit current) and find R_{th} by $R_{th} = v_{oc}/i_{sc}$. Fig. 7 shows the circuit obtained by shorting the output terminals.

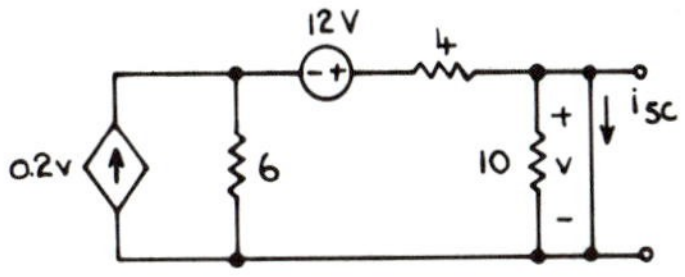

Fig 7

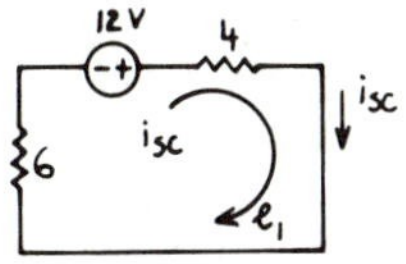

Fig. 8

Now since v = 0,

A KVL equation around loop ℓ_1 gives i_{sc}.

$$12 = i_{sc}\ (6 + 4)$$

$$i_{sc} = \frac{12}{6+4} = 1.2\ A$$

$$R_{th} = \frac{v_{oc}}{i_{sc}} = \frac{15}{1.2} = 12.5\ \Omega.$$

Fig. 9 shows the resulting Thevenin equivalent circuit.

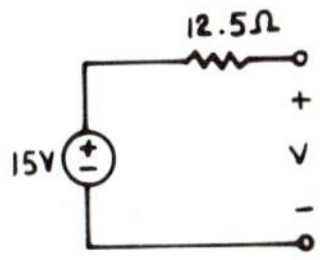

Fig 9

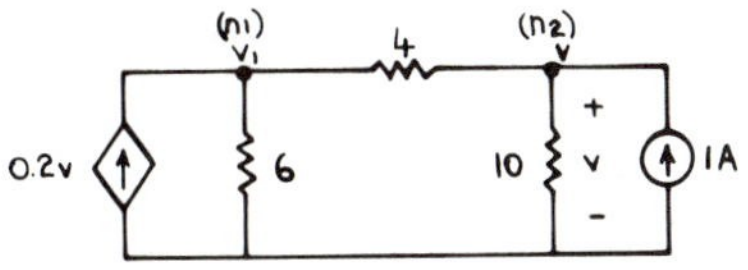

Fig. 10

(c) Since no independent voltage or current sources are given i_{sc} and v_{oc} can't be found to obtain R_{th} by v_{oc}/i_{sc}. If a 1-A current source is connected as shown in Fig. 10, one can find R_{th} by finding the voltage v.

Writing a KCL equation for node n_1 gives

$$\frac{1}{5}v = \frac{v_1}{6} + \frac{v_1 - v}{4}$$

Solving for v_1 in terms of v yields

$$\left(\frac{1}{5} + \frac{1}{4}\right) v = v_1 \left(\frac{1}{6} + \frac{1}{4}\right)$$

$$v_1 = v \frac{\left(\frac{1}{5} + \frac{1}{4}\right)}{\left(\frac{1}{6} + \frac{1}{4}\right)} = \frac{.45}{.4167} = 1.08v.$$

Writing a KCL equation for node n_2 gives

$$1 = \frac{v}{10} + \frac{v - v_1}{4}$$

$$1 = \frac{v}{10} + \frac{v\ (1 - 1.08)}{4}$$

$$R_{th} = \frac{v}{1} = \frac{1}{\frac{1 - 1.08}{4} + \frac{1}{10}} = \frac{1}{0.08} = 12.5\ \Omega$$

The entire network is dependent, hence acts passively and supplies no voltage so $v_{th} = 0$.

• PROBLEM 2-43

> The circuit shown in Fig. 1 is a linear equivalent circuit of a transistor circuit. The source v is an ideal independent voltage source, but $h_{21}i_1$ and $h_{12}v_2$ are dependent current and voltage sources, respectively. Determine Thevenin's equivalent circuit with respect to terminals a - b.

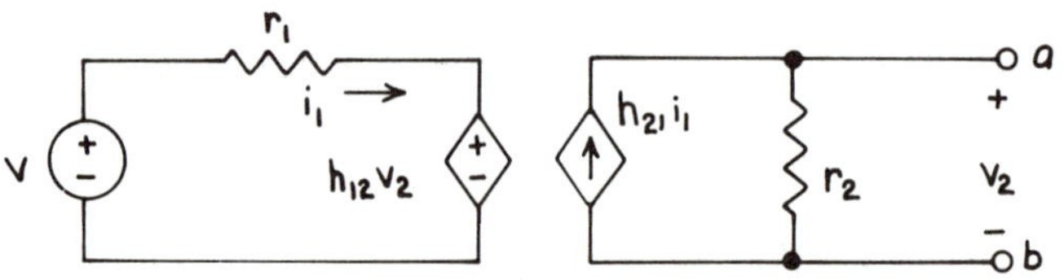

FIG. 1

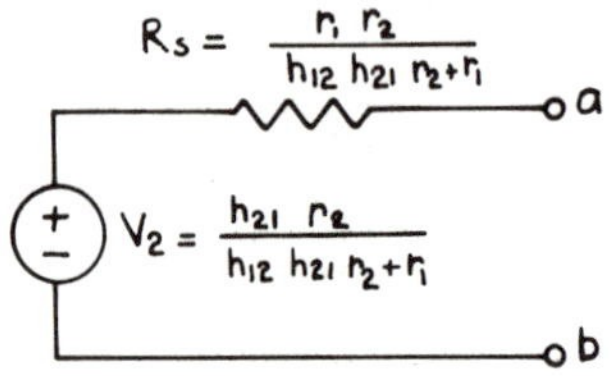

FIG. 2

Solution: The first step in finding the Thevenin equivalent of a circuit is to find the open circuit voltage, $(v_2)_{oc}$ which replaces all the sources in the Thevenin equivalent circuit.

In the right half of the circuit of Fig. 1 note that

$$(v_2)_{oc} = r_2\ h_{21}i_1 \tag{1}$$

where i_1 is found from the left half of the circuit

$$i_1 = \frac{v - h_{12}i_2}{r_1}\ . \tag{2}$$

Hence, substituting Eq. (2) into (1) yields

$$(v_2)_{oc} = r_2\ h_{21} \left(\frac{v - h_{12}v_2}{r_1}\right) \tag{3}$$

But since $v_2 = (v_2)_{oc}$, solve for $(v_2)_{oc}$,

$$(v_2)_{oc} = r_2\ h_{21} \left(\frac{v - h_{12}(v_2)_{oc}}{r_1}\right)$$

$$(v_2)_{oc} = \frac{r_2\ h_{21}\ v}{r_1} - \frac{r_2\ h_{21}\ h_{12}\ (v_2)_{oc}}{r_1}$$

$$(v_2)_{oc}\left[1 + \frac{r_2\ h_{21}\ h_{12}}{r_1}\right] = \frac{r_2\ h_{21}\ v}{r_1}$$

$$(v_2)_{oc} = \frac{\dfrac{r_2\ h_{21}\ v}{r_1}}{\left[1 + \dfrac{r_2\ h_{21}\ h_{12}}{r_1}\right]}$$

obtaining,

$$(v_2)_{oc} = \frac{\dfrac{r_2\ h_{21}\ v}{r_1}}{\dfrac{r_1 + r_2\ h_{21}\ h_{12}}{r_1}} = \frac{r_2\ h_{21}\ v}{r_1 + r_2\ h_{21}\ h_{12}} \quad .$$

In order to find the Thevenin impedance Z_s, short circuit terminals a - b and find the current i_s flowing from a to b which yields

$$Z_s = \frac{(v_2)_{oc}}{i_s}$$

i_s is to be the current produced by the dependent current source $h_{21}\ i_1$.

Since it has already been established that

$$i_1 = \frac{v - h_{12}\ v_2}{r_1} \quad ,$$

i_1 must be $\frac{v}{r_1}$ when the terminals a - b are short-circuited ($v_2 \equiv 0$). Hence,

$$i_s = \frac{h_{21}\ v}{r_1} \qquad \text{and}$$

$$Z_s = \frac{(v_2)_{oc}}{i_s} = \frac{r_1\ r_2}{r_1 + r_2\ h_{21}\ h_{12}} \quad .$$

Since h_{21} and h_{12} are real positive numbers Z_s is a pure resistance, R_s.

The Thevenin equivalent circuit is the Thevenin source $(v_2)_{oc}$ in series with the Thevenin resistance R_s shown in Fig. 2.

R_s is called the output impedance and the demensionless quantity $\frac{h_{21}\ r_2}{h_{12}\ h_{21}\ r_2 + r_1}$ is called the voltage gain.

• PROBLEM 2-44

Find the equivalent resistor R and equivalent capacitor C as shown in Fig. 1, where μF denotes a microfarad, equivalent to 10^{-6} farad.

FIG. 1

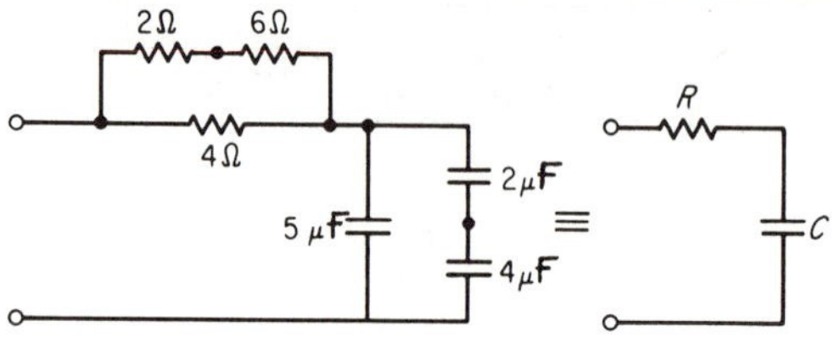

Solution: First, examine the resistors. Note that the 2-Ω and 6-Ω resistors are in series and can be replaced by a single resistance of 6 + 2 = 8Ω. Now the 8-Ω and 4-Ω resistors are in parallel. Combine them to get a single resistance R given by

$$\frac{1}{R} = \frac{1}{8} + \frac{1}{4}$$

$$R = \frac{8 \cdot 4}{8 + 4} = \frac{32}{12} = \frac{8}{3}\ \Omega\ .$$

For the capacitors, the 2μF and 4μF capacitors are in series and they can be combined to give a capacitor $\frac{2 \cdot 4}{2 + 4}$ μF or $\frac{4}{3}$ μF. This will be in parallel with a 5μF capacitor. Hence

$$C = 5 + \frac{4}{3} = \frac{19}{3}\ \mu F.$$

• PROBLEM 2-45

Find R, L, and C for the networks shown.

FIG. 1

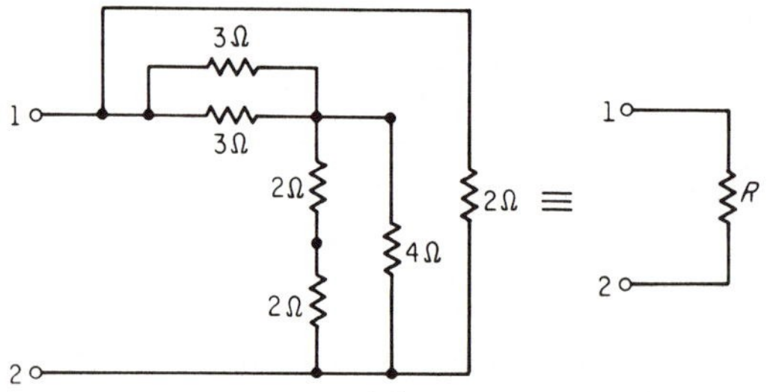

(a)

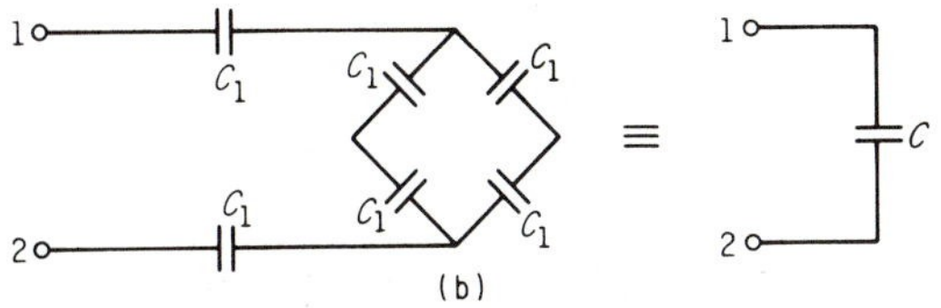

(b)

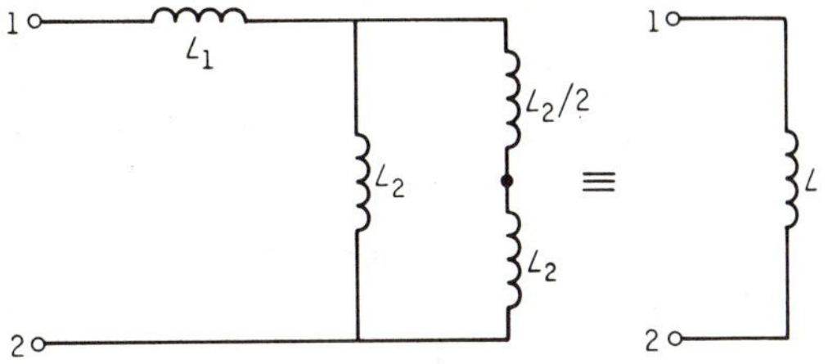

(c)

Solution: (a) First, combine the parallel resistances, the two 3-Ω resistors and the two 4-Ω resistors.

Finally; (1.5 + 2) Ω ||2 Ω

$$R = \frac{3.5\ (2)}{5.5} = 1.272\ \Omega$$

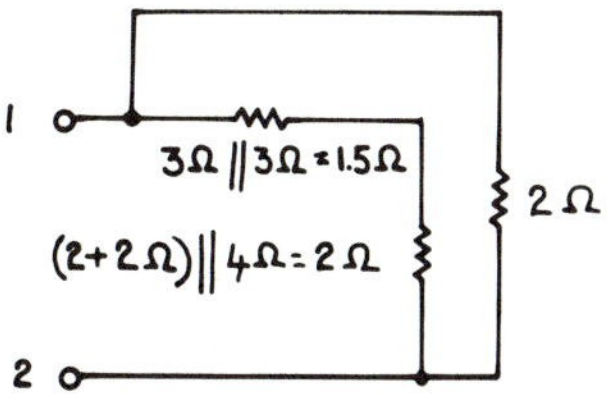

FiG. 2

(b) Since capacitors combine the conductances,

and finally,

$$C = \frac{1}{\frac{1}{C_1} + \frac{1}{C_1} + \frac{1}{C_1}} = \frac{1}{\frac{3}{C_1}}$$

$$C = \frac{C_1}{3}\ \text{F}.$$

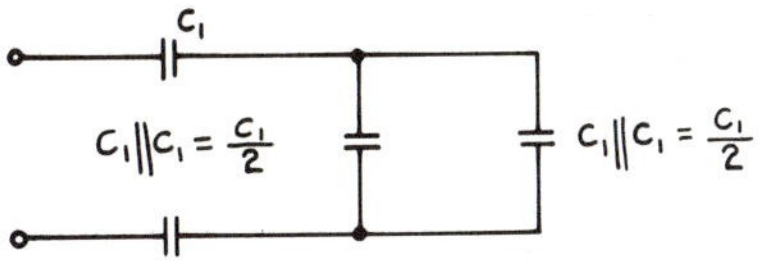

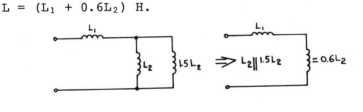

FiG. 3

(c) Since inductors combine like resistances

$L = (L_1 + 0.6L_2)$ H.

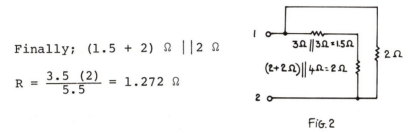

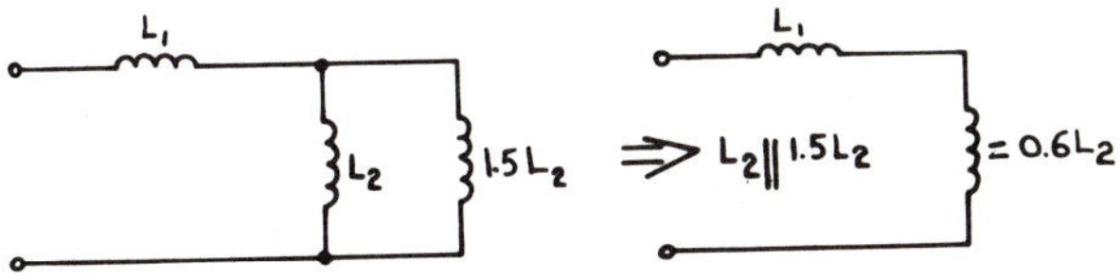

FiG. 4

• PROBLEM 2-46

(a) Use superposition to determine the magnitude and direction of the current in the 4-Ω resistor of Fig. 1.
(b) How would the above result change if there were also a 40-V source, positive reference to the right, in series with the 4-Ω resistor?

Solution: (a) First, open-circuit the current source and combine resistances to obtain the simplified circuit in Fig. 2.

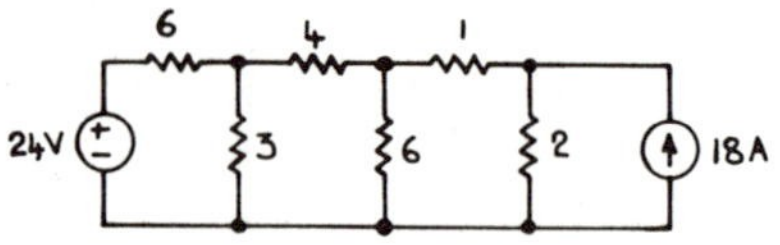

Fig. 1

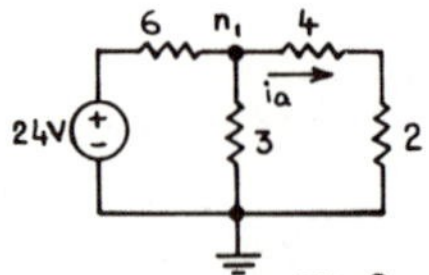

Fig. 2

By using the concept of voltage division, the voltage at node n_1 is found to be

$$v_{n_1} = \frac{24 \dfrac{3(4+2)}{3+(4+2)}}{6 + \dfrac{3(4+2)}{3+(4+2)}} = \frac{24(2)}{6+2} = 6\text{V}.$$

The current i_a is

$$i_a = \frac{v_{n_1}}{4+2} = \frac{6}{6} = 1\text{A}.$$

Now, short-circuit the 24-V source and combine resistances to obtain the simplified circuit in Fig. 3.

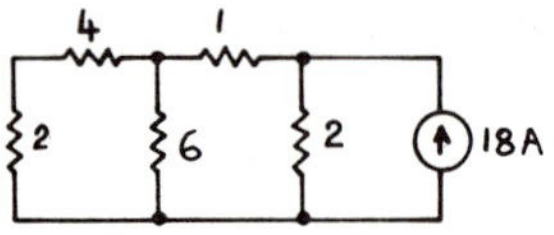

Fig. 3

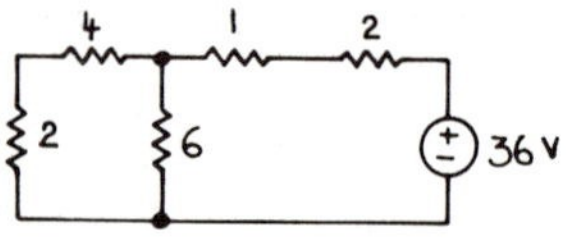

Fig. 4

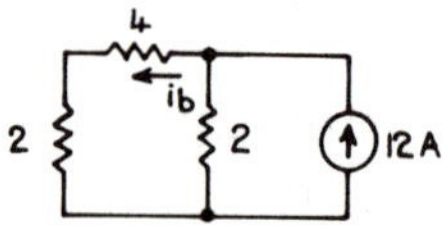

Fig. 5

Source transformations follow in Figs. 4 - 5.

Using the concept of current division, find the current i_b,

$$i_b = \frac{12(2)}{4+2+2} = \frac{24}{8} = 3\text{A}.$$

The total current through the 4-Ω resistor is illustrated in Fig. 6.

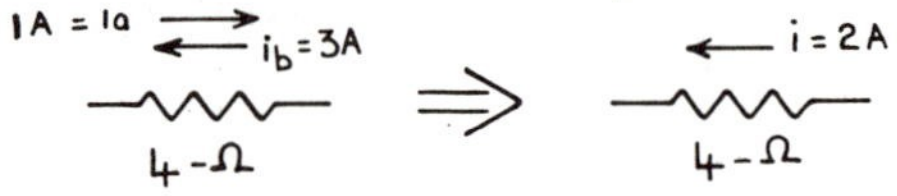

Fig. 6

(b) With the 40-V source in series with the 4-Ω resistor, the current through the 4-Ω resistor would receive i_a and i_b found in part (a) plus an additional current, i_c, due to the 40-V source alone.

Short circuit the 24-V source and open-circuit the 18-A source and then simplify the resulting circuit to give the circuit in Fig. 7.

i_c is found to be

$$i_c = \frac{40}{8} = 5\text{A}$$

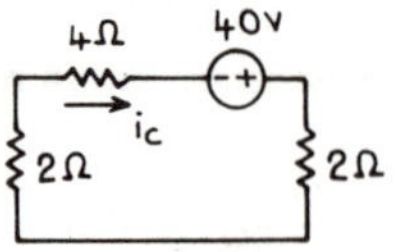

Fig. 7

so the total current is

3-A to the right.

• PROBLEM 2-47

> Use the principles of superposition and homogeneity to show that the network below is linear. The input is the voltage v and the output is the current i.

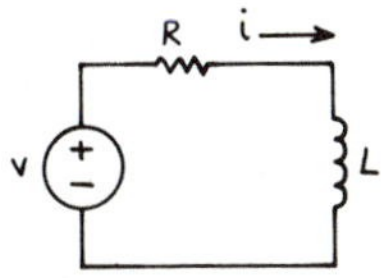

Solution: A network is said to be linear if it satisfies either superposition or homogeneity conditions.

Using superposition, KVL yields

$$v_1 = i_1R + L \frac{di_1}{dt} \tag{1}$$

when $v = v_1$, $i = i_1$ and $i(0) = 0$

$$v_2 = i_2R + L \frac{di_2}{dt} \tag{2}$$

when $v = v_2$, $i = i_2$ and $i(0) = 0$.

Hence, $$v_1 + v_2 = R\,(i_1 + i_2) + L \frac{d}{dt}(i_1 + i_2) \tag{3}$$

is the sum of the two conditions stated above. Note that an input of $v_1 + v_2$ gives an output of $i_1 + i_2$. Hence, the principle of superposition is satisfied. Next, apply the principle of homogeneity letting $v = kv_1$, $i = i_4$ and $i(0) = 0$. KVL gives,

$$kv_1 = i_4R\ L \frac{di_4}{dt} \tag{4}$$

multiplying equ. (1) by k gives

$$kv_1 = ki_1R + kL \frac{di_1}{dt} \quad .$$

Comparing with eq. (4) shows that $i_4 = ki_1$. Hence, the response to kv_1 is ki_1. The principle of homogeneity is satisfied. Therefore, the network is linear.

It is intuitively obvious that any combination of linear resistors, capacitors and inductors will result in a linear network. There are two important classes of networks that may be considered as nonlinear. First: networks, capacitors and inductors, but for which some of the initial conditions are nonzero; second: networks that contain one or more nonlinear elements.

• **PROBLEM 2-48**

Use superposition to find v in each of the circuits shown in Fig. 1.

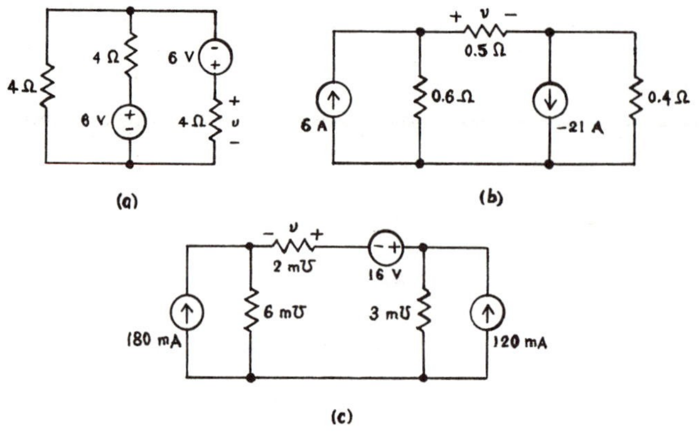

Fig. 1

Solution: (a) Replace the 6-V source on the right with a short circuit and find the currents through the simplified circuit in Fig. 2.

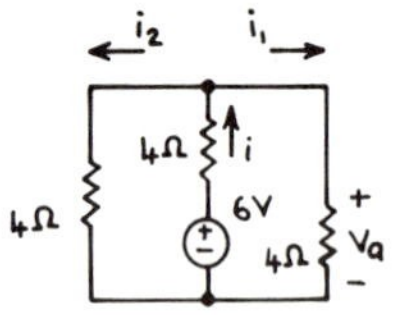

Fig. 2

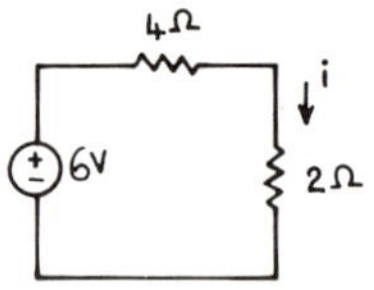

Fig. 3

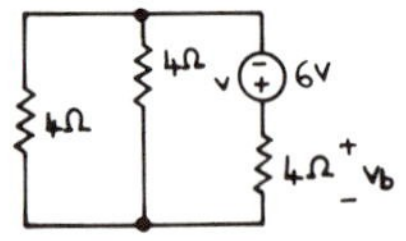

Fig. 4

Note that $i = i_1 + i_2$ and, since both i_1 and i_2 flow through a 4-Ω resistor,

$$i_1 = i_2 \qquad \text{and} \qquad i = \frac{i_2}{2} = \frac{i_1}{2} .$$

Combining the two 4-Ω resistors yields the circuit in Fig. 3.

$$i = \frac{6}{4 + 2} = 1 \text{ A}$$

$$i_1 = \tfrac{1}{2} \text{ A} , \qquad v_a = (\tfrac{1}{2})\, 4 = 2\text{V}.$$

Go back to the original circuit in Fig. 1(a) and replace the 6-V source on the left with a short circuit, and solve for the currents in the simplified circuit shown in Fig. 4.

From the circuit in Fig. 4, the circuit in Fig. 5 can be derived. Solving for i_3, we have

$$i_3 = \frac{6}{4 + 2} = 1 \text{ A;} \qquad v_b = (1)(4) = 4\text{V}.$$

By adding the currents, obtained by analyzing the effect of one source at a time on the circuit the current, when all sources are connected, can be found.

Thus, $(i_1 + i_3)\ 4 = v$

$(\frac{1}{2} + 1)\ 4 = 6V.$

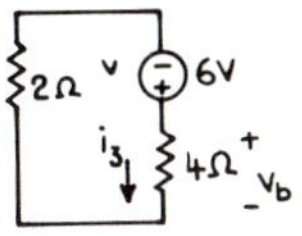

Fig. 5

(b) Applying the theorem of superposition yields the two circuits in Figs. 6 and 7.

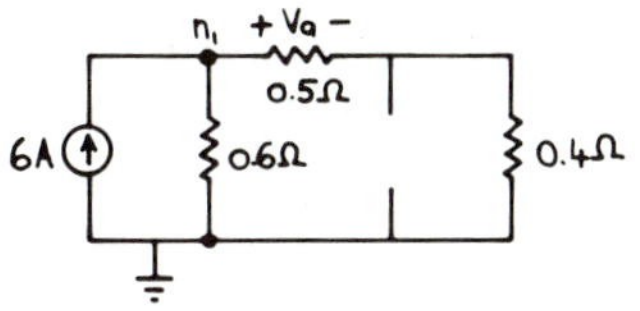

Fig. 6

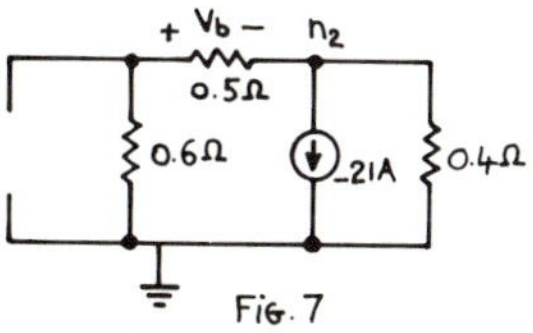

Fig. 7

In the circuit of Fig. 6, apply KCL to the voltage at node n_1.

$$6 = v_{n_1}\left(\frac{1}{.5 + .4} + \frac{1}{.6}\right)$$

$$v_{n_1} = \frac{6}{\frac{1}{.5 + .4} + \frac{1}{.6}} = \frac{6}{1.111 + 1.667} = 2.16V.$$

The voltage v_a, due to 6-A source alone is

$$v_a = i(0.5) = \left(\frac{v_{n_1}}{.5 + .4}\right)(0.5)$$

$$v_a = \left(\frac{2.16}{.9}\right)(0.5) = 1.2V.$$

In the circuit of Fig. 7, apply KCL to find the voltage at node n_2

$$-(-21) = v_{n_2}\left(\frac{1}{.5 + .6} + \frac{1}{.4}\right)$$

$$v_{n_2} = \frac{21}{\frac{1}{.5 + .6} + \frac{1}{.4}} = \frac{21}{.909 + 2.5} = 6.16\ V.$$

The voltage v_b, due to the - 21 A source alone is

$$v_b = i\ (0.5) = \left(-\frac{v_{n_1}}{.5 + .6}\right)(0.5)$$

$$v_b = \left(-\frac{6.16}{1.1}\right)(0.5) = -2.8\ V.$$

Obtain v by adding the two voltages

$$v = v_a + v_b = 1.2 - 2.8 = -1.6\text{V}.$$

(c) Applying the superposition theorem, obtain three circuits from Fig. 1(c) shown in Figs. 8 - 10.

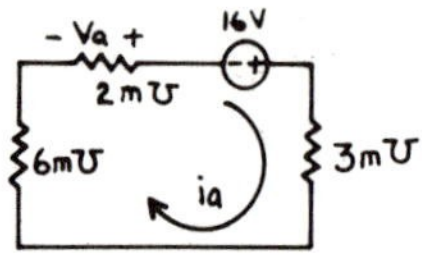

180 mA and 120 mA sources open circuited

Fig. 8

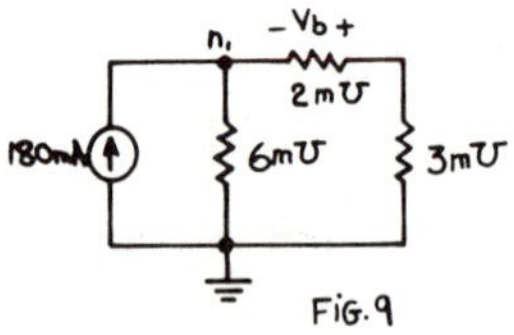

16V source short circuited and 120 mA source open circuited

Fig. 9

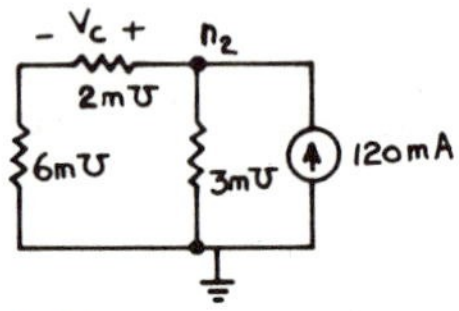

16V source short circuited and 180 mA source open circuited

Fig. 10

Superposition gives $v = v_a + v_b + v_c$

Note that conductance values are given.

Writing the KVL equation around the loop in Fig. 8 yields

$$16 = i_a \left(\frac{1}{2} + \frac{1}{6} + \frac{1}{3}\right) \times \frac{1}{10^{-3}}$$

$$i_a = 16 \text{ mA}$$

$$v_a = -\frac{16 \text{ mA}}{2 \text{ m}} = -8\text{V}.$$

Writing the KCL equation for node n_1 in Fig. 9 yields

$$180 \text{ mA} = v_{n_1} \left(\frac{2 \cdot 3}{2 + 3} + 6\right) \times 10^{-3}$$

$$v_{n_1} = \frac{180}{\left(\frac{6}{5} + 6\right)} = 25\text{V}$$

$$v_b = -\frac{i_b}{2\text{m}} = -\frac{v_{n_1}\left(\frac{6}{5}\right)}{2} = -15.0\text{V}.$$

Writing the KCL equation for node n_2 in Fig. 10 gives

$$120\text{ mA} = v_{n2}\left(\frac{2 \cdot 6}{2 + 6} + 3\right) \times 10^{-3}$$

$$v_{n2} = \frac{120}{\frac{12}{8} + 3} = 26.667\text{ V}$$

$$v_c = \frac{i_c}{2m} = \frac{v_{n2}\left(\frac{12}{8}\right)}{2} = 20\text{V}.$$

Applying superposition

$$v = v_a + v_b + v_c$$

$$v = (-8) + (-15) + (20) = -3\text{V}.$$

• PROBLEM 2-49

First find v_x in the circuit of Fig. 1 by using the superposition principle and then determine the power generated by: (a) the 5-A source; (b) the 6-V source; (c) the dependent source.

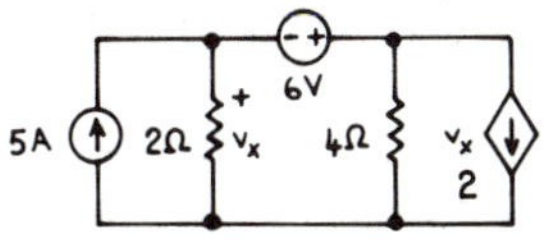

Fig. 1

Solution: Find v_x before attempting to find the power values for parts (a), (b) and (c) because the voltage v_x makes it possible to find the power absorbed or supplied by every element using simple algebraic manipulation.

Finding v_x by superposition simplifies our task since there are two independent sources and one dependent current source.

Use superposition to obtain the two circuits in Figs. 2 and 3. Since dependent sources are always active, include them in each circuit.

The KCL equation for node n_1 in Fig. 2:

$$5 - \frac{v_a}{2} = v_a\left(\frac{1}{4} + \frac{1}{2}\right)$$

$$1.25\, v_a = 5$$

$$v_a = 4\text{V}.$$

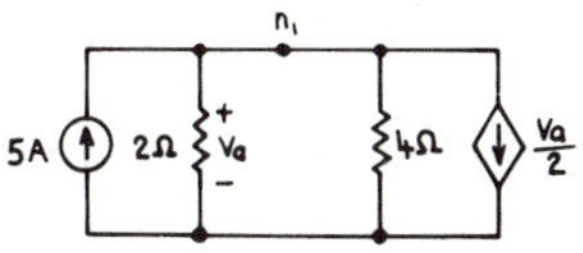

Fig. 2

The KCL equation for Fig. 3:

$$-\frac{v_b}{2} = v_b\left(\frac{1}{2}\right) + (v_b + 6)\,\frac{1}{4}$$

$$-\frac{v_b}{2} = v_b\left(\frac{3}{4}\right) + 1.5$$

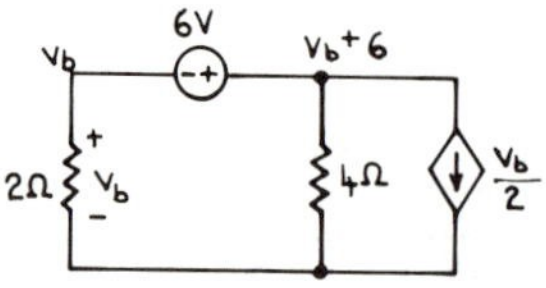

Fig.3

$$1.25\ v_b = -\ 1.5$$

$$v_b = -\ 1.2V.$$

Since $v_x = v_b + v_a$ then $v_x = 4 + (-\ 1.2)$

$$v_x = 2.8V.$$

(a) Since v_x is the voltage across the 5-A source, one can find the power it generates.

$$p = vi = (2.8)(5) = 14W$$

(b) Find the current through the 6-V source by applying KCL at the node n_3 in Fig. 4.

$$5A - i_2 = i_1\ ; \quad i_2 = \frac{2.8V}{2\Omega} = 1.4A$$

$$5 - 1.4 = i_1 = 3.6$$

$$p = vi = (6)(3.6) = 21.6$$

(c) Find the voltage across the dependent current source $\frac{v_x}{2}$ by applying KVL around the loop ℓ_1 in Fig. 4 to find the voltage v_y. Then apply KVL to loop ℓ_2 to prove that the voltage across the dependent current source is $-\ v_y$.

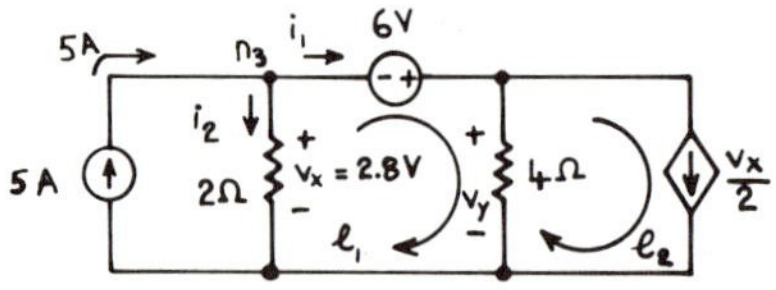

Fig. 4

$$v_x + 6 = v_y$$

$$2.8 + 6 = v_y = 8.8.$$

$\frac{v_x}{2}$ current source supplies negative power.

$$p = vi = (-\ 8.8)\left(\frac{v_x}{2}\right) = -\ (8.8)\left(\frac{2.8}{2}\right) = -\ 12.32W.$$

CHAPTER 3

MATRIX METHODS

MATRIX MATH

• PROBLEM 3-1

Let

$$A = \begin{bmatrix} 1 & -1 & 2 \\ 3 & 2 & 3 \end{bmatrix} \text{ and } B = \begin{bmatrix} 2 & 8 & 3 \\ 1 & 5 & 1 \end{bmatrix}, \text{ find } A + B \text{ and } A - B.$$

Solution: Since the two matrices are of the same dimension, we can add or subtract each corresponding element. Hence,

$$A + B = \begin{bmatrix} 1+2 & -1+8 & 2+3 \\ 3+1 & 2+5 & 3+1 \end{bmatrix} = \begin{bmatrix} 3 & 7 & 5 \\ 4 & 7 & 4 \end{bmatrix}$$

$$A - B = \begin{bmatrix} 1-2 & -1-8 & 2-3 \\ 3-1 & 2-5 & 3-1 \end{bmatrix} = \begin{bmatrix} -1 & -9 & -1 \\ 2 & -3 & 2 \end{bmatrix}$$

• PROBLEM 3-2

Obtain the following products and state the dimension of the new matrix. Can the order of multiplication be reversed?

(a) $$\begin{bmatrix} 2 & 4 & -5 \\ -7 & -1 & -2 \\ -3 & 0 & -1 \end{bmatrix} \begin{bmatrix} 1 & -2 \\ 3 & 3 \\ 1 & 5 \end{bmatrix}$$

(b) $$\begin{bmatrix} 1 & -3 & -1 \\ 5 & 2 & 7 \end{bmatrix} \begin{bmatrix} 1 \\ 2 \\ 3 \end{bmatrix}$$

(c) $$\begin{bmatrix} 6 & 1 & -1 \\ -1 & 4 & 7 \\ -3 & 3 & 1 \end{bmatrix} \begin{bmatrix} 2 & -4 & -6 \\ 4 & -1 & 0 \\ 6 & 3 & -2 \end{bmatrix}$$

(d) $\begin{bmatrix} 1 & 2 & 1/2 \\ 3/2 & -1/2 & 5/2 \\ -1 & -2 & 3 \end{bmatrix} \begin{bmatrix} 1 & 7 & 11 \\ 14 & 3 & 2 \\ -7 & 5 & -6 \end{bmatrix}$

(e) $\begin{bmatrix} 3 & -3 \\ 1 & 1 \end{bmatrix} \begin{bmatrix} x_1 \\ x_2 \end{bmatrix} + \begin{bmatrix} x_1 & -3 \\ -2 & x_2 \end{bmatrix} \begin{bmatrix} 1 \\ 2 \end{bmatrix}$

Solution: (a) Matrix multiplication is defined as

$$AB = \sum_{k=1}^{r} a_{ik} b_{kj} \qquad \begin{matrix} i = 1,2,\ldots,m \\ j = 1,2,\ldots,n \end{matrix}$$

where $A = (a_{jj})_{m \times r}$ $B = (b_{ij})_{r \times n}$

Hence

$$\begin{bmatrix} 2 & 4 & -5 \\ -7 & -1 & -2 \\ -3 & 0 & -1 \end{bmatrix} \begin{bmatrix} 1 & -2 \\ 3 & 3 \\ 1 & 5 \end{bmatrix} =$$

$$\begin{bmatrix} (1)(2) + 3(4) + (1)(-5) & (-2)(2) + (3)(4) + (5)(-5) \\ (1)(-7) + 3(-1) + (1)(-2) & (-2)(-7) + (3)(-1) + (5)(-2) \\ (1)(-3) + 3(0) + (1)(-1) & (-2)(-3) + (3)(0) + (5)(-1) \end{bmatrix}$$

Hence $\begin{bmatrix} 9 & -17 \\ -12 & 1 \\ -4 & 1 \end{bmatrix}$ is the product of the two matrices.

A 3 × 2 matrix

(b) $\begin{bmatrix} 1 & -3 & -1 \\ 5 & 22 & 7 \end{bmatrix} \begin{bmatrix} 1 \\ 2 \\ 3 \end{bmatrix} = \begin{bmatrix} (1)(1) + (2)(-3) + (3)(-1) \\ (1)(5) + (2)(2) + (3)(7) \end{bmatrix} = \begin{bmatrix} -8 \\ 30 \end{bmatrix}$

A 2 × 1 matrix

(c) $\begin{bmatrix} 6 & 1 & -1 \\ -1 & 4 & 7 \\ -3 & 3 & 1 \end{bmatrix} \begin{bmatrix} 2 & -4 & -6 \\ 4 & -1 & 0 \\ 6 & 3 & -2 \end{bmatrix} =$

$$\begin{bmatrix} (2)(6)+(4)(1)+(6)(-1) & (-4)(6)+(-1)(1)+(3)(-1) & (-6)(6)+(0)(1)+(-2)(-1) \\ (2)(-1)+(4)(4)+(6)(7) & (-4)(-1)+(-1)(4)+(3)(7) & (-6)(-1)+(0)(4)+(-2)(7) \\ (2)(-3)+(4)(3)+(6)(1) & (-4)(-3)+(-1)(3)+(3)(1) & (-6)(-3)+(0)(3)+(-2)(1) \end{bmatrix}$$

(1st column) (2nd column) (3rd column)

$$= \begin{bmatrix} 10 & -28 & -34 \\ 56 & 21 & -8 \\ 12 & 12 & 16 \end{bmatrix}$$

The product of two square matrices of the same dimension square matrix.

(d) $$\begin{bmatrix} 1 & 2 & 1/2 \\ 3/2 & -1/2 & 5/2 \\ -1 & -2 & 3 \end{bmatrix} \begin{bmatrix} 1 & 7 & 11 \\ 14 & 3 & 2 \\ -7 & 5 & -6 \end{bmatrix} =$$

$$\begin{bmatrix} (1)(1)+(14)(2)+(-7)(\frac{1}{2}) & (7)(1)+(3)(2)+(5)(\frac{1}{2}) & (11)(1)+(2)(2)+(-6)(\frac{1}{2}) \\ (1)(\frac{3}{2})+(14)(-\frac{1}{2})+(-7)(\frac{5}{2}) & (7)(\frac{3}{2})+(3)(-\frac{1}{2})+(5)(\frac{5}{2}) & (11)(\frac{3}{2})+(2)(-\frac{1}{2})+(-6)(\frac{5}{2}) \\ (1)(-1)+(14)(-2)+(-7)(3) & (7)(-1)+(3)(-2)+(5)(3) & (11)(-1)+(2)(-2)+(-6)(3) \end{bmatrix}$$

(1st column) (2nd column) (3rd column)

$$= \begin{bmatrix} \frac{51}{2} & \frac{31}{2} & 12 \\ -23 & \frac{43}{2} & \frac{1}{2} \\ -50 & 2 & -33 \end{bmatrix}$$

(e) $$\begin{bmatrix} 3 & -3 \\ 1 & 1 \end{bmatrix} \begin{bmatrix} x_1 \\ x_2 \end{bmatrix} + \begin{bmatrix} x_1 & -3 \\ -2 & x_2 \end{bmatrix} \begin{bmatrix} 1 \\ 2 \end{bmatrix}$$

$$= \begin{bmatrix} 3x_1 - 3x_2 \\ x_1 + x_2 \end{bmatrix} + \begin{bmatrix} x_1 - 6 \\ -2 + 2x_2 \end{bmatrix} = \begin{bmatrix} 4x_1 - 3x_2 - 6 \\ x_1 + 3x_2 \quad 2 \end{bmatrix}$$

• PROBLEM 3-3

Let $A = \begin{bmatrix} 2 & 2 \\ 1 & 2 \\ 4 & 2 \end{bmatrix}$, $B = \begin{bmatrix} 3 & 1 \\ 1 & 2 \end{bmatrix}$. Find AB.

Solution: The resulting matrix AB will be of dimension 3×2. Note that BA does not exist.

$$AB = \begin{bmatrix} 3 \cdot 2 + 1 \cdot 2 & 1 \cdot 2 + 2 \cdot 2 \\ 3 \cdot 1 + 1 \cdot 2 & 1 \cdot 1 + 2 \cdot 2 \\ 3 \cdot 4 + 1 \cdot 2 & 1 \cdot 4 + 2 \cdot 2 \end{bmatrix} = \begin{bmatrix} 8 & 6 \\ 5 & 5 \\ 14 & 8 \end{bmatrix}$$

• PROBLEM 3-4

Find $\overline{P} = \overline{AB}$ if

$$\overline{A} = \begin{bmatrix} 4 & 3 & 2 & 1 \\ 2 & 1 & 1 & 1 \\ 0 & 4 & 2 & 1 \\ 1 & 3 & 2 & 2 \end{bmatrix}, \quad \overline{B} = \begin{bmatrix} 1 & 1 & 1 & 1 \\ 2 & 2 & 3 & 1 \\ 3 & 1 & 1 & 1 \\ 1 & 1 & 1 & 2 \end{bmatrix}.$$

Solution: If $A = (a_{ij})_{m\times r}$ and $\overline{B} = (b_{ij})_{r\times n}$ then $\overline{P} = \overline{A}\,\overline{B} = (\sum_{k=1}^{r} a_{ik}b_{kj})$; $i = 1,2,\ldots, m$, $j = 1, 2, \ldots, n$. In the above example, $\overline{A} = (a_{ij})_{4\times 4}$, $\overline{B} = (b_{ij})_{4\times 4}$ and $\overline{P} = \overline{A}\,\overline{B} = (\sum_{k=1}^{4} a_{ik}b_{kj})$; $i = 1, 2, 3, 4$; $j = 1, 2, 3, 4$. Hence,

$$p_{11} = a_{11}b_{11} + a_{12}b_{21} + a_{13}b_{31} + a_{14}b_{41}$$
$$= 4(1) + 3(2) + 2(3) + 1(1) = 17$$

$$p_{21} = a_{21}b_{11} + a_{22}b_{21} + a_{23}b_{31} + a_{24}b_{41}$$
$$= 2(1) + 1(2) + 1(3) + 1(1) = 8$$

$$p_{12} = a_{11}b_{12} + a_{12}b_{22} + a_{13}b_{32} + a_{14}b_{42}$$
$$= 4(1) + 3(2) + 2(1) + 1(1) = 13$$

$$p_{32} = a_{31}b_{13} + a_{32}b_{23} + a_{33}b_{33} + a_{34}b_{43}$$
$$= 0(1) + 4(2) + 2(1) + 1(1) = 11$$

etc. The resulting matrix, $\bar{P}$, is

$$\bar{P} = \begin{bmatrix} 17 & 13 & 16 & 11 \\ 8 & 6 & 7 & 6 \\ 15 & 11 & 15 & 8 \\ 15 & 11 & 14 & 10 \end{bmatrix}$$

• **PROBLEM 3-5**

Find the determinate of matrix A

$$A = \begin{bmatrix} 2 & 3 \\ 1 & 2 \end{bmatrix}.$$

Solution: To evaluate the determinant, proceed as follows: Select any row j or column k, multiply each element in that row or column by its minor and by $(-1)^{j+k}$, then add the products.

Note: The minor of the element appearing in both row j and column k is the determinant obtained when row j and column k are removed.

For this example:

1. Choose $a_{11} = 2$ (row 1, column 1 value)
2. eliminate all other elements in row 1 and column 1, $a_{22} = 2$ is the minor.

$$(-1^{j+k})(a_{11} \times a_{22}) = (2 \times 2)(-1^2) = (2 \times 2)$$

3. choose $a_{12} = 2$ (row 1, column 2 value)
4. eliminate all other elements in row 1 and column 2, $a_{21} = 1$ is the minor.

$$(-1^{j+k})(a_{12} \times a_{21}) = (3 \times 1)(-1^3) = -(3 \times 1)$$

5. add to get the final solution

$$\Delta = \text{Determinant} = (2 \cdot 2) - (3 \cdot 1) = 4 \cdot 3 = 1$$

• **PROBLEM 3-6**

Evaluate the determinant

$$\Delta = \begin{vmatrix} 1 & -4 & -5 \\ 1 & 2 & 3 \\ -3 & 1 & -2 \end{vmatrix}$$

Solution: Find the cofactors for any row or column and sum the products of each corresponding element to find the determinant of a square matrix.

Taking the first row, the cofactors are (2)(-2) - (3)(1), -[(1)(-2) - (3)(-3)], and (1)(1) - (-3)(2) for each element in the row left to right. Hence the determinant is

$$\Delta = 1[(2)(-2) - (3)(1)] - (-4)[(1)(-2) - (3)(-3)] - 5[(1)(1) - (-3)(2)]$$

$$\Delta = 1[-7] + 4[7] - 5[7] = -14$$

• PROBLEM 3-7

(a) Evaluate the determinants of the following matrices, first by cofactors and then by simplification.

$$A = \begin{bmatrix} 3 & 2 & 3 \\ -2 & 2 & 1 \\ 9 & 7 & -3 \end{bmatrix}, \quad B = \begin{bmatrix} 1 & 5 & -4 \\ -2 & 1 & 0 \\ 1 & 0 & 3 \end{bmatrix}, \quad C = \begin{bmatrix} 2 & 5 & -1 \\ 1 & -1 & 2 \\ 3 & 1 & 4 \end{bmatrix}$$

(b) Find A^{-1}, B^{-1}, and C^{-1}.

Solution: By using the definition of determinant for a given 3×3 matrix

$$|\Delta| = \begin{vmatrix} a_{11} & a_{12} & a_{13} \\ a_{21} & a_{22} & a_{23} \\ a_{31} & a_{32} & a_{33} \end{vmatrix} = a_{11}\begin{vmatrix} a_{22} & a_{23} \\ a_{32} & a_{33} \end{vmatrix} - a_{12}\begin{vmatrix} a_{21} & a_{23} \\ a_{31} & a_{33} \end{vmatrix} + a_{13}\begin{vmatrix} a_{21} & a_{22} \\ a_{31} & a_{32} \end{vmatrix}$$

and knowing that

$$\begin{vmatrix} a_{11} & a_{12} \\ a_{21} & a_{22} \end{vmatrix} = a_{11}a_{22} - a_{21}a_{12}$$

yields the following determinant equation for a 3 × 3 matrix.

$$|\Delta| = [a_{11}a_{22}a_{33}] + [a_{12}a_{23}a_{31}] + [a_{13}a_{21}a_{32}] - [a_{13}a_{22}a_{31}] - [a_{11}a_{23}a_{32}] - [a_{12}a_{21}a_{33}]$$

Apply this to matrix $\overline{A}$ where

$$a_{11} = 3 \quad a_{12} = 2 \quad a_{13} = 3$$

$a_{21} = -2 \quad a_{22} = 2 \quad a_{23} = 1$

$a_{31} = 9 \quad a_{32} = 7 \quad a_{33} = -3$

$$|\bar{A}| = [(3)(2)(-3)] + [(2)(1)(9)] + [(3)(-2)(7)] - [(3)(2)(9)] - [(3)(1)(7)] - [(2)(-2)(-3)]$$

$$= -18 + 18 - 42 - 54 - 21 - 12$$

$$= -129$$

Similarly

$$|\bar{B}| = [(1)(1)(3)] + [(5)(0)(1)] + [(-4)(-2)(0)] - [(-4)(1)(1)] - [(1)(0)(0)] - [(5)(-2)(3)]$$

$$= 3 + 0 + 0 + 4 - 0 + 30 = 37$$

and $\bar{C} = [(2)(-1)(4)] + [(5)(2)(3)] + [(-1)(1)(1)] - [(-1)(-1)(3)] - [(2)(2)(3)] - [(5)(1)(4)]$

$$= -8 + 30 -1 - 3 - 12 -20 = -14.$$

(b) Given a matrix $\bar{A}$, the inverse of $\bar{A}$ is obtained as

$$\bar{A}^{-1} = \begin{bmatrix} \frac{A_{11}}{|A|} & \frac{A_{21}}{|A|} & \frac{A_{31}}{|A|} & \cdots & \frac{A_{n1}}{|A|} \\ \frac{A_{12}}{|A|} & \cdots & \cdots & \cdots & \frac{A_{n2}}{|A|} \\ \frac{A_{1n}}{|A|} & \cdots & \cdots & \cdots & \frac{A_{nn}}{|A|} \end{bmatrix}$$

where A_{ij} is the cofactor of a_{ij} and $|\bar{A}|$ is the determinant of matrix $\bar{A}$ the technique of obtaining the inverse of $\bar{A}$ therefore is

1) Replace each element a_{ij} with its cofactor A_{ij} in matrix $\bar{A}$. This new matrix is then (A_{ij}).
2) Find the transpose of the new matrix which is $(A_{ij})T = (A_{ji})$.
3) Divide each element of this matrix by determinant of $\bar{A}$,

the resulting matrix is the inverse of matrix A.

1) Finding the cofactors given

$$\text{Matrix } \bar{A} = \begin{bmatrix} a_{11} & a_{12} & a_{13} \\ a_{21} & a_{22} & a_{23} \\ a_{31} & a_{32} & a_{33} \end{bmatrix}$$

Cofactor $\bar{A}_C$ =

$$\begin{bmatrix} (+)\begin{vmatrix} a_{22} & a_{23} \\ a_{32} & a_{33} \end{vmatrix} = A_{11} & (-)\begin{vmatrix} a_{21} & a_{23} \\ a_{31} & a_{33} \end{vmatrix} = A_{12} & (+)\begin{vmatrix} a_{21} & a_{22} \\ a_{31} & a_{32} \end{vmatrix} = A_{13} \\ (-)\begin{vmatrix} a_{12} & a_{13} \\ a_{32} & a_{33} \end{vmatrix} = A_{21} & (+)\begin{vmatrix} a_{11} & a_{13} \\ a_{31} & a_{33} \end{vmatrix} = A_{22} & (-)\begin{vmatrix} a_{11} & a_{12} \\ a_{31} & a_{32} \end{vmatrix} = A_{23} \\ (+)\begin{vmatrix} a_{12} & a_{13} \\ a_{22} & a_{23} \end{vmatrix} = A_{31} & (-)\begin{vmatrix} a_{11} & a_{13} \\ a_{21} & a_{23} \end{vmatrix} = A_{32} & (+)\begin{vmatrix} a_{11} & a_{12} \\ a_{21} & a_{22} \end{vmatrix} = A_{33} \end{bmatrix}$$

Applying these results to matrix $\bar{A}$

$A_{11} = [(2)(-3) - (7)(1)](+1) = -13$

$A_{12} = [(-2)(-3) - (9)(1)](-1) = +3$

$A_{13} = [(-2)(7) - (9)(2)](+1) = -32$

$A_{21} = [(2)(-3) - (7)(3)](-1) = +15$

$A_{22} = [(3)(-3) - (9)(3)](+1) = -36$

$A_{23} = [(3)(7) - (9)(2)](-1) = -3$

$A_{31} = [(2)(1) - (2)(3)](+1) = -4$

$A_{32} = [(3)(1) - (-2)(3)](-1) = -9$

$A_{33} = [(3)(2) - (-2)(2)](+1) = 10$

$$\bar{A}_C = \begin{bmatrix} -13 & +3 & -32 \\ +15 & -36 & -3 \\ -4 & -9 & +10 \end{bmatrix}$$

and

$$\bar{A}^{-1} = \begin{bmatrix} -13/-129 & +15/-129 & -4/-129 \\ +3/-129 & -36/-129 & -9/-129 \\ -32/-129 & -3/-129 & +10/-129 \end{bmatrix}$$

Similarly

$$\bar{B}_C = \begin{bmatrix} 3 & 6 & -1 \\ -15 & 7 & 5 \\ 4 & 8 & 11 \end{bmatrix} \quad \bar{B}^{-1} = \begin{bmatrix} 3/37 & -15/37 & 4/37 \\ 6/37 & 7/37 & 8/37 \\ -1/37 & 5/37 & 11/37 \end{bmatrix}$$

and

$$\bar{C}_C = \begin{bmatrix} -6 & 2 & 4 \\ -21 & 11 & 13 \\ 9 & -5 & -7 \end{bmatrix} \quad \bar{C}^{-1} = \begin{bmatrix} -6/-14 & -21/-14 & 9/-14 \\ 2/-14 & 11/-14 & -5/-14 \\ 4/-14 & 13/-14 & -7/-14 \end{bmatrix}$$

• **PROBLEM 3-8**

Solve for x_1, x_2, x_3.

$$x_1 - x_2 + x_3 = 1$$

$$2x_1 + x_2 + 2x_3 = 5$$

$$-x_1 + 2x_2 + 3x_3 = 7$$

Solution: From the set of equations given, form the augmented matrix

$$\left[\begin{array}{ccc|c} 1 & -1 & 1 & 1 \\ 2 & 1 & 2 & 5 \\ -1 & 2 & 3 & 7 \end{array}\right]$$

which can be simplified to

$$\left[\begin{array}{ccc|c} 1 & 0 & 0 & x_1 \\ 0 & 1 & 0 & x_2 \\ 0 & 0 & 1 & x_3 \end{array}\right]$$

by Gaussian elimination.

Adding the first row to the third, and subtracting twice the first from the second gives,

$$\left[\begin{array}{ccc|c} 1 & -1 & 1 & 1 \\ 0 & 3 & 0 & 3 \\ 0 & 1 & 4 & 8 \end{array}\right].$$

Divide the second row by 3 and add it to the first and subtract it from the second to obtain

$$\left[\begin{array}{ccc|c} 1 & 0 & 1 & 2 \\ 0 & 1 & 0 & 1 \\ 0 & 0 & 4 & 7 \end{array}\right].$$

Dividing the third row by 4 and subtracting from the first gives

$$\left[\begin{array}{ccc|c} 1 & 0 & 0 & \frac{1}{4} \\ 0 & 1 & 0 & 1 \\ 0 & 0 & 1 & \frac{7}{4} \end{array}\right]$$

where $x_1 = \frac{1}{4}$, $x_2 = 1$, and $x_3 = \frac{7}{4}$.

• PROBLEM 3-9

Find the eigenvalues of matrices

(a) $\begin{bmatrix} 3 & 1 \\ 1 & 3 \end{bmatrix}$ (b) $\begin{bmatrix} 2 & 2 & 1 \\ 1 & 3 & 1 \\ 1 & 2 & 2 \end{bmatrix}$ (c) $\begin{bmatrix} 2 & -1 & 0 \\ 9 & 4 & 6 \\ -8 & 0 & -3 \end{bmatrix}$

Solution: The eigenvalue λ of a square matrix $\bar{A}$ is obtained from the equation,

$$|\bar{A} - \lambda\bar{I}| = 0$$

where $\bar{I}$ is an identity matrix of the same order as $\bar{A}$.

$$\bar{A} - \lambda\bar{I} = \begin{bmatrix} 3 & 1 \\ 1 & 3 \end{bmatrix} - \lambda\begin{bmatrix} 1 & 0 \\ 0 & 1 \end{bmatrix}$$

$$= \begin{bmatrix} 3-\lambda & 1 \\ 1 & 3-\lambda \end{bmatrix}$$

Taking the determinant of the matrix above yields

$$g(\lambda) = (3 - \lambda)^2 - 1 = \lambda^2 - 6\lambda + 8$$

$$= (\lambda - 4)(\lambda - 2)$$

$$\lambda = 4, 2$$

(b)

$$\bar{A} - \lambda\bar{I} = \begin{bmatrix} 2 & 2 & 1 \\ 1 & 3 & 1 \\ 1 & 2 & 2 \end{bmatrix} - \lambda\begin{bmatrix} 1 & 0 & 0 \\ 0 & 1 & 0 \\ 0 & 0 & 1 \end{bmatrix}$$

$$= \begin{bmatrix} 2-\lambda & 2 & 1 \\ 1 & 3-\lambda & 1 \\ 1 & 2 & 2-\lambda \end{bmatrix}$$

$$|\bar{A} - \lambda\bar{I}| = (2-\lambda)[(3-\lambda)(2-\lambda)-2]-[2(2-\lambda)-2]+[2-(3-\lambda)]$$

$$= (2-\lambda)(\lambda^2 - 5\lambda + 4) - (2 - 2\lambda) + (-1 + \lambda)$$

$$= -\lambda^3 + 7\lambda^2 - 14\lambda + 8 - 2 + 2\lambda - 1 + \lambda$$

$$g(\lambda) = -\lambda^3 + 7\lambda^2 - 11\lambda + 5$$

We find one root to be 1.0 by trial and error. Now obtain a quadratic by dividing

$$\begin{array}{r} -\lambda^2 + 6\lambda - 5 \\ (\lambda - 1)\overline{)-\lambda^3 + 7\lambda^2 - 11\lambda + 5} \\ \underline{-\lambda^3 + \lambda^2} \qquad\qquad \\ 6\lambda^2 - 11\lambda \qquad \\ \underline{6\lambda^2 - 6\lambda} \qquad \\ -5\lambda + 5 \\ \underline{-5\lambda + 5} \end{array}$$

Applying the quadratic formula yields

$$\lambda_2, \lambda_3 = \frac{-b \pm \sqrt{b^2 - 4ac}}{2a}$$

$$\lambda_2, \lambda_3 = \frac{-6 \pm \sqrt{36 - 4 -1)(-5)}}{-2}$$

$$\lambda_2, \lambda_3 = 3 \pm \frac{\sqrt{36 - 20}}{-2} = 3 \pm \frac{\sqrt{16}}{-2} = 3 \pm (-2)$$

$\lambda_2 = 1$; $\lambda_3 = 5$. The eigenvalues are 1, 1, and 5.

(c)

$$\bar{A} - \lambda\bar{I} = \begin{bmatrix} 2 & -1 & 0 \\ 9 & 4 & 6 \\ -8 & 0 & -3 \end{bmatrix} - \lambda \begin{bmatrix} 1 & 0 & 0 \\ 0 & 1 & 0 \\ 0 & 0 & 1 \end{bmatrix}$$

$$= \begin{bmatrix} 2-\lambda & -1 & 0 \\ 9 & 4-\lambda & 6 \\ -8 & 0 & -3-\lambda \end{bmatrix}$$

$$|\bar{A} - \lambda\bar{I}| = (2-\lambda)(4-\lambda)(-3-\lambda) - 9[-1(-3-\lambda)] - 8(-6)$$

$$g(\lambda) = -\lambda^3 + 3\lambda^2 + 10\lambda - 24 - 27 - 9\lambda + 48$$

$g(\lambda) = -\lambda^3 + 3\lambda^2 + \lambda - 3$. We find one of the three roots to be 1.0 by trial and error. Dividing gives

$$\begin{array}{r} -\lambda^2 + 2\lambda + 3 \\ (\lambda - 1)\overline{)-\lambda^3 + 3\lambda^2 + \lambda - 3} \\ \underline{-\lambda^3 + \lambda^2} \qquad\qquad \\ 2\lambda^2 + \lambda \qquad \\ \underline{2\lambda^2 - 2\lambda} \qquad \\ 3\lambda - 3 \\ \underline{3\lambda - 3} \end{array}$$

Using the quadratic formula gives the remaining two roots:

$$\lambda_2, \lambda_3 = \frac{-b \pm \sqrt{b^2 - 4ac}}{2a}$$

$$\lambda_2, \lambda_3 = \frac{-2 \pm \sqrt{4 - 4(-1)(3)}}{-2}$$

$$\lambda_2, \lambda_3 = 1 \pm \frac{\sqrt{16}}{-2} = 1 \pm (-2)$$

$\lambda_2 = -1$; $\lambda_3 = 3$. The eigenvalues are 1, -1, and 3.

• PROBLEM 3-10

For the matrix

$$\bar{A} = \begin{bmatrix} -1 & 0 \\ 2 & -1 \end{bmatrix}$$

compute $e^{\bar{A}t}$.

Solution: We compute $e^{\bar{A}t}$ in two steps. First solve $\det(\bar{A}-\lambda\bar{I}) = 0$ for values of λ. If these values are not distinct, then we solve the following set of equations:

Type 1:

$$\alpha_0 + \alpha_1\lambda_1 + \alpha_2\lambda_1^2 + \ldots + \alpha_{n-1}\lambda_1^{n-1} = e^{\lambda_1 t}$$

$$\frac{d}{d\lambda_1}(\alpha_0 + \alpha_1\lambda_1 + \alpha_2\lambda_1^2 + \ldots + \alpha_{n-1}\lambda_1^{n-1}) = \frac{d}{d\lambda_1} e^{\lambda_1 t}$$

$$\frac{d^2}{d\lambda_1^2}(\alpha_0 + \alpha_1\lambda_1 + \alpha_2\lambda_1^2 + \ldots + \alpha_{n-1}\lambda_1^{n-1}) = \frac{d^2 e^{\lambda_1 t}}{d\lambda_1^2}$$

$$\frac{d^{m-1}}{d\lambda_1^{m-1}}(\alpha_0 + \alpha_1\lambda_1 + \alpha_2\lambda_1^2 + \ldots + \alpha_{n-1}\lambda_1^{n-1}) = \frac{d^{m-1}}{d\lambda_1^{m-1}} e^{\lambda_1 t}$$

Type 2:

$$\alpha_0 + \alpha_1\lambda_{m+1} + \alpha_2\lambda^2_{m+1} + \ldots + \alpha_{n-1}\lambda^{n-1}_{m+1} = e^{\lambda_{m+1}t}$$

$$\vdots$$

$$\alpha_0 + \alpha_1\lambda_n + \alpha_2\lambda_n^2 + \ldots + \alpha_{n-1}\lambda_n^{n-1} = e^{\lambda_n t}$$

where m is the root multiplicity of $\det(\bar{A} - \lambda\bar{I}) = 0$, which has n-m distinct roots. The total number of equations solved is n, where m is the number of type 1 equations and n-m is the number of type 2 equations. After obtaining $(\alpha_0 \ldots \alpha_n)$ we substitute into

$$e^{\bar{A}t} = \alpha_0(t)\bar{I} + \alpha_1(t)\bar{A} + \alpha_2(t)\bar{A}^2 + \ldots + \alpha_{n-1}(t)\bar{A}^{n-1}. \tag{1}$$

In our example

$$\bar{A} = \begin{bmatrix} -1 & 0 \\ 2 & -1 \end{bmatrix}$$ and the eigenvalues of A are the solution of

$$\det(\bar{A} - \lambda\bar{I}) = \begin{vmatrix} -\lambda-1 & 0 \\ 2 & -\lambda-1 \end{vmatrix} = (\lambda + 1)^2 = 0.$$

Since A has non-distinct eigenvalues, we solve for α_0 and α_1. In the two equations obtained from the general form above, note that both are type 1 equations.

$$\alpha_0 + \alpha_1\lambda_1 = e^{\lambda_1 t}$$

$$\frac{d}{d\lambda_1}(\alpha_0 + \alpha_1\lambda_1) = \frac{d}{d\lambda_1} e^{\lambda_1 t}.$$

Then, since $\lambda_1 = -1$

$$\alpha_0 - \alpha_1 = e^{-t}$$

$$\alpha_1 = te^{-t}$$

hence

$$\alpha_0(t) = e^{-t} + te^{-t} \text{ and } \alpha_1(t) = te^{-t}.$$

Substituting into Eq(1) gives

$$e^{\bar{A}t} = \begin{bmatrix} e^{-t} + te^{-t} & 0 \\ 0 & e^{-t} + te^{-t} \end{bmatrix} + \begin{bmatrix} -te^{-t} & 0 \\ 2te^{-t} & -te^{-t} \end{bmatrix}$$

$$e^{\bar{A}t} = \begin{bmatrix} e^{-t} & 0 \\ 2te^{-t} & e^{-t} \end{bmatrix}$$

• **PROBLEM 3-11**

For the matrix

$$\bar{A} = \begin{bmatrix} -2 & 1 \\ 0 & -1 \end{bmatrix}$$

compute $e^{\bar{A}t}$ for all $t \geq 0$.

Solution: We compute $e^{\bar{A}t}$ in two steps. First, find $\det(\bar{A} - \lambda\bar{I})$ and solve $\det(\bar{A} - \lambda\bar{I}) = 0$ for values of λ. If the eigenvalues of $\bar{A}$ are distinct, we can solve the following n-linear algebraic equations:

$$\alpha_0 + \alpha_1\lambda_1 + \alpha_2\lambda_1^2 + \ldots + \alpha_{n-1}\lambda_1^{n-1} = e^{\lambda_1 t}$$

$$\alpha_0 + \alpha_1\lambda_2 + \alpha_2\lambda_2^2 + \ldots + \alpha_{n-1}\lambda_2^{n-1} = e^{\lambda_2 t}$$

$$\vdots \qquad \vdots \qquad \vdots$$

$$\alpha_0 + \alpha_1\lambda_n + \alpha_2\lambda^2{}_n + \ldots + \alpha_{n-1}\lambda_n^{n-1} = e^{\lambda_n t}$$

for $\alpha_0 - \alpha_n$.

Second, substitute all values found for $\alpha_0 - \alpha_n$ into

$$e^{\bar{A}t} = \alpha_0(t)\bar{I} + \alpha_1(t)\bar{A} + \alpha_2(t)\bar{A}^2 + \ldots + \alpha_{n-1}(t)\bar{A}^{n-1}$$

In our case, since n = 2, we use $e^{\bar{A}t} = \alpha_0(t)\bar{I} + \alpha_1(t)\bar{A}$ to obtain $e^{\bar{A}t}$.

Eigenvalues of $\bar{A}$ are the solution of

$$\det(\bar{A} - \lambda\bar{I}) = \begin{vmatrix} -\lambda-2 & 1 \\ 0 & -\lambda-1 \end{vmatrix}$$

$$= (\lambda + 1)(\lambda + 2) = 0$$

therefore, $\lambda_1 = -1$, $\lambda_2 = -2$.

Substituting into the general form for n-linear algebraic equations for distinct eigenvalues yields

$$\alpha_0 - \alpha_1 = e^{-t}$$

$$\alpha_0 - 2\alpha_1 = e^{-2t}$$

Solving for α_0 and α_1 we obtain

$$\alpha_0 = \frac{\begin{vmatrix} e^{-t} & -1 \\ e^{-2t} & -2 \end{vmatrix}}{\begin{vmatrix} 1 & -1 \\ 1 & -2 \end{vmatrix}} = \frac{-2e^{-t} + e^{-2t}}{-2 + 1} = 2e^{-t} - e^{-2t}$$

$$\alpha_1 = \frac{\begin{vmatrix} 1 & e^{-t} \\ 1 & e^{-2t} \end{vmatrix}}{\begin{vmatrix} 1 & -1 \\ 1 & -2 \end{vmatrix}} = \frac{e^{-2t} - e^{-t}}{-2 + 1} = e^{-t} - e^{-2t}$$

Substituting into $e^{\bar{A}t} = \alpha_0 I + \alpha_1 A$ yields

$$e^{\bar{A}t} = (2e^{-t} - e^{-2t})\begin{bmatrix} 1 & 0 \\ 0 & 1 \end{bmatrix} + (e^{-t} - e^{-2t})\begin{bmatrix} -2 & 1 \\ 0 & -1 \end{bmatrix}$$

$$e^{\bar{A}t} = \begin{bmatrix} 2e^{-t}-e^{-2t} & 0 \\ 0 & 2e^{-t} - e^{-2t} \end{bmatrix} + \begin{bmatrix} -2e^{-t}+2e^{-2t} & e^{-t}-e^{-2t} \\ 0 & -e^{-t}+e^{-2t} \end{bmatrix}$$

$$e^{\bar{A}t} = \begin{bmatrix} e^{-2t} & e^{-t}-e^{-2t} \\ 0 & e^{-t} \end{bmatrix}$$

• **PROBLEM 3-12**

Solve for a, b, c, and d using Gauss Elimination.

$$a + b + c - d = 5$$

$$2a + 3b + 2c = 1$$

$$a - 2b + 4c + 3d = 4$$

$$a - b + c + d = -1$$

Solution: Form the augmented matrix from the four equations.

$$\begin{matrix} \text{row 1} \\ \text{row 2} \\ \text{row 3} \\ \text{row 4} \end{matrix} \quad \begin{bmatrix} 1 & 1 & 1 & -1 & 5 \\ 2 & 3 & 2 & 0 & 1 \\ 1 & -2 & 4 & 3 & 4 \\ 1 & -1 & 1 & 1 & -1 \end{bmatrix}$$

After the following operations are performed

-2(row 1) + (row 2) → row 2

(row 3) - (row 1) → row 3

(row 4) - (row 1) → row 4

The matrix becomes

$$\begin{matrix} \text{row 1} \\ \text{row 2} \\ \text{row 3} \\ \text{row 4} \end{matrix} \quad \begin{bmatrix} 1 & 1 & 1 & -1 & 5 \\ 0 & 1 & 0 & 2 & -9 \\ 0 & -3 & 3 & 4 & -1 \\ 0 & -2 & 0 & 2 & -6 \end{bmatrix} .$$

The next operations that are performed are

3(row 2) + (row 3) → row 3

2(row 2) + (row 4) → row 4

(row 1) - (row 2) → row 1 .

This yields the matrix

$$\begin{matrix} \text{row 1} \\ \text{row 2} \\ \text{row 3} \\ \text{row 4} \end{matrix} \quad \begin{bmatrix} 1 & 0 & -1 & -3 & 14 \\ 0 & 1 & 0 & 2 & -9 \\ 0 & 0 & 3 & 10 & -28 \\ 0 & 0 & 0 & 6 & -24 \end{bmatrix}.$$

After the following two operations are performed

$$\tfrac{1}{3}(\text{row 3}) \rightarrow \text{row 3}$$

$$(\text{row 1}) - (\text{row 3}) \rightarrow \text{row 1}$$

the matrix becomes

$$\begin{matrix} \text{row 1} \\ \text{row 2} \\ \text{row 3} \\ \text{row 4} \end{matrix} \quad \begin{bmatrix} 1 & 0 & 0 & -\frac{19}{3} & \frac{70}{3} \\ 0 & 1 & 0 & 2 & -9 \\ 0 & 0 & 1 & \frac{10}{3} & -\frac{28}{3} \\ 0 & 0 & 0 & 6 & -24 \end{bmatrix}$$

The next operation that is performed

$$\tfrac{1}{6}(\text{row 4}) \rightarrow \text{row 4}$$

yields

$$\begin{matrix} \text{row 1} \\ \text{row 2} \\ \text{row 3} \\ \text{row 4} \end{matrix} \quad \begin{bmatrix} 1 & 0 & 0 & -\frac{19}{3} & \frac{70}{3} \\ 0 & 1 & 0 & 2 & -9 \\ 0 & 0 & 1 & \frac{10}{3} & -\frac{28}{3} \\ 0 & 0 & 0 & 1 & -4 \end{bmatrix}.$$

Finally, the following operations

$$\tfrac{19}{3}(\text{row 4}) + (\text{row 1}) \rightarrow \text{row 1}$$

$$-2(\text{row 4}) + (\text{row 2}) \rightarrow \text{row 2}$$

$$-\tfrac{10}{3}(\text{row 4}) + (\text{row 3}) \rightarrow \text{row 3}$$

yield

$$\begin{matrix} \text{row 1} \\ \text{row 2} \\ \text{row 3} \\ \text{row 4} \end{matrix} \quad \begin{bmatrix} 1 & 0 & 0 & 0 & -2 \\ 0 & 1 & 0 & 0 & -1 \\ 0 & 0 & 1 & 0 & 4 \\ 0 & 0 & 0 & 1 & -4 \end{bmatrix}.$$

From this matrix it is seen that

$a = -2$

$b = -1$

$c = 4$

$d = -4$

KCL WITH MATRICES

• PROBLEM 3-13

Consider the network shown in Fig. 1. Solve for voltages at nodes n_1 and n_2.

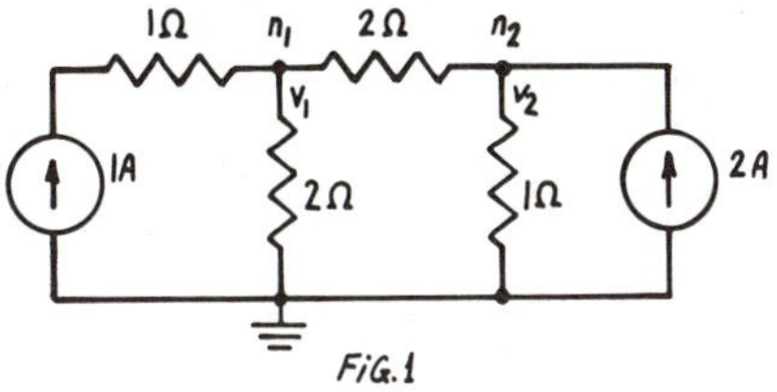

FiG.1

Solution: Using the two nodes and respective voltages indicated one can write two KCL equations.

Hence,

$$\text{node } n_1: \quad \frac{v_1}{2} + \frac{v_1 - v_2}{2} = 1$$

$$\text{node } n_2: \quad \frac{v_2}{1} + \frac{v_2 - v_1}{2} = 2.$$

Multiplying both equations through by two, and writing them in matrix form gives,

$$\begin{bmatrix} 2 & -1 \\ -1 & 3 \end{bmatrix} \begin{bmatrix} v_1 \\ v_2 \end{bmatrix} = \begin{bmatrix} 2 \\ 4 \end{bmatrix}$$

or

$$\bar{G}\,\bar{v} = \bar{i}$$

In order to obtain $\bar{v}$, we need $\bar{G}^{-1}$, so that $\bar{v} = \bar{G}^{-1}\bar{i}$ where $\bar{G}^{-1} = \dfrac{G_c^{-T}}{|\bar{G}|}$ and

$$\bar{G}_c = \begin{bmatrix} 3 & 1 \\ 1 & 2 \end{bmatrix}, \quad G_c^{T} = G_c \text{ and}$$

$|\bar{G}| = 2(3) - (-1)(-1) = 5$, thus

$$\bar{G}^{-1} = \frac{1}{5}\begin{bmatrix} 3 & 1 \\ 1 & 2 \end{bmatrix}.$$

Hence

$$\begin{bmatrix} v_1 \\ v_2 \end{bmatrix} = \frac{1}{5}\begin{bmatrix} 3 & 1 \\ 1 & 2 \end{bmatrix}\begin{bmatrix} 2 \\ 4 \end{bmatrix} = \begin{bmatrix} 2 \\ 2 \end{bmatrix} V,$$

$v_1 = 2V$, $v_2 = 2V$.

• PROBLEM 3-14

Find the node voltages of the circuit of fig. 1 by writing matrices for self-conductances and mutual conductances directly from an inspection of the network.

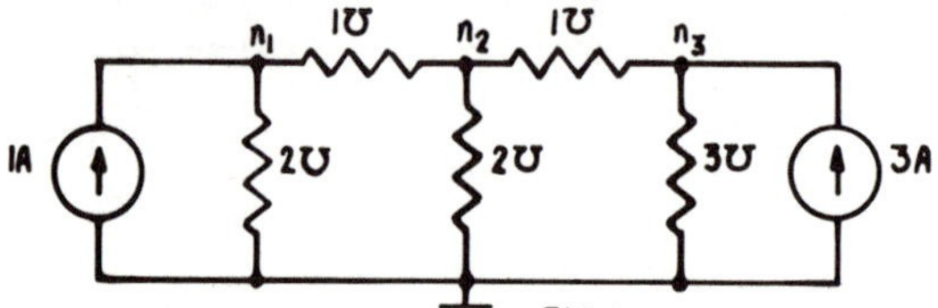

FIG. 1

Solution: Self-conductance can be found at each node by summing all the conductances in all the branches comprising that node. Mutual conductance is the conductance of the branch connecting two nodes. It is always negative in nodal analysis since a positive voltage is assigned at each node and the branches connecting these nodes have voltages which are opposite each other when taken with respect to each node. The self conductances for nodes n_1, n_2, and n_3 are written as G_{11}, G_{22}, and G_{33} respectively. There are six mutual conductances which are written G_{12}, G_{13}, G_{21}, G_{23}, G_{31}, and G_{23}. Note that $G_{12} = G_{21}$, $G_{23} = G_{32}$, and $G_{31} = G_{13}$. Also, note that since there is no branch connecting nodes n_1 and n_3, $G_{31} = G_{13} = 0$. The two connecting branches in this circuit contain 1mho conductances so that,

$$G_{12} = G_{21} = G_{23} = G_{32} = -1 \text{ mho}.$$

The self conductances are

$$G_{11} = 3,\ G_{22} = 4, \text{ and } G_{33} = 4 \text{ mho}.$$

Arranging the self and mutual conductances in matrix form gives

$$\bar{G} = \begin{bmatrix} G_{11} & G_{12} & G_{13} \\ G_{21} & G_{22} & G_{23} \\ G_{31} & G_{23} & G_{33} \end{bmatrix} = \begin{bmatrix} 3 & -1 & 0 \\ -1 & 4 & -1 \\ 0 & -1 & 4 \end{bmatrix}.$$

In order to find the node voltages v_1, v_2, and v_3 we must solve three equations written in matrix notation as $\bar{G}\,\bar{V} = \bar{I}$, where $\bar{G}$ is the matrix of self and mutual conductances found above, and $\bar{I}$ is a one column three-element matrix representing the algebraic sum of all the current sources in the branches comprising each node. In our circuit we have,

$$\bar{I} = \begin{bmatrix} 1 \\ 0 \\ 3 \end{bmatrix}.$$

$\bar{V}$ is also a 1x3 matrix which is our node voltage matrix,

$$\bar{V} = \begin{bmatrix} v_1 \\ v_2 \\ v_3 \end{bmatrix}.$$

We can write the complete set of equations,

$$\bar{G}\,\bar{V} = \bar{I}$$

$$\begin{bmatrix} 3 & -1 & 0 \\ -1 & 4 & -1 \\ 0 & -1 & 4 \end{bmatrix} \begin{bmatrix} v_1 \\ v_2 \\ v_3 \end{bmatrix} = \begin{bmatrix} 1 \\ 0 \\ 3 \end{bmatrix}.$$

There are various techniques at our disposal for solving $\bar{V}$. We can use determinants, Gaussian elimination or the matrix inversion method.

Using the matrix in version method solve for v by

$$\bar{V} = \bar{G}^{-1}\bar{I}$$

$$\bar{G}^{-1} = \begin{bmatrix} \frac{-1(16-1)}{\Delta} & \frac{-1(-4-0)}{\Delta} & \frac{1(1-0)}{\Delta} \\ \frac{-1(-4-0)}{\Delta} & \frac{1(12-0)}{\Delta} & \frac{-1(-3-0)}{\Delta} \\ \frac{1(1-0)}{\Delta} & \frac{-1(-3-0)}{\Delta} & \frac{1(12-1)}{\Delta} \end{bmatrix}$$

where $\Delta = |\bar{G}|$ (the determinant of $\bar{G}$)

$$\Delta = 3(16 - 1) - (-1)(-4-0) = 41$$

giving

$$\bar{G}^{-1} = \begin{bmatrix} \frac{15}{41} & \frac{4}{41} & \frac{1}{41} \\ \frac{4}{41} & \frac{12}{41} & \frac{3}{41} \\ \frac{1}{41} & \frac{3}{41} & \frac{11}{41} \end{bmatrix}$$

Finally, we obtain

$$\bar{V} = \bar{G}^{-1}\bar{I} \quad \begin{bmatrix} \frac{15}{41} & \frac{4}{41} & \frac{1}{41} \\ \frac{4}{41} & \frac{12}{41} & \frac{3}{41} \\ \frac{1}{41} & \frac{3}{41} & \frac{11}{41} \end{bmatrix} \begin{bmatrix} 1 \\ 0 \\ 3 \end{bmatrix} = \begin{bmatrix} \frac{15}{41} + \frac{3}{41} \\ \frac{4}{41} + \frac{9}{41} \\ \frac{1}{41} + \frac{33}{41} \end{bmatrix}$$

thus,

$$v_1 = \frac{18}{41}, \; v_2 = \frac{13}{41}, \; \text{and } v_3 = \frac{34}{41} \text{ V}.$$

KVL WITH MATRICES

• PROBLEM 3-15

In the resistance network of Fig. 1 find the currents in all resistances as a function of the source functions $v_A(t)$ and and $i_B(t)$.

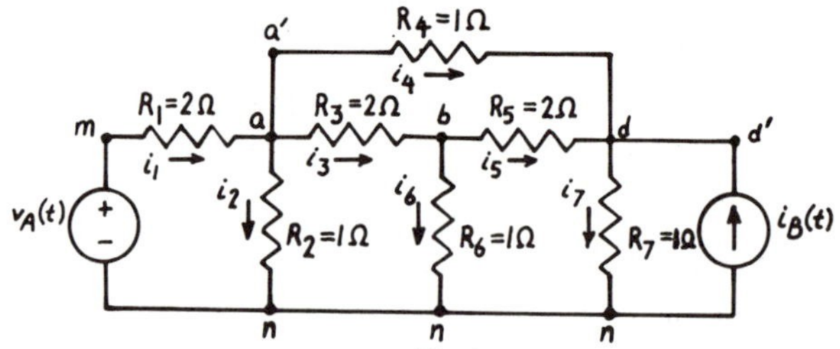

FiG.1

Solution: In order to solve for the currents indicated in fig. 1, it would be necessary to solve five KVL equations. If we use source transformations to reduce the number of loops in the circuit to 3, the work would be greatly reduced since only 3 KVL equations would be needed. We can transform the resistive current source into a voltage source and series resistor. This will eliminate one loop. Two source transformations of the resistive voltage source, V_A and R_1, R_2, will eliminate another loop, giving 3 loops. Figs. 2 and 3 show how the current source is transformed, and figs 4 - 6 show how the voltage source is transformed. Fig. 6 is the final 3-loop circuit showing loop currents arbitrarily assigned.

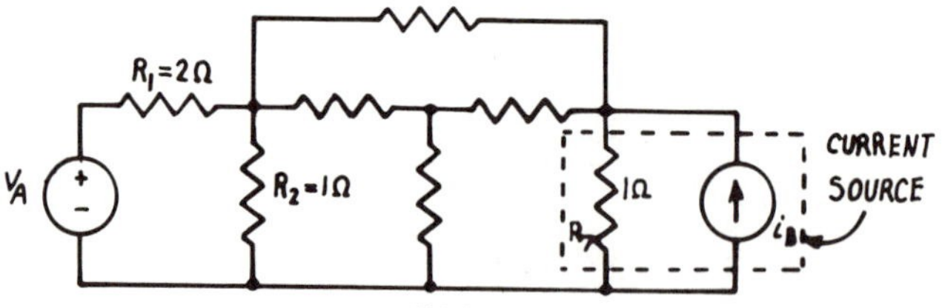

FiG.2

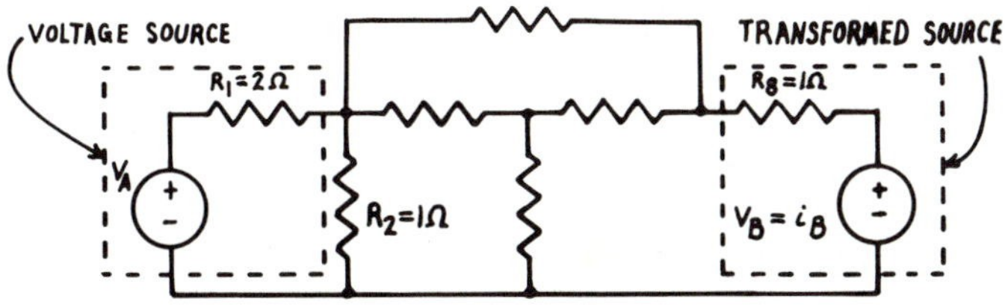

FiG.3

Note that resistors R_4, R_3, R_6, and R_5 remain unchanged; thus, the current through them can be expressed in terms of the new loop currents.

$$i_3 = i_C - i_E \qquad (1)$$

$$i_5 = -(i_E + i_D) \qquad (2)$$

$$i_4 = i_E \qquad (3)$$

$$i_6 = i_C + i_D \qquad (4)$$

The currents i_1, i_2, and i_7 can be found by algebraic manipulations in terms of i_C, i_D, and i_E.

Hence, from fig. 1,

$$i_7 = i_4 + i_5 + i_B.$$

Substituting yields $i_7 = i_E - i_E - i_D + i_B = i_B - i_D$. (5)

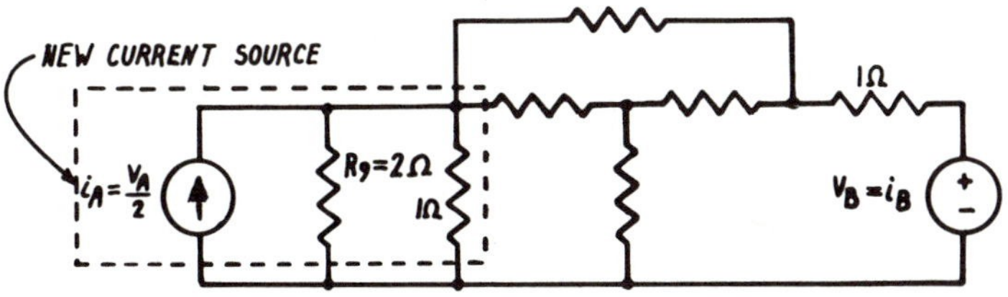

FiG.4

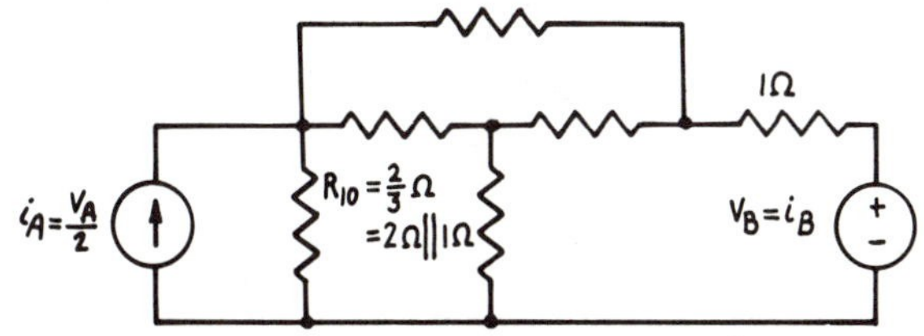

FiG.5

Find i_2 in terms of the new loop currents by writing the voltage v_{an} (see fig. 6) in terms of the new loop currents. v_{an} is the voltage across R_2, thus, the current i_2 is $\frac{v_{an}}{1} = v_{an}$.

Hence,

$$v_{an} = R_3(i_C - i_E) + R_6(i_C + i_D)$$

$$v_{an} = 2(i_C - i_E) + i_C + i_D$$

$$v_{an} = 3i_C - 2i_E + i_D$$

$$i_2 = \frac{v_{an}}{1} = 3i_C - 2i_E + i_D. \qquad (6)$$

Now find i_1 by summing the currents at node a (fig. 1).

Hence, $i_1 = i_2 + i_3 + i_4$

$$i_1 = 3i_C - 2i_E + i_D + i_C - i_E + i_E$$

$$i_1 = 4i_C - 2i_E + i_D \qquad (7)$$

Now proceed to solve for i_C, i_D, and i_E by writing 3 KVL equations for the circuit in fig. 6.

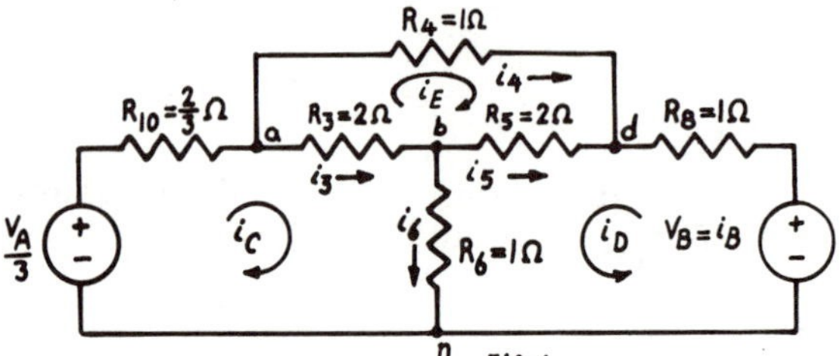

FIG. 6

loop C: $\frac{v_A}{3} = i_C\left(\frac{2}{3} + 3\right) + i_D - 2i_E$

loop D: $i_B = i_C + 4i_D + 2i_E$

loop E: $0 = -2i_C + 2i_D + 5i_E.$

In matrix form,

$$\begin{bmatrix} \frac{11}{3} & 1 & -2 \\ 1 & 4 & 2 \\ -2 & 2 & 5 \end{bmatrix} \begin{bmatrix} i_C \\ i_D \\ i_E \end{bmatrix} = \begin{bmatrix} \frac{v_A}{3} \\ i_B \\ 0 \end{bmatrix},$$

$$\bar{R}\,\bar{I} = \bar{v}.$$

We can find, $\bar{I} = \bar{R}^{-1}\bar{v}$, by finding $\bar{R}^{-1}$.

Since $\bar{R}^{-1} = \frac{\bar{R}_C^T}{|\bar{R}|}$, the cofactor of $\bar{R}$, $\bar{R}_C$, can be found.

Hence,

$$\bar{R}_C = \begin{bmatrix} (4)(5)-(2)(2) & -[(1)(5)-(2)(2)] & (1)(2)-(4)(-2) \\ -[(1)(5)-(-2)(2)] & \left[\left(\frac{11}{3}\right)(5)-(-2)(-2)\right] & -\left[\frac{11}{3}(2)-(1)(-2)\right] \\ (1)(2)-(-2)(4) & -\left(\frac{11}{3}\right)(2)-(-2)(1) & \left(\frac{11}{3}\right)(4)-(1)(1) \end{bmatrix}$$

$$\bar{R}_C = \begin{bmatrix} 16 & -9 & 10 \\ -9 & \frac{43}{3} & -\frac{28}{3} \\ 10 & -\frac{28}{3} & \frac{41}{3} \end{bmatrix}$$

With this matrix $\bar{R}_C = \bar{R}_C^T$ and $|\bar{R}| =$

$$\frac{11}{3}(16) + (-9) - 2(10) = \frac{176}{3} - \frac{87}{3} = \frac{89}{3}.$$

Therefore,

$$\begin{bmatrix} i_C \\ i_D \\ i_E \end{bmatrix} = \frac{3}{89} \begin{bmatrix} 16 & -9 & 10 \\ -9 & \frac{43}{3} & -\frac{28}{3} \\ 10 & \frac{28}{3} & \frac{41}{3} \end{bmatrix} \begin{bmatrix} \frac{v_A}{3} \\ i_B \\ 0 \end{bmatrix}$$

$$\begin{bmatrix} i_C \\ i_D \\ i_E \end{bmatrix} = \begin{bmatrix} \frac{48}{89} & \frac{-27}{89} & -\frac{30}{89} \\ \frac{-27}{89} & \frac{43}{89} & \frac{-28}{89} \\ \frac{30}{89} & \frac{-28}{89} & \frac{41}{89} \end{bmatrix} \begin{bmatrix} \frac{v_A}{3} \\ i_B \\ 0 \end{bmatrix}$$

$$\begin{bmatrix} i_C \\ i_D \\ i_E \end{bmatrix} = \begin{bmatrix} \frac{48}{89}\cdot\frac{v_A}{3} - \frac{27}{89} i_B \\ \frac{-27}{89}\cdot\frac{v_A}{3} + \frac{43}{89} i_B \\ \frac{30}{89}\cdot\frac{v_A}{3} - \frac{28}{89} i_B \end{bmatrix} = \begin{bmatrix} \frac{16v_A - 27i_B}{89} \\ \frac{-9v_A + 43i_B}{89} \\ \frac{10v_A - 28i_B}{89} \end{bmatrix}$$

Using equations (1) through (7) one finds $i_1 - i_7$. Hence, the currents in each resistor, $i_1 = 4i_C - i_E - i_D$

$$i_1 = 4\left(\frac{16v_A - 27i_B}{89}\right) - 2\left(\frac{10v_A - 28i_B}{89}\right) + \left(\frac{-9v_A + 43i_B}{89}\right)$$

$$i_1 = \left(\frac{64v_A - 108i_B}{89}\right) + \left(\frac{-29v_A + 99i_B}{89}\right)$$

$$i_1 = \frac{35v_A - 9i_B}{89}$$

$$i_2 = i_1 - i_C = \frac{35v_A - 9i_B}{89} - \frac{16v_A - 27i_B}{89}$$

$$i_2 = \frac{19v_A + 15i_B}{89}$$

$$i_3 = i_C - i_E = \frac{16v_A - 27i_B}{89} - \frac{10v_A - 28i_B}{89}$$

$$i_3 = \frac{6v_A + i_B}{89}$$

$$i_4 = i_E = \frac{10v_A - 28i_B}{89}$$

$$i_5 = -(i_E + i_D) = -\frac{(10v_A - 28i_B)}{89} + \frac{(-9v_A + 43i_B)}{89}$$

$$i_5 = \frac{-v_A - 15i_B}{89}$$

$$i_6 = i_C + i_D = \frac{16v_A - 27i_B}{89} + \frac{-9v_A + 43i_B}{89}$$

$$i_6 = \frac{7v_A + 16i_B}{89}$$

$$i_7 = i_B - i_D = i_B - \frac{(-9v_A + 43i_B)}{89}$$

$$i_7 = \frac{9v_A - 42i_B}{89}$$

• **PROBLEM** 3-16

Solve for loop currents for networks shown. Use matrix methods.

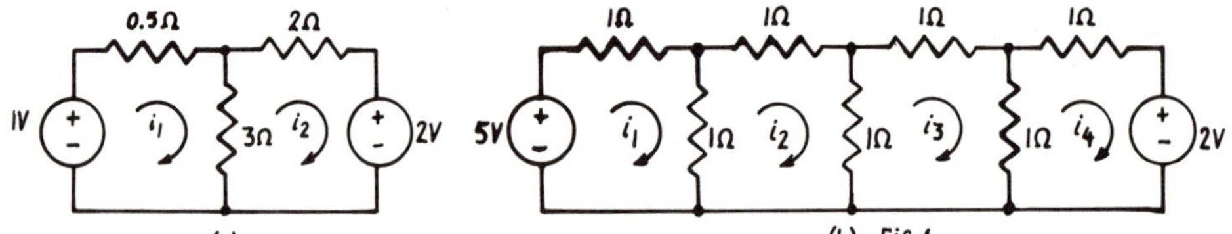

FIG.1

Solution: (a) Writing the loop equations for the two current loops shown in fig. 1 (a) yields

$$1 = 0.5i_1 + 3(i_1 - i_2)$$

$$-2 = 3(i_2 - i_1) + 2i_2$$

Simplifying

$$1 = 3.5i_1 - 3i_2$$

$$-2 = -3i_1 + 5i_2$$

Using determinants to solve the above yields

$$i_1 = \frac{\begin{vmatrix} 1 & -3 \\ -2 & 5 \end{vmatrix}}{\begin{vmatrix} 3.5 & -3 \\ -3 & 5 \end{vmatrix}} = \frac{5 - 6}{17.5 - 9} = -\frac{1}{8.5} = -.118.$$

$$i_2 = \frac{\begin{vmatrix} 3.5 & 1 \\ -3 & -2 \end{vmatrix}}{8.5} = \frac{(-7+3)}{8.5} = -\frac{4}{8.5} = -.471$$

b) We write the loop equations as follows,

$$5 = 2i_1 - i_2$$
$$0 = -i_1 + 3i_2 - i_3$$
$$0 = \quad - i_2 + 3i_3 - i_4$$
$$-2 = \quad - i_3 + 2i_4$$

In matrix form

$$\begin{bmatrix} 2 & -1 & 0 & 0 \\ -1 & 3 & -1 & 0 \\ 0 & -1 & 3 & -1 \\ 0 & 0 & -1 & 2 \end{bmatrix} \begin{bmatrix} i_1 \\ i_2 \\ i_3 \\ i_4 \end{bmatrix} = \begin{bmatrix} 5 \\ 0 \\ 0 \\ -2 \end{bmatrix}$$

First, form the augmented matrix. Then, use Gaussian elimination to solve for the four currents i_1, i_2, i_3 and i_4.

$$\begin{bmatrix} 2 & -1 & 0 & 0 & 5 \\ -1 & 3 & -1 & 0 & 0 \\ 0 & -1 & 3 & -1 & 0 \\ 0 & 0 & -1 & 2 & -2 \end{bmatrix}$$

(1) Divide the first row by 2 and add to the second row.

$$\begin{bmatrix} 1 & -\frac{1}{2} & 0 & 0 & \frac{5}{2} \\ 0 & 3-\frac{1}{2} & -1 & 0 & \frac{5}{2} \\ 0 & -1 & 3 & -1 & 0 \\ 0 & 0 & -1 & 2 & -2 \end{bmatrix}$$

(2) Divide the second row by 5 and add to the first row.

$$\begin{bmatrix} 1 & 0 & -\frac{1}{5} & 0 & \frac{6}{2} \\ 0 & \frac{1}{2} & -\frac{1}{5} & 0 & \frac{1}{2} \\ 0 & -1 & 3 & -1 & 0 \\ 0 & 0 & -1 & 2 & -2 \end{bmatrix}$$

(3) Multiply the second row by 2 and add to the third row.

$$\begin{bmatrix} 1 & 0 & -\frac{1}{5} & 0 & \frac{6}{2} \\ 0 & 1 & -\frac{2}{5} & 0 & 1 \\ 0 & 0 & 3-\frac{2}{5} & -1 & 1 \\ 0 & 0 & -1 & 2 & -2 \end{bmatrix}$$

(4) Divide the third row by 13 and add to the first row.

$$\begin{bmatrix} 1 & 0 & 0 & -\frac{1}{13} & \frac{6}{2}+\frac{1}{13} \\ 0 & 1 & -\frac{2}{5} & 0 & 1 \\ 0 & 0 & \frac{1}{5} & -\frac{1}{13} & \frac{1}{13} \\ 0 & 0 & -1 & 2 & -2 \end{bmatrix}$$

(5) Multiply the third row by 2 and add it to the second.

$$\begin{bmatrix} 1 & 0 & 0 & -\frac{1}{13} & \frac{6}{2}+\frac{1}{13} \\ 0 & 1 & 0 & -\frac{2}{13} & 1+\frac{2}{13} \\ 0 & 0 & \frac{2}{5} & -\frac{2}{13} & \frac{2}{13} \\ 0 & 0 & -1 & 2 & -2 \end{bmatrix}$$

(6) Multiply the third row by 2.5 and add it to the fourth row.

$$\begin{bmatrix} 1 & 0 & 0 & -\frac{1}{13} & \frac{6}{2}+\frac{1}{13} \\ 0 & 1 & 0 & -\frac{2}{13} & 1+\frac{2}{13} \\ 0 & 0 & 1 & -\frac{5}{13} & \frac{5}{13} \\ 0 & 0 & 0 & 2-\frac{5}{13} & -2+\frac{5}{13} \end{bmatrix}$$

(7) Divide the fourth row by 21 and add it to the first row.

$$\begin{bmatrix} 1 & 0 & 0 & 0 & \frac{6}{2}+\frac{1}{13}-\frac{2}{21}+\frac{5}{273} \\ 0 & 1 & 0 & -\frac{2}{13} & 1+\frac{2}{13} \\ 0 & 0 & 1 & -\frac{5}{13} & \frac{5}{13} \\ 0 & 0 & 0 & \frac{1}{13} & -\frac{2}{21}+\frac{5}{273} \end{bmatrix}$$

(8) Multiply the fourth row by 2, adding it to the second.

$$\begin{bmatrix} 1 & 0 & 0 & 0 & \frac{6}{2}+\frac{1}{13}-\frac{2}{21}+\frac{5}{273} \\ 0 & 1 & 0 & 0 & 1+\frac{2}{13}-\frac{4}{21}+\frac{10}{273} \\ 0 & 0 & 1 & -\frac{5}{13} & \frac{5}{13} \\ 0 & 0 & 0 & \frac{2}{13} & -\frac{4}{21}+\frac{10}{273} \end{bmatrix}$$

(9) Multiplying the fourth row by 2.5 and adding it to the third row gives,

$$\begin{bmatrix} 1 & 0 & 0 & 0 & \frac{6}{2}+\frac{1}{13}-\frac{2}{21}+\frac{5}{273} \\ 0 & 1 & 0 & 0 & 1+\frac{2}{13}-\frac{4}{21}+\frac{10}{273} \\ 0 & 0 & 1 & 0 & \frac{5}{13}-\frac{10}{21}+\frac{25}{273} \\ 0 & 0 & 0 & \frac{5}{13} & -\frac{10}{21}+\frac{25}{273} \end{bmatrix}$$

Finally, multiply the fourth row by $\frac{13}{5}$ to obtain an identity matrix in columns 1-4 and the values of i_1, i_2, i_3 and i_4 in the 5th column.

$$\begin{bmatrix} 1 & 0 & 0 & 0 \\ 0 & 1 & 0 & 0 \\ 0 & 0 & 1 & 0 \\ 0 & 0 & 0 & 1 \end{bmatrix} \begin{bmatrix} \frac{6}{2} + \frac{1}{13} - \frac{2}{21} + \frac{5}{273} \\ 1 + \frac{2}{13} - \frac{4}{21} + \frac{10}{273} \\ \frac{5}{13} - \frac{10}{21} + \frac{25}{273} \\ -\frac{130}{105} + \frac{325}{1365} \end{bmatrix} = \begin{bmatrix} i_1 \\ i_2 \\ i_3 \\ i_4 \end{bmatrix}$$

Adding the fractions gives, $i_1 = 3A$

$i_2 = 1A$

$i_3 = 0A$

$i_4 = -1A$

• **PROBLEM** 3-17

In the circuit of fig. 1, write the KVL equations for currents i_1 through i_5 in matrix form. Solve for the currents by Gaussian elimination.

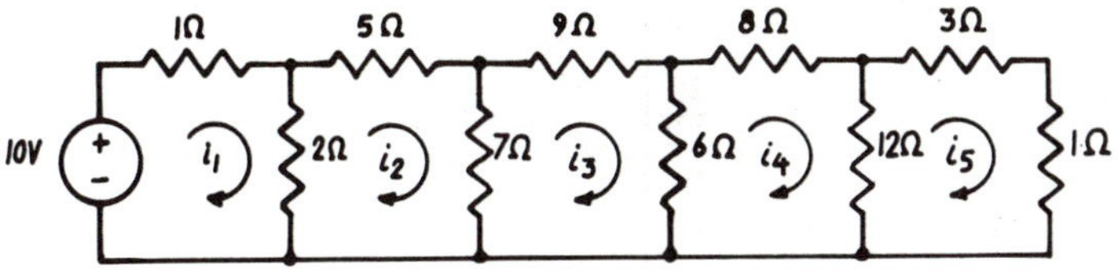

FiG. 1

Solution: Writing KVL equations for the five loops yields

$$10 = 3i_1 - 2i_2$$

$$0 = -2i_1 + 14i_2 - 7i_3$$

$$0 = -7i_2 + 22i_3 - 6i_4$$

$$0 = -6i_3 + 26i_4 - 12i_5$$

$$0 = -12i_4 + 16i_5 .$$

The equations are in the general form, $\bar{A}\,\bar{I} = \bar{V}$; where $\bar{A}$ is a 5 by 5 square matrix, $\bar{I}$ and $\bar{V}$ are 5 element vectors (single column matrix).

Writing out the matrices gives,

$$\begin{bmatrix} 3 & -2 & 0 & 0 & 0 \\ -2 & 14 & -7 & 0 & 0 \\ 0 & -7 & 22 & -6 & 0 \\ 0 & 0 & -6 & 26 & -12 \\ 0 & 0 & 0 & -12 & 16 \end{bmatrix} \begin{bmatrix} i_1 \\ i_2 \\ i_3 \\ i_4 \\ i_5 \end{bmatrix} = \begin{bmatrix} 10 \\ 0 \\ 0 \\ 0 \\ 0 \end{bmatrix}.$$

For Gaussian elimination, our composite matrix becomes $\bar{A}$ combined with $\bar{V}$. Our aim is to eliminate certain rows to obtain a composite matrix in the form

$$\begin{bmatrix} 1 & 0 & 0 & 0 & 0 & x_1 \\ 0 & 1 & 0 & 0 & 0 & x_2 \\ 0 & 0 & 1 & 0 & 0 & x_3 \\ 0 & 0 & 0 & 1 & 0 & x_4 \\ 0 & 0 & 0 & 0 & 1 & x_5 \end{bmatrix}$$

where x_1 through x_5 are the values of the currents $i_1 - i_5$.

$$\begin{bmatrix} 3 & -2 & 0 & 0 & 0 & 10 \\ -2 & 14 & -7 & 0 & 0 & 0 \\ 0 & -7 & 22 & -6 & 0 & 0 \\ 0 & 0 & -6 & 26 & -12 & 0 \\ 0 & 0 & 0 & -12 & 16 & 0 \end{bmatrix}$$

Multiply the first row by 2/3 and add it to the second row, then divide the first row by 3 to obtain,

$$\begin{bmatrix} 1 & -\frac{2}{3} & 0 & 0 & 0 & \frac{10}{3} \\ 0 & 14-\frac{4}{3} & -7 & 0 & 0 & \frac{20}{3} \\ 0 & -7 & 22 & -6 & 0 & 0 \\ 0 & 0 & -6 & 26 & -12 & 0 \\ 0 & 0 & 0 & -12 & 16 & 0 \end{bmatrix}.$$

Multiply the second row by $\frac{1}{19}$ and add it to the first row. Multiply the second row by $\frac{21}{38}$ and add it to the third. Finally, divide the second row by $\frac{38}{3}$.

Our new composite matrix becomes

$$\begin{bmatrix} 1 & 0 & -\frac{7}{19} & 0 & 0 & \frac{10}{3}+\frac{20}{57} \\ 0 & 1 & -\frac{21}{38} & 0 & 0 & \frac{20}{38} \\ 0 & 0 & \frac{689}{38} & -6 & 0 & \frac{20}{3}\cdot\frac{21}{38} \\ 0 & 0 & -6 & 26 & -12 & 0 \\ 0 & 0 & 0 & -12 & 16 & 0 \end{bmatrix}$$

Multiply row three by $\frac{14}{689}$, $\frac{21}{689}$, and $\frac{228}{689}$ and add to rows one, and add to rows one, two, and four, respectively. Finally, multiply row three by $\frac{38}{689}$ to obtain,

$$\begin{bmatrix} 1 & 0 & 0 & -\frac{84}{689} & 0 & \frac{10}{3}+\frac{20}{57}+\left[\frac{20\cdot 21}{114}\cdot\frac{14}{689}\right] \\ 0 & 1 & 0 & -\frac{126}{689} & 0 & \frac{20}{38}+\left[\frac{20\cdot 21}{114}\cdot\frac{21}{689}\right] \\ 0 & 0 & 1 & -\frac{228}{689} & 0 & \frac{20}{3}\cdot\frac{21}{38}\cdot\frac{38}{689} \\ 0 & 0 & 0 & \frac{16546}{689} & -12 & \frac{20\cdot 21}{114}\cdot\frac{228}{689} \\ 0 & 0 & 0 & -12 & 16 & 0 \end{bmatrix}$$

Multiply row four by $\frac{84}{16546}$, $\frac{126}{16546}$, $\frac{228}{16546}$, and $\frac{8268}{16546}$, then add to rows one, two, three, and five respectively. Finally, multiply row four by $\frac{689}{16546}$ to obtain,

$$\begin{bmatrix} 1 & 0 & 0 & 0 & -\frac{1008}{16546} & \frac{10}{3}+\frac{20}{57}+\left[\frac{20\cdot 21}{114}\cdot\frac{14}{689}\right]+\left[\frac{84}{16546}\cdot\frac{840}{689}\right] \\ 0 & 1 & 0 & 0 & -\frac{1512}{16546} & \frac{20}{38}+\left[\frac{20\cdot 21}{114}\cdot\frac{21}{689}\right]+\left[\frac{126}{16546}\cdot\frac{840}{689}\right] \\ 0 & 0 & 1 & 0 & -\frac{2736}{16546} & \left[\frac{20}{3}\cdot\frac{21}{38}\cdot\frac{38}{689}\right]+\left[\frac{228}{16546}\cdot\frac{840}{689}\right] \\ 0 & 0 & 0 & 1 & -\frac{8268}{16546} & \frac{840}{16546} \\ 0 & 0 & 0 & 0 & \frac{165520}{16546} & \frac{840}{689}\cdot\frac{8268}{16546} \end{bmatrix}$$

Multiply row five by $\frac{1008}{165520}$, $\frac{1512}{165520}$, $\frac{2736}{165520}$, and $\frac{8268}{165520}$ and add to rows one, two, three and four respectively. Finally, multiply row five by $\frac{16546}{165520}$ to obtain the required composite matrix.

$$\begin{bmatrix} 1\;0\;0\;0\;0 & \frac{10}{3}+\frac{20}{57}+\left[\frac{20\cdot 21}{114}\cdot\frac{14}{689}\right]+\left[\frac{84}{16546}\cdot\frac{840}{689}\right]+\left[\frac{1008}{165520}\cdot\frac{840}{689}\cdot\frac{8268}{16546}\right] \\ 0\;1\;0\;0\;0 & \frac{20}{38}+\left[\frac{20\cdot 21}{114}\cdot\frac{21}{689}\right]+\left[\frac{126}{16546}\cdot\frac{840}{689}\right]+\left[\frac{1512}{165520}\cdot\frac{840}{689}\cdot\frac{8268}{16546}\right] \\ 0\;0\;1\;0\;0 & \left[\frac{20}{3}\cdot\frac{21}{38}\cdot\frac{38}{689}\right]+\left[\frac{228}{16546}\cdot\frac{840}{689}\right]+\left[\frac{2736}{165520}\cdot\frac{840}{689}\cdot\frac{8268}{16546}\right] \\ 0\;0\;0\;1\;0 & \frac{840}{16546}+\left[\frac{8268}{16546}\cdot\frac{840}{689}\cdot\frac{8268}{16546}\right] \\ 0\;0\;0\;0\;1 & \frac{840}{689}\cdot\frac{8268}{16546}\cdot\frac{16546}{165520} \end{bmatrix}$$

Combining the fractions gives,

$i_1 = 3.76897$

$i_2 = 0.65346$

$i_3 = 0.23006$

$i_4 \quad = 0.08120$

$i_5 \quad = 0.06090$

• PROBLEM 3-18

Find the currents i_1, i_2, and i_3 for network shown in Fig. 1 if $R_1 = R_2 = R_3 = 1\ \Omega$ and $v_1 = v_2 = v_3 = 1$ volt. Assume R_4 to be a negative resistance of $-5\ \Omega$.

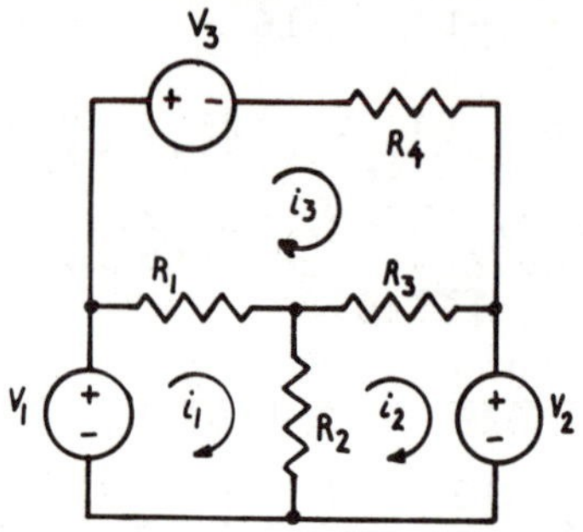

FIG. 1

Solution: Since the current loop directions have already been chosen, we can write three KVL equations for the voltage drops around each loop,

$$\text{loop 1: } V_1 = i_1(R_1 + R_2) - i_2(R_2) - i_3(R_1)$$

$$\text{loop 2: } -V_2 = -i_1(R_2) + i_2(R_2 + R_3) - i_3(R_3)$$

$$\text{loop 3: } -V_3 = -i_1(R_1) - i_2(R_3) + i_3(R_1 + R_3 + R_4).$$

Substituting the values given for the resistors and the voltage sources gives,

$$1 = 2i_1 - i_2 - i_3$$

$$-1 = -i_1 + 2i_2 - i_3$$

$$-1 = -i_1 - i_2 - 3i_3 .$$

These three equations may be written in matrix form as follows:

$$\begin{bmatrix} 2 & -1 & -1 \\ -1 & 2 & -1 \\ -1 & -1 & -3 \end{bmatrix} \begin{bmatrix} i_1 \\ i_2 \\ i_3 \end{bmatrix} = \begin{bmatrix} 1 \\ -1 \\ -1 \end{bmatrix}$$

To find $\bar{I}$, we must solve for $\bar{R}^{-1}$ in the equation $\bar{R}^{-1}\,\bar{v} = \bar{I}$. The inverse of matrix $\bar{R}$ is the cofactor of matrix $\bar{R}$ transposed, divided by the determinant of $\bar{R}$. Hence

$$\bar{R}^{-1} = \frac{\bar{R}_C^T}{|\bar{R}|}$$

first, $\bar{R}_C$ =

$$\begin{bmatrix} (-2)(-3)-(-1)(-1) & -[(-1)(-3)-(-1)(-1)] & (-1)(-1)-(-1)(2) \\ -[(-1)(-3)-(-1)(-1)] & (2)(-3)-(-1)(-1) & -[(2)(-1)-(-1)(-1)] \\ (-1)(-1)-(2)(-1) & -[(2)(-1)-(-1)(-1)] & (2)(2)-(-1)(-1) \end{bmatrix}$$

$$\bar{R}_C = \begin{bmatrix} -7 & -2 & 3 \\ -2 & -7 & 3 \\ 3 & 3 & 3 \end{bmatrix}$$

then, transpose the cofactor of $\bar{R}$ (exchange rows and columns) to obtain

$$\bar{R}_C^T = \begin{bmatrix} -7 & -2 & 3 \\ -2 & -7 & 3 \\ 3 & 3 & 3 \end{bmatrix}$$

To find the determinant of $\bar{R}$, sum the products of the elements of any corresponding row or columns in the matrices $\bar{R}_C$ and $\bar{R}$.

Taking the first row of each matrix gives,

$$|\bar{R}| = (2)(-7) + (-1)(-2) + (-1)(3)$$

$$|\bar{R}| = -14 + 2 - 3 = -15.$$

Thus $\bar{R}^{-1} = \dfrac{\bar{R}_C^T}{-15}$

$$\bar{R}^{-1} = \begin{bmatrix} \frac{7}{15} & \frac{2}{15} & -\frac{3}{15} \\ \frac{2}{15} & \frac{7}{15} & -\frac{3}{15} \\ -\frac{3}{15} & -\frac{3}{15} & -\frac{3}{15} \end{bmatrix}$$

and

$$\bar{I} = \bar{R}^{-1}\bar{V}$$

$$\begin{bmatrix} i_1 \\ i_2 \\ i_3 \end{bmatrix} = \begin{bmatrix} \frac{7}{15} & \frac{2}{15} & -\frac{3}{15} \\ \frac{2}{15} & \frac{7}{15} & -\frac{3}{15} \\ -\frac{3}{15} & -\frac{3}{15} & -\frac{3}{15} \end{bmatrix} \quad \begin{bmatrix} 1 \\ -1 \\ -1 \end{bmatrix}$$

$$\begin{bmatrix} i_1 \\ i_2 \\ i_3 \end{bmatrix} = \begin{bmatrix} \frac{7}{15} - \frac{2}{15} + \frac{3}{15} \\ \frac{2}{15} - \frac{7}{15} + \frac{3}{15} \\ -\frac{3}{15} + \frac{3}{15} + \frac{3}{15} \end{bmatrix}$$

Hence
$$\begin{bmatrix} i_1 \\ i_2 \\ i_3 \end{bmatrix} = \begin{bmatrix} \frac{8}{15} \\ -\frac{2}{15} \\ \frac{3}{15} \end{bmatrix} \text{ A}$$

• PROBLEM 3-19

> For the ladder network shown in Fig. 1, find the power dissipated in resistor R.

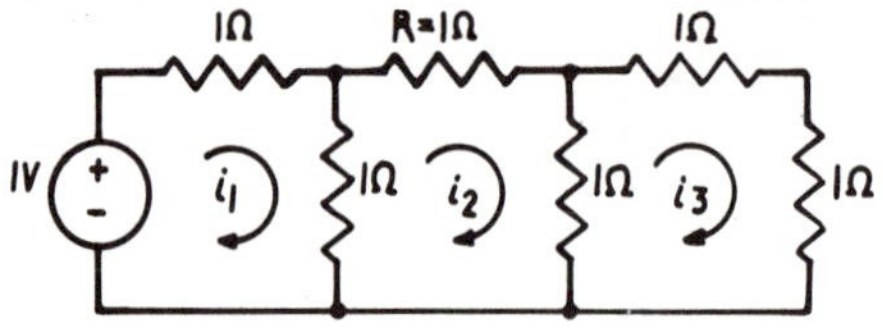

FIG. 1

Solution: Since the current loops are given, write the KVL equations as

loop 1: $1 = 2i_1 - i_2$

loop 2: $0 = -i_1 + 3i_2 - i_3$

loop 3: $0 = \quad i_2 + 3i_3.$

In matrix form, write

$$\begin{bmatrix} 2 & -1 & 0 \\ -1 & 3 & -1 \\ 0 & -1 & 3 \end{bmatrix} \begin{bmatrix} i_1 \\ i_2 \\ i_3 \end{bmatrix} = \begin{bmatrix} 1 \\ 0 \\ 0 \end{bmatrix}.$$

Since we need to know only i_2 in order to find the power dissipated in Resistor R, we write

$$i_2 = \frac{\begin{vmatrix} 2 & 1 & 0 \\ -1 & 0 & -1 \\ 0 & 0 & 3 \end{vmatrix}}{\begin{vmatrix} 2 & -1 & 0 \\ -1 & 3 & -1 \\ 0 & -1 & 3 \end{vmatrix}} = \frac{-(-1)[3]}{(2)[(3)(3)-(-1)(-1)]-(-1)[(-1)(3)]}$$

$$i_2 = \frac{3}{13} \text{ A}$$

Hence, the power dissipated by $R = 1\ \Omega$ is

$$p = i_2^2 R = i_2^2 = \frac{9}{169} \text{ W}.$$

Find the current in each branch in the circuit of fig. 1.

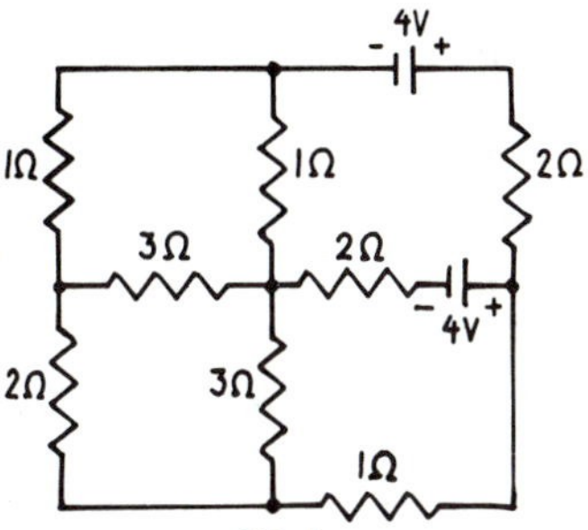

FiG. 1

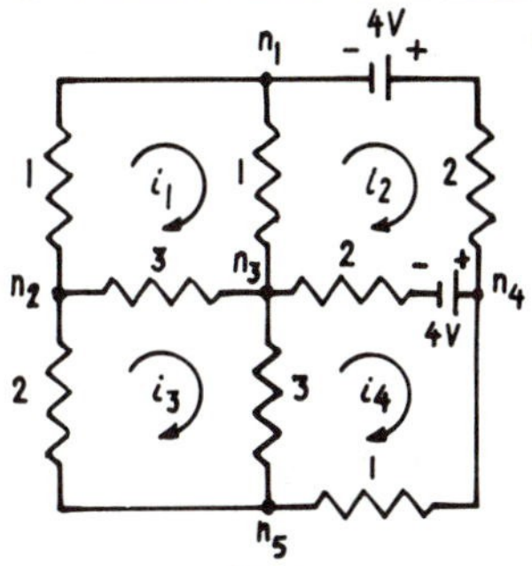

FIG.2

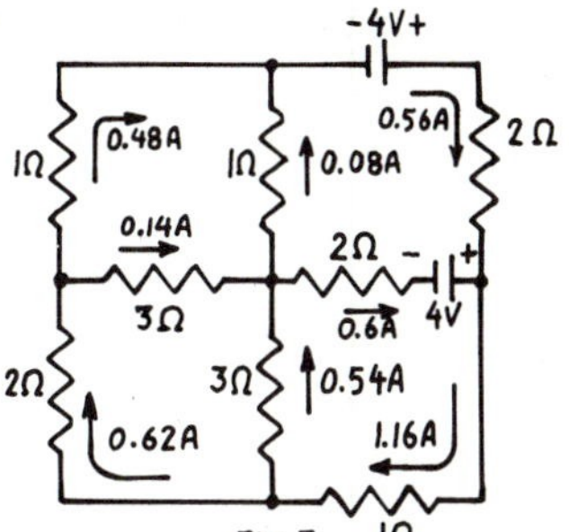

FiG.3

Solution: We assign four loop currents as shown in fig. 2. The current in each of the branches can be expressed in terms of these four loop currents. Writing the loop equations,

$$\ell_1: \quad 0 = 5i_1 - i_2 - 3i_3$$

$$\ell_2: \quad 0 = -i_1 + 5i_2 \qquad -2i_4$$

$$\ell_3: \quad 0 = -3i_1 \qquad +8i_3 \quad -3i_4$$

$$\ell_4: \quad 4 = \qquad -2i_2 \; -3i_3 \quad +6i_4 \; .$$

In matrix form,

$$\begin{bmatrix} 5 & -1 & -3 & 0 \\ -1 & 5 & 0 & -2 \\ -3 & 0 & 8 & -3 \\ 0 & -2 & -3 & 6 \end{bmatrix} \begin{bmatrix} i_1 \\ i_2 \\ i_3 \\ i_4 \end{bmatrix} = \begin{bmatrix} 0 \\ 0 \\ 0 \\ 4 \end{bmatrix} .$$

Solving by use of Gaussian elimination, we multiply the first row of the augmented matrix by $\frac{1}{5}$ and add it to the second row, thus eliminating the second element in the first column, obtaining

$$\begin{bmatrix} 5 & -1 & -3 & 0 & 0 \\ 0 & \frac{24}{5} & -\frac{3}{5} & -2 & 0 \\ -3 & 0 & 8 & -3 & 0 \\ 0 & -2 & -3 & 6 & 4 \end{bmatrix}$$

Multiplying the first row by $\frac{3}{5}$ and adding it to the third row eliminates the third element in the first column. We have reduced the first column. The second column can be reduced by multiplying the second row by the proper constant in order to eliminate the 3rd and 4th elements in the second column. The matrix can be reduced in this way to the lower triangular form

$$\begin{bmatrix} a_{11} & a_{12} & a_{13} & \cdots & & a_{1m} & b_1 \\ 0 & a_{22} & a_{23} & \cdots & & a_{2m} & b_2 \\ 0 & 0 & a_{33} & \cdots & & a_{3m} & b_3 \\ \vdots & & & \ddots & & \vdots & \vdots \\ 0 & 0 & & & a_{n-1\,m-1} & a_{n-1m} & b_{n-1} \\ 0 & 0 & & \cdots & 0 & a_{nm} & b_n \end{bmatrix}$$

The solution can be calculated by noting that

$$i_n = \frac{b_n}{a_{nm}}.$$

Then this can be substituted into the equation,

$$a_{n-1\;m-1}i_{n-1} + a_{n-1\;m}i_n = b_{n-1},$$

to solve for i_{n-1}, and so on.

The lower triangular form of the augmented matrix is

$$\begin{bmatrix} 5 & -1 & -3 & 0 & 0 \\ 0 & 4.8 & -0.6 & -2 & 0 \\ 0 & 0 & 6.13 & -3.25 & 0 \\ 0 & 0 & 0 & 3.44 & 4 \end{bmatrix}.$$

Hence

$$i_4 = \frac{4}{3.44} = 1.16 \text{ A}, \quad 6.13i_3 - 3.25(1.16) = 0,$$

$$i_3 = \frac{3.77}{6.13} = 0.62 \text{ A}. \quad \text{Similarly } i_2 = .56 \text{ A, and}$$

$i_1 = 0.48$ A

Fig. 3 shows the currents found in each branch.

• PROBLEM 3-21

Consider the network shown in fig. 1. Given $i_1 = 1.1$ A, $i_2 = 2.1$ A and $i_3 = 0.7$ A. Find v_1, v_2, and v_3.

Solution: Writing the loop equations for the loops shown in fig. 1 gives three equations in the form,

$$\bar{R}\,\bar{I} = \bar{V}$$

where $\bar{v}$ is unknown.

The loop equations are:

$$\ell_1: \quad v_1 = 6.2i_1 - 3.9i_2 - 2.3i_3 \qquad (1)$$

$$\ell_2: \quad -v_2 = -3.9i_1 + 5.7i_2 - 1.8i_3 \qquad (2)$$

$$\ell_3: \quad -v_3 = -2.3i_1 - 1.8i_2 + 4.7i_3 \qquad (3)$$

Writing the equations above in matrix form and multiplying eqs (2) and (3) by (-1) yields:

$$\begin{bmatrix} 6.2 & -3.9 & -2.3 \\ 3.9 & -5.7 & 1.8 \\ 2.3 & 1.8 & -4.7 \end{bmatrix} \begin{bmatrix} i_1 \\ i_2 \\ i_3 \end{bmatrix} = \begin{bmatrix} v_1 \\ v_2 \\ v_3 \end{bmatrix}$$

Since $\bar{I} = \begin{bmatrix} 1.1 \\ 2.1 \\ 0.7 \end{bmatrix}$, $\bar{v} = \begin{bmatrix} v_1 \\ v_2 \\ v_3 \end{bmatrix}$ is the product of the current and resistance matrix.

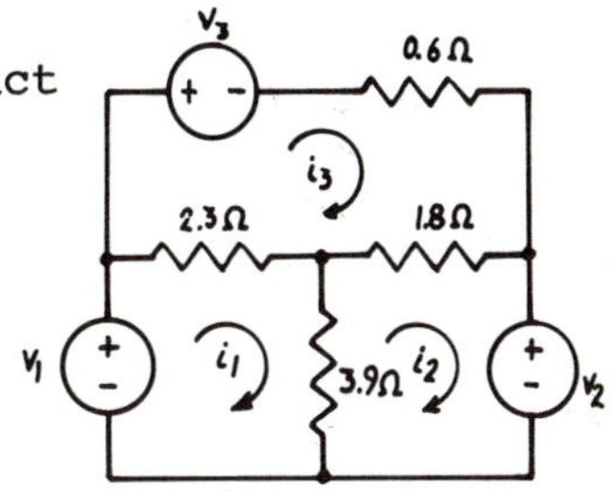

FIG. 1

Hence,

$$\bar{v} = \begin{bmatrix} 6.2 & -3.9 & -2.3 \\ 3.9 & -5.7 & 1.8 \\ 2.3 & 1.8 & -4.7 \end{bmatrix} \begin{bmatrix} 1.1 \\ 2.1 \\ 0.7 \end{bmatrix}$$

$$= \begin{bmatrix} 6.2(1.1) - 3.9(2.1) - 2.3(0.7) \\ 3.9(1.1) - 5.7(2.1) + 1.8(0.7) \\ 2.3(1.1) + 1.8(2.1) - 4.7(0.7) \end{bmatrix}$$

$$\bar{v} = \begin{bmatrix} -2.98 \\ -6.42 \\ 3.02 \end{bmatrix}, \quad v_1 = -2.98, \; v_2 = -6.42, \text{ and } v_3 = 3.02 \text{ V.}$$

• **PROBLEM 3-22**

Write the equations for currents $i_1 - i_4$, in the network shown, in matrix form

$$\bar{R}\,\bar{I} = \bar{V}.$$

Solution: The network in fig. 1 has 20 branches and 12 nodes for which we can write 20-(12-1) = 9 linear independent loop equations. Writing the 9 loop equations yields

$$\ell_1 : \quad (6 + 2 + 10 + 13)i_1 - 13\,i_2 - 10i_6 = -25 + 10$$

ℓ_2: $-13i_1 + (13+11+9+2)i_2 - 9i_3 - 11i_5 = -10 + 17 + 20$

ℓ_3: $-9i_2 + (9 + 10 + 7 + 5)i_3 - 10i_4 = -17 - 6$

ℓ_4 : $-10i_3 + (10+30+9+30)i_4 - 30i_5 - 9i_9 = 0$

ℓ_5 : $-11i_2 - 30i_4 + (11+7+5+30)i_5 - 7i_6 - 5i_8 = -20$

ℓ_6 : $-10i_1 - 7i_5 + (10+30+7)i_6 - 30i_7 = 12$

ℓ_7 : $-30i_6 + (30 + 1 + 10)i_7 = -7$

ℓ_8 : $-5i_5 + (5+20+2)i_8 - 2i_9 = 7$

ℓ_9 : $-9i_4 - 2i_8 + (9 + 2 + 8 + 10)i_9 = -10$

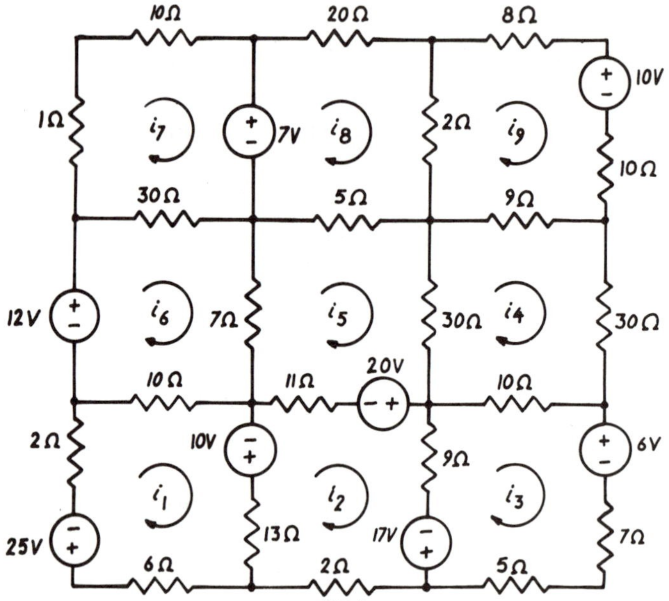

FIG. 1

Writing the equations in matrix form $\bar{R}\,\bar{I} = \bar{V}$ we obtain,

$$\begin{bmatrix} 31 & -13 & 0 & 0 & 0 & -10 & 0 & 0 & 0 \\ -13 & 35 & -9 & 0 & -11 & 0 & 0 & 0 & 0 \\ 0 & -9 & 31 & -10 & 0 & 0 & 0 & 0 & 0 \\ 0 & 0 & -10 & 79 & -30 & 0 & 0 & 0 & -9 \\ 0 & -11 & 0 & -30 & 53 & -7 & 0 & -5 & 0 \\ -10 & 0 & 0 & 0 & -7 & 47 & -30 & 0 & 0 \\ 0 & 0 & 0 & 0 & 0 & -30 & 41 & 0 & 0 \\ 0 & 0 & 0 & 0 & -5 & 0 & 0 & 27 & -2 \\ 0 & 0 & 0 & -9 & 0 & 0 & 0 & -2 & 29 \end{bmatrix} \begin{bmatrix} i_1 \\ i_2 \\ i_3 \\ i_4 \\ i_5 \\ i_6 \\ i_7 \\ i_8 \\ i_9 \end{bmatrix} = \begin{bmatrix} -15 \\ 27 \\ -23 \\ 0 \\ -20 \\ 12 \\ -7 \\ 7 \\ -10 \end{bmatrix}$$

the required matrix form equation .

Solve for the loop currents $i_1 - i_4$ in the network of fig. 1 using matrix methods.

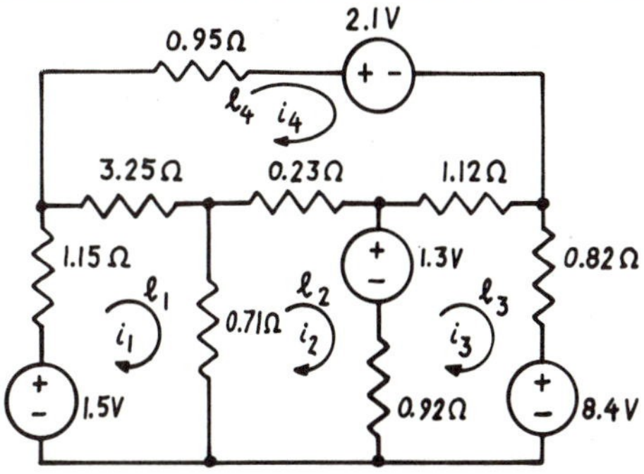

FIG. 1

Solution: There are four loops in the network, therefore, we can write four loop equations,

$$\ell_1 : \quad 1.5 = 5.11i_1 - 0.71i_2 - 3.25i_4$$

$$\ell_2 : \quad -1.3 = -0.71i_1 + 1.86i_2 - 0.92i_3 - 0.23i_4$$

$$\ell_3 : \quad 1.3-8.4 = \quad - 0.92i_2 + 2.86i_3 - 1.12i_4$$

$$\ell_4 : \quad -2.1 = -3.25i_1 - 0.23i_2 - 1.12i_3 + 5.55i_4$$

In matrix form we write,

$$\begin{bmatrix} 5.11 & -0.71 & 0 & -3.25 \\ -0.71 & 1.86 & -0.92 & -0.23 \\ 0 & -0.92 & 2.86 & -1.12 \\ -3.25 & -0.23 & -1.12 & 5.55 \end{bmatrix} \begin{bmatrix} i_1 \\ i_2 \\ i_3 \\ i_4 \end{bmatrix} = \begin{bmatrix} 1.5 \\ -1.3 \\ -7.1 \\ -2.1 \end{bmatrix}$$

Using Gaussian elimination to solve for the loop currents, form the augment matrix,

$$\left[\begin{array}{cccc|c} 5.11 & -0.71 & 0 & -3.25 & 1.5 \\ -0.71 & 1.86 & -0.92 & -0.23 & -1.3 \\ 0 & -0.92 & 2.86 & -1.12 & -7.1 \\ -3.25 & -0.23 & -1.12 & 5.55 & -2.1 \end{array}\right]$$

Multiply the first row by $\frac{0.71}{5.11}$ and add it to the second row, also multiply the first row by $\frac{3.25}{5.11}$ and add it to the fourth row to obtain the partially reduced augment matrix

$$\left[\begin{array}{cccc|c} 5.11 & -0.71 & 0 & -3.25 & 1.5 \\ 0 & 1.76 & -0.92 & -0.68 & -1.09 \\ 0 & -0.92 & 2.86 & -1.12 & -7.1 \\ 0 & -0.68 & -1.12 & 3.48 & -1.15. \end{array}\right].$$

Multiplying the second row by $\frac{0.92}{1.76}$ and $\frac{0.68}{1.76}$ and adding to the third and fourth rows respectively gives,

$$\left[\begin{array}{cccc|c} 5.11 & -0.71 & 0 & -3.25 & 1.5 \\ 0 & 1.76 & -0.92 & -0.68 & -1.09 \\ 0 & 0 & 2.38 & -1.48 & -7.67 \\ 0 & 0 & -1.48 & 3.22 & -1.57 \end{array}\right]$$

Similar reduction gives the matrix in reduced lower triangular form

$$\left[\begin{array}{cccc|c} 5.11 & -0.71 & 0 & -3.25 & 1.5 \\ 0 & 1.76 & -0.92 & -0.68 & -1.09 \\ 0 & 0 & 2.38 & -1.48 & -7.67 \\ 0 & 0 & 0 & 2.30 & -6.34 \end{array}\right]$$

Hence,

$2.30i_4 = -6.34$, $i_4 = -2.75$ A and $2.38i_3 - 1.48(-2.75)$ $= -7.67$, $i_3 = \frac{-7.67 - 4.07}{2.38} = -4.93$A and $1.76i_2 - 0.92(-4.93)$ $-0.68(-2.75) = -1.09$, $i_2 = \frac{-1.09 - 4.54 - 1.87}{1.76} = -4.26$ A,

finally $5.11i_1 - 0.71(-4.26) - 3.25(-2.75) = 1.5$,

$i_1 = \frac{1.5 - 3.02 - 8.94}{5.11} = -2.04$ A.

• PROBLEM 3-24

Solve for i_1, i_2, and i_3 using matrix methods.

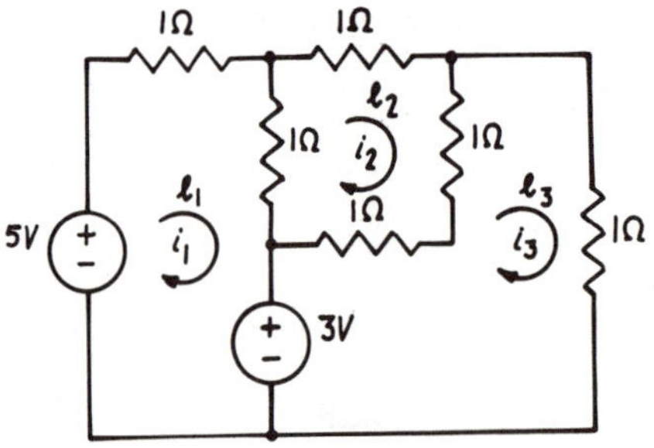

FIG. 1

Solution: There are three loops; therefore we can write three KVL loop equations for the loop currents indicated. The KVL equations for the three loops are

$$\ell_1 : 5 - 3 = 2i_1 - i_2$$

$$\ell_2 : 0 = -i_1 + 4i_2 - 2i_3$$

$$\ell_3 : 3 = \quad - 2i_2 + 3i_3.$$

Writing the loop equations in the matrix form

$$\bar{R}\,\bar{I} = \bar{V}$$

yields

$$\begin{bmatrix} 2 & -1 & 0 \\ -1 & 4 & -2 \\ 0 & -2 & 3 \end{bmatrix} \begin{bmatrix} i_1 \\ i_2 \\ i_3 \end{bmatrix} = \begin{bmatrix} 2 \\ 0 \\ 3 \end{bmatrix}$$

There are numerous techniques available to solve the simultaneous linear equations. We will use the method of determinants. Hence,

$$i_1 = \frac{\begin{vmatrix} 2 & -1 & 0 \\ 0 & 4 & -2 \\ 3 & -2 & 3 \end{vmatrix}}{\begin{vmatrix} 2 & -1 & 0 \\ -1 & 4 & -2 \\ 0 & -2 & 3 \end{vmatrix}} = \frac{2[12-4] + 3[2]}{2[12-4] + 1[-3]} = \frac{22}{13} = 1.69 \text{ A}$$

$$i_2 = \frac{\begin{vmatrix} 2 & 2 & 0 \\ -1 & 0 & -2 \\ 0 & 3 & 3 \end{vmatrix}}{13} = \frac{2[6] + 1[6]}{13} = \frac{18}{13} = 1.39 \text{ A}$$

$$i_3 = \frac{\begin{vmatrix} 2 & -1 & 2 \\ -1 & 4 & 0 \\ 0 & -2 & 3 \end{vmatrix}}{13} = \frac{2[12] + 1[-3+4]}{13} = \frac{25}{13} = 1.92 \text{ A}$$

• **PROBLEM** 3-25

Consider the resistive network shown in Fig. 1. Represent the resulting network equations in the matrix form.

Solution: Writing the three KVL equations for the voltage drops around the loops indicated, gives

loop 1: $i_1(R_1 + R_2) - i_2(R_2) - i_3(R_1) = v_1$

loop 2: $-i_1(R_2) + i_2(R_2 + R_3) - i_3(R_3) = -v_2$

loop 3: $-i_1(R_1) - i_2(R_3) + i_3(R_1 + R_3 + R_4) = -v_3$

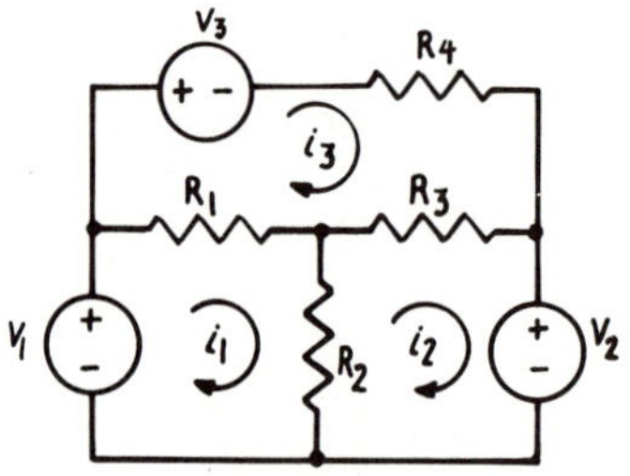

FIG. 1

We can write the above equations in matrix form, $\bar{R}\,\bar{I} = \bar{v}$ or

$$\begin{bmatrix} (R_1 + R_2) & -R_2 & -R_1 \\ -R_2 & (R_2 + R_3) & -R_3 \\ -R_1 & -R_3 & (R_1 + R_3 + R_4) \end{bmatrix} \begin{bmatrix} i_1 \\ i_2 \\ i_3 \end{bmatrix} = \begin{bmatrix} v_1 \\ -v_2 \\ -v_3 \end{bmatrix}$$

CHAPTER 4

INDUCTORS AND CAPACITORS

VOLTAGE-CURRENT RELATIONSHIPS FOR INDUCTORS

• PROBLEM 4-1

The current through an inductor with inductance $L = 10^{-3}$ henry is given as

$$i_L(t) = 0.1 \sin 10^6 t.$$

Find the voltage $\nu_L(t)$ across this inductor.

Solution: We apply the definition for voltage across an inductor

$$V_L(t) = L \frac{di_L(t)}{dt}$$

$$V_L(t) = L \frac{d}{dt}(0.1 \sin 10^6 t)$$

$$V_L(t) = 10^{-3} 10^6 (.1) \cos 10^6 t$$

$$V_L(t) = 100 \cos 10^6 t.$$

• PROBLEM 4-2

The inductance of linear time-varying inductor is given by

$$L(t) = L_0 (t + \tanh t)$$

and the current through it is given by

$$i(t) = \cos \omega t$$

Find the voltage across this inductor.

Solution: Since $\phi = L(t)\ i(t)$ and

$\nu(t) = \frac{d\phi}{dt}$, then $\nu(t) = \frac{d}{dt} L(t)\ i(t)$.

$$\nu(t) = \frac{d}{dt} L_o (t + \tanh t) \cos \omega t$$

$$U(t) = \frac{d}{dt} \quad U(t) \leftrightarrow V(t) = U(t)\frac{dv(t)}{dt} + v(t)\frac{du(t)}{dt}$$

$$\frac{d}{dt}(\tanh u) = \frac{du}{dt} \quad \text{sech}^2 u$$

$$v(t) = L_o\ (1 + \text{sech}^2 t)\cos\omega t - \omega L_o\ (t + \tanh t)\sin\omega t.$$

• PROBLEM 4-3

Find the voltage across an inductor, shown in Fig. 1, whose inductance is given by

$$L(t) = te^{-t} + 1$$

and the current through it is given by

$$i(t) = \sin \omega t$$

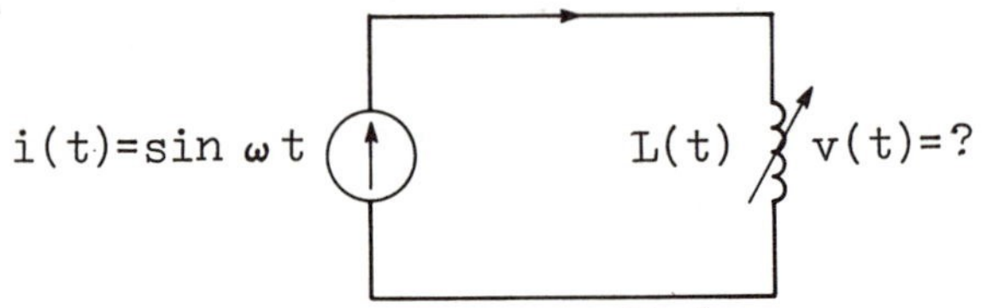

Fig 1

Solution: The voltage across an inductor is defined as $v(t) = \frac{d\phi}{dt}$ and $\phi = L\ i(t)$. In this problem L is a time-varying inductance L(t).

$$\phi(t) = L(t)\ i(t) = (te^{-t} + 1)(\sin \omega t)$$

the voltage becomes

$$v(t) = \frac{d}{dt}\left[\ (te^{-t} + 1)(\sin \omega t)\ \right]$$

$$v(t) = (1 + te^{-t})(\omega\cos \omega t) + (\sin \omega t)$$
$$(e^{-t} - te^{-t})$$

$$v(t) = \omega\cos \omega\ (1 + te^{-t}) + e^{-t} \sin \omega t\ (1 - t)$$

• PROBLEM 4-4

Consider a nonlinear inductor which is defined by

$$\phi = i + \tanh(i)$$

let the current through this inductor be

$$i(t) = \sin t.$$

Find the voltage across it (see Fig. 1).

Solution: We are given $\phi(i)$ and $i(t)$. To obtain the voltage $v(t)$, we find the product of the derivatives $\frac{d\phi}{di} \cdot \frac{di}{dt}$.

First, $$\phi = i + \tanh(i)$$

$$\frac{d\phi}{di} = 1 + \text{sech}^2(i)$$

also, $$i(t) = \sin t$$

$$\frac{di}{dt} = \cos t$$

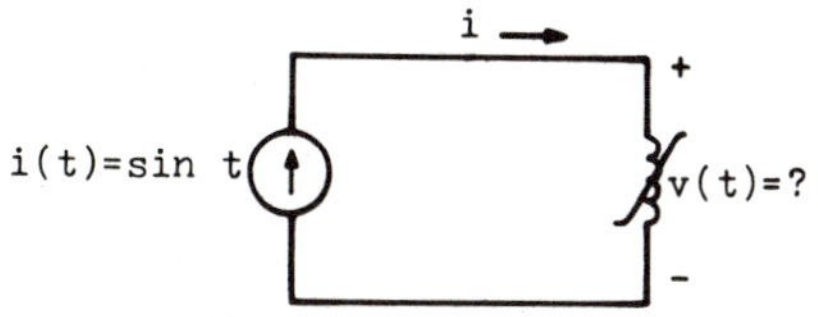

Figure 1

finally, $$v(t) = [\, 1 + \text{sech}^2(i)\,] \cos t$$

since $i(t) = \sin t$

$$v(t) = [\, 1 + \text{sech}^2(\sin t)\,] \cos t.$$

• PROBLEM 4-5

Assuming the passive sign convention, find the current through a 0.1 H inductor at t = 2s if (a) $V_L = 0.5\ t$ V for $0 \le t \le 3$s, $v_L = 0$ for $t < 0$ and $t > 3$, and $i_L = 0$ at $t = -3$; (b) $V_L = 0.5\ (t+2)$ V for $-2 \le t \le 1$ s, $v_L = 0$ for $t<-2$ and $t > 1$, and $i_L = 0$ at $t = -3$; (c) $v_L = \frac{2}{t^2+4}$ V for all t, $i_L\ (-\infty) = 0$.

Solution: (a) To find the current through an inductor we solve $v_L = L \frac{di}{dt}$ for i, which gives

$$i(t) = i(t_o) + \frac{1}{L} \int_{t_o}^{t} v_L\ (\tau)\ d\tau\ .$$

An initial value of the current is given at $t_o = -3$ s as $i_L(t_o) = 0$ but $v_L(t)$ is equal to zero in this region, so that if $V_L = 0.5\ t$ V then

$$i(t) = \frac{1}{0.1} \int_0^2 0.5\ t\ \ dt$$

$$i(t) = \frac{1}{0.1} \left. \frac{0.5t^2}{2} \right|_0^2 = 10 \text{ A}$$

(b) We are also given,

$$i_L = 0 \quad \text{at} \quad t = -3$$

but $V_L = 0$ for $t < -2$, so that if $V_L = 0.5\ (t + 2)$ V then

$$i(t) = \frac{1}{0.1} \int_{-2}^{1} 0.5\ (t + 2)\ dt$$

$$i(t) = \frac{1}{0.1} \left| _{-2}^{1} \frac{0.5(t+2)^2}{2} \right.$$

$$i(t) = \frac{1}{0.1} [2.25 - 0] = 22.5\ A.$$

The current in the inductor remains at 22.5 A even after V_L goes to zero. This is the dual analogy to a capacitor which, when charged, remains at the last value of voltage just before the current goes to zero.

(c) If $V_L = \frac{2}{t^2+4}$ V for all t and $i_L\ (-\infty) = 0$ then

$$i(t) = \frac{1}{0.1} \int_{-\infty}^{2} \frac{2}{t^2+4}\ dt$$

$$i(t) = \frac{2}{0.1} \int_{-\infty}^{2} \frac{dt}{t^2+\ 4}$$

$$i(t) = \frac{2}{0.1} \left| _{-\infty}^{2} \frac{1}{2} \arctan \frac{t}{2} \right.$$

$$i(t) = \frac{2}{0.1} \left[\frac{1}{2} \left(\frac{\pi}{4} - \left[- \frac{\pi}{2} \right] \right) \right]$$

$$i(t) = 23.6\ A$$

• PROBLEM 4-6

For the voltage waveform shown in Fig. 1, calculate the current through an inductor if L = 1/2 henry and $i_L(0) = -2$ amperes.

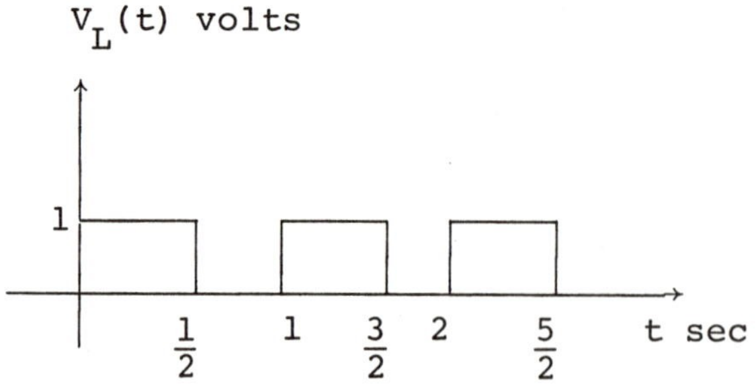

Fig. 1.

Solution: We employ the relation

$$i_L(t) = \frac{1}{L} \int_{t_o}^{t} v_L(\tau)\, d\tau + i(t_o)$$

with $t_o = 0$, $1/L = 2$ and $i(t_o) = -2$ A

$$i_L(t) = 2 \int_{0}^{t} v_L(\tau)\, d\tau \; -2.$$

We recognize that the integral of $V_L(t)$ is the area under the wave form in Fig. 1.

For the first pulse in the wave form we have,

$$i_L(1/2) = 2 \int_0^{1/2} 1\, dt \; -2$$

$$i_L(1/2) = 2 \left[t \right]_0^{1/2} -2$$

$$i_L(1/2) = -1 \text{ A}.$$

Fig. 2. shows that the current starts at -2 A and the results of our integration show that i_L is -1 A at $t = 1/2$ s.

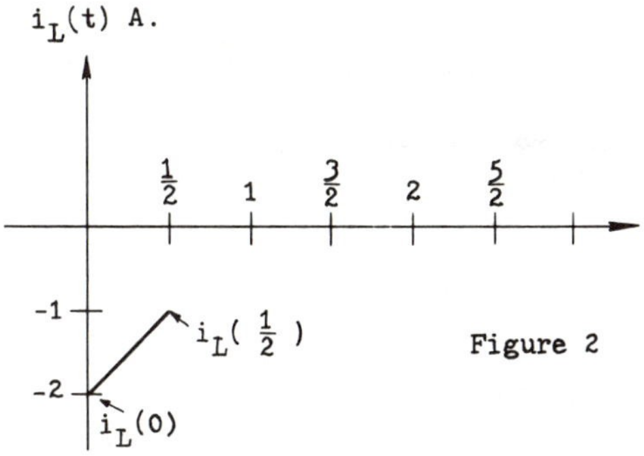

Figure 2

When the voltage across an inductor goes to zero, the current remains at the last value just before zero volts.

For $1/2 < t < 1$, i_L remains at -1 A. For the next pulse, our integration limits become $t = 1$ to $t = 3/2$ and our new initial current is -1 A.

Thus

$$i_L(3/2) = 2 \int_1^{3/2} 1\, dt \; - 1$$

$$i_L(3/2) = 0 .$$ (see Fig. 3).

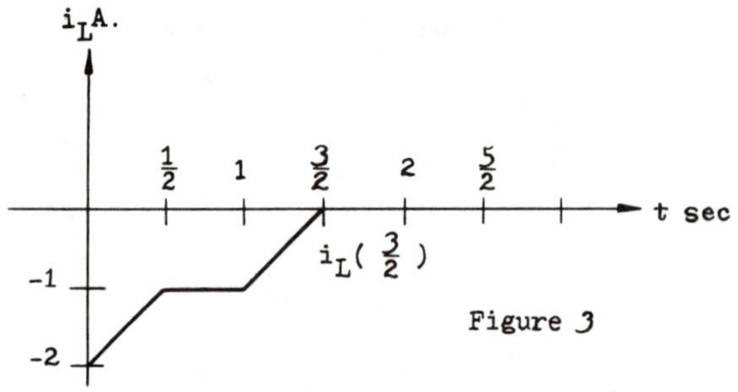

Figure 3

We can continue this way for the entire wave form. The results are in Fig. 4.

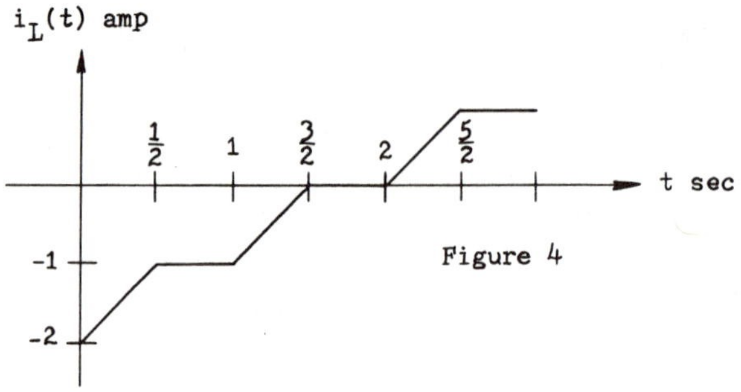

Figure 4

• PROBLEM 4-7

The voltage across a nonlinear inductor is given by

$$\nu = \tanh i$$

Show that this resistor is both voltage-controlled and current controlled. Find i in terms of ν.

Solution: When the voltage is written in terms of the current,

$$v = \tanh i,$$

note that all values of i produce single values of v, thus the inductor is current controlled.

Using the definition for tanh i, solve for i in terms of ν.

$$\nu = \tanh i = \frac{e^{i} - e^{-i}}{e^{i} + e^{-i}}\,.$$

First, multiply the right side of equation by $\frac{e^{i}}{e^{i}}$

$$\nu = \frac{e^{i}}{e^{i}}\left(\frac{e^{i} - e^{-i}}{e^{i} + e^{-i}}\right)$$

$$\nu = \frac{e^{2i} - e^{o}}{e^{2i} + e^{o}} = \frac{e^{2i} - 1}{e^{2i} + 1}\,.$$

Algebraic manipulation yields

$$\nu(e^{2i} + 1) = (e^{2i} - 1)$$

$$\nu e^{2i} + \nu = e^{2i} - 1.$$

Rearranging terms,

$$-\nu e^{2i} + e^{2i} = 1 + \nu$$

$$e^{2i}(1 - \nu) = 1 + \nu$$

$$e^{2i} = \frac{1 + \nu}{1 - \nu}$$

and taking the natural log of each side gives

$$\ln e^{2i} = 2i = \ell n\left(\frac{1 + \nu}{1 - \nu}\right).$$

Finally,

$$i = \frac{1}{2}\ln\left(\frac{1 + \nu}{1 - \nu}\right)$$

is a single valued function if $\nu^2 < 1$ thus, the resistor is both voltage and current controlled.

• PROBLEM 4-8

The current through a 10 henry inductor is observed to be the function plotted here.

Plot the function $V_L(t)$ and indicate the values for (a) $t = 0^+$, (b) $t = 2^-$, (c) $t = 2^+$, (d) $t = 5.5$, (e) $t = 5.5^+$, (f) $t = 6$, (g) $t = 6.5$, (h) $t = 8^-$.

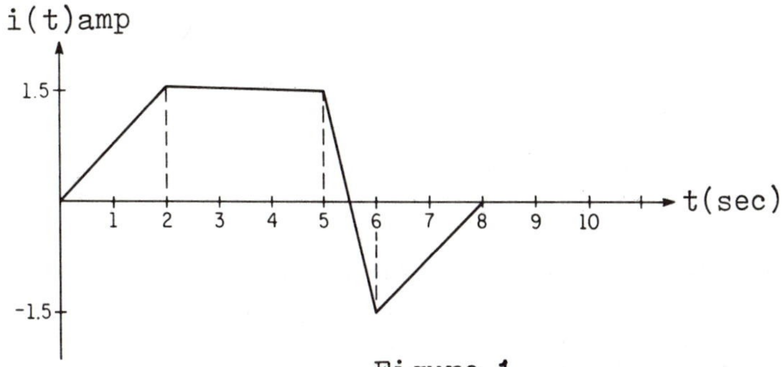

Figure 1

Solution: We recognize that the voltage across an inductor is defined as

$$V(t) = L\,\frac{di}{dt}$$

which means that V(t) would be the product of inductance and the slope of the wave form in Fig. 1.

By piece-wise analysis we proceed to find the slope of each section of the wave form in Fig. 1.

For $0 < t < 2$ slope = 0.75 A/S

$2 < t < 5$ slope = 0 A/S

$5 < t < 6$ slope = -3 A/S

$6 < t < 8$ slope = 0.75 A/S

By multiplying each of the slopes by the inductance, we obtain the plot of V(t) in Fig. 2.

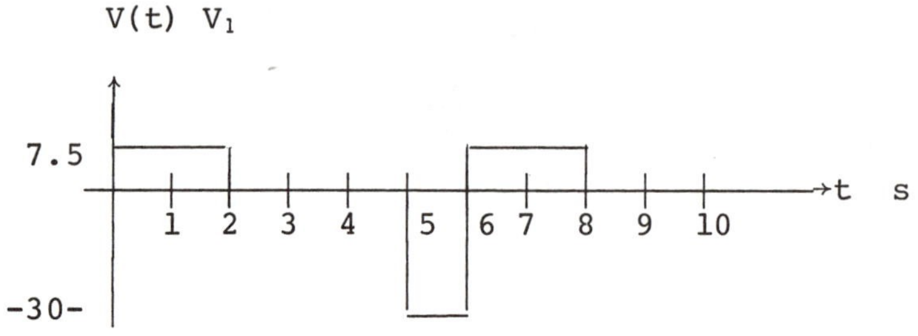

Fig. 2.

To find the values of time indicated, we read off the graph in Fig. 2.

(a) $t = 0^+$ $v(t) = 7.5$ V

(b) $t = 2^-$ $v(t) = 7.5$ V

(c) $t = 2^+$ $v(t) = 0$ V

(d) $t = 5.5$ $v(t) = -30$ V

(e) $t = 5.5^+$ $v(t) = -30$ V

(f) $t = 6$ v(t) is undefined

at $t = 6^-$ $v(t) = -30$ V at

$t = 6^+$ $v(t) = 7.5$ V

(g) $t = 6.5$ $v(t) = 7.5$ V

(h) $t = 8^-$ $v(t) = 7.5$ V

• **PROBLEM** 4-9

The variation of current through a 20-mH inductor as a function of time is shown in Fig. 1. Find the inductor voltage at t = : (a) 0.5; (b) 2.7; (c) 4.01 ms.

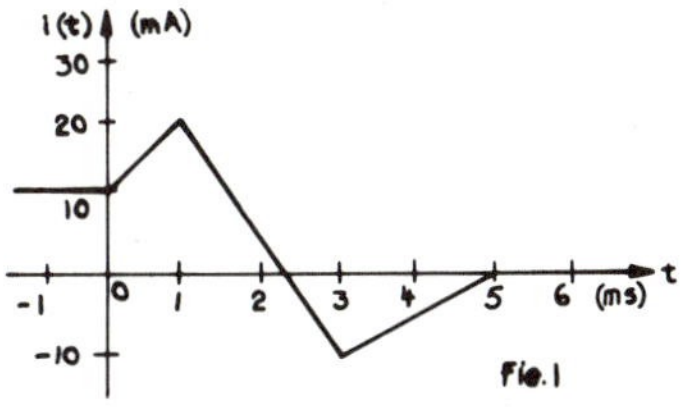

Fig. 1

Solution: In order to find the voltage across the inductor at specific times we must do a graphical piece-wise differentiation of Fig. 1. In order to do this we remember that the derivative of a function is the slope of that function at any point.

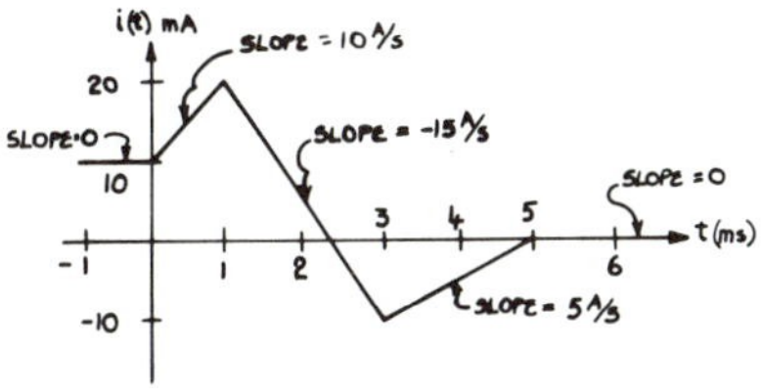

FiG. 2(a)

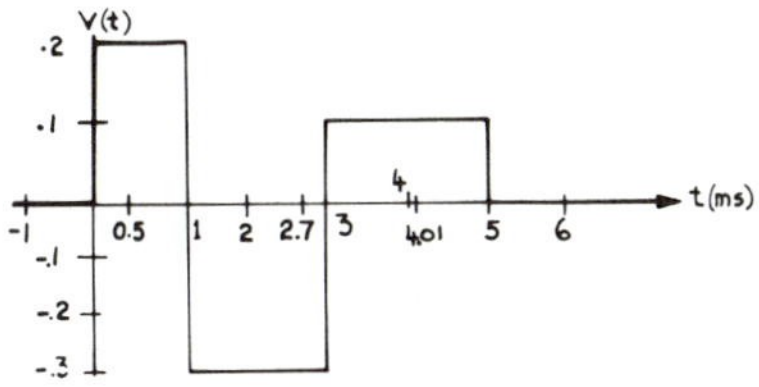

FiG. 2(b)

Fig. 2 (a) shows the function i(t) and the slope for each section of the graph.

Fig. 2 (b) shows the function $v(t) = L\frac{di}{dt}$.

Below are the values of v(t) for each section of the graph.

(a) $v(t) = 0.02\ (10) = 0.2\ V$

(b) $v(t) = 0.02\ (-15) = -0.3\ V$

(c) $v(t) = 0.02\ (5) = 0.1\ V$

• PROBLEM 4-10

The flux through a nonlinear inductor is given in terms of current through it by

$$\phi(t) = \frac{2i(t)}{1 + |i(t)|}.$$

(a) Find i(t) in terms of $\phi(t)$.
(b) Show that this inductor is both flux controlled and current controlled.

Solution: (a) First write the $\phi(t)$ expression as a composite function:

$$\phi(t) = \frac{2i(t)}{1 + |i(t)|} = \begin{cases} \frac{2i(t)}{1 + i(t)}, & i \geq 0 \\ \frac{2i(t)}{1 - i(t)}, & i < 0 \end{cases}$$

Next, separately invert the expressions to solve for i.

For $i \geq 0$, $2i = (1+i)\phi$; $(2-\phi)i = \phi$; $i = \frac{\phi}{2-\phi}$ provided $0 \leq \phi < 2$

For $i < 0$, $2i = (1-i)\phi$; $(2+\phi)i = \phi$; $i = \frac{\phi}{2+\phi}$ provided $-2 < \phi \leq 0$

The provisions on ϕ are necessary to place i in the specified range. We have

$$i(t) = \begin{cases} \dfrac{\phi(t)}{2-\phi(t)}, & 0 \leq \phi < 2 \\ \dfrac{\phi(t)}{2+\phi(t)}, & -2 < \phi \leq 0 \end{cases} = \frac{\phi(t)}{2-|\phi(t)|}, \quad |\phi| < 2.$$

(b) An inductor is said to be current controlled if its flux can be expressed as an explicit function of its current. It is said to be flux controlled if its current can be expressed as an explicit function of flux. We have explicit equations for current as a function of flux and flux as a function of current, so this nonlinear inductor is both flux controlled and current controlled.

VOLTAGE-CURRENT RELATIONSHIPS FOR CAPACITORS

• **PROBLEM** 4-11

Given a voltage source $v_{ab}(t) = 10\ tu(t)$ as shown in Fig. 1, calculate and sketch the current i_{ab} through an element connected to v_{ab} if the element is (a) a 2-ohm resistance; (b) a 2-henry inductance with $i(0) = 3$; (c) a 2-farad capacitance.

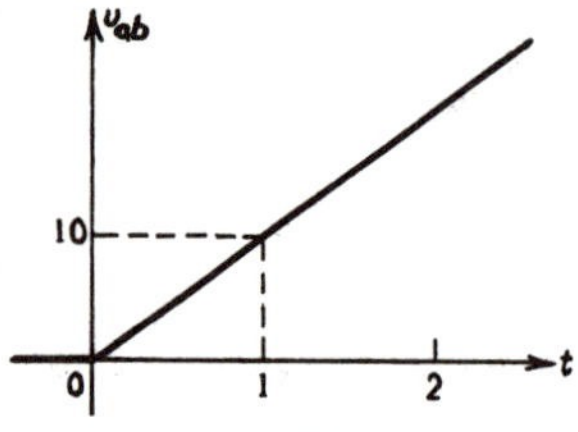

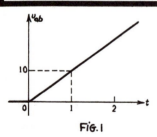

FIG. 1

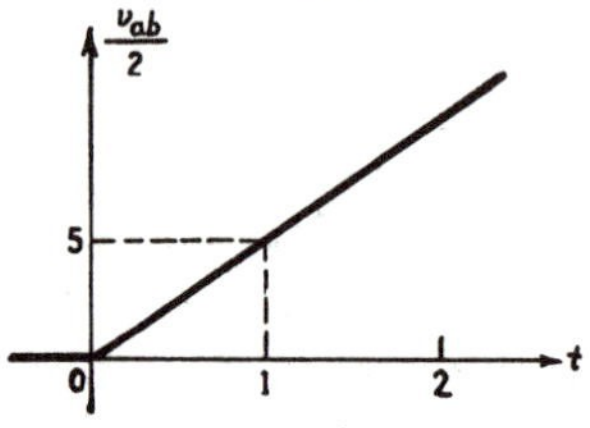

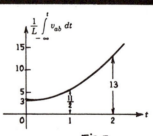

FIG. 2

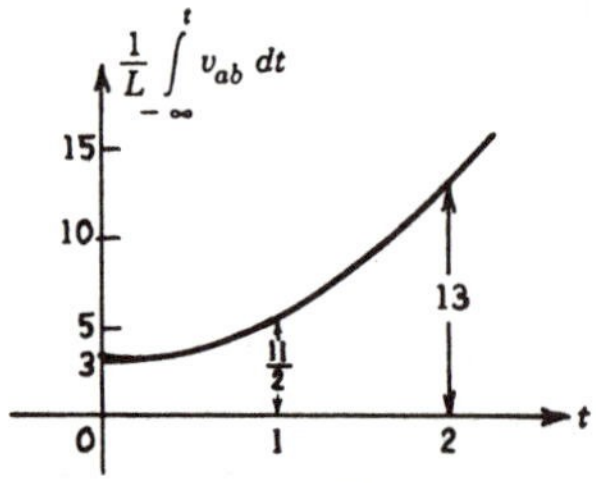

FIG. 3

Solution: (a) The current through a 2-ohm resistance is

$$\frac{v_{ab}}{2} = \frac{10t}{2}\ u(t) = 5t\ u(t).$$

A ramp with a slope of 5, as shown in Fig. 2.

(b) The current through an inductor is given as

$$i_L = i_L(0) + \frac{1}{L}\int_0^t v_L\, dT$$

where $i_L(0) = 3$, $v_L = 10t$, and $L = 2$ H. Hence,

$$i_L = 3 + \frac{1}{2}\int_0^t 10\ T\ dT$$

$$i_L = 3 + \frac{1}{2}\left[\frac{10\ T^2}{2}\right]_0^t$$

$$i_L = 3 + \frac{5}{2}\ t^2\ u(t)\ \text{A}$$

a parabola as shown in Fig. 3.

(c) The current through a capacitor is given as

$$i_c = C\frac{d\ v_c}{dt} = 2\ \frac{d}{dt}\ 10t\ u(t)$$

$$i_c = 20\ u(t)\ \text{A}$$

a step function as shown in Fig. 4.

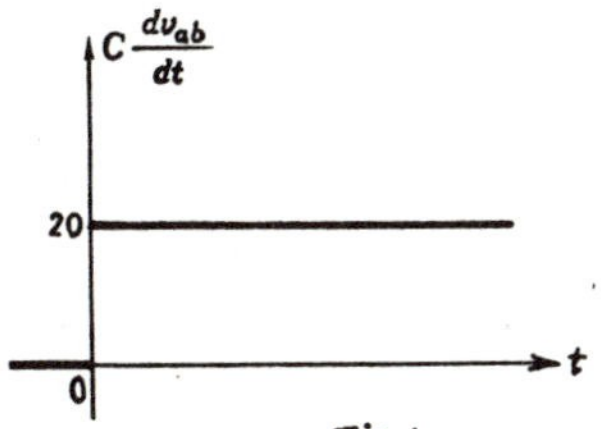

FIG.4

• PROBLEM 4-12

In Fig. 1. $v_c = \sin 2\pi T$. Find an expression for i and calculate i at the instants (a) t = 0, (b) t = 1/4s, (c) t = 1/2s.

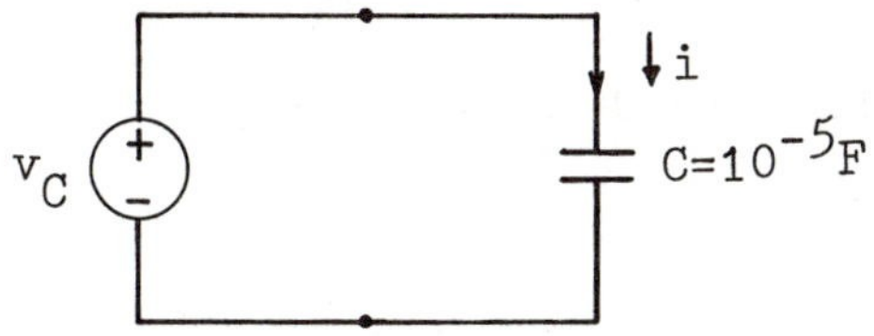

Figure 1

Solution: Current through a capacitor is defined as $i = C\frac{dv}{dt}$.

Since $v_c = \sin 2\pi t$

$$i = C\ 2\pi\ \cos 2\pi t$$

(a) at $t = 0$

$$i = 10^{-5} 2\pi\ (1) = 2\pi \times 10^{-5}\ \text{A}$$

(b) at $t = 1/4\text{s}$

$$i = 10^{-5}\ 2\pi\ \cos\ \frac{2\pi}{4}$$

$$i = 10^{-5}\ 2\pi\ (0) = 0\ \text{A}$$

(c) at $t = 1/2\text{s}$

$$i = 10^{-5}\ 2\pi\ \cos\ \frac{2\pi}{2}$$

$$i = 10^{-5}\ 2\pi\ (-1) = -2\pi \times 10^{-5}\ \text{A.}$$

• PROBLEM 4-13

Find the current through a 0.01 μF capacitor at $t = 0$ if the voltage across it is (a) $2 \sin 2\pi\ 10^6 t$ V; (b) $-2e^{-10^7 t}$ V; (c) $-2e^{-10^7 t} \sin 2\pi\ 10^6 t$ V.

Solution: (a) We use the definition

$$i(t) = C\frac{d\nu}{dt}\ ; \qquad \nu(t) = 2 \sin 2\pi\ 10^6 t$$

$$i(t) = 0.01 \times 10^{-6} \frac{d}{dt}\ 2 \sin 2\pi\ 10^6 t$$

$$i(t) = 0.01 \times 10^{-6} (2)(2\pi\ 10^6) \cos 2\pi\ 10^6 t$$

$$i(0) = 0.1256\ \text{A.}$$

(b) $$v(t) = -2e^{-10^7 t}\ \text{V}$$

$$i(t) = C\frac{dv}{dt}$$

$$i(t) = 0.01 \times 10^{-6} \frac{d}{dt} - 2e^{-10^7 t}$$

$$i(t) = (0.01 \times 10^{-6})(-10^7)\ -2e^{-10^7 t}$$

$$i(0) = 0.2\ \text{A}$$

(c) $$v(t) = -2e^{-10^7 t} \sin 2\pi\ 10^6 t\ \text{V}$$

$$i(t) = 0.01 \times 10^{-6} \frac{d}{dt} (-2e^{-10^7 t} \sin 2\pi\ 10^6 t)$$

$$i(t) = 0.01 \times 10^{-6} \Big[-2e^{-10^7 t} (2\pi\ 10^6) \cos 2\pi\ 10^6 t\ +$$

$$(-10^7)\ -2e^{-10^7 t} \sin 2\pi\ 10^6 t \Big]$$

$$i(0) = 0.01 \times 10^{-6} [\ -1.256 \times 10^7 + 0\ ]$$

$i(0) = -0.1256$ A

• PROBLEM 4-14

Consider the capacitor shown in Fig. 1. The capacitance C(t) is given by

$$C(t) = C_0(1 + 0.5 \sin t)$$

The voltage across this capacitor is given by

$$\nu(t) = 2 \sin \omega t$$

Find the current through the capacitor.

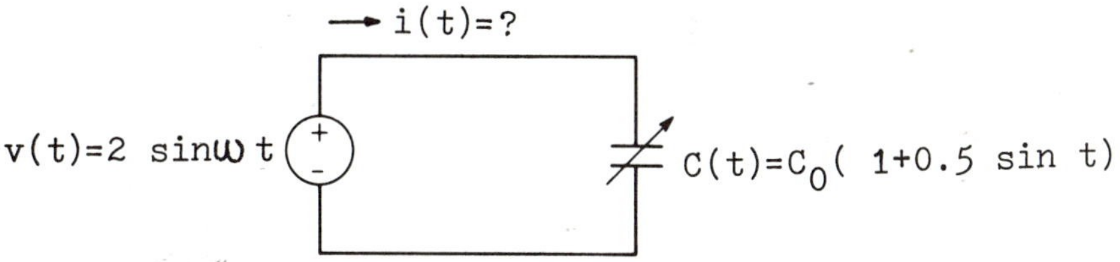

Figure 1

Solution: We can find the charge on the capacitor q(t) by using the definition $q(t) = C\,\nu(t)$. In this problem C is a time varying function C (t).

$$q(t) = C(t)\ \nu(t)$$

$$q(t) = C_0\ (1 + 0.5 \sin t)(2 \sin \omega t).$$

Since $i(t) = \frac{dq}{dt}$,

we have

$$i(t) = \frac{d}{dt}[\, C_0\ (1 + 0.5 \sin t)(2 \sin \omega t)$$

$$= (2 \sin \omega t)(0.5\, C_0 \cos t) +$$

$$C_0\ (1 + 0.5 \sin t)(2\omega \cos \omega t).$$

$$i(t) = C_0 \sin \omega t \cos t + 2\omega\, C_0 \cos \omega t\ (1 + 0.5 \sin t).$$

• PROBLEM 4-15

Consider a nonlinear capacitor which is defined by

$$q = \nu + \frac{1}{3}\nu^3$$

Let the voltage across this capacitor be

$$\nu(t) = \sin t$$

Find the current through this capacitor (see Fig. 1.)

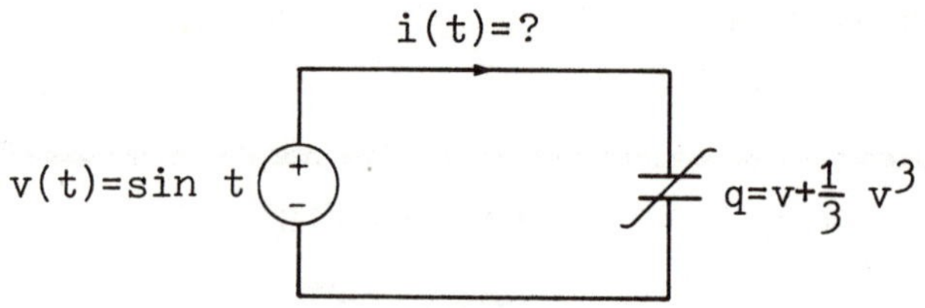

Figure 1

<u>Solution</u>: We are given $q(\nu)$ and $\nu(t)$. If we find the product of $\frac{dq}{d\nu}$ and $\frac{d\nu}{dt}$ we will obtain the current through capacitor $i(t) = \frac{dq}{dt}$.

First,
$$q = \nu + \frac{1}{3}\nu^3$$
$$\frac{dq}{d\nu} = 1 + \nu^2$$
and
$$\nu = \sin t$$
$$\frac{d\nu}{dt} = \cos t$$
finally
$$i(t) = \frac{dq}{dt} \cdot \frac{d\nu}{dt} = \frac{dq}{dt}$$
$$i(t) = (1 + \nu^2)\cos t$$
by substituting sin t for ν we have
$$i(t) = (1 + \sin^2 t)\cos t$$
using trig identity
$$\sin^2 t = 1/2\,(1 - \cos 2t)$$
$$i(t) = 1 + 5\cos t - 1/2\cos^2 t\cos t$$
and,
$$\cos A \cos B = 1/2\,[\cos(A - B) + \cos(A + B)]$$
Thus,
$$i(t) = 1.5\cos t - .25\cos t - .25\cos 3t$$
$$= 1.25\cos t - .25\cos 3t$$

• PROBLEM 4-16

The nonlinear characteristic of a capacitor is given by
$$q(t) = 0.5\nu^2(t)$$
The voltage across this capacitor is given by
$$\nu(t) = 1 + 0.5\sin t$$
Find the current through this capacitor.

Solution: If we find $\frac{dq}{d\nu}$ from

$$q(t) = 0.5\nu^2 (t)$$

and $\frac{d\nu}{dt}$ from

$$\nu(t) = 1 + 0.5 \sin t$$

their product gives $\frac{dq}{d\nu}\frac{d\nu}{dt} = \frac{dq}{dt} = i(t)$.

Hence,

$$\frac{dq}{d\nu} = \nu(t) \;; \qquad \frac{d\nu}{dt} = 0.5 \cos t$$

since we are given $\nu(t) = 1 + 0.5 \sin t$, and

$$\frac{dq}{d\nu}\frac{d\nu}{dt} = i(t) = (1 + 0.5 \sin t)(0.5 \cos t)$$

$$= .5 \cos t + .25 \sin t \cos t$$

Using the trig identity,

$$\sin A \cos B = 1/2 [\sin (A + B) + \sin (A - B)]$$

we obtain,

$$i(t) = .5 \cos t + \frac{.25}{2} [\sin 2t]$$

$$= .5 \cos t + .125 \sin 2t.$$

• PROBLEM 4-17

For the voltage waveform $\nu_c(t)$ shown in Fig. 1, find the charge and current for a capacitor of value C = 1/10 farads.

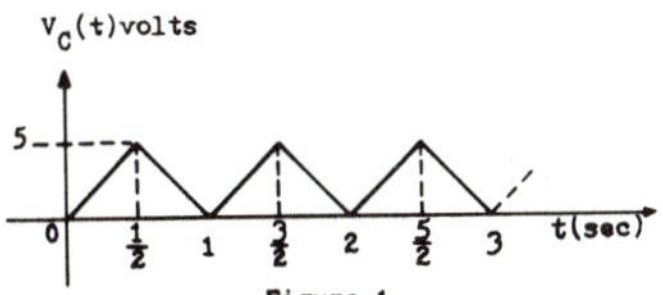

Figure 1

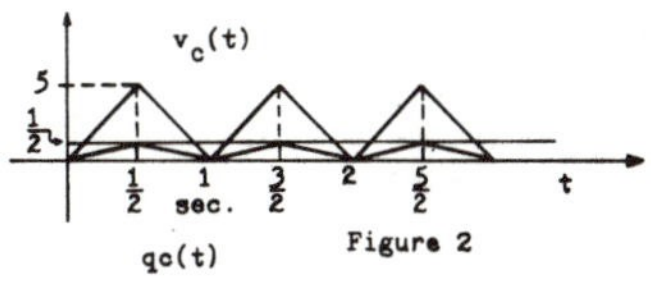

Figure 2

Solution: First we note the two relations for the capacitor

1) charge; $q(t) = C\,\nu(t)$

2) current; $i(t) = \frac{dq}{dt} = C\,\frac{d\nu}{dt}$,

C given as $\frac{1}{10}$ farads. Therefore, $q_c(t) = \frac{V_c(t)}{10}$

(see Fig. 2). Since

$$i_c(t) = C\,\frac{d\nu_c(t)}{dt}$$

we note that $\frac{d\nu_c(t)}{dt}$ is the slope of $\nu_c(t)$.

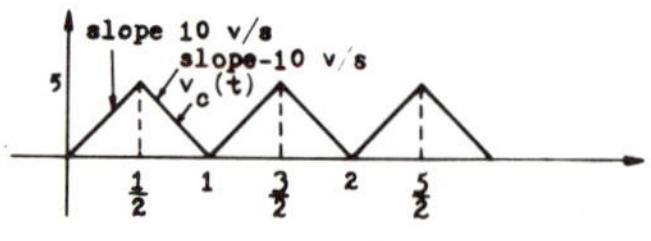

Figure 3

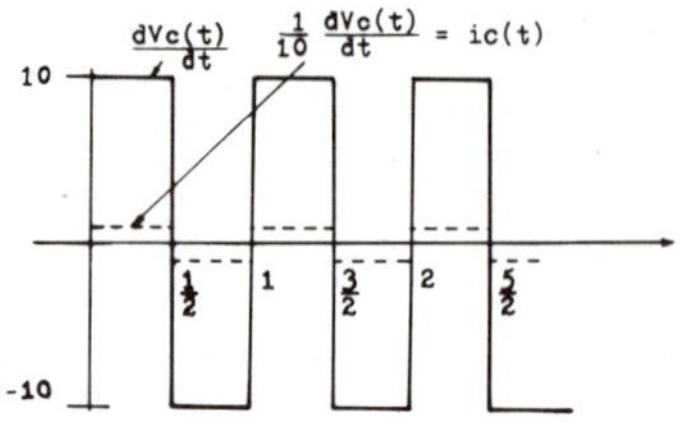

Figure 4

The functions $\frac{d\nu_c(t)}{dt}$ and $i_c(t) = \frac{1}{10}\,\frac{d\nu_c(t)}{dt}$ are shown in Fig. 4.

• PROBLEM 4-18

> Consider a capacitor with capacitance $C = 10^{-6}$ farad. Assume that initial voltage across this capacitor is $\nu_c(0) = 1$ volt. Find the voltage $\nu_c(t)$ at time $t \geq 0$ on this capacitor if the current through it is $i_c(t) = \cos(10^6 t)$.

Solution: We use the definition,

$$i = C\,\frac{d\nu}{dt}$$

solving for ν

$$\frac{1}{C}\int_{-\infty}^{t} i(\tau)\;d\tau = \nu(t).$$

If we have an initial voltage at time t_o, $-\infty < t_o < t$, we may state that

$$\frac{1}{C}\int_{-\infty}^{t_o} i(\tau)\,d\tau + \frac{1}{C}\int_{t_o}^{t} i(\tau)\,d\tau = \nu(t)$$

$$\nu(t_o) + \frac{1}{C}\int_{t_o}^{t} i(\tau)\;d\tau = \nu(t). \qquad (1)$$

In this problem we are given $\nu(t_o) = \nu_c(o) = 1$ volt,

$C = 10^{-6}$ farad, and $i_c(t) = \cos(10^6 t)$. We are asked to find $\nu_c(t)$ at time $t \geq 0$.

Substituting the above conditions into eq. (1)

$$\nu_c(t) = 1 + \frac{1}{10^{-6}} \int_0^t \cos(10^6 \tau)\, d\tau$$

$$\nu_c(t) = 1 + \frac{1}{10^6 \cdot 10^{-6}} \Big[\sin(10^6 \tau) \Big]_0^t$$

$$\nu_c(t) = 1 + \sin(10^6 t).$$

• PROBLEM 4-19

In Fig. 1. let $i = \sin 2\pi t$. Calculate ν_c at the instants (a) $t = 0$, (b) $t = 1/4$, (c) $t = 1/2$. Assume $\nu_c(o) = 0$.

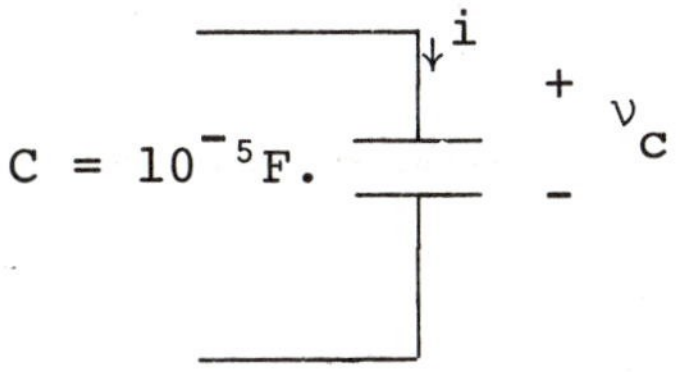

Fig. 1.

Solution: (a) Since we are told that $\nu_c(o) = 0$, at $t = 0$ the capacitor voltage ν_c is 0.

(b) We apply the relation

$$\nu(t) = \frac{1}{C} \int_{t_o}^{t} i \, d\tau + \nu(t_o)$$

$i(t) = \sin 2\pi t$

$$\nu(t) = \frac{1}{10^{-5}} \int_0^{1/4} \sin 2\pi t \quad + \quad 0$$

$$\nu(t) = \frac{1}{10^{-5}} \left(\frac{1}{2\pi} \right) \Big[-\cos 2\pi t \Big]_0^{1/4}$$

$$\nu(t) = \frac{1}{2\pi\ 10^{-5}} \Big[0 - (-1) \Big]$$

$$\nu(1/4) = 1.59 \times 10^4 \text{ V.}$$

(c) To find the voltage for the same capacitor at t = 1/2, we change the limits of integration.

$$\nu(t) = \frac{1}{10^{-5}} \int_0^{1/2} \sin 2\pi t \quad + \quad 0$$

$$\nu(t) = \frac{1}{10^{-5}} \left(\frac{1}{2\pi}\right) \left[- \cos 2\pi t \right]_0^{1/2}$$

$$\nu(t) = \frac{1}{2\pi\ 10^{-5}} \left[-(-1) - (-1) \right] = \frac{2}{2\pi\ 10^{-5}}$$

$$\nu(1/2) = 3.18 \times 10^4 \text{ V.}$$

• PROBLEM 4-20

Find the voltage across a 0.01 μF capacitor at t = 0.25 μs if $\nu(o) = 0.6$ V and the current through it is: (a) $0.2 \sin 2\pi\ 10^6\ t$ A, (b) $0.2e^{-10^7 t}$ A, (c) 0.2 A; $0 \le t \le 3$ μs.

Solution: (a) We make use of the equation

$$V(t) = \frac{1}{C} \int_{t_o}^{t} i(\tau)\ d\tau + V(t_o).$$

Substituting the values above into the equation we get

$$V(t) = \frac{1}{0.01\ 10^{-6}} \int_0^{0.25\ 10^{-6}} 0.2 \sin 2\pi\ 10^6 t \quad + 0.6$$

$$V(t) = \frac{0.2}{0.01 \times 10^{-6}} \left(\frac{1}{2\pi\ 10^6}\right) \left| - \cos 2\pi\ 10^6 t \right|_0^{0.25\times10^{-6}} + 0.6$$

$$V(t) = 3.183 \left(- \cos \frac{\pi}{2} + \cos 0 \right) + 0.6$$

$$V(t) = 3.183\ (\ 1\ ) \quad + \quad 0.6 \quad = 3.78 \text{ V.}$$

(b) $i(t) = 0.2e^{-10^7 t}$ A

$$V(t) = \frac{1}{C} \int_{t_o}^{t} i(\tau) \quad d\tau \quad + \quad V(t_o)$$

$$V(t) = \frac{1}{0.01\times10^{-6}} \int_{0}^{0.25\times10^{-6}} 0.2e^{-10^7 t}\, dt \quad + 0.6$$

$$V(t) = -\frac{0.2}{(.01\times10^{-6})10^7} \left[e^{-10^7 t} \right]_{0}^{0.25\times10^{-6}} + 0.6$$

$$V(t) = -2 \left[e^{-(.25\times10^{-6})10^7} - e^{0} \right] \quad + 0.6$$

$$V(t) = -2\ (-.918) + 0.6 = 2.44\ V.$$

(c) In this case $i(t) = 0.2A$ is a constant over the interval $0 \le t \le 3\ \mu s$ which covers the range of time values in which we will be integrating.

$$V(t) = \frac{1}{0.01\times10^{-6}} \int_{0}^{0.25\times10^{-6}} 0.2\ dt \ + \ 0.6$$

$$V(t) = \frac{0.2}{0.01\times10^{-6}} \left[t \right]_{0}^{0.25\times10^{-6}} \quad + 0.6$$

$$V(t) = 5 + 0.6 = 5.6\ V$$

ENERGY, CHARGE, & POWER

• PROBLEM 4-21

The charge (in C) that has entered a circuit element since $t = -\infty$ is numerically equal to 50 times the current (in A) at that point for every instant of time. (a) If the current is 4A at $t = 10s$ find $q(t)$ (b) If the total charge is 5C at $t = -20s$, find $i(t)$.

Solution: We are given the relation

$$q = 50i$$

also, we are told that $i(10) = 4$.

We are asked to find a relation for charge with respect to time.

First, we substitute the definition of current $i = \frac{dq}{dt}$ into $q = 50i$ to obtain

$$q = 50\,\frac{dq}{dt}\ .$$

We proceed to solve for q in terms of t.

$$dt = 50 \frac{dq}{q}$$

$$\int dt = 50 \int \frac{1}{q}\, dq$$

$$t + C_1 = 50 \ln q + C_2$$

$$t + (C_1 - C_2) = 50 \ln q, \quad \text{let } C_1 - C_2 = K$$

$$t + K = 50 \ln q$$

$$\frac{t}{50} + \frac{K}{50} = \ln q, \qquad \text{we use the relation}$$

$x = e^{\ln x}$ to obtain,

$$e^{\left[\frac{t}{50} + \frac{K}{50}\right]} = e^{\ln q}$$

$$e^{\frac{K}{50}} e^{\frac{t}{50}} = q, \qquad \text{Let } e^{\frac{K}{50}} = K'$$

$$K' e^{\frac{t}{50}} = q.$$

Substituting the given relation $q = 50i$,

$$K' e^{\frac{t}{50}} = 50\, i$$

since $i(t) = i(10) = 4$

$$K' e^{\frac{10}{50}} = 50\ (4)$$

$$K' = \frac{50\ (4)}{e^{\frac{10}{50}}} = 163.7$$

our final result is,

$$q = 163.7\, e^{\frac{t}{50}}\ \text{C}.$$

(b) From part (a) we found that

$$q = K' e^{\frac{t}{50}}$$

and it is given that $q = 50\, i$

$$50\, i = K' e^{t/50}$$

$$i = \frac{K'}{50} e^{t/50}$$

in part (b) we are told that q(t)

$$= q(-20) = 5 . \qquad i = \frac{q}{50} = \frac{5}{50} = .1$$

$$0.1 = \frac{K^1}{50} e^{\frac{-20}{50}}$$

$$K' = \frac{(.1)50}{e^{\frac{-20}{30}}} = 7.46$$

our final result is

$$i = \frac{7.46}{50} e^{t/50} = 0.149\, e^{t/50}\ \text{A}.$$

• PROBLEM 4-22

A voltage source $\nu(t)$ is connected to a capacitor C = 2f. Find the energy stored in the capacitor from t = 0 to t = 10 sec if

(a) $\nu(t) = t^2 e^{-2t}$ (b) $\nu(t) = t \sin t$

(c) $\nu(t) = (\sin t) e^{-t}$

Solution: We make use of the equation for energy and the relations for a capacitor.

$$E = \int_{t_o}^{t} vi\, dt ; \qquad \text{since } i = C \frac{d\nu}{dt}$$

$$E = C \int_{t_o}^{t} \nu \frac{d\nu}{dt} dt = C \int_{\nu(t_o)}^{\nu(t)} \nu \, d\nu$$

$$E = C\, 1/2\, [\nu(t)^2 - \nu(t_o)^2]$$

(a) $\nu(t) = t^2 e^{-2t}$

$$\nu(10) = 10^2 e^{-20} = 2.06 \times 10^{-7}$$

$$\nu(o) = 0$$

$$E = 2\, 1/2\, [(2.06 \times 10^{-7})^2 - 0^2]$$

$$E = 4.3 \quad 10^{-14}\ \text{J} \cong 0^+\ \text{J}$$

(b) $\nu(t) = t \sin t$

$$\nu(10) = 10 \sin 10 = 10(-.544) = -5.44$$

$$\nu(o) = 0$$

$$E = 2\ 1/2\ [(-5.44)^2 - 0^2]$$

$$E = 29.5 \text{ J}$$

(c) $$\nu(t) = (\sin t)e^{-t}$$

$$\nu(10) = (\sin 10)e^{-10} = (-.544)(4.54 \times 10^{-5})$$

$$\nu(o) = 0$$

$$E = 2\ 1/2\ [(-2.469 \times 10^{-5})^2 - 0^2]$$

$$E = 6.09 \times 10^{-10} \text{ J} \cong 0^+ \text{ J}$$

• PROBLEM 4-23

(a) If $q = -10^{-7}e^{-10^5 t}$C, find the average current during the time from $t = -5\mu s$ to $+5\mu s$. Given $i = 12 \cos(1000t + \frac{\pi}{6})$A; (b) Find the average current during the interval, $0 < t < \pi/3$ ms; (c) Determine the total charge transferred between $t = 0$ and $t = 2\pi/3$ ms.

Solution: The average of a function over a given interval $(t_2 - t_1)$ can be found by integrating that function over the interval and dividing by the interval. Hence,

$$i_{avg} = \frac{1}{t_2 - t_1} \int_{t_1}^{t_2} i\ dt.$$

Since, $$i = \frac{dq}{dt}$$

then $$i_{avg} = \frac{1}{t_2 - t_1} \int_{t_1}^{t_2} dq$$

$$i_{avg} = \frac{1}{t_2 - t_1} \left[q(t_2) - q(t_1) \right].$$

Hence, $t_1 = -5\mu s$, $t_2 = 5\mu s$ and

$$i_{avg} = 10^5 \left[-10^{-7}e^{-10^5(5\times10^{-6})} + 10^{-7}e^{-10^5(-5\times10^{-6})} \right]$$

$$i_{avg} = 10^5 \left[-10^{-7}(.607) + 10^{-7}(1.649) \right]$$

$$i_{avg} = 10.42 \times 10^{-3}\text{A} = 10.42\text{mA}.$$

(b) For $i = 12\cos(1000t + \frac{\pi}{6})A$;

$$i_{avg} = \frac{1}{t_2 - t_1}\int_{t_1}^{t_2} i\,dt$$

$$i_{avg} = \frac{1}{t_2 - t_1}\left[\frac{12}{1000}\sin(1000t + \frac{\pi}{6})\right]_{t_1}^{t_2}$$

where $t_1 = 0$ and $t_2 = \frac{\pi}{3}$ ms;

$$i_{avg} = 955\left[\frac{12}{1000}(\sin\frac{\pi}{2} - \sin\frac{\pi}{6})\right] = 5.73A.$$

(c) The charge transferred over a time interval $(t_2 - t_1)$ is

$$q = \int_{t_1}^{t_2} i\,dt = \int_0^{\frac{2\pi}{3}\times 10^{-3}} 12\cos(1000t + \frac{\pi}{6})dt$$

$$q = \left[\frac{12}{1000}\sin(1000t + \frac{\pi}{6})\right]_0^{\frac{2\pi}{3}\times 10^{-3}}$$

$$q = \left[\frac{12}{1000}\sin\frac{5\pi}{6} - \frac{12}{1000}\sin\frac{\pi}{6}\right]$$

$$q = \frac{12}{1000} - \frac{12}{1000} = 0\ C.$$

• **PROBLEM 4-24**

Find the energy stored in an inductor from t = 0 to t = 1 whose ϕ-i characteristic is given by

$$\phi(t) = 1/3i^3\ (t)$$

and the current through it is given by

$$i(t) = 2\sin t.$$

Solution: We can obtain an expression for $v(t)$ for the inductor by solving

$$\frac{d\phi}{di}\ \frac{di}{dt} = v(t)$$

since, $\frac{d\phi}{di} = i^2 (t)$

and $\frac{di}{dt} = 2 \cos t$

$$\nu(t) = (2 \sin t)^2 (2 \cos t).$$

The energy stored is found by integrating power over the time interval.

$$E = \int_{t_o}^{t} p(t)dt$$

$$E = \int_0^1 \nu i \, dt = \int_0^1 (2 \sin t)^3 (2 \cos t) \, dt$$

$$u = 2 \sin t$$

$$\frac{du}{dt} = 2 \cos t \Rightarrow \quad du = 2 \cos t \, dt$$

then,

$$\int_0^1 (2 \sin t)^3 \, 2 \cos t \, dt = \int_0^1 u^3 du$$

$$E = \frac{u^4}{4} \Bigg|_0^1$$

$$E = \Bigg|_0^1 \frac{(2 \sin t)^4}{4} = \frac{(1.682)^4}{4}$$

$$E = 2.00 \text{ J}$$

• PROBLEM 4-25

Find the magnitude of the voltage across an inductor: (a) of 30 mH, if the current through it is increasing at the rate 20 mA/ms (b) of 0.4 mH at t = 0, if the current in it is $50e^{-10^4 t}$ mA; (c) at t = 0, if the power entering it is given by 12 cos 100 πt mW and the current is 150 mA at t = 0.

Solution: (a) If the current through an inductor is increasing at the rate of

$$20 \frac{mA}{ms} \quad \text{then} \quad \frac{di(t)}{dt} = 20 \text{ A/s}$$

and if L = 0.03 H then

$$V = L\frac{di}{dt} = (0.03)(20) \text{ V} = .6 \text{ V}$$

(b) Again,
$$V = L\frac{di}{dt}$$
$$V = 0.4 \times 10^{-3}\frac{d}{dt}50 \times 10^{-3}e^{-10^4t}$$
$$V = 0.4 \times 10^{-3}[-500\ e^{-10^4t}]$$

at t = 0
$$|V| = 0.2\ V$$

(c) Since $p = \nu i = L\frac{di}{dt}i$

we have $p = L\ i\frac{di}{dt} = 12 \cos 100\ \pi t$ mW

also given is i(o) = 150 mA.

We now find the voltage to be
$$V = \frac{L\ i\ \frac{di}{dt}}{i} = \frac{p}{i} = \frac{12 \cos 100\ \pi t}{150}$$

or
$$|V| = 0.08\ V; \quad \text{at } t = 0$$

• PROBLEM 4-26

Find the energy stored in a nonlinear capacitor with characteristics given by
$$g(t) = 0.5\ \nu^2(t)$$
and
$$\nu(t) = 1 + 0.5 \sin t$$
from t = 0 to t = 1 sec.

Solution: Energy or power over a time interval is defined as,
$$E = \int_{t_o}^{t} p\ dt$$

since $p = \nu i$;
$$E = \int_{t_o}^{t} \nu i\ dt$$

$$i(t) = \frac{dq}{d\nu}\frac{d\nu}{dt} = \frac{dq}{dt} = (1 + 0.5 \sin t)(0.5 \cos t)$$
$$p = \nu i = (1 + 0.5 \sin t)^2(0.5 \cos t)$$
$$E = \int_{o}^{1}(1 + 0.5 \sin t)^2(0.5 \cos t)\ dt$$

Let $u = (1 + 0.5 \sin t)$
$$du = 0.5 \cos t\ dt$$

$$E = \int_0^1 u^2\,du = \frac{u^3}{3}\Bigg|_0^1 = \frac{(1 + 0.5\sin t)^3}{3}\Bigg|_0^1$$

$$E = [\ .956 - .333\ ] = 0.623\ J$$

• **PROBLEM 4-27**

The current through a 4-μF capacitor is an infinite sequence of 100-mA pulses of 1-ms duration and 3-ms separation. That is, the period is 4ms. The first pulse begins at t = 0 when the capacitor has no voltage across it. (a) When does the capacitor voltage reach 1000 V? (b) What is the capacitor charge (in C) at t = 12-ms? (c) If the pulse amplitude is reduced to zero after t = 0.1s, what is the energy stored in the capacitor at t = 0.2s?

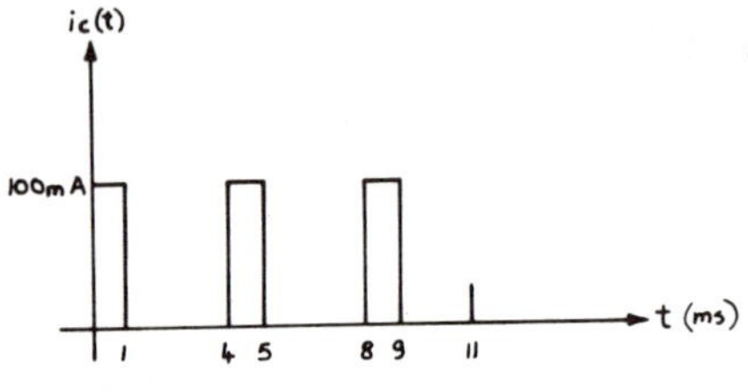

Fig. 1

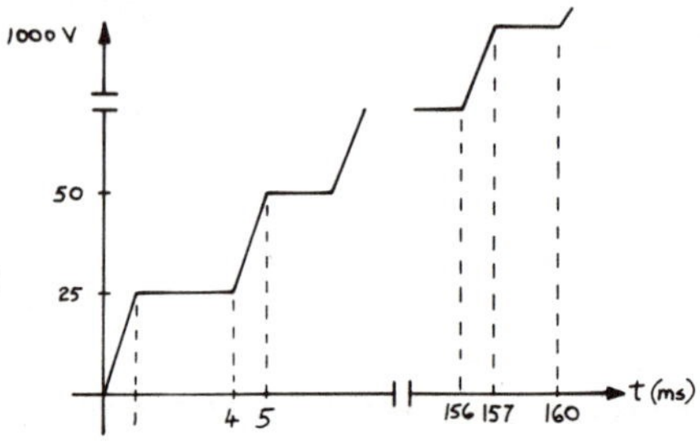

Fig. 2

<u>Solution:</u> (a) The applied waveform shown in Fig. 1 can be integrated graphically to obtain the resultant voltage waveform.

The voltage across the capacitor is given by

$$v_C(t) = v_C(t_0) + \frac{1}{C}\int_{t_0}^{t} i\,dt$$

and it is given that $v_C(0) = 0$.

Then $$v_C(1ms) = \frac{1}{4 \times 10^{-6}}\int_0^{.001} 100 \times 10^{-3} dt$$

$$v_C(1ms) = 25V.$$

Referring to Fig. 2, it can be seen that the voltage reaches successive multiples of 25V every (4n+1)ms, where n = 0,1,2... (i.e. at t = 1,5,9,...). Furthermore, whenever t = 4n+1, the voltage has made (n+1) jumps of 25V and is thus given by v = 25(n+1). (e.g. at t = 5ms, n = 1 and v = 50V). Thus v reaches 1000V when 25(n+1) = 1000 or when n = 39. Then t = 4n+1, or t = 157ms, as shown in Fig. 2.

(b) To find v_C at t = 12ms, note that at t = 9ms (corresponding to n = 2). The voltage has just reached 25(n+1) or 75V, and will stay at this voltage for the next 3ms. Thus at t = 12-ms v_C = 75V. Hence,

$q = Cv$

$q = (4 \times 10^{-6})(75) = 300 \times 10^{-6}C.$

(c) For t = .1 sec = 100ms: The voltage is at a value that it reached when t = 97ms (4n+1 = 97, n = 24). Then v = 25(24+1) = 625V. At t = .1s v_C = 625V hence,

$$E_C = \frac{1}{2}Cv^2$$

$$= \frac{1}{2}(4 \times 10^{-6})(625)^2 = 0.781J.$$

• PROBLEM 4-28

A 5-μf capacitor($C = 5 \times 10^{-6}$) has an initial voltage of V_0 volts at t = 0 when the current waveform shown is applied. (a) Find and sketch $\nu_c(t)$ for (i) $V_0 = 0$, (ii) $V_0 = -5$, (iii) $V_0 = 5$. (b) Plot the power p(t) for each condition i, ii, and iii. (c) Evaluate the energy stored at t = 1, 2, ∞ for each condition i, ii, and iii.

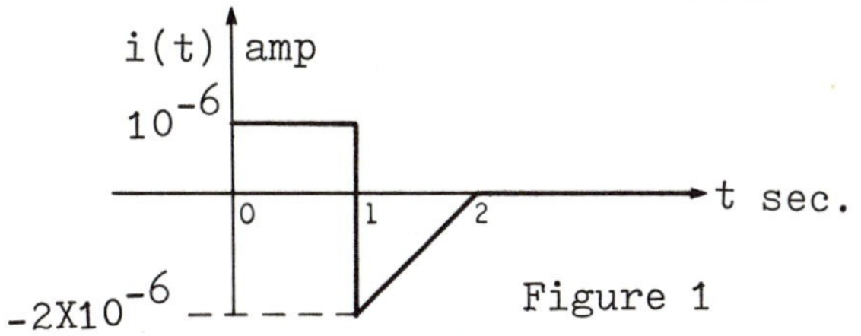

Figure 1

Solution: Define i(t) as follows:

$$i(t) = \begin{cases} 10^{-6} \text{ A} & 0 \le t < 1 \\ 2\times10^{-6}(t-2) \text{ A} & 1 \le t < 2 \\ 0 \text{ A} & 2 \le t \end{cases} \quad \text{[From Fig(1)]}$$

we solve $i = c\frac{d\nu}{dt}$ for $\nu(t)$ and obtain

$$\nu(t) = \frac{1}{c} \int_{t_0}^{t} i(\tau)d\tau + \nu(t_0) \qquad (1)$$

with initial condition $\nu(0) = V_0$ given. We find $\nu(t)$ with a piece-wise analysis of the graph in Fig. 1.

First solve $\nu(t)$ for the constant current portion of the graph, $0 \le t < 1$,

$$\nu_a(t) = \frac{1}{c}\int_0^t 10^{-6}\, d\tau + V_0$$

$$= \frac{1}{5\times10^{-6}}\Big[10^{-6}\,\tau\Big]_0^t + V_0$$

$$\therefore \quad \nu_a(t) = \frac{t}{5} + V_o \qquad (2)$$

Next we find the equation for the ramp portion of the graph, $1 \leq t < 2$.

The slope, $m = 2\times10^{-6}$ and

t-axis intercept $k = 2$.

Using point slope representation for a straight line,

$$i = m(t - k), \quad \text{where} \quad m = \text{slope}$$
$$k = \text{t-axis intercept}$$

$$i_b(t) = 2\times10^{-6}(t-2) \; . \qquad 1 \leq t < 2 \qquad (3)$$

Substituting into Eq (1)

$$\nu_b(t) = \frac{1}{c}\int_1^t 2\times10^{-6}(\tau-2)\,d\tau + \nu_a(1)$$

$$= \frac{1}{5 \times 10^{-6}}\left[\frac{2\times10^{-6}\tau^2}{2} - 4\times10^{-6}\tau\right]_1^t + \nu_a(1)$$

$$\nu_b(t) = \frac{t^2}{5} - \frac{4t}{5} + \frac{3}{5} + \left(\frac{1}{5} + V_o\right) \qquad 1 \leq t < 2$$

For $t > 2$, $\nu(t)$ would become V_o.

Therefore,

$$\nu(t) = \begin{cases} \frac{t}{5} + V_o & o \leq t < 1 \\ \frac{t^2}{5} - \frac{4}{5}t + \frac{4}{5} + V_o & 1 \leq t < 2 \\ V_o & 2 \leq t < \infty \end{cases}$$

Note: for cases (i), (ii) and (iii), insert the proper value of V_o.

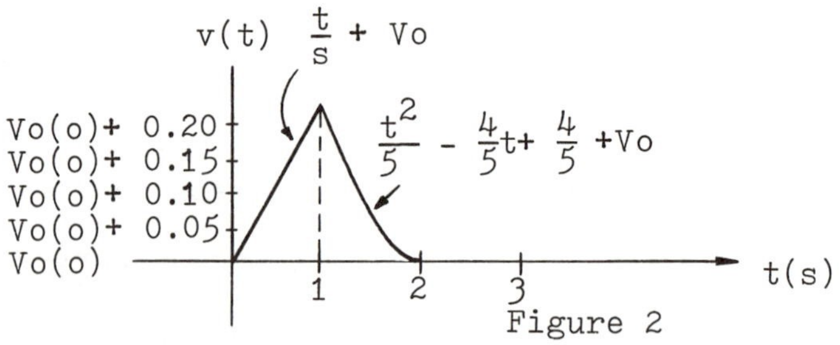

Figure 2

Fig. 2 shows a plot of the voltage $\nu(t)$, across the 5μF capacitor. This graph is generalized for any initial voltage V_o.

To obtain a plot of the power we take the product of the functions in the graphs of Figs. 1 and 2. Instantaneous power is defined as

$$p(t) = \nu(t)\ i(t) \quad W$$

Therefore;

$$P(t) = \begin{cases} t/5 + V_o & \times \quad 10^{-6} & o \le t < 1 \\ \frac{t^2}{5} - \frac{4}{5}t + \frac{4}{5} + V_o & \times \quad 2\times10^{-6}(t-2) & 1 \le t < 2 \\ V_o & \times \quad o & 2 \le t \end{cases}$$

Multiplying yields

$$P(t) = \begin{cases} 10^{-6}\left(\frac{t}{5} + V_o\right) W & o < t < 1 \\ 10^{-6}\left[\frac{2}{5}t^2 (t-6) + t\left(\frac{24}{6} + 2V\right) - \left(\frac{16}{5}+4V_o\right)\right] W & 1 \le t < 1 \\ o & 2 < t \end{cases}$$

The cubic equation for $1 \le t < 2$ may not be represented on a generalized graph since the shape will change for various values of V_0. Fig. 3 is a plot of the power for (i) $V_0 = 0$. After two seconds the current $i(t) = 0$, hence the power is zero.

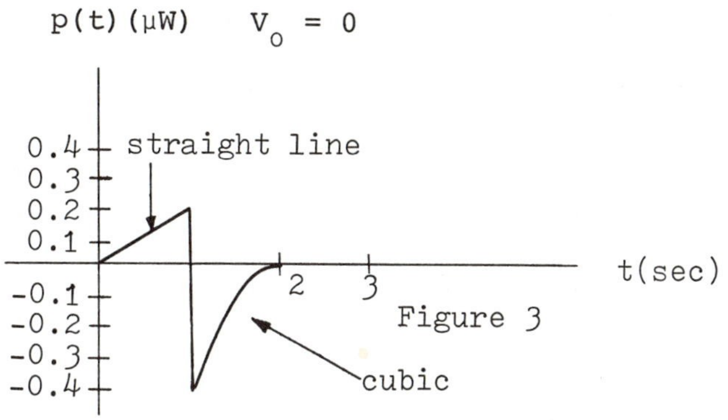

Figure 3

The energy stored is defined as power over a time interval.

$$E = \int_{t_o}^{t} P(t)d\,t = \int_{t_o}^{t} V_i \quad dt$$

Since $\quad i(t) = C\,\frac{d\nu}{dt}$

$$E = \int_{t_o}^{t} C\,\frac{d\nu}{dt}\ \nu dt = \int_{V(t_o)}^{\nu(t)} C\nu\, d\nu$$

$$E\,(t) = \frac{1}{2}\,C\,[\ \nu(t)^2 - \nu(t_o)^2\ ]$$

$$E(t) = 2.5 \times 10^{-6} [\nu(t)^2 - \nu(o)^2]$$

For $t = 1$, $\nu(t) = \frac{1}{5} + V_o$

$$\therefore E(t) = 2.5 \times 10^{-6} [(1/5 + V_o)^2 - V_o^2]$$

$$= 2.5 \times 10^{-6} \left[\frac{1}{25} + \frac{2}{5} V_o\right]$$

for $t = 2$ $\nu(t) = \frac{4}{5} - \frac{8}{5} + \frac{4}{5} + V_o$

$\nu(2) = V_o$ $\nu^2(2) = V_o^2$

$$E(t) = 2.5 \quad 10^{-6} [V_o^2 - V_o^2] = 0$$

for $t \geq 2$ $\nu(t) = V_o$ so again

$$E(t = \infty) = 2.5 \times 10^{-6} [V_o^2 - V_o^2] = 0$$

• PROBLEM 4-29

(a) Find the value of an inductor which carries 2 A. of current and in which 20 J of energy is stored.
(b) Find the value of a capacitor with 500 volts across it, in which 20 J of energy is stored.

Solution: (a) Since $E_L = 1/2\, Li^2$ we can solve for the unknown L.

$$L = \frac{2 E_L}{i^2} = \frac{40}{(2)^2} = 10 \text{ H}$$

(b) Similarly, $E_C = 1/2\, C \nu^2$

$$C = \frac{2 E_C}{\nu^2} = \frac{40}{(500)^2} = 1.6 \times 10^{-4} \text{ F.} = 160 \ \mu\text{F.}$$

• PROBLEM 4-30

Let the current i_1 enter the positive terminal of a circuit element across which the voltage v_1 is present. Find the power being absorbed by the element at $t = 10s$ if $i_1 = 2e^{-0.1t}A$ and (a) $v_1 = 6 \frac{di_1}{dt}$; (b) $v_1 = \frac{1}{6} \int_o^t i_1 dt^1 + 2 \text{ V}$

Solution: The power being absorbed by the element at any instant of time is $p(t) = i_1(t) \quad \nu_1(t)$.

(a) $$p(t) = i_1(t)\nu_1(t) = 2e^{-0.1t}\, 6\,\frac{di_1}{dt}$$

$$= 2e^{-0.1t}\,(6)(-0.1)(2)e^{-0.1t}$$

$$p(t) = -2.4e^{-0.1t}\, e^{-0.1t}$$

$$= -2.4e^{-0.2t}$$

We are asked to find the power absorbed at t = 10s. Hence,

$$p(t) = -2.4e^{-0.2(10)} = -0.325 \text{ W}.$$

(b) $$p(t) = \nu_1(t)\; i_1\;(t) = \left[\frac{1}{6}\int_0^t i_1\; dt^1 + 2V\right] 2e^{-0.1t}$$

$$p(t) = \left[\frac{1}{6}\int_0^t 2e^{-0.1t}\; dt^1 + 2V\right] 2e^{-0.1t}$$

$$p(t) \left[\frac{2}{(-0.1)6}\left[e^{-0.1t^1}\right]_0^t + 2\right] 2e^{-0.1t}$$

When t = 10s

$$p(10) = \left[\frac{-2}{0.6}(e^{-1} - e^0)+2\right] 2e^{-1} = [4.107]\; 0.736$$

$$= p(10) = 3.02 \text{ W}.$$

• PROBLEM 4-31

A 1000-pF capacitor, a 1-mH inductor, and a current source of $2 \cos 10^6 t$ mA are in series. The capacitor voltage is zero at t = 0. Find the energy stored (a) in the inductor at t = 0; (b) in the capacitor at t = 1.571 μs; (c) in both inductor and capacitor together at t = 1 μs.

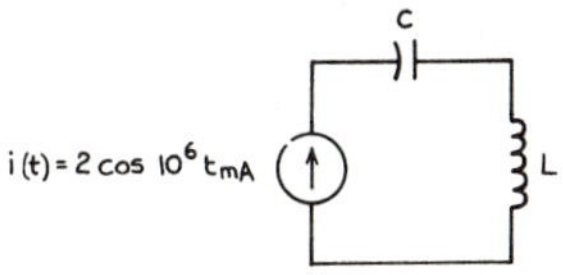

FIG.1

Solution: (a) To find the energy stored in the inductor we use

$$E = 1/2\; L\; i^2$$

$$E(t) = \frac{1}{2}(1 \times 10^{-3})[(2 \cos 10^6 t) \times 10^{-3}]^2$$

$$E(o) = (0.5 \times 10^{-3})(4 \times 10^{-6}) = 2 \times 10^{-9} \text{ J}$$

(b) In order to find the energy stored in a capacitor we must first find the voltage across it.

$$v(t) = \frac{1}{c} \int_{t_o}^{t} i \, d\tau + v(t_o)$$

$v(t_o) = 0$, so

$$v(1.571\ 10^{-6}) = \frac{1 \times 10^{-3}}{1000 \times 10^{-12}} \int_{o}^{1.571 \times 10^{-6}} 2 \cos 10^6 t \ dt.$$

We must remember that the current source is in mA. This is the reason we multiply the equation by 10^{-3} in order to get our results in Joules.

$$v(t) = \frac{2 \times 10^{-3}}{1000 \times 10^{-12}(10^6)} \left| \sin 10^6 t \right|_{o}^{1.571 \times 10^{-6}}$$

$$v(t) = 2 [\sin 1.571] = 2 \text{ V}.$$

Sin 1.571 is evaluated for 1.571 radians (1 rad = 57.3°).

Now we find the energy stored by

$$E(t) = 1/2\ Cv^2(t)$$

$$E = \frac{1000 \times 10^{-12}}{2} (2)^2 = 2 \times 10^{-9} \text{ J}.$$

(c) Using the same methods as those in both parts (a) and (b) we find the total energy stored in both inductor and capacitor at $t = 1\ \mu s$.

First the inductor:

$$E_L = (1/2) L\ i^2$$

$$E_L = \frac{1}{2}(1 \times 10^{-3})\ [(2 \cos 1) \times 10^{-3}]^2$$

$$E_L = 5.839 \times 10^{-10}$$

The voltage across the capacitor:

$$v(t) = \frac{1}{c} \int_{t_o}^{t} i(\tau)\, d\tau + v(t_o)$$

$$v(t) = \frac{1 \times 10^{-3}}{1000\ 10^{-12}} \int_0^{1 \times 10^{-6}} 2 \cos 10^6 t \ dt .$$

$$v(t) = \frac{2 \times 10^{-3}}{1000 \times 10^{-12}(10^6)} \ (\sin 1)$$

$$v(t) = 1.683 \text{ V.}$$

We find the energy stored in the capacitor:

$$E_c = (1/2) C\, v^2$$

$$E_c = \frac{1000 \times 10^{-12}}{2} (1.683)^2 = 1.416 \times 10^{-9} \text{ J.}$$

The total energy becomes

$$E_T = E_L + E_c = 5.839 \times 10^{-10} + 1.416 \times 10^{-9} = 2 \times 10^{-9} \text{J.}$$

• PROBLEM 4-32

The element in the terminal pair a-b in the figure is either an inductor or a capacitor. If i = cos 3t and v = sin 3t, what is the element and its value if

(a) $v = v_{ab}$; (b) $v = v_{ba}$?

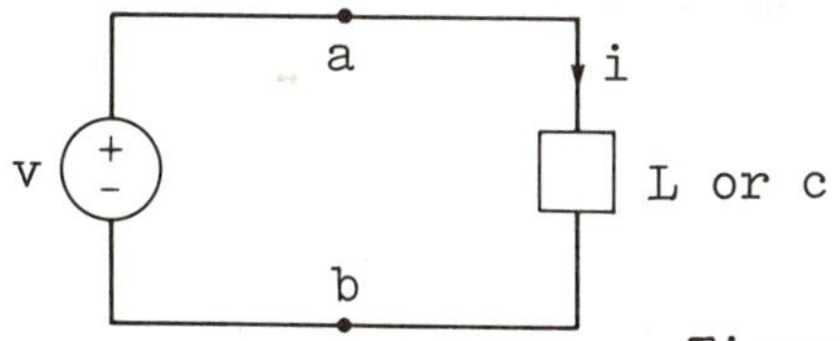

Figure 1

Solution: We know that the dual relations

$$v = L \frac{di}{dt} \quad (1)$$

$$i = C \frac{dv}{dt} \quad (2)$$

define the parameters for the unknown element in (a) $v = v_{ab}$. We notice that $\frac{dv}{dt} = 3 \cos 3t$ but $\frac{di}{dt} = -3 \sin 3t$. Since both sides of equation (2) remain positive, the unknown element in part (a) must be a capacitor. In (b) however, the voltage across the unknown element is the opposite of the passive sign convention which is illustrated in Fig. 2.

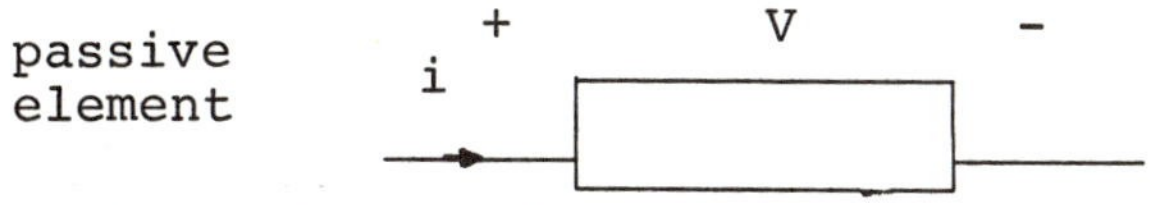

Fig. 2.

Equation (1) satisfies this condition, therefore, the element must be an inductor.

CHAPTER 5

NATURAL RESPONSE OF RL AND RC CIRCUITS

RL CIRCUITS

• **PROBLEM** 5-1

For the circuit shown in Fig. 1, find i and v as functions of time for $t > 0$.

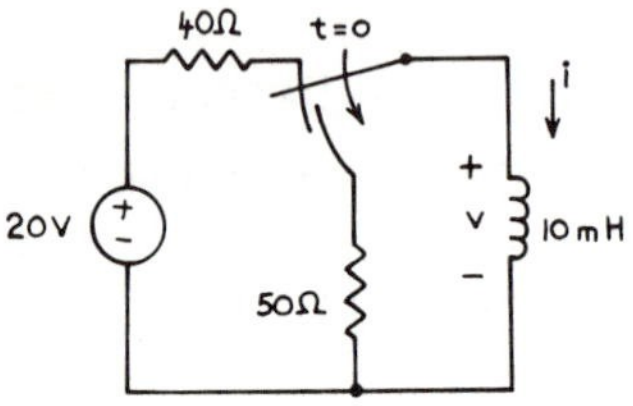

Fig. 1

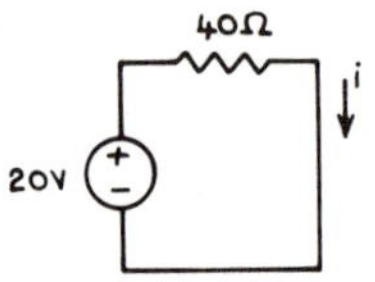

Fig. 2

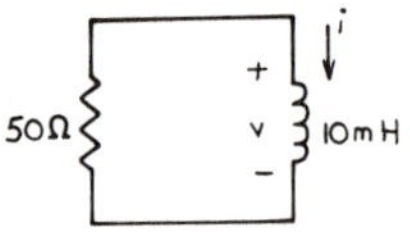

Fig. 3

Solution: First, find the current through the inductor i, just before the switch is thrown. Fig. 2 shows the circuit at $t = 0^-$. In Fig. 2, $i = \frac{20v}{40\,\Omega} = 0.5A$. Fig. 3 shows the circuit in Fig. 1 at $t = 0^+$.
In Fig. 3, in order for the voltage drops to sum to zero around loop, the voltage v must be $-(50\Omega)(0.5A) = -25V$. The time constant is found to be

$$\frac{L}{R} = \frac{10mH}{50\,\Omega} = \frac{1}{5000} .$$

Write the response,

$$i(t) = 0.5e^{-5000t} A \; ; \; t > 0$$
$$v(t) = -25e^{-5000t} V \; ; \; t > 0 .$$

• **PROBLEM** 5-2

A 30-mH inductor is in series with a 400-Ω resistor. If the energy stored in the coil at $t = 0$ is 0.96 μJ, find the magnitude of the current at (a) $t = 0$; (b) $t = 100\,\mu s$; (c) $t = 300\,\mu s$.

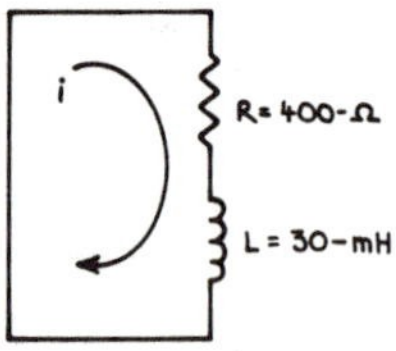

Solution: (a) Find the initial current $(i(0))$ by making use of the energy relationship for an inductor

$$E = \frac{1}{2} Li^2 .$$

Since we are given E and asked to find i,

$$i = \sqrt{\frac{2E}{L}}$$

$$i = \sqrt{\frac{2(0.96 \times 10^{-6})}{0.03}}$$

$$i = \sqrt{\frac{1.92 \times 10^{-6}}{3 \times 10^{-2}}} = \sqrt{64 \times 10^{-6}}$$

$$i = 8 \times 10^{-3} = 8\text{mA}$$

(b) After $t = 0$ the current through the inductor is governed by the response of the series RL circuit.

$$i(t) = I_o e^{-Rt/L} .$$

To find i at 100 μs ,

$$i(100\ \mu s) = (.008)\exp\left[\frac{-400(100 \times 10^{-6})}{0.03}\right]$$

$$i(100\ \mu s) = (.008)(2.64) = 2.11\text{mA} .$$

(c) To find i at 300 μs ,

$$i(300\ \mu s) = (.008)\exp\left[\frac{-400(300 \times 10^{-6})}{0.03}\right]$$

$$i(300\ \mu s) = (.008)e^{-4} = (.008)(.018) = 0.15\text{mA} .$$

• PROBLEM 5-3

Each circuit shown in Fig. 1 has been in the condition shown for an extremely long time. Determine $i(0)$ in each circuit.

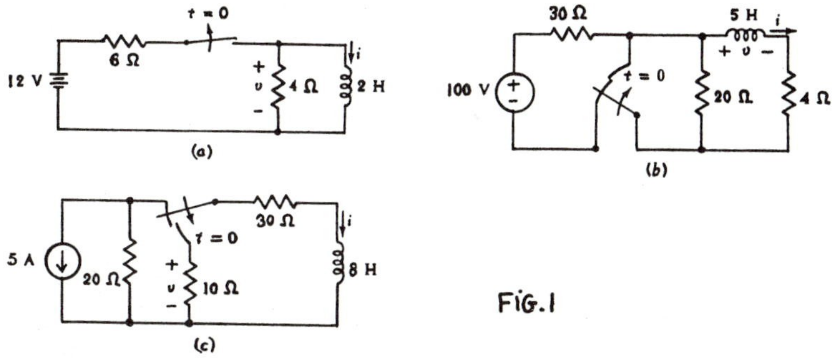

FIG. 1

Solution: Since the current through an inductor cannot change instantaneously, the current through each inductor in Fig. 1 is the same for $t = 0^-$ as for $t = 0$. (a) An inductor behaves as a short circuit in the D.C. steady state condition.

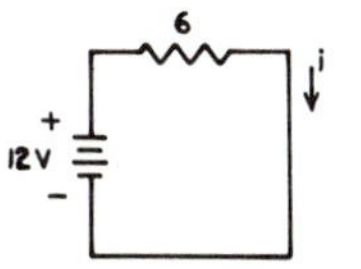

FIG. 2

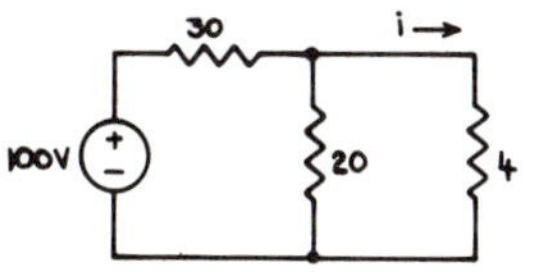

FIG. 3

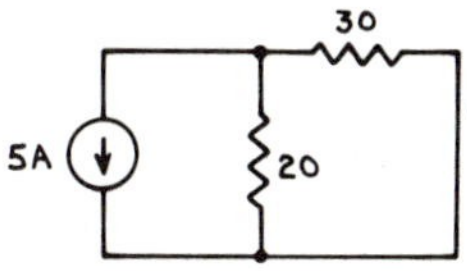

FIG. 4

Redrawing the circuit in Fig. 1(a) gives the D.C. steady state circuit of Fig. 2

$$i = \frac{12V}{6\Omega} = 2A$$

$$i(0) = 2A$$

b) The D.C. steady state circuit is shown in Fig. 3.

Using the voltage division concept, obtain the voltage across the 4-Ω resistor, so that the current i is found.

$$i = \frac{v}{R} = \frac{100(20||4)}{30 + (20||4)} \Big/ 4$$

$$i = \frac{100(20(4)/(20+4))}{[30 + (20(4)/(20+4))]4} = \frac{100(3.33)}{[33.33]4} = 2.5A$$

$$i(0) = 2.5A.$$

c) Again the D.C. steady state circuit is shown in Fig. 4. Using the current division concept yields

$$i = -\frac{5(20)}{20 + 30} = -2A$$

$$i(0) = -2A.$$

• PROBLEM 5-4

At the instant just after the switches are thrown in the circuits of Fig. 1, find v.

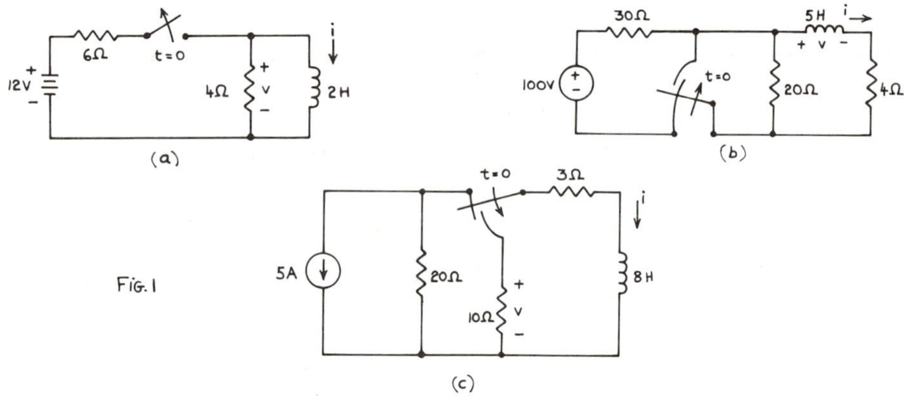

Fig. 1

Solution: In order to find the voltages indicated just after the switches are thrown, find the currents in the inductors at $t = 0^-$. Since the current in an inductor cannot change instantaneously and remains the same before, during, and after the switch is thrown, we can find v in each case by simple loop methods. For each of the three parts of the problem Fig. 2 shows the circuit just before the switch is thrown in D.C. steady state condition. Fig. 3 shows the circuit just after the switch is thrown.

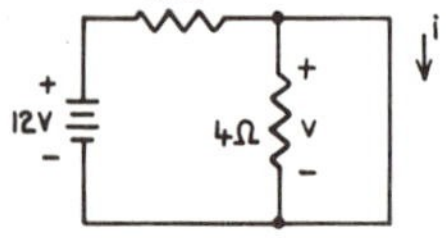

Fig. 2 (a)

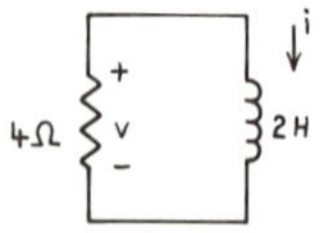

Fig. 3 (a)

a) In Fig. 2(a), i is found to be $12v/6\Omega = 2A$. In Fig. 3(a), if $i = 2A$, then $v = iR = -2(4) = -8V$.

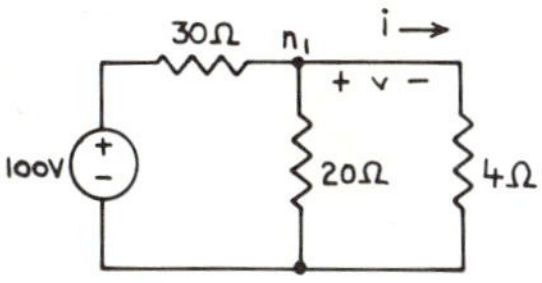

Fig. 2 (b)

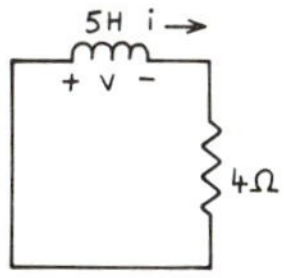

Fig. 3 (b)

b) In Fig. 2(b) by KCL find the voltage at node n_1 and i.

$$0 = \frac{v-100}{30} + \frac{v}{20} + \frac{v}{4}$$

$$v = \frac{100}{30(\frac{1}{30} + \frac{1}{20} + \frac{1}{4})} = \frac{100}{10} = 10V$$

$$i = \frac{v}{4} = \frac{10}{4} = 2.5A$$

We can find the voltage across the 4-Ω resistor in Fig. 3(b). $v_{4-\Omega} = 4(2.5) = 10V$ and since the voltage drops around the loop must sum to zero, v must be -10V.

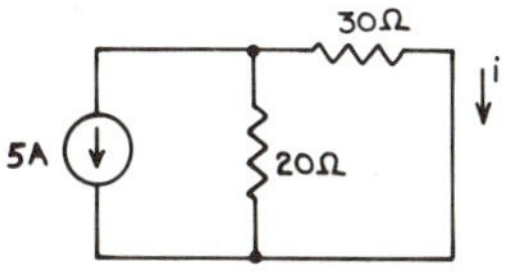

Fig. 2 (c)

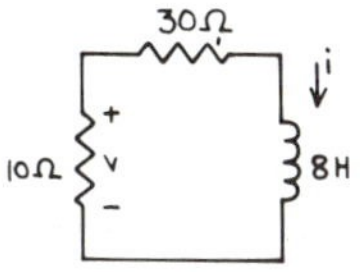

Fig. 3 (c)

c) The current i in Fig. 2(c) by current division is

$$i = -\frac{5(20)}{20 + 30} = -2A$$

The voltage v in Fig. 3(c) becomes $v = iR = -(-2)(10) = 20V$.

• PROBLEM 5-5

After $t = 0$, each of the circuits in Fig. 1 is source-free. Find expressions for i and v in each case for $t > 0$.

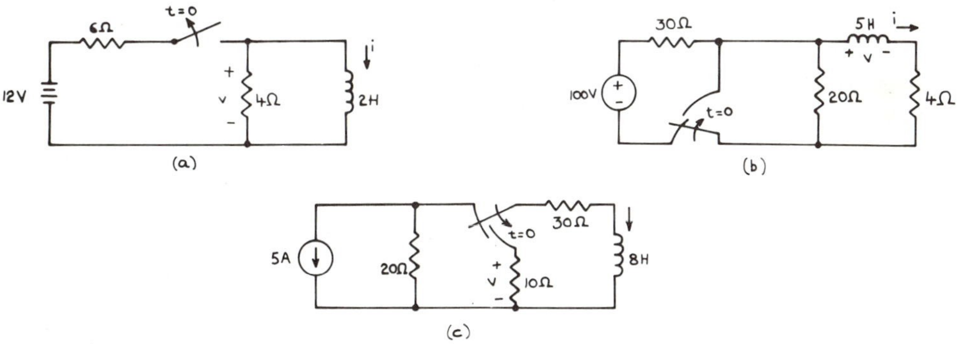

Fig. 1

Solution: The response of each circuit for $t > 0$ will be exponential in the form

$$i(t) = i(0^+)e^{-\frac{t}{\tau}}$$

$$v(t) = v(0^+)e^{-\frac{t}{\tau}}$$

where $\tau = L/R$ for each circuit we must find $i(0^+)$, $v(0^+)$ and R/L. Use the values of $i(0^+)$ and $v(0^+)$ found in the preceeding problems.

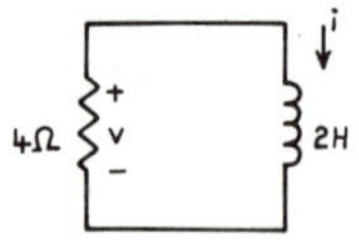

Fig. 2

a)
$$\tau = L/R = 2/4 = 0.5s$$

$$i(t) = 2e^{-2t}A; \quad t > 0$$

$$v(t) = -8e^{-2t}V; \quad t > 0$$

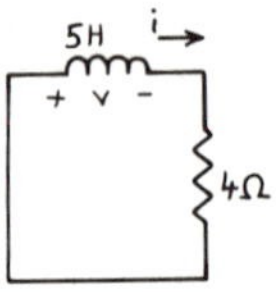

Fig. 3

b)
$$t = L/R = 5/4 = 1.25$$

$$i(t) = 2.5e^{-0.8t}A; \quad t > 0$$

$$v(t) = 10e^{-0.8t}A; \quad t > 0$$

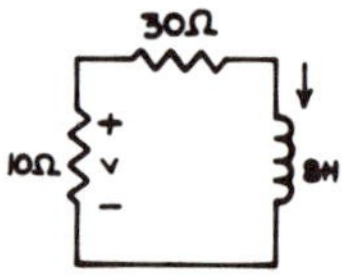

Fig. 4

c)
$$\tau = \frac{1}{R} = \frac{8}{30 + 10} = 0.2s$$

$$i(t) = -2e^{-5t}A; \quad t > 0$$

$$v(t) = 20e^{-5t}V; \quad t > 0$$

• PROBLEM 5-6

The left switch in the circuit of Fig. 1 is closed at $t = 0$. (a) Find $i_L(t)$ and $i_1(t)$ for $0 < t < 0.1s$. (b) The right switch is then closed at $t = 0.1s$. Find $i_L(t)$ and $i_1(t)$ for $t > 0.1s$.

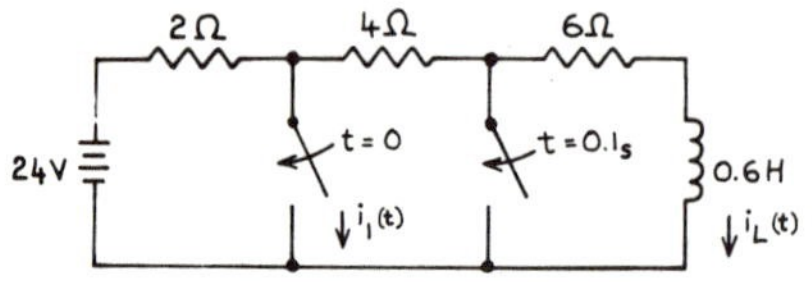

Fig. 1

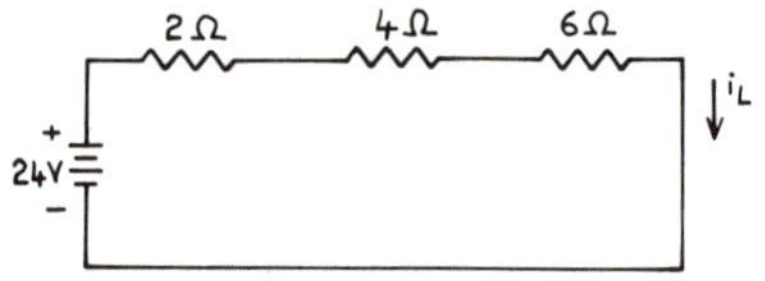

Fig. 2

Solution: First, find the current i_L at $t = 0^-$. Fig. 2 shows the circuit at $t = 0^-$.

$$i_L = \frac{24}{2+4+6} = 2A$$

after the switch on the left is closed at $t = 0$, we have the RL circuit shown in Fig. 3.

Note that $i_A = i_1 + i_L$ and that $i_1 = i_A - i_L$; $i_A = \frac{24v}{2\Omega} = 12A$ and i_L is an exponential response. The time constant is $L/R = 0.6/4+6 = 0.06$ so that i_L is

$$i_L(t) = 2e^{-t/0.06}A \; ; \; 0 < t < 0.1s \qquad (1)$$

$$i_1(t) = 12 - 2e^{-t/0.06}A \; ; \; 0 < t < 0.1s \qquad (2)$$

When the second switch is closed at $t = 0.1s$, the current i_L has already decayed to a specific value given by Eq. (1) above

$$i_L(0.1) = 2e^{-.1/0.06} = 2e^{-1/0.6} = 2(.189)$$

$$i_L(0.1) = .378A \quad .$$

Fig.3

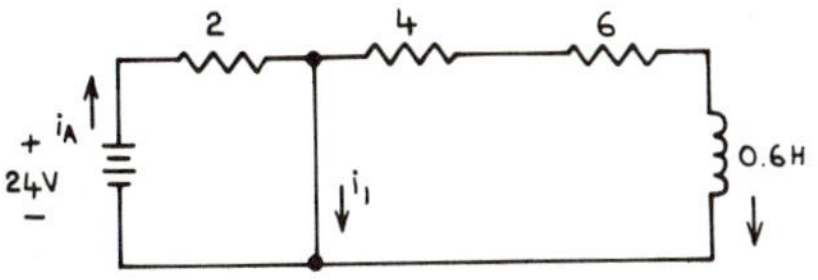

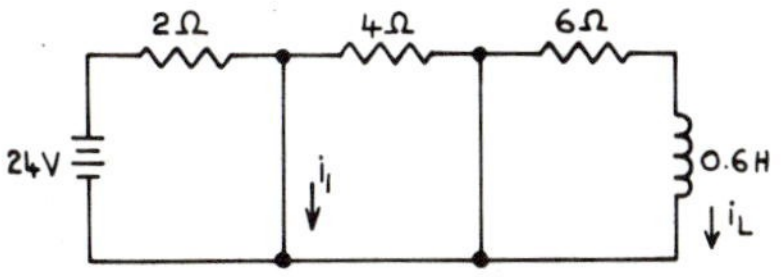

Fig.4

Fig. 4 shows the circuit when the switch is closed at $t = 0.1s$. The time constant now becomes

$$\frac{L}{R} = \frac{0.6}{6} = \frac{1}{10}$$

and for $t > 0.1s$

$$i_L(t) = .378e^{-10(t-0.1)}$$

$$i_1(t) = 12 - 2e^{-.1/.06} = 11.62A.$$

• PROBLEM 5-7

The two switches in the circuit of Fig. 1 are thrown simultaneously at $t = 0$. (a) Find $i_1(0^+)$, $i_2(0^+)$. (b) Find L_{eq} and τ. (c) Write $i(t)$ for $t > 0$. (d) To find $i_1(t)$ or $i_2(t)$, it is necessary to include a possible constant current present in the inductive loop. This may be done by finding the voltage across the 6-Ω resistor, which is the voltage across each coil, and integrating the inductor voltage to find the current; the known initial value is used as the constant of integration. Find $i_1(t)$ and $i_2(t)$. (e) Show that the sum of the energies remaining in the two coils as $t \to \infty$ plus that dissipated since $t = 0$ in the resistor is equal to the sum of the inductor energies at $t = 0$.

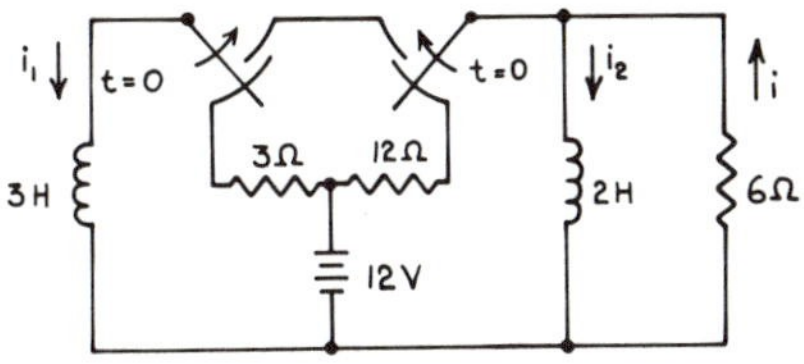

Fig. 1

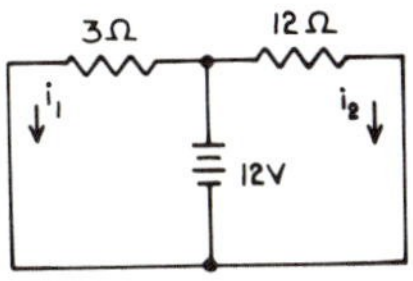

Fig. 2

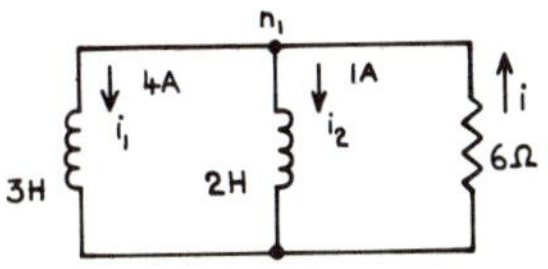

Fig. 3

Solution: (a) Just before the two switches are thrown, the currents i_1 and i_2 can be found from Fig. 2.

$$i_1 = \frac{12v}{3\Omega} = 4A$$

$$i_2 = \frac{12v}{12\Omega} = 1A \quad .$$

Since these two currents are inductor currents, they remain at the same value at $t = 0^+$. The circuit shown in Fig. 3 is for $t = 0^+$. An equation for node n, gives $i = 4A + 1A = 5A$ so for $t = 0^+$ we have $i_1 = 4A$, $i_2 = 1A$ and $i = 5A$.

(b) Fig. 3 shows that L_{eq} is the parallel combination of the two inductors

$$L_{eq} = \frac{3\times 2}{3+2} = \frac{6}{5} = 1.2H$$

and the time constant is

$$\frac{L_{eq}}{R} = \frac{1.2}{6} = 0.2s$$

(c) From the information obtained in parts (a) and (b) we can write an equation for $i(t)$; it > 0 ; $i(t)=5e^{-5t}$ A; $t>0$

(d) Note that the voltage across the 6-Ω resistor can be written, $(6)(5)e^{-5t}V$. Paying attention to the current directions shown in Fig.3, observe that the voltage across each of the inductors is $-30e^{-5t}V$. By using the relation,

$$i(t) = \frac{1}{L}\int_{t_0}^{t} v(\tau)d\tau + i(t_0) \quad , \qquad (1)$$

we can find i_1 and i_2 for $t > 0$. Substituting into eq. (1) for i_1, yields

$$i_1(t) = \frac{1}{3}\int_0^t -30e^{-5\tau}d\tau + 4$$

$$i_1(t) = \frac{-30}{3(-5)} e^{-5\tau}\Big|_0^t + 4$$

$$i_1(t) = 2(e^{-5t} - 1) + 4$$

$$i_1(t) = 2 + 2e^{-5t}A \quad . \qquad (2)$$

Substituting into eq. (1) for i_2, yields

$$i_2(t) = \frac{1}{2}\int_0^{t} -30e^{-5\tau}d\tau + 1$$

$$i_2(t) = \frac{-30}{2(-5)} e^{-5\tau}\Big|_0^t + 1$$

$$i_2(t) = 3(e^{-5t} - 1) + 1$$

$$i_2(t) = -2 + 3e^{-5t}A \quad . \qquad (3)$$

Note that $i_1 + i_2 = 5e^{-5t} = i$ and that $i_1(0^+)$ and $i_2(0^+)$ still equal 4A and 1A respectively.

(e) The energy stored in the inductors at $t = 0$ is $\frac{1}{2}L[i(0)]^2$. The energy stored at $t \to \infty$ is $\frac{1}{2}L[i(\infty)]^2$ and the energy dissipated in the resistor since $t = 0$ is $\int_0^{\infty} P_R\, dt$. We are asked to prove

$$\frac{1}{2}L_1[i_1(\infty)]^2 + \frac{1}{2}L_2[i_2(\infty)]^2 + \int_0^{\infty} P_R\, dt$$

$$= \frac{1}{2}L_1[i_1(0)]^2 + \frac{1}{2}L_2[i_2(0)]^2 \qquad (4)$$

Evaluate:

$$i_1(0),\ i_2(0),\ i_1(\infty),\ \text{and}\ i_2(\infty) .$$

Eqs. (2) and (3) yields, $i_1(0) = 4A$, $i_2(0) = 1A$, $i_1(\infty) = 2A$, $i_2(\infty) = -2A$, and $P_R(t) = (30e^{-5t})(5e^{-5t}) = 150e^{-10t}$. Substituting into eq. (4),

$$\frac{1}{2}(3)(2)^2 + \frac{1}{2}(2)(-2)^2 + \int_0^{\infty} 150e^{-10t}\, dt$$

$$= \frac{1}{2}(3)(4)^2 + \frac{1}{2}(2)(1)^2$$

$$6 + 4 + \frac{150}{-10}\, e^{-10t}\Big|_0^{\infty} = 24 + 1$$

$$6 + 4 + \frac{150}{-10}\, [0 - 1]_0 = 24 + 1$$

$$[6 + 4 + 15 = 24 + 1]\ J$$

$$25 = 25$$

RC CIRCUITS

• PROBLEM 5-8

After having been closed for a long time, the switch in the network of Fig. 1 is opened at $t = 0$. Find $v_C(t)$ for $t > 0$.

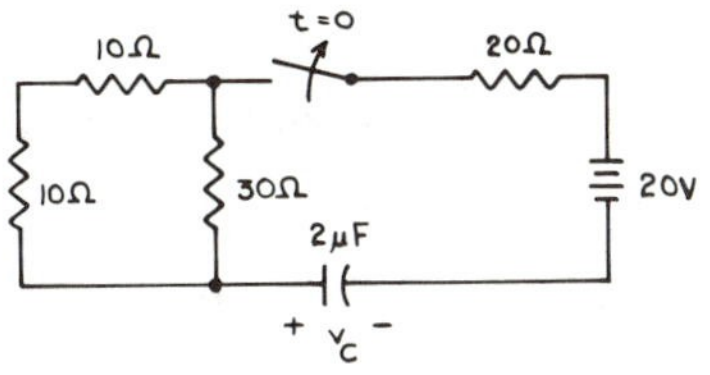

Fig. 1

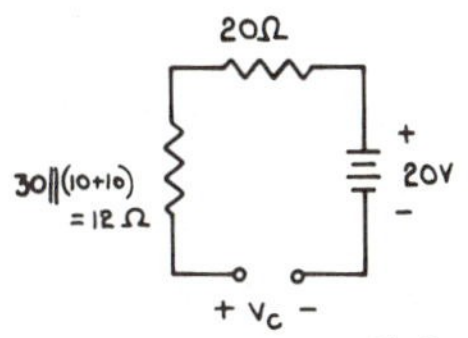

Fig. 2

Solution: The voltage across the capacitor v_C at steady state $(t = 0^-)$ can be found in Fig. 2. to be 20V. After the switch is opened, the capacitor is no longer in an RC loop. Therefore the voltage v_C remains at 20V for $t > 0$.

• PROBLEM 5-9

In the circuit of Fig. 1, $v(0^+) = 10$. Solve for $i(t)$ directly, without finding $v(t)$.

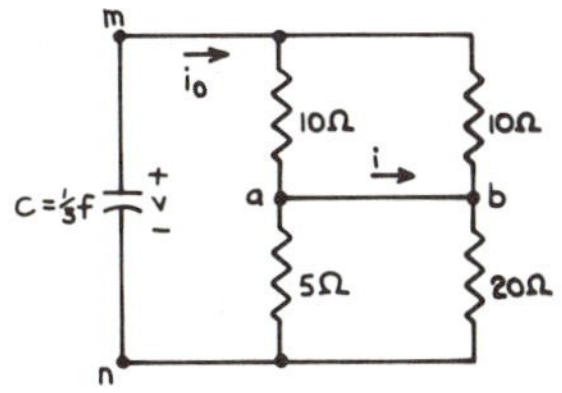

Fig. 1

Solution: Directly write a relation for i_0

$$i_0 = \frac{v_0}{R_{eq}}\, e^{-t/R_{eq}C}$$

where R_{eq} is the equivalent resistance across the capacitor.

$$R_{eq} = \frac{10(10)}{20} + \frac{20(5)}{25} = 9\Omega$$

$$i_0 = \frac{10}{9} e^{-t/(9)(1/3)} = \frac{10}{9} e^{-t/3} \text{ A} .$$

Write a relation for i in terms of i_0. By using the concept of current division, we have

$$i_{ma} = \frac{i_0(10)}{20}$$

$$i_{mb} = \frac{i_0(10)}{20}$$

$$i = \frac{i_{ma} 5}{5+20} - \frac{i_{mb} 20}{5+20}$$

$$i = i_0 \frac{1}{2}\left(\frac{5}{25} - \frac{20}{25}\right) = i_0 \frac{1}{2}\left(-\frac{15}{25}\right)$$

$$i = i_0\left(-\frac{15}{50}\right) = -\frac{3}{10} i_0$$

thus

$$i = -\frac{3}{10}\frac{10}{9} e^{-t/3} \text{A} = -\frac{1}{3} e^{-t/3} \text{A} \; ; \; t > 0$$

• PROBLEM 5-10

In the circuit of Fig. 1, $v_C(0^+) = 10$. Find v_{ab} for all $t \geq 0^+$.

Solution: Write a relationship for v_{ab} in terms of v_C. By using voltage division concepts, note that,

$$v_{mn} = \frac{v_C[(4+8)||(3+1)]}{9 + [(4+8)||(3+1)]}$$

and

$$v_{an} = \frac{v_{mn}(8)}{4+8}$$

$$v_{bn} = \frac{v_{mn}(1)}{3+1}$$

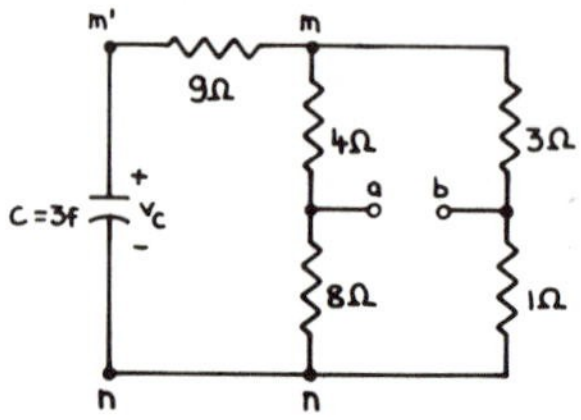

Fig.1

Finally

$$v_{ab} = v_{an} - v_{bn} = v_{mn}\left(\frac{8}{12}\right) - v_{mn}\left(\frac{1}{4}\right)$$

$$v_{ab} = \frac{v_C[(4+8)||(3+1)]}{9 + [(4+8)||(3+1)]}\left(\frac{8}{12} - \frac{1}{4}\right)$$

$$v_{ab} = \frac{v_C(3)}{12}\left(\frac{1}{2.4}\right) = v_C \frac{3}{28.8} = \frac{v_C}{9.6} .$$

Since we have a source free RC circuit, the response will be in the form

$$v_C = V_0 e^{-t/RC} .$$

For this circuit, R is the equivalent resistance across the capacitor.

$$R = 9 + [(4+8)||(3+1)] = 12\Omega$$

$$\tau = RC = 12(3) = 36\text{s}$$

and since

$$V_0 = 10\text{V} \quad \text{and} \quad v_{ab} = \frac{v_C}{9.6}$$

we have,
$$v_{ab} = \frac{10e^{-t/36}}{9.6} = \frac{25}{24}e^{-t/36}\text{V} \ ; \ t \geq 0^{+} \ .$$

• PROBLEM 5-11

A certain precision 1-μF capacitor has very high resistance material used between its conducting surfaces. The capacitor is charged to 1V at t = 0 and disconnected from the source. It is found that the voltage drops to 0.9V in 100 hr. Find the insulation resistance.

Solution: We realize that the parameters of the transient response of an RC circuit are given since we are told that the capacitor has a resistance between its conducting surfaces. Solve for R in our relationship for RC circuits.

$$v(t) = V_0 e^{-t/RC} \ .$$

Taking the natural log. of both sides and

$$\ln v(t) = \ln v_0 + \left(\frac{-t}{RC}\right) \ .$$

Solving for R yields,

$$R = \frac{-t}{[\ln(v(t)) - \ln(v_0)]C} \qquad (1)$$

In the problem it is given that $V_0 = 1$ V ,

$$t = 100 \text{ hr}(60 \text{ min/hr})(60 \text{ sec/min}) = 3.6 \times 10^5 \text{s} \ ,$$

$v(t) = 0.9$V and $C = 1 \times 10^{-6}$F . Substituting the above parameters into equation (1) yields,

$$R = \frac{-3.6 \times 10^5}{(\ln 0.09 - \ln 1)1 \times 10^{-6}}$$

$$R = \frac{-3.6 \times 10^5}{(-.1054 - 0)1 \times 10^{-6}} =$$

$$R = \frac{-3.6 \times 10^5}{-1.054 \times 10^{-7}} = 3.42 \times 10^{12} - \Omega \ .$$

• PROBLEM 5-12

After being closed for a long time, the switch in the circuit of Fig. 1 is opened at t = 0 . Find i(t) for t > 0 .

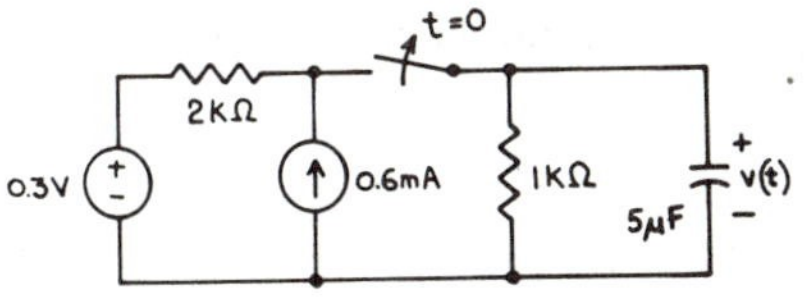

Fig. 1

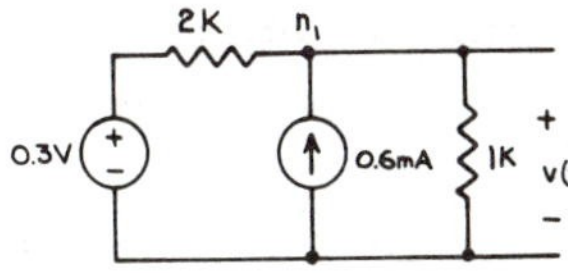

Fig. 2

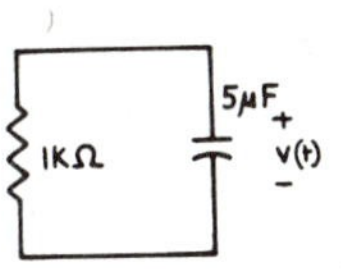

Fig. 3

Solution: Fig. 2 shows the circuit in Fig. 1 at $t = 0^-$. Find $v(0^-)$ by writing a KCL equation for node n_1 .

$$0.6(10^{-3}) = \left(\frac{v-0.3}{2} + \frac{v}{1}\right)\frac{1}{10^3}$$

$$0.6(10^{-3}) + \frac{0.3}{2}(10^{-3}) = \left(\frac{v}{2} + \frac{1}{1}\right) 10^{-3}$$

$$\frac{3}{2} v = 0.75$$

$$v(0^-) = \frac{2(.75)}{3} = 0.5V .$$

We know that $v(0^-) = v(0^+)$ since the voltage across the capacitor cannot change instantaneously. At $t = 0^+$ we have the RC circuit shown in Fig. 3. The time constant is $RC = (1K)(5\mu F) = 1/200$. Write v(t) as:

$$v(t) = 0.5e^{-200t} V \; ; \; t > 0 .$$

• PROBLEM 5-13

In the circuit of Fig. 1 the switch is open for $0 < t < t_1$ and closed for $t > t_1$. If $v_{ab}(0+) = V_0$, formulate $v_{ab}(t)$ and $i(t)$ for all $t \geq 0^+$.

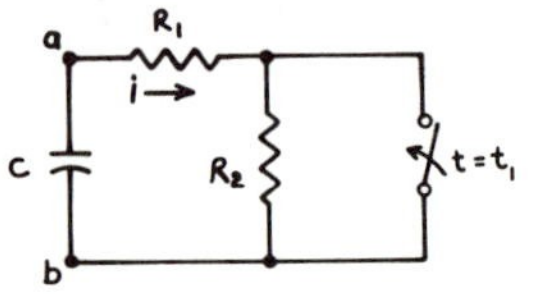

FIG.1

Solution: We have the solution for the interval $t = 0$ to $t = t_1$:

$$v_{ab} = V_0 \, e^{-t/(R_1+R_2)C} \tag{1}$$

$$i = \frac{V_0}{R_1+R_2} e^{-t/(R_1+R_2)C} \tag{2}$$

At $t = t_1$, the switch shorts the resistor R_2. This changes the time constant for the circuit and the current i. But since all currents are finite (no impulse currents), the voltage v_{ab} just before t_1 and just after t_1 must be equal. Using eqs.(1) and (2) write the relation for voltage v_{ab} and current i at t_1^-.

$$v_{ab}(t_1^-) = V_0 \, e^{-t_1/(R_1+R_2)C} \; ; \tag{3}$$

$$i(t_1^-) = \frac{V_0}{R_1+R_2} e^{-t_1/(R_1+R_2)C} \; ; \tag{4}$$

By calling the new initial voltage $v_{ab}(t_1^-)$, V_{t_1}, we have the new response for $t \geq t_1$

$$v_{ab}(t) = V_{t_1} \, e^{-\frac{(t-t_1)}{R_1 C}} \tag{5}$$

Since the current at t_1 is discontinuous $i(t_1^-) \neq i(t_1^+)$

$$i(t_1^+) = \frac{V_0}{R_1} e^{-t_1/(R_1+R_2)C} = \frac{v_{ab}(t_1^-)}{R_1} .$$

So that, for $t \geq t_1$

$$i(t) = \frac{V_{t_1}}{R_1} e^{-(t-t_1)/R_1C}$$

• PROBLEM 5-14

The independent source in the circuit of Fig. 1 is 140V for $t < 0$ and 0 for $t > 0$. Find $i(t)$ and $v_0(t)$ for $t > 0$.

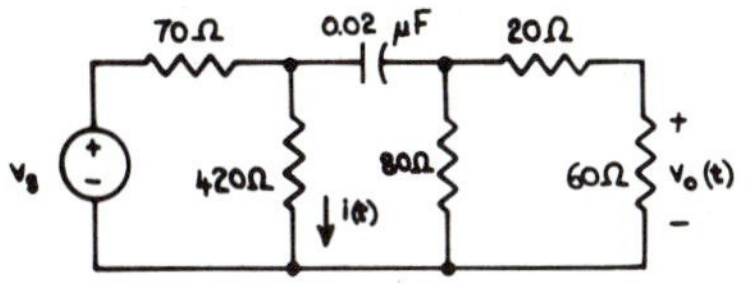

Fig. 1

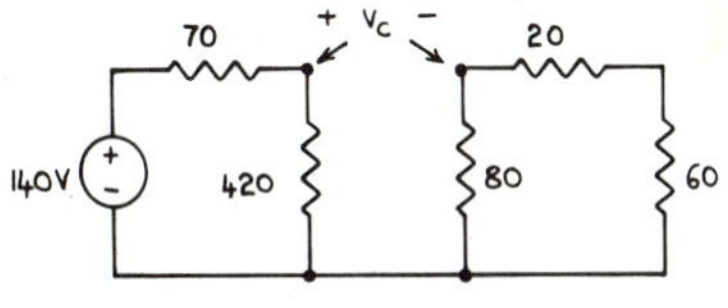

Fig. 2

Solution: In order to find the current $i(t)$ and voltage $v_0(t)$, first find the voltage across the capacitor. Since the voltage across a capacitor cannot change instantaneously, we can find it at $t = 0^-$. Fig. 2 shows the circuit in Fig. 1 at $t = 0^-$.

Note that the voltage v_C is the voltage across the 420-Ω resistor. By voltage division,

$$v_C = \frac{140(420)}{420+70} = 120V .$$

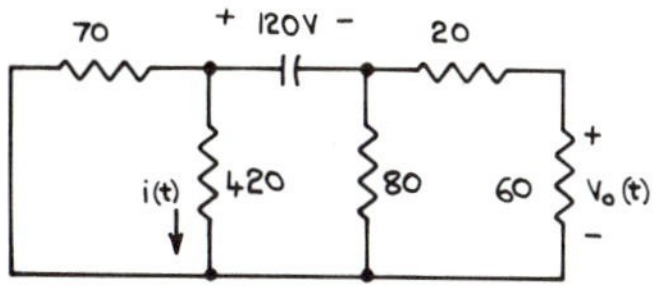

Fig. 3

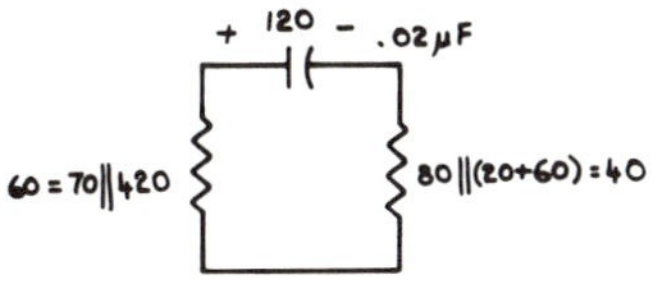

Fig. 4

The circuit in Fig. 3 is at $t = 0^+$. In order to find $i(t)$ and $v_0(t)$, further reduce the above circuit. In Fig. 4, find the voltage drop across each of the resistors.

$$v_{60\text{-}\Omega} \text{ resistor} = \frac{120(60)}{100} = 72V$$

$$v_{80\text{-}\Omega} \text{ resistor} = \frac{120(40)}{100} = 48V .$$

Now, one can find

$$i(0^+) = \frac{72V}{420\Omega} = .1714A$$

$$v_0(0^+) = \frac{48}{80}(60) = 36V .$$

From Fig. 4 the time constant is found to be

$$\tau = RC = (60 + 40)(0.02\mu F) = \frac{1}{500{,}000} s .$$

We can now write

$$i(t) = .1714e^{-500{,}000t} A \; ; \; t > 0$$

$$v_0(t) = 36e^{-500{,}000t} V \; ; \; t > 0 .$$

With reference to the circuit shown in Fig. 1, let $v(0) = 9V$. Find $i(t)$ for $t > 0$.

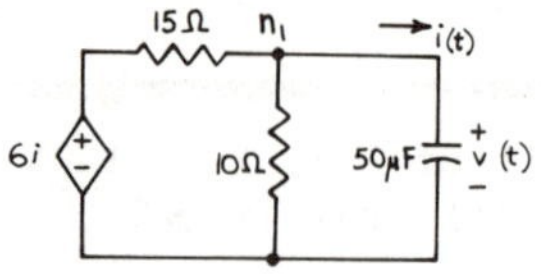

FIG. 1

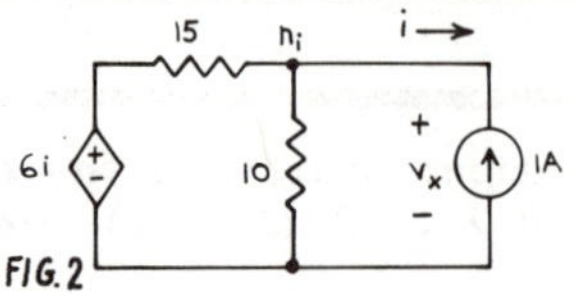

FIG. 2

Solution: By writing a KCL equation at node n_1, $i(0)$ can be found.

$$0 = \frac{v - 6i}{15} + \frac{v}{10} + i$$

since $v(0) = 9V$, and we must find $i(0)$ we have,

$$0 = \frac{9 - 6i(0)}{15} + \frac{9}{10} + i$$

$$-\frac{6}{15} i(0) + i(0) = -\left(\frac{9}{10} + \frac{9}{15}\right)$$

$$0.6i(0) = -1.5$$

$$i(0) = \frac{-1.5}{0.6} = -2.5A .$$

Find R_{eq} across the terminals of the capacitor. By connecting a 1-A source in place of the capacitor as shown in Fig. 2, we find

$$R_{eq} = \frac{v_x}{1} .$$

Writing a KCL equation at node n_1 yields

$$1 = \frac{v_x - 6i}{15} + \frac{v_x}{10}$$

since $i = -1A$,

$$1 = \frac{v_x + 6}{15} + \frac{v_x}{10}$$

$$\left(\frac{1}{15} + \frac{1}{10}\right) v_x = 1 - \frac{6}{15}$$

$$R_{eq} = \frac{v_x}{1} = \frac{1 - 6/15}{\left(\frac{1}{15} + \frac{1}{10}\right)} = 3.6\Omega .$$

The time constant can now be found

$$\tau = R_{eq} C = 3.6(50\mu F) = 1.8 \times 10^{-4} s .$$

$i(t)$ becomes

$$i(t) = -2.5e^{-t/1.8 \times 10^{-4}} A ; \; t > 0 .$$

• PROBLEM 5-16

For each of the circuits shown in Fig. 1, determine the transient response.

Solution: In order to find the transient response for $t > 0$, for each circuit first find v at $t > 0^-$ and then i at $t = 0^+$. We know that the voltage across a capacitor cannot change instantaneously so

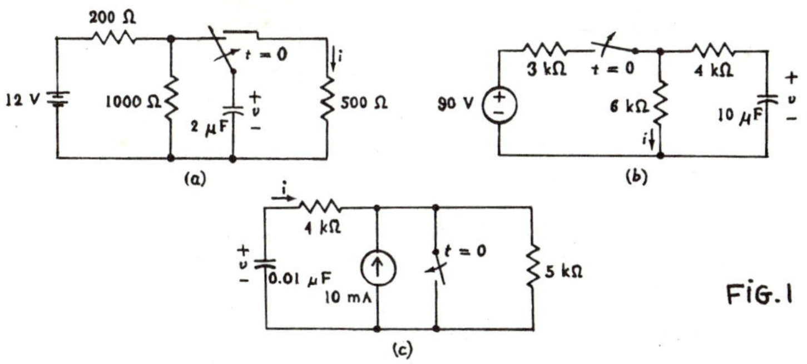

FiG.1

that $v(0^-) = v(0^+)$. Finally, find the time constant for the RC circuit (1/RC) and write the response in the form:

$$v(t) = v(0^+)e^{-t/RC} \quad ; \quad t > 0$$

$$i(t) = i(0^+)e^{-t/RC} \quad ; \quad t > 0$$

For each part of the problem, Fig. 2 shows the circuit at $t = (0^-)$ and Fig. 3 shows the circuit at $t = (0^+)$. At $t = (0^-)$, the circuit is in D.C. steady state so that the capacitor is an open circuit.

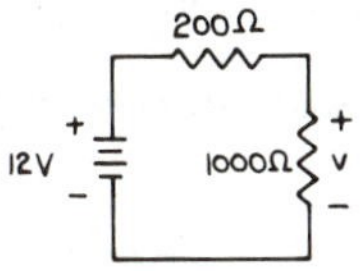

FiG.2 (a)

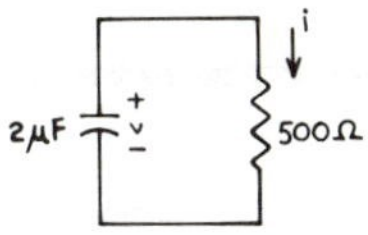

FiG.3 (a)

(a) In Fig. 2(a) by voltage division, we have

$$v = \frac{12(1000)}{200 + 1000} = 10V .$$

In Fig. 3(a), the current i becomes $i = \frac{v}{R} = \frac{10}{500} = 20mA$. The time constant is $RC = 500(2\mu F) = 500(2 \times 10^{-6}) = .001s$. The responses are:

$$v(t) = 10e^{-1000t} V \; ; \; t > 0$$

$$i(t) = 20e^{-1000t} mA \; ; \; t > 0$$

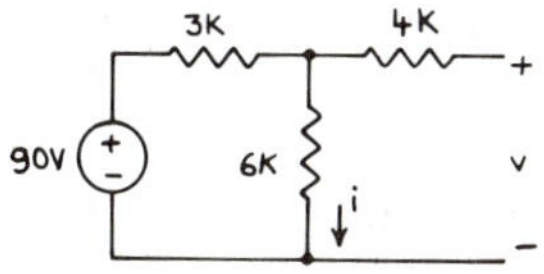

FiG.2 (b)

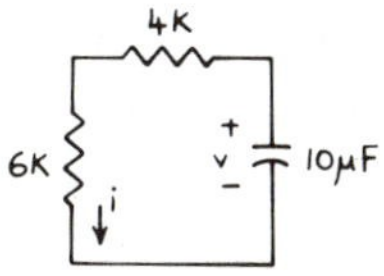

FiG.3 (b)

(b) In Fig. 2(b), find v by voltage division

$$v = \frac{90(6K)}{(3+6)K} = 60V .$$

In Fig. 3(b) the voltage across the 6K-Ω resistor can be found by voltage division.

$$v_{6-\Omega} = \frac{60(6K)}{6K + 4K}$$

$$i = \frac{v_{6-\Omega}}{6K} = \frac{60}{6K + 4K} = 6mA .$$

The time constant is $RC = (6K + 4K)(10\mu F) = .1s$. Write the response

$$v(t) = 60e^{-10t} V \; ; \; t > 0$$

$$i(t) = 6e^{-10t} mA \; ; \; t > 0 .$$

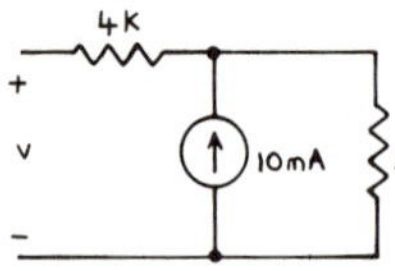

FiG. 2 (c)

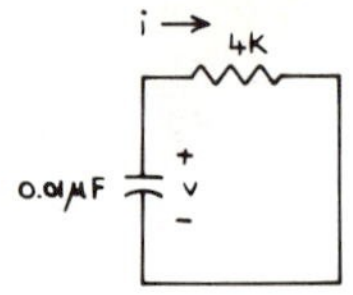

FiG.3 (c)

(c) In Fig. 2(c) the voltage v is the voltage across the current source. Since for a D.C. current a capacitor acts as an open circuit, the current through the $4K\Omega$ resistor, $i_{4K\Omega} = 0$. In order for the voltage around the loop to be zero, the voltage v must be (10mA)(5K) = 50V. In Fig. 3(c) the current i must be $v/R = 50/4K = 12.5mA$ in order for the voltage drops around the loop to sum to zero. The time constant is $RC = (4K)(.01\mu F) = 4 \times 10^{-5}$. Write the response

$$v(t) = 50e^{-25,000t}\,V \;;\; t > 0$$

$$i(t) = 12.5e^{-25,000t}\,mA \;;\; t > 0 \quad .$$

• PROBLEM 5-17

For the circuit shown in Fig. 1, find $v_1(t)$ and $v_2(t)$ for $t > 0$.

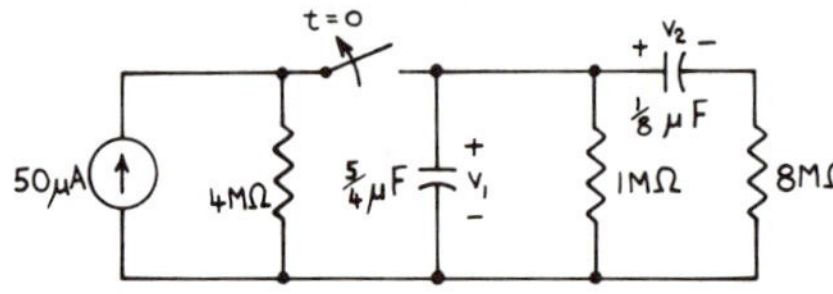

FiG. I

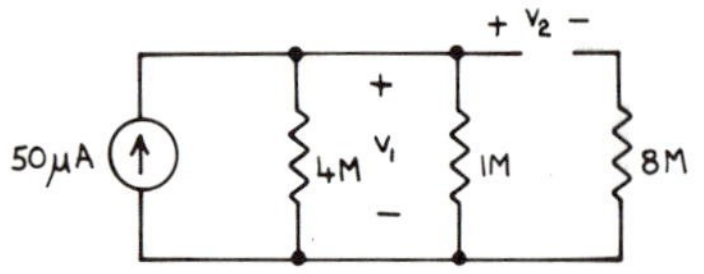

FiG. 2

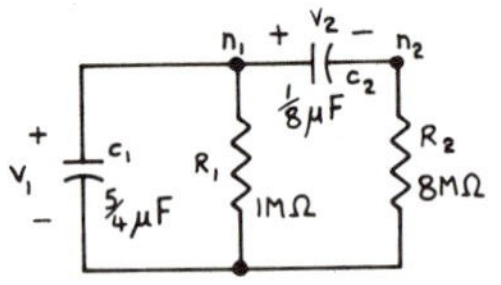

FiG. 3

Solution: First, find the initial voltages for the two capacitors. Do this by finding $v_1(0^-)$ and $v_2(0^-)$. Fig. 2 shows the circuit at $t = 0^-$. By current division

$$v_1 = \frac{50\,\mu A\,(1M)}{1M + 4M}\,(4M) = 40V \quad .$$

Since the $8M\text{-}\Omega$ resistor is not part of the "live" circuit voltage, $v_2 = 40V$. The initial conditions are,

$$v_1(0) = v_2(0) = 40V \;.$$

When the switch is opened at $t = 0$, we have the circuit in Fig. 3.

Notice that in Fig. 3 we cannot combine capacitors or resistors. Solve for v_1 and v_2 by differential equations. Calling the voltage at node n_1 , v_1 and that at node n_2 , $v_1 - v_2$, write the KCL equations for these two nodes.

$$n_1:\; 0 = C_1 \frac{dv_1}{dt} + \frac{v_1}{R_1} + C_2 \frac{d}{dt}\,[v_1 - (v_1 - v_2)]$$

$$n_2:\; 0 = C_2 \frac{d}{dt}\,[(v_1 - v_2) - v_1] + \frac{v_1 - v_2}{R_2}$$

Simplifying the above equations yields

$$C_1 \frac{dv_1}{dt} + C_2 \frac{dv_2}{dt} + \frac{v_1}{R_1} = 0 \qquad (1)$$

$$-C_2 \frac{dv_2}{dt} + \frac{v_1}{R_2} - \frac{v_2}{R_2} = 0 \qquad (2)$$

Knowing that the solution to the above set of differential equations must be in the form,

$$C_1 e^{s_1 t} + C_2 e^{s_2 t} \quad .$$

Let,

$$v_1 = Ae^{s_1 t} + Be^{s_1 t} \qquad (3)$$

and

$$v_2 = Ce^{s_1 t} + De^{s_1 t} \quad . \qquad (4)$$

Substituting eqs. (3) and (4) and the capacitor and resistor values into equations (1) and (2) yields

$$\frac{5}{4}(As_1 e^{s_1 t} + Bs_2 e^{s_2 t}) + \frac{1}{8}(Cs_1 e^{s_1 t} + Ds_2 e^{s_2 t}) + (Ae^{s_1 t} + Be^{s_2 t}) = 0$$

$$-\frac{1}{8}(Cs_1 e^{s_1 t} + Ds_2 e^{s_2 t}) + \frac{1}{8}(Ae^{s_1 t} + Be^{s_2 t}) - \frac{1}{8}(Ce^{s_1 t} + De^{s_2 t}) = 0 \ .$$

Rearranging the terms yields

$$(\frac{5}{4} As_1 + \frac{1}{8} Cs_1 + A)e^{s_1 t} + (\frac{5}{4} Bs_2 + \frac{1}{8} Ds_2 + B)e^{s_2 t} = 0 \qquad (5)$$

$$(-\frac{1}{8} Cs_1 - \frac{1}{8} C + \frac{1}{8} A)e^{s_1 t} + (-\frac{1}{8} Ds_2 - \frac{1}{8} D + \frac{1}{8} B)e^{s_2 t} = 0 \qquad (6)$$

In order for equations (5) and (6) to be zero, each term in parenthesis must be zero. This enables us to write four equations:

$$(\frac{5}{4} s_1 + 1)A + \frac{1}{8} Cs_1 = 0 \qquad (7)$$

$$(\frac{5}{4} s_2 + 1)B + \frac{1}{8} Ds_2 = 0 \qquad (8)$$

$$-\frac{1}{8}[(s_1 + 1)C - A] = 0 \qquad (9)$$

$$-\frac{1}{8}[(s_2 + 1)D - B] = 0 \qquad (10)$$

Dividing equation (7) by $(5/4\ s_1 + 1)$, and equation (9) by $(-1/8)$, and adding them yields

$$\frac{1/8\ s_1 C}{(5/4\ s_1 + 1)} + (s_1 + 1)C = 0 \qquad (11)$$

eliminating C, we obtain

$$\frac{1/8\ s_1}{(5/4\ s_1 + 1)} = (-s_1 - 1) \quad .$$

Simplifying

$$\frac{1}{8} s_1 = (\frac{5}{4} s_1 + 1)(-s_1 - 1)$$

$$\frac{1}{8} s_1 = - \frac{5}{4} s_1^{\ 2} - [(1 + \frac{5}{4}) s_1] - 1$$

$$0 = - \frac{5}{4} s_1^{\ 2} - [(1 + \frac{5}{4} + \frac{1}{8}) s_1] - 1$$

Multiplying both sides of the equation by -1 and combining fractions, we have

$$0 = \frac{10}{8} s_1^{\ 2} + \frac{19}{8} s_1 + \frac{8}{8}$$

$$0 = 10s_1^{\ 2} + 19s_1 + 8 \qquad (12)$$

Solving the quadratic equation (12) we obtain two values for s_1 .

$$s_1 = \frac{-b \pm \sqrt{b^2 - 4ac}}{2a}$$

$$s_1 = \frac{-19 \pm \sqrt{19^2 - 4(10)(8)}}{20}$$

$$s_1 = \frac{-19 \pm \sqrt{41}}{20} = \frac{-19 \pm 6.4}{20}$$

$$s_1 = \frac{-19 + 6.4}{20} = 0.63$$

$$s_1 = \frac{-19 - 6.4}{20} = -1.27$$

If we were to solve equations (8) and (10) for s_2 , the same results would be obtained since the same equation as eq. (12) would result when we eliminate the variables C and D . Since

$$s_1 = -0.63, - 1.27$$

$$s_2 = -0.63, -1.27$$

and

$$v_1 = Ae^{s_1 t} + Be^{s_2 t} \qquad (13)$$

$$v_2 = Ce^{s_1 t} + De^{s_2 t} \qquad (14)$$

We can choose s_1 and s_2 to be any of the above values as long as they are different. $s_1 = -0.63$ and $s_2 = -1.27$ were chosen. At $t = 0$ we found $v_1 = v_2 = 40V$, eqs. (13) and (14) become

$$40 = A + B \qquad (15)$$
$$40 = C + D \qquad (16)$$

since $e^0 = 1$. Taking eqs. (7)-(10) and substituting s_1 and s_2 , yields

$$.213A - .079C = 0 \qquad (17)$$
$$-.588B - .159D = 0 \qquad (18)$$
$$.370C - 1.000A = 0 \qquad (19)$$
$$-.270D - 1.000B = 0 \qquad (20)$$

Solving eqs. (15) and (16) for A and C respectively and substituting into eq. (19) produces,

$$B - .37D = 25.2 \qquad (21)$$

Solving eqs. (15) and (16) for B and D respectively, and substituting into eq. (20), yields

$$A + .27C = 50.8 \tag{22}$$

By using determinants, take eq. (21) and (18) and solve for B and D.

$$B = \frac{\begin{vmatrix} 0 & -.159 \\ 25.2 & -.370 \end{vmatrix}}{\begin{vmatrix} -.588 & -.159 \\ 1 & -.360 \end{vmatrix}} = \frac{-(-4.00)}{.218 - (-159)} = 10.6$$

$$D = \frac{\begin{vmatrix} -.588 & 0 \\ 1 & 25.2 \end{vmatrix}}{.377} = \frac{-14.82}{.377} = -39.3$$

By using determinants, take eqs. (22) and (17) and solve for A and C.

$$A = \frac{\begin{vmatrix} 0 & -.079 \\ 50.8 & +.27 \end{vmatrix}}{\begin{vmatrix} .213 & -.079 \\ 1 & +.27 \end{vmatrix}} = \frac{-(-4.01)}{(.058) - (-.079)} = 29.4$$

$$C = \frac{\begin{vmatrix} .213 & 0 \\ 1 & 50.8 \end{vmatrix}}{0.137} = \frac{10.82}{0.137} = 79.3$$

The final results are obtained by substituting A, B, C, D, s_1 and s_2 into eqs. (3) and (4), yielding

$$v_1(t) = 29.4e^{-.630t} + 10.6e^{-1.270t} \quad ; \quad t > 0$$

$$v_2(t) = 79.3e^{-.630t} - 39.3e^{-1.270t} \quad ; \quad t > 0$$

• PROBLEM 5-18

A voltage source, $10e^{-4t}$V, is connected to the series combination of 8Ω and 3H. (a) Show that the circuit equations are satisfied by a current, $i = -2.5e^{-4t}$A. (b) Construct the exact dual of the above circuit. (c) Find the current through the capacitor in the dual circuit.

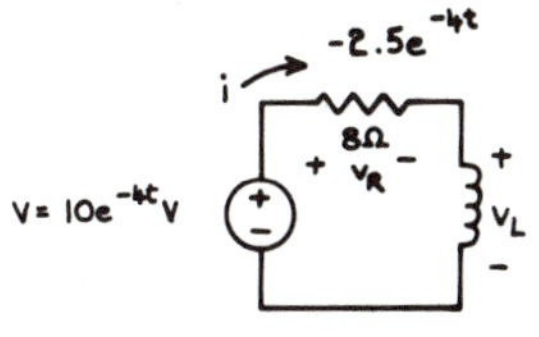

FIG. 1

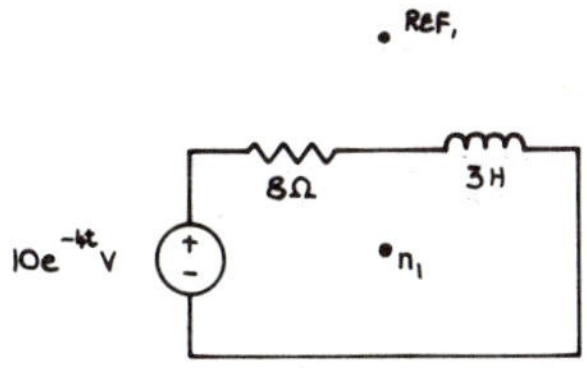

FIG. 2

Solution: (a) Fig. 1 shows the circuit described. In order to prove that the above voltage and current satisfy one another, prove that the KVL equation below is an equality.

$$10e^{-4t} \stackrel{?}{=} v_R + v_L$$

$$10e^{-4t} \stackrel{?}{=} iR + L\frac{di}{dt}$$

$$10e^{-4t} \stackrel{?}{=} (-2.5e^{-4t})8 + (3)\frac{d(-2.5e^{-4t})}{dt}$$

$$10e^{-4t} \stackrel{?}{=} -20e^{-4t} + (3)(-2.5)(-4)e^{-4t}$$

$$10e^{-4t} \stackrel{?}{=} -20e^{-4t} + 30e^{-4t}$$

$$10e^{-4t} \equiv 10e^{-4t} .$$

(b) We can construct a dual of the circuit in Fig. 1 directly from the diagram. In Fig. 2 redraw the circuit in Fig. 1 indicate the reference node outside the loop and the second node (n_1) inside the loop. Now join the two nodes by 3 branches passing through each element in the circuit. Each branch will contain the dual of that element. The $10e^{-4t}$V voltage source will become a $10e^{-4t}$A current source, the 8Ω resistor will become an 8℧ conductance, and the 3-H inductor will become a 3-F capacitor. From Fig. 3, obtain the dual circuit shown in Fig. 4.

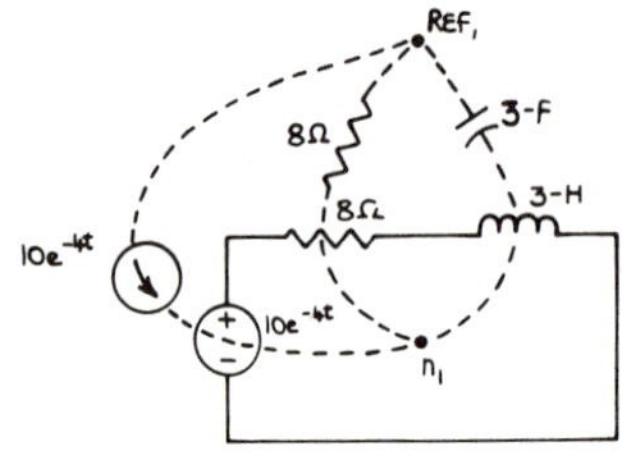

Fig. 3

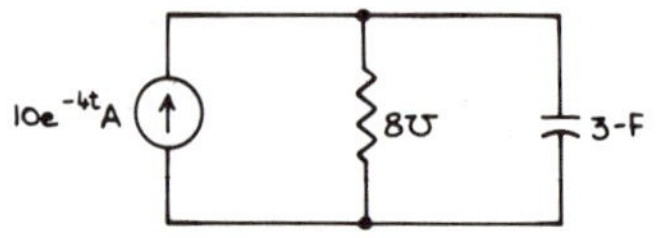

Fig. 4

(c) Write a KCL equation for the pair of nodes

$$10e^{-4t} = v(8) + C\frac{dv}{dt}$$

$$10e^{-4t} = 3\frac{dv}{dt} + 8v \qquad (1)$$

Remember that v must be in the form Ae^{-4t}. Substituting into the differential equation, yields

$$10e^{-4t} = 3\frac{dAe^{-4t}}{dt} + 8Ae^{-4t}$$

$$10e^{-4t} = 3(-4)Ae^{-4t} + 8Ae^{-4t}$$

dividing both sides by e^{-4t} yields

$$10 = 3(-4)A + 8A$$

$$10 = -12A + 8A$$

$$10 = -4A$$

$$A = -\frac{10}{4} = -2.5$$

so that $v = -2.5e^{-4t}$ this is the dual of the current in Fig. 1. The current through the capacitor is $C\frac{dv}{dt}$.

$$i_C = C\frac{dv}{dt} = 3\frac{d}{dt}(-2.5e^{-4t})$$

$$i_C = e(-2.5)(-4)e^{-4t}$$

$$i_C = 30e^{-4t}A$$

CHAPTER 6

FORCED RESPONSE OF RL AND RC CIRCUITS

THE UNIT STEP FUNCTION

• PROBLEM 6-1

Develop an analytical expression using step function notation for the signal g(t) shown in Figure 1.

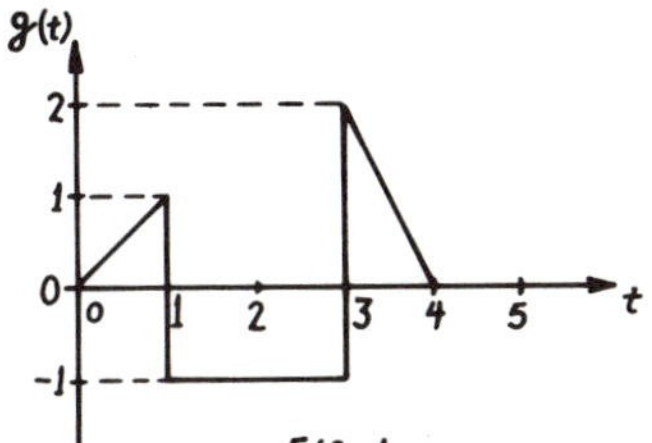

FiG. 1

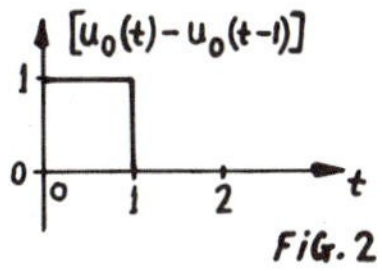

FiG. 2

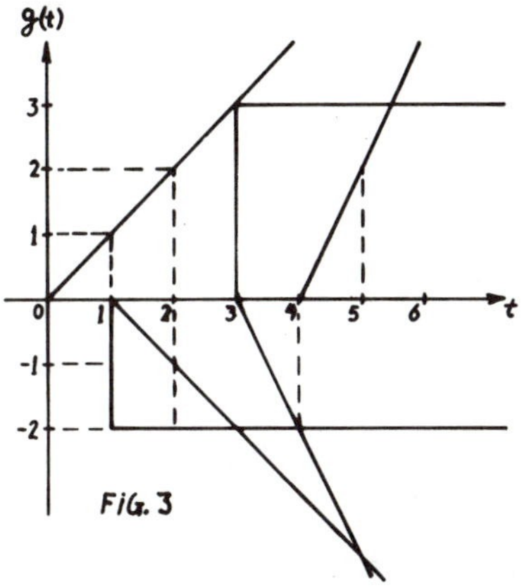

FiG. 3

Solution: As the function appears to consist of three distinct regions, it is convenient to separate g(t) into three components.

$$g(t) = g_1(t) + g_2(t) + g_3(t), \tag{1}$$

where $g_1(t) = g(t)$ in the interval $(0 \le t < 1)$; $g_2(t) = g(t)$ in the interval $(1 \le t < 3)$; $g_3(t) = g(t)$ in the interval $(3 \le t < 4)$.

Begin by combining step functions to form a new function, one which is equal to unity in the interval of interest and equal to zero elsewhere. For example, in the first interval $(0 \le t < 1)$ we sum

$$U_0(t) = \begin{cases} 0, & t < 0 \\ +1, & t \ge 0 \end{cases} \tag{2}$$

with

$$-U_0(t-1) = \begin{cases} 0, & t < 1 \\ -1, & t \ge 1 . \end{cases} \tag{3}$$

The result is the function

$$[U_0(t) - U_0(t-1)] = \begin{cases} +1, & (0 \le t < 1) \\ 0, & \text{elsewhere} \end{cases}, \tag{4}$$

which has the form shown in Figure 2

The product of this function with t gives $g_1(t)$, a function that linearly increases from $g_1(0) = 0$ to $g_1(1) = 1$ and is zero elsewhere:

$$g_1(t) = t[U_0(t) - U_0(t-1)]. \tag{5}$$

Now, combine step functions to form a function which is equal to one within the second interval, $(1 \le t < 3)$, and is zero elsewhere. This new function is

$$[U_0(t-1) - U_0(t-3)] = \begin{cases} +1, & (1 \le t < 3) \\ 0, & \text{elsewhere} \end{cases}. \tag{6}$$

The function $g_2(t)$ is equal to -1 in this interval and, therefore,

$$g_2(t) = -[U_0(t-1) = U_0(t-3)]. \tag{7}$$

For the third interval, we again form a function which is unity between t = 3 and t = 4 and is zero elsewhere:

$$[U_0(t-3) - U_0(t-4)] = \begin{cases} +1, & (3 \le t < 4) \\ 0, & \text{elsewhere} \end{cases}. \tag{8}$$

The function $g_3(t)$ falls linearly from $g_3(3) = 2$ to $g_3(4) = 0$, and we must multiply (8) by a function that exhibits this behavior. The function (4-t) goes from one to zero in this interval and, therefore, 2(4-t), or -2(t-4), satisfies our requirements. The function $g_3(t)$ is then

$$g_3(t) = -2(t-4)[U_0(t-3) - U_0(t-4)]. \tag{9}$$

Hence, the complete expression for g(t) is, from (1),

$$g(t) = t[U_0(t) - U_0(t-1)] - [U_0(t-1) - U_0(t-3)]$$
$$-2(t-4)[U_0(t-3) - U_0(t-4)].$$

A second approach to this problem would be to break the function up into a series of step and ramp functions as shown in fig. 3.

The reasoning behind this construction is as follows: For $0 < t < 1$ a unit ramp function is required, so draw a line through the origin, with slope = 1, and extend it out infinitely. Next consider the interval $1 < t < 3$. At t=1 we need to level off the function and then pull it down to -1, where it will remain until t=3. This can be accomplished by adding to the unit ramp function another ramp function with slope = -1. Note that this will level off g(t) at the value 1 (since the slopes cancel each other, the function neither rises nor falls, but remains fixed at the value g(t) = 1). What is now needed is a step function of magnitude -2 to pull g(t) down to the value -1. Next consider what happens at t=3. The function is pulled up to a value g(t) = 2. Since, for $1 < t < 3$, it has been "sitting" at g(t) = -1, a step function of magnitude 3 acting at t=3 will pull g(t) up to the value 2. Next a ramp function of slope -2 is needed, again applied at t = 3. For $t \geq 4$ we need g(t) = 0. Note that the function at this point is at t = 0 (at t = 3, g(t) = 2 and the ramp function of slope -2 has pulled it down to g(t) = 0 by t = 4). However, it is necessary to cancel out the ramp function so that g(t) remains at 0. This is easily accomplished with a ramp function of slope +2 starging at t = 4.

Now it is easy to read off g(t), which is simply the sum of all the ramp and step functions, from fig. 3:

$$g(t) = tU(t) - (t-1)U(t-1) - 2U(t-1) + 3U(t-3) -$$

$$- 2(t-3)U(t-3) + 2(t-4)U(t-4)$$

Just as a check, note what happens for given intervals:

$t < 1$: $g(t) = t$

$1 < t < 3$: $g(t) = t - (t-1) - 2$; $g(t) = -1$

$3 < t < 4$: $g(t) = t - (t-1) - 2 + 3 - 2(t-3)$; $g(t) = -2t + 8$

$t > 4$: $g(t) = t-(t-1) - 2 + 3 - 2(t-3) + 2(t-4)$; $g(t)=0$

Also note that this solution is exactly the same as the previous one.

• PROBLEM 6-2

Develop an analytical expression for the waveform shown.

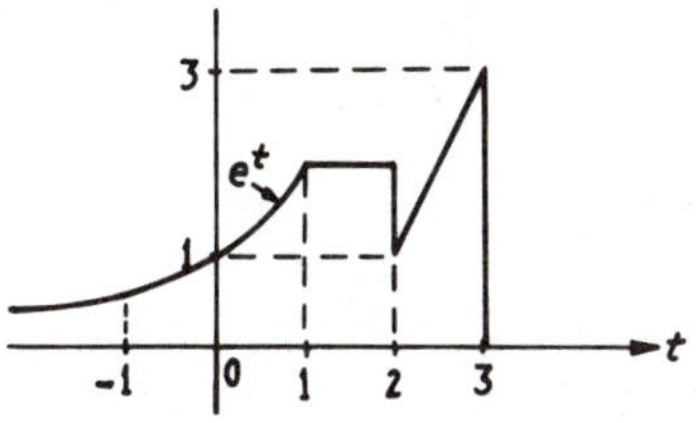

Solution: As seen from the figure, the given curve can be considered to consist of the following three functions:

$$f_1(t) \equiv e^t u(-t) \qquad \text{for } t < 0 \qquad (1)$$

$$f_1(t) = e^t[u(t) - u(t-1)] \quad \text{for } 1 > t > 0 \qquad (2).$$

In this equation, $[u(t) - u(t-1)]$ represents a rectangularly shaped function of unit magnitude which "turns on" at t = 0 and "turns off" at t = 1.

$$f_2(t) \equiv e[u(t-1) - u(t-2)] \text{ for } 1 < t < 2 \qquad (3).$$

In this equation, the function $f_2(t) = e$ "turns on" at $t = 1$ and "turns off" at $t = 2$. The last straight line portion of the curve shown is represented by

$$f_3(t) \equiv (2t-3)[u(t-2)-u(t-3)] \text{ for } 2 < t < 3. \tag{4}$$

In this equation, $[u(t-2)-u(t-3)]$ represents a rectangular pulse with a magnitude of 1 which turns on at $t = 2$ and turns off at $t = 3$. Therefore, the total response is

$$f(t) = f_1(t) + f_2(t) + f_3(t) = e^t[u(-t) + u(t) - u(t-1)]$$

$$+ e[u(t-1) + u(t-2)] + (2t-3)[u(t-2) - u(t-3)]. \tag{5}$$

The function $[u(-t) + u(t) - u(t-1)]$ represents a unit step function which was "turned on" at $t = -\infty$ and "turned off" at $t = 1$. The above function is the same as $u(t+1)$ if t is replaced by $-t$.

The resulting function, $u(1-t)$, is a unit step function of unity magnitude as long as $1-t \geq 0$, which means the entire range $-\infty \leq t \leq +1$. So,

$$u(1-t) = [u(-t) + u(t) - u(t-1)] \tag{6}$$

Then,

$$f(t) = e^t u(1-t) + e[u(t-1) + u(t-2)] + (2t-3)[u(t-2)$$

$$-u(t-3)] \tag{7}$$

• PROBLEM 6-3

Evaluate (i) $t\delta(t - 1)$; (ii) $t\delta^{(1)}(t - 1)$.

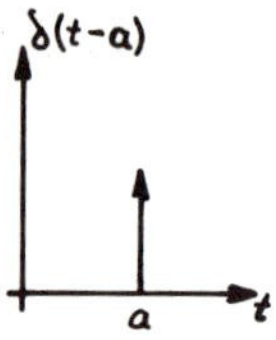

FiG. 1

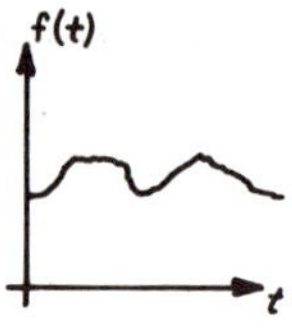

FiG. 2

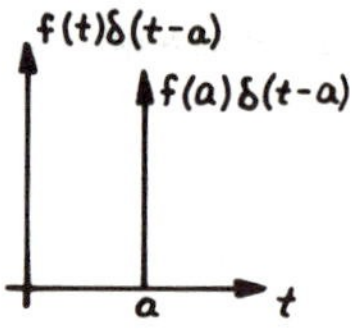

FiG. 3

Solution: In this problem δ is an impulse function. An impulse function is defined as a function that has infinite value at one point and zero value at all other points. Expressed mathematically:

$$\delta(t - a) = 0 \text{ for } t \neq a$$

$$= \infty \text{ for } t = a$$

Fig. 1 is the graphical expression of an impulse function.

(i) From Figs 2 and 3 it is seen that

$$f(t)\ \delta(t - a) = f(a)\,\delta(t - a)$$

In this problem $f(t) = t$ and $a = 1$; therefore, $f(1) = 1$.

So

$$t\delta(t-1) = 1\cdot\delta(t-1) = \delta(t-1)$$

(ii) The result can be found by using the sampling property which states that

$$f(t)\ \delta^{(1)}(t-a) = f(a)\delta^{(1)}(t-a) - f^{(1)}(a)\delta(t-a)$$

where $\delta^{(1)}(t-a) = \frac{d}{dt}[\delta(t-a)]$

and $\quad f^{(1)}(a) = \frac{d}{dt}[f(a)]$

In this case $f(t) = t$, $a = 1$ and $f^{(1)}(a) = \left.\frac{d}{dt}t\right|_{t=a} = 1$

Therefore,

$$t\ \delta^{(1)}(t-1) = 1\cdot\delta^{(1)}(t-1) - 1\cdot\delta(t-1)$$
$$= \delta^{(1)}(t-1) - \delta(t-1).$$

• PROBLEM 6-4

Write an expression for the wave forms shown in fig. 1 using unit step function notation.

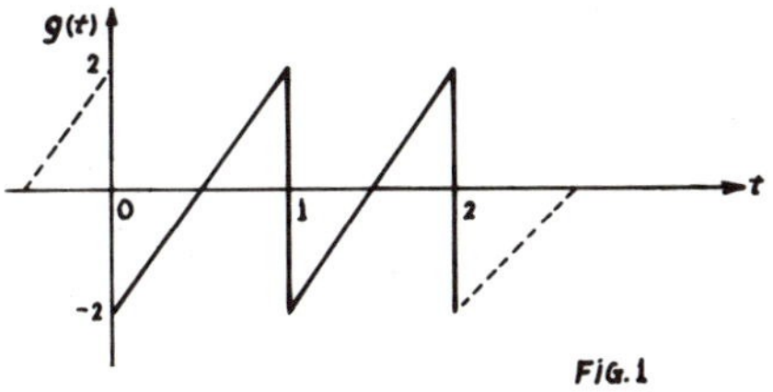

FIG. 1

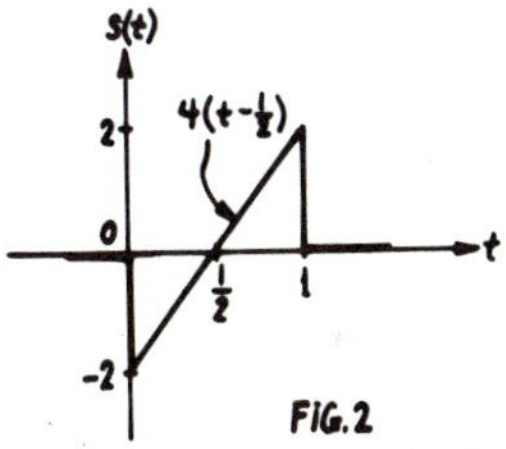

FIG. 2

Solution: The waveform g(t) is periodic with period T = 1s. By writing an expression for the waveform shown in fig. 2 and then shifting it by T left and right and summing we can obtain g(t).

$$s(t) = 4(t - \tfrac{1}{2})[u(t) - u(t-1)] \tag{1}$$

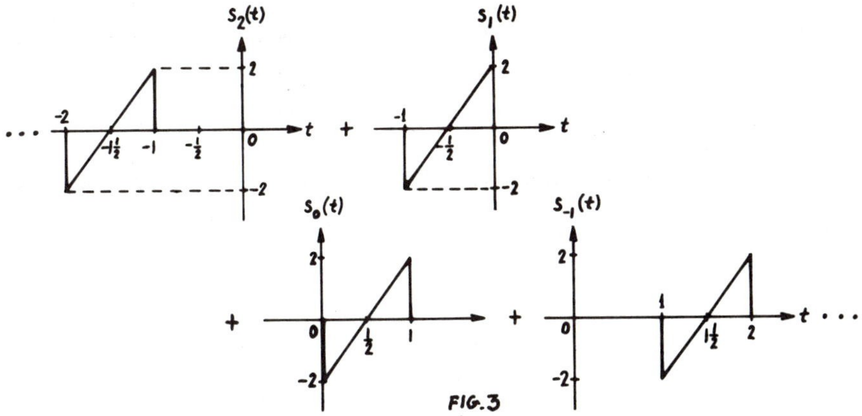

FIG. 3

Hence, g(t) can be formed by summing the shifted waveforms shown in fig. 3.

$$g(t) = \sum_{n=-\infty}^{\infty} s_n(t + n) ; \qquad (2)$$

$$= \ldots s_2(t) + s_1(t) + s_0(t) + s_{-1}(t) + s_{-2}(t) \ldots$$

Substituting (1) into (2) and replacing t by t + n gives

$$g(t) = \sum_{n=-\infty}^{\infty} 4(t + n - \tfrac{1}{2})[u(t + n) - u(t + n - 1)].$$

• PROBLEM 6-5

Evaluate:

(a) $\sin t\ \delta[t-(\pi/2)]$; (b) $\cos t\ \delta(t - \pi)$;

(c) $\sin^2 t\ \ \delta^{(1)}(t)$.

Solution: (a) $\delta[t - \left(\frac{\pi}{2}\right)] = \begin{cases} 0, & t \neq \frac{\pi}{2} \\ \delta[t - \left(\frac{\pi}{2}\right)], & t = \frac{\pi}{2} \end{cases}$

$$\sin \frac{\pi}{2} = 1$$

Hence, $\sin t\ \delta[t - \left(\frac{\pi}{2}\right)] = \sin \frac{\pi}{2}\delta[t - \frac{\pi}{2}] = \delta(t - \frac{\pi}{2})$.

(b) Similarly, $\delta(t - \pi) = \begin{cases} 0, & t \neq \pi \\ \delta(t-\pi), & t = \pi \end{cases}$.

Therefore, $\cos t\ \delta(t-\pi) = \cos \pi\delta(t-\pi) = -\delta(t-\pi)$.

(c) $\delta^{(1)}(t) = 0, \ t \neq 0$

$$\sin^2 t\ \ \delta^{(1)}(t) = \sin^2 0\ \delta^{(1)}(t) = 0\delta^{(1)}(t) = 0.$$

RL CIRCUITS

• PROBLEM 6-6

Use the impulse-train response to evaluate the steady-state component in the RL network shown. Assume that the input is $u_T(t) = \sin t$.

Solution: The impulse response (h(t)) in an RL circuit is given by the current i:

$$i(t) = h(t) = e^{At}bu_0(t) \quad \text{where } A = -\frac{R}{L} \text{ and } b = \frac{1}{L}.$$

Therefore, $h(t) = \frac{1}{L}\, e^{-(R/L)t}u_0(t)$. (1)

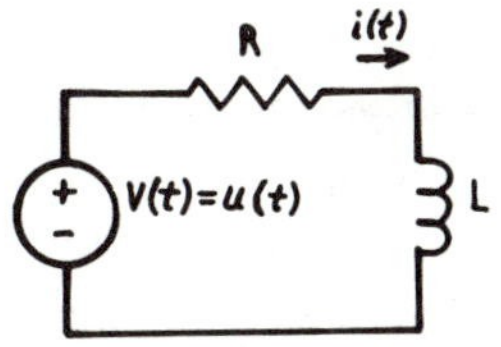

This result implies that if a voltage impulse function is applied to the RL series network, a current with exponential decay will flow in the network in the direction shown.

And so, when the voltage u(t) is the driving function and the current i(t) through the inductor is the desired response, the impulse response is h(t) (eq (1)).

When the impulse response is in the form $e^{-at}u_0(t)$, the impulse-train response is given as,

$$\frac{e^{-at}}{1 - e^{-at}} . \tag{2}$$

Substituting eq. (1) for e^{-at} in eq. (2) the impulse-train response becomes

$$h_T(t) = \frac{(1/L)e^{-(R/L)t}}{1 - e^{-(R/L)T}} \qquad 0 < t < T. \tag{3}$$

Use h_T to find the steady-state response ("finite-convolution") integral:

$$y_T(t) = \int_b^{b+T} u_T(\tau)h_T(t-\tau)\,d\tau \tag{4}$$

or $$y_T(t) = \int_b^{b+T} U_T(t-\tau)h_T(\tau)\,d\tau \tag{5}$$

where constant b is arbitrary.

Substituting equation (3) into equation (5) where $T = 2\pi$ and b is chosen to be 0, yields,

$$y_T(t) = \int_0^{2\pi} \left(\frac{1/L}{1 - e^{-(R/L)2\pi}}\right) e^{-(R/L)\tau} \sin(t-\tau)\,d\tau$$

evaluation of this integral yields,

$$i_T(t) = \frac{1/L}{1 + (R/L)^2}\left[\frac{R}{L}\sin t - \cos t\right]$$

and

$$i_T(t) = \frac{1}{\sqrt{R^2 + L^2}} \sin(t-\theta)$$

where $\tan\theta = (L/R)$.

NOTE: For a sinusoidal input, the steady state response is also sinusoid of the same frequency but with a phase shift and different magnitude.

Consider applying a unit ramp voltage source to a series RL circuit as shown in fig. 1. Compute the voltages $v_R(t)$ and $v_L(t)$ with zero initial condition for (a) L = 0.1H; (b) L = 1H; (c) L = 10H.

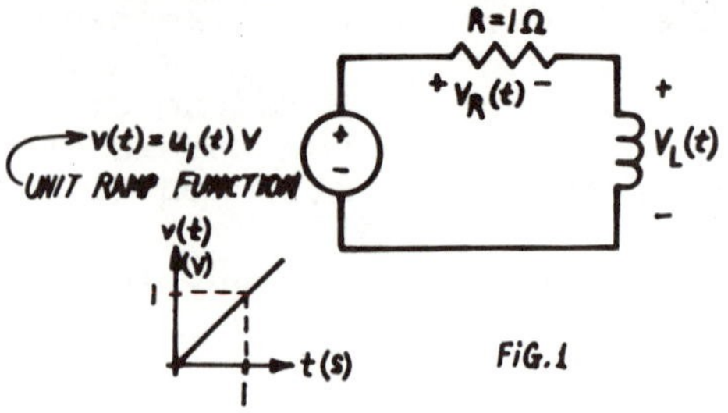

FIG. 1

Solution: The solution to this problem requires solving a first order linear differential equation. The voltages of interest in the circuit are expressed as,

$$v(t) = t \quad \text{(unit ramp function)}$$

$$V_R = iR$$

$$V_L = L\frac{di}{dt}$$

where i is the current in the loop. Using KVL we write

$$v(t) = V_R + V_L$$

hence, the differential equation is found to be

$$t = iR + L\,\frac{di}{dt}\,. \tag{1}$$

First we solve the homogeneous differential equation

$$iR + L\,\frac{di}{dt} = 0.$$

The equation can be written as

$$-\,\frac{R}{L}\,dt = \frac{di}{i}$$

Integrating both sides of the equation,

$$-\,\frac{R}{L}\int dt = \int \frac{1}{L}\,di$$

$$-\,\frac{R}{L}t = \ln\, i$$

and taking the exponential of both sides gives the homogeneous solution,

$$e^{-\frac{R}{L}t} = e^{\ln i}$$

$$K_0\,e^{-\frac{R}{L}t} = i_n(t) \tag{2}$$

The particular solution is guessed to be of the form

$$i_p(t) = K_1 t + K_2 . \tag{3}$$

Substituting $i_p(t)$ into the differential equation [Eq(1)] gives

$$t = R[K_1 t + K_2] + L[K_1]$$

$$t = RK_1 t + RK_2 + LK_1$$

Equating coefficients of like terms yields

$$RK_1 = 1; \quad K_1 = \frac{1}{R}$$

and $RK_2 + LK_1 = 0$

Hence, $K_2 = -\frac{L}{R^2}$

The total solution is the sum of the particular and homogeneous solutions, hence

$$i_T(t) = i_p(t) + i_h(t)$$

$$i_T(t) = \frac{1}{R}t - \frac{L}{R^2} + K_0 e^{-\frac{R}{L}t}$$

Find the constant K_0 by applying the initial condition $i_T(0) = 0$.

$$i_T(0) = 0 = -\frac{L}{R^2} + K_0$$

$$K_0 = \frac{L}{R^2}$$

Then, the total current and the voltages v_R and v_L are given by

$$i(t) = \frac{1}{R}t + \frac{L}{R^2}(e^{-\frac{R}{L}t} - 1)$$

$$v_R(t) = i(t)R = t + \frac{L}{R}(e^{-\frac{R}{L}t} - 1)$$

$$v_L(t) = L\frac{di(t)}{dt} = \frac{L}{R}(1 - e^{-\frac{R}{L}t})$$

Finally, since $R = 1\ \Omega$, we have

$$i(t) = t + L(e^{-t/L} - 1)$$

$$v_R(t) = t + L(e^{-t/L} - 1)$$

$$v_L(t) = L(1 - e^{-t/L})$$

(a) When L = 0.1H

$$V_R = t + \frac{1}{10}(e^{-10t} - 1)V; \quad V_L = \frac{1}{10}(1 - e^{-10t})V$$

(b) When L = 1H

$V_R = t + (e^{-t} - 1)V; \quad V_L = (1 - e^{-t})V$

(c) When L = 10 H

$V_R = t + 10(e^{-\frac{t}{10}} - 1)V; \quad V_L = 10(1 - e^{-\frac{t}{10}})V$

• PROBLEM 6-8

> A current source of 0.2 u(t) A, a 100-Ω resistor, and a 0.4-H inductor are in parallel. Find the magnitude of the inductor current (a) as $t \to \infty$; (b) at $t = 0^+$; (c) at t = 4 ms.

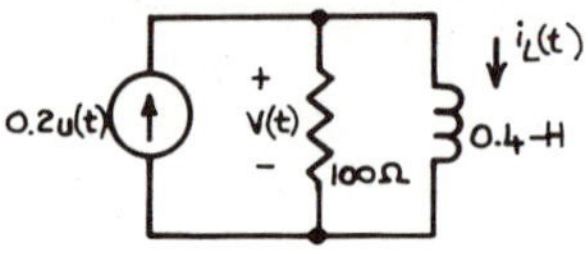

FIG.1

Solution: (a) The circuit described above is shown in Fig. 1 below.

We know that as $t \to \infty$, the inductor becomes a short circuit, so that the full current, 0.2A, from the source must flow through it: $i_L(\infty) = 12A$.

(b) At $t = 0^-$ the current source is "off", the entire circuit is "dead". At t = 0 the current source is turned on. The current through the inductor cannot change instantaneously for a finite voltage. Thus, $i_L(0^+) = 0$.

(c) Using the relation

$$i_L(t) = i_0 + \frac{1}{L}\int_{t_0}^{t} v \, dT,$$

find the current through the inductor given the initial current. We know that $i_0 = 0$, and v(t) must be in the form $V_0 e^{-\frac{R}{L}t}$. We can find V_0 because it is known that at $t = 0^+$ no current was flowing through the inductor, thus the current through the 100-Ω resistor must have been 0.2A. Therefore, $V_0 = 100(0.2) = 20V$. We can now solve the integral

$$i_L(t) = \frac{1}{0.4}\int_0^t 20 \, e^{-\frac{100}{.4}T} dT$$

$$i_L(t) = \frac{20}{0.4\ (-250)}\Big[e^{-250T} \Big]_0^t$$

$$i_L(t) = -0.2\left(e^{-250t} - 1\right)$$

$$i_L(t) = 0.2\left[1 - e^{-250t}\right] \qquad (1)$$

$$i_L(.004) = 0.2\left[1 - e^{-250(.004)}\right] = 0.2(1 - 3.68)$$

$$i_L(.004) = 0.126 \text{ A.}$$

Equation (1) can be written in a more general form:

$$i_L(t) = \frac{V_0}{R}\left[1 - e^{-\frac{R}{L}t}\right] u(t).$$

• **PROBLEM** 6-9

A current source, 8U(t)A, a 5 Ω resistor, and a 20 mH inductor are in parallel. At t = 3ms, find the power absorbed by the: (a) resistor; (b) inductor; (c) source.

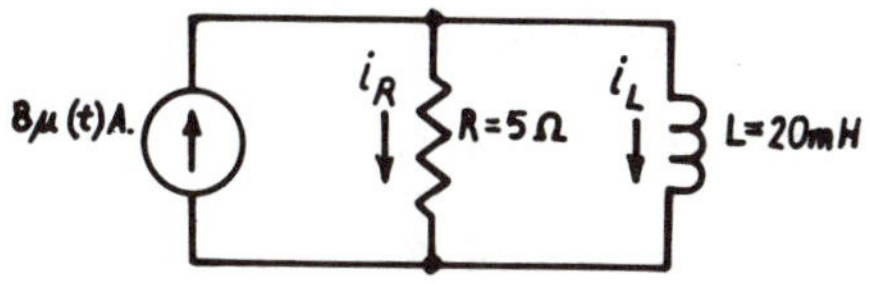

FiG.1

Solution: The parallel arrangement is shown in fig. 1 with currents i_L and i_R as indicated.

The current source is an 8 A step function so that for all $t > 0$,

$$i_R + i_L = 8 \quad \text{by application of Kirchoff's Current Law.}$$

Since all branches are in parallel, the current i_R can be expressed in terms of the branch voltage.

Thus, $i_R = \frac{1}{R} L\left(\frac{di_L}{dt}\right)$

and $\frac{L}{R}\frac{di_L(t)}{dt} + i_L(t) = 8$ is the governing equation for $t > 0$.

The complementary solution is found by solving the homogeneous differential equation, $\frac{L}{R}\frac{di_L}{dt} + i_L = 0$,

assuming the solution

$$i(t) = Ae^{-t/\tau} = Ae^{-\frac{R}{L}t}$$

Since the forcing function (the current source) is time invariant after t = 0, the particular solution must be one in which all circuit variables are constant in time. Thus,

$$\frac{di_L}{dt} = 0 \text{ and } i(t) = 8 \text{ A for } t >> 0.$$

The total solution is:

$$i(t) = 8 + Ae^{-\frac{R}{L}t} \qquad (t > 0)$$

At t = 0+, just after the source comes on, $i_L(t = 0+) = 0$ since the current in the inductor cannot change at once. Using this information in the total solution gives the value of the constant A at t = 0+: $i(t) = 0 = 8 + Ae^{-0}$

$$A = -8,$$

and the solution is, with all values inserted,

$$i_L(t) = 8(1 - e^{-250t})\ A;\ t > 0$$

At t = .003 seconds, all necessary information can be found.

$$i_L = 8(1 - e^{-.75}) = 4.22\ A$$

$$i_R = 8 - i_L = 3.78\ A$$

$$V_R = V_L = 5(3.78) = 18.9\ V,$$

and the power in each branch is (at t = .003s)

$$P_R = 18.9(3.78) = 71.44\ W$$

$$P_L = 18.9(4.22) = 79.76\ W$$

$$P_S = -8(18.9) = -151.2\ W$$

• **PROBLEM** 6-10

The zero-state response of a circuit is the response with zero initial conditions. The zero-input response is the response to nonzero initial conditions when the driving function is set equal to zero. It is true in general, that the total response of a linear circuit is the sum of the zero-state and zero-input responses. Show that this is true for the circuit in fig. 1. Let the input be the current source i(t) and the response be the current through the inductor $i_L(t)$. Assume $i_L(0) = I_0$.

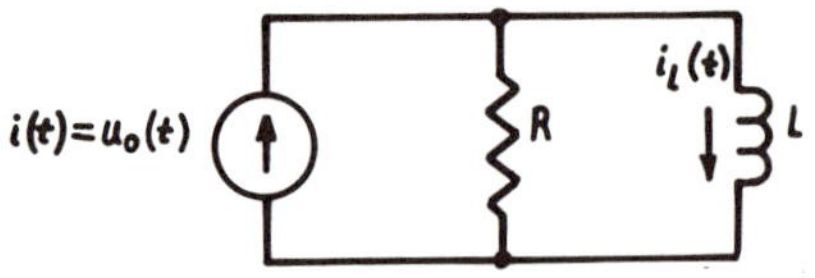

FiG.1

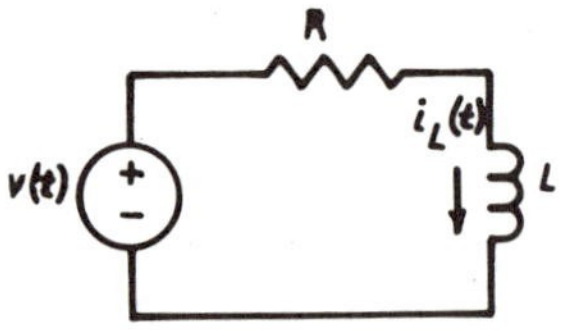

FiG.2

Solution: Since the inductor current is of interest, it is convenient to Thevenin-transform the current source and resistor to obtain the network shown in fig. 2, where v(t) = RV for t ≥ 0. Writing a loop equation,,

$$L\dot{i}_L + Ri_L = R$$

So the differential equation to be solved is

$$\dot{i}_L + \left(\frac{R}{L}\right) i_L = \frac{R}{L}, \quad i_L(0) = I_0$$

The natural response is the solution of the homogeneous equation $(\dot{i}_L)_{nat'l} + \left(\frac{R}{L}\right)(i_L)_{nat'l} = 0$, and is $Ae^{-(R/L)t}$, where A is a constant. The zero-input response is the natural response evaluated with actual initial conditions, and is

$$(i_L)_{z-i} = I_0 e^{-\frac{R}{L}t} A, \quad t \geq 0$$

The forced response is that component of the response that is of the form of the driving function and its derivatives, and is seen to be $(i_L)_{forced} = 1$, $t \geq 0$. The complete response is the superposition of the forced and natural responses, $i_L(t) = 1 + Ae^{-(R/L)t}$, and the zero-state response is the complete response evaluated for initial conditions, so

$$(i_L)_{z-s} = 1 - e^{-\frac{R}{L}t} A, \quad t \geq 0.$$

The complete response evaluated with actual initial conditions is

$$i_L(t) = 1 + (I_0 - 1)e^{-\frac{R}{L}t} A, \quad t \geq 0,$$

obtainable either as $(i_L)_{z-i} + (i_L)_{z-s}$ or from a direct evaluation of the complete response expression.

• PROBLEM 6-11

Write an expression for the inductor current $i_L(t)$ for $t \geq 0$ in the parallel RL circuit in fig. 1 by writing and solving the differential equation for the circuit.

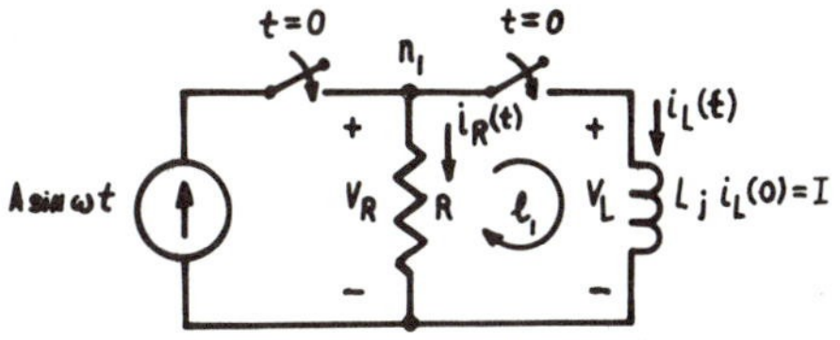

FiG.1

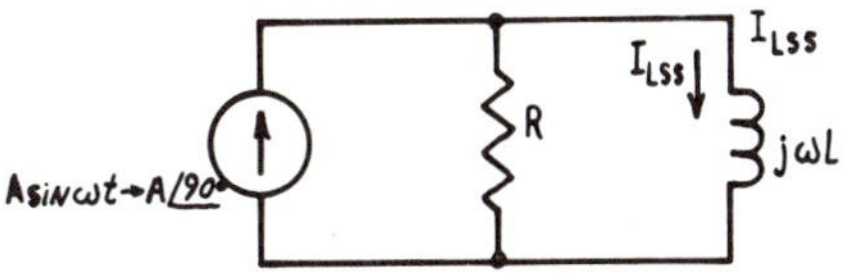

FiG. 2

Solution: We can start to write one of many differential equations for the circuit in fig. 1 by applying KCL at node n_1 giving

$$i_R(t) + i_L(t) = A \sin \omega t \tag{1}$$

Also applying KVL around the loop l_1 gives

$$V_R = V_L \tag{2}$$

Solve for $i_L(t)$ in terms of the circuit parameter constants, A, ω, R, and initial condition I. Find a way of expressing

$i_R(t)$ in Eq (1) in terms of the above constants.

Note that the branch voltage current relationship for the resistor is

$$i_R(t) = \frac{1}{R} V_R$$

and since $V_R = V_L = L \frac{di_L(t)}{dt}$, substitute $\frac{L}{R} \frac{di_L(t)}{dt}$ for $i_R(t)$ in Eq (1) giving a first-order linear differential equation

$$\frac{L}{R} \frac{di_L(t)}{dt} + i_L(t) = A \sin \omega t \qquad (3)$$

where $i_L(0) = I$.

The total solution to this differential equation, $i_L(t)$, can be found by summing the homogeneous and particular solutions of Eq (3). The homogeneous solution is found by solving

$$\frac{L}{R} \frac{d}{dt} [i_L(t)]_h + [i_L(t)]_h = 0 \qquad (4)$$

Proceed as follows,

$$\frac{L}{R} \frac{d}{dt}[i_L(t)]_h = -[i_L(t)]_h$$

$$\frac{d}{dt} [i_L(t)]_h = -\frac{R}{L}[i_L(t)]_h$$

$$\frac{d[i_L(t)]_h}{[i_L(t)]_h} = -\frac{R}{L} dt$$

integrating both sides gives

$$\int_{[i_L(0)]_h}^{[i_L(t)]_h} \frac{d[i_L(t)]_h}{[i_L(t)]_h} = -\frac{R}{L} \int_0^t dt$$

$$\ln[i_L(t)]_h - \ln[i_L(0)]_h = -\frac{R}{L}t \qquad (5)$$

Note that since the total solution is the sum of the homogeneous and particular solutions

$$i_L(t) = [i_L(t)]_h + [i_L(t)]_p$$

then $[i_L(t)]_h = i_L(t) - [i_L(t)]_p$ or more specifically,

$[i_L(0)]_h = i_L(0) - [i_L(0)]_p$, thus we can substitute into eq (5) to obtain

$$\ln[i_L(t)]_h - \ln(i_L(0) - [i_L(0)]_p) = -\frac{R}{L}t .$$

Taking the exponential of both sides

$$\frac{[i_L(t)]_h}{i_L(0)-[i_L(0)]_p} = e^{-\frac{R}{L}t}$$

since $i_L(0) = I$, the homogeneous solution becomes

$$[i_L(t)]_h = (I - I_p)\, e^{-\frac{R}{L}t} \qquad (6)$$

where $I_p = [i_L(0)]_p$, the initial condition of the particular solution.

We note that the general homogeneous solution, $ke^{-\frac{R}{L}t}$, in mathematics is equivalent to the natural response in circuit analysis.

In the same way the particular solution is equivalent to the forced response of our circuit. In our case the forcing function is a sinusoid so the forced response will also be a sinusoid; more specifically, the forced response will be the sinusoidal steady-state solution.

We can however find the particular solution by conventional mathematical methods by substituting the general particular solution

$$[i_L(t)]_p = k_1 \cos \omega t + k_2 \sin \omega t \qquad (7)$$

for $i_L(t)$ in Eq (3).

We note that $[i_L(0)]_p = I_p = k_1$. Substituting (7) for $i_L(t)$ in Eq (3) gives: $\frac{L}{R}\frac{d}{dt}(k_1 \cos \omega t + k_2 \sin \omega t) + k_1 \cos \omega t$

$$+ k_2 \sin \omega t = A \sin \omega t$$

$$\frac{L}{R}\omega(-k_1 \sin \omega t + k_2 \cos \omega t) + k_1 \cos \omega t + k_2 \sin \omega t$$

$$= A \sin \omega t$$

Collecting terms gives

$$(-\frac{L\omega}{R}k_1 + k_2) \sin \omega t + (k_1 + k_2 \frac{L\omega}{R}) \cos \omega t = A \sin \omega t.$$

Since this relation must hold true for all $t \geq 0$ we can substitute $t = 1$ and $t = 0$ to obtain the two equations

$$\frac{-L\omega}{R}k_1 + k_2 = A$$

$$k_1 + \frac{L\omega}{R}k_2 = 0$$

which we now solve for k_1 and k_2.

$$I_p = k_1 = \frac{\begin{vmatrix} A & 1 \\ 0 & \frac{L\omega}{R} \end{vmatrix}}{\begin{vmatrix} \frac{-L\omega}{R} & 1 \\ 1 & \frac{L\omega}{R} \end{vmatrix}} = -\frac{A\left(\frac{L\omega}{R}\right)}{1 + \left(\frac{L\omega}{R}\right)^2}$$

$$k_2 = \frac{\begin{vmatrix} \frac{-L\omega}{R} & A \\ 1 & 0 \end{vmatrix}}{\begin{vmatrix} \frac{-L\omega}{R} & 1 \\ 1 & \frac{L\omega}{R} \end{vmatrix}} = \frac{A}{1 + \left(\frac{L\omega}{R}\right)^2}.$$

Substituting k_1 and k_2 into Eq (7) gives our particular solution (forced response)

$$[i_L(t)]_p = \frac{A}{1 + \left(\frac{L\omega}{R}\right)^2}\left(\sin \omega t - \frac{L\omega}{R} \cos \omega t\right).$$

Finally our total solution is the sum of homogeneous and particular solutions

$$i_L(t) = \left(I + \frac{A\left(\frac{L\omega}{R}\right)}{1 + \left(\frac{L\omega}{R}\right)^2}\right) e^{-\frac{R}{L}t} + \frac{A}{1+\left(\frac{L\omega}{R}\right)^2}\left\{\sin \omega t - \frac{L\omega}{R} \cos \omega t\right\};$$

$$t \geq 0.$$

We can prove that the particular solution, $[i_L(t)]_p$, is the sinusoidal steady-state inductor current.

Using current division theory in the steady-state transformed circuit in fig. 2, we obtain an expression for I_{LSS} directly,

$$I_{LSS} = \frac{R(A \angle 90^\circ}{R + j\omega L} = \frac{-jRA}{R + j\omega L} \quad .$$

In polar form we have

$$\vec{I}_{LSS} = \frac{RA \angle -90^\circ}{\sqrt{R^2 + (\omega L)^2} \angle \tan^{-1}\left(\frac{\omega L}{R}\right)}$$

$$\vec{I}_{LSS} = \frac{A}{\sqrt{1 + \left(\frac{\omega L}{R}\right)^2}} \angle -90^\circ - \tan^{-1}\left(\frac{\omega L}{R}\right)$$

and transforming from phasor form into the conventional time form we get

$$I_{LSS} = \frac{A}{\sqrt{1 + \left(\frac{\omega L}{R}\right)^2}} \cos\left(\omega t - 90° - \tan^{-1}\left(\frac{\omega L}{R}\right)\right)$$

Since $\cos(\theta - 90°) = \sin(\theta)$ we have

$$I_{LSS} = \frac{A}{\sqrt{1 + \left(\frac{\omega L}{R}\right)^2}} \sin\left(\omega t - \tan^{-1}\left(\frac{\omega L}{R}\right)\right) \qquad (8)$$

If we let $\phi = \tan^{-1}\left(\frac{\omega L}{R}\right)$

then $\sin\phi = \dfrac{\frac{\omega L}{R}}{\sqrt{1 + \left(\frac{\omega L}{R}\right)^2}}$ and $\cos\phi = \dfrac{1}{\sqrt{1 + \left(\frac{\omega L}{R}\right)^2}}$

Using the trigonometric angle-difference relation to expand Eq (8) we get

$$I_{LSS} = \frac{A}{\sqrt{1 + \left(\frac{\omega L}{R}\right)^2}} [\sin\omega t \cos\phi - \cos\omega t \sin\phi]$$

Substituting for $\sin\phi$ and $\cos\phi$ we obtain

$$I_{LSS} = \frac{A}{\sqrt{1 + \left(\frac{\omega L}{R}\right)^2}} \left(\frac{1}{\sqrt{1 + \left(\frac{\omega L}{R}\right)^2}} \sin\omega t - \frac{\left(\frac{\omega L}{R}\right)}{\sqrt{1 + \left(\frac{\omega L}{R}\right)^2}} \cos\omega t\right)$$

simplifying gives

$$I_{LSS} = \frac{A}{1 + \left(\frac{\omega L}{R}\right)^2} \left(\sin\omega t - \frac{\omega L}{R} \cos\omega t\right)$$

the same result as the particular solution.

• PROBLEM 6-12

For the circuit shown in fig. 1, set up a first-order differential equation that can be solved for the current i(t). Assume that $i_{L_1}(0) = 1A$ and $i_{L_2}(0) = 0$.

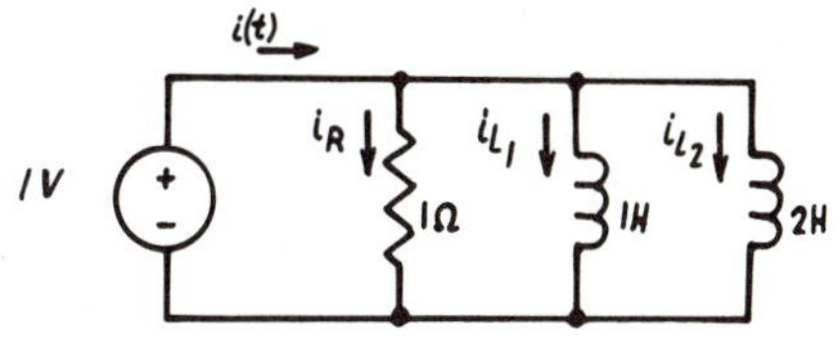

FIG.1

Solution: Using KCL,

$$i = i_R + i_{L_1} + i_{L_2}$$

$$i = 1 + i_{L_1}(0) + \frac{1}{L_1}\int_0^t V_L \, dt + i_{L_2}(0) + \frac{i}{L_2}\int_0^t V_L \, dt$$

$$i = 1 + 1 + \int_0^t V_L \, dt + 0 + \frac{1}{2}\int_0^t V_L \, dt$$

$$i = 2 + \frac{3}{2}\int^t V_L \, dt$$

The above relation can be looked at as an expression for the inductor current of an inductor $L = \frac{2}{3}$ with initial condition $i(0^+) = 2$, hence,

$$V_L = \frac{2}{3}\frac{di}{dt} \; ; \quad i(0^+) = 2A$$

since V_L is always 1V.

$$\frac{di}{dt} = \frac{3}{2} \; ; \quad i(0^+) = 2A.$$

• PROBLEM 6-13

Both switches in the circuit of Fig. 1 are closed at t = 0. Find $i_1(t)$, $i_2(t)$, and $i_3(t)$ for $t > 0$.

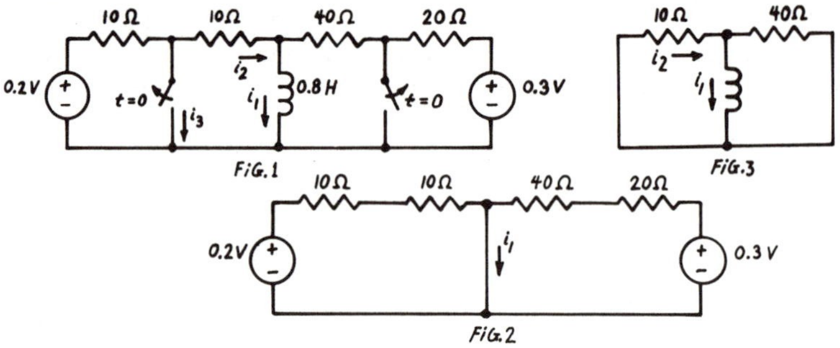

Solution: Since the switches have been open a long time, the inductor is a short circuit in the steady state. The current i_1 is the superposition of currents from each source which operates independently because of the short. The circuit could be redrawn as in Fig. 2.

$$i_1 = \frac{0.2}{10 + 10} + \frac{.3}{40 + 20} = 15 \text{ mA}$$

When the switches are closed at t = 0, the inductor is isolated from the sources and shows its natural response through the parallel combination of a 10 Ω and a 40 Ω resistor. The circuit for t > 0 could be drawn as in Fig. 3 where the inductor is concerned. The time constant is

$$L/R_{eq} = \frac{.8}{\frac{10(40)}{10 + 40}} = .1 \text{ sec}^{-1}$$

$i_1(t)$ starts at 15mA and decays to 0.

$$i_1(t) = 15\ e^{-10t}\text{mA}.$$

$i_2(t)$ is found by current division.

$$i_2(t) = +\frac{40}{10 + 40}\ i_1(t) = +12\ e^{-10t}\ \text{mA}$$

However, the branch in which i_3 is designated carries $-i_2(t)$ plus a contribution of 0.2/10 from the source.

$$i_3(t) = \frac{.2}{10} - i_2(t) = 20 - 12e^{-10t}\text{mA}$$

• PROBLEM 6-14

In the circuit shown in Fig. 1, the switch has been in the position shown for a long time. At t = 0 it is moved to the left. Find and sketch $i_L(t)$.

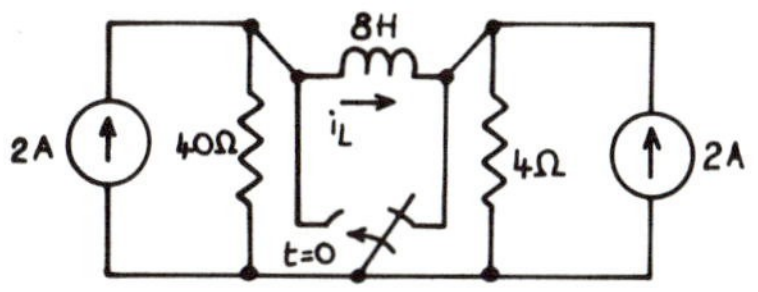

FiG.1

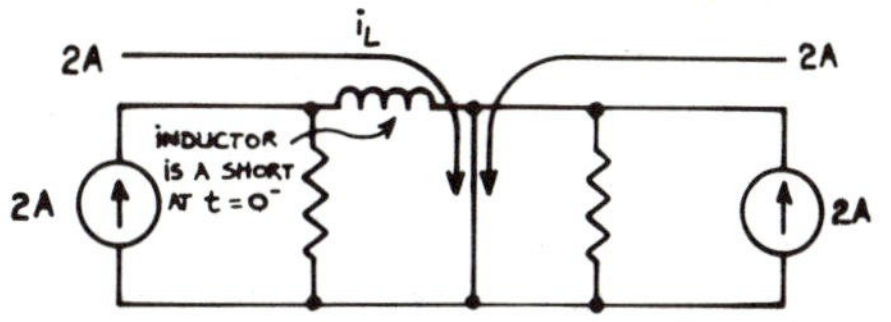

Fig. 2

Solution: We must find the two initial conditions $i_L(0^+)$ and $v_L(0^+)$. Note that $i_L(0^+) = i_L(0^-)$ and $i_L(0^-) =$ 2A as shown in the circuit at $t = 0^-$ in Fig. 2.

Also, $t = 0^+$, as shown in the new circuit in Fig. 3.

We see from Fig. 3 that $i_R = 4A$ thus $v_L(0^+) = i_R\ 4\Omega =$ 16V.

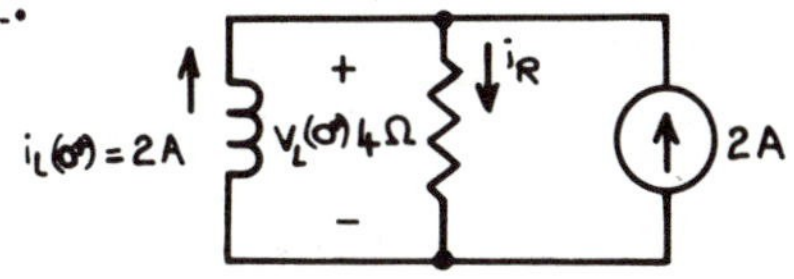

FiG.3

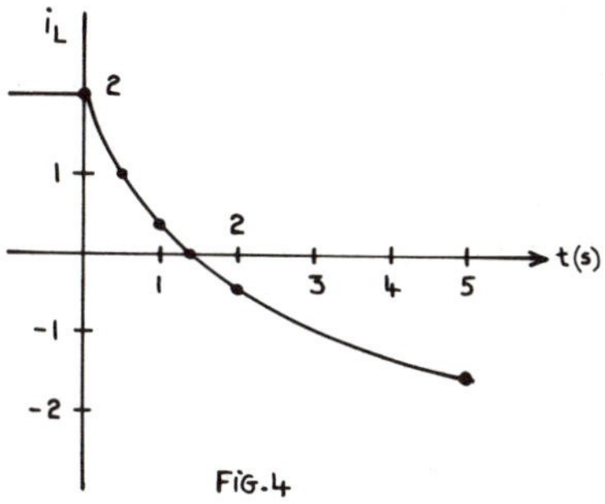

FiG.4

From Fig. 3 we can immediately write an expression for $v_L(t)$

$$v_L(t) = 16\ e^{-\frac{1}{2}t}\ u(t)\ \text{V}.$$

By using the relation

$$i_L(t) = i_0 + \frac{1}{L}\int_{t_0}^{t} v\ dT \qquad (1)$$

we can find i_L.

Note that the current direction indicated in the problem is opposite the voltage and current convention adopted for inductors, capacitors and resistors. In order to find i_L in the direction indicated by the problem, use $-16\,e^{-\frac{1}{2}t}$ for v(t) in equation (1).

$$i_L(t) = i_0 + \frac{1}{8}\int_0^t -16\,e^{-\frac{1}{2}T}\,dT\,.$$

Since $i_0 = 2A$,

$$i_L(t) = 2 + \frac{-16\,(-2)}{8}\left[e^{-\frac{1}{2}T}\right]_0^t$$

$$i_L(t) = 2 + 4\,(e^{-\frac{1}{2}t} - 1)\,u(t)\ A.$$

$$i_L(t) = 2 + (-4 + 4\,e^{-\frac{1}{2}t})\,u(t)\ A.$$

Fig. 4 shows the plot of the function $i_L(t)$.

• PROBLEM 6-15

For the circuit shown in Fig. 1, find $i_L(t)$.

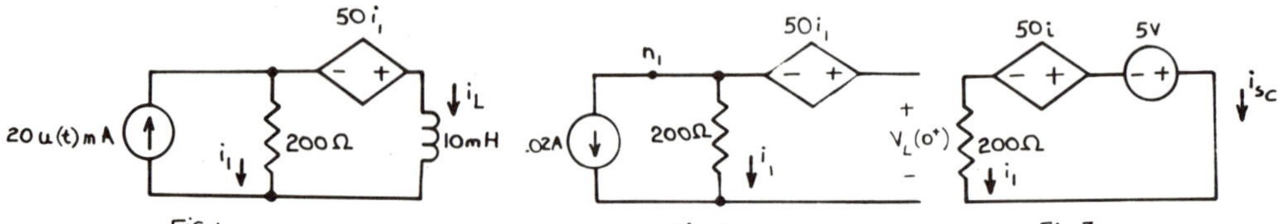

FIG.1 FIG.2 FIG.3

Solution: Note that $i_L(0) = 0$ and that the open circuit voltage v_{oc} across the inductor is $v_L(0^+)$. This voltage also gives the Thevenin equivalent circuit that will enable us to find R_{eq} across the inductor.

Writing the KCL equation for node n_1 in Fig. 2,

gives $i_1 = 0.02$ A

$$v_L(0^+) = 200\,(.02) + 50(.02) = 5\ V.$$

If we open circuit the 2-A current source and place a 5-V source in series with the 50i-V dependent source, we have the Thevenin equivalent circuit. Short circuiting the output yields, i_{sc} and

$$\frac{v_{oc}}{i_{sc}} = \frac{v_L\ (0^+)}{i_{sc}} = R_{eq}\ .$$

A KVL equation around the loop in Fig. 3 gives

$$50\ i_1 + 5V = i_{sc}(200).$$

Since $i_1 = -\ i_{sc}$,

$$5 = i_{sc}\ (200 + 50)$$

$$i_{sc} = \frac{5}{250}$$

$$R_{eq} = 5\left(\frac{250}{5}\right) = 250\ \Omega.$$

Write the relation for the voltage across the inductor.

$$v_L(t) = 5\ e^{-\frac{250}{.01}t} = 5\ e^{-\ 25{,}000t}$$

using, $i_L(t) = i_0 + \frac{1}{L}\int_{t_0}^{t} v_L(T)\ dT$

give $\quad i_L(t) = 0 + \frac{1}{.01}\int_0^t 5\ e^{-\ 25{,}000T}\ dT$

$$i_L(t) = \frac{5}{.01}\ \frac{1}{(-\ 25{,}000)}\left[\ e^{-\ 25{,}000T}\right]_0^t$$

$$i_L(t) = -\ .02\ (e^{-\ 25{,}000t} - 1)\ u(t)\ A.$$

$$i_L(t) = .02\ (1 - e^{-\ 25{,}000t})\ u(t)\ A.$$

• PROBLEM 6-16

For the circuit shown in Fig. 1, find: (a) i_L; (b) v_L.

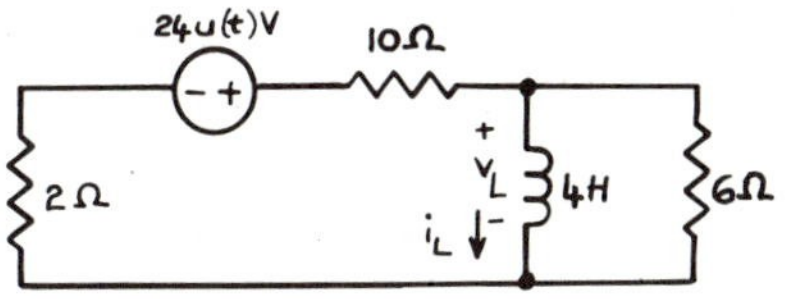

FiG.1

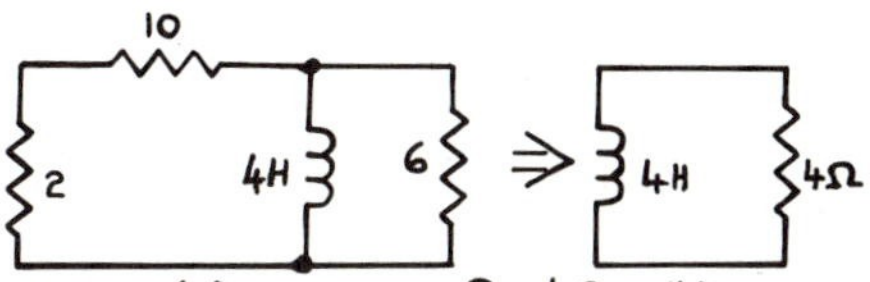

FiG.2 (a) Req=4Ω (b)

Solution: We can determine the complete response $i_L(t)$ by finding the sum of the natural and forced responses.

In the circuit of Fig. 1, the natural response would

be the response which occurs when the 24-V source is "off" or short-circuited.

The natural response for the circuit in Fig. 2(b) is, therefore,

$$i_n = A\,e^{-t}$$

The forced response is the response that would occur at $t = \infty$. Fig. 3 shows the circuit for $t = \infty$, with the inductor short circuited.

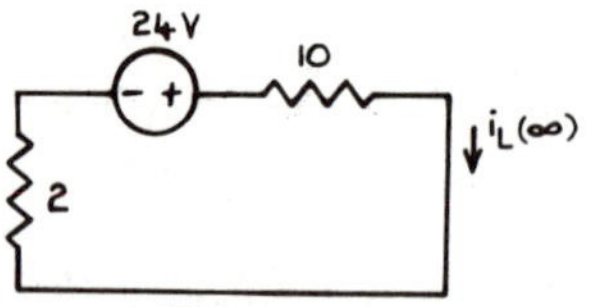

Fig. 3

$$i_f = i_L(\infty) = \frac{24\text{ V}}{12\ \Omega} = 2\text{A}$$

The complete response becomes

$$i_L(t) = i_n + i_f = Ae^{-t} + 2,$$

since the current through an inductor cannot change instantaneously and $i(0^-) = 0$. Thus,

$$i(0^+) = 0$$

$$i_L(0^+) = A + 2 = 0$$

giving, $A = -2$

$$i_L(t) = 2(1 - e^{-t})\,u(t)\text{ A.}$$

Find the voltage v_L in the same way.

$$v_L = v_n + v_f$$

We already know that $v_f = v_L(\infty) = 0$ so that

$$v_L = v_n + 0 = V_0\,e^{-t}.$$

V_0 is the initial voltage across the inductor. We can find this voltage by observing that at $t = 0^+$, $i_L = 0$ so that the voltage across the 6-Ω resistor is equal to $v_L(0^+)$. Since all the current in the loop flows through the 6-Ω resistor

$$v_L(0^+) = \frac{24}{6 + 2 + 10}(6) = 8\text{V}$$

thus $v_L(t) = 8\,e^{-t}\,u(t)$ V.

For the circuit of Fig. 1, let $i_s = 2u(t)$ A and find; (a) i_L; (b) i_1.

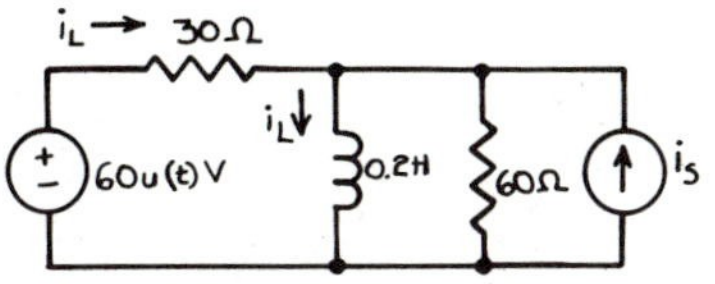

FiG. 1

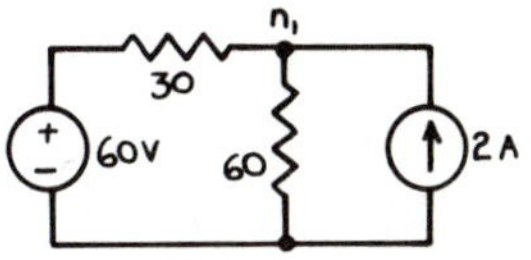

FiG. 2

Solution: Since both sources turn "on" at t = 0, $i_L(0) = 0$, by finding the voltage across the 60-Ω resistor at $t = 0^+$, we find the voltage across the inductor. Writing a KCL equation for node n_1 in Fig. 2 gives $v_L(0^+)$.

$$2 = \frac{v_L - 60}{30} + \frac{v_L}{60}$$

$$\left(\frac{1}{30} + \frac{1}{60}\right) v_L = 2 + \frac{60}{30}$$

$$v_L(0^+) = \frac{4}{\left(\frac{1}{30} + \frac{1}{60}\right)} = \frac{4}{.05} = 80\text{V}$$

The RL circuit which gives the natural response of the circuit in Fig. 1 is shown in Fig. 3.

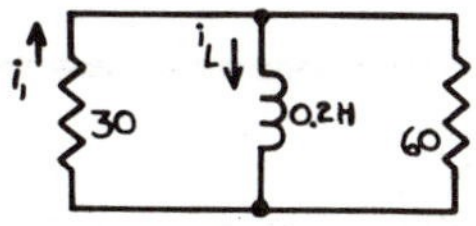

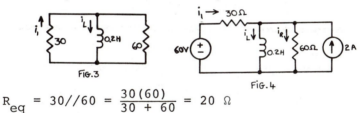

FiG. 3

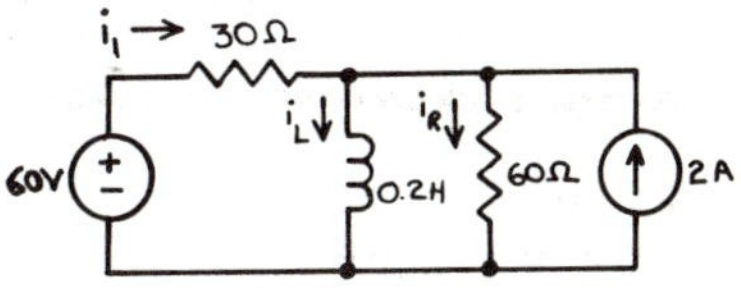

FiG. 4

$$R_{eq} = 30//60 = \frac{30(60)}{30 + 60} = 20\ \Omega$$

We can now write

$$v_L(t) = 80\, e^{-\frac{20}{0.2}t} = 80\, e^{-100t}\, u(t)\ \text{V}.$$

Substituting $v_L(t)$ into the following equation gives $i_L(t)$

$$i_L(t) = i_0 + \frac{1}{L}\int_0^t v_L(T)\,dT$$

$$i_0 = 0,$$

$$i_L(t) = \frac{1}{0.2}\int_0^t 80\, e^{-100T}\, dT$$

$$i_L(t) = \frac{80}{0.2}\left(\frac{1}{-100}\right)[e_{0}^{t}{}^{-100T}]$$

$$i_L(t) = -4\,(e^{-100t} - 1)A.$$

$$i_L(t) = 4 - 4\,e^{-100t}\,u(t)\ A.$$

Find i_1 by summing the currents at node n_1 in Fig. 4.

$$2 + i_1 = i_L + i_R$$

$$i_R = \frac{v_L}{60} = \frac{80}{60}\ e^{-100t} = \frac{4}{3}\ e^{-100t}$$

$$2 + i_1 = 4 - 4\,e^{-100t} + \frac{4}{3}\,e^{-100t}$$

$$i_1 = -2 + 4 - 4\,e^{-100t} + \frac{4}{3}\,e^{-100t}$$

$$i_1(t) = 2 - \frac{8}{3}\,e^{-100t}\,u(t)\ A.$$

• PROBLEM 6-18

Consider the circuit shown in Fig. 1. Assume that $i_{L1}(0) = I_1$ and $i_{L2}(0) = I_2$, and find $i_R(t)$ for $t \geq 0$.

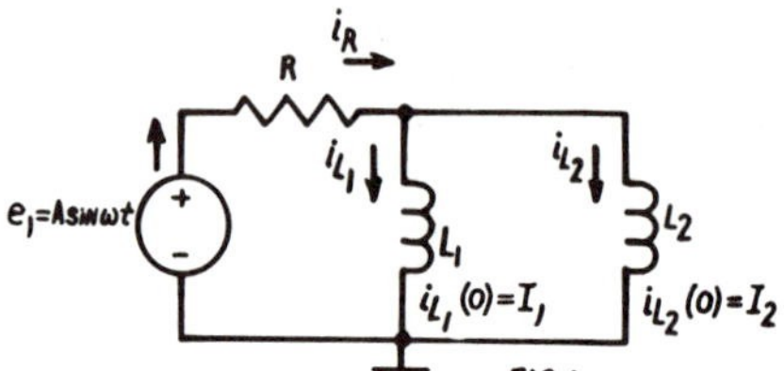

FIG. 1

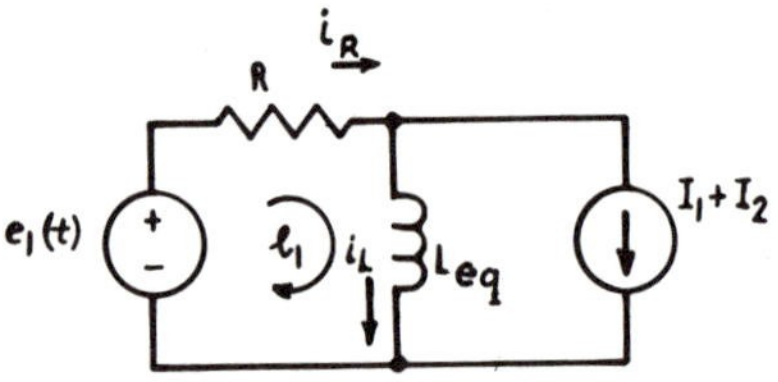

FIG. 2

Solution: Since the inductor currents at t = 0 are constants they can be represented by independent current sources. Thus combining L_1 and L_2 to their equivalent we get the circuit shown in fig. 2.

where, $i_L(0) = 0$

and $L_{eq} = \frac{L_1 L_2}{L_1 + L_2}$

writing loop and nodal equations yields

$$i_R(t) = i_{L_{eq}}(t) + (I_1 + I_2) \tag{1}$$

$$e_1(t) = i_R(t)R + L\,\frac{di_L(t)}{dt} \tag{2}$$

Immediately solve for

$$i_R(0) = I_1 + I_2 \qquad \text{since } i_{L_{eq}}(0) = 0$$

Differentiating both sides of eq (1) gives

$$\frac{di_R(t)}{dt} = \frac{di_L(t)}{dt}$$

and substituting this result into equation 2, it is clear that;

$$A \sin \omega t = Ri_R(t) + L\frac{d}{dt}i_R(t) \qquad (3)$$

Assume a solution of the form

$$i_R(t) = i_R(0)e^{-kt}$$

Substituting into eq (3) yields

$$A \sin \omega t = (R - L_{eq}K)\, i_R(0)e^{-kt}\ ;$$

for t = 0

$$0 = (R - L_{eq}K)\, i_R(0)$$

therefore

$$R - L_{eq}K = 0$$

$$K = \frac{R}{L_{eq}}$$

thus $i_R(t) = i_R(0)e^{-\frac{R}{L_{eq}}t} = (I_1 + I_2)e^{(-R/L_{eq})t}$

• **PROBLEM** 6-19

For the RL circuit calculate the current $i_L(t)$. The initial condition is $i_L(0^-) = -1$ A.

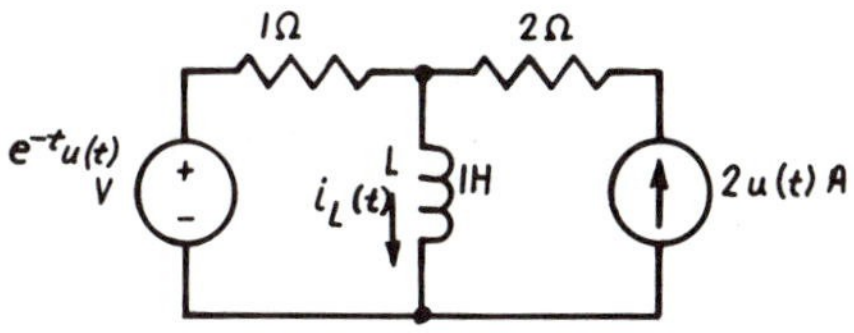

FiG. 1

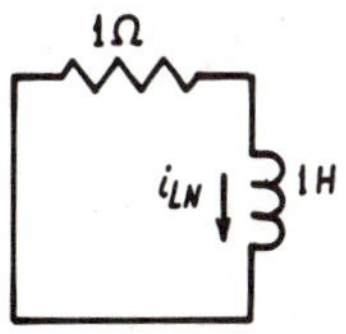

FiG. 2

Solution: The complete response is the sum of the natural and forced responses. The natural response i_{LN} can be calculated from the network of figure 2, in which both sources have been set to zero value (Note that the 2Ω resistor, being in series with the current source, has no effect on i_L).

Thus, $i_{LN}(t) = Ae^{-\frac{R}{L}t} = Ae^{-t}$ for $t \geq 0$.

The forced response can be obtained by superposition, using the differential equation approach. The forced response due to the current source is obtained from figure 3. The forced response is a dc current, so the inductor is a short circuit.

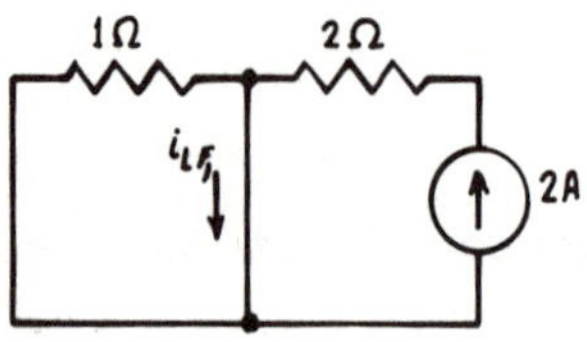

FIG.3

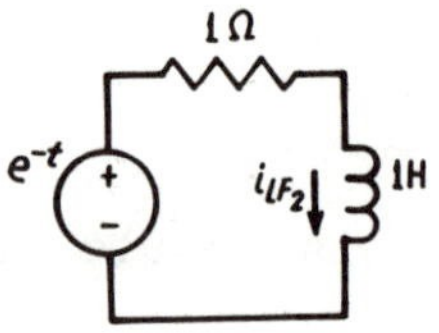

FIG.4

The forced component due to the current source is $i_{LF_1} = 2A$. Figure 4 is the equivalent source for calculation of the forced response due to the exponential voltage source.

Now $\frac{di_{LF_2}}{dt} + i_{LF_2} = e^{-t}$ has the characteristic equation

$$s + 1 = 0$$

which has the root $s = -1$. Since this is the complex frequency associated with e^{-t}, we must solve the differential equation directly. First, multiply by an integrating factor ϕ (to be determined):

$$\phi \frac{di_{LF_2}}{dt} + \phi i_{LF_2} = \phi e^{-t}$$

Imposing the condition that $\frac{d\phi}{dt} = \phi$ leads to $\phi(t) = e^t$.

Hence,

$$\frac{d}{dt}(e^t i_{LF_2}) = e^t \cdot e^{-t} = 1$$

so

$$e^t i_{LF_2}(t) = t + c$$

where c is a constant to be determined. Thus,

$$i_{LF_2}(t) = te^{-t} + ce^{-t}$$

Now since the forced response is merely <u>any one</u> solution of the nonhomogeneous equation, we can pick $c = 0$. Therefore,

$$i_L(t) = i_{F_1} + i_{F_2} + Ae^{-t}\ ; \quad t \geq 0$$

or

$$i_L(t) = 2 + te^{-t} + Ae^{-t}\ ; \quad t \geq 0$$

In order to evaluate A, we can use the initial condition, $i_L(0+) = i_L(0-) = -1 = 2 + A$. Thus $A = -3$, and

$$i_L(t) = (2 + te^{-t} - 3e^{-t})\, U_{-1}(t)$$

RC CIRCUITS

• **PROBLEM** 6-20

For the network of fig. 1

(a) Choose i(t) as the unknown of the circuit and write a first-order differential equation in i(t).

(b) Give the homogeneous solution to the equation obtained in (a).

(c) Give the particular solution to the equation obtained in (a).

(d) Write down the total solution of the equation obtained in (a) and determine the unknown constant using the initial condition.

(e) Sketch the plot of i(t) versus t.

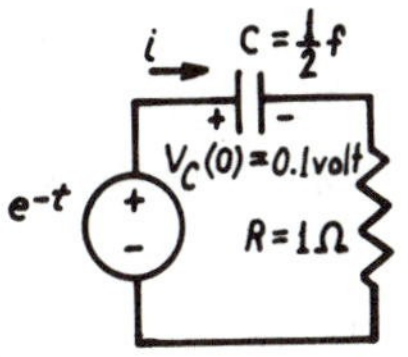

FIG. 1

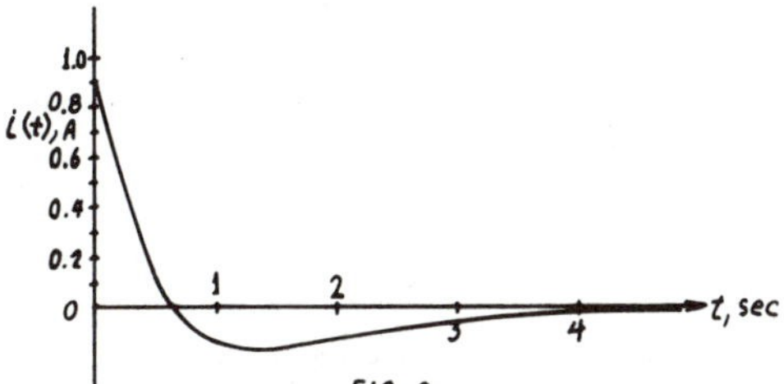

FIG. 2

Solution: (a) In general, for a series RC circuit,

$$\frac{1}{C}\int i(t)\,dt + Ri(t) = v(t).$$

For this problem we have

$$2\int i\,dt + i = e^{-t}.$$

A differential equation is obtained by differentiating the above equation with respect to t. Thus

$$2i + \frac{di}{dt} = -e^{-t}.$$

(b) The homogeneous equation is

$$2i + \frac{di}{dt} = 0.$$

Assume a solution of the form $i = e^{mt}$.

Substituting this into the differential equation gives the characteristic equation,

$$2 + m = 0$$

whose solution is m = 2. Hence

$$i_h = Ke^{-2t} \text{ A.}$$

(c) The particular solution is found by assuming a current of the same form as the driving function. $i = ae^{-t}$. Substituting this into the differential equation gives

$$2ae^{-t} - ae^{-t} = -e^{-t}.$$

Solving for a

$$2a - a = -1$$

or

$$a = -1.$$

The particular solution is

$$i_p = -e^{-t} \text{ A.}$$

(d) The total solution is

$$i = i_h + i_p = Ke^{-2t} - e^{-t}.$$

At t = 0 the total voltage across R is $e^{-0} - 0.1 = 0.9$ volts. The current through R is 0.9/1 = 0.9 amps. Evaluating i for t = 0 gives

$$i(0) = K - 1 = 0.9 \text{ A.}$$

Solving for K we get K = 1.9.

The final solution is

$$i(t) = 1.9e^{-2t} - e^{-t} \text{A.}$$

(e) A plot of i(t) is shown in fig. 2.

• PROBLEM 6-21

Find the unit impulse response of the network of fig. 1. Assume zero initial conditions and $v(t) = \delta(t)$.

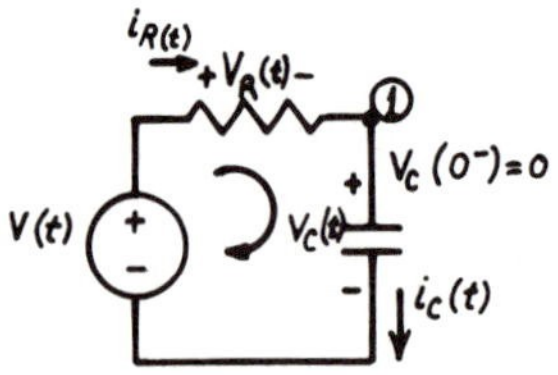

FiG.1

Solution: Writing Kirchoff's voltage law around the loop gives

$$v_R(t) + v_C(t) = v(t) \tag{1}$$

Using voltage current relationship of the capacitor and Kirchoff's current law at node 1 we get;

$$i_R(t) = i_C(t) = C\frac{dv_C}{dt} \qquad (2)$$

Using $v_R(t) = i_R(t)R$ and combining eqs (1) and (2) yields the differential equation defining the response $v_C(t)$:

$$RC\frac{dv_C(t)}{dt} + v_C(t) = \delta(t) \qquad v_C(0^-) = 0$$

to get the homogeneous solution of this equation we set the equation equal to zero and use definitions given below.

NOTE: The diff. input function is given by

$$a_1\frac{dx(t)}{dt} + a_0x(t) = 0$$

then the homogeneous solution is

$$x_h(t) = ke^{-(a_0/a_1)t}$$

The homogeneous solution is therefore;

$$[v_C(t)]_h = ke^{-t/RC}$$

The particular solution to a differential equation is any solution which satisfies the equation:

$$[v_C(t)]_p = \frac{1}{RC}e^{-t/RC}u_0(t)$$

The total solution is the addition of homogeneous and particular solutions.

$$V_C(t) = [v_C(t)]_h + [v_C(t)]_p$$

Since the initial conditions are zero the homogeneous solution, which describes the response of the circuit before the application of any forcing function, will be zero and the total solution is

$$v_C(t) = [v_C(t)]_p = \frac{1}{RC}e^{-t/RC}u_0(t).$$

• PROBLEM 6-22

For the circuit shown in fig. 1 set up a first order differential equation that can be solved for the current i. Specify the initial conditions by replacing the initial capacitor voltages by independent voltage sources.

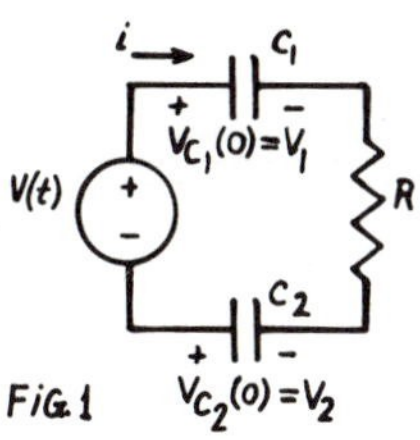

FiG.1

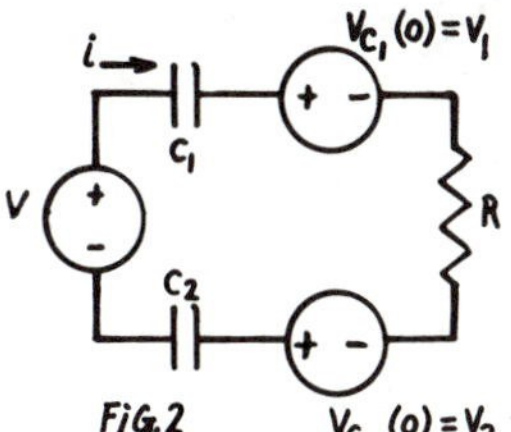

FiG.2

Solution: The initial capacitor voltages can be replaced by independent voltage sources in fig. 1 by noting that the definition of capacitor voltage

$$V_C(t) = V_C(0) + \frac{1}{C}\int_0^t i_C \, dt$$

can be looked at as an independent voltage source $V_1 = V_C(0)$ in series with a capacitor with zero initial condition. Writing the KVL equation for the single loop circuit in fig. 2

$$V = V_{C_1}(0) + \frac{1}{C_1}\int_0^t i \, dt + iR - V_{C_2}(0) + \frac{1}{C_2}\int_0^t i \, dt$$

Hence,

$$V - V_1 + V_2 = \frac{1}{C_1}\int_0^t i \, dt + \frac{1}{C_2}\int_0^t i \, dt + iR$$

differentiating with respect to time gives

$$\frac{d(V - V_1 + V_2)}{dt} = \frac{1}{C_1} i + \frac{1}{C_2} i + \frac{di}{dt} R$$

Let $V_s = V - V_1 + V_2$

$$\frac{di}{dt} + \frac{1}{R}\left(\frac{1}{C_1} + \frac{1}{C_2}\right) i = \frac{1}{R}\frac{dV_s}{dt} .$$

• PROBLEM 6-23

In the circuit of fig. 1 the switch has been closed for a long time and is opened at $t = t_0$. Find the voltage $V_C(t)$ for $t \geq t_0$. How does the phase angle ϕ affect voltage $V_C(t)$?

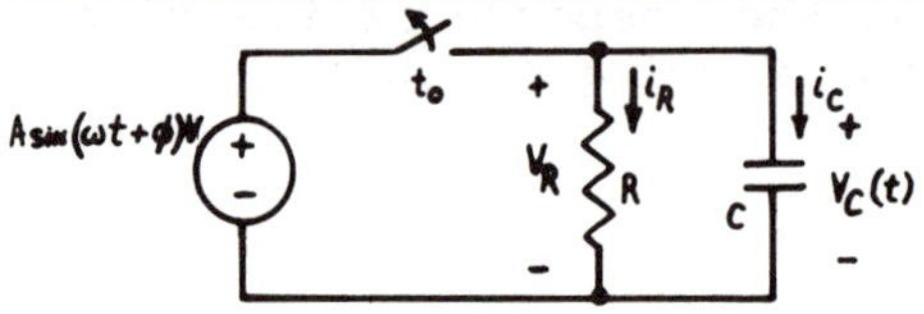

FIG.1

Solution: We write the first-order homogeneous differential equation by removing the voltage source and writing KVL and KCL equations for the remaining RC circuit.

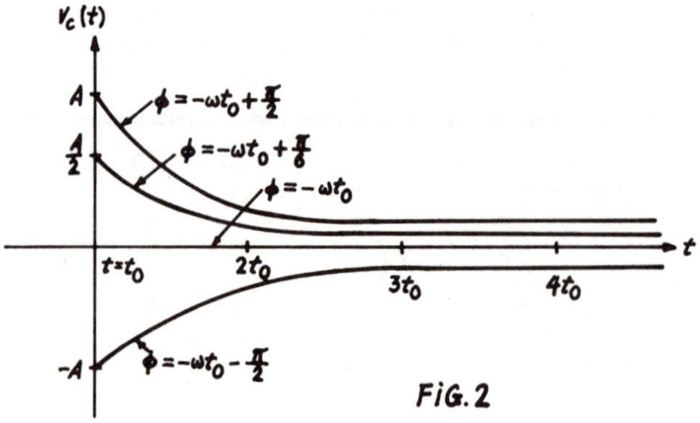

FIG.2

The KVL equation is

$V_C(t) = V_R(t)$, and the KCL equation is

$i_R(t) = -\, i_C(t) \quad t \geq t_0$.

Since $i_C = C\,\frac{dV_C}{dt}$ and $i_R(t) = \frac{V_R(t)}{R}$, substitute into the KCL equation above to obtain

$$\frac{V_R(t)}{R} + C\frac{dV_C}{dt} = 0.$$

Substitute $V_C(t)$ for $V_R(t)$ from the KVL equation, obtaining the differential equation.

$$RC\,\frac{dV_C(t)}{dt} + V_C(t) = 0 \qquad t \geq t_0$$

Recognize that the solution of this equation is of the form

$$V_C(t) = ke^{-t/RC} \qquad t \geq t_0 \qquad (1)$$

where k must be determined from the initial conditions.

Find the initial condition $V_C(t_0^+)$ by noting that $V_C(t_0^-) = V_C(t_0^+)$, since the voltage across a capacitor cannot change instantaneously. Noting that $V_C(t_0^-)$ is the source voltage just before the switch is opened,

$$V_C(t_0^-) = V_C(t_0^+) = A \sin\,(\omega t_0 + \phi), \qquad (2)$$

gives an equation for k. From the solution of the homogeneous differential equation Eq (1) at $t = t_0$ we have

$$V_C(t_0) = ke^{-t_0/RC} = A \sin\,(\omega t_0 + \phi)$$

solving for k gives

$$k = \frac{A \sin\,(\omega t_0 + \phi)}{e^{-t/RC}} = A \sin\,(\omega t_0 + \phi)\,e^{t/RC}$$

Substituting this equation for k into our solution Eq (1) gives

$$V_C(t) = A \sin\,(\omega t_0 + \phi)\,e^{t_0/RC}e^{-t/RC}$$

$$V_C(t) = A \sin\,(\omega t_0 + \phi)\,e^{-(t-t_0)/RC}.$$

We note that the phase angle ϕ affects the amplitude B of $V_C(t)$ where $B = A \sin\,(\omega t_0 + \phi)$.

We note that $B = A$ when $\omega t_0 + \phi = \frac{\pi}{2}, \frac{-3\pi}{2}, \quad \frac{5\pi}{2}, -\frac{7\pi}{2} \ldots$.

(see fig. 2)

For the circuit shown in Fig. 1, find $v_c(t)$ and $i_c(t)$ if i_s = (a) 25u(t) mA; (b) 10 + 15u(t) mA.

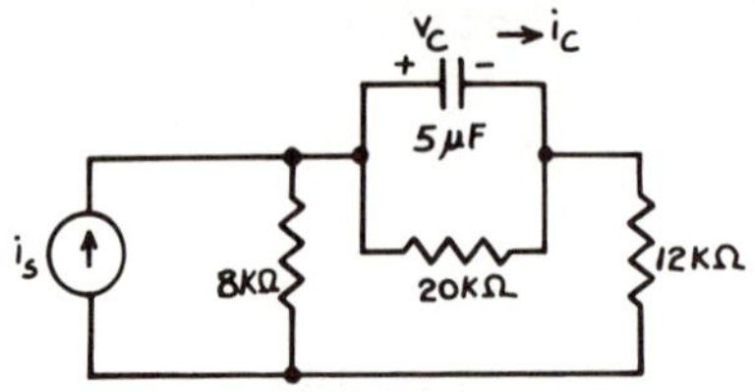

FIG. 1

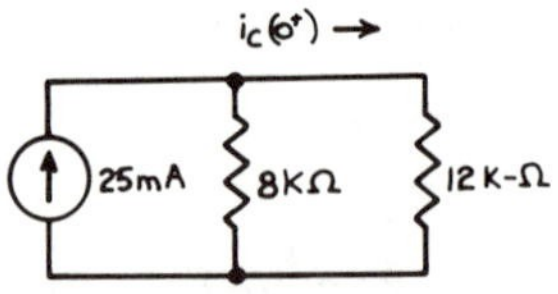

FIG. 2

Solution: (a) Note that $v_c(0^+) = 0$, this means that at $t = 0^+$ the capacitor short-circuits the 20k-Ω resistor, Fig. 2 shows the circuit used to find $i_c(0^+)$.

Using current division,

$$i_c(0^+) = \frac{25\ (8)}{20} = 10 \text{ mA.}$$

The voltage across the capacitor can be found using the relation,

$$v_c(t) = V_0 + \frac{1}{C}\int_{t_0}^{t} i(T)\, dT$$

$$v_c(t) = 0 + \frac{1}{5\mu F}\int_0^t 10 \times 10^{-3}\, e^{-\frac{T}{R_{eq}C}}\, dT$$

R_{eq} can be found by open circuiting the current source.

$$R_{eq} = ((8 + 12)||20)k = 10k\ \Omega$$

$$R_{eq}C = 5\mu F\ (10k) = .05$$

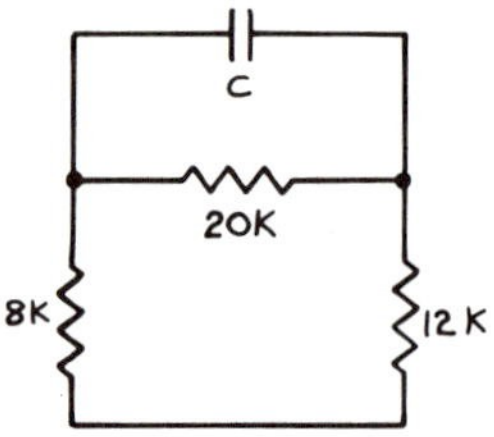

FIG. 3

$$v_c(t) = \frac{10 \times 10^{-3}}{5 \times 10^{-6}}\left(-\frac{1}{20}\right)\left[e^{-20T}\right]_0^t$$

$$v_c(t) = 100\ (e^{-20t} - 1)\ u(t)V$$

$$i_c(t) = 10\ e^{-20t}\ u(t) \text{ mA.}$$

(b) Fig. 4 shows a plot of i_s for this problem.

Knowing that $v_c(0^+) = v_c(0^-)$, find $v_c(0^+)$ for the circuit when it is in steady state condition with i_s = 10 mA. Fig. 5 shows that the voltage across the 20-kΩ resistor is

$v_c(0^-) = v_c(0^+)$.

By current division,

$$v_c(0^+) = v_c(0^-) = \frac{10\ (8)}{40}\ (20) = 40\ \text{V}.$$

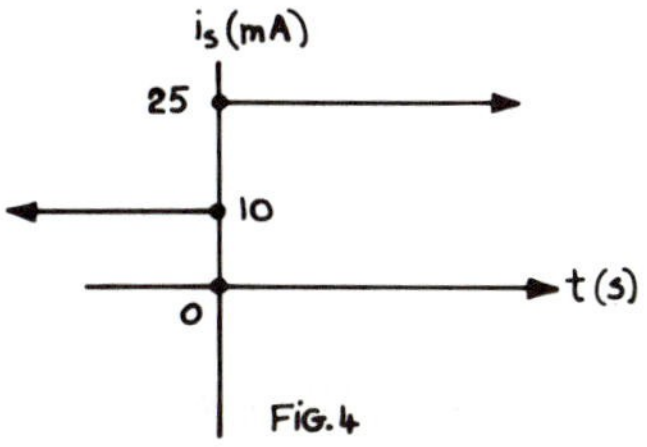

FIG.4

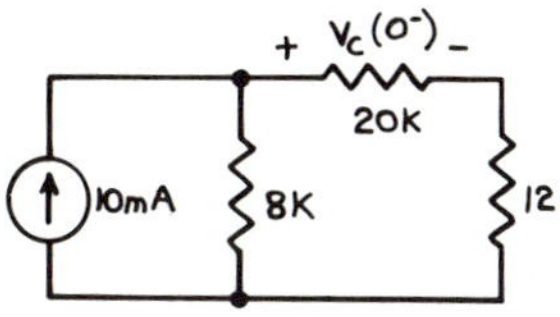

FIG.5

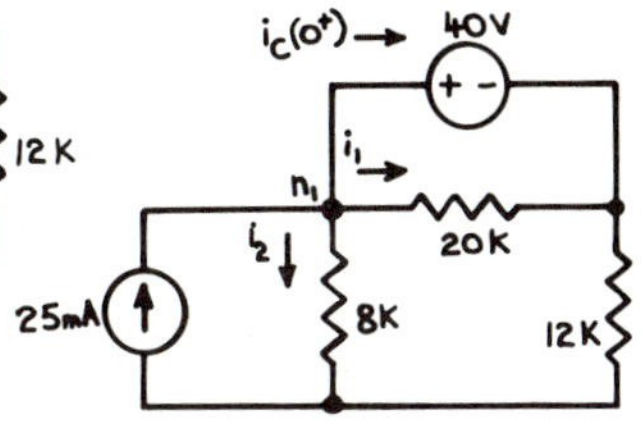

FIG.6

One can now find $i_c(0^+)$ by replacing the capacitor with a 40-V source in the circuit for $t = 0^+$ shown in Fig. 6.

By summing the currents at node n_1 we have

$$25\ \text{mA} = i_1 + i_2 + i_c(0^+)$$

$$i_1 = \frac{40\text{v}}{20\text{k}} = 2\ \text{mA}.$$

Find i_2 by using the superposition theorem on the circuit in Fig. 6.

$$i_a = \frac{25\ (12)}{20} = 15\ \text{mA}$$

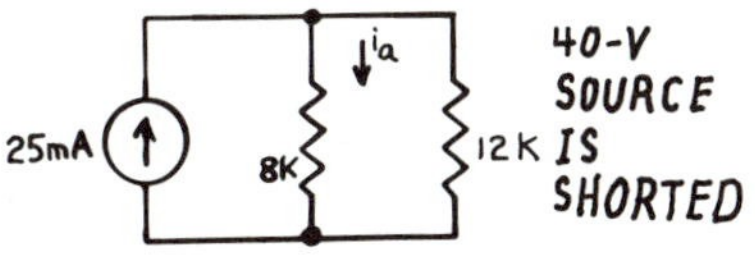

FIG.7

$$i_b = \frac{40\text{v}}{20\text{k}} = 2\ \text{mA}$$

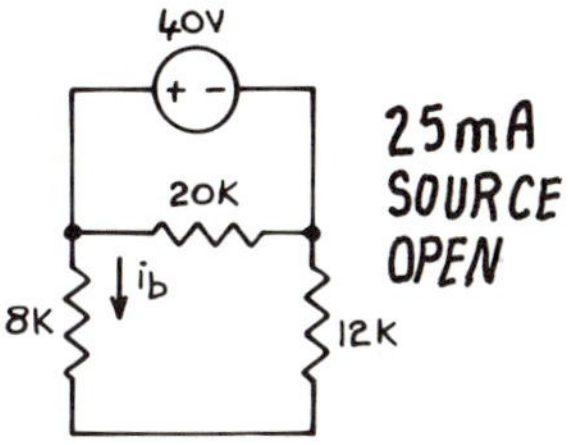

FIG.8

$$i_2 = i_a + i_b = (15 + 2)\ \text{mA} = 17\ \text{mA}$$

$$i_c(0^+) = (25 - 2 - 17)\ \text{mA} = 6\ \text{mA}$$

Knowing $i_c(0^+) = 6$ mA and $v_c(0^+) = 40$ V makes it possible to find $i_c(t)$ and $v_c(t)$.

$$i_c(t) = 6\ e^{-\ 20t}\ u(t)\ \text{mA}$$

$$v_c(t) = v_0 + \frac{1}{C}\int_{t_0}^{t} i(T)\ dT$$

$$v_c(t) = 40 + \frac{1}{5\mu F}\int_0^t 6\, e^{-20T}\ \text{mA}\ dT$$

$$v_c(t) = 40 + \frac{6 \times 10^{-3}}{5 \times 10^{-6}}\left(-\frac{1}{20}\right)\left[e^{-20T}\right]_0^t$$

$$v_c(t) = 40 - 60\,(e^{-20t} - 1)\, u(t)\ \text{V}.$$

$$v_c(t) = 40 + 60(1 - e^{-20t})\, u(t)\ \text{V}.$$

• PROBLEM 6-25

With reference to the circuit shown in Fig. 1, let $v_s = 30u(t)$ V and determine: (a) $v_c(t)$; (b) $v_1(t)$.

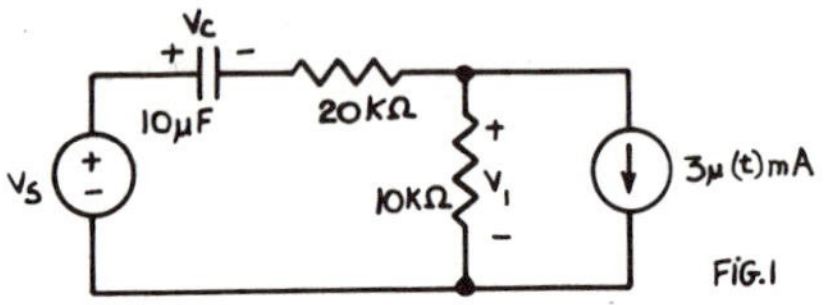

FiG.1

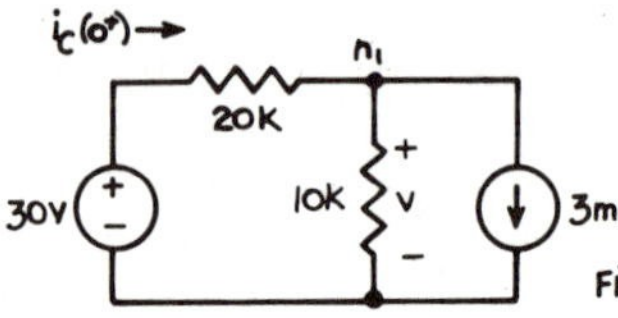

FiG.2

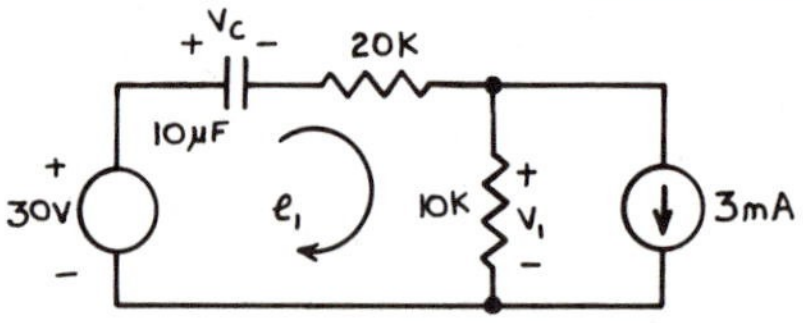

FiG.3

Solution: (a) Since all sources are "dead" for $t < 0$, $v_c(0^+) = 0$. Replacing the capacitor with a short circuit yields the circuit for $t = 0^+$ in Fig. 2.

Writing a KCL equation for node n_1

$$-3\ \text{mA} = \frac{v - 30}{20k} + \frac{v}{10k}$$

gives the initial conditions

$$v = \frac{-3\ \text{mA} + \frac{3}{2}\ \text{mA}}{\left(\frac{1}{20k} + \frac{1}{10k}\right)} = -10\ \text{V}$$

$$i_c(0^+) = -\frac{v - 30}{20k} = -\frac{-10 - 30}{20k} = 2\ \text{mA}.$$

One can use

$$v_c(t) = V_0 + \frac{1}{C}\int_{t_0}^t i(T)\, dT$$

given $i(t) = 2\, e^{-\frac{t}{R_{eq} C}}$ mA,

$R_{eq} = 30\text{k}\Omega$, and $T = 0.3$ s, to find the capacitor voltage.

Hence, $$v_c(t) = 0 + \frac{1}{10 \times 10^{-6}} \int_0^t 2 \times 10^{-3}\, e^{-\frac{T}{0.3}}\, dT$$

$$v_c(t) = \frac{2 \times 10^{-3}}{10 \times 10^{-6}} (-0.3) \left[e^{-\frac{T}{0.3}} \right]_0^t$$

$$v_c(t) = -60 \left(e^{-\frac{t}{0.3}} - 1 \right) u(t)\ \text{V}.$$

v_1 can be found by summing the voltages around the loop ℓ_1 in Fig. 3.

$$30 = v_c + v_2 + v_1$$

$$30 = 60 - 60\, e^{-\frac{t}{0.3}} + i_c(t)(20\text{k}) + v_1$$

$$i_c(t) = 2\, e^{-\frac{t}{0.3}}\ \text{mA}$$

$$v_1 = 30 - 60 + 60\, e^{-\frac{t}{0.3}} - 40\, e^{-\frac{t}{0.3}}$$

$$v_1 = \left[-30 + 20\, e^{-\frac{t}{0.3}} \right] u(t)\ \text{V}$$

Note that $v\ (0) = [-30 + 20]\text{V} = -10\text{V}$, corresponds to the initial condition found earlier.

• PROBLEM 6-26

For the circuit shown in Fig. 1, find: (a) $v_1(t)$; (b) $v_2(t)$; (c) $i_3(t)$.

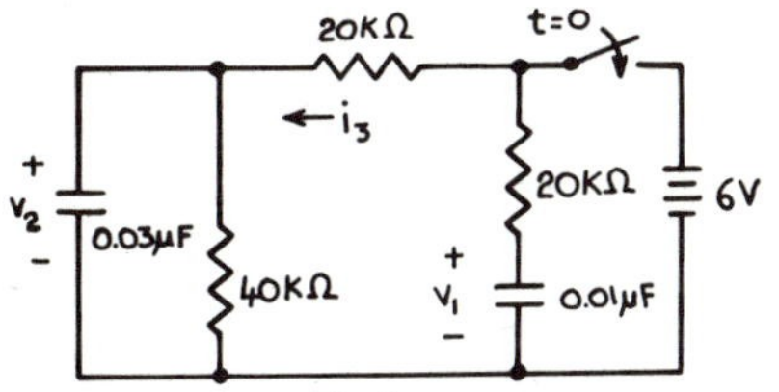

FIG. 1

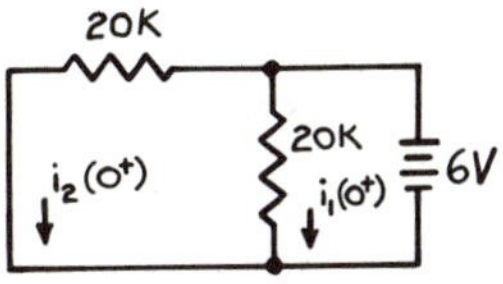

FIG. 2

Solution: Recognizing that $v_1(0^+) = 0$ and $v_2(0^+) = 0$, $i_1(0^+)$ and $i_2(0^+)$ can be derived from the circuit in Fig. 2.

By current division,

$$i_1(0^+) = i_2(0^+) = \frac{6V}{20k} = 3.0 \times 10^{-4}\ A$$

To find the natural response for this circuit, draw the circuit in Fig. 1 for t > 0 with the 6-V source shorted.

Observe from Fig. 3 that there are two separate RC circuits.

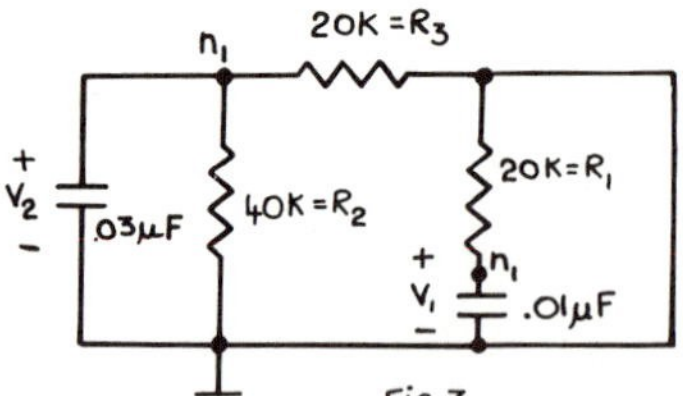

FiG.3

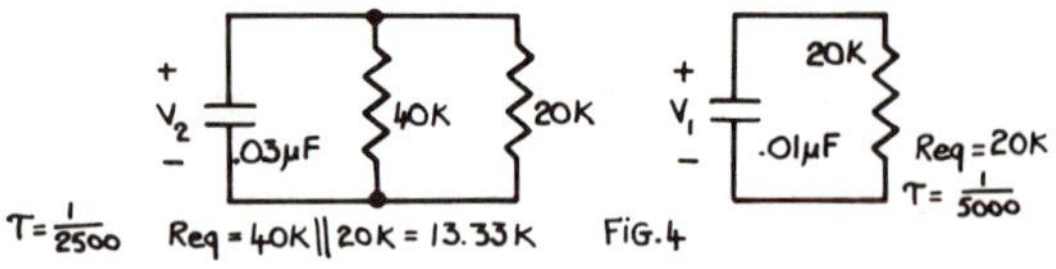

FiG.4

By writing KCL equations for nodes n_1 and n_2 with the reference ground as indicated in Fig. 3, it becomes even more apparent that there are two separate RC circuits. The two KCL equations are,

$$n_1: \quad C_1 \frac{dv_1}{dt} + \frac{v_1}{R_1} = 0$$

$$n_2: \quad C_2 \frac{dv_2}{dt} + \frac{v_2}{R_2} + \frac{v_2}{R_3} = 0$$

Notice that the two equations above are independent of each other.

Fig. 4 shows these two circuits separated.

We utilize

$$v_C(t) = V_0 + \frac{1}{C}\int_{t_0}^{t} i(T)\, dT$$

$$v_1(t) = 0 + \frac{1}{.01 \times 10^{-6}}\int_0^t 3.0 \times 10^{-4}\, e^{-5000T}\, dT$$

$$v_1(t) = \frac{3.0 \times 10^{-4}}{.01 \times 10^{-6}}\left(\frac{1}{-5000}\right)\Big[e^{-5000T}\Big]_0^t$$

$$v_1(t) = -6(e^{-5000t} - 1)\, u(t)\ V$$

$$v_1(t) = (6 - 6e^{-5000t})\, u(t)\ V.$$

$$v_2(t) = 0 + \frac{1}{.03 \times 10^{-6}}\int_0^t 3.0 \times 10^{-4}\, e^{-2500T}\, dT$$

$$v_2(t) = \frac{3.0 \times 10^{-4}}{.03 \times 10^{-6}} \left(\frac{1}{-2500}\right) \left[e^{-2500T} \right]_0^t$$

$$v_2(t) = -4\,(e^{-2500T} - 1)\,u(t)\ \text{V}.$$

$$v_2(t) = (4 - 4e^{-2500t})\,u(t)\ \text{V}.$$

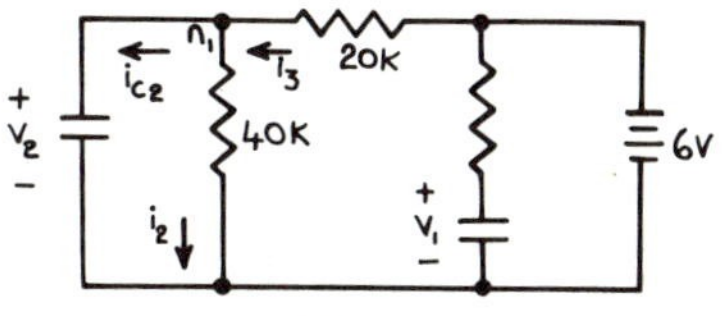

FiG.5

To find i_3, sum the currents at node n_1 in Fig. 4.

$$i_3 = i_{C_2} + i_2$$

$$i_3 = 3 \times 10^{-4}\, e^{-2500t} + \frac{\left(4 - 4e^{-2500t}\right)}{40k}$$

$$i_3 = (0.1 + 0.2e^{-2500t})\,u(t)\ \text{mA}.$$

• **PROBLEM** 6-27

Find $i_1(0^+)$, $i_2(0^+)$, and $i_3(0^+)$ in the circuit shown in Fig. 1 if i_s = : (a) 0; (b) 3 A; (c) 3u(t) A.

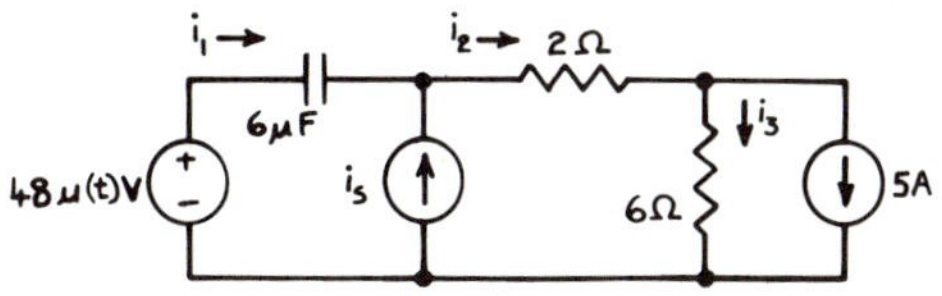

FiG.1

FiG.2

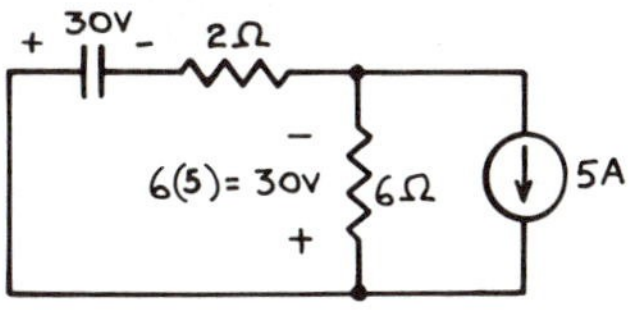

Solution: (a) In order to find the currents indicated, find the voltage across the capacitor. Since $v_c(0^-) = v_c(0^+)$, we can find $v_c(0^+)$ by using the circuit in Fig. 2 for $t = 0^-$.

Observe in Fig. 2 that the 2-Ω resistor and the capacitor are both in the same branch and since no current flows in this branch $v_c = 30$ V.

At $t = 0^+$, the 48-V source becomes "active." Fig. 3 shows the circuit at $t = 0^+$.

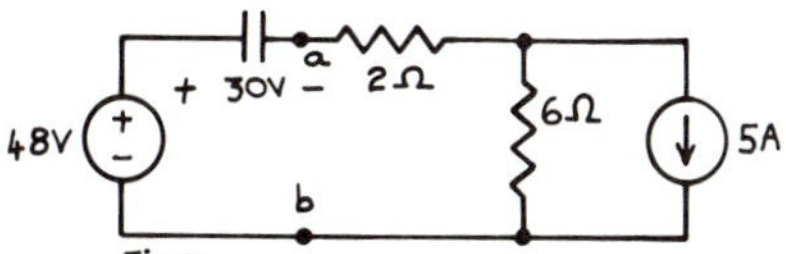

FiG.3

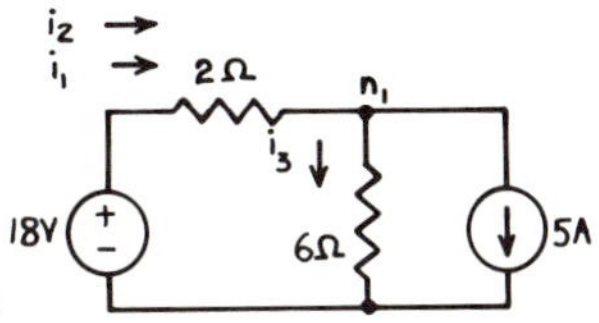

FiG.4

Note that $v_{ab} = 48 - 30 = 18$ V. Fig. 3 is changed into Fig. 4 by connecting an 18-V source between ab.

By writing a KCL equation for node n_1 one can solve for i_1, i_2 and i_3.

$$-5 = \frac{v}{6} + \frac{v - 18}{2}$$

$$\left(\frac{1}{6} + \frac{1}{2}\right) v = -5 + 9$$

$$v = \frac{4}{\left(\frac{1}{6} + \frac{1}{2}\right)} = 6 \text{ V}$$

$$i_1 = i_2 = \frac{6 - 18}{2} = 6 \text{ A}$$

$$i_3 = \frac{6}{6} = 1 \text{ A.}$$

(b) Again, find $v_c(0^+)$ by finding v_c in the circuit in Fig. 5 at $t = 0^-$.

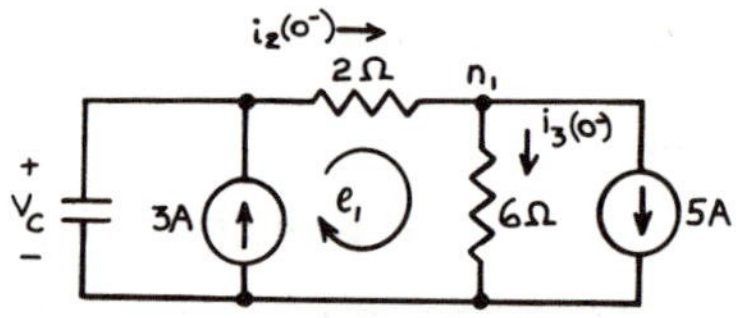

FIG.5

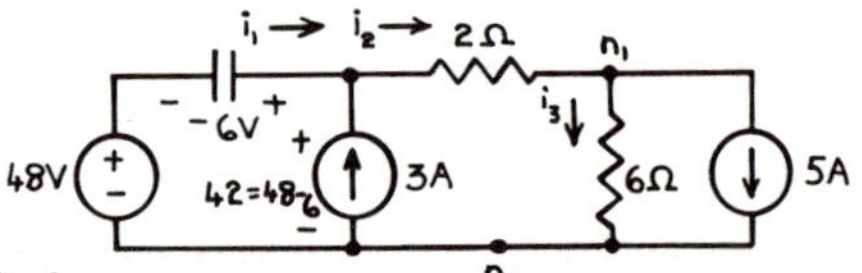

FIG.6

By summing the currents at node n_1 in Fig. 5, we find $i_2(0^-)$ and $i_3(0^-)$.

$$i_2(0^-) = i_3(0^-) + 5$$

$$i_2(0^-) = 3\text{A}$$

$$i_3(0^-) = 5 - 3 = 2\text{A}$$

Summing voltages around the loop ℓ_1 gives

$$v_c(0^-) = v_c(0^+) = i_2(0^-)2\Omega - i_3(0^-)6\Omega$$

$$v_c(0^+) = 3(2) - 2(6) = -6\text{V.}$$

The circuit in Fig. 6 is for $t = 0^+$.

Since the voltage across the 3-A source is 42V a KCL equation can be written for node n_1 and thus one can find $i_1(0^+)$, $i_2(0^+)$ and $i_3(0^+)$.

$$-5 = \frac{v - 42}{2} + \frac{v}{6}$$

$$\left(\frac{1}{2} + \frac{1}{6}\right) v = -5 + 21$$

$$v = \frac{16}{\left(\frac{1}{2} + \frac{1}{6}\right)} = 24V$$

$$i_3(0^+) = \frac{24V}{6\Omega} = 4A$$

$$i_2(0^+) = -\frac{(24 - 42)V}{2\Omega} = 9A$$

Writing an equation for the sum of the currents at node n_2 in Fig. 6 yields

$$i_1(0^+) + 3A = i_3(0^+) + 5A$$

$$i_1(0^+) = i_3(0^+) + 5A - 3A = 4 + 5 - 3 = 6A\ .$$

(c) The circuit for $t = 0^-$ in this part of the problem is the same as that in part (a) thus $v_c(0^+) = 30$ V.

The circuit at $t = 0^+$ is shown in Fig. 7.

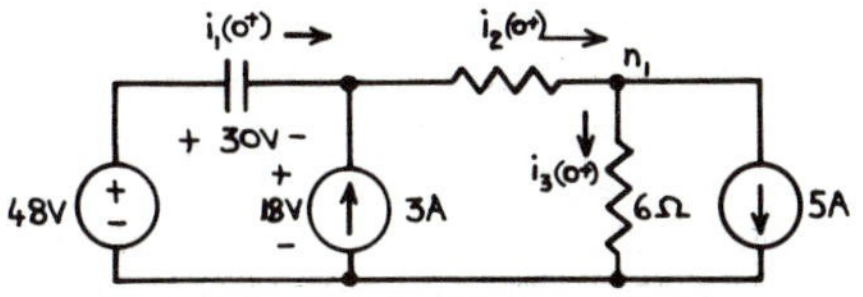

Fig.7

Writing a KCL equation for node n_1 in Fig. 7 yields

$$-5 = \frac{v - 18}{2} + \frac{v}{6}$$

$$\left(\frac{1}{2} + \frac{1}{6}\right) v = -5 + 9$$

gives $$v = \frac{4}{\left(\frac{1}{2} + \frac{1}{6}\right)} = 6V.$$

$$i_3(0^+) = \frac{6V}{6\Omega} = 1A$$

$$i_2(0^+) = -\frac{(6 - 18)V}{2\Omega} = 6A$$

$$i_1(0^+) + 3 = i_3(0^+) + 5$$

$$i_1(0^+) = i_3(0^+) + 5 - 3 = 1 + 5 - 3 = 3A.$$

Find the transient and steady-state current if the waveforms shown in fig. 1 are applied to

(i) series RC network (R = C = 1).

(ii) series RL network (R = L = 1).

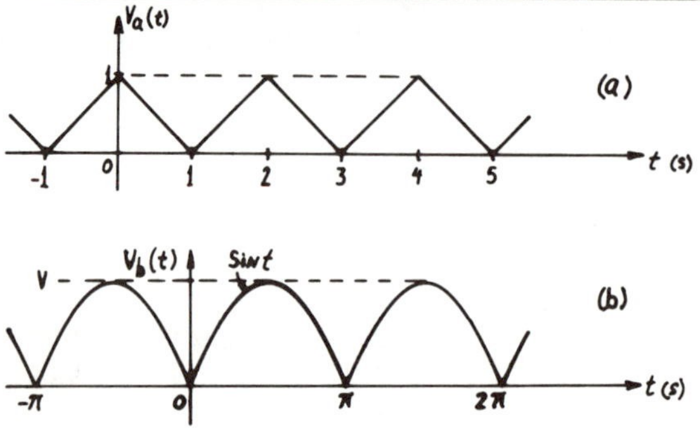

FIG. 1

Solution: In each part of this problem, the procedure will be to solve a first order linear differential equation for the particular integral solution, and then, choosing the appropriate constraints, solve for the integration constants. The transient solution is found second because its integration constant is determined from the steady-state solution.

(i) The differential equation for this circuit is found from the voltage equation,

$$V(t) = Ri + \frac{1}{C}\int i\ dt.$$

Differentiating, we get

$$\frac{dV(t)}{dt} = R\frac{di}{dt} + \frac{i}{C}\ i = \frac{di}{dt} + i \qquad (1)$$

where we have used R = C = 1.

Waveform (a):

There is no equation for a triangular wave so we divide the wave into the two sections which we can then treat separately. $0 \leq t \leq 1$: The voltage is $V_a(t) = 1 - t$.

Substitute this into equation (1) to get

$$-1 = \frac{di_1}{dt} + i_1$$

The solution is found using the integration factor $e^{\int dt} = e^t$. Thus

$$-1\ e^t = e^t\ \frac{di_1}{dt} + e^t\ i_1 = \frac{d}{dt}\ (i_1 e^t).$$

Integrating,

$$-e^t + k_1 = ie^t$$

or

$$i_1 = k_1 e^{-t} - 1. \tag{2}$$

$1 \le t \le 2$: The voltage is $V_a(t) = t - 1$.

Substituting $V_a(t)$ into equation (1) we get

$$1 = \frac{di_2}{dt} + i_2.$$

Solving for i_2 as above gives

$$i_2 = k_2 e^{-t} + 1. \tag{3}$$

Next we solve for k_1 and k_2 by applying constraints which tie the two solutions together. The first is

$$i_1(0) = i_2(2)$$

which is necessary for periodicity.

Placing equations (2) and (3) into this constraint gives

$$k_1 - 1 = 1 + k_2 e^{-2}$$

or

$$k_2 = (k_1 - 2)\, e^2 \tag{4}$$

The second constraint is

$$i_1(1) = i_2(1).$$

Again using equations (2) and (3) gives

$$-1 + k_1 e^{-1} = 1 + k_2 e^{-1}$$

or $$k_2 = k_1 - 2e. \tag{5}$$

Equating equations (4) and (5), and solving for k_1, yields

$$k_1 = 1.462 \tag{6}$$

Putting this into equation (5) we find

$$k_2 = 1.462 - 2e. \tag{7}$$

The final step is to place the values of k_1 and k_2 into their respective equations, (2) and (3). The results are

$$i_1(t) = 1.462\, e^{-t} - 1. \qquad 0 \le t \le 1$$

and $$i_2(t) = (1.462 - 2e)\, e^{-t} + 1$$

$$= 1.462e^{-t} - 2e^{-(t-1)} + 1 \qquad 0 \le t \le 2,$$

for the steady-state solution.

The transient solution is found from the homogeneous part of equation (1) which is

$$\frac{di}{dt} + i = 0$$

The characteristic equation is $m + 1 = 0$ so the solution is

$$i_{tr} = Ae^{-t}.$$

The total current for $0 \leq t \leq 1$ is the sum of the transient and steady-state solutions,

$$i(t) = Ae^{-t} + 1.462e^{-t} - 1 \qquad 0 \leq t \leq 1.$$

The voltage is

$$V(t) = i + \frac{di}{dt} = 0.462 + A - t$$

which is found by substituting i(t) into the equation. At $t = 0$, the applied voltage is 1. Hence

$$V(0) = 1 = 0.462 + A$$

and $A = 0.538$.

The complete transient solution is

$$i_{tr}(t) = 0.538e^{-t} \qquad t \geq 0$$

or $$i_{tr}(t) = 0.538e^{-t}u(t).$$

Waveform (b):

The solution for i proceeds the same as for waveform (a), except in this case it is easier since we do not have to divide the wave into two segments. Again use equation (1).

$$\frac{dV_b(t)}{dt} = \frac{di(t)}{dt} + i(t).$$

Setting $V_b(t) = \sin t$ we get

$$\cos t = \frac{di}{dt} + i. \qquad 0 \leq t \leq \pi$$

Multiplying by the integration factor and solving,

$$e^t \cos t = e^t \frac{di}{dt} + e^t i = \frac{d}{dt}(e^t i)$$

$$\frac{1}{2} e^t (\cos t + \sin t) + k_3 = i e^t$$

or $$i(t) = \frac{1}{2} (\cos t + \sin t) + k_3 e^{-t}$$

To solve for k_3 use $i(0) = i(\pi)$. Then

$$\frac{1}{2} + k_3 = -\frac{1}{2} + k_3 e^{-\pi}$$

or $$k_3 = -1.045.$$

The steady-state solution is

$$i(t) = \frac{1}{2}(\cos t + \sin t) - 1.045e^{-t}$$

or $\quad i(t) = 0.707 \sin(t + 45°) - 1.045e^{-t} \qquad 0 \le t \le \pi$

The form of the transient solution is the same as the triangular wave case,

$$i_{tr}(t) = Be^{-t}.$$

The total solution is

$$i(t) = Be^{-t} + 0.707 \sin(t + 45°) - 1.045\, e^{-t}.$$

At t = 0, the applied voltage is zero so the initial current is zero. Hence

$$i(0) = 0 = B + 0.5 - 1.045.$$

Thus $B = 0.545$.

The transient solution is

$$i_{tr}(t) = 0.545\, e^{-t}\, u(t).$$

(ii) RL network.

The differential equation for this circuit is

$$V(t) = Ri(t) + L\frac{di(t)}{dt} = i + \frac{di}{dt}. \tag{8}$$

Waveform (a):

Time interval $0 \le t \le 1$:

The applied voltage is $V(t) = 1 - t$. Then equation (8) gives

$$1 - t = i_1 + \frac{di_1}{dt}$$

whose solution is

$$i_1(t) = 2 - t + k_4e^{-t}. \tag{9}$$

Time interval $1 \le t \le 2$:

The applied voltage is $V(t) = t - 1$. Using this equation (8) becomes

$$t - 1 = i_2 + \frac{di_2}{dt}$$

whose solution is

$$i_2(t) = t - 2 + k_5e^{-t}. \tag{10}$$

Two constraints used to solve for k_4 and k_5 are

$$i_1(0) = i_2(2)$$

$$i_1(1) = i_2(1).$$

Applying these constraints to equations (9) and (10) gives

$$k_5 = (2 + k_4)e^2$$

and $$k_5 = 2e + k_4.$$

Solving for the constants, we get

$$k_4 = -1.462$$

$$k_5 = 2e - 1.462.$$

Placing the constants into equations (9) and (10) gives the final steady-state solution,

$$i_1(t) = 2 - t - 1.462\, e^{-t} \qquad 0 \le t \le 1$$

$$i_2(t) = t - 2 - 1.462\, e^{-t} + 2e^{-(t-1)} \qquad 1 \le t \le 2$$

To find the transient solution use the homogeneous form of equation (8),

$$i + \frac{di}{dt} = 0$$

which has a solution of the form ce^{-t}. The total current in the region $0 \le t \le 1$ is

$$i(t) = ce^{-t} + 2 - t - 1.462e^{-t}.$$

The inductor prohibits current from flowing at $t = 0$, so

$$i(0) = 0 = C + 2 - 1.462$$

and $$C = -0.538.$$

The transient solution is

$$i_{tr}(t) = -0.538e^{-t}\, u(t).$$

Waveform (b):

In the time interval $0 \le t \le \pi$ the differential equation with applied voltage sin t is

$$\sin t = i + \frac{di}{dt}.$$

Solving with the integration factor e^t,

$$e^t \sin t = \frac{d}{dt}(e^t i)$$

so $$i(t) = \frac{1}{2}(\sin t - \cos t) + k_6 e^{-t}$$

Determine k_6 from the condition $i(0) = i(\pi)$. Then

$$-\frac{1}{2} + k_6 = \frac{1}{2} + k_6 e^{-\pi}$$

and $k_6 = 1.045$

The steady-state solution is

$$i(t) = \frac{1}{2}(\sin t - \cos t) + 1.045\, e^{-t}$$

or

$$i(t) = 0.707 \sin(t - 45°) + 1.045e^{-t} \quad 0 \le t \le \pi$$

The form of the transient solution is

$$i_{tr}(t) = De^{-t}$$

and the total current is

$$i(t) = De^{-t} + 0.707 \sin(t - 45°) + 1.045\, e^{-t}.$$

At $t = 0$, the current is zero so

$$i(0) = 0 = D - 0.5 + 1.045$$

and

$$D = -0.545.$$

The transient solution is

$$i_{tr}(t) = -0.545e^{-t}\, u(t)$$

DIFFERENTIAL EQUATIONS

• PROBLEM 6-29

Given the first-order differential equation

$$4\frac{dx(t)}{dt} + 2x(t) = 3e^{-t}$$

find: (a) the homogeneous solution, (b) the particular solution, and (c) the total solution if $x(0) = 1$.

Solution: (a) The homogeneous solution is the solution found when the input function is zero. To find the homogeneous solution set the differential equation equal to zero and find $x(t)$.

$$4\frac{dx(t)}{dt} + 2x(t) = 0 \qquad (1)$$

Exchanging terms:

$$4\frac{dx(t)}{dt} = -2x(t)$$

$$4\,dx(t) = -2x(t)\,dt$$

$$\frac{dx(t)}{x(t)} = -\frac{2}{4}\,dt \qquad (2)$$

Integrating both sides gives:

$$\ln[x(t)] = -\frac{1}{2}t + k \qquad (3)$$

where k is a constant of integration. Taking the exponential of each side gives:

$$x(t) = \exp[-\tfrac{1}{2}t + k]$$

$$= e^{-\frac{1}{2}t}e^{k}$$

$$= Ke^{-\frac{1}{2}t} \qquad (4)$$

where $K = e^k$ is a non-zero constant which depends on the initial conditions.

The homogeneous solution is

$$x_h(t) = Ke^{-(1/2)t} \qquad (5)$$

The value of K will be found in part C.

(b) The particular solution, $x_p(t)$, is found when the input function, $3e^{-t}$, is set equal to the differential equation. So

$$4\frac{dx_p(t)}{dt} + 2x_p(t) = 3e^{-t}. \qquad (6)$$

The method of undetermined coefficients is used to find $x_p(t)$. With this method the form of $x_p(t)$ is guessed and then the undetermined coefficients are found by substitution into the differential equation. The particular solution depends on the input function; therefore the particular solution has the form:

$$x_p(t) = k_1e^{-t}.$$

Substituting this form into equation (6) gives

$$4\frac{d}{dt}[k_1e^{-t}] + 2k_1e^{-t} = 3e^{-t}$$

or $$-4K_1e^{-t} + 2k_1e^{-t} = 3e^{-t}.$$

Now solve for k_1 by dividing both sides by e^{-t}.

$$-4k_1 + 2k_1 = 3$$

$$k_1 = -\frac{3}{2}$$

The particular solution is

$$x_p(t) = -\frac{3}{2}e^{-t}.$$

(c) The total solution of any linear differential equation is the sum of the homogeneous solution and the particular solution. So

$$x(t) = x_h(t) + x_p(t)$$

$$x(t) = ke^{-(1/2)t} - \frac{3}{2}e^{-t}$$

The constant k from the homogeneous equation is now found from the given conditions that at t = 0 x(t) = 1. Then

$$x(0) = 1 = ke^{-(1/2)0} - \frac{3}{2}e^{(0)} = k - \frac{3}{2}$$

Therefore

$$k - \frac{3}{2} = 1$$

$$k = \frac{5}{2}.$$

The total solution is

$$x(t) = \frac{5}{2}e^{-(1/2)t} - \frac{3}{2}e^{-(t)}.$$

• PROBLEM 6-30

For the following differential equation

(a) Find the homogeneous solution.

(b) Find the particular solution.

(c) Write down the total solution and determine the unknown constant.

$$-2\frac{dv(t)}{dt} + 2v(t) = t^2 + 2t + 1; \quad v(0) = 1$$

Solution: (a) The homogeneous solution, $v_h(t)$, is found by setting the input function equal to zero. In this case the input function is on the right side. Hence $v_h(t)$ can be solved for the equation

$$-2\frac{dv_h(t)}{dt} + 2v_h(t) = 0. \qquad (1)$$

Exchanging terms:

$$\frac{d\,v_h(t)}{v_h(t)} = 1\;dt$$

Integrating both sides gives

$$\ln[v_h(t)] = t + k$$

where k is a constant of integration. Taking the exponent of each side gives

$$v_h(t) = e^{t+k} = e^t e^k = Ke^t$$

where $K = e^k$ is a non-zero constant which depends on the initial condition. This is the "unknown constant" that will be determined in part (c). The answer to part (a) then is

$$v_h(t) = Ke^t. \qquad (2)$$

(b) The particular solution, $v_p(t)$, is solved from the complete equation

$$-2\,\frac{dv_p(t)}{dt} + 2v_p(t) = t^2 + 2t + 1. \qquad (3)$$

The method of undetermined coefficients will be used to solve this equation. In this case, the particular solution has the form of the forcing function, that is,

$$v_p(t) = At^2 + Bt + C. \qquad (4)$$

Hence, $$\frac{dv_p(t)}{dt} = 2At + B. \qquad (5)$$

Substituting (4) and (5) into (3) and rearranging gives

$$(2A)t^2 + (2B - 4A)t + (2C - 2B) = t^2 + 2t + 1 \qquad (6)$$

Now A, B, and C must be found. From equation (6) it is seen that:

$$2A = 1\;;\quad 2B - 4A = 2\;;\quad 2C - 2B = 1.$$

$$A = \frac{1}{2}\;;\quad B = 2\;;\quad C = \frac{5}{2}\,.$$

The particular solution is

$$v_p(t) = \frac{1}{2}t^2 + 2t + \frac{5}{2}\,. \qquad (7)$$

(c) The total solution is the sum of the homogeneous and particular solutions. Combining equations (2) and (7) gives

$$v(t) = \frac{1}{2}t^2 + 2t + \frac{5}{2} + ke^t. \qquad (8)$$

Now the unknown constant k must be found. It is given that the total solution at t = 0 is 1. Hence equation (8) can be rewritten for t = 0.

$$v(0) = \frac{1}{2}(0)^2 + 2(0) + \frac{5}{2} + ke^{(0)}$$

$$1 = \frac{5}{2} + k$$

Hence the unknown constant is $1 - \frac{5}{2} = -\frac{3}{2}$.

The total solution then is

$$v(t) = \frac{1}{2}t^2 + 2t + \frac{5}{2} - \frac{3}{2}e^t.$$

• PROBLEM 6-31

For the following differential equation

(a) Find the homogeneous solution.

(b) Find the particular solution.

(c) Write the total solution and determine the unknown constant

$$\frac{d\,x(t)}{dt} = e^{-t} \sin 2t; \quad x(0) = 1$$

Solution: A homogeneous differential equation is a differential equation with no forcing function or driving function. In the equation given, the homogeneous solution is a general solution of the following ordinary differential equation

$$\frac{d\,x(t)}{dt} = 0 \tag{1}$$

Integrating both sides

$$x_h(t) = K \tag{2}$$

where $x_h(t)$ is the homogeneous solution of x(t) and K is an integrating constant.

(b) An integral of the given differential equation

$$\frac{d\,x(t)}{dt} = e^{-t} \sin 2t \tag{3}$$

is a particular solution of Eq. (3).

$$x_p(t) = \int e^{-t} \sin 2t\, dt \tag{4}$$

When Eq. (4) is integrated, the following integral formula is employed.

$$\int u(t)v'(t)dt = u(t)v(t) - \int u'(t)v(t)dt \tag{5}$$

This method is known as integration by parts. In this

equation,

$$\text{let } u(t) \equiv \sin 2t \tag{6}$$

$$v'(t) \equiv e^{-t} \tag{7}$$

Then, from (6), $u'(t) = 2\cos 2t$ (8)

and from (7) $v(t) = -e^{-t}$ (9)

So, substituting Eqs (6) - (9) into Eq (5),

$$\int \sin 2t \cdot e^{-t}dt = \sin 2t \cdot (-e^{-t}) - \int 2\cos 2t(-e^{-t})dt$$

$$= -e^{-t}\sin 2t + 2\int e^{-t}\cos 2t\, dt \tag{10}$$

Applying Eq (5) for the last integral by letting

$$u(t) = \cos 2t \tag{11}$$

$$v'(t) = e^{-t} \tag{12}$$

Then $u'(t) = -2\sin 2t$ (13)

and $v(t) = -e^{-t}$ (14)

Substituting Eqs (11) - (14) into (5),

$$\int \cos 2t \cdot e^{-t}\, dt = \cos 2t\ (-e^{t}) - \int (-2\sin 2t)(-e^{-t})\, dt$$

$$= -e^{t}\cos 2t - 2\int e^{-t}\sin 2t\, dt \tag{15}$$

Substituting Eq. (15) into Eq. (10)

$$\int e^{-t}\sin 2t\, dt$$

$$= -e^{-t}\sin 2t + 2(-e^{t}\cos 2t - 2\int e^{-t}\sin 2t\, dt)$$

$$= -e^{-t}\sin 2t - 2e^{t}\cos 2t - 4\int e^{-t}\sin 2t\, dt \tag{16}$$

Collecting terms of $\int e^{-t}\sin 2t\, dt$,

$$5\int e^{-t}\sin 2t\, dt = -e^{-t}\sin 2t - 2e^{-t}\cos 2t$$

$$\int e^{-t}\sin 2t\, dt = \frac{-1}{5}e^{-t}\sin 2t - \frac{2}{5}e^{-t}\cos 2t$$

$$= 0.2e^{-t}\sin 2t - 0.4\, e^{-t}\cos 2t \tag{17}$$

Substituting Eq (17) into (4),

$$x_p(t) = -0.2e^{-t}\sin 2t - 0.4e^{-t}\cos 2t \tag{18}$$

This is the particular solution.

(c) The total solution $x(t)$ is a sum of the homogeneous solution $x_h(t)$ and the particular solution $x_p(t)$.

$$x(t) = x_h(t) + x_p(t) \tag{19}$$

Substituting Eq. (2) and Eq. (18) into Eq. (19)

$$x(t) = K - 0.2e^{-t} \sin 2t - 0.4 e^{-t} \cos 2t \qquad (20)$$

The unknown constant K is determined from the initial condition given for t = 0,

$$x(0) = 1 \qquad (21)$$

In Eq. (20), for t = 0,

$$x(0) = K - 0 - 0.4 \qquad (22)$$

Equating Eqs (21) and (22)

$$1 = K - 0.4.$$

$$K = 1 + 0.4 = 1.4 \qquad (23)$$

Now the unknown constant has been determined. Substituting (23) into Eq. (20), the total solution is

$$x(t) = 1.4 - 0.2 e^{-t} \sin 2t - 0.4 e^{-t} \cos 2t \qquad (24)$$

CHAPTER 7

RLC CIRCUITS

SERIES RLC

• **PROBLEM** 7-1

Find the homogeneous solution to a differential equation for the voltage $v_C(t)$ across a $\frac{1}{2}$ F capacitor in a series RLC circuit where $R = 4\Omega$, $L = 4H$, $i_L(0) = \frac{1}{4}$ A, and $v_C(0) = 0$.

Solution: Since we are interested in the homogeneous solution of an RLC circuit, start the formulation of the differential equation by writing

$$v_L + v_R + v_C = 0 \quad \text{where}$$

$v_L = L\frac{di_L}{dt}$ and, $v_R = i_L R$. Substituting $C\frac{dv_C}{dt}$ for i_L gives the homogeneous second order differential equation in terms of v_C.

$$LC\frac{d^2v_C}{dt^2} + RC\frac{dv_C}{dt} - v_C = 0$$

It is known that the solution must be of the form $e^{-\alpha t}(A\cos\omega_d t + B\sin\omega_d t)$ when $\alpha^2 = \left(\frac{R}{2L}\right)^2 < \omega_0{}^2 = \frac{1}{LC}$. Since $\alpha^2 = \frac{1}{4}$ and $\omega_0{}^2 = \frac{1}{2}$ this is the proper solution form for our circuit. Find ω_d by solving

$$\omega_d = \sqrt{\omega_0{}^2 - \alpha^2}$$

$$\omega_d = \sqrt{\frac{1}{2} - \frac{1}{4}}$$

$$\omega_d = \sqrt{\frac{1}{4}} = \frac{1}{2}$$

Hence, $\qquad v_C(t) = e^{-\frac{1}{2}t}\left(A\cos\frac{1}{2}t + B\sin\frac{1}{2}t\right)$

At $t = 0$ $\qquad v_C(0) = A = 0$ and, since

$$\frac{i_L}{C} = \frac{dv_C}{dt} = -\frac{1}{2}e^{-\frac{1}{2}t} A\cos\frac{1}{2}t - \frac{1}{2}e^{-\frac{1}{2}t} A\sin\frac{1}{2}t$$

$$-\frac{1}{2}e^{-\frac{1}{2}t} B\sin\frac{1}{2}t + \frac{1}{2}e^{-\frac{1}{2}t} B\cos\frac{1}{2}t$$

$$\frac{i_L(0)}{C} = \frac{dv_C(0)}{dt} = -\frac{1}{2}A + \frac{1}{2}B$$

$$\frac{i_L(0)}{C} = \frac{\frac{1}{4}}{\frac{1}{2}} = \frac{1}{2} = -\frac{1}{2}A + \frac{1}{2}B \quad \text{since } A = 0 \quad B = 1.$$

Therefore, the homogeneous solution is $e^{-\frac{1}{2}t}\sin\frac{1}{2}t$ V.

• PROBLEM 7-2

Find the homogeneous solution to a differential equation for the voltage, $v_C(t)$ across a $\frac{4}{3}$ F. capacitor in a series RLC circuit where $R = 1\Omega$, $L = \frac{1}{4}$ H, $i_L(0) = 4A$, and $v_C(0) = 0$.

Solution: Since the homogeneous solution of a RLC circuit is of interest, start the formulation of the differential equation by writing

$$v_L + v_R + v_C = 0$$

where $v_L = L\frac{di_L}{dt}$ and, $v_R = i_L R$. Substituting $C\frac{dv_C}{dt}$ for i_L gives a homogeneous second order differential equation in terms of v_C

$$LC\frac{d^2v_C}{dt^2} + RC\frac{dv_C}{dt} + v_C = 0.$$

It is known that the solution must be in the form $Ae^{s_1t} + Be^{s_2t}$, when $\alpha^2 = \left(\frac{R}{2L}\right)^2 > \omega_0{}^2 = \frac{1}{LC}$. Since $\alpha^2 = 4$ and $\omega^2 = 3$, this is the proper solution form for our circuit. Find s_1 and s_2 by determining

$$s_1, s_2 = -\alpha \pm \sqrt{\alpha^2 \quad \omega_0{}^2}$$

$$s_1, s_2 = -2 \pm \sqrt{4 - 3}$$

$$s_1, s_2 = -2 \pm 1$$

$$s_1 = -1$$

$$s_2 = -3$$

Hence, the solution is in the form $v_C(t) = Ae^{-t} + Be^{-3t}$.

At $t = 0$, $v_C(0) = 0 = A + B$ and since

$$\frac{i_L}{C} = \frac{dv_C}{dt} = -Ae^{-t} - 3\,Be^{-3t}$$

$$\frac{i_L(0)}{C} = \frac{dv_C(0)}{dt} = -A - 3\,B$$

$\dfrac{4}{\frac{4}{3}} = 3 = -A - 3\,B$. We can now solve these two equations for A and B,

$$A = \frac{\begin{vmatrix} 0 & 1 \\ 3 & -3 \end{vmatrix}}{\begin{vmatrix} 1 & 1 \\ -1 & -3 \end{vmatrix}} = \frac{-3}{-3+1} = \frac{3}{2}$$

$$B = \frac{\begin{vmatrix} 1 & 0 \\ -1 & 3 \end{vmatrix}}{-2} = -\frac{3}{2}$$

Therefore the homogeneous solution is found to be

$\frac{3}{2}(e^{-t} - e^{-3t})$ V.

• PROBLEM 7-3

For the circuit shown in Fig. 1, calculate the total response, $v_C(t)$, as the sum of the particular and homogeneous solutions of the second order differential equation describing the network. (Assume zero initial conditions).

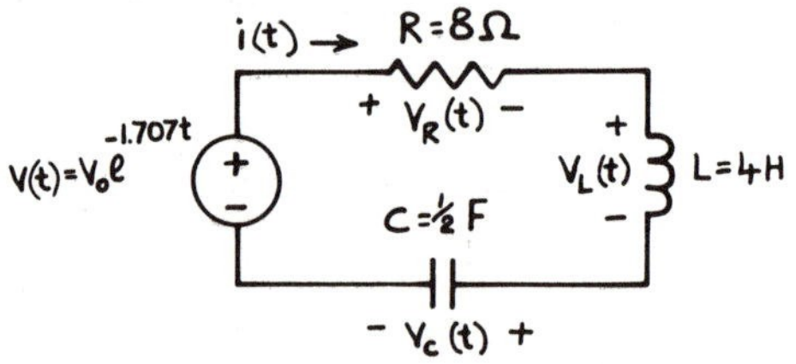

FiG.1

Solution: Writing a KVC equation for the single loop circuit yields

$$v(t) = v_R(t) + v_L(t) + v_C(t) \tag{1}$$

write,

$$v(t) = R\,i(t) + L\,\frac{di(t)}{dt} + v_C(t) \tag{2}$$

and eliminate i(t) by noting that

$$i(t) = C \frac{d\, v_C(t)}{dt} .$$

Hence, $v(t) = RC \frac{d\, v_C}{dt} + LC \frac{d^2 v_C}{dt^2} + v_C$ is the second order differential equation describing the circuit in terms of the desired response $v_C(t)$. Substituting values given for R, C, L and v(t) yields

$$\underbrace{v_0\, e^{-1.707t}}_{\text{driving function}} = \underbrace{2 \frac{d^2 v_C}{dt^2} + 4 \frac{d\, v_C}{dt} + v_C}_{\text{differential equation}} \qquad (3)$$

Setting the driving function to zero gives the homogeneous differential equation

$$0 = 2 \frac{d^2 v_C}{dt^2} + 4 \frac{d\, v_C}{dt} + v_C$$

which can be solved for the homogeneous solution. It has already been shown that for a series RLC circuit, the homogeneous solution is in the form

$$K_1 e^{s_1 t} + K_2 e^{s_2 t}$$

if $\frac{R}{2L} > \frac{1}{\sqrt{LC}}$ where $s_{1,2} = -\frac{R}{2L} \pm \sqrt{\left(\frac{R}{2L}\right)^2 - \frac{1}{LC}}$. Hence, we obtain $\frac{R}{2L} = 1$, $\frac{1}{LC} = \frac{1}{2}$ and $s_1 = -0.293$, $s_2 = -1.707$. The homogeneous solution is

$$K_1 e^{-0.293t} + K_2 e^{-1.707t}.$$

Using the method of undetermined coefficients, assume the particular solution is in the form $K_3 te^{-1.707t}$. This is suggested by the driving function, $v_0 e^{-1.707t}$. Substituting $v_C(t) = K_3 te^{-1.707t}$ in Eq. (3) gives K_3

$$v_0 e^{-1.707t} = 2 \frac{d^2}{dt^2} \left[K_3 te^{-1.707t}\right]$$

$$+ 4 \frac{d}{dt} \left[K_3 te^{-1.707t}\right] + K_3 te^{-1.707t}$$

$$v_0 e^{-1.707t} = 2\left[- K_3 1.707e^{-1.707t} - K_3 1.707e^{-1.707t}\right.$$

$$\left. + K_3 2.914te^{-1.707t}\right] + 4\left[K_3 e^{-1.707t} - K_3 1.707te^{-1.707t}\right]$$

$$+ K_3 te^{-1.707t}.$$

Combining terms and solving for K_3 gives,

$$K_3\left[- 2.828e^{-1.707}\right] = v_0 e^{-1.707t}$$

$$K_3 = - \frac{v_0 e^{-1.707t}}{2.828e^{-1.707t}} = - v_0\ 0.354.$$

Hence, the total solution is the sum of the homogeneous solution and the particular solution.

$$v_C(t) = K_1 e^{-0.293t} + K_2 e^{-1.707t} - v_0 0.354te^{-1.707t}$$

The constants K_1 and K_2 can be determined from initial conditions. Since $v_C(0) = 0$, one can substitute $t = 0$ into the total solution $v_C(t)$ above to obtain

$$v_C(0) = K_1 + K_2 = 0. \qquad (4)$$

The second initial condition relates to the inductor current $i_L(t)$. Since the inductor and capacitor are in series combination, it must be true that

$$i_L(t) = i_C(t) = C\,\frac{d\ v_C}{dt}$$

for all instants of time. In particular at $t = 0^+$, the inductor current is:

$$i_L(0^+) = C\,\frac{d\ v_C}{dt}\bigg|_{t=0^+}$$

$$= \frac{1}{2}\ [K_1\ (-0.243) + K_2(-1.707) - 0.354\ v_0] = 0$$

$$0.177\ v_0 = - K_1\ 0.146 - K_2\ 0.854 \qquad (5)$$

Solving Eqs. (4) and (5) simultaneously gives

$$K_1 = \frac{\begin{vmatrix} 0 & 1 \\ 0.177\ v_0 & -\ 0.854 \end{vmatrix}}{\begin{vmatrix} 1 & 1 \\ -0.146 & -\ 0.854 \end{vmatrix}} = \frac{-\ 0.177\ v_0}{-\ 0.854 + 0.146}$$

$$K_1 = 0.25\ v_0.$$

$$K_2 = \frac{\begin{vmatrix} 1 & 0 \\ -\ 0.146 & 0.177\ v_0 \end{vmatrix}}{-\ 0.708} = \frac{0.177\ v_0}{-\ 0.708} = -\ 0.25\ v_0$$

so that the total solution is given by

$$v_C(t) = 0.25\ v_0 e^{-0.293t} - 0.25\ v_0 e^{-1.707t} - 0.345\ v_0 te^{-1.707t}.$$

• PROBLEM 7-4

In the series RLC circuit it is known that $v_C(0) = 12.5$ V and $i(0) = -0.6$ A. Find $v_C(t)$ and $i(t)$ for all $t \geq 0$.

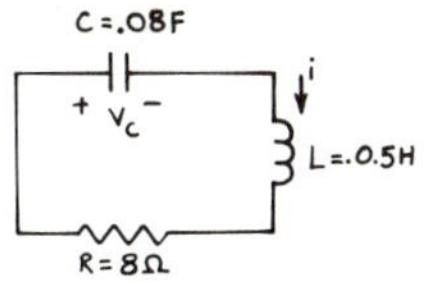

FIG.1

Solution: This problem requires solving for the natural response of a series RLC circuit with given initial conditions. The first step is to determine if the circuit is overdamped, critically damped or underdamped. Since we have shown that if $\left(\frac{R}{2L}\right)^2 > \frac{1}{LC}$, a series RCL circuit is underdamped with a response in the form

$$K_1 e^{s_1 t} + K_2 e^{s_2 t}, \text{ where}$$

$s_{1,2} = -\ \frac{R}{2L} \pm \sqrt{\left(\frac{R}{2L}\right)^2 - \frac{1}{LC}}$. Hence, for the circuit of Fig. 1,

$$\left(\frac{R}{2L}\right)^2 = \left(\frac{8}{(2)\ (0.5)}\right)^2 = 64, \text{ and } \frac{1}{LC} = 25$$

indicating that the circuit is underdamped with $s_{1,2} = -\ 8 \pm \sqrt{64 - 25}$, $s_1 = -\ 1.755$, and $s_2 = -\ 14.245$. Hence the response (in general form) for both $v_C(t)$ and $i(t)$ is

$$v_C(t) = K_1 e^{-1.755t} + K_2 e^{-14.245t} \quad ; \quad t \geq 0$$

$$i(t) = K_3 e^{-1.755t} + K_4 e^{-14.245t} \quad ; \quad t \geq 0.$$

We can determine the value of the constants K_1 and K_2 from the given initial conditions. First, $v_C(0) = 12.5$ V, thus

$$K_1 + K_2 = 12.5 \quad \text{at } t = 0 \qquad (1)$$

Second $i(0) = -0.6 = C \dfrac{d\, v_C(0)}{dt}$, thus

$$\frac{d\, v_C}{dt} = -1.755\, K_1 e^{-1.755t} - 14.245\, K_2 e^{-14,245t}$$

and $$\frac{d\, v_C(0)}{dt} = \frac{-0.6}{0.08} = -7.5 = -1.755\, K_1 - 14.245\, K_2 \qquad (2)$$

There are now two equations (Eqs. (1) and (2)) in two unknowns. Hence, by method of determinants,

$$K_1 = \frac{\begin{vmatrix} 12.5 & 1 \\ -7.5 & -14.245 \end{vmatrix}}{\begin{vmatrix} 1 & 1 \\ -1.755 & -14.245 \end{vmatrix}} = \frac{-178.063 + 7.5}{-14.245 + 1.755} = 13.65$$

and $$K_2 = \frac{\begin{vmatrix} 1 & 12.5 \\ -1.755 & -7.5 \end{vmatrix}}{-12.49} = -1.15$$

Hence, the voltage response across the capacitor is

$$v_C(t) = 13.65e^{-1.755t} - 1.15e^{-14.245t}$$

The determination of constants K_3 and K_4 for the current response is not necessary since the current can be obtained from the relation

$$i(t) = C \frac{d\, v_C}{dt}$$

$$= 0.08\ [13.65\ (-1.755)e^{-1.755t}$$

$$-1.15\ (-14.245)e^{-14.245t}]$$

$$i(t) = -1.91e^{-1.755t} + 1.31e^{-14.245t}$$

In the circuit of Fig. 1, the switch is thrown from position d to position a' at t = o. It is known that at $t = 0^-$, $v_C = 12.5$ V and $i(0^-) = -0.6$ A. Solve for $v_C(t)$ for $t \geq 0^+$.

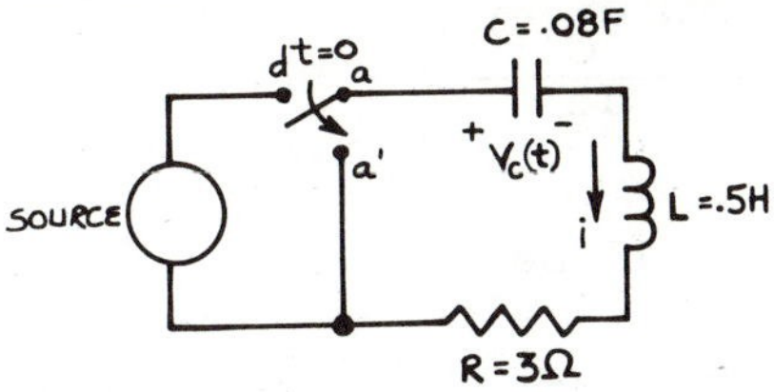

FIG.1

Solution: In this problem it is required to find the natural response of a series RLC circuit with given initial conditions. The first step is to determine if this RLC circuit is overdamped, critically damped, or underdamped. To determine this, use the following tests;

1) If $\left(\frac{R}{2L}\right)^2 > \frac{1}{LC}$, then the circuit is overdamped and the response is in the form $K_1e^{s_1t} + K_2e^{s_2t}$: where $s_{1,2} = -\frac{R}{2L} \pm \sqrt{\left(\frac{R}{2L}\right)^2 - \frac{1}{LC}}$.

2) If $\left(\frac{R}{2L}\right)^2 = \frac{1}{LC}$ then the circuit is critically damped and the response is in the form $K_1e^{st} + K_2te^{st}$ where $s = -\frac{R}{2L}$.

3) If $\left(\frac{R}{2L}\right)^2 < \frac{1}{LC}$, then the circuit is underdamped and the response is in the form $K_1e^{st}\cos\omega_d t + K_2e^{st}\sin\omega_d t$

where $\omega_d = \sqrt{\frac{1}{LC} - \left(\frac{R}{2L}\right)^2}$ and $s = -\frac{R}{2L}$.

Since $\left(\frac{R}{2L}\right)^2 = \left(\frac{3}{2\ (0.5)}\right)^2 = 9$ is less than $\frac{1}{LC} = \frac{1}{(0.5)(0.08)}$ $= 25$, the circuit is underdamped and the response desired, $v_C(t)$, is in the form

$$K_1e^{st}\cos\omega_d t + K_2e^{st}\sin\omega_d t$$

since $\omega_d = \sqrt{25 - 9} = \sqrt{16} = 4$ and $s = -\frac{R}{2L} = -3$. Hence,

$$v_C(t) = K_1e^{-3t}\cos 4t + K_2e^{-3t}\sin 4t. \qquad (1)$$

The constants K_1 and K_2 can be found by applying the initial conditions. Since we are told that $v_C(0) = 12.5$ V,

$$K_1 = 12.5$$

from Eq. (1).

K_2 can be determined by applying the initial condition $i(0) = -0.6$ A. Since

$$i(0) = C\frac{d\,v_C(0)}{dt} = -0.6,$$

$\frac{d\,v_C(t)}{dt}$ will be

$$-3K_1e^{-3t}\cos 4t - 4K_1e^{-3t}\sin 4t - 3K_2e^{-3t}\sin 4t$$

$$+\ 4K_2e^{-3t}\cos 4t.$$

Hence, $\quad C\frac{d\,v_C(0)}{dt}$ is

$0.08\,[-3K_1 + 4K_2] = -0.6$. Substituting $K_1 = 12.5$ gives

$$-3 + 0.32\,K_2 = -0.6$$

$$0.32\,K_2 = 2.4$$

$$K_2 = 7.5.$$

Hence, the desired response is

$$v_C(t) = 12.5e^{-3t}\cos 4t + 7.5e^{-3t}\sin 4t.$$

PARALLEL RLC

• PROBLEM 7-6

Determine values α, ω_0 and s_1 for a parallel RLC circuit in which $C = 0.1\ \mu F$, $R = 1k\Omega$ and $L =$: (a) 0.625 H (b) 0.4 H; (c) 0.256 H.

Solution: For a parallel RLC circuit, write a KCL equation

$$\frac{v}{R} + \frac{1}{L}\int_0^t v\,dt - i_0 + C\frac{dv}{dt} = 0$$

given $v(0^+) = v_0$; $i(0^+) = I_0$. Differentiating with respect to time gives a linear, second-order, homogeneous differential equation.

$$C \frac{dv^2}{dt^2} + \frac{1}{R} \frac{dv}{dt} + \frac{1}{L} v = 0$$

Assuming that the solution is in the form

$v = Ae^{st}$, we have

$$CAs^2 e^{st} + \frac{1}{R} Ase^{st} + \frac{1}{L} Ae^{st} = 0$$

$$Ae^{st} \left(Cs^2 + \frac{1}{R} s + \frac{1}{L}\right) = 0.$$

The characteristic equation

$$Cs^2 + \frac{1}{R} s + \frac{1}{L} = 0$$

can be solved for s using the quadratic formula.

The solution

$$s_1, s_2 = \frac{-b \pm \sqrt{b^2 - 4aC}}{2a}$$

can be written as;

$$s_1, s_2 = \frac{-b}{2a} \pm \sqrt{\left(\frac{b}{2a}\right)^2 - \frac{C}{a}}$$

where $\quad b = \frac{1}{R}; \; a = C; \; C = \frac{1}{L}$

$$s_1, s_2 = -\frac{1}{2RC} \pm \sqrt{\left(\frac{1}{2RC}\right)^2 - \frac{1}{LC}}. \qquad (1)$$

Now, $\frac{1}{2RC}$ is called the exponential damping coefficient α, and $\frac{1}{\sqrt{LC}}$ the resonant frequency ω_0. Substituting these terms into Eq. (1), yields

$$s_1 = -\alpha + \sqrt{\alpha^2 - \omega_0^2}$$

$$s_2 = -\alpha - \sqrt{\alpha^2 - \omega_0^2}$$

a) $$\alpha = \frac{1}{2RC} = \frac{1}{(2)(1 \times 10^3)(.1 \times 10^{-6})} = 5 \times 10^3$$

$$\omega_0 = \frac{1}{\sqrt{LC}} = \frac{1}{\sqrt{.625\,(.1 \times 10^{-6})}} = 4 \times 10^3$$

$$s_1 = -\alpha + \sqrt{\alpha^2 - \omega_0^{\,2}} = -5 \times 10^3 + \sqrt{25 \times 10^6 - 16 \times 10^6}$$

$$= -5 \times 10^3 + 3 \times 10^3$$

$$s_1 = -2 \times 10^3$$

b) $$\alpha = \frac{1}{2RC} = \frac{1}{(2)(1 \times 10^3)(.1 \times 10^{-6})} = 5 \times 10^3$$

$$\omega_0 = \frac{1}{\sqrt{LC}} = \frac{1}{\sqrt{(.4)\ .1 \times 10^{-6}}} = 5 \times 10^3$$

$$s_1 = -\alpha + \sqrt{\alpha^2 - \omega_0^{\,2}} = 5 \times 10^3 + \sqrt{25 \times 10^6 - 25 \times 10^6}$$

$$s_1 = -5 \times 10^3$$

c) $$\alpha = \frac{1}{2RC} = 5 \times 10^3$$

$$\omega_0 = \frac{1}{\sqrt{LC}} = \frac{1}{\sqrt{.256\ (.1 \times 10^{-6})}} = 6.25 \times 10^3$$

$$s_1 = -\alpha + \sqrt{\alpha^2 - \omega_0^{\,2}} = -5 \times 10^3 + \sqrt{25 \times 10^6 - 39.06 \times 10^6}$$

$$= -5 \times 10^3 + \sqrt{-14.06 \times 10^6}$$

$$s_1 = -5 \times 10^3 + j\ 3.75 \times 10^3$$

• PROBLEM 7-7

Find the homogeneous solution to a differential equation for the voltage $v_C(t)$ across a $\frac{1}{2}$ F capacitor in a parallel RLC circuit where R = 1Ω, L = 4H, $i_L(0) = \frac{1}{4}$ A, and $v_C(0) = 0$.

Solution: Since the homogeneous solution of an RLC circuit is of interest start the formulation of the differential equation by writing

$$i_L + i_R + i_C = 0, \text{ where}$$

$$i_C = C\frac{dv_C}{dt},\ i_R = \frac{v_C}{R} \text{ and } i_L = v_C(0) + \frac{1}{L}\int_0^t v_C\, d\tau$$

Hence, $\frac{1}{L}\int_0^t v_C \, d\tau + \frac{v_C}{R} + C\frac{dv_C}{dt} = 0$

can be put in the form of a second order differential equation by taking its derivative with respect to time, thus

$$\frac{v_C}{L} + \frac{1}{R}\frac{dv_C}{dt} + C\frac{d^2Vc}{dt^2} = 0$$

is our homogeneous equation for v_C. This solution must be in the form $Ae^{s_1 t} + Be^{s_2 t}$ when $\alpha^2 = \left(\frac{1}{2RC}\right)^2 > \omega_0^{\,2} = \frac{1}{LC}$. Since $\alpha^2 = 1$ and $\omega_0^{\,2} = \frac{1}{2}$ this is the proper solution form for our circuit. s_1 and s_2 are

$$s_1, s_2 = -\alpha \pm \sqrt{\alpha^2 - \omega_0^{\,2}}$$

$$s_1, s_2 = -1 \pm \sqrt{1 - \frac{1}{2}}$$

$$s_1, s_2 = -1 \pm \frac{1}{\sqrt{2}}$$

$$s_1 = -1 + \frac{1}{\sqrt{2}}$$

$$s_2 = -1 - \frac{1}{\sqrt{2}}$$

Hence, $v_C(t) = Ae^{(-1 + \frac{1}{\sqrt{2}})t} + Be^{(-1 - \frac{1}{\sqrt{2}})t}$.

At $t = 0$ $v_C(0) = A + B = 0$ and since

$$\frac{i_L}{C} = \frac{dv_C}{dt} = (-1 + \frac{1}{\sqrt{2}})\,Ae^{(-1 + \frac{1}{\sqrt{2}})t}$$

$$+ (-1 - \frac{1}{\sqrt{2}})\,B^{(-1 - \frac{1}{\sqrt{2}})t}$$

$$\frac{i_L(0)}{C} = \frac{dv_C(0)}{dt} - (-1 + \frac{1}{\sqrt{2}})\,A - (1 + \frac{1}{\sqrt{2}})\,B$$

$\frac{\frac{1}{4}}{\frac{1}{2}} = \frac{1}{2} = (-1 + \frac{1}{\sqrt{2}})\,A - (1 + \frac{1}{\sqrt{2}})\,B$, we can solve the two equations above for A and B.

$$A = \frac{\begin{vmatrix} 0 & 1 \\ \frac{1}{2} & -(1 + \frac{1}{\sqrt{2}}) \end{vmatrix}}{\begin{vmatrix} 1 & 1 \\ (-1 + \frac{1}{\sqrt{2}}) & -(1 + \frac{1}{\sqrt{2}}) \end{vmatrix}} = \frac{-\frac{1}{2}}{-(1 + \frac{1}{\sqrt{2}}) - (-1 + \frac{1}{\sqrt{2}})}$$

$$= \frac{1}{2\sqrt{2}}$$

$$B = -\frac{1}{2\sqrt{2}}$$

Therefore, the homogeneous solution is found to be

$$\frac{1}{2\sqrt{2}} e^{(-1 + \frac{1}{\sqrt{2}})t} - \frac{1}{2\sqrt{2}} e^{(-1 - \frac{1}{\sqrt{2}})t}.$$

• PROBLEM 7-8

A parallel circuit is comprised of the elements; L = 10H, R = 320Ω, C = $\frac{125}{8}$ μF. The initial capacitor voltage is $v_C(0^+) = -160$ V. The initial value of the capacitor current is $i_C(0^+) = 0.7$ A, where i_C and v_C are related by the passive sign convention. (a) Find the initial energy storage in the inductor and capacitor. (b) At what value of t is $v_C = 0$? (c) At what value of t is v_C a positive maximum?

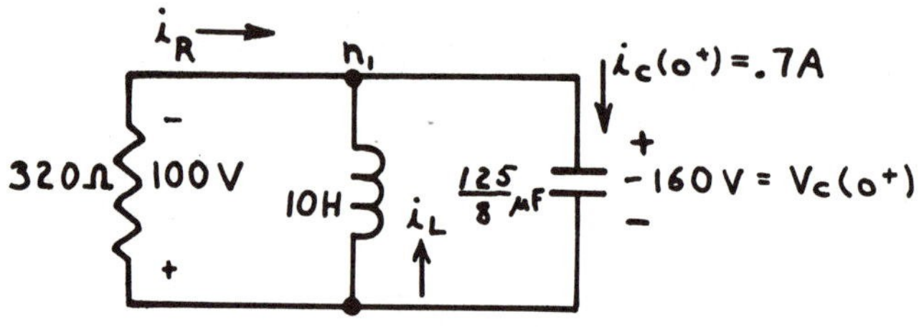

FIG. 1

Solution: Fig. 1 shows the RLC circuit described with $i_C(0^+)$ and $v_C(0^+)$ related by the passive sign convention.

Since the voltage across the capacitor is given, the current in the resistor can be found.

$$i_R = \frac{160 \text{ V}}{360 \ \Omega} = 0.5 \text{ A}$$

Summing the currents at node n_1 yields

$$i_L = - i_R + i_C = - 0.5 + 0.7 = 0.2 \text{ A.}$$

We find $\alpha = \frac{1}{2RC} = \frac{1}{2(320)(\frac{125}{8} \times 10^{-6})} = 100$

$$\omega_0 = \frac{1}{\sqrt{LC}} = \frac{1}{\sqrt{10(\frac{125}{8} \times 10^{-6})}} = 80$$

$$s_1 = - \alpha + \sqrt{\alpha^2 - \omega_0^2} = - 100 + 60 = - 40$$

$$s_2 = - \alpha - \sqrt{\alpha^2 - \omega_0^2} = - 100 - 60 = - 160$$

giving the general form of the response for an overdamped parallel RLC circuit

$$v(t) = A_1 e^{-40t} + A_2 e^{-160t}$$

$$v(0) = - 160 = A_1 + A_2 \qquad (1)$$

$$\frac{dv(0)}{dt} = \frac{i_C(0^+)}{C} = \frac{0.7}{\frac{125}{4} \times 10^{-6}} = - 40 A_1 - 160 A_2$$

$$4.48 \times 10^4 = - 40 A_1 - 160 A_2 \quad (2)$$

Solving Eqs. (1) and (2)

$$A_1 = \frac{\begin{vmatrix} -160 & 1 \\ 4.48 \times 10^4 & -160 \end{vmatrix}}{\begin{vmatrix} 1 & 1 \\ -40 & -160 \end{vmatrix}} = \frac{25600 - 4.48 \times 10^4}{- 160 + 40}$$

$$= \frac{- 1.92 \times 10^4}{- 120}$$

$$A_1 = 160$$

$$A_2 = \frac{\begin{vmatrix} 1 & -160 \\ -40 & 4.48 \times 10^4 \end{vmatrix}}{\begin{vmatrix} 1 & 1 \\ 40 & -160 \end{vmatrix}} = \frac{4.48 \times 10^4 - 6400}{- 120}$$

$$A_2 = - 320$$

$$v(t) = 160 e^{-40t} - 320 e^{-160t} \qquad (3)$$

(a) Since $E_0 = \frac{1}{2} C(v_0)^2$ and $v_C(0^+) = - 160$ V

$$E_0 = \frac{1}{2}\left(\frac{125}{8} \times 10^{-6}\right)(-160)^2 = 0.2\ J$$

$$E_0 = \frac{1}{2} L (i_0)^2 \text{ and } i_L\ (0^+) = 0.2\ A$$

$$E_0 = \frac{1}{2}(10)\ (.2)^2 = 0.2\ J$$

b) Set v(t) = 0 in equation (3), and solve for t.

$$0 = 160\ e^{-40t} - 320\ e^{-160t}$$

$$320\ e^{-160t} = 160\ e^{-40t}$$

taking the natural log of both sides,

$$\ell n\ 320 - 160t = \ell n 160 - 40t$$

$$-40t + 160t = \ell n\ 320 - \ell n 160$$

$$t = \frac{\ell n\ 320 - \ell n 160}{120} = \frac{\ell n \frac{320}{100}}{120}$$

$$t = \frac{\ell n 2}{120} = \frac{.693}{120} = 5.78\ \text{ms}$$

c) To find the maximum of a function, set the derivative of the function equal to zero.

$$\frac{dv}{dt} = 0 = (-40)\ (160)e^{-40t} - 320\ (-160)e^{-160t}$$

Solving for t

$$6400\ e^{-40t} = 51200\ e^{-160t}$$

$$-40t + 160t = \ell n 51200 - \ell n 6400$$

$$120t = \ell n \frac{51200}{6400}$$

$$t = \frac{\ell n\ 8}{120} = \frac{2.08}{120} = 17.3\ \text{ms.}$$

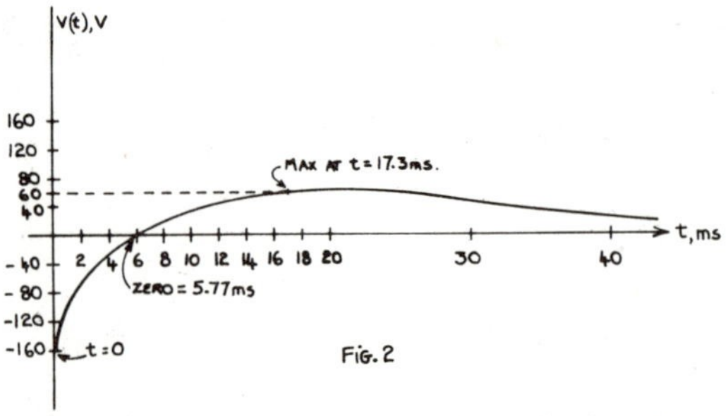

FIG. 2

Fig. 2 shows the response v(t) and critical points.

(a) The circuit shown on the response curve of Fig. 1 is modified by increasing the size of the inductor to 1400/23H. If the initial conditions remain unchanged, find and sketch v(t); compare with the response of the original circuit.
(b) What will the response become as L approaches infinity?

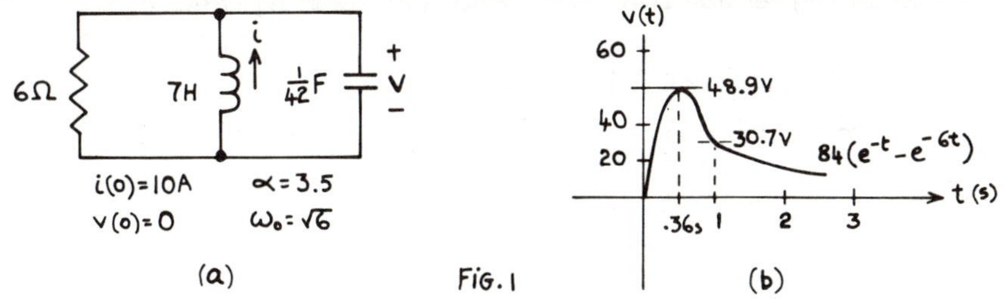

(a) FIG. 1 (b)

Solution: α remains the same, α = 3.5

$$\omega_0 = \frac{1}{\sqrt{LC}} = \frac{1}{\sqrt{\frac{1400}{23}\left(\frac{1}{42}\right)}} = \sqrt{.69}$$

giving: $s_1 = -0.1$

$s_2 = -6.9$

Then the solution is of the form

$$v(t) = A_1 e^{-0.1t} + A_2 e^{-6.9t}$$

and $\quad v(0) = 0 = A_1 + A_2 \qquad (1)$

to obtain another equation in A_1 and A_2, we apply KCL at n_1 in Fig. 2.

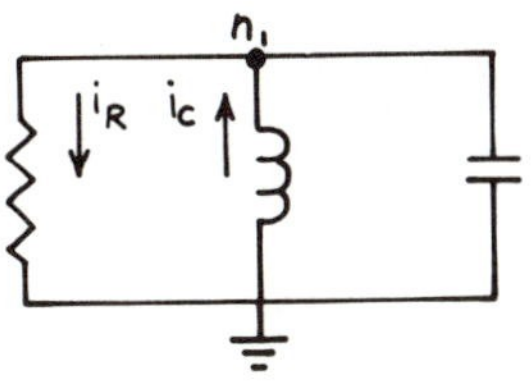

FIG. 2

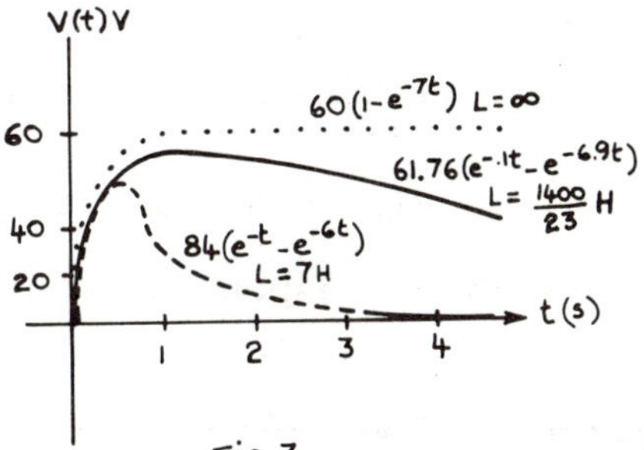

FIG. 3

$$i_R + i_C = i_L$$

$$i_C = i_L - i_R$$

$$i_C(0^+) = i_L(0^+) + v(0^+)\ R$$

$$i_C(0^+) = 10 - 0$$

$$\frac{i_C(0^+)}{C} = \frac{dv(0^+)}{dt} = -0.1\,A_1 - 6.9\,A_2$$

$$\frac{i_C(0^+)}{C} = \frac{10}{\frac{1}{42}} = 420 = -0.1\,A_1 - 6.9\,A_2 \quad (2)$$

Solving equation (1), and (2) for A_1 and A_2.

$$A_1 = \frac{\begin{vmatrix} 420 & -6.9 \\ 0 & 1 \end{vmatrix}}{\begin{vmatrix} -.1 & -6.9 \\ 1 & 1 \end{vmatrix}} = \frac{420}{-.1 + 6.9} = 61.76$$

$$A_2 = \frac{\begin{vmatrix} -.1 & 420 \\ 1 & 0 \end{vmatrix}}{6.8} = \frac{-420}{6.8} = -61.76$$

$$v(t) = 61.76\,(e^{-0.1t} - e^{-6.9t})$$

We find

$$v_{C\,max}(.622) = 57.2\text{ V.}$$

When $L \to \infty$, $\omega_0 \to 0$ so that

$$s_1 = -\alpha + \alpha = 0$$

$$s_2 = -\alpha - \alpha = -2\alpha = -2(3.5) = -7$$

$$v(t) = A_1 + A_2 e^{-7t}$$

$$v(0) = 0 = A_1 + A_2$$

$$\frac{dv(0)}{dt} = 420 = -7A_2\ ;\ A_2 = -\frac{420}{7} = -60;\ A_1 = 60$$

$$v(t) = 60\,(1 - e^{-7t})$$

The three responses found are shown in Fig. 3.

• PROBLEM 7-10

For the circuit shown in Fig. 1, calculate the total response, $v_C(t)$, as the sum of the particular and homogeneous solutions of the second order differential equation describing the network. The initial conditions are, $i_L(0) = 3.5$A and $v_C(0) = 10$V.

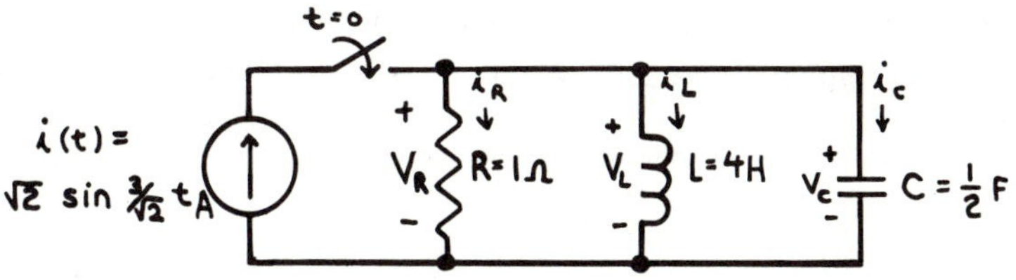

Fig. 1

Solution: Applying KCL at the top node yields

$$i(t) = i_R(t) + i_L(t) + i_C(t).$$

We can write,

$$i(t) = \frac{1}{R} v_R(t) + i_L(0) + \frac{1}{L}\int_0^t v_L(t)\,dt + C\,\frac{d\,v_C(t)}{dt} \qquad (1)$$

but since $v_C(t) = v_L(t) = v_R(t)$, replace them with v(t) in the integro-differential equation (1) above. Differentiating both sides of Eq. (1) once with respect to t eliminates the integral and gives the required second order equation in terms of the desired response v(t).

$$\frac{di(t)}{dt} = \frac{1}{L} v(t) + \frac{1}{R}\frac{dv(t)}{dt} + C\,\frac{d^2v(t)}{dt^2}. \qquad (2)$$

The associated homogeneous equation is

$$0 = \frac{1}{L} v(t) + \frac{1}{R}\frac{dv(t)}{dt} + C\,\frac{d^2v(t)}{dt^2}.$$

It has already been shown that for a parallel RLC circuit, the homogeneous solution is in the form

$$K_1 e^{s_1 t} + K_2 e^{s_2 t}$$

if $\frac{1}{2RC} > \frac{1}{\sqrt{LC}}$ where $s_{1,2} = -\frac{1}{2RC} \pm \sqrt{\left(\frac{1}{2RC}\right)^2 - \frac{1}{LC}}$.

Hence, $\frac{1}{2RC} = 1$, $\frac{1}{LC} = \frac{1}{2}$ and $s_1 = -0.293$, $s_2 = -1.707$ are obtained. The homogeneous solution is

$$K_1 e^{-0.293t} + K_2 e^{-1.707t} \qquad (3)$$

Using the method of undetermined coefficients, assume that the particular solution is in the form

$$K_3 \sin\frac{3}{\sqrt{2}}t + K_4 \cos\frac{3}{\sqrt{2}}t. \qquad (4)$$

(This is suggested by the sinusoidal driving function $\sqrt{2}\sin\frac{3}{\sqrt{2}}t$.) Substituting $v(t) = K_3\sin\frac{3}{\sqrt{2}}t + K_4\cos\frac{3}{\sqrt{2}}t$ into Eq. (2) enables us to determine the coefficients. Hence,

$$\frac{d}{dt}\left[\sqrt{2}\sin\frac{3}{\sqrt{2}}t\right] = \frac{1}{2}\frac{d^2}{dt^2}\left[K_3\sin\frac{3}{\sqrt{2}}t + K_4\cos\frac{3}{\sqrt{2}}t\right]$$

$$+ \frac{d}{dt}\left[K_3\sin\frac{3}{\sqrt{2}}t + K_4\cos\frac{3}{\sqrt{2}}t\right] + \frac{1}{4}\left[K_3\sin\frac{3}{\sqrt{2}}t\right.$$

$$\left.+ K_4\cos\frac{3}{\sqrt{2}}t\right]$$

$$3\cos\frac{3}{\sqrt{2}}t = \frac{1}{2}\left[-\frac{9}{2}K_3\sin\frac{3}{\sqrt{2}}t - \frac{9}{2}K_4\cos\frac{3}{\sqrt{2}}t\right]$$

$$+ \left[K_3\frac{3}{\sqrt{2}}\cos\frac{3}{\sqrt{2}}t - K_4\frac{3}{\sqrt{2}}\sin\frac{3}{\sqrt{2}}t\right]$$

$$+ \frac{1}{4}\left[K_3\sin\frac{3}{\sqrt{2}}t + K_4\cos\frac{3}{\sqrt{2}}t\right].$$

Combining terms gives

$$3\cos\frac{3}{\sqrt{2}}t = \left(-2K_3 - \frac{3}{\sqrt{2}}K_4\right)\sin\frac{3}{\sqrt{2}}t$$

$$+ \left(\frac{3}{\sqrt{2}}K_3 - 2K_4\right)\cos\frac{3}{\sqrt{2}}t.$$

Note that in order for the above equation to hold true,

$$-2K_3 - \frac{3}{\sqrt{2}}K_4 = 0$$

and $$\frac{3}{\sqrt{2}}K_3 - 2K_4 = 3.$$

Therefore,

$$K_3 = \frac{\begin{vmatrix} 0 & -\frac{3}{\sqrt{2}} \\ 3 & -2 \end{vmatrix}}{\begin{vmatrix} -2 & -\frac{3}{\sqrt{2}} \\ \frac{3}{\sqrt{2}} & -2 \end{vmatrix}} = \frac{\frac{9}{\sqrt{2}}}{4 + \frac{9}{2}} = \frac{9}{17}\sqrt{2}$$

$$K_4 = \frac{\begin{vmatrix} -2 & 0 \\ \frac{3}{\sqrt{2}} & 3 \end{vmatrix}}{\frac{17}{2}} = \frac{-6}{\frac{17}{2}} = -\frac{12}{17}$$

Hence, the particular solution is

$$\frac{9}{17}\sqrt{2}\sin\frac{3}{\sqrt{2}}t - \frac{12}{17}\cos\frac{3}{\sqrt{2}}t. \qquad (5)$$

The total solution is obtained by summing the homogeneous and particular solutions Eq. (3) and (5) respectively. The total solution is

$$v(t) = K_1 e^{-0.293t} + K_2 e^{-1.707t}$$

$$+ \frac{9}{17}\sqrt{2}\sin\frac{3}{\sqrt{2}}t - \frac{12}{17}\cos\frac{3}{\sqrt{2}}t.$$

The constants K_1 and K_2 can be determined from initial conditions. Since $v_C(0) = v(0) = 10$

$$v(0) = K_1 + K_2 - \frac{12}{17} = 10$$

or $$K_1 + K_2 = \frac{182}{17}. \qquad (6)$$

A second relationship between K_1 and K_2 must be determined by making use of the initial condition on the inductor. Since $i_L(0) = i_L(0^+) = 3.5$

and $i_L(0+) = -i_R(0^+) - i_C(0^+) + i(0^+) = 3.5$

from the KCL equation. Solve for $i_C(0^+) = i(0^+) - i_R(0^+) - 3.5$

where $i(0^+) = 0$,

and $i_R(0^+) = \frac{v_R(0^+)}{R} = \frac{v_C(0^+)}{R} = 10.$

Hence, $i_C(0^+) = -10 - 3.5 = -13.5$ A.

Now, $$i_C(0^+) = C\frac{d\,v_C(0^+)}{dt} = \frac{1}{2}\left[-0.293K_1 - 1.707K_2 + \frac{9\sqrt{2}}{17}\,\frac{3}{\sqrt{2}}\right] = -13.5$$

gives the second relationship,

$$- K_1 0.293 - K_2\ 1.707 = - \frac{486}{17}. \tag{7}$$

Solving Eqs. (6) and (7) simultaneously gives

$$K_1 = \frac{\begin{vmatrix} \frac{182}{17} & 1 \\ -\frac{486}{17} & -1.707 \end{vmatrix}}{\begin{vmatrix} 1 & 1 \\ -0.293 & -1.707 \end{vmatrix}} = \frac{\frac{182}{17\sqrt{2}} + \frac{304}{17}}{\sqrt{2}}$$

$$K_1 = \frac{91}{17} - \frac{304}{17\sqrt{2}} = -7.29$$

$$K_2 = \frac{\begin{vmatrix} 1 & \frac{182}{17} \\ -0.293 & -\frac{486}{17} \end{vmatrix}}{-\sqrt{2}} = \frac{-\frac{182}{17\sqrt{2}} - \frac{304}{17}}{-\sqrt{2}}$$

$$K_2 = \frac{91}{17} + \frac{304}{17\sqrt{2}} = 18.$$

Hence, the total response is given by,

$$v(t) = -7.29e^{-0.293t} + 18e^{-1.707t} + \frac{9}{17}\sqrt{2}\sin\frac{3}{\sqrt{2}}t - \frac{12}{17}\cos\frac{3}{\sqrt{2}}t \text{ V}.$$

COMBINATION

• PROBLEM 7-11

Calculate the current $i_L(t)$ and the voltage $v_C(t)$ in the parallel RLC circuit in Fig. 1 in terms of the circuit elements R, C, L, and initial conditions $v_C(t_0) = V$, $i_L(t_0) = I$. Solve for the conditions:

a) $\left(\frac{1}{2RC}\right)^2 > \frac{1}{LC}$ overdamped,

b) $\left(\frac{1}{2RC}\right)^2 = \frac{1}{LC}$ critically damped, and

c) $\left(\frac{1}{2RC}\right)^2 < \frac{1}{LC}$ underdamped.

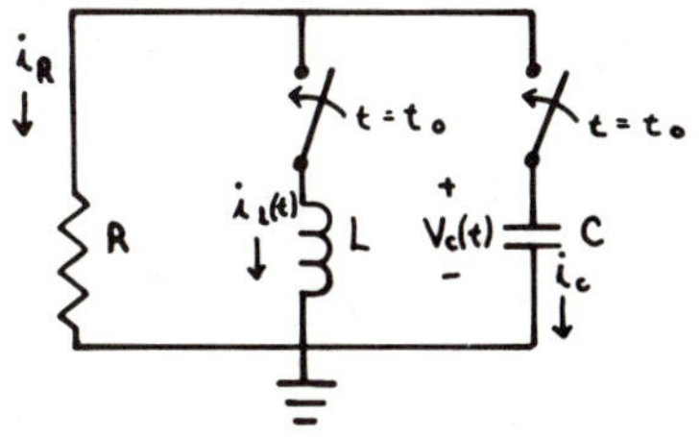

Fig.1

Solution: The solution requires finding the natural response of a parallel RLC circuit with given initial values under the conditions (a), (b), and (c). These, we shall see, are the overdamped, critically damped, and underdamped responses respectively.

Our first step is to write the differential equation governing the circuit. Using KCL, write the sum of the currents in the three branches.

$$i_R + i_L + i_C = 0$$

$$\frac{v_C(t)}{R} + I + \frac{1}{L}\int_{t_0}^{t} v_C(t)\,dt + C\,\frac{dv_C(t)}{dt} = 0. \quad (1)$$

Differentiating with respect to t yields

$$C\,\frac{d^2v_C}{dt} + \frac{1}{R}\,\frac{dv_C}{dt} + \frac{1}{L}\,v_C = 0. \quad (2)$$

Since $v_C(t) = L\,\frac{di_L(t)}{dt}$, one can write Eq. (1) as

$$LC\,\frac{d^2i_L}{dt} + \frac{L}{R}\,\frac{di_L}{dt} + i_L = 0. \quad (3)$$

Dividing Eq. (2) by C and Eq. (3) by LC gives

$$\frac{d^2v_C}{dt^2} + \frac{1}{RC}\,\frac{dv_C}{dt} + \frac{1}{LC}\,v_C = 0 \quad (4)$$

$$\frac{d^2i_L}{dt^2} + \frac{1}{RC}\,\frac{di_L}{dt} + \frac{1}{LC}\,i_L = 0. \quad (5)$$

For condition (a), $\left(\frac{1}{2RC}\right)^2 > \frac{1}{LC}$, since

$$s_1 = -\frac{1}{2RC} + \sqrt{\left(\frac{1}{2RC}\right)^2 - \frac{1}{LC}} \quad \text{and}$$

$$s_2 = -\frac{1}{2RC} - \sqrt{\left(\frac{1}{2RC}\right)^2 - \frac{1}{LC}}$$

then, the overdamped response must be in the form

$$K_1 e^{s_1(t - t_0)} + K_2 e^{s_2(t - t_0)}.$$

The constants K_1 and K_2 can be evaluated from the initial conditions $v_C(t_0) = V$, $i_L(t_0) = I$. If,

$$i_L(t) = K_1 e^{s_1(t - t_0)} + K_2 e^{s_2(t - t_0)}; \; t \geq t_0$$

then $i_L(t_0) = I = K_1 + K_2$

and $\dfrac{di_L(t_0)}{dt} = s_1 K_1 + s_2 K_2 = \dfrac{V}{L}.$

Hence,

$$K_1 = \frac{\begin{vmatrix} I & 1 \\ \frac{V}{L} & s_2 \end{vmatrix}}{\begin{vmatrix} 1 & 1 \\ s_1 & s_2 \end{vmatrix}} = \frac{s_2 I - \frac{V}{L}}{s_2 - s_1}$$

and

$$K_2 = \frac{\begin{vmatrix} 1 & I \\ s_1 & \frac{V}{L} \end{vmatrix}}{s_2 - s_1} = \frac{\frac{V}{L} - s_1 I}{s_2 - s_1}$$

The inductor current for condition (a) is

$$i_L(t) = \frac{s_2 I - \frac{V}{L}}{s_2 - s_1} e^{s_1(t - t_0)} + \frac{\frac{V}{L} - s_1 I}{s_2 - s_1} e^{s_2(t - t_0)}. \quad (6)$$

For the capacitor voltage, if

$$v_C(t) = K_3 e^{s_1(t - t_0)} + K_4 e^{s_2(t - t_0)}$$

then $v_C(t_0) = V = K_3 + K_4$

and $C\dfrac{dv_C(t_0)}{dt} = i_C(t_0)$, since

$$i_C(t_0) + i_R(t_0) + i_L(t_0) = 0$$

$$i_C(t_0) = -i_R(t_0) - I \quad \text{where } i_R(t_0) = \frac{v_C(t_0)}{R} = \frac{V}{R}.$$

Hence, $\dfrac{dv_C(t_0)}{dt} = -\dfrac{I}{C} - \dfrac{V}{RC} = s_1 K_3 + s_2 K_4.$

Solving for K_3 and K_4 by method of determinants gives

$$K_3 = \frac{\begin{vmatrix} V & 1 \\ -\frac{I}{C} - \frac{V}{RC} & s_2 \end{vmatrix}}{\begin{vmatrix} 1 & 1 \\ s_1 & s_2 \end{vmatrix}} = \frac{Vs_2 + \frac{I}{C} + \frac{V}{RC}}{s_2 - s_1}$$

and $$K_4 = \frac{\begin{vmatrix} 1 & V \\ s_1 & -\frac{I}{C} - \frac{V}{RC} \end{vmatrix}}{s_2 - s_1} = \frac{-\left[\frac{I}{C} + \frac{V}{RC} + s_1 V\right]}{s_2 - s_1}.$$

Therefore, the capacitor voltage for condition (a) is

$$v_C(t) = \frac{Vs_2 + \frac{I}{C} + \frac{V}{RC}}{s_2 - s_1} e^{s_1(t - t_0)} - \frac{Vs_1 + \frac{I}{C} + \frac{V}{RC}}{s_2 - s_1} e^{s_2(t - t_0)} \qquad (7)$$

For condition (b) the response must be of the form

$$K_5 e^{s(t - t_0)} + K_6 (t - t_0) e^{s(t - t_0)}$$

where $s = -\frac{1}{2RC}$.

We can determine the constants K_5 and K_6 by the given initial conditions.

For the inductor current, if

$$i_L(t) = K_5 e^{s(t - t_0)} + K_6 (t - t_0) e^{s(t - t_0)},$$

then $i_L(t_0) = I = K_5$.

Since $$L \frac{di_L(t_0)}{dt} = v_C(t_0) = V$$

$$\frac{V}{L} = \frac{di_L(t_0)}{dt} = s\,K_5 + K_6.$$

Hence, $$K_6 = \frac{V}{L} - s\,K_5 = \frac{V}{L} - s\,I.$$

The inductor current for condition (b) is

$$i_L(t) = \left[I + \left(\frac{V}{L} - s\,I \right) (t - t_0) \right] e^{s(t - t_0)}; \quad t \geq t_0. \quad (8)$$

For the capacitor voltage under condition (b) note that

if $\quad v_C(t) = K_7 e^{s(t - t_0)} + K_8 (t - t_0) e^{s(t - t_0)},$

then $v_C(t_0) = V = K_7.$

Since $i_C(t_0) = C \dfrac{d\, v_C(t_0)}{dt}$ and

$$i_C(t_0) = - i_R(t_0) - i_L(t_0) = - \frac{v_C(t_0)}{R} - i_L(t_0) = - \frac{V}{R} - I$$

then, $\dfrac{d\, v_C(t_0)}{dt} = - \dfrac{V}{RC} - \dfrac{I}{C} = s\,K_7 + K_8.$

Hence, $K_8 = - \dfrac{V}{RC} - \dfrac{I}{C} - s\,K_7 = - \left[\dfrac{V}{RC} + \dfrac{I}{C} + s\,V \right].$

The capacitor voltage for condition (b) is

$$v_C(t) = \left\{ V - \left[\frac{V}{RC} + \frac{I}{C} + s\,V \right] (t - t_0) \right\} e^{s(t - t_0)}; \; t \geq t_0. \quad (9)$$

For condition (c) (underdamped case) the response must be in the form

$$K_9 e^{s(t - t_0)} \cos\,[\omega_d (t - t_0)] + K_{10} e^{s(t - t_0)} \sin\,[\omega_d (t - t_0)]$$

where $s = - \dfrac{1}{2RC}$ and $\omega_d = \sqrt{\dfrac{1}{LC} - \left(\dfrac{1}{2RC} \right)^2}$. Determining the constants K_9 and K_{10} for the inductor current response follows a procedure similar to the previous. If

$$i_L(t) = K_9 e^{s(t - t_0)} \cos\,[\omega_d (t - t_0)]$$

$$+ K_{10} e^{s(t - t_0)} \sin\,[\omega_d (t - t_0)]$$

then $i_L(t_0) = I = K_9$

and $\dfrac{di_L(t_0)}{dt} = \dfrac{V}{L} = s\,K_9 + \omega_d\,K_{10}.$

Hence, $K_{10} = \left(\dfrac{V}{\omega_d L} - \dfrac{sK_9}{\omega_d} \right) = \dfrac{V}{\omega_d L} - \dfrac{sI}{\omega_d}.$

The inductor current under condition (c) is

$$i_L(t) = Ie^{s(t - t_0)}\cos[\omega_d(t - t_0)]$$

$$+ \left[\frac{V}{\omega_d L} - \frac{sI}{\omega_d}\right] e^{s(t - t_0)}\sin[\omega_d(t - t_0)];\ t \geq t_0 \qquad (10)$$

For the capacitor voltage under condition (c), it can be said that, if

$$v_C(t) = K_{11}e^{s(t - t_0)}\cos[\omega_d(t - t_0)]$$

$$+ K_{12}e^{s(t - t_0)}\sin[\omega_d(t - t_0),$$

then $v_C(t_0) = V = K_{11}$.

$$\text{Since, } C\frac{dv_C(t_0)}{dt} = i_C(t_0) = -i_R(t_0) - i_L(t_0)$$

$$= -\frac{v_C(t_0)}{R} - i_L(t_0) = -\frac{V}{R} - I$$

$$\frac{dv_C(t_0)}{dt} = -\frac{V}{RC} - \frac{I}{C} = s\,K_{11} + \omega_d\,K_{12}.$$

Hence,
$$K_{12} = -\left(\frac{V}{RC\omega_d} + \frac{I}{C\omega_d} + \frac{sK_{11}}{\omega_d}\right)$$

$$K_{12} = -\left(\frac{V}{RC\omega_d} + \frac{I}{C\omega_d} + \frac{sV}{\omega_d}\right).$$

The capacitor voltage under condition (c) is

$$v_C(t) = Ve^{s(t - t_0)}\cos[\omega_d(t - t_0)] - \left[\frac{V}{RC\omega_d} + \frac{I}{C\omega_d} + \frac{sV}{\omega_d}\right]$$

$$e^{s(t - t_0)}\sin[\omega_d(t - t_0)];\quad t \geq t_0 \qquad (11)$$

• PROBLEM 7-12

Calculate $i_L(t)$ for the circuit in Fig. 1 using differential equations.

Solution: Start by writing a KCL equation for the single node circuit in Fig. 2.

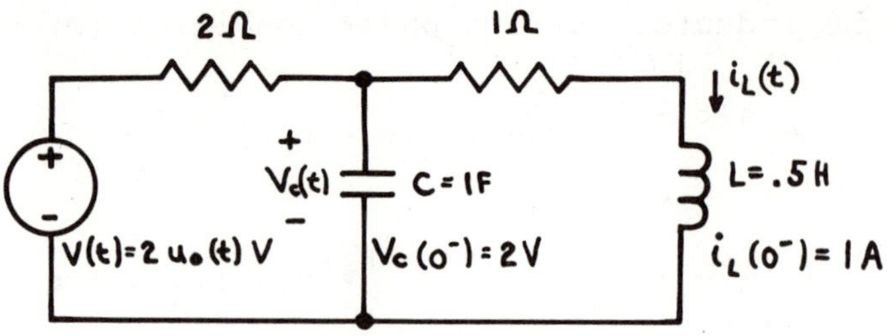

Fig. 1

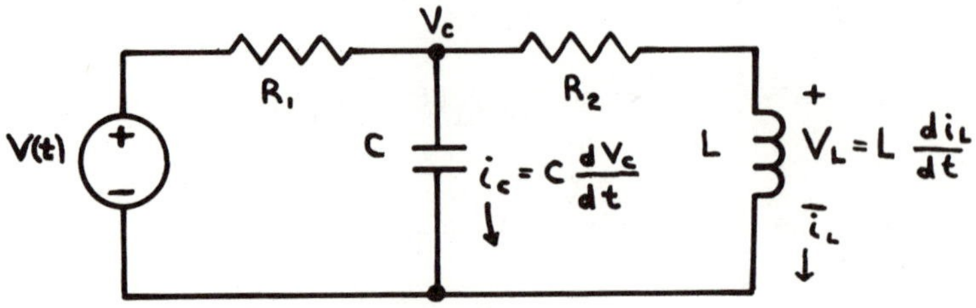

Fig. 2

$$\frac{v_C - v(t)}{R_1} + i_C + i_L = 0$$

$$\frac{v_C - v(t)}{R_1} + C \frac{dv_C}{dt} + \frac{v_C - v_L}{R_2} = 0$$

$$\frac{v_C - v(t)}{R_1} + C \frac{dv_C}{dt} + \frac{v_C - L \frac{di_L}{dt}}{R_2} = 0. \quad (1)$$

We can eliminate v_C from the equation above by writing a KVL equation for the loop shown in Fig. 3.

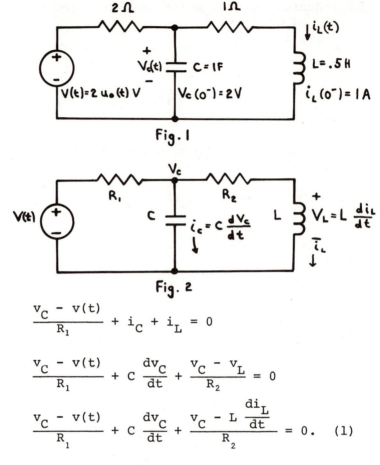

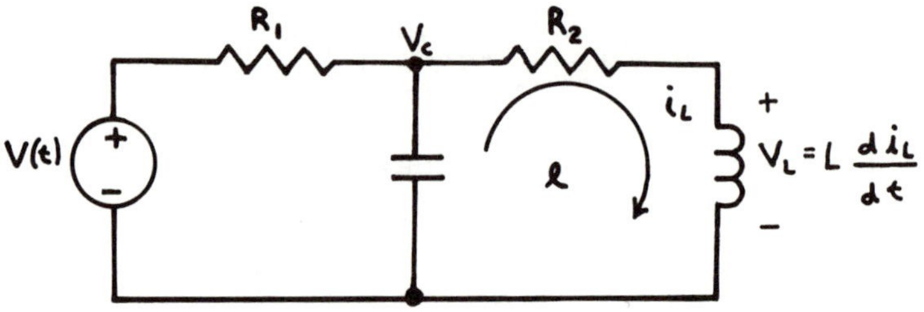

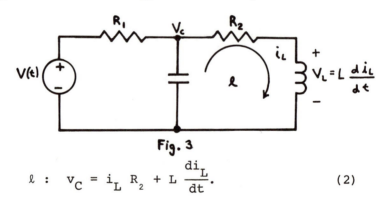

Fig. 3

$$\ell : \quad v_C = i_L R_2 + L \frac{di_L}{dt}. \quad (2)$$

Differentiating Eq. (2) with respect to time yields

$$\frac{dv_C}{dt} = R_2 \frac{di_L}{dt} + L \frac{d^2 i_L}{dt^2}. \quad (3)$$

Substituting (2) and (3) into (1) gives the second order differential equation for i_L,

$$v(t) = \frac{d^2 i_L}{dt^2} C L R_1 + \frac{di_L}{dt} (L + CR_1 R_2) + i_L (R_2 + R_1). \quad (4)$$

Substituting the values for R_1, R_2, C and L in (4) yields,

$$\frac{d^2 i_L}{dt^2} + \frac{5}{2} \frac{di_L}{dt} + 3 i_L = v(t). \quad (5)$$

The solution to the homogeneous second order differential equation is in the form

$$i_{Lh} = K_1 e^{s_1 t} + K_2 e^{s_2 t} \quad (6)$$

where s_1 and s_2 are found by solving the characteristic equation

$$s^2 + \frac{5}{2} s + 3 = 0$$

hence, $s_1 = -\frac{5}{4} + j\,1.2$ and $s_2 = -\frac{5}{4} - j\,1.2$.

The solution (6) can now be written as

$$i_{Lh} = Ae^{-\frac{5}{4}t} \sin 1.2t + Be^{-\frac{5}{4}t} \cos 1.2t$$

where K_1 and K_2 are chosen so that A and B are real.

Using the identities

$$\sin t = \frac{e^{jt} - e^{-jt}}{2j}$$

$$\cos t = \frac{e^{jt} + e^{-jt}}{2}$$

Since the applied voltage v(t) is a constant voltage, 2V for t > 0, the particular solution i_{Lp} must also be a constant current for t > 0. Hence, we assign

$$i_{Lp} = C. \quad (7)$$

Substituting (7) into (5) yields

$$3C = 2$$

hence, $C = \frac{2}{3}$.

The total solution is the sum of the homogeneous and particular solutions;

$$i_L(t) = \frac{2}{3} + Ae^{-\frac{5}{4}t} \sin 1.2t + Be^{-\frac{5}{4}t} \cos 1.2t. \quad (8)$$

The remaining constants A and B can be found by application of the given initial conditions, $i_L(0) = 1$ A and $v_C(0) = 2$ V. Substituting $t = 0$ into Eq. (8) gives the first initial condition, hence the equation

$$1 = \frac{2}{3} + B$$

and
$$B = \frac{1}{3}.$$

Apply the second initial condition by substituting (8) into (2) to find v_C as follows

$$v_C = i_L + \frac{1}{2}\frac{di_L}{dt}$$

$$v_C = \frac{2}{3} + Ae^{-\frac{5}{4}t} \sin 1.2t + \frac{1}{3} e^{-\frac{5}{4}t} \cos 1.2t$$

$$- \frac{5}{8} Ae^{-\frac{5}{4}t} \sin 1.2t + 0.6\, Ae^{-\frac{5}{4}t} \cos 1.2t$$

$$- \frac{5}{24} e^{-\frac{5}{4}t} \cos 1.2t - 0.2\, e^{-\frac{5}{4}t} \sin 1.2t.$$

Applying the initial condition,

$$v_C(0) = 2 = \frac{2}{3} + \frac{1}{3} + 0.6\, A - \frac{5}{24}$$

gives
$$A = \frac{1 + \frac{5}{24}}{0.6} = 2.014.$$

Hence the total solution is,

$$i_L(t) = \frac{2}{3} + 2.014\, e^{-\frac{5}{4}t} \sin 1.2t + 0.333\, e^{-\frac{5}{4}t} \cos 1.2t.$$

In order to simplify this result slightly, consider the terms

$$2.014e^{-\frac{5}{4}t} \sin 1.2t + .333e^{-\frac{5}{4}t} \cos 1.2t$$

$$= e^{-\frac{5}{4}t} [2.014 \sin 1.2t + .333 \cos 1.2t]$$

$$= e^{-\frac{5}{4}t} [\, 2.014 \underline{/- 90°} + .333 \underline{/0}]$$

$$= e^{-\frac{5}{4}t} [.333 - j\, 2.014]$$

$$= e^{-\frac{5}{4}t}\left[\sqrt{(.333)^2 + (2.014)^2}\ \underline{/\text{arctan}}\left(-\frac{2.014}{.333}\right)\right]$$

$$= 2.04e^{-\frac{5}{4}t}\ \underline{/-\ 80.6°}$$

$$= 2.04e^{-\frac{5}{4}t}\ (\cos 1.2t - 80.6°)$$

Then the total solution may be written as

$$i_L(t) = \frac{2}{3} + 2.04e^{-\frac{5}{4}t} \cos (1.2t - 80.6°) \text{ A} \quad \text{for } t > 0.$$

• PROBLEM 7-13

For the circuit in Fig. 1 calculate i(t), assume zero initial conditions.

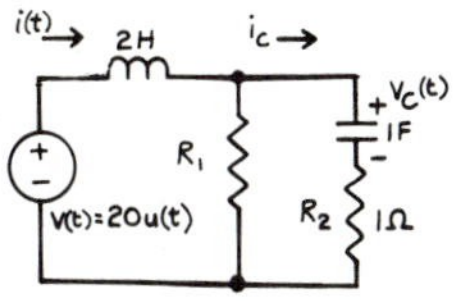

Fig. 1

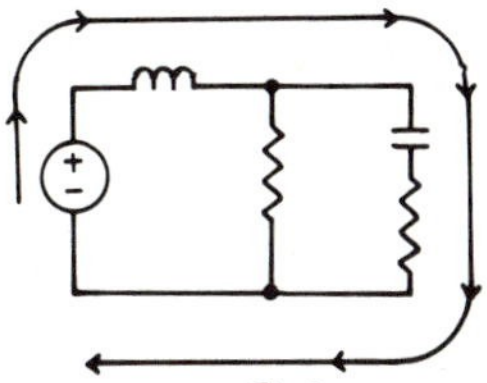

Fig. 2

Solution: First search for the describing differential equation for i(t). Since i(t) is flowing through an inductance, and the voltage source is given, Kirchoff's Voltage Law is useful. Choosing the loop shown in Fig. 2

$$20 = 2\frac{di(t)}{dt} + v_C + i_C \tag{1}$$

where i_C is the charging current of the IF capacitor.

By Kirchoff's Current Law, the current through R_1 is $\{i(t) - i_C\}$ (A). Therefore, the voltage across R_1 is $\{i(t) - i_C\}$. This voltage must be equal to $(v_C + i_C)$ because R_1 is in parallel with the series combination of R_2 and the capacitor. Hence

$$i(t) - i_C = v_C + i_C \tag{2}$$

Substituting Eq. (2) into Eq. (1)

$$20 = 2\frac{di(t)}{dt} + i(t) - i_C \tag{3}$$

$$i_C = 2\frac{di(t)}{dt} + i(t) - 20 \tag{4}$$

Differentiating Eq. (1) with respect to time

$$0 = 2\frac{d^2i(t)}{dt^2} + \frac{dv_C}{dt} + \frac{di_C}{dt} \qquad (5)$$

the term $\frac{dv_C}{dt}$ in Eq. (5) represents i_C going through a capacitance of 1F.

So, $$i_C = \frac{dv_C}{dt} \qquad (6)$$

With Eqs. (5) and (6),

$$0 = 2\frac{d^2i(t)}{dt^2} + i_C + \frac{di_C}{dt} \qquad (7)$$

From Eq. (4)

$$\frac{di_C}{dt} = 2\frac{d^2i(t)}{dt^2} + \frac{di(t)}{dt}. \qquad (8)$$

Substituting Eqs. (4) and (8) into Eq. (7),

$$0 = 2\frac{d^2i(t)}{dt^2} + 2\frac{di(t)}{dt} + i(t) - 20 + 2\frac{d^2i(t)}{dt^2} + \frac{di(t)}{dt}$$

$$4\frac{d^2i(t)}{dt^2} + 3\frac{di(t)}{dt} + i(t) = 20$$

$$\frac{d^2i(t)}{dt^2} + \frac{3}{4}\frac{di(t)}{dt} + \frac{1}{4}(t) = 5. \qquad (9)$$

This is the describing differential equation. The characteristic equation is then

$$s^2 + \frac{3}{4}s + \frac{1}{4} = 0 \qquad (10)$$

$$s = \frac{1}{2}\left(-\frac{3}{4} \pm \sqrt{\frac{9}{16} - 1}\right)$$

$$= \frac{1}{2}\left(-\frac{3}{4} \pm j\frac{\sqrt{7}}{4}\right) = -\frac{3}{8} \pm j\frac{\sqrt{7}}{8}. \qquad (11)$$

So, the natural response is

$$i_n(t) = e^{-\frac{3}{8}t}\left[A\cos\frac{\sqrt{7}}{8}t + B\sin\frac{\sqrt{7}}{8}t\right] \qquad (12)$$

where A and B are constants, respectively.

The forced response is obtained by letting $t \to \infty$.

By that time, the capacitance is fully charged and does not accept any more current. Thus it acts like an open circuit. The inductor now acts like a short circuit. So, the current flows through R_1 at $t \to \infty$ due only to the d.c. voltage source of 20 V. Therefore $i_f(t) = \frac{20}{1} = 20$ (A) (13)

Then the total response is

$$i(t) = i_n(t) + i_f(t) = e^{-\frac{3}{8}t}\left[A \cos \frac{\sqrt{7}}{8}t + B \sin \frac{\sqrt{7}}{8}t\right] + 20 \quad (14)$$

at $t = 0$, $i(0) = 0$ (15)

$$0 = A + 20 \quad (16)$$

$$A = -20 \quad (17)$$

at $t = 0^+$ $v_C = 0$ (18)

and $i_C = 0$ (19)

Substituting Eqs. (18) and (19) in Eq. (1),

$$20 = 2\,\frac{di(0^+)}{dt} \quad \text{or} \quad \frac{di(0^+)}{dt} = 0 \quad (20)$$

In Eq. (14)

$$\frac{di(t)}{dt} = -\frac{3}{8}\,e^{-\frac{3}{8}t}\left[A \cos \frac{\sqrt{7}}{8}t + B \sin \frac{\sqrt{7}}{8}t\right]$$

$$+ e^{-\frac{3}{8}t}\left[-A\,\frac{\sqrt{7}}{8} \sin \frac{\sqrt{7}}{8} + \frac{\sqrt{7}}{8} B \cos \frac{\sqrt{7}}{8}t\right] \quad (21)$$

Then, $$\frac{di(0)}{dt} = -\frac{3}{8} A + \frac{\sqrt{7}}{8} B \quad (22)$$

Combining Eqs. (20) and (22),

$$10 = -\frac{3}{8} A + \frac{\sqrt{7}}{8} B \quad (23)$$

$$80 = -3A + \sqrt{7}\,B \quad (24)$$

Substituting Eq. (17) in Eq. (24),

$$80 = +60 + \sqrt{7}\,B \quad (25)$$

$$B = \frac{20}{\sqrt{7}} \quad (26)$$

Substituting both Eqs. (17) and (26) into Eq. (14), the complete response is,

$$i(t) = e^{-\frac{3}{8}t}\left[-20\cos\frac{\sqrt{7}}{8}t + \frac{20}{\sqrt{7}}\sin\frac{\sqrt{7}}{8}t\right] + 20$$

$$= 20\left[1 - e^{-\frac{3}{8}t}\left\{\cos\frac{\sqrt{7}}{8}t - \frac{1}{\sqrt{7}}\sin\frac{\sqrt{7}}{8}t\right\}\right]\mu(t)\ \text{A}$$

• PROBLEM 7-14

The switch in the circuit shown in Fig. 1 is closed at t = 0. Find v at t = : (a) 0; (b) 2.5 ms; (c) 5π/3 ms.

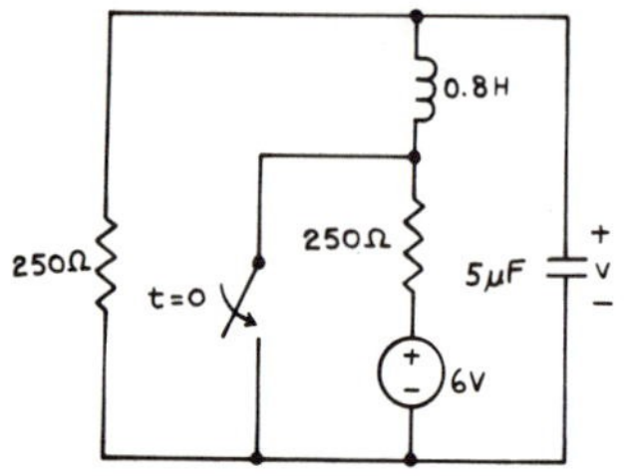

FiG. I

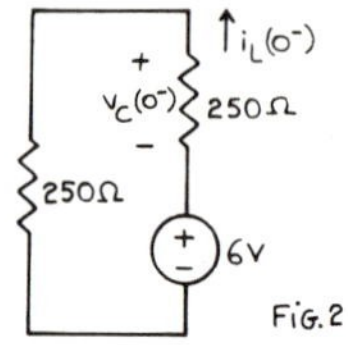

FiG. 2

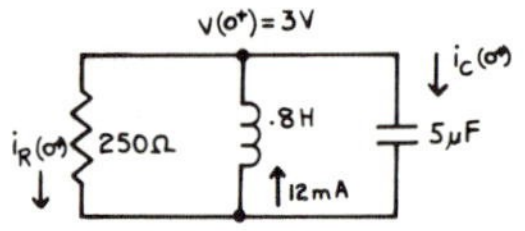

FiG. 3

Solution: First all initial conditions, (i.e., $i_L(0^+)$, $i_C(0^+)$, $i_R(0^+)$, $v(0^+)$) must be found. It is known that $i_L(0^-) = i_L(0^+)$, and that $v_C(0^-) = v(0^+)$, giving $i_R(0^+)$ and $i_C(0^+)$. Fig. 2 shows the circuit at $t = 0^-$.

Note that, under the assumption that the circuit has been operating for a long time, the capacitor is an open circuit and the inductor is a short circuit.

By voltage division

$$v_C(0^-) = v(0^+) = 6 - \frac{6\ (250)}{250 + 250} = 3\ \text{V}$$

and
$$i_L(0^-) = \frac{6\ \text{v}}{250 + 250} = 12\ \text{mA} = i_L(0^+).$$

Fig. 3 shows the RLC circuit at $t = 0+$.

We obtain $i_R(0^+) = \dfrac{3\ \text{v}}{250\ \Omega} = 12$ mA, and at the top node $i_C(0^+) = 0$.

Now list the initial conditions:

$$i_L(0^+) = 12\ \text{mA}$$

$$i_C(0^+) = 0$$

$i_R(0^+) = 12$ mA

$v(0^+) = 3$ V (our solution to part (a))

Now find α, ω_0 and $s_{1,2}$. First, test to see if this RLC circuit is underdamped, critically damped or overdamped.

$$\alpha = \frac{1}{2\ RC} = \frac{1}{2\ (250)(5 \times 10^{-6})} = 400$$

$$\omega_0 = \frac{1}{\sqrt{LC}} = \frac{1}{\sqrt{.8\ (5 \times 10^{-6})}} = 500$$

Since $\alpha < \omega_0$, the circuit is underdamped and response will be in the form,

$$v = e^{-\alpha t}(B_1 \cos \omega_d t + B_2 \sin \omega_d t)$$

where $\omega_d = \sqrt{\omega_0^2 - \alpha^2} = \sqrt{500^2 - 400^2} = 300$

giving $v = e^{-400t}(B_1 \cos 300t + B_2 \sin 300t)$. (1)

The initial condition $v(0) = 3$, can be substituted into Eq. (1) to give

$$3 - B_1.$$

Since $i_C = C \frac{dv}{dt}$ and $\frac{i_C(0)}{C} = \frac{dv(0)}{dt}$, we can find the value of of B_2 by finding the derivative of Eq. (1).

$$\frac{dv}{dt} = -\ 300\ B_1 e^{-400t} \sin 300t - 400\ B_1 e^{-400t} \cos 300t$$

$$+\ 300\ B_2 e^{-400t} \cos 300t - 400\ B_2 e^{-400t} \sin 300t$$

substituting $t = 0$

$$\frac{dv(0)}{dt} = \frac{i_C(0)}{C} = -\ 400\ B_1 + 300\ B_2$$

since it has been determined that $i_C(0) = 0$ and $B_1 = 3$, we have $B_2 = \frac{1200}{300} = 4$ giving

$$v(t) = e^{-400t}(3 \cos 300t + 4 \sin 300t)$$

(b) $v(2.5 \times 10^{-3}) = e^{-1}(3(.732) + 4\ (.682))$

$v(2.5 \times 10^{-3}) = .368\ (2.196 + 2.728)$

$$v(2.5 \times 10^{-3}) = .368(4.924) = 1.81 \text{ V}$$

(c) $$v\left(\frac{5\pi}{3} \times 10^{-3}\right) = e^{-2.09}(3(0) + 4(1))$$

$$v\left(\frac{5\pi}{3} \times 10^{-3}\right) = .123(4) = 0.493 \text{ V}$$

• PROBLEM 7-15

Note that the current source in the circuit of Fig. 1 goes to zero at t = 0, and find $i(0^+)$ and $i'(0^+)$ if R = : (a) 500 Ω; (b) 400 Ω; (c) 320 Ω.

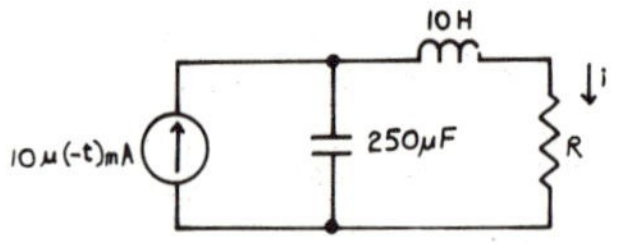

Fig. 1

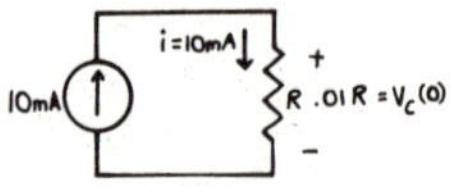

Fig. 2

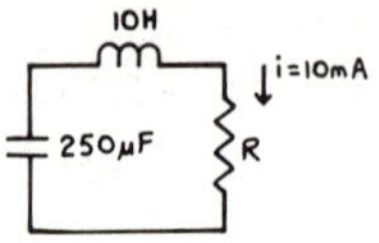

Fig. 3

Solution: For t < 0 assuming that the circuit has been operating for a long time, the circuit shown in Fig. 2 applies. Note that the capacitor is an open circuit while the inductor is a short circuit.

At $t = 0^+$, as shown in Fig. 3, i must remain at 10 mA no matter what value of R is placed in the circuit, since the inductor current cannot change instantaneously for a finite voltage change. Since the voltage across a capacitor is subject to the same restrictions,

$$v_L(0^-) = v_L(0^+) = 0 \quad \text{thus} \quad \frac{di(0^+)}{dt} = \frac{v_L(0^+)}{L} = \frac{0}{10} = 0$$

for all values of R.

• PROBLEM 7-16

For the circuit shown in Fig. 1, determine values at $t = 0^-$ and $t = 0^+$ for (a) v_1; (b) v_2; (c) v_3.

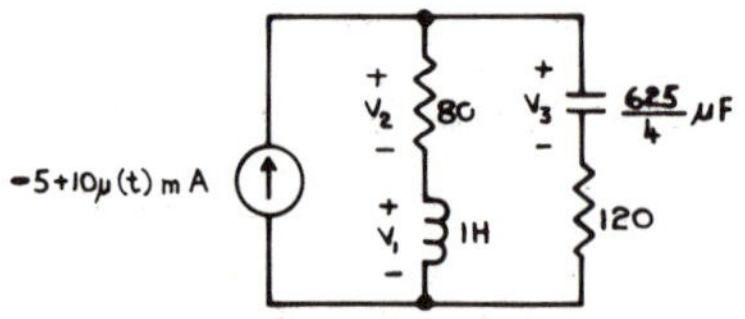

Fig. 1

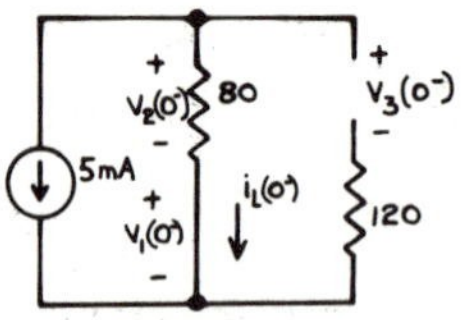

Fig. 2

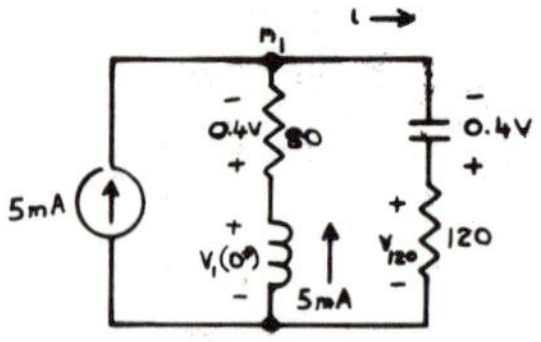

Fig. 3

Solution: Find $i_L(0^-)$ and $v_3(0^-)$ in the circuit for $t = 0^-$ shown in Fig. 2.

It is immediately apparent that $v_1(0^-) = 0$ V and that

$i_L(0^-) = -5$ mA; $v_2(0^-) = v_3(0^-) = i_L(0^-)$ (80) = - .4 V (there is no voltage drop across the 120 Ω resistor since no current flows in it).

For $t = 0^+$, it is known that $v_3(0^-) = v_3(0^+) = -.4$ V and that $i_L(0^-) = i_L(0^+)$. This means that $v_2(0^-) = v_2(0^+)$ = - .4 V.

Fig. 3 shows the circuit for $t = 0^+$.

Applying KCL at n_1, we find that, i = 10 mA thus the voltage across the 120 Ω resistor is 1.2 V.

Since the voltage across each branch must be equal, $v_1(0^+) = v_{120\ \Omega} = 1.2$ V.

• PROBLEM 7-17

For the circuit of Fig. 1, find values at $t = 0^-$ and $t = 0^+$ for: (a) $\frac{dv_1}{dt}$; (b) $\frac{dv_2}{dt}$; (c) $\frac{dv_3}{dt}$.

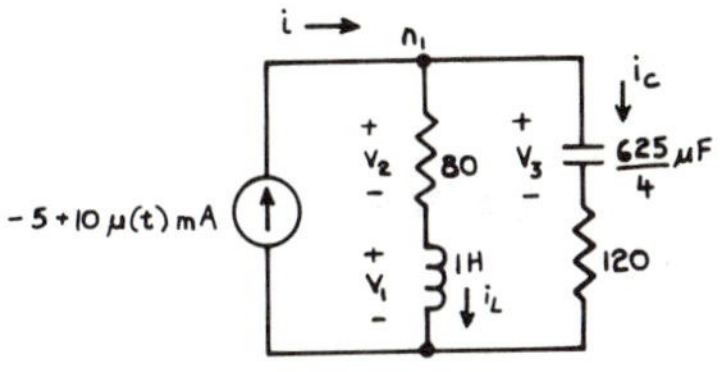

Fig. 1

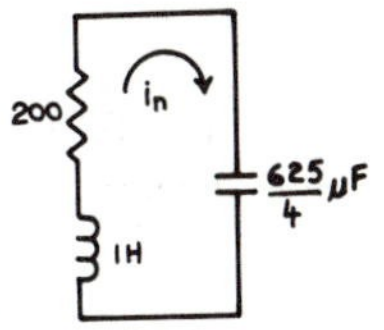

Fig. 2

Solution: In the previous problem all voltages were constant for t < 0. Hence, $\frac{dv_1}{dt} = \frac{dv_2}{dt} = \frac{dv_3}{dt} = 0$. It was determined that $i_C(0^+) = 10$ mA and that $v_1(0^+) = 1.2$ V. Fig. 2 shows the circuit used to obtain the natural response.

$$\alpha = \frac{R}{2L} = 100 \text{ rad/s}$$

$$\omega_0 = \frac{1}{\sqrt{LC}} = 80 \text{ rad/s}$$

$$s_1 = -\alpha + \sqrt{\alpha^2 - \omega_0^2} = -40$$

$$s_2 = -\alpha - \sqrt{\alpha^2 - \omega_0^2} = -160$$

The natural response is

$$i_n = A_1 e^{-40t} + A_2 e^{-160t}$$

In Fig. 1 at $t = 0^+$ at node n_1 we have $i_L + i_C = 5$ mA so that $i_L = 5\text{ mA} - i_C$

$$i_C = 5\text{ mA} - i_L.$$

It is also known that the total response is the sum of the forced response and the natural response. Therefore,

$$i_L = 5\text{ mA} + (-\,i_n)$$

$$i_C = 0 + i_n.$$

Giving $i_C = A_1e^{-40t} + A_2e^{-160t}$

$$i_L = .005 - A_1e^{-40t} - A_2e^{-160t}$$

Since $i_C = C\dfrac{dv_3}{dt}$ and $\dfrac{dv_3}{dt} = \dfrac{i_C}{C}$

$$\frac{dv_3}{dt} = \frac{i_C}{C} = \frac{A_1e^{-40t} + A_2e^{-160t}}{\frac{625}{4} \times 10^{-6}} \qquad (1)$$

$$\frac{dv_3(0^+)}{dt} = \frac{i_C(0^+)}{C} = \frac{10\text{ mA}}{\frac{625}{4} \times 10^{-6}} = 64\text{ V/s}$$

Substituting t = 0 into Eq. (1) yields

$$64\text{ V/s} = 6400\,A_1 + 6400\,A_2. \qquad (2)$$

Since $v_1 = L\dfrac{di_L}{dt}$ and $\dfrac{dv_1}{dt} = L\dfrac{d^2i_L}{dt^2}$.

$$\frac{dv_1}{dt} = L\left(\frac{d^2i_L}{dt^2}\right) = L\left(\frac{d^2\,[.005 + (-\,i_n)]}{dt^2}\right)$$

$$\frac{dv_1}{dt} = -\,(1)\left(40^2A_1e^{-40t} + 160^2A_2e^{-160t}\right) \qquad (3)$$

In order to find $\dfrac{dv_1(0^+)}{dt}$, solve for A_1 and A_2.

A second equation is obtained that can be used with Eq. (2) to solve for A_1 and A_2.

Since $v_1 = L \frac{di_L}{dt}$: $\frac{di_L}{dt} = \frac{v_1}{L}$

$$\frac{di_L(0^+)}{dt} = \frac{v_1(0^+)}{L} = \frac{1.2}{1} \text{ A/s}$$

$$\frac{di_L}{dt} = \frac{d}{dt}[.005 - i_n]$$

$$\frac{di_L}{dt} = 40\, A_1 e^{-40t} + 160\, A_2 e^{-40t}$$

Substituting t = 0 into the above equation gives

$$1.2 = 40\, A_1 + 160\, A_2 \qquad (4)$$

Eq. (4) and (2) are solved simultaneously.

$$A_1 = \frac{\begin{vmatrix} 1.2 & 160 \\ 64 & 6400 \end{vmatrix}}{\begin{vmatrix} 40 & 160 \\ 6400 & 6400 \end{vmatrix}} = \frac{7680 - 10240}{256000 - 1024000} = \frac{-2560}{-768000}$$

$A_1 = .0033$

$$A_2 = \frac{\begin{vmatrix} 40 & 1.2 \\ 6400 & 64 \end{vmatrix}}{\begin{vmatrix} 40 & 160 \\ 6400 & 6400 \end{vmatrix}} = \frac{2560 - 7680}{-768000} = \frac{-5120}{-768000}$$

$A_2 = .0067$

Substituting t = 0, A_1 = .0033 and A_2 = .0067 into Eq. (3) gives,

$$\frac{dv_1(0^+)}{dt} = -[40^2 (.0033) + 160^2 (.0067)]$$

$$\frac{dv_1(0^+)}{dt} = -176 \text{ V/s}$$

We can find $\frac{dv_2}{dt}(0^+) = 80 \frac{di_L}{dt}(0^+)$

$$\frac{dv_2}{dt}(0^+) = 80\,(40\,(.0033) + 160\,(.0067)) = 96 \text{ V/s.}$$

For the circuit of Fig. 1; for t > 0 find: (a) $v_1(t)$; (b) $v_2(t)$; (c) $v_3(t)$.

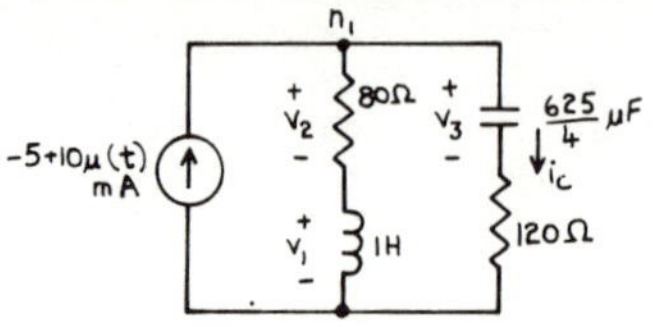

Fig. 1

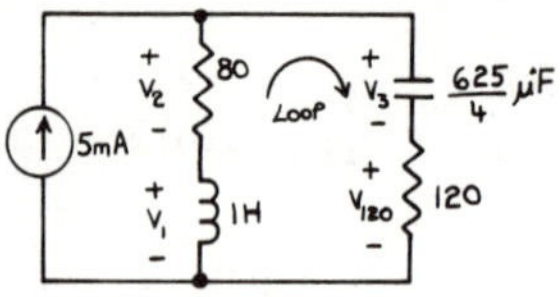

Fig. 2

Solution: In the previous problem it was found that

$$i_C = .0033e^{-40t} + .0067e^{-160t} \text{ A.}$$

$$i_L = .005 = .0033e^{-40t} - .0067e^{-160t} \text{ A.}$$

Find v_1 by $v_1 = L\frac{di_L}{dt}$

$$v_1 = (1)\left[\frac{d}{dt}(.005 - .0033e^{-40t} - .0067e^{-160t})\right]$$

$$v_1 = 0.133e^{-40t} + 1.067e^{-160t} \text{ V.}$$

$$v_2 = i_L(80) = 0.4 - .267e^{-40t} - 0.533e^{-160t} \text{ V.}$$

V_3 is found by writing an equation for the loop in Fig. 2.

$$v_2 + v_1 = v_3 + v_{120}$$

$$v_3 = v_2 + v_1 - v_{120}\ ;\ v_{120} = i_C(120)$$

$$v_3 = 0.4 - .267e^{-40t} - 0.533e^{-160t} + .133e^{-40t}$$

$$+ 1.067e^{-160t} - [120](.0033e^{-40t} + .0067e^{-160t})$$

$$v_3 = 0.4 - .134e^{-40t} + .534e^{-160t} - .4e^{-40t}$$

$$- .8e^{-160t}.$$

$$v_3 = 0.4 - 0.533e^{-40t} - 0.267e^{-160t}$$

Check the last answer by applying

$$v_3 = \frac{1}{C}\int_0^t i_C(\tau)\,d\tau + v_C(0^+)$$

$$v_3 = \frac{1}{\frac{625}{4} \times 10^{-6}} \int_0^t (.0033e^{-40t} + .0067e^{-160t})$$

$$+ (-.4)$$

$$v_3 = \frac{.0033}{\frac{625}{4} \times 10^{-6}(-40)} \left[e^{-40t} \right]_0^t$$

$$+ \frac{.0067}{\frac{625}{4} \times 10^{-6}(-160)} \left[e^{-160t} \right]_0^t + (-0.4)$$

$$v_3 = -.533 (e^{-40t} - 1) - .267 (e^{-160t} - 1) - 0.4$$

$$v_3 = -.533e^{-40t} + .533 - .267e^{-160t} + .267 - 0.4$$

$$v_3 = 0.4 - 0.533e^{-40t} - 0.267e^{-160t} \text{ V.}$$

• PROBLEM 7-19

Use differential equations to calculate the voltage response $v_C(t)$ for the network in Fig. 1. Assume zero initial conditions. Is the circuit underdamped, critically damped, or overdamped.

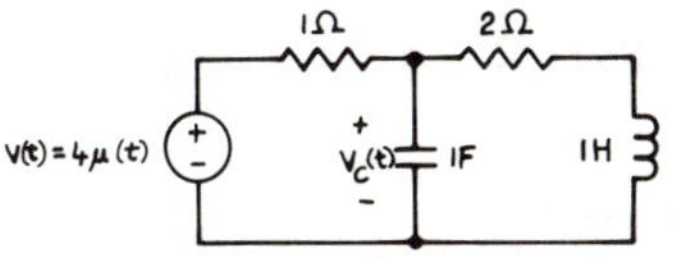

FIG. 1

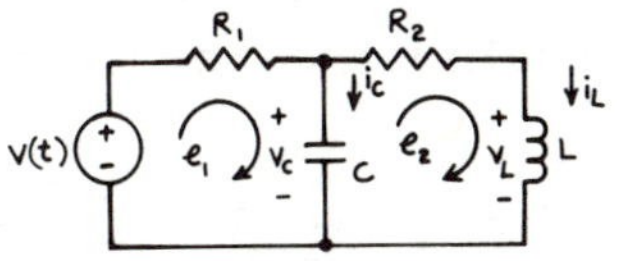

FIG. 2

Solution: In order to write a differential equation, assign the voltages and currents shown in Fig. 2. This facilitates writing two KVL equations around the loops.

$$\ell_1: v(t) = R_1(i_C + i_L) + v_C$$

$$\ell_2: 0 = -v_C + R_2 i_L + v_L$$

Since $i_C = C\frac{dv_C}{dt}$ and $v_L = L\frac{di_L}{dt}$, they can be substituted into the above loop equations to obtain:

$$v(t) = R_1 C \frac{dv_C}{dt} + R_1 i_L + v_C \qquad (1)$$

$$0 = -v_C + R_2 i_L + L\frac{di_L}{dt} \qquad (2)$$

Solving Eq. (1) for i_L yields,

$$i_L = \frac{v(t)}{R_1} - C\frac{dv_C}{dt} - \frac{v_C}{R_1} \qquad (3)$$

Substituting Eq. (3) into Eq. (2) gives

$$0 = -v_C + R_2\left[\frac{v(t)}{R_1} - C\frac{dv_C}{dt} - \frac{v_C}{R_1}\right] + L\frac{d}{dt}\left[\frac{v(t)}{R_1} - C\frac{dv_C}{dt} - \frac{v_C}{R_1}\right]$$

$$0 = -v_C + \frac{R_2}{R_1}v(t) - CR_2\frac{dv_C}{dt} - \frac{R_2}{R_1}v_C + \frac{L}{R_1}\frac{dv(t)}{dt} - LC\frac{d^2v_C}{dt^2}$$

$$-\frac{L}{R_1}\frac{dv_C}{dt}.$$

Combining terms,

$$\frac{R_2}{R_1}v(t) + \frac{L}{R_1}\frac{dv(t)}{dt} = LC\frac{d^2v_C}{dt^2} + \left[R_2C + \frac{L}{R_1}\right]\frac{dv_C}{dt} + \left[1 + \frac{R_2}{R_1}\right]v_C.$$

Since v(t) is a constant for t > 0, we obtain,

$$v(t) = \left[\frac{R_1LC}{R_2}\right]\frac{d^2v_C}{dt^2} + \left[R_1C + \frac{L}{R_2}\right]\frac{dv_C}{dt} + \left[1 + \frac{R_1}{R_2}\right]v_C, \qquad (4)$$

a second order differential equation.

Substituting the values for R,L,C,and v(t) yields

$$4 = \frac{1}{2}\frac{d^2v_C}{dt^2} + \frac{3}{2}\frac{dv_C}{dt} + \frac{3}{2}v_C.$$

By multiplying through by 2;

$$8 = \frac{d^2v_C}{dt} + 3\frac{dv_C}{dt} + 3v_C. \qquad (5)$$

We can find the homogeneous solution of

$$0 = \frac{d^2v_C}{dt^2} + 3\frac{dv_C}{dt} + 3v_C \qquad (6)$$

by writing the operation $\frac{d}{dt}$ as s and $\frac{d^2}{dt^2}$ as s^2. Hence,

$$0 = s^2 + 3s + 3$$

$$s_{1,2} = \frac{-3 \pm \sqrt{9-12}}{2}$$

$$s_1 = \frac{-3}{2} + j\frac{\sqrt{3}}{2} \qquad (7)$$

$$s_2 = -\frac{3}{2} - j\frac{\sqrt{3}}{2}. \qquad (8)$$

The homogeneous solution is in the form

$$K_1 e^{s_1 t} + K_2 e^{s_2 t}.$$

The particular solution must be a constant K_3 since the driving function $4\mu(t)$ is a constant for $t > 0$. Find the value of K_3 by substituting $v_C(t) = K_3$ into Eq. (5). Hence,

$$8 = \frac{d^2}{dt^2}K_3 + 3\frac{d}{dt}K_3 + 3K_3$$

$$8 = 3K_3$$

$$K_3 = \frac{8}{3}.$$

The total solution is the sum of the particular and homogeneous solutions. Hence,

$$v_C(t) = \frac{8}{3} + K_1 e^{s_1 t} + K_2 \, {}^{s_2 t} \qquad (9)$$

The constants K_1 and K_2 can be determined from the initial conditions, $i_L(0) = 0$ and $v_C(0) = 0$. Hence,

$$v_C(0) = 0 = \frac{8}{3} + K_1 + K_2, \text{ thus}$$

$$-\frac{8}{3} = K_1 + K_2. \qquad (10)$$

Since $i_C(0) = C\frac{dv_C(0)}{dt}$, and $i_C(0) = i_{R_1}(0) - i_L(0)$ as shown in Fig. 3.

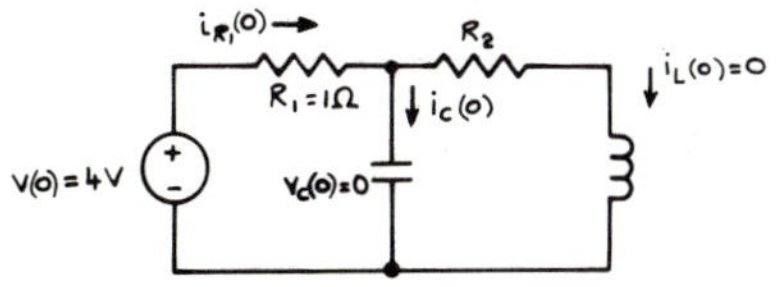

FIG. 3

From Fig. 3, we can show that

$$i_C(0) = i_{R_1}(0) = \frac{v(0)}{R_1} = 4.$$

Hence, $$\frac{dv_C(0)}{dt} = \frac{4}{C} = 4 = s_1K_1 + s_2K_2. \tag{11}$$

Solving for K_1 and K_2 in Eqs. (10) and (11) by the method of determinants yields

$$K_1 = \frac{\begin{vmatrix} -\frac{8}{3} & 1 \\ 4 & s_2 \end{vmatrix}}{\begin{vmatrix} 1 & 1 \\ s_1 & s_2 \end{vmatrix}} = \frac{-\frac{8}{3}s_2 - 4}{s_2 - s_1}.$$

Substituting for s_1 and s_2 from Eq. (7) and (8),

$$K_1 = \frac{-\frac{8}{3}\left(-\frac{3}{2} - j\frac{\sqrt{3}}{2}\right) - 4}{-j\sqrt{3}} = \frac{j\frac{4\sqrt{3}}{3}}{-j\sqrt{3}} = -\frac{4}{3}$$

$$K_2 = \frac{\begin{vmatrix} 1 & -\frac{8}{3} \\ s_1 & 4 \end{vmatrix}}{s_2 - s_1} = \frac{4 + \frac{8}{3}s_1}{s_2 - s_1}$$

$$K_2 = \frac{4 + \frac{8}{3}\left(-\frac{3}{2} + j\frac{\sqrt{3}}{2}\right)}{-j\sqrt{3}} = \frac{j\frac{4\sqrt{3}}{3}}{-j\sqrt{3}} = -\frac{4}{3}$$

Substituting K_1, K_2, s_1, and s_2 into Eq. (9) yields

$$v_C(t) = \frac{8}{3} - \frac{4}{3}\left[e^{\left(-\frac{3}{2} + j\frac{\sqrt{3}}{2}\right)t} + e^{\left(-\frac{3}{2} - j\frac{\sqrt{3}}{2}\right)t}\right]$$

which can be rewritten

$$v_C(t) = \frac{8}{3} - \frac{8}{3}e^{-\frac{3}{2}t}\left[\frac{e^{+j\frac{\sqrt{3}}{2}t} + e^{-j\frac{\sqrt{3}}{2}t}}{2}\right].$$

Since, $\cos \omega t = \dfrac{e^{j\omega t} + e^{-j\omega t}}{2}$, the complete solution for $v_C(t)$ can be written as

$$v_C(t) = \frac{8}{3}\left[1 - e^{-\frac{3}{2}t}\cos\frac{\sqrt{3}}{2}t\right] \text{ V}; \quad t \geq 0.$$

The circuit is obviously underdamped, since a step excitation produces a damped sinusoid.

The circuit shown in Fig. 1 is source-free for $t > 0$. Find $v_C(t)$.

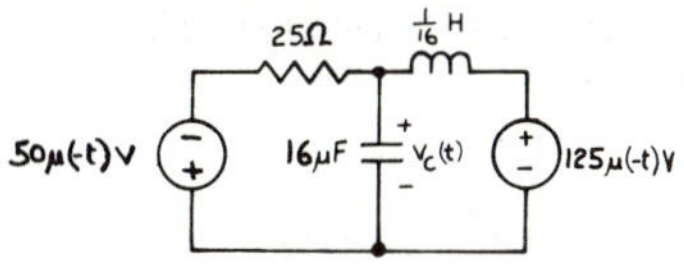

FIG. 1

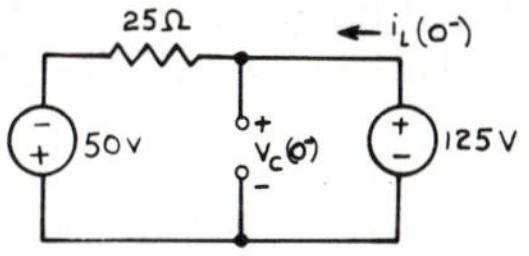

FIG. 2

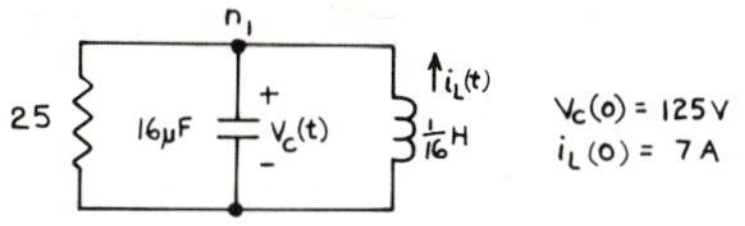

FIG. 3

Solution: At $t = 0^-$ both voltage sources are "on" and the circuit is in the state as shown in Fig. 2.

Knowing that the capacitor voltage cannot change instantaneously, we know the initial condition $v_C(0^-) = v_C(0^+) = 125$ V. Knowing that the current through the inductor cannot change instantaneously, we know the second initial condition, $i_L(0^-) = i_L(0^+) = \frac{50 + 125}{25} = 7$ A. At $t = 0^+$ both voltage sources are "turned-off" giving the RLC circuit shown in Fig. 3.

Using KCL at node n_1

$$\frac{v_C(t)}{R} + C\frac{dv_C(t)}{dt} + I(0) + \frac{1}{L}\int_0^t v(t)\,dt = 0.$$

Taking the derivative,

$$\frac{d^2v}{dt^2} + \frac{1}{RC}\frac{dv}{dt} + \frac{1}{LC} = 0.$$

The solution to this second order differential equation is in the form

$$v_C(t) = Ae^{s_1 t} + Be^{s_2 t},$$

where $s_1, s_2 = -\frac{1}{2RC} \pm \sqrt{\left(\frac{1}{2RC}\right)^2 - \frac{1}{LC}}$. It is found that,

$$s_1 = -1250 + \sqrt{562500} = -500$$

$$s_2 = -1250 - \sqrt{562500} = -2000$$

hence, $$v_C(t) = Ae^{-500t} + Be^{-2000t}. \qquad (1)$$

The constants are found by applying the initial conditions. Note that at $t = 0$ $v_C(0) = 125$ V and Eq. (1) becomes

$$125 = A + B. \qquad (2)$$

Also, since $i_C = C\frac{dv_C}{dt}$, then $i_C(0) = C\frac{dv_C(0)}{dt}$, $i_C(0)$ can be found from Fig. 3 and $C \frac{dv_C}{dt}$ is $(-500 Ae^{-500t} - 2000 Be^{-2000t})\ 16 \times 10^{-6}$.

In Fig. 3 we find the current through the resistor to be 5 A hence, the current through the capacitor (using the passive sign convention) must be 2 A. Hence,

$$2 = C \frac{dv_C(0)}{dt} = (-500\ A - 2000\ B)\ 16 \times 10^{-6}. \quad (3)$$

Eqs. 2 and 3 can be solved simultaneously to give A = 250 and B = - 125.

Hence, $\quad v_C(t) = 250e^{-500t} - 125e^{-2000t}$ V; $\quad t \geq 0$.

• PROBLEM 7-21

For the circuit in Fig. 1, solve analytically for the capacitor voltage $v_C(t)$. Establish the initial conditions from the circuit relationships.

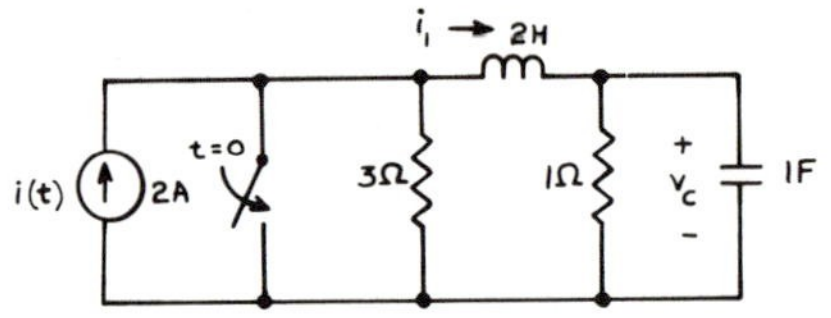

Fig. 1

Solution: This solution is the natural response of a simple circuit formed by connecting R, L, and C in parallel with C initially charged to some voltage. The derivation may be found in circuit textbooks and will not be repeated. The circuit may be under-damped, critically-damped, or over-damped depending on the values of R, L, and C. In this case, as will be seen, the response is oscillatory as it decays.

We need to know two initial conditions since there are two arbitrary constants. These are $v_C(0^-)$ and $\left.\frac{dv_C}{dt}\right|_{t=0^-}$, where $t = 0^-$ is the time just before the switch is closed. The source is 2 A d.c. and therefore the inductance will act as a short-circuit, placing the 1Ω and 3Ω resistors in parallel with the capacitor with an initial voltage, $V_C = I\ R_{eq} = 2 \frac{(3)\ (1)}{3 + 1} = 1.5$ V.

Since the initial voltage across the capacitor is d.c., then $\left.\frac{dv_C}{dt}\right|_{t=0^-} = 0$, with the switch closed, at t = 0, the 3Ω resistor is shorted and R = 1Ω, L = 1H and C = 1F with an initial charge of 1.5 V.

The exponential form of the solution is

$$v_C(t) = A_1 e^{s_1 t} + A_2 e^{s_2 t}$$

where $s_{1,2} = -\alpha \pm \sqrt{\alpha^2 - \omega_0^2}$

$$\alpha = \frac{1}{2RC} = 0.5 \text{ rad/s}$$

$$\omega_0 = \frac{1}{\sqrt{LC}} = \frac{\sqrt{2}}{2} \text{ rad/s}$$

A_1 and A_2 are constants depending on the initial conditions.

We see that $\omega_0^2 > \alpha^2$. This indicates underdamping and with $j = \sqrt{-1}$

$$\sqrt{\alpha^2 - \omega_0^2} = j\sqrt{\omega_0^2 - \alpha^2}$$

Let $\omega_d = \sqrt{\omega_0^2 - \alpha^2} = 0.5$ rad/s be the natural resonant frequency.

Then, we can write

$$v_C(t) = e^{-\alpha t}(A_1 e^{j\omega_d t} + A_2 e^{-j\omega_d t}).$$

But $\cos \omega_d t = \dfrac{e^{j\omega_d t} + e^{-j\omega_d t}}{2}$

$$\sin \omega_d t = \frac{e^{j\omega_d t} - e^{-j\omega_d t}}{j2}.$$

We can then write

$$v_C(t) = e^{-\alpha t}[(A_1 + A_2)\cos \omega_d t + j(A_1 - A_2)\sin \omega_d t] \quad (1)$$

To evaluate A_1 and A_2:

$$v_C(t) = A_1 e^{s_1 t} + A_2 e^{s_2 t}$$

$$v_C(t = 0^-) = A_1 e^0 + A_2 e^0 = A_1 + A_2 = 1.5$$

or $A_1 + A_2 = 1.5$ (2)

$$\frac{dv_C}{dt} = A_1 s_1 e^{s_1 t} + A_2 s_2 e^{s_2 t} = A_1 s_1 e^{0} + A_2 s_2 e^{0} = A_1 s_1 + A_2 s_2 = 0$$

or $\quad A_1 s_1 + A_2 s_2 = 0. \qquad (3)$

From Equation (3),

$$A_2 = -\frac{s_1}{s_2} A_1 .$$

Substituting in Equation (2)

$$A_1 - \frac{s_1}{s_2} A_1 = 1.5$$

or $\quad A_1 = \dfrac{1.5 s_2}{s_2 - s_1}$

and $\quad A_2 = 1.5 \dfrac{s_1}{s_1 - s_2}.$

Then $A_1 + A_2 = 1.5$

$$A_1 - A_2 = 1.5 \left(\frac{s_2}{s_2 - s_1} - \frac{s_1}{s_1 - s_2} \right) = 1.5 \left(\frac{s_2 + s_1}{s_2 - s_1} \right)$$

$$= (1.5) \frac{(-1)}{(-j)}$$

or $\quad A_1 - A_2 = \dfrac{1.5}{j} = -j\,1.5.$

Substituting all values into Equation (1), we have

$$v_C(t) = e^{-0.5t} [1.5 \cos 0.5t - j^2\, 1.5 \sin 0.5t]$$

where $j^2 = -1$

This can be simplified to

$$v_C(t) = 1.5e^{0.5t} (\cos 0.5t + \sin 0.5t).$$

This Equation is equivalent to

$$v_C(t) = 1.5e^{-0.5t} [\sqrt{2} \cos (0.5t + 45°) + 2 \sin t].$$

Since, by the trigonometric identity

$\cos(A + B) = \cos A \cos B - \sin A \sin B$

$$v_C(t) = 1.5e^{-0.5t}(\sqrt{2}\cos 45° \cos 0.5t$$

$$-\sqrt{2}\sin 45° \sin 0.5t + 2\sin 0.5t)$$

$$= 1.5e^{-0.5t}(\cos 0.5t - \sin 0.5t + 2\sin 0.5t)$$

$$= 1.5^{-0.5t}(\cos 0.5t + \sin 0.5t) \text{ V.}$$

• PROBLEM 7-22

For the network in Fig. 1, let $V_C(0) = 5$ V and $i_L(0) = 1$ A. Find the complete response v(t).

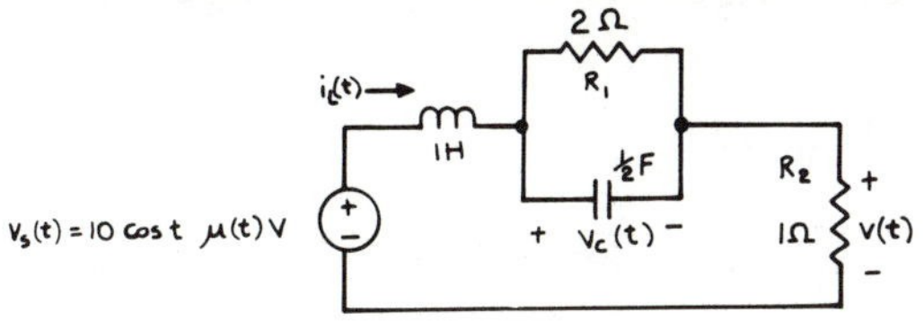

FIG.1

Solution: The complete response v(t) is found by solving for $i_L(t)$, since $i_L(t)(1) = v(t)$. We find that $i_L(t)$ is the sum of the current in the 2Ω resistor and the capacitor. Hence,

$$i_L(t) = i_R + i_C$$

$$i_L(t) = \frac{v_C}{R_1} + C\frac{dv_C}{dt} \tag{1}$$

To find $i_L(t)$, $v_C(t)$ must be solved first. Using KVL,

$$v_s(t) = v_L(t) + v_C(t) + v(t)$$

$$v_s(t) = L\frac{di_L}{dt} + v_C + i_L R_2 \tag{2}$$

Substituting $i_L = \frac{v_C}{R_1} + C\frac{dv_C}{dt}$ into Eq. (2) yields

$$v_s(t) = L\left(\frac{1}{R_1}\frac{dv_C}{dt} + C\frac{d^2v_C}{dt^2}\right) + V_C + \frac{R_2}{R_1}v_C + R_2C\frac{dv_C}{dt}$$

$$v_s(t) = \frac{d^2v_C}{dt^2}LC + \frac{dv_C}{dt}\left(\frac{L}{R_1} + R_2C\right) + v_C\left(1 + \frac{R_2}{R_1}\right)$$

substituting $R_1 = 2\Omega$, $R_2 = 1\Omega$, $L = 1$H and $C = \frac{1}{2}$F yields,

$$10 \cos t = v_s(t) = \frac{1}{2}\frac{d^2 v_C}{dt^2} + \frac{dv_C}{dt} + \frac{3}{2} v_C. \qquad (3)$$

The homogeneous solution $v_{Ch}(t)$ is found by solving

$$0 = \frac{1}{2}\frac{d^2 v_C}{dt} + \frac{dv_C}{dt} + \frac{3}{2} v_C$$

hence, $$v_{Ch}(t) = K_A e^{s_1 t} + K_B e^{s_2 t}$$

where s_1 and s_2 are the roots of $s^2 + 2s + 3$. This yields

$$v_{Ch}(t) = K_A e^{-1t + j\sqrt{2}t} + K_B e^{-1t - j\sqrt{2}t}$$

which can be written as

$$v_{Ch}(t) = K_1 e^{-t} \sin\sqrt{2}t + K_2 e^{-t} \cos\sqrt{2}t. \qquad (4)$$

Since the source is a sinusoid of frequency $\omega = 1$, the particular solution must also be a sinusoid of frequency $\omega = 1$ of the form

$$v_{Cp}(t) = K_3 \sin t + K_4 \cos t. \qquad (5)$$

Substituting (5) into Eq. (3) yields,

$$10 \cos t = -\frac{1}{2} K_3 \sin t - \frac{1}{2} K_4 \cos t + K_3 \cos t$$

$$- K_4 \sin t + \frac{3}{2} K_3 \sin t + \frac{3}{2} K_4 \cos t$$

$$10 \cos t = K_3 (\sin t + \cos t) + K_4 (-\sin t + \cos t)$$

letting $t = 0$ $\qquad 10 = K_3 + K_4$

letting $t = \frac{\pi}{2}$ $\qquad 0 = K_3 - K_4$

hence $\qquad K_3 = K_4 = 5.$

The total solution $v_C(t)$ is the sum of the homogeneous and particular solutions.

$$v_C(t) = 5 \sin t + 5 \cos t + K_1 e^{-t} \sin \sqrt{2}t$$

$$+ K_2 e^{-t} \cos \sqrt{2}t \qquad (6)$$

Setting $t = 0$ in Eq. (6) gives the initial condition $v_C(0) = 5$, hence,

$$v_C(0) = 5 = 5 + K_2;\ K_2 = 0$$

Substituting $v_C(t) = 5 \sin t + 5 \cos t + K_1 e^{-t} \sin \sqrt{2}t$ into Eq. (1) yields

$$i_L(t) = 5 \cos t - \frac{K_1}{2} e^{-t} \sin \sqrt{2}t + \frac{K_1\sqrt{2}}{2} e^{-t} \cos \sqrt{2}t. \quad (7)$$

The constant K_1, is found by setting $t = 0$ in Eq. (7) hence,

$$i_L(0) = 1 = 5 + K_1 \frac{\sqrt{2}}{2}$$

$$K_1 = -4\sqrt{2}$$

The complete solution is

$$i_L = 5 \cos t + \frac{4}{\sqrt{2}} e^{-t} \sin \sqrt{2}t - 4e^{-t} \cos \sqrt{2}t \text{ A.}$$

• PROBLEM 7-23

The switch in the circuit of Fig. 1 has been closed during a laboratory experiment. Seeing a maximum source voltage of only 10 V in the circuit, student D carelessly lets his hands roam all over it; in particular, across the inductor at $t = 0$ just as his newest enemy opens the switch. What is the maximum voltage to which he will be subjected? Sketch $v_L(t)$ vs. t.

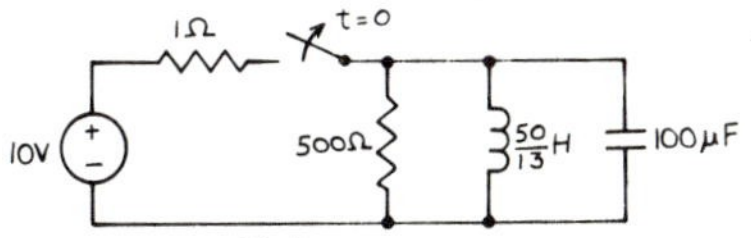

Fig.1

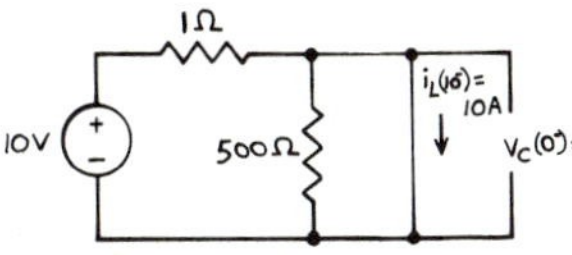

Fig.2

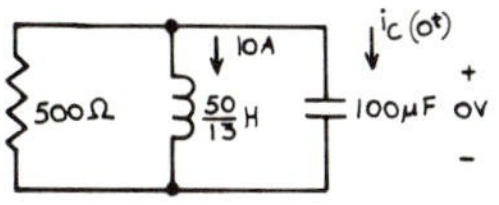

Fig.3

Solution: Find the initial conditions $i_L(0)$ and $v_C(0)$ for the circuit in Fig. 2 at $t = 0^-$.

Since these two initial conditions do not change instantaneously with time, apply the initial conditions to the circuit in Fig. 3 at $t = 0^+$.

Since the voltage across the node pair is zero, $i_C(0^+) = -10A$.

Find the parameters α and ω_0

$$\alpha = \frac{1}{2RC} = \frac{1}{(2)(500)(100 \times 10^{-6})} = \frac{1}{.1} = 10 \text{ rad/s}$$

$$\omega_0 = \frac{1}{\sqrt{LC}} = \frac{1}{\sqrt{\frac{50}{13}\, 100 \times 10^{-6}}} = \frac{1}{.0196} = 51 \text{ rad/s}$$

Since $\omega_0 > \alpha$, the RLC circuit is underdamped, then the response will be an exponentially damped sinusoidal voltage.
In the form

$$v(t) = B_1 e^{-10t} \cos 50t + B_2 e^{-10t} \sin 50t.$$

Knowing that v(0) = 0, we find

$$v(0) = B_1 = 0, \text{ and}$$

$$\frac{i_C(0^+)}{C} = \frac{dv(0)}{dt} = 50\ B_2 e^{-10(0)} \cos 50(0)$$

$$- 10\ B_2 e^{-10(0)} \sin 50(0)$$

$$\frac{i_C(0^+)}{C} = \frac{-10}{100 \times 10^{-6}} = 50\ B_2$$

$$\text{giving } B_2 = \frac{-10}{(50)(100 \times 10^{-6})} = -2000$$

$$B_1 = 0.$$

Our response becomes

$$v(t) = -2000\ e^{-10t} \sin 50t \text{ V}.$$

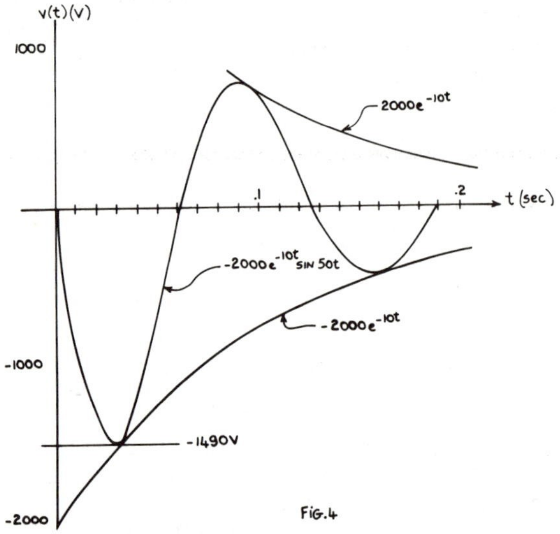

Fig.4

V_{max} is found graphically (see Fig. 4) to be - 1490 V.

• PROBLEM 7-24

> The circuit of Fig. 1 has been in the configuration shown for a long time. At t = 0, the 100Ω resistor is connected into the circuit as the switch opens. Find i(t) for all t.

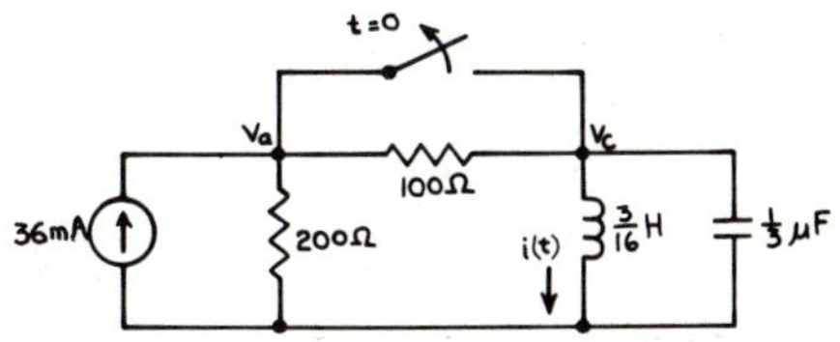

FIG. 1

Solution: This type of problem can be solved by obtaining a describing or governing differential equation for i(t). To find the governing differential equation for i(t), Kirchoff's Current Law is applied at the node of potential v_C in Fig. 1, for t > 0.

$$\frac{v_a - v_C}{100} = i(t) + \frac{1}{3} \times 10^{-6} \frac{dv_C}{dt} \tag{1}$$

and $$v_C = \frac{3}{16} \frac{di(t)}{dt}. \tag{2}$$

Substituting Eq. (2) into Eq. (1)

$$\frac{v_a}{100} - \frac{3}{1600} \frac{di(t)}{dt} = i(t) + \frac{1}{3} \times 10^{-6} \times \frac{3}{16} \frac{d^2i(t)}{dt^2}$$

$$\frac{v_a}{100} = \frac{1}{16} \times 10^{-6} \frac{d^2i(t)}{dt^2} + \frac{3}{1600} \frac{di(t)}{dt} + i(t) \tag{3}$$

If Kirchoff's Current Law is applied at the node v_a for t > 0,

$$36 \times 10^{-3} = \frac{v_a}{200} + \frac{v_a - v_C}{100}$$

$$= \left(\frac{1}{200} + \frac{1}{100}\right) v_a - \frac{v_C}{100}$$

$$= \frac{3}{200} v_a - \frac{v_C}{100} \tag{4}$$

Then $$v_a = \frac{200}{3} \left(36 \times 10^{-3} + \frac{v_C}{100}\right) \tag{5}$$

Combining both equations (2) and (5),

$$v_a = \frac{200}{3} \left(36 \times 10^{-3} + \frac{3}{1600} \frac{di(t)}{dt}\right) \tag{6}$$

Substituting Eq. (6) into Eq. (3),

$$\frac{2}{3} \left(36 \times 10^{-3} + \frac{3}{1600} \frac{di(t)}{dt}\right) = \frac{1}{16} \times 10^{-6} \frac{d^2i(t)}{dt^2}$$

$$+ \frac{3}{1600} \frac{di(t)}{dt} + i(t)$$

$$\frac{1}{16} \times 10^{-6} \frac{d^2 i(t)}{dt^2} + \frac{1}{1600} \frac{di(t)}{dt} + i(t) = 24 \times 10^{-3}$$

multiplying through by 1600

$$10^{-4} \frac{d^2 i(t)}{dt^2} + \frac{di(t)}{dt} + 1600\ i(t) = 24 \times 16 \times 10^{-1}$$

multiplying through by 10^4

$$\frac{d^2 i(t)}{dt^2} + 10^4 \frac{di(t)}{dt} + 16 \times 10^6\ i(t) = 24 \times 16 \times 10^3 \quad (7)$$

The Characteristic Equation of Eq. (7) is

$$s^2 + 10^4 s + 16 \times 10^6 = 0 \quad (8)$$

$$s = \frac{1}{2}(-10^4 \pm \sqrt{10^8 - 4 \times 16 \times 10^6})$$

$$= \frac{1}{2}(-10^4 \pm \sqrt{36 \times 10^3})$$

$$= \frac{10^3}{2}(-10 \pm 6) = -2000, -8000 \quad (9)$$

The Natural Response should be

$$i_n(t) = A^+ e^{-2000t} + A^- e^{-8000t} \quad (10)$$

Both A^+ and A^- are constants to be determined from the initial condition given.

At $t = 0^-$, the switch is closed and 36 mA flows through the inductor. So

$$i(0^-) = 36 \times 10^{-3} \text{ (A)} \quad (11)$$

But before doing this, the forced response of the given circuit must be otained by letting $t \to \infty$. At $t = \infty$, the capacitor is fully charged and accepts no more current. It can be considered open. The voltage across the inductor is zero or $v_C = 0$ at $t = \infty$.

So, by the current division principle,

$$i_f = i(\infty) = 36 \times 10^{-3} \frac{200}{100 + 200} = 24 \times 10^{-3}. \quad (12)$$

The total response of the current i(t) is then

$$i(t) = i_n + i_f = \left(A^+ e^{-2000t} + A^- e^{-8000t} + 24 \times 10^{-3}\right)\mu(t) \quad (13)$$

Using condition (11)

$$36 \times 10^{-3} = A^{+} + A^{-} + 24 \times 10^{-3}$$

$$A^{+} + A^{-} = 12 \times 10^{-3} \tag{14}$$

$$v_C(0^-) = 0 \tag{15}$$

then from Eq. (2) $\frac{di(0)}{dt} = 0.$ (16)

From Eq. (13)

$$\frac{di(t)}{dt} = -2000 A^{+}e^{-2000t} - 8000 A^{-}e^{-8000t} \tag{17}$$

So, $\frac{di(0)}{dt} = -2000 A^{+} - 8000 A^{-}.$ (18)

Combining Eq. (16) and (18)

$$-2000 A^{+} - 8000 A^{-} = 0$$

$$-A^{+} - 4 A^{-} = 0 \tag{19}$$

Adding Eq. (14) and (19), $-3 A^{-} = 12 \times 10^{-3}$

$$A^{-} = -4 \times 10^{-3} \tag{20}$$

Substituting Eq. (20) into Eq. (14)

$$A^{+} = 16 \times 10^{-3} \tag{21}$$

Using Eq. (20) and (21) in Eq. (13),

$$i(t) = (16 \times 10^{-3} e^{-2000t} - 4 \times 10^{-3} e^{-8000t} + 24 \times 10^{-3})\, \mu(t) \tag{22}$$

This solution is only for $t \geq 0$ (23)

but the problem is asking for a solution for all t. When $t > 0^-$ the switch is closed and the 200Ω resistor is short-circuited, or 36 mA is flowing in the inductor. So

$$i(-t) = 36 \times 10^{-3} \mu(-t). \tag{24}$$

Therefore the complete response for all t is

$$i(t) = 36 \times 10^{-3} \mu(-t) + (16 \times 10^{-3}e^{-2000t} - 4 \times 10^{-3}e^{-8000t} + 24 \times 10^{-3})\, \mu(t)\,(A) \tag{25}$$

or $i(t) = 36\, \mu(-t) + (16e^{-2000t} - 4e^{-8000t} + 24)\, \mu(t)$ (mA) (26)

After being closed for 1s, the switch in the circuit of Fig. 1 is opened at t = 0. Find $i_1(t)$ for all t > 0.

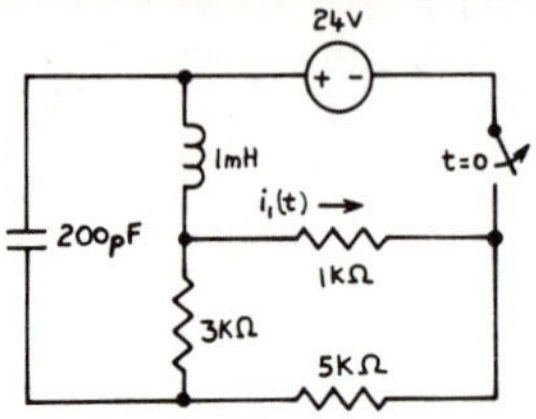

Fig.1

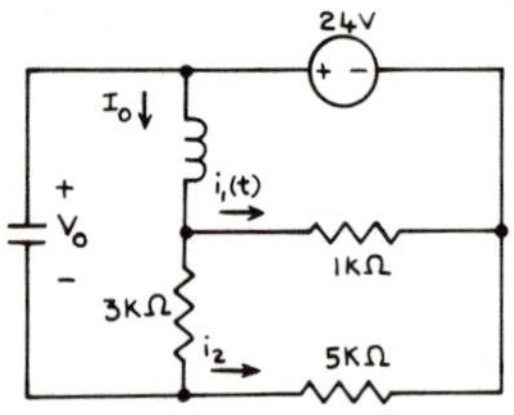

Fig.2 (Situation at t = 0-)

Solution: It is assumed that when the switch is closed for 1s, all circuit variables are in the steady state, or time invariant, since the 24V source is time invariant. It is necessary, then, to find the inductor current I_0 and capacitor voltage V_0 as shown in Figure 2. Keep in mind that at $t = 0^-$, just before the switch is opened, the inductor is a short circuit ($di_L/dt = 0$) and the capacitor is an open circuit ($dv_C/dt = 0$).

As just stated, the voltage across the inductor at this time is 0v. The capacitor is in parallel with the 3kΩ resistor, and the inductor current will be the total current out of the source or, $I_0 = i_1 + i_2$, as shown.

$$i_1 = \frac{24}{1k} = 24 \text{ mA}$$

$$i_2 = \frac{24}{3k + 5k} = 3 \text{ mA}$$

$$\left.\begin{aligned} V_0 &= 3 \text{ mA } (3k) = 9 \text{ V} \\ I_0 &= 24 + 3 = 27 \text{ mA} \end{aligned}\right\}$$

These are the values at $t = 0^-$ and at $t = 0^+$ just after the switch is opened, since neither the inductor current nor the capacitor voltage can change at once. With the switch open, the circuit becomes a series loop with initial stored energy in both inductor and capacitor. The equivalent circuit is shown in Figure 3a and 3b. The current i(t) in Figure 3b can be found as for any RLC series loop, and then $i_1(t)$ can be otained by current division.

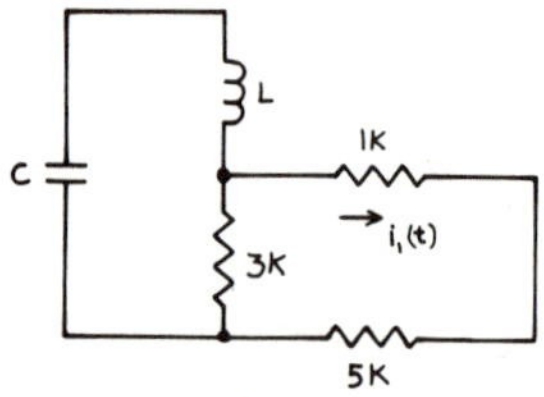

Fig.3a

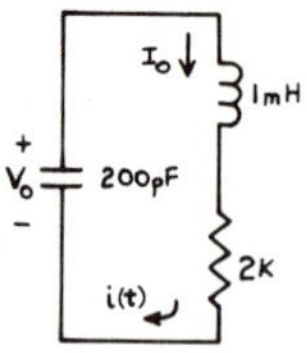

Fig.3b

The governing differential equation for the series RLC is

$$L\frac{d^2i}{dt^2} + R\frac{di}{dt} + \frac{i}{C} = 0.$$

The damped frequency is

$$\omega_d = \sqrt{\omega_0^2 - \alpha^2} \quad \text{where} \quad \omega_0^2 = \frac{1}{LC}$$

$$\alpha^2 = \left(\frac{R}{2L}\right)^2,$$

and these are evaluated to give

$\alpha = 10^6$ rad/s; $\omega_0 = 2.24 \times 10^6$ rad/s; $\omega_d = 2 \times 10^6$ rad/s.

Since $\omega_0^2 > \alpha^2$, the circuit is underdamped and an appropriate solution to assume is

$$i(t) = e^{-\alpha t}(B_1 \cos \omega_d t + B_2 \sin \omega_d t).$$

From the initial conditions above, it is known that $i(0^+) = 27$ mA

$$i(0^+) = 27 = (1)\, B \cos (0); \quad B_1 = 27 \text{ mA}$$

Again, looking at Figure 3b, note that the inductor voltage at $t = 0^+$ is the difference between the 9 volts on the capacitor and the 54 volts produced by the $I_0 = 27$ mA in the 2kΩ resistor. This will be used to find the constant B_2.

$$\frac{di}{dt} = B_1 \left[-\alpha e^{-\alpha t} \cos \omega_d t - e^{-\alpha t} \omega_d \sin \omega_d t \right]$$

$$+ B_2 \left[-\alpha e^{-\alpha t} \sin \omega_d t + e^{-\alpha t} \omega_d \cos \omega_d t \right]$$

at $t = 0^+$. This is evaluated at B_1 - 27 mA giving

$$\left.\frac{di}{dt}\right|_{t=0} = -.027\alpha + B_2 \omega_d \text{ and}$$

$$L \frac{di}{dt} = -45 = (.001)(-.027)(10^6) + 2 \times 10^6 (.001) B_2$$

$$B_2 = -\frac{45 + 27}{2000} = -.018 = -18 \text{ mA}.$$

It is apparent from Figure 3a that, by current division, only $\frac{1}{3}$ of the current in the loop is found in the 1k and 5k resistors. Thus, the final solution is

$$i(t \geq 0) = e^{-10^6 t} [9 \cos (2 \times 10^6 t)$$

$$- 3 \sin (2 \times 10^6)\, t] \text{ mA}.$$

Write the single equation required to describe the circuit of Fig. 1, using i_C as the variable. By comparing this equation with the basic equations for the series and parallel RLC circuit, determine equivalent values for R, L, and C. Find $i_C(t)$ for $t > 0$.

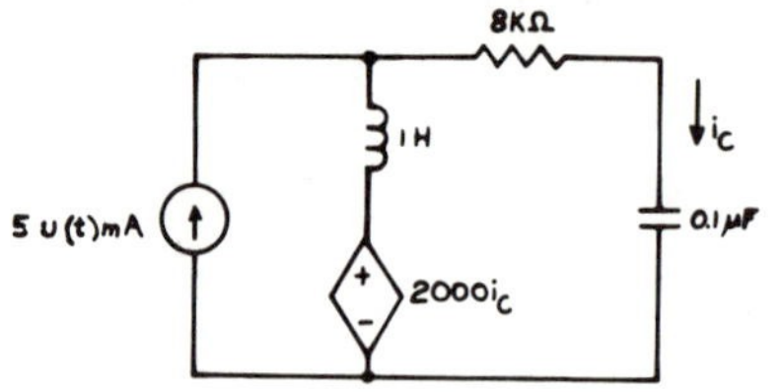

FiG.1

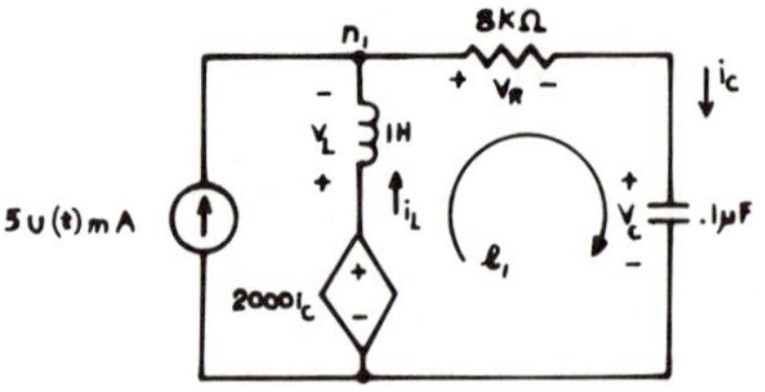

FiG.2

Solution: In Fig. 2, write a KVL equation around the loop shown, noting that $i_L = i_C - .005A$.

$$0 = -\ 2000i_C + 8000i_C + L\frac{d}{dt}(i_C - .005) + \frac{1}{C}\int_0^t i_C\, dt$$
$$+\ v_C(0^+)$$

$L = 1H$, $C = .1 \times 10^{-6}F$ $v_C(0^+) = 0$ since the current source is "dead" for $t < 0$.

$$0 = 6000i_C + \frac{d}{dt}(-\ .005 + i_C) + 10^7\int_0^t i_C\, dt$$

$$0 = 6000i_C = \frac{di_C}{dt} + 10^7\int_0^t i_C\, dt \qquad (1)$$

Eq. (1) describes the series RLC circuit containing a 6kΩ resistor, a 1-H inductor, and a .1-μF capacitor.

Since $i_L(0^-) = 0$ and $v_C(0^-) = 0$, then $i_L(0^+) = 0$ and $v_C(0^+) = 0$. If $i_L(0^+) = 0$, then all the supply current must flow through the 8-kΩ resistor and the capacitor. The voltage across the resistor capacitor branch would be $8000(.005) + v_C(0^+) = 40V + 0 = 40V$.

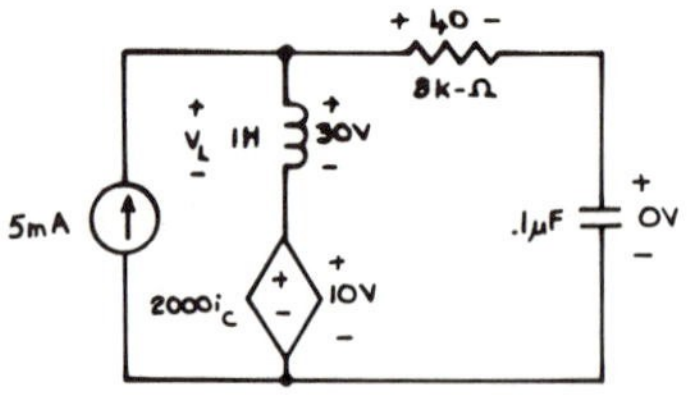

FiG.3

Fig. 3 shows that the dependent voltage source is

$2000(.005)v = 10V$ and $v_L(0^+) = -30V + 40V$

It is now established:

$i_C(0) = .005A \qquad i_L(0) = 0A$

$v_C(0) = 0V \qquad v_L(0) = -30V$

Find the parameters α and ω_0

$$\alpha = \frac{R}{2L} = \frac{6000}{2(1)} = 3000$$

$$\omega_0 = \frac{1}{\sqrt{LC}} = \frac{1}{\sqrt{.1 \times 10^{-6}}} = 3162.3$$

Since $\omega_0 > \alpha$, the response will be an exponentially damped sinusoidal current.

$$\omega_d = \sqrt{\omega_0^2 - \alpha^2} = 1000$$

$$i_C(t) = B_1 e^{-3000t} \cos 1000t + B_2 e^{-3000t} \sin 1000t \quad (2)$$

$$i_C(0) = .005 = B_1 e^{-3000(0)} \cos 1000(0) = B_1$$

Find B_2 by taking the derivative of i_C

$$\frac{di_C}{dt} = -3000\, B_1 e^{-3000t} \cos 1000t$$

$$- 1000\, B_1 e^{-3000t} \sin 1000t - 3000\, B_2 e^{-3000t} \sin 1000t$$

$$+ 1000\, B_2 e^{-3000t} \cos 1000t$$

$$\frac{v_L(0^+)}{L} = \frac{di(0)}{dt} = -300\, B_1 + 1000\, B_2$$

$$\frac{v_L(0^+)}{L} = -30 = -3000\, B_1 + 1000\, B_2; \quad B_1 = .005$$

$$B_2 = \frac{-30 + 3000\,(.005)}{1000} = -.015$$

Substituting the values found for B_1 and B_2 into Eq. (2):

$$i_C(t) = e^{-3000t}(5 \cos 1000t - 15 \sin 1000t) \text{ mA}; \; t > 0$$

CHAPTER 8

RMS VALUES, PHASORS, AND POWER

RMS VALUES

• PROBLEM 8-1

Find the rms and average values of the waveform of fig. 1.

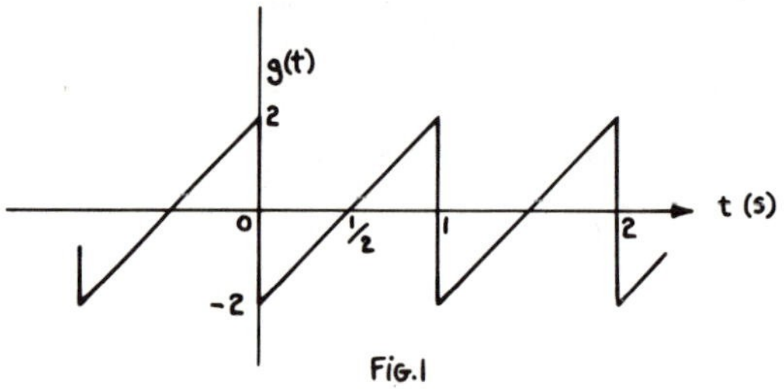

Fig.1

Solution: The average of a periodic waveform is:

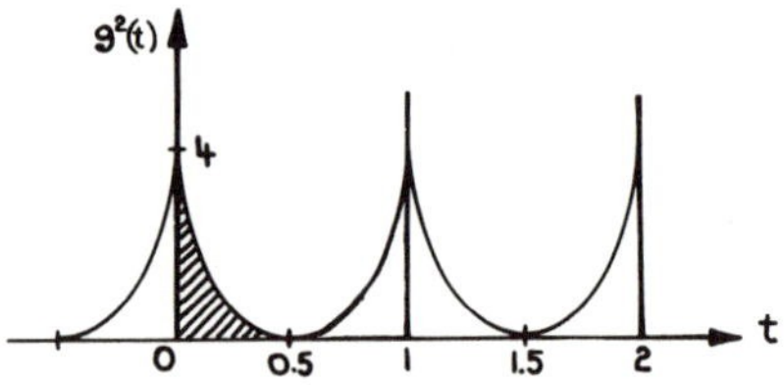

$$\text{Average of } g(t) = \frac{1}{T}\int_0^T g(t)\,dt$$

$$= \left(\frac{1}{T}\right)\left(\text{net area over one cycle;}\right) \qquad (1)$$

the rms value of a periodic waveform is:

$$\text{rms value of } g(t) = \sqrt{\frac{1}{T}\int_0^T g^2(t)\,dt} = \sqrt{\text{average of } g^2(t)}. \qquad (2)$$

The simple geometric form of the given graph suggests use of the area method rather than actual integration, i.e., the net area is equal to the value of the integral.

(a) Find the net area of each half-cycle and combine the two areas algebraically. First, for t = 0 to t = 0.5 seconds

$$\text{Area}_1 = \frac{1}{2}(\Delta t)(-2) = \frac{1}{2}\left(\frac{1}{2}\right)(-2) = -\frac{1}{2}, \quad (3)$$

using the fact that the

$$\text{area of a triangle} = \left(\frac{1}{2}\right)(\text{base})(\text{altitude}) \quad (4)$$

Similarly, for the time interval t = 0.5 to t = 1 seconds,

$$\text{Area}_2 = \frac{1}{2}(\Delta t)(+2) = \frac{1}{2}\left(\frac{1}{2}\right)(+2) = +\frac{1}{2}. \quad (5)$$

Finding the area from t = 0 to t = 1.0 seconds,

$$\text{net area} = \text{area}_1 + \text{area}_2 = -\frac{1}{2} + \frac{1}{2} = \text{zero}. \quad (6)$$

Substituting this into equation (1),

$$\text{Average of } g(t) = \frac{1}{T} \times \text{net area} = \text{zero}. \quad (7)$$

(b) To obtain the rms value by the geometric method, first square the ordinates of the curve, obtaining a new plot for $g^2(t)$. This is seen in the accompanying figure. Due to the squaring, the curve's ordinates are all positive.

Finding the area for the same time intervals used previously, we use the fact that the squared ordinates form a parabola, and

$$\text{Area under a parabolic curve} = \frac{1}{3} \times \text{base} \times \text{altitude}. \quad (8)$$

Thus, the area under the cross-hatched area is

$$\text{Area}_1 = \frac{1}{3}(\Delta t)(+4) = \frac{1}{3}\left(\frac{1}{2}\right)(+4) = +\frac{2}{3} \quad (9)$$

for the interval t = 0 to t = 0.5. Similarly, the area from t = 0.5 to t = 1 is also + 2/3, so the net area over a period of one cycle is

$$\text{Net area} = \text{area}_1 + \text{area}_2 = \frac{2}{3} + \frac{2}{3} = \frac{4}{3}. \quad (10)$$

Now we can substitute the area into equation (1) to find the average of $g^2(t)$.

$$\text{Average of } g^2(t) = \frac{1}{T} \times \text{net area} = \frac{1}{1} \times \frac{4}{3} = \frac{4}{3}. \quad (11)$$

Insert this quantity into equation (2):

$$\text{rms value} = \sqrt{\text{average of } g^2(t)} = \sqrt{\frac{4}{3}} = \frac{2}{\sqrt{3}} = 1.15. \quad (12)$$

(c) Another method would be analytically expressing the straight line from t = 0 to t = 1 as

$$g\ (t) = 4\ t - 2 \tag{13}$$

and squaring it:

$$g^2\ (t) = 16\ t^2 - 16\ t + 4. \tag{14}$$

Then, we proceed to integrate in accordance with equations (1) and (13):

$$\text{Average} = \frac{1}{T}\int_0^T (4t - 2)\ d\ t = \frac{1}{T}\left(\frac{4t^2}{2} - 2t\right) = 2T - 2T = 0; \tag{15}$$

substituting equation (14) into equation (2),

$$\text{rms} = \sqrt{\frac{1}{T}\int_0^T (16t^2 - 16t + 4)\ dt} = \sqrt{\frac{1}{T}\left[\frac{16t^3}{3} - \frac{16t^2}{2} + 4t\right]_0^T} \tag{16}$$

$$\text{rms} = \sqrt{\frac{16}{3} - 8 + 4} = \sqrt{\frac{4}{3}} = \frac{2}{\sqrt{3}} = 1.15 \tag{17}$$

for T = 1.

• PROBLEM 8-2

Using the sawtooth wave of fig.1, (a) calculate the rms value over one period; (b) calculate the rms value over the interval (0, 9.5 T) and compare your answer with part (a). (c) Does the interval affect the rms value in this case?

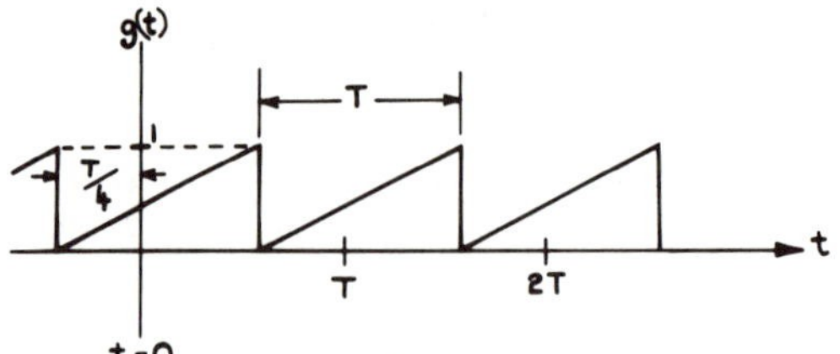

Fig.1

Solution: (a) A convenient region over which to calculate the rms value is $-\frac{1}{4}T < t < \frac{3}{4}t$. The equation of the line is

$$g\ (t) = \frac{1}{T}\ t + \frac{1}{4}.$$

The rms value squared is

$$g_{rms}^2 = \frac{1}{T} \int_{-\frac{1}{4}T}^{\frac{3}{4}T} g^2(t)\, dt$$

$$= \frac{1}{T} \int_{-\frac{1}{4}T}^{\frac{3}{4}T} \left(\frac{1}{T} t + \frac{1}{4} \right)^2 dt$$

$$= \frac{1}{T} \left(\frac{t^3}{3T^2} + \frac{t^2}{4T} + \frac{t}{16} \right) \Bigg|_{-\frac{1}{4}T}^{\frac{3}{4}T}$$

$$= 0.3333.$$

Then, $g_{rms} = 0.577$.

(b) The integral from 0 to 9.5 T can be simplified because the integral over each periodic interval is the same.

Again, the rms value squared is

$$g_{rms}^2 = \frac{1}{9.5T} \int_0^{9.5T} g^2(t)\, dt$$

$$= \frac{1}{9.5T} \left[\int_0^{\frac{3}{4}T} + \int_{\frac{3}{4}T}^{1\frac{3}{4}T} + \dots + \int_{7\frac{3}{4}T}^{8\frac{3}{4}T} + \int_{8\frac{3}{4}T}^{9.5T} \right]$$

$$= \frac{1}{9.5T} \left[\int_0^{\frac{3}{4}T} g^2(t)\, dt + 8 \text{ integrals over one period} + \int_{8\frac{3}{4}T}^{9.5T} g^2(t)\, dt \right].$$

Taking advantage of the calculations in part a, we can write

$$g_{rms}^2 = \frac{1}{9.5T} \left[\left(\frac{t^3}{3T^2} + \frac{t^2}{4T} + \frac{t}{16} \right) \Bigg|_0^{\frac{3}{4}T} + 8\,(0.3333) \right.$$

$$+ \left(\frac{t^3}{3T^2} + \frac{t^2}{4T} + \frac{t}{16} \right) \Bigg|_{-\frac{1}{4}T}^{\frac{1}{2}T} \Bigg] .$$

The latter interval can be justified by noting that the area under the curve is the same as that of the corresponding integral in the next to the last step. Evaluating the expressions, we get

$$g_{rms}^2 = \frac{1}{9.5} (0.328 + 2.666 + 0.141).$$

Hence, $g_{rms} = 0.574$

(c) Yes, the interval affects the rms value because in the second case the interval was not an integral number of periods.

• PROBLEM 8-3

Evaluate the rms value of the periodic waveform of Fig. 1.

Assume T = 1.

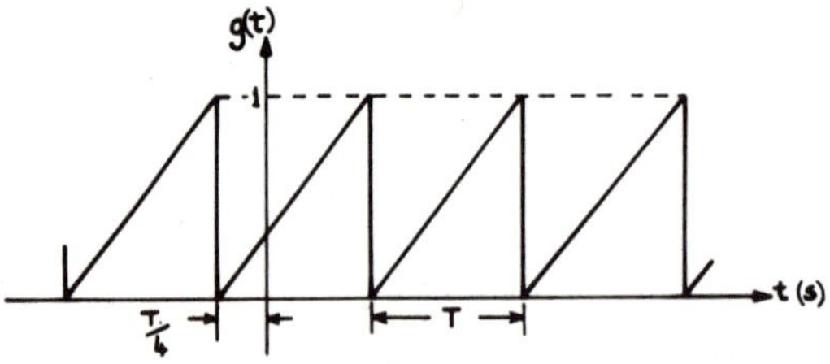

FIG. 1

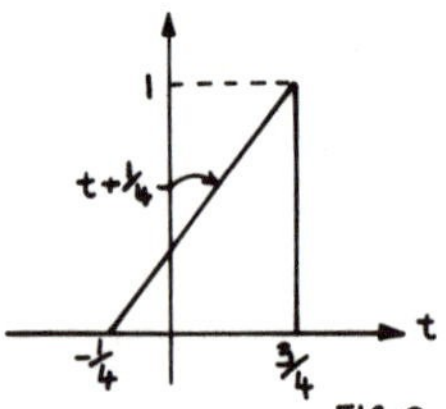

FIG. 2

Solution: The rms value of a periodic function is given by

$$\sqrt{\frac{1}{T} \int_0^T [g(t)]^2 \, dt} .$$

In our example we can integrate over the period shown in Fig. 2.

Hence,

$$G_{rms} = \sqrt{\frac{1}{1} \int_{-\frac{1}{4}}^{\frac{3}{4}} \left(t + \frac{1}{4} \right)^2 dt}$$

$$G_{rms} = \sqrt{\int_{-\frac{1}{4}}^{\frac{3}{4}} \left(t^2 + \frac{1}{2}t + \frac{1}{16}\right) dt}$$

$$G_{rms} = \sqrt{\left[\frac{t^3}{3} + \frac{t^2}{4} + \frac{t}{16}\right]_{-\frac{1}{4}}^{\frac{3}{4}}}$$

$$G_{rms} = \sqrt{\left[\frac{9}{64} + \frac{9}{64} + \frac{3}{64} + \frac{1}{192} - \frac{1}{64} + \frac{1}{64}\right]}$$

$$G_{rms} = \sqrt{\frac{27 + 27 + 9 + 1 - 3 + 3}{192}}$$

$$G_{rms} = \sqrt{\frac{64}{192}} = \frac{1}{\sqrt{3}} = 0.57735.$$

• PROBLEM 8-4

Find the rms value of the waveform of fig. 1 over the time intervals:

(a) (0, 5); (b) (0, 10)

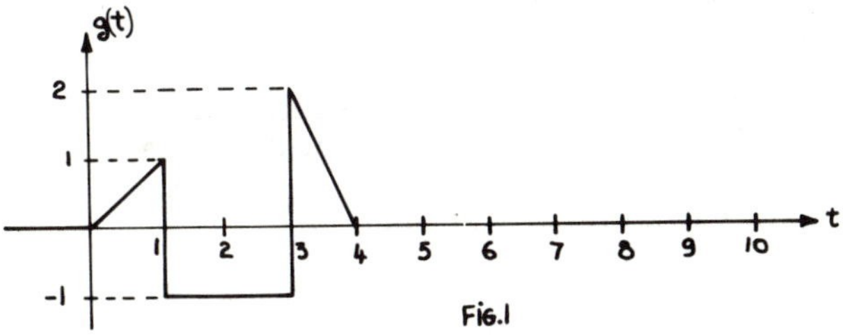

Fig.1

Solution: (a) The solution will be found by integrating over each type of geometric interval separately.

For each interval we want to calculate

$$\int g^2(t)\, dt$$

as follows:

Interval I: $0 \leq t \leq 1$

$$\int_0^1 t^2\, dt = \frac{1}{3}$$

Interval II: $1 \leq t \leq 3$

$$\int_1^3 (-1)^2\, d\, t = 2$$

Interval III: This integral is easier to calculate if the same curve is considered to go from t = 0 to t = 1. Thus,

$$\int_0^1 (-2\, t + 2)^2\, d\, t = \frac{4}{3}.$$

The square of the rms value is:

$$g_{rms}^2 = \frac{1}{5}\left[\int_0^1 + \int_1^3 + \int_3^4 + \int_4^5\right]$$

$$= \frac{1}{5}\left[\frac{1}{3} + 2 + \frac{4}{3} + 0\right].$$

Hence, $g_{rms} = 0.86$.

(b) The rms of g (t) can be evaluated easily by adding more terms for each of the intervals to cover the interval (0, 10).

Thus, we have the following:

$$g_{rms}^2 = \frac{1}{10}\left[\int_0^1 + \int_1^3 + \int_3^4 + \int_4^{10}\right]$$

$$= \frac{1}{10}\left[\frac{1}{3} + 2 + \frac{4}{3} + 0\right]$$

which gives an rms of $g_{rms} = 0.6$.

Let $i = 4 - 8\,[u\,(t - 2) - u\,(t - 3)] + t\,u\,(t - 2)$ for one period over the interval $0 \leq t \leq 5$. For this current, determine: (a) the positive and negative peak values; (b) the average value; (c) the effective value.

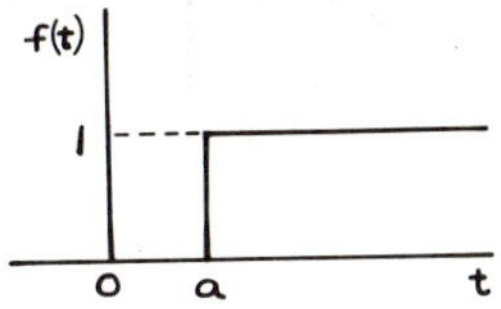

Fig.1

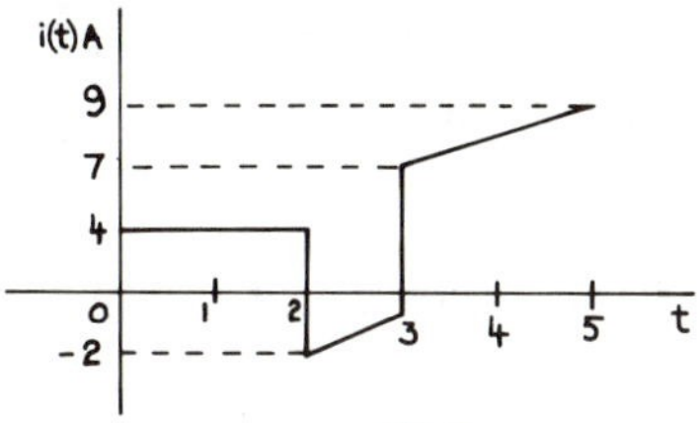

Fig.2

Solution: In order to understand the problem, it is desirable to review the properties of the unit step function, $f\,(t) = u\,(t - a)$ as sketched in Fig. 1.

This function has the value of zero for all time less than a and the value of one for all time greater than a. Note that at a, the function is discontinuous. While no physical quantities can change in a zero time interval, in several cases they may approximate such a change.

We see therefore that we may write the equation for the current in three time segments within the period interval.

These are:

$0 < t < 2$

$i\,(t) = 4 + 0 + 0 = 4\text{ A}$

$2 < t < 3$

$i\,(t) = 4 - 8\,(1) + 8\,(0) + t\,(1) = t - 4\text{ A}$

$3 < t \leq 5$

$i\,(t) = 4 - 8\,(1) + 8\,(1) + t\,(1) = t + 4\text{ A}.$

This current is shown in Fig. 2.

From Fig. 2 we see that the positive peak occurs at $t = 5$ and is 9 A. The negative peak is at $t = 2$ and is $- 2$ A. Therefore (a) $I_{pos.\ peak} = + 9\text{ A}$; $I_{neg.\ peak} = - 2\text{ A}$.

To obtain the average value, we can use the defining equation,

$$I_{ave} = \frac{1}{T}\int_0^T i\,(t)\,dt.$$

To integrate the function in this case we must integrate separately over the three intervals as follows, where

T = 5, the period:

$$I_{ave} = \frac{1}{5}\left[\int_0^2 4\,d\,t + \int_2^3 (t-4)\,d\,t + \int_3^5 (t+4)\,d\,t\right]$$

$$I_{ave} = \frac{1}{5}\left[4\,t\Big|_0^2 + \frac{(t-4)^2}{2}\Big|_2^3 + \frac{(t+4)^2}{2}\Big|_3^5\right]$$

$$I_{ave} = \frac{1}{5}\left[8 - 0 + \frac{(-1)^2}{2} - \frac{(-2)^2}{2} + \frac{9^2}{2} - \frac{7^2}{2}\right].$$

(b) $$I_{ave} = \frac{1}{5}\left[8 + \frac{1}{2} - 2 + \frac{81}{2} - \frac{49}{2}\right] = 4.5\text{ A}.$$

The defining equation of the effective value of current is $I_{eff} = \sqrt{\frac{1}{T}\int_0^T i\,(t)^2\,dt}$. It is easiest to find I_{eff}^2 first, and then perform the square root operation. In this example,

$$I_{eff}^2 = \frac{1}{5}\left[\int_0^2 4^2\,d\,t + \int_2^3 (t-4)^2\,d\,t + \int_3^5 (t+4)^2\,d\,t\right]$$

Performing the integration yields:

$$I_{eff}^2 = \frac{1}{5}\left[16\,t\Big|_0^2 + \frac{(t-4)^3}{3}\Big|_2^3 + \frac{(t+4)^3}{3}\Big|_3^5\right]$$

$$I_{eff}^2 = \frac{1}{5}\left[32 + \frac{(-1)^3}{3} - \frac{(-2)^3}{3} + \frac{9^3}{3} - \frac{7^3}{3}\right]$$

$$= \frac{1}{5}\left[32 - \frac{1}{3} + \frac{8}{3} + \frac{729}{3} - \frac{343}{3}\right]$$

$$= \frac{1}{5}\left[\frac{489}{3}\right] = \frac{489}{15} = 32.6\text{ A}^2.$$

(c) $$I_{eff} = \sqrt{32.6} = 5.71\text{ A}$$

• **PROBLEM 8-6**

(a) Find the rms value of the waveform of fig. 1 over the time interval (-1,3). (b)If the waveform of part (a) represents a current, find the energy delivered to a 5 - Ω resistor by the current over the interval (- 1, 3).

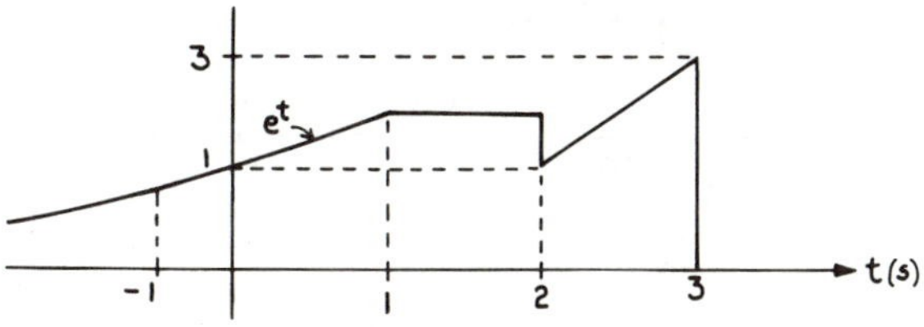

FiG.1

Solution: (a) the region is divided into three sections for which the area under the squared curve can be computed as follows:

Section I $\quad f(t) = e^t \quad -1 < t < 1$

$$\text{Area} = \int_{-1}^{1} e^{2t}\,dt = \frac{1}{2} e^{2t}\Big|_{-1}^{1} = \frac{1}{2}\left[e^2 - e^{-2}\right] = 3.63$$

Section II $\quad f(t) = e^1 = 2.72 \quad 1 < t < 2$

$$\text{Area} = (2.72)^2 \times 1 \text{ sec} = 7.40$$

Section III $\quad f(t) = 1(t - 1) \quad 2 < t < 3$

$$f^2(t) = t^2 - 2t + 1$$

$$\text{Area} = \int_{2}^{3} (t^2 - 2t + 1)\,dt = \left[\frac{1}{3}t^3 - t^2 + t\right]\Big|_2^3 = 4.33.$$

The sum of the areas under squared functions is, then, 3.63 + 7.40 + 4.33 = 15.36.

The average of this quantity is $\frac{15.36}{4 \text{ secs}} = 3.84$.

The root of this value is 1.96 which is the rms value desired.

(b) The power is $I^2_{rms}R$ for the 4 second interval for which the rms calculation is valid. Consequently, the energy is $(1.96^2 \times 5)\,\frac{\text{Joules}}{\text{sec}} \times 4 \text{ sec} = 76.8$ joules.

Find the effective (rms) values of the three periodic current waveforms shown in Fig. 1.

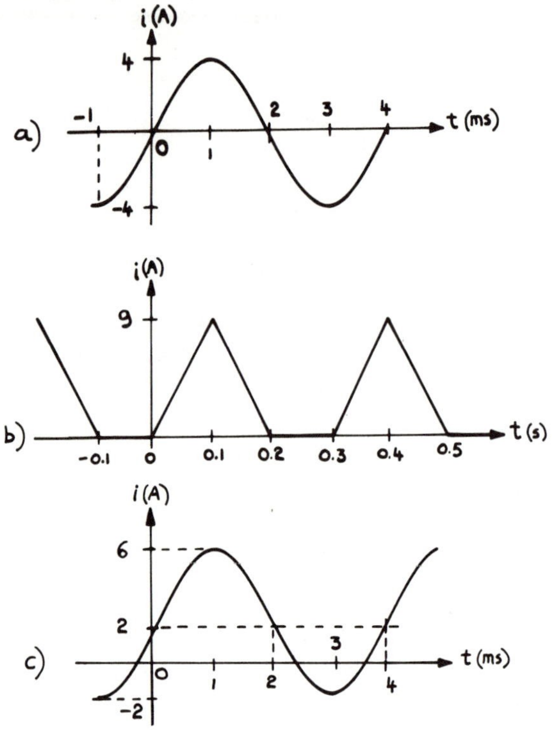

FIG. 1

Solution: The effective value of any periodic waveform can be evaluated using

$$x_{eff} = \sqrt{\frac{1}{T}\int_0^T [x(t)]^2\, dt} \qquad (1)$$

where x (t) is the periodic function defined over the period T.

a) The current waveform in Fig. 1 (a) is i (t) $= 4 \sin \omega t$ where $\omega = \frac{2\pi}{T}$ and T = 4 ms. Substituting into equation (1) gives

$$i_{eff} = \sqrt{\frac{1}{.004}\int_0^{0.004} 16 \sin^2\left(\frac{2\pi}{0.004}t\right) dt.}$$

Substituting $\sin^2 x = \frac{1}{2}(1 - \cos 2x)$ yields:

$$i_{eff} = \sqrt{\frac{16}{0.004}\left[\int_0^{0.004} \frac{1}{2}\,dt - \int_0^{0.004} \frac{1}{2}\cos\left(\frac{4\pi}{0.004}t\right)dt\right]}$$

$$i_{eff} = \sqrt{\left[\frac{16}{0.004}\left[\frac{t}{2} - \frac{0.004}{8\pi}\sin\left(\frac{4\pi}{0.004}t\right)\right]\right]_0^{0.004}}$$

$$i_{eff} = \sqrt{\frac{16}{0.004}\left(\frac{0.004}{2} - \frac{0.004}{8\pi}\sin 4\pi\right)};$$

since $\sin 4\pi = 0$

$$i_{eff} = \sqrt{\frac{16}{2}} = \sqrt{8} = 2.83 \text{ A}.$$

The result could have been easily obtained by taking the peak voltage and dividing it by $\sqrt{2}$, $i_{eff} = \frac{4}{\sqrt{2}} = 2.83$ A. This is applicable only to sinusoids with a zero D.C. component.

(b) The waveform of Fig. 1 (b) can be represented over its period T = 0.3 s by the following functions:

$$i(t) = \begin{cases} 90t & 0 \le t < 0.1 \\ -90t + 18 & 0.1 \le t < 0.2 \\ 0 & 0.2 \le t < 0.3 \end{cases}$$

The square of the function i(t) is

$$[i(t)]^2 \begin{cases} 8100t^2 & 0 \le t < 0.1 \\ 8100t^2 - 3240t + 324 & 0.1 \le t < 0.2 \\ 0 & 0.2 \le t < 0.3. \end{cases}$$

We can now substitute into eq. (1):

$$i_{eff} = \sqrt{\frac{1}{0.3}\left[\int_0^{0.1}(8100t^2)dt + \int_{0.1}^{0.2}(8100t^2 - 3240t + 324)dt\right]}$$

$$i_{eff} = \sqrt{\frac{8100}{0.3}\int_0^{0.1}(t^2)\,dt + \frac{324}{0.3}\int_{0.1}^{0.2}(25t^2 - 10t + 1)\,dt}$$

$$i_{eff} = \sqrt{\frac{8100}{0.3}\left[\frac{t^3}{3}\right]_0^{0.1} + \frac{324}{0.3}\left[\frac{25t^3}{3} - \frac{10t^2}{2} + t\right]_{0.1}^{0.2}}$$

$$i_{eff} = \sqrt{\frac{8100}{0.3}\left(\frac{0.001}{3}\right) + \frac{324}{0.3}\left(\frac{0.2}{3} - \frac{0.4}{2} + 0.2 - \frac{0.025}{3} + \frac{0.1}{2} - 0.1\right)}$$

$$i_{eff} = \sqrt{\frac{8.1}{0.9} + \frac{324}{0.3}\left(\frac{1}{120}\right)} = \sqrt{18} = 4.24 \text{ A.}$$

(c) The waveform of Fig. 1 (c) is the same as that of Fig. 1 (a), except that it has a + 2V d. c. component. Hence

$$i(t) = 2 + 4 \sin \omega t$$

$$[i(t)]^2 = (2 + 4 \sin \omega t)(2 + 4 \sin \omega t)$$

$$= 4 + 16 \sin \omega t + 16 \sin^2 \omega t.$$

Substituting into Eq. (1) yields

$$i_{eff} = \sqrt{\frac{1}{0.004}\int_0^{0.004}\left(4 + 16 \sin \frac{2\pi}{0.004}t + 8 - 8 \cos \frac{4\pi}{0.004}t\right) dt}$$

$$i_{eff} = \sqrt{\frac{1}{0.004}\left[4t - \frac{0.032}{\pi}\cos\frac{2\pi}{0.004}t + 8t - \frac{0.008}{\pi}\sin\frac{4\pi}{0.004}t\right]_0^{0.004}}$$

$$i_{eff} = \sqrt{\frac{1}{0.004}\left(0.016 - \frac{0.032}{\pi}\cos 2\pi + \frac{0.032}{\pi}\cos 0 + 0.032\right.}$$

$$\left. \frac{0.008}{\pi} \sin 4\pi \right]$$

$$i_{eff} = \sqrt{\frac{1}{0.004} \left(\frac{0.048}{1}\right)} = \sqrt{12} = 3.46 \text{ A.}$$

• PROBLEM 8-8

Find the average power delivered to a 12-Ω resistor by each of the three periodic current waveforms displayed in Fig. 1.

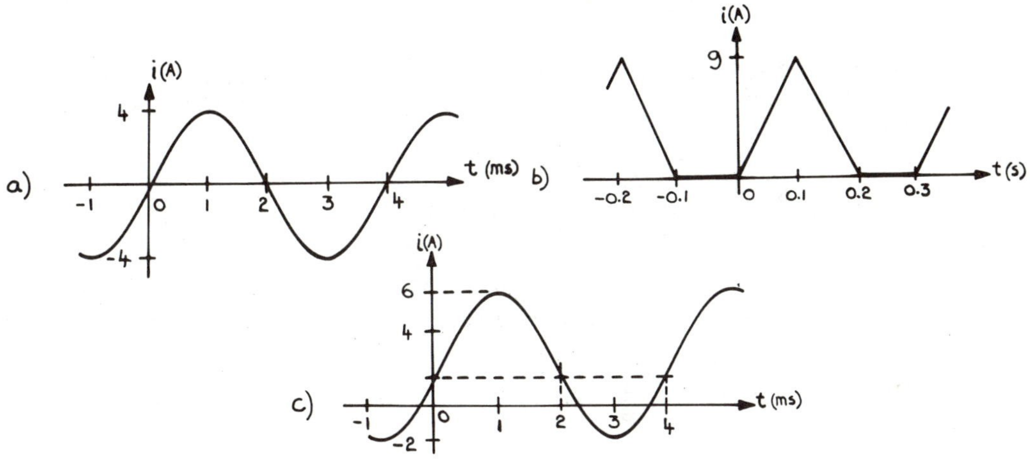

FIG. 1

Solution: Since power is the time rate of energy absorption, average power for a periodic current or voltage waveform is obtained by first finding the energy absorbed in one complete cycle of current and then dividing this energy by the length of one cycle (T). To find the energy we integrate the instantaneous power over the period (one cycle):

$$\text{Energy per cycle} = \int p\,(t)\,dt = \int_0^T Ri^2\,dt, \quad (1)$$

from which we obtain

$$P = \text{average power} = \frac{\text{Energy per cycle}}{\text{period}} = \frac{1}{T}\int_0^T Ri^2\,dt. \quad (2)$$

(a) (1) the first waveform is seen to be a pure sine wave, which can be expressed analytically as

$$i\,(t) = A \sin \omega t = A \sin \frac{2\pi}{T} t \quad (3)$$

where $\omega = 2\,\pi\,f$ and $f = 1/T$. (4)

Now, substituting the value for i (t) stated in equation (3) into equation (2) we obtain

$$P = \frac{1}{T} \int_0^T R\,A^2 \sin^2 \omega\, t\, dt\,. \qquad (5)$$

Next modify it to

$$P = \frac{R\,A^2}{T} \int_0^T \left(\frac{1}{2} - \frac{1}{2} \cos 2\,\omega\, t\right) d\,t \qquad (6)$$

by making use of the trigonometric identity

$$\sin^2 \theta = \frac{1}{2} - \frac{1}{2} \cos 2\,\theta \qquad (7)$$

in order to facilitate integration. Now, performing the integration indicated in equation (6), term by term, we obtain

$$P = \frac{R\,A^2}{T} \left[\frac{t}{2} - \left(\frac{1}{2}\right) \frac{1}{2\omega} \sin 2\,\omega\, t\right]_0^T, \qquad (8)$$

where we have made use of the integrating relationship

$$\int \cos a\,t\,d\,t = \frac{1}{a} \sin a\,t. \qquad (9)$$

Next we must evaluate the integral over the specified limits:

$$P = \frac{R\,A^2}{T} \left[\frac{T}{2} - 0 - \frac{1}{4\omega} (\sin 4\,\pi - \sin 0)\right] = \frac{R\,A^2}{2}, \qquad (10)$$

since $\sin 4\,\pi = 0$ and $\sin 0 = 0$.

To obtain the final solution, substitute the known values of R and A into equation (10):

$$\text{Average power} = P = \frac{R\,A^2}{2} = \frac{12 \times 4^2}{2} = 96 \text{ watts.} \qquad (11)$$

(2) There is another way of doing this problem which involves taking a "short cut." We start with:

$$\text{Average power} = \text{Average of } Ri^2 = \text{Avg. of } R\,A^2 \sin^2 \omega\, t, \qquad (12)$$

which becomes (by using the trigonometric identity of equation (6) again):

$$P = \text{Avg. of } R\,A^2 \left(\frac{1}{2} - \frac{1}{2}\cos 2\,\omega\,t\right). \qquad (13)$$

Now, the average of any pure sine or cosine is zero. This leads us to the solution

$$P = R\,A^2 \left(\frac{1}{2} - 0\right) = \frac{R\,A^2}{2} = 96 \text{ watts} \qquad (14)$$

without going through the actual process of integration.

(b) The straight-line components of the waveform in (6) suggest the use of a geometric approach rather than mathematical integration. That is, since integration can be equated with finding the net area under a curve, it is sometimes convenient to look at the problem graphically.

We begin by squaring ordinates of the current curve, which results in the parabolic curves shown in Fig. 2.

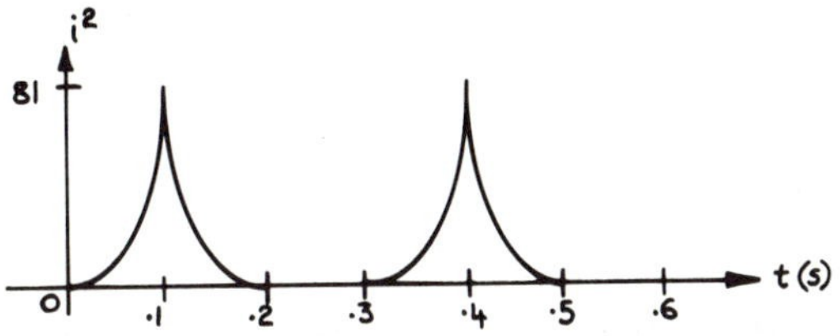

FIG. 2

Now the area under the i^2 curve can be found using the following information:

$$\text{Area under a parabola} = \frac{1}{3} \times \text{base} \times \text{altitude}. \qquad (15)$$

Making the appropriate substitutions in this formula, we have

$$\int i^2\,dt = \text{Area under } i^2 \text{ curve} = \frac{1}{3} \times 0.2 \times 81$$

$$= 5.4 \text{ amps}^2 - \text{sec}. \qquad (16)$$

Substituting this result into equation (2) yields

$$P = \frac{1}{T} R \int_0^T i^2\,d\,t = \frac{1}{0.3} \times 12 \times 5.4 = 216 \text{ watts}. \qquad (17)$$

This method is quicker than putting the curve into analytical form, squaring and integrating, and can be used for waveforms such as triangles or rectangles

(c) The only difference between the curve in (c) and the curve in (a) is that a d-c component of 2 amperes has been added. We have seen that obtaining the square of the current is a key step in this type of problem.

To obtain the square of the current first express the current analytically,

$$i(t) = 2 + 4 \sin \omega t, \tag{18}$$

and then square the current,

$$i^2(t) = 2^2 + 2 (2)(4 \sin \omega t) + 4^2 \sin^2 \omega t. \tag{19}$$

Proceed as in the short-cut shown in (a):

$$R i^2 = 12 \left[4 + 16 \sin \omega t + 16 \left(\frac{1}{2} - \frac{1}{2} \cos 2 \omega t\right)\right]. \tag{20}$$

We have again used the trigonometric expansion of equation (6).

Proceeding to find the average we obtain

$$P = \text{avg. of } R i^2 = 12 \left[4 + 0 + 16 \left(\frac{1}{2} - 0\right)\right] = 144 \text{ watts}, \tag{21}$$

recognizing again that the average value of a pure sine or cosine is zero.

• PROBLEM 8-9

A resultant current wave is made up of two components, a 5-A d-c component and a 60-Hz a-c component which is of sinusoidal waveform and which has a max value of 4-A.

a) Draw a sketch of the resultant current wave.

b) Write the analytical expression for the resultant wave, choosing the t = 0 reference where the a-c component is zero and where $\frac{di}{dt}$ is positive.

c) What is its average value for one cycle?

d) What is the effective value of the current?

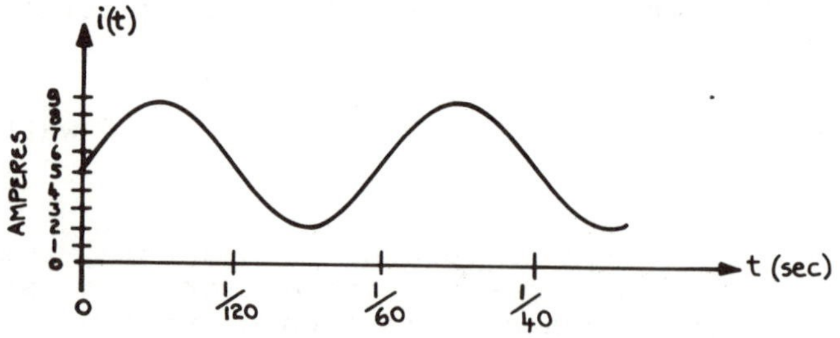

FIG. 1 GRAPH OF RESULTANT CURRENT

Solution: Picture this voltage as a sine wave of 4 volts amplitude which is displaced vertically in the positive direction by 5 volts. Likewise, the analytical expression is the sum of the two voltages. The average value of a periodic waveform is

$$\text{Average of } i\ (t) = \frac{1}{T} \int_0^T i\ dt\ , \tag{1}$$

and its effective value is

$$\text{effective (or rms) value of } i\ (t) = \sqrt{\frac{1}{T} \int_0^T i^2\ dt}\ . \tag{2}$$

(a) The current in question is seen to be a sine wave "riding" on a positive d-c value. The peak value of the current occurs when the sine wave is at its maximum value of 4; at that instant the current is 5 + 4 = 9 amperes. Similarly, when the sine wave is at its negative maximum value of -4, the current has a minimum value of 5 - 4 = 1 amperes. The waveform is, therefore, pictured as a sinusoidal wave form varying between the limits of 9 and 1 amperes, as pictured in Figure 1.

(b) To write the analytical expression for this waveform, add the two components algebraically:

$$i\ (t) = \text{d-c component} + \text{a-c component}. \tag{3}$$

The angular frequency of the sine wave is

$$\omega = \frac{2\pi}{T} = 2\ \pi\ f = 2\ \pi\ (60) \quad \text{rad/sec}. \tag{4}$$

Thus, we write

$$i\ (t) = 5 + 4 \sin\ (120\ \pi)\ t\ A \tag{5}$$

(c) To find the average value, substitute equation (5) into equation (1).

$$\text{Average of } i\ (t) = \frac{1}{T} \int_0^T (5 + 4 \sin\ (2\pi/T)\ t)\ d\ t. \tag{6}$$

Perform the indicated integration:

$$\text{Avg. of } i\ (t) = \frac{1}{T} \left[5\ t + \frac{4}{2\pi/T} \left(- \cos \frac{2\pi}{T}\ t \right) \right]_0^T \tag{7}$$

Then, evaluate for the stated limits:

$$\text{Avg. of } i\ (t) = \frac{1}{T} \left[5\ T + \frac{2\ T}{\pi}\ (- \cos\ 2\ \pi + \cos\ 0) \right] \tag{8}$$

$$\text{Avg. of } i(t) = 5 + \frac{2}{\pi}(-1 + 1) = 5 \text{ A}. \qquad (9)$$

(d) To find the effective (rms) value, first perform the squaring of i (t) as called for in equation (2):

$$i^2(t) = 25 + 40 \sin \omega t + 16 \sin^2 \omega t. \qquad (10)$$

To make the third term easier to integrate, replace it with its equivalent according to the trigonometric identity

$$\sin^2 x = \frac{1}{2} - \frac{1}{2} \cos 2x \qquad (11)$$

which changes equation (10) to

$$i^2(t) = 25 + 40 \sin \omega t + 8 - 8 \cos 2 \omega t. \qquad (12)$$

Substituting equation (12) into equation (2) results in

rms value of i (t) =

$$\sqrt{\frac{1}{T} \int_0^T \left[33 + 40 \sin \frac{2\pi}{T} t - 8 \cos \frac{4\pi}{T} t\right] dt}. \qquad (13)$$

Performing the indicated integration,

$$\text{rms value} = \sqrt{\frac{1}{T}\left[33t + \left(\frac{40}{2\pi/T}\right)\left(-\cos \frac{2\pi}{T} t\right) - \left(\frac{8}{4\pi/T}\right)\left(\sin \frac{4\pi}{T} t\right)\right]_0^T}, \qquad (14)$$

and evaluating,

$$\text{rms value} = \sqrt{\frac{1}{T}\left[33T + \frac{20T}{\pi}(-\cos 2\pi + \cos 0) - \frac{2T}{\pi}(\sin 4\pi - \sin 0)\right]} \qquad (15)$$

$$= \sqrt{33 + \frac{20}{\pi}(-1 + 1) - \frac{2}{\pi}(0 - 0)} = \sqrt{33}. \qquad (16)$$

The rms value of i (t) = 5.75 A. (17)

• **PROBLEM 8-10**

Write the equivalent polar form of the phasor 3 + j4 where the numbers refer to unit lengths. Illustrate the phasor by means of a diagram.

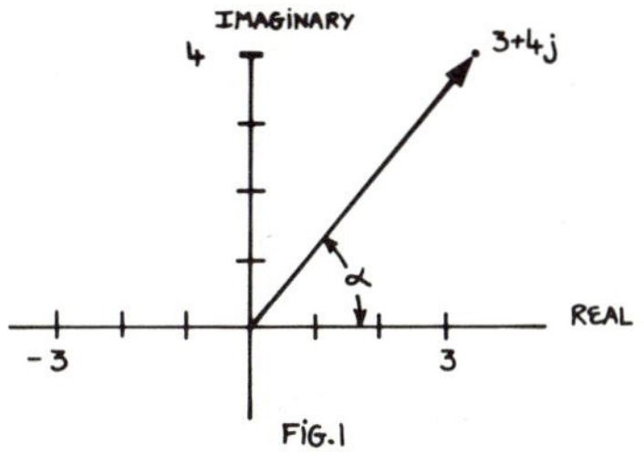

FIG.1

Solution: A phasor in the form a + jb can be written

$\sqrt{a^2 + b^2}\, e^{j \tan^{-1} \frac{b}{a}}$, or in shorthand notation,

$\sqrt{a^2 + b^2} \angle \tan^{-1} \frac{b}{a}$.

We have $\sqrt{a^2 + b^2} = \sqrt{3^2 + 4^2} = \sqrt{25} = 5$

$$\tan^{-1} \frac{b}{a} = \tan^{-1} \frac{4}{3} = 53.1°;$$

therefore $3 + j4 = 5e^{j53.1°} = 5\angle 53.1°$.

The solution can be obtained graphically by finding the point (3, j4) on a real-imaginary axis.

In Fig. 1, draw a vector from the origin to the point 3 + j4 and note that the length of the vector is $\sqrt{3^2 + 4^2} = 5$ and the angle is $\alpha = \tan^{-1} \frac{4}{3} = 53.1°$.

• **PROBLEM 8-11**

A phasor is given in the form of $10e^{-j120°}$. Write the symbolic polar and cartesian forms of the phasor, and illustrate, by means of a phasor diagram, the magnitude and phase position of the phasor.

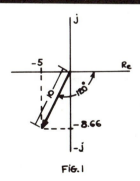

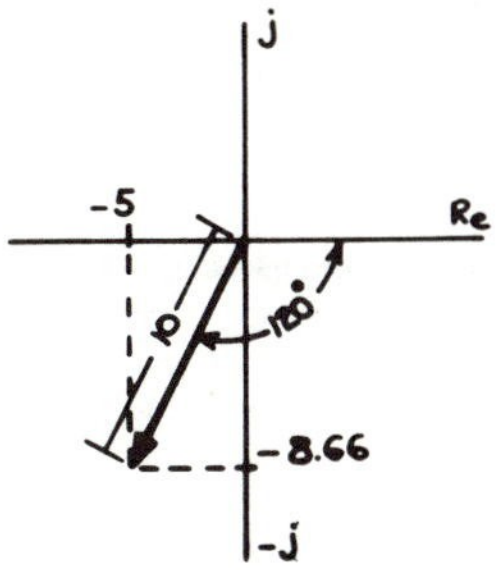

FIG.1

Solution: The phasor $10e^{-j120°}$ is written as $10\underline{/-120°}$ in symbolic polar form. In order to write it in the cartesian form, we must make use of Euler's identity, which is stated as:

$$Ce^{j\alpha} = C\cos\alpha + jC\sin\alpha$$

where $C = \sqrt{a^2 + b^2}$

and $a = C\cos\alpha = 10\cos(-120°) = -5$

$b = jC\sin\alpha = 10\sin(-120°) = -8.66.$

This gives $10e^{-j120°} = 10\underline{/-120°} = -5 - j8.66.$

Fig. 1 shows a graphical solution.

• PROBLEM 8-12

Transform each of the following currents to the time domain: (a) 6 - j8 A; (b) - 8 + j6 A; (c) - j10 A

Solution: (a) 6 - j8 in polar form is $\sqrt{6^2 + 8^2}\underline{/\tan^{-1} -\frac{8}{6}}$

or $10\underline{/-53.1°}$. Therefore, the time domain representation is:

$$i(t) = 10\cos(\omega t - 53.1°)\text{ A}.$$

(b) - 8 + j6 conversion gives:

$$\sqrt{8^2 + 6^2}\underline{/\tan^{-1}\frac{6}{-8}} = 10\underline{/+143.1}$$

so that $i(t) = 10\cos(\omega t + 143.1)$ A.

(c) $-j10 = 0 - j10 = 10\underline{/-90°}$

$$i(t) = 10\cos(\omega t - 90°).$$

• PROBLEM 8-13

Give each voltage in phasor form:

(a) $20\cos(\omega t + 40°)$; (b) $12\cos(\omega t + 360°) + 10\cos(\omega t + 70°)$; (c) $-20\sin(\omega t - 10°) + 20\cos(\omega t - 50°)$ V.

Solution: In each case the sinusoidal function, v (t), must be expressed in terms of a magnitude R and an angle θ. The basic identity that will be used is

$$\vec{V} = R\cos(\omega t + \theta) = R\underline{/\theta}\text{ V}. \quad (1)$$

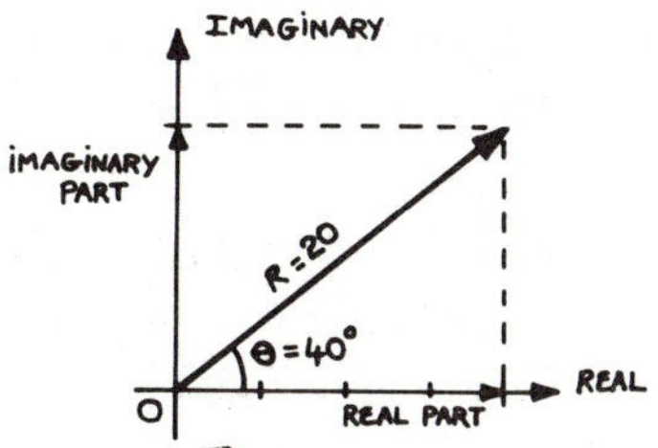

FIG. 1

(a) In this case the magnitude is $R = 20$ and the angle is $\theta = 40°$. Hence, the phasor voltage is expressed $\vec{V} = 20 \underline{/40°}$. It is often desirable to express the phasor quantity in terms of real and imaginary parts. This is easily done with the use of Fig. 1. The real part is $20 \cos 40° = 15.3$. The imaginary part is $20 \sin 40° = 12.8$.

Hence, the complex vector can be written as $\vec{V} = 15.3 + j\ 12.8$ V.

(b) In this case the phasor can be expressed as the sum of two phasors $\vec{V}_1$ and $\vec{V}_2$ each having magnitudes R_1 and R_2 and angles θ_1 and θ_2 respectively.

$\vec{V}_1$ will be the phasor expression for $12 \cos (\omega t + 360°)$. For this first expression $R_1 = 12$ and $\theta = 360°$; hence,

$$\vec{V}_1 = 12 \underline{/360°} = 12 \underline{/0°}\ .$$

$\vec{V}_2$ will be the phasor expression for $10 \cos (\omega t + 70°)$. For $\vec{V}_2$ $R_2 = 10$ and $\theta = 70°$; hence

$$\vec{V}_2 = 10 \underline{/70°}.$$

The total phasor then is

$$\vec{V} = \vec{V}_1 + \vec{V}_2 = 12 \underline{/0°} + 10 \underline{/70°}\ \text{V}.$$

It is seen that $\vec{V}_1$ has no imaginary part because it lies entirely on the x-axis. $\vec{V}_2$ has a real part $10 \cos 70° = 3.4$ and an imaginary part $10 \sin 70° = 9.4$. Hence the total complex vector can be written as

$$\vec{V} = 12 + 3.4 + j\ 9.4\ \text{V}$$

$$\vec{V} = 15.4 + j\ 9.4\ \text{V}.$$

(c) Once again the phasor will be expressed as two phasors $\vec{V}_1$ and $\vec{V}_2$, each having magnitudes R_1 and R_2 and angles θ_1 and θ_2 respectively. In this case, however, the first expression uses a sine function instead of the cosine function. In order to use the identity in equation (1), the sine function must be expressed as a cosine function. Note that

$\sin(\omega t + \theta) = \cos(\omega t + \theta - 90°)$;

hence $-20 \sin(\omega t - 10°) = -20 \cos(\omega t - 100°)$,

$R_1 = -20$, $\theta_1 = -100°$, and $\vec{V}_1 = -20\angle -100°$. The second expression $20 \cos(\omega t - 50°)$ already uses a cosine function so there is no need to convert from sine to cosine.

$R_2 = 20$, $\theta_2 = -50°$ and $\vec{V}_2 = 20\angle -50°$.

The total phasor then is

$$\vec{V} = \vec{V}_1 + \vec{V}_2$$

$$\vec{V} = -20\angle -100° + 20\angle -50° \text{ V}$$

$\vec{V}_1$ has an imaginary part $-20 \sin -100° = 19.7$ and a real part $-20 \cos -100° = 3.5$. $\vec{V}_2$ has an imaginary part $20 \sin -50° = 15.3$ and a real part $20 \cos -50° = 12.9$. Hence, the total complex expression is

$$\vec{V} = 3.5 + 12.9 + j(19.7 - 15.3) \text{ V}$$

$$\vec{V} = 16.4 + j\,4.4 \text{ V}.$$

• PROBLEM 8-14

Perform the following phasor operations: (a) $\vec{A} + \vec{B}$ where $\vec{A} = 10\angle 36.9°$ and $\vec{B} = 6\angle 120°$; (b) $\vec{A} + \vec{B}$ where $\vec{A} = 14\angle 60°$ and $\vec{B} = 20\angle 15°$; (c) $\vec{A} + \vec{B} + \vec{C}$ where $\vec{A} = 40e^{j120°}$ $\vec{B} = 20\angle -40°$ and $\vec{C} = 26.46 + j0$.

Solution: (a) Given two phasors in polar form, add them by first writing them in rectangular form. Hence,

$$10\angle 36.9° = 10 \cos(36.9°) + j\,10 \sin(36.9°) = 8 + j6$$

and

$$6\angle 120° = 6 \cos(120°) + j6 \sin(120°) = -3 + j5.2.$$

Adding the two phasors component by component gives

$$\begin{array}{r} 8 + j6 \\ \underline{-3 + j5.2} \\ 5 + j11.2 \end{array}$$

in rectangular form and

$$\sqrt{5^2 + 11.2^2} \angle \tan^{-1} \frac{11.2}{5} = 12.27\angle 65.9° \text{ in polar form.}$$

(b) Convert to rectangular form and add as in part (a).

Hence,

$$14\angle 60° = 14 \cos (60°) + j14 \sin (60°)$$
$$= 7 + j12.12$$
$$20\angle 15° = 20 \cos (15°) + j20 \sin (15°)$$
$$= 19.3 + j5.18$$
$$7 + j12.12 + 19.3 + j5.18 = 26.3 + j17.3$$

which is $\sqrt{26.3^2 + 17.3^2} \angle \tan^{-1} \frac{17.3}{26.3} = 31.5\angle 33.3°$ in polar form.

(c) Note that the three given phasors are written in three different forms: $40e^{j120°}$ is written in complex polar form, $20\angle -40°$ is written in "shorthand" polar form, and 26.46 + j0 is written in rectangular form. Converting the first two phasors in polar form to rectangular form and adding them together with the third gives

$$\begin{array}{rl} -20 & + j34.6 \\ 15.3 & - j12.9 \\ 26.46 & + j0 \\ \hline 21.76 & + j21.7, \end{array}$$

which is $30.8\angle 45°$ in polar form.

• PROBLEM 8-15

Add the following current waves as phasors:

$$i_1 = 5 \sin \omega t$$

$$i_2 = 10 \sin (\omega t + 60°)$$

Solution: Since a phasor is defined as Re $\{e^{j\omega t}\} = \cos \omega t \Rightarrow 1\angle 0°$,

$$i_1 = 5 \sin \omega t \Rightarrow 5\angle -90°$$

$$i_2 = 10 \sin (\omega t + 60°) \Rightarrow 10\angle -90° + 60° = 10\angle -30°.$$

In rectangular form, $i_1 = 5 \cos (-90°) + j\, 5 \sin (-90°)$ $i_1 = 0 - j5$, $i_2 = 10 \cos (-30°) + j\, 10 \sin (-30°) = 8.66 - j5$ the sum of i_1 and i_2 gives $i_0 = 8.66 - j10$ which is written as

$$\vec{I}_0 = \sqrt{8.66^2 + 10^2}\,\underline{/\tan^{-1} - \frac{10}{8.66}} = 13.23\underline{/-49.1°}$$

in polar form. Transforming the phasor back to a wave gives

$$13.23 \cos (\omega t - 49.1°),$$

which can also be written as,

$$13.23 \sin (\omega t - 49.1° + 90°) =$$

$$13.23 \sin (\omega t + 40.9°).$$

• PROBLEM 8-16

Find $\vec{A} - \vec{B}$ for the phasors

$$\vec{A} = 30\underline{/60°}$$

$$\vec{B} = 21 (\cos 160° - j \sin 160°).$$

Solution: Converting the polar expression $30\underline{/60°}$ to rectangular form gives $30 (\cos 60° + j \sin 60°)$, the same form as the expression for $\vec{B}$.

Hence, $\vec{A} = 30 (\cos 60° + j \sin 60°) = 15 + j26$

$$\vec{B} = 21 (\cos 160° - j \sin 160°) = - 19.73 - j7.18$$

$$\vec{A} - \vec{B} = 34.73 + j33.18,$$

which is $48\underline{/43.6°}$ in polar form.

• PROBLEM 8-17

Given the phasors $\vec{V}_{12} = 6\underline{/30°}$, $\vec{V}_{23} = 2\underline{/0°}$, $\vec{V}_{24} = 6\underline{/-60°}$, and $\vec{V}_{15} = 4\underline{/90°}$, find: (a) $\vec{V}_{14}$; (b) $\vec{V}_{43}$; (c) $\vec{V}_{35}$.

Solution: The double subscript notation gives potential differences on a relative basis. To find the desired point-to-point voltage, we must add or subtract the given quantities in the appropriate order.

(a) $\vec{V}_{14} = \vec{V}_{12} + \vec{V}_{24} = 6\underline{/30°} + 6\underline{/-60°}$

$$\vec{V}_{14} = 5.2 + j3 + 3 - j5.2 = 8.2 - j2.2 = 8.49\underline{/-15°}$$

(b) To find $\vec{V}_{43}$,

$$\vec{V}_{43} = \vec{V}_{42} - \vec{V}_{32} = - \vec{V}_{24} - (-\vec{V}_{23}) = -6\underline{/-60°} + 2\underline{/0°}$$

$$\vec{V}_{43} = - 3 + j5.2 + 2 + j0 = - 1 + j5.2 = 5.3\underline{/100.9°}\ \text{V}$$

(c) To find $\vec{V}_{35}$, given $\vec{V}_{15}$, we first find $\vec{V}_{13}$.

$$\vec{V}_{13} = \vec{V}_{12} + \vec{V}_{23} = 5.2 + j3 + 2 + j0 = 7.2 + j3$$

Now, $\vec{V}_{51} = -\,(4\underline{/90°}) = 4\underline{/-90°} = 0 - j4$

$$\vec{V}_{31} = -\,\vec{V}_{13} = -\,7.2 - j3$$

then, $\vec{V}_{53} = \vec{V}_{51} - \vec{V}_{31} = 0 - j4 - (-7.2 - j3) = 7.2 - j1.$

Finally, $\vec{V}_{35} = -\,\vec{V}_{53} = -7.2 + j1 = 7.27\underline{/172.1°}.$

• PROBLEM 8-18

Find the product of the two phasors

2 (cos 40° + j sin 40°) and 3 (cos 100° + j sin 100°).

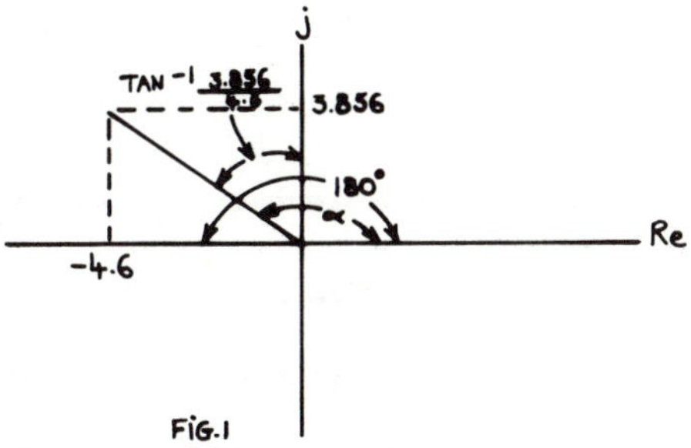

FIG. 1

Solution: Evaluate each of the expressions, 2 (cos 40° - j sin 40°) = 1.532 + j1.286 and 3 (cos 100° + j sin 100°) = - 0.521 + j2.954.

Multiplying the two expressions found above, gives

$$\begin{array}{r} 1.532 + j1.286 \\ \underline{-0.521 + j2.954} \\ j4.526 - 3.800 \\ \underline{-0.800 - j0.670 \quad\quad} \\ -4.6 \quad + j3.856 \end{array}$$

Writing -4.6 + j3.856 in polar form gives

$\sqrt{4.6^2 \times 3.856^2}\underline{/\tan^{-1} - \frac{3.856}{4.6} + 180°}$. Add 180° since the vector -4.6 + j3.856 is in the second quadrant as shown in Fig. 1.

Hence, $-4.6 + j3.856 = 6\underline{/140°}$.

• PROBLEM 8-19

Raise the phasor (8.66 + j5.0) to the second power; to the fifth power.

Solution: Change the phasor to polar form:

$$8.66 + j5.0 = \sqrt{8.66^2 + 5^2}\,\angle \tan^{-1}\frac{5}{8.66}$$

$$= 10\angle 30°.$$

Raise it to the second power:

$$(8.66 + j5.0)^2 = 10^2\angle 2 \times 30° = 100\angle 60° \;.$$

Raise it to the 5th power:

$$(8.66 + j5.0)^5 = 10^5\angle 5 \times 30° = 100{,}000\angle 150° \;.$$

• PROBLEM 8-20

Find $\frac{\vec{A}}{\vec{B}}$, when A = 10 + j17.3 and B = 4.33 + j2.5, by the method of complex conjugates and by conversion into polar form.

Solution: First write

$$\frac{A}{B} = \frac{10 + j17.3}{4.33 + j2.5}$$

then multiply both numerator and denominator by the complex conjugate of the denominator.

Hence,
$$\frac{A}{B} = \frac{10 + j17.3}{4.33 + j2.5}\,\frac{(4.33 - j2.5)}{(4.33 - j2.5)}$$

$$\frac{A}{B} = \frac{(43.3 + 43.3) + j\,(75 - 25)}{4.33^2 + 2.5^2}$$

$$\frac{A}{B} = 3.465 + j2,$$

or $4\angle 30°$ in polar form.

A second method is to convert A and B to polar form and divide, as follows:

$$A = 10 + j17.3 = \sqrt{10^2 + 17.3^2} \angle \tan^{-1} \frac{17.3}{10} = 20\angle 60°$$

$$B = 4.33 + j2.5 = \sqrt{4.33^2 + 2.5^2} \angle \tan^{-1} \frac{2.5}{4.33} = 5\angle 30°.$$

$$\text{Hence,} \quad \frac{A}{B} = \frac{20\angle 60°}{5\angle 30°} = \frac{20}{5} \angle 60° - 30° = 4\angle 30°,$$

which is the same result obtained above.

PHASORS

• PROBLEM 8-21

> Let it be required to find the square roots of A where A = 3.08 + j8.455.

Solution: A has two roots which can be found by converting A to polar form:

$$A = \sqrt{3.08^2 + 8.455^2} \angle \tan^{-1} \frac{8.455}{3.08} = 9.0\angle 70°$$

$$\text{Then the first root is} \quad \sqrt{9.0} \angle \frac{70°}{2} = 3\angle 35°$$

$$\text{and the second root is} \quad \sqrt{9.0} \angle \frac{70° + 360°}{2} = 3\angle 215°.$$

Note that either root squared yields the phasor A.

• PROBLEM 8-22

> Find the cube roots of the phasor (8 + j0), and draw a complete phasor diagram of the phasor and its three roots.

Solution: The phasor 8 + j0 has three cube roots which can be found by first converting 8 + j0 to polar form. Hence,

$$8 + j0 = 8 \angle 0°$$

$$\text{The first root is} \quad \sqrt[3]{8} \angle \frac{0°}{3} = 2\angle 0°.$$

$$\text{The second root is} \quad \sqrt[3]{8} \angle \frac{0° + 360°}{3} = 2\angle 120°.$$

The third root is $\sqrt[3]{8} \angle \frac{0° + 2\,(360°)}{3} = 2\angle 240°$.

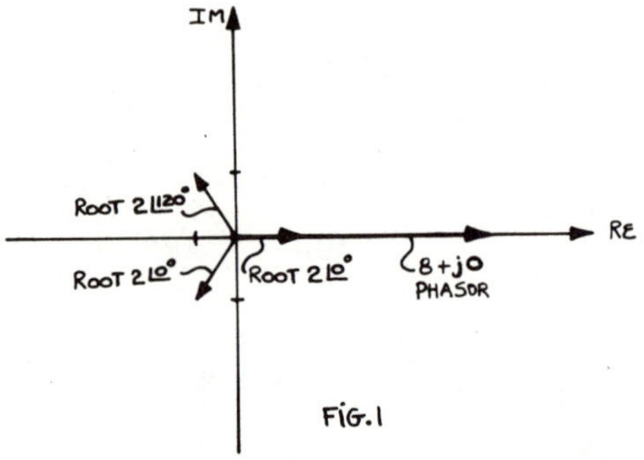

FIG. 1

Note that any of the above roots when cubed equals $8\angle 0°$. Fig. 1 shows the phasor and its three roots.

• PROBLEM 8-23

Find the phasor current $\vec{I}$.

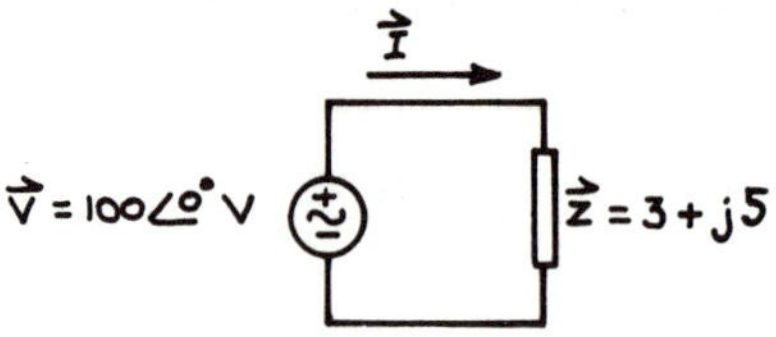

Solution: Using Ohm's Law we find the phasor current to be

$$\vec{I} = \frac{\vec{V}}{\vec{Z}} = \frac{100\angle 0°}{3 + j5} = \frac{100}{3 + j5}\ .$$

Multiplying both numerator and denominator by the complex conjugate of 3 + j5 yields

$$\vec{I} = \frac{100}{3 + j5}\ \frac{(3 - j5)}{(3 - j5)} = \frac{300 - j500}{3^2 + 5^2}$$

$$\vec{I} = \frac{300 - j500}{34} = 8.82 - j\ 14.71\ \text{A}$$

Expressed in polar form:

$$\vec{I} = A\angle\theta = \sqrt{8.82^2 + 14.71^2}\ \angle \tan^{-1} - \frac{14.71}{8.82}$$

$$\vec{I} = 17.15\ \angle -59.1°\ \text{A}.$$

• PROBLEM 8-24

Find the voltage source necessary to produce the response in Fig. 1.

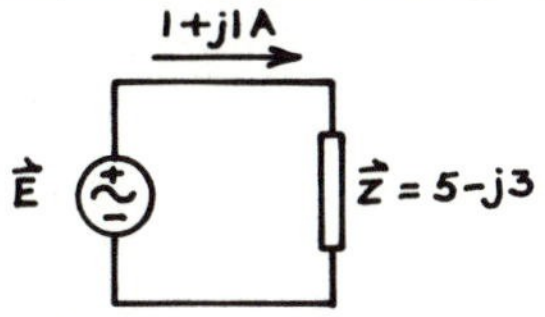

FIG.1

Solution: Using Ohm's Law we find

$$\vec{E} = \vec{I}\,\vec{Z} = (1 + j)\,(5 - j3)$$

multiplying out to simplify,

$$\vec{E} = 5 + j5 - j3 + 3$$

$$\vec{E} = 8 + j2 \text{ V.}$$

Expressing $\vec{E}$ in polar form:

$$\vec{E} = \sqrt{8^2 + 2^2}\ \underline{/\tan^{-1}\frac{2}{8}}$$

$$\vec{E} = 8.25\ \underline{/14°}\text{ V.}$$

• PROBLEM 8-25

Find the voltage source necessary to produce the response in Fig. 1.

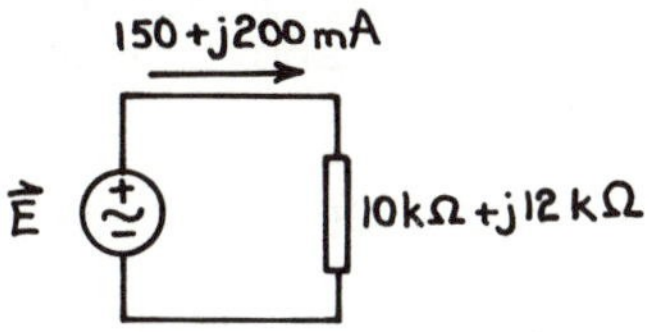

FIG.1

Solution: Using Ohm's Law we find

$$\vec{E} = \vec{I}\,\vec{Z} = (150 + j200)\,(10 + j12)\text{ V}$$

multiplying out to simplify,

$$\vec{E} = 1500 + j2000 + j1800 - 2400$$

$$\vec{E} = -\,900 + j3800\text{ V.}$$

Expressing $\vec{E}$ in polar form

$$\vec{E} = \sqrt{900^2 + 3800^2}\ \angle \tan^{-1} - \frac{3800}{400} + 180°$$

$$\vec{E} = 3905\ \angle - 76.7 + 180°$$

$$\vec{E} = 3905\ \angle 103.3°\ \text{V}.$$

• PROBLEM 8-26

A box containing only passive elements possesses four terminals, labeled 1, 2, 3, and 4. If a current source, $i_{in,1} = 10 \cos (1000\,t - 20°)$ A, is applied between terminals 1 and 2, then the response $i_{out,3} = 2 \sin (1000\,t + 50°)$ A, is obtained at terminals 3 and 4. Find $i_{out,3}$ if $i_{in,1}$ = (a) $5 \sin (1000\,t - 30°)$; (b) $6\,e^{j\,1000\,t}$; (c) $j\,5\,e^{j\,(1000\,t + 20°)}$; (d) $(8 + j\,6)\,e^{j\,(1000\,t + 30°)}$ A.

Solution: The input current can be written in polar form,

$$\vec{i}_{in,1} = 10\angle -20°\ \text{A}$$

and the output $\vec{i}_{out,3} = 2\ \angle -40°$ A.

Note that the amplitude of the current is reduced $\frac{1}{5}$th and a phase shift of - 20° is added. Since the box contains passive elements any sinusoidal input current at the same frequency as the test current above should produce the same changes in amplitude and phase.

(a) input: $5 \sin (1000\,t - 30°) \Rightarrow 5\ \angle -120°$

output: $1\ \angle -140° \Rightarrow \cos (1000\,t - 140°)$ or

$\sin (1000\,t - 50°)$

(b) input: $6e^{j\,1000\,T} \Rightarrow 6\ \angle 0°$

output: $\frac{6}{5}\ \angle -20° \Rightarrow 1.2\,e^{j\,(1000\,t - 20°)}$

(c) input: $j\,5\,e^{j\,(1000\,t + 20°)} \Rightarrow 5\ \angle 110°$

output: $1\ \angle 90° \Rightarrow e^{j\,(1000\,t + 90°)}$

(d) input: $(8 + j\,6)\,e^{j\,(1000\,t + 30°)} \Rightarrow (10\,\underline{/52.87°})(1\,\underline{/30°})$

$= 10\,\underline{/82.87°}$

output: $2\,\underline{/62.87°} \Rightarrow 2\,e^{j\,(1000\,t + 62.87°)}$

POWER

• PROBLEM 8-27

> With reference to the circuit of Fig. 1, what should be the value of R_L to absorb a maximum power? Find the value of this maximum power.

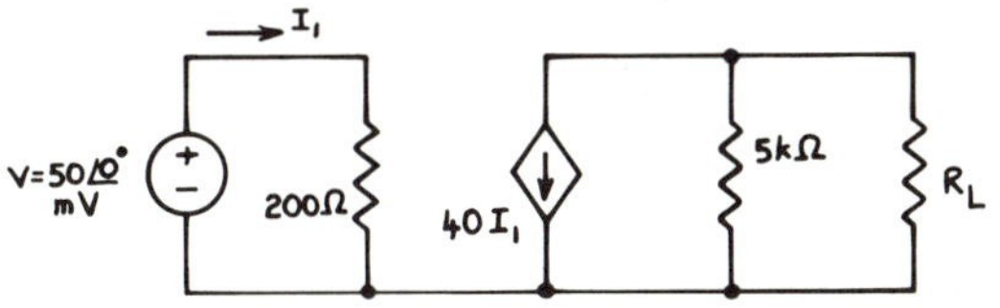

Fig.1

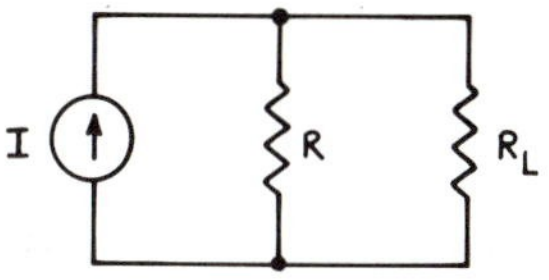

Fig.2

Solution: To absorb maximum power in R_L in a circuit as shown in Fig. 2, $R = R_L$.

$I = 5\,\underline{/0°}$ mA

$i_{rms} = \frac{5}{\sqrt{2}}$ mA

In the circuit of Fig. 1 R_L must equal the impedance Z looking back into the circuit.

Hence, as shown in Fig. 3, $R_L = 5\ K\ \Omega$ for maximum power transfer. When $V = 50$ m V, $I_1 = \dfrac{.05\,\underline{/0°}}{200} = 0.25\,\underline{/0°}$ m A; hence we obtain the circuit shown in Fig. 4.

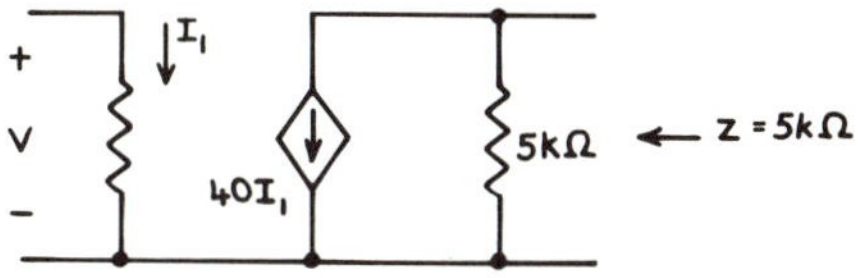

Fig.3

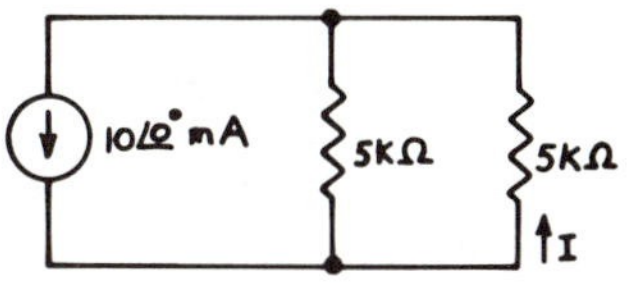

Fig.4

Using the relation $p = i_{rms}^{\ 2}\,R$ the maximum power absorbed by R_L is $\left(\dfrac{.005}{\sqrt{2}}\right)^2 (5 \times 10^3) = 62.5$ m W.

• PROBLEM 8-28

> In the circuit shown, determine the following: (a) currents i_1, i_2, and i_3; (b) peak and average values of total power dissipated; (c) peak and average values of power dissipated in each resistance.

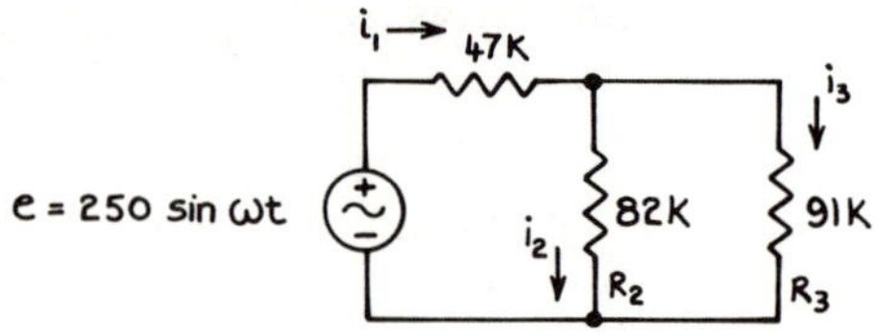

Solution: Use Ohm's Law to solve for the current i_1, and then use the current divider principle to solve for i_2 and i_3. Next, find the power delivered by the source, $p(t) = ei$ Lastly, use Joule's Law, $p = i^2 R$, to find the power in each resistance.

(a) Compute the total resistance of the network as seen by the source.

$$R = R_1 + \frac{1}{\frac{1}{R_2} + \frac{1}{R_3}} = 47K + \frac{1}{\frac{1}{82K} + \frac{1}{91K}} = 90.1\ K\Omega. \quad (1)$$

Then use it to calculate i_1:

$$i_1 = \frac{e}{R} = \frac{250 \sin \omega t}{90{,}100} = 2.77 \sin \omega t \text{ mA} . \quad (2)$$

Apply the current divider principle to solve for i_2 and i_3

$$i_2 = i_1 \frac{R_3}{R_2 + R_3} = 2.77(\sin \omega t) \times \frac{91}{82 + 91} = 1.46 \sin \omega t \text{ mA} \quad (3)$$

$$i_3 = i_1 \frac{R_2}{R_2 + R_3} = 2.77(\sin \omega t) \times \frac{82}{82 + 91} = 1.31 \sin \omega t \text{ mA} \quad (4)$$

Multiply the amplitudes of these currents by 0.707 to obtain the rms values:

$$I_1 = 1.96 \text{ m A},\ I_2 = 1.03 \text{ m A},\ I_3 = 0.93 \text{ m A}. \quad (5)$$

(b) The total power dissipated is the power being supplied by the source.

$$\text{Total power} = p\,(t) = e\,i \quad (6)$$

$$p\,(t) = (250 \sin \omega t)(2.77 \times 10^{-3} \sin \omega t) = 693 \sin^2 \omega t \text{ mW} . \quad (7)$$

The peak power is the highest instantaneous value of $p\,(t)$ which occurs when $\sin \omega t = 1.0$.

$$\text{Peak power} = 693\ (1)^2 = 693 \text{ m W}. \quad (8)$$

Solving for the average power is aided by using the trigonometric identity

$$\sin^2 a = \frac{1}{2} - \frac{1}{2} \cos 2a \quad (9)$$

which converts the instantaneous power to

$$p\ (t) = \frac{693}{2} - \frac{693}{2} \cos 2\ \omega\ t \text{ mW} . \qquad (10)$$

Now, the average of any periodic function is:

$$\text{Average of } x\ (t) = \frac{1}{T} \int_0^T x\ (t)\ dt \qquad (11)$$

where T is the period of the waveform and

$$\omega = \frac{2\pi}{T}. \qquad (12)$$

Applying this principle to the second term of equation (10), we have

$$\text{Average of } - \frac{693}{2} \cos 2\,\omega\ t = - \frac{693}{2} \frac{1}{T} \int_0^T \cos\ 2\ \omega\ t\ dt \qquad (13)$$

$$\text{Average} = - \frac{693}{2} \frac{1}{T} \frac{1}{2\omega} [\sin\ 2\ \omega\ t]_0^T \qquad (14)$$

$$\text{Average} = - \frac{693}{8\pi} [\sin\ 4\ \pi - \sin\ 0] = 0\text{W} \qquad (15)$$

where use has been made of equation (12). A shorter way is recognition of the fact that the average of any sinusoidal wave is zero over one or more integral number of periods since the area above the time axis is equal to the area below the axis.

Returning to equation (10), solve for the average power.

$$\text{Average of } p\ (t) = \frac{693}{2} - 0 = 347 \text{ m W}.$$

(c) The power in each resistance can be found by applying Joule's Law,

$$p\ (t) = R\ i^2 = R\ (I_m \sin\ \omega\ t)^2 = R\ I_m{}^2 \sin^2\ \omega\ t \text{ watts} \qquad (16)$$

$$\text{and } p\ (t) = R\ I_m{}^2 \left(\frac{1}{2} - \frac{1}{2} \cos\ 2\ \omega\ t\right) \qquad (17)$$

From equation (16), it is evident that

$$\text{peak power in each resistor} = R\ I_m{}^2 , \qquad (18)$$

and from equation (17), we obtain

Average power in each resistor $= R\, I_m^{\,2}/2 = \frac{1}{2}$ (peak power). (19)

Substituting appropriate values for R and I_m in equations (18) and (19), we are able to calculate:

$$\text{Peak power in } R_1 = (47{,}000)\ (2.77 \times 10^{-3})^2 = 361 \text{ m W} \quad (20)$$

$$\text{Peak power in } R_2 = (82{,}000)\ (1.46 \times 10^{-3})^2 = 175 \text{ m W} \quad (21)$$

$$\text{Peak power in } R_3 = (91{,}000)\ (1.31 \times 10^{-3})^2 = 156 \text{ m W} \quad (22)$$

Check: 361 + 175 + 156 = 692 mW (23)

$$\text{Average power in } R_1 = \frac{1}{2}\ (361) = 181 \text{ m W} \quad (24)$$

$$\text{Average power in } R_2 = \frac{1}{2}\ (175) = 88 \text{ m W} \quad (25)$$

$$\text{Average power in } R_3 = \frac{1}{2}\ (156) = 78 \text{ m W} \quad (26)$$

Check: 181 + 88 + 78 = 347 m W (27)

• PROBLEM 8-29

Find the input impedance, Z, at 60 Hz.

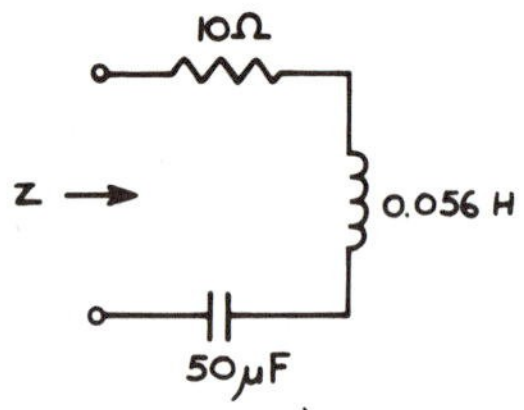

FIG.1

Solution: The input impedance, $\vec{Z}$, can be found by computing

$Z_R + Z_L + Z_C$, where

$Z_R = R$, $Z_L = j\omega L$, and $Z_C = \dfrac{1}{j\omega C}$,

$$\omega = 2\pi \quad = 2\pi(60)\ \frac{\text{rad}}{\text{s}} = 377\ \frac{\text{rad}}{\text{s}}.$$

$$\text{Hence,}\quad \vec{Z} = 10\ \Omega + j\ (377)\ (0.56) + \frac{1}{j377\ (50 \times 10^{-6})}$$

$$\vec{Z} = 10 + j21.1 - j\left(\frac{1}{0.019}\right)$$

$$\vec{Z} = 10 + j21.1 - j52.6$$

$\vec{Z} = 10 - j31.5\ \Omega$ in rectangular form.

$\vec{Z}$ in polar form can be found by

$$A + jB = \sqrt{A^2 + B^2} \angle \tan^{-1} \frac{B}{A} .$$

Hence, $\vec{Z} = \sqrt{10^2 + 31.5^2} \angle \tan^{-1} - \frac{31.5}{10}$

$$\vec{Z} = 33.05\angle -72.39° \ \Omega .$$

• PROBLEM 8-30

A 120- pF capacitor is connected in series with a 15-mH inductor whose series resistance is 500Ω. What is the impedance of this circuit at a frequency of 115-kHz? at 120 kHz?

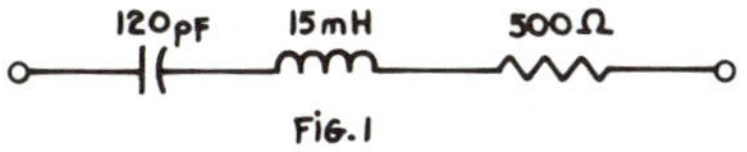

Fig. 1

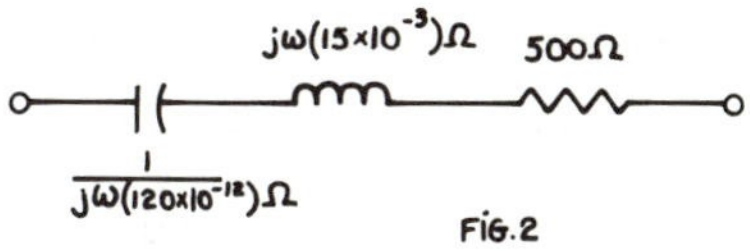

Fig. 2

Solution: Figure 1 shows the given connection in the time domain. In Fig. 2, we have transformed the network into the frequency domain.

The impedance is given by

$$\vec{Z} = - jX_C + jX_L + R$$

$$= - j \left(\frac{1}{\omega (120 \times 10^{-12})} \right) + j (\omega (15 \times 10^{-3}) + 500 \ \Omega .$$

a) The frequency, f, is given as 115 kHz. Then

$$\omega = 2\pi f = 723 \times 10^3 \text{ rad/sec}$$

and $$\vec{Z} = - j \left(\frac{1}{(723 \times 10^3)(120 \times 10^{-12})} \right)$$

$$+ j (723 \times 10^3 \times 15 \times 10^{-3}) + 500$$

$$= - j (1.15 \times 10^4) + j (1.08 \times 10^4) + 500$$

$$\vec{Z} = 500 - j700 \ \Omega$$

$$= \sqrt{(500)^2 + (700)^2} \angle \arctan \left(\frac{-700}{500} \right)$$

$$Z = 860 \angle -54.5° \ \Omega .$$

b) When f = 120 kHz, $\omega = 2\pi\ (120) = 754 \times 10^3$ rad./sec.

Then, $$Z = -j\left(\frac{1}{754 \times 10^3\ (120 \times 10^{-12})}\right)$$

$$+ j\ (754 \times 10^3 \times 15 \times 10^{-3}) + 500$$

$$= -j\ (1.11 \times 10^4) + j\ (1.13 \times 10^4) + 500$$

$$\vec{Z} = 500 + j200$$

$$= \sqrt{(500)^2 + (200)^2} \ \angle \arctan\left(\frac{200}{500}\right)$$

$$\vec{Z} = 539 \angle 21.8° \ \Omega .$$

• PROBLEM 8-31

> At 2000 rad/s find the admittance of (a) a 5 - Ω resistor in series with a 100 - μF capacitor; (b) the parallel combination of a 2.5 m H inductor and the network of (a) above; (c) the series combination of a 100 - uF capacitor and the network of (b) above.

Solution: (a) A series impedance Z of a series connection of a resistor of R Ohm and a capacitance C Farad is given by

$$Z = R + \frac{1}{j\ \omega\ C} \qquad (1)$$

where ω is an angular frequency in rad/s. This can be further modified to

$$Z = \frac{j\ \omega\ C\ R + 1}{j\ \omega\ C} \qquad (2)$$

Admittance is the reciprocal of the impedance. Therefore

$$Y = \frac{1}{Z} = \frac{j\ \omega\ C}{j\ \omega\ C\ R + 1}\ . \qquad (3)$$

Rationalizing the above equation yields

$$Y = \frac{j\ \omega\ C}{1 + j\ \omega\ C\ R} \cdot \frac{1 - j\ \omega\ C\ R}{1 - j\ \omega\ C\ R}$$

$$= \frac{j\ \omega\ C + \omega^2 C^2\ R}{1 + \omega^2 C^2\ R^2}$$

$$= \frac{\omega^2 C^2 R}{1 + \omega^2 C^2 R^2} + j \frac{\omega C}{1 + \omega^2 C^2 R^2}. \qquad (4)$$

Now, $\omega = 2000$ rad/s

$$\left.\begin{aligned} R &= 5\ \Omega \\ C &= 100\ \mu F = 100 \times 10^{-6}\ F \\ &= 10^{-4}\ F \end{aligned}\right\} \qquad (5)$$

Substituting values given in (5) into (4) we obtain

$$Y = \frac{(2x10^3)^2 x (10^{-4})^2 x\ 5}{1 + (2x10^3)^2 x (10^{-4})^2 x\ 5^2} + j \frac{2x10^3\ x10^{-4}}{1 + (2x10^3)^2 x (10^{-4})^2 x\ 5^2}$$

$$\vec{Y} = 0.1 + j\ 0.1 \text{ mho}. \qquad (6)$$

(b) A parallel admittance of a combination of an inductor of L Henry and an admittance of Y_a is given by

$$Y_b = \frac{1}{j \omega L} + Y_a. \qquad (7)$$

If Y_a is represented by a conductance G_a and a susceptance B_a as

$$\vec{Y}_a = G_a + j B_a \qquad (8)$$

then $Y_b = \frac{1}{j \omega L} + G_a + j B_a$

$$= G_a + j B_a - j \frac{1}{\omega L}$$

$$\vec{Y}_b = G_a + j \left(B_a - \frac{1}{\omega L}\right). \qquad (9)$$

From Eq. (6),

$$\vec{Y}_a = G_a + j B_a = 0.1 + j\ 0.1 \qquad (10)$$

or
$$\left.\begin{aligned} G_a &= 0.1 \\ B_a &= 0.1 \end{aligned}\right\} \qquad (11)$$

It is also given that

$$\left.\begin{aligned} \omega &= 2 \times 10^3 \text{ rad/s} \\ L &= 2.5 \text{ m H} = 2.5 \times 10^{-3}\ H \end{aligned}\right\} \qquad (12)$$

Substituting Eqs. (11) and (12) in Eq. (9) yields

$$Y_b = 0.1 + j\left(0.1 - \frac{1}{2 \times 10^3 \times 2.5 \times 10^{-3}}\right)$$

$$= 0.1 + j\,(-\,0.1)$$

$$\vec{Y}_b = 0.1 - j\,0.1.\ \text{mho} \tag{13}$$

(c) Impedance of series combination of a capacitor of C Farad and an admittance of Y_b mho is given by

$$Z_C = \frac{1}{j\,\omega\,C} + \frac{1}{Y_b}\ . \tag{14}$$

The admittance Y_b can be expressed by the conductance G_b and the susceptance B_b, as

$$\vec{Y}_b = G_b + j\,B_b \tag{15}$$

Substituting Eq. (15) in Eq. (14) yields

$$\vec{Z}_C = \frac{1}{j\,\omega\,C} + \frac{1}{G_b + j\,B_b}$$

$$= \frac{G_b + j\,B_b + j\,\omega\,C}{j\,\omega\,C\,(G_b + j\,B_b)}$$

$$= \frac{G_b + j\,(B_b + \omega\,C)}{-\omega\,C\,B_b + j\,\omega\,C\,G_b}\ . \tag{16}$$

The admittance is a reciprocal of the impedance. Therefore

$$\vec{Y}_C = \frac{1}{Z_C} = \frac{-\,\omega C\,B_b + j\,\omega\,C\,G_b}{G_b + j\,(B_b + \omega\,C)}. \tag{17}$$

Rationalizing, we obtain

$$\vec{Y}_C = \frac{-\,\omega C\,B_b + j\,\omega\,C\,G_b}{G_b + j\,(B_b + \omega\,C)} \cdot \frac{G_b - j\,(B_b + \omega\,C)}{G_b - j\,(B_b + \omega\,C)}$$

$$= \frac{-\omega C B_b G_b + \omega C G_b\,(B_b + \omega\,C) + j\{\omega C G_b{}^2 + \omega\,C B_b\,(B_b + \omega\,C)\}}{G_b{}^2 + (B_b + \omega\,C)^2}$$

$$= \frac{\omega^2 C^2 G_b + j\omega C (G_b{}^2 + B_b{}^2 + \omega C B_b)}{G_b{}^2 + (B_b + \omega C)^2}. \quad (18)$$

$$\left.\begin{aligned} \text{Now} \quad \omega &= 2 \times 10^3 \quad \text{rad/s} \\ C &= 10^{-4} \quad \text{F} \\ G_b &= 0.1 \quad \text{mho} \\ B_b &= -0.1 \quad \text{mho} \end{aligned}\right\} \quad (19)$$

Substitute (19) in (18):

$$\vec{Y}_C = \frac{(2x10^3)^2 x (10^{-4})^2 x\ 0.1 + j2x10^3 x\ 10^{-4}(0.1^2 + 0.1^2 - 2x10^3 x\ 10^{-4} x0,1)}{(0.1)^2 + (-0.1 + 2x10^3 x\ 10^{-4})^2}$$

$$\vec{Y}_C = 0.2 + j\,0 = 0.2 \text{ mho}. \quad (20)$$

Important steps used in solving this type of problem are:

1. Obtain series impedance first, then admittance.
2. Hold rationalization until the last moment.
3. Simplify the final equation as much as possible before substituting given values.

The widely accepted international scientific unit for mho at present is Siemens.

• PROBLEM 8-32

If the remaining two terminals are left open-circuited and ω = 1 k rad/s , Find the input impedance of the network below at terminals: (a) a-b; (b) c-d; (c) a-c; (d) Find Z_{in} at c-d if a and b are short-circuited.

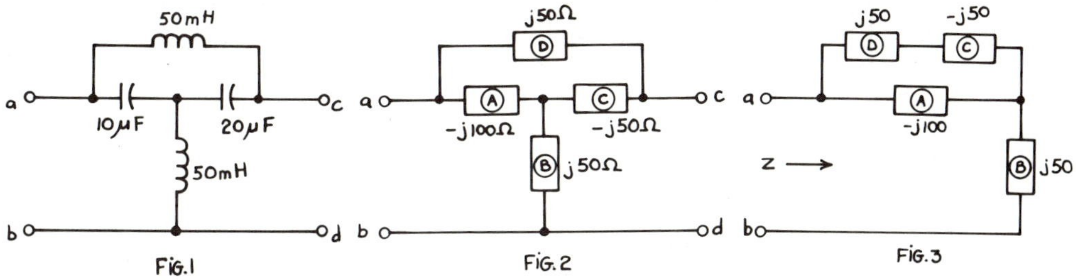

Solution: Before proceeding, compute the impedance of each element in the circuit of Fig. 1 and draw the new circuit shown in Fig. 2.

$$Z_L = j\omega L$$

$$Z_C = -\frac{j}{\omega C}$$

(a) The impedance across the terminals a-b can be found by combining impedances in the circuit shown in Fig. 3.

Note Fig. 3 is the same circuit as Fig. 2.

Since impedances D and C add to zero they short circuit impedance A and the impedance at terminals a-b is B = j50Ω.

(b) Re-drawing Fig. 2 to find the impedance at terminals c-d yields:

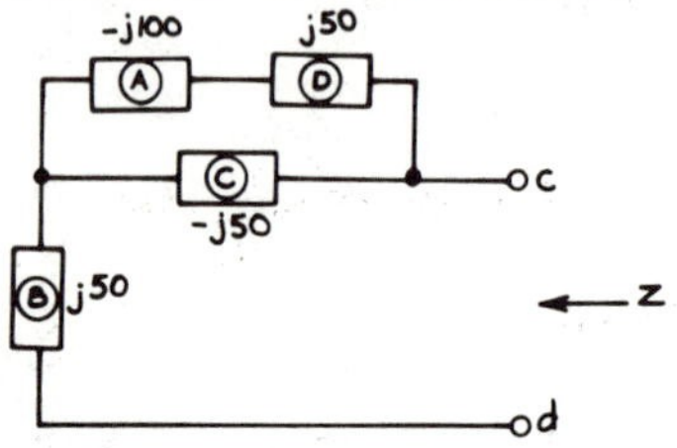

Fig.4

$$Z = \frac{(Z_A + Z_D)\, Z_C}{Z_A + Z_D + Z_C} + Z_B$$

$$Z = \frac{(-j50)\ j50}{j100} + j50 = j25\Omega .$$

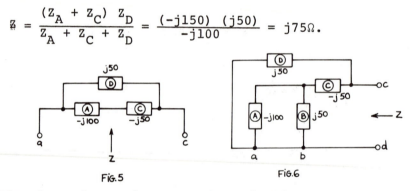

(c) Note that impedance B does not affect the impedance at terminals a-c. The impedance is shown in Fig. 5.

$$Z = \frac{(Z_A + Z_C)\, Z_D}{Z_A + Z_C + Z_D} = \frac{(-j150)\ (j50)}{-j100} = j75\Omega .$$

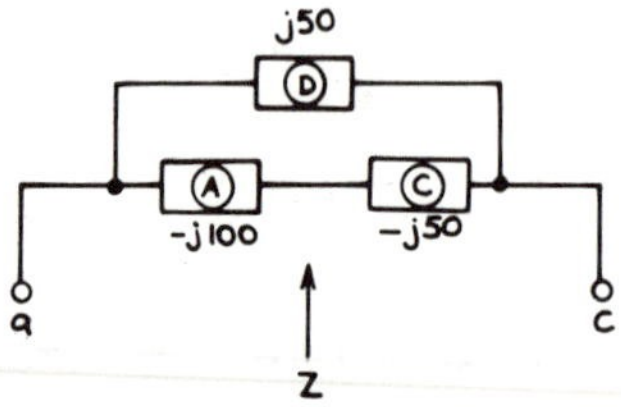

Fig.5

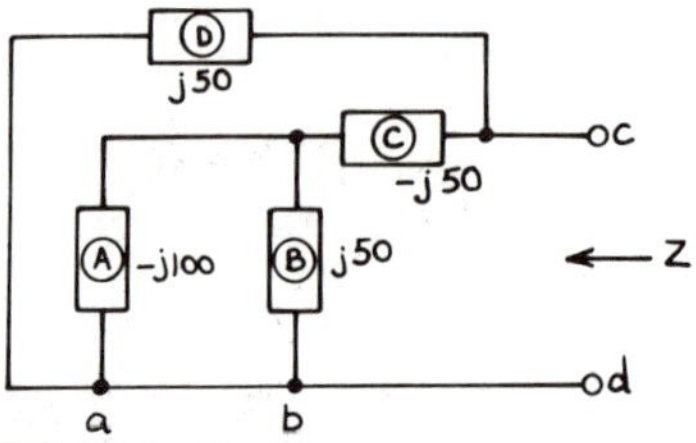

Fig.6

(d) Fig. 2 can be re-drawn as shown in Fig. 6

The impedance at terminals c-d is

$$Z = \frac{\left[\frac{Z_A\ Z_B}{Z_A + Z_B}\right] + Z_C \quad Z_D}{\frac{Z_A\ Z_B}{Z_A + Z_B} + Z_C + Z_D}$$

$$Z = \frac{\left(\frac{(-j100)\ (j50)}{-j50} - j50\right)\ j50}{\frac{(-j100)\ (j50)}{-j50} - j50 + j50}$$

$$Z = \frac{(j100 - j50)\ j50}{j100}$$

$$Z = \frac{(j50)\ (j50)}{j100} = \frac{-50^2}{j100}$$

$$Z = j25\Omega.$$

• PROBLEM 8-33

Determine the power being dissipated in the circuit shown in fig.1.

Solution: The power dissipation of a circuit is the power absorbed by purely resistive elements. Hence the power dissipated is

$$P_\Omega = I_1^2 R_1 + I_2^2 R_2 + \ldots + I_n^2 R_n.$$

Where $I_1, I_2, \ldots, I_n$ is the current in each branch and $R_1, R_2, R_3, \ldots, R_n$ is the pure resistance in each branch.

In the above circuit there is one resistive element ($R = 100\Omega$). Thus we must find I_{rms} of the circuit, and then calculate $P_\Omega = I_{rms}^2\ (100)$

To calculate I_{rms}, notice that:

$I_{rms} = \frac{V_{rms}}{|Z|}$ where V_{rms} is given (100 V_{rms}). The total impedance is

$$Z = R + Z_L \text{ where } Z_L = j\ \omega\ L = j\ 2\pi(60)\ 200 \times 10^{-3}$$

$$Z_L = j\ 75.4\ \Omega.$$

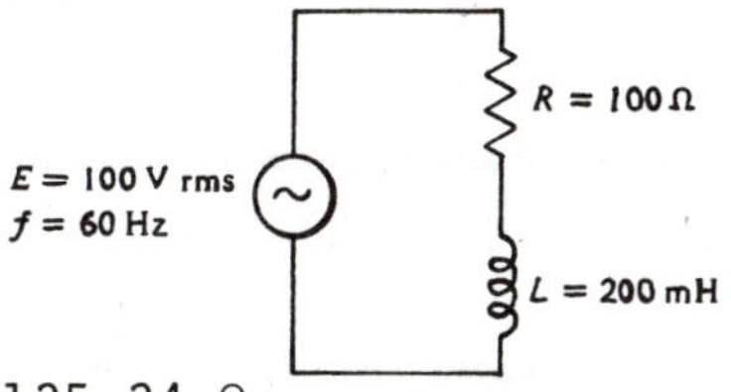

FiG.1

Hence $\vec{Z} = 100 + j\ 75.4$

$$|Z| = \sqrt{100^2 + 75.4^2} = \sqrt{15685.16} = 125.24\ \Omega.$$

$$\text{Thus}\quad I_{rms} = \frac{100}{|Z|} = \frac{100}{125.24} = 0.798$$

$$P_\Omega = I_{rms}^2\ (100) = (0.748)^2\ (100) = 63.7\ W.$$

• PROBLEM 8-34

A resistance R and reactance X in series are connected to a 115-V 60-Hz voltage supply. Good laboratory instruments are used to show that the reactor voltage is 75V and the total power supplied to the circuit is 190 W. Find R, X, and the rms current.

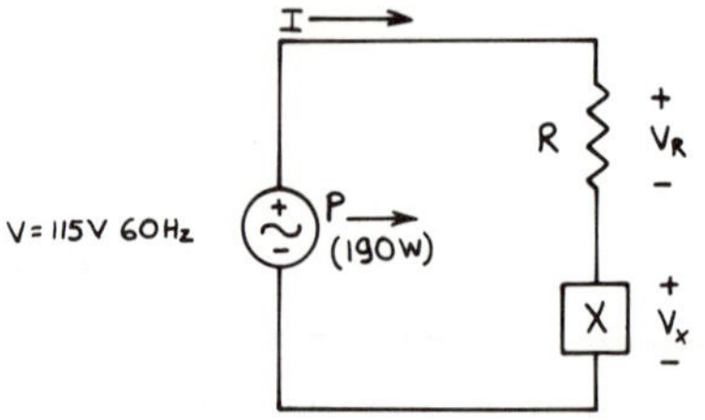

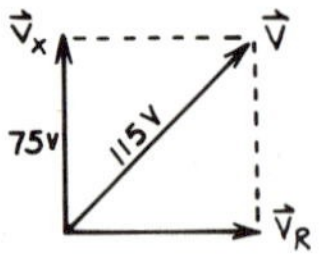

FIG. 1

Solution: Since $V = V_R + V_X$ and since V_R and V_X have a 90° phase difference (see the accompanying figure), then

$$|V|^2 = |V_R|^2 + |V_X|^2 \tag{1}$$

where $|V|$ indicates the effective or rms value of V. Thus from equation (1)

$$|V_R| = \sqrt{|V|^2 - |V_X|^2} = \sqrt{(115)^2 - (75)^2} = 87.2 \text{ V}.$$

The reactor (inductance or capacitance) cannot absorb real power (watts), hence $P = 190 \text{ W} = |V_R|^2/R = (87.2)^2/R$. Thus $R = 40\Omega$. The current is then $|I| = |V_R|/R = 2.18$ A and the reactance is $X = \pm|V_X|/|I| = \pm 75/2.18 = \pm 34.4\Omega$. If the reactor is an inductor, the plus sign applies; for a capacitor, the negative sign applies.

• PROBLEM 8-35

In the circuit shown in fig. 1, determine the current and power dissipated in each branch.

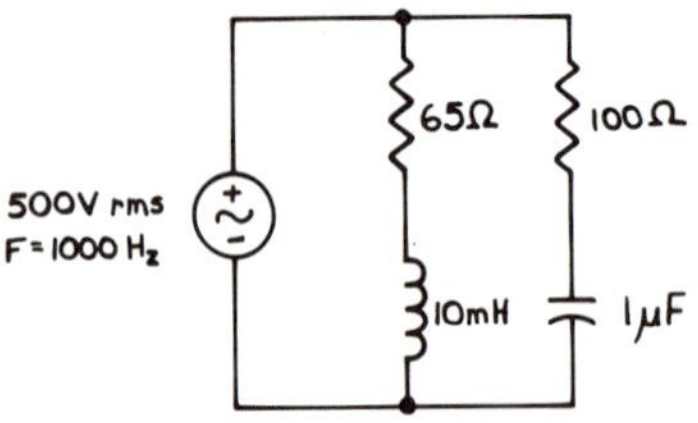

FIG. 1

Solution: At f = 1000 Hz , ω = 6280 rad/sec

$$Z_C = \frac{-j}{\omega C} = \frac{-j}{6280 \times 1 \times 10^{-6}} = -\,j\,159.2\ \Omega$$

$$Z_L = j\,\omega\,L = +\,j\,62.8\Omega.$$

In the R-C branch,

$$\vec{I}_C = \frac{500\angle 0^\circ}{100 - j\,159.2} = \frac{500\angle 0^\circ}{188\angle -57.87^\circ} = 2.66\angle +57.87^\circ$$

and using P = V I cos θ watts, P = 500 (2.66) cos 57.87° = 707.4 watts.

In the R-L branch

$$\vec{I}_L = \frac{500\angle 0^\circ}{65 - j\,62.8} = \frac{500}{90.4\angle 44^\circ} = 5.53\angle -44^\circ\ A_{rms},$$

and P = 500 (5.53) cos 44° = 1989 watts.

• PROBLEM 8-36

In the circuit of Fig. 1, find the average power received by the following: (a) resistor; (b) dependent source; (c) independent source.

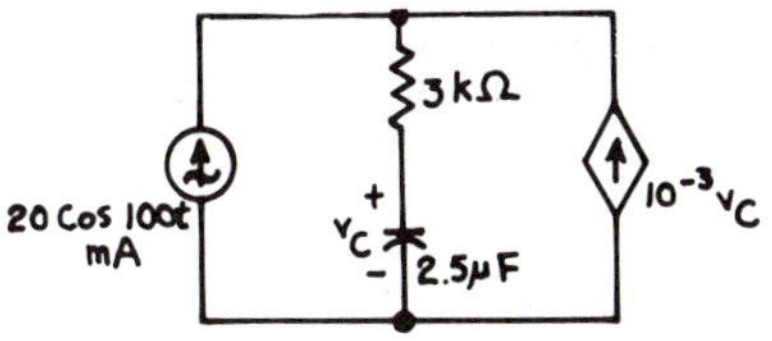

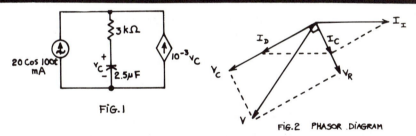

FIG. 1

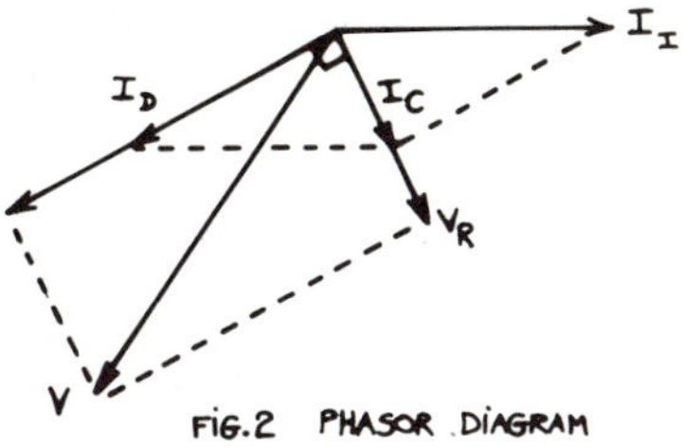

FIG. 2 PHASOR DIAGRAM

Solution: The key to this problem is finding the current in the middle branch. First, this will allow the determination of the capacitor voltage which leads to finding the value of the dependent-source current. In addition, the middle-branch current is used in finding the resistor power and in solving for the voltage between junctions, the latter being required in order to evaluate power in each of the two sources.

(a) Assign a current i_C to the capacitor branch (directed downward) and apply Kirchhoff's current law at either top or bottom junction. Transform all voltages and currents to their rms phasor equivalents to facilitate calculations.

$$\vec{I}\text{ (independent source)} + \vec{I}\text{ (dependent source)} - \vec{I}\text{ (capacitor branch)} = 0 \quad (1)$$

$$(20 \times 10^{-3})\ (0.707)\ \underline{/0°} + 10^{-3}\ \vec{V}_C - \vec{I}_C = 0 \qquad (2)$$

Next, write the voltage-current relationship for the capacitive reactance:

$$\vec{V}_C = -j\left(\frac{1}{\omega C}\right)\vec{I}_C \quad \text{rms volts} \qquad (3)$$

which becomes, numerically,

$$\vec{V}_C = -j\,\frac{1}{100\ (2.5 \times 10^{-6})}\,\vec{I}_C. \qquad (4)$$

It is now possible to substitute equation (4) into equation (2) in order to solve for the current in the capacitive branch:

$$14.14 \times 10^{-3}\ \underline{/0°} + 10^{-3}\left(-j\,\frac{1}{0.25 \times 10^{-3}}\right)\vec{I}_C - \vec{I}_C = 0 \quad (5)$$

Now, combining terms,

$$\vec{I}_C\ (1 + j\,4) = 14.14 \times 10^{-3}\underline{/0°} \qquad (6)$$

and solving,

$$\vec{I}_C = \frac{14.14 \times 10^{-3}\underline{/0°}}{4.12\underline{/76.0°}} = 3.43 \times 10^{-3}\underline{/-76°} \text{ rms amperes.} \qquad (7)$$

(b) Solve for power received by the resistor by using this current since resistor current and capacitor current are the same:

$$P\ (\text{resistor}) = I^2\,R = (3.43 \times 10^{-3})^2 \times 3000 = 35 \text{ milliwatts.} \ (8)$$

(c) Solve for capacitor voltage V_C in order to evaluate current in the dependent source by substituting equation (7) into equation (4):

$$\vec{V}_C = -j\left(\frac{1}{0.25 \times 10^{-3}}\right)(3.43 \times 10^{-3}\underline{/-76°})$$

$$= 13.7\underline{/-90° - 76°}\ V_{rms}. \qquad (9)$$

Now, evaluate the dependent-source current:

$$\vec{I}_D = 10^{-3}\vec{V}_C = 13.7 \times 10^{-3}\underline{/-166°}\ A_{rms}. \qquad (10)$$

(d) Before the powers in the two sources can be evaluated, it is necessary to solve for the voltage between junctions, V, with the positive polarity at the top. In accordance with the usual convention in an impedance,

$$\vec{V} = \vec{I}_C \vec{Z} = (3.43 \times 10^{-3} \angle{-76°}) \left[3000 - j\left(\frac{1}{0.25 \times 10^{-3}}\right)\right] \quad (11)$$

$$\vec{V} = (3.43 \times 10^{-3})\ (5000)\ \angle{-76° - 53.1°}$$

$$= 17.2\ \angle{-129.1°} \text{ rms volts.} \quad (12)$$

(e) Turning our attention to the dependent-source power,

$$P \text{ (dependent source)} = -\ V\ I_D \cos\theta \text{ W} \quad (13)$$

where θ is the angle between V and I_D. The minus sign is included because the current direction is opposite to the direction of voltage drop. Substituting the proper values into equation (13) yields

$$P \text{ (dependent source)} = -(17.2)(13.7 \times 10^{-3}) \cos(-129.1° - (-166°)) \quad (14)$$

$$P_D = -\ 188 \text{ m W} \quad (15)$$

The minus sign indicates that power is being delivered from the dependent source.

(f) Finally, power received by the independent source is

$$P \text{ (independent source)} = -\ V\ I_I \cos\phi \quad (16)$$

$$P_I = (17.2)(14.14 \times 10^{-3}) \cos(-129.1° - 0°) = 153 \text{ m W.} \quad (17)$$

(g) Checks:

$$\vec{I}_I + \vec{I}_D - \vec{I}_C = 14.1\angle{0°} + 13.7\angle{-166°} - 3.43\angle{-76} \text{ m A} \quad (18)$$

$$\Sigma\ \vec{I} = 14.1 + j\ 0 - 13.3 - j\ 3.3 - 0.8 + j\ 3.3 = 0 \quad (19)$$

$$\Sigma\ P = P_I + P_D + P_R = 153 - 188 + 35 = 0 \quad (20)$$

• PROBLEM 8-37

A circuit draws 4 A at 25 V_{rms}, and dissipates 50 W. Find:

(a) apparent power; (b) reactive power; (c) power factor and phase angle; and (d) impedance in both polar and rectangular forms.

Solution: (a) The apparent power is $|S| = V_{eff}\ I_{eff} = (4)\ (25) = 100$ V A. (b) Since $\vec{S} = P + j\ Q$ and $|S| = \sqrt{P^2 + Q^2} = 100$ V A we can find Q because we are given P (dissipated power) = 50 W.

Hence $Q = \sqrt{|S|^2 - P^2}$

$$Q = \sqrt{100^2 - 50^2} = \sqrt{7500} = 86.6 \text{ V A}.$$

(c) The power factor is defined as the ratio of dissipated power to apparent power. Hence,

$$pf = \frac{P}{|S|} = \frac{50}{100} = 0.5.$$

The phase angle is the $\cos^{-1} pf$, thus $\phi = \cos^{-1} 0.5 = 60°$.

(d) the magnitude of the impedance can be found from

$$|Z| = \frac{V_{eff}}{I_{eff}} = \frac{25}{4} = 6.25 \ \Omega.$$

We can therefore write Z in polar form:

$$\vec{Z} = |Z| \underline{/\phi}$$

$$\vec{Z} = 6.25 \underline{/60°}.$$

In rectangular form we must write

$\vec{Z} = 3.125 \pm j\ 5.41\ \Omega$ since we are not given any information to determine the polarity of the phase angle (that is, whether V lags or leads I). Thus the j term could be either plus or minus.

• PROBLEM 38

Determine the power factor (p.f.) associated with a load that:

(a) consists of a 6.25 - μF capacitor in parallel with the series combination of 400 - Ω resistance and 1 H at ω = 400 rad/s;

(b) is inductive and draws 33 A rms and 6.9 k W at 230 V rms;

(c) is composed of parallel loads, one of which draws 10 kVA at 0.8 p.f. lagging and the other 8 kVA at 0.9 p.f. leading.

Solution: The power factor (p.f.) of a load is the cosine of the phase angle of the load impedance.

(a) The impedance of a 6.25 μF capacitor at a frequency ω = 400 rad/sec is $1/j\omega C = -j400\Omega$, and the impedance of a 1 H inductor is $j\omega L = j400\Omega$.

The total impedance Z for the specified network is then

$$\vec{Z} = \frac{(-j400)\ (400 + j400)}{400 + j400 - j400} = 400\ (1 - j1)$$

$= 400\sqrt{2}\ \underline{/-45°}.$

The phase angle is -45° and the p.f. is cos 45° = 0.707 leading.

(b) Since an inductive load implies a lagging p.f., we can use $P = |V||I| \cos\theta$ to find $\cos\theta$, the magnitude of the p.f.:

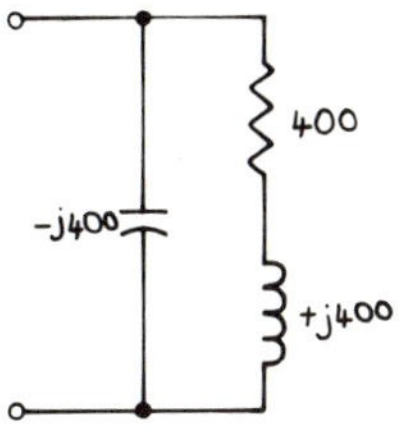

$$\cos\theta = \frac{P}{|V|\ |I|} = \frac{6900}{(230)\ (33)} = 0.909.$$

The p.f. is thus 0.909 lagging.

(c) A load of 10 kVA at 0.8 p.f. lagging is equivalent to a draw of (10 k) (0.8) = 8 k w, (10 k) (0.6) = -6 Kvars where $\cos\theta = 0.8$ lagging has been used to determine $\theta = -\cos^{-1} 0.8$. Therefore $\sin\theta = -0.6$.

Similarly, a 8 kVA of 0.9 p.f. leading is equivalent to

$(8\text{ k})(0.9) = 7.2$ k W, $(8\text{ k})\sqrt{1 - (0.9)^2} = (8\text{ k})(.436) = +3.488$ Kvars. The combination of the two loads draws a real power of (8 + 7.2) k W = 15.2 k W and a reactive power of (-6 + 3.488) Kvars = -2.512 Kvars.

Since the net reactive power is negative, the p.f. will be lagging, and its magnitude will be given by $\cos\theta$ for θ obtained from the sketch shown.

$$\cos\theta = \frac{15.2}{\sqrt{(2.512)^2 + (15.2)^2}} = 0.987$$

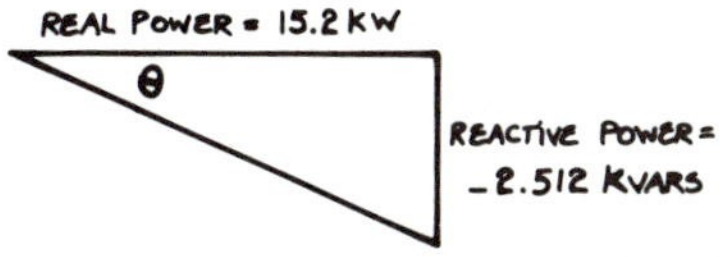

The p.f. is 0.987 lagging.

• PROBLEM 8-39

Calculate the current, the voltage drops V_1, V_2, and V_3, the power consumed by each impedance, and the total power taken by the circuit with the constants shown in Fig. 1. The impressed voltage will be taken along the reference axis.

Solution: The source voltage V is a real quantity and it can be expressed as

$$\vec{V} = 100 + j\,0.$$

The entire impedance, Z, is the sum of the individual impedances:

$$Z = 4 + j\,3 + 6 - j\,8 + 2 = 12 - j\,5.$$

The current is now found by using Ohm's Law:

$$\vec{I} = \frac{\vec{V}}{Z} = \frac{100 + j\,0}{12 - j\,5} = \frac{(100 + j\,0)(12 + j\,5)}{(12 - j\,5)\,(12 + j\,5)} = \frac{1200 + j\,500}{169}$$

$$\vec{I} = 7.1 + j\,2.96 \text{ amperes.}$$

The voltage drops $\vec{V}_1$, $\vec{V}_2$, and $\vec{V}_3$ are also found from Ohm's Law:

$$\vec{V}_1 = \vec{I}\,\vec{Z}_1 = (7.1 + j\,2.96)\,(4 + j\,3)$$

$$= 28.4 + j\,33.14 + j^2\,8.88.$$

However, $j^2 = -1$. Therefore:

$$\vec{V}_1 = 28.4 - 8.88 + j\,33.14 = 19.52 + j\,33.14 \text{ volts.}$$

Similarly,

$$\vec{V}_2 = I\,Z_2 = (7.1 + j\,2.96)\,(6 - j\,8)$$

$$= 66.2 - j\,39.04 \text{ volts}$$

$$\vec{V}_3 = I\,Z_3 = (7.1 + j\,2.96)\,2$$

$$= 14.2 + j\,5.92 \text{ volts.}$$

The power consumption of each impedance is the product of the square of the magnitude of the current, and the real part of each impedance. The magnitude of the current is:

$$|I| = \sqrt{(7.1)^2 + (2.96)^2}.$$

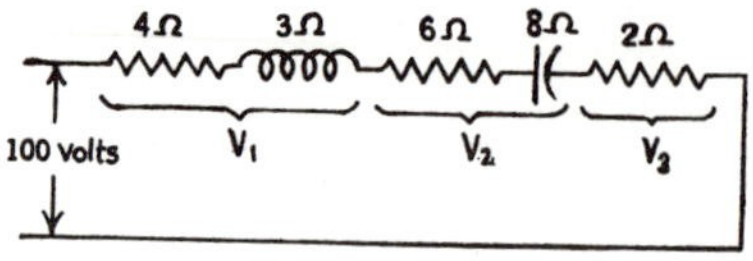

FIG. 1

Thus, $|I|^2 = (7.1)^2 + (2.96)^2 = 59.17$

Therefore, $P_1 = R_1\,|I|^2 = 4\,(59.17) = 237$ watts

$$P_2 = R_2\,|I|^2 = 6\,(59.17) = 355 \text{ watts}$$

$$P_3 = R_3\,|I|^2 = 2\,(59.17) = 118 \text{ watts.}$$

The total power delivered to the circuit is the sum of the powers consumed by each impedance:

$$P_T = P_1 + P_2 + P_3 = 237 + 355 + 118 = 710 \text{ watts.}$$

In order to check this result, recall that the total power delivered by the source is given by $P_T = V\,I\cos\theta$ where θ is the angle of impedance. Then

$$P_T = (100)\,(7.69)\cos[\arctan -5/12]$$

$$P_T = 710 \text{ V.}$$

The terminals of an a-c generator, which has an internal resistance of 2 ohms and an equivalent internal inductive reactance of 6 ohms, are connected to a particular RLC series branch, the R of which is 10 ohms, the ωL of which is 20 ohms, and the 1/ωC of which is 40 ohms. If the magnitude of the internally generated emf is 500 volts, find the current that flows in the series circuit and the terminal voltage of the generator.

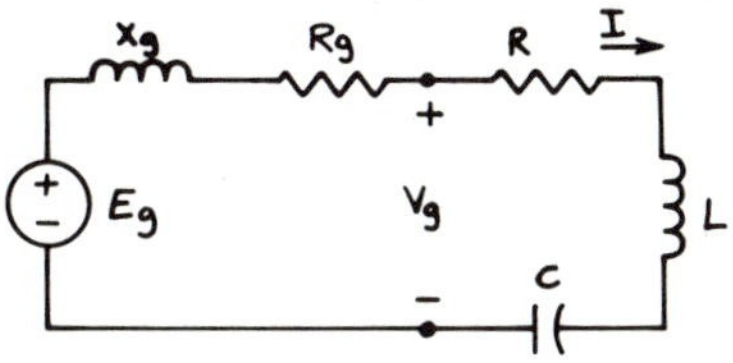

Fig. 1

Solution: The circuit described is shown in Fig. 1 below, where $X_g = 6\Omega$, $R_g = 2\Omega$, $R = 10\Omega$, $\omega L = 20\Omega$, $\frac{1}{\omega C} = 40\Omega$, and $|E_g| = 500$ V. The generated emf, E_g is arbitrarily chosen to coincide with the reference axis. Therefore,

$$\vec{E}_g = 500 + j0 = 500\underline{/0°} \text{ V.}$$

Using KVL, the current I is found by solving the loop equation:

$$\vec{E}_g = \vec{I}\ (j6 + 2 + 10 + j20 - j40)$$

$$\vec{I} = \frac{500\underline{/0°}}{12 - j14} = \frac{500\underline{/0°}}{18.44\underline{/-49.4°}}$$

$$\vec{I} = 27.1\underline{/49.9°} \text{ A.}$$

The generator voltage $\vec{V}_g$ is

$$\vec{E}_g - (R_g + jX_g)\ \vec{I}$$

$$(500 + j0) - (2 + j6)\ 27.1\underline{/49.4°}$$

$$(500 + j0) - (6.32\underline{/71.57°})\ (27.1\underline{/49.4°})$$

$$(500 + j0) - 171.27\underline{/120.97°}$$

$$(500 + j0) - (-88.1 + j146.9).$$

$$V_g = 588.1 - j146.9 = 606\underline{/-14°} \text{ V.}$$

Three passive loads Z_1, Z_2, and Z_3, are receiving the complex power values 2 + j3, 3 - j1, and 1 + j2 V A, respectively. What total complex power is received if the three loads are: (a) in series with a voltage source $100\underline{/30°}$ V; (b) in parallel with a current source $1200\underline{/-17°}$ A?

Solution: (a) Under $100\underline{/30°}$ V, the first passive load Z_1 is receiving complex power value

$$\vec{S}_1 = 2 + j3 \text{ V A};$$

the second passive load Z_2 is receiving

$$\vec{S}_2 = 3 - j1 \text{ V A};$$

and the third passive load Z_3 is receiving

$$\vec{S}_3 = 1 + j2 \text{ V A}$$

with a series connection to the voltage source.

Therefore, the total complex power is

$$\vec{S} = \vec{S}_1 + \vec{S}_2 + \vec{S}_3 = (2 + 3 + 1) + j\,(3 - 1 + 2)$$

$$= 6 + j4 \text{ V A}.$$

(b) In parallel with a current source $1200\underline{/-17°}$ A, the first passive load Z_1 is receiving complex power value

$$\vec{S}_1 = 2 + j3 \text{ V A}.$$

For the second passive load Z_2,

$$\vec{S}_2 = 3 - j1 \text{ V A},$$

and the third passive load Z_3,

$$\vec{S}_3 = 1 + j2 \text{ V A}.$$

So, the total complex power received is

$$\vec{S} = \vec{S}_1 + \vec{S}_2 + \vec{S}_3 = (2 + 3 + 1) + j\,(3 - 1 + 2)$$

$$= 6 + j4 \text{ V A}.$$

In this problem the voltage $100\underline{/30°}$ V, the current source of $1200\underline{/-17°}$ and the circuit configurations described are so called "dummies" because they do not affect the answers to this problem. An important principle to know is the principle of superposition. Total phasor power is the summation of the phasor power of each component in the circuit.

Generally, the load impedance of Z_1, Z_2, Z_3 in part (a) of this problem is different from the load impedance Z_1, Z_2 and Z_3 of part (b) of this problem.

• PROBLEM 8-42

The circuit in fig. 1 is excited by v(t). Find the steady-state output voltage and plot it approximately. Also find its root mean square value.

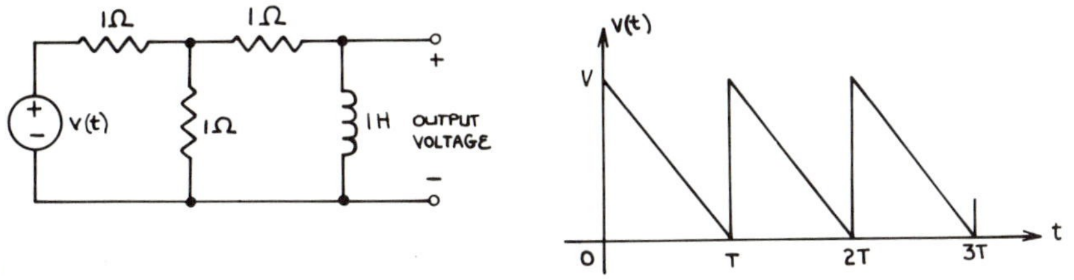

FIG. 1

Solution: The problem can be simplified by replacing the source and resistors by a Thevenin equivalent, producing the circuit shown in fig. 2.

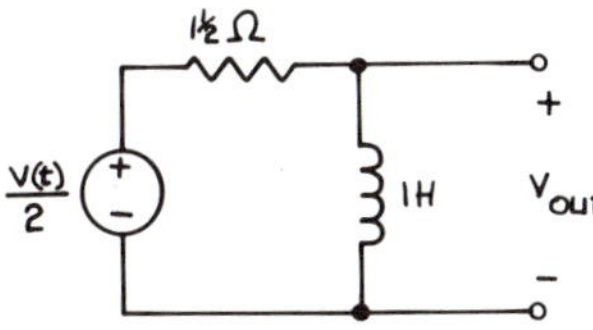

FIG. 2

There are two approaches to solving the problem. One is to replace the given v (t) by a Fourier series expansion, with a subsequent phasor solution for each sinusoidal component and a final superposition. The other approach is to formulate a differential equation and solve it. Using the latter approach, a KCL equation at the upper output node is

$$\int v_{out}\, dt + \frac{v_{out} - v(t)/2}{1\frac{1}{2}} = 0.$$

Differentiating through and rearranging yields

$$v_{out} + \frac{2}{3}\dot{v}_{out} - \frac{1}{3}\dot{v} = 0\ ;\ \dot{v}_{out} + \frac{3}{2}v_{out} = \frac{1}{2}\dot{v}.$$

Rather than solve this equation directly, solve a simpler system, then use linear operations and superposition to arrive at the actual solution. First solve

$$\dot{y} + \frac{3}{2}y = u(t)$$

to obtain $y(t) = \frac{2}{3}\left(1 - e^{-\frac{3}{2}t}\right) u(t)$.

In order to use this result as a building block, the given v (t) must be expressed as a superposition of step functions or linearly related quantities (impulses, ramps). We can express the given v (t) as

$$v\,(t) = \int \left[-\frac{V}{T}\,u\,(t) + \sum_{n=0}^{\infty} V\,\delta\,(n\,T) \right] dt.$$

The actual excitation is $\frac{1}{2}\,\dot{v} = -\frac{V}{2T}\,u\,(t) + \frac{V}{2}\sum_{n=0}^{\infty} \delta\,(\,n\,T)$.

Since the solution to $\dot{y} + \frac{3}{2}\,y = u\,(t)$ is $y\,(t)$, the solution to $\dot{y}_1 + \frac{3}{2}\,y_1 = \delta\,(t)$ is $y_1\,(t) = \frac{d\,y}{d\,t}$

$$= \frac{d}{d\,t}\left[\frac{2}{3}\,(1 - e^{-\frac{3}{2}t})\;u\,(t)\right]$$

$$= \frac{2}{3}\left[(1 - {}^{\frac{3}{2}t})\;\delta\,(t) \;+ u\,(t)\,\frac{3}{2}\,e^{-\frac{3}{2}t}\right]$$

$$= e^{-\frac{3}{2}t}\,u\,(t),$$

and the solution to $\dot{v}_{out} + \frac{3}{2}\,v_{out} = \frac{1}{2}\,\dot{v}$ is

$$v_{out}\,(t) = -\frac{V}{3T}\,(1 - e^{-\frac{3}{2}t})\;u\,(t) + \frac{V}{2}\sum_{n=0}^{\infty} e^{-\frac{3}{2}(t - n\,T)}\,u\,(t - n\,T).$$

At steady-state, the initial transient will have subsided, with the result that

$$\left[v_{out}\,(t)\right]_{\text{steady-state}} = -\frac{V}{3T} + \frac{V}{2}\sum_{n=0}^{\infty} e^{-\frac{3}{2}(t - nT)}\;u\,(t - n\,T).$$

A sketch of the steady-state output is in fig. 3, for the particular input voltage period T = 1 second:

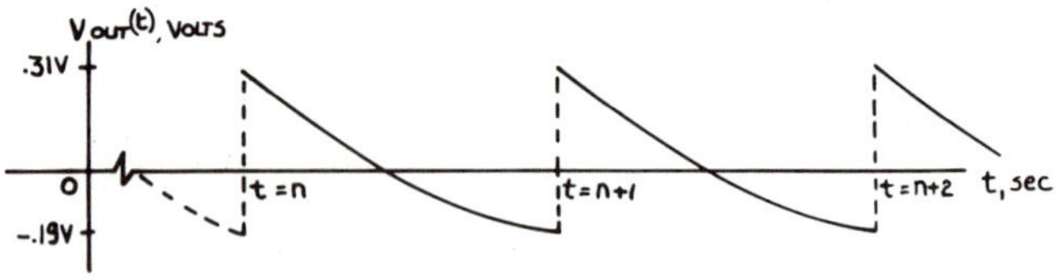

FIG. 3

Note that steady-state output values involve the cumulative effect of the sequence of decaying exponentials.

Within any one period, say from t = 0 to t = 1, the steady-state output is given by

$$\frac{V}{3}\left[1.93\, e^{-\frac{3}{2}t} - 1\right].$$

The rms value is calculated from

$$v_{rms}^{2} = \frac{1}{1}\int_0^1 \left[\frac{V}{3}\left(1.93\, e^{-\frac{3}{2}t} - 1\right)\right]^2 d\,t$$

$$= \frac{V^2}{9}\int_0^1 \left[3.725\, e^{-3t} - 3.86\, e^{-\frac{3}{2}t} + 1\right] d\,t$$

$$= \frac{V^2}{9}\left[\frac{3.725}{-3}\, e^{-3t} + \frac{3.86}{1.5}\, e^{-\frac{3}{2}t} + t\right]_0^1$$

$$= \frac{V^2}{9}\;[\;1.2417\;(1 - 0.0498) - 2.573\;(1 - 0.223) + 1\;]$$

$$= \frac{V^2}{9}\;(0.1806).$$

Therefore $v_{rms} = 0.142$ V.

CHAPTER 9

STEADY-STATE ANALYSIS

IMPEDANCE-ADMITTANCE CALCULATIONS

• PROBLEM 9-1

Calculate the total impedance of the circuit in Fig. 1.

Solution: Note that circuit elements form two branch impedances, $Z_1 = R_1 + X_1$ and $Z_2 = R_2 + X_2$, in parallel.

Hence,

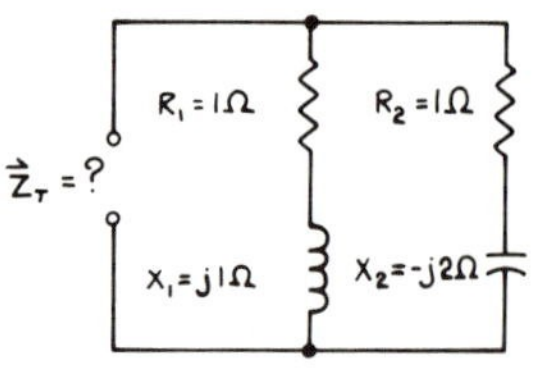

Fig. 1

$$\vec{Z}_T = \frac{Z_1 Z_2}{Z_1 + Z_2} = \frac{(R_1+X_1)(R_2+X_2)}{(R_1+R_2)+(X_1+X_2)}$$

$$\vec{Z}_T = \frac{(1+j)(1-j^2)}{(2-j)} = \frac{3-j}{2-j}$$

$$\vec{Z}_T = \frac{(3-j)(2+j)}{2^2+1} = \frac{7+j}{5} = \frac{7}{5} + j\frac{1}{5}$$

In polar coordinates we have

$$Z_T = \sqrt{\left(\frac{7}{5}\right)^2 + \left(\frac{1}{5}\right)^2} \underline{/\tan^{-1}\frac{1}{7}}$$

$$Z_T = \sqrt{\frac{49}{25} + \frac{1}{25}}\ \underline{/8.1°}$$

$$Z_T = \sqrt{\frac{50}{25}}\underline{/8.1°}$$

$$Z_T = \sqrt{2}\underline{/8.1°}\ \Omega.$$

• PROBLEM 9-2

What is the impedance of a circuit if $100\underline{/10°}$ V causes a current 7.07 + j7.07 A?

Solution: Using Ohm's Law;

$$\vec{Z} = \frac{\vec{V}}{\vec{I}} = \frac{100\underline{/10°}}{7.07 + j7.07} = \frac{100\underline{/10°}}{\sqrt{2(7.07)^2}\underline{/\tan^{-1} 1}}$$

$$\vec{Z} = \frac{\vec{V}}{\vec{I}} = \frac{100\underline{/10°}}{\sqrt{100}\underline{/45°}} = 10\underline{/10° - 45°} = 10\underline{/-35°}\ \Omega.$$

In rectangular form;

$$\vec{Z} = 10 \cos(-35°) + j10 \sin(-35°)$$

$$\vec{Z} = 8.2 - j5.7\ \Omega.$$

• PROBLEM 9-3

A phasor diagram is constructed for the circuit shown in fig. 1. The scale used is 20 V/cm and 0.4 A/cm. If $\vec{V}_1$ is represented by an arrow 1.6 cm long at an angle of -90°, $\vec{V}_2$ by a 2.0 cm long arrow at an undesignated angle, and $\vec{I}_s$ by a 1.8 cm long arrow at an angle of -90°, find the impedance of the: (a) capacitor; (b) resistor; (c) inductor.

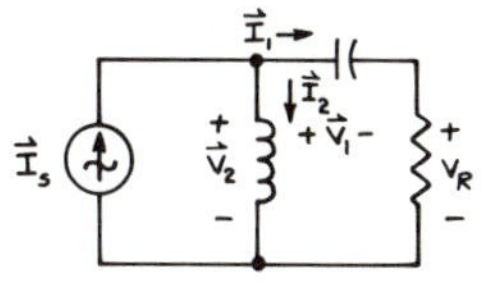

FIG. 1

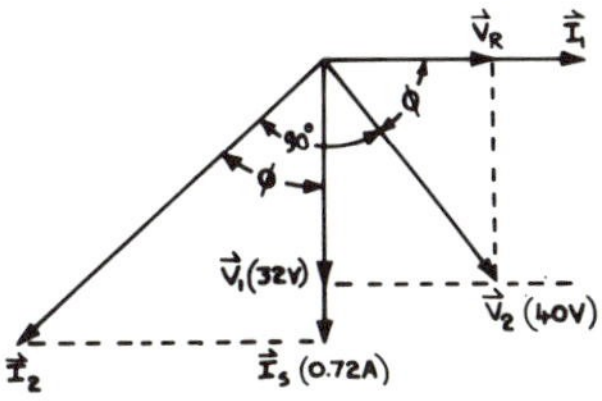

FIG. 2

Solution: Using the scale factors, the magnitudes of $\vec{V}_1$, $\vec{V}_2$ and $\vec{I}_s$ are calculated as 32V, 40V and 0.72A, respectively. The angle of $\vec{I}_1$ is zero since the current leads the voltage by 90 for a capacitor. The resistor voltage $\vec{V}_R$ is in phase with $\vec{I}_1$. The vector sum of $\vec{V}_1$ and $\vec{V}_R$ is equal to $\vec{V}_2$ (see Fig. 2). Hence

$$|V_R| = \sqrt{|V_2|^2 - |V_1|^2} = \sqrt{40^2 - 32^2} = 24V$$

And $$\phi = \tan^{-1}\left(\frac{32}{24}\right) = 53.13°.$$

Since $\vec{I}_s = \vec{I}_1 + \vec{I}_2$ and $\vec{I}_1$ is real, the imaginary parts of $\vec{I}_2$ and $\vec{I}_s$ are equal. That is

$$|I_2| = |I_s|/\cos\phi = 1.2A.$$

Likewise, since $\vec{I}_s$ is imaginary, the real parts of $\vec{I}_1$ and $\vec{I}_2$ must be negatives of each other, or

$$|I_1| = |I_2|\sin\phi = 0.96A.$$

The three impedances are now calculated as

$$R = \frac{V_R}{I_1} = 25\Omega$$

$$\vec{Z}_L = j\frac{|V_2|}{I_2} = j33.3\Omega$$

$$\vec{Z}_C = -j\frac{|V_1|}{I_1} = -j33.3\Omega$$

• PROBLEM 9-4

Find the equivalent impedance of the network to the right of the terminals a,b in fig. 1.

Solution: The total impedance of this series RLC network is the sum of the individual impedances. That is,

$$\vec{Z}_{ab} = \vec{Z}_R + \vec{Z}_L + \vec{Z}_C$$

or

$$\vec{Z}_{ab} = R + j\omega L + \frac{1}{j\omega C}$$

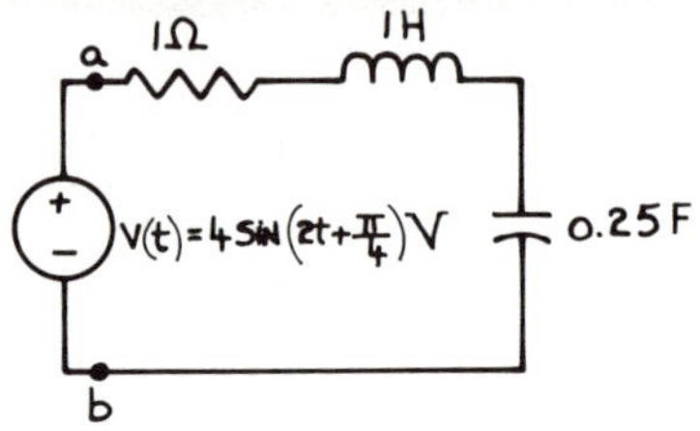

FIG. 1

where ω is given in the expression for v(t) as $2\frac{rad}{s}$. Hence

$$\vec{Z}_{ab} = 1 + j2 + \frac{1}{j0.5}$$

$$\vec{Z}_{ab} = 1 + j2 - j2$$

$$\vec{Z}_{ab} = 1\Omega.$$

• PROBLEM 9-5

What is the admittance of a circuit having an impedance of 2 + j3 Ω?

Solution: If $\vec{Z} = 2 + j3\ \Omega$, then the admittance is $\vec{Y} = \frac{1}{\vec{Z}} = \frac{1}{2 + j3}$ mhos. Multiplying numerator and denominator by the conjugate of the denominator gives

$$\vec{Y} = \frac{1}{2 + j3}\,\frac{(2 - j3)}{(2 - j3)} = \frac{2 - j3}{2^2 + 3^2} = \frac{2 - j3}{13}$$

$$\vec{Y} = \frac{2}{13} - j\frac{3}{13} = 0.154 - j0.221 \text{ mhos.}$$

• **PROBLEM** 9-6

> The following impedances are connected in series: $\vec{Z}_1 = 10\underline{/30°}\ \Omega$; $\vec{Z}_2 = 7\underline{/60°}\ \Omega$. What is the total series impedance?

Solution: In order to add the two impedances $\vec{Z}_1$ and $\vec{Z}_2$, first convert them from their given polar form to rectangular form.

Since $\vec{C} = C\underline{/\theta} = C \cos\theta + jC \sin\theta$,

$$\vec{Z}_1 = 10\underline{/30°} = 10 \cos 30° + j10 \sin 30°$$

$$\vec{Z}_1 = 8.66 + j5.00\ \Omega$$

and $\vec{Z}_2 = 7\underline{/60°} = 7 \cos 60° + j7 \sin 60°$

$$\vec{Z}_2 = 3.5 + j6.06\ \Omega.$$

Hence $\vec{Z}_T = \vec{Z}_1 + \vec{Z}_2 = 8.66 + j5.00$

$$\underline{\ 3.50 + j6.06}$$

$$\vec{Z}_T = 12.16 + j11.06$$

To convert back to polar form, use $A + jB = \sqrt{A^2 + B^2}\underline{/\tan^{-1} \frac{B}{A}}$, thus,

$$\vec{Z}_T = 12.16 + j11.06 = \sqrt{12.16^2 + 11.06^2}\underline{/\tan^{-1} \frac{11.06}{12.16}}$$

$$\vec{Z}_T = 16.44\underline{/42.29°}\ \Omega.$$

EQUIVALENT CIRCUITS

• **PROBLEM** 9-7

> For a series RLC circuit with the elements, R = 20Ω, L = 8mH, and C = 2.22μF, deduce the equivalent circuits consisting of two elements at the frequencies (a) f_1 = 796 Hz; (b) f_2 = 1194 Hz; (c) f_3 = 1592 Hz.

Solution: (a) At the frequency f_1 = 796 Hz, the total impedance of the series RLC circuit is

$$\vec{Z}_T = \vec{Z}_R + \vec{Z}_L + \vec{Z}_C$$

$$\vec{Z}_T = R + j\omega L + \frac{1}{j\omega C}$$

$$\vec{Z}_T = 20 + j2\pi(796)(8\times10^{-3}) + \frac{1}{j2\pi(796)(2.22\times10^{-6})}$$

$$\vec{Z}_T = 20 + j40 - j90$$

$$\vec{Z}_T = 20 - j50\Omega.$$

The RLC circuit can be represented by the impedance 20-j50 at F = 796 Hz or a series combination of a 20Ω resistance and a 50Ω capacitive reactance. The value of the capacitor needed is

$$C = \frac{1}{50\ 2\pi(796)} = 4 \times 10^{-6}F = 4\mu F.$$

(b) For f_2 = 1194 Hz, the total impedance is

$$\vec{Z}_T = 20 + j\ 2\pi(1194)(8\times10^{-3}) + \frac{1}{j2\pi(1194)(2.22\times10^{-6})}$$

$$\vec{Z}_T = 20 + j60 - j60$$

$$\vec{Z}_T = 20\Omega.$$

Hence the entire RLC circuit can be replaced by a 20Ω resistor at this frequencey.

(c) For f_3 = 1592 Hz the total impedance is

$$\vec{Z}_T = 20 + j2\ (1592)(8\times10^{-3}) + \frac{1}{j2\pi(1592)(2.22\times10^{-6})}$$

$$\vec{Z}_T = 20 + j80 - j45$$

$$\vec{Z}_T = 20 + j35\Omega.$$

In this case, the RLC circuit can be replaced by a series combination of a resistor and, because of the positive reactance, an inductor. The value of the inductor is found from the reactance. Hence

$$L = \frac{35}{2\pi(1592)} = 3.5mH.$$

• PROBLEM 9-8

For the circuit in fig. 1, determine the Thévenin equivalent circuit with respect to terminals a-b.

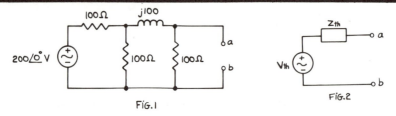

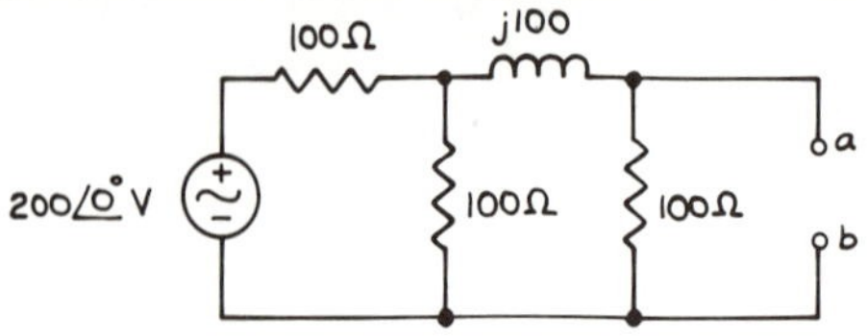

FiG.1

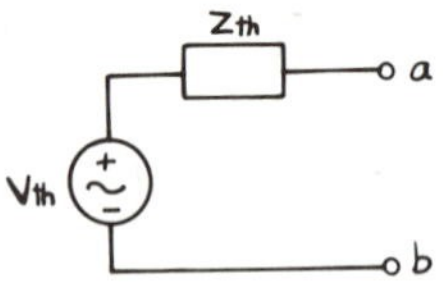

FiG.2

Solution: To find a Thévenin equivalent circuit, one has to find the Thevenin equivalent impedance and Thévenin equivalent voltage source as shown in fig. 2. Z_{th} is found by setting all independent sources equal to zero (i.e. replacing all voltage sources with short circuits and all current sources with open circuits) and then finding the impedance looking into terminals a-b.

In our circuit, Z_{th} is found from the circuit in fig. 3 where the voltage source has been replaced by a short circuit.

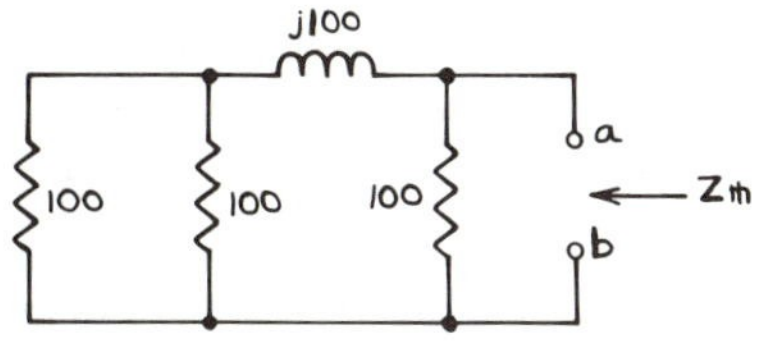

FiG.3

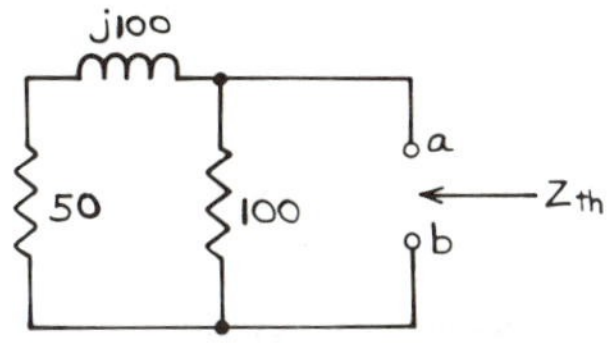

FiG.4

$$\vec{Z}_{th} = 50 + j100 \| 100$$

$$\vec{Z}_{th} = \frac{(50+j100)100}{150 + j100}$$

$$\vec{Z}_{th} = \frac{100(50+j100)(150-j100)}{150^2 + 100^2}$$

$$\vec{Z}_{th} = \frac{100(17500+j10000)}{32500} = \frac{17500+j10000}{325}$$

$$\vec{Z}_{th} = 53.84 + j30.76\Omega$$

V_{th} is found by solving for the open circuit voltage (the voltage across terminals a-b).

V_{th} can be found in the circuit of fig. 5 by using voltage division concepts.

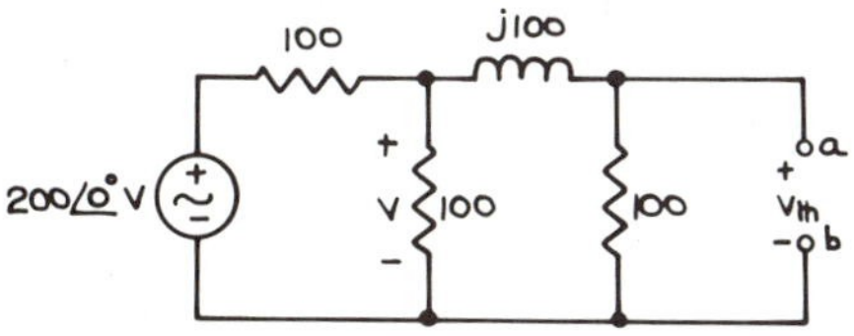

FiG.5

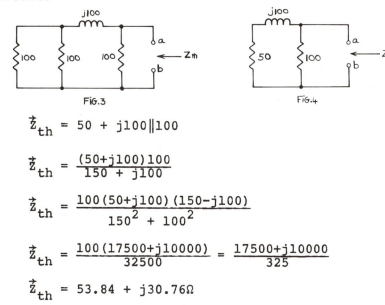

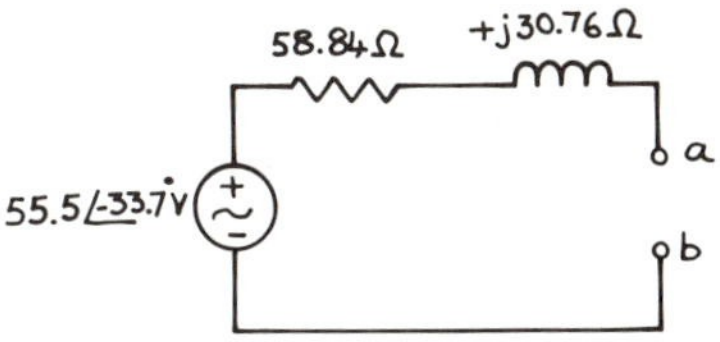

FiG.6

$$\vec{V} = \frac{[100 \| 100+j100]}{100+[100 \| 100+j100]}(200\underline{/0°}) \text{ where } 100 \| 100+j100$$

$$= \frac{100(100+j100)}{200 + j100}$$

$$= 60 + j20$$

$$V_{th} = \frac{100(V)}{100+j100}$$

$$\vec{V}_{th} = \frac{(60+j20)(200)(100)}{(160+j20)(100+j100)} = \frac{20(60+j20)}{14 + j18}$$

$$\vec{V}_{th} = \frac{20(60+j20)(14-j18)}{14^2 + 18^2} = \frac{200}{52}(12-j8)$$

$$\vec{V}_{th} = 3.85(12 - j8) = 55.5\underline{/-33.7° \text{ V}}.$$

Hence our Thévenin equivalent circuit is shown in fig. 6.

• PROBLEM 9-9

When the three load elements, shown below, are successively placed across the linear network terminals, the magnitude of the element voltages are $|V_1| = 25V$, $|V_2| = 100V$ and $|V_3| = 50V$. Determine the Thévenin's equivalent of the network.

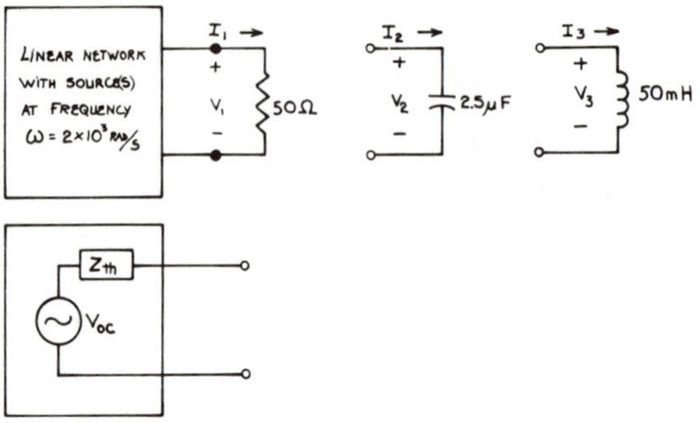

Solution: The three load impedances are readily calculated as;

$$\vec{Z}_1 = 50 , \vec{Z}_2 = -j\left[\frac{1}{(2\times10^3)(2.5\times10^{-6})}\right] = -j200\Omega,$$

and $\vec{Z}_3 = j(2\times10^3)(5\times10^{-2}) = j100\Omega.$

The magnitudes of the three load currents are therefore

$$|I_1| = \frac{25V}{|Z_1|} = 0.5A, \quad |I_2| = \frac{100V}{|Z_2|} = 0.5A \quad \text{and}$$

$$|I_3| = \frac{50V}{|Z_3|} = 0.5A.$$

Since $|I_1| = |I_2| = |I_3|$, the magnitude of the total impedance, including Z_{th}, must be equal in all three cases. Letting $\vec{Z}_{th} = R + jX$, this means that:

$$(R+50)^2 + X^2 = R^2 + (X-200)^2 = R^2 + (X+100)^2.$$

Equating the last two of these means that

$$X - 200 = -(X+100) \quad \text{or} \quad X = 50\Omega,$$

and equating the first two (with X = 50Ω) gives,

$$(R+50)^2 + (50)^2 = R^2 + (150)^2, \quad \text{or} \quad R = 175\Omega.$$

Therefore $\vec{Z}_{th} = 175 + j50\Omega$. The magnitude of the total impedance is then

$$|Z_{th}+50| = \left[(175+50)^2 + (50)^2\right]^{1/2} = 230.5\Omega$$

and $V_{oc} = |I_1|(230.5\Omega) = 0.5(230.5) = 115.25V$

• PROBLEM 9-10

Find the Thevenin and Norton equivalent circuits that would represent the circuit shown in fig. 1.

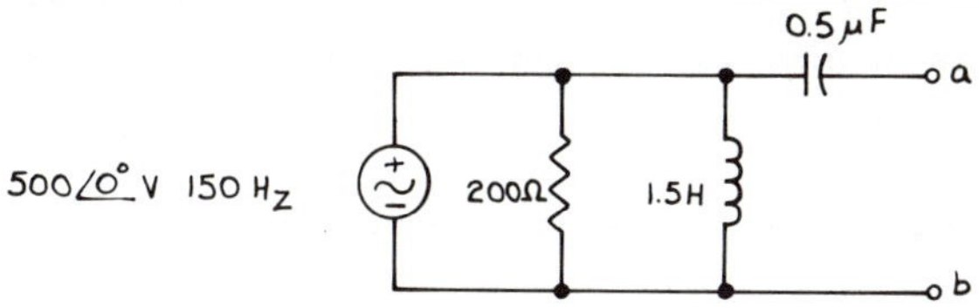

Fig. 1

Solution: Since we are given $\omega = 2\pi f = 2\pi 150 = 942.5 \frac{\text{rad}}{\text{s}}$ we can transform the circuit in fig. 1 to its steady-state equivalent.

Hence the 1.5 H inductor becomes $j\omega(1.5) = j(942.5)(1.5) = j1413.75\Omega$, $0.5\mu F \Rightarrow \frac{1}{j(0.5\times10^{-2})(942.5)} = -j2122.25\Omega$

We obtain the circuit in fig. 2.

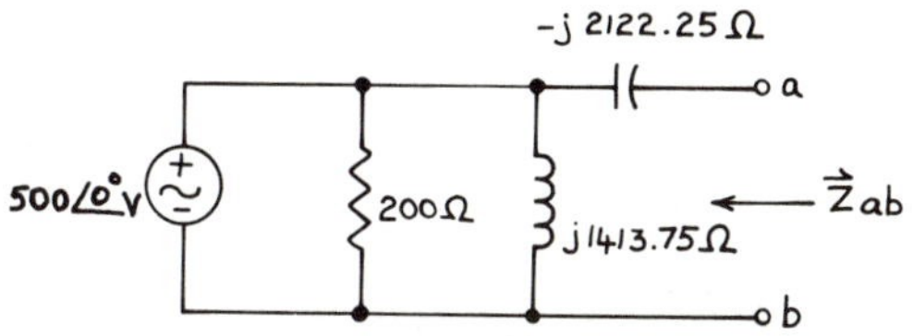

Fig. 2

We can now proceed to find the Thévenin and Norton equivalent circuits. First we find the Thevenin impedance $\vec{Z}_{ab}$ looking into terminals ab.

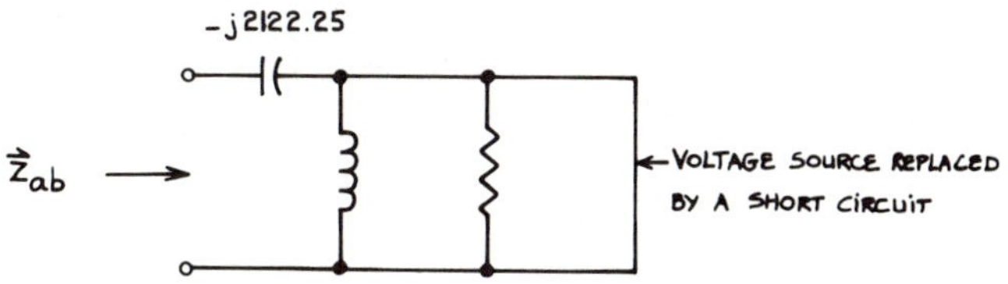

Fig. 3

Hence, from fig. 3, $\vec{Z}_{ab} = -j2122.25\Omega$ is the Thévenin impedance. The Thévenin voltage source would be the voltage across terminals ab, which by inspection is found to be $500\underline{/0^\circ}$V.

The Thévenin equivalent circuit is shown below in fig. 4.

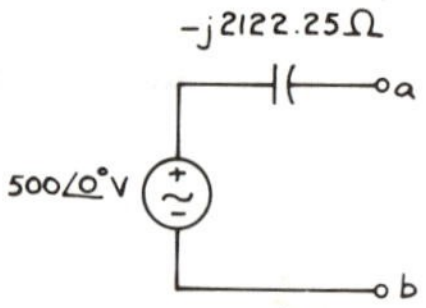

FIG. 4

The Norton equivalent would be the source transformed circuit of the Thévenin circuit above, thus

$$\vec{I} = \frac{500}{-j2122.25}\text{ A}$$

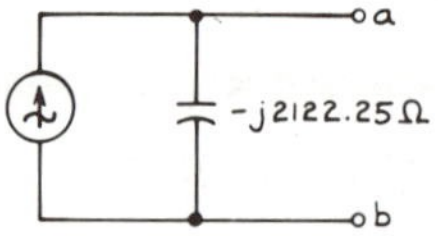

where, $\vec{I} = 0.236\underline{/90^\circ}$ A.

• **PROBLEM** 9-11

Find the Thévenin and Norton equivalents of the circuit shown in fig. 1.

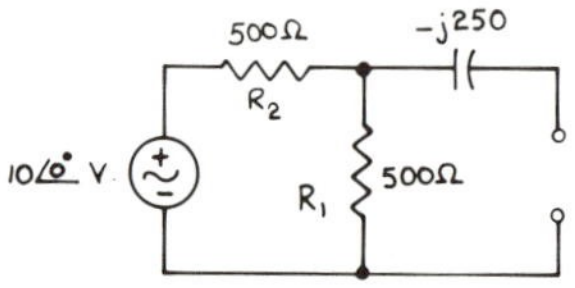

FIG. 1

Solution: Thévenin's equivalent voltage is the open circuit voltage shown in fig. 1. If the terminals are open, there will be no current flowing in the capacitive reactance of -j250Ω and therefore no voltage drop across it. The open circuit voltage is equal to the voltage across the 500Ω resistor R_1. Using the voltage division principle, the source voltage $10\underline{/0^\circ}$V will be divided into the voltages across two 500Ω resistors. Therefore the Thevenin equivalent voltage v_{th} is given by

$$\vec{V}_{th} = 10\underline{/0^\circ}\,\frac{500}{500+500} = 5\underline{/0^\circ}\text{V} \qquad (1)$$

The Norton equivalent internal impedance can be obtained by turning off the $10\underline{/0^\circ}$V source and short circuiting the source terminals. Then the internal impedance of the network presented across the output terminals is the series impedance of the capacitive reactance of -j250Ω, and the parallel resistance of two 500 resistors. Therefore, the Norton equivalent internal impedance

$$Z_{no} = \frac{500}{2} - j250 = 250 - j250 = 250\,(1-j1) = 250\underline{/-45^\circ}\Omega. \qquad (2)$$

RESPONSE

A black box containing only passive elements possesses four terminals, labeled A, B, C, and D. If a voltage source $v_{AB} = 100e^{j(4t+30°)}V$ is applied, the response, $v_{CD} = 40e^{j(4t-120°)}V$, is obtained. Find v_{CD} if v_{AB} =: (a) $50e^{j(4t+30°)}$; (b) $50e^{j(4t-30°)}$; (c) 20 cos 4t; (d) 30 sin(4t-45°); (e) $100e^{j(7t+30°)}V$.

Solution: The solution to this problem relies heavily on superposition. Define the transfer function

$$H(j\omega) = \frac{\vec{V}_{CD}(j\omega)}{\vec{V}_{AB}(j\omega)}$$

to show that

$$\vec{H}(j4) = \frac{40\angle -120°}{100\angle 30°} = 0.4\angle -150°.$$

Hence, as long as the excitation frequency is ω = 4 rad/sec, we get $\vec{V}_{CD} = H(j4)V_{AB}$ (in phasor form). Thus,

(a) $\vec{V}_{CD} = (.4\angle -150°)(50\angle 30°) = 20\angle -120°V$, so

$$v_{CD}(t) = 20e^{j(4t-120\ )}V.$$

(b) $\vec{V}_{CD} = (.4\angle -150°)(50\angle -30\ ) = 20\angle -180\ V$, so

$$v_{CD}(t) = 20e^{j(4t-180\ )}V.$$

(c) $\vec{V}_{CD} = (.4\angle -150°)(20\angle 0°) = 8\angle -180°V$, so

$$v_{CD}(t) = 8\cos(4t - 150°)V.$$

(d) $\vec{V}_{CD} = (.4\angle -150°)(30\angle -45°) = 12\angle -195°V$, so

$$v_{CD}(t) = 12\sin(4t - 195°)V.$$

(e) Since v_{AB} has a frequency which is different from 4 rad/sec, we have no knowledge of $H(j\omega)$, hence we cannot solve the problem.

Note that in parts (a) and (b), the inputs were complex exponentials, in part (c) we must take the real part of $v_{CD}e^{j4t}$ to evaluate the response to 20 cos 4t, and in part (d) we must take the imaginary part.

Find the angle by which v_2 leads v_1 if

$$v_1 = 4\cos(1000t - 40°)\text{V} \quad \text{and} \quad v_2 = ;$$

(a) $3\sin(1000t - 40°)$V; (b) $-2\cos(1000t - 120°)$V;
(c) $5\sin(1000t - 180°)$V.

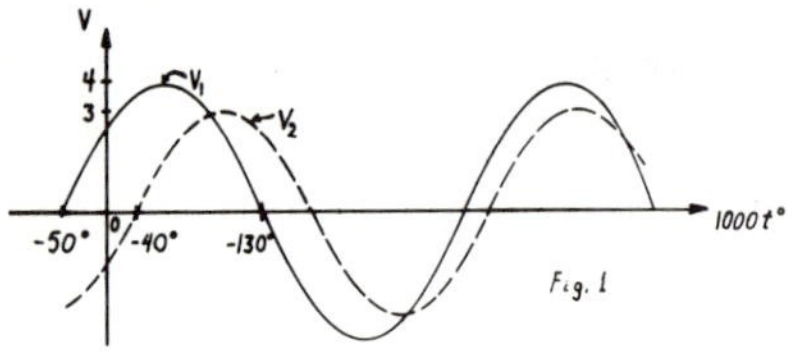

Fig. 1

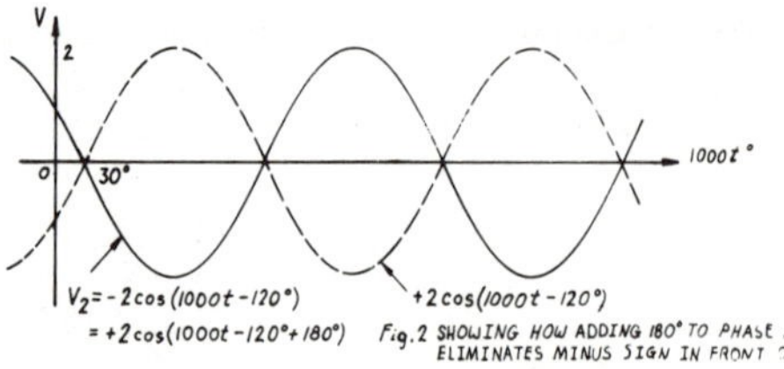

Fig. 2 SHOWING HOW ADDING 180° TO PHASE ANGLE ELIMINATES MINUS SIGN IN FRONT OF COSINE

Solution: Although sines and cosines are actually two ways of describing the same wave of sinusoidal shape, the determination of phase angles between two sinusoids is less apt to result in mistakes if we first convert sines to cosines. Secondly, since we are asked specifically for the angle by which v_2 leads v_1 subtract the phase angle of v_1 from the phase angle of v_2. Thus, if v_2 actually lags v_1 the calculated angle will be negative.

(a) As a preliminary step, convert the sine function to a cosine function by subtracting 90 degrees from its phase angle. This yields

$$v_2 = 3\sin(1000t - 40°) = 3\cos(1000t - 40° - 90°)\text{V} \quad (1)$$

Since both voltages are cosines, subtract the angle of v_1 from the angle of v_2:

$$\theta_2 - \theta_1 = (-130°) - (-40°) = -90°. \quad (2)$$

The negative sign means that v_2 actually lags v_1. It is left in this form because the original statement of the problem asks for the amount by which v_2 leads v_1.

The relationship of these two sinusoids is shown in Fig. 1.

As a check, note that v_2 reaches its peak value when $1000t = 130°$ as sketched, and since a sine wave reaches its positive peak when its total angle is 90°, i.e.

$$v_2 = 3 \text{ when } \sin(1000t - 40°) = \sin 90° = 1, \quad (3)$$

$$\text{then } 1000t - 40° = 90° \quad (4)$$

$$\text{and } 1000t = 130°. \quad (5)$$

(b) Conversion is unnecessary this time, but the minus sign in front of the cosine function should be interpreted as an additional 180° phase angle (either + or −, it doesn't matter). That is, the v_2 wave shown in Fig. 2 could be interpreted mathematically as the mirror image of 2 cos(1000t - 120°) (the minus sign does that), or it could also be interpreted as the 2 cos(1000t - 120°) shifted right or left by 180°.

Once v_2 has been modified, proceed as before to subtract phase angles:

$$v_2 = 2\cos(1000t - 120° + 180°) = 2\cos(1000t + 60°)\text{V} \quad (6)$$

$$\theta_2 - \theta_1 = (+60°) - (-40°) = 100°. \quad (7)$$

(c) Again it is necessary to convert the sine function to a cosine by subtracting 90 degrees, thus

$$v_2 = 5\sin(1000t - 180°) = 5\cos(1000t - 270°)\text{V} \quad (8)$$

Subtracting phase angles gives

$$\theta_2 - \theta_1 = -270° - (-40°) = -230°. \quad (9)$$

Although numerically correct, our answer will be modified by adding 360 because angles of lead or lag are customarily expressed as equal to or less than 180 in electrical engineering applications,

$$\theta_2 - \theta_1 = -230° + 360° = 130°. \quad (10)$$

• PROBLEM 9-14

A voltage of $25\angle 45°$ V is applied to an impedance of $4 - j2\,\Omega$. Determine the current in both rectangular and polar forms.

Solution: Using Ohm's Law, the current is found to be

$$\vec{I} = \frac{\vec{V}}{\vec{Z}} = \frac{25\angle 45°}{4-j2} = \frac{25\angle 45°}{\sqrt{4^2 + 2^2}\angle \tan^{-1} - \frac{1}{2}}$$

$$\vec{I} = \frac{25\angle 45°}{\sqrt{20}\angle -26.6°} = \frac{25}{\sqrt{20}}\angle 45° + 26.6°$$

$$\vec{I} = 2.5\sqrt{5}\angle 71.6°\text{A}$$

in polar form.

$\vec{I}$ could have been found in rectangular form directly by converting the voltage given in polar form to rectangular form and by solving for I.

Hence,

$$\vec{I} = \frac{\vec{V}}{\vec{Z}} = \frac{25\angle 45}{4 - j2} = \frac{\frac{25}{\sqrt{2}} + j\frac{25}{\sqrt{2}}}{4 - j2}$$

$$\vec{I} = \frac{\left(\frac{25}{\sqrt{2}} + j\frac{25}{\sqrt{2}}\right)(4 + j2)}{4^2 + 2^2} = \frac{5}{2\sqrt{2}} + j\frac{15}{2\sqrt{2}}$$

$$\vec{I} = 1.25\sqrt{2} + j3.75\sqrt{2}\text{A}.$$

To check the results, note that

$$\sqrt{(1.25\sqrt{2})^2 + (3.75\sqrt{2})^2} \text{ should equal } 2.5\sqrt{5}$$

$$\sqrt{\left(\frac{5}{4}\sqrt{2}\right)^2 + \left(\frac{15}{4}\sqrt{2}\right)^2} \overset{?}{=} 2.5\sqrt{5}$$

$$\sqrt{\frac{50}{16} + \frac{450}{16}} \overset{?}{=} 2.5\sqrt{5}$$

$$\sqrt{\frac{500}{16}} = \frac{\sqrt{500}}{4} \overset{?}{=} 2.5\sqrt{5}$$

$$\frac{10\sqrt{5}}{4} = 2.5\sqrt{5}.$$

• **PROBLEM** 9-15

Find the current in the circuit shown in Fig. 1.

Solution: Using KVL, write the equation

$$100\underline{/0°} = \vec{I}(j6 + 2 + 3 - j4)$$

solving for $\vec{I}$:

$$100\underline{/0°} = \vec{I}(5 + j2)$$

$$\vec{I} = \frac{100}{5 - j2} = \frac{100(5 - j2)}{5^2 + 2^2}$$

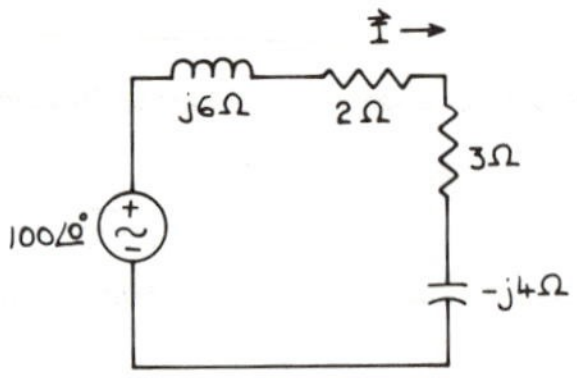

FIG.1

$$\vec{I} = \frac{100}{29}(5 - j2) = 17.24 - j6.9 \text{ A}.$$

In polar form, write

$$\vec{I} = \sqrt{17.24^2 + 6.9^2}\underline{/\tan^{-1} - \frac{6.9}{17.24}}$$

$$\vec{I} = 18.57\underline{/-21.8°} \text{ A}.$$

• **PROBLEM** 9-16

A voltage source 60 cos 1000t V is in series with a 2-kΩ resistor and a 1 - uF capacitor. Find i_{forced} .

Solution: In this case we use the phasor method since we want only the forced (steady-state) current.

Express the instantaneous voltage as a phasor. The reference angle is zero and the frequency is suppressed until we convert the phasor current into its instantaneous value. Thus, v = 60 cos 1000t V, becomes the phasor voltage, $\vec{V} = 60\underline{/0°}$ V.

The impedance of the circuit, $\vec{Z} = R - j\frac{1}{\omega C}$ is

$$\vec{Z} = 2000 - j\frac{1}{(1000)(1\text{x}10^{-6})} = 2000 - j1000°.$$

It is convenient to express in polar-form and also in KΩ. Therefore $\vec{Z} = 2.24\underline{/-26.6°}$ KΩ. Now apply Ohm's Law to find the phasor current,

$$I = \frac{\vec{V}}{\vec{Z}} = \frac{60\underline{/0°}}{2.24\underline{/-26.6°}} = 26.8/26.6° \text{ mA}.$$

The final step is to convert this phasor current into its equivalent instantaneous value.

The angle indicates that the current leads by 26.6° and the amplitude of the current is 26.8 mA. The frequency must be the same as the frequency of the voltage, ω = 1000 rad/s. Thus, i_{forced} = 26.8 cos(1000t + 26.6°)mA. Now use a trigonometric identity,

$$\cos(A + B) = \cos A \cos B - \sin A \sin B$$

$$i_{forced} = 26.8(\cos 26.6° \cos 1000t - \sin 26.6° \sin 1000t)\text{mA}$$

Carrying out the operations and rounding off the results,

$$i_{forced} = 24 \cos 1000t - 12 \sin 1000t \text{ mA}.$$

• PROBLEM 9-17

With $V_C(t)$ as the desired response, find the sinusoidal steady-state transfer function H(jω) for the circuit of Fig. 1.

Solution: The transfer function H(jω) is defined as

$$\frac{V_C(j\omega)}{V(j\omega)}.$$

Since $V_C(j\omega)$ can be found by voltage division as

$$V_C(j\omega) = \frac{V(j\omega)Z_C(j\omega)}{Z_R(j\omega) + Z_L(j\omega) + Z_C(j\omega)},$$

the transfer function is

$$H(j\omega) = \frac{V_C(j\omega)}{V(j\omega)} = \frac{Z_C(j\omega)}{Z_R(j\omega) + Z_L(j\omega) + Z_C(j\omega)}.$$

Hence,

$$H(j\omega) = \frac{\frac{1}{j\omega 0.25}}{1 + j\omega + \frac{1}{j\omega 0.25}}$$

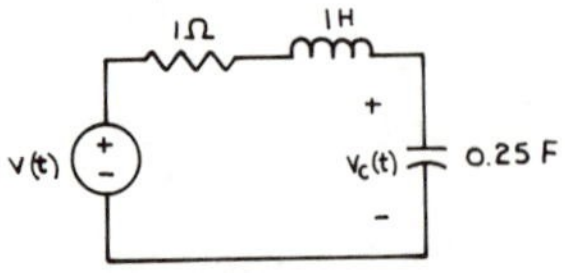

Fig.1

$$H(j\omega) = \frac{\frac{4}{j\omega}}{1 + j\omega + \frac{4}{j\omega}}$$

$$H(h\omega) = \frac{4}{(j\omega)^2 + j\omega + 4} = \frac{4}{4-\omega^2 + j\omega}.$$

• PROBLEM 9-18

> In the series circuit shown in fig. 1, calculate i(t) in the steady-state and draw a voltage phasor diagram.

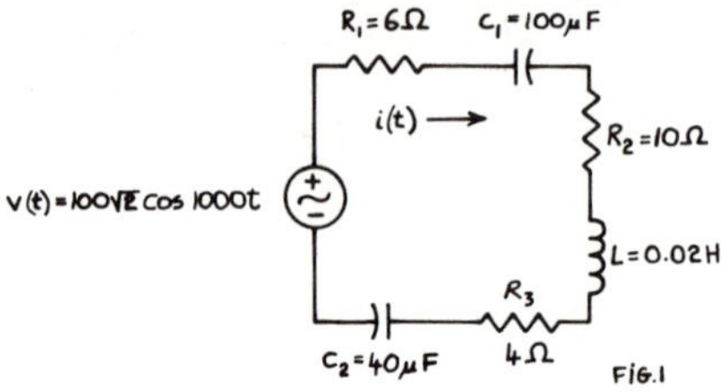

Fig.1

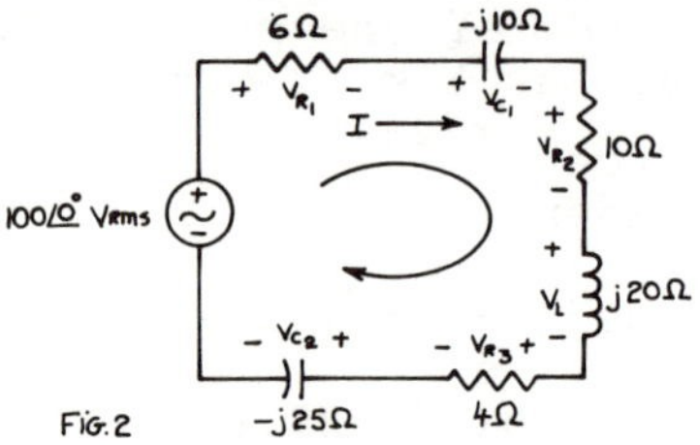

Fig.2

Solution: First, find the steady-state equivalent of the circuit by transforming, element-by-element;

$$100\sqrt{2}\cos 1000t \text{ V} \Rightarrow 100\underline{/0^\circ} \text{ V rms}$$

$$C_1 = 100\mu F \Rightarrow \frac{1}{j\omega C} = \frac{1}{j(1000)(100\times 10^{-6})} = -j10\Omega$$

$$L = 0.02H = \quad j\omega L = j1000(0.02) = j20$$

$$C_2 = 40\mu F \Rightarrow \frac{1}{j\omega C} = \frac{1}{j1000(40\times 10^{-6})} = -j25\Omega.$$

Hence, we have the circuit shown in fig. 2.
Using KVL around the loop in fig. 2, we find

$$100\underline{/0^\circ} = \vec{I}(6 - j10 + 10 + j20 + 4 - j25)$$

the phasor current is,

$$\vec{I} = \frac{100}{20 - j15}$$

$$\vec{I} = \frac{100(20 + j15)}{20^2 + 15^2} = \frac{20 + j15}{6.25}$$

$$\vec{I} = 3.2 + j2.4 \text{ A}$$

in polar form the current is,

$$\vec{I} = 4\underline{/36.87^\circ} \text{ A}.$$

One can now calculate the voltage drop across each element,

$$\vec{V}_{R_1} = 6(4\angle 36.87°) = 24\angle 36.87° \text{ Vrms}$$

$$\vec{V}_{C_1} = -j10(4\angle 36.87°)$$

$$= (10\angle -90°)(4\angle 36.87°) = 40\angle -53.13° \text{ Vrms}$$

$$\vec{V}_{R_2} = 10(4\angle 36.87°) = 40\angle 36.87° \text{ Vrms}$$

$$\vec{V}_L = j20(4\angle 36.87°)$$

$$= (20\angle 90°)(4\angle 36.87°) = 80\angle 126.87° \text{ Vrms}$$

$$\vec{V}_{R_3} = 4(4\angle 36.87°) = 16\angle 36.87° \text{ Vrms}$$

$$\vec{V}_{C_2} = -j25(4\angle 36.87°)$$

$$= (25\angle -90°)(4\angle 36.87°) = 100\angle -53.13° \text{ Vrms.}$$

In rectangular form,

$$\vec{V}_{R_1} = 19.2 + j14.4 \text{ V}$$

$$\vec{V}_{C_1} = 24.0 - j32.0 \text{ V}$$

$$\vec{V}_{R_2} = 32.0 + j24.0 \text{ V}$$

$$\vec{V}_L = -48.0 + j64.0 \text{ V}$$

$$\vec{V}_{R_3} = 12.8 + j9.6 \text{ V}$$

$$\vec{V}_{C_2} = 60.0 - j80.0 \text{ V}$$

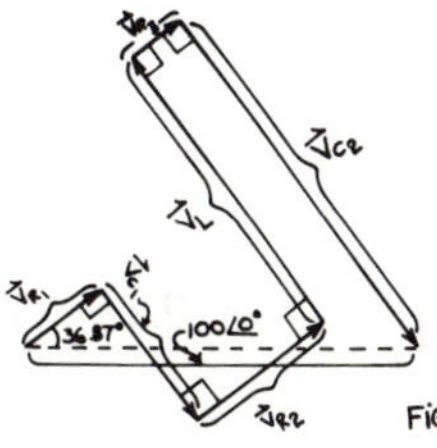
FIG.3

Adding the voltages in rectangular form gives 100 + j0 V, the expected resulting source voltage.

Using the voltage in polar form, add the phasor vectorially using the source voltage as a reference, Fig. 3 shows the vector addition of the voltage drops.

• **PROBLEM** 9-19

In the circuit shown in fig. 1, the following magnitudes are known:

$$|V_{ab}| = 100 \text{ V}$$

$$|V_{a'b}| = 200 \text{ V}$$

$$|I| = 20 \text{ A.}$$

In addition, it is known that $X_L = 2R$. Solve for $\vec{Z}_{a'a}$ and $\vec{V}_{a'a}$.

Solution: Since $Z_{ab} = R_1$ and $\vec{V}_{ab}$ and $\vec{I}$ are in phase (current through a resistor must always be in phase with the voltage across it), then,

$$\vec{Z}_{ab} = R_1 = \frac{|V_{ab}|}{|I|} = \frac{100}{20} = 5\Omega.$$

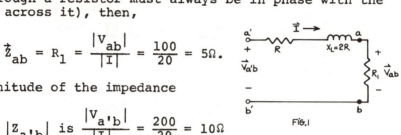

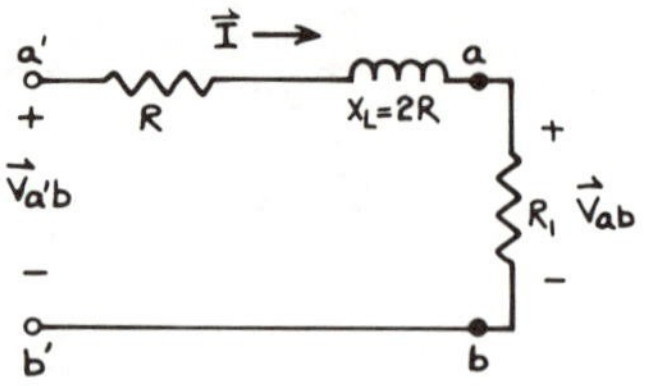

FIG. 1

The magnitude of the impedance

$$|Z_{a'b}| \text{ is } \frac{|V_{a'b}|}{|I|} = \frac{200}{20} = 10\Omega$$

and, since,

$$|Z_{a'b}| = \sqrt{(R + R_1)^2 + (X_L)^2} = 10$$

then,

$$\sqrt{(R + 5)^2 + (2R)^2} = 10$$

$$\sqrt{R^2 + 10R + 25 + 4R^2} = 10$$

$$\sqrt{5R^2 + 10R + 25} = 10$$

Squaring both sides, and solving for R:

$$5R^2 + 10R + 25 = 100$$

$$R^2 + 2R + 5 = 20$$

$$R^2 + 2R - 15 = 0$$

$$R = \frac{-2 \pm \sqrt{4 - 4(-15)}}{2}$$

$$R = -1 \pm \frac{\sqrt{64}}{2}$$

$$R = -1 \pm 4.$$

Hence, $R = 3\Omega$, (rejecting solution $R = -5\Omega$)

$$X_L = 6\Omega, \quad \text{and} \quad \vec{Z}_{a'a} = 3 + j6\Omega.$$

We can find the magnitude of the voltage

$$|V_{a'a}| = |I||Z_{a'a}|$$

$$= 20\sqrt{3^2 + 6^2}$$

$$|V_{a'a}| = 20(6.71) = 134.2 \text{ V.}$$

Since $$|V_{a'a}| = \sqrt{V_R^2 + V_L^2}$$

where $$V_R = |I|R = (3)(20) = 60 \text{ V.}$$

Solve for V_L:

$$134.2 = \sqrt{60^2 + V_L^2}$$

$$134.2^2 = 60^2 + V_L^2$$

$$\sqrt{134.2^2 - 60^2} = V_L^2$$

$$\sqrt{14410} = V_L^2$$

$$\pm 120 \text{ V} = V_L.$$

We chose V_L = +120 V since the current through the inductor is positive and

$$X_L = 6 = \frac{120}{|I|} = \frac{120}{20} = 6, \quad \text{is also positive.}$$

One can now write

$$V_{a'a} = 60 + j120 \text{ V}$$

$$Z_{a'a} = 3 + j6\Omega\ .$$

• PROBLEM 9-20

A sinusoidal driving function V sin ωt is applied to a series RL network. Find the behavior of the current under steady state conditions. Sketch the magnitude and phase versus ω.

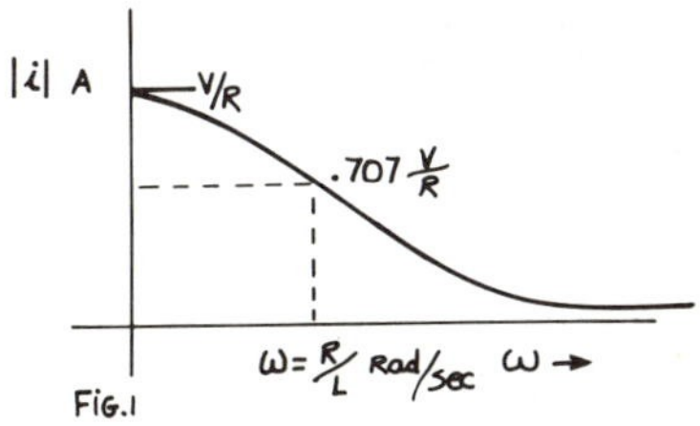

FIG.1

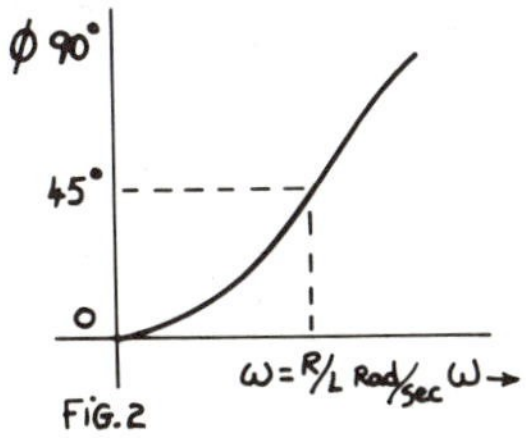

FIG.2

Solution: The differential equation which results from KVL applied to the R-L circuit is:

$$L\frac{di}{dt} + iR = V \sin \omega t$$

Since the response i(t) must assume the time dependence of the forcing function, and since one term contains a derivative (with 90° shift), it would seem logical to assume:

$$i(t) = I_1 \sin\omega t + I_2 \cos \omega t \tag{1}$$

where I_1 and I_2 are real constants which must depend on R, L, ω and V. A less cumbersome form results if we recognize that an equivalent form using a single amplitude constant and phase factor will give the same result.

$$i(t) = A \sin(\omega t + \phi) \tag{2}$$

Putting (2) into the differential equation gives:

$$L[\omega A \cos(\omega t + \phi)] + R[A \sin(\omega t + \phi)] = V \sin \omega t.$$

Expanding the trigonometric forms gives:

$$A\omega L[\cos \omega t \cos \phi - \sin \omega t \sin \phi]$$

$$+ AR[\sin \omega t \cos \phi + \cos \omega t \sin \phi] = V \sin \omega t. \tag{3}$$

Now equate the coefficients of cos ωt:

$$A\omega L \cos \phi + AR \sin \phi = 0$$

from which

$$\tan \phi = \frac{\sin \phi}{\cos \phi} = \frac{-\omega L}{R}.$$

Also, from this result, it can be seen that

$$\sin \phi = \frac{-\omega L}{\sqrt{R^2 + \omega^2 L^2}} \quad \text{and} \quad \cos \phi = \frac{R}{\sqrt{R^2 + \omega^2 L^2}} \tag{4}$$

Now equate the coefficients of sin ωt in (3)

$$-A\omega L \sin \phi + AR \cos \phi = V.$$

Substituting the results of (4),

$$\frac{\omega^2 L^2 A}{\sqrt{R^2 + \omega^2 L^2}} + \frac{R^2 A}{\sqrt{R^2 + \omega^2 L^2}} = V$$

From which it is easy to obtain

$$A = \frac{\sqrt{R^2+\omega^2L^2}}{R^2+\omega^2L^2} V = \frac{V}{R^2+\omega^2L^2}$$

The solution is then written:

$$i(t) = \frac{V}{\sqrt{R^2 + \omega^2 L^2}} \sin(\omega t + \tan^{-1} - \frac{\omega L}{R})$$

Figures 1 and 2 are sketches of magnitude phase of i versus ω.

The complete response of the circuit in fig. 1 was found to be $v_C(t) = \frac{1}{2}[e^{-t} - \cos t + \sin t]u(t)V$ for an input of $v(t) = \sin t\ u(t)V$. Use sinusoidal steady state methods to find the steady state voltage $V_{C_{ss}}(t)$ when the input is, $v(t) = \sin(t)V$. Compare the results with the expression for complete response, $v_C(t)$, above.

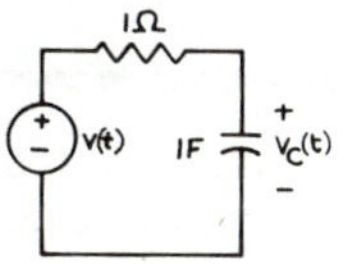

Fig.1

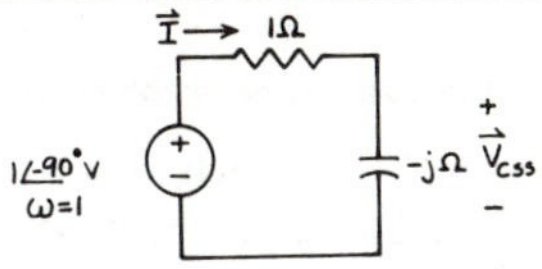

Fig.2

Solution: In order to perform steady state analysis, recall the convention for phasor representation of sinusoidals. It was stated previously that

$$v(t) = A\cos(\omega t + \phi) = \text{Re}(\text{Im}\ e^{j(\omega t+\phi)}),$$

is represented by, $V = A\ e^{j\phi} = A\underline{/\phi}$, the definition of the phasor.

Since, $\sin t = \cos(t - 90°)$, the phasor input becomes, $V = 1\underline{/-90°}$ V. Fig. 2 shows the steady state phasor representation of the circuit in fig. 1.
Writing a KVL equation yields,

$$1\underline{/-90°} = \vec{I}(1 - j)$$

$$\vec{V}_{C_{ss}} = \vec{I}(-j) = \frac{1\underline{/-90°}}{1 - j}(-j)$$

in polar form:

$$\vec{V}_{C_{ss}} = \frac{(1\underline{/-90°})(1\underline{/-90°})}{\sqrt{2}\underline{/-45°}} = \frac{1}{\sqrt{2}}\underline{/-135°}\ \text{V}$$

Using the convention stated above change the phasor $\vec{V}_{C_{ss}}$ to $v_{C_{ss}}(t)$ sinusoidal.

$$v_{C_{ss}}(t) = \frac{1}{\sqrt{2}}\cos(t - 135°)\text{V}. \qquad (1)$$

By using the angle difference relation,

$$\cos(\alpha-\beta) = \cos\alpha\cos\beta + \sin\alpha\sin\beta,$$

expand the expression for $V_{C_{ss}}$ into the sum of two sinusoids

$$v_{C_{ss}}(t) = \frac{1}{\sqrt{2}}\cos(t-135°) = \frac{1}{\sqrt{2}}[\cos t\cos(135°) + \sin t\sin(135°)]$$

$$V_{c_{ss}}(t) = \frac{1}{\sqrt{2}} \cos(t-135°) = -\frac{1}{2} \cos t + \frac{1}{2} \sin t \text{ V.} \qquad (2)$$

Explain the meaning of the results in equation (1) by comparing them to the complete response $V_c(t)$. Since the complete response contains both transient and steady state responses, the steady state response can be found by allowing, $t \to \infty$. At $t = \infty$ the term $\frac{1}{2}e^{-t}$ becomes zero and the steady state response becomes $-\frac{1}{2} \cos t + \frac{1}{2} \sin t$.

• PROBLEM 9-22

Evaluate the sinusoidal steady-state current for the circuit shown in Fig. 1 by replacing the circuit by its sinusoidal steady state equivalent.

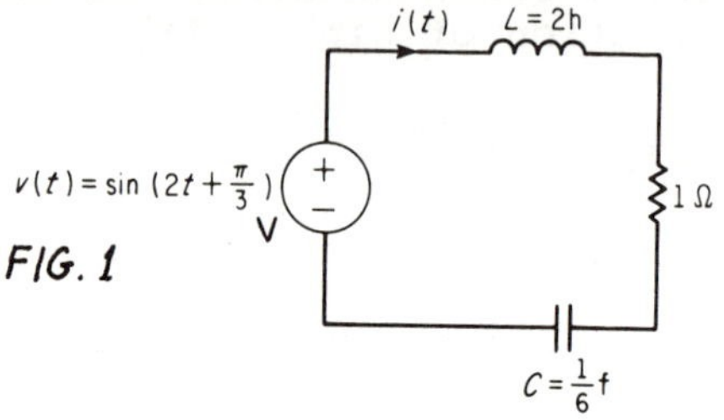

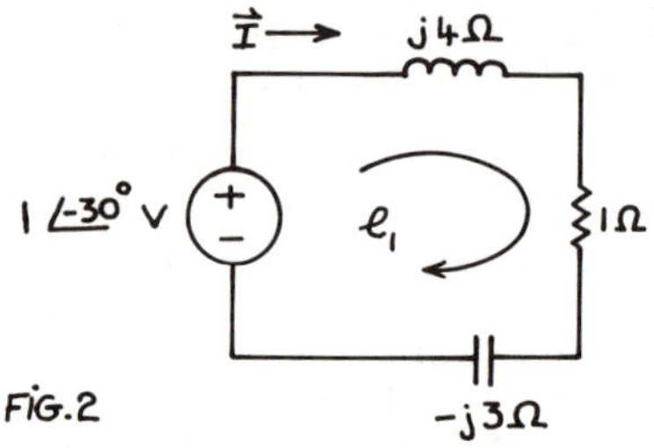

Solution: In order to replace the circuit in Fig. 1 by its sinusoidal steady state equivalent, note that from $v(t) = \sin(2t + \frac{\pi}{3})$ that $\omega = 2$ and that $v(t) = \sin(2t + 60°)$ $= \cos(2t+60 -90°) = \cos(2t-30°) = \text{Re}\left[e^{j(2t-30°)}\right]$. This can be written in phasor form as $1\angle -30°$ V.

$$\vec{Z}_L = j\omega 2 = j2(2) = j4\Omega$$

$$\vec{Z}_R = 1\Omega$$

$$\vec{Z}_C = \frac{1}{j\omega(\frac{1}{6})} = \frac{1}{j2(\frac{1}{6})} = \frac{-j}{\frac{1}{3}} = -j3\Omega.$$

Replace each of the elements in Fig. 1 by those found above. Fig. 2 shows the sinusoidal steady-state equivalent of the circuit in Fig. 1. Using KVL around the loop ℓ, solve for $\vec{I}$.

$$1\angle 30° = \vec{I}(j4 + 1 - j3)$$

$$\vec{I} = \frac{1\angle 30°}{1 + j1} = \frac{1\angle 30°}{\sqrt{2}\angle 45°}$$

$$I = \frac{1}{\sqrt{2}}\angle -75° = \frac{1}{\sqrt{2}} e^{j(2t-75°)} = \frac{1}{\sqrt{2}} \cos(2t - 75°)$$

$$I = \frac{1}{\sqrt{2}} \sin(2t - 75° + 90°) = \frac{1}{\sqrt{2}} \sin(2t + 15°)$$

$$I = \frac{1}{\sqrt{2}} \sin(2t + \frac{\pi}{12}) A.$$

• PROBLEM 9-23

A 200-Ω resistor, a 0.04 H inductor, and a 0.25-μF capacitor are connected in series. Find the phasor voltage across the combination if the phasor current, $30\angle 45°$ mA, is applied at ω =: (a) 8000 rad/s; (b) 10,000 rad/s; (c) 12,500 rad/s.

Solution: To find the required voltage, we must use the phasor form of Ohm's Law:

$$\vec{V} = \vec{I}\vec{Z} \tag{1}$$

This is sometimes called the AC form or complex form of Ohm's Law.

First, find the impedance Z of the series circuit which is the sum of the individual impedances.

$$\vec{Z} = \vec{Z}_R + \vec{Z}_L + \vec{Z}_C \tag{2}$$

Using $\vec{Z}_R = R$, $\vec{Z}_L = j\omega L$, and $\vec{Z}_C = -j/\omega C$, we have

$$Z = R + j(\omega L - 1/\omega C). \tag{3}$$

The magnitude of this impedance is then found from

$$|Z| = \sqrt{R^2 + (\omega L - 1/\omega C)^2}, \tag{4}$$

and the phase angle from

$$\theta = \tan^{-1}[(\omega L - 1/\omega C)/R]. \tag{5}$$

(a) For ω = 8000 rad/s,

$$\vec{Z} = 200 + j(8000 \times 0.04 - (1/8000 \times 0.25 \times 10^{-6}))$$

$$= 200 + j(320 - 500)$$

$$= 200 + j(-180)\Omega$$

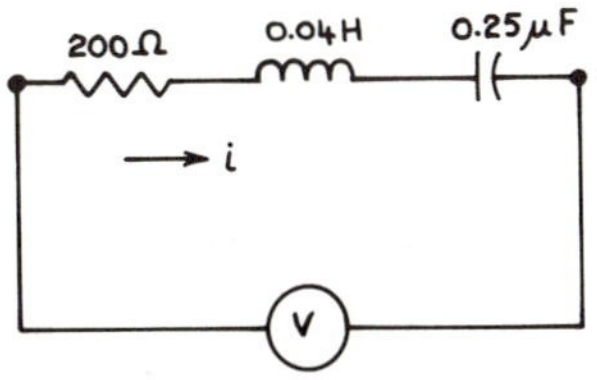

FIG.1

which, from (4) and (5), gives

$$|Z| = 269.1\Omega$$

$$\theta = -42.0°.$$

This results are often expressed as

$$\vec{Z} = 269.1\angle -42° \ \Omega$$

in phasor notation.

To use the phasor form of Ohm's Law, we multiply the magnitudes of the phasor quantities I and Z and add the phase angles.

$$\vec{V} = (0.03\underline{/45^\circ})(269.1\underline{/-42^\circ})$$

and $$\vec{V} = 8.07\underline{/3^\circ} \text{ V.}$$

(b) for $= 10{,}000$ rad/s,

$$\vec{Z} = 200 + j(400 - 400)$$

$$Z = 200 + j(0)\Omega.$$

Therefore,

$$|Z| = 200,$$

$$\theta = 0^\circ,$$

and $$\vec{Z} = 200\underline{/0^\circ} \ \Omega.$$

From the phasor form of Ohm's Law,

$$\vec{V} = (0.03\underline{/45^\circ})(200\underline{/0^\circ})$$

$$\vec{V} = 6.0\underline{/45^\circ} \text{ V}$$

(c) For $\omega = 12{,}500$ rad/s,

$$\vec{Z} = 200 + j(500 - 320)$$

$$= 200 + j(180)\Omega.$$

This gives

$$|Z| = 269.1$$

$$\theta = 42.0^\circ$$

or, $$\vec{Z} = 269.1\underline{/42^\circ} \ \Omega.$$

Therefore, from Ohm's Law, the voltage is

$$\vec{V} = (0.03\underline{/45^\circ})(269.1\underline{/42^\circ})$$

$$\vec{V} = 8.07\underline{/87^\circ} \text{ V.}$$

• PROBLEM 9-24

Find the current that will flow in a series circuit containing a 5-kΩ resistor, a 470-ρF capacitor, a 150-mH inductor, and a 0.001μF capacitor if the exciting voltage is 65V at 25-kHz.

Solution: It is convenient to assume that the applied voltage has zero angle. That is, $\vec{V} = 65\underline{/0^\circ}$ V. The equivalent capacitance of the two series capacitors is

$$\frac{1}{C_{eq}} = \frac{1}{C_1} + \frac{1}{C_2} = \frac{1}{4.75\times10^{-10}} + \frac{1}{10^{-9}} = 3.13\times10^{9}$$

$$C_{eq} = 3.2 \times 10^{-10}F.$$

This capacitance is effectively in series with the resistor and inductor, so the three impedances are added to get the equivalent impedance across the source.

$$Z_{eq} = R + j\omega L - j\left(\frac{1}{\omega C_{eq}}\right)$$

$$= 5\times10^{3} + j2\pi(2.5\times10^{4})(0.15) - j\left[\frac{1}{2\pi(2.5\times10^{4})(3.2\times10^{-10})}\right]$$

$$= 5\times10^{3} + j3.668\times10^{3} = 6.2\times10^{3}\underline{/36.8°}\,\Omega.$$

Then $I = \frac{\vec{V}}{\vec{Z}_{eq}} = 65/6.2\times10^{3}\underline{/36.3°} = 10.5\underline{/-36.3°}$mA.

• **PROBLEM** 9-25

In a series circuit with a 20-kΩ resistor, a 10-kΩ capacitive reactance, and a 5-kΩ inductive reactance, what is the current through the inductor when 1500V is placed across the circuit?

Solution: Figure 1 is a schematic diagram of the network associated with the problem
The problem is to determine I. Now

$$\vec{I} = \frac{\vec{V}}{\vec{Z}} = \frac{1500\underline{/0°}}{(20-j10+j5)\times10^{3}} = \frac{1.5\underline{/0°}}{20 - j5}$$

$$= \frac{0.3\underline{/0°}}{4 - j} = \frac{0.3\underline{/0°}}{4.12\underline{/-14°}} = .0728\underline{/14}\ \ H$$

$$= 72.8\underline{/14°}\text{ mA}.$$

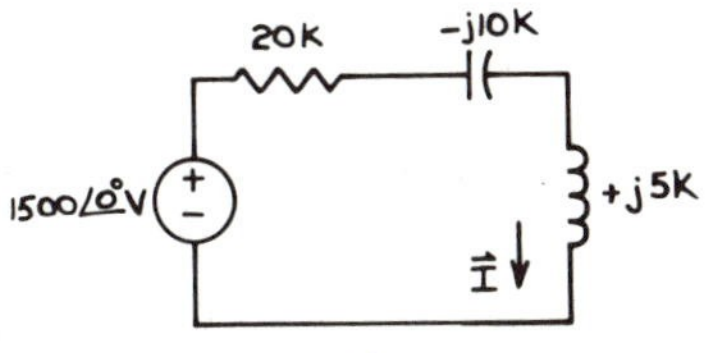

FIG.1

• **PROBLEM** 9-26

Find the steady-state current i(t) in the circuit shown in fig. 1. Assume

$$v(t) = 4\sin(2t + \frac{\pi}{4})V.$$

Solution: Steady-state current i(t) is expressed in phasor form as

$$\vec{I} = I\,e^{j\phi_i} \qquad (1)$$

where I is the rms value of i(t), ϕ_i is the initial phase angle of i(t), and j is equal to $\sqrt{-1}$. Steady-state voltage v(t) is expressed in a phasor form as

$$\vec{V} = V e^{j\phi_n} \tag{2}$$

where V is the rms value of v(t) and ϕ_v is the initial phase angle of v(t). Both $\vec{I}$ and $\vec{V}$ are related to each other by Ohm's law as follows

$$\vec{I} = \frac{\vec{V}}{\vec{Z}} \tag{3}$$

Here $\vec{Z}$ is the impedance of the circuit. If the circuit is formed of a series connection of a resistance R, inductance L and capacitance C then

$$\vec{Z} = R + j(\omega L - \frac{1}{\omega C})$$

$$= \sqrt{R^2 + (\omega L - \frac{1}{\omega C})^2} \exp\left[j \tan^{-1} \frac{L - \frac{1}{\omega C}}{R}\right] \tag{4}$$

Now, $v(t) = 4 \sin(2t + \frac{\pi}{4})$

$$= 4 \cos(2t + \frac{\pi}{4}) - \frac{\pi}{2} = 4 \cos(2t - \frac{\pi}{4}) \tag{5}$$

$$\vec{V} = \frac{4}{\sqrt{2}} e^{j(-\frac{\pi}{4})} V \tag{6}$$

Now R = 1, L = 1, $C = \frac{1}{4}$ and $\omega = 2$ so from Eq. (4)

$$\vec{Z} = \sqrt{1^2 + \left(2 \times 1 - \frac{1}{2\times\frac{1}{4}}\right)^2} \exp\left[j \tan^{-1} \frac{2 \times 1 - \frac{1}{2 \times \frac{1}{4}}}{1}\right]$$

$$\vec{Z} = 1e^{j \tan^{-1} 0} = 1e^{j0}\Omega. \tag{7}$$

So, by Eq. (3),

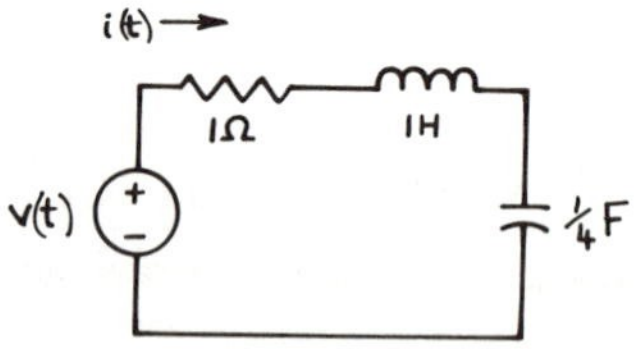

FIG.1

$$\vec{I} = \frac{\frac{4}{2}e^{-j\frac{\pi}{4}}}{1e^{j0}} = \frac{4}{\sqrt{2}} e^{-j\frac{\pi}{4}} A$$

$i_{ss}(t) = \sqrt{2}\, I \cos(2t - \frac{\pi}{4})$

$$= 4 \sin(2t - \frac{\pi}{4} + \frac{\pi}{2}) = 4 \sin(2t + \frac{\pi}{4})A \qquad (8)$$

This problem can be solved by introducing a concept of power factor, or current lags voltage by an angle of

$$\theta = \tan^{-1} \frac{\omega L - \frac{1}{\omega C}}{R} \qquad (9)$$

Then,

$$i_{ss}(t) = \frac{4}{\sqrt{R^2 + \left(\omega L - \frac{1}{\omega C}\right)^2}} \sin(2t + \frac{\pi}{4} - \theta)$$

$$= \frac{4}{1} \sin(2t + \frac{\pi}{4} - 0)$$

$$= 4 \sin(2t + \frac{\pi}{4})A \qquad (10)$$

Note that all calculations using phasors here could have been done using only peak values given, then converting the answer if rms values were desired.

• PROBLEM 9-27

The R-L circuit shown has been connected to a sinusoidal source

$$v(t) = V_m \cos(\omega t + \phi)V$$

for a sufficiently long time so that the steady state is reached. At t = 0 the circuit is deenergized, solve for i(t) for all $t \geq 0^+$.

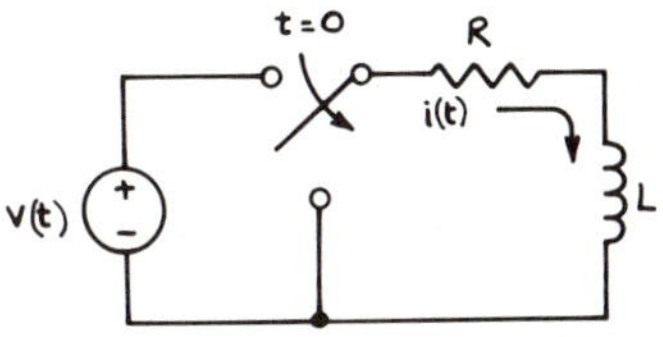

Solution: The first step of the solution is to find the steady-state current through the inductor which will then be the initial condition for the de-energized circuit.

The steady-state current with the switch up is found simply by

$$\frac{V_m \cos(\omega t + \phi)}{R + j\omega L}$$

when the switch is down the circuit equation is

$$Ri + L\frac{di}{dt} = 0$$

which has a solution of the form

$$i(t) = Ke^{-t/\tau}$$

where $\tau = L/R$.
The integration constant, K, is evaluated by setting the current equal to the original steady-state current for $t = 0$. Thus

$$K = \frac{V_m \cos \phi}{R + j\omega L}.$$

The solution is

$$i(t) = \frac{V_m \cos \phi}{R + j\omega L} e^{-t/\tau}.$$

Another form of the solution is obtained by setting $R + j\omega L = Z\underline{/\theta}$. In complex notation we have

$$\frac{V_m e^{j\phi}}{Z e^{j\theta}} e^{-t/\tau} = \frac{V_m}{Z} e^{j(\phi - \theta)} e^{-t/\tau}.$$

Taking the real part gives

$$i(t) = \left[\frac{V_m}{Z}\cos(\phi - \theta)\right] e^{-t/\tau}.$$

• **PROBLEM** 9-28

Calculate the steady-state voltage response of the network shown in Fig. 1.

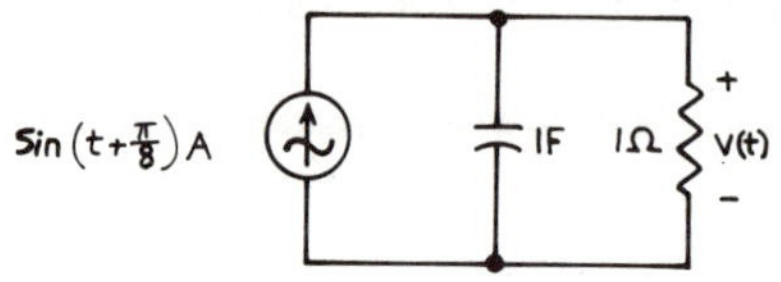

FIG. 1

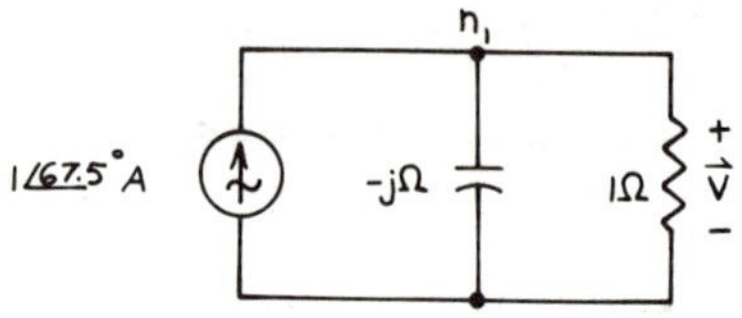

FIG. 2

Solution: Using phasor concepts, transform the circuit into its steady-state equivalent.
Doing so,

$$\sin(t + \tfrac{\pi}{8})A \Rightarrow 1\underline{/-\tfrac{\pi}{2} + \tfrac{\pi}{8}} = 1/67.5 \quad A,$$

$$1F \Rightarrow -j\Omega,$$

and $1\Omega \Rightarrow 1\Omega$, thus, we have the circuit shown in Fig. 2.
Using KCL at node n_1 we obtain an equation which can be solved for V, the phasor branch voltage. Hence,

$$1\underline{/67.5°} = \frac{\vec{V}}{-j} + \frac{\vec{V}}{1},$$

solving for $\vec{V}$,

$$1\underline{/67.5°} = \vec{V}\left(\frac{1}{-j} + 1\right)$$

$$1\underline{/67.5°} = \vec{V}(j + 1)$$

$$\vec{V} = \frac{1\underline{/67.5}}{1 + j}$$

$$\vec{V} = \frac{1\underline{/67.5}}{\sqrt{2}\underline{/45}}$$

gives,

$$\vec{V} = \frac{1}{\sqrt{2}}\underline{/22.5\ V}.$$

Converting from phasor form to time form yields

$$v(t) = \frac{1}{\sqrt{2}} \cos\left(t + \frac{\pi}{8}\right) V.$$

• PROBLEM 9-29

Calculate the steady-state current response of the network shown in fig. 1 below.

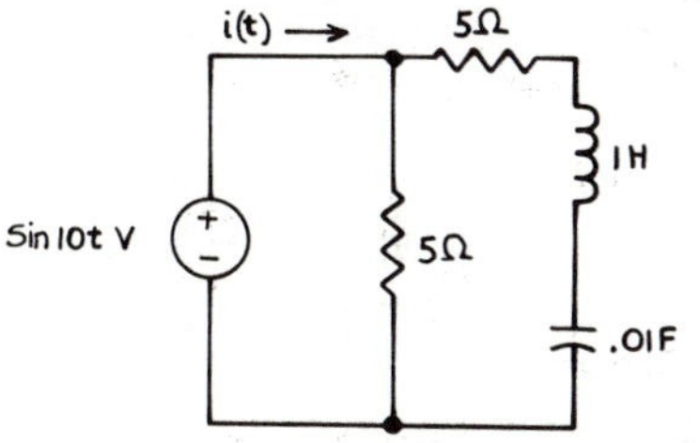

FIG. 1

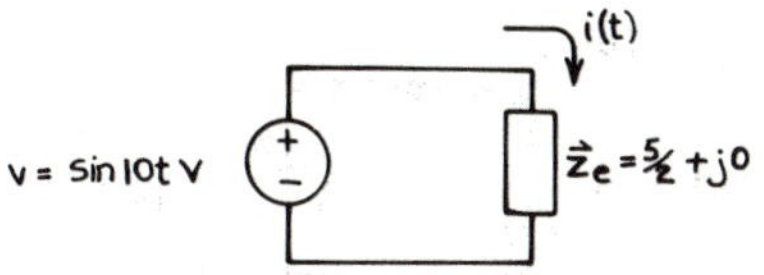

FIG. 2

Solution: Since the response we are asked to find does not involve an individual branch we can start by finding the equivalent impedance of the parallel branches.

$$Z_e = 5 \,\|\left[j\omega(1) + \frac{1}{j\omega(0.01)}\right]$$

Since the input is sin 10t where $\omega = 10$.

$$Z_e = 5\|\left[5 + j10(1) - \frac{j}{10(.01)}\right]$$

$$= 5\,\|\left[5 + j10 - j10\right]$$

or $\qquad Z_e = 5\|5 = \frac{25}{10}$

$$\vec{Z}_e = \frac{5}{2} + j0\Omega.$$

We now have the circuit of fig. 2.

Therefore,

$$i(t) = \frac{2}{5}\sin 10t \text{ A (by Ohm's Law).}$$

• PROBLEM 9-30

Analyze the circuit shown in Fig. 1 by constructing a reasonably accurate phasor diagram with the voltage $\vec{V}_1$ as the reference phasor. Let the arrow representing $\vec{V}_1$ be 2 in. long at an angle of 0°, and let the $\vec{I}_1$ arrow be 3 in. long. What is the length of the arrow for: (a) $\vec{I}_3$; (b) $\vec{I}_2$; (c) $\vec{V}_s$?

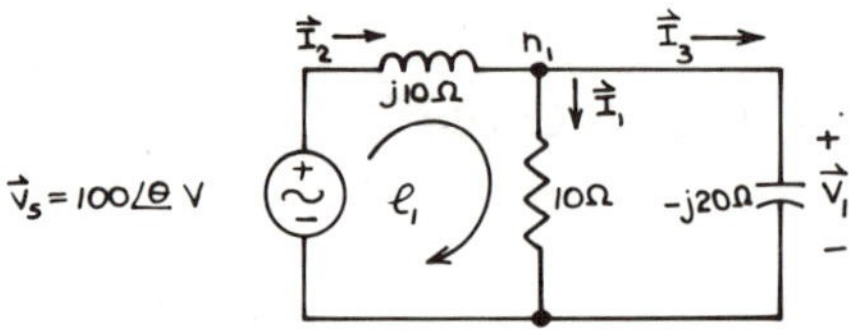

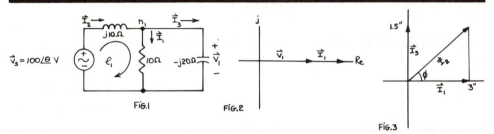

Fig.1

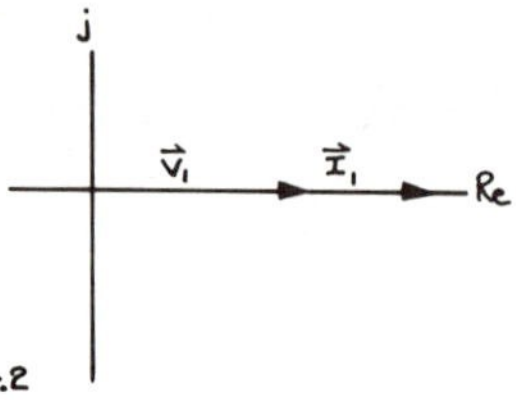

Fig.2

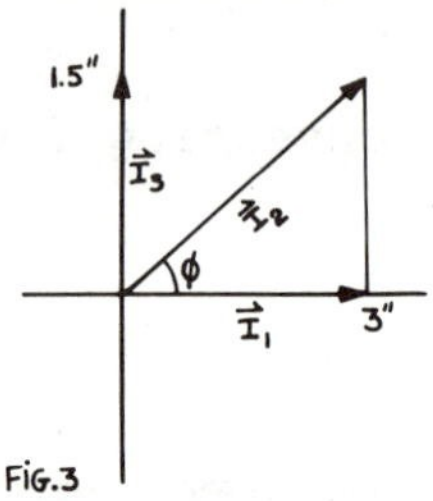

Fig.3

<u>Solution:</u> Writing a KCL equation for node n_1 yields

$$\vec{I}_1 + \vec{I}_3 = \vec{I}_2$$

$$\frac{\vec{V}_1}{10} + \frac{\vec{V}_1}{-j20} = \frac{\vec{V}_1 - \vec{V}_s}{j10}.$$

It is established that on a phasor diagram $\vec{V}_1$ is represented by an arrow 2" long at an angle of 0° as shown in fig. 2. $\vec{I}_1$ which is equal to $\frac{\vec{V}_1}{10}$ and 3" long, has an angle which must also be zero.

Note that

$$\vec{I}_3 = \frac{\vec{V}_1}{-j20} = \frac{2"\angle 0°}{20\angle -90°} = \frac{2"}{20}\angle 90°$$

and I_1 was found to be $3"\angle 0°$

$$I_1 = \frac{V_1}{10\Omega} = \frac{2"\angle 0°}{10} = C\frac{2"}{10}\angle 0° = 3"\angle 0°$$

a conversion factor, C, is used to arrive at $3"\angle 0°$.

$$C = 10\frac{3"}{2"} = 15\Omega.$$

This yields

$$I_3 = C\frac{2"}{20}\angle 90° = 1.5"\angle 90°.$$

Fig. 3 shows the two phasors, $\vec{I}_1$, $\vec{I}_3$ and their sum $\vec{I}_2$.

$$|I_2| = \sqrt{1.5^2 + 3^2} = \sqrt{11.25} = 3.35"$$

$$\phi = \tan^{-1}\frac{1.5}{3} = 26.57°.$$

Write a KVL equation around the loop ℓ_1 in fig. 1.

$$\vec{V}_s = \vec{I}_2(j10 + 10||-j20)$$

$$\vec{V}_s = \vec{I}_2(j10 + \frac{-j200}{10 - j20})$$

$$\vec{V}_s = \vec{I}_2\left(10\underline{/90°} + \frac{200\underline{/-90°}}{22.36\underline{/-63.43°}}\right)$$

$$\vec{V}_s = \vec{I}_2(10\underline{/90°} + 8.94\underline{/-26.57°})$$

solving for I_2, yields

$$\vec{I}_2 = \frac{V_s}{10\underline{/90°} + 8.94\underline{/-26.57°}}$$

$$\vec{I}_2 = \frac{100\underline{/\theta}}{j10 + 8 - j4} = \frac{100\underline{/\theta}}{8 + j6}$$

$$\vec{I}_2 = \frac{100\underline{/\theta}}{10\underline{/36.87}} = 10\underline{/\theta - 36.87}\text{ A}$$

Now, find V_1

$$\vec{V}_1 = \vec{I}_2(8.94\underline{/-26.57°})$$

$$\vec{V}_1 = (10\underline{/\theta - 36.87})(8.94\underline{/-26.57°})$$

$$\vec{V}_1 = 89.4\underline{/\theta - 36.87 - 26.57}$$

$$\vec{V}_1 = 89.4\underline{/\theta - 63.44°}\text{ V}$$

It is now possible to form a ratio of real voltage values and length in inches.

$$\frac{100}{89.4} = \frac{X}{2"}$$

$$X = \frac{2"(100)}{89.4} = 2.24".$$

Now we have

$$|V_s| = 2.24";\quad \vec{V}_s = 2.24"\underline{/\theta}$$

To find θ note that $V_1 = 89.4\underline{/\theta - 63.44°}$ or, on the graph, $V_1 = 2"\underline{/0°}$ so that

$$0° = \theta - 63.44°$$

$$\theta = 63.44°$$

$$\vec{V}_s = 100/63.44° \text{ V}; \quad 2.24"\underline{/64.34°}$$

$$\vec{V}_1 = 89.4\underline{/0°} \text{ V}; \quad 2.00"\underline{/0°}\ .$$

• PROBLEM 9-31

For the circuit shown in Fig. 1, calculate i(t). The source voltage v(t) is given as

$$v(t) = 156 \cos 377t \text{ V.}$$

The element values are R = 100Ω, L = 0.20 H, C = 20 μF.

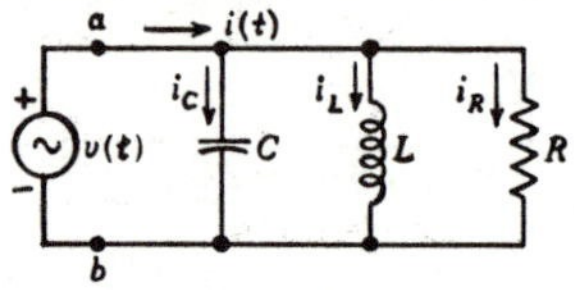

FIG.1

Solution: To find the current i(t), i_C, i_L, and i_R must be found. Thus, $i(t) = i_C + i_L + i_R$, where

$$i_C = \frac{v(t)}{Z_C}, \quad i_L = \frac{v(t)}{Z_L}, \quad \text{and} \quad i_R = \frac{v(t)}{R}.$$

We are given v(t) = 156 cos 377t, thus ω = 377 and in phasor form $V = \frac{156}{\sqrt{2}}\underline{/0}$ Vrms. Hence,

$$\vec{I}_C = \frac{V}{\frac{1}{j\omega C}} = \frac{\frac{156}{\sqrt{2}}\underline{/0°}}{\frac{1}{j377}(20\times10^{-6})}$$

$$\vec{I}_C = \frac{\frac{156}{\sqrt{2}}\underline{/0°}}{132.6\underline{/-90°}} = 0.832\ \underline{/90°} \text{ Arms.}$$

Similarly,

$$\vec{I}_L = \frac{\vec{V}}{j\omega L} = \frac{\frac{156}{\sqrt{2}}\underline{/0°}}{j(377)(.20)} = \frac{\frac{156}{\sqrt{2}}\underline{/0°}}{75.4\underline{/-90°}}$$

$$\vec{I}_L = 1.463\underline{/-90°} \text{ A rms}$$

$$\vec{I}_R = \frac{\vec{V}}{R} = \frac{\frac{156}{\sqrt{2}}\underline{/0°}}{100} = 1.103 \text{ A rms.}$$

Hence,

$$\vec{I} = I_C + I_L + I_R$$

$$\vec{I} = 0.832\underline{/90^\circ} + 1.463\underline{/-90^\circ} + 1.103\underline{/0^\circ}$$

$$\vec{I} = j0.832 - j1.463 + 1.103$$

$$\vec{I} = 1.103 - j0.631$$

$$\vec{I} = 1.271 \underline{/-30^\circ} \text{A rms}$$

$$i(t) = 1.271(\sqrt{2})\cos(377t - 30^\circ)$$

$$i(t) = 1.8 \cos(377t - 30^\circ)$$

• PROBLEM 9-32

Use phasor techniques to find $v_L(t)$ in the circuit below.

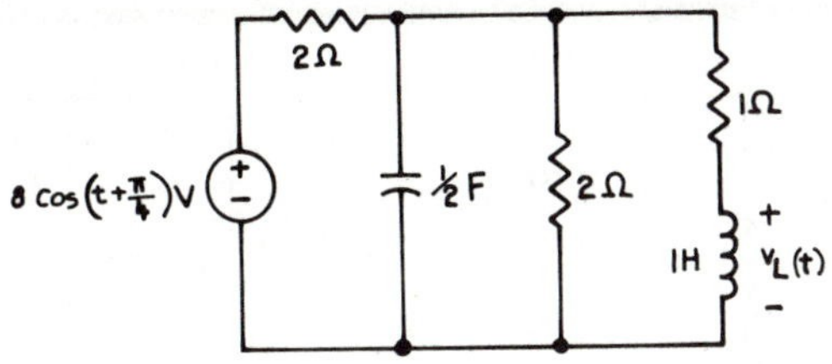

Solution: The impedance of the R-L branch is

$$1 + j1\Omega \; (\omega = 1)$$

and its admittance is, therefore,

$$\frac{1}{\sqrt{2}\underline{/90^\circ}} = .707\underline{/-90} = \frac{1}{2} - j\frac{1}{2} \text{ mho.}$$

The capacitor and resistor parallel branches have admittances

$$\vec{Y}_C = \omega C = 0 + j\frac{1}{2} \text{ mho}$$

$$\vec{Y}_R = \frac{1}{2} + j0 \text{ mho.}$$

The equivalent admittance of the parallel branches is

$$\vec{Y}_p = (\frac{1}{2} + j0) + (0 + j\frac{1}{2}) + (\frac{1}{2} - j\frac{1}{2}) = 1 + j0 \text{ mho,}$$

and, consequently, the total impedance driven by the source is

$$\vec{Z}_t = 2 + j0 + \frac{1}{Y_p} = 2 + j0 + 1 + j0 = 3 + j0\Omega.$$

The current out of the source is

$$\vec{I}_t = \frac{8\angle +45}{3\angle 0^\circ} = \frac{8}{3}\angle +45^\circ \text{ A.}$$

This current produces a voltage across the parallel branches of

$$\vec{V}_p = \frac{8}{3}\angle 45 \; (1\angle 0^\circ) = \frac{8}{3}\angle 45^\circ \text{ V.}$$

By voltage division, then,

$$\vec{V}_L = \frac{j}{1+j1}\,\frac{8}{3}\angle 45^\circ = \frac{1\angle 90^\circ \frac{8}{3}\angle 45}{\sqrt{2}\angle 45^\circ} = \frac{8}{3\sqrt{2}}\angle 90^\circ \text{ V.}$$

In the time domain $v_L(t) = 1.89 \cos(t + 90^\circ)$V.

• PROBLEM 9-33

Using the complex plane, plot all the voltages and currents as phasors for the circuit shown in fig. 1.

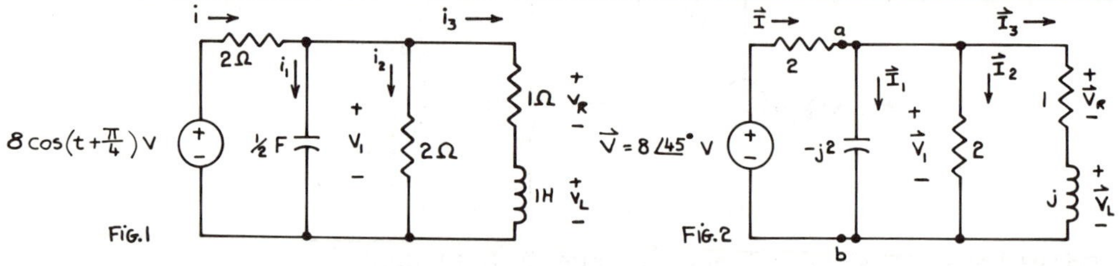

FiG.1 FiG.2

Solution: Each of the currents and voltages in fig. 1 can be plotted on the complex plane, provided they are expressed as phasors in polar form. The steady-state equivalent circuit representing the circuit in fig. 1 is shown in fig. 2.

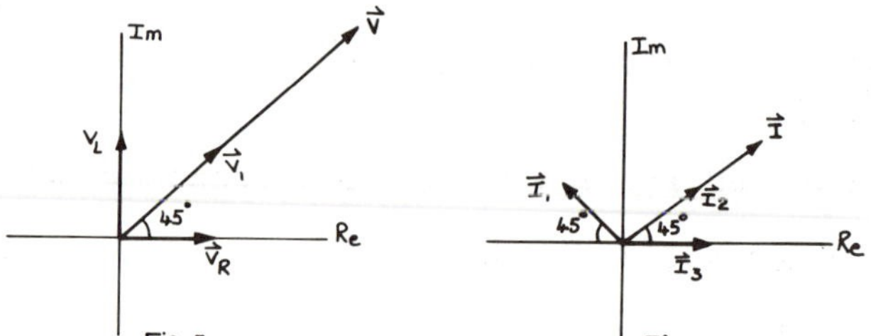

FiG.3 FiG.4

We can find the phasor voltage $\vec{V}_1$ by use of the voltage division concept,

$$\vec{V}_1 = \frac{\vec{V}[(j+1\|2)\| - j2]}{2+[(j+1\|2)\| - j2]}$$

where $(j+1\|2)\| - j2$ is the equivalent impedance to the right of terminals a-b. Hence,

$$[(j+1\|2)] - j2 = \frac{\dfrac{-j4(j+1)}{j+3}}{-j2 + \dfrac{2(j+1)}{j+3}}$$

$$= \frac{\frac{-j4(j+1)}{j+3}}{\frac{-j2(j+3)+2(j+1)}{j+3}} = \frac{-j4(j+1)}{-j2(j+3)+2(j+1)}$$

$$= \frac{4-j4}{4-j4} = 1$$

$$\vec{V}_1 = \frac{V}{3} = \tfrac{8}{3}\underline{/45^\circ}\ \text{V.}$$

$\vec{V}_1$ and the source voltage V are "in-phase". The current $\vec{I}$ is

$$\frac{\vec{V}}{3} = \tfrac{8}{3}\underline{/45^\circ}\ \text{A.}$$

The current $\vec{I}_1$ is

$$\frac{\vec{V}_1}{-j2} = \frac{\frac{8}{3}\underline{/45^\circ}}{2\underline{/-90^\circ}} = \tfrac{4}{3}\underline{/135^\circ}\ \text{A.}$$

The current $\vec{I}_L$ is

$$\frac{\vec{V}_1}{2} = \tfrac{4}{3}\underline{/45^\circ}\ \text{A.}$$

$\vec{I}_3$ is obtained from

$$\vec{I} - \vec{I}_1 - \vec{I}_2 = \tfrac{8}{3}\underline{/45^\circ} - \tfrac{4}{3}\underline{/135^\circ} - \tfrac{4}{3}\underline{/45^\circ}$$

$$\vec{I}_3 = \tfrac{4}{3}\underline{/45^\circ} - \tfrac{4}{3}\underline{/135^\circ}$$

$$\vec{I}_3 = 0.943 + j0.943 - (-0.943 + j0.943)$$

$$\vec{I}_3 = 1.886 = \frac{4\sqrt{2}}{3}\underline{/0^\circ}\ \text{A.}$$

The resistor and inductor voltages are

$$\vec{V}_R = \vec{I}_3(1) = \tfrac{4}{3}\sqrt{2} = \tfrac{4}{3}\sqrt{2}\underline{/0^\circ}\ \text{V}$$

$$\vec{V}_L = jI_3 = (1\underline{/90^\circ})(\tfrac{4}{3}\sqrt{2}) = \tfrac{4}{3}\sqrt{2}\underline{/90^\circ}\ \text{V}$$

Figs. 2 and 3 show plots of the voltage and current phasors respectively.

• **PROBLEM** 9-34

For the circuit in fig. 1, calculate the current in each branch and the total impedance $\vec{Z}_{ab}$.

<u>Solution:</u> Given is: $\omega = 377$, thus

$$\vec{Z}_c = \frac{1}{j\omega C} = \frac{1}{j377(10\times10^{-6})} = -j265\Omega$$

and

$$\vec{Z}_L = j\omega L = j377(1) = j377\Omega.$$

We can draw a circuit for the steady state, shown in fig. 2.

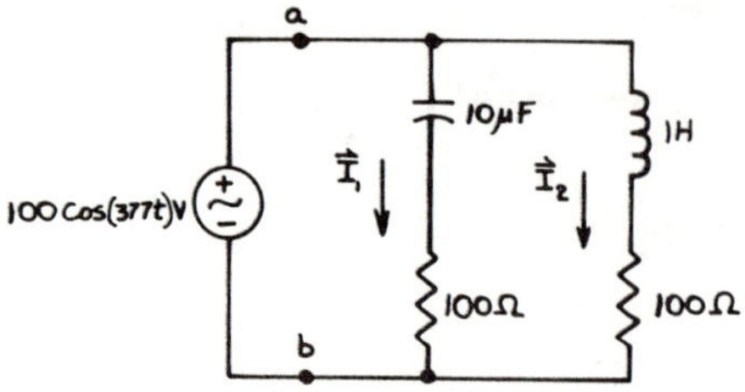

FIG.1   FIG.2

From this circuit, calculate both I_1 and I_2 by dividing the voltage, $100\underline{/0^\circ}$, across the two branches by the impedance of each branch. Hence,

$$\vec{I}_1 = \frac{100}{100 - j265} = \frac{100(100 + j265)}{100^2 + 265^2}$$

$$\vec{I}_1 = \frac{100}{80225}(100 + j265) = 0.125 + j0.330 \text{ A}$$

and

$$\vec{I}_2 = \frac{100}{100 + j377} = \frac{100(100 - j377)}{100^2 + 377^2}$$

$$\vec{I}_2 = \frac{100}{152129}(100 - j377) = 0.066 - j0.248 \text{ A}.$$

Since $\vec{I}_1$ and $\vec{I}_2$ have already been found in rectangular form they can be added to obtain the total current produced by the voltage source. The total impedance, $\vec{Z}_{ab}$, is the voltage source divided by the total current.

$$\vec{Z}_{ab} = \frac{V}{I_1 + I_2} = \frac{100}{0.191 + j0.082}$$

$$\vec{Z}_{ab} = \frac{100(0.191 - j0.082)}{0.191^2 + 0.082^2}$$

$$\vec{Z}_{ab} = \frac{100}{0.0432}(0.191 - j0.082)$$

$$\vec{Z}_{ab} = 2315(0.191 - j0.082)$$

$$Z_{ab} = 442 - j190\Omega.$$

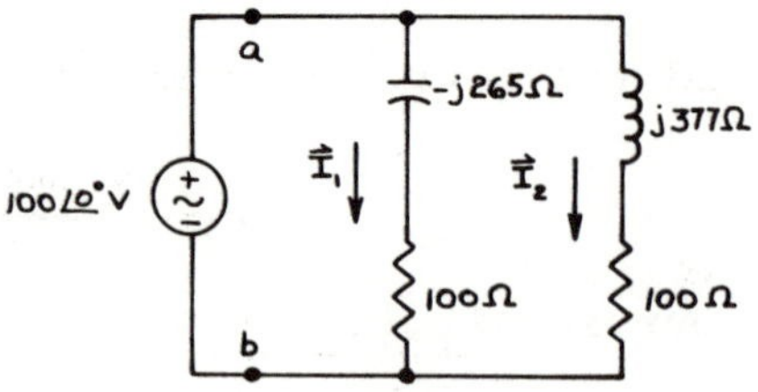

• PROBLEM 9-35

Determine the total current and the current in each branch for the circuit in fig. 1.

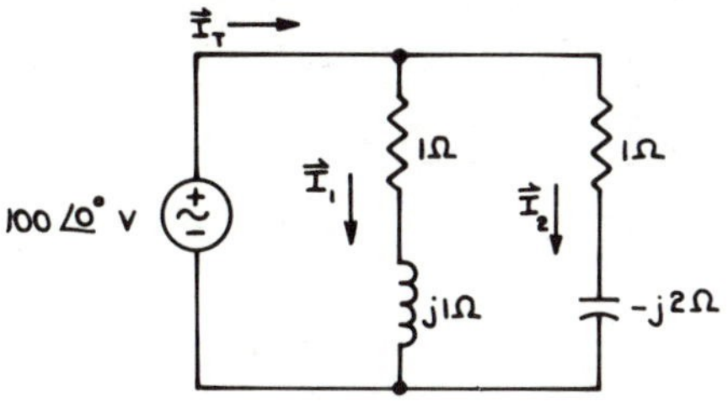

Fig. 1

Solution: To find the current in each branch, divide the voltage across the branches, $100\angle 0°$ V, by the impedance of each branch.

Hence,

$$\vec{I}_1 = \frac{100}{1+j} = \frac{100(1-j)}{2} = 50 - j50 \text{ A}$$

and $$\vec{I}_2 = \frac{100}{1+j2} = \frac{100(1+j2)}{5} = 20 + j40 \text{ A.}$$

The total current is found by summing

$$\vec{I}_T = \vec{I}_1 + \vec{I}_2 = 70 - j10 \text{ A.}$$

In polar form, this is

$$\vec{I}_T = \sqrt{70^2 + 10^2}\angle \tan^{-1}(-\frac{10}{70})$$

$$\vec{I}_T = \sqrt{5000}\angle 8.1°$$

$$\vec{I}_T = \sqrt{2 \times 2500}\angle 8.1° = \sqrt{2}(50)\angle 8.1° \text{ A}$$

• PROBLEM 9-36

Calculate the steady-state voltage, $v_c(t)$, for the network shown in Fig. 1.

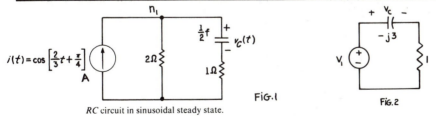

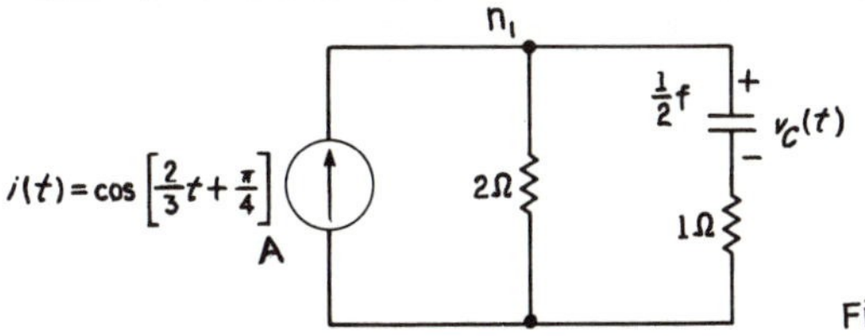

FIG. 1

RC circuit in sinusoidal steady state.

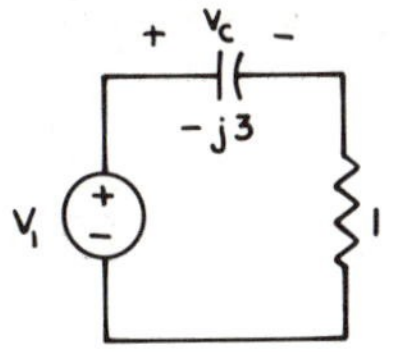

Fig. 2

Solution: Write a KCL equation for node n_1 in Fig. 1:

$$i(t) = \frac{V_1}{2} + \frac{V_1}{1 + \frac{1}{j\omega C}} \tag{1}$$

One can write

$$i(t) = \cos(\tfrac{2}{3}t + \tfrac{\pi}{4}) = \mathrm{Re}\left[e^{j(\frac{2}{3}t+\frac{\pi}{4})}\right]$$

in polar form,

$$\vec{I} = 1\angle\tfrac{\pi}{4} = 1\angle 45^\circ \text{ A}$$

also, $\omega = \frac{2}{3}$ rad/s and $C = \frac{1}{2}$F. Substituting these values into Eq. (1) gives

$$1\angle 45^\circ = \frac{\vec{V}_1}{2} + \frac{\vec{V}_1}{\dfrac{1}{1 + j(\frac{2}{3})(\frac{1}{2})}}$$

$$1\angle 45^\circ = \vec{V}_1(.5) + \frac{\vec{V}_1}{1 - \dfrac{j}{\frac{1}{3}}}$$

$$1\angle 45^\circ = \vec{V}_1(.5 + \frac{1}{1 - j3})$$

Solving for V_1

$$\vec{V}_1 = \frac{1\angle 45^\circ}{.5 + \dfrac{1}{\sqrt{10}\angle -71.57}} = \frac{1\angle 45^\circ}{.5 + \frac{\sqrt{10}}{10}\angle 71.57}$$

$$\vec{V}_1 = \frac{1\angle 45^\circ}{.5 + .1 + j.3} = \frac{1\angle 45^\circ}{.6 + j.3}$$

$$\vec{V}_1 \quad \frac{1\angle 45^\circ}{\sqrt{0.45}\angle 26.57^\circ}.$$

Using the concept of voltage division as illustrated in fig. 2,

$$V_C = \frac{V_1\ (-j3)}{1 - j3}$$

$$\vec{V}_C = \frac{\dfrac{1\angle 45}{\sqrt{0.45}/26.57}\ 3\angle -90^\circ}{\sqrt{10}\angle -71.56^\circ}$$

$$V_C = \frac{3}{\sqrt{0.45}}\left(\frac{1}{\sqrt{10}}\right)\angle 45 - 26.57 - 90 + 71.56$$

$$V_C = \sqrt{\frac{9}{0.45(10)}}\angle 0^\circ$$

$$V_C = \sqrt{\frac{9}{4.5}}\angle 0^\circ = \sqrt{2}\angle 0^\circ \text{ V}$$

$$V_C = \sqrt{2}\cos\left(\frac{2}{3}t\right)$$

• PROBLEM 9-37

A 0.5μF capacitor, an 80-mH inductor, and a 500-Ω resistor are in parallel with the voltage source, 12-j6V. Find the source current as a phasor if ω =: (a) 500 rad/s; (b) 5000 rad/s; (c) 50,000 rad/s.

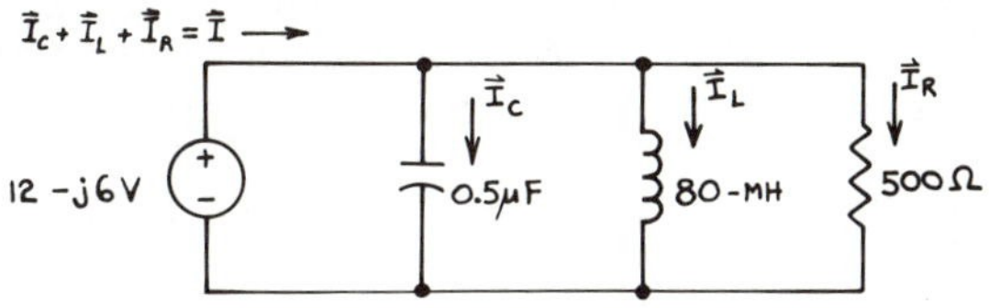

FiG.1

Solution: We can see in fig. 1 that the source current is equal to the sum of the currents in each of the other branches. Hence

$$I = \frac{\vec{V}}{\vec{Z}_C} + \frac{\vec{V}}{\vec{Z}_L} + \frac{\vec{V}}{\vec{Z}_R}$$

$$\vec{I} = \vec{V}(\vec{Y}_D + \vec{Y}_L + \vec{Y}_R)$$

$$\vec{I} = 12\text{-}j6\left\{j\omega(0.5\times 10^{-6}) - \frac{j}{\omega(80\times 10^{-3})} + \frac{1}{500}\right\}$$

for ω = 500 rad/s,

$$\vec{I} = 12 - j6(j0.25\times 10^{-3} - j25\times 10^{-3} + 2\times 10^{-3})$$

$$\vec{I} = (12 - j6)(2 - j24.75)$$

$$\vec{I} = 24 - 148.5 - j12 - j297$$

$$\vec{I} = -124.5 - j309 \text{ mA}$$

in polar form $\vec{I} = 333\angle -112^\circ$ mA.

For ω = 5000 rad/s

$$\vec{I} = (12 - j6)(j2.5\times 10^{-3} - j2.5\times 10^{-3} + 2\times 10^{-3})$$

$$\vec{I} = (12 - j6)(2)$$

$\vec{I} = 24 - j12$ mA, in polar form

$$\vec{I} = 26.8\angle -26.6^\circ \text{ mA.}$$

For ω = 50,000 rad/s,

$$\vec{I} = (12 - j6)(j2.5 \times 10^{-2} - j2.5 \times 10^{-4} + 2 \times 10^{-3})$$

$$\vec{I} = (12 - j6)(2 + j24.75)$$

$$\vec{I} = 24 - j12 + j297 + 148.5$$

$\vec{I} = 172.5 + j285$ mA in polar form

$\vec{I} = 333\underline{/58.8^\circ}$ mA.

• PROBLEM 9-38

A sinusoidal current source, 10 cos 1000t A, is in parallel both with a 20-Ω resistor and the series combination of a 10-Ω resistor and a 10-mH inductor. Remembering the usefulness of Thévenin's theorem, find the voltage across the following: (a) inductor; (b) 10-Ω resistor; (c) current source.

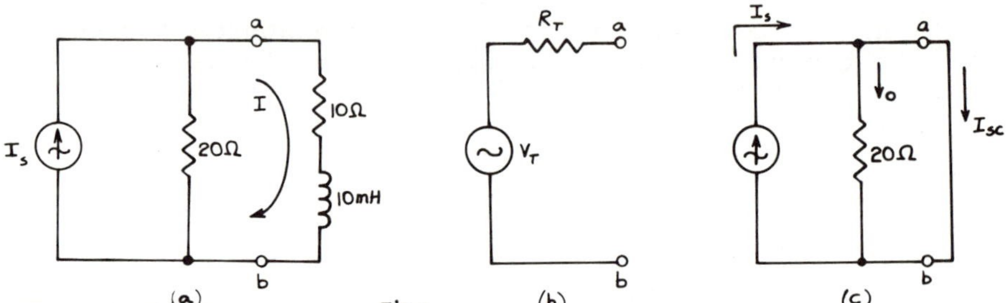

FIG. 1

Solution: Following the suggestion, use Thevenin's theorem to convert the current source and 20-ohm resistor into a voltage source as shown in Figure 1(b). The resultant circuit shown in Figure 2 can now be solved for current and the current used to find the required voltages.

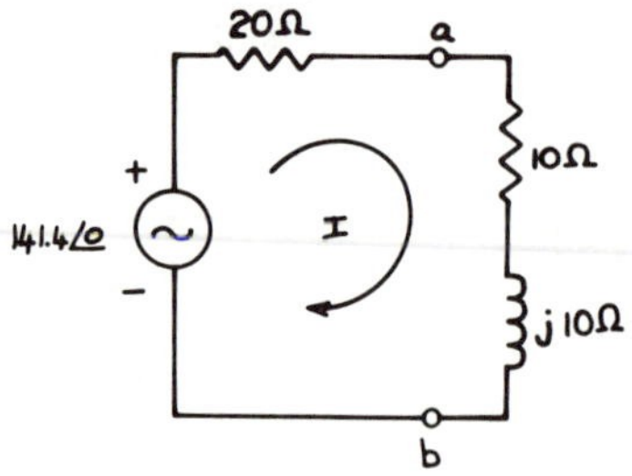

FIG. 2 CIRCUIT AS MODIFIED

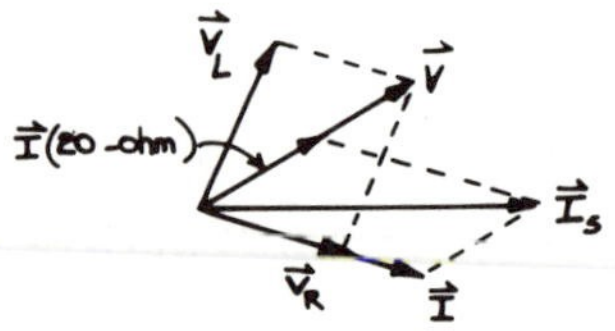

FIG. 3 PHASOR DIAGRAM

The original circuit as shown in Figure 1(a) will be modified by converting the current source and 20-ohm resistor to the left of a-b to an equivalent voltage source as shown in Figure 1(b). This is acomplished by first separating this part of the circuit from the remaining part and then determining the open-circuit voltage between a and b.

$$\text{Open-Circuit Voltage} = \vec{V}_t = 20\,\vec{I}_s \qquad (1)$$

where I_s and V_t are phasor quantities (rms),

$$\vec{V}_t = 20 \times 10(.707)\underline{/0^\circ} = 141.4\underline{/0^\circ}\text{ V.} \quad (2)$$

Next, determine the current through a hypothetical short-circuit between points a and b:

$$\text{Short-circuit Current } \vec{I}_{sc} = \vec{I}_s = 7.07\underline{/0^\circ}\text{ A.} \quad (3)$$

Finally, the Thevenin resistance R_t is found at the quotient

$$R_t = \frac{V_{ab}(\text{open circuit})}{I_{sc}(\text{short circuit})} = \frac{141.4\underline{/0^\circ}}{7.07\underline{/0^\circ}} = 20\Omega. \quad (4)$$

The resultant modified circuit is shown in Figure 2. Now we are ready to solve this simpler circuit.

First, solve for the inductive reactance

$$X_L = \omega L = 1000 \times .010 = 10\Omega. \quad (5)$$

and determine current with Kirchhoff's law:

$$\vec{I} = \frac{141.4\underline{/0^\circ}}{(10+20)+j10} = \frac{141.4\underline{/0^\circ}}{31.62\underline{/18.4^\circ}} = 4.47\underline{/-18.4^\circ}\text{ A.} \quad (6)$$

Converted to an instantaneous value, the current is

$$i = 4.47\sqrt{2}\cos(1000t - 18.4^\circ) = 6.32\cos(1000t - 18.4^\circ)\text{A} \quad (7)$$

where $\sqrt{2}$ is included to convert the rms value to peak value.

The current can now be used to find out the voltage across the inductor:

$$\vec{V}_L = j\omega L\,\vec{I} = 10\underline{/90^\circ} \times 4.47\underline{/-18.4^\circ} = 44.7\underline{/71.6^\circ}\text{ V} \quad (8)$$

which converts to

$$v_L = 63.2\cos(1000t + 71.6^\circ)\text{V.} \quad (9)$$

(d) The voltage across the 10-ohm resistor is

$$\vec{V}_R = \vec{I}\vec{R} = 4.47\underline{/-18.4^\circ} \times 10 = 44.7\underline{/-18.4^\circ}\text{ V} \quad (10)$$

which converts to

$$v_R = 63.2\cos(1000t - 18.4^\circ)\text{V.} \quad (11)$$

(e) Finally, voltage across the current source is seen to be the voltage across a-b in Figure 1(a) which is

$$\vec{V} = \vec{I}\vec{Z} = (4.47)\underline{/-18.4^\circ}(10+j10) = (4.47)(14.14)\underline{/-18.4^\circ+45^\circ} \quad (12)$$

$$\vec{V} = 63.2\underline{/26.6^\circ}\text{ V.} \quad (13)$$

This converts to

$$v = 89.4 \cos(1000t + 26.6°)\text{V}. \qquad (14)$$

The phasor diagram for this circuit is shown in Figure 3.

• PROBLEM 9-39

> Solve for the steady-state current $i_L(t)$ in the circuit shown in fig. 1.

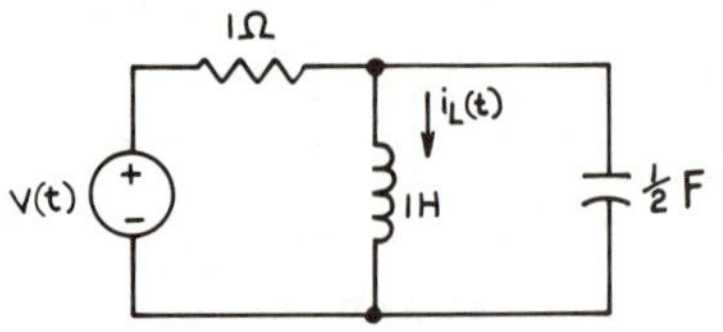

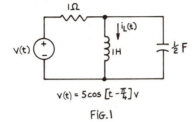

$v(t) = 5\cos\left[t - \frac{\pi}{4}\right]$ V

FIG. 1

Solution: The phasor representation of the voltage v(t) is:

$$\vec{V} = 5\angle -45° \text{ V}$$

with a frequency of 1 rad/sec. The impedance of each element is

$$\vec{Z}_R = 1 + j0 = 1\angle 0° \ \Omega$$

$$\vec{Z}_L = 0 + j\omega L = 0 + j1 = 1\angle 90° \ \Omega$$

$$\vec{Z}_C = \frac{1}{j\omega C} = \frac{-j}{1\frac{1}{2}} = 0 - j2 = 2\angle -90° \ \Omega.$$

Since the capacitor and inductor are in parallel, the equivalent impedance is:

$$Z_P = \frac{Z_C Z_L}{Z_L + Z_C} = \frac{1\angle 90° \ 2\angle -90°}{+j1 - j2} = \frac{2\angle 0°}{1\angle -90°} = 2\angle 90° = 0 + j2\Omega \ .$$

The total impedance is then:

$$\vec{Z}_T = \vec{Z}_R + \vec{Z}_P = 1 + j0 + 0 + j2 = 1 + j2 = \sqrt{5}\angle 63.43° \ \Omega$$

It follows that the total current is

$$\vec{I}_T = \frac{\vec{V}}{\vec{Z}_T} = \frac{5\angle -45°}{\sqrt{5}\angle 63.43°} = \sqrt{5}\angle -108.43° \text{ A.}$$

The voltage across the parallel branches is

$$\vec{V}_P = \vec{I}_T\vec{Z}_P = \sqrt{5}\angle -108.43° \ 2\angle 90°$$

$$= 2\sqrt{5}\angle -18.43° \text{ V.}$$

The inductor current must be

$$\vec{I}_L = \frac{\vec{V}_P}{\vec{Z}_L} = \frac{2\sqrt{5}\angle{-18.43^\circ}}{1\angle{+90}} = 2\sqrt{5}\angle{-108.43^\circ}\ \text{A}.$$

In the time domain representation this is

$$i_L(t) = 2\sqrt{5}\cos(t - 108.43^\circ) = 2\sqrt{5}\cos(t - 45 - 63.43^\circ)\text{A}.$$

• PROBLEM 9-40

Write an expression for $v_c(t)$ in the circuit shown in fig. 1 by using a sinusoidal steady-state equivalent network.

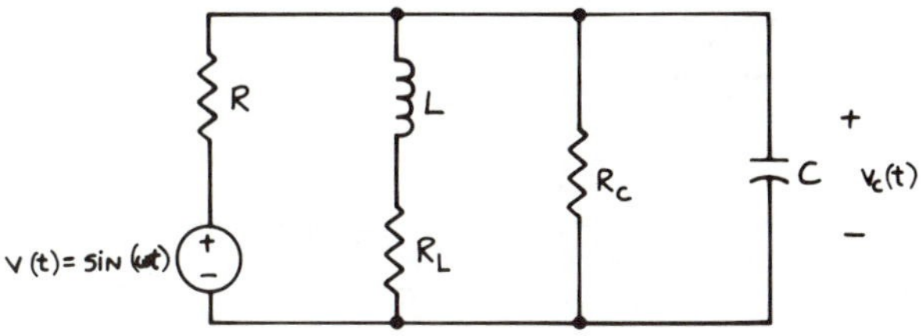

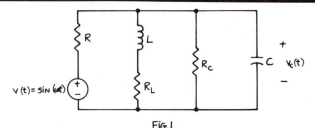

FIG. 1

Solution: Replacing each element by its sinusoidal steady-state impedance produces the circuit in fig. 2.

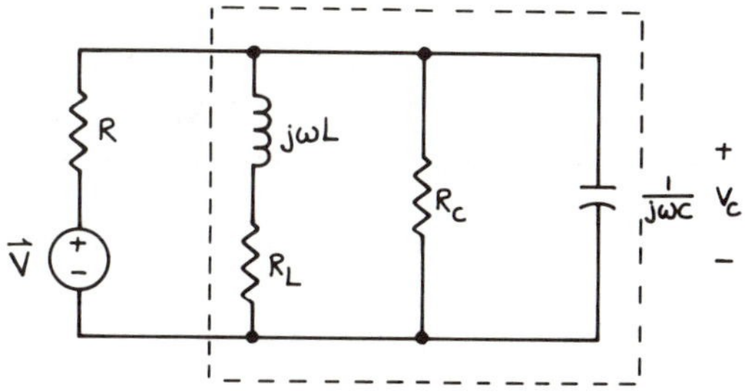

FIG. 2

By replacing the part of the circuit in fig. 2 which is enclosed by the dotted line, the voltage division concept can be applied to obtain V_C.

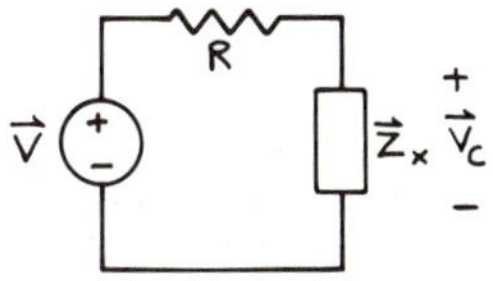

FIG. 3

$$\vec{Z}_X = (j\omega L + R_L) \| R_C \| \frac{1}{j\omega C}$$

$$\vec{Z}_X = \left[\frac{(j\omega L + R_L)R_C}{j\omega L + R_L + R_C}\right] \| \frac{1}{j\omega C}$$

$$\vec{Z}_X = \frac{\dfrac{(j\omega L + R_L)R_C}{(j\omega L + R_L + R_C)j\omega C}}{\dfrac{(j\omega L + R_L)R_C}{j\omega L + R_L + R_C} + \dfrac{1}{j\omega C}}$$

Simplifying gives,

$$\vec{Z}_X = \frac{\dfrac{j\omega LR_C + R_LR_C}{j^2\omega^2LC + j\omega CR_L + j\omega CR_C}}{\dfrac{(j\omega L + R_L)j\omega CR_C + j\omega L + R_L + R_C}{j^2\omega^2LC + j\omega CR_L + j\omega CR_C}}$$

$$\vec{Z}_X = \frac{j\omega LR_C + R_LR_C}{(j\omega L + R_L)(j\omega CR_C + 1) + R_C}$$

$$\vec{Z}_X = \frac{R_C(j\omega L + R_L)}{(j\omega L + R_L)(j\omega CR_C + 1) + R_C}$$

since $\vec{V}_C = \dfrac{V(Z_X)}{(R + Z_X)}$

$$\frac{\vec{Z}_X}{R + \vec{Z}_X} = \frac{\dfrac{R_C(j\omega L + R_L)}{(j\omega L + R_L)(j\omega CR_C + 1) + R_C}}{R + \dfrac{R_C(j\omega L + R_L)}{(j\omega L + R_L)(j\omega CR_C + 1) + R_C}}$$

Simplifying,

$$= \frac{\dfrac{R_C(j\omega L + R_L)}{(j\omega L + R_L)\qquad(j\omega CR_C + 1) + R_C}}{\dfrac{R(j\omega L + R_L)(j\omega CR_C + 1) + RR_C + R_C(j\omega L + R_L)}{(j\omega L + R_L)(j\omega CR_C + 1) + R_C}}$$

and dividing through by $R_C(j\omega L + R_L)$

$$= \frac{1}{\dfrac{R(j\omega CR_C + 1)}{R_C} + \dfrac{R}{j\omega L + R_L} + 1}$$

finally,

$$\vec{V}_c = \frac{V}{j\omega CR + \dfrac{R}{R_C} + \dfrac{R}{j\omega L + R_L} + 1}$$

• **PROBLEM 9-41**

Find the steady-state current through the RL branch in the circuit shown in fig. 1.

Solution: The network can be viewed as a current source in parallel with four admittance elements as shown in the steady state equivalent circuit of fig. 2. Since the voltage $V(j\omega)$ across the four parallel branches can be found by

$$\vec{V}(j\omega) = \frac{\vec{I}(j\omega)}{\vec{Y}_1 + \vec{Y}_2 + \vec{Y}_3 + \vec{Y}_4},$$

the current $\vec{I}_L$ through the RL branch is

$$\vec{I}_L = \vec{V}(j\omega)\vec{Y}_2.$$

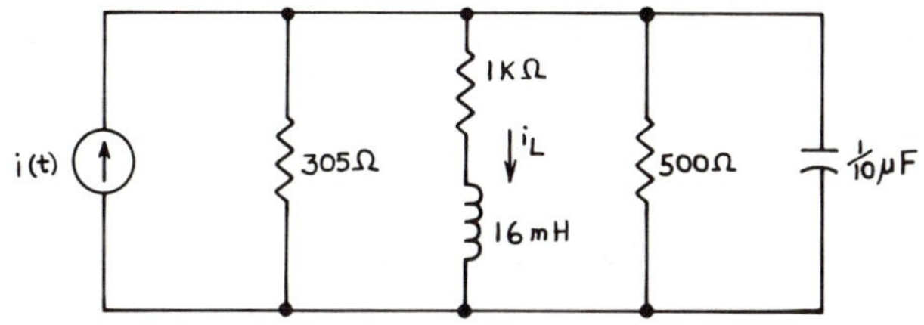

FIG. 1 $\quad i(t) = \sin((2\pi \times 10^4)t + 45^\circ)$

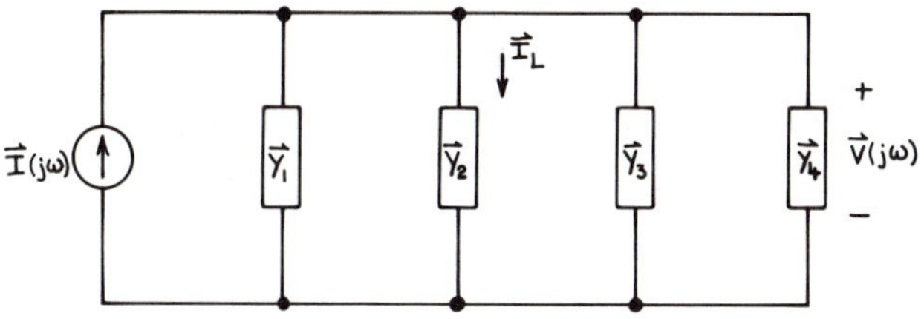

FIG. 2

Hence, the relationship

$$\vec{I}_L = \vec{I}(j\omega)\,\frac{\vec{Y}_2}{\vec{Y}_1 + \vec{Y}_2 + \vec{Y}_3 + \vec{Y}_4} \tag{1}$$

describes the concept of current division. The current in any particular branch is the total current multiplied by the ratio of the admittance of the particular branch to the total admittance in the parallel combination. Substituting the values given in fig. 1 into Eq. (1) yields

$$\vec{I}_L = 1\underline{/-45^\circ}\left(\frac{\dfrac{1}{1000 + j1000}}{3.28\times10^{-3} + \dfrac{1}{1000 + j1000} + 2\times10^{-3} + j6.28\times10^{-3}}\right)$$

$$\vec{I}_L = 1\underline{/-45^\circ}\left(\frac{0.5 - j0.5}{3.28 + 0.5 - j0.5 + 2 + j6.28}\right)$$

$$\vec{I}_L = 1\underline{/-45^\circ}\left(\frac{0.5 - j0.5}{5.78 + j5.78}\right)$$

$$\vec{I}_L = 1\underline{/-45^\circ}\left(\frac{0.707\underline{/-45^\circ}}{8.174\underline{/45^\circ}}\right)$$

$$\vec{I}_L = 1\underline{/-45^\circ}\,(0.086\underline{/-90\ })$$

$$\vec{I}_L = 0.086\underline{/-135^\circ}\text{ A}$$

$$i_L(t) = 0.086 \cos((2\pi\times 10^4)t - 135°)A$$

$$i_L(t) = 86 \cos((2\pi\times 10^4)t - 135°)\ mA$$

• PROBLEM 9-42

Find $v_c(t)$ steady state in the circuit of fig. 1.

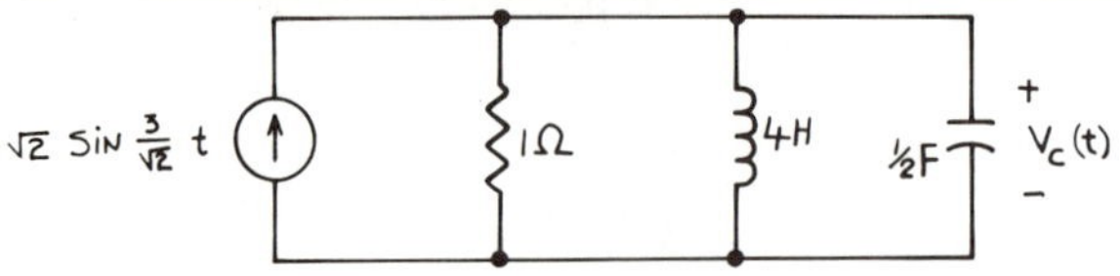

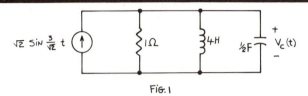

FIG. 1

Solution: Transforming the above circuit to its phasor equivalent gives the circuit of fig. 2. Note that cos $\sqrt{\frac{3}{2}}$ t is used as a reference. Note also that instead of impedance values, conductance values are used to facilitate nodal analysis.

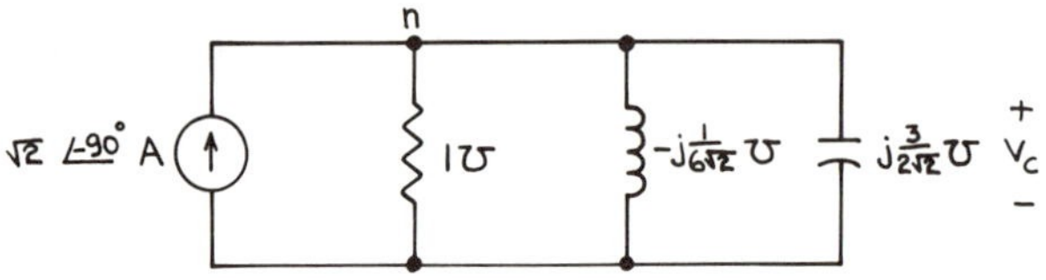

FIG. 2

Now the application of KCL at n yields

$$\sqrt{2}\underline{/-90°} = -j\sqrt{2} = \vec{V}_C\left(1 - j\frac{1}{6\sqrt{2}} + j\frac{3}{2\sqrt{2}}\right)$$

$$\vec{V}_C = \frac{-j\sqrt{2}}{1 + j\frac{4}{3\sqrt{2}}}$$

$$\vec{V}_C = \frac{-j\sqrt{2}}{1 + j\frac{4}{3\sqrt{2}}} \cdot \frac{1 - j\frac{4}{3\sqrt{2}}}{1 - j\frac{4}{3\sqrt{2}}}$$

$$\vec{V}_C \quad \frac{-\frac{4}{3} - j\sqrt{2}}{\frac{17}{9}} = -\frac{12}{17} - j\frac{9\sqrt{2}}{17}$$

$$\vec{V}_C = 1.029\underline{/133.3°}\ V$$

$$v_c(t) = 1.029 \cos\left(\frac{3}{\sqrt{2}}t + 133.3°\right)V$$

For the ladder network shown in Fig. 1, find V_{out} if $V_{in} = 100\angle 0°$ mV.

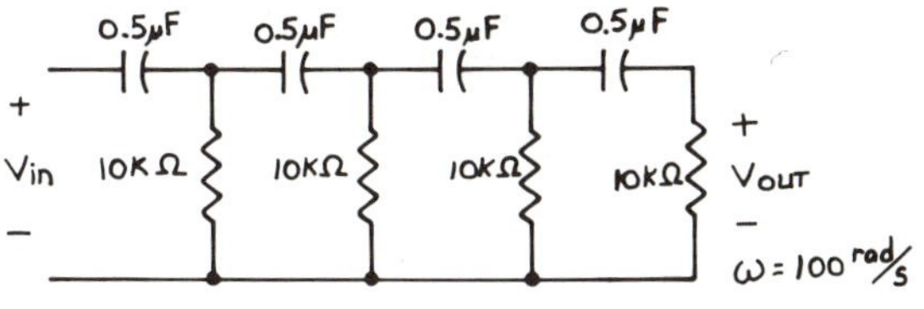

Fig. 1

Solution: The method used is to find the current through the 10k load resistor solving for the mesh current, I_4, with the mesh currents as indicated in Figure 2. This can be found by the method of determinants. Then $V_{out} = I_4 \times 10k\Omega V$.

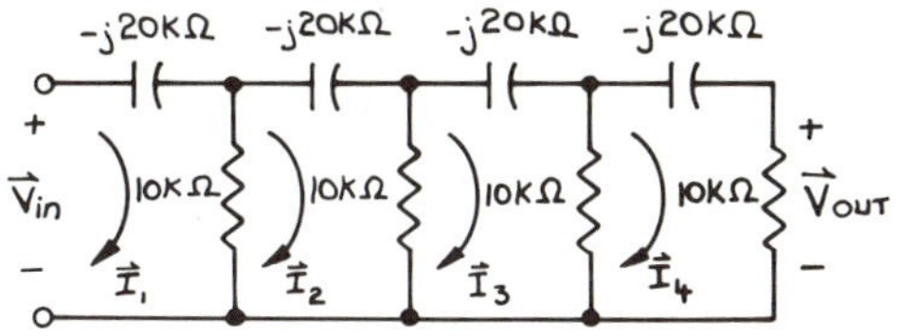

Fig. 2

First find

$$X_C = -j\frac{1}{\omega C} = -j\frac{1}{(100)(0.5\times 10^{-6})} = -j20k.$$

$$\vec{Z}_{11} = 10k - j20k = 10\sqrt{5}\angle -63° \; k\Omega$$

$$\vec{Z}_{22} = \vec{Z}_{33} = \vec{Z}_{44} = 20k - j20k = 20\sqrt{2}\angle -45 \; k\Omega$$

$$\vec{Z}_{12} = \vec{Z}_{21} = \vec{Z}_{23} = \vec{Z}_{32} = \vec{Z}_{34} = \vec{Z}_{43} = -10k\Omega$$

$$\vec{Z}_{13} = \vec{Z}_{31} = \vec{Z}_{14} = \vec{Z}_{41} = \vec{Z}_{24} = \vec{Z}_{42} = 0\Omega$$

Therefore, by the method of determinants

$$I_4 = \frac{\begin{vmatrix} 10\sqrt{5}\angle -63° & -10 & 0 & \frac{V_{in}}{10^3} \\ -10 & 20\sqrt{2}\angle -45° & -10 & 0 \\ 0 & -10 & 20\sqrt{2}\angle -45° & 0 \\ 0 & 0 & -10 & 0 \end{vmatrix}}{\begin{vmatrix} 10\sqrt{5}\angle -63° & -10 & 0 & 0 \\ -10 & 20\sqrt{2}\angle -45° & -10 & 0 \\ 0 & -10 & 20\sqrt{2}\angle -45° & -10 \\ 0 & 0 & -10 & 20\sqrt{2}\angle -45° \end{vmatrix}}$$

Expand the numerator determinant along the fourth column gives

$$-\frac{V_{in}}{10^3}\begin{vmatrix} -10 & 20\sqrt{2}\underline{/-45^\circ} & -10 \\ 0 & -10 & 20\sqrt{2}\underline{/-45^\circ} \\ 0 & 0 & -10 \end{vmatrix} = \frac{1000\ V_{in}}{10^3} = V_{in}$$

Expand the denominator determinant along the first column gives

$$10\sqrt{5}\underline{/-63^\circ}\begin{vmatrix} 20\sqrt{2}\underline{/-45^\circ} & -10 & 0 \\ -10 & 20\sqrt{2}\underline{/-45^\circ} & -10 \\ 0 & -10 & 20\sqrt{2}\underline{/-45^\circ} \end{vmatrix}$$

$$+10\begin{vmatrix} -10 & 0 & 0 \\ -10 & 20\sqrt{2}\underline{/-45^\circ} & -10 \\ 0 & -10 & 20\sqrt{2}\underline{/-45^\circ} \end{vmatrix}$$

Further expansion of the third-order determinants gives

$$10\sqrt{5}\,\underline{/63^\circ}\,[(20\sqrt{2}\underline{/-45^\circ})^3 - 2000\sqrt{2}\underline{/-45^\circ} - 2000\sqrt{2}\underline{/-45^\circ}]$$

$$+ 10[-8000\underline{/-90^\circ} + 1000]$$

$$= 505,964\underline{/162^\circ} - 126,491\underline{/-108^\circ} - 80,000\underline{/-90^\circ} + 1000$$

$$= (-481,200 + j156,351) - (-39,088 - j120,300) + j80,000 + 1000$$

$$= -442,012 + j356,651 = 567,956\underline{/141^\circ}$$

$$\vec{I}_4 = \frac{V_{in}}{567,956\underline{/141^\circ}}$$

$$\vec{V}_{out} = I_4 \times 10 \times 10^3$$

$$\vec{V}_{out} \quad \frac{V_{in} \times 10 \times 10^3}{567,956\underline{/141^\circ}} = \frac{100 \times 10^{-3} \times 10 \times 10^3}{567,956\underline{/141^\circ}}$$

$$\vec{V}_{out} = 1.76 \times 10^{-3}\underline{/-141^\circ}$$

$$\vec{V}_{out} = 1.76\underline{/-141^\circ}\ \text{mV}$$

• **PROBLEM 9-44**

Show that the currents, $\vec{I}_1$ in Fig. 1 and $\vec{I}_2$ in Fig. 2, are equal.

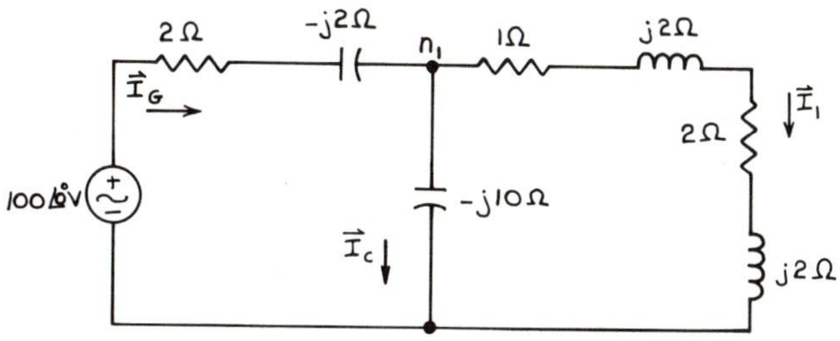

Fig.1

Solution: Using KCL at node n_1 in Fig. 1, write an equation for the currents at the node,

$$\vec{I}_G - \vec{I}_C = \vec{I}_1 \qquad \text{where}$$

$$\vec{I}_G = \frac{100\angle 0^\circ - V_{n1}}{2 - j2}$$

$$\vec{I}_C = \frac{\vec{V}_{n1}}{-j10}$$

$$\vec{I}_1 = \frac{\vec{V}_{n1}}{(1 + j2) + (2 + j2)}. \tag{1}$$

Hence, the KCL equation becomes

$$\frac{100\angle 0^\circ - \vec{V}_{n1}}{2 - j2} - \frac{\vec{V}_{n1}}{-j10} = \frac{\vec{V}_{n1}}{3 + j4}. \tag{2}$$

Solve Eq. (2) for $\vec{V}_{n1}$ and substitute into (1) to find $\vec{I}_1$.

$$\frac{100\angle 0^\circ}{2 - j2} = \vec{V}_{n1}\left[\frac{1}{2 - j2} + \frac{1}{-j10} + \frac{1}{3 + j4}\right]$$

$$\vec{V}_{n1} = \frac{\dfrac{100\angle 0}{2 - j2}}{\left(\dfrac{1}{2 - j2} + \dfrac{1}{-j10} + \dfrac{1}{3 + j4}\right)}$$

Simplifying the expression gives

$$\vec{V}_{n1} = \frac{\dfrac{100}{2 - j2}}{\left(\dfrac{2 + j2}{8} + \dfrac{j10}{100} + \dfrac{3 - j4}{25}\right)}$$

$$\vec{V}_{n1} = \frac{\dfrac{100}{2 - j2}}{\dfrac{12.5(2 + j2) + j10 + 4(3 - j4)}{100}}$$

$$\vec{V}_{n1} = \frac{100^2}{(37 + j19)(2 - j2)}$$

$$\vec{V}_{n1} = \frac{100^2}{112 - j36} = \frac{100^2(112 + j36)}{13840}$$

$$\vec{V}_{n1} = 0.723(112 + j36) = 81 + j26 \text{ V.}$$

Substituting V_{n1} into Eq. (1) yields

$$\vec{I}_1 = \frac{81 + j26}{3 + j4} = \frac{(81 + j26)(3 - j4)}{25}$$

$$\vec{I}_1 = \frac{347 - j246}{25} = 13.88 - j9.84 \text{ A.}$$

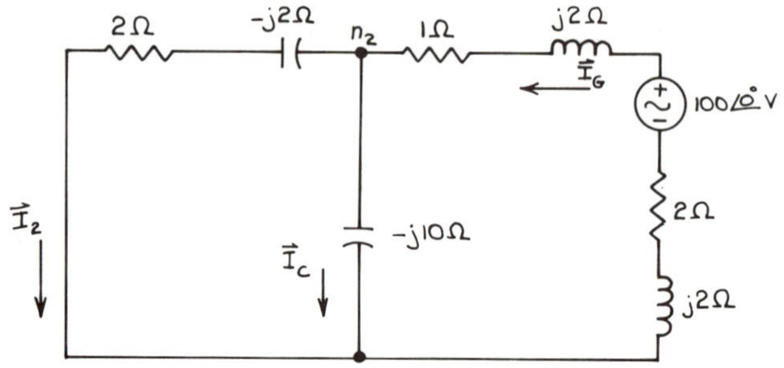

FIG.2

Using KCL at node n_2 in Fig. 2, we can write an equation for the currents at the node, which is,

$$\vec{I}_G - \vec{I}_C = \vec{I}_2 \qquad \text{where}$$

$$\vec{I}_G = \frac{100\angle 0 - V_{n2}}{(1 + j2) + (2 + j2)}$$

$$\vec{I}_C = \frac{V_{n2}}{-j10}$$

$$\vec{I}_2 = \frac{V_{n2}}{2 - j2} \tag{3}$$

Hence, the KCL equation becomes

$$\frac{100\angle 0 - V_{n2}}{3 + j4} - \frac{V_{n2}}{-j10} = \frac{V_{n2}}{2 - j2} \tag{4}$$

Solve Eq. (4) for V_{n2} and substitute into Eq. (3) to find I_2.

$$\frac{100\angle 0}{3 + j4} = V_{n2}\left(\frac{1}{3 + j4} + \frac{1}{-j10} + \frac{1}{2 - j2}\right)$$

$$\vec{V}_{n2} = \frac{\frac{100}{3 + j4}}{\left(\frac{1}{3 + j4} + \frac{1}{-j10} + \frac{1}{2 - j2}\right)}$$

$$\vec{V}_{n2} = \frac{100^2}{(37 + j19)(3 + j4)}$$

$$\vec{V}_{n2} = \frac{100^2}{35 + j205} = \frac{100^2(35 - j205)}{43250}$$

$$\vec{V}_{n2} = 0.231(35 - j205) = 8.09 - j47.4 \text{ V.}$$

Substituting $\vec{V}_{n2}$ into Eq. (3) yields

$$\vec{I}_2 = \frac{8.09 - j47.4}{2 - j2} = \frac{(8.09 - j47.4)(2 + j2)}{8}$$

$$\vec{I}_2 = \frac{110.18 - j78.62}{8} = 13.8 - j9.84 \text{ A.}$$

Hence, $\vec{I}_1 = \vec{I}_2 = 13.8 - j9.84$ A, this problem illustrates the reciprocity theorem.

• PROBLEM 9-45

Find all currents and voltage drops (phasor form) in the a.c. steady state circuit shown in Fig. 1. Make sure that all are specified clearly for both magnitude and phase.

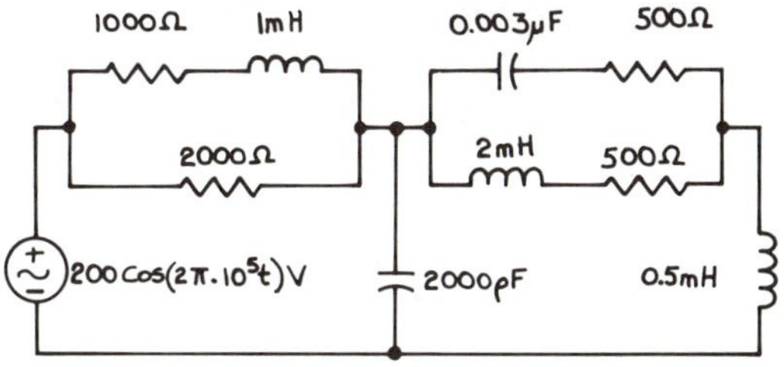

FIG.1

Solution: First replace the circuit by a phasor equivalent, the circuit with all L, C replaced by their respective impedances, $Z_L = j\omega L$ and $Z_C = 1/j\omega C$. Also, assign a label and reference convention to each voltage and current of the circuit.

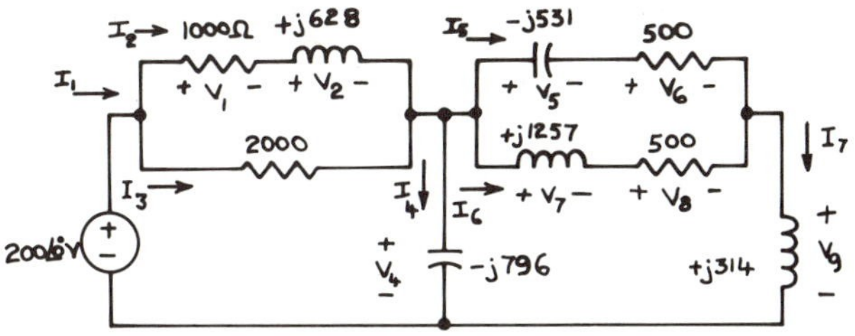

FIG.2

There are many ways the solution can proceed, and all involve a considerable amount of labor. One approach is to simplify the circuit by combining impedances in series and parallel. The top left leg impedances is

$$\frac{(1000 + j628)(2000)}{3000 + j628} = 723 + j267\Omega.$$

The top right impedance is

$$\frac{(500 - j531)(500 + j1257)}{1000 + j726} = \frac{729\angle -46.72^\circ \; 1353\angle +68.31^\circ}{1236\angle +35.98^\circ}$$

$$= 798\angle -14.39^\circ = 773 - j198\Omega.$$

The $-j796\Omega$ and $+j314\Omega$ impedances combine with the one above to form

$$\frac{(773 - j198 + j314)(-j796)}{773 - j680} = \frac{782\angle +8.53^\circ \;\; 796\angle -90^\circ}{1030\angle -41.34^\circ}$$

$$= 604\angle -40.13^\circ = 462 - j389\Omega.$$

With the use of Fig. 3, we then solve for $\vec{I}_1$, $\vec{V}_3$, and $\vec{V}_4$.

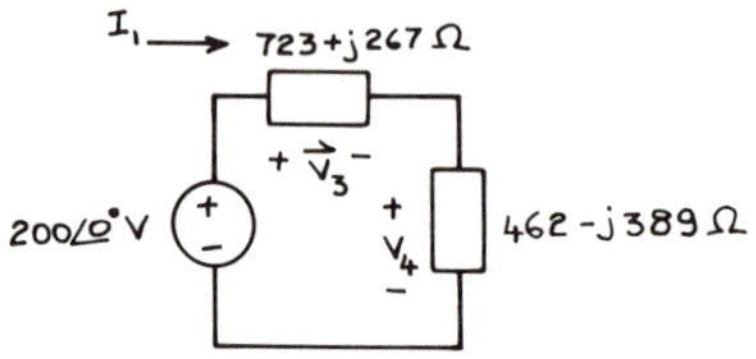

FiG.3

$$\vec{I}_1 = \frac{200}{(723 + j267) + (462 - j389)} = \frac{200}{1185 - j122} = 0.168\angle 5.88^\circ \text{ A}$$

$$\vec{V}_3 = \vec{I}_1(723 + j267) = (0.168\angle 5.88^\circ)(771\angle 20.27^\circ)$$

$$= 129.5\angle 26.15^\circ \text{ V}$$

$$\vec{V}_4 = \vec{I}_1(462 - j389) = (0.168\angle 5.88^\circ)(604\angle -40.13^\circ)$$

$$= 101.5\angle -34.25^\circ \text{ V.}$$

By Ohm's Law, $\vec{I}_3 = \frac{V_3}{2000} = 0.0648\angle 26.15^\circ$ A.

By KCL, $\vec{I}_2 = \vec{I}_1 - \vec{I}_3 = (0.167 + j.017) - (0.0582 + j.0286)$

$$= 0.109 - j0.0116 = 0.110\angle -6.07^\circ \text{ A}$$

By Ohm's Law, $\vec{V}_1 = 100\vec{I}_2 = 110\angle -6.07^\circ$ V

and $\vec{V}_2 = j628\vec{I}_2 = (628\angle 90^\circ)(.110\angle -6.07^\circ) = 69.1\angle 83.93^\circ$ V

and $\vec{I}_4 = \frac{\vec{V}_4}{-j796} = \frac{101.5\angle -34.25^\circ}{796\angle -90^\circ} = 0.128\angle 55.75^\circ$ A

and $\vec{I}_7 = \frac{\vec{V}_4}{(773 - j198) + j314} = \frac{101.5\angle -34.25^\circ}{781.66\angle 8.51^\circ} = 0.130\angle -42.76^\circ$ A

and $\vec{V}_9 = j314\vec{I}_7 = (314\angle 90^\circ)(0.130\angle -42.76^\circ) = 40.82\angle 47.24^\circ$ V.

By KVL and Ohm's Law,

$$\vec{I}_5 = \frac{\vec{V}_4 - \vec{V}_9}{500 - j531} = \frac{(83.90 - j57.12) - (27.7 + j29.97)}{500 - j531}$$

$$= \frac{56.19 - j87.09}{500 - j531} = \frac{103.64\angle -57.17°}{729.36\angle -46.72°}$$

$$= 0.142\angle -10.45° \text{ A}$$

Similarly,

$$\vec{I}_6 = \frac{\vec{V}_4 - \vec{V}_9}{500 + j1257} = \frac{103.64\angle -57.15°}{1352.8\angle 68.31°} = 0.0766\angle -125.48° \text{ A}$$

By Ohm's Law:

$$\vec{V}_5 = -j531\vec{I}_5 = (531\angle -90°)(.142\angle -10.45°) = 75.4\angle -100.45° \text{ V}$$

$$\vec{V}_6 = 500\vec{I}_5 = (500)(.142\angle -10.45°) = 71.0\angle -10.45° \text{ V}$$

$$\vec{V}_7 = j1257\vec{I}_6 = (1257\angle 90°)(.0766\angle -125.48°) = 96.29\angle -35.48° \text{ V}$$

$$\vec{V}_8 = 500\vec{I}_6 = (500)(.0766\angle -125.48°) = 38.3\angle -125.48° \text{ V}$$

• PROBLEM 9-46

Find the Thévenin equivalent circuit of the network shown in Fig. 1 and use it to find I_{ab} through the 30Ω resistor.

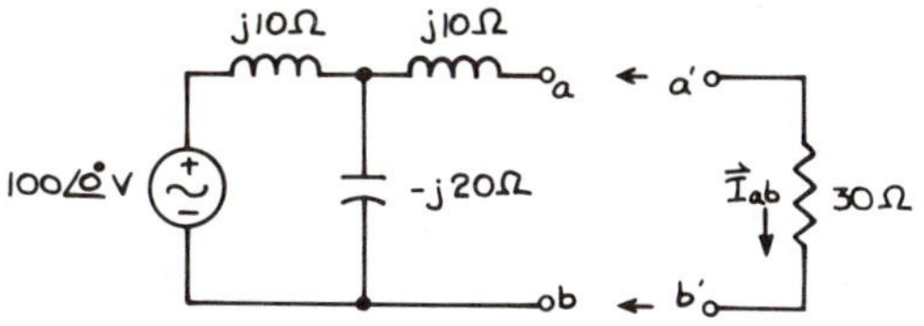

FiG.1

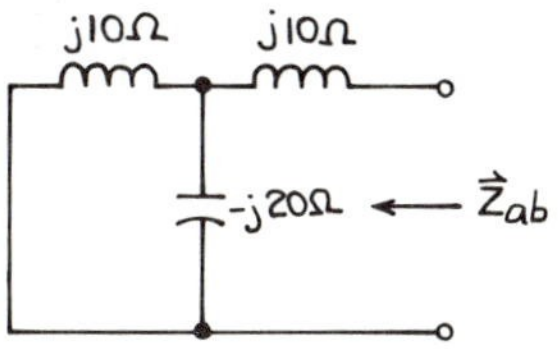

FiG.2

Solution: A Thévenin equivalent circuit is found by locating the open circuit voltage across the terminals of the desired network. This yields the Thévenin equivalent voltage source. The Thévenin equivalent impedance is found by determining the impedance across the terminals of the network when all independent voltage and/or current sources are "off." (i.e. voltages sources short circuited and current sources open circuited). The Thévenin equivalent circuit becomes the series combination of these two parameters.

In this example, the open circuit voltage $\vec{V}_{ab}$ is found by determining the voltage across the capacitor.

Hence,

$$\vec{V}_{ab} = \frac{(100\angle 0°)(-j20)}{j10 - j20} = \frac{-j2000}{-j10}$$

$$\vec{V}_{ab} = 200\underline{/0^\circ}\ \text{V}.$$

The impedance across terminals ab is shown in Fig. 2.

$$\vec{Z}_{ab} = \frac{(j10)(-j20)}{j10 - j20} + j10$$

$$\vec{Z}_{ab} = \frac{200}{-j10} + j10$$

$$\vec{Z}_{ab} = \frac{j2000}{100} + j10$$

$$\vec{Z}_{ab} = j20 + j10 = j30\Omega.$$

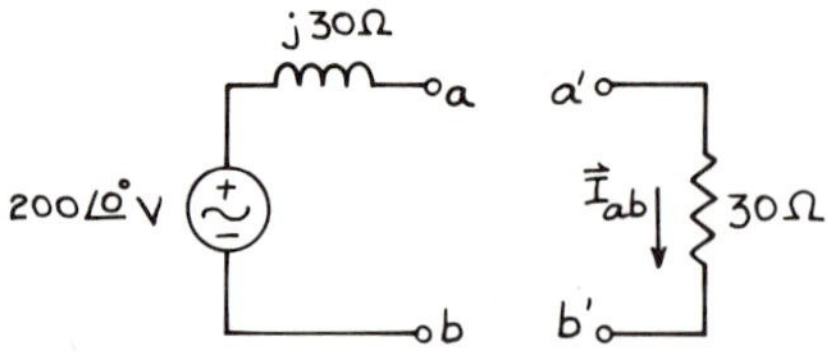

FIG.3

Hence, the Thévenin equivalent circuit is shown in Fig. 3. The current $\vec{I}_{ab}$ is found by appling KVL:

$$200\underline{/0^\circ} = \vec{I}_{ab}(j30 + 30)$$

$$\vec{I}_{ab} = \frac{200}{30 + j30} = \frac{200\underline{/0^\circ}}{42.4\underline{/45^\circ}}$$

$$\vec{I}_{ab} = 3.33 - j3.33\ \text{A}$$

$$\vec{I}_{ab} = 4.7\underline{/-45^\circ}\ \text{A}.$$

• PROBLEM 9-47

Find the steady-state voltage $v_c(t)$ in the circuit of fig. 1.

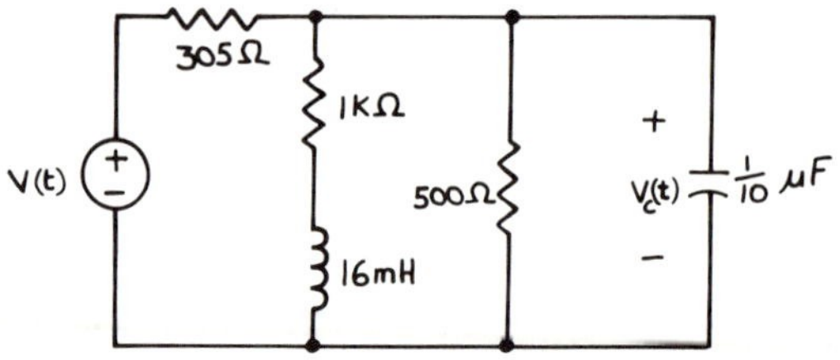

FIG.1 V(t) = sin ((2π × 10⁴)t + 45°) V

Solution: The equivalent steady-state network can be represented as three parallel admittances in series with an impedance and voltage source, as shown in Fig. 2.

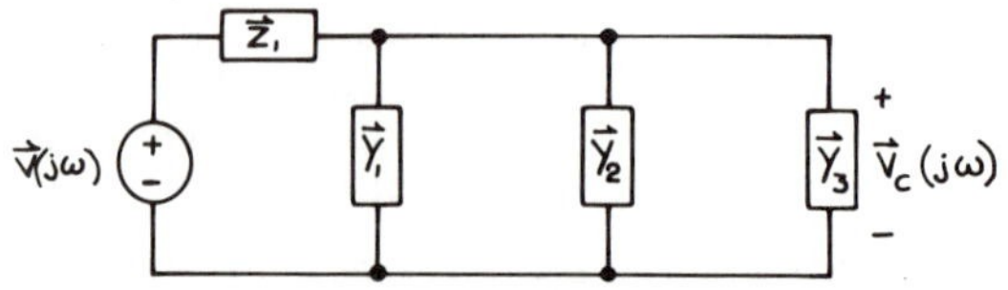

FIG.2

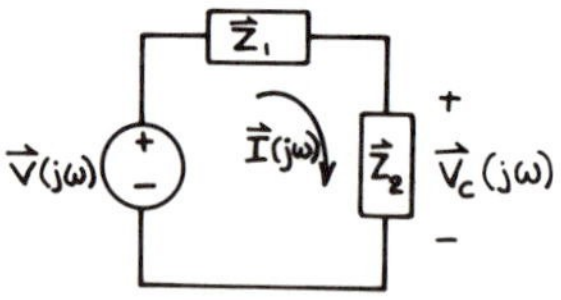

FIG.3

By combining the three parallel admittances, the total admittance is obtained

$$\vec{Y}_T = \vec{Y}_1 + \vec{Y}_2 + \vec{Y}_3$$

This admittance can be represented by a single impedance, $\frac{1}{Y_T} = Z_2$, shown in the equivalent circuit of Fig. 3.

The current around the single loop in the circuit of Fig. 3 is

$$\vec{I}(j\omega) = \frac{\vec{V}(j\omega)}{\vec{Z}_1 + \vec{Z}_2}$$

and $$\vec{V}_C(j\omega) = \vec{I}(j\omega)Z_2 = \vec{V}(j\omega)\frac{\vec{Z}_2}{\vec{Z}_1 + \vec{Z}_2} \qquad (1)$$

Eq. (1) describes the concept of voltage division. In any series combination of impedances, the voltage across any particular element is the ratio of the impedance of the element to the total impedance in the series combination multiplied by the applied voltage.

Substituting the values given in Fig. 1 into Eq. (1) yields

$$\vec{V}_C = 1\underline{/-45^\circ}\left(\frac{\dfrac{1}{\dfrac{1}{1000+j1000} + 2\times 10^{-3} + j6.28\times 10^{-3}}}{305 + \dfrac{1}{\dfrac{1}{1000+j1000} + 2\times 10^{-3} + j6.28\times 10^{-3}}}\right)$$

$$\vec{V}_C = 1\underline{/-45^\circ}\left(\frac{\left(\dfrac{1}{0.5 - j0.5 + 2 + j6.28}\right)\dfrac{1}{10^{-3}}}{305 + \left(\dfrac{1}{0.5 - j0.5 + 2 + j6.28}\right)\dfrac{1}{10^{-3}}}\right)$$

$$\vec{V}_C = 1\underline{/-45^\circ}\left(\frac{\left(\dfrac{1}{2.5 + j5.78}\right)\dfrac{1}{10^{-3}}}{305 + \left(\dfrac{1}{2.5 + j5.78}\right)\dfrac{1}{10^{-3}}}\right)$$

$$\vec{V}_C = 1\underline{/-45^\circ}\left(\frac{\left(\dfrac{2.5 - j5.78}{6.3}\right)\dfrac{1}{10^{-3}}}{305 + \left(\dfrac{2.5 - j5.78}{6.3}\right)\dfrac{11}{10^{-3}}}\right)$$

$$\vec{V}_C = 1\underline{/-45^\circ}\left(\frac{3.97 - j917}{305 + 397 - j917}\right)$$

$$\vec{V}_C = 1\underline{/-45^\circ}\,\frac{1000\underline{/-66.6}}{1154.9\underline{/-52.6}}$$

$$\vec{V}_C = 0.866\underline{/-59^\circ}\text{ V}$$

$$v_c(t) = 0.866\cos((2\pi \times 10^4) - 59^\circ)\text{V}.$$

• **PROBLEM 9-48**

Find the voltage across R_2 due to the two sources in Fig. 1.

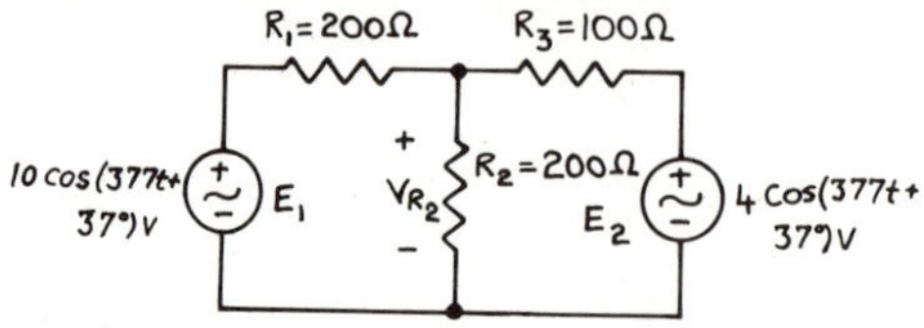

Fig. 1

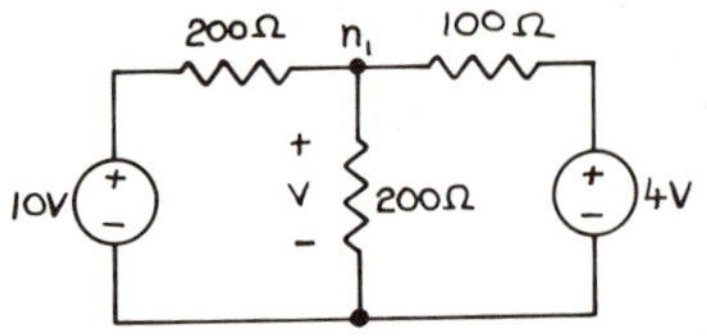

Fig. 2

Solution: Since both voltage sources have the same frequency of 377 rad/s and the same phase angle of 37°, it is established that all voltages and currents will remain in-phase, thus, $V_{R_2} = V\underline{/37°}$, where V can be found by simple D.C. methods.

Fig. 2 shows the circuit converted to D. C. V can be found by writing a KCL equation for node n_1.

$$\frac{10 - V}{200} + \frac{4 - V}{100} = \frac{V}{200}$$

$$\frac{10}{200} + \frac{4}{100} = V\left(\frac{1}{200} + \frac{1}{200} + \frac{1}{100}\right)$$

$$V = \frac{\frac{10}{200} + \frac{4}{100}}{\frac{1}{200} + \frac{1}{200} + \frac{1}{100}} = \frac{18}{9} = 4.5 \text{ V}$$

Hence the voltage across R_2 in our A. C. circuit is $4.5\underline{/37°} = 4.5 \cos(377t + 37°)$V.

• **PROBLEM 9-49**

Use the superposition theorem to find the current through R_2 in Fig. 1. Both sources are of the same frequency.

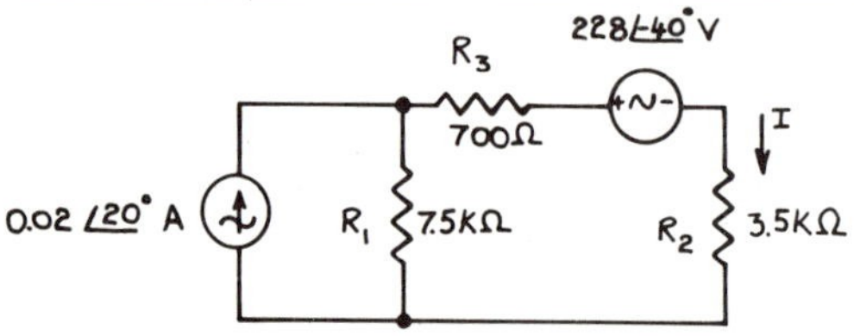

Fig. 1

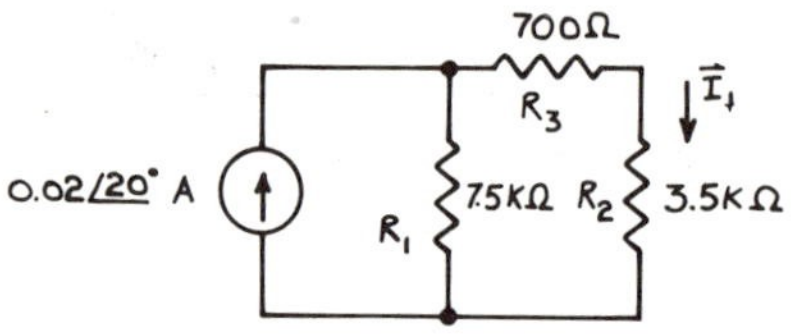

Fig. 2

Solution: First, consider the effect of the current source. With the voltage source "off," the circuit shown in Fig. 2 is obtained.
Using current division concepts produces

$$\bar{I}_1 = \frac{R_1}{R_1 + R_3 + R_2}(0.02\underline{/20°})$$

$$\vec{I}_1 = \frac{7.5K\Omega}{7.5K\Omega + 700\Omega + 3.5K\Omega}(0.02\underline{/20°})$$

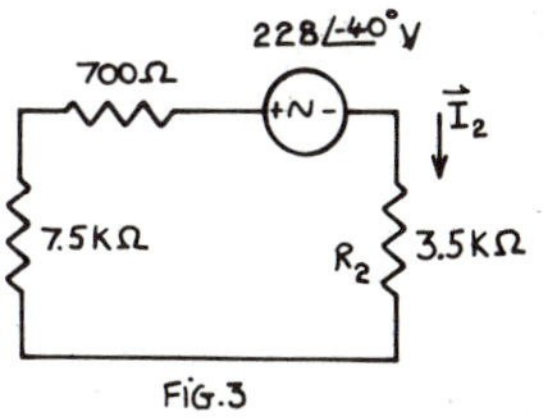

FiG.3

$$\vec{I}_1 = \frac{7.5}{11.7}(0.02\underline{/20°})$$

$$\vec{I}_1 = 12.8\underline{/20°}\text{ mA.}$$

Next, consider the effect of the voltage source. With the current source "off," the circuit shown in Fig. 3 is obtained.
Using KVL in the single loop circuit of Fig. 3, write the equation

$$-228\underline{/-40°} = \vec{I}_2(7.5K\Omega + 700\Omega + 3.5K\Omega)$$

where, $\vec{I}_2 = \dfrac{-228\underline{/-40°}}{11.7 \times 10^3}$

$$\vec{I}_2 = -19.5\underline{/-40°}\text{ mA.}$$

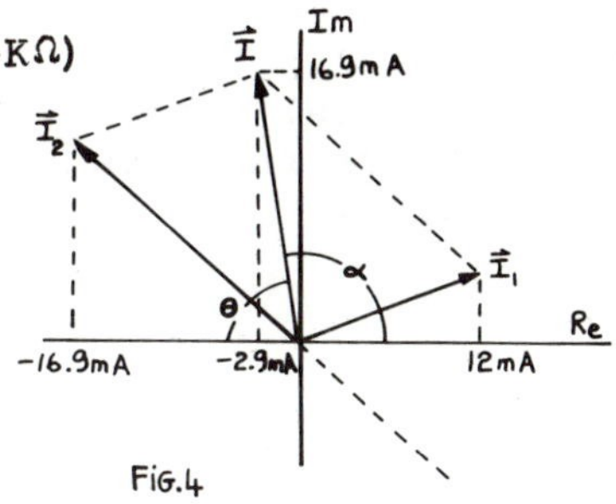

FiG.4

By superposition, we have

$$\vec{I} = \vec{I}_1 + \vec{I}_2 = (12.8\underline{/20°} - 19.5\underline{/-40°})\text{mA.}$$

Change the currents to rectangular form and add to obtain the current I.

$$\vec{I} = 12.8 \cos 20° - 19.5 \cos(-40°)$$
$$+ j(12.8 \sin 20° - 19.5 \sin(-40°))$$

$$\vec{I} = (12 - 14.9) + j(4.4 + 12.5)$$

$$\vec{I} = -2.9 + j16.9 \text{ mA}$$

$$\vec{I} = \sqrt{2.9^2 + 16.9^2}\underline{/-\tan^{-1}\frac{16.9}{2.9} + 180°} = 17.1\underline{/99.7°}\text{ mA.}$$

Fig. 4 shows vectoriol representation of the addition of vectors $\vec{I}_1$ and $\vec{I}_2$.

• PROBLEM 9-50

Calculate the voltage $v_c(t)$ in the circuit shown in Fig. 1.

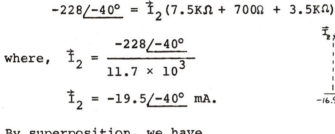

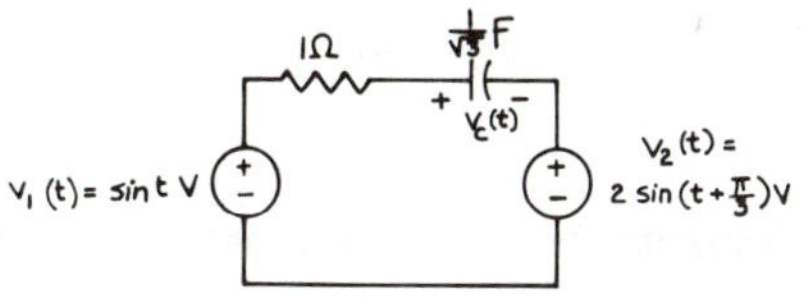

FiG.1

Solution: First, find the net voltage across the impedance and apply Ohm's law to determine the current. Once the current is known, the product of this current and the impedance of the capacitor will equal the voltage required.

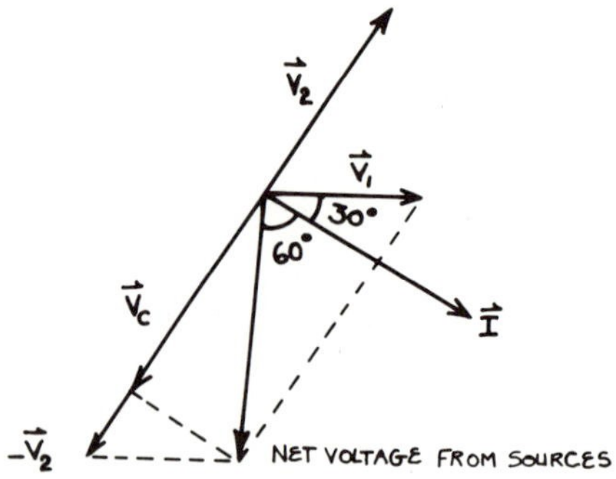

FIG. 2 PHASOR DIAGRAM

(a) Convert the two sinusoidal voltages to their phasor forms and obtain the complex impedance of the capacitor.

$$\vec{V}_1 = 1\angle 0^\circ \text{ V} ; \quad V_2 = 2\angle 60^\circ = 1 + j\sqrt{3} \text{ V} \tag{1}$$

$$\vec{Z}_C = \frac{1}{j\omega C} = -j\frac{1}{(1)(1/\sqrt{3})} = -j\sqrt{3} \tag{2}$$

(b) Write Kirchhoff's voltage law around the loop in the clockwise direction:

$$\vec{V}_1 - \vec{I}(R - j\frac{1}{\omega C}) - V_2 = 0 \tag{3}$$

$$(1 + j0) - \vec{I}(1 - j\sqrt{3}) = (1 + j\sqrt{3}) = 0 \tag{4}$$

(c) Solve for $\vec{I}$:

$$I = \frac{-j\sqrt{3}}{1-j\sqrt{3}} = \frac{\sqrt{3}\angle -90^\circ}{2\angle -60} = \frac{\sqrt{3}}{2}\angle -30^\circ \text{ A} \tag{5}$$

(b) The desired voltage in phasor form is

$$V_C = IZ_C = \frac{\sqrt{3}}{2}\angle -30^\circ \quad \sqrt{3}\angle -90^\circ = \frac{3}{2}\angle -120^\circ \text{ V} \tag{6}$$

This is converted back to the trigonometric form as

$$v_c(t) = \frac{3}{2}\sin(t - 120^\circ) = \frac{3}{2}\sin(t - \frac{4}{3}\pi)\text{V} \tag{7}$$

(c) The phasor diagram is shown in Fig. 2. Note that $\vec{I}$ leads $\vec{V}_C$ by 90° as it should.

• PROBLEM 9-51

Calculate the phasor current I when $\vec{E}_a$ is (a) $100\angle 0^\circ$ V; (b) 0 V, in Fig. 1.
Assume each voltage source has an internal impedance of

(1 + j3)Ω for $\vec{E}_a$ and (1 + j5)Ω for $\vec{E}_b$.

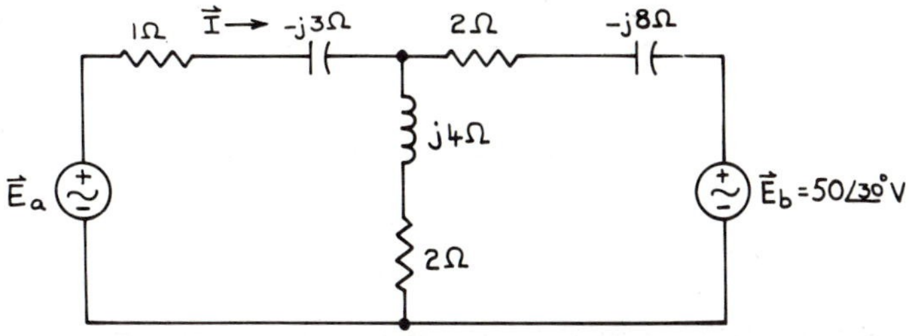

FIG. 1

Solution: (a) Re-draw the circuit of Fig. 1 to include the internal impedances of the sources. KCL at node n_1 yields an equation for $\vec{V}$.

$$\vec{I} + \frac{50\underline{/30°} - \vec{V}}{(2 - j8) + (1 + j5)} = \frac{\vec{V}}{2 + j4}$$

where
$$I = \frac{100\underline{/0°} - \vec{V}}{(1 + j3) + (1 - j3)}.$$

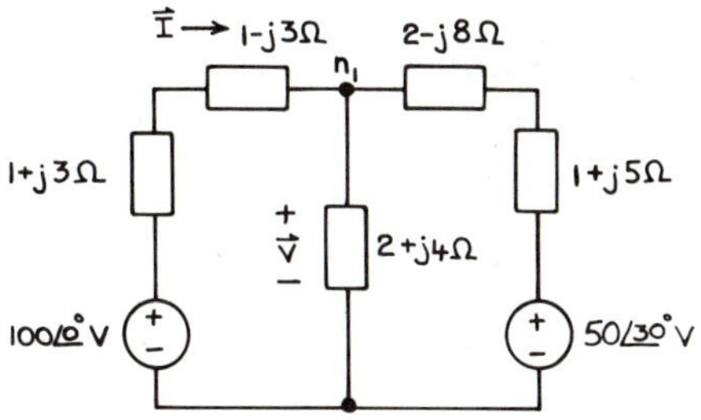

FIG. 2

Hence,

$$\vec{V}\left(\frac{1}{2 + j4} + \frac{1}{3 - j3} + \frac{1}{2}\right) = \frac{50\underline{/30°}}{3 - j3} + \frac{100\underline{/0°}}{2}$$

$$\vec{V} = \frac{\dfrac{50\underline{/30°}}{3 - j3} + 50}{\dfrac{1}{2 + j4} + \dfrac{1}{3 - j3} + \dfrac{1}{2}}$$

and
$$\vec{I} = \frac{100\underline{/0°} - \vec{V}}{2}, \quad \text{thus}$$

$$\vec{I} = 50\underline{/0°} - \frac{1}{2}V$$

$$\vec{I} = 50 - \frac{1}{2}\left[\frac{\dfrac{50\underline{/30°}}{3 - j3} + 50}{\dfrac{1}{2 + j4} + \dfrac{1}{3 - j3} + \dfrac{1}{2}}\right]$$

$$\vec{I} = 50 - \left[\frac{11.79\underline{/75°} + 50}{\dfrac{4 - j8}{20} + \dfrac{6 + j6}{18} + 1}\right]$$

$$\vec{I} = 50 - \left[\frac{3.05 + j11.39 + 50}{1.533 - j0.066}\right]$$

$$\vec{I} = 50 - \frac{54.26\underline{/12.12}}{1.53\underline{/-2.50}}$$

$$\vec{I} = 50 = 35.46\underline{/14.62}$$

$$\vec{I} = 15.7 - j8.95 \text{ A.}$$

• PROBLEM 9-52

Find V_0 in fig. 1 if $\omega = 5$ Mrad/s.

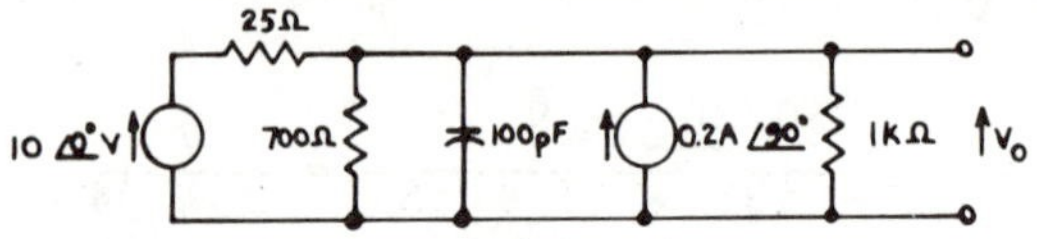

FiG.1

Solution: The method used here will be to convert the voltage source plus the 25Ω series resistance into a Norton equivalent. Thus, there will be two current sources in parallel. These can be replaced by one current source across the output impedance.

The circuit in Fig. 1 is transformed by Norton's theorem in Fig. 2.

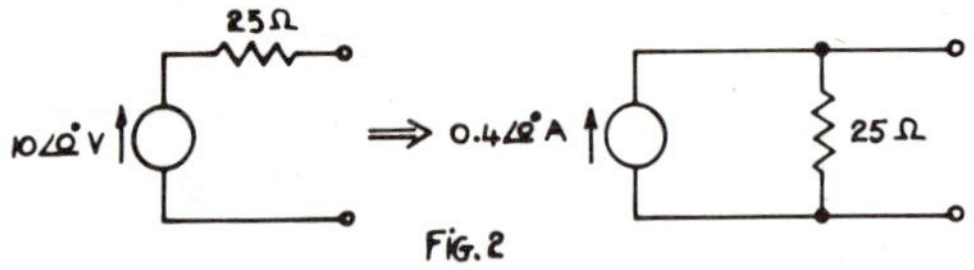

FiG.2

The network is then as shown in Fig. 3.

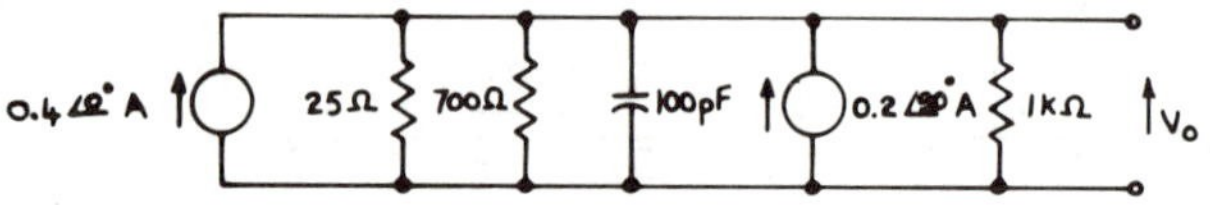

FiG.3

Combine the two current sources

$$\vec{I} = 0.4\underline{/0°} + 0.2\underline{/90°} = 0.447\underline{/26.6°} \text{ A}$$

$$\omega = 5 \times 10^6 \text{ rad/s.}$$

Admittance of capacitive branch, Y_C, is

$$\vec{Y}_C = j\omega C = j(5 \times 10^6)(100 \times 10^{-12}) = j5 \times 10^{-4} \text{ mho.}$$

The admittance of the resistive branches, Y_R, is

$$\vec{Y}_R = \frac{1}{25} + \frac{1}{700} + \frac{1}{1000} = 4.24 \times 10^{-2} \text{ mho.}$$

The total admittance across the output is

$$\vec{Y} = 4.24 \times 10^{-2} + j5 \times 10^{-4} = 4.24 \times 10^{-2}\underline{/0.7°}$$

Then,

$$\vec{V}_0 = \frac{\vec{I}}{\vec{Y}} = \frac{0.447\underline{/26.6^\circ}}{4.24 \times 10^{-2}\underline{/0.7^\circ}} = 10.5\underline{/25.9^\circ}\ \text{V}$$

• PROBLEM 9-53

Find the phasor voltage $\vec{V}_C$ and phasor currents $\vec{I}_1$, $\vec{I}_2$ and $\vec{I}_3$ in the circuit of Fig. 1.

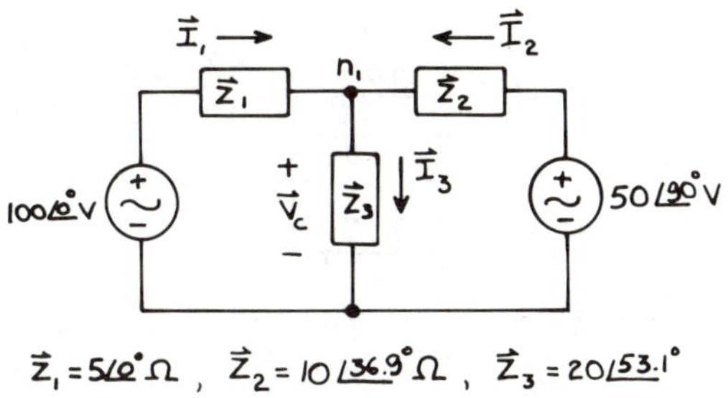

FIG. 1

Solution: Using KCL at node n_1 gives the equation

$$\vec{I}_1 + \vec{I}_2 = \vec{I}_3 \qquad \text{where,}$$

$$\vec{I}_1 = \frac{100\underline{/0^\circ} - \vec{V}_C}{5\underline{/0^\circ}}$$

$$\vec{I}_2 = \frac{50\underline{/90^\circ} - \vec{V}_C}{10\underline{/36.9^\circ}},$$

and

$$\vec{I}_3 = \frac{\vec{V}_C}{20\underline{/53.1^\circ}},$$

thus

$$\frac{100\underline{/0^\circ} - \vec{V}_C}{5\underline{/0^\circ}} + \frac{50\underline{/90^\circ} - \vec{V}_C}{10\underline{/36.9^\circ}} = \frac{\vec{V}_C}{20\underline{/53.1^\circ}}.$$

We can solve the above KCL equation for $\vec{V}_C$ and then find $\vec{I}_1$, $\vec{I}_2$, and $\vec{I}_3$.

Solving for $\vec{V}_C$ produces

$$\frac{100\underline{/0^\circ} - \vec{V}_C}{5\underline{/0^\circ}} + \frac{50\underline{/90^\circ} - \vec{V}_C}{10\underline{/36.9^\circ}} = \frac{\vec{V}_C}{20\underline{/53.1}}$$

$$\frac{100\underline{/0^\circ}}{5\underline{/0^\circ}} + \frac{50\underline{/90^\circ}}{10\underline{/36.9^\circ}} = V_C\ \frac{1}{20\underline{/53.1^\circ}} + \frac{1}{5\underline{/0^\circ}} + \frac{1}{10\underline{/36.9^\circ}}$$

$$\vec{V}_C = \frac{\dfrac{100\angle 0^\circ}{5\angle 0^\circ} + \dfrac{50\angle 90^\circ}{10\angle 36.9^\circ}}{\dfrac{1}{20\angle 53.1^\circ} + \dfrac{1}{5\angle 0^\circ} + \dfrac{1}{10\angle 36.9^\circ}}$$

$$\vec{V}_C = \frac{20\angle 0^\circ + 5\angle 53.1^\circ}{\frac{1}{20}\angle -53.1^\circ + \frac{1}{5}\angle 0 + \frac{1}{10}\angle -36.9^\circ}$$

$$\vec{V}_C = \frac{20 + 3 + j4}{\frac{3}{100} - j\frac{4}{100} + \frac{20}{100} + \frac{8}{100} - j\frac{6}{100}}$$

$$\vec{V}_C = \frac{23 + j4}{\frac{31}{100} - j\frac{10}{100}} = \frac{23.3\angle 9.87^\circ}{0.326\angle -17.88^\circ}$$

$$\vec{V}_C = 71.6\angle 27.76^\circ \text{ V.}$$

Hence, $I_1 = \dfrac{100\angle 0^\circ - V_C}{5\angle 0^\circ} = \dfrac{100\angle 0^\circ - 71.6\angle 27.76^\circ}{5\angle 0^\circ}$

$$I_1 = \frac{100 - 63.36 - j33.35}{5}$$

$$I_1 = 7.33 - j6.67 = 9.9\angle -42.3^\circ \text{ A}$$

$$\vec{I}_2 = \frac{50\angle 90^\circ - \vec{V}_C}{10\angle 36.9^\circ} = \frac{50\angle 90^\circ - 71.6\angle 27.76^\circ}{10\angle 36.9^\circ}$$

$$\vec{I}_2 = \frac{j50 - 63.36 - j33.35}{10\angle 36.9^\circ} = \frac{-63.36 + j16.65}{10\angle 36.9^\circ}$$

$$\vec{I}_2 = \frac{65.5\angle 165.3^\circ}{10\angle 36.9^\circ} = 6.55\angle 128.4^\circ \text{ A}$$

and

$$\vec{I}_3 = \frac{\vec{V}_C}{20\angle 53.1^\circ} = \frac{71.6\angle 27.76^\circ}{20\angle 53.1^\circ} = 3.58\angle -25.34^\circ \text{ A.}$$

• PROBLEM 9-54

Find the current and voltage for each impedance in Fig. 1. Redraw the diagram with the appropriate polarities.

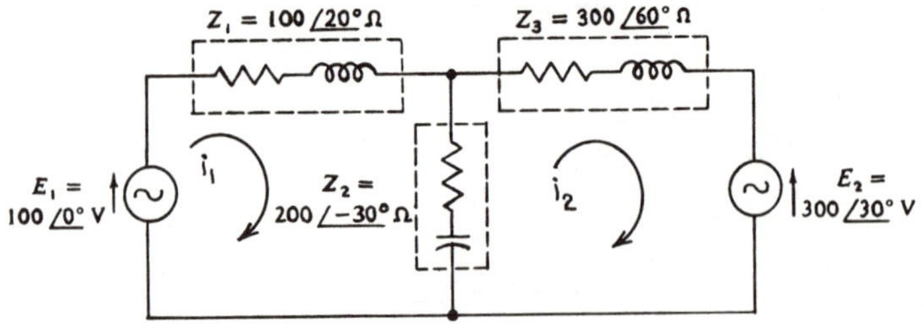

FIG. 1

Solution: $\vec{V}_{Z2}$ can be found directly from the nodal method.

$$\frac{\vec{V}_{Z2} - \vec{E}_1}{\vec{Z}_1} + \frac{\vec{V}_{Z2}}{\vec{Z}_1} + \frac{\vec{V}_{Z2} - \vec{E}_2}{\vec{Z}_3} = 0$$

$$\vec{V}_{Z2}\left(\frac{1}{\vec{Z}_1} + \frac{1}{\vec{Z}_2} + \frac{1}{\vec{Z}_3}\right) = \frac{\vec{E}_1}{\vec{Z}_1} + \frac{\vec{E}_2}{\vec{Z}_3}$$

$$\vec{V}_{Z2}\left[\frac{\vec{Z}_2\vec{Z}_3 + \vec{Z}_1\vec{Z}_3 + \vec{Z}_1\vec{Z}_2}{\vec{Z}_1\vec{Z}_2\vec{Z}_3}\right] = \frac{\vec{Z}_1\vec{E}_2 + \vec{Z}_3\vec{E}_1}{\vec{Z}_1\vec{Z}_3}$$

$$\vec{V}_{Z2} = \frac{\vec{Z}_1\vec{Z}_2\vec{E}_2 + \vec{Z}_2\vec{Z}_3\vec{E}_1}{\vec{Z}_2\vec{Z}_3 + \vec{Z}_1\vec{Z}_3 + \vec{Z}_1\vec{Z}_2}$$

$$= \frac{60\underline{/20^\circ} + 6\underline{/30^\circ}}{60\underline{/30^\circ} + 30\underline{/80^\circ} + 20\underline{/-10^\circ}} \times 10^3$$

$$= \frac{10.8 + j5.1}{77 + j56.1} \times 10^3$$

$$= \frac{11.9\underline{/25.3^\circ}}{95.3\underline{/36^\circ}}$$

$$\vec{V}_{Z2} = 125\underline{/-10.8^\circ}\ \text{V}$$

$\vec{V}_{Z1}$ and $\vec{V}_{Z3}$ are found by KVL.

$$\vec{V}_{Z1} = \vec{V}_{Z2} - \vec{E}_1$$

$$= 125\underline{/-10.8^\circ} - 100\underline{/0^\circ}$$

$$= 22.8 - j23.4$$

$$\vec{V}_{Z1} = 32.7\underline{/-45.7^\circ}\ \text{V}$$

$$\vec{V}_{Z3} = \vec{V}_{Z2} - \vec{E}_2$$

$$= 125\underline{/-10.8^\circ} - 300\underline{/30^\circ}$$

$$= -137.2 - j173.4$$

$$\vec{V}_{Z3} = 221.1\underline{/-128.3}\ \text{V}$$

I_{Z1}, I_{Z2} and I_{Z3} are found by Ohm's Law

$$\vec{I}_{Z1} = \frac{\vec{V}_{Z1}}{\vec{Z}_1}$$

$$= \frac{32.7\underline{/-45.7°}}{100\underline{/20°}}$$

$$\vec{I}_{Z1} = .327\underline{/-65.7°} \text{ A}$$

$$\vec{I}_{Z2} = \frac{\vec{V}_{Z2}}{Z_2}$$

$$= \frac{125\underline{/-10.8°}}{200\underline{/-30°}}$$

$$\vec{I}_{Z2} = .75\underline{/19.2°} \text{ A}$$

$$\vec{I}_{Z3} = \frac{\vec{V}_{Z3}}{\vec{Z}_3}$$

$$= \frac{221.1\underline{/-128.3°}}{300\underline{/60°}}$$

$$\vec{I}_{Z3} = 0.73\underline{/-188.3°} \text{ A.}$$

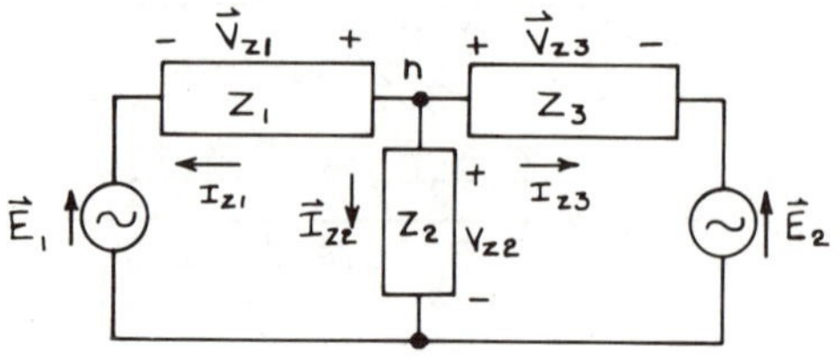

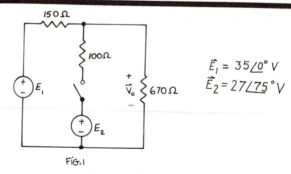

FiG.2

Fig. 2 shows the voltage and current polarities.

• PROBLEM 9-55

Find the voltage V_0 resulting from the action of source E_1 in Fig. 1. What does V_0 become after the switch is closed?

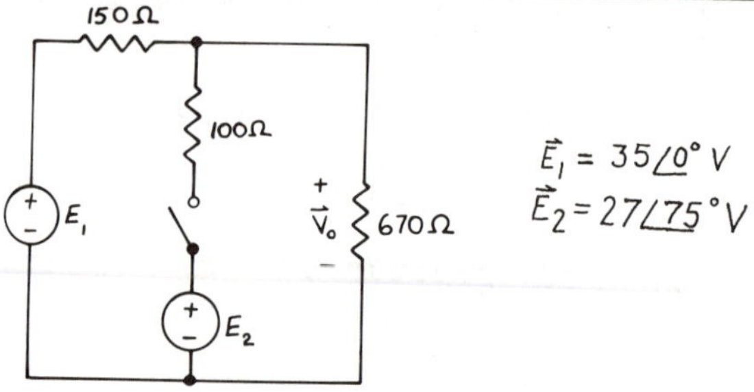

FiG.1

Solution: The voltage resulting from source E_1 can be found by voltage division

$$\vec{V}_0 = \frac{670}{150 + 670}E_1 = 28.6\underline{/0°}$$

When the switch is closed, V_0 can be found by superposition as shown in Figure 2.

From Figure 2(a), $\vec{V}_{0_1} = \frac{100\Omega \| 670\Omega}{150\Omega + 100\Omega \| 670\Omega}E_1 = \frac{87}{237}E_1$

$$= 12.85\underline{/0°} \text{ V}$$

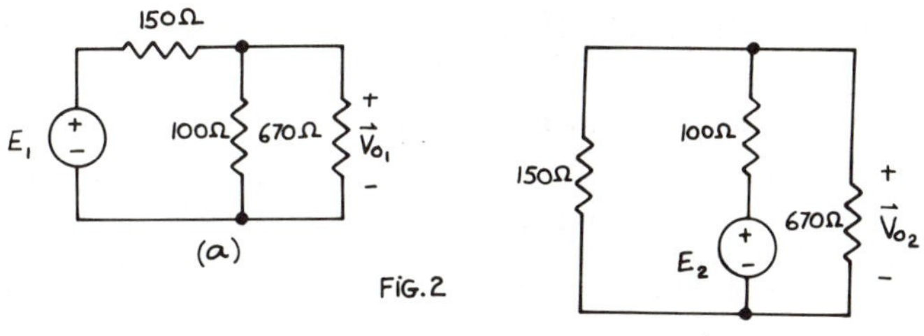

(a) FIG. 2 (b)

From Figure 2(b), $\vec{V}_{0_2} = \dfrac{150\,||\,670}{100\Omega + 150\Omega\,||\,670\Omega} E_2 = \dfrac{122.6}{222.6} E_2 =$

$$= 14.87\underline{/75°} \text{ V.}$$

Thus, $\vec{V}_0 = \vec{V}_{0_1} + \vec{V}_{0_2} = 12.85\underline{/0°} + 14.87\underline{/75°} = 12.85 + j0$

$$+ 3.85 + j14.36$$

$$= 22.03\underline{/40°} \text{ V.}$$

• PROBLEM 9-56

Find i_1 and v_2 in the circuit shown in Fig. 1.

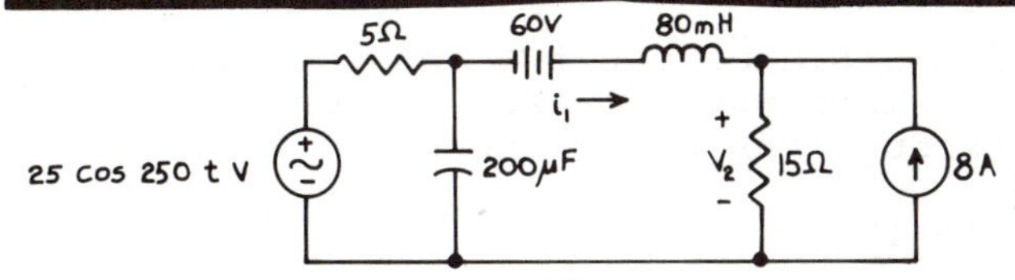

Fig. 1

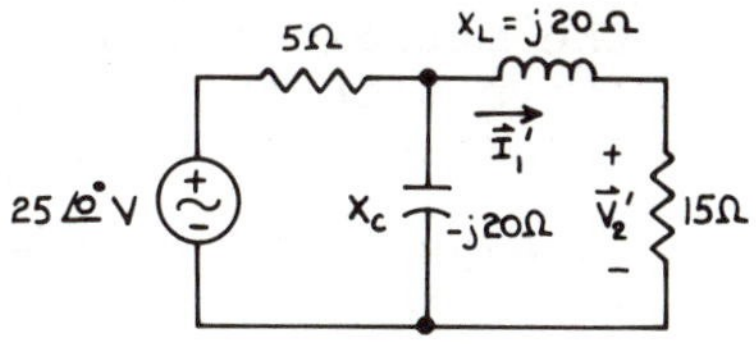

Fig. 2

Solution: We apply the superposition theorem to this problem. This theorem states that, for a linear network, the current and voltage for any branch will be the algebraic sum of the currents and voltages found with only one source applied at a time. Furthermore, when voltage sources are omitted they are replaced by short circuits with internal impedances left in the network. Current sources are open-circuited when omitted with internal admittances remaining. Also inductances act as short-circuits and capacitances are open-circuited for d.c.

Solve for one component with an a.c. voltage source. This is shown in Fig. 2. The network is solved by the phasor method and then converted to instantaneous values. Also this is a two-mesh network and we can use the method of determinants to immediately solve for the desired current in phasor form and find the component of voltage using Ohm's Law.

$$X_L = j\omega L = j(250)(80 \times 10^{-3}) = j20 \quad \text{(inductive reactance)}$$

$$X_C = -j\frac{1}{\omega C} = -j\frac{1}{(250)(200 \times 10^{-6})} = -j20 \quad \text{(capacitive reactance)}$$

$\vec{I}_1'$ is the phasor current through the inductance due to the a.c. voltage source.
$\vec{V}_2'$ is the phasor voltage across the 15Ω resistor. For convenience, assume the mesh current in the mesh consisting of the voltage source, 5Ω resistor and the capacitor is clockwise. The mesh current in the mesh consisting of the capacitor, inductance and 15Ω-resistor is the same as $\vec{I}_1'$ and clockwise.

The determinant solution for $\vec{I}_1'$ is then of the form

$$\vec{I}_1' = \frac{\begin{vmatrix} \vec{E}_1 & \vec{Z}_{12} \\ \vec{E}_2 & \vec{Z}_{22} \end{vmatrix}}{\begin{vmatrix} \vec{Z}_{11} & \vec{Z}_{12} \\ \vec{Z}_{21} & \vec{Z}_{22} \end{vmatrix}} \qquad (1)$$

$\vec{E}_1 = 0$ V is the total electromotive force (emf) in mesh with mesh current, $\vec{I}_1'$. $\vec{E}_2 = 25\underline{/0}$ is the emf in second mesh. Note that the sign is positive because the current enters at the negatively assigned terminal of the voltage soruce.

$$\vec{Z}_{12} = \vec{Z}_{22} = -(-j20) = j20 = 20\underline{/90^\circ}\ \Omega$$

is the mutual (common) impedance of the two meshes. The negative sign is required since the two mesh currents are in opposite directions through the capacitance.

$$\vec{Z}_{11} = -j20 + j20 + 15 = 15\underline{/0^\circ}\ \Omega$$

is the self-impedance of the mesh with mesh-current, I_1'.

$$\vec{Z}_{22} = 5 - j20 = 20.6\underline{/-76.0^\circ}\ \Omega$$

is the self-impedance of the other mesh.

Substituting in equation (1),

$$\vec{I}_1' = \frac{\begin{vmatrix} 0 & 20\underline{/90^\circ} \\ 25\underline{/0^\circ} & 20.6\underline{/-76.0^\circ} \end{vmatrix}}{\begin{vmatrix} 15\underline{/0^\circ} & 20\underline{/90^\circ} \\ 20\underline{/90^\circ} & 20.6\underline{/-76.0^\circ} \end{vmatrix}}$$

Expanding the determinants and performing the operations

$$\vec{I}_1' = \frac{-500\underline{/90^\circ}}{309\underline{/-76.0^\circ} - 400\underline{/180^\circ}} = \frac{500\underline{/-90^\circ}}{74.8 - j300 + 400} = \frac{500\underline{/-90^\circ}}{474.8 - j300}$$

$$\vec{I}_1' = \frac{500\underline{/-90^\circ}}{561.6\underline{/-32.3^\circ}} = 0.890\underline{/-57.7^\circ}\ \text{A}$$

By Ohm's Law,

$\vec{V}_2' = \vec{I}_1'\vec{Z}_2 = (0.890\underline{/-57.7°})(15) = 13.3\underline{/-57.7°}$ V

$\vec{Z}_2 = 15$ is the impedance of the resistor across which we wish to determine the voltage, V_2. Now convert the phasor values to their instantaneous values.

$$i_1' = 0.890 \cos(250t - 57.7°)\text{A}$$

$$v_2' = 13.3 \cos(250t - 57.7°)\text{V}.$$

Place the 60V d.c. source in the network with the other two sources removed. The network is shown in Fig. 3.

$I_1'' = \dfrac{V}{R_1 + R_2}$ is the current component of i_1' due to the 60-v source and V = 60 V is the total emf. Note that the voltage is positive for assumed current direction. $R_1 = 5\Omega$ and $R_2 = 15\Omega$ are the series resistances.

$$I_1'' = \frac{60}{5 + 15} = 3\text{A}.$$

By Ohm's Law,

$V_2'' = (3)(15) = 45$ V. This is positive since the current enters the positive terminal of the resistor as marked. Now, consider the network with only the current source. This is shown in Fig. 4.

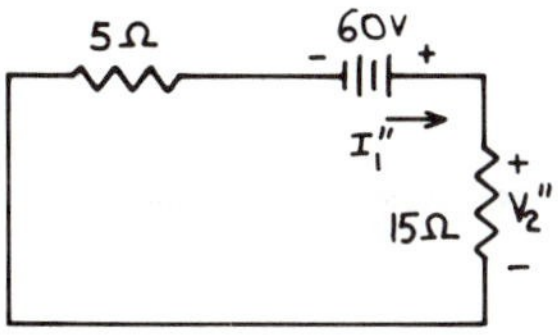

FIG.3

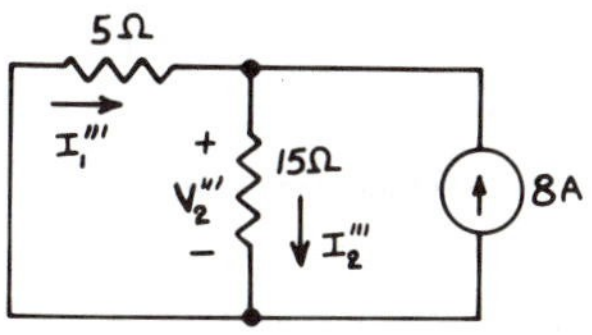

FIG.4

By Kirchhoff's Current Law (KCL),

$$I_1''' + 8 = I_2 \qquad (2)$$

By Kirchhoff's Voltage Law (KVL),

$$5I_1''' + 15I_2''' = 0$$

or, $$I_2''' = -\frac{1}{3}I_1''' . \qquad (3)$$

Substituting, (3) in (2),

$$I_1''' + \frac{1}{3}I_1''' = \frac{4}{3}I_1''' = -8$$

or $\quad I_1''' = -6\ A \quad$ (current is opposite to assumed direction.)

and substituting in (3)

$$I_2''' = (-\frac{1}{3})(-6) = 2\ A$$

$$V_2''' = (2)(15) = 30\ V.$$

We now use the superposition process to find the solutions with all sources in the network. Thus,

$$i_1 = i_1' + I_1'' + I_1'''$$

$$i_1 = 0.890 \cos(250t - 57.7^\circ) + 3 - 6\ A$$

or,

$$i_1 = -3 + 0.890 \cos(250t - 57.7^\circ) A$$

and,

$$v_2 = v_2' + V_2'' + V_2'''$$

$$v_2 = 13.3 \cos(250t - 57.7^\circ) + 45 + 30\ V$$

or,

$$v_2 = 75 + 13.3 \cos(250t - 57.7^\circ) V$$

• PROBLEM 9-57

In the circuit shown in Fig. 1 determine: a) I_{nN}; (b) I_{AN}; (c) I_{AB}.

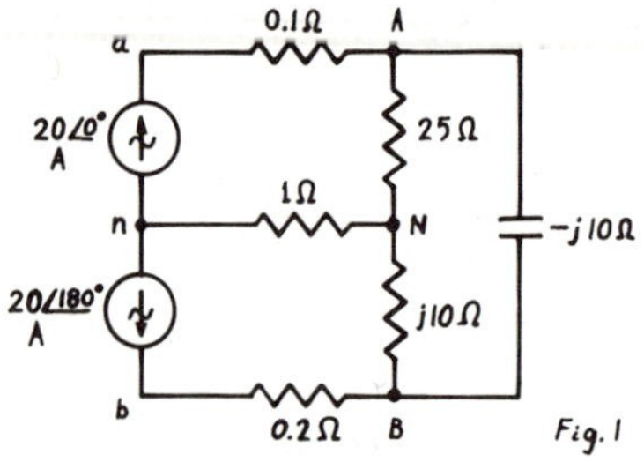

Fig. 1

Solution: One method of solving a circuit such as this is by loop analysis. Thus, one must only write an equation for the right-hand mesh since the other two loop currents are known. Assuming a loop current I in the clockwise direction in that mesh, yields

$$(25 + j10 - j10)\vec{I} - 25(20\underline{/0^\circ}) + j10(20\underline{/180^\circ}) = 0$$

or $\quad 25\vec{I} - 500 - j200 = 0$

or $\vec{I} = \frac{500 + j200}{25} = 20 + j8 = 4(5 + j2)A.$

Thus,

a) $\vec{I}_{AN} = 20\underline{/0^\circ} + 20\underline{/180^\circ} = 0$ A

b) $\vec{I}_{AN} = 20\underline{/0^\circ} - I = 20 - 20 - j8 = 8\underline{/-90^\circ}$ A

c) $\vec{I}_{AB} = I = 4(5 + j2) = 21.5\underline{/21.8^\circ}$ A

• PROBLEM 9-58

Use nodal analysis on the circuit shown in Fig. 1 to evaluate the phasor voltage: (a) $\vec{V}_1$; (b) $\vec{V}_2$; (c) $\vec{V}_3$.

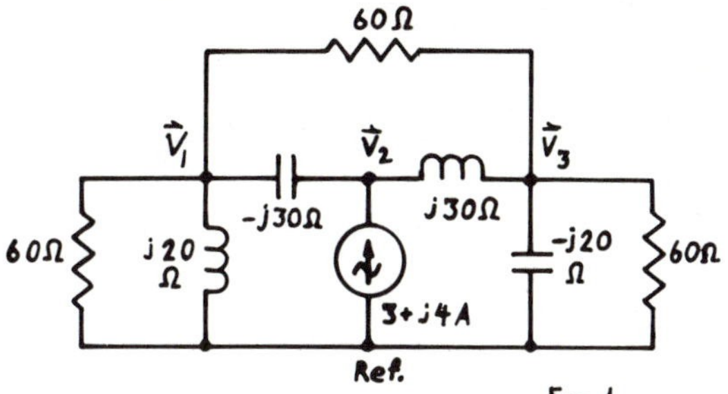

Fig. 1

Solution: Writing a KCL equation for each node yields three linear independent equations.

$$0 = \frac{\vec{V}_1}{60} + \frac{\vec{V}_1}{j20} + \frac{\vec{V}_1 + \vec{V}_2}{-j30} + \frac{\vec{V}_1 - \vec{V}_3}{60}$$

$$3 + j4 = \frac{\vec{V}_2 - \vec{V}_1}{-j30} + \frac{\vec{V}_1 - V_2}{j30}$$

$$0 = \frac{\vec{V}_3}{-j20} + \frac{\vec{V}_3}{60} + \frac{\vec{V}_3 - \vec{V}_2}{j30} + \frac{\vec{V}_3 - \vec{V}_1}{60}$$

Collecting like terms and factoring gives

$$0 = \vec{V}_1\left(\frac{1}{60} + \frac{1}{60} + \frac{1}{j20} + \frac{1}{-j30}\right) + V_2\left(\frac{1}{j30}\right) + V_3\left(-\frac{1}{60}\right)$$

$$3 + j4 = V_1\left(\frac{1}{j30}\right) + V_2\left(\frac{1}{-j30} + \frac{1}{j30}\right) + V_3\left(-\frac{1}{j30}\right)$$

$$0 = V_1\left(-\frac{1}{60}\right) + V_2\left(-\frac{1}{j30}\right) + V_3\left(\frac{1}{60} + \frac{1}{60} + \frac{1}{-j20} + \frac{1}{j30}\right).$$

Simplifying the above equations gives

$$0 = V_1\left(\frac{1}{30} - j\frac{1}{60}\right) + V_2\left(-j\frac{1}{30}\right) + V_3\left(-\frac{1}{60}\right)$$

$$3 + j4 = \vec{V}_1(-j\tfrac{1}{30}) + \vec{V}_2(0) + \vec{V}_3(j\tfrac{1}{30})$$

$$0 = \vec{V}_1(-\tfrac{1}{60}) + \vec{V}_2(j\tfrac{1}{30}) + \vec{V}_3(\tfrac{1}{30} + j\tfrac{1}{60}).$$

Changing each of the terms to polar form and writing them in matrix form gives

$$\begin{bmatrix} \frac{1}{26.83}\underline{/-26.57^\circ} & \frac{1}{30}\underline{/-90^\circ} & \frac{11}{60}\underline{/180^\circ} \\ \frac{1}{30}\underline{/-90^\circ} & 0 & \frac{1}{30}\underline{/90^\circ} \\ \frac{1}{60}\underline{/180^\circ} & \frac{1}{30}\underline{/90^\circ} & \frac{1}{26.83}\underline{/26.57^\circ} \end{bmatrix} \begin{bmatrix} \vec{V}_1 \\ \vec{V}_2 \\ \vec{V}_3 \end{bmatrix} = \begin{bmatrix} 0 \\ 5\underline{/53.13^\circ} \\ 0 \end{bmatrix}$$

Solving the matrix by use of determinants

$$\Delta = \frac{1}{26.83}\underline{/-26.57}\left[-(\tfrac{1}{30})^2\underline{/180^\circ}\right] - \frac{1}{30}\underline{/-90^\circ}\left[(\tfrac{1}{30})(\tfrac{1}{26.83})\underline{/-63.43^\circ} - (\tfrac{1}{60})(\tfrac{1}{30})\underline{/270^\circ}\right] + \frac{1}{60}\underline{/180^\circ}\left[(\tfrac{1}{30})^2\underline{/0^\circ}\right]$$

$$\Delta = -\left(\frac{1}{30^2}\right)\left(\frac{1}{26.83}\right)\underline{/153.43^\circ} - \left(\frac{1}{30^2}\right)\left(\frac{1}{26.83}\right)\underline{/-153.43^\circ} + \left(\frac{1}{30^2}\right)\left(\frac{1}{60}\right)\underline{/180^\circ} + \left(\frac{1}{30^2}\right)\left(\frac{1}{60}\right)\underline{/180^\circ}$$

$$\Delta = -4.14 \times 10^{-5}\underline{/153.43^\circ} - 4.14 \times 10^{-5}\underline{/-153.43^\circ} + 3.704 \times 10^{-5}\underline{/180^\circ}.$$

Changing each term back to cartesian form

$$\Delta = 3.704 \times 10^{-5} - j1.852 \times 10^{-5} + 3.704 \times 10^{-5} + j1.852 \times 10^{-5} - 3.704 \times 10^{-5}$$

$$\Delta = 3.704 \times 10^{-5}$$

Solving for $\vec{V}_1$

$$\vec{V}_1 = \frac{\begin{vmatrix} 0 & \frac{1}{30}\underline{/-90^\circ} & \frac{1}{60}\underline{/180^\circ} \\ 5\underline{/53.13^\circ} & 0 & \frac{1}{30}\underline{/90^\circ} \\ 0 & \frac{1}{30}\underline{/90^\circ} & \frac{1}{26.83}\underline{/26.57^\circ} \end{vmatrix}}{\Delta}$$

$$\vec{V}_1 = \frac{-5\underline{/53.13^\circ}\left[\left(\frac{1}{30}\right)\left(\frac{1}{26.83}\right)\underline{/-63.43^\circ} - \left(\frac{1}{30}\right)\left(\frac{1}{60}\right)\underline{/270^\circ}\right]}{3.704 \times 10^{-5}\underline{/0}}$$

$$\vec{V}_1 = \frac{\frac{-5}{30(26.83)}\underline{/-10.3^\circ} + \frac{5}{(60)(30)}\underline{/323.13^\circ}}{3.704 \times 10^{-5}\underline{/0^\circ}}$$

$$\vec{V}_1 = \frac{-6.21 \times 10^{-3}\underline{/-10.3^\circ} + 2.78 \times 10^{-3}\underline{/323.13^\circ}}{3.704 \times 10^{-5}\underline{/0^\circ}}$$

$$\vec{V}_1 = \frac{-6.11 \times 10^{-3} + j1.11 \times 10^{-3} + 2.22 \times 10^{-3} - j1.67 \times 10^{-3}}{3.704 \times 10^{-5}\underline{/0^\circ}}$$

$$\vec{V}_1 = \frac{-3.89 \times 10^{-3} - j0.56 \times 10^{-3}}{3.704 \times 10^{-5}\underline{/0^\circ}} = -105 - j15$$

$$\vec{V}_1 = \frac{3.93 \times 10^{-3}\underline{/188.19}}{3.704 \times 10^{-5}\underline{/0^\circ}}$$

$$\vec{V}_1 = 106.1\underline{/188.19^\circ} = -105 - j15 \text{ V.}$$

Solving for $\vec{V}_2$

$$\vec{V}_2 = \frac{\begin{vmatrix} \frac{1}{26.83}\underline{/-26.57^\circ} & 0 & \frac{1}{60}\underline{/180^\circ} \\ \frac{1}{30}\underline{/-90^\circ} & 5\underline{/53.13^\circ} & \frac{1}{30}\underline{/90^\circ} \\ \frac{1}{60}\underline{/180^\circ} & 0 & \frac{1}{26.83}\underline{/26.57^\circ} \end{vmatrix}}{3.704 \times 10^{-5}\underline{/0^\circ}}$$

$$\vec{V}_2 = \frac{5\underline{/53.13^\circ}\left[\left(\frac{1}{26.83}\right)^2\underline{/0^\circ} - \left(\frac{1}{60}\right)^2\underline{/0^\circ}\right]}{3.704 \times 10^{-5}\underline{/0^\circ}}$$

$$\vec{V}_2 = \frac{5\angle 53.13°(1.11 \times 10^{-3}\angle 0°}{3.704 \times 10^{-5}\angle 0°}$$

$$\vec{V}_2 = (30)5\angle 53.13° = 150\angle 53.13°$$

$$\vec{V}_2 = 90 + j120 \text{ V.}$$

Solving for V_3:

$$\vec{V}_3 = \frac{\begin{vmatrix} \frac{1}{26.83}\angle -26.57° & \frac{1}{30}\angle -90° & 0 \\ \frac{1}{30}\angle -90° & 0 & 5\angle 53.13° \\ \frac{1}{60}\angle 180° & \frac{1}{30}\angle 90° & 0 \end{vmatrix}}{3.704 \times 10^{-5}\angle 0}$$

$$\vec{V}_3 = \frac{-5\angle 53.13° \left[\frac{1}{26.83(30)}\angle 63.43° - \frac{1}{(30)(60)}\angle 90°\right]}{3.704 \times 10^{-5}\angle 0°}$$

$$V_3 = \frac{\frac{-5}{26.83(30)}\angle 116.56 + \frac{5}{(30)(60)}\angle 143.13}{3.704 \times 10^{-5}\angle 0°}$$

$$V_3 = \frac{-6.21 \times 10^{-3}\angle 116.56° + 2.78 \times 10^{-3}\angle 143.13°}{3.704 \times 10^{-5}\angle 0°}$$

$$\vec{V}_3 = \frac{2.78 \times 10^{-3} - j5.55 \times 10^{-3} - 2.22 \times 10^{-3} + j1.67 \times 10^{-3}}{3.704 \times 10^{-5}}$$

$$\vec{V}_3 = \frac{0.56 \times 10^{-3} - j3.89 \times 10^{-3}}{3.704 \times 10^{-5}} = 15 - j105 \text{ V.}$$

• PROBLEM 9-59

In the circuit shown in Fig. 1 find: (a) $\vec{I}_1$; (b) $\vec{I}_2$; (c) $\vec{I}_3$.

Solution: Writing 2 KVL equations around the loops shown in Fig. 2 yields

$$\ell_1: \quad 0 = \vec{I}_1\left(\frac{1}{j\omega\frac{25}{6} \times 10^{-6}}\right) - (\vec{I}_2)(j\omega 5 \times 10^{-3})$$

$$- (\vec{I}_2 + \vec{I}_3)(j\omega 10 \times 10^{-3})$$

ℓ_2: $0 = \vec{I}_2(j\omega 5 \times 10^{-3}) + (\vec{I}_1 + \vec{I}_2)(10) - \vec{I}_3\left(\frac{1}{j\omega 25 \times 10^{-6}}\right)$

A third equation can be written by summing the currents at node n_1 in Fig. 2.

$$2 = \vec{I}_1 + \vec{I}_2 + \vec{I}_3.$$

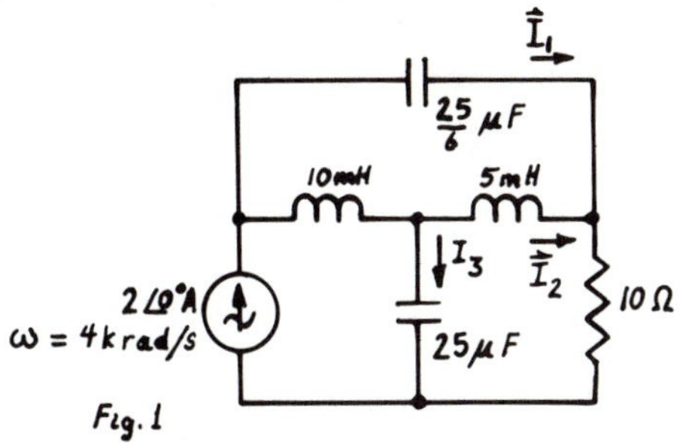

Fig. 1

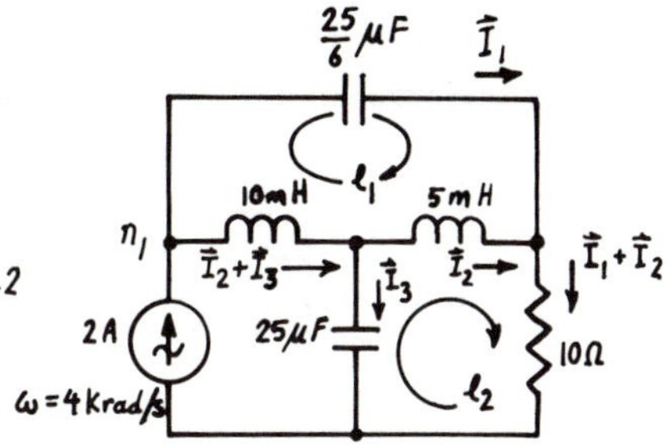

Fig. 2

Any variable chosen can be eliminated from the two loop equations.

Substituting $2 - \vec{I}_1 - \vec{I}_2$ for I_3 in the two loop equations yields

$$0 = \vec{I}_1(-j60) - \vec{I}_2(j20) - (2 - \vec{I}_1)(j40)$$

$$0 = \vec{I}_2(j20) + (\vec{I}_1 + \vec{I}_2)(10) - (2 - \vec{I}_1 - \vec{I}_2)(-j10).$$

Simplifying gives

$$j80 = \vec{I}_1(-j20) + \vec{I}_2(-j20)$$

$$-j20 = \vec{I}_1(10 - j10) + \vec{I}_2(10 + j10)$$

Solving for I_1

$$\vec{I}_1 = \frac{\begin{vmatrix} j80 & -j20 \\ -j20 & 10+j10 \end{vmatrix}}{\begin{vmatrix} -j20 & -j20 \\ 10-j10 & 10+j10 \end{vmatrix}} = \frac{j80(10+j10) - (j^2)20^2}{-j20(10+j10) + j20(10-j10)}$$

$$\vec{I}_1 = \frac{j800 - 800 + 400}{-j200 + 200 + j200 + 200} = \frac{-400 + j800}{400} = (-1 + j2)\text{A}$$

Solving for $\vec{I}_2$

$$\vec{I}_2 = \frac{\begin{vmatrix} -j20 & j80 \\ 10-j10 & -j20 \end{vmatrix}}{400} = \frac{j^2 20^2 - j80(10 - j10)}{400}$$

$$\vec{I}_2 = \frac{-400 - j800 - 800}{400} + \frac{-1200 - j800}{400} = (-3 - j2)\text{A}$$

$\vec{I}_3$ can be found

$$\vec{I}_3 = 2 - I_1 - I_2 = 2 + 1 - j2 + 3 + j2 = 6A.$$

• PROBLEM 9-60

Use Thévenin's theorem to calculate the phasor $\vec{V}_{ab}$ in the circuit in Fig. 1.

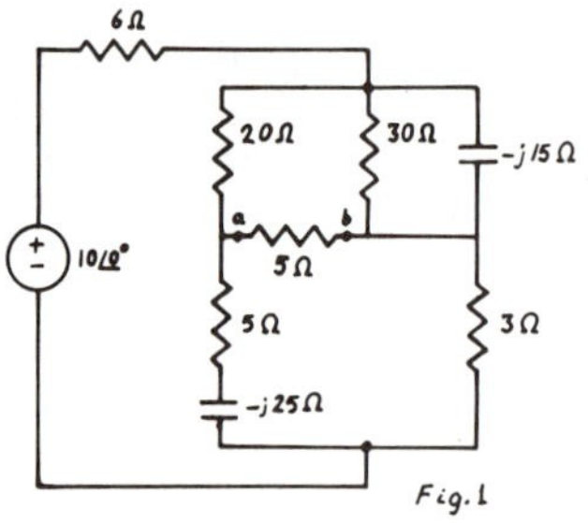

Fig. 1

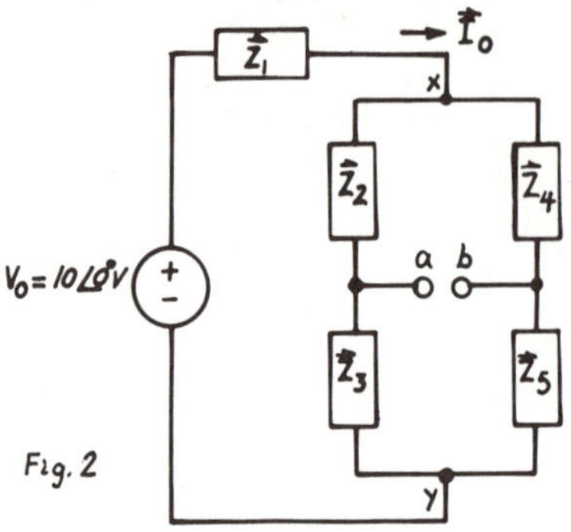

Fig. 2

Solution: The plan is to remove the 5-Ω resistance between a and b and proceed to solve for the open-circuit voltage between these two points. This is the voltage source in our equivalent circuit. Next, connect points a and b together (a short circuit) and calculate the current in the short circuit. Dividing the open-circuit voltage by the short-circuit current will give us the impedance of the Thevenin equivalent circuit. Finally, place the 5-Ω resistance between terminals a and b of the equivalent circuit. The complete circuit now consists of the Thévenin voltage, the Thévenin impedance and the 5-Ω resistance in series. Solve for the current in this series circuit and multiply this current by 5-Ω to obtain the voltage $\vec{V}_{ab}$.

First, convert the parallel combination consisting of the 30-Ω resistance and the 15-Ω capacitive reactance to a series impedance; we shall call this impedance $\vec{Z}_4$.

$$\vec{Z}_4 = \frac{1}{\frac{1}{30} + \left[\frac{1}{-j15}\right]} = \frac{30\underline{/0°} \times 15\underline{/-90°}}{30 - j15} \tag{1}$$

$$\vec{Z}_4 = 13.4\underline{/-63.4°} = 6 - j12\Omega \tag{2}$$

Next, remove the 5-Ω resistance between a and b; this leaves the network as shown in Figure 2. Now solve for $\vec{V}_{ab}$.

The first step in finding $\vec{V}_{ab}$, the open-circuit voltage, is to solve for the total impedance of the network. From Figure 1, we see that

$$\vec{Z}_0 = \vec{Z}_1 + \frac{(\vec{Z}_2+\vec{Z}_3)(\vec{Z}_4+\vec{Z}_5)}{(\vec{Z}_2+\vec{Z}_3)(\vec{Z}_4+\vec{Z}_5)} \tag{3}$$

$$\vec{Z}_0 = 6 + j0 + \frac{(25 - j25)(9 - j12)}{(25 - j25) + (9 - j12)} \quad (4)$$

$$\vec{Z}_0 = 12.7 - j8.2 = 15.1\underline{/-32.8^\circ}\ \Omega \quad (5)$$

This enables us to calculate $\vec{I}_0$:

$$\vec{I}_0 = \frac{\vec{V}_0}{\vec{Z}_0} = \frac{10\underline{/0}}{15.1\underline{/-32.8^\circ}} = 0.663\underline{/32.8^\circ}\ \text{A} \quad (6)$$

With this solution we can calculate $\vec{V}_{xy}$:

$$\vec{V}_{xy} = \vec{V}_0 - I_0Z_1 = 10 + j0 - (0.663\underline{/32.8^\circ})(6\underline{/0}) \quad (7)$$

$$\vec{V}_{xy} = 6.65 - j2.15 = 7.00\underline{/-17.9^\circ}\ \text{V} \quad (8)$$

By finding $\vec{V}_{xy}$, it is possible to apply the voltage divider principle in order to solve for $\vec{V}_{ay}$ and $\vec{V}_{by}$ because

$$\vec{V}_{ab} = \vec{V}_{ay} - \vec{V}_{by} \quad (9)$$

around the lower loop.

Solving for $\vec{V}_{ay}$,

$$\vec{V}_{ay} = \vec{V}_{xy}\frac{\vec{Z}_3}{\vec{Z}_3+\vec{Z}_2} = 7.00\underline{/-17.9^\circ}\ \frac{(5 - j25)}{(25 - j25)} \quad (10)$$

$$\vec{V}_{ay} = 5.04\underline{/-51.6} = 3.13 - j3.95\ \text{V}. \quad (13)$$

Solving for $\vec{V}_{by}$, (11)

$$\vec{V}_{by} = V_{xy}\ \frac{\vec{Z}_5}{\vec{Z}_5+\vec{Z}_4} = 7.00\underline{/-17.9^\circ}\ \frac{(3+j0)}{(9-j12)}$$

$$\vec{V}_{by} = 1.40\underline{/35.2^\circ} = 1.14 + j0.81\ \text{V}. \quad (12)$$

Finally, determining $\vec{V}_{ab}$,

$$\vec{V}_{ab}(\text{open-circuit}) = \vec{V}_{ay} - \vec{V}_{by} = (3.13 - j3.95) - (1.14 + j0.81) \quad (14)$$

$$\vec{V}_{ab}(\text{open-circuit}) = 1.99 - j4.76 = 5.16\underline{/-67.3^\circ}\ \text{V}. \quad (15)$$

We now have the voltage in the Thévenin equivalent circuit. It now remains to determine the equivalent impedance.

The first step in finding the equivalent impedance is finding the current in a short circuit joining a and b.

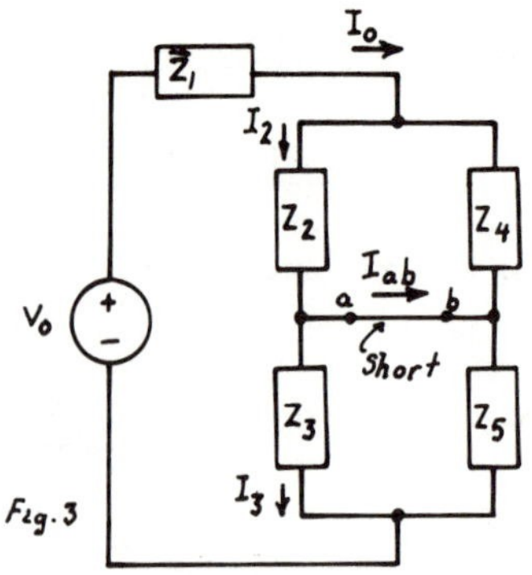

Fig. 3

Redraw the network as shown in Figure 3, showing this condition. We begin by finding the impedance of Z_2 and Z_4 in parallel:

$$\vec{Z}_{24} = \frac{\vec{Z}_2\,\vec{Z}_4}{\vec{Z}_2+\vec{Z}_4} = \frac{(20\underline{/0})(13.4\underline{/-63.4\ })}{(20 + j0) + (6 - j12)} \tag{16}$$

$$\vec{Z}_{24} = 9.37\underline{/-38.6^\circ} = 7.32 - j5.85\Omega\ , \tag{17}$$

and, similarly, the impedance of $\vec{Z}_3$ and $\vec{Z}_5$ in parallel is

$$\vec{Z}_{35} = \frac{\vec{Z}_3\vec{Z}_5}{\vec{Z}_3+\vec{Z}_5} = \frac{(5 - j25)(3 + j0)}{(5 - j25)(3 + j0)} \tag{18}$$

$$\vec{Z}_{35} = 2.91\underline{/-6.43} = 2.90 - j0.33\ \Omega. \tag{19}$$

Now we are able to combine the impedances into the total impedance seen by the source. Hence,

$$\vec{Z}_0 = \vec{Z}_1 + \vec{Z}_{24} + \vec{Z}_{35} \tag{20}$$

$$\vec{Z}_0 = (6 + j0) + (7.32 - j5.85) + (2.90 - j0.33). \tag{21}$$

Substituting equations (17) and (19) into (20) results in

$$Z_0 = 16.2 - j6.18 = 17.4\underline{/-20.9^\circ}\ \Omega. \tag{22}$$

Now we are ready to find the <u>source</u> current under the conditions specified (a short-circuit between a and b).

$$\text{Source current} = \vec{I}_0 = \frac{\vec{V}_0}{\vec{Z}_0} = \frac{10\underline{/0}}{17.4\underline{/-20.9^\circ}} \tag{23}$$

$$\vec{I}_0 = 0.576\underline{/20.9^\circ}\ \text{A}. \tag{24}$$

Knowing the source current, it is possible to apply the current divider principle to find $\vec{I}_2$ and $\vec{I}_3$ (in Figure 2):

$$\vec{I}_2 = \vec{I}_0 \frac{\vec{Z}_4}{\vec{Z}_4+\vec{Z}_2} = 0.576\underline{/20.9} \times \frac{13.4\underline{/-63.4^\circ}}{(6 - j12) + (20 + j0)} \quad (25)$$

$$\vec{I}_2 = 0.270\underline{/-17.8^\circ} = 0.257 - j0.082 \text{ A.} \quad (26)$$

Similarly,

$$\vec{I}_3 = I_0 \frac{\vec{Z}_5}{\vec{Z}_5+\vec{Z}_3} = 0.576\underline{/20.9^\circ} \times \frac{3\underline{/0}}{(3 + j0) + (5 - j25)} \quad (27)$$

$$\vec{I}_3 = 0.066\underline{/93.1^\circ} = -.004 + j.066 \text{ A.} \quad (28).$$

The short-circuit current is found by applying Kirchhoff's current law to the node between $\vec{Z}_2$ and $\vec{Z}_3$ in Figure 2:

$$\vec{I}_{ab} = \vec{I}_2 - \vec{I}_3, \quad (29)$$

and, substituting equations (26) and (28) into equation (29),

$$\vec{I}_{ab} = (0.257 - j0.082) - (-0.004 + j0.066) \quad (30)$$

$$\vec{I}_{ab} = 0.261 - j0.148 = 0.300\underline{/-29.6^\circ} \text{ A.} \quad (31)$$

Finally, the Thévenin equivalent impedance is found from the defining equation,

$$\vec{Z}_{TH} = \frac{\vec{V}_{ab}(\text{open circuit})}{\vec{I}_{ab}(\text{short circuit})} \quad (32)$$

into which we substitute values from equations (15) and (31).

$$\vec{Z}_{TH} = \frac{5.16\underline{/-67.3^\circ}}{0.300\underline{/-29.6}} = 17.20\underline{/-37.75} = 13.6 - j10.5\ \Omega. \quad (33)$$

At this point, a circuit diagram should be drawn to clarify things as they now stand. This is shown in Figure 4. From this diagram, we can apply Kirchhoff's voltage law and solve for the current.

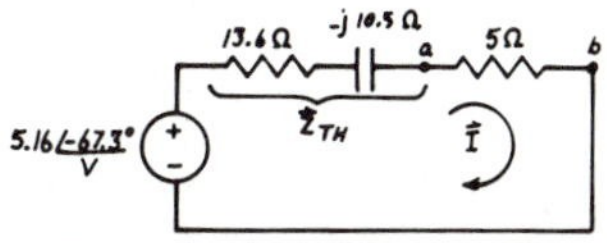

FIG. 4. ORIGINAL CIRCUIT AFTER BEING REDUCED TO 5OHM RESISTANCE BETWEEN a AND b AND THE THEVENIN EQUIVALENT FOR THE REMAINDER OF THE NETWORK.

$$\vec{I} = \frac{\vec{V}}{\vec{Z}} = \frac{5.16\underline{/-67.3^\circ}}{(13.6 - j10.5) + 5} \quad (34)$$

$$\vec{I} = 0.24\underline{/-37.8°}\ A, \tag{35}$$

and, finally, the voltage across the 5-ohm resistor is

$$\vec{V}_{ab} = IR = (0.24\underline{/-37.8°})(5\underline{/0°}) = 1.2\underline{/-37.8°}\ V. \tag{36}$$

• PROBLEM 9-61

> Find the voltage gain, $\frac{E_{out}}{E_{in}}$, under steady-state conditions for the equivalent circuit of the Field-effect Transistor amplifier in Fig. 1. The constant g_m for the dependent current source is 200×10^{-5} mho, also $E_{in} = \sqrt{2} \sin 3770t$ V, and $Rg = R_1$.

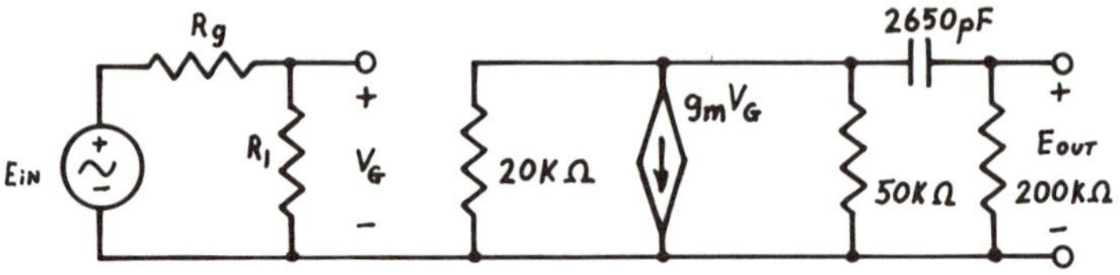

Fig.1

Solution: Transform the entire circuit into its steady-state equivalent. Hence,

$$\sqrt{2} \sin 3770t\ V \Rightarrow 1\underline{/-90°}\ V_{rms}$$

$$2650\rho F = \frac{1}{j\omega(2650 \times 10^{-12})}$$

$$2650\rho F \Rightarrow \frac{1}{j3770(2650 \times 10^{-12})}$$

$$2650\rho F \Rightarrow \frac{1}{j(1 \times 10^{-5})}$$

$$2650\rho F \Rightarrow -j(1 \times 10^{5}).$$

Fig. 2 shows the equivalent circuit.

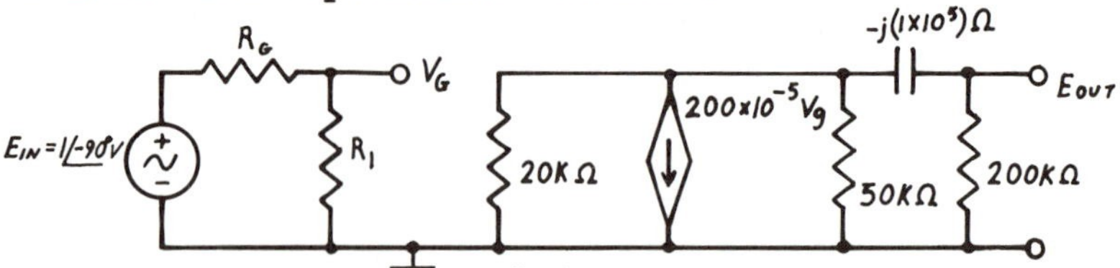

Fig.2

Note that since $R_G = R_1$, the portion of the circuit on the left forms a voltage divider,

$$\frac{\vec{E}_{IN}\ R_1}{R_G + R_1} = \vec{V}_G$$

since $R_1 = R_G = R$

$$\frac{\vec{E}_{IN}\,R}{2R} = \frac{\vec{E}_{IN}}{2} = \vec{V}_G$$

Note that all voltage and current values are rms values.

e.g. $\vec{E}_{IN} = \sqrt{2}\underline{/-90°}\ V_{pp}$

$$\vec{E}_{IN} = \frac{\sqrt{2}}{\sqrt{2}}\ \underline{/-90°} = 1\underline{/-90°}\ V_{rms}$$

$$\vec{V}_G = \frac{1\underline{/-90°}}{2} = 0.5\underline{/-90°}.$$

The value of the dependent current source can now be found

$$200 \times 10^{-5}(0.5\underline{/-90°}) = 1\underline{/-90°}\ mA.$$

At this pint, use Kirchoff's laws to write network equations for the steady-state output voltage in the circuit of Fig. 3. Combine the 20KΩ and the 50KΩ resistors, and then transform the current source to a voltage source.

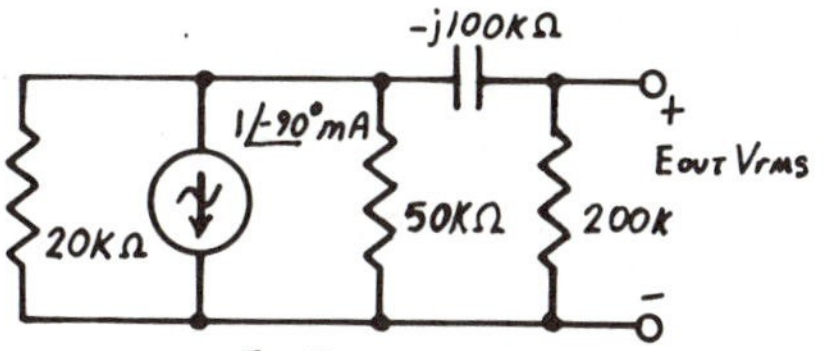

Fig. 3

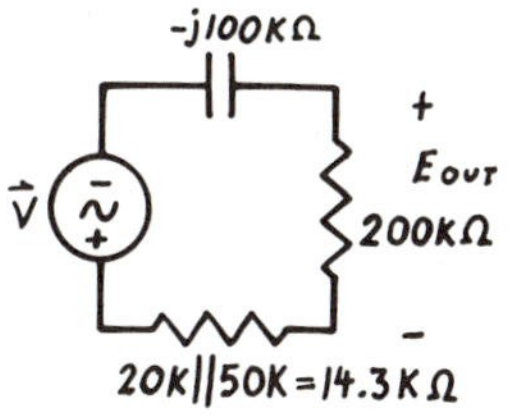

Fig. 4

Hence, we have the circuit of Fig. 4

where $\vec{V} = (20K||50K)1\underline{/-90°}$ mA

$\vec{V} = 14.3\underline{/-90°}$ V

Using KVL,

$$14.3\underline{/-90°} = \vec{I}(214.3 \times 10^3 - j100 \times 10^3)$$

$$\vec{I} = \frac{14.3\underline{/-90°}}{214.3 - j100}\ mA$$

$$\vec{E}_{OUT} = -I(200 \times 10^3)V_{rms}$$

$$\vec{E}_{OUT} = \frac{-2860\underline{/-90°}}{214.3 - j100}\ V_{rms}$$

$$\vec{E}_{OUT} = \frac{-2860\underline{/-90°}}{236.5\underline{/-25°}}$$

$$\vec{E}_{OUT} = -12.09\underline{/-65°}\ V_{rms}.$$

Hence, the gain is

$$\frac{\vec{E}_{OUT}}{\vec{E}_{OUT}} = \frac{-12.09\underline{/-65^\circ}}{1\underline{/-90^\circ}} = -12.09\underline{/25^\circ}$$

and

$$-12.09\underline{/25^\circ} = 12.09\underline{/-155^\circ}$$

• PROBLEM 9-62

Using superposition to find $\vec{I}$ in the circuit shown in Fig. 1, what is the partial (phasor) response produced by the: (a) left source? (b) right source? (c) upper source?

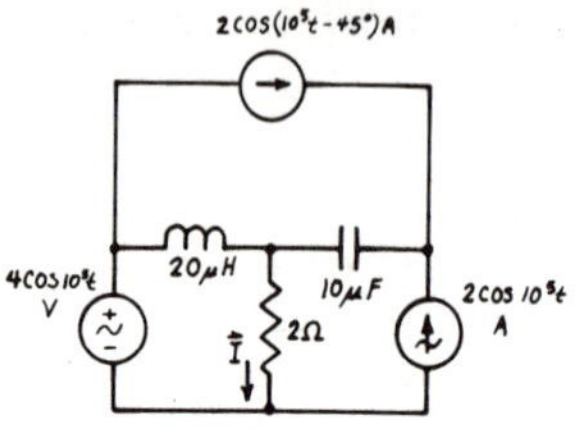

Fig. 1

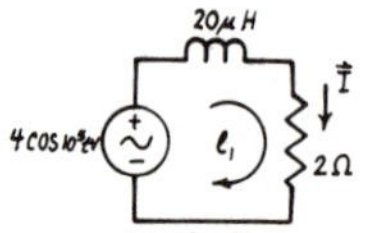

Fig. 2

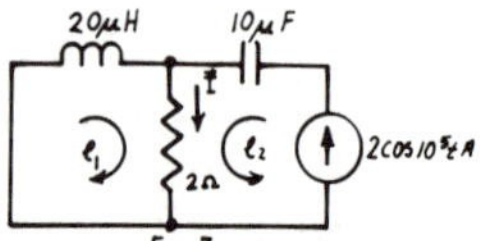

Fig. 3

Solution: (a) To find the partial phasor response produced by the left source, open-circuit the two current sources and obtain the circuit shown in Fig. 2. Since

$$V_m \cos(\omega t + \phi) = \text{Re}(V_m e^{j(\omega t+\phi)}),$$

the phasor $\vec{V}$ is written $V_m e^{j\phi}$, or $V_m\underline{/\phi^\circ}$ so that

$$4 \cos 10^5 t = 4\underline{/0^\circ}$$

and $$\omega = 10^5 \text{ rad/s}$$

Writing an equation around the loop in Fig. 2 yields

$$4 \cos 10^5 t - \vec{I}(j\omega 20 \times 10^{-6} + 2)$$

substituting $4\underline{/0^\circ}$ for $4 \cos 10^5 t$ and 10^5 for ω

$$4\underline{/0^\circ} = \vec{I}(j2 + 2)$$

$$\vec{I} = \frac{4\underline{/0^\circ}}{j2 + 2} = \frac{4\underline{/0^\circ}}{\sqrt{2^2 + 2^2}\underline{/\tan^{-1} \frac{2}{2}}} = \frac{4\underline{/0^\circ}}{\sqrt{8}\underline{/45^\circ}}$$

$$\vec{I} = \frac{4\underline{/0^\circ}}{2\sqrt{2}\underline{/45^\circ}} = \frac{2\underline{/0^\circ}}{\sqrt{2}\underline{/45^\circ}} = \sqrt{2}\underline{/-45^\circ} \text{ A}$$

or $\vec{I} = 1 - j1$ A.

(b) The circuit shown in Fig. 3 shows the partial

response due to the right source. Re-drawing the circuit in Fig. 3 to include the current in each branch and the impedance of each element, yields the results seen in Fig. 4.

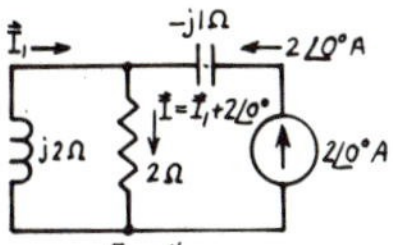

Fig. 4

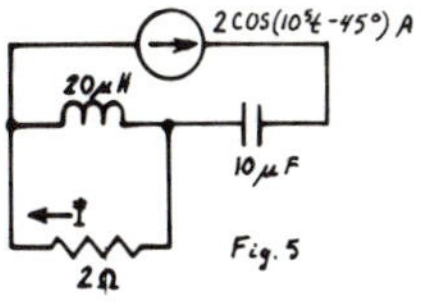

Fig. 5

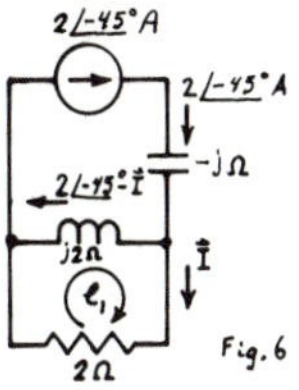

Fig. 6

Writing an equation for loop ℓ_1 gives

$$0 = \vec{I}_1(j2) + (\vec{I}_1 + 2\angle 0°)(2)$$

since $$\vec{I}_1 = \vec{I} - 2\angle 0°$$

$$0 = (\vec{I} - 2\angle 0°)(j2) + \vec{I}(2)$$

$$0 = \vec{I} - 2\angle 0°(2\angle 90°) + 2\vec{I}$$

$$4\angle 90° = \vec{I}(j2 + 2)$$

$$\vec{I} = \frac{4\angle 90°}{2 + j2} = \frac{4\angle 90°}{2\sqrt{2}\angle 45°} = \sqrt{2}\angle 45° \text{ A}$$

or $\vec{I} = 1 + j1$ A.

(c) The circuit in Fig. 5 shows the partial response due to the upper source.

Fig. 6 is the circuit in Fig. 5 redrawn to show the current in each branch,

An equation around the loop ℓ_1 in Fig. 6 yields

$$0 = \vec{I}(2) - (2\angle -45° - \vec{I})(j2)$$

$$0 = \vec{I}(2) + (2\angle -45° - \vec{I})(2\angle 90°)$$

$$0 = 2\vec{I} - 4\angle 45° + \vec{I}(j2)$$

$$+4\angle 45° = \vec{I}(2 + j2)$$

$$\vec{I} = \frac{+4\angle 45°}{2 - j2} = \frac{4\angle 45°}{\sqrt{8}\angle -45°} = \sqrt{2}\angle 0° \text{ A}$$

or $\vec{I} = 1.414$ A.

The total current is

$$\vec{I} = 1.414 + 1 - j1 + 1 + j1 = 3.414 \text{ A.}$$

• PROBLEM 9-63

Find $v_x(t)$ in the circuit shown in Fig. 1.

Solution: Since the unknown quantity is a voltage, the nodal method is favored, because we can assign the middle node (at the bottom end of the capacitor) as the common node and then solve for v_X directly. Both the mesh method and the nodal method require the solution of simultaneous equations.

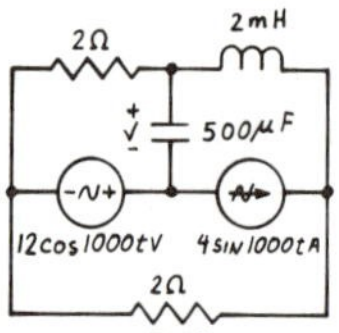

Fig. 1

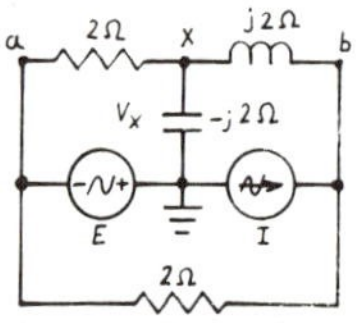

Fig. 2

The first step is to convert the inductance and capacitance to complex impedances:

$$\vec{Z}_C = \frac{1}{j\omega C} = \frac{1}{j(1000)500 \times 10^{-6}} = -j2\ \Omega, \tag{1}$$

$$\vec{Z}_L = j\omega L = j(1000)2 \times 10^{-3} = j2\ \Omega. \tag{2}$$

The second step is to assign the letters E and I to the two sources for algebraic convenience, that is

$$\vec{E} = 12\underline{/0^\circ}\ \text{V} \quad \text{and} \quad \vec{I} = 4\underline{/-90^\circ}\ \text{A}, \tag{3}$$

where the current has been first converted to a cosine:

$$i(t) = 4 \sin 1000t = 4 \cos(1000t - 90^\circ). \tag{4}$$

In addition, we shall assign symbols to each node, as shown in Fig. 2, a, X and b, with the ground symbol indicating the chosen common node. Now we are ready to write a current equation at each node:

Currents out of the node - Currents into the node = 0, (5)

where the current in each passive circuit element is designated as the difference between the potentials at each end of the impedance divided by the value of the impedance. Proceeding on this basis, the equations for each node are:

<u>Node X</u>: $$(\vec{V}_X - \vec{V}_a)(\tfrac{1}{2}) + \vec{V}_X(-\tfrac{1}{j2}) + (\vec{V}_X - \vec{V}_b)(\tfrac{1}{j2}) = 0 \tag{6}$$

<u>Node b</u>: $$(\vec{V}_b - \vec{V}_X)(\tfrac{1}{j2}) - \vec{I} + (\vec{V}_b - \vec{V}_a)(\tfrac{1}{2}) = 0 \tag{7}$$

<u>Node a</u>: $$\vec{V}_a = -\vec{E}. \tag{8}$$

The equation for Node a is not a current equation; it identifies the potential at that node in terms of one of the known quantities. Now we have three equations with three unknowns, $\vec{V}_a$, $\vec{V}_b$ and $\vec{V}_X$, and the next step is to solve them for V_X. But first these equations should be written again in a more consolidated form:

$$\vec{V}_X(\tfrac{1}{2} - \tfrac{1}{j2} + \tfrac{1}{j2}) - \vec{V}_a(\tfrac{1}{2}) - \vec{V}_b(\tfrac{1}{j2}) = 0 \tag{9}$$

$$V_X\left(-\frac{1}{j2}\right) - \vec{V}_a\left(\frac{1}{2}\right) + \vec{V}_b\left(\frac{1}{j2} + \frac{1}{2}\right) - \vec{I} = 0. \qquad (10)$$

Now, simplifying by multiplying each equation by j2 and substituting $(-\vec{E})$ for $\vec{V}_a$,

$$j\vec{V}_X + jE - \vec{V}_b = 0 \qquad (11)$$

$$-\vec{V}_X + j\vec{E} + (1 + j1)\vec{V}_b - j2\vec{I} = 0 \qquad (12)$$

Since we are only interested in $\vec{V}_X$, eliminate $\vec{V}_b$ by first multiplying equation (11) by $(1 + j1)$,

$$(1 + j1)j\vec{V}_X + (1 + j1)j\vec{E} - (1 + j1)\vec{V}_b = 0, \qquad (13)$$

and then summing equations (12) and (13):

$$-(2 - j1)\vec{V}_X - (1 - j2)\vec{E} - j2\vec{I} = 0. \qquad (14)$$

Now solving equation (14) for $\vec{V}_X$

$$\vec{V}_X = \frac{-j2I}{2-j1} - \frac{(1 - j2)E}{(2 - j1)} \qquad (15)$$

and further simplifying plus substituting for E and I from equation (3):

$$\vec{V}_X = \frac{1\underline{/-90^\circ}}{\sqrt{5}\underline{/-26.6^\circ}}(2)(4\underline{/-90^\circ}) - \frac{\sqrt{5}\underline{/-63.4^\circ}}{\sqrt{5}\underline{/-26.6^\circ}}(12\underline{/0^\circ}). \qquad (16)$$

Simplifying:

$$\vec{V}_X = 3.58\underline{/-153.4} - 12\underline{/-36.8} \qquad (17)$$

$$\vec{V}_X = -3.2 - j1.6 - (9.6 - j7.2) \qquad (18)$$

$$\vec{V}_X = -12.8 + j5.6 = 13.97\underline{/156.4^\circ}\ \text{V}. \qquad (19)$$

Finally we convert the polar form to the trigonometric form:

$$v_X(t) = 13.97 \cos(1000t + 156.4^\circ)\text{V} \qquad (20)$$

POWER

• PROBLEM 9-64

> Given the circuit in Fig. 1, find $\vec{I}$, $\vec{I}_1$, $\vec{I}_2$, and the total power consumed.

Solution: First find I_1 and I_2 by use of Ohm's law (phasor version).

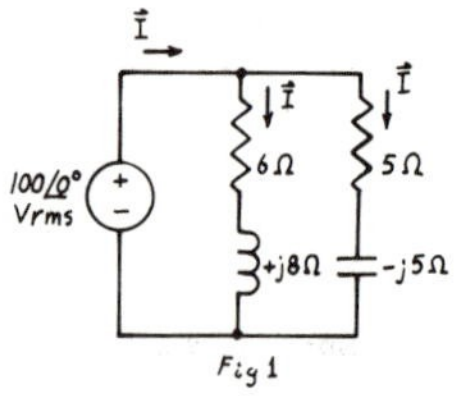

Fig 1

$$\vec{I}_1 = \frac{100\angle 0^\circ}{6 + j8}\left(\frac{6 - j8}{6 - j8}\right) = \frac{600 - j800}{36 + 64} = 6 - j8 \text{ A rms},$$

or, in magnitude-phase form, $I_1 = 10\underline{/-53.1^\circ}$ A rms

$$\vec{I}_2 = \frac{100\angle 0^\circ}{5 - j5}\left(\frac{5 + j5}{5 + j5}\right) = \frac{500 + j500}{25 + 25} = 10 + j10 \text{ A rms},$$

or, in magnitude-phase form, $\vec{I}_2 = 14.14\underline{/+45^\circ}$ A rms.
$\vec{I}$ is found by application of Kirchoff's Current Law.

$$\vec{I} = \vec{I}_1 + \vec{I}_2 = (6 - j8) + (10 + j10) = 16 + j2$$

$$= 16.125\underline{/7.125^\circ} \text{ A rms.}$$

The total power consumed will be that dissipated in the resisters, so

$$P = |I_1|^2 6 + |I_2|^2 5$$

$$= (10)^2 6 + (14.14)^2 5 = 600 + 1000 = 1600 \text{ w.}$$

Check on power: An alternate expression for the power delivered to the circuit is

$$P = |V||I|\cos(\theta),$$

where θ is the angle between the current and voltage phasors, so

$$P = (100)(16.125)\cos(7.125^\circ) = 1600\text{w.}$$

• PROBLEM 9-65

In the network of Fig. 1, find the instantaneous power, the average power and the reactive power at the terminals ab. What is the energy delivered to the network over a time interval of 5s.

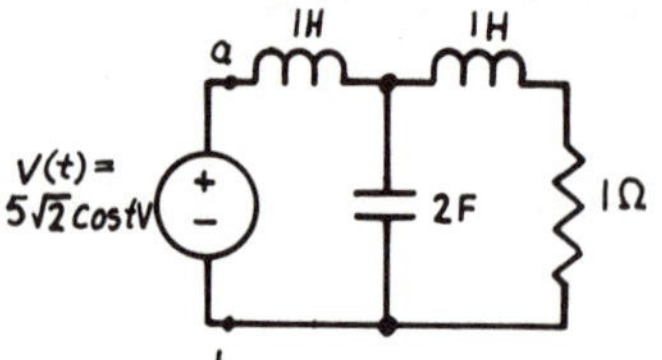

Fig. 1

Solution: Find the impedance of the series-parallel network between a and b. Then use Ohm's law to determine the current in the source. Instantaneous power is the product of this current and the source voltage; from this expression we can find the specific powers asked for.

Convert the inductances and capacitance to reactances, where ω is obtained from the expression for the source voltage, $V_m \cos \omega t$.

Inductances:

$$j\omega L = j1; \tag{1}$$

Capacitance:

$$-j\frac{1}{\omega C} = -j\frac{1}{2} \,. \tag{2}$$

The total impedance may now be determined:

$$\vec{Z} = j1 + \frac{(1 + j1)(-j\frac{1}{2})}{1 + j1 - j\frac{1}{2}} = j1 + \frac{\frac{1}{2} - j\frac{1}{2}}{1 + j\frac{1}{2}}, \tag{3}$$

$$\vec{Z} = 0.20 + j0.40 = 0.447\underline{/63.4^\circ} \quad . \tag{4}$$

With this we can now solve for phasor current.

$$\vec{I} = \frac{\vec{V}}{\vec{Z}} = \frac{5\underline{/0^\circ}}{0.447\underline{/63.4^\circ}} = 11.2\underline{/-63.4^\circ}\ A_{rms}. \tag{5}$$

Converting to instantaneous current,

$$i(t) = 11.2\sqrt{2}\cos(t - 63.4^\circ)A; \tag{6}$$

calculating instantaneous power,

$$p(t) = vi = (5\sqrt{2}\cos t)(11.2\sqrt{2}\cos(t - 63.4^\circ))W \tag{7}$$

$$p(t) = 112 \cos t \cos(t - 63.4^\circ)W. \tag{8}$$

Average power may be found from

$$P = |V||I|\cos\theta \tag{9}$$

where $|V|$ and $|I|$ are rms values and θ is the angle of the impedance.

$$P = 5 \times 11.2 \times \cos 63.4^\circ = 25W. \tag{10}$$

Reactive power may be found from

$$P_X = VI \sin\theta = 5 \times 11.2 \times \sin 63.4^\circ = 50 \text{ vars}. \tag{11}$$

Since power is the rate at which energy is being consumed, energy is the integral of power over a period of time:

$$\text{Energy} = \int_0^t p\,dt \quad \text{joules.} \tag{12}$$

Before proceeding with the actual integration, divide the instantaneous power into two terms by using the trigonometric identity

$$\cos A \cos B = \frac{1}{2}\cos(A + B) + \frac{1}{2}\cos(A - B), \tag{13}$$

$$p(t) = \frac{112}{2}\cos(2t - 63.4°) + \frac{112}{2}\cos 63.4° \text{W} \tag{14}$$

where A = t and B = (t - 63.4°) in equation (8). Now we can integrate in two parts over the first 5 seconds:

$$\text{Energy} = \int_0^5 56\cos(2t - 63.4°)dt + \int_0^5 56\cos 63.4°\,dt \text{ joules} \tag{15}$$

Carrying out the integration results in

$$\text{Energy} = 56(\tfrac{1}{2})[\sin(2t - 63.4°)]_0^5 + 56\cos 63.4°[t]_0^5 \tag{16}$$

$$\text{Energy} = 28[\sin(10 - 1.11) - \sin(0 - 63.4°)] + 25(5 - 0) \tag{17}$$

where 63.4° has been changed from degrees to radians by using the conversion

$$\text{radians} = \text{degrees} \times \pi/180. \tag{18}$$

Carrying out the numerical evaluation,

$$\text{Energy} = 28[0.51 - (-.89)] + 125 \tag{19}$$

$$\text{Energy} = 164 \text{ joules.} \tag{20}$$

• PROBLEM 9-66

For the circuit shown in Fig. 1:

(a) Draw a phasor diagram indicating all voltages and currents.

(b) Using the diagram, calculate the average and reactive powers.

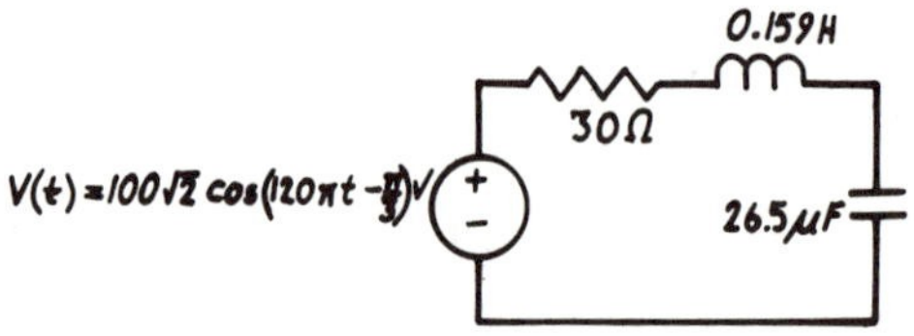

Fig. 1

Solution: Note that the source voltage is a sinusoid of frequency 60 Hz, ω = 2π(60) = 377 rad/s, and can be expressed as a phasor $\vec{V} = 100\angle{-60°}$ Vrms. The impedance of each passive element in the circuit can be represented as a phasor. Hence,

$$\vec{Z}_R = 30\underline{/0^\circ}\ \Omega$$

$$\vec{Z}_L = \omega L\underline{/90^\circ} = 60\underline{/90^\circ}\ \Omega$$

$$\vec{Z}_C = \frac{1}{C}\underline{/-90^\circ} = 100\underline{/-90^\circ}\ \Omega$$

The current in the single loop circuit is

$$\vec{I} = \frac{\vec{V}}{\vec{Z}_R + \vec{Z}_L + \vec{Z}_C} = \frac{100\underline{/-60^\circ}}{30\underline{/0^\circ} + 60\underline{/90^\circ} + 100\underline{/-90^\circ}}$$

$$\vec{I} = \frac{100\underline{/-60^\circ}}{30 - j40} = \frac{100\underline{/-60^\circ}}{50\underline{/-53^\circ}} = 2\underline{/-7^\circ}\ \text{Arms.}$$

We can now find the voltage across each element

$$\vec{V}_R = \vec{I}\,\vec{Z}_R = 60\underline{/-7^\circ}\ \text{Vrms}$$

$$\vec{V}_L = \vec{I}\,\vec{Z}_L = 120\underline{/83^\circ}\ \text{Vrms}$$

$$\vec{V}_C = \vec{I}\,\vec{Z}_C = 200\underline{/-97^\circ}\ \text{Vrms.}$$

(a) The phasor voltages and current is shown in a diagram in Fig. 2.

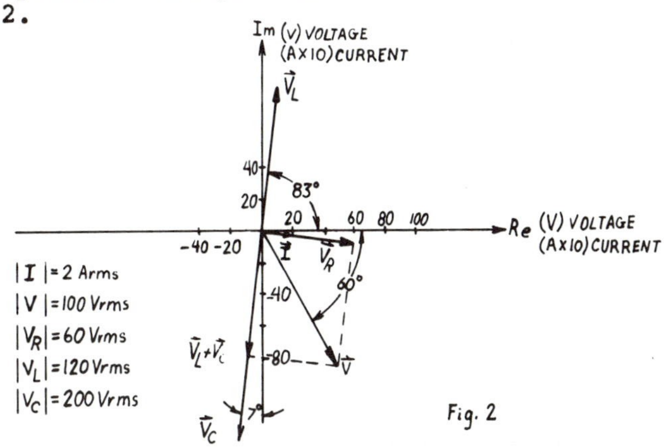

Fig. 2

(b) To find the average power P and the reactive power Q, we can use the formula,

$$P = \frac{1}{2}|\vec{V}|\,|\vec{I}|\ \cos\phi$$

and

$$Q = \frac{1}{2}|\vec{V}|\,|\vec{I}|\ \sin\phi$$

where

$$\phi = -60^\circ - (-7^\circ) = -53,$$

but it is much simpler to find the average and reactive powers graphically. The quantity $|\vec{V}|\cos\phi$ is the projection of the phasor $\vec{V}$ onto $\vec{I}$ (or V_R as shown in Fig. 2) and since $|\vec{I}| = 2$ Arms then $P = |\vec{V}| \cos\phi = 60$ watts (from Fig. 2). Similarly, $|\vec{V}| \sin\phi$ is the projection of the phasor $\vec{V}$ onto $\vec{V}_L + \vec{V}_C$. Hence, $Q = -80$ vars.

• PROBLEM 9-67

In the circuit shown $\vec{I}_1 = 2e^{-j\pi/3}$A, $\vec{I}_2 = 5e^{j\pi/3}$, $\vec{I}_3 = 1$A.

(a) Draw a phasor diagram indicating all voltages and currents in the circuit.

(b) Is the circuit inductive, capacitive or resonant?

(c) Find the average and reactive power in the circuit.

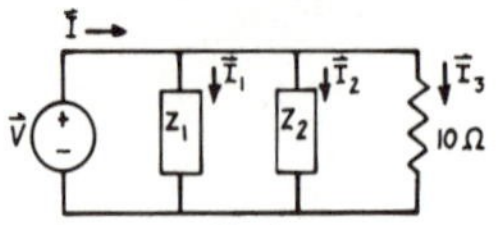

Fig. 1

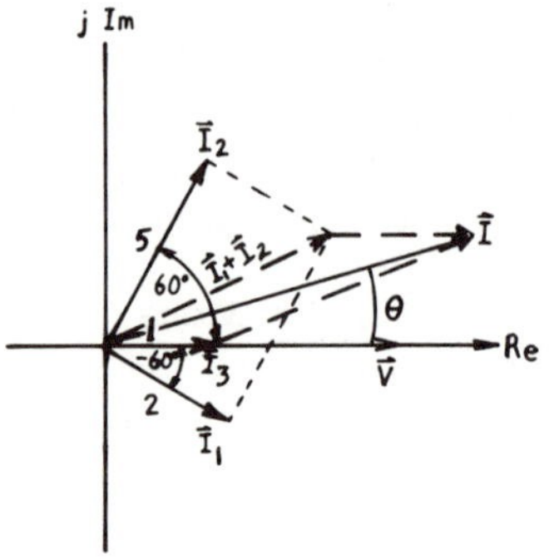

Solution: (a) This is a parallel circuit, so each element has the same voltage as any other. Thus,

$$\vec{V} = 10I_3 = 10 = 10\angle 0^\circ \text{ V},$$

$$\vec{I} = \vec{I}_1 + \vec{I}_2 + \vec{I}_3 = 2\left(\frac{1}{2} - \frac{j\sqrt{3}}{2}\right) + 5\left(\frac{1}{2} + \frac{j\sqrt{3}}{2}\right) + 1 = 5.2\angle 30^\circ \text{ A}.$$

Hence, $\theta = 30°$.

(b) It is clear that the circuit is capacitive since the voltage lags the current.

(c) The power is given by $P = |\vec{V}|\cdot|\vec{I}| \cos(-\theta) = 10 \times 5.2 \cos 30°$ so $P = 45$ W.

The reactive power is given by $Q = |\vec{V}|\cdot|\vec{I}| \sin(-\theta) = -26$ vars.

• PROBLEM 9-68

A variable capacitor is connected in series with an inductor and resistor as shown.

(a) The power is observed to be 360 W. Find the value of C and the power factor.

(b) The capacitor is readjusted and the power is observed to be 200 W. Find the new value of C and the new power factor.

Solution: The power absorbed by the circuit is defined as the apparent power times the power factor,

$$P = \overbrace{|V||I|}^{\text{apparent power}} \underbrace{\cos\theta}_{\text{power factor}}$$

where $\theta = \theta_v - \theta_i$ and $|\vec{V}|$ and $|\vec{I}|$ are rms values.

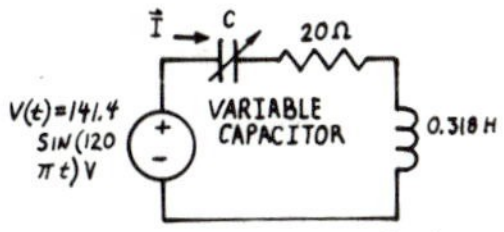

Fig. 1

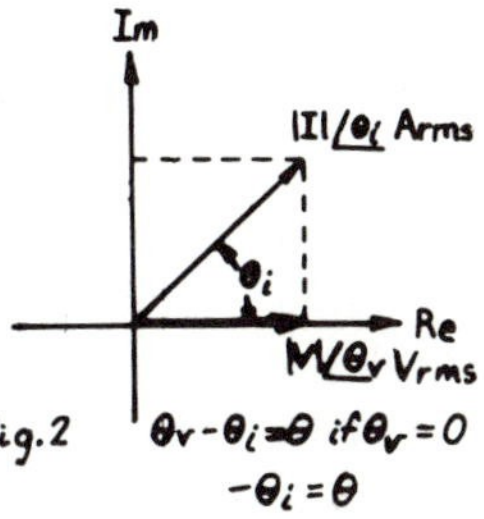

Fig. 2

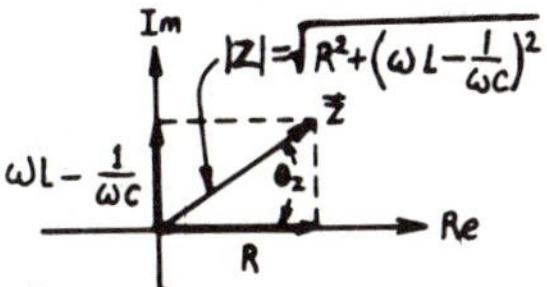

Fig. 3

(a) We are given the power P = 360W, the voltage $\vec{V} = \frac{141.4}{\sqrt{2}}\angle 0°$ Vrms at $\omega = 120\pi$ rad/s and the elements, $R = 20\Omega$, $L = 0.318H$. From this information we are asked to find the capacitance, C and the power factor, $\cos\theta$.

Since,

$$\vec{I} = I\angle\theta_i = \frac{V\angle\theta_v}{Z\angle\theta_z}$$

and

$$Z\angle\theta_t = \frac{1}{j\omega C} + R + j\omega L = R + j\left(\omega L - \frac{1}{\omega C}\right)$$

$$Z\angle\theta_z = \sqrt{R^2 + \left(\omega L - \frac{1}{C}\right)^2}\angle\tan^{-1}\frac{\omega L - \frac{1}{\omega C}}{R} \tag{2}$$

then,

$$\vec{I} = I\angle\theta_i = \frac{|V|}{\sqrt{R^2 + \left(\omega L - \frac{1}{\omega C}\right)^2}}\angle -\tan\frac{\omega L - \frac{1}{\omega C}}{R} \tag{3}$$

Figs. 2 and 3 show the phasors and angles used in computation.

Hence,

$$|I| = \frac{|V|}{\sqrt{R^2 + \left(\omega L - \frac{1}{\omega C}\right)^2}}$$

$$\theta = -\theta_i = \theta_z = \tan^{-1}\frac{\omega L - \frac{1}{\omega C}}{R}$$

and

$$P = \frac{|V|^2}{\sqrt{R^2 + \left(\omega L - \frac{1}{\omega C}\right)^2}}\cos\underbrace{\left[\tan^{-1}\frac{\omega L - \frac{1}{\omega C}}{R}\right]}_{\theta} \tag{4}$$

Since $\theta_Z = \theta$ we can find $\cos\theta$ from the triangle formed in Fig. 3, hence

$$\cos\theta = \frac{R}{\sqrt{R^2 + (\omega L - \frac{1}{\omega C})^2}} \,. \qquad (5)$$

Substituting into the equation for power (Eq. (4)) yields

$$P = \frac{|V|^2 R}{R^2 + (\omega L - \frac{1}{\omega C})^2} \,. \qquad (6)$$

Solving for the capacitance

$$C = \frac{1}{\omega\left[\omega L - \sqrt{\frac{|V|^2 R}{P} - R^2}\right]} \qquad (7)$$

$C = 24.7 \times 10^{-6}$ F for $P = 360$ W.

An equation for the power factor can be found directly from Eq. (6), hence

since, $$\cos\theta = \frac{1}{\sqrt{R^2 + (\omega L - \frac{1}{\omega C})^2}}$$

and $$P = \frac{|V|^2 R}{R^2 + (\omega L - \frac{1}{\omega C})^2}$$

then $$\cos\theta = \frac{\sqrt{PR}}{|V|} = \frac{\sqrt{360.20}}{100} = 0.849$$

$$\theta = 31.9°.$$

(b) For $P = 200$W

$$\cos\theta = \frac{\sqrt{200 \times 20}}{100} = 0.632$$

$$\theta = 50.8°$$

and from Eq. (7)

$$C = 27.8 \times 10^{-6} \text{ F}.$$

• PROBLEM 9-69

Find R_1 and $|I_S|$ in the circuit shown in Fig. 1 if the average power delivered to each resistor is 1.5W.

Solution: The solution is obtained by using the power dissipation requirement to determine the current through the inductor branch and then the voltage drop across the branch. This voltage and the power dissipation specifies the resistance of R_1 and its current. Finally the currents through the two branches are summed to find I_S.

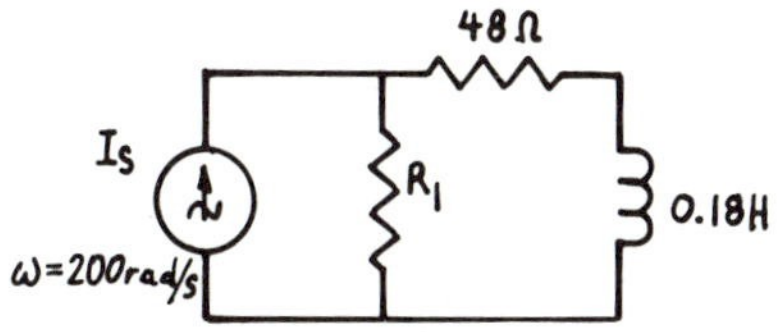

Fig. 1

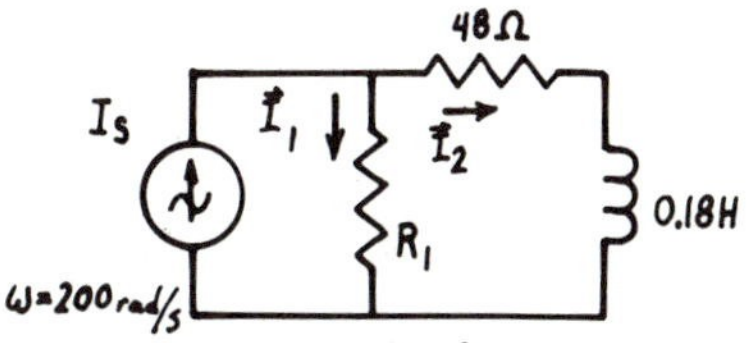

Fig 2

Fig. 1 is redrawn in Fig. 2 with labeled currents. The inductor is lossless so the power dissipated in branch 2 is given by

$$|I_{2rms}|^2 48 = 1.5$$

or $$|I_{2rms}| = \frac{1}{4\sqrt{2}} = 0.1768 \text{ A}$$

where $|I_{2rms}|$ designates the rms value of the current in branch 2. The voltage across branch 2 is

$$\vec{V}_{rms} = 48\vec{I}_{2rms} + j\omega(0.18)\vec{I}_{2rms}$$

or $$\vec{V}_{rms} = [48 + j200(0.18)]\vec{I}_{2rms}.$$

Substituting for $\vec{I}_{2rms}$ and solving gives

$$\vec{V}_{rms} = (48 + j36)\frac{1}{4\sqrt{2}} = \frac{1}{\sqrt{2}}(12 + j9)\text{V}.$$

Next the power equation for branch 1 produces

$$\frac{|V_{rms}|^2}{R_1} = 1.5.$$

Solve for R_1 and substitute for $\vec{V}_{rms}$. Note that $|V_{rms}|^2 = \vec{V}_{rms}\vec{V}^*_{rms}$, where the asterisk denotes complex conjugate.

$$R_1 = \frac{|V_{rms}|^2}{1.5}$$

$$R_1 = \frac{1}{\sqrt{2}}(12 + j9)\frac{1}{\sqrt{2}}(12 - j9)\frac{1}{1.5}$$

$$R_1 = 75\Omega.$$

The current through branch 1 is

$$\vec{I}_{1rms} = \frac{\vec{V}_{rms}}{R_1} = \frac{12 + j9}{75\sqrt{2}} = 0.1414\underline{/36.87°}\ A.$$

The current, I_S, is the sum of the currents in branches 1 and 2. Hence,

$$\vec{I}_{S_{rms}} = \vec{I}_{1rms} + \vec{I}_{2rms}$$

$$= 0.1414\underline{/36.87°} + 0.1768$$

$$= 0.3021\underline{/16.31°}\ A.$$

To get the peak value multiply by $\sqrt{2}$. Then

$$\vec{I}_S = 0.427\underline{/16.31°}\ A$$

and $$|I_S| = 0.427\ A.$$

• PROBLEM 9-70

In Fig. 1, find the average power delivered to the (a) 10-Ω load (b) 100-Ω load (c) 20 + j10-Ω load.

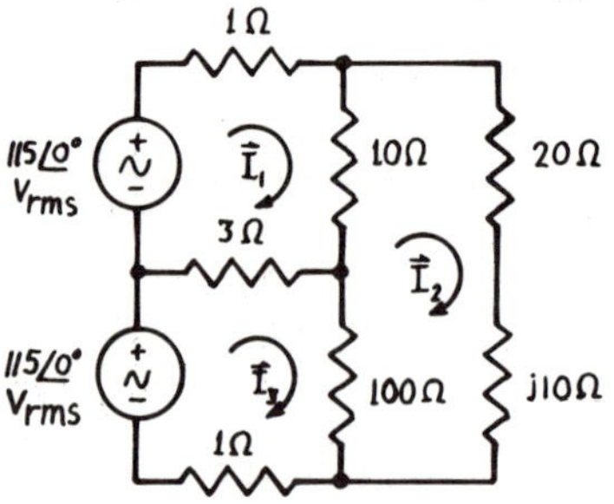

Fig. 1

Solution: Voltage equations are written around the three meshes, designated by $\vec{I}_1$, $\vec{I}_2$ and $\vec{I}_3$.

$$115 = \vec{I}_1 + 10(\vec{I}_1 - \vec{I}_2) + 3(\vec{I}_1 - \vec{I}_3)$$

$$0 = 10(\vec{I}_2 - \vec{I}_1) + (20 + j10)\vec{I}_2 + 100(\vec{I}_2 - \vec{I}_3)$$

$$115 = 3(\vec{I}_3 - \vec{I}_1) + 100(\vec{I}_3 - \vec{I}_2) + \vec{I}_3.$$

The terms are collected in these equations to give

$$14I_1 - 10I_2 - 3I_3 = 115$$

$$-10I_1 + (130 + j10)I_2 - 100I_3 = 0$$

$$-3I_1 - 100I_2 + 104I_3 = 115.$$

The characteristic determinant Δ is that, whose elements are the coefficients of the current phasors.

$$\Delta = \begin{vmatrix} 14 & -10 & -3 \\ -10 & 130 + j10 & -100 \\ -3 & -100 & 104 \end{vmatrix} = 31{,}710 + j14470 = 34{,}855\underline{/24.528°}$$

The three currents are then

$$\mathbf{I}_1 = \frac{\begin{vmatrix} 115 & -10 & -3 \\ 0 & 130 + j10 & -100 \\ 115 & -100 & 104 \end{vmatrix}}{\Delta} = 16.2 - j3.51 \text{ A.}$$

$$\mathbf{I}_2 = \frac{\begin{vmatrix} 14 & 115 & -3 \\ -10 & 0 & -100 \\ 3 & 115 & 104 \end{vmatrix}}{\Delta} = 9.139\underline{/-24.528°} = 8{,}314 - j3.794 \text{ A}$$

$$I_3 = \frac{\begin{vmatrix} 14 & -10 & 115 \\ -10 & 130 + j10 & 0 \\ 3 & -100 & 115 \end{vmatrix}}{\Delta} = 9.567 - j3.749 \text{ A.}$$

The magnitude of the currents through the 10Ω, 100Ω and 20 + j10Ω loads are

$$|I_{10\Omega}| = |I_1 - I_2| = |7.886 + j0.284| = 7.891 \text{ A}$$

$$|I_{100\Omega}| = |I_3 - I_2| = |1.253 + j0.045| = 1.254 \text{ A}$$

$$|I_{20+j10\Omega}| = |I_2| = 9.139 \text{ A}$$

respectively. The required powers are then

$$P_{10\Omega} = (7.891)^2(10) = 623\text{W}$$

$$P_{100\Omega} = (1.254)^2(100) = 157\text{W}$$

$$P_{20+j10\Omega} = (9.139)^2(20) = 1670\text{W}$$

• PROBLEM 9-71

Find the value of an impedance which: (a) absorbs a complex power of $4600\underline{/30°}$ VA at 230 V rms; (b) absorbs a complex power of $5000\underline{/45°}$ VA when the rms current through it is 12.5 A; (c) requires -1500 vars at 230 V rms and 10 A rms.

Solution: The solution of these problems rely on the knowledge that complex power is directly related to complex impedance. That is,

$$|Z|\underline{/\theta} = R + jX\ \Omega, \tag{1}$$

but multiplying this equation by the square of the rms current converts it to complex power,

$$|I|^2|Z|\underline{/\theta} = |I|^2R + j|I|^2X\ \text{VA}. \tag{2}$$

The corresponding units in this equation are:

$$\text{VA} = \text{watts} + j\ \text{vars}, \tag{3}$$

where "vars" are "volt-amperes reactive" because this component of the complex power is associated with the reactive component of the impedance. In addition, the complex power can be expressed as

$$|V||I|\underline{/\theta} = |V||I|\cos\theta + j|V||I|\sin\theta \tag{4}$$

since $\quad |I|^2|Z| = |I|(|I||Z|) = |V||I|.$ (5)

(a) In this problem we are given the complex power in the polar form, $4600\underline{/30°}$, as it is in the left side of Equation (4). We are also given the rms voltage, so we see a way to find the rms current,

$$|I| = \frac{|V||I|}{|V|} = \frac{4600}{230} = 20\ \text{amperes}. \tag{6}$$

With this additional information it is now possible to solve for the magnitude of the impedance, by Ohm's law,

$$|Z|\frac{|V|}{|I|} = \frac{230}{20} = 11.5\ \Omega. \tag{7}$$

Since Equation (1) was merely multiplied by a constant magnitude $|I|^2$ to obtain Equation (2), it follows that the angle of the complex power is identical to the angle on the impedance, hence,

$$\vec{Z} = |Z|\underline{/\theta} = 11.5\underline{/30°} = 9.95 + j5.75\ \Omega. \tag{8}$$

(b) The solution to this problem is similar to that of part (a) except that the current is given and the voltage is to be found, so

$$|V| = \frac{|V||I|}{|I|} = \frac{5000}{12.5} = 400\ \text{V} \tag{9}$$

and therefore,

$$|Z| = \frac{|V|}{|I|} = \frac{400}{12.5} = 32.0\ \Omega. \tag{10}$$

Now, supplying the angle from the complex power,

$$\vec{Z} = |Z|\underline{/\theta} = 32.0\underline{/45°} = 22.6 + j22.6\ \Omega. \tag{11}$$

(c) The approach to this problem differs because both |V| and |I| are known but the angle is not given. If we search our stock of equations, number (4) appears to be helpful because it includes |V| and |I| and the angle. We can see by referring to Equation (3) that the vars component is known, and therefore, by inspection,

$$|V||I| \sin\theta = -1500 \text{ vars} \tag{12}$$

from which we can solve for θ, since

$$\sin\theta = \frac{-1500}{230 \times 10} = -0.652 \tag{13}$$

which leads us to

$$\theta = \sin^{-1}(-0.652) = -40.7°. \tag{14}$$

Now this inofrmation can be used to find the complex impedance, since

$$\vec{Z} = |Z|\underline{/\theta} = \frac{|V|}{|I|}\underline{/\theta} = \frac{230}{10}\underline{/-40.7°} = 23\underline{/-40.7°}\ \Omega \tag{15}$$

which can be converted to the rectangular form,

$$\vec{Z} = 23\underline{/-40.7°} = 17.4 - j15.0\ \Omega. \tag{16}$$

• PROBLEM 9-72

An ordinary household outlet of 110 V rms at 60 Hz is connected to a series circuit consisting of an 8 Ω resistance, 0.0531-H inductance, and 189.7-μF capacitance. Calculate the current, effective power, complex power, power factor, vars, and apparent power. Find the voltage drop across each element.

Solution: To find the current through the circuit we must find the total impedance and divide the voltage by this impedance. Finding the impedance of each element:

$$\vec{Z}_R = 8\ \Omega$$

$$\vec{Z}_L = j\omega L = j2\pi 60(.0531) = j20\ \Omega$$

$$\vec{Z}_C = \frac{1}{j\omega C} = \frac{-j}{2\pi 60\left(189.7 \times 10^{-6}\right)} = -j14\ \Omega.$$

Summing each of the impedances above gives the total impedance for the circuit.

$$\vec{Z}_T = \vec{Z}_R + \vec{Z}_L + \vec{Z}_C = 8 + j20 - j14$$

$$\vec{Z}_T = 8 + j6\ \Omega.$$

Expressing the impdeance in polar form

$$\vec{Z}_T = 8 + j6 = \sqrt{8^2 + 6^2}\angle\tan^{-1}\frac{6}{8} = 10\angle 36.87^\circ\ \Omega.$$

The current is found by dividing the voltage across the elements by the total impedance.

$$\vec{I} = \frac{\vec{V}}{\vec{Z}_T} = \frac{110\angle 0^\circ}{10\angle 36.87^\circ} = 11\angle -36.87^\circ \text{ Arms.}$$

Notice that the magnitude of the current here is the effective value, I_{eff}, since the voltage is given as the rms or effective value. We can find the effective power by use of $P_{eff} = I_{eff}V_{eff}\cos\theta_{v-i}$ where $\theta_{v-i} = 0^\circ - (-36.87^\circ)$ $P_{eff} = 11(110)\cos(36.87^\circ) = 968$ W. The complex power is found by applying the definition

$$\vec{P} = \vec{V}_{eff}\vec{I}^*_{eff}.$$

We must first find $\vec{I}^*_{eff}$ ($\vec{I}_{eff}$ conjugate).

$$\vec{I}_{eff} = 11\angle -36.87^\circ = 8.8 - j6.6 \text{ A}$$

$$\vec{I}^*_{eff} = 8.8 + j6.6 \text{ A}$$

so that

$$\vec{P} = 110(8.8 + j6.6) = 968 + j726 \text{ W}$$

in polar form $\vec{P} = 1210\angle 36.86$ W. The power factor is defined as the cosine of the phase difference between voltage and current.

$$\text{pf} = \cos(\theta_V - \theta_I).$$

Since in our circuit $\theta_V = 0^\circ$ and $\theta_I = -36.86^\circ$

$$\text{pf} = \cos(36.86^\circ) = 0.8.$$

The unit vars refers to reactive power; this is defined as the imaginary part of the complex power.

Since $\vec{P} = 968 + j726$ W

reactive power = 726 vars.

The apparent power is defined as

$$P_{apparent} = |\vec{V}_{eff}||\vec{I}_{eff}| = (110)11 = 1210 \text{ VA.}$$

Notice that the real part of the complex power is the effective power and this term can be found by multiplying the sum of all the purely resistive elements times the magnitude of the effective current squared.

$$P_{eff} = |I_{eff}|^2 R_T = (11)^2 8 = 968 \text{ W.}$$

The reactive power (Q) can be found by multiplying the sum of the reactive elements by the magnitude of the effective current squared.

$$Q = |I_{eff}|^2 X_T = (11)^2 6 = 726 \text{ vars.}$$

And the apparent power is the vectorial sum of effective power and reactive power.

$$P_{apparent} = \sqrt{968^2 + 726^2} = 1210 \text{ VA.}$$

The voltage drop found across each element is,

$$\vec{V}_R = \vec{I}_{eff} R = 8(8.8 - j6.6) = 70.4 - j52.8 \text{ V}$$

$$\vec{V}_L = \vec{I}_{eff} \vec{Z}_L = j20(8.8 - j6.6) = 132 + j176 \text{ V}$$

$$\vec{V}_C = \vec{I}_{eff} \vec{Z}_C = -j14(8.8 - j6.6) = -92.4 - j123.2 \text{ V}$$

$$\vec{V}_R = 70.4 - j52.8 = 88\underline{/-36.87^\circ} \text{ V}$$

$$\vec{V}_L = 132.0 + j176.0 = 220\underline{/+53.13^\circ} \text{ V}$$

$$\vec{V}_C = -92.4 - j123.2 = 154\underline{/233.13^\circ} \text{ V}$$

$$\text{sum} = \vec{V} = 110 + j10$$

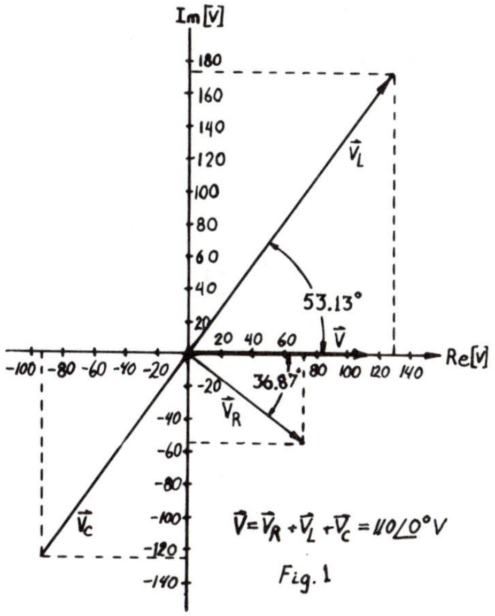

Fig. 1 illustrates the vectorial sum of the voltage drops. Notice that the voltage drops in polar form appear to sum to more than 110 V, but when vectorially added they sum to $110\underline{/0^\circ}$ V.

• PROBLEM 9-73

(a) Find the average and reactive power for the circuit shown in Fig. 1.
(b) Find the energy dissipated in the resistors over the interval $(0, 10\pi)$.

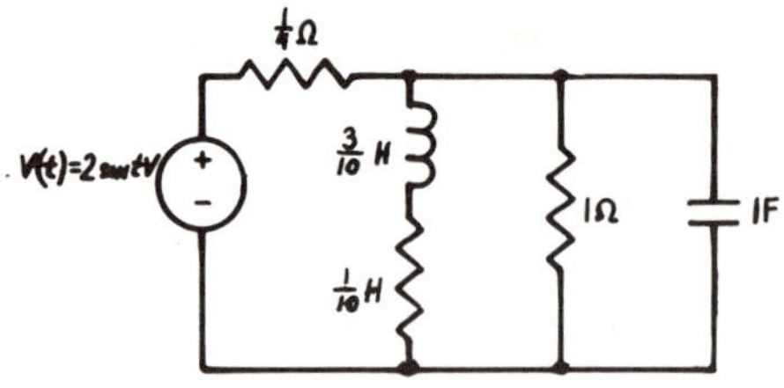

Fig. 1

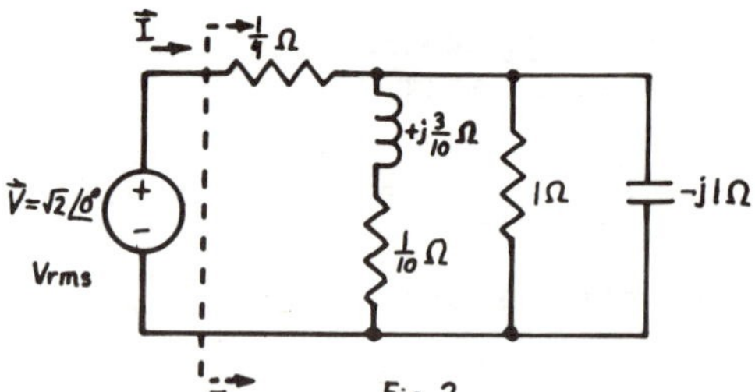

Fig. 2

Solution: Replace the circuit by a phasor equivalent, as shown in Fig. 2. Choose to use rms-sine phasors, so a time function $A\sin(\omega t + \theta)$ is represented as $(A/\sqrt{2})\underline{/\theta^\circ}$ in the phasor domain. The capacitive impedance of a 1F capacitor at the specified frequency of $\omega = 1$ is $1/(j\omega C) = -j1\ \Omega$, and the inductive impedance of a 3/10 H inductor is $j\omega L = +j3/10\ \Omega$.

The impedance Z seen by the voltage source is obtained by parallel and series combinations of the elemental impedances.

$$\vec{Z} = \frac{1}{4} + \frac{1}{\dfrac{1}{\frac{1}{10} + j\frac{3}{10}} + \dfrac{1}{1} + \dfrac{1}{-j1}} = \frac{1}{4} + \frac{1}{\dfrac{10}{1+j3} + 1 + j1}$$

$$= \frac{1}{4} + \frac{1+j3}{10 + (1+j1)(1+j3)} = \frac{1}{4} + \frac{1+j3}{8+j4} = \frac{1}{4}\left[1 + \frac{1+j3}{2+j1}\right]$$

$$= \frac{1}{4}\left[\frac{2 + j1 + 1 + j3}{2+j1}\right] = \frac{1}{4}\left[\frac{3+j4}{2+j1}\right].$$

Next solve for the current delivered by the source.

$$\vec{I} = \frac{\vec{V}}{\vec{Z}} = \sqrt{2}\ 4\left[\frac{2+j1}{3+j4}\right]\left(\frac{3-j4}{3-j4}\right) = 4\sqrt{2}\ \frac{(6+4) + j(3-8)}{9+16}$$

$$\vec{I} = 1.6\sqrt{2} - j0.8\sqrt{2}\ \text{A rms}.$$

(a) The average real power P is the product of V and the component of I in phase with V, so

$$P = |V| \times \text{Re}(I) = \sqrt{2}(1.6\sqrt{2}) = 3.2\ \text{W}.$$

The reactive power Q is the product of V and the magnitude of the component of I out of phase with V, so

$$Q = |V| \times \text{Im}(I) = \sqrt{2}(0.8\sqrt{2}) = 1.6\ \text{vars}.$$

(b) The period of the given voltage source function is 2π seconds, so the interval $(0,10\pi)$ consists of 5 full periods. The rate energy is dissipated in the resistors of the circuit varies over a cycle, but the average rate over a complete cycle is the power P calculated in (a) to be 3.2 watts. Since a watt is a joule/second,

$$\text{Energy dissipated over } (0,10\pi) = (3.2)(10\pi) = 32\pi$$

$$= 100.53 \text{ joules.}$$

• PROBLEM 9-74

A power company provides 660-VAC power to a certain customer who has connected three loads in parallel. The average power and power factor of each load is specified as follows:

Load 1, 15KW, 0.6 pf lagging

Load 2, 20KW, 0.707 pf leading

Load 3, 10KW, 0.4 pf lagging.

The loads are connected to a substation by means of cables having a total resistance of 1.1 ohms with negligible reactance. Calculate the voltage at the substation. Compare the power supplied by the substation to the power paid for by the customer.

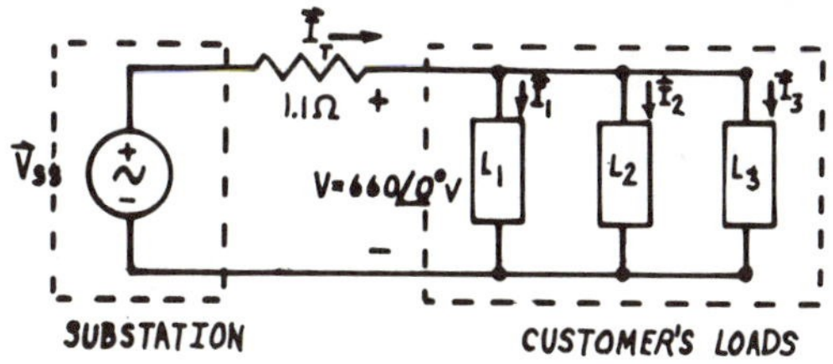

Fig. 1

Solution: A schematic diagram of the system is drawn in Fig. 1.
It is not necessary to calculate the impedances of each of the loads since the three loads can be combined into one equivalent load by noting that the complex power in each load is, $\vec{V}\vec{I}_1^*$, $\vec{V}\vec{I}_2^*$, and $\vec{V}\vec{I}_3^*$. But noting that

$$\vec{I}_T^* = \vec{I}_1^* + \vec{I}_2^* + \vec{I}_3^* = (\vec{I}_1 + \vec{I}_2 + \vec{I}_3)^*$$

we can find the total power dissipated in all three loads

$$\vec{P}_T = \vec{P}_1 + \vec{P}_2 + \vec{P}_3 = VI_T^* = V(I_1 + I_2 + I_3)^*.$$

Complex power is written

$$\vec{V}\vec{I}^* = P + jQ = \sqrt{P^2 + Q^2}\angle\tan^{-1}\frac{Q}{P}; \text{ where } \tan^{-1}\frac{Q}{P} = \theta.$$

Since we are given P for each load and pf = cos θ we can solve for Q, for each load and sum all the complex power terms to obtain a single value for complex power.

$$\text{Since } \tan^{-1}\frac{Q}{P} = \theta$$

$$\tan\theta = \frac{Q}{P} \qquad (1)$$

and $\quad pf = \cos\theta$ so that $\quad \theta = \cos^{-1}(pf)$

Substituting $\cos^{-1}(pf)$ for θ in equation (1) and solving for Q gives

$$Q = P\tan(\cos^{-1}[pf]).$$

For each load we have

$$Q_1 = 15\tan(\cos^{-1} 0.6) = 20K \text{ vars}$$

$$Q_2 = -20\tan(\cos^{-1} 0.707) = -20K \text{ vars}$$

$$Q_3 = 10\tan(\cos^{-1} 0.4) = 22.9K \text{ vars}.$$

We adopt the convention; when current lags voltage we have an inductive load, thus inductive loads produce lagging pf's and when current leads the voltage we have a capacitive load thus, capacitive loads produce leading pf's.

Since $\quad \vec{P} = \vec{V}\vec{I}^*$

for an inductor $\quad \vec{P} = V\underline{/90°}(I\underline{/0°})^* = P\underline{/90°}$

$\qquad = P + jQ;\ pf = 1$ (lagging)

for a capacitor $\quad \vec{P} = V\underline{/0°}(I\underline{/90°})^* = P\underline{/-90°}$

$\qquad \vec{P} = P - jQ;\ pf = 1$ (leading).

For each load

$$\vec{P}_1 = P_1 + jQ_1 = (15 + j20)\text{KW}$$

$$\vec{P}_2 = P_2 + jQ_2 = (20 - j20)\text{KW}$$

$$\vec{P}_3 = P_3 + jQ_3 = \underline{(10 + j22.9)\text{KW}}$$

$$\vec{P}_T = \vec{P}_1 + \vec{P}_2 + \vec{P}_3 = (45 + j22.9)\text{KW}$$

The customer pays for an apparent power of $\sqrt{45^2 + 22.9^2}$ = 50.5KVA. To find the current I_T;

$$\vec{I}_T = \left(\frac{\vec{P}_T}{\vec{V}}\right)^* = \left(\frac{\vec{V}\vec{I}_T^*}{\vec{V}}\right)^*$$

$$\vec{I}_T = \left(\frac{\vec{P}_T}{\vec{V}}\right)^* = \left[\frac{(45 + j22.9) \times 10^3}{660}\right]^*$$

$$\vec{I}_T = \left[\frac{50.5\angle 26.98° \text{ KW}}{660\angle 0° \text{ V}}\right]^* = [76.5\angle 26.98°]^*$$

$$\vec{I}_T = [68.2 + j34.7]^* = 68.2 - j34.7$$

$$\vec{I}_T = 76.5\angle -26.98° \text{ A.}$$

The voltage drop due to the cable resistance is

$$\vec{V}_C = R\vec{I}_T$$

$$\vec{V}_C = (1.1)(76.5\angle -26.98\ ) = 75 - j38.2 \text{ V.}$$

Now $\vec{V}_{SS} = \vec{V}_C + \vec{V} = 75 - j38.2 + 660$ V

$$\vec{V}_{SS} = 735 - j38.2 = 736\angle -3° \text{ V.}$$

The power supplied by the substation is

$$\vec{P}_{SS} = V_{SS}I_T^* = (736\angle -3°)(76.5\angle +26.98°)$$

$$\vec{P}_{SS} = 56.3\angle 23.98° \text{ KW.}$$

The customer's power, $\vec{P}_T = 50.5\angle 26.98°$ KW, is 89.7% of the total power dissipated.

• PROBLEM 9-75

For the circuit shown in Figure 1, find the value of C in terms of the other circuit elements which will yield a unity power factor.

Solution: Recalling that the power factor is given by pf = cos θ, where $\theta = \tan^{-1}\left(\frac{\text{Im}Z}{\text{Re}Z}\right)$, it is clear that we must calculate the impedance. Therefore,

$$Z(j\omega) = \frac{(R_L + j\omega L)\left(R_C + \frac{1}{j\omega C}\right)}{R_L + R_C + j\omega L + \frac{1}{j\omega C}}$$

so $$Z(j\omega) = \frac{(R_L + j\omega L)(1 + j\omega R_C C)}{(1 - \omega^2 LC) + j\omega(R_L + R_C)C}$$

and $$\theta(j\omega) = \tan^{-1}\left(\frac{\omega L}{R_L}\right) + \tan^{-1}(\omega R_C C) - \tan^{-1}\left[\frac{\omega(R_L + R_C)C}{1 - \omega^2 LC}\right].$$

For unity power factor (pf = 1), $\theta(j\omega) = 0$. Thus, we get

$$\underbrace{\tan^{-1}\left(\frac{\omega L}{R_L}\right)}_{A} + \underbrace{\tan^{-1}(\omega R_C C)}_{B} = \tan^{-1}\left[\frac{\omega(R_L + R_C)C}{1 - \omega^2 LC}\right]$$

Hence, $$\tan(A + B) = \frac{\omega(R_L + R_C)C}{1 - \omega^2 LC}.$$

Recalling that $\tan(A + B) = \frac{\tan A + \tan B}{1 - \tan A \tan B}$, we get

$$\frac{\frac{\omega L}{R_C} + \omega R_C C}{1 - \frac{\omega^2 LR_C C}{R_L}} = \frac{\omega(R_L + R_C)C}{1 - \omega^2 LC}$$

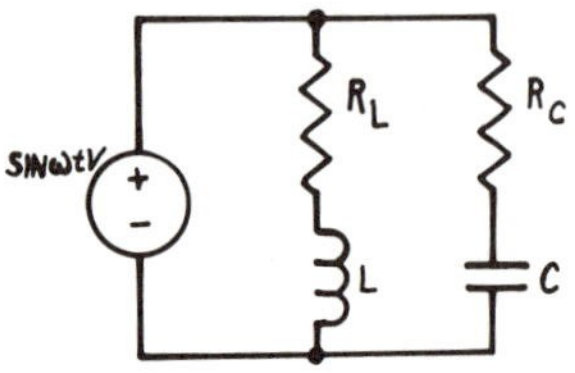

Fig. 1

A little algebraic manipulation yields the desired value of C:

$$\left[\frac{L}{R_L} + R_C C\right][1 - \omega^2 LC] = [R_L + R_C]C\left[1 - \frac{\omega^2 LR_C C}{R_L}\right]$$

or

$$-\omega^2 R_C LC^2 + \left(R_C - \frac{\omega^2 L}{R_L}\right)C + \frac{L}{R_L} = -\frac{\omega^2 LR_C}{R_L}\left[R_L + R_C\right]C^2 + (R_L + R_C)C.$$

Hence $$\frac{\omega^2 LR_C^2}{R_L}C^2 - \left[\frac{\omega^2 L}{R_L} + R_L\right]C + \frac{L}{R_L} = 0$$

or $$\omega^2 LR_C^2 C^2 = \left[\omega^2 L + R_L^2\right]C + L = 0.$$

Using the quadratic formula, we get

$$C = \frac{R_L^2 + \omega^2 L^2 \pm \sqrt{(R_L^2 + \omega^2 L^2)^2 - 4L^2\omega^2 R_C^2}}{2\omega^2 LR_C^2}.$$

CHAPTER 10

THREE-PHASE CIRCUITS

Y CONNECTED

• **PROBLEM** 10-1

A balanced three phase Y connected source has, using a positive phase sequence, $\vec{V}_{an} = 390\underline{/30°}$ Vrms. Find: (a) $\vec{V}_{cn}$; (b) $\vec{V}_{bc}$; (c) $\vec{V}_{ac}$.

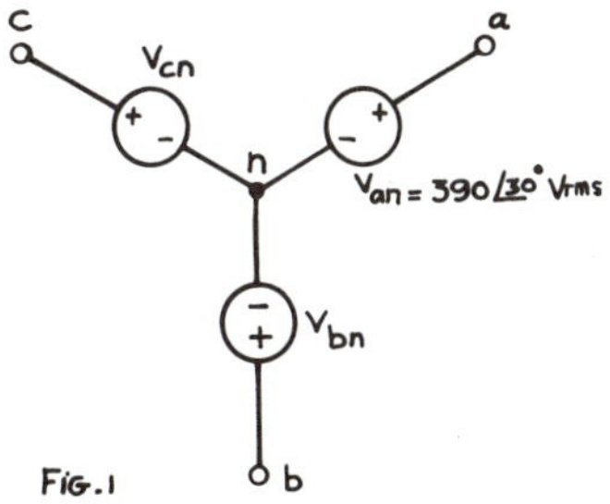

FIG. 1

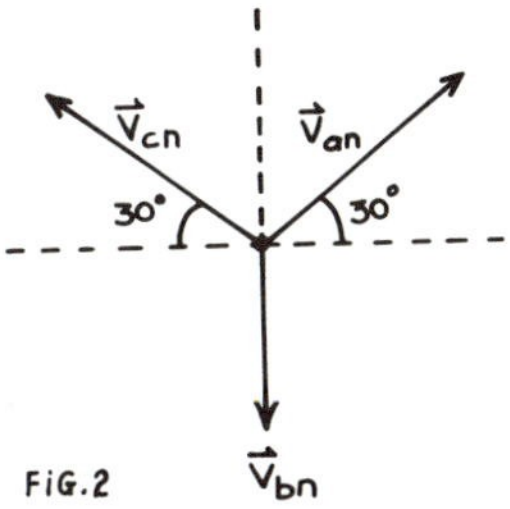

FIG. 2

Solution: Since the source is balanced it is known that $|V_{an}| = |V_{bn}| = |V_{cn}|$. Since we are using positive phase sequence it is known that $\vec{V}_{bn}$ lags $\vec{V}_{an}$ by 120° and $\vec{V}_{cn}$ lags $\vec{V}_{bn}$ by 120°. Fig. 1 shows the circuit diagram.

(a) It is known that $\vec{V}_{bn}$ lags $\vec{V}_{an}$ by 120° and that $|\vec{V}_{bn}| = |\vec{V}_{an}|$. Therefore:

$$\vec{V}_{bn} = |V_{an}|\underline{/30°-120°} = 390\underline{/-90°} \text{ Vrms}$$

similarly,

$$\vec{V}_{cn} = |V_{bn}|\underline{/-90°-120°} = 390\underline{/-210°} \text{ Vrms}$$

(b) Now that the values of $\vec{V}_{bn}$ and $\vec{V}_{cn}$ are known, a vector diagram is drawn in Fig. 2. $\vec{V}_{bc}$ can now be found by vector addition.

$$\vec{V}_{bc} = \vec{V}_{bn} - \vec{V}_{cn}. \qquad (1)$$

Fig. 3 shows the position of $\vec{V}_{bc}$.

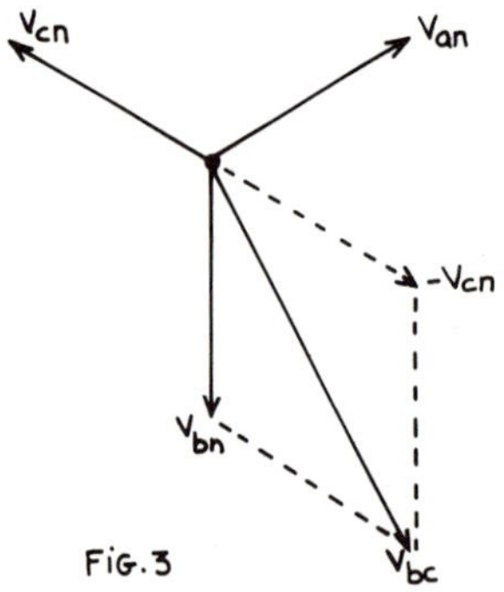

FiG.3

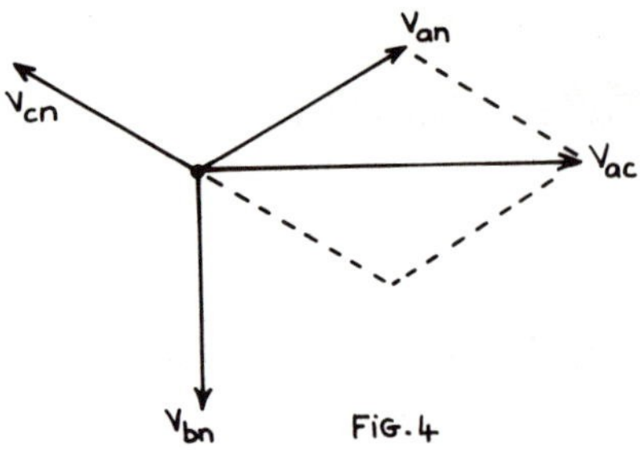

FiG.4

$\vec{V}_{bc}$ along the y-axis is

$$\vec{V}_{bcy} = \vec{V}_{bn} - \vec{V}_{cn} \sin 30° \qquad (2)$$

$$= -390 - 390(\tfrac{1}{2})$$

$$= -585 \text{ Vrms}$$

$\vec{V}_{bc}$ along the x-axis is

$$\vec{V}_{bcx} = -V_{cnx} \qquad (3)$$

$$= 390 \cos 30°$$

$$= 337 \text{ Vrms}$$

and $$|V_{bc}| = \sqrt{|V_{bcy}{}^2| + |V_{bcx}{}^2|}$$

$$= \sqrt{585^2 + 337^2} = 675 \text{ Vrms.}$$

Notice that

$$|V_{bc}| = |V_{line}| = \sqrt{3}\,|V_{an}|. \qquad (4)$$

This is true for all balanced Y connected circuits. The angle for $\vec{V}_{bc}$ is

$$\theta = \tan^{-1}\frac{V_{bcx}}{V_{bcy}} = \tan^{-1}\frac{-585}{337} = -60°.$$

So $\vec{V}_{bc} = 675\underline{/-60°}$ Vrms.

(c) This time we will use equation (4) with $\vec{V}_{ac}$ substituted for $\vec{V}_{bc}$.

$$|V_{ac}| = \sqrt{3}\,|V_{an}|$$

So $$|V_{ac}| = 675 \text{ Vrms}$$

Fig. 4 shows the position of $\vec{V}_{ac}$. It is seen from Fig. 4

that V_{ac} lags V_{an} by 30°. Therefore

$$\vec{V}_{ac} = |V_{ac}| \underline{/30°-30°} = 675\underline{/0°} \text{ Vrms.}$$

• PROBLEM 10-2

A 220-V (line-to-neutral) three-phase source is connected to three 10Ω resistance loads with both source and load in the Y-configuration. Determine the current in each phase the phase of each current, and the current in the neutral line.

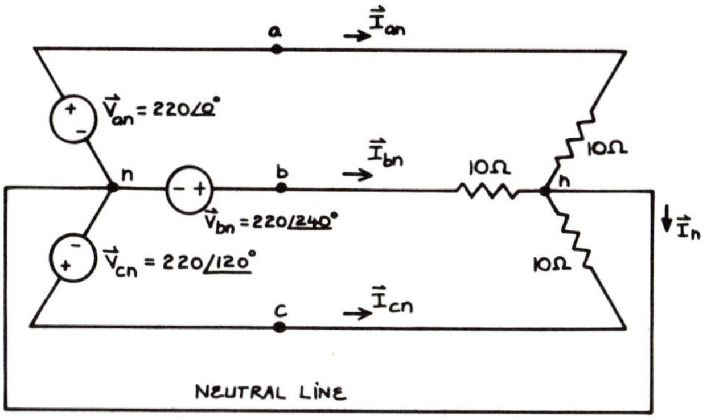

FIG.1

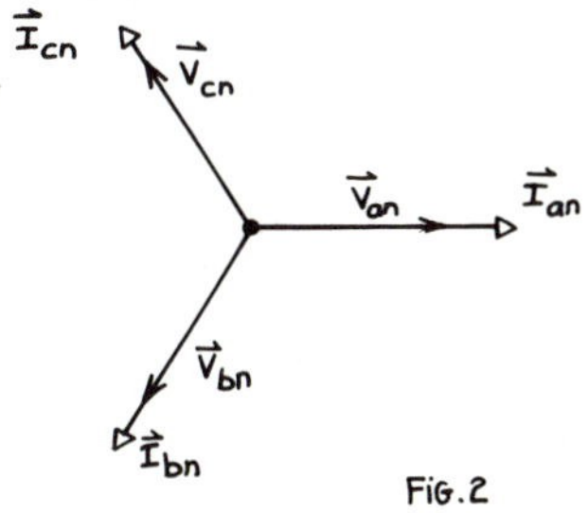

FIG.2

Solution: First, draw a circuit diagram as shown in Fig. 1. Then proceed to solve for the current in each phase of the load (line to neutral). Finally, add the three complex currents in the neutral line.

The three phases of the source have been identified and assigned angles in accordance with a positive phase sequence ($\vec{V}_{bn}$ lagging $\vec{V}_{an}$, $\vec{V}_{cn}$ lagging $\vec{V}_{bn}$, $\vec{V}_{an}$ lagging $\vec{V}_{cn}$), which is the usual system.

Next, compute the current in each impedance, utilizing the phase relationship.

$$\text{Phase current} = \frac{\text{Phase voltage (line to neutral)}}{\text{Phase impedance}} \tag{1}$$

which, for phase a, becomes

$$\vec{I}_{an} = \frac{\vec{V}_{an}}{\vec{Z}_{an}} = \frac{220\underline{/0°}}{10\underline{/0°}} = 22\underline{/0°} \text{ A.} \tag{2}$$

Similarly, for the other phases,

$$\vec{I}_{bn} = \frac{\vec{V}_{bn}}{\vec{Z}_{bn}} = \frac{220\underline{/240°}}{10\underline{/0°}} = 22\underline{/240°} \text{ A,} \tag{3}$$

and $$\vec{I}_{cn} = \frac{\vec{V}_{cn}}{\vec{Z}_{cn}} = \frac{220\underline{/120°}}{10\underline{/0°}} = 22\underline{/120°} \text{ A}. \tag{4}$$

These currents and the given voltages are displayed in the phasor diagram in Fig. 2.

If the impedances had been other than resistive, each current would lead or lag its corresponding voltage by the same angle, and, hence, the three currents would maintain their 120 degree symmetry.

Studying Fig. 1, it can be seen that the neutral current is the sum

$$\vec{I}_n = \vec{I}_{an} + \vec{I}_{bn} + \vec{I}_{cn}. \tag{5}$$

Carrying out this addition of phasor currents,

$$\vec{I}_n = 22\underline{/0°} + 22\underline{/240°} + 22\underline{/120°} \tag{6}$$

which simplifies to

$$\vec{I}_n = 22(1\underline{/0°} + 1\underline{/240°} + 1\underline{/120°}). \tag{7}$$

It is then expanded to

$$\vec{I}_n = 22((1+jo) + (-0.5 - j0.866) + (-0.5 + j0.866)). \tag{8}$$

We find these components add to zero:

$$I_n = 22((1 - 0.5 - 0.5) + j(0.866 - 0.866)) - 0. \tag{9}$$

This should not be surprising **since** the three current phasors in Fig. 2 lie in a symmetrical pattern. We can generalize on this information by making the following statement: As long as the three impedances are identical in magnitude and angle, the neutral line will carry no current.

• PROBLEM 10-3

A 220 V, three-phase source is connected to equal impedances, $10\underline{/45°}$ Ω, with both connected in a Y-configuration. Determine the current in each phase, the phase of each current, and the current in the neutral line.

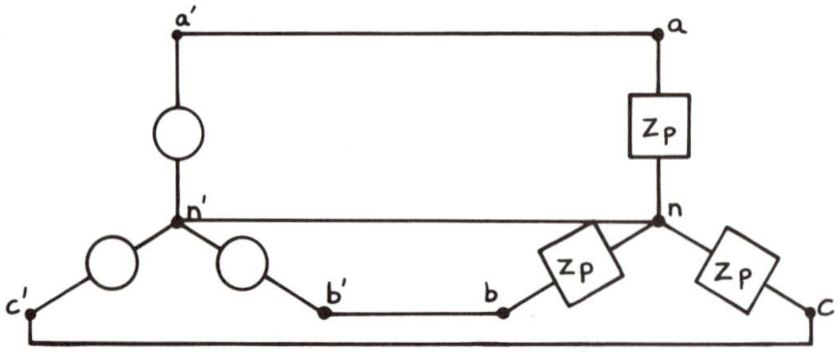

Solution: A balanced 3-phase system results in complete symmetry of response. Each phase current ($\vec{I}_p$) can be obtained by dividing the phase voltage ($\vec{V}_p$) by phase impedance ($\vec{Z}_p$); therefore,

$$\vec{I}_p = \frac{\vec{V}_p}{\vec{Z}_p}$$

taking $\vec{V}_{an}$ as the reference point gives

$$\vec{I}_{an} = \frac{220\underline{/0°}}{10\underline{/45°}} = (220/10)\underline{/0°-45°} = 22\underline{/-45°}\ \text{A}$$

$$\vec{I}_{bn} = \frac{220\underline{/-120°}}{10\underline{/45°}} = (220/10)\underline{/-120°-45°} = 22\underline{/-165°}$$

$$= 22\underline{/-165°+360°} = 22\underline{/+195°}\ \text{A}$$

$$\vec{I}_{cn} = \frac{220\underline{/-240°}}{10\underline{/+45°}} = (220/10)\underline{/-240-45°} = 22\underline{/-285°}$$

$$= 22\underline{/-285°+360°} = 22\underline{/+75°}\ \text{A}$$

The current in the neutral line $\vec{I}_{nn'} = \vec{I}_{an} + \vec{I}_{bn} + \vec{I}_{cn}$, converting the complex results into rectangular form:

$$\vec{I}_{nn'} = 15.55 - j15.15 - 21.25 - j5.69 + 5.69 + j21.25 = 0$$

The result is 0, as expected for a balanced system.

• PROBLEM 10-4

A 440-V rms (line voltage) three-phase three-wire system feeds two balanced Y-connected loads. One load is an induction motor which may be represented by an impedance of 10 + j5Ω per phase. The other is a lighting load equivalent to 15Ω per phase. Find the average power: (a) delivered to the lighting load; (b) delivered to the induction motor; (c) provided by one phase of the source.

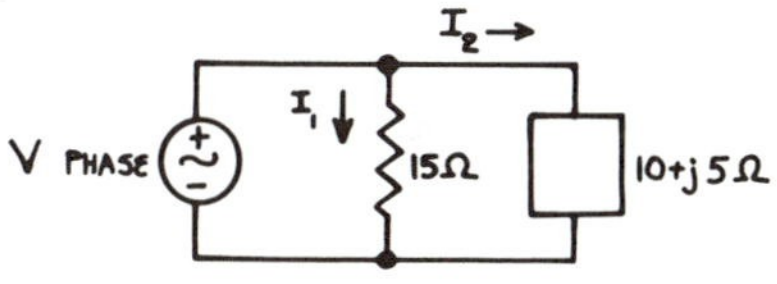

FIG.1

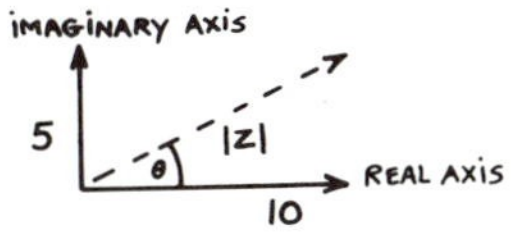

FIG.2

Solution: This problem is easier to solve when it is reduced to a single phase problem. Fig. 1 shows one of the three phases.

(a) V_{line} is given, so V_{phase} can be calculated:

$$V_{phase} = \frac{V_{line}}{\sqrt{3}} = \frac{440}{\sqrt{3}} = 254 \text{ Vrms} \qquad (1)$$

Now the current for the lighting load can be found by using Ohm's Law.

$$I_1 = \frac{V_{phase}}{15\Omega} = \frac{254}{15} = 16.9 \text{ A.} \qquad (2)$$

From (1) and (2) we can find the power per phase of the lighting load.

$$P_1 = V_{phase}\, I_1 = 254 \times 16.9 = 4.3 \text{ kW} \qquad (3)$$

The average power delivered to the lighting load is

$$3 \cdot P_1 = 12.9 \text{ kW.}$$

(b) The same process is used to find the average power to the motor. The magnitude of the motor impedance is found by using Fig. 2.

$$|Z| = \sqrt{5^2 + 10^2} = 11.2$$

$$\theta = \tan^{-1}\frac{5}{10} = 26.57° \therefore \vec{Z} = 11.2\underline{/26.57°}$$

The current through the motor can now be found:

$$\vec{I}_2 = \frac{\vec{V}_{phase}}{\vec{Z}} = \frac{254\underline{/0°}}{11.2\underline{/26.57}} = 22.68\underline{/-26.57} \text{ A}$$

$$= 20.28 - j10.14.$$

Now the power per phase to the motor can be found:

$$P = V_{phase} \times \text{Re}[\vec{I}_2] = 254 \times 20.28 = 5.15 \text{ kW.}$$

The average power to the motor is:

$$3P = 15.4 \text{ kW.}$$

(c) The power delivered to both loads (per phase) is the sum of the average per phase power delivered to the lighting load and the motor load.

$$P = 5.1 \text{ kW} + 4.3 \text{ kW} = 9.4 \text{ kW.}$$

• PROBLEM 10-5

Calculate the magnitude of a line current in the circuit shown in Fig. 1.

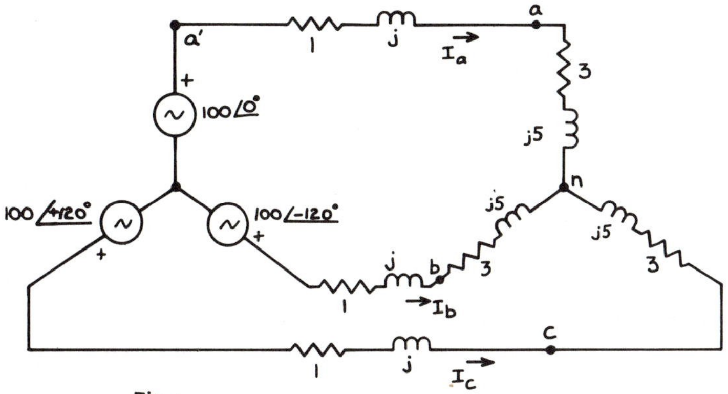

FIG. 1

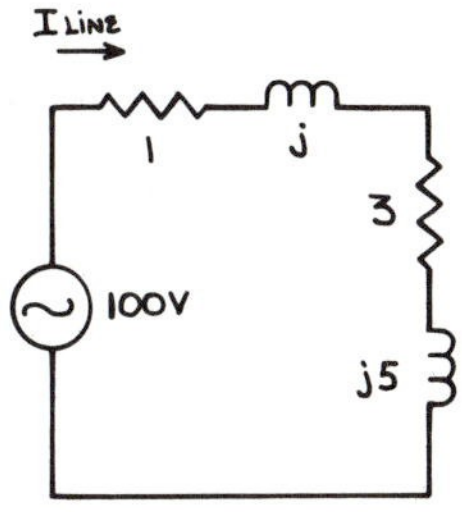

FIG. 2

Solution: The circuit is balanced because the load for each phase is the same, the magnitude of the source for each phase is the same, and the angle for each phase is displaced by 120°. Since the circuit is balanced, the magnitude of the line current in each phase is the same. To find the line current, Fig. 1 is redrawn in Fig. 2 to include only one phase.

The total impedance of one phase is 1 + j + 3 + j5 = 4 + j6Ω. The magnitude of the line current is

$$|I_{line}| = \left|\frac{100}{4+j6}\right| = \frac{100}{\sqrt{4^2+6^2}} = \frac{100}{7.21} = 13.85 \text{ A.}$$

• PROBLEM 10-6

A three-phase, Y-connected system with 220V per phase is connected to three loads: 10Ω, $20\underline{/20°}$ Ω, and $12\underline{/-35°}$ Ω to phases 1, 2 and 3, respectively. Find the current in each line and in the neutral line.

Solution: This three-phase, Y-connected system has a conductor between the generator neutral and the load neutral. Because of this the current in each line can be calculated as if we had single phase circuits.

The generator terminals will be unprimed and the load terminals primed. The line currents are (assume the phasing is 1-3-2)

$$\vec{I}_{11'} = \frac{\vec{V}_{1'n'}}{\vec{Z}_1} = \frac{220\underline{/0°}}{10\underline{/0°}} = 22.0\underline{/0°} \text{ A}$$

$$\vec{I}_{22'} = \frac{\vec{V}_{2'n'}}{\vec{Z}_2} = \frac{220\underline{/120°}}{20\underline{/20°}} = 11.0\underline{/100°} \text{ A}$$

$$\vec{I}_{33'} = \frac{\vec{V}_{3'n'}}{\vec{Z}_3} = \frac{220\underline{/-120°}}{12\underline{/-35°}} = 18.3\underline{/-85°} \text{ A}$$

The current in the neutral line is found by summing the above three currents.

$$\vec{I}_{nn'} = \vec{I}_{11'} + \vec{I}_{22'} + \vec{I}_{33'} = 22.9\underline{/-18.8°} \text{ A}$$

• PROBLEM 10-7

A balanced set of three-phase voltages is connected to an unbalanced set of Y-connected impedances as shown in Fig. 1. The following values are known:

$\vec{V}_{ab} = 212\underline{/\ \ 90°}$ V $\qquad \vec{Z}_{an} = 10 + j0\ \Omega$

$\vec{V}_{bc} = 212\underline{/-150°}$ V $\qquad \vec{Z}_{bn} = 10 + j10\Omega$

$\vec{V}_{ca} = 212\underline{/\ -30°}$ V $\qquad \vec{Z}_{cn} = 0 - j20\Omega$

Find the line currents $\vec{I}_{a'a}$, $\vec{I}_{b'b}$ and $\vec{I}_{c'c}$.

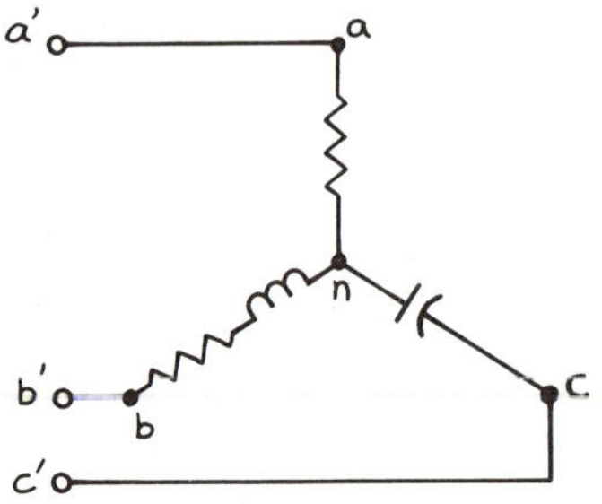

Fig. 1

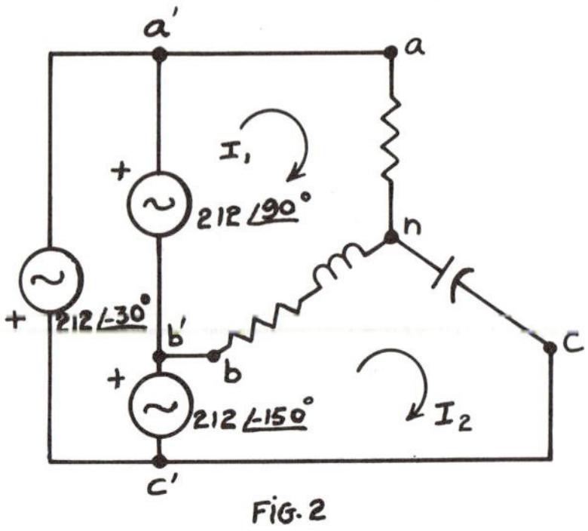

Fig. 2

Solution: In the unbalanced case, any of the usual methods of analysis can be applied. If we draw the circuit with the sources attached, as in Fig. 2, we can see that two mesh current equations can be solved and, thus, all three line currents are determined.

The equations are:

$$(10+j0)\vec{I}_1 + (10+j10)\vec{I}_1 - (10+j10)\vec{I}_2 = 212\underline{/90°}$$

$$-(10+j10)\vec{I}_1 + (10+j10)\vec{I}_2 + (0-j20)\vec{I}_2 = 212\underline{/-150}$$

Solving by determinants,

$$\vec{I}_1 = \frac{\begin{vmatrix} 0+j212 & -10-j10 \\ -183.6-j106 & 10-j10 \end{vmatrix}}{\begin{vmatrix} 20+j10 & -10-j10 \\ -10-j10 & 10-j10 \end{vmatrix}} = 3.66\underline{/15°} \text{ A}$$

$$\vec{I}_2 = \frac{\begin{vmatrix} 20+j10 & 0+j212 \\ -10-j10 & -183.6+j100 \end{vmatrix}}{\text{same}} = 11.96\underline{/-113.8°} \text{ A}$$

Now, $\vec{I}_{aa'} = \vec{I}_1$ and $\vec{I}_{cc'} = -\vec{I}_2 = 11.96\underline{/66°}$.

The third line current is found from KCL applied at n:

$$\vec{I}_{bb'} = -\vec{I}_{aa'} + \vec{I}_{c'c} = -\vec{I}_1 + \vec{I}_2$$

$$\vec{I}_{bb'} = -(3.54+j.95) + -4.83 - j10.94$$

$$= -8.37 - j11.89 = 14.54\underline{/-125.1°} \text{ A.}$$

• PROBLEM 10-8

For the circuit shown find: (a) $\vec{I}_{aA}$; (b) $\vec{I}_{bB}$; (c) $\vec{I}_{cC}$.

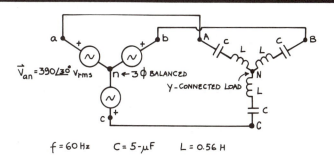

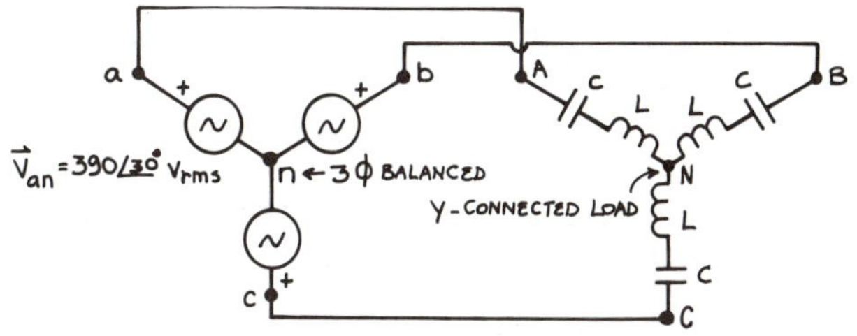

f = 60 Hz C = 5-μF L = 0.56 H

Solution: Since the load is balanced, there is no potential difference between n and N. Hence, each line current can be calculated as if we had three separate single phase circuits. For a frequency of 60 Hz the load impedance on each phase is

$$\vec{Z} = j\omega L - \frac{j}{\omega C} = j\left[2\pi(60)(0.56) - \frac{1}{2\pi(60)(5\text{x}10^{-6})}\right]$$

$$= -j319.4 \ \Omega$$

The currents are

$$\vec{I}_{aA} = \frac{\vec{V}_{an}}{\vec{Z}} = \frac{390\underline{/30°}}{319.4\underline{/-90°}} = 1.22\underline{/120°} \text{ A rms}$$

$$\vec{I}_{bB} = \frac{\vec{V}_{bn}}{\vec{Z}} = \frac{390\underline{/-90°}}{319.4\underline{/-90°}} = 1.22\underline{/0°} \text{ A rms}$$

$$\vec{I}_{cC} = \frac{\vec{V}_{cn}}{\vec{Z}} = \frac{390\underline{/-210°}}{319.4\underline{/-90°}} = 1.22\underline{/-120°} \text{ A rms}$$

The results can be checked to see if

$$\vec{I}_{aA} + \vec{I}_{bB} + \vec{I}_{cC} = 0.$$

• PROBLEM 10-9

$\vec{Z}_{na} = \vec{Z}_{nb} = \vec{Z}_{nc} = 2 + j8\ \Omega$

Find $\vec{V}_{a'b'}$, $\vec{V}_{b'c'}$, and $\vec{V}_{c'a'}$.

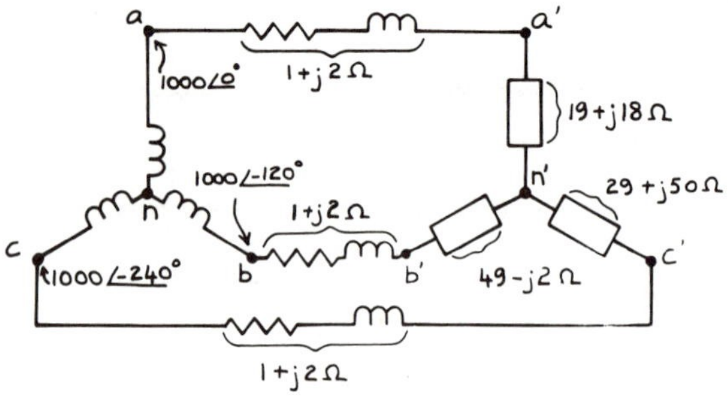

Solution: The procedure in this problem will be to determine the current through each line and then find the voltage drop across each load.

To find the current, it is necessary to know the voltage drop from n' to n. This is given by

$$\vec{V}_{n'n} = \vec{Z}_p \left[\frac{1000\underline{/0°}}{\vec{Z}_{na}+\vec{Z}_{aa'}+\vec{Z}_{a'n'}} + \frac{1000\underline{/-120°}}{\vec{Z}_{nb}+\vec{Z}_{bb'}+\vec{Z}_{b'n'}} + \frac{1000\underline{/-240°}}{\vec{Z}_{nc}+\vec{Z}_{cc'}+\vec{Z}_{c'n'}} \right]$$

where $\vec{Z}_p$ is the parallel combination of $(\vec{Z}_{na}+\vec{Z}_{aa'}+\vec{Z}_{a'n'})$, $(\vec{Z}_{nb}+\vec{Z}_{bb'}+\vec{Z}_{b'n'})$, and $(\vec{Z}_{nc}+\vec{Z}_{cc'}+\vec{Z}_{c'n'})$.

Substituting the numerical values into the equation, we find

$$\frac{1}{\vec{Z}_p} = \frac{1}{(2+j8)+(1+j2)+(19+j18)} + \frac{1}{(2+j8)+(1+j2)+(49-j2)}$$

$$+ \frac{1}{(2+j8)+(1+j2)+(29+j50)}$$

or, $\vec{Z}_p = 17.42\underline{/41.39°}\ \Omega$

Then,

$$\vec{V}_{n'n} = 17.42\underline{/41.39}\left[\frac{1000\underline{/0°}}{22+j28} + \frac{1000\underline{/-120°}}{52+j8} + \frac{1000\underline{/-240°}}{32+j60}\right]$$

$$= 484.21\underline{/-20.15°}\ \text{V.}$$

The currents through each line are

$$\vec{I}_{aa'} = \frac{1000\underline{/0°} - V_{n'n}}{\vec{Z}_{na}+\vec{Z}_{aa'}+\vec{Z}_{a'n'}} = \frac{1000\underline{/0°} - 484.21\underline{/-20.15°}}{22+j28}$$

$$= 16.02\underline{/-34.83°}\ \text{A.}$$

$$\vec{I}_{bb'} = \frac{1000\underline{/-120°} - V_{n'n}}{Z_{nb}+Z_{bb'}+Z_{bn'}} = \frac{1000\underline{/-120°} - 484.21\underline{/-20.51°}}{52+j8}$$

$$= 22.48\underline{/-152.52°}\ \text{A.}$$

$$\vec{I}_{cc'} = \frac{1000\underline{/-240°} - \vec{V}_{n'n}}{\vec{Z}_{nc}+\vec{Z}_{cc'}+\vec{Z}_{c'n'}} = \frac{1000\underline{/-240°} - 484.21\underline{/-20.15°}}{32+j60}$$

$$= 20.67\underline{/70.81°}\ \text{A.}$$

Checking, we find that $I_{aa'} + I_{bb'} + I_{cc'} = 0$ as required. The voltage drop across each load is found by multiplying the load impedance by the above current. Thus,

$$\vec{V}_{a'n'} = \vec{I}_{aa'}\ \vec{Z}_{a'n'} = (16.02\underline{/-34.83°})(19+j18)$$

$$= 419.28\underline{/8.62°} = 414.55 + j62.86\ \text{V.}$$

$$\vec{V}_{b'n'} = \vec{I}_{bb'}\ \vec{Z}_{b'n'} = (22.48\underline{/-152.52°})(49-j2)$$

$$= 1102.44\underline{/-154.86°} = -997.98 - j468.40\ \text{V.}$$

$$\vec{V}_{c'n'} = \vec{I}_{cc'}\,\vec{Z}_{c'n'} = (20.67\underline{/70.81°})(29+j50)$$

$$= 1194.75\underline{/130.70°} = -779.04 + j905.84 \text{ V.}$$

The last step is to combine the load voltages appropriately to get the desired results.

$$\vec{V}_{a'b'} = \vec{V}_{a'n'} + \vec{V}_{n'b'} = 414.55 + j62.86 - (-997.98$$

$$-j468.40) = 1509\underline{/20.6°} \text{ V.}$$

$$\vec{V}_{b'c'} = \vec{V}_{b'n'} + \vec{V}_{n'c'} = -997.98 - j468.40 - (-779.04$$

$$+ j905.84) = 1392\underline{/-99.0°} \text{ V.}$$

$$\vec{V}_{c'a'} = \vec{V}_{c'n'} + \vec{V}_{n'a'} = -779.04 + j905.84 - (414.55$$

$$+ j62.86) = 1461\underline{/144.8°} \text{ V.}$$

• PROBLEM 10-10

A balanced three-phase three-wire system supplies two balanced Y-connected loads. The first draws 6-kW at 0.8PF lagging while the other requires 12kW at 0.833PF leading. If the current in each line is 8A rms, find the current in the: (a) first load; (b) second load; (c) source phase.

Solution: It does not matter how the load is connected. For any balanced 3-phase system,

$$P = |V|\,|I|\cos(\theta)$$

where P is the average power, $|V|$ is the rms voltage magnitude, $|I|$ is the rms current magnitude, and $\cos(\theta)$ is the power factor. θ is the angle by which the voltage leads the current in each line.

(a) For $P_1 = 6\text{kW}$, $\cos(\theta_1) = 0.8$ lagging (that is, θ is negative),

$$\theta_1 = -\cos^{-1}(0.8) = -36.87°$$

$$|I_1| = \frac{P_1}{|V|\cos(\theta_1)} = \frac{(6000)}{|V|(0.8)} = \frac{7500}{|V|} \text{ A rms.}$$

For the second load, $P_2 = 12\text{kW}$, $\cos(\theta_2) = 0.833$ leading

$$\theta_2 = +\cos^{-1}(0.833) = 33.59°$$

$$|I_2| = \frac{P_2}{|V|\cos(\theta_2)} = \frac{12{,}000}{|V|\,(0.833)} = \frac{14{,}406}{|V|} \text{ A rms.}$$

We are also given that $|I_1 + I_2| = 8$ A rms, so

$$\frac{1}{|V|}\left|7500\underline{/-36.87°} + 14{,}406\underline{/+33.59°}\right| = 8$$

$$\frac{1}{|V|}\left|6000 - j4500 + 12{,}000 + j7979\right| = 8$$

$$\frac{1}{|V|}\left|18{,}000 + j3470\right| = 8$$

$$|V| = \frac{18{,}331}{8} = 2291 \text{ V rms.}$$

Finally, $|I_1| = \frac{7500}{2291} = 3.273$ A rms.

(b) From (a) $|I_2| = \frac{14{,}406}{|V|} = \frac{14{,}406}{2291} = 6.288$ A rms.

(c) The source phase current is the same as the line current, so it is 8 A rms.

Δ CONNECTED

• PROBLEM 10-11

Find the total power in the delta-connected load shown.

$\vec{V}_{ba} = 220\underline{/0°}$ V

$\vec{V}_{cb} = 220\underline{/-120°}$ V

$\vec{V}_{ac} = 220\underline{/\ 120°}$ V

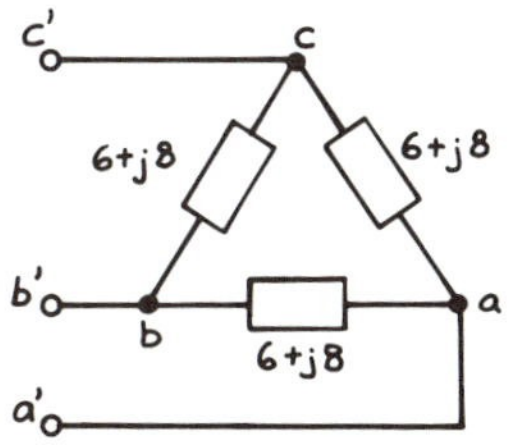

Solution: For a delta-connected load, the phase voltage $\vec{V}_P$ equals the line voltage $\vec{V}_L$ in each case. Therefore, each phase current can be found by dividing the phase voltage by phase impedance.

$$\vec{I}_P = \frac{\vec{V}_P}{\vec{Z}_P} = \frac{220\underline{/\phi}}{10\underline{/53°}}$$

Phase impedance $\vec{Z}_P$ is converted from rectangular form of x+jy to polar form $Z\underline{/\phi}$ by using the following equations.

1) x+jy - Rectangular form

2) $Z\underline{/\phi}$ - Polar form

where $Z = \sqrt{x^2+y^2}$ and $\phi = \tan^{-1}\frac{y}{x}$.

Therefore, using x+jy = 6+j8,

$$Z = \sqrt{6^2+8^2} = 10$$

$$\phi = \tan^{-1}\frac{y}{x} = \tan^{-1}\frac{8}{6} = 53°$$

Then, using complex division, $\vec{I}_P = 22\underline{/\theta-53°}$

The power per phase is given by $P_P = |\vec{V}_P||I_P|\cos\phi$ where ϕ = angle of phase impedance.

Therefore, $P_P = |V_P||I_P|\cos = (220)(22)(\cos 53°) = 2912$, and the total power is $P_T = 3|V_P||I_P|\cos\phi = 3(P_P)$

$$= (3)(2912) = 8736W.$$

• PROBLEM 10-12

In the balanced Δ-connected generator shown in Fig. 1,

$$|V_a| = |V_b| = |V_c| = 173 \text{ V and}$$

$$\vec{V}_a + \vec{V}_b + \vec{V}_c = 0.$$

Calculate the absolute value of the current phasor $\vec{I}$ in the 6Ω resistor.

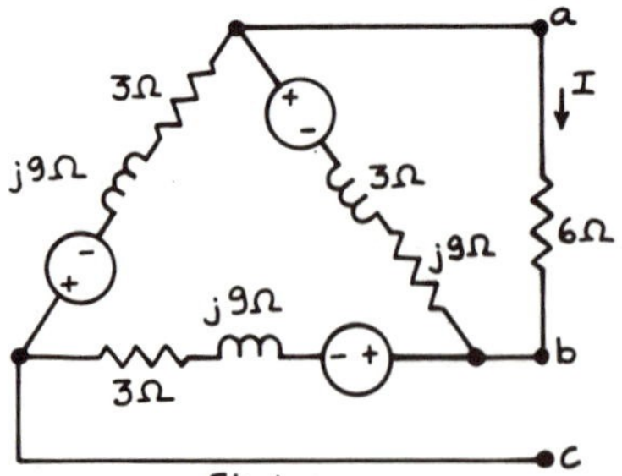

Fig.1

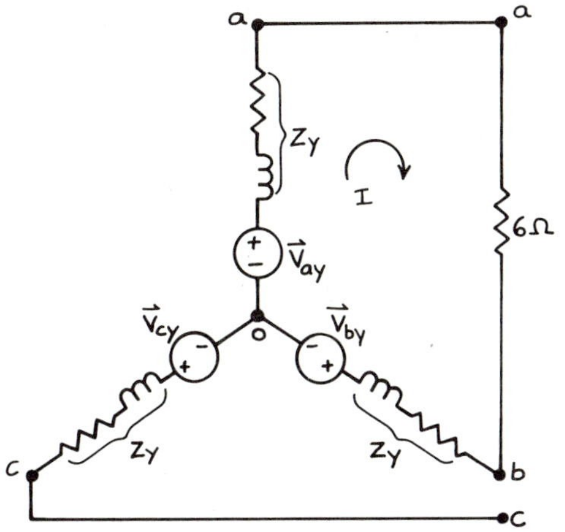

FiG.2 DELTA GENERATOR CONVERTED TO AN EQUIVALENT WYE GENERATOR

Solution: Always bear in mind that a three-phase delta network can be converted into a wye which often makes it easier to solve. In this particular problem, we can reduce a two-mesh network to a single loop by so doing. Comparing the converted network of Fig. 2 with the original circuit will make this obvious.

The conversion from delta to wye can be accomplished by the use of Thévenin's theorem. First, obtain the open circuit voltage between any two terminals of the delta, removing the 6Ω resistance temporarily. This is known to be $100\sqrt{3}$, as given. Now, in the corresponding wye of Fig. 1, we know that

$$|V_{ay}| = |V_{by}| = |V_{cy}| = \frac{|V_{ab}|}{\sqrt{3}} = 100 \text{ V} \tag{1}$$

from basic three-phase theory.

The next step in applying Thévenin's theorem is to short circuit between terminals a-b, b-c and c-a of the delta and calculate the line current as seen in Fig. 3.

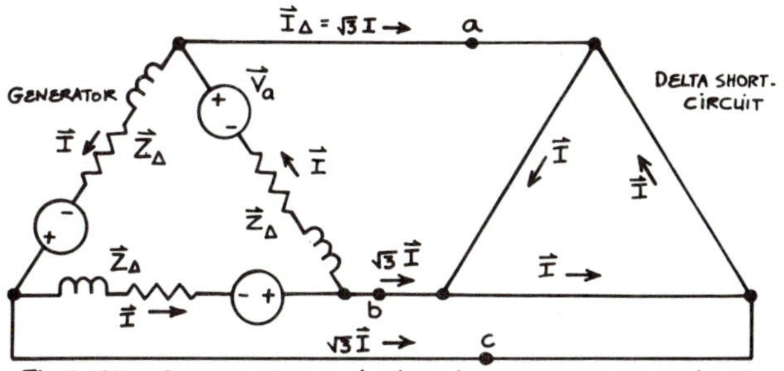

FiG.3 THREE-PHASE DELTA SHORT-CIRCUIT APPLIED TO THE GENERATOR TERMINALS.

Due to symmetry, all phase currents will be alike. Thus, let us choose one set to concentrate on: the a-b phase. Since the voltage across a-b is zero,

$$\vec{V}_{ab} = \vec{V}_a - \vec{I}\,\vec{Z}_\Delta = 0, \tag{2}$$

and $$|I| = \left|\frac{V_a}{Z_\Delta}\right| = \frac{100\sqrt{3}}{|Z_\Delta|} . \qquad (3)$$

The line current must be $\sqrt{3}|I|$ from basic three-phase theory for a delta network, so

$$|I_\Delta| = |\sqrt{3}\ I| = \frac{300}{|Z_\Delta|} . \qquad (4)$$

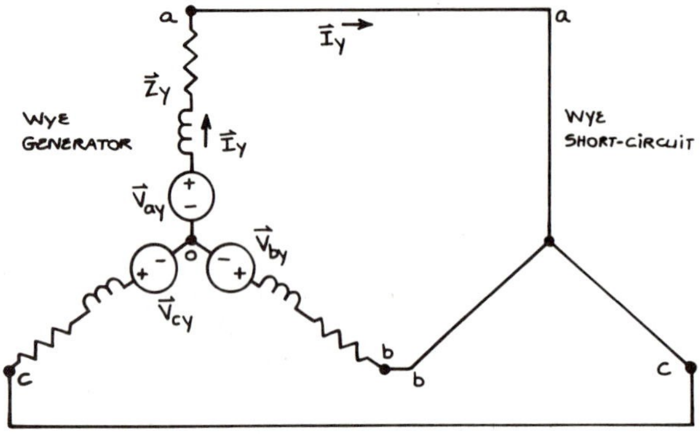

FIG. 4 THREE-PHASE WYE SHORT CIRCUIT APPLIED TO THE GENERATOR TERMINALS.

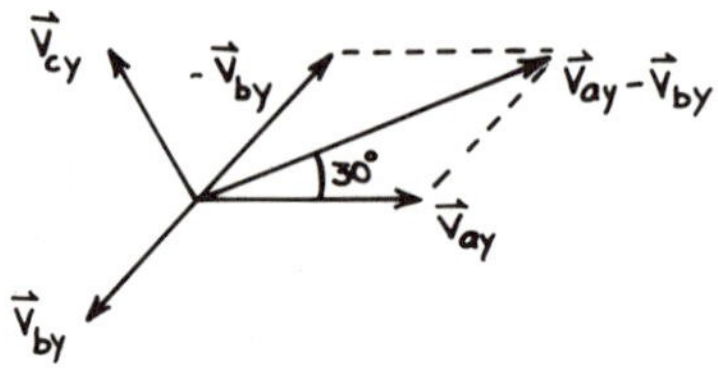

FIG. 5 PHASOR VOLTAGES IN THE WYE GENERATOR.

Now, turning to the equivalent wye, we short its terminals as shown in Fig. 4 and solve for the line current which equals phase current. Under symmetrical short circuit conditions,

$$\vec{V}_{oa} = \vec{V}_{ay} - \vec{I}_y\vec{Z}_y = 0 \qquad (5)$$

and $$|Z_y| = \left|\frac{V_{ay}}{I_y}\right| . \qquad (6)$$

But, for the delta and wye generators to be equivalent, both line currents must be the same:

$$|I_y| = |I_\Delta| = \left|\frac{300}{Z_\Delta}\right| , \qquad (7)$$

so, substituting equations (1) and (7) into equation (6), we have

$$|Z_y| = \left|\frac{V_{ay}}{I_y}\right| = \left|\frac{V_{ay}}{I_\Delta}\right| = \left|\frac{100}{300}\ Z_\Delta\right| . \qquad (8)$$

Therefore,

$$\vec{Z}_y = \frac{1}{3}\vec{Z}_\Delta = \frac{1}{3}(3+j9) = 1+j3\ \Omega. \qquad (9)$$

Now we are ready to solve the circuit of Fig. 1. Because terminal c is not connected to any outside impedance, the current in branch c of the wye is zero; hence, the current in branches a and b are the same. Traversing the loop in the direction of current $\vec{I}$ and applying Kirchhoff's voltage law,

$$(\vec{V}_{ay} - \vec{V}_{by}) - \vec{I}(\vec{Z}_y + \vec{Z}_y + 6) = 0. \qquad (10)$$

By constructing the phasor diagram of Fig. 5, it becomes evident that

$$\left|\vec{V}_{ay} - \vec{V}_{by}\right| = \sqrt{3}\left|V_{ay}\right| = \sqrt{3}\ 100. \qquad (11)$$

Thus, substituting equations (9) and (11) into equation (10), we are able to solve for I, the current in the 6-ohm resistance:

$$|I| = \left|\frac{V_{ay} - V_{by}}{Z_y + Z_y + 6}\right| = \left|\frac{100\sqrt{3}}{8+j6}\right| = \frac{100\sqrt{3}}{10} \text{ A}. \qquad (12)$$

• PROBLEM 10-13

A 3-phase motor takes 10 kva at 0.6 pf lagging from a source of 220-V. It is in parallel with a balanced delta-load having 16Ω resistance and -j12Ω reactance in series in each phase. Find the total volt-amperes, power, line current and power factor of the combination.

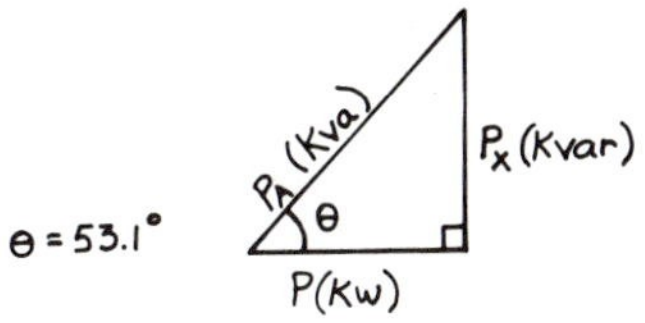

Fig.1

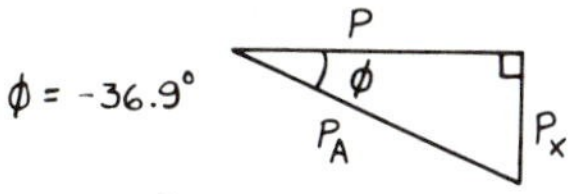

Fig.2

Solution: The approach to three-phase problems involving more than one load is to determine P, the active power (kilowatts), and P_x, the reactive power (kilovars), in each load; then find the sum of the active powers and the net reactive power. From this result, P_A, the apparent power (kva), current and power factor of the combination can be calculated.

For the motor, sketch the power triangle as shown in Fig. 1. Since the power factor is known to be equal to the cosine θ,

$$\theta = \cos^{-1}0.6 = 53.1°. \qquad (1)$$

From this we can proceed to evaluate the unknown sides of the triangle:

$$P = P_A \cos\theta = 10 \cos 53.1° = 6 \text{ kw} \qquad (2)$$

$$P_x = P_A \sin\theta = 10 \sin 53.1° = 8 \text{ kvar.} \quad (3)$$

These results will be used later.

For the delta load,

$$\text{Delta phase voltage} = \vec{V}_P = \text{Line voltage} = 220 \text{ V.} \quad (4)$$

$$\text{Delta phase impedance} = \vec{Z}_P = 16 - j12 = 20\underline{/-36.9°}\ \Omega. \quad (5)$$

$$\text{Delta phase current} = V_P/Z_P = 220/20 = 11 \text{ A.} \quad (6)$$

$$\text{Delta Line current} = I_L = \sqrt{3}\, I_P = 19.05 \text{ A.} \quad (7)$$

The power triangle for the delta load is sketched in Fig. 2 with the P_x leg drawn in the downward direction because the reactive component of the impedance is designated with a minus sign. This indicates capacitive reactance which is associated with current leading the voltage and, hence, "leading kvars." It is in contrast with the motor's power triangle, shown in Fig. 1, which indicates "lagging" vars in accordance with the given information.

Now, solving for the delta load's apparent, active and reactive powers,

$$P_A = \sqrt{3}\, V_L\, I_L = (\sqrt{3})(220)(19.05) = 7.26 \text{ kvar} \quad (8)$$

$$P = P_A \cos\phi = 7.26 \cos(-36.9°) = 5.81 \text{ kw} \quad (9)$$

$$P_x = P_A \sin\phi = 7.26 \sin(-36.9°) = -4.36 \text{ kvar} \quad (10)$$

Having solved for the constituent powers in each load, we now combine them.

$$\text{Total active power} = P_1 + P_2 = 6 + 5.81 = 11.81 \text{ kw} \quad (11)$$

$$\text{Net reactive power} = P_{x1} + P_{x2} = 8 - 4.36 = 3.64 \text{ kvar} \quad (12)$$

Since these are two sides of a right triangle, the hypotenuse (apparant power) can be found by the Pythagorean theorem:

$$P_A = \sqrt{11.81^2 + 3.64^2} = 12.36 \text{ kva.} \quad (13)$$

From this, the current can be calculated:

$$I_L = \frac{P_A}{\sqrt{3}V_L} = \frac{12,360}{\sqrt{3}(220)} = 32.4 \text{ A.}$$

Power factor can be determined from the defining equation:

$$\text{power factor} = \frac{P}{P_A} = \frac{11.81}{12.36} = 0.955 \text{ lagging.} \qquad (14)$$

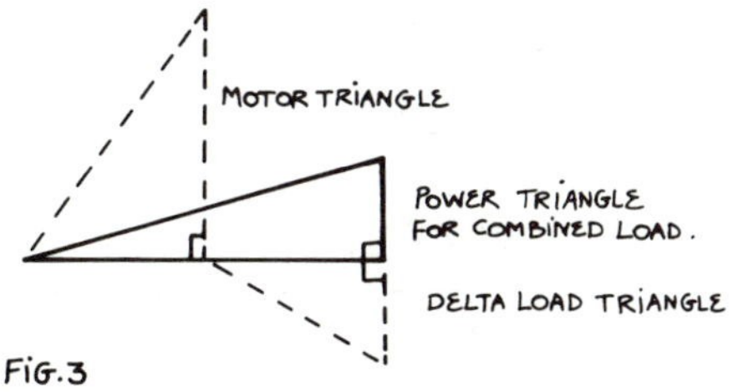

FIG. 3

A graphical representation of the situation is sketched in Fig. 3.

• PROBLEM 10-14

A balanced three-phase three-wire system has a Δ-connected load with a 50Ω resistor, a 5 μF capacitor, and a 0.56 H inductor in series in each phase. Using positive phase sequence with $\vec{V}_{an} = 390\underline{/30°}$ V rms and ω = 500 rad/s, find the magnitude and phases of all line currents.

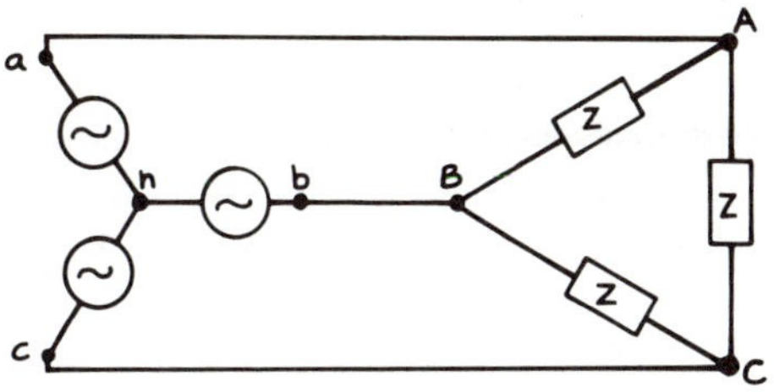

Solution: The circuit described is as shown. Since $\vec{V}_{an} = 390\underline{/30°}$ V rms and a positive phase sequence is specified, $\vec{V}_{bn} = 390\underline{/-90°}$ V rms and $\vec{V}_{cn} = 390\underline{/+150°}$ V rms.

The line voltages are:

$$\vec{V}_{AB} = \vec{V}_{an} - \vec{V}_{bn} = 390\underline{/30°} - 390\underline{/-90°} = 390\sqrt{3}\underline{/60°} \text{ V rms}$$

$$\vec{V}_{BC} = \vec{V}_{bn} - \vec{V}_{cn} = 390\underline{/-90°} - 390\underline{/+150°} = 390\sqrt{3}\underline{/-60°} \text{ Vrms}$$

$$\vec{V}_{CA} = \vec{V}_{cn} - \vec{V}_{an} = 390\underline{/+150°} - 390\underline{/30°} = 390\sqrt{3}\underline{/180°} \text{ V rms.}$$

For a frequency ω = 500 rad/secs the load impedance in each phase is

$$\vec{Z} = 50+j\left[(500)(0.56) - \frac{1}{(500)(5\text{x}10^{-6})}\right] = 50-j120 = 130\underline{/-67.4°}\ \Omega$$

The load phase currents are then

$$\vec{I}_{AB} = \frac{\vec{V}_{AB}}{\vec{Z}} = \frac{390\sqrt{3}\underline{/60°}}{130\underline{/-67.4°}} = 3\sqrt{3}\underline{/127.4°} \text{ A rms}$$

$$\vec{I}_{BC} = \frac{\vec{V}_{BC}}{\vec{Z}} = \frac{390\sqrt{3}\underline{/-60°}}{130\underline{/-67.4°}} = 3\sqrt{3}\underline{/7.4°} \text{ A rms}$$

$$\vec{I}_{CA} = \frac{\vec{V}_{CA}}{\vec{Z}} = \frac{390\sqrt{3}\underline{/180°}}{130\underline{/-67.4°}} = 3\sqrt{3}\underline{/-112.6°} \text{ A rms}$$

Finally the line currents are

$$\vec{I}_{aA} = \vec{I}_{AB} - \vec{I}_{CA} = 3\sqrt{3}\underline{/127.4°} - 3\sqrt{3}\underline{/-112.6°} = 9\underline{/97.4°} \text{ A rms}$$

$$\vec{I}_{bB} = \vec{I}_{BC} - \vec{I}_{AB} = 3\sqrt{3}\underline{/7.4°} - 3\sqrt{3}\underline{/127.4°} = 9\underline{/-22.6°} \text{ A rms}$$

$$\vec{I}_{cC} = \vec{I}_{CA} - \vec{I}_{BC} = 3\sqrt{3}\underline{/-112.6°} - 3\sqrt{3}\underline{/7.4°} = 9\underline{/-142.6°} \text{ A rms}$$

COMBINATION OF Y & Δ

• PROBLEM 10-15

A 120 V per phase, three-phase, Y connected source delivers power to the following delta-connect-d load:

Phase 1: $40\underline{/0°}$ Ω

Phase 2: $20\underline{/-60°}$ Ω

Phase 3: $15\underline{/45°}$ Ω.

Determine the phase currents, line currents, and line voltages. Show that the line currents add up to zero.

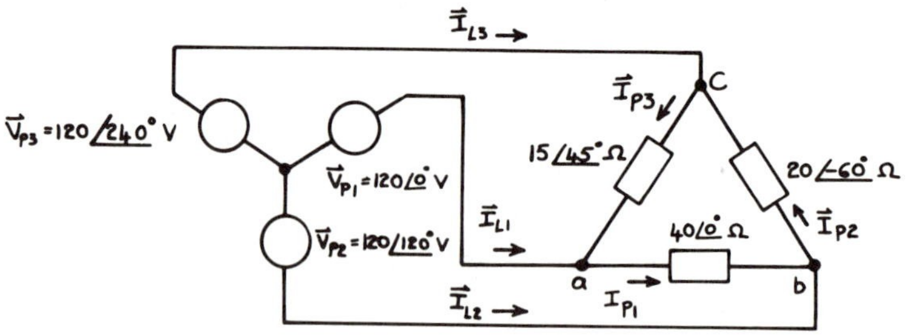

FIG. 1

Solution: The circuit diagram for this problem is indicated in Fig. 1. The line voltages are found from the formula:

$$|V_L| = \sqrt{3}|V_P|$$

Once the line voltages are found, the phase currents can be found by Ohm's Law. The line currents are then found by summing the proper phase currents. Notice that the line current cannot be found from the formula

$$I_L = \sqrt{3}\, I_P$$

because the delta-connected load is not balanced.

Since the Y-connected source is balanced, the line voltage is

$$V_L = \sqrt{3}\, 120 = 208 \text{ V.}$$

This is the magnitude of the voltage appearing across each load. V_{ab} will be adopted as the reference voltage, hence

$$\vec{V}_{L1} = \vec{V}_{ab} = 208\underline{/0°} \text{ V}$$

$$\vec{V}_{L2} = \vec{V}_{bc} = 208\underline{/120°} \text{ V}$$

$$\vec{V}_{L3} = \vec{V}_{ca} = 208\underline{/240°} \text{ V.}$$

Now the phase current can be found by Ohm's Law.

$$\vec{I}_{P1} = \frac{\vec{V}_{L1}}{40\underline{/0°}} = \frac{208\underline{/0°}}{40\underline{/0°}} = 5.2\underline{/0°} \text{ A}$$

$$\vec{I}_{P2} = \frac{\vec{V}_{L2}}{20\underline{/-60°}} = \frac{208\underline{/120°}}{20\underline{/-60°}} = 10.4\underline{/180°} \text{ A}$$

$$\vec{I}_{P3} = \frac{\vec{V}_{L3}}{15\underline{/45°}} = \frac{208\underline{/240°}}{15\underline{/45°}} = 13.8\underline{/195°} \text{ A.}$$

From Fig. 1 it is seen that

$$\vec{I}_{L1} = \vec{I}_{p1} - \vec{I}_{p3}.$$

Converting the phase currents to rectangular form we have

$$\vec{I}_{L1} = 5.2 + j0 - (-13.33 - j3.57) = 18.5 + j3.57 \text{ A.}$$

Similarly,

$$\vec{I}_{L2} = \vec{I}_{P2} - \vec{I}_{P1}$$

$$= -10.4 + j0 - (5.2 + j0) = -15.6 + j0 \text{ A}$$

$$\vec{I}_{L3} = \vec{I}_{P3} - \vec{I}_{P2}$$

$$= -13.33 - j3.57 - (-10.4+j0) = -2.9 - j3.57 \text{ A.}$$

When the three line currents are added it is seen that

$$\vec{I}_{L1} + \vec{I}_{L2} + \vec{I}_{L3} = 0 + j0 \text{ A.}$$

This must be true because there is no neutral line in a delta-connected load.

• PROBLEM 10-16

A balanced Δ-connected load contains 8 + j4Ω per phase, while a balanced Y-connected load consists of 2 - j1Ω per phase. Both loads are connected in parallel to a three-phase three-wire system in which V_L = 120 V rms at the loads. The three lines extending from the loads to the source each have a resistance of 0.2Ω. Find the total power: (a) delivered to the Δ-connected load; (b) delivered to the Y-connected load; (c) lost in the wires.

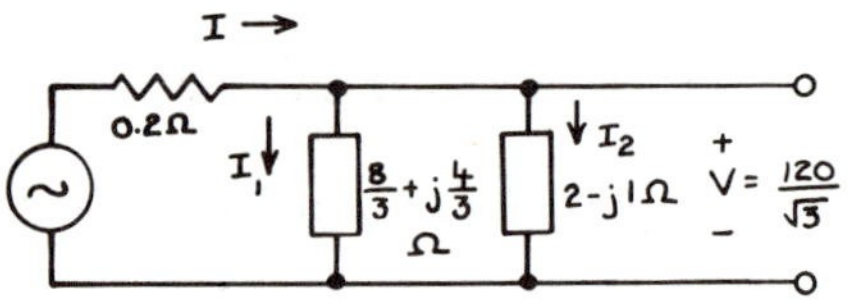

Solution: There are several ways this problem can be solved. One of the simplest is to convert everything to a line-to-neutral basis.

A 120 V rms line voltage is equivalent to a $\frac{120}{\sqrt{3}}$ V rms line-to-neutral voltage.

The delta-wye transformation for a balanced system is $Z_{wye} = \frac{1}{3} Z_{delta}$, so the 8+j4Ω per-phase delta load is equivalent to $\frac{8}{3} + j\frac{4}{3}$ Ω per-phase for a wye-connected load.

The original problem is now reduced to a single phase problem, as shown.

(a) The current in one phase of the transformed delta load is

$$\vec{I}_1 = \frac{V}{\frac{8}{3}+j\frac{4}{3}} = \frac{120}{\sqrt{3}} \frac{3}{4(2+j1)} = \frac{30\sqrt{3}(2-j1)}{(2+j1)(2-j1)} = 12\sqrt{3}-j6\sqrt{3} \text{ A rms}$$

The power delivered to one phase of this load is $|I_1|^2 \times \frac{8}{3}$, and the total power to this load is 3 times this, so

$$P_{delta} = 3\left|I_1\right|^2 \frac{8}{3} = 8(6\sqrt{3})^2|2-j1|^2 = (8)(36)(3)(5)$$

$$= 4.32 \text{ kW}$$

(b) The current in one phase of the original wye-connected load is

$$\vec{I}_2 = \frac{V}{2-j1} = \frac{120}{\sqrt{3}(2-j1)} \frac{2+j1}{2+j1} = \frac{24}{\sqrt{3}}(2+j1) = 8\sqrt{3}(2+j1) \text{ A rms}$$

Following the procedure as in (a),

$$P_{wye} = 3\left|I_2\right|^2 2 = (3)(64)(3)(5)(2) = 5.76 \text{ kW.}$$

(c) The line current is

$$\vec{I} = \vec{I}_1+\vec{I}_2 = 6\sqrt{3}(2-j1)+8\sqrt{3}(2+j1) = 28\sqrt{3}+j2\sqrt{3}$$

$$= 2\sqrt{3}(14+j1)$$

The power lost in all 3 wires is

$$P_{wires} = 3|I|^2 0.2 = (3)(4)(3)(197)(0.2) = 1.42 \text{ kW.}$$

• PROBLEM 10-17

The balanced three-phase load shown in Fig. 1 draws 20 kw at a lagging power factor of 0.4. The frequency is 60 hertz. (a) Calculate the total kvar rating of three capacitances which, when connected in delta or in wye in parallel with the load, will raise the power factor of the combination to 0.9 lagging. (b) If the line voltage at the load terminals is 1,300 volts, calculate the value of each capacitance of (a) for (1) delta connection; (2) wye connection.

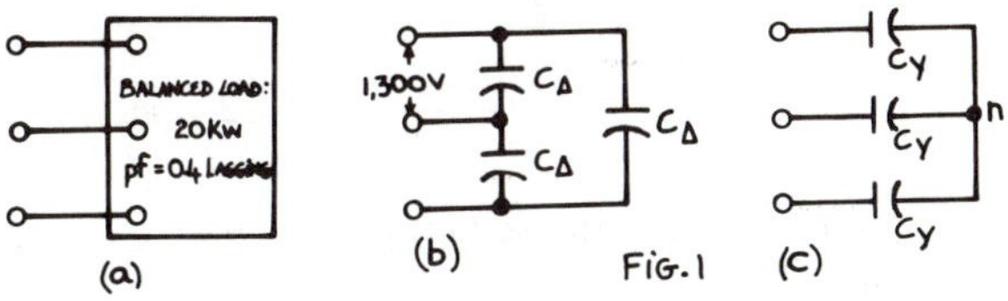

Fig. 1

Solution: (a) In this problem, the imaginary power rating Q at a capacitive load is to be found if the capacitive load is in parallel with another load with a given Q. The balanced load has a real power rating P_B = 20 kw and an imaginary power rating Q_B. Since the power factor for the balanced load is given, Q_B can be found from Fig. 2(a).

$$\theta_a = \text{Cos}^{-1}(0.4) = 66.4°.$$

Hence, $Q_B = -P_B \tan\theta_a = -20$ kva tan 66.4° = -45.8 kvar.

When the capacitive load is in parallel with the load of Fig. 1(a), P for the combined load remains the same, while Q for the combined load changes. The new power factor is 0.9. Hence, the angle θ_b changes to

$$\theta_b = \text{Cos}^{-1}(0.9) = 25.8°.$$

θ_b decreased, therefore Q must have increased to

$$Q_{BC} = -P_B \tan(25.8°) = -9.7 \text{ kvar}.$$

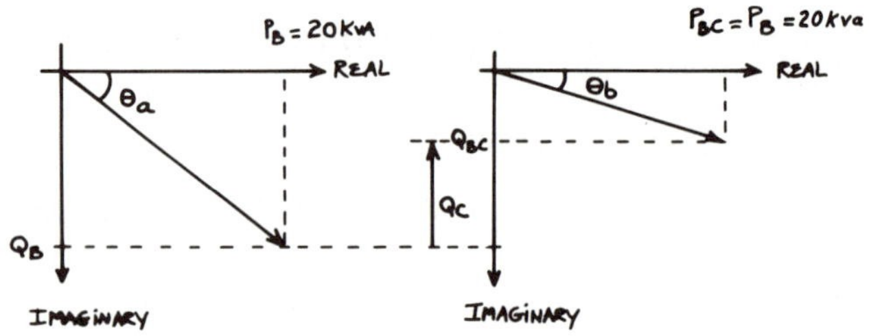

FIG. 2 REAL AND IMAGINARY POWER RATINGS FOR
a) LOAD OF FIG 1(a)
b) LOAD OF FIG 1(a) IN PARALLEL WITH CAPACITIVE LOAD

Since Q_B increased, the imaginary power for the capacitive load, Q_C, must be positive.

$$Q_C = Q_B - Q_{BC} = -45.8 - (-9.7) - 36.1 \text{ kvar}$$

Q for each capacitor is

$$\frac{Q_C}{3} = 12 \text{ kvar}.$$

(b) If the capacitors are delta connected, the voltage across each capacitor is 1300V and the power consumed by each is 12 kvar (found in part (a)). Hence, the current through each capacitor I_C is

$$I_C = \frac{Q}{V} = \frac{12,000}{1,300} = 9.23 \text{ amps}.$$

The impedance of each capacitor is $\frac{1}{\omega C}$. Since f = 60 H_z; $\omega = 2\pi f = 377$.

Hence,

$$Z_{C\Delta} = \frac{1}{377C} = \frac{V}{I} = \frac{1,300}{9.23} = 141\ \Omega$$

$$C_\Delta = \frac{1}{141 \cdot 377} = 18.8\ \mu f.$$

If the capacitors are Y connected, then the voltage across each capacitor is $V_C = \frac{1300}{\sqrt{3}} = 751$ V. The power consumed by each capacitor is still 12 kva. Hence, I_C is

$$I_C = \frac{Q}{V} = \frac{12,000}{751} = 16 \text{ amps}$$

Therefore,

$$Z_{cy} = \frac{1}{377C} = \frac{V}{I} = \frac{751}{16} = 47\ \Omega$$

$$C_y = \frac{1}{47 \cdot 377} = 56.5\ \mu f.$$

Notice that C_y could have been found by multiplying C_Δ by 3.

• PROBLEM 10-18

Two balanced three-phase loads are connected in parallel (Fig. 1). Load 1 draws 20 kw at a lagging power factor of 0.85; load 2 draws 15 kw at a lagging power factor of 0.5. The line voltage at the terminals of the load is 440 volts. The load terminals are connected to the source terminals by means of cables which have an impedance of 1 + j3 ohms each. Calculate (a) the line current drawn by the combination of the two loads; (b) the line voltage at the terminals of the source; (c) the power factor at the load terminals and at the source terminals.

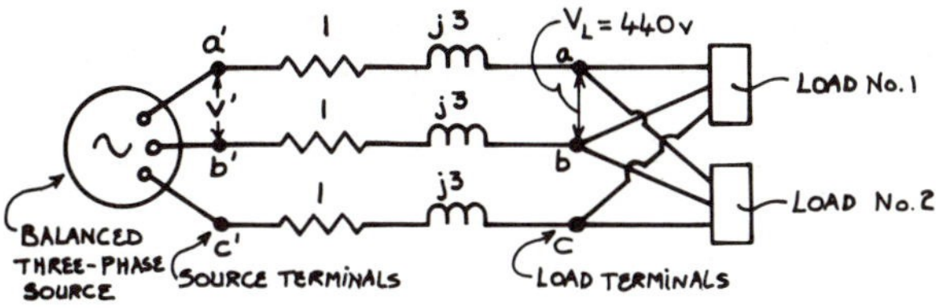

FIG. 1

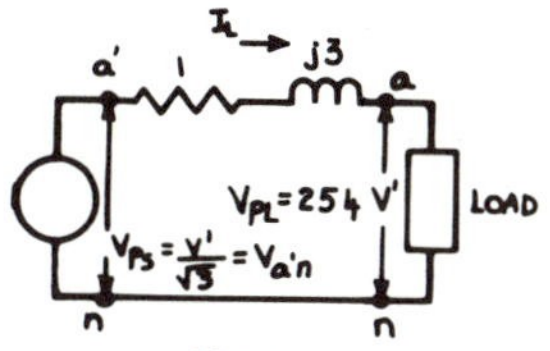

FIG. 2

Solution: (a) The line current can be found from the equation P = VI. In order to use this equation, the combined power of the two loads must be found. For load 1, the real power quantity, P_1, is 20 kw and θ is $\cos^{-1}(.85) = 31.8°$. From P and θ the imaginary power Q_1 is found to be

$$Q_1 = -P_1 \tan\theta_1 = (-20 \text{ kw}) \quad \tan(31.8) = -12.4 \text{ kvar.}$$

For load 2, the real power quantity, P_2, is 15 kw and θ_2 is $\cos^{-1}(.5) = 60°$. The imaginary power Q_2 is found to be

$$Q_2 = -P_2 \tan\theta_2 = -15 \tan(60°) = -26 \text{ kvar.}$$

The combination of loads 1 and 2 draw

$$P = P_1 + P_2 = 20 + 15 = 35 \text{ kw}$$

$$Q = Q_1 + Q_2 = -12.4 - 26 = -38.4 \text{ kvar.}$$

Now the single phase representation is made. Assuming a wye connected source and load, each phase draws

$$P_P = \frac{P}{3} = 11.7 \text{ kW} \tag{1}$$

$$Q_P = \frac{Q}{3} = -12.8 \text{ kvar.} \tag{2}$$

Hence, the magnitude of the power drawn per-phase is

$$\left|P_P\right| = \sqrt{\left(\frac{P}{3}\right)^2 + \left(\frac{Q}{3}\right)^2} = \sqrt{(11.7)^2 + (12.8)^2} = 17.3 \text{ kva.}$$

The load voltage per-phase is

$$V_{PL} = \frac{V_{Line}}{\sqrt{3}} = \frac{440}{\sqrt{3}} = 254 \text{ V.} \tag{3}$$

Using P = VI, the line current is calculated

$$I_L = \frac{P}{V_{PL}} = \frac{17.3 \times 10^3}{254} = 68 \text{ A.}$$

(b) To find the line voltage at the terminal of the source, one of the three phases is drawn in Fig. 2. Once the voltage across the single phase source is found, the line voltage can be found. Applying KVL around the loop yields

$$\vec{V}_{a'n} = \vec{V}_{a'a} + \vec{V}_{an} . \tag{4}$$

If I_{an} is chosen to be at 0° (the reference axis) then $\vec{I}_{an} = I_L\underline{/0°} = 68\underline{/0°}$ amps. $\vec{V}_{an}$ is then $254\underline{/\theta}$ where θ is determined by the power factor of the combined loads. It is seen that

$$\theta = \text{Tan}^{-1}\frac{-Q_P}{P_P} = \tan^{-1}\frac{12.8}{11.7} = 47.6°.$$

Hence, $\vec{V}_{an} = 254\underline{/47.6°}$. (5)

$V_{a'a}$ is now calculated to be

$$\vec{V}_{a'a} = \vec{I}_{an}\,(1+j3) = 68(1+j3) = 68 + j204 \text{ V.} \tag{6}$$

Putting (5) and (6) into (4) gives

$$\vec{V}_{a'n} = 254\underline{/47.6°} + 68 + j204$$

$$\vec{V}_{a'n} = 254 \text{ Cos } 47.6 + j254 \text{ Sin } 47.6 + 68 + j204$$

$$= 171 + j187 + 68 + j204 = 239 + j391 \text{ V}$$

$$= |V_{a'n}| \underline{/\theta_{a'n}} \text{ V}$$

where $|V_{a'n}| = \sqrt{(239)^2 + (391)^2} = 458$

and $\theta_{a'n} = \tan^{-1} \frac{391}{239} = 58.5°.$

Therefore

$$V_{a'n} = 458\underline{/58.5°} \text{ V.}$$

The line voltage then is

$$V' = \sqrt{3}|V_{a'n}|$$

$$V' = 458\sqrt{3} = 793 \text{ V.}$$

(c) The power factor for the load is the cosine of the angle between $\vec{V}_{an}$ and $\vec{I}_{an}$, found in part (b).

$$pf = \cos 47.6° = 0.67 \text{ lagging}$$

Similarly, the power factor for the source is the cosine of the angle between $\vec{I}_{a'n}$ and $\vec{V}_{a'n}$.

$$pf = \cos 58.5° = 0.52 \text{ lagging}$$

CHAPTER 11

LAPLACE TRANSFORM TECHNIQUES

SIMPLE TIME FUNCTIONS

• **PROBLEM 11-1**

Write an expression for $V_C(t)$ in Fig. 1 using the unit ramp function, defined as

$$Au_1(t-T) = \begin{cases} At \; ; \; (t-T) \geq 0 \\ 0 \; ; \; (t-T) < 0 \end{cases}$$

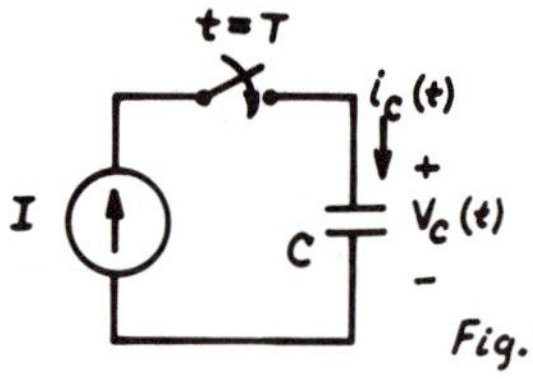

Fig.1

Solution: The voltage $V_C(t)$ is

$$v_C(t) = \frac{1}{C}\int_{-\infty}^{t} i_C(t)dt.$$

However, $\quad i_C(t) = Iu(t-T).$

Hence

$$v_C(t) = \frac{1}{C}\int_{-\infty}^{t} Iu(t-T)dt$$

since

$$u(t-T) = \begin{cases} 1 \; ; \; t \geq T \\ 0 \; ; \; t < T \end{cases}$$

Then

$$v_C(t) = \begin{cases} \frac{1}{C}\int_{-\infty}^{T} 0 \quad dt + \frac{1}{C}\int_{T}^{t} I \, dt \; ; \; T \geq 0 \\ \frac{1}{C}\int_{-\infty}^{t} 0 \cdot dt \qquad\qquad ; \; T < 0 \end{cases}$$

Hence

$$v_C(t) = \frac{I}{C}(t-T)u(t-T)$$

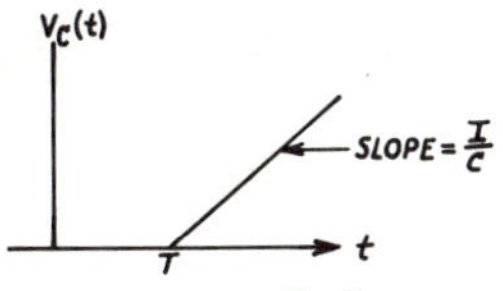

Fig. 2

Using unit ramp function notation, we can write

$$v_C(t) = \frac{I}{C} u_1(t-T)$$

analogous with the unit step-function.

• PROBLEM 11-2

Develop an analytical expression for the signal shown using step-function notation.

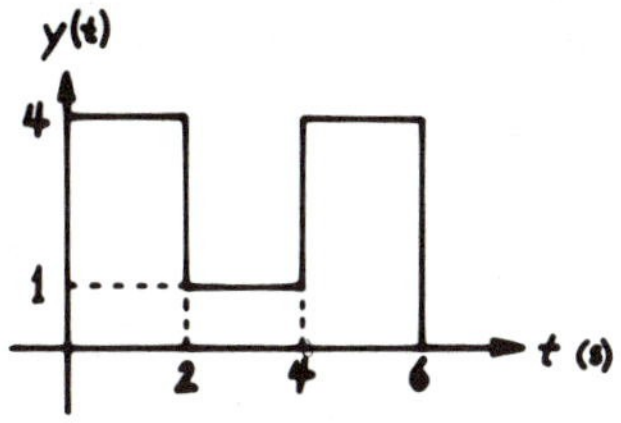

Solution: The appropriate step function notation is

$$u(t) = \begin{cases} 1, & t \geq 0 \\ 0, & t < 0 \end{cases}$$

Any piece-wise constant composite function can be written as a superposition of suitably delayed steps multiplied by appropriate constants. By inspection,

$$y(t) = 4\, u(t) - 3\, u(t-2) + 3\, u(t-4) - 4\, u(t-6)$$

• PROBLEM 11-3

Write the signal f(t) shown in Fig. 1 using the step functions.

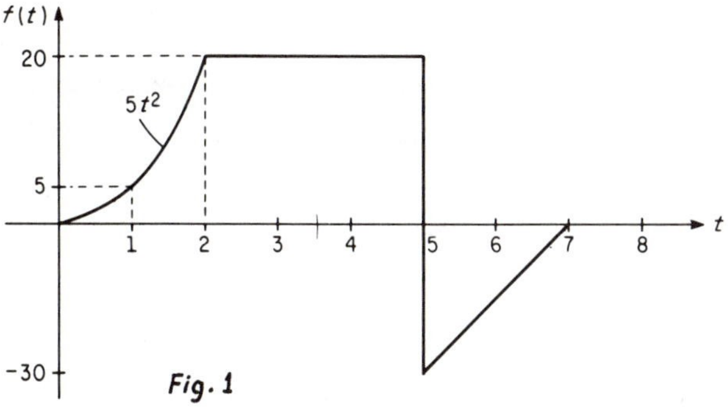

Fig. 1

Solution: From Fig. 1 we can write a piece-wise representation of the signal f(t) as

$$f(t) = \begin{cases} 0 & t < 0 \\ 5t^2 & 0 \le t < 2 \\ 20 & 2 \le t < 5 \\ 15(t-7) & 5 \le t < 7 \\ 0 & 7 \le t \end{cases}$$

Write the function f(t) as the sum of the three functions $f_1(t)$, $f_2(t)$, and $f_3(t)$ shown below.

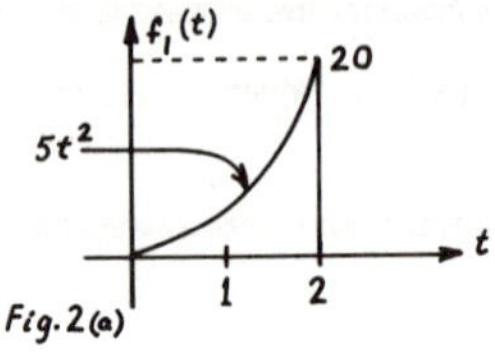

Fig. 2(a)

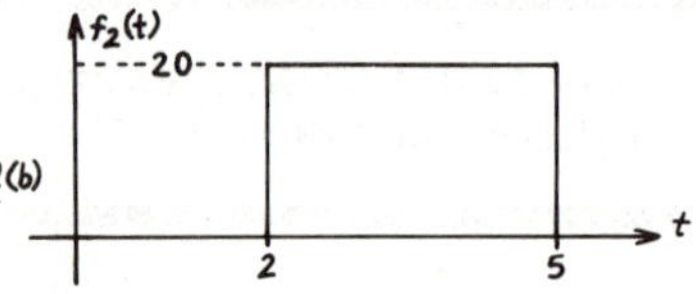

Fig. 2(b)

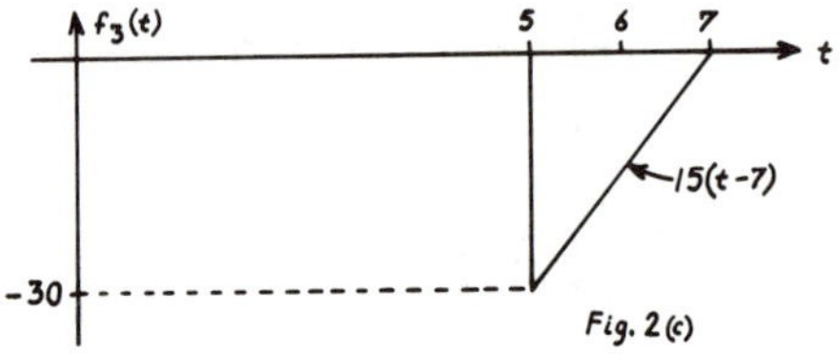

Fig. 2(c)

We can write in step function notation for each waveform in Figs. 2a, b, and c

$$f_1(t) = 5t^2 \, [u(t) - u(t-2)]$$

$$f_2(t) = 20 \, [u(t-2) - u(t-5)]$$

$$f_3(t) = 15(t-7) \, [u(t-5) - u(t-7)]$$

respectively.

Since $f(t) = f_1(t) + f_2(t) + f_3(t)$

$$f(t) = 5t^2[u(t)-u(t-2)] + 20[u(t-2)-u(t-5)] + 15(t-7)[u(t-5)-u(t-7)]$$

• **PROBLEM 11-4**

Sketch each of the following waveforms as functions of time and calculate its value at t = 2s:

(a) $u(t+3) - u(t-3)$; (b) $u(3+t) - u(3-t)$;

(c) $t\,u(t) + u(t-1.5) - (t-3)\,u(t-3)$;

(d) $e^{-t}\,u(t) - e^{-1}\,u(t-1)$; (e) $u(t-t^2)$

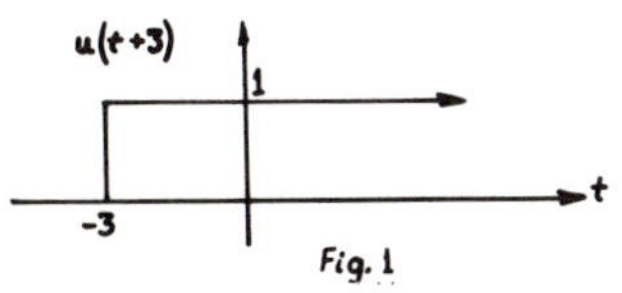

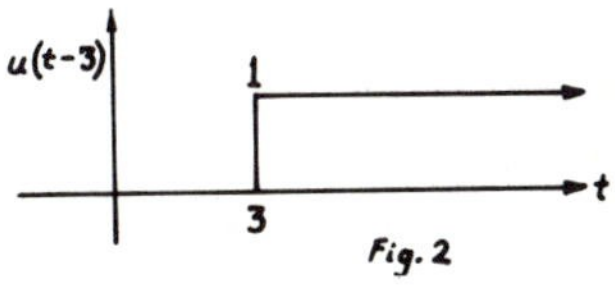

Solution: (a) If we draw the graph for each step function and its argument and subtract them, the graph of the entire function is obtained.

If we subtract the function in Fig. 2 from that in Fig. 1 the function u(t+3) - u(t-3) in Fig. 3 is obtained.

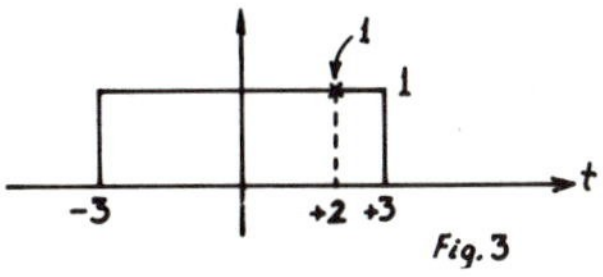

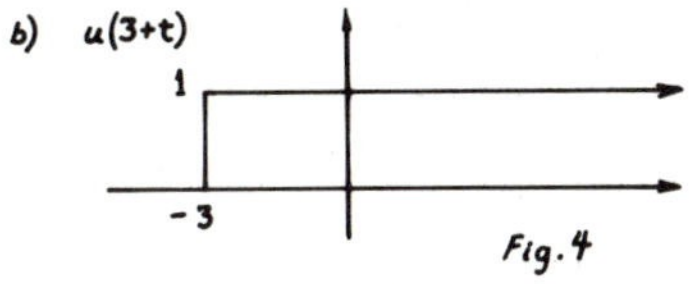

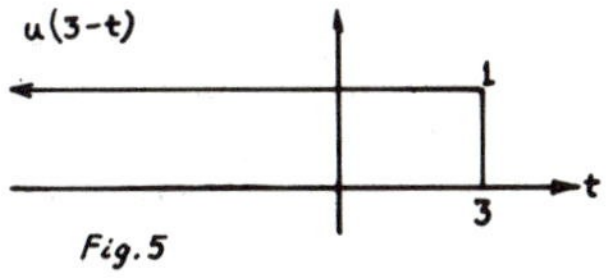

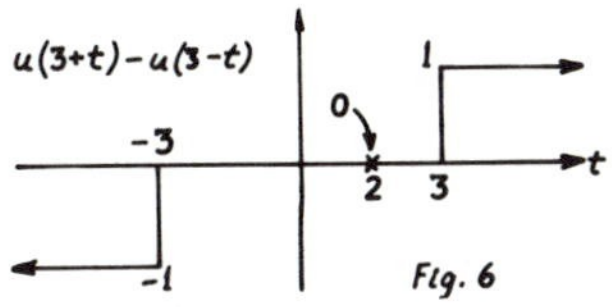

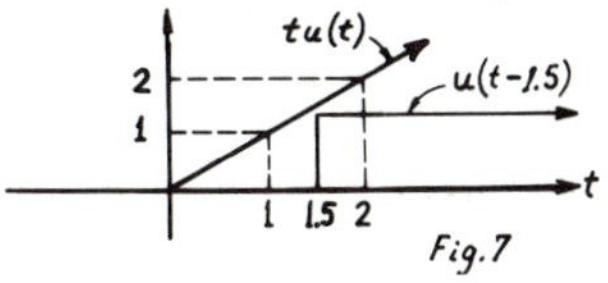

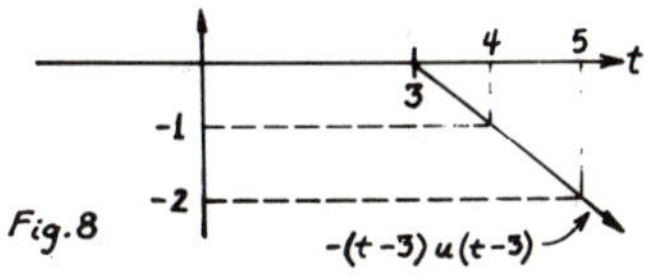

(c) The three functions and their arguments are shown in Figs. 7 and 8; if they are added we obtain the plot in Fig. 9.

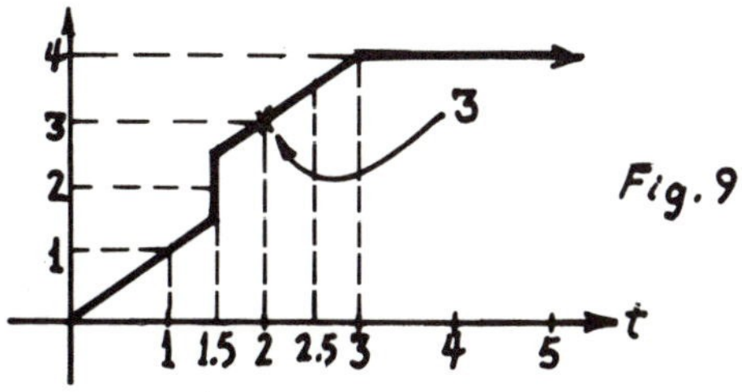

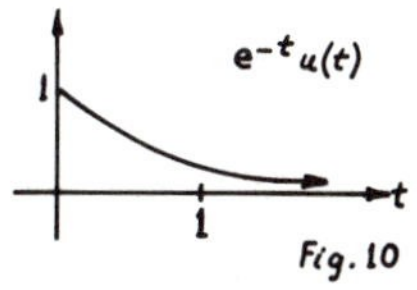

(d) The function e^{-t} has a value of 1 at t = 0. Starting at t = 1, subtract e^{-1} = .368. The value of the function e^{-t} at t = 1 is e^{-1}, thus, the function $e^{-t}u(t)$ - $e^{-1}(t-1)$ is zero at t = 0. The function is plotted in Fig. 11.

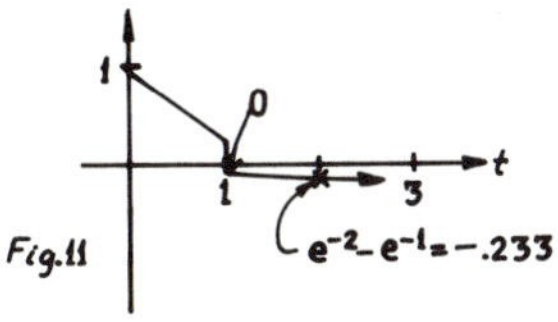

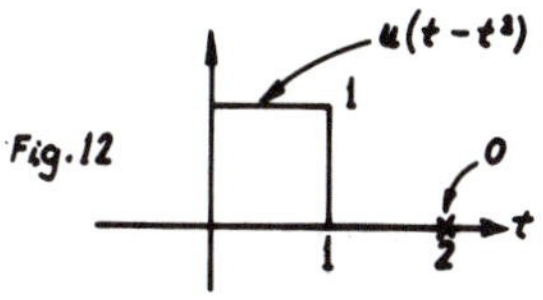

(e) Only, 0 < t < 1, values of t give a positive argument.

• PROBLEM 11-5

Use step-function notation

$$A\,u(t-K) = \begin{cases} A \; ; & t \geq K \\ 0 \; ; & t < K \end{cases}$$

to write an expression for the output $v_a(t)$ when the switch in Fig. 1 is closed at $t = T$. $(T > 0)$

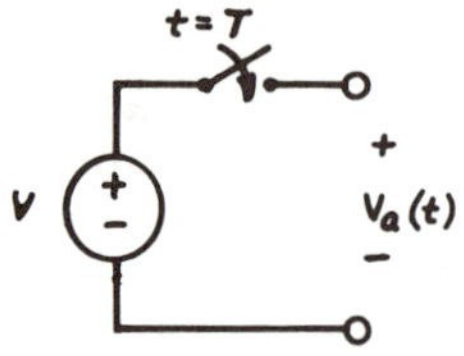

Fig. 1

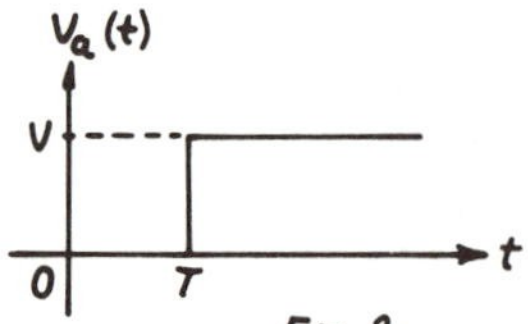

Fig. 2

Solution: It is obvious that the output $v_a(t)$ will be a step function shown in Fig. 2. Hence, we need to find the value of K for

$$V\,u(t-K) = \begin{cases} V & ; \ t \geq T \\ 0 & ; \ t < T \end{cases}$$

Since the value (t-K) will be positive or zero when the first condition above is satisfied, namely $t \geq T$, we can establish that:

$$(t-K) \geq 0, \quad \text{when } t \geq T \text{ and the output is } V.$$

Taking $t \geq T$ and subtracting T from both sides of the inequality, yields

$$t - T \geq 0, \quad \text{hence } t - T = t - K \text{ and } K = T.$$

Our expression for $v_a(t)$ can be written as

$$v_a(t) = V\,u(t-T).$$

• PROBLEM 11-6

Find the first derivative of the function f(t) shown below.

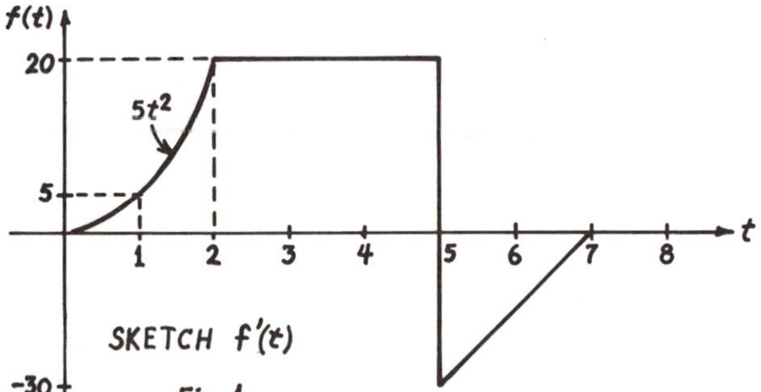

Fig. 1

Solution: We can represent f(t) piecewise as

$$f(t) = \begin{cases} 0 & ; \ t < 0 \\ 5t^2 & ; \ 0 \leq t < 2 \\ 20 & ; \ 2 \leq t < 5 \\ 15(t-7) & ; \ 5 \leq t < 7 \\ 0 & ; \ 7 \leq t \end{cases}$$

where we define,

$$f_1(t) = 5t^2; \; 0 \le t < 2$$

$$f_2(t) = 20 \; ; \; 2 \le t < 5$$

$$f_3(t) = 15(t-7); \; 5 \le t < 7.$$

Using unit step function notation,

$$f_1(t) = 5t^2[u(t) - u(t-2)]$$

$$f_2(t) = 20[u(t-2) - u(t-5)]$$

$$f_3(t) = 15(t-7)[u(t-5) - u(t-7)].$$

Hence,

$$f(t) = f_1(t) + f_2(t) + f_3(t)$$

and $$f'(t) = \frac{df(t)}{dt} = \frac{d}{dt} f_1(t) + \frac{d}{dt} f_2(t) + \frac{d}{dt} f_3(t).$$

Taking the derivative of each function yields

$$\frac{d}{dt} f_1(t) = 10t[u(t)-u(t-2)] + 5t^2[\delta(t)-\delta(t-2)],$$

since $\frac{d\,u(t)}{dt} = \delta(t)$ and $\frac{d}{dt} u(t-T) = \delta(t-T)$

$$\frac{d}{dt} f_2(t) = 0[u(t-2)-u(t-5)] + 20[\delta(t-2)-\delta(t-5)]$$

and $$\frac{d}{dt} f_3(t) = 15[u(t-5)-u(t-7)] + 15(t-7)[\delta(t-5)-\delta(t-7)]$$

Make use of the sampling property which states

$$f(t)\delta(t) = f(0)\delta(t).$$

Hence,

$$\frac{d}{dt} f_1(t) = 10t[u(t)-u(t-2)] + 0\delta(t) - 20\delta(t-2)$$

$$\frac{d}{dt} f_2(t) = 20\delta(t-2) - 20\delta(t-5)$$

and

$$\frac{df_3(t)}{dt} = 15[u(t-5) - u(t-7)] - 30\delta(t-5).$$

Hence,

$$\frac{d}{dt} f(t) = 10t[u(t)-u(t-2)] - 50\delta(t-5) + 15[u(t-5)-u(t-7)]$$

A sketch of $\frac{d}{dt} f(t)$ is shown in Fig. 2.

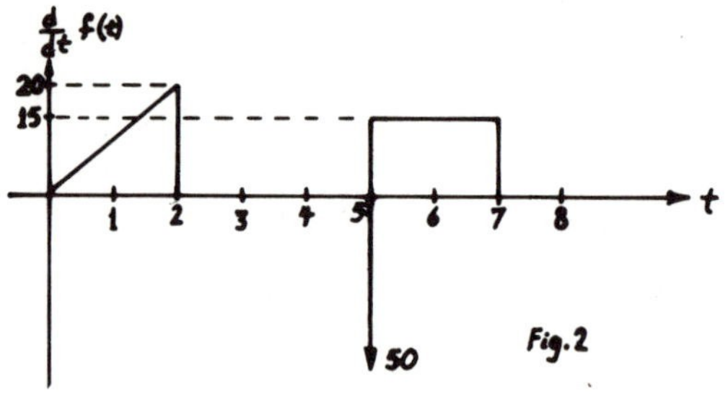

Fig.2

LAPLACE TRANSFORM OF TIME FUNCTIONS

• PROBLEM 11-7

Determine the Laplace transform of the functions of time depicted in Fig. 1.

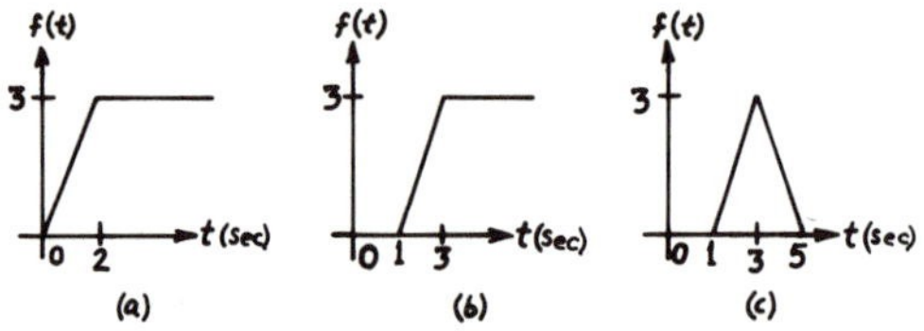

Fig. 1

Solution: (a) In order to solve this problem, the function in Fig. 1a is broken down into two functions. The two functions are then transformed into their Laplace identities and then recombined to form the Laplace transform of the entire function. The function of Fig. 1a can be considered as the sum of $f_2(t)$ and $f_3(t)$ in Figs. 2 and 3.

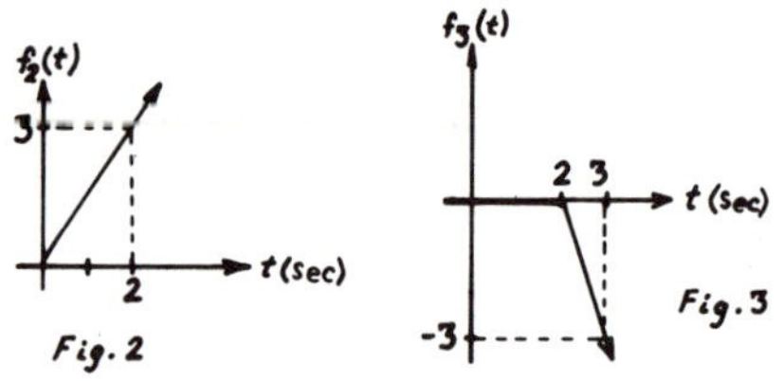

Fig. 2 is a ramp function, $f_2(t) = 1.5t(u(t))$. The Laplace transform of this function is $F_2(s) = \frac{1.5}{s^2}$. Fig. 3 is the same ramp function inverted and shifted in time by 2 seconds to the right. $f_3(t) = -1.5(t-2)(u(t-2))$.

The Laplace transform of $f_3(t)$ is found by the shifting theorem, which states that when a function f(t) is shifted on the t axis by an amount, a, the new function, f(t-a), has the Laplace transform $e^{-as}F(s)$ where $F(s) = L[f(t)]$. The function in Fig. 3 is shifted by 2 seconds; therefore a = 2 and

$$F_3(s) = \frac{-1.5e^{-2s}}{s^2} .$$

The Laplace transform of the function in Fig. 1a is

$$F(s) = F_2(s) + F_3(s)$$

$$F(s) = \frac{1.5(1-e^{-2s})}{s^2} .$$

(b) Fig. 1(b) is the function in Fig. 1(a) shifted by 1 second. Using the shifting theorem, the Laplace transform of the function is $e^{-s}F(s)$ where F(s) was found in (a). Hence, the Laplace transform for this function is

$$\frac{1.5e^{-s}(1-e^{-2s})}{s^2} .$$

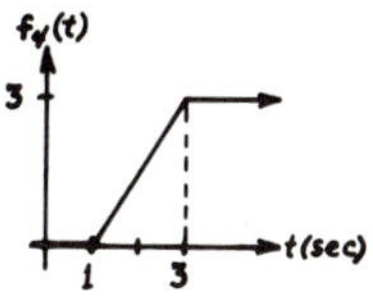

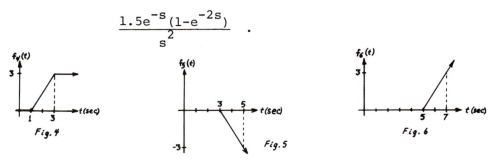

Fig. 4

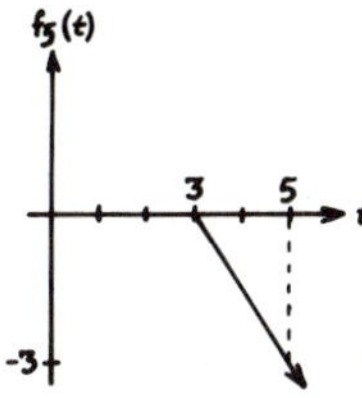

Fig. 5

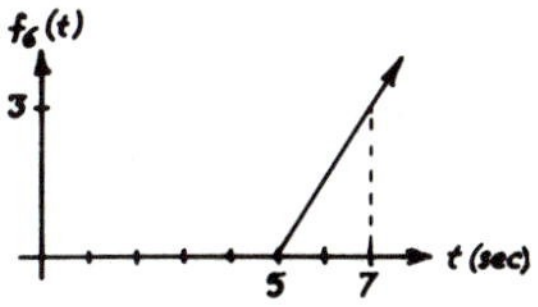

Fig. 6

(c) The function in Fig. 1(c) can be considered as the sum of the functions in Figs. 4, 5, and 6. The Laplace of $f_4(t)$ was found in part (b) to be

$$F_4(s) = \frac{1.5e^{-s}(1-e^{-2s})}{s^2} .$$

$f_5(t)$ is $f_3(t)$ shifted by 1 sec; therefore,

$$F_5(s) = F_3(s)e^{-s} = \frac{-1.5e^{-3s}}{s^2} .$$

Similarly, $f_6(t)$ is $f_2(t)$ shifted by 5 seconds; hence,

$$F_6(s) = F_2(s)e^{-5s} = \frac{1.5e^{-5s}}{s^2} .$$

The Laplace transform for the function in Fig. 1(c) is

$$F(s) = F_4(s) + F_5(s) + F_6(s)$$

$$F(s) = \frac{1.5}{s^2}(e^{-s} - e^{-3s} - e^{-3s} + e^{-5s})$$

$$F(s) = \frac{1.5}{s^2}(e^{-s} - 2e^{-3s} + e^{-5s}).$$

Find the Laplace transform of the modified ramp function $v_a(t)$ shown in Fig. 1(a), and use the result to find the Laplace transform of the trapezoidal pulse shown in Fig. 2.

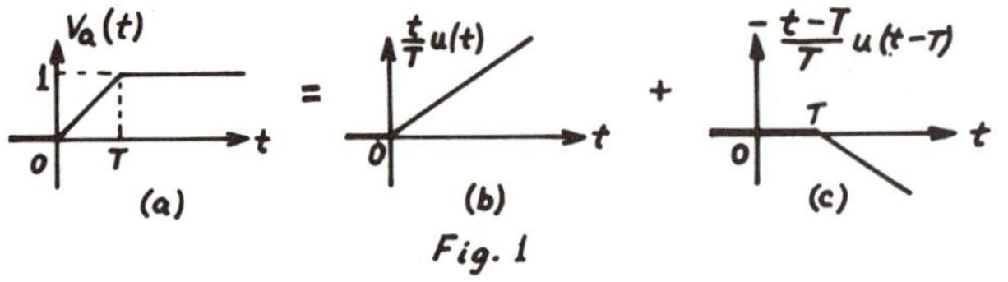

Fig. 1

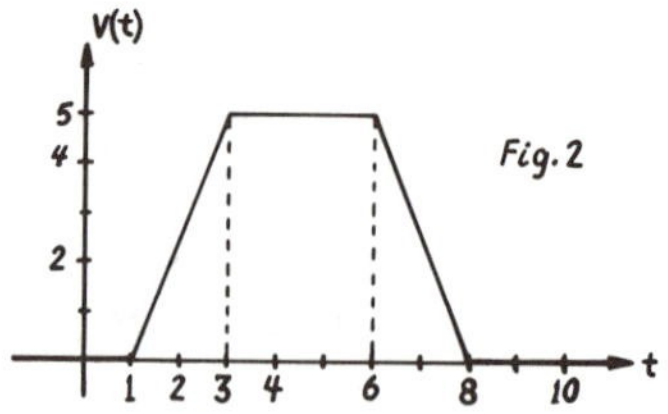

Fig. 2

Solution: Taking the Laplace transform of the ramp in Fig. 1(b) gives $L\{\frac{1}{T}\ tu(t)\} = \frac{1}{T}\ L\{tu(t)\} = \frac{1}{T}\ \left(\frac{1}{s^2}\right)$. By using the time-axis translation property of the Laplace transform expressed as:

$$L\{f(t-t_o)u(t-t_o)\} = e^{-st_o}F(s),$$

the ramp is transformed in Fig. 1(c) by multiplying $L\{-\frac{1}{T}\ tu(t)\}$ by e^{-st}.

Therefore, $$L\{-\frac{t-T}{T}\ u(t-T)\} = -\frac{1}{T}\ (\frac{1}{s^2})\ e^{-st}.$$

By adding the transform of Fig. 1(b) to that of Fig. 1(c) the transform of the modified ramp v_a is obtained.

$$v_a(s) = \left[\frac{1}{T}\quad \frac{1}{s^2} - \frac{1}{s^2}\ e^{-st}\right]$$

The trapezoidal pulse can be broken down into four ramp functions which add to form the trapezoidal pulse. Fig. 3 (a-d) shows the four ramps.

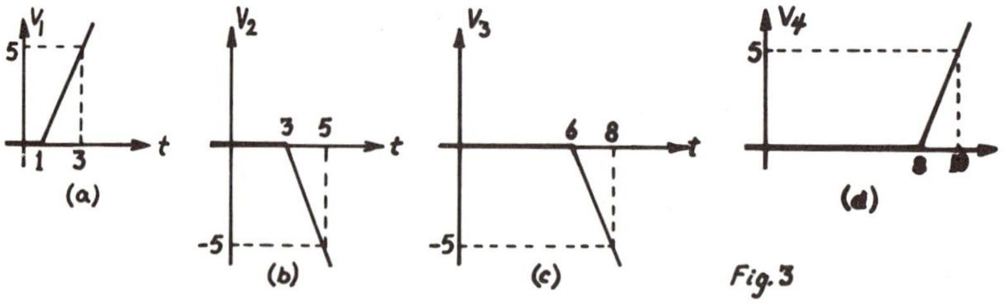

Fig. 3

$$v(t) = v_1 + v_2 + v_3 + v_4$$

Note that here, T = 2.

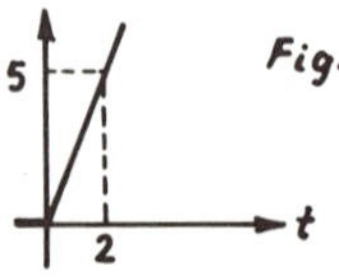

Fig. 4

Since the ramp in Fig. 4 can be represented as $\frac{5}{2}t\ u(t)$ and its Laplace transform as $L\left[\frac{5}{2}t\ u(t)\right] = \frac{5}{2}\frac{1}{s^2}$, we get the Laplace transform for all the ramps in Figs. 3 (a-d) by multiplying by e^{-st_o} for v_1, $t_o = 1$; v_2, $t_o = 3$; v_3, $t_o = 6$; and v_4, $t_o = 8$. Note that v_2 and v_3 have negative slopes. The following can be written directly

$$v(s) = v_1(s) + v_2(s) + v_3(s) + v_4(s)$$

$$v(s) = \frac{5}{2}\frac{1}{s^2}e^{-s} - \frac{5}{2}\frac{1}{s^2}e^{-3s} - \frac{5}{2}\frac{1}{s^2}e^{-6s} + \frac{5}{2}\frac{1}{s^2}e^{-8s}$$

• PROBLEM 11-9

Find the Laplace transform of the periodic waveform shown in Fig. 1.

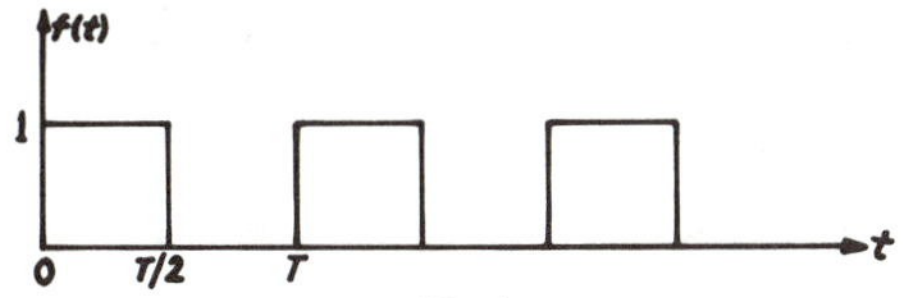

Fig. 1

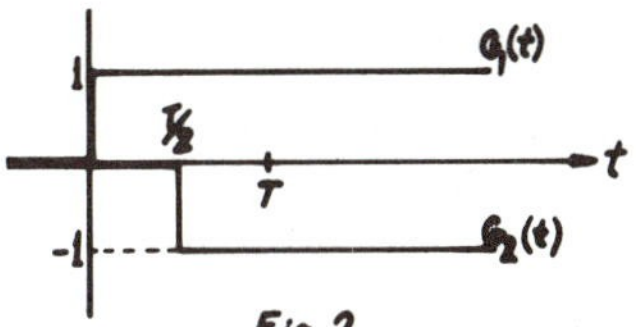

Fig. 2

Solution: Express the Laplace transform of a semi-periodic function by writing it in the form,

$$F(s) = \frac{G(s)}{1 - e^{-sT}}, \quad \text{where T is the period of}$$

the function f(t) and G(s) is the Laplace transform of the function for $0 < t < T$.

G(t) is mathematically expressed as,

$$G(t) \begin{cases} 1\ ; & 0 < t < T/2 \\ 0\ ; & T/2 < t \leq T . \end{cases}$$

But, in order to illustrate the transformation of G(t) to G(s), write

$$G(t) = G_1(t) + G_2(t) \quad \text{where}$$

$$G_1(t) = \quad 1\ ;\ t > 0$$

$$G_2(t) = -1\ ;\ t > T/2$$

Fig. 2 shows that $G_1(t) + G_2(t) = G(t)$ thus $G_1(s) + G_2(s) = G(s)$. We find $G_1(s)$ to be $\frac{1}{s}$ and $-G_2(t)$ to be the same function as $G_1(t)$ delayed by T/2. A time shift in the time domain is equivalent to multiplying by $e^{-s\tau}$ where τ is the time shift interval, thus $G_2(s) = \frac{-e^{-s\,T/2}}{s}$ and

$$G(s) = G_1(s) + G_2(s)$$

$$G(s) = \frac{1 - e^{-s\,T/2}}{s}.$$

The Laplace transform of the periodic function is

$$F(s) = \frac{G(s)}{1-e^{-sT}} = \frac{1-e^{-s\,T/2}}{s(1-e^{-sT})}$$

• PROBLEM 11-10

Find the Laplace transforms of the periodic functions illustrated in Fig. 1.

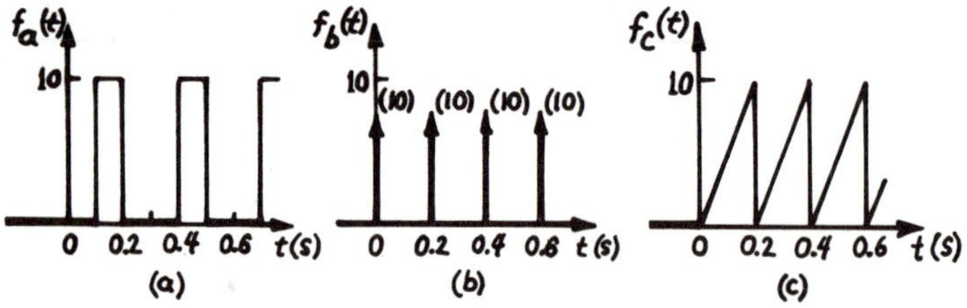

Fig 1

Solution: We first develop a general theorem for the Laplace transform of a periodic function. The definition for the LaPlace transform of a function f(t) is

$$L\{f(t)\} = \int_0^\infty f(t)e^{-st}dt. \tag{1}$$

The right hand side can be expanded as

$$L\{f(t)\} = \int_0^\tau f(t)e^{-st}dt + \int_\tau^{2\tau} f(t)e^{-st}dt + \int_{2\tau}^{3\tau} f(t)e^{-st}dt + \ldots \tag{2}$$

Now change variables to replace t by t + τ in the 2d term, t + 2τ in the 3d term, t + 3τ in the 4th term, etc.

$$L\{f(t)\} = \int_0^\tau f(t)e^{-st}dt + \int_0^\tau f(t+\tau)e^{-s(t+\tau)}d(t+\tau)$$

$$+ \int_0^\tau f(t+2\tau)e^{-s(t+2\tau)}d(t+2\tau) + \ldots \quad (3)$$

$$= \int_0^\tau f(t)e^{-st}dt + e^{-s\tau}\int_0^\tau f(t)e^{-st}dt$$

$$+ e^{-2s\tau}\int_0^\tau f(t)e^{-st}dt + \ldots \quad (4)$$

Where we have made use of the periodicity to set

$$f(t) = f(t+\tau) = f(t+2\tau) = f(t+3\tau) = \ldots \quad (5)$$

provided τ is a period of f(t). Next, factor $\int_0^\tau f(t)e^{-st}dt$ from (4) and use the identity

$$\frac{1}{1-e^{-s\tau}} = 1 + e^{-s\tau} + e^{-2s\tau} + e^{-3s\tau} + \ldots, \quad (6)$$

which can be verified by long division, to write

$$L\{f(t)\} = \frac{\int_0^\tau f(t)e^{-st}dt}{1 - e^{-s\tau}} \quad . \quad (7)$$

This is a general result, applicable to any Laplace-transformable periodic function.

(a) The function shown graphically has a period of 0.3 seconds, and over the interval $0 \le t < 0.2$,

$$f_a(t) = \begin{cases} 0, & 0 \le t < 0.1 \text{ sec} \\ 10, & 0.1 \le t < 0.2 \text{ sec} \\ 0, & 0.2 \le t < 0.3 \text{ sec} \end{cases}$$

By (7),

$$L\{f_a(t)\} = \frac{\int_{0.1}^{0.2} 10e^{-st}dt}{1 - e^{-0.3s}} \quad . \quad (8)$$

Evaluating the integral,

$$\int_{0.1}^{0.2} 10e^{-st}dt = 10\left[\frac{e^{-st}}{-s}\right]_{0.1}^{0.2} = \frac{10}{-s}\left(e^{-0.2s} - e^{-0.1s}\right), \tag{9}$$

and substituting (9) into (8) produces the final result:

$$L\left\{f_a(t)\right\} = \frac{10(e^{-0.1s} - e^{-0.2s})}{s(1 - e^{-0.3s})} .$$

(b) The function $f_b(t)$, shown graphically, has a period of $\tau = 0.2$ seconds. Over the interval $0 \leq t < 0.2$ seconds the function may be written

$$f_b(t) = \begin{cases} 10\delta(t), & t = 0 \text{ sec} \\ 0, & 0 < t < 0.2 \text{ sec} , \end{cases}$$

where $\delta(t)$ represents a unit impulse.

By (7),

$$L\{f_b(t)\} = \frac{\int_{0-}^{0+} 10\delta(t)e^{-st}dt}{1 - e^{-0.2s}} = \frac{10}{1 - e^{-0.2s}}$$

(c) The function $f_c(t)$, shown graphically, has a period of $\tau = 0.2$ seconds. Over the interval $0 \leq t < 0.2$ seconds the function may be written

$$f_c(t) = 50t, \; 0 \leq t < 0.2 \text{ seconds.}$$

By (7),

$$L\{f_c(t)\} = \frac{\int_0^{0.2} 50te^{-st}dt}{1 - e^{-0.2s}} = \frac{50}{1 - e^{-0.2s}} \int_0^{0.2} t\, e^{-st}dt. \tag{10}$$

The integral may be evaluated by integrating by parts, using the relation

$$\int_a^b udv = uv\Big|_a^b - \int_a^b vdu. \tag{11}$$

Setting $u = t$, $dv = e^{-st}dt$, we have $du = dt$ and $v = \frac{1}{-s}e^{-st}$, and

$$\int_0^{0.2} te^{-st}dt = \frac{t}{-s}\, e^{-st}\Big|_0^{0.2} - \int_0^{0.2} \frac{1}{-s}\, e^{-st}dt$$

$$= \frac{0.2}{-s}\, e^{-0.2s} - \frac{1}{s^2}\, e^{-st}\Big|_0^{0.2}$$

$$= \frac{0.2}{-s}\, e^{-0.2s} - \frac{1}{s^2}\, [e^{-0.2s} - 1] \tag{12}$$

Substituting (12) into (10) and rearranging terms produces the final result.

$$L\{f_c(t)\} = \frac{50[(0.2s+1)(-e^{-0.2s})+1]}{(1-e^{-0.2s})s^2}$$

• PROBLEM 11-11

Determine the Laplace transform of the time functions shown in: (a) Fig. 1(a); (b) Fig. 1(b).

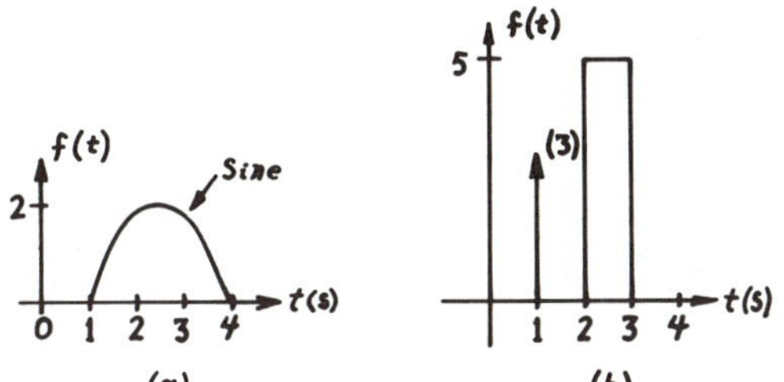

Fig. 1

Solution: In both cases the functions in Fig. 1 will be broken down into several simpler functions whose Laplace transform can be found in a table.

(a) The sinusoidal function in Fig. 1(a) can be considered as the sum of the functions in Fig. 2.

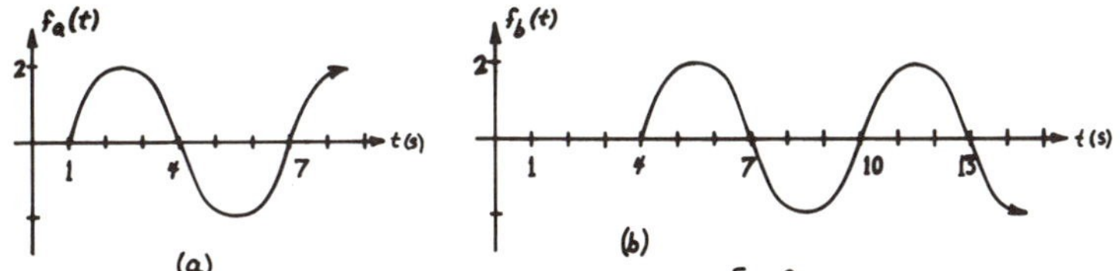

Fig. 2

The function in Fig. 2(a) is

$$f_a(t) = 2\text{Sin}\ \frac{\pi}{3}\ (t-1)\ u(t-1).$$

The function in Fig. 2(b) is

$$f_b(t) = 2\text{Sin}\ \frac{\pi}{3}\ (t-4)\ u(t-4).$$

Hence,

$$f(t) = f_a(t) + f_b(t)\ ,$$

$$f(t) = 2\text{Sin}\ \frac{\pi}{3}\ (t-1)u(t-1) + 2\text{Sin}\ \frac{\pi}{3}\ (t-4)u(t-4).$$

Using the shifting theorem, the Laplace transform of f(t) is found to be

$$F(s) = 2L[\text{Sin}\tfrac{\pi}{3}(t-1)u(t-1)] + 2L[\text{Sin}\tfrac{\pi}{3}(t-4)u(t-4)]$$

$$F(s) = (e^{-s} + e^{-4s})\ 2L[\text{Sin}\tfrac{\pi}{3}t].$$

It is found in a Laplace transform table that

$$L[\text{Sin}\omega t] = \frac{\omega}{s^2+\omega^2} .$$

In this case $\omega = \frac{\pi}{3}$, hence,

$$F(s) = \frac{2\frac{\pi}{3}}{s^2 + \frac{\pi^2}{9}} [e^{-s} + e^{-4s}] .$$

(b) The function in Fig. 1(b) can be considered as the sum of the functions in Fig. 3.

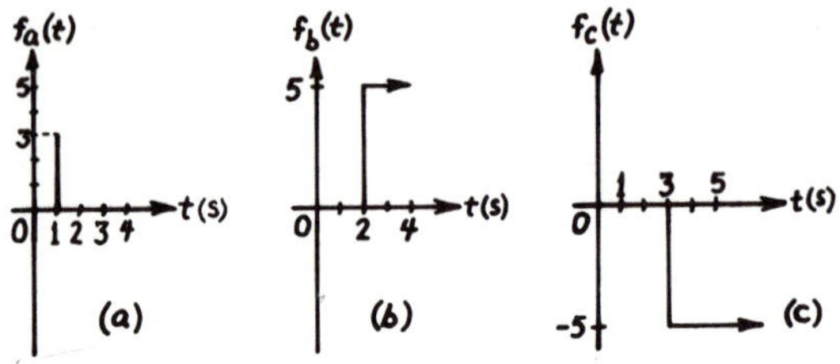

Fig. 3

The function in Fig. 3(a) is

$$f_a(t) = 3\delta(t-1).$$

The function in Fig. 3(b) is

$$f_b(t) = 5u(t-2).$$

The function in Fig. 3(c) is

$$f_c(t) = -5u(t-3).$$

Hence,

$$f(t) = f_a(t) + f_b(t) + f_c(t)$$

$$f(t) = 3\delta(t-1) + 5u(t-2) - 5u(5-3)$$

Using the shifting theorem the Laplace transform is found to be

$$F(s) = 3e^{-s} + \frac{5}{s}(e^{-2s} - e^{-3s}).$$

• PROBLEM 11-12

Use the time-shift theorem to obtain the Laplace transform of the: (a) non-periodic staircase function shown in Fig. 1(a); (b) non-periodic sawtooth function of Fig. 1(b).

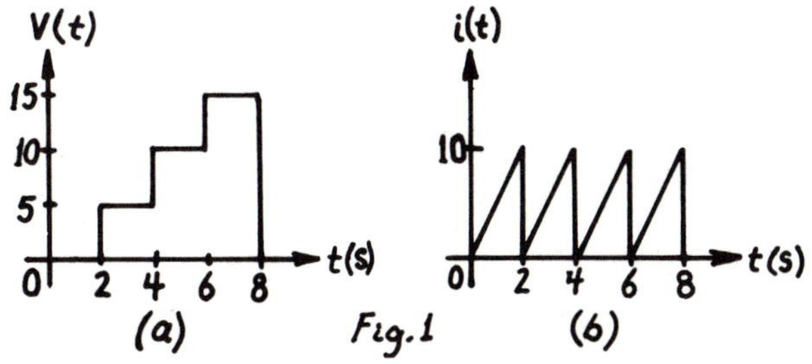

Fig. 1

Solution: The functions in Fig. 1 will be broken down into several simpler functions.

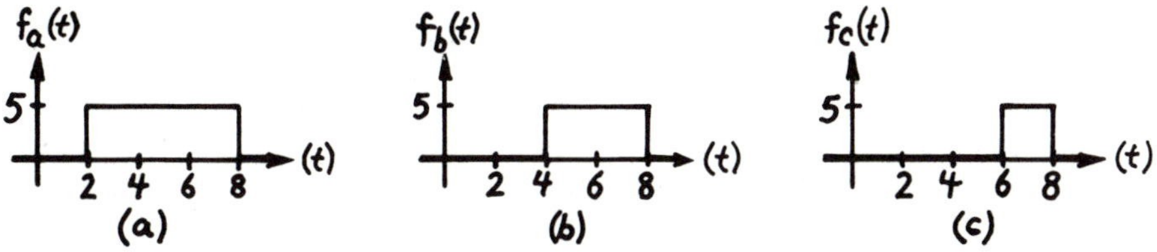

Fig. 2

(a) The step function in Fig. 1(a) can be considered as the sum of the three functions in Fig. 2.

The function in Fig. 2(a) is $f_a(t) = 5[U(t-2) - U(t-8)]$.

The function in Fib. 2(b) is $f_b(t) = 5[U(t-4) - U(t-8)]$.

The function in Fig. 2(c) is $f_c(t) = 5[U(t-6) - U(t-8)]$.

Hence,

$$v(t) = f_a(t) + f_b(t) + f_c(t)$$

$$v(t) = 5[U(t-2) + U(t-4) + U(t-6) - 3U(t-8)].$$

Using the shifting theorem, the Laplace transform is found to be

$$V(s) = \frac{5}{s}\,[e^{-2s} + e^{-4s} + e^{-6s} - 3e^{-8s}].$$

(b) In this case the function in Fig. 1(b) can be separated into four identical ramp functions each one shifted by 2 sec. The first ramp can be seen as the sum of the functions in Fig. 3.

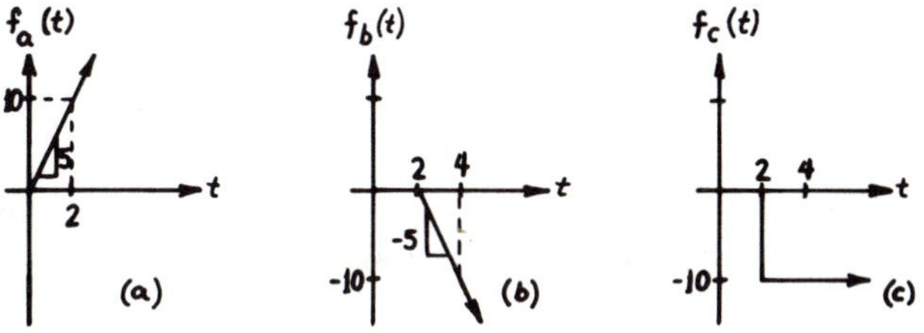

Fig. 3

The function in Fig. 3(a) is $f_a(t) = 5tU(t)$.

The function in Fig. 3(b) is $f_b(t) = -5(t-2)U(t-2)$.

The function in Fig. 3(c) is $f_c(t) = -10U(t-2)$.

Hence, the function for the first ramp is

$$f_1(t) = f_a(t) + f_b(t) + f_c(t)$$

$$= 5tU(t) - 5(t-2)U(t-2) - 10U(t-2).$$

The function for the second ramp, $f_2(t)$, is $f_1(t)$ shifted by 2 sec. Similarly, the function for the third ramp, $f_3(t)$, is $f_2(t)$ shifted by 2 sec. and the function for the fourth ramp, $f_4(t)$ is $f_3(t)$ shifted by 2 sec. Hence,

$$f_2(t) = 5(t-2)U(t-2) - 5(t-4)U(t-4) - 10U(t-4)$$

$$f_3(t) = 5(t-4)U(t-4) - 5(t-6)U(t-6) - 10U(t-6)$$

$$f_4(t) = 5(t-6)U(t-6) - 5(5-8)U(t-8) - 10U(t-8).$$

The total function then is the sum of the four individual ramp functions.

$$i(t) = f_1(t) + f_2(t) + f_3(t) + f_4(t)$$

$$i(t) = 5tU(t) - 10U(t-2) - 10U(t-4) - 10U(t-6)$$
$$- 10U(t-8) - 5(t-8)U(t-8).$$

Using the shifting theorem, the Laplace transform of i(t) is found to be:

$$I(s) = \frac{5}{s^2} - \frac{10}{s}e^{-2s} - \frac{10}{s}e^{-4s} - \frac{10}{s}e^{-6s} - \frac{10}{s}e^{-8s}$$
$$- \frac{5}{s^2}e^{-8s}$$

rearranging terms:

$$I(s) = \frac{5}{s^2}(1-e^{-8s}) - \frac{10}{s}e^{-2s}(1 + e^{-2s} + e^{-4s} + e^{-6s}).$$

LAPLACE TRANSFORM PROPERTIES

• **PROBLEM 11-13**

The periodic waveform shown in Fig. 1 is applied to a series RL network (R = L = 1). Find the transient and steady-state current using the Laplace transform.

Solution: The waveform given can be represented as a series of positive and negative step functions

$$v(t) = u(t) - u(t-T/2) + u(t-T) - u(t-\frac{3T}{3}) + u(t-2T) \ . \ . \ .$$

For a step function applied to an R-L with R/L = 1, the response is easy to find.

$$V(s) = 1/s \qquad H(s) = R + sL = 1 + s$$

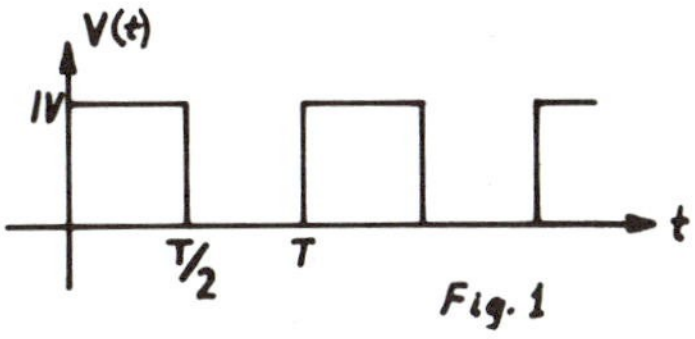

Fig. 1

$$I(s) = \frac{V(s)}{H(s)} = \frac{1/s}{1+s} = \frac{1}{s(1+s)} \ .$$

We can set

$$\frac{A}{s} + \frac{B}{1+s} = \frac{1}{s(1+s)}$$ and see that A + As + Bs = 1. This is possible only if A = 1 and B = -A = -1.

Therefore, $$I(s) = \frac{1}{s} - \frac{1}{1-s}$$

and $$i(t) = u(t) - e^{-t}u(t) = (1-e^{-t})u(t). \qquad (1)$$

Using the time shifting theorem of Laplace transforms, we can immediately write the response for each term in Eq. (1).

$$i(t) = (1-e^{-t})u(t) - (1-e^{-(t-T/2)})u(t-T/2) + (1-e^{-(t-T)})u(t-T)$$

$$- (1-e^{-(t-\frac{3T}{2})})u(t-\frac{3T}{2}) \ . \ . \ . \qquad (2)$$

or, $$i(t) = \sum_{m=o}^{\infty} (1-e^{-(t-mT)})u(t-mT) - \sum_{n=1}^{\infty} (1-e^{-(t-\frac{mT}{2})})u(t-\frac{mT}{2})$$

We should, in order to get a closed form answer, look at a finite sum for some number of full periods of the impressed waveform. Even then there are two cases to consider.

Case I: $NT \leq t \leq NT + T/2$ (t is a time during pulse "on"). The response due to positive steps is then

$$i_+ = N + 1 - \sum_{n=o}^{N} e^{-(t-nT)} \ , \qquad (3)$$

and the response to negative steps is

$$i_- = -N + \sum_{n=o}^{N+1} e^{-(t-T/2-nT)} \ .$$

From past experience with such sums (or by reference to the appropriate handbook), we know that

$$\sum_{n=o}^{N} e^{nT} = \frac{1 - e^{(N+1)T}}{1 - e^{T}} \qquad (4)$$

and $$\sum_{n=o}^{n-1} e^{nT} = \frac{1 - e^{NT}}{1 - e^{T}} \tag{5}$$

Since e^{-t} is a common factor in each term, we can write

$$i(t) = i_+ + i_- = 1 - e^{-t}\left[\frac{1-e^{(N+1)T}}{1-e^{T}}\right] + e^{-t}e^{T/2}\left[\frac{1-e^{NT}}{1-e^{T}}\right]$$

$$= 1 - \underbrace{e^{-t}\left[\frac{1-e^{T/2}}{1-e^{T}}\right]}_{\text{Transient}} - e^{-t}\left[\frac{e^{(N+\frac{1}{2})T} - e^{(N+1)T}}{1-e^{T}}\right]$$

Case II: $NT + T/2 \le t \le (N+1)T$ (t is a time during pulse "off"). For this case,

$$i_+ = N + 1 - \sum_{n=o}^{N} e^{-(t-nT)}$$

$$i_- = -(N+1)\sum_{n=o}^{N} e^{-(t-\frac{T}{2}-nT)} .$$

Then,

$$i(t) = -e^{-t}(1-e^{T/2})\sum_{n=o}^{N} e^{nT}$$

Using Eq.(4) again,

$$i(t) = -e^{-t}(1-e^{T/2})\left[\frac{1-e^{(N+1)T}}{1-e^{T}}\right]$$

$$= \underbrace{\frac{-e^{-t}(1-e^{T/2})}{1-e^{T}}}_{\text{Transient}} + \frac{e^{-t}(1-e^{T/2})}{1-e^{T}}e^{(N+1)T}$$

• PROBLEM 11-14

Calculate the Laplace transform of the function (a) $e^{-at}u_o(t)$, (b) $e^{-a(t-1)}u_o(t-1)$, and (c) $e^{-a(t-1)}u_o(t)$.

Solution: (a) The Laplace transform is defined as

$$F(s) = L[f(t)] = \int_{o^-}^{\infty} f(t)e^{-st}dt. \tag{1}$$

In this case $f(t) = e^{-at}u_o(t) = e^{-at}$ for $t \geq 0$.

Putting f(t) into (1) gives

$$F(s) = L[e^{-at}u_o(t)] = \int_{0^-}^{\infty} e^{-at}e^{-st}dt$$

$$F(s) = \int_{0^-}^{\infty} e^{-t(s+a)}dt = \left.\frac{e^{-(s+a)t}}{-(s+a)}\right|_{0^-}^{\infty} .$$

Hence,

$$F(s) = \frac{e^{-(s+a)\infty}}{-(s+a)} - \frac{e^{-(s+a)o}}{-(s+a)}$$

$$F(s) = \frac{1}{s+a} .$$

(b) The Laplace transform can be found from the shifting property, which states that

$$L[f(t-t_o)u_o(t-t_o)] = e^{-st_o}F(s),$$

where $F(s) = L[f(t)]$.

In this case $t_o = 1$ and F(s) was found in part (a) to be $\frac{1}{s+a}$. Hence,

$$L[e^{-a(t-1)}u_o(t-1)] = \frac{e^{-s}}{s+a} .$$

(c) As in part a, the general definition of the Laplace transform is used.

In this case $f(t) = e^{-a(t-1)}u_o(t) = e^{-a(t-1)}$ for $t \geq 0$.

Putting f(t) into (1) gives

$$F(s) = L[e^{-a(t-1)}u_o(t)] = \int_{-o}^{\infty} e^{-a(t-1)}e^{-st}$$

$$F(s) = e^{a}\int_{o}^{\infty} e^{-t(s+a)} .$$

Hence,

$$F(s) = \left.\frac{e^{a}}{-(s+a)}\right|_{0^-}^{\infty}$$

$$F(s) = \frac{e^{a}}{s+a} .$$

• PROBLEM 11-15

Sketch f(t) if $F(s) = (1/s)/(1 - e^{-s})$.

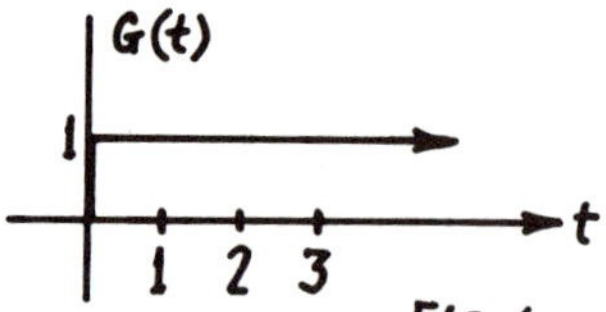

FIG. 1

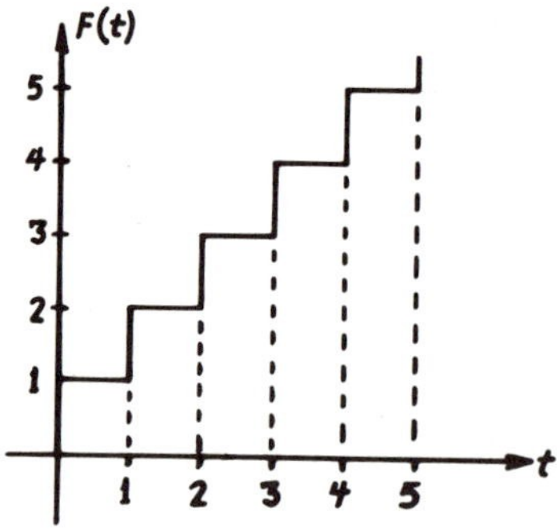

FIG. 2

Solution: Remember that $F(t) = L\{F(s)\} = L \dfrac{G(s)}{1-e^{-st}}$ is a semiperiodic function where T is the period provided, G(s) represents a pulse that is zero for t > T.

In this example, T = 1 and $G(s) = \frac{1}{s}$. $G(t) = L\{G(s)\}$ is shown in fig. 1.

Note that for t > 1 G(t) is not zero, thus the pulse will be "modified" periodically over period T.

F(t) is sketched in fig. 2.

• PROBLEM 11-16

For $f(t) = e^{-2t}u(t)$, find the initial value and slope both by letting $t = 0^+$ and from the initial-value theorem.

Solution: Substituting $t = 0^+$ into f(t), results in $e^{-o} = 1$, and in $\frac{df}{dt} = -2e^{-2t}$, $\frac{df(0^+)}{dt} = -2$.

The initial value theorem states:

$$f(0^+) = \lim_{s\to\infty} s\,F(s) \qquad \text{and}$$

$$\frac{df(0^+)}{dt} = \lim_{s\to\infty} [s^2F(s) - sF(0^+)]$$

Thus

$$L\{e^{-2t}\} = F(s) = \frac{1}{s+2}$$

$$f(0^+) = \lim_{s\to\infty} \frac{s}{s+2}$$

$$f(0^+) = \lim_{s\to\infty} \frac{1}{s+\frac{2}{s}} = 1$$

$$\frac{df(0^+)}{dt} = \lim_{s\to\infty} \frac{s^2}{s+2} - s$$

$$\frac{df(0^+)}{dt} = \lim_{s\to\infty} \frac{-2s}{s+2}$$

$$\frac{df(0^+)}{dt} = \lim_{s\to\infty} \frac{-2}{1+\frac{2}{s}} = -2.$$

• PROBLEM 11-17

Find $v(0^+)$ and $(dv/dt)_{0^+}$ if

$$V(s) = \frac{-36s^2 - 24s + 2}{12s^3 + 17s^2 + 6s}$$

Solution: Using the initial value theorem,

$$v(0^+) = \lim_{s\to\infty} [sV(s)]$$

$$v(0^+) = \lim_{s\to\infty} \left[\frac{-36s^2-24s+2}{12s^2+17s+6}\right]$$

$$v(0^+) = \lim_{s\to\infty} \left[\frac{-36+\frac{12}{s}+\frac{2}{s^2}}{12+\frac{17}{s}+\frac{6}{s^2}}\right] = \frac{-36}{12} = -3$$

and,

$$\frac{dv(0^+)}{dt} = \lim_{s\to\infty} \left[s^2V(s) - sV(0^+)\right]$$

$$= \lim_{s\to\infty}\left[\frac{-36s^3-24s^2+2s}{12s^2+17s+6} + 3s\right]$$

$$= \lim_{s\to\infty} \left[\frac{-36s^3-24s^2+2s+3s(12s^2+17s+6)}{12s^2+17s+6}\right]$$

$$\frac{dv(0^+)}{dt} = \lim_{s\to\infty} \left[\frac{27s^2+20s}{12s^2+12s+6}\right]$$

$$\frac{dv(0^+)}{dt} = \lim_{s\to\infty} \left[\frac{27+\frac{20}{s}}{12+\frac{12}{s}+\frac{6}{s^2}}\right] = \frac{27}{12} = 2.25.$$

• PROBLEM 11-18

Given the initial conditions $y(0^-) = y_o$, $y'(0^-) = y_1$, and $x(0^-) = 0$ solve for $Y(s)$ from the relationship

$$\frac{d^2y}{dt^2} + a_1 \frac{dy}{dt} + a_o y = b_1 \frac{dx}{dt} + b_o x.$$

Solution: View the differential equation as representing the operation of an electrical network. y(t) is then the network response to a known driving function x(t). Since we want to solve for Y(s), begin by transforming the differential equation. This yields

$$L\left[\frac{d^2y}{dt^2}\right] + a_1 L\left[\frac{dy}{dt}\right] + a_o L[y] = b_1 L\left[\frac{dx}{dt}\right] + b_o L[x]$$

Since $L\left(\frac{d^n}{dt^n} f(t)\right) =$

$$s^n F(s) - s^{n-1} f(0^-) - s^{n-2} f^1(0^-) - \ldots - f^{(n-1)}(0^-)$$

we can transform the differential equation obtaining,

$$\underbrace{s^2Y(s) - sy_o - y_1}_{L\left|\frac{d^2y}{dt}\right|} + \underbrace{a_1 sY(s) - a_1 y_o}_{a_1 L\left|\frac{dy}{dt}\right|} + \underbrace{a_o Y(s)}_{a_o L[y]}$$

$$= \underbrace{b_1 sX(s) - 0}_{b_1 L\left|\frac{dx}{dt}\right|} + \underbrace{b_o X(s)}_{b_o L[x]}$$

Combining terms yields,

$$[s^2+a_1 s+a_o]Y(s) = [sb_1+b_o]X(s) + [s+a_1]y_o + y_1$$

Solving for Y(s) gives

$$Y(s) = \frac{b_1 s+b_o}{s^2+a_1 s+a_o} X(s) + \frac{s+a_1}{s^2+a_1 s+a_o} y_o + \frac{1}{s^2+a_1 s+a_o} y_1$$

This example shows that differential equations can be transformed into simple algebraic equations while simultaneously introducing the necessary initial conditions.

• PROBLEM 11-19

Calculate the Laplace transform of $t^n u(t)$.

Solution: To determine the Laplace transform of $t^n u(t)$, first calculate the transform of t u(t). Calling u(t) a function f(t) and using the definition for the unilateral Laplace transform,

$$L\{f(t)\} = F(s) = \int_{0^-}^{\infty} f(t)e^{-st}dt,$$

we obtain

$$L\{tf(t)\} = \int_{0^-}^{\infty} tf(t)e^{-st}dt.$$

Substitute

$$-\frac{d}{ds}(e^{-st})$$

for te^{-st} in the above integral to obtain

$$L\{tf(t)\} = -\frac{d}{ds}\int_{0^-}^{\infty} f(t)e^{-st}dt.$$

Noticing that this is

$$L\{tf(t)\} = -\frac{d}{ds}F(s) = -\frac{d}{ds}L\{f(t)\},$$

formulate a general expression for $L\{t^n f(t)\}$ showing that the substitution:

$$-\frac{d}{ds}e^{-st}$$

for te^{-st} would be true for n = 1. A substitution of

$$\frac{d^2}{ds^2}e^{-st}$$

for t^2e^{-st} would be true for n = 2, etc. Hence,

$$L\{t^n f(t)\} = (-1)^n \frac{d^n}{ds^n} F(s),$$

where $F(s) = L\{f(t)\} = L\{u(t)\}$.

Since $u(t) = 1$ for $t > 0$ in the unilateral transform, we can write

$$L\{u(t)\} = \int_{0^-}^{\infty} e^{-st}\, dt = \left.\frac{e^{-st}}{-s}\right|_{0^-}^{\infty} = \frac{1}{s}$$

Hence,

$$L\{t^n u(t)\} = (-1)^n \frac{d^n}{ds^n} \frac{1}{s}$$

Note that $L\{t^n u(t)\}$ is

$$\frac{1}{s^2},\quad \frac{2}{s^3},\quad \frac{6}{s^4},\quad \frac{24}{s^5},\quad \frac{120}{s^6} \qquad \text{for}$$

n = 1, 2, 3, 4, and 5 respectively which facilitates writing:

$$L\{t^n u(t)\} = \frac{n!}{s^{n+1}}\,.$$

CONVOLUTION

• PROBLEM 11-20

Find $f_1(t)*f_2(t)$ if $f_1(t) = 2e^{-4t}u(t)$ and $f_2(t) = 5\cos 3t\ u(t)$.

Solution: The convolution is represented by $f_1(t)*f_2(t)$ or "$f_1(t)$ asterisk $f_2(t)$" and it means

$$f_1(t)*f_2(t) = \int_{-\infty}^{+\infty} f_2(\tau)f_1(t-\tau)d\tau = \int_{-\infty}^{+\infty} f_1(\tau)f_2(t-\tau)d\tau \qquad (1)$$

In this problem

$$f_1(t) = 2e^{-4t}u(t)$$

$$f_2(t) = 5\cos 3t\ u(t).$$

A function g(t) when multiplied by the unit step function u(t) means that g(t) "turns on" at t = 0, or

$$f_1(t) = 0 \qquad \text{for } t < 0$$

$$f_1(t) = 2e^{-4t} \qquad \text{for } t \geq 0$$

$$f_2(t) = 0 \qquad \text{for } t < 0$$

$$f_2(t) = 5 \cos 3t \quad \text{for } t \geq 0$$

Now if the convolution of $f_1(t)$ and f_2(tP is computed using (1),

$$f_1(t)*f_2(t) = \int_{-\infty}^{\infty} 5 \cos 3\tau \; u(\tau) \cdot 2e^{-4(t-\tau)} u(t-\tau) d\tau \qquad (2)$$

and

$$u(t-\tau) = 1 \quad \text{for } (t-\tau) \geq 0 \quad \text{or} \quad \tau \leq t \qquad (3)$$

$$u(t-\tau) = 0 \quad \text{for } (t-\tau) < 0 \quad \text{or} \quad \tau > t \qquad (4)$$

$$u(\tau) = 1 \quad \text{for} \quad \tau \geq 0 \qquad (5)$$

$$u(\tau) = 0 \quad \text{for} \quad \tau < 0 \qquad (6).$$

Therefore, combining (2) through (6)

$$\begin{aligned} f_1(t)*f_2(t) &= \int_0^t 5 \cos 3\tau \cdot 2e^{-4(t-\tau)} d\tau \\ &= 10e^{-4t} \int_0^t e^{4\tau} \cos 3\tau d\tau \\ &= \frac{10e^{-4t}}{16+9} \left| e^{4\tau}(4 \cos 3\tau + 3 \sin 3\tau) \right|_0^t \\ &= 1.6 \cos 3t + 1.2 \sin 3t - 1.6e^{-4t} \end{aligned}$$

or

$$f_1(t)*f_2(t) = (1.6 \cos 3t + 1.2 \sin 3t - 1.6e^{-4t})u(t)$$

• PROBLEM 11-21

Find f(t)*g(t) in Fig.1(a) and f(t)*g(t) in Fig.1(b).

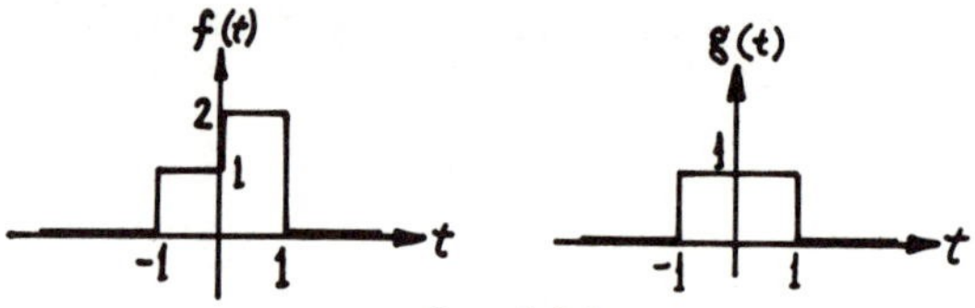

Fig. 1(a)

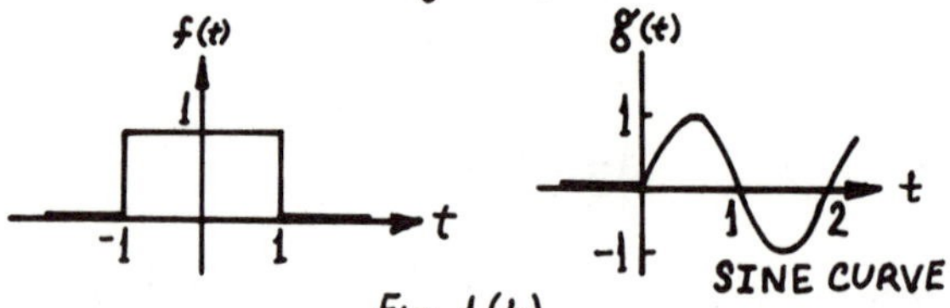

Fig. 1(b)

Solution: The definition of convolution of two functions is given as follows

$$f(t)*g(t) = \int_{-\infty}^{+\infty} f(u)g(t-u)du \tag{1}$$

(a) By inspection of the given graph, it is known that

$f(u) = 0$	for	$-\infty < u < -1$
$f(u) = 1$	for	$-1 \leq u < 0$
$f(u) = 2$	for	$0 \leq u < 1$
$f(u) = 0$	for	$1 < u$
$g(t-u) = 0$	for	$-\infty < (t-u) < -1$
$g(t-u) = 1$	for	$-1 \leq (t-u) \leq +1$
$g(t-u) = 0$	for	$+1 < (t-u)$

Therefore, when Eq.(1) is calculated, by inspection there are following significant check points.

$t = -2$	$g(t-u)$ touches $f(u)$.
$t = -1$	a half of $g(t-u)$ overlaps with $f(u)$.
$t = 0$	entire range of $g(t-u)$ coincides with $f(u)$
$t = +1$	a left half of $g(t-u)$ overlaps with $f(u)$
$t = +2$	the left edge of $g(t-u)$ touches with the right edge of $f(u)$

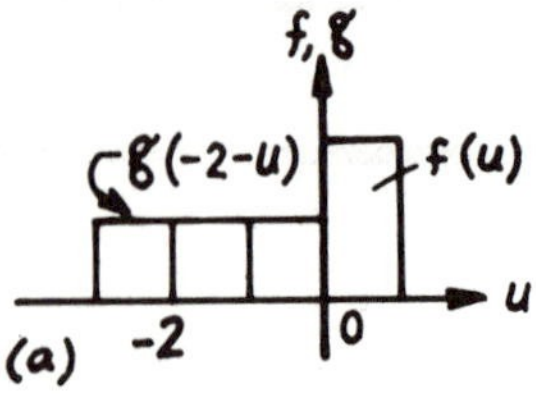

FIG. 2 (a)

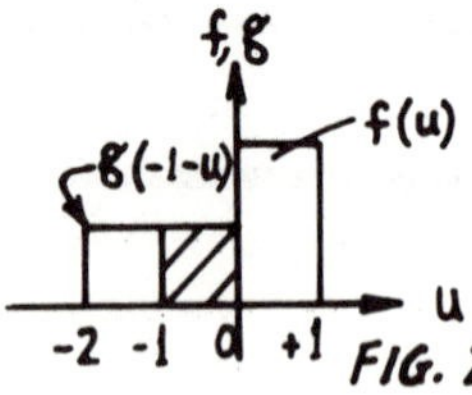

FIG. 2 (b)

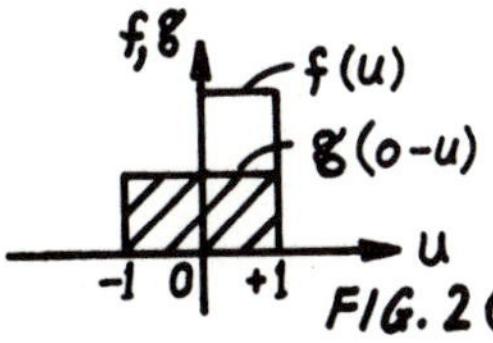

FIG. 2(c)

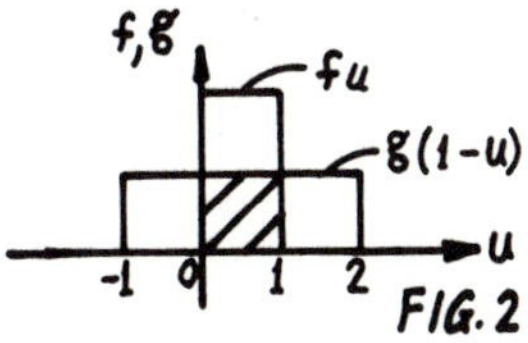

FIG. 2(d)

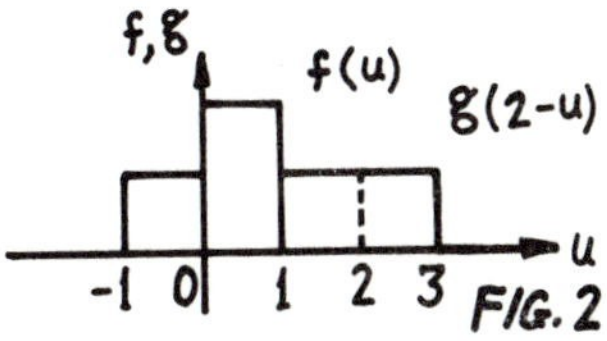

FIG. 2(e)

So at t = -2,

$$f(-2)*g(-2) = \int_{-\infty}^{+\infty} f(u)g(-z-u)du = 0 \quad \text{(See Fig. 2(a).)}$$

at t = -1,

$$f(-1)*g(-1) = \int_{-\infty}^{+\infty} f(u)g(-1-u)du = \int_{-1}^{0} 1 \times 1\, du = [u]_{-1}^{0} = 1$$

(See Fig. 2(b).)

at t = 0,

$$f(0)*g(0) = \int_{\infty}^{\infty} f(u)g(0-u)du = \int_{-1}^{0} 1 \times 1\, du + \int_{0}^{1} 2 \times 2\, du$$

(See Fig. 2(c).)

$$= [u]_{-1}^{0} + [2u]_{0}^{1} = +1 + 2 = 3$$

at t = 1,

$$f(1)*g(1) = \int_{-\infty}^{\infty} f(u)g(1-u)du = \int_{0}^{1} 2 \times 1\, du = [2u]_{0}^{1} = 2$$

(See Fig. 2(d).)

at t = 2,

$$f(2)*g(2) = \int_{-\infty}^{+\infty} f(u)g(2-u)du = 0$$

(See Fig. 2(e).)

So the plot is given by

f(-2)*g(-2)	. . .	0
f(-1)*g(-1)	. . .	1
f(0)*g(0)	. . .	3
f(+1)*g(+1)	. . .	2
f(+2)*g(+2)	. . .	0

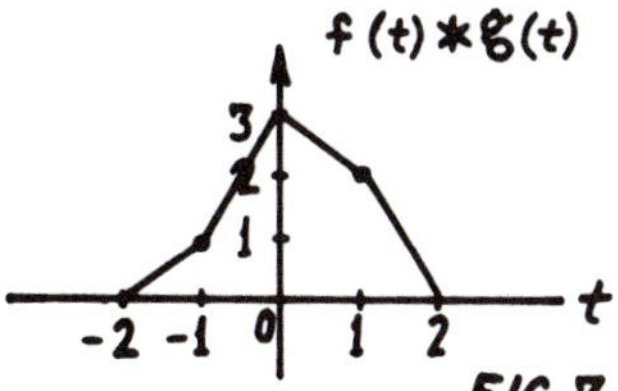

FIG. 3

(See Fig. 3)

(b) By definition, the convolution of f(t) and g(t) is

$$f(t)*g(t) \equiv \int_{-\infty}^{+\infty} f(u)g(t-u)du \tag{2}$$

By inspection of Fig. 1 (b)

$$f(u) = 0 \quad \text{for} \quad -\infty \leq u < -1, \quad 1 < u \leq +\infty \tag{3}$$

$$f(u) = 1 \quad \text{for} \quad -1 \leq u \leq +1 \tag{4}$$

$$g(t-u) = 0 \quad \text{for} \quad -\infty \leq (t-u) \leq 0 \tag{5}$$

$$g(t-u) = \text{Sin}\pi(t-u) \quad \text{for} \quad 0 \leq (t-u) \leq +\infty \tag{6}$$

The situation may be illustrated by Fig. 4. Therefore

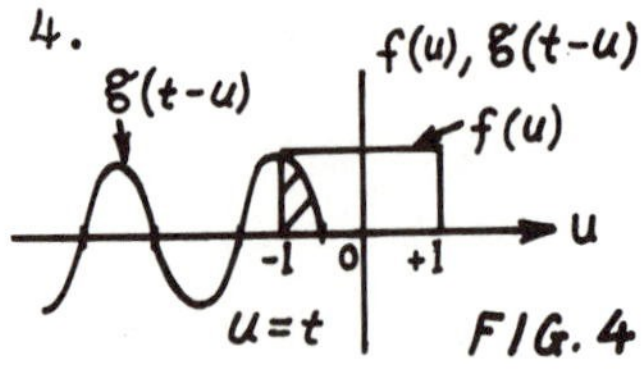

FIG. 4

$$f(t)*g(t) = \int_{-\infty}^{+\infty} f(u)g(t-u)du$$

$$= \int_{-\infty}^{-1} f(u)g(t-u)du + \int_{-1}^{+1} f(u)g(t-u)du + \int_{+1}^{+\infty} f(u)g(t-u)du$$

$$= \int_{-\infty}^{1} 0\cdot\text{Sin}\pi(t-u)du + \int_{-1}^{+t} 1\cdot\text{Sin}\pi(t-u)du + \int_{+1}^{+\infty} 0\cdot\text{Sin}\pi(t-u)du$$

$$= 0 + \left[\frac{-\text{Cos}\pi(t-u)}{-1}\right]_{-1}^{+t} + 0$$

$$= 0 + \frac{1}{\pi}\left[\text{Cos}\pi(t-t) - \text{Cos}\pi(t+1)\right] + 0 \tag{7}$$

The convolution is zero for $t<-1$ and $t>1$ but in the range $-1\leq t<+1$

$$f(t)*g(t) = \frac{1}{\pi}\left[1 - \text{Cos}\pi(t+1)\right]$$

This is an even function of t

t	$\text{Cos}\pi(t+1)$	$f(t)*g(t)$
0	-1	$\frac{2}{\pi}$
1	+1	0

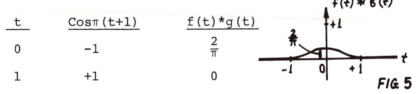

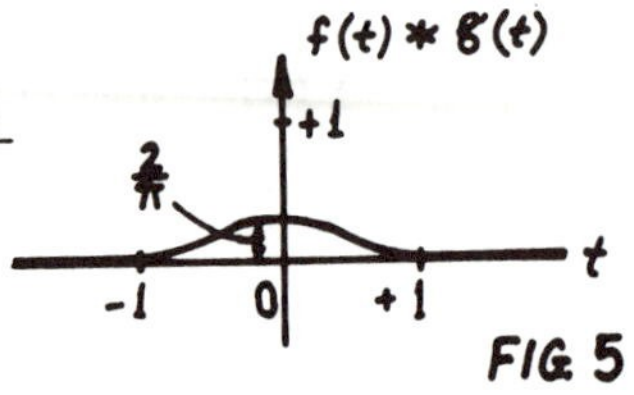

FIG 5

The sketch of f(t)*g(t) is shown in Fig. 5.

• PROBLEM 11-22

Use the convolution integral to find the inverse Laplace transform of $F_1(s)F_2(s)$ =: (a) $(\frac{1}{s})(\frac{1}{s})$; (b) $(\frac{1}{s})(\frac{1}{s})^2$; (c) $(\frac{1}{s})(\frac{1}{s+1})$.

Solution: The convolution integral can be written as

$$\int_0^t f_1(\tau)f_2(t-\tau)d\tau = L^{-1}[F_1(s)F_2(s)]$$

where the operator L^{-1} indicates the inverse Laplace transform and

$$F_1(s) = L[f_1(t)]$$

$$F_2(s) = L[f_2(t)].$$

(a) $F_1(s)F_2(s) = \frac{1}{s}\frac{1}{s}$

Let

$$F_1(s) = \frac{1}{s}$$

$$F_2(s) = \frac{1}{s}.$$

The inverse transform of $\frac{1}{s}$ is 1 so

$$f_1(t) = 1 \quad \text{for} \quad t > 0$$

$$f_2(t) = 1 \quad \text{for} \quad t > 0$$

Then

$$L^{-1}[F_1(s)F_2(s)] = \int_0^t f_1(\tau)f_2(t-\tau)d\tau$$

$$L^{-1}[\frac{1}{s}\frac{1}{s}] = \int_0^t (1)(1)d\tau = \tau\Big|_0^t = t \quad \text{for } t > 0$$

or using step function notation, $u(t)$,

$$L^{-1}[\frac{1}{s}\frac{1}{s}] = t\,u(t) \quad \text{for all } t.$$

(b) $F_1(s)F_2(s) = \frac{1}{s}\left(\frac{1}{s}\right)^2$

Let

$$F_1(s) = \frac{1}{s}$$

$$F_2(s) = \frac{1}{s^2}$$

Then

$$f_1(t) = 1$$

$$f_2(t) = t.$$

The inverse transform of $\frac{1}{s}(\frac{1}{s})^2$ is

$$L^{-1}[\frac{1}{s}(\frac{1}{s})^2] = \int_o^t f_1(\tau)f_2(t-\tau)d\tau$$

$$= \int_o^t (1)(t-\tau)d\tau$$

$$= (t\tau - \frac{\tau^2}{2})\Big|_o^t$$

$$= t^2 - \frac{1}{2}t^2$$

$$= \frac{1}{2}t^2 \text{ for } t > 0$$

$$= \frac{1}{2}t^2 u(t) \quad \text{for all } t.$$

(c) $F_1(s)F_2(s) = \frac{1}{s}\frac{1}{s+1}$.

Let

$$F_1(s) = \frac{1}{s}$$

$$F_2(s) = \frac{1}{s+1} .$$

Then

$$f_1(t) = 1$$

$$f_2(t) = e^{-t}.$$

The inverse transform of $\frac{1}{s}\frac{1}{s+1}$ is

$$L^{-1}[\frac{1}{s}\frac{1}{s+1}] = \int_o^t f_1(\tau)f_2(t-\tau)d\tau$$

$$= \int_o^t (1)e^{-(t-\tau)}d\tau$$

$$= e^{-t}\int_o^t e^{\tau}d\tau$$

$$= e^{-t}e^{\tau}\Big|_o^t$$

$$= e^{-t}(e^{t}-1)$$

$$= 1 - e^{-t} \quad \text{for } t > 0$$

$$= (1 - e^{-t})u(t) \quad \text{for all } t.$$

EXPANSION BY PARTIAL FRACTIONS

• PROBLEM 11-23

Find the partial-fraction expansion of

$$F(s) = \frac{8s + 2}{(s^2 + 3s + 2)}$$

Solution: F(s) is a rational function, since it is in the form, $\frac{N(s)}{D(s)}$, where the degree of D(s) is greater than that of N(s), thus write a partial-fraction expansion in the form

$$F(s) = \frac{A}{s+1} + \frac{B}{s+2} = \frac{8s+2}{s^2+3s+2}$$

where $A = \left|(s+1)f(s)\right|_{s=-1} = \left|\frac{8s+2}{s+2}\right|_{s=-1} = -6$

and $B = \left|(s+2)F(s)\right|_{s=-2} = \left|\frac{8s+2}{s+1}\right|_{s=-2} = 14$

Therefore,

$$F(s) = \frac{-6}{s+1} + \frac{14}{s+2}.$$

• PROBLEM 11-24

Find the partial-fraction expansion of

$$F(s) = \frac{s + 3}{(s + 1)(s^2 + 4s + 8)}$$

Solution: Factor the denominator of F(s) by finding the roots of the quadratic $s^2 + 4s + 8$.

Hence,

$$s_1 = \frac{-4 + \sqrt{16-32}}{2} \ ; \ s_2 = \frac{-4 - \sqrt{16-32}}{2}$$

$$s_1 = -2 + j2 \qquad s_2 = -2 - j2 \ ,$$

thus by partial fraction expansion

$$F(s) = \frac{s+3}{(s+1)(s+2-j2)(s+2+j2)}$$

$$= \frac{A}{s+1} + \frac{B}{s+2-j2} + \frac{B^*}{s+2+j2}$$

where, $A = (s+1)F(s)\Big|_{s=-1} = \frac{s+3}{s^2+4s+8}\Bigg|_{s=-1} = \frac{2}{5}$

and $\quad B = (s+2-j2)F(s)\Big|_{s=-2+j2} = \frac{s+3}{(s+1)(s+2+j2)}\Bigg|_{s=-2+j2}$

$$B = \frac{1+j2}{(-1+2j)(j4)} = \frac{1+j2}{-8-j4}$$

$$B = \frac{1+j2}{-8-j4} \cdot \frac{-8+j4}{-8+j4} = \frac{-16-j12}{80} = -\frac{1}{5} - j\frac{3}{20}$$

$$B^* = -\frac{1}{5} + j\frac{3}{20} \ .$$

Hence,

$$F(s) = \frac{\frac{2}{5}}{s+1} + \frac{-\frac{1}{5}-\frac{3}{20}j}{s+2-j2} + \frac{-\frac{1}{5}+\frac{3}{20}j}{s+2+j2}$$

The last two terms can be combined to obtain,

$$F(s) = \frac{\frac{2}{5}}{s+1} + \frac{(-\frac{1}{5}-\frac{3}{20}j)(s+2+j2) + (-\frac{1}{5}+\frac{3}{20}j)(s+2-j2)}{(s+2-j2)(s+2+j2)}$$

$$F(s) = \frac{\frac{2}{5}}{s+1} + \frac{\frac{1}{5}(2s+1)}{s^2+4s+8} \ .$$

• PROBLEM 11-25

Find the partial-fraction expansion of

$$F(s) = \frac{s+3}{(s+1)(s^2+4s+8)}$$

Solution: The rational fraction F(s) can be expanded into the form

$$F(s) = \frac{s+3}{(s+1)(s^2+4s+8)} = \frac{A}{s+1} + \frac{Bs+C}{s^2+4s+8} .$$

Multiplying both sides by $(s+1)(s^2+4s+8)$ gives

$$(s+3) = A(s^2+4s+8) + (Bs+C)(s+1)$$

$$s+3 = As^2 + 4As + 8A + Bs^2 + Bs + Cs + C$$

Factoring by powers of s gives,

$$s+3 = s^2(A+B) + s^1(4A+B+C) + s^0(8A+C)$$

Note that since there are no s^2 terms on the left side of the equation A+B=0, and similarly, 4A+B+C=1, and 8A+C=3. Solving the three equations simultaneously gives

$$A = \frac{2}{5}$$

$$B = -\frac{2}{5}$$

$$C = -\frac{1}{5}$$

thus

$$F(s) = \frac{\frac{2}{5}}{s+1} - \frac{\frac{2}{5}s + \frac{1}{5}}{s^2 + 4s + 8}$$

• PROBLEM 11-26

Obtain the partial-fraction expansion of

$$F(s) = \frac{s + 2}{(s + 1)^2(s + 3)}$$

Solution: The rational function F(s) has a multiple root in its denominator. Thus, expand accordingly

$$F(s) = \frac{s+2}{(s+1)^2(s+3)} = \frac{A_1}{(s+1)^2} + \frac{A_2}{s+1} + \frac{B}{s+3}$$

where $A_1 = \left|(s+1)^2F(s)\right|_{s=-1} = \left|\frac{s+2}{s+3}\right|_{s=-1} = \frac{1}{2}$

$$A_2 = \left|\frac{1}{(2-1)!}\frac{d}{ds}[(s+1)^2F(s)]\right|_{s=-1} = \left|\frac{d}{ds}\frac{s+2}{s+3}\right|_{s=-1}$$

$$A_3 = \left|\frac{(s+3) - (s+2)}{(s+3)^2}\right|_{s=-1} = \frac{1}{4}$$

$$B = \left|(s+3)F(s)\right|_{s=-3} = \left|\frac{s+2}{(s+1)^2}\right|_{s=-3} = -\frac{1}{4}.$$

Thus,

$$F(s) = \frac{\frac{1}{2}}{(s+1)^2} + \frac{\frac{1}{4}}{(s+1)} - \frac{\frac{1}{4}}{(s+3)}.$$

• PROBLEM 11-27

Obtain the partial-fraction expansion of

$$F(s) = \frac{s^2 + 3s + 1}{(s + 1)^3 (s + 2)^2}$$

Solution: The rational function F(s) has multiple roots in its denominator. Expand as follows:

$$F(s) = \frac{A_1}{(s+1)^3} + \frac{A_2}{(s+1)^2} + \frac{A_3}{(s+1)} + \frac{B_1}{(s+2)^2} + \frac{B_2}{(s+2)}$$

where the constants can be found by use of the general formula for a multiple root $(s + a)^n$

$$K_c = \left|\frac{1}{(c-1)!} \frac{d^{c-1}}{ds^{c-1}} [(s+a)^n F(s)]\right|_{s=-a.} \quad c = 1,2 \ldots n$$

Therefore,

$$A_1 = \left|\frac{1}{0!} \frac{d^\circ}{ds^\circ} [(s+1)^3 F(s)]\right|_{s=-1} = \left|\frac{s^2+3s+1}{(s+2)^2}\right|_{s=-1}$$

$$A_1 = -1$$

$$A_2 = \left|\frac{1}{(s-1)!} \frac{d}{ds} [(s+1)^3 F(s)]\right|_{s=-1}$$

$$A_2 = \left|\frac{(s+2)^2(2s+3) - 2(s^2+3s+1)(s+2)}{(s+2)^4}\right|_{s=-1}$$

$$A_2 = \left|\frac{s+4}{(s+2)^3}\right|_{s=-1} = 3$$

$$A_3 = \left| \frac{1}{(3-1)!} \frac{d^2}{ds^2} [(s+1)^3 F(s)] \right|_{s=-1}$$

$$= \left| \frac{1}{2} \left[\frac{1}{(s+2)^3} - \frac{3(s+4)}{(s+2)^4} \right] \right|_{s=-1}$$

$$A_3 = \left| \frac{(s+2) - 3(s+4)}{2(s+2)^4} \right|_{s=-1} = \left| \frac{-s-5}{(s+2)^4} \right|_{s=-1} = -4$$

Similarly,

$$B_1 = \left| (s+2)^2 F(s) \right|_{s=-2} = \left| \frac{s^2+3s+1}{(s+1)^3} \right|_{s=-2} = 1$$

$$B_2 = \left| \frac{1}{(2-1)!} \frac{d}{ds} [(s+2)^2 F(s)] \right|_{s=-2}$$

$$= \left| \frac{2s+3}{(s+1)^3} - \frac{3(s^2+3s+1)}{(s+1)^4} \right|_{s=-2}$$

$$B_2 = \left| \frac{-s^2 - 4s}{(s+1)^4} \right|_{s=-2} = 4.$$

Hence,

$$F(s) = \frac{-1}{(s+1)^3} + \frac{3}{(s+1)^2} - \frac{4}{s+1} + \frac{1}{(s+2)^2} + \frac{4}{s+2} .$$

INVERSE LAPLACE TRANSFORMS

• PROBLEM 11-28

By expansion into partial fractions, find the inverse Laplace transform of: (a) $1/[s(s+3)]$; (b) $1/[s^2(s+3)]$; (c) $1/[s(s+1)(s+3)]$.

Solution: By using partial fraction expansions, these Laplace equations can be simplified into sums of smaller equations. Once simplified, the inverse Laplace transform can be found.

(a) Expand the equation into partial fractions:

$$\frac{1}{s(s+3)} = \frac{A}{s} + \frac{B}{s+3} . \qquad (1)$$

From this equation we see that:

$$A(s+3) + Bs = 1, \tag{2}$$

therefore, $$As + 3A + Bs = 1. \tag{3}$$

This equation gives two more equations

$$A + B = 0 \qquad \text{and} \qquad 3A = 1.$$

From these two equations it is found that $A = \frac{1}{3}$ and $B = -\frac{1}{3}$. Now equation (1) is rewritten with the values of A and B included.

$$\frac{1}{s(s+3)} = \frac{1}{3s} - \frac{1}{3(s+3)} \tag{4}$$

Equation (1) has been expanded to equation (4). Now the inverse Laplace is found for the individual sums.

$$L^{-1}[\frac{1}{3s}] = \frac{1}{3} L^{-1} \frac{1}{s} = \frac{1}{3}$$

and

$$L^{-1}[\frac{1}{3(s+3)}] = \frac{1}{3} L^{-1} \frac{1}{s+3} = \frac{1}{3} e^{-3t} .$$

So,

$$L^{-1}\left(\frac{1}{s(s+3)}\right) = \frac{1}{3} - \frac{1}{3} e^{-3t} = \frac{1}{3} (1 - e^{-3t}).$$

(b) First expand the equation into partial fractions.

$$\frac{1}{s^2(s+3)} = \frac{As+b}{s^2} + \frac{C}{s+3} . \tag{5}$$

From this equation, we see that

$$(As + b)(s + 3) + Cs^2 = 1,$$

therefore, $$s^2(A+C) + s(3A+B) + 3B = 1. \tag{6}$$

From this equation we can get three more equations:

$$A + C = 0, \quad 3A + B = 0, \quad \text{and } 3B = 1.$$

From these three equations it is found that

$$C = \frac{1}{9}, \quad B = \frac{1}{3}, \quad A = -\frac{1}{9} .$$

Now these values are substituted into equation (5),

$$\frac{1}{s^2(s+3)} = \frac{-\frac{1}{9} s + \frac{1}{3}}{s^2} + \frac{1}{9} \cdot \frac{1}{(s + 3)} .$$

Expanding further,

$$\frac{1}{s^2(s+3)} = -\frac{1}{9} \cdot \frac{1}{s} + \frac{1}{3} \cdot \frac{1}{s^2} + \frac{1}{9} \cdot \frac{1}{s+3} .$$

Now the inverse Laplace is found for the individual sums.

$$L^{-1}[-\frac{1}{9}\cdot\frac{1}{s}] = -\frac{1}{9}L^{-1}[\frac{1}{s}] = -\frac{1}{9}$$

$$L^{-1}[\frac{1}{3}\cdot\frac{1}{s^2}] = \frac{1}{3}L^{-1}[\frac{1}{s^2}] = \frac{1}{3}\ t$$

$$L^{-1}[\frac{1}{9}\cdot\frac{1}{s+3}] = \frac{1}{9}L^{-1}[\frac{1}{s+3}] = \frac{1}{9}\ e^{-3t}\ .$$

So,

$$L^{-1}[\frac{1}{s^2(s+3)}] = -\frac{1}{9} + \frac{1}{3}\ t + \frac{1}{9}\ e^{-3t}\ .$$

(c) First expand the equation into partial fractions.

$$\frac{1}{s(s+1)(s+3)} = \frac{A}{s} + \frac{B}{s+1} + \frac{C}{s+3}\ . \qquad (7)$$

From this equation it is seen that

$$A(s+1)(s+3) + Bs(s+3) + Cs(s+1) = 1,$$

therefore $\quad s^2(A+B+C) + s(4A+3B+C) = 3A = 1. \qquad (8)$

From (8) it is seen that

$$A + B + C = 0 \qquad 4A + 3B + C = 0 \qquad 3A = 1.$$

From these three equations it is found that

$A = \frac{1}{3}$, $B = -\frac{1}{2}$, and $C = \frac{1}{6}$. Now equation (7) is rewritten with the values of A, B, and C,

$$\frac{1}{s(s+1)(s+3)} = \frac{1}{3}\cdot\frac{1}{s} - \frac{1}{2}\cdot\frac{1}{s+1} + \frac{1}{6}\cdot\frac{1}{s+3}\ .$$

Now the inverse Laplace is found for the individual sums.

$$L^{-1}[\frac{1}{3}\cdot\frac{1}{s}] = \frac{1}{3}L^{-1}[\frac{1}{s}] = \frac{1}{3}$$

$$L^{-1}[\frac{1}{2}\cdot\frac{1}{s+1}] = \frac{1}{2}L^{-1}[\frac{1}{s+1}] = \frac{1}{2}\ e^{-t}$$

$$L^{-1}[\frac{1}{6}\cdot\frac{1}{s+3}] = \frac{1}{6}L^{-1}[\frac{1}{s+3}] = \frac{1}{6}\ e^{-3t}$$

So,

$$L^{-1}[\frac{1}{s(s+1)(s+3)}] = \frac{1}{3} - \frac{1}{2}\ e^{-t} + \frac{1}{6}\ e^{-3t}\ .$$

Evaluate the inverse Laplace transform of F(s) if $F(s) = \frac{2s + 4}{s^2 + 2s + 5}$. Write a generalized Laplace transform pair which can be added to a transform table to transform functions of this type.

Solution: Note that the quadratic in the denominator of F(s) can be factored into complex roots only. Thus,

$$F(s) = \frac{2s + 4}{[s + (x+zj)][s + (x-zj)]}$$

The partial fraction expansion of F(s) will be in the form:

$$F(s) = \frac{A}{s + (x+zj)} + \frac{A^*}{s + (x-zj)},$$ where A and A* will

be complex. Finally, f(t) will take the form:

$$f(t) = Ae^{-x-zj} + A^*e^{-x+zj},$$

if $A = x' + z'j$, and $A^* = x' = z'j$.

Then

$$f(t) = (x' + z'j)e^{-x-zj} + (x' - z'j)e^{-x+zj}$$

which can be written as:

$$f(t) = 2x'e^{-x}\,\frac{e^{zj}+e^{-zj}}{2} - e^{-x}z'j(2j)\,\frac{e^{zj}-e^{-zj}}{2j}$$

giving,

$$f(t) = 2x'e^{-x}(\cos zt) + 2z'e^{-x}(\sin zt) \qquad (1)$$

an exponentially damped sinusoid. Note that the damping (x) and the frequency (z) are the same for both cosine and sine terms.

Knowing that the function F(s) transforms to an exponentially damped sinusoid, check the Laplace table to find possible Laplace transform pairs which can transform F(s).

Most Laplace transform tables list the following two transform pairs:

$$c_1e^{-at}\cos bt \Rightarrow \frac{c_1(s+a)}{(s+a)^2 + b^2}$$

$$c_2e^{-at}\sin bt \Rightarrow \frac{c_2(b)}{(s+a)^2 + b^2}$$

Note that the function cannot be manipulated in any of these forms since

$$F(s) = \frac{2s+4}{s^2+2s+5} = \frac{2(s+2)}{(s+1)^2+4} = \frac{2(s+1)+2}{(s+1)^2+4} .$$

We can however, combine the two transform pairs since it is already known that f(t) must be in the form indicated in Eq.(1).

$$c_1 e^{-at}\cos bt + c_2 e^{-at}\sin bt \Rightarrow \frac{c_1(s+a) = c_2 b}{(s+a)^2 + b^2}$$

$$c_1 e^{-at}\cos bt + c_2 e^{-at}\sin bt \Rightarrow \frac{c_1 s + c_1 a + c_2 b}{(s+a)^2 + b^2}$$

Note that the damping a, and frequency b can be found readily from the above form.

$$F(s) = \frac{2s + 4}{(s+1)^2 + 4^2} ; \quad a = 1, \; b = 2$$

c_1 is found to be 2, c_2 must be solved. Allowing $c_1 a + c_2 b = K_2$, solve for c_2

$$c_2 = \frac{K_2 - c_1 a}{b} .$$

Thus, write a generalized expression for the Laplace transform pair as

$$c_1 e^{-at}\cos bt + \frac{K_2 - c_1 a}{b} e^{-at}\sin bt \Rightarrow \frac{c_1 s + K_2}{(s+a)^2 + b^2}$$

Using F(s) in our example:

F(t)

$$F(s) = \frac{2s+4}{(s+1)^2+4} \Rightarrow 2e^{-t}\cos 2t + \frac{4-2}{2} e^{-t}\sin 2t$$

$$f(t) = 2e^{-t}\cos 2t + e^{-t}\sin 2t.$$

• PROBLEM 11-30

Evaluate f(t) if

$$F(s) = \frac{3s^2 + 17s + 47}{(s + 2)(s^2 + 4s + 29)}$$

Solution: Since the degree of the numerator polynomial is less than the degree of the denominator polynomial, factor the denominator and expand the rational fraction by partial fractions.

Thus,

$$F(s) = \frac{3s^2 + 17s + 47}{(s + 2)(s + r_1)(s + r_2)}$$

where

$$r_1 = \frac{-4+\sqrt{16-116}}{2} \quad ; \qquad r_2 = \frac{-4-\sqrt{16-116}}{2}$$

$$r_1 = -4 + j5 \quad ; \qquad r_2 = -4 - j5$$

Expanding by partial fractions gives

$$F(s) = \frac{K_1}{s+2} + \frac{K_2}{s+2+j5} + \frac{K_2^*}{s+2-j5}$$

where

$$K_1 = \left|(s+2)\,F(s)\right|_{s=-2} = \left|\frac{3s^2+17s+47}{s^2+4s+29}\right|_{s=-2} = \frac{25}{25} = 1$$

and $$K_2 = \left|(s+2+j5)F(s)\right|_{s=-2-j5} = \left|\frac{3s^2+17s+47}{(s+2)(s+2-j5)}\right|_{s=-2-j5}$$

$$= \frac{3(-21+j20)-34-j85+47}{(-j5)(-j10)} = \frac{-50-j25}{-50} = 1 + j0.5$$

$$K_2^* = 1 - j0.5.$$

Substituting K_1, K_2, and K_2^* into the partial fraction expansion gives

$$F(s) = \frac{1}{s+2} + \frac{1+j0.5}{s+2+j5} + \frac{1-j0.5}{s+2-j5}\,.$$

Taking the inverse Laplace transform of F(s) gives

$$f(t) = e^{-2t} + (1+j0.5)e^{-2t-j5t} + (1-j0.5)e^{-2t+j5t}$$

Multiplying out:

$$f(t) = e^{-2t} + e^{-2t}e^{-j5t} + j0.5e^{-2t}e^{-j5t} + e^{-2t}e^{j5t} - j0.5e^{-2t}e^{j5t}$$

factoring to obtain the form:

$$\cos \omega t = \frac{e^{j\omega t} + e^{-j\omega t}}{2} \qquad \sin \omega t = \frac{e^{j\omega t} - e^{-j\omega t}}{2j}$$

$$f(t) = e^{-2t} + 2e^{-2t}\left|\frac{e^{j5t} + e^{-j5t}}{2}\right| + (2j)(-j0.5e^{-2t})\left|\frac{e^{j5t} - e^{-j5t}}{2j}\right|$$

$$f(t) = [e^{-2t} + 2e^{-2t}\cos 5t + e^{-2t}\sin 5t]\ u(t).$$

We can combine 2 cos 5t + sin 5t into a single sinusoidal expression by converting each term into a phasor and adding

$$\begin{array}{rcll} 2\cos 5t & \Rightarrow & 2\underline{/0^\circ} & = 2 \\ +\quad \sin 5t & \Rightarrow & 1\underline{/-90^\circ} & = -j \\ \hline 2\cos 5t + \sin 5t & \Rightarrow & 2\underline{/0^\circ} + 1\underline{/90^\circ} & = 2-j \end{array}$$

$$|2-j| = \sqrt{2^2+1^2}\underline{/\tan^{-1}-\tfrac{1}{2}} = 2.24\underline{/-26.6^\circ}$$

$2.24\underline{/-26.6^\circ} \Rightarrow 2.24\cos(5t-26.6^\circ)$.

The final result is

$$f(t) = [e^{-2t} + 2.24e^{-2t}\cos(5t-26.6^\circ)]u(t)$$

• PROBLEM 11-31

Obtain the time function corresponding to

$$F(s) = \frac{s^2 + 3s + 1}{(s + 1)^3(s + 2)^2}$$

Solution: Use partial fraction techniques to expand F(s) since $F(s) = \frac{N(s)}{D(s)}$ and the degree of D(s) is greater than N(s).

Thus,

$$F(s) = \frac{A_1}{(s+1)^3} + \frac{A_2}{(s+1)^2} + \frac{A_3}{s+1} + \frac{B_1}{(s+2)^2} + \frac{B_2}{s+2}$$

where constants are found to be

$$A_1 = \left|(s+1)^3F(s)\right|_{s=-1} = \left|\frac{s^2+3s+1}{(s+2)^2}\right|_{s=-1}$$

$$A_1 = -1$$

$$A_2 = \left|\frac{1}{(2-1)!}\frac{d}{ds}(s+1)^3 F(s)\right|_{s=-1}$$

$$A_2 = \left|\frac{d}{ds}\frac{s^2+3s+1}{(s+2)^2}\right|_{s=-1} = \left|\frac{(s+2)^2(2s+3)-(s^2+3s+1)2(s+2)}{(s+2)^4}\right|_{s=-1}$$

$$A_2 = 3$$

$$A_3 = \left|\frac{1}{(3-1)!}\frac{d^2}{ds^2}\frac{s^2+3s+1}{(s+2)^2}\right|_{s=-1}$$

$$A_3 = \left|\frac{1}{2}\frac{[2(s+2)(2s+3)+(s+2)^2 2-[(2s+3)2(s+2)+(s^2+3s+1)2]](s+2)^4}{(s+2)^8}\right.$$

$$\left. - \frac{[(s+2)^2(2s+3)-(s^2+3s+1)2(s+2)]4(s+2)^3}{(s+2)^8}\right|_{s=-1}$$

$$A_3 = \frac{[(2+2)-(2-2)]-[1+2]4}{2} = \frac{4-12}{2} = -4$$

$$B_1 = \left|(s+2)^2 F(s)\right|_{s=-2} = \left|\frac{s^2+3s+1}{(s+1)^3}\right|_{s=-2}$$

$$B_1 = 1$$

$$B_2 = \left|\frac{1}{(2-1)!}\frac{d}{ds}(s+2)^3 F(s)\right|_{s=-2}$$

$$= \left|\frac{(s+1)^3(2s+3)-(s^2+3s+1)3(s+1)^2}{(s+1)^6}\right|_{s=-2}$$

$$B_2 = 4$$

Substituting the above values for the constants into the partial fraction expansion gives,

$$F(s) = \frac{-1}{(s+1)^3} + \frac{3}{(s+1)^2} - \frac{4}{(s+1)} + \frac{1}{(s+2)^2} + \frac{4}{(s+2)}$$

Looking into a table of Laplace transforms yields

$$\frac{A\,n!}{(s+a)^{n+1}} \Leftrightarrow At^n e^{-at}$$ but

the terms of the partial fraction expansion exist in the form

$\frac{A}{(s+a)^n}$. Find the time-domain of this function by subtracting 1 from each n in the transform pair found in the tables above, thus,

$$\frac{A(n-1)!}{(s+a)^n} \Leftrightarrow At^{n-1}e^{-at}$$

Finally, divide each side of the pair by (n-1)! to give the form desired.

$$\frac{A}{(s+a)^n} \Leftrightarrow \frac{A}{(n-1)!}t^{n-1}e^{-at}$$

Applying this transform pair to partial fraction gives,

$$F(t) = -\frac{1}{2}t^2e^{-t} + 3te^{-t} - 4e^{-t} + te^{-2t} + 4e^{-2t}.$$

• PROBLEM 11-32

Evaluate f(t) if

$$F(s) = \frac{2s^2 + 13s + 17}{(s+1)(s+3)}$$

Solution: Use partial fraction expansion on F(s) provided that the function is rational, thus the degree on the numerator polynomial must be lower than the degree of the denominator polynomial.

If the function is not rational, as in this example, then divide the numerator by the denominator in order to obtain a rational function.

$$F(s) = \frac{2s^2 + 13s + 17}{(s+1)(s+3)} = \frac{2s^2 + 13s + 17}{s^2 + 4s + 3}$$

Dividing gives,

$$\begin{array}{r|l} & 2 \\ s^2+4s+3 & 2s^2+13s+17 \\ & \underline{2s^2+\ 8s+\ 6} \\ & 5s+11 \end{array}$$

Thus,

$$F(s) = 2 + \frac{5s + 11}{s^2 + 4s + 3}$$

Now, expand the second term by partial fractions:

$$F(s) = 2 + \frac{A}{s + 1} + \frac{B}{s + 3}$$

where, $A = \left|\frac{5s+11}{s+3}\right|_{s=-1} = 3$, and

$$B = \left|\frac{5s+11}{s+1}\right|_{s=-3} = 2.$$

The function F(s) becomes:

$$F(s) = 2 + \frac{3}{s+1} + \frac{2}{s+3} \cdot$$

Since, $\{K\}^{-L} = K\delta(t)$,

and $\frac{K}{s+a}^{-L} = Ke^{-at}$, transform F(s) to the time domain.

$$F(t) = 2\delta(t) + (3e^{-t} + 2e^{-3t})u(t).$$

• PROBLEM 11-33

Find the inverse Laplace transform of

$$F(s) = \frac{1 - 2e^{-s} + c^{-2s}}{s^2(1 - e^{-2s})}$$

Solution: Recall that $F(t) = L\{F(s)\}$ in a semiperiodic function given in the form $L\left\{\frac{G(s)}{1-e^{-Ts}}\right\}$, where T is the period, provided G(s) represents a pulse that is zero for $t > T$.

In the example, $T = 2$ and $G(s) = \frac{1-2e^{-s}+e^{-2s}}{s^2}$, which can be written as

$$\frac{1}{s^2} - \frac{2}{s^2}e^{-s} + \frac{1}{s^2}e^{-2s} = G_1(s) + G_2(s) + G_3(s)$$

Find G(t) by transforming each term,

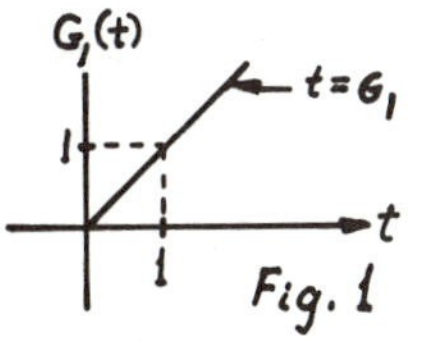

Fig. 1

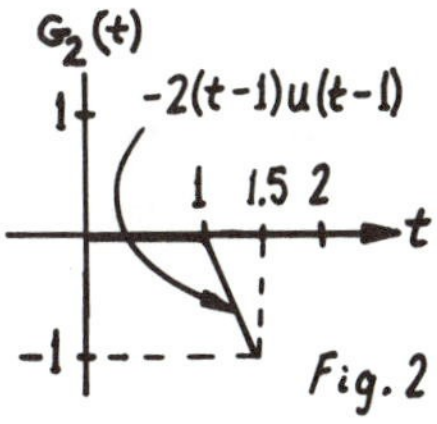

Fig. 2

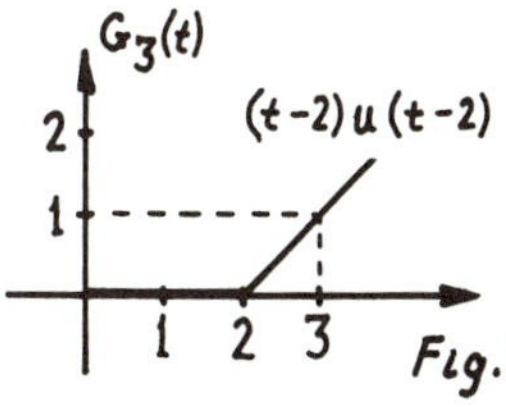

Fig. 3

$$G_1(t) = L^{-1}\left(\frac{1}{s^2}\right) = t\ u(t) \qquad \text{(fig. 1)}$$

Note in $G_2(t)$, e^{-s} is the time displacement term where in $e^{-\tau s}$, τ is the number of time units by which the signal is delayed.

Thus,

$$G_2(t) = L^{-1}\left(-\frac{2}{s^2}\right) \text{ delayed by } 1 = -2(t-1)u(t-1) \quad \text{(fig. 2)}$$

Similarly,

$$G_3(t) = L^{-1}\left(\frac{1}{s^2}\right) \text{ delayed by } 2 = (t-2)u(t-2) \qquad \text{(fig. 3)}$$

In order to visualize the entire pulse $G(t)$, add the three waveforms found (fig. 4)

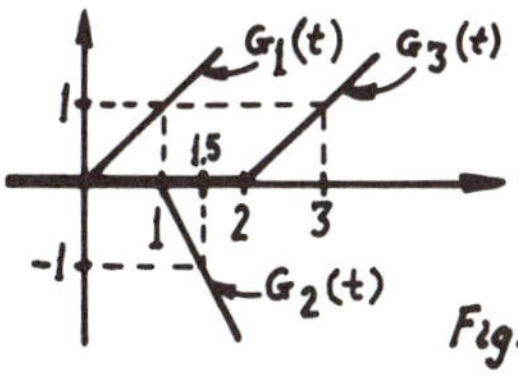

Fig. 4

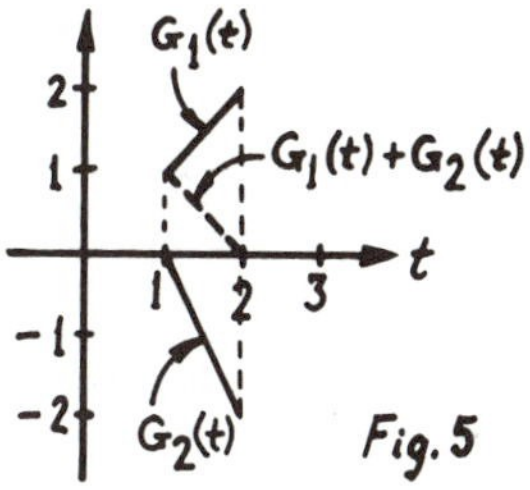

Fig. 5

For $t < 0$, $G(t) = 0$; for $0 < t < 1$, $G(t) = G_1(t)$ since $G_2(t) = G_3(t) = 0$; for $1 < t < 2$, $G(t) = G_1(t) + G_2(t)$ since $G_3(t) = 0$. Thus $G_1(t) + G_2(t)$ is shown in fig. 5.

For $t > 2$, $G(t) = G_1(t) + G_2(t) + G_3(t) = 0$ as shown in fig. 6.

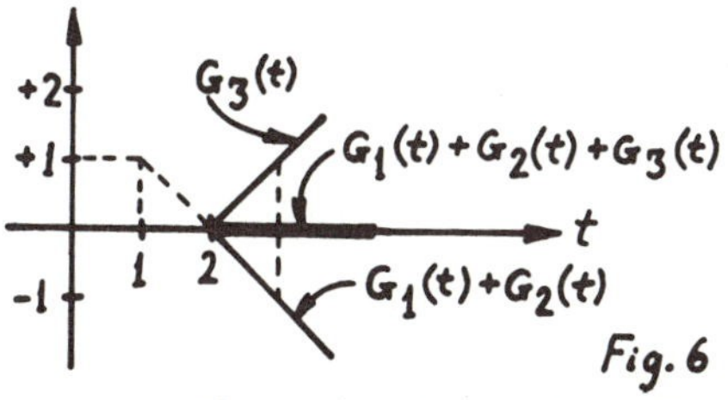

Fig. 6

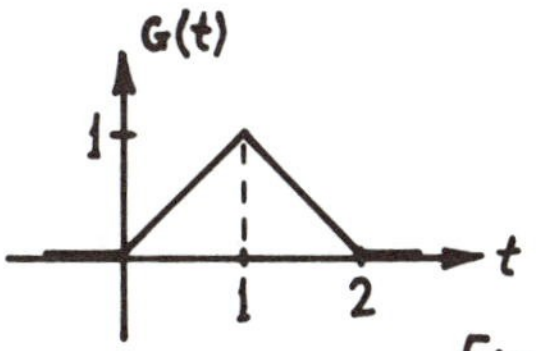

Fig. 7

The single pulse $G(t)$ is shown in fig. 7.

Note that for $t > 2$, $G(t) = 0$. Thus, we can say that $F(t)$ is periodic with period $T = 2$. $F(t)$ is shown in fig. 8.

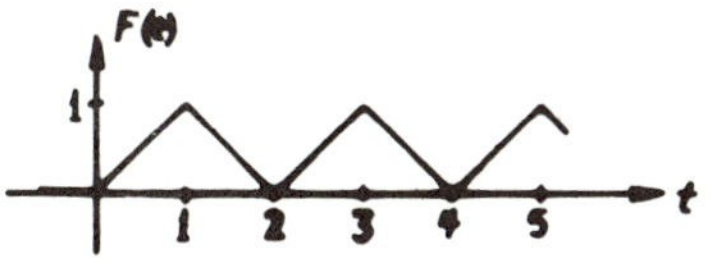

Fig. 8

In a certain network the response y(s) is related to the source $\phi(s)$ by the equation

$$y(s) = \frac{s+2}{(s+1)(s^2+3s+5)}\,\phi(s).$$

Use Laplace transform techniques to calculate the component of the response due to each of the following sources: (a) $\phi(t) = 10\,u(t)$; (b) $\phi(t) = 10t\,u(t)$; (c) $\phi(t) = 10e^{-2t}u(t)$; (d) $\phi(t) = 10e^{-4t}u(t)$; (e) $\phi(t) = e^{-t}u(t)$.

Solution: (a) We can transform $\phi(t)$ to $\phi(s)$ and substitute it into the equation for the network response. Thus,

$$\phi(s) = L[10u(t)] = \frac{10}{s}$$

yields

$$y(s) = \frac{10(s+2)}{s(s+1)(s^2+2s+5)}.$$

Using partial fraction expansion we obtain

$$y(s) = \frac{A}{s} + \frac{B}{s+1} + \frac{C}{s+r_3} + \frac{D}{s+r_4}.$$

We find that the component of the response due to the source is $\frac{A}{s}$ by comparing the above expanded function with the given network function,

$$\frac{s+2}{(s+1)(s^2+2s+5)}.$$

By partial fractions,

$$A = s\,y(s)\Big|_{s=0} = \frac{10(0+2)}{(0+1)(0+0+5)} = \frac{20}{5} = 4$$

thus, the component of the response due to the source is

$$\frac{4}{s} \Rightarrow 4\,u(t).$$

(b) Using the same method as in (a) yields

$$\phi(s) = L\{\phi(t)\} = L\{10t\,u(t)\} = \frac{10}{s^2}.$$

Thus,

$$y(s) = \frac{10(s+2)}{s^2(s+1)(s^2+2s+5)}$$

and $$y(s) = \frac{A}{s} + \frac{B}{s^2} + \frac{C}{s+1} + \frac{D}{s+r_4} + \frac{E}{s+r_5}\ .$$

The component due to the source is

$$\frac{A}{s} + \frac{B}{s^2}$$

where $$B = s^2\ \widehat{y(s)}\Big|_{s=0} = 4$$ and

$$A = \frac{1}{(2-1)!}\ \frac{d}{ds}\ [(s^2 y(s)]\Big|_{s=0} = \frac{d}{ds}\ \frac{10(s+2)}{(s+1)(s^2+2s+5)}\Bigg|_{s=0}$$

$$A = \frac{10(s+1)(s^2+2s+5) - 10(s+2)[(s^2+2s+5) + (s+1)(2s+2)]}{[(s+1)(s^2+2s+5)]^2}\Bigg|_{s=0}$$

$$A = \frac{10(1)(5) - 10(2)[5 + (1)(2)]}{[(1)(5)]^2} = \frac{50 - 140}{25}$$

$A = -3.6.$

Our component response becomes

$$\frac{-3.6}{s} + \frac{4}{s^2} \Rightarrow (-3.6 + 4t)u(t)$$

(c) $\phi(t) = 10e^{-2t}u(t) \Rightarrow \phi(s) = \dfrac{10}{s+2}$, gives

$$\underline{y}(s) = \frac{10(s+2)}{(s+2)(s+1)(s^2+2s+5)}\ .$$ Thus, the component of the response due to the source is zero.

(d) $\phi(t) = 10e^{-4t} \Rightarrow \phi(s) = \dfrac{10}{s+4}$, gives

$$\underline{y}(s) = \frac{10(s+2)}{(s+4)(s+1)(s^2+2s+5)}\ .$$ By partial fractions we obtain

$$y(s) = \frac{A}{s+4} + \frac{B}{s+1} + \frac{C}{s+r_3} + \frac{D}{s+r_4}\ .$$

The component due to the source is

$$\frac{A}{s+4}$$

where $$A = (s+4)y(s)\Big|_{s=-4} = \frac{10(-4+2)}{(-4+1)(16-8+5)}$$

$$A = \frac{20}{39}.$$

Thus, our component is $\frac{\frac{20}{39}}{s+4} \Rightarrow \frac{20}{39}e^{-4t}u(t)$.

(e) $\phi(t) = e^{-t}u(t) \Rightarrow \frac{1}{s+1}$ gives

$$y(t) = \frac{(s+2)}{(s+1)^2(s^2+2s+5)}.$$ By partial fractions we have

$$y(t) = \frac{A}{s+1} + \frac{B}{(s+1)^2} + \frac{C}{s+r_3} + \frac{D}{s+r_4}$$

where the component due to the source is $\frac{B}{(s+1)^2}$ because the given network function $\frac{s+2}{(s+1)(s^2+2s+5)}$ can be expanded to $y(s) = \frac{E}{s+1} + \frac{F}{s+r_3} + \frac{G}{s+r_4}$. Thus, the response due to the source must be $\frac{B}{(s+1)^2}$, where

$$B = (s+1)^2 y(s)\Big|_{s=-1} = \frac{1}{4}.$$

The response is $\frac{\frac{1}{4}}{(s+1)^2} \Rightarrow \frac{1}{4}te^{-t}u(t)$.

• PROBLEM 11-35

A response transform $V_{ab}(s)$ is related to a source transform $V(s)$ by

$$V_{ab}(s) = \frac{3V(s) + 3s + 27}{s^2 + 6s + 8}$$

Find $v_{ab}(t)$ if $v(t) = 16e^{-3t}u(t)$.

Solution: Find $L\{v(t)\} = V(s)$ and substitute into $V_{ab}(s)$.

Thus, $$L\{16e^{-3t}\} = V(s) = \frac{16}{s+3}$$

$$V_{ab}(s) = \frac{\frac{(3)(16)}{s+3} + 3s + 27}{s^2 + 6s + 8} = \frac{48 + 3s(s+3) + 27(s+3)}{(s+3)(s^2+6s+8)}$$

$$V_{ab}(s) = \frac{3s^2 + 36s + 129}{(s+3)(s^2+6s+8)} = \frac{3s^2 + 36s + 129}{(s+3)(s+2)(s+4)}$$

Using partial fraction expansion, write:

$$V_{ab}(s) = \frac{A}{s+3} + \frac{B}{s+2} + \frac{C}{s+4}$$

where $A = \left| (s+3)V_{ab}(s) \right|_{s=-3} = \left| \frac{3s^2+36s+129}{(s+2)(s+4)} \right|_{s=-3} =$

$$A = \frac{27 - 108 + 129}{(-3+4)(-3+2)} = \frac{48}{-1} = -48$$

$$B = \left| (s+2)V_{ab}(s) \right|_{s=-2} = \left| \frac{3s^2+36s+129}{(s+3)(s+4)} \right|_{s=-2}$$

$$B = \frac{12 - 72 + 129}{(-2+3)(-2+4)} = \frac{69}{2}$$

$$C = \left| (s+4)V_{ab}(s) \right|_{s=-4} = \left| \frac{3s^2+36s+129}{(s+2)(s+4)} \right|_{s=-4}$$

$$C = \frac{48 - 144 + 129}{(-4+2)(-4+3)} = \frac{33}{2}$$

Finally, we obtain $V_{ab}(t)$ by taking the inverse Laplace transform of $V_{ab}(s)$,

$$V_{ab}(s) = \frac{-48}{s+3} + \frac{\frac{69}{2}}{s+2} + \frac{\frac{33}{2}}{s+4}$$

$$v_{ab}(t) = -48e^{-3t} + \frac{69}{2}e^{-2t} + \frac{33}{2}e^{-4t}$$

• **PROBLEM 11-36**

If $v(t) = 10e^{-4t}u(t)$, and $V_{ab}(s)$ is related to $V(s)$ by

$$V_{ab}(s) = \frac{3V(s) + 3s + 27}{s^2 + 6s + 8},$$

formulate $v_{ab}(t)$.

Solution: Substituting $L[v(t)] = \frac{10}{s+4}$ for $V(s)$ in $V_{ab}(s)$ gives

$$V_{ab}(s) = \frac{\frac{30}{s+4} + 3s + 27}{s^2 + 6s + 8}$$

$$V_{ab}(s) = \frac{3s^2 + 39s + 138}{(s+4)(s^2+6s+8)}$$

factoring the denominator,

$$V_{ab}(s) = \frac{3s^2 + 39s + 138}{(s+4)^2(s+2)} .$$

By partial fraction expansion:

$$V_{ab}(s) = \frac{A}{(s+4)^4} + \frac{B}{s+4} + \frac{C}{s+2}$$

where $A = \left| (s+4)^2 V_{ab}(s) \right|_{s=-4} = \left| \frac{3s^2 + 39s + 138}{s + 2} \right|_{s=-4}$

$A = -15$

$$B = \frac{1}{(2-1)!} \left| \frac{d}{ds} (s+4)^2 V_{ab}(s) \right|_{s=-4}$$

$$B = \left| \frac{(s+2)(6s+39) - (3s^2+39s+138)}{(s+2)^2} \right|_{s=-4} = -15$$

$$C = \left| (s+2)V_{ab}(s) \right|_{s=-2} = \left| \frac{3s^2+39s+138}{(s+4)^2} \right|_{s=-2} = 18$$

Hence,

$$V_{ab}(s) = \frac{-15}{(s+4)^2} + \frac{-15}{s+4} + \frac{18}{s+2}$$

the inverse Laplace transform of $V_{ab}(s)$ gives

$$v_{ab}(t) = -15te^{-4t} - 15e^{-4t} + 18e^{-2t} .$$

• PROBLEM 11-37

Given the transfer function,

$$\frac{V_o}{V_i} = \frac{s^2 + 2s + 1}{s^2 + 2s + 2} ,$$

find the component of the response due to the source for: (a) $V_i(t) = 20u(t)$ and (b) $V_i(t) = 20t^2u(t)$, knowing that the component of the response due to the source, $v_i(t) = 10t\ u(t)$, is $[v_o(t)]_{source} = [5+5t]u(t)$.

Solution: (a) It is given that the source $v_i(t) = 10t\ u(t)$, causes a response $[v_o(t)]_s = [5+st]u(t)$. When transformed, $V_i(s) = \frac{10}{s^2}$ and $[V_o(s)]_s = \frac{5}{s} + \frac{5}{s^2}$. Find the response due to the new source $v_i(t) = 20u(t)$ by writing it as a product;

$$L\{v_i(t)\} = V_i(s) = \frac{20}{s} = \frac{10}{s^2} A$$

This would mean that $[V_o(s)]_s = A(\frac{5}{s} + \frac{5}{s^2})$.

Find A by solving

$$\frac{20}{s} = \frac{10}{s^2} (A)$$

or

$$A = \frac{20}{s} \cdot \frac{s^2}{10} = 2s$$

Remember that s represents the operation $\frac{d}{dt}$, so that

$$[v_o(t)]_s = 2 \frac{d}{dt} (5+5t)$$

$$[v_o(t)]_s = 10$$

(b) Using the same procedure as in part (a), find $L\{v_i(t)\} = V_i(s) = \frac{40}{s^3} = \frac{10}{s^2} (A)$

Hence,

$$A = \frac{40}{s^3} \cdot \frac{s^2}{10} = \frac{4}{s}.$$

Remember that $\frac{1}{s}$ can represent the operation of integration $\int_o^t$, thus

$$[v_o(t)]_s = 4\int (5+5t)dt = 20t + 10t^2 + C$$

where C is the constant of integration.

Find C by substituting the values of $V_i(s)$ and $[v_o(t)]_s$ for $V_i(s)$ and $v_o(t)$ respectively in the transfer equation.

Hence,

$$\frac{L\{20t + 10t^2 + C\}}{\frac{40}{s^2}} = \frac{s^2 + 2s + 1}{s^2 + 2s + 2}$$

$$\frac{\frac{20}{s^2} + \frac{20}{s^3} + \frac{C}{s}}{\frac{40}{s^3}} = \frac{s^2 + 2s + 1}{s^2 + 2s + 2}$$

multiplying the left side by $\frac{s^3}{s^3}$ gives,

$$\frac{Cs^2 + 20s + 20}{40} = \frac{s^2 + 2s + 1}{s^2 + 2s + 2}$$

multiplying through gives,

$$s^4(c) + s^3(2C+20) + s^2(2C+60) + 80s + 40 = 40s^2 + 80s + 40$$

Since C must be a non-zero constant, and $[V_1(s)]_s$ does not represent the complete response, disregard the s^4 and s^3 terms.

Hence, $2C + 60 = 40$

$$C = \frac{40 - 60}{2} = -10$$

and $[v_o(t)]_s = 20t + 10t^2 - 10$

• PROBLEM 11-38

Find $F(s) = L\{e^{-4t}\cos(10t-30°)u(t)\}$ by using the concept of Phasors to expand the sinusoid. Use the Laplace operation of frequency displacement to obtain the result.

Solution: Since cos(10t-30°) is displaced by a phase angle of 30°, we cannot find its Laplace transform directly. It must be expanded. Using Euler's identity, it can be said that, since $e^{j\omega t} = \cos \omega t + j \sin \omega t$,

then $\mathrm{Re}\{e^{j\omega t}\} = \cos \omega t$. This is the basis of the Phasor concept. Thus, we can write $\cos(10t-30°) = \mathrm{Re}\{e^{j(10t-30°)}\}$. Using Euler's identity to expand $e^{j(10t-30°)}$ obtains

$$e^{j(10t-30°)} = e^{j10t} \cdot e^{-j30°}$$

$$= (\cos 10t + j\sin 10t)(\cos 30° - j\sin 30°)$$

$$= \cos 30°\cos 10t + j\cos 30°\sin 10t$$

$$- j\sin 30°\cos 10t + \sin 30°\sin 10t$$

Taking the real part yields

$$\text{Re}\{e^{j(10t-30°)}\} = \cos(10t-30°) = \cos 30° \cos 10t + \sin 30° \sin 10t$$

$$= 0.866\cos 10t + 0.5\sin 10t$$

Now use the following two Laplace transform pairs to transform the result.

$$\frac{As}{s^2 + \omega^2} \Leftrightarrow A\cos\omega t$$

$$\frac{B\omega}{s^2 + \omega^2} \Leftrightarrow B\sin\omega t$$

Since $F_1(s) + F_2(s) \Leftrightarrow F_1(t) + F_2(t)$ is permissible, combine the two transforms to obtain

$$\frac{As + B\omega}{s^2 + \omega^2} \Leftrightarrow A\cos\omega t + B\sin\omega t.$$

Thus,

$$L\{0.866\cos 10t + 0.5\sin 10t\} = \frac{0.866s + 5}{s^2 + 100}$$

Using the frequency displacement Laplace operation, which is stated as:

$$F(s+a) \Leftrightarrow e^{-at}f(t),$$

obtains

$$e^{-4t}\cos(10t-30°) \Leftrightarrow \frac{0.866(s+4) + 5}{(s+4)^2 + 100},$$

the desired result.

• **PROBLEM 11-39**

The impulse response of a network is given by sin t, $0 < t < \pi$ and zero otherwise. Find the response due to a pulse of height 1 and duration π sec.

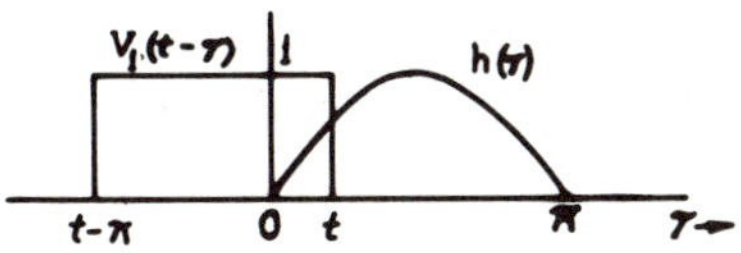

Solution: The Laplace transform of the impulse response is the transfer function of the network which must be obtained first.

$$H(s) = L[h(t)] = \int_0^{\pi} \sin t \; e^{-st} \, dt$$

where the upper limit of integration is π since h(t) has zero value beyond this. To facilitate integration, sin t is replaced by its exponential form to give

$$H(s) = \frac{1}{2j}\int_0^{\pi} [e^{(-s+j)t} - e^{(-s-j)t}]dt$$

$$= \frac{1}{2j}\left[\frac{e^{(-s+j)t}}{(-s+j)} - \frac{e^{(-s-j)t}}{(-s-j)}\right]_0^{\pi}$$

$$= \frac{1}{2j}\left[\frac{e^{(-s+j)\pi} - 1}{(-s+j)} - \frac{e^{(-s-j)\pi} - 1}{(-s-j)}\right]$$

Since $e^{\pm j\pi} = -1$, H(s) becomes

$$H(s) = \frac{1}{2j}\left[\frac{e^{-\pi s}+1}{s-j} - \frac{e^{-\pi s}+1}{s+j}\right] = \frac{e^{-\pi s}+1}{2j}\left[\frac{1}{s-j} - \frac{1}{s+j}\right]$$

$$= \frac{e^{-\pi s}+1}{s^2+1}$$

An alternate way to derive this is to note that h(t) is the sum of sin(t)u(t) and sin(t-π)u(t-π). The respective transforms of these are $1/(s^2+1)$ and $e^{-s\pi}/(s^2+1)$ where the shifting theorem was applied to obtain the latter transform. The sum of these two transforms gives the required H(s).

The input pulse is the sum of u(t) and -u(t-π). Applying the shifting theorem again, the pulse transform is

$$V_1(s) = \frac{1}{s}(1-e^{-\pi s}).$$

The transform of the response is, thus,

$$V_2(s) = H(s)V_1(s) = \frac{1}{s(s^2+1)}[1-e^{-2\pi s}]$$

Consider first the fraction $V_2'(s) = 1/s(s^2+1)$. The inverse transform is readily obtained by partial fraction expansion (or tables) as:

$$V_2'(t) = (1-\cos t)u(t).$$

Thus, the response $V_2(t)$ consists of $V_2'(t)$ minus $V_2'(t-\pi)$ since the factor $e^{-2\pi s}$ corresponds to a delay of 2π seconds in the inverse transform. Thus,

$$V_2(t) = (1-\cos t)u(t) - [1-\cos(t-2\pi)]u(t-2\pi)$$

$$= (1-\cos t)[u(t) - u(t-2\pi)]$$

The response can be obtained even more simply by using convolution. That is,

$$V_2(t) = \int_0^t h(\tau)V_1(t-\tau)d\tau = \int_0^t \sin(\tau)d\tau = 1 - \cos t$$

It is apparent from the figure that $V_1(t-\tau)$ and $h(\tau)$ only overlap for $0 < t < 2\pi$. Thus, $V_2(t)$ is non-zero over this range only and can be written more properly as

$$V_2(t) = (1-\cos t)[u(t) - u(t-2\pi)]$$

CHAPTER 12

LAPLACE TRANSFORM APPLICATIONS

SERIES CIRCUITS

• PROBLEM 12-1

For the circuit of Fig. 1 calculate i(t) when (a) v(t) = u(t), (b) v(t) = δ(t). Assume zero initial conditions.

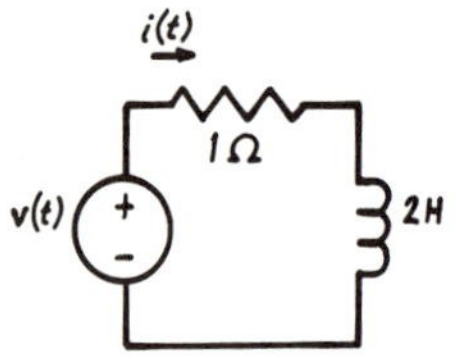

FiG.1

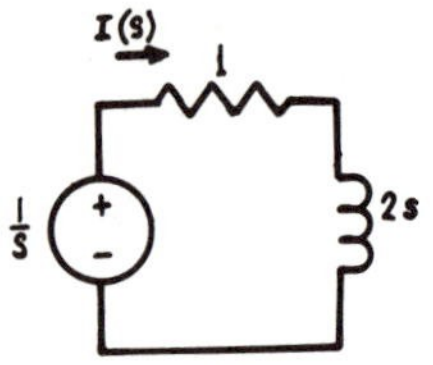

FiG.2

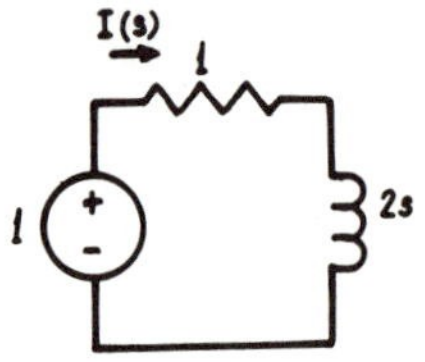

FiG.3

Solution: Figure 2 shows the given network transformed to the s domain, with the assumption of zero initial conditions.

Recall that L {u(t)} = $\frac{1}{s}$. Writing a loop equation,

$$\frac{1}{s} = I(s)\ (1 + 2s)$$

$$I(s) = \frac{1}{s(2s + 1)}$$

$$I(s) = \frac{1}{2s\left(s + \frac{1}{2}\right)}$$

$$I(s) = \frac{\frac{1}{2}}{s\left(s + \frac{1}{2}\right)}$$

expanding into partial fractions in order to take the inverse Laplace,

$$I(s) = \frac{\frac{1}{2}}{s\left(s + \frac{1}{2}\right)} = \frac{A}{s} + \frac{B}{s + \frac{1}{2}}$$

$$A = (I(s))(s)\Bigg|_{s=0} = \frac{\frac{1}{2}}{s + \frac{1}{2}}\Bigg|_{s=0} = 1$$

$$B = (I(s))\left(s + \frac{1}{2}\right)\Bigg|_{s=-\frac{1}{2}} = \frac{\frac{1}{2}}{s}\Bigg|_{s=-1/2} = -1$$

so $$I(s) = \frac{1}{s} - \frac{1}{s + \frac{1}{2}}$$

and $$i(t) = u(t) - e^{-\frac{t}{2}}u(t)$$

$$i(t) = (1 - e^{-\frac{t}{2}})\,u(t).$$

For part (b) (Figure 3) we recall that $L\{\delta(t)\} = 1$. Then writing a loop equation,

$$1 = I(s)\,(1 + 2s)$$

$$I(s) = \frac{1}{1 + 2s} = \frac{\frac{1}{2}}{s + \frac{1}{2}}$$

and $$i(t) = \frac{1}{2}e^{-\frac{t}{2}}u(t).$$

• **PROBLEM 12-2**

Calculate the response $v_C(t)$.

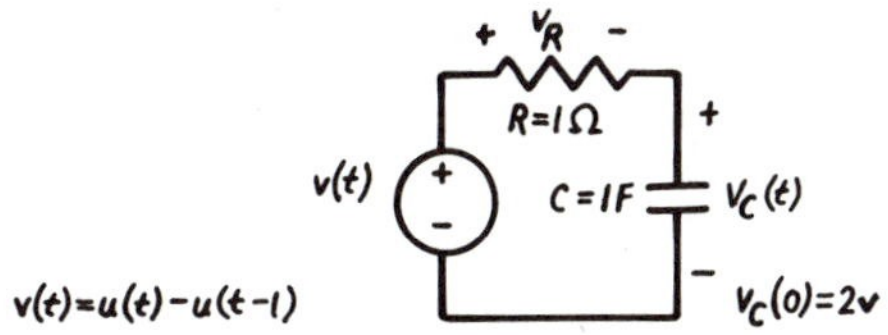

Solution: This problem is solved by first writing $v_C(t)$ as a differential equation using the nodal relationship for currents. The differential equation is then solved using Laplace transform techniques.

The nodal equation at the node between R and C is

$$\frac{v_C(t) - [u(t) - u\,(t - 1)]}{R} + C\,\frac{dv_C(t)}{dt} = 0$$

$R = C = 1.$

Hence, $$v_C(t) + \frac{dv_C(t)}{dt} = u(t) - u(t - 1).$$

Taking the Laplace transform and using the time-shift function,

$$V_C(s) + s\,V_C(s) - V_C(0) = \frac{1}{s} - \frac{1}{s}e^{-s}.$$

Substituting $V_C(0) = 2$ and rearranging,

$$V_C(s) = \frac{1}{s(s+1)} - \frac{1}{s(s+1)}e^{-s} + \frac{2}{s+1}.$$

Using partial fraction expansion,

$$V_C(s) = \left(\frac{1}{s} - \frac{1}{s+1}\right) - \left(\frac{1}{s} - \frac{1}{s+1}\right)e^{-s} + \frac{2}{s+1}.$$

Taking the inverse Laplace transform,

$$v_C(t) = (1 - e^{-t})\,u(t) - (1 - e^{-(t-1)})\,u(t-1) + 2e^{-t}\,u(t).$$

Collecting terms,

$$v_C(t) = (1 + e^{-t})\,u(t) - (1 - e^{-(t-1)})\,u(t-1).$$

But $u(t) = \begin{cases} 0, & t < 0 \\ 1, & t > 0 \end{cases}$

$u(t-1) = \begin{cases} 0, & t < 1 \\ 1, & t > 1 \end{cases}$

therefore, $v_C(t) = \begin{cases} 1 + e^{-t}\ \text{V} & 1 > t > 0 \\ e^{-(t-1)} + e^{-t}\ \text{V} & t > 1 \end{cases}.$

• PROBLEM 12-3

For the circuit of Fig. 1: (a) let $v_s(t) = 3t\,u(t)$ and find $i(t)$; (b) Let $v_s(t) = 3\delta(t)$ and find $v(t)$.

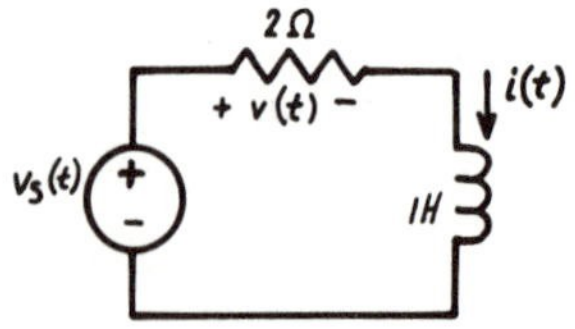

FIG.1

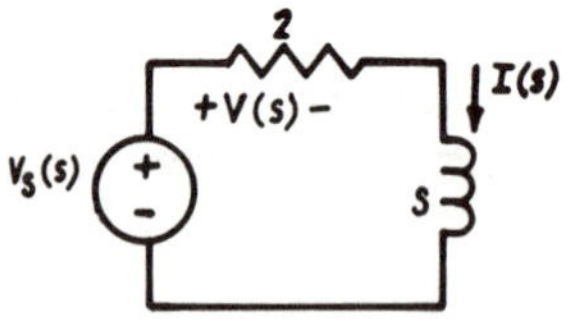

FIG. 2

Solution: This problem can be solved using Laplace transforms. First transform the circuit into the s-domain as indicated in Figure 2.

From the Laplace transform pairs:

$$L\{v_s(t)\} = L\{3t\ u(t)\} = \frac{3}{s^2} = V_s(s)$$

and $$I(s) = \frac{V_s(s)}{Z(s)} = \frac{\left(\frac{3}{s^2}\right)}{s+2} = \frac{3}{s^2(s+2)}.$$

By the partial fraction expansion theorem,

$$I(s) = \frac{3}{s^2(s+2)} = \frac{K_1}{s+2} + \frac{N_1}{s} + \frac{N_2}{s^2}$$

where $$K_1 = \left.\frac{3}{s^2}\right|_{s=-2} = \frac{3}{4}$$

$$N_1 = \frac{d}{ds}\left.\left(\frac{3}{s+2}\right)\right|_{s=0} = -\left.\frac{3}{(s+2)^2}\right|_{s=0} = -\frac{3}{4}$$

$$N_2 = \left.\frac{3}{s+2}\right|_{s=0} = \frac{3}{2}.$$

Hence, $$I(s) = \frac{3}{4(s+2)} - \frac{3}{4s} + \frac{3}{2s^2}.$$

Taking the inverse Laplace from a table,

$$i(t) = L^{-1}\{I(s)\} = \frac{3}{4}e^{-2t}u(t) - \frac{3}{4}u(t) + \frac{3}{2}t\ u(t),$$

or rearranging and factoring,

$$i(t) = \left(-\frac{3}{4} + \frac{3}{2}t + \frac{3}{4}e^{-2t}\right)u(t)\ \text{A}.$$

For part (b),

$$V_s(s) = L\{3\,\delta(t)\} = 3$$

$$I(s) = \frac{V_s(s)}{Z(s)} = \frac{3}{s+2}$$

$$i(t) = L^{-1}\{I(s)\} = 3e^{-2t}u(t)$$

$$v(t) = i(t)R = 6e^{-2t}u(t)\ \text{V}.$$

Find the current i(t) when a step input u(t) is applied to the RL network shown in Fig. 1. Use the convolution integral and then compare with Laplace transform techniques.

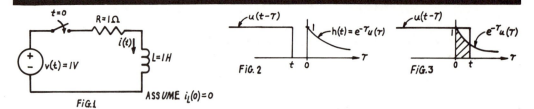

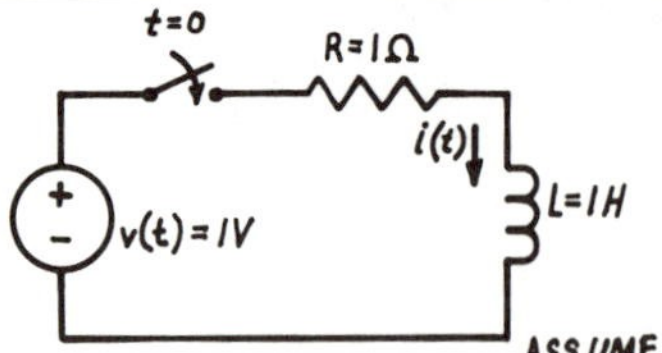

ASSUME $i_L(0)=0$

FIG.1

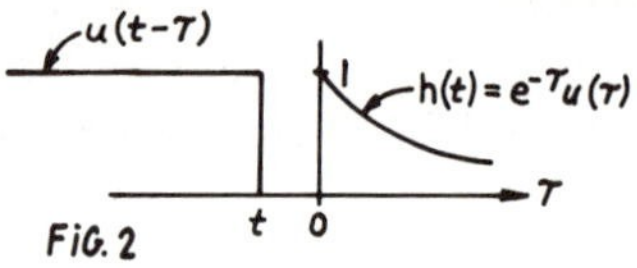

FIG. 2

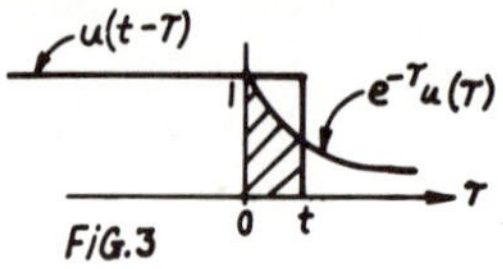

FIG.3

Solution: Find i(t) by use of the convolution integral

$$g(t) = f_1(t) * f_2(t) = \int_{-\infty}^{\infty} f_1(t - \tau)\, f_2(\tau)\, d\tau$$

where g(t) is the output response i(t), $f_1(t)$ is the input driving function u(t), and $f_2(t)$ is the impulse response of the circuit h(t).

The impulse response can be found by writing the differential equation(s) that describe(s) the circuit in the form

$$\dot{X} = \bar{A}X + \bar{b}u(t)$$

then taking $e^{\bar{A}t}\,\bar{b}u(t)$ which equals $\bar{h}(t)$.

In this example write the single differential equation

$$L\frac{di}{dt} + Ri = u(t).$$

In the form mentioned above,

$$\dot{X} = -\frac{R}{L}X + \frac{1}{L}u(t).$$

Hence $\quad A = -\frac{R}{L}$ and $b = \frac{1}{L}$ the impulse response is

$$e^{At}bu(t) = \frac{1}{L}e^{-\frac{R}{L}t}u(t) = e^{-t}u(t) \text{ where } R = L = 1.$$

Therefore, the convolution integral gives

$$i(t) = \int_{-\infty}^{\infty} u(t - \tau)\, e^{-\tau}\, u(\tau)\, d\tau$$

Figs. 2 and 3 show a graph of the function $u(t - \tau)$ for $t < 0$ and $t > 0$ respectively.

Note from these figures that the convolution integral will give values of zero for i(t) when $t \leq 0$. When $t > 0$, however, we obtain the area shown in cross-hatched lines in Fig. 3. Hence, integrate $e^{-t}u(t)$ for $t > 0$

$$i(t) = \int_0^t e^{-\tau}\,d\tau = 1 - e^{-t}\ \text{A},\quad t > 0$$

Using Laplace transform, check the results by noting that multiplication of two functions in the frequency domain is analogous to convolving two functions in the time domain.

Hence, $I(s) = V(s)\,H(s)$ where $V(s) = L\{v(t)\} = L\{u(t)\} = \frac{1}{s}$, and $H(s) = L\{e^{-t}u(t)\} = \frac{1}{s+1}$.

We obtain $I(s) = \frac{1}{s}\left(\frac{1}{s+1}\right) = \frac{1}{s(s+1)}$, using partial fraction expansion

$$\frac{1}{s(s+1)} = \frac{A}{s} + \frac{B}{s+1},\ \text{where } A = \left.\frac{1}{s+1}\right|_{s=0} = 1 \text{ and}$$

$$B = \left.\frac{1}{s}\right|_{s=-1} = -1,\ \text{giving } I(s) = \frac{1}{s} - \frac{1}{s+1}$$

which can be transformed back to the time domain noting that $\frac{1}{s} \overset{L^{-1}}{\Rightarrow} u(t)$, and $-\frac{1}{s+1} \overset{L^{-1}}{\Rightarrow} -e^{-t}$. Hence,

$$i(t) = 1 - e^{-t}\ \text{A},\quad t > 0.$$

• PROBLEM 12-5

The voltage shown in Fig. 1 is applied to the RC circuit shown in Fig. 2. (a) Find the capacitor voltage $v_C(t)$. (b) Show that the RC circuit "delays" the input by $\tau = RC$ seconds.

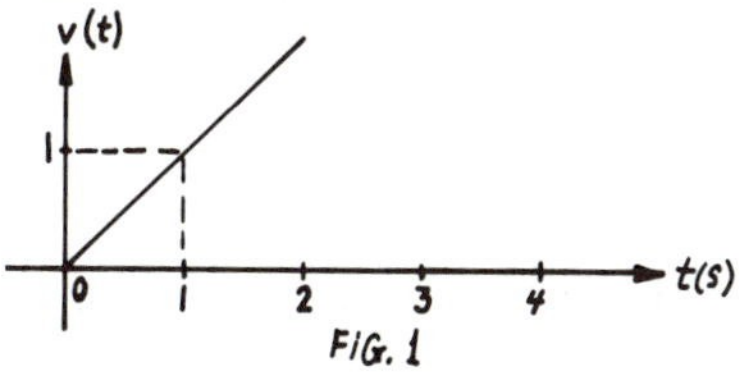

FIG. 1

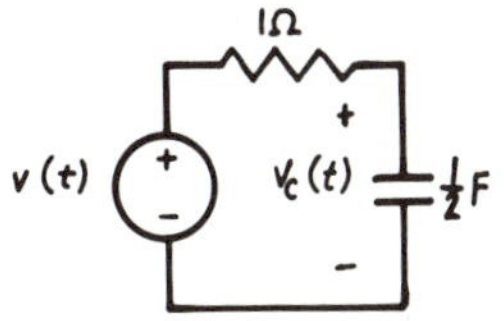

FIG. 2

Solution: (a) Using Laplace transforms we can quickly find the transfer function

$$H(s) = \frac{V_C(s)}{V(s)} = \frac{\frac{2}{s}}{1 + \frac{2}{s}} = \frac{2}{s+2}.$$

The input ramp is transformed to $V(s) = \frac{1}{s^2}$. Hence, the output $V_C(s)$ is $H(s)\,V(s)$,

$$V_C(s) = \frac{2}{s^2\,(s + 2)}.$$

Using partial fraction expansion we can write

$$V_C(s) = \frac{A}{s^2} + \frac{B}{s} + \frac{C}{s + 2}$$

when $A = \left| V_C(s)\, s^2 \right|_{s=0} = \left| \frac{2}{s + 2} \right|_{s=0} = 1$

$$B = \left| \frac{d}{ds} V_C(s)\, s^2 \right|_{s=0} = \left| \frac{-2}{(s + 2)^2} \right|_{s=0} = -\frac{1}{2}$$

$$C = \left| V_C(s)\,(s + 2) \right|_{s=-2} = \left| \frac{2}{s^2} \right|_{s=-2} = \frac{1}{2}.$$

Hence, $\qquad V_C(s) = \frac{1}{s^2} - \frac{\frac{1}{2}}{s} + \frac{\frac{1}{2}}{s + 2}$

can be inverse-Laplace transformed to

$$v_C(t) = \left[t - \frac{1}{2} + \frac{1}{2} e^{-2t} \right] u(t) \quad V.$$

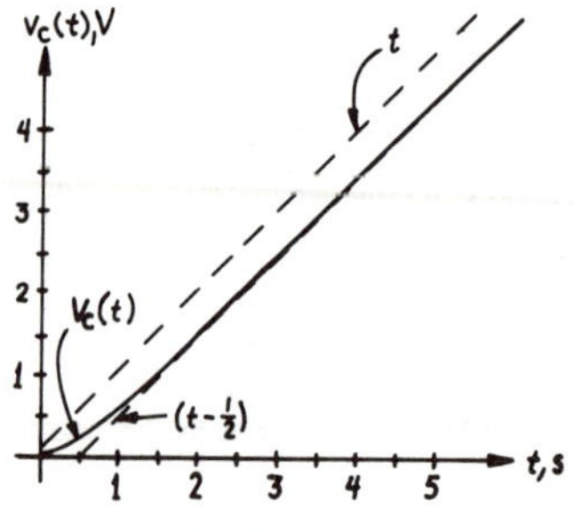

FIG.3

(b) We note that the output of the RC circuit is $\left(t - \frac{1}{2}\right)$ plus the exponential $\frac{1}{2} e^{-2t}$. The output is zero at $t = 0$ and follows the function $t - \frac{1}{2}$ as $t \to \infty$. The function $t - \frac{1}{2}$ is nothing more than the ramp input delayed by $\frac{1}{2}$ s. The plot of the output in Fig. 3 shows that the input is essentially delayed by $\tau = RC = \frac{1}{2}$ when passed through the RC circuits.

Write a generalized expression for $v_C(t)$, in the RC circuit shown in Fig. 1, in terms of v_s, v_i, R, and C by using Laplace transforms.

Can the expression for $v_C(t)$ be further generalized by expressing it in terms of $v_C(\infty) = V_{ss}$ (steady state voltage) instead of v_s?

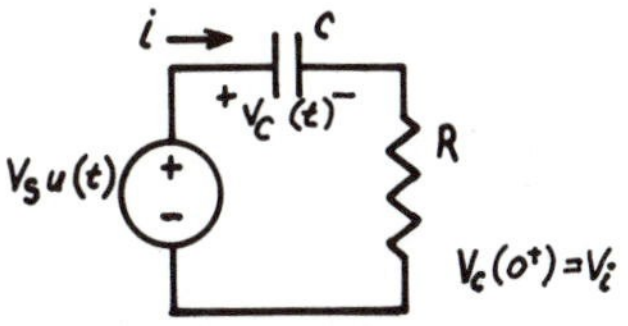

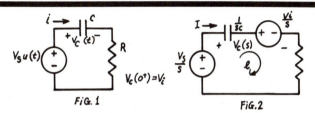

FiG. 1

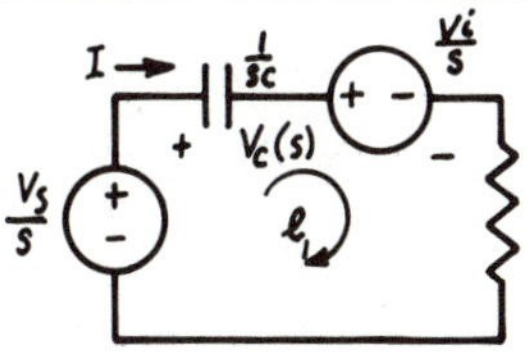

FiG. 2

Solution: Using Laplace transforms enables us to find an expression for $v_C(t)$ by simple algebraic manipulations.

We can transform each element in the circuit to form a new transformed circuit as follows

$$V_s\ u(t) \Rightarrow \frac{V_s}{s}$$

$$v_C(t) = v_C(0^+) + \frac{1}{C}\int_0^t i\ dt \overset{L}{\Rightarrow} \frac{v_C(0^+)}{s} + \frac{I}{sC}$$

$$R \Rightarrow R$$

giving the circuit in Fig. 2.

We must now use partial fraction expansion to write

$$\frac{\frac{1}{RC}(V_s - V_i)}{s\left(s + \frac{1}{RC}\right)} = \frac{A}{s} + \frac{B}{\left(s + \frac{1}{RC}\right)}$$

$$\text{where } A = \left.\frac{\frac{1}{RC}(V_s - V_i)}{\left(s + \frac{1}{RC}\right)}\right|_{s=0} = V_s - V_i,$$

$$\text{and} \quad B = \left.\frac{\frac{1}{RC}(V_s - V_i)}{s}\right|_{s=-\frac{1}{RC}} = -(V_s - V_i).$$

We can now write $V_C(s)$ in expanded form

$$V_C(s) = \frac{V_i}{s} + \frac{V_s - V_i}{s} - \frac{V_s - V_i}{\left(s + \frac{1}{RC}\right)}$$

transforming this expression to the time-domain gives our generalized expression.

In order to transform back to the time-domain ($V_C(s) \overset{-L}{\Rightarrow} v_C(t)$) we must write $V_C(s)$ in the form containing the factored terms

$$\frac{K}{(s + r_1)\ (s + r_2)\ \ldots}.$$

The first term $\frac{V_i}{s}$ is already in factored form $\frac{V_i}{(s + 0)}$. The second term must be factored as follows

$$= \frac{V_s - V_i}{s^2\ C\ \left(\frac{1}{sC} + R\right)}$$

$$= \frac{V_s - V_i}{(s + s^2\ RC)}$$

$$= \frac{V_s - V_i}{s\ (s\ RC + 1)},\ \text{finally obtaining}$$

$$= \frac{\frac{1}{RC}\ (V_s - V_i)}{s\ \left(s + \frac{1}{RC}\right)}$$

Using KVL around loop ℓ_1 yields the equation

$$\frac{V_s - V_i}{s} = I(s)\ \left[\frac{1}{sC} + R\right].$$

Solving for I(s) yields

$$I(s) = \frac{V_s - V_i}{s\ \left(\frac{1}{sC} + R\right)}.$$

Since we have already found the transform of $v_C(t)$ we can substitute I(s) directly to obtain

$$V_C(s) = \frac{V_i}{s} + \frac{I}{sC}$$

$$V_C(s) = \frac{V_i}{s} + \frac{V_s - V_i}{s^2\, C\left(\frac{1}{sC} + R\right)}.$$

$$v_C(t) = V_i + V_s - V_i - (V_s - V_i)\, e^{-\frac{t}{RC}}$$

$$v_C(t) = V_s - (V_s - V_i)\, e^{-\frac{t}{RC}} \text{ V} \quad t > 0.$$

We note that in our circuit, V_{ss} (steady state voltage) is equal to the source voltage V_s. In other words, $v_C(\infty) = V_s = V_{ss}$. For any RC circuit, all we need to know is $v_C(0^+)$ and $v_C(\infty)$ in order to find $v_C(t)$ for $t \geq 0$. Our generalized expression for an RC circuit is

$$v_C(t) = v_C(\infty) - [v_C(\infty) - v_C(0^+)]\, e^{-\frac{t}{RC}}; \; t > 0.$$

A dual expression can be obtained for a parallel RL circuit

$$i_L(t) = i_L(\infty) - [i_L(\infty) - i_L(0^+)]\, e^{-\frac{t}{L/R}}. \text{ A}$$

We leave this for the reader to verify.

• PROBLEM 12-7

Find the current i(t) in the circuit of Fig. 1 below for the input shown.

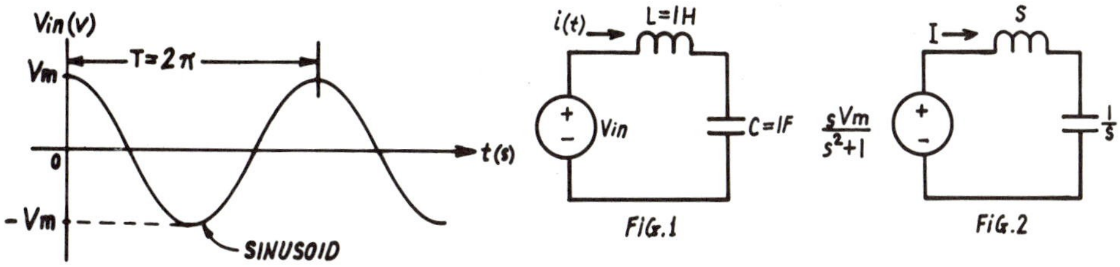

Solution: The input voltage V_{in} can be written $V_m \cos \omega t\, u(t)$ where $\omega = \frac{2\pi}{T} = 1$, thus $V_{in} = V_m \cos t\, u(t)$.

The best way to solve for i(t) in this problem is to use Laplace transforms. Transform the circuit in Fig. 1 and obtain the circuit in Fig. 2.

Writing a KVL equation yields

$$\frac{sV_m}{s^2 + 1} = I\left(s + \frac{1}{s}\right).$$

Solving for I gives

$$I = \frac{sV_m}{(s^2 + 1)\ (s + \frac{1}{s})}.$$

Multiplying by $\frac{s}{s}$ yields

$$I = \frac{s^2 V_m}{(s^2 + 1)^2}.$$

In order to use partial fraction expansion, find the roots of $(s^2 + 1)^2 = (s^2 + 1)(s^2 + 1)$ to be $+ j$, $- j$, $+ j$, and $- j$.

$$I = \frac{s^2 V_m}{(s - j)(s + j)(s - j)(s + j)} = \frac{s^2 V_m}{(s - j)^2 (s + j)^2}$$

By partial fraction expansion:

$$I = \frac{A}{(s - j)^2} + \frac{A^*}{(s + j)^2} + \frac{B}{(s - j)} + \frac{B^*}{(s + j)}$$

where, $A = \left|(s - j)^2\ I\right|_{s=j} = \left|\frac{s^2 V_m}{(s + j)^2}\right|_{s=j} = \frac{j^2 V_m}{(2j)^2} = \frac{V_m}{4}$

$$B = \left|\frac{1}{(2 - 1)!}\ \frac{d}{ds}\ (s - j)^2\ I\right|_{s=j}$$

$$= \left|\frac{(s + j)^2\ (V_m 2s) - V_m s^2 (2)(s + j)}{(s + j)^4}\right|_{s=j}$$

$$B = -\frac{V_m}{4}\ j$$

$$A^* = \frac{V_m}{4} \quad \text{and } B^* = +\frac{V_m}{4}\ j$$

Substituting into the expansion yields

$$I(s) = \frac{\frac{V_m}{4}}{(s - j)^2} + \frac{\frac{V_m}{4}}{(s + j)^2} + \frac{-\frac{V_m}{4} j}{s - j} + \frac{\frac{V_m}{4} j}{s + j}.$$

$L^{-1}\ \{I(s)\} = I(t)$ gives

$$I(t) = \frac{V_m}{4}\ t\ e^{+jt} + \frac{V_m}{4}\ t\ e^{-jt} - \frac{V_m}{4}\ j\ e^{+jt} + \frac{V_m}{4}\ j\ e^{-jt}$$

$$I(t) = \frac{V_m}{4}\left[t\left(e^{jt} + e^{-jt}\right) - j\left(e^{jt} - e^{-jt}\right)\right]$$

$$I(t) = \frac{V_m}{4}\left[2t\left(\frac{e^{jt} + e^{-jt}}{2}\right) - j(2j)\left(\frac{e^{jt} - e^{-jt}}{2j}\right)\right]$$

Since $\dfrac{e^{jt} + e^{-jt}}{2} = \cos t$ and $\dfrac{e^{jt} - e^{-jt}}{2j} = \sin t$,

$$I(t) = \left\{\frac{V_m}{2}\,[t \cos t + \sin t]\right\} u(t).$$

• PROBLEM 12-8

Consider the network of Fig. 1. Use Laplace transform techniques to find the output voltage $v_C(t)$ when the input applied is $v(t) = \sin t$, $0 < t < \pi$ and, $v(t) = 0$ for all other time intervals.

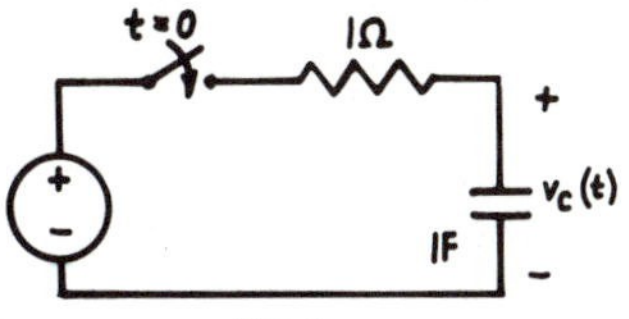

FiG. 1

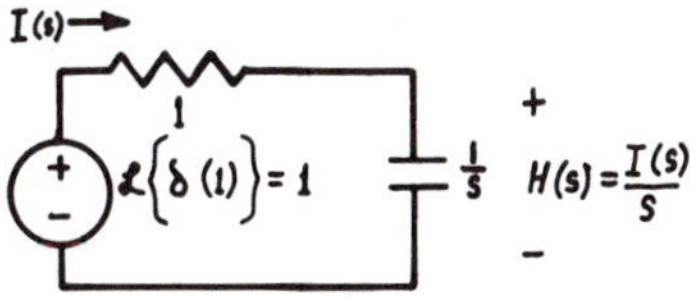

FiG. 2

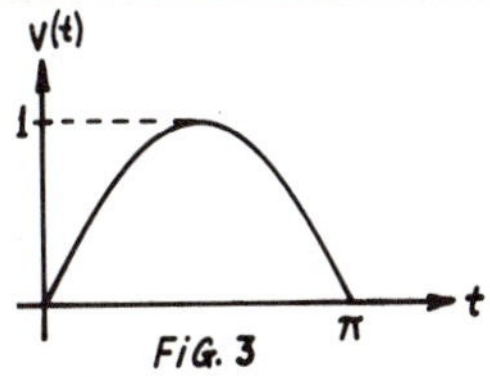

FiG. 3

Solution: Determine the transfer function for $\frac{V_C}{V}(s)$ by finding the impulse response of the transformed circuit in Fig. 2.

A KVL equation gives:

$$1 = I(s)\left(1 + \frac{1}{s}\right) = I(s)\left(\frac{s+1}{s}\right)$$

$$I(s) = \frac{s}{s+1};\quad H(s) = \frac{I(s)}{s} = \frac{1}{s+1}.$$

The voltage being applied is $v(t) = \sin t$ for $0 < t < \pi$, and $v(t) = 0$ for all other times as sketched in Fig. 3.

This can be represented by the sum of the two sine waves in Figs. 4 (b, c).

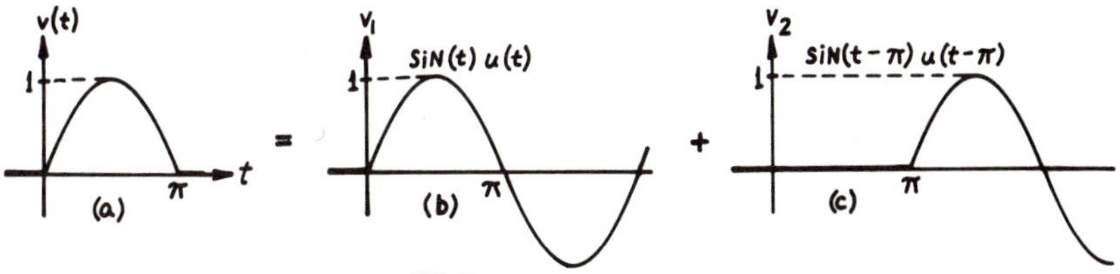

FiG. 4

The Laplace transform of sin(t) u(t) is L {sin(t) u(t)} $= \frac{1}{s^2 + 1}$, so that the transform for the wave in Fig. 4(c) becomes L {sin (t − π) u(t − π)} $= \frac{e^{-s\pi}}{s^2 + 1}$. The applied voltage v(t) is transformed to V(s)

$$V(s) = V_1(s) + V_2(s) = \frac{1}{s^2 + 1}\,(1 + e^{-s\pi})$$

$V_C(s)$ can be found by multiplying the applied voltage, V(s), with the transfer function H(s).

$$V_C(s) = V(s)\,H(s) = \left(\frac{1}{s^2 + 1}\right)\left(\frac{1}{s + 1}\right)(1 + e^{-s\pi})$$

Note that $V_C(s)$ has two parts to its response

$$V_C(s) = \frac{1}{(s^2 + 1)\,(s + 1)} + \frac{e^{-s\pi}}{(s^2 + 1)\,(s + 1)} \qquad (1)$$

illustrating the principle of superposition. The first part is the response due to the wave in Fig. 4(b) and the second part is the response due to the wave in Fig. 4(c).

Using partial fraction expansion, express Eq. (1) as two partial fractions

$$V_C(s) = \left(\frac{A}{S + 1} + \frac{B}{s + j} + \frac{B^*}{s - j}\right) +$$

$$\left(\frac{D}{s + 1} + \frac{E}{s + j} + \frac{E^*}{s - j}\right)$$

where $$A = (s + 1)\,\frac{1}{(s + 1)\,(s^2 + 1)}\Bigg|_{s=-1} = \frac{1}{2}$$

$$B = (s + j)\,\frac{1}{(s + 1)\,(s - j)\,(s + j)}\Bigg|_{s=-j} = \frac{1}{(1 - j)\,(-\,2j)}$$

$$= \frac{1}{-\,2 - j2}$$

$$B = \frac{1}{\sqrt{8}\underline{/-\,135°}} = \frac{1}{\sqrt{8}}\,\underline{/135°} = -\,\frac{1}{4} + j\,\frac{1}{4}$$

$$B^* = -\frac{1}{4} - j\frac{1}{4}$$

$$D = \left.\frac{e^{-s\pi}(s+1)}{(s+1)(s^2+1)}\right|_{s=-1} = \frac{e^{\pi}}{2}$$

$$E = \left.\frac{e^{-s\pi}(s+j)}{(s+1)(s+j)(s-j)}\right|_{s=-j} = \frac{e^{j\pi}}{(1-j)(-j2)}$$

$$E = -\frac{1}{4}e^{j\pi} + j\frac{1}{4}e^{j\pi}$$

$$E^* = -\frac{1}{4}e^{-j\pi} - j\frac{1}{4}e^{-j\pi}$$

Substituting into the partial fraction gives,

$$V_C(s) = \left(\frac{\frac{1}{2}}{s+1} + \frac{-\frac{1}{4} + j\frac{1}{4}}{s+j} + \frac{-\frac{1}{4} - j\frac{1}{4}}{s-j}\right) +$$

$$\left(\frac{\frac{1}{2}e^{\pi}}{s+1} + \frac{-\frac{1}{4}e^{j\pi} + j\frac{1}{4}e^{j\pi}}{s+j} + \frac{-\frac{1}{4}e^{-j\pi} - j\frac{1}{4}e^{-j\pi}}{s-j}\right)$$

Note that when the inverse Laplace transform of $V_C(s)$ is taken,

$$L^{-1}\{V_C(s)\} = v_C(t) = L^{-1}\{V_0(s)\} + L^{-1}\{V_{\pi}(s)\}$$

where $$V_0(s) = \left(\frac{\frac{1}{2}}{s+1} + \frac{-\frac{1}{4} + j\frac{1}{4}}{s+j} + \frac{-\frac{1}{4} - j\frac{1}{4}}{s-j}\right)$$

and $$V_{\pi}(s) = \left(\frac{\frac{1}{2}e^{\pi}}{s+1} + \frac{-\frac{1}{4}e^{j\pi} + j\frac{1}{4}e^{j\pi}}{s+j} + \frac{-\frac{1}{4}e^{-j\pi} - j\frac{1}{4}e^{-j\pi}}{s-j}\right)$$

$$v_C(t) = v_0(t)\,u(t) + v_{\pi}(t)\,u(t-\pi)$$

But $v_{\pi}(t) = v_0(t-\pi)$; so that

$$v_C(t) = v_0(t)\,u(t) + v_0(t-\pi)\,u(t-\pi)$$

All that is needed is to find the inverse transform of

$V_0(s)$, and then to replace all t's by $t - \pi$ in $v_0(t)$ to obtain $v_0(t - \pi)$.

$$v_0(t) = \frac{1}{2} e^{-t} + \left(-\frac{1}{4} + j\frac{1}{4}\right) e^{-jt} + \left(-\frac{1}{4} - j\frac{1}{4}\right) e^{jt}$$

$$v_0(t) = \frac{1}{2} e^{-t} - \frac{1}{4}\left(e^{jt} + e^{-jt}\right) - j\frac{1}{4}\left(e^{jt} - e^{-jt}\right)$$

Note the forms:

$$\cos t = \frac{e^{jt} + e^{-jt}}{2}, \text{ and}$$

$$\sin t = \frac{e^{jt} - e^{-jt}}{2j}, \text{ in the above equation}$$

$$v_0(t) = \frac{1}{2} e^{-t} - \frac{1}{2}\left(\frac{e^{jt} + e^{-jt}}{2}\right) + \frac{1}{2}\left(\frac{e^{jt} - e^{-jt}}{2j}\right)$$

$$v_0(t) = \left[\frac{1}{2} e^{-t} - \frac{1}{2}\cos t + \frac{1}{2}\sin t\right] u(t)$$

replacing all t's by $t - \pi$ gives

$$v_0(t - \pi) = \left[\frac{1}{2} e^{-(t-\pi)} - \frac{1}{2}\cos(t - \pi) + \frac{1}{2}\sin(t - \pi)\right] u(t - \pi)$$

The resulting output is

$$v_C(t) = \frac{1}{2}\Big[(e^{-t} - \cos t + \sin t)\, u(t) + (e^{-(t-\pi)} - \cos(t - \pi) + \sin(t - \pi))\, u(t - \pi)\Big]$$

• PROBLEM 12-9

> For the series RLC network of Fig. 1, calculate (a) input impedance, Z(s), (b) current i(t).

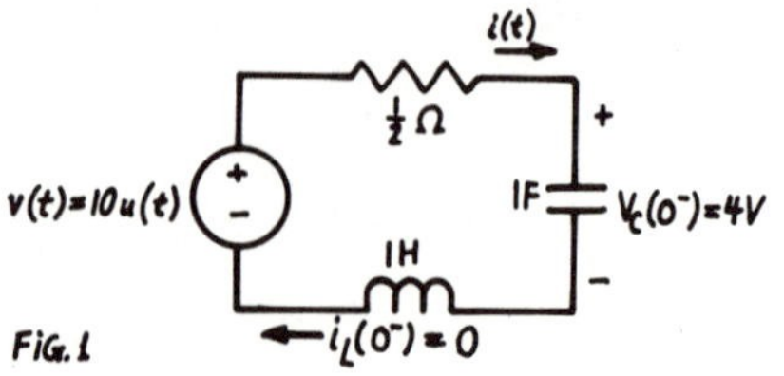

FiG. 1

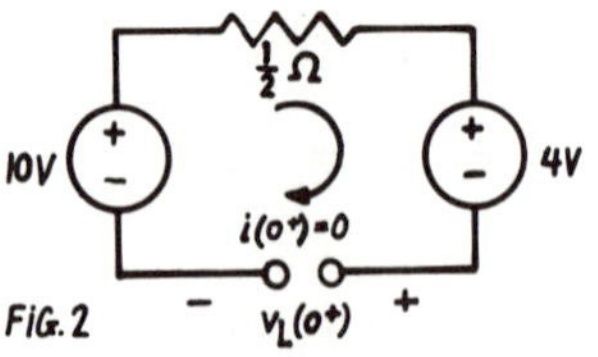

FiG. 2

Solution: (a) The input impedance is merely the sum of the three element impedances,

$$Z(s) = \frac{1}{2} + \frac{1}{s} + s = \frac{2s^2 + s + 2}{2s}.$$

(b) The forced component of the response is given by

$$i_F(t) = Y(0)\cdot 10 = 0.$$

The natural response is given by the poles of the admittance function,

$$Y(s) = \frac{2s}{2s^2 + s + 2} = \frac{2s}{\left(s + \frac{1}{4} - j\frac{\sqrt{15}}{4}\right)\left(s + \frac{1}{4} + j\frac{\sqrt{15}}{4}\right)}$$

Hence, the natural and, in this case, the total response is given by

$$i(t) = K_1 e^{\left(-\frac{1}{4} + j\frac{\sqrt{15}}{4}\right)t} + K_2 e^{\left(-\frac{1}{4} - j\frac{\sqrt{15}}{4}\right)t}$$

$$= K_1' e^{-\frac{1}{4}t} \cos\left(\frac{\sqrt{15}}{4}t + \phi\right) u(t).$$

Now $i(0^-) = i(0^+) = 0$ from the inductor initial condition. Thus, $0 = K_1' \cos\phi$, so $\phi = \pm\frac{\pi}{2}$ ($K_1' = 0$ gives a solution which is identically zero). Now another constraint can be derived from the initial conditions of Figure 2.

Using the inductor relation and KVL,

$$\left.\frac{di}{dt}\right|_{t=0^+} = v_L(0^+) = 10 - 4 = 6V.$$

Thus, $$6 = K_1'\left[-\frac{1}{4}\cos\phi - \frac{\sqrt{15}}{4}\sin\phi\right].$$

Now, picking $\phi = -\frac{\pi}{2}$ $\left(+\frac{\pi}{2}\text{ will lead to the same result}\right)$

$$6 = +\frac{\sqrt{15}}{4}K_1'$$

so $$K_1' = \frac{24}{\sqrt{15}}$$

and $$i(t) = \frac{24}{\sqrt{15}} e^{-\frac{1}{4}t} \cos\left(\frac{\sqrt{15}}{4}t - \frac{\pi}{2}\right) u(t)$$

or $$i(t) = 6.26\, e^{-0.25t} \sin(.968\, t)\, u(t)\ A.$$

A voltage wave form represented in Fig. 1(a) is applied to the RC network in Fig. 1(b). Find the transient as well as steady-state current I(t) using Laplace transforms.

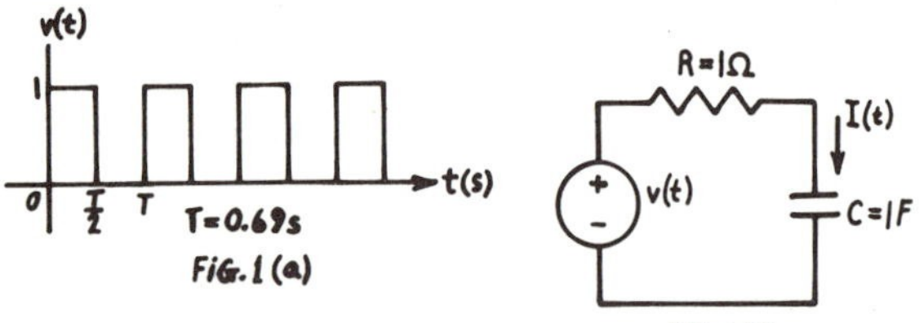

Fig. 1 (a)

Fig. 1 (b)

Solution: Represent $L\{v(t)\} = V(s)$ as $\dfrac{G(s)}{1 - e^{-st}}$ where G(t) is one pulse which is zero for $t > T$.

G(t) is represented by the sum of two waveforms.

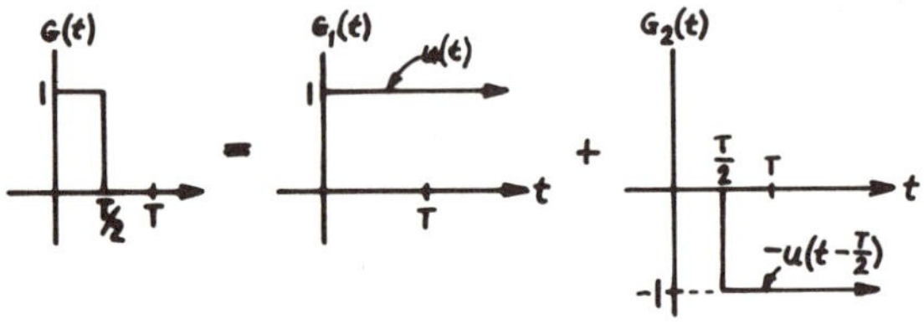

Thus,

$$G(t) = G_1(t) + G_2(t)$$

$$G(t) = u(t) - u(t - T/2)$$

$$L\{G(t)\} = G(s) = \frac{1}{s} - \frac{1}{s} e^{-sT/2} = \frac{1 - e^{-sT/2}}{s}.$$

Hence, $$V(s) = \frac{G(s)}{1 - e^{-sT}} = \frac{1 - e^{-sT/2}}{s(1 - e^{-sT})}.$$

The equivalent transformed circuit of Fig. 1(b) is shown in Fig. 2.

$$V(s) = \frac{1 - e^{-sT/2}}{s(1 - e^{-sT})}$$

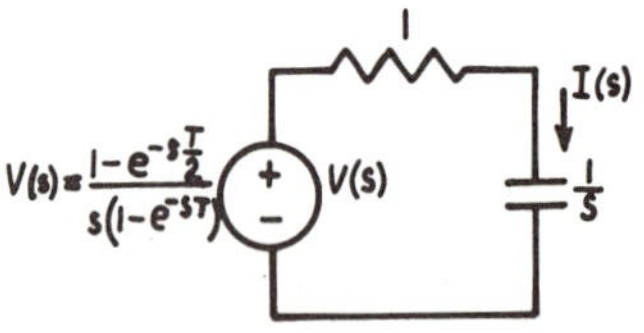

Fig. 2

where: $$V(s) = I(s)\left[1 + \frac{1}{s}\right]$$

$$V(s) = I(s)\left[\frac{s + 1}{s}\right]$$

$$I(s) = \frac{sV(s)}{s + 1} = \frac{1 - e^{-sT/2}}{(s + 1)(1 - e^{-sT})}$$

I(s) can be represented in two parts, a transient part called $I_t(s)$ and a periodic part called $I_p(s)$ where

$$I(s) = I_t(s) + I_p(s).$$

Since $I_p(s)$ must be in the form $\frac{H(s)}{1 - e^{-sT}}$ and I_t in the form $\frac{J(s)}{(s + 1)}$, write

$$I(s) = \frac{J(s)}{s + 1} + \frac{H(s)}{1 - e^{-sT}}$$

Using partial fraction methods yields

$$J(s) = \left|(s + 1)\, I(s)\right|_{s=-1} = \left|\frac{1 - e^{-sT/2}}{1 - e^{-sT}}\right|_{s=-1}$$

$$J(s) = \frac{1 - e^{T/2}}{1 - e^{T}}$$

Hence, $$I_t(s) = \frac{H(s)}{1 - e^{-sT}} = \frac{1 - e^{-sT/2}}{(s + 1)(1 - e^{-sT})} - \frac{\frac{1 - e^{T/2}}{1 - e^{T}}}{s + 1}$$

$$H(s) = \frac{1 - e^{-sT/2}}{s + 1} - \left(\frac{1 - e^{T/2}}{1 - e^{T}}\right) \frac{1 - e^{-sT}}{s + 1}$$

H(t), remember, represents one pulse of the entire periodic waveform,

$$H(s) = \frac{1}{s + 1} - \frac{e^{-sT/2}}{s + 1} - \frac{K}{s + 1} + \frac{Ke^{-sT}}{s + 1}$$

where $K = \frac{1 - e^{T/2}}{1 - e^{T}}$.

Transforming H(s) obtains

$$H(t) = e^{-t}\, u(t) - e^{-(t - T/2)} u(t - T/2)$$
$$- Ke^{-t}\, u(t) + Ke^{-(t - T)}\, u(t - T).$$

Substituting T = 0.69s obtains

$$H(t) = e^{-t}\, u(t) - e^{-(t - .35)}\, u(t - .35)$$

$$- .41\,e^{-t}\,u(t) + .41e^{-(t - .69)}\,u(t - .69)$$

$$H(t) = 0.59e^{-t}\,u(t) - e^{-(t - .35)}u(t - .35)$$

$$+ .41e^{-(t - .69)}\,u(t - .69)$$

Sketch H(t) by drawing each wave form.

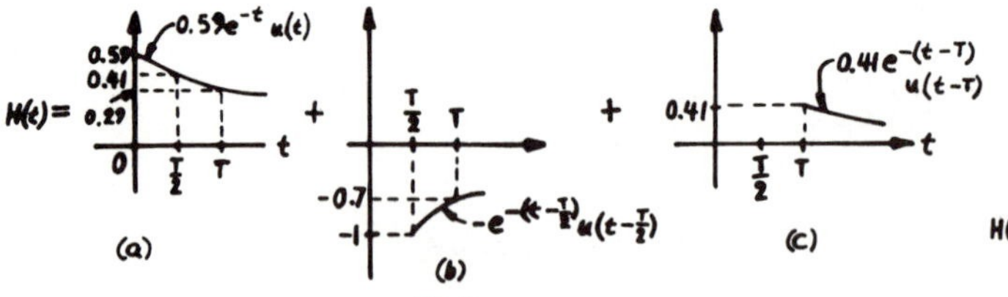

FIG.3

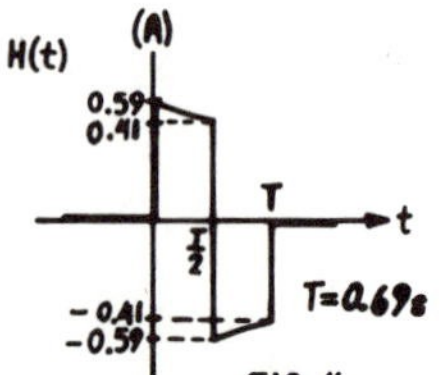

FIG. 4

$I_t(s)$ (the transient response) $= \dfrac{J(s)}{s + 1}$ where J(s) was found to be $K = \dfrac{1 - e^{T/2}}{1 - e^{T}} = 0.41$. Thus $I_t(t) = 0.41e^{-t}$.

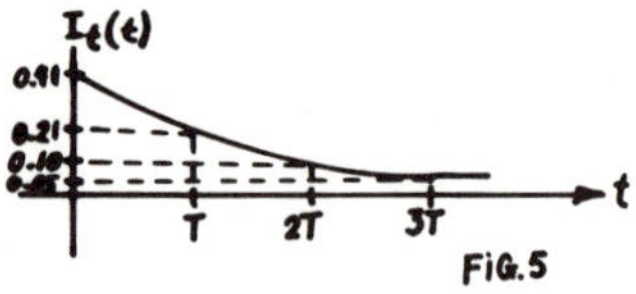

FIG.5

Sketch the entire response by noting that H(s) represents one pulse of the periodic response and $I_t(t) + I_p(t)$ is the total response.

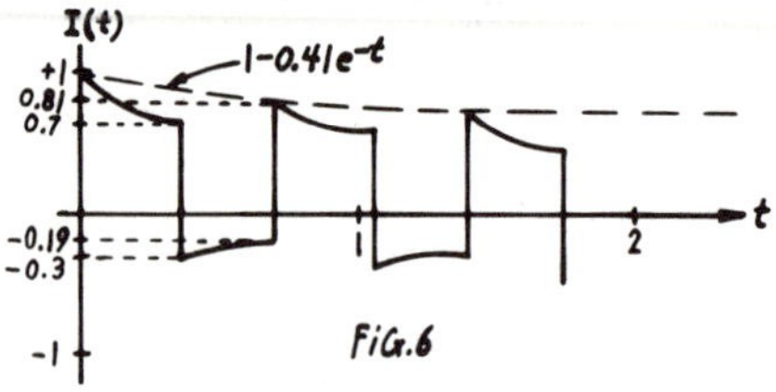

FIG.6

Also, note that at $t = \infty$, $I_t(t) = 0$ and the total response is just $I_p(t)$. This is also called the steady-state response.

• PROBLEM 12-11

> Calculate the current at $t = 0^+$ in the circuit of Fig. 1. The current at $t = 0^-$ is zero.

Solution: This problem can be solved using two methods. In the first method, note that the series combination of the impulse-function source and the inductance can be interpreted as an inductance with initial-energy storage. This interpretation is difficult to see intuitively, but by analizing the inductor and given initial condition representation, this interpretation becomes clear.

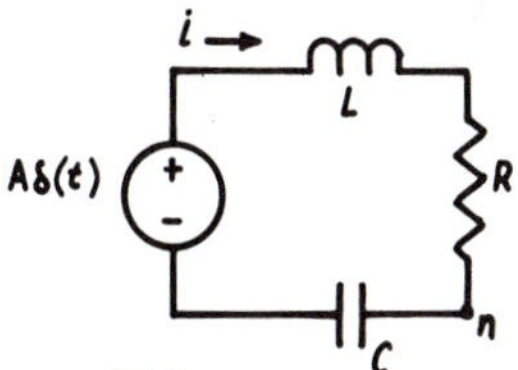

Fig.1

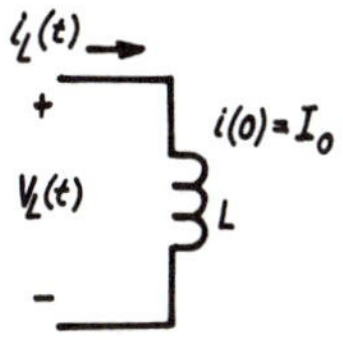

Fig.2

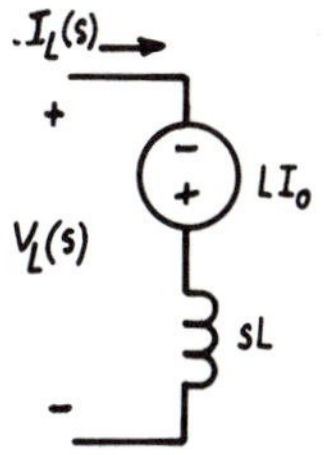

Fig.3

The second method relies on manipulation of the differential equation which describes the circuit, and interpretation of the relationship between charge and current in the capacitor during the interval $t = 0^-$ and $t = 0^+$.

Using the first method, we analyze the inductive element L with initial current I_0 shown in Fig. 2.

$$i_L(t) = I_0 + \frac{1}{L}\int_0^t V_L(t)\, dt \qquad (1)$$

By using Laplace transforms, the relationship in Eq. (1) can be described as

$$I_L(s) = \frac{I_0}{s} + \frac{1}{sL} V_L(s). \qquad (2)$$

Solving Eq. (2) above for $V_L(s)$ yields

$$V_L(s) = sLI_L(s) - LI_0 \qquad (3)$$

which can be represented as a voltage source LI_0 in series with the inductor L shown in Fig. 3.

In the time domain equation, (3) is written

$$v_L(t) = L\frac{di_L(t)}{dt} - LI_0\delta(t). \qquad (4)$$

This equation is equivalent to saying,

$$v_L(t) = L\frac{di_L(t)}{dt};\ i(0) = I_0. \qquad (5)$$

Hence, in the RLC network of Fig. 1, we have the same inductor-impulse source combination where $A = LI_0$ and $I_0 = \frac{A}{L}$. Since it is established that $i_L(0^-) = 0$, $i_L(0^+)$ must equal $I_0 = \frac{A}{L}$ because the only other contributing term to the current, $i_L(t)$, is $\frac{1}{L}\int_0^t v_L(t)\,dt$ from Eq. (1) is zero at $t = 0^+$. But, since it has already been shown that the inductor-impulse source can be represented by an inductor with initial condition I_0, note that $i_L(0^+)$ must equal I_0, because the current through the inductor cannot change instantaneously in the interval $(0^-, 0^+)$.

The second method involves writing the KVL differential equation for the RLC circuit,

$$L\frac{di}{dt} + Ri + \frac{1}{C}\int_{-\infty}^{t} i(t)\,dt = A\,\delta(t) \qquad (6)$$

noting that $\int_{-\infty}^{t} i(t) = q$, is the charge in the capacitor.

Hence,

$$L\frac{di}{dt} + Ri + \frac{1}{C}q = A\,\delta(t) \qquad (7)$$

since by definition, $\int_{0^-}^{0^+} A\,\delta(t) = A$, then Eq. (7) becomes

$$L\int_{0^-}^{0^+}\frac{di}{dt}\,dt + R\int_{0^-}^{0^+} i\,dt + \frac{1}{C}\int_{0^-}^{0^+} q\,\delta t = A \qquad (8)$$

integrating,

$$L\int_{0^-}^{0^+} di + R\int_{0^-}^{0^+} i\,dt + \frac{1}{C}\int_{0^-}^{0^+} q\,dt = A$$

$$L\,[i(0^+) - i(0^-)] + R\,[q(0^+) - q(0^-)] + \frac{1}{C}\int_{0^-}^{0^+} q\,dt = A \qquad (9)$$

Since the current in the circuit is finite at all times, it follows that a finite current cannot transfer non-zero charge to the capacitance in zero time. Hence,

$q(0^+) = q(0^-)$, $\quad R\,[q(0^+) - q(0^-)] = 0$ and

$$\int_{0^-}^{0^+} q\,dt = 0.$$

If the differential equation is to be satisfied at $t = 0$, the current i must experience a jump. Hence, Eq. (9) becomes

$$L\,[i(0^+) - i(0^-)] = A$$

and since $i(0^-)$ was given as zero, $i(0^+) = \frac{A}{L}$.

• PROBLEM 12-12

A series RLC circuit contains a 2 μF capacitor, a $\frac{1}{2}$ H inductor, and a 2600Ω resistor. If the inductor current and voltage $\left(v_L = L\frac{di_L}{dt}\right)$ at $t = 0$ are $v_L = 30V$ and $i_L = 50mA$, find $i_L(t)$ for $t > 0$.

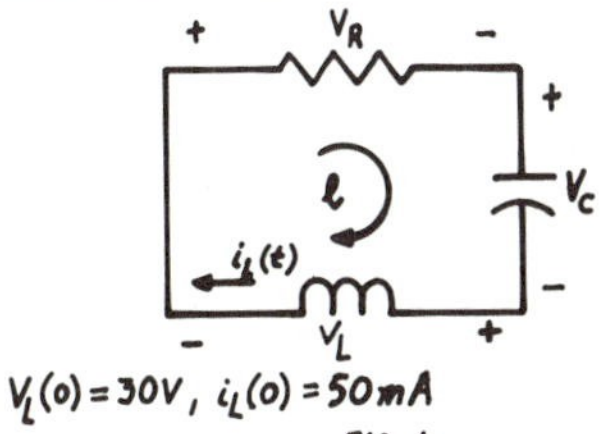

FIG. 1

Solution: Fig. 1 shows the series RLC circuit and initial conditions.

Writing the KVL equation around the loop in Fig. 1 gives

$$V_R + V_C + V_L = 0 \quad \text{where}$$

$$V_R = i_L(t)\ R$$

$$V_C = \frac{1}{C}\int_0^t i_L(\tau)\ d\tau$$

$$V_L = L\,\frac{di_L(t)}{dt}.$$

We can substitute these voltage current relationships into the loop equation and differentiate once with respect to t to obtain the differential equation

$$L\,\frac{d^2 i_L}{dt^2} + R\,\frac{di_L}{dt} + \frac{1}{C}\,i_L = 0. \qquad (1)$$

If we substitute s for $\frac{d}{dt}$, s^2 for $\frac{d^2}{dt^2}$ and divide by i_L, we obtain the quadratic equation

$$Ls^2 + Rs + \frac{1}{C} = 0$$

where $$s_1, s_2 = -\frac{R}{2L} \pm \sqrt{\left(\frac{R}{2L}\right)^2 - \left(\frac{1}{\sqrt{LC}}\right)^2}$$

if $s_1 \neq s_2$ and $\frac{R}{2L} > \frac{1}{\sqrt{LC}}$. The solution to our differential equation is in the form

$$i_L(t) = Ae^{s_1t} + Be^{s_2t}.$$

Hence, $$\frac{R}{2L} > \frac{1}{\sqrt{LC}}$$

$$\frac{2600}{2\left(\frac{1}{2}\right)} > \frac{1}{\sqrt{\left(\frac{1}{2}\right) 2 \times 10^{-6}}}$$

$$2600 > 1000$$

thus, $$s_1 = -2600 + 2400 = -200$$

and $$s_2 = -2600 - 2400 = -5000.$$

Using the initial condition information given, we can find the coefficients A and B in our solution

$$i_L(t) = Ae^{-200t} + Be^{-5000t}.$$

Since $i_L(0) = .05A$ we can write

$$i_L(0) = .05 = Ae^0 + Be^0$$

$$i_L(0) = .05 = A + B \qquad (2)$$

also, since $V_L(0) = 30V = L\frac{di_L(0)}{dt}$ we can write

$$v_L(0) = 30 = L(-200\,Ae^0 - 5000\,Be^0)$$

$$v_L(0) = 30 = -100A - 2500B. \qquad (3)$$

Eqs. (2) and (3) can be solved simultanesouly for A and B,

$$
\begin{array}{rcl}
[\ 0.05 & = & A + B]\ 100 \\
30.00 & = & -100A - 2500B \\
\hline
35 & = & -2400B
\end{array}
$$

$$B = -\frac{7}{480}$$

$$A = \frac{1}{20} + \frac{7}{480} = \frac{31}{480}.$$

Hence, the solution is

$$i_L(t) = \frac{1}{480}\left(31e^{-200t} - 7e^{-5000t}\right) \text{ A.}$$

• PROBLEM 12-13

> Given the initial conditions $i_L(0^-)$ and $V_C(0^-)$, find the Laplace transform of the series RLC network shown in Fig. 1 and derive an expression for the Laplace transform of the current i(t).

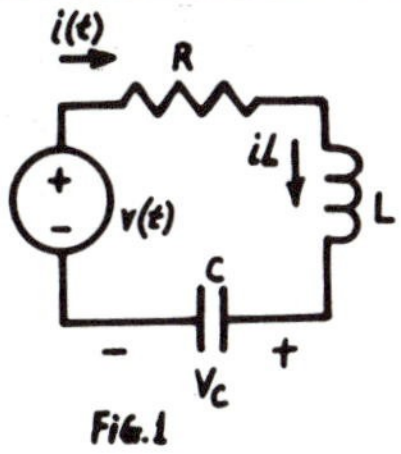

Fig.1

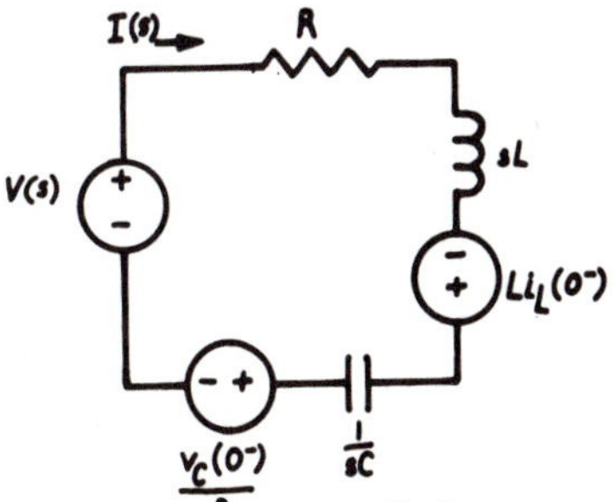

Fig.2

Solution: First, write an integro-differential equation for the voltage drops around the loop using KVL.

$$\text{Hence, } v(t) = i(t)\ R + L\frac{di_L}{dt} + v_C(0^-) + \frac{1}{C}\int_0^t i(\tau)\ d\tau$$

Taking the Laplace transform of each term in the above loop equation yields

$$V(s) = I(s)\ R + sL\ I(s) - L\ i_L(0^-) + \frac{v_C(0^-)}{s} + I(s)\ \frac{1}{sC}$$

Note that besides the three voltage drops for the three circuit elements $I(s)\ R$, $sLI(s)$, and $I(s)\ \frac{1}{sC}$

one can represent the initial conditions in the transformed circuit by voltage sources

$Li_L(0^-)$ and $\frac{v_C(0^-)}{s}$, shown in Fig. 2.

From the transformed loop equation, solve for I(s) to obtain

$$I(s) = \frac{1}{R + sL + \frac{1}{sC}} \left[V(s) + Li_L(0^-) - \frac{v_C(0^-)}{s} \right]$$

• PROBLEM 12-14

Solve for the current response, i(t), for a unit step voltage applied to the series RLC network of Fig. 1 under zero initial conditions.

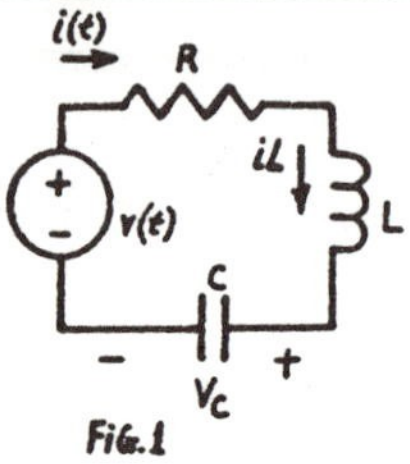

Fig.1

Solution: Since the applied voltage is a unit step, V(s) $= \frac{1}{s}$, also $v_C(0^-) = 0$ and $i_L(0^-) = 0$. Hence, from the equation derived in the last problem, we obtain

$$I(s) = \frac{\frac{1}{s}}{R + sL + \frac{1}{sC}}$$

$$I(s) = \frac{1}{s^2L + sR + \frac{1}{C}} = \frac{\frac{1}{L}}{s^2 + s\frac{R}{L} + \frac{1}{LC}}$$

$$I(s) = \frac{\frac{1}{L}}{(s + a)(s + b)} \tag{1}$$

where $a, b = -\frac{R}{2L} \pm \sqrt{\left(\frac{R}{2L}\right)^2 - \frac{1}{LC}}$

Note that there are three possible solutions for I(s);

(A) both a and b are real and $a \neq b$

(B) both a and b are real and $a = b$

(C) a and b are complex and $a = b^*$.

For case A: $a \neq b$, by partial fraction expansion

$$I(s) = \frac{\frac{1}{L}}{(s + a)(s + b)} = \frac{D}{s + a} + \frac{E}{s + b}$$

where, $$D = \left.\frac{\frac{1}{L}}{(s + b)}\right|_{s=-a} = \frac{1}{L\ (b - a)}$$

and $$E = \left.\frac{\frac{1}{L}}{(s + a)}\right|_{s=-b} = \frac{1}{L\ (a - b)},$$ thus

$$I(s) = \frac{\frac{1}{L}}{(s + a)(s + b)} = \frac{\frac{1}{L\ (b - a)}}{(s + a)} + \frac{\frac{1}{L\ (a - b)}}{(s + b)} \quad (2)$$

when $a \neq b$

For case B: $a = b$, substitute $c = a = b$ into equation (2) above to obtain

$$I(s) = \frac{\frac{1}{L}}{(s + C)^2} \text{ when } a = b = c$$

For case C: $\left(\frac{R}{2L}\right)^2 < \frac{1}{LC}$ and $a = b^*$

Define $a = -\alpha_0 + j\omega_d$ and $b = -\alpha_0 - j\omega_d$ where $\alpha_0 = \frac{R}{2L}$, and $\omega_d = \sqrt{\frac{1}{LC} - \left(\frac{R}{2L}\right)^2}$. Using partial fraction expansion, obtains

$$I(s) = \frac{\frac{1}{L}}{(s + \alpha_0 - j\omega_d)(s + \alpha_0 + j\omega_d)}$$

$$= \frac{F}{(s + \alpha_0 - j\omega_d)} + \frac{F^*}{(s + \alpha_0 + j\omega_d)}$$

where $$F = (s + \alpha_0 - j\omega_d)\ I(s)\Big|_{s=-\alpha_0 + j\omega_d}$$

$$F = \left.\frac{\frac{1}{L}}{(s + \alpha_0 + j\omega_d)}\right|_{s=-\alpha_0 + j\omega_d} = \frac{1}{j\ 2\omega_d L}$$

$$F^* = -\frac{1}{j\ 2\omega_d L}$$

Hence $$I(s) = \frac{\frac{1}{j\ 2\omega_d L}}{(s + \alpha_0 - j\omega_d)} + \frac{\frac{1}{j\ 2\omega_d L}}{(s + \alpha_0 + j\omega_d)}$$

for $a = b^*$

Taking the inverse Laplace transform of each of the three cases obtains

A: $$\frac{\frac{1}{L(b-a)}}{s+a} + \frac{\frac{1}{L(a-b)}}{s+b} \overset{L^{-1}}{\Longrightarrow} \left[\frac{1}{L(b-a)} e^{-at} + \frac{1}{L(a-b)} e^{-bt}\right] u(t)$$

Therefore $$i(t) = \left[\frac{1}{L(b-a)} e^{-at} + \frac{1}{L(a-b)} e^{-bt}\right] u(t)$$

for $a \neq b$

B: $$\frac{\frac{1}{L}}{(s+c)^2} \overset{L^{-1}}{\Longrightarrow} \frac{1}{L} t e^{-ct} u(t)$$

Therefore $i(t) = \frac{1}{L} t e^{-ct} u(t)$ for $a = b = c$

C: $$\frac{\frac{1}{j2\omega_d L}}{s+\alpha_0 - j\omega_d} + \frac{\frac{1}{j2\omega_d L}}{s+\alpha_0 + j\omega_d} \overset{L^{-1}}{\Longrightarrow}$$

$$\frac{1}{j2\omega_d L} e^{-\alpha_0 t} e^{j\omega_d t} - \frac{1}{j2\omega_d L} e^{-\alpha_0 t} e^{-j\omega_d t}$$

which can be written

$$i(t) = \frac{e^{-\alpha_0 t}}{\omega_d L} \underbrace{\left(\frac{e^{j\omega_d t} - e^{-j\omega_d t}}{2j}\right)}_{\sin \omega_d t}$$

$$i(t) = \left[\frac{e^{-\alpha_0 t}}{\omega_d L} \sin \omega_d t\right] u(t) \quad \text{for } a = b^* = -\alpha_0 + j\omega_d$$

• PROBLEM 12-15

In the circuit of Fig. 1 $v_C(0^-) = 10V$. The source voltage is given in Fig. 2. Use Laplace transform techniques to obtain the response $v_R(t)$.

Solution: Represent the pulse waveform in Fig. 2 as the sum of two unit step functions as illustrated in Fig. 3.

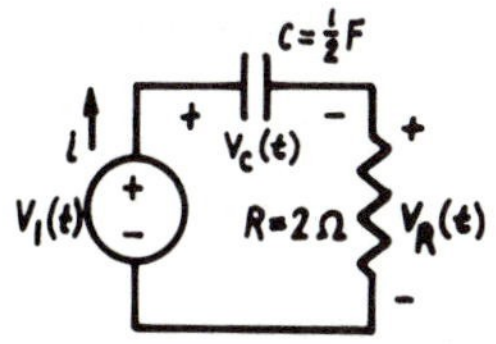

FiG.1

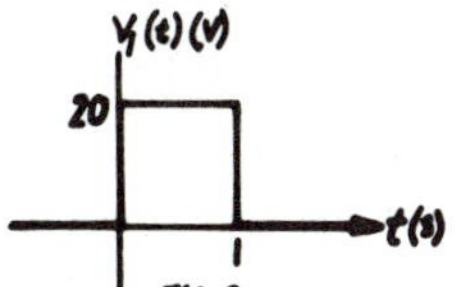

FiG.2

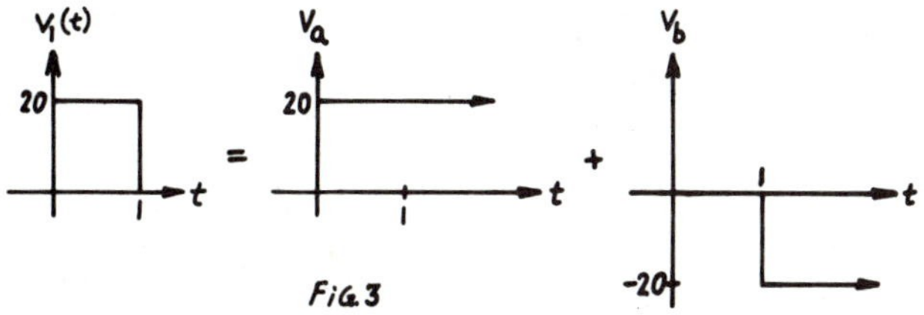

FiG 3

$v_a = 20\ u(t);\ v_b = -\ 20\ u(t - 1)$

giving $v_1(t) = 20\ [u(t) - u(t - 1)]$

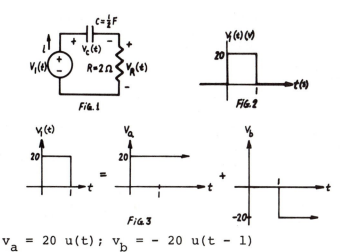

Note, that when transforming a capacitor with an initial condition;

$$V_C(s) = \frac{v_C(0^-)}{s} + \frac{1}{sC}\ I_C(s).$$

The transformed circuit is shown in Fig. 4.

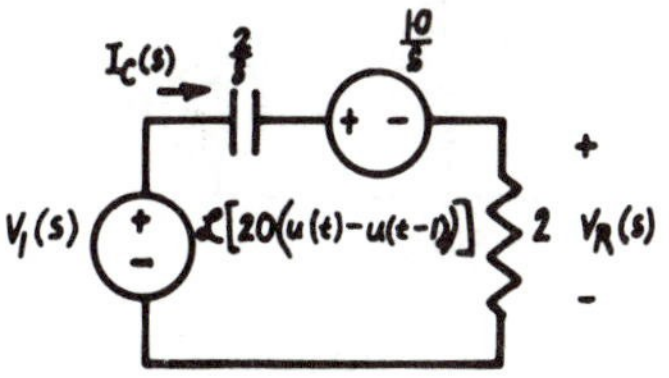

FiG.4

We find $L\ [20\ (u(t) - u(t - 1))]$

$$= L\ [20\ u(t)] - L\ [20\ u(t - 1)]$$

recalling that $f\ (t - b) \Rightarrow e^{-sb}\ F(s)$ where $f(t) \Rightarrow F(s)$ and $f(t) = 0$ for $t < 0$. Time axis translation in the time domain is equivalent to multiplying by e^{-sb}, where $-\ b$ is the translation time, in the frequency domain. Therefore,

$$V_1(s) = L\ [20\ (u(t) - u(t - 1))] = \frac{20}{s} - e^{-s}\ \frac{20}{s}$$

Writing a KVL equation around the single loop yields

$$\frac{20}{s} - e^{-s}\ \frac{20}{s} - \frac{10}{s} = I_C(s)\ \left[\frac{2}{s} + 2\right]$$

$$\frac{10}{s} - e^{-s}\ \frac{20}{s} = 2\ I_C(s)\ \left(\frac{1}{s} + 1\right)$$

Note that $2 I_C(s) = V_R(s)$. The following can be written:

$$V_R(s) = \frac{\frac{10}{s} - e^{-s}\frac{20}{s}}{\frac{1}{s} + 1} = \frac{\frac{10 - e^{-s} 20}{\not{s}}}{\frac{1 + s}{\not{s}}}$$

$$V_R(s) = \frac{10}{s + 1} - e^{-s} \frac{20}{s + 1} \qquad (1)$$

Recalling that $L^{-1}\left(\frac{1}{s + 1}\right) = e^{-t}$ and $L^{-1}(e^{-s} F(s)) = F(t - 1)$.

Transform Eq. (1) into the time domain to obtain the desired result:

$$v_R(t) = 10e^{-t} u(t) - 20e^{-(t - 1)} u(t - 1)$$

• PROBLEM 12-16

In the circuit of Fig. 1 use Laplace transforms to draw the transformed circuit. Find $v_C(t)$ for $t > 0$.

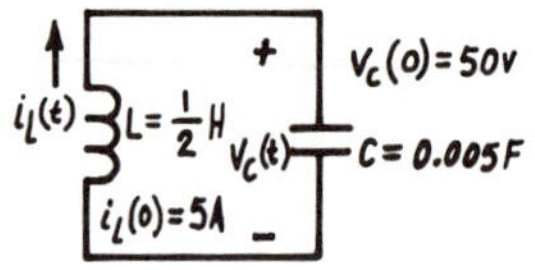

FiG. 1

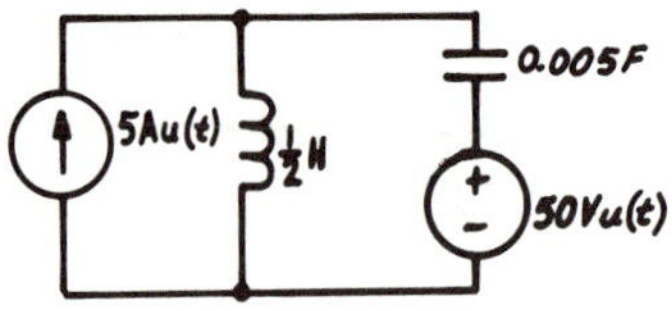

FiG. 2

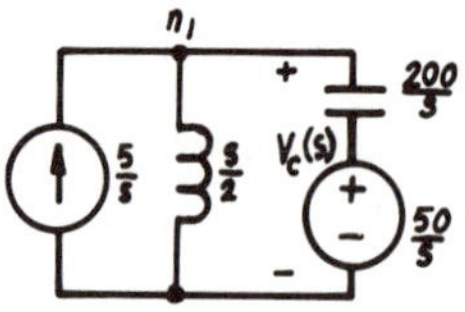

FiG. 3

Solution: In order to draw a complete transformed circuit we must take into account the initial conditions in the circuit and represent them in the transformed circuit.

We can represent the initial conditions in our time-domain circuit, shown in Fig. 2., by a voltage source in series with the capacitor for $v_C(0)$ and a current source in parallel with the inductor for $i_L(0)$.

Since the two initial condition sources are specified for $t = 0$, they can be represented by unit step functions.

We can now transform each of the two sources and two elements in the circuit of Fig. 2 to obtain the frequency domain circuit of Fig. 3.

Writing a KCL equation for node n_1 we can solve for $V_C(s)$,

$$\frac{5}{s} = \frac{V_C(s)}{\frac{s}{2}} + \frac{V_C(s) - \frac{50}{s}}{\frac{200}{s}}$$

$$\frac{5}{s} = V_C(s)\left[\frac{2}{s} + \frac{s}{200}\right] - \frac{1}{4}$$

$$V_C(s) = \frac{\frac{1}{4} + \frac{5}{s}}{\frac{2}{s} + \frac{s}{200}}.$$

Multiplying top and bottom by s gives

$$V_C(s) = \frac{\frac{s}{4} + 5}{\frac{s^2}{200} + 2}.$$

We can write $V_C(s)$ as

$$V_C(s) = \frac{\frac{s}{4}}{\frac{s^2}{200} + 2} + \frac{5}{\frac{s^2}{200} + 2}.$$

Recognizing that we can write $V_C(s)$ in the form

$$V_C(s) = K_1\left[\frac{s}{s^2 + a^2}\right] + K_2\left[\frac{a^2}{s^2 + a^2}\right]$$

where $v_C(t) = [K_1 \cos at + K_2 \sin at]\ u(t)$
we multiply through by 200 and factor to get,

$$V_C(s) = 50\left[\frac{s}{s^2 + 20^2}\right] + 50\left[\frac{20^2}{s^2 + 20^2}\right].$$

Transforming back to the time domain gives

$$v_C(t) = [50 \cos 20t + 50 \sin 20t]\ u(t).$$

Using phasor notation we can combine the two sinusoids into one expression.

$$\vec{v}_C(t) = 50\underline{/0°} + 50\underline{/-90°}$$

$$= 50 - j\,50$$

$$= \sqrt{50^2 + 50^2}\underline{/\tan^{-1}(-1)} = 70.7\underline{/-45°}\ \text{V}$$

$$v_C(t) = [70.7 \cos(20t - 45°)]\ u(t)\ \text{V}$$

A battery V_0, an open switch that closes at t = 0, and an impedance $Z(s) = \frac{2(s+2)}{(s+4)}$ Ω are in series. (a) Determine the appropriate form for the current in the circuit after t = 0. (b) Evaluate the unknown amplitudes by making use of the information that $i(0^+) = 6A$ and $\frac{di(0^+)}{dt} = 12\ \frac{A}{s}$. (c) Now assume that the switch, after being closed for a long time, is opened at t = 0. Determine v(t), the voltage across the impedance, if $v(0^+) = -12V$.

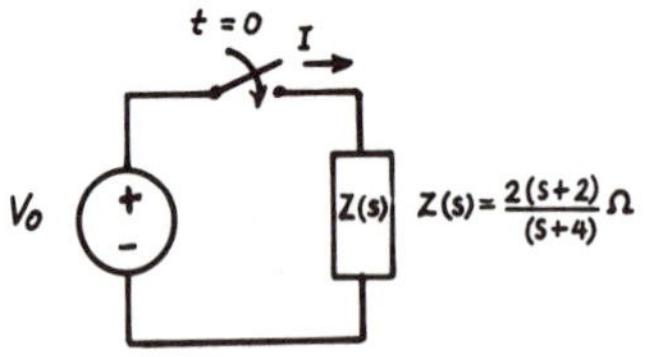

FiG. 1

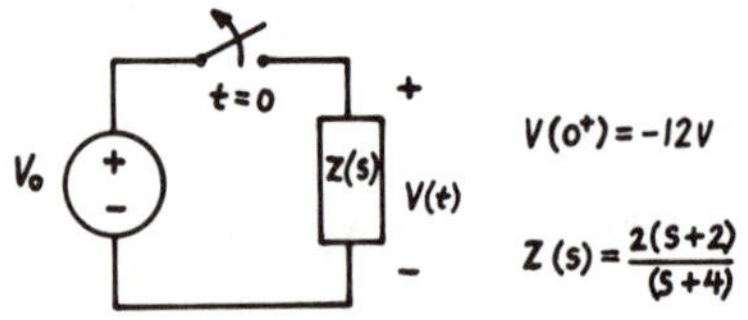

FiG. 2

Solution: (a) Fig. 1 shows the circuit described.

Since the current is found from

$$I(s) = \frac{V(s)}{Z(s)} = \frac{\frac{V_0}{s}}{\frac{2(s+2)}{(s+4)}} = \frac{V_0(s+4)}{2s(s+2)}$$

the denominator of I(s) tells us that the response must be composed of a DC term (since there is a pole at s = 0) and an exponential term (since there is a pole at s = - 2). Hence,

$$i(t) = AV_0 + Be^{-2t}. \tag{1}$$

(b) Given the initial condition $i(0^+) = 6A$ we set t in equation (1) equal to zero and obtain the equation

$$i(0^+) = 6 = AV_0 + B.$$

The initial condition $\frac{di(0^+)}{dt} = 12\ \frac{A}{s}$ is applied by differentiating Eq. (1), hence

$$\frac{di(0^+)}{dt} = 12 = -2B; \quad B = -6$$

and $AV_0 = 12$. The output current is

$$i(t) = 12 - 6e^{-2t}\ A.$$

(c) We now must find the voltage across the impedance after the impedance is brought to steady state at $v(0^+) = -12V$ as shown in Fig. 2.

The denominator term of the impedance tells us that response must be exponential with initial condition - 12V (since there is a pole at s = -4). Hence,

$$v(t) = -12e^{-4t}\text{ V.}$$

• PROBLEM 12-18

Consider the network of Fig. 1. Use the convolution integral to find the output voltage $v_C(t)$ when the input applied is $v(t) = \sin t$, $0 < t < \pi$, and $v(t) = 0$ otherwise. Sketch the output voltage and find its maximum value.

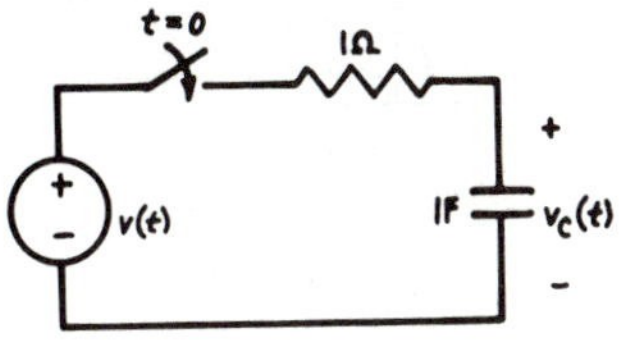

FiG.1

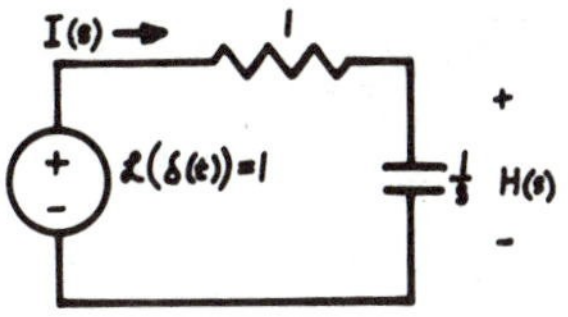

FiG.2

Solution: The input applied to the circuit convolved with the impulse response gives the desired output response of the circuit.

$$v_C(t) = v(t) * h(t) = \int_{-\infty}^{\infty} v(t - \tau)\, h(\tau)\, d\tau \qquad (1)$$

In order to determine the limits of the convolution integral, find the impulse response and the input voltage and plot them both.

First, find the impulse response

Writing a KVL equation for the circuit in Fig. 2 gives h(s), the impulse response in the frequency domain.

$$1 = I(s)\left[1 + \frac{1}{s}\right] = I(s)\left[\frac{s+1}{s}\right]$$

since $I(s)\left(\frac{1}{s}\right) = H(s)$

$$I(s) = \frac{s}{s+1}$$

$$H(s) = \left(\frac{s}{s+1}\right)\left(\frac{1}{s}\right) = \frac{1}{s+1}$$

The impulse response in the time domain becomes

$$h(t) = L\,[h(s)] = L\left[\frac{1}{s+1}\right] = e^{-t}$$

Plots of the input voltage v(t) and the impulse response h(t) are shown in Fig. 3.

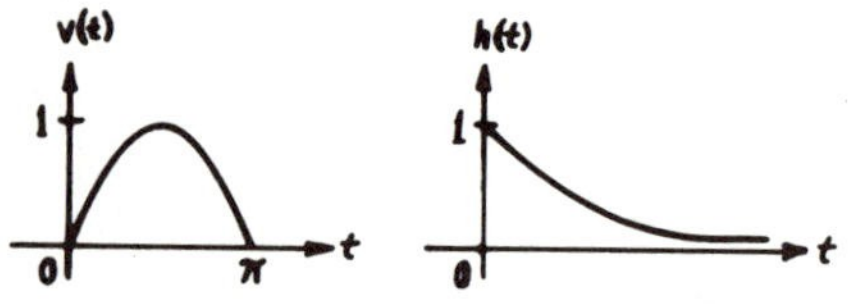

Fig. 3

Plots of $v(t - \tau)$ and $e^{-\tau}$ are shown in Fig. 4 at $t = 0$.

Fig. 4 shows that for $t \leq 0$, $v_C = 0$ because the product of the two functions is 0 for this interval.

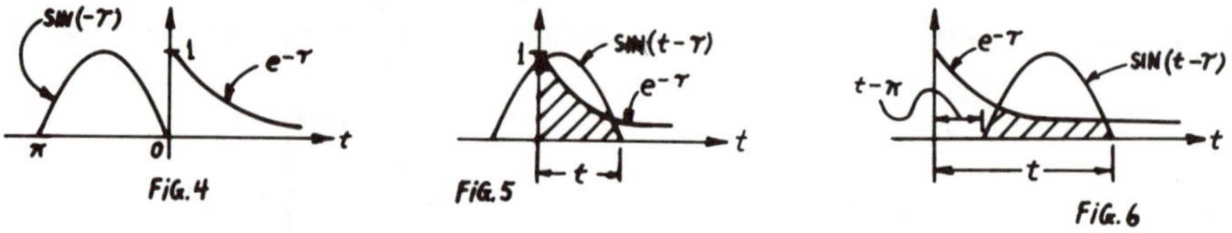

Fig. 5 shows the convolution in the interval $0 < t < \pi$.

$$v_C(t) = \int_0^t \sin(t - \tau)\, e^{-\tau}\, d\tau, \quad 0 < t < \pi \qquad (2)$$

The next time interval of interest is shown in Fig. 6 for $\pi < t < \infty$.

$$v_C(t) = \int_{t-\pi}^t \sin(t - \tau)\, e^{-\tau}\, d\tau; \quad \pi < t < \infty \qquad (3)$$

Now turn the attention to evaluating these integrals.

First, note that $\sin(t - \tau)$ can be expressed as

$\sin t \cos \tau - \cos t \sin \tau$.

This allows writing the two convolution integrals (Eqs. (2), (3)) as

$$v_C(t) = \sin t \int_0^t (\cos \tau)\, e^{-\tau}\, d\tau - \cos t \int_0^t (\sin \tau)\, e^{-\tau} d\tau;$$

$$0 < t < \pi \qquad (4)$$

$$v_C(t) = \sin t \int_{t-\pi}^t (\cos \tau)\, e^{-\tau}\, d\tau - \cos t \int_{t-\pi}^t (\sin \tau)\, e^{-\tau} d\tau;$$

$$\pi < t < \infty \qquad (5)$$

Let the first term of Eq. (4) be called (a), and the second term (b). Using the exponential form (by Euler's Identity), transform each term of Eq. (4);

For term (a):

$$\frac{e^{jt} - e^{-jt}}{2j} \int_0^t \frac{e^{j\tau} + e^{-j\tau}}{2} (e^{-\tau})\, d\tau$$

$$= \frac{e^{jt} - e^{-jt}}{2j} \int_0^t \frac{e^{j\tau-\tau} + e^{-j\tau-\tau}}{2}\, d\tau$$

$$= \frac{e^{jt} - e^{-jt}}{2j} \left[\frac{e^{j\tau-\tau}}{2\,(j - 1)} \Bigg|_0^t + \frac{e^{-j\tau-\tau}}{2\,(- j - 1)} \Bigg|_0^t \right]$$

$$\text{term (a)} = \frac{e^{jt} - e^{-jt}}{2j} \left[\frac{e^{jt-t}}{2j - 2} + \frac{e^{-jt-t}}{-2j - 2} - \frac{1}{2j - 2} - \frac{1}{- 2j - 2} \right] \qquad (6)$$

For term (b):

$$\frac{e^{jt} + e^{-jt}}{2} \int_0^t \frac{e^{j\tau} - e^{-j\tau}}{2j}\, e^{-\tau}\, d\tau$$

$$= \frac{e^{jt} + e^{-jt}}{2} \int_0^t \frac{e^{j\tau-\tau} + e^{-j\tau-\tau}}{2j}\, d\tau$$

$$= \frac{e^{jt} + e^{-jt}}{2} \left[\frac{e^{j\tau-\tau}}{2\,(j - 1)} \Bigg|_0^t + \frac{e^{-j\tau-\tau}}{2\,(- j - 1)} \Bigg|_0^t \right]$$

$$\text{term (b)} = \frac{e^{jt} + e^{-jt}}{2} \left[\frac{e^{-jt-t}}{2j - 2} + \frac{e^{-jt-t}}{- 2j - 2} - \frac{1}{2j - 2} - \frac{1}{- 2j - 2} \right] \qquad (7)$$

After performing the subtraction indicated by $v_C(t)$

= a - b, Eq. (6) and (7), combine terms and rewrite the result in trigonometric form to yield

$$v_C(t) = \frac{e^{-t}\sin^2 t + \sin t}{2} + \frac{e^{-t}\cos^2 t - \cos t}{2} \quad (8)$$

Substituting the trigonometric identities,

$$\sin^2 t = \frac{1}{2}(1 - \cos 2t)$$

$$\cos^2 t = \frac{1}{2}(1 + \cos 2t)$$

into Eq. (8) yields

$$v_C(t) = \frac{e^{-t}}{2} + \frac{\sin t}{2} - \frac{\cos t}{2}; \quad 0 < t < \pi$$

Evaluating Eq. (5), using the procedure outlined for Eq. (4) gives the following result:

$$v_C(t) = \frac{e^{-t}}{2}\sin^2 t + \frac{e^{-t}}{2}\cos^2 t + \frac{e^{-(t-\pi)}}{2};$$

since $\sin^2 t + \cos^2 t = 1$,

$$v_C(t) = \frac{e^{-t}}{2} + \frac{e^{-(t-\pi)}}{2}; \quad \pi < t < \infty$$

Note that the two functions

$$v_C(t) = \frac{e^{-t}}{2} + \frac{\sin t}{2} - \frac{\cos t}{2}; \quad 0 < t < \pi \quad (9)$$

$$v_C(t) = \frac{e^{-t}}{2} + \frac{e^{-(t-\pi)}}{2}; \quad \pi < t < \infty \quad (10)$$

are equal at $t = \pi$

$$\frac{e^{-\pi}}{2} + \frac{\sin \pi}{2} - \frac{\cos \pi}{2} = \frac{e^{-\pi}}{2} + \frac{e^{-(\pi-\pi)}}{2}$$

$$\frac{e^{-\pi}}{2} + 0 - \left[-\frac{1}{2}\right] = \frac{e^{-\pi}}{2} + \frac{1}{2}$$

$$\frac{e^{-\pi}}{2} + \frac{1}{2} \equiv \frac{e^{-\pi}}{2} + \frac{1}{2} = 0.5216$$

Eq. (9) is equal to 0 at $t = 0$, and Eq. (10) is a monotonically decreasing function whose value at $t = \infty$ is 0.

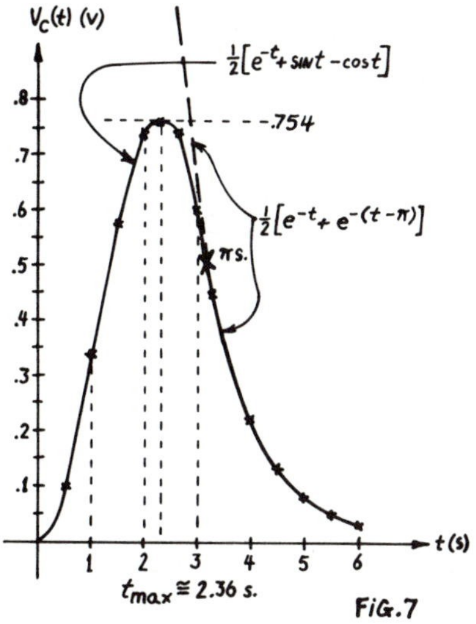

FIG.7

Fig. 7 shows a plot of the two functions. The maximum value was obtained graphically to be approximately .754 at $t = 2.36s \approx \frac{3\pi}{4}$.

PARALLEL CIRCUITS

• PROBLEM 12-19

Use Laplace transforms to find i(t) for t > 0 if: (a) $2\frac{di}{dt} + 5i = 3u(t)$, i(0) = 4; (b) $\frac{d^2i}{dt^2} + 4\frac{di}{dt} + 3i = -14\delta(t)$, i'(0) = 2 i(0) = 3; (c) i(t) is the current indicated in the figure.

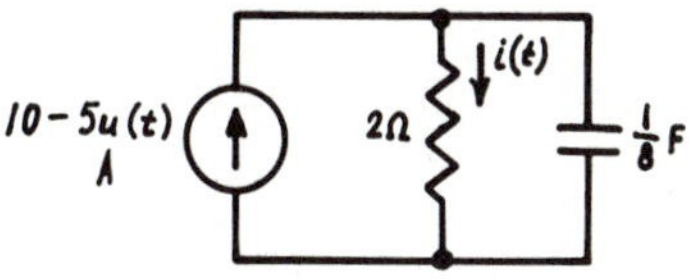

Solution: This problem is good practice in the use of Laplace transforms of derivatives and other simple functions.

(a) $$2\frac{di}{dt} + 5\,i = 3u(t),\; i(0) = 4 \tag{1}$$

$$L\left[\frac{di}{dt}\right] = sI(s) - i(0). \tag{2}$$

In this particular case, i(0) = 4 (3)

so, $$L\left[\frac{di}{dt}\right] = sI(s) - 4 \tag{4}$$

$$L\,[i(t)] = I(s). \tag{5}$$

It is known that

$$L\,[u(t)] = \frac{1}{s} \qquad (6)$$

substituting (4), (5) and (6) into Eq. (1) and Laplace transforming both sides of Eq. (1) yields

$$2\,\{sI(s) - 4\} + 5\,I(s) = 3\,\frac{1}{s} \qquad (7)$$

$$(2s + 5)\,I(s) - 8 = 3\,\frac{1}{s}. \qquad (8)$$

Dividing through by 2

$$(s + 2.5)\,I(s) - 4 = 1.5\,\frac{1}{s} \qquad (9)$$

$$(s + 2.5)\,I(s) = 4 + 1.5\,\frac{1}{s}$$

$$I(s) = \frac{4}{s + 2.5} + \frac{1.5}{s\,(s + 2.5)}$$

$$= \frac{4s + 1.5}{s\,(s + 2.5)}. \qquad (10)$$

Using the partial fraction expansion technique, let

$$I(s) = \frac{A}{s} + \frac{B}{s + 2.5} \qquad (11)$$

$$= \frac{A\,(s + 2.5) + Bs}{s\,(s + 2.5)}$$

$$= \frac{(A + B)s + 2.5A}{s\,(s + 2.5)}. \qquad (12)$$

Comparing Eq. (12) with Eq. (10),

$$A + B = 4 \qquad (13)$$

$$2.5A = 1.5 \qquad (14)$$

From Eq. (14), $A = \frac{1.5}{2.5} = \frac{3}{5} = 0.6.$ (15)

Substituting Eq. (15) into Eq. (13),

$$0.6 + B = 4, \quad B = 3.4. \qquad (16)$$

Substituting Eqs. (15) and (16) into Eq. (11)

$$I(s) = \frac{0.6}{s} + \frac{3.4}{s + 2.5} \qquad (17)$$

$$= 0.6\left(\frac{1}{s}\right) + 3.4\left(\frac{1}{s + 2.5}\right). \qquad (18)$$

Remembering that

$$L^{-1}\left(\frac{1}{s}\right) = u(t) \tag{19}$$

$$L^{-1}\left(\frac{1}{s+a}\right) = e^{-at}\, u(t) \tag{20}$$

$$L^{-1}[I(s)] = 0.6\, L^{-1}\left(\frac{1}{s}\right) + 3.4\, L^{-1}\left(\frac{1}{s+2.5}\right)$$

$$i(t) = 0.6\, u(t) + 3.4\, e^{-2.5t}\, u(t)$$

$$= (0.6 + 3.4\, e^{-2.5t})\, u(t)\ A \tag{21}$$

(b) $$\left.\begin{aligned} &\frac{d^2 i}{dt^2} + 4\frac{di}{dt} + 3i = -14\delta(t) \\ &i'(0) = 2, \quad i(0) = 3 \end{aligned}\right\} \tag{22}$$

Remembering that

$$L\left[\frac{di}{dt}\right] = sI(s) - i(0) \tag{23}$$

$$L\left[\frac{d^2 i}{dt^2}\right] = s^2 I(s) - si(0) - i'(0) \tag{24}$$

$$L[i] = I(s) \tag{25}$$

$$L[\delta(t)] = 1. \tag{26}$$

In this case,

$$L\left[\frac{di}{dt}\right] = sI(s) - 3 \tag{27}$$

$$L\left[\frac{d^2 i}{dt^2}\right] = s^2 I(s) - 3s - 2. \tag{28}$$

Therefore, the Laplace transform of Eq. (22) is,

$$s^2 I(s) - 3s - 2 + 4sI(s) - 12 + 3I(s) = -14. \tag{29}$$

Arranging in descending order of s,

$$(s^2 + 4s + 3)\, I(s) = 3s \tag{30}$$

$$(s+1)\,(s+3)\, I(s) = 3s \tag{31}$$

$$I(s) = \frac{3s}{(s+1)(s+3)}. \tag{32}$$

Again, using partial fraction expansion

$$I(s) = \frac{A}{s + 1} + \frac{B}{s + 3} \tag{33}$$

$$= \frac{A(s + 3) + B\ (s + 1)}{(s + 1)\ (s + 3)}$$

$$= \frac{(A + B)s + 3A + B}{(s + 1)\ (s + 3)}. \tag{34}$$

Comparing Eqs. (34) and (32),

$$A + B = 3 \tag{35}$$

$$3A + B = 0. \tag{36}$$

Subtracting Eq. (35) from (36),

$$2A = -3$$

$$A = -1.5. \tag{37}$$

Substituting Eq. (37) into Eq. (35),

$$-1.5 + B = 3 \tag{38}$$

$$B = 4.5. \tag{39}$$

Substituting Eqs. (37) and (39) into Eq. (33)

$$I(s) = \frac{-1.5}{s + 1} + \frac{4.5}{s + 3} \tag{40}$$

$$i(t) = L^{-1}\ [I(s)]$$

$$= -1.5\ L^{-1}\left[\frac{1}{s + 1}\right] + 4.5\ L^{-1}\left[\frac{1}{s + 3}\right]. \tag{41}$$

Remembering Eq. (20),

$$i(t) = -1.5e^{-t}\ u(t) + 4.5e^{-3t}\ u(t)$$

$$= (4.5e^{-3t} - 1.5e^{-t})\ u(t)$$

$$= 1.5\ (3e^{-3t} - e^{-t})\ u(t)\ A \tag{43}$$

(c) Applying Kirchhoff's Current Law for $t > 0$

$$(10 - 5)u(t) = i(t) + \frac{1}{8}\ \frac{dv}{dt} \tag{44}$$

where v is the voltage across the $\frac{1}{8}$F capacitor. This is equal to the voltage across the 2Ω resistor which is connected in parallel to the capacitor

$$v = 2\ i(t). \tag{45}$$

Combining Eq. (44) and (45),

$$5\ u(t) = i(t) + \frac{1}{8} \cdot 2\ \frac{di(t)}{dt} \tag{46}$$

Now, the differential equation of i(t) is introduced.

$$\frac{1}{4}\ \frac{di(t)}{dt} + i(t) = 5\ u(t). \tag{47}$$

Remembering Eqs. (2), (5), and (6), the Laplace transform of Eq. (47) is

$$\frac{1}{4}\ [sI(s) - i(0)] + I(s) = 5\ \frac{1}{s}. \tag{48}$$

According to the given initial condition,

$$i(0^-) = 10\ A \tag{49}$$

is already flowing through the 2Ω resistor due to the 10 A current source which has been on for $t < 0$. So, for $t < 0$, the $\frac{1}{8}$F capacitor is considered to be fully charged and accepts no more current. Combining Eqs. (48) and (49),

$$\frac{1}{4}\ sI(s) - 2.5 + I(s) = 5\ \frac{1}{s}$$

$$sI(s) - 10 + 4\ I(s) = 20\ \frac{1}{s}$$

$$(s + 4)\ I(s) = 10 + 20\ \frac{1}{s}$$

$$I(s) = \frac{10}{s + 4} + \frac{20}{s\ (s + 4)}$$

$$= \frac{10s + 20}{s\ (s + 4)}. \tag{50}$$

Using the principle of partial fractions

$$I(s) = \frac{A}{s} + \frac{B}{s + 4} \tag{51}$$

$$= \frac{A\ (s + 4) + Bs}{s\ (s + 4)} \tag{52}$$

$$= \frac{(A + B)s + 4A}{s\ (s + 4)}. \tag{53}$$

Comparing Eqs. (50) and (53)

$$\left.\begin{aligned} A + B &= 10 \\ 4A &= 20 \end{aligned}\right\} \quad (54)$$

$$\text{so,} \quad \left.\begin{aligned} A &= 5 \quad \text{and} \\ B &= 5 \end{aligned}\right\} \quad (55)$$

Combining Eqs. (55), and (51),

$$I(s) = \frac{5}{s} + \frac{5}{s + 4}$$

$$i(t) = L^{-1} [I(s)] = 5 L^{-1} \left[\frac{1}{s}\right] + 5 L^{-1} \left[\frac{1}{s + 4}\right]. \quad (56)$$

Remembering Eq. (6) and (20)

$$i(t) = 5 u(t) + 5 e^{-4t} u(t)$$

$$= 5 (1 + e^{-4t}) u(t) \text{ A.} \quad (57)$$

• PROBLEM 12-20

Write a generalized expression for $v_C(t)$, in the RC circuit shown in Fig. 1, in terms of I_s, V_i, R, and C using Laplace transforms.

Show that

$$v_C(t) = v_C(\infty) - \left[v_C(\infty) - v_C(0^+)\right] e^{-\frac{t}{RC}}$$

holds for this circuit.

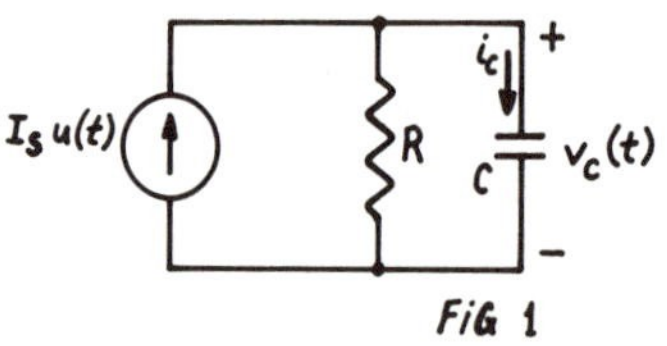

FiG 1

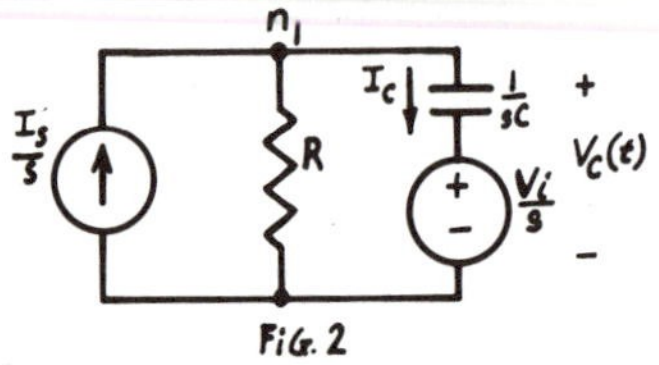

FiG. 2

Solution: We transform each element in the circuit to form a new transformed circuit as follows

$$I_s u(t) \Rightarrow \frac{I_s}{s}$$

$$R \Rightarrow R$$

$$v_C(t) = v_C(0^+) + \frac{1}{C}\int_0^t i_C dt \Rightarrow \frac{V_i}{s} + \frac{I_C}{sC}$$

giving the circuit of Fig 2.

Using KCL at node n_1 we write

$$\frac{I_s}{s} = \frac{V_C}{R} + \frac{V_C - \frac{V_i}{s}}{\frac{1}{sC}}$$

solving for $V_C(s)$ yields

$$\frac{I_s}{s} = V_C \left(\frac{1}{R} + sC\right) - CV_i$$

$$V_C(s) = \frac{I_s}{s\left(\frac{1}{R} + sC\right)} + \frac{CV_i}{\left(\frac{1}{R} + sC\right)}.$$

In order to transform the expression back to the time domain ($V_C(s) \Rightarrow v_C(t)$) by partial fraction expansion, we must write each term of $V_C(s)$ in the form

$$\frac{K}{(s + r_1)\ (s + r_2)\ \ldots}$$

To do this, we multiply the numerator and denominator of each term by 1/C to obtain the desired form.

$$V_C(s) = \frac{\frac{1}{C} I_s}{s\left(s + \frac{1}{RC}\right)} + \frac{V_i}{\left(s + \frac{1}{RC}\right)}.$$

We can now use partial fraction expansion to expand the first term of the above expression as follows

$$V_C(s) = \frac{A}{s} + \frac{B}{\left(s + \frac{1}{RC}\right)} + \frac{V_i}{\left(s + \frac{1}{RC}\right)}$$

where $$A = \left|\frac{\frac{1}{C} I_s}{s + \frac{1}{RC}}\right|_{s=0} = RI_s,$$

and $$B = \left|\frac{\frac{1}{C} I_s}{s}\right|_{s=-\frac{1}{RC}} = -RI_s.$$

We write the complete expression

$$V_C(s) = \frac{RI_s}{s} - \frac{RI_s}{\left(s + \frac{1}{RC}\right)} + \frac{V_i}{\left(s + \frac{1}{RC}\right)}$$

and transform it to the time-domain to obtain our generalized expression for $v_C(t)$.

$$v_C(t) = RI_s - (RI_s - V_i)\, e^{-\frac{t}{RC}}.$$

We note that RI_s is the steady state voltage across the the capacitor $v_C(\infty)$ and $V_i = v_C(0^+)$. Thus, our expression is

$$v_C(t) = v_C(\infty) - [v_C(\infty) - v_C(0^+)]\, e^{-\frac{t}{RC}}.$$

• PROBLEM 12-21

A single sinusoidal current pulse, $i_s(t) = 5 \sin 2t\,[u(t) - u(t - \pi/2)]$ A, is applied to the parallel combination of a $\frac{1}{8}$ H inductor and a $\frac{1}{2}$ F capacitor. Find the capacitor voltage v(t) for t > 0.

Solution: Figure 1 below illustrates the problem.

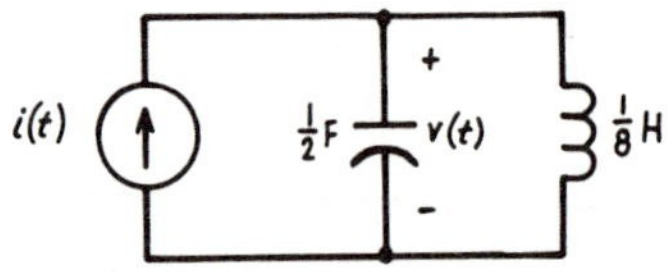

FIG. 1

Since the problem involves transients, it is convenient to use the Laplace Transform. Now V(s) = Z(s) I(s), while

$$Z(s) = \frac{1}{\frac{1}{2}s + \frac{8}{s}} = \frac{2s}{s^2 + 16}$$

and $I(s) \int_0^{\pi/2} 5 \sin 2t\, e^{-st}\, dt$

$$u = 5 \sin 2t \qquad dv = e^{-st}\, dt$$

$$du = 10 \cos 2t\, dt \qquad v = -\frac{1}{s} e^{-st}$$

$$I(s) = -\frac{5}{s} e^{-st} \sin 2t \Big|_0^{\pi/2} + \frac{1}{s}\int_0^{\pi/2} 10 \cos 2t\, e^{-st}\, dt$$

$$u = 10 \cos 2t \qquad dv = e^{-st}\, dt$$

$$du = -20 \sin 2t\, dt \qquad v = -\frac{1}{s} e^{-st}$$

or $$I(s) = \frac{1}{s}\left[-\frac{10}{s}\cos 2t\, e^{-st}\Big|_0^{\frac{\pi}{2}} - \frac{1}{s}\int_0^{\frac{\pi}{2}} 20 \sin 2t\, e^{-st} dt\right]$$

so $$I(s) = -\frac{10}{s^2}\left[-e^{-s\frac{\pi}{2}} - 1\right] - \frac{4}{s^2} I(s)$$

Hence, $$I(s) = 10\,\frac{1 + e^{-s\frac{\pi}{2}}}{s^2 + 4}$$

Therefore $$V(s) = \frac{2s}{s^2 + 16} \times \frac{10\,(1 + e^{-s\frac{\pi}{2}})}{s^2 + 4} = F(s) + F(s)e^{-s\frac{\pi}{2}}$$

where $$F(s) = \frac{20s}{(s^2 + 16)(s^2 + 4)} = \frac{As + B}{s^2 + 16} + \frac{Cs + D}{s^2 + 4}$$

Thus $$20s = (As + B)\;(s^2 + 4) + (Cs + D)\;(s^2 + 16)$$

Matching coefficients, we get $A + C = 0$, $B + D = 0$, $4A + 16C = 20$, and $4B + 16D = 0$. Solving simultaneously, we get $C = -A = \frac{5}{3}$ and $B = D = 0$.

Hence, $$F(s) = \frac{-\frac{5}{3}s}{s^2 + 16} + \frac{\frac{5}{3}s}{s^2 + 4}$$

so $$f(t) = \left[-\frac{5}{3}\cos 4t + \frac{5}{3}\cos 2t\right] u(t)$$

$$= \frac{5}{3}\,[\cos 2t - \cos 4t]\; u(t)$$

and $$v(t) = L^{-1}\{F(s)\, e^{-s\frac{\pi}{2}}\} = f(t) + f\left(t - \frac{\pi}{2}\right)$$

or $$v(t) = \frac{5}{3}\,[\cos 2t - \cos 4t]\; u(t) + \frac{5}{3}\left[\cos 2\left(t - \frac{\pi}{2}\right) - \cos 4\left(t - \frac{\pi}{2}\right)\right] u\left(t - \frac{\pi}{2}\right)\; V$$

MIXED CIRCUITS

• PROBLEM 12-22

Find the equivalent impedance of the network shown in Fig. 1.

Solution: The equivalent impedance can be found by two methods. One method, which requires few calculations, finds the impedance by successive combinations of series and parallel impedances. In the other method, which requires more calculations, the voltage V(s) and the current I(s) are found and then the equivalent impedance is found by Ohm's Law. The method requiring fewer calculations is done first. The Laplace transformed network is shown in Fig. 2.

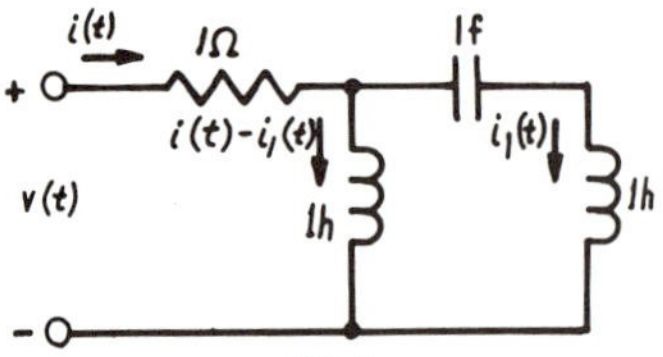

FiG.1

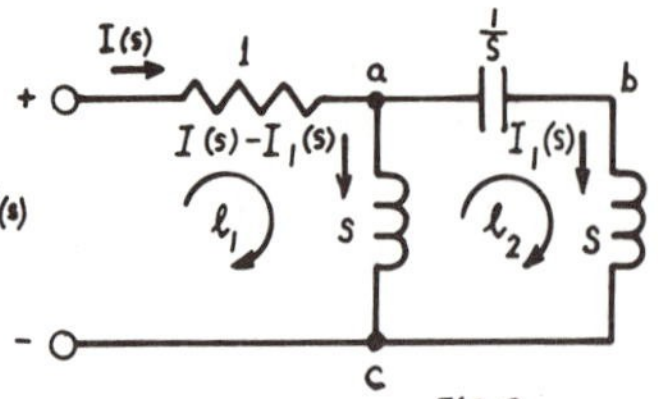

FiG.2

The impedance Z_{abc} is the impedance of the capacitor and inductor in series.

$$Z_{abc}(s) = \frac{1}{s} + s$$

The impedance Z_{ac} is Z_{abc} in parallel with the inductor.

$$Z_{ac} = \frac{(Z_{abc})s}{Z_{abc} + s}.$$

$$= \frac{\left(\frac{1}{s} + s\right)s}{s + s + \frac{1}{s}}$$

$$= \frac{s^2 + 1}{2s + \frac{1}{s}}$$

$$Z_{ac} = \frac{s^3 + s}{2s^2 + 1}$$

The equivalent impedance is Z_{ac} in series with the resistor.

$$Z_{eq}(s) = 1 + Z_{ac}(s)$$

$$= 1 + \frac{s^3 + s}{2s^2 + 1}$$

$$Z_{eq}(s) = \frac{s^3 + 2s^2 + s + 1}{2s^2 + 1}$$

Now the method requiring more calculations is done.

Applying KVL around loops ℓ_1 and ℓ_2 gives

ℓ_1: $\quad V(s) = I(s) + s\,(I(s) - I_1(s))$

$$V(s) = I(s) + sI(s) - sI_1(s) \qquad (1)$$

ℓ_2: $$0 = \frac{1}{s}I_1(s) + sI_1(s) - s\,(I(s) - I_1(s))$$

$$0 = I_1(s)\left[\frac{1}{s} + 2s\right] - sI(s) \qquad (2)$$

From equation (2) $I_1(s)$ is solved in terms of $I(s)$ and substituted into equation (1).

$$I_1(s) = \frac{s}{\frac{1}{s} + 2s} I(s) = \frac{s^2}{2s^2 + 1} I(s)$$

$$V(s) = I(s) + sI(s) - \frac{s^3}{2s^2 + 1} I(s)$$

$$= I_s\left[\frac{s^3 + 2s^2 + s + 1}{2s^2 + 1}\right]$$

Applying Ohm's Law yields $Z_{eq}(s)$

$$Z_{eq}(s) = \frac{V(s)}{I(s)} = \frac{s^3 + 2s^2 + s + 1}{2s^2 + 1}.$$

• PROBLEM 12-23

Use Laplace transform techniques to calculate the voltage response $v_C(t)$ for the network shown in Fig. 1. Assume zero initial conditions.

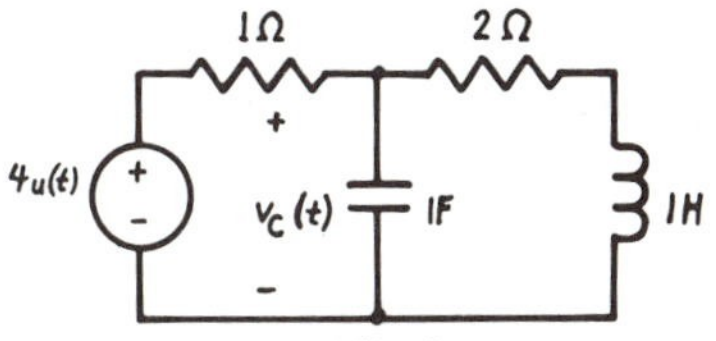

FIG. 1

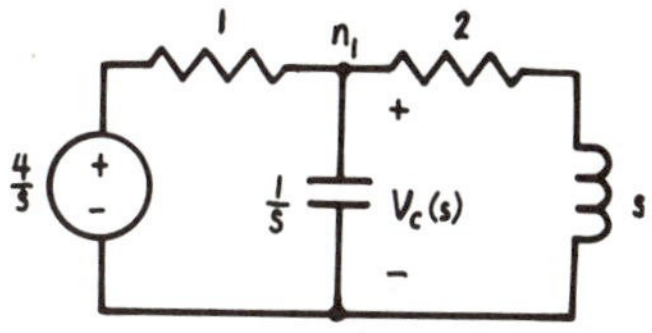

FIG. 2

Solution: Transforming the network of Fig. 1 by use of Laplace transforms produces the network shown in Fig. 2.

Using KCL at node n_1 write the equation,

$$\frac{\frac{4}{s} - V_C(s)}{1} = \frac{V_C(s)}{\frac{1}{s}} + \frac{V_C(s)}{s + 2}.$$

Solving for $V_C(s)$ yields

$$\frac{4}{s} = V_C(s)\left[1 + s + \frac{1}{s + 2}\right]$$

$$V_C(s) = \frac{4}{s\left[1 + s + \frac{1}{s+2}\right]}$$

$$V_C(s) = \frac{4\ (s + 2)}{s\ (s^2 + 3s + 3)}$$

Factor $s^2 + 3s + 3$ by finding the roots of the equation $s^2 + 3s + 3 = 0$.

Hence,

$$s_{1,2} = \frac{-3 \pm \sqrt{9 - 12}}{2}$$

$$s_1 = -\frac{3}{2} + j\frac{\sqrt{3}}{2}$$

$$s_2 = -\frac{3}{2} - j\frac{\sqrt{3}}{2}$$

$$V_C(s) = \frac{4\ (s + 2)}{s\left(s + \frac{3}{2} - j\frac{\sqrt{3}}{2}\right)\left(s + \frac{3}{2} + j\frac{\sqrt{3}}{2}\right)}$$

Using partial fraction expansion yields

$$\frac{4\ (s + 2)}{s\ (s^2 + 3s + 3)} = \frac{A}{s} + \frac{B}{\left(s + \frac{3}{2} - j\frac{\sqrt{3}}{2}\right)} + \frac{B^*}{\left(s + \frac{3}{2} + j\frac{\sqrt{3}}{2}\right)}$$

where $A = s\ V_C(s)\Big|_{s=0} = \frac{4\ (s + 2)}{(s^2 + 3s + 3)}\Big|_{s=0} = \frac{8}{3}$

$$B = \left(s + \frac{3}{2} - j\frac{\sqrt{3}}{2}\right) V_C(s)\Big|_{s=-\frac{3}{2} + j\frac{\sqrt{3}}{2}}$$

$$B = \frac{4\ (s + 2)}{s\left(s + \frac{3}{2} + j\frac{\sqrt{3}}{2}\right)}\Bigg|_{s=-\frac{3}{2} + j\frac{\sqrt{3}}{2}} = \frac{4\left(\frac{1}{2} + j\frac{\sqrt{3}}{2}\right)}{\left(-\frac{3}{2} + j\frac{\sqrt{3}}{2}\right)\left(j\sqrt{3}\right)}$$

$$B = \frac{2 + j\ 2\sqrt{3}}{-\frac{3}{2} - j\frac{3\sqrt{3}}{2}} = \frac{2 + j\ 2\sqrt{3}}{-\frac{3}{2} - j\frac{3\sqrt{3}}{2}} \cdot \frac{\left(-\frac{3}{2} + j\frac{3\sqrt{3}}{2}\right)}{\left(-\frac{3}{2} + j\frac{3\sqrt{3}}{2}\right)}$$

$$B = \frac{-3 + j\,3\sqrt{3} - j\,3\sqrt{3} - 9}{\frac{9}{4} + \frac{27}{4}} = -\frac{4}{3}$$

and $B^* = -\frac{4}{3}$.

Hence,
$$V_C(s) = \frac{\frac{8}{3}}{s} - \frac{\frac{4}{3}}{\left(s + \frac{3}{2} - j\frac{\sqrt{3}}{2}\right)} - \frac{\frac{4}{3}}{\left(s + \frac{3}{2} + j\frac{\sqrt{3}}{2}\right)}$$

can be inversely transformed to become

$$V_C(t) = \frac{8}{3} - \frac{4}{3}\left(e^{\left(-\frac{3}{2} + j\frac{\sqrt{3}}{2}\right)t} + e^{\left(-\frac{3}{2} - j\frac{\sqrt{3}}{2}\right)t}\right).$$

This can be written as

$$V_C(t) = \frac{8}{3} - \frac{8}{3}e^{-\frac{3}{2}t}\left[\frac{e^{j\frac{\sqrt{3}}{2}t} + e^{-j\frac{\sqrt{3}}{2}t}}{2}\right].$$

Recognizing that,

$$\cos\omega t = \frac{e^{+j\omega t} + e^{-j\omega t}}{2},$$

facilitates writing

$$V_C(t) = \frac{8}{3}\left(1 - e^{-\frac{3}{2}t}\cos\frac{\sqrt{3}}{2}t\right); \quad t \geq 0.$$

• PROBLEM 12-24

In the circuit of Fig. 1, derive the network function relating v_1 to v by voltage division.

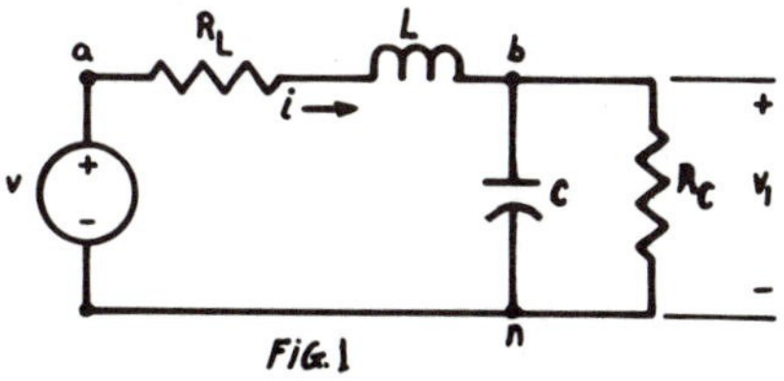

FiG.1

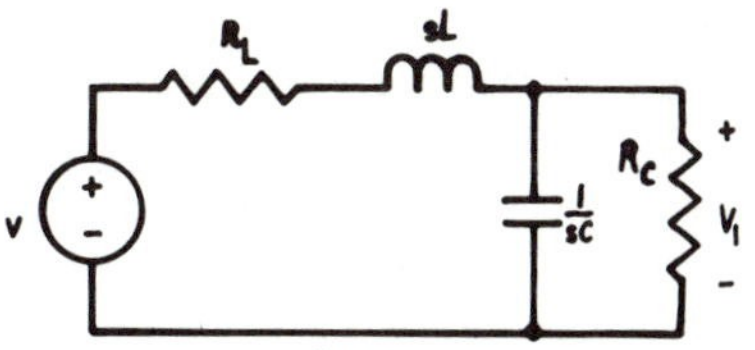

FiG.2

Solution: We use complex frequency s, as shown in Fig. 2, to express each impedance.

Combining $\frac{1}{sC}$ and R_C we obtain the circuit in Fig. 3.

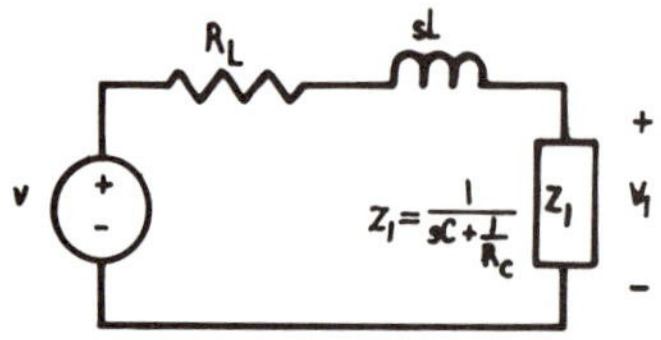

Fig. 3

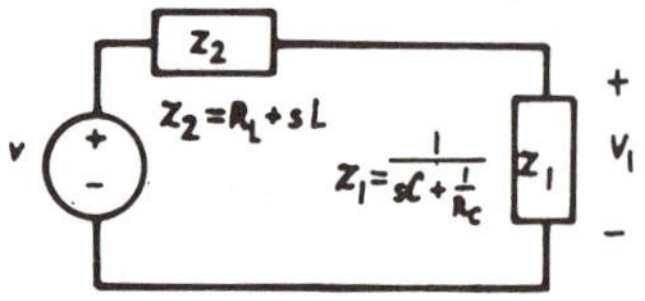

Fig. 4

Combining R_L and sL we obtain the circuit in Fig. 4.

Now by voltage division,

$$V_1 = \frac{Z_1 V}{Z_1 + Z_2}$$

thus,

$$V_1 = \frac{\dfrac{V}{sC + \dfrac{1}{R_C}}}{R_L + sL + \dfrac{1}{sC + \dfrac{1}{R_C}}}$$

simplifying,

$$V_1 = \frac{\dfrac{V R_C}{s R_C C + 1}}{R_L + sL + \dfrac{R_C}{s R_C C + 1}}$$

$$V_1 = \frac{\dfrac{V R_C}{s R_C C + 1}}{\dfrac{R_L (s R_C C + 1) + sL (s R_C C + 1) + R_C}{s R_C C + 1}}$$

yields

$$\frac{V_1}{V} = H(s) = \frac{R_C}{s^2 (R_C L C) + s (R_L R_C C + L) + R_L R_C}$$

which is the desired result.

• PROBLEM 12-25

Find the impulse response of the circuit of Fig. 1 by constructing the frequency-domain circuit, finding the appropriate transfer function, and evaluating the inverse Laplace transform. The output is (a) $v_A(t)$; (b) $v_B(t)$; (c) $v_C(t)$.

Solution: We obtain the frequency-domain network using the Laplace transforms. Using this network, we obtain the required transfer function, H(s). Taking the inverse Laplace will give the required impulse response, h(t). The mesh currents, $I_C(s)$ and $I_B(s)$ are obtained first using the determinant method. These currents are the same as the branch currents through the output impedances. Hence, we can find the output voltages and transfer functions. The current, $I_A(s)$ through the impedance for $V_A(s)$ can be found by Kirchhoff's Current Law. The frequency-domain network together with the transformed currents and voltages is shown in Fig. 2.

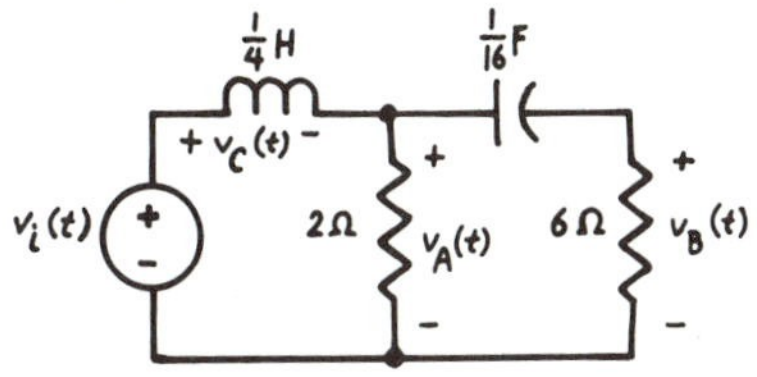

FiG. 1

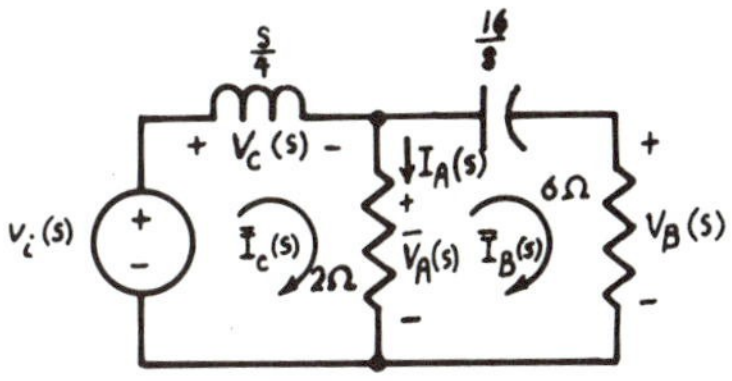

FiG. 2

First find the mesh current, $I_C(s)$ using determinants

$$I_C(s) = \frac{\begin{vmatrix} V_i(s) & -2 \\ 0 & \frac{16}{s} + 8 \end{vmatrix}}{\begin{vmatrix} \frac{2}{4} + 2 & -2 \\ -2 & \frac{16}{s} + 8 \end{vmatrix}} = \frac{\left(\frac{16}{s} + 8\right) V_i(s)}{\left(\frac{s}{4} + 2\right)\left(\frac{16}{s} + 8\right) - 4}.$$

Clearing fractions and multiplying

$$I_C(s) = \frac{32s + 64}{8s^2 + 80s + 128 - 16s} V_i(s) = \frac{32s + 64}{8s^2 + 64s + 128} V_i(s).$$

Now divide out the common factor in the numerator and denominator. We see that the denominator has a multiple root at $s = -4$.

$$I_C(s) = \frac{4s + 8}{s^2 + 8s + 16} V_i(s) = \frac{4s + 8}{(s + 4)^2} V_i(s).$$

Now, $V_C(s) = I_C(s)\ Z_C(s) = \dfrac{4s + 8}{(s + 4)^2} \left(\dfrac{s}{4}\right) V_i(s)$

$$V_C(s) = \frac{s^2 + 2s}{(s + 4)^2} V_i(s).$$

Therefore, $H_C(s) = \dfrac{V_C(s)}{V_i(s)} = \dfrac{s^2 + 2s}{(s + 4)^2}.$

To find the impulse response in the time domain, we have to express H(s) in terms of partial fractions such that the inverse of each fraction can be found from a table of Laplace transform pairs. The expansion method will not be covered in detail here, but will be used in this case with a pair of roots, s = - 4.

Expand $H_C(s)$ as follows:

$$H_C(s) = \frac{A(s)}{B(s)} = \frac{s^2 + 2s}{(s + 4)^2} = \frac{N_1}{s + 4} + \frac{N_2}{(s + 4)^2}$$

$$N_1 = \frac{d}{ds}(s^2 + 2s)\Big|_{s=-4} = 2s + 2\Big|_{s=-4} = -8 + 2 = -6$$

$$N_2 = s^2 + 2s\Big|_{s=-4} = 16 - 8 = 8.$$

Therefore, $H_C(s) = \dfrac{-6}{s + 4} + \dfrac{8}{(s + 4)^2}.$

Taking the inverse from a table

(c) $h_C(t) = \delta(t) - (6e^{-4t} - 8te^{-4t})\, u(t).$

Note the $\delta(t)$ term appears in the response since this output is in series with the input which is a unit impulse function. The other two terms found from the table do not contain $\delta(t)$ and if not included, there would be no response mathematically for the other two outputs.

Solve for mesh current, $I_B(s)$

$$I_B(s) = \frac{\begin{vmatrix} \frac{s}{4} + 2 & V_i(s) \\ -2 & 0 \end{vmatrix}}{\begin{vmatrix} \frac{s}{4} + 2 & -2 \\ -2 & \frac{16}{s} + 8 \end{vmatrix}} = \frac{2\, V_i(s)}{\left(\frac{s}{4} + 2\right)\left(\frac{16}{s} + 8\right) - 4}$$

$$I_B(s) = \frac{s}{s^2 + 8s + 16} V_i(s) = \frac{s}{(s + 4)^2} V_i(s)$$

$$V_B(s) = I_B(s)\, Z_B(s) = \frac{6s}{(s + 4)^2} V_i(s)$$

$$H_B(s) = \frac{V_B(s)}{V_i(s)} = \frac{6s}{(s+4)^2} = \frac{N_1}{s+4} + \frac{N_2}{(s+4)^2}$$

$$N_1 = \frac{d}{ds}(6s)\bigg|_{s=-4} = 6$$

$$N_2 = 6s\bigg|_{s=-4} = -24.$$

Therefore, $H_B(s) = \dfrac{6s}{(s+4)^2} = \dfrac{6}{s+4} - \dfrac{24}{(s+4)^2}$

(b) $h_B(t) = L^{-1}\{H_B(s)\} = L^{-1}\left\{\dfrac{6}{s+4} - \dfrac{24}{(s+4)^2}\right\}$

$$h_B(t) = (6e^{-4t} \quad 24te^{-4t})\, u(t).$$

Find $I_A(s)$ using Kirchhoff's Current Law.

$$I_A(s) = I_C(s) - I_B(s)$$

$$I_A(s) = \left[\frac{4s+8}{(s+4)^2} - \frac{s}{(s+4)^2}\right] V_i(s)$$

$$I_A(s) = \frac{3s+8}{(s+4)^2} V_i(s).$$

Then $V_A(s) = I_A(s)\ Z_A(s) = \dfrac{(3s+8)\,2}{(s+4)^2} V_i(s)$

$$H_A(s) = \frac{V_A(s)}{V_i(s)} = \frac{6s+16}{(s+4)^2} = \frac{N_1}{s+4} + \frac{N_2}{(s+4)^2}$$

$$N_1 = \frac{d}{ds}(6s+16)\bigg|_{s=-4} = 6$$

$$N_2 = (6s+16)\bigg|_{s=-4} = -24 + 16 = -8.$$

Hence, $H_A(s) = \dfrac{6}{s+4} - \dfrac{8}{(s+4)^2}$

$$L^{-1}\{H_A(s)\} = h_A(t) = L^{-1}\left\{\frac{6}{s+4} - \frac{8}{(s+4)^2}\right\}.$$

From a table of Laplace transform pairs

(a) $h_A(t) = (6e^{-4t} - 8te^{-4t})\ u(t).$

• PROBLEM 12-26

Use Laplace transforms to find $v_2(t)$ if $v_s(t)$ =:

(a) 12 u(t);

(b) 12 cos t u(t);

(c) $12\ e^{-t}\ u(t)$.

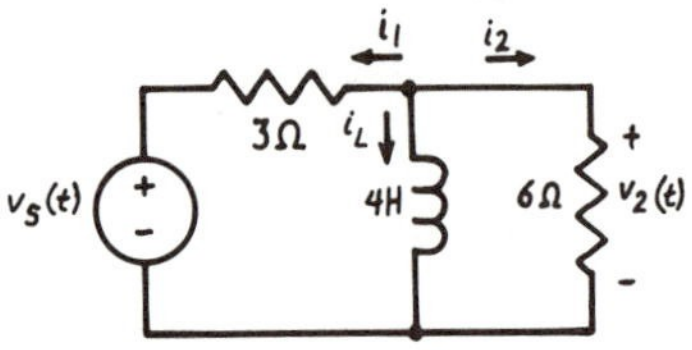

FIG. 1 $i_L(0^-) = 2A$

Solution: The transform of $v_2(t)$, in terms of the initial inductor current and the source voltage transform $V_s(s)$, will be obtained first. Then specific transforms for $V_s(s)$ can be substituted as required.

Nodal equations will be used to solve for $V_2(s)$. The current out of the node above the inductor is

$$i_1 + i_L + i_2 = 0$$

or $$\frac{v_2(t) - v_s(t)}{3\Omega} + \frac{1}{4}\int_0^t v_2(t)\ dt + 2A + \frac{v_2(t)}{6\Omega} = 0.$$

The Laplace transform is then taken.

$$\frac{V_2(s)}{3} - \frac{V_s(s)}{3} + \frac{V_2(s)}{4s} + \frac{2}{s} + \frac{V_2(s)}{6} = 0.$$

Solving algebraically for $V_2(s)$,

$$V_2(s) = \frac{\frac{2}{3}s\ V_s(s)}{s + \frac{1}{2}} - \frac{4}{s + \frac{1}{2}} = V_2'(s) - \frac{4}{s + \frac{1}{2}}$$

(a) $V_s(s) = \frac{12}{s}$ and $V_2(s)$ becomes

$$V_2(s) = \frac{8}{s + \frac{1}{2}} - \frac{4}{s + \frac{1}{2}} = \frac{4}{s + \frac{1}{2}}.$$

Hence, $v_2(t) = 4e^{-t/2}\,u(t)$ V

(b) $V_s(s) = \frac{12s}{s^2 + 1}$ and $V_2{}'(s)$ is

$$V_2{}'(s) = \frac{8s^2}{\left(s^2 + 1\right)\left(s + \frac{1}{2}\right)}$$

This is represented in partial fractions as

$$V_2{}'(s) = \frac{A}{s + \frac{1}{2}} + \frac{B}{s + j} + \frac{B^*}{s - j}$$

where B* is the complex conjugate of B. The three constants are evaluated as

$$A = \left(s + \frac{1}{2}\right) V_2{}'(s)\Big|_{s=-\frac{1}{2}} = \frac{8}{5}$$

$$B = (s + j)\, V_2{}'(s)\Big|_{s=-j} = \frac{8}{5}\,(2 + j)$$

$$B^* = \frac{8}{5}\,(2 - j).$$

These values are substituted into the partial fraction expansion of $V_2{}'(s)$ and added to $\frac{-4}{s + \frac{1}{2}}$ in order to obtain $V_2(s)$

$$V_2(s) = \frac{\frac{8}{5} - 4}{s + \frac{1}{2}} + \frac{8}{5}\left[\frac{2 - j}{s + j} + \frac{2 + j}{s - j}\right]$$

$$= \frac{-2.4}{s + \frac{1}{2}} + \frac{16}{5}\left[\frac{2s - 1}{s^2 + 1}\right]$$

$$= \frac{-2.4}{s + \frac{1}{2}} + \frac{6.4s}{s^2 + 1} - \frac{3.2}{s^2 + 1}.$$

The inverse transform is then

$$v_2(t) = (-2.4e^{-t/2} + 6.4 \cos t - 3.2 \sin t)\, u(t) \text{ V}$$

(c) $V_s(s) = \dfrac{12}{s+1}$ and $V_2'(s)$ is $\dfrac{8s}{(s+1)\left(s+\frac{1}{2}\right)}$, which expanded in partial fractions becomes

$$V_2'(s) = \frac{A}{s+\frac{1}{2}} + \frac{B}{s+1}.$$

The two constants are

$$A = \left(s+\frac{1}{2}\right) V_2'(s)\Big|_{s=-\frac{1}{2}} = -8$$

$$B = (s+1)\, V_2'(s)\Big|_{s=-1} = 16.$$

A and B are substituted into the expansion, and again added to $\dfrac{-4}{s+\frac{1}{2}}$, in order to obtain $V_2(s)$.

$$V_2(s) = \frac{-12}{s+\frac{1}{2}} + \frac{16}{s+1}$$

and $v_2(t) = (-12e^{-t/2} + 16e^{-t})\, u(t)$ V.

• PROBLEM 12-27

Use Laplace transforms to find the complete response v(t) in the circuit of Fig. 1.

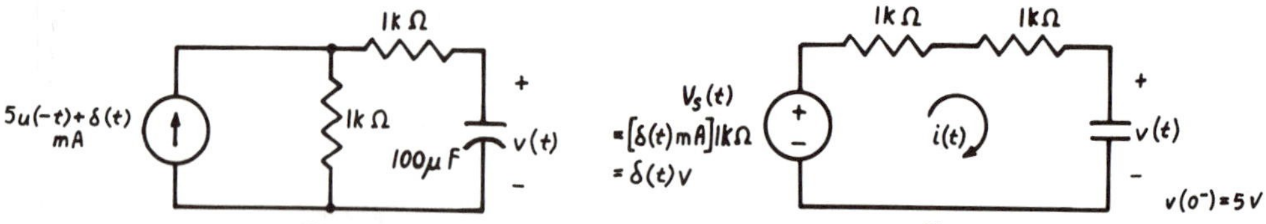

FIG. 1 FIG. 2

Solution: Before jumping into this problem with Laplace transforms, one needs to consider the effect of the 5 u(-t) mA term of the current source. This has a value of 5 mA for t < 0 and a value of zero for t ≥ 0. The Laplace transform of this term is therefore zero. That is

$$5 \text{ mA} \int_{0^-}^{\infty} u(-t)\, e^{-st}\, dt = 0$$

We cannot ignore this term however, since it represents a long-time steady (or dc) current of 5 mA flowing into the circuit until t = 0. The effect of this current will be to develop a voltage of (5 mA)(1kΩ) = 5V across the resistor, shunting the source terminals. The capacitor voltage v(t) will likewise have charged up to 5V long before t = 0. Thus the 5 u(t) mA term of the current source can be replaced by an initial voltage of $v(t = 0^-) = 5V$ across the capacitor.

The remaining current source, δ(t) mA, and the resistor, shunting the source terminals, are converted into a Thevenins equivalent circuit, as shown in Fig. 2. Summing voltages, we obtain

$$\delta(t) = (2k\Omega)\, i(t) + \frac{1}{100\mu F} \int_0^t i(t)\, dt + 5$$

The Laplace transform of this equation is

$$1 = 2k\Omega\; I(s) + V(s) = 2k\Omega\; I(s) + \frac{10^4\, I(s)}{s} + \frac{5}{s}$$

I(s) is solved for algebraically to obtain

$$I(s) = 5 \times 10^{-4}\, \frac{s - 5}{s + 5}$$

The transformed capacitor voltage is therefore

$$V(s) = \frac{10^4\, I(s)}{s} + \frac{5}{s} = \frac{5(s - 5) + 5(s + 5)}{s(s + 5)} = \frac{10}{s + 5}$$

and $v(t) = 10e^{-st}\, u(t)$ V.

• PROBLEM 12-28

Find $v_C(t)$ and $i_L(t)$ for $t \geq 0$ in Fig. 1, assume $v_C(0) = i_L(0) = 0$.

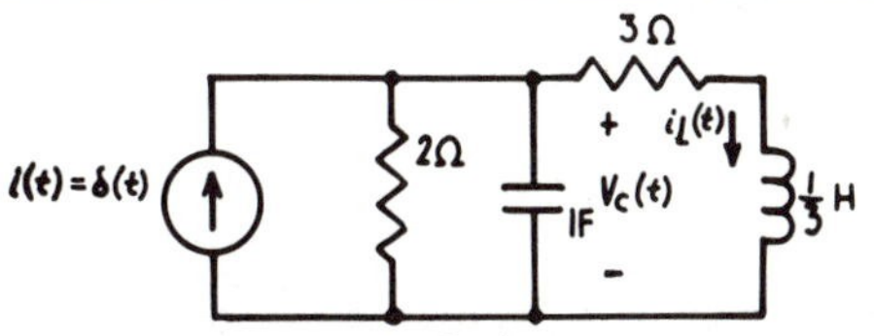

FIG.1

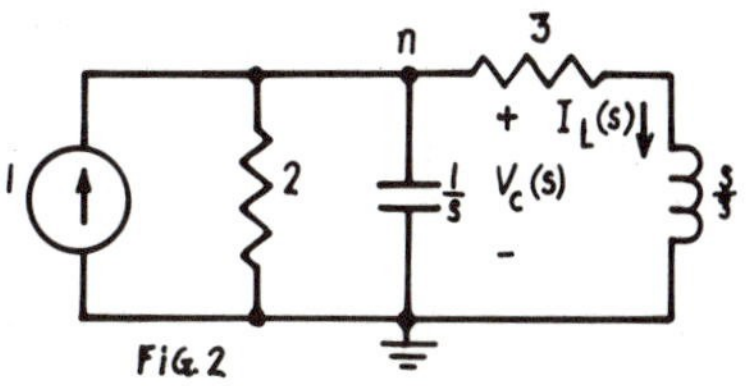

FIG.2

Solution: Obtain the Laplace transformed circuit shown in Fig. 2.

Note that L {δ(t)} = 1.

Writing a KCL equation for node n in Fig. 2 yields,

$$1 = \frac{V_C(s)}{2} + \frac{V_C(s)}{\frac{1}{s}} + \frac{V_C(s)}{3 + \frac{s}{3}}$$

Note that $\dfrac{V_C(s)}{3 + \frac{s}{3}} = I_L(s)$.

Solving the above KCL equation,

$$1 = V_C(s)\left[\frac{1}{2} + s + \frac{1}{3 + \frac{s}{3}}\right]$$

$$V_C(s) = \frac{1}{\frac{1}{2} + s + \frac{1}{3 + \frac{s}{3}}}$$

$$V_C(s) = \frac{2\,(s + 9)}{(s + 9) + 2s\,(s + 9) + 6}$$

yields, $V_C(s) = \dfrac{s + 9}{s^2 + 9.5s + 7.5}$.

$V_C(s)$ can be expanded into partial fractions and inverse transformed as follows

$$V_C(s) = \frac{s + 9}{(s + 8.63)\,(s + 0.87)}$$

$$V_C(s) = \frac{A}{s + 8.63} + \frac{B}{s + 0.87}$$

where $A = V_C(s)\,(s + 8.63)\Big|_{s=-8.63} = \dfrac{0.37}{-7.76} = -0.048$

$B = V_C(s)\,(s + 0.87)\Big|_{s=-0.87} = \dfrac{8.13}{7.76} = 1.048$

hence, $v_C(t) = 1.048e^{-0.87t} - 0.048e^{-8.63t}$ V

To find $i_L(t)$, we find the inverse transform of $\dfrac{V_C(s)}{3 + \frac{s}{3}}$.

Hence, $I_L(s) = \dfrac{V_C(s)}{3 + \frac{s}{3}} = \dfrac{\frac{s + 9}{s + 9.5s + 7.5}}{\frac{9 + s}{3}}$

$$I_L(s) = \frac{3\ (s + 9)}{(s + 8.63)\ (s + 0.87)\ (s + 9)}$$

$$I_L(s) = \frac{3}{(s + 8.63)\ (s + 0.87)}$$

$$I_L(s) = \frac{A}{s + 8.63} + \frac{B}{s + 0.87}$$

where $$A = \left| I_L(s)\ (s + 8.63) \right|_{s=-8.63} = -\ 0.39$$

$$B = \left| I_L(s)\ (s + 0.87) \right|_{s=-0.87} = +\ 0.39$$

and $i_L(t) = 0.39e^{-0.87t} - 0.39e^{-8.63t}$ A

• PROBLEM 12-29

Use an impulse input for $v_1(t)$ in the circuit of Fig. 1, to find the transfer function $\frac{V_2}{V_1}(s)$. Use the transfer function and Laplace transform techniques to find the component of the response for $t \geq 0^+$ due to the sources (a) $v_1 = 10\ u(t)$; (b) $v_1 = 10t\,u(t)$; (c) $v_1 = (10 + 10t)\ u(t)$; (d) $v_1 = 5e^{-3t}\ u(t)$; (e) $v_1 = (10t + 5e^{-3t})\ u(t)$;

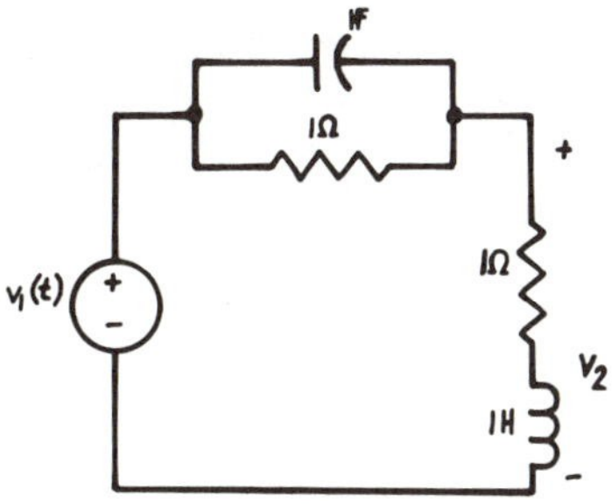

FiG.1

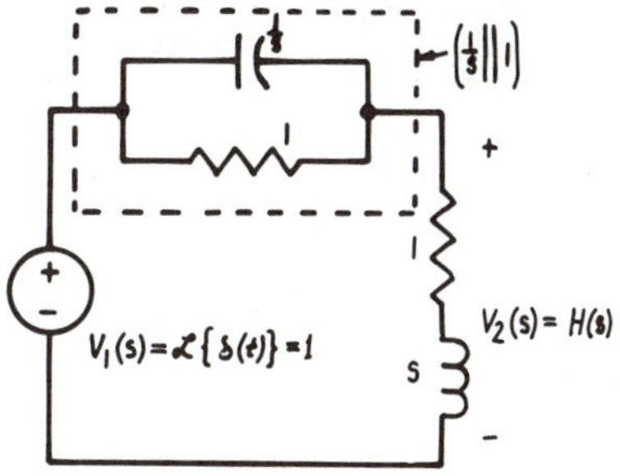

FiG.2

Solution: In order to find the transfer function, remember that the impulse response will give the transfer function directly.

Writing an equation for voltage division in the transformed circuit in Fig. 2 gives the impulse response, H(s).

$$H(s) = \frac{V_1\ (s)\ (s + 1)}{\left(\frac{1}{s}\ ||\ 1\right) + \left(s + 1\right)}$$

$$H(s) = \frac{(1)\ (s+1)}{\dfrac{\frac{1}{s}}{\frac{1}{s}+1} + (s+1)} = \frac{(s+1)}{\dfrac{\frac{1}{\not s}}{\frac{(s+1)}{\not s}} + (s+1)}$$

$$H(s) = \frac{(s+1)}{\dfrac{1}{(s+1)} + (s+1)}$$

multiplying numerator and denominator by (s + 1) gives

$$H(s) = \frac{(s+1)\ (s+1)}{(s+1)\ (s+1)+1} = \frac{s^2+2s+1}{s^2+2s+2}$$

It is now possible to find the response due to any input by finding the product of the input and the transfer function.

The block diagram in Fig. 3 illustrates the process. Remember that the operation described must take place in the frequency domain. In other words, the functions must be in terms of complex frequency, s. In the time domain the operation would be convolution, as illustrated in Fig. 4.

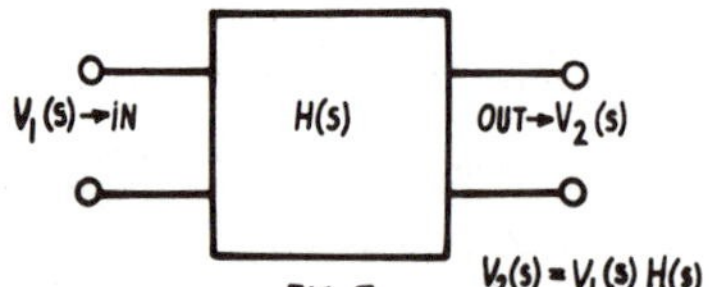

FIG. 3

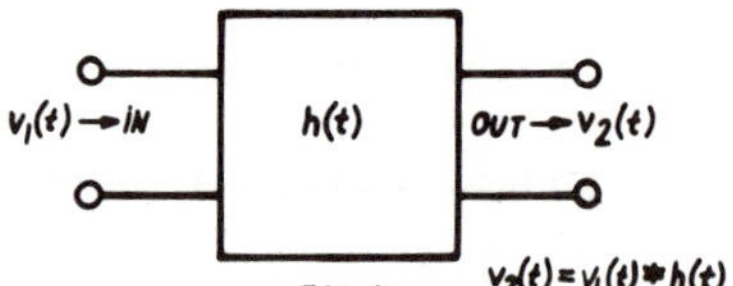

FIG. 4

To find the component of the response due to the input, multiply the transformed input with the transfer function and proceed with partial fractions to find the inverse Laplace of the response. Find the coefficients for those terms that are due to the input source. Realize that the component of the response due to the input must be a function that is the same as that of the input. For example, a linear input must produce a linear output response and an exponential input gives the same component output response.

(a) The input is given as $v_1(t) = 10\ ut$, transforming this input gives $V_1(s) = \frac{10}{s}$. The output response becomes,

$$V_2(s) = \frac{10}{s}\left(\frac{(s+1)(s+1)}{s^2+2s+2}\right).$$

This can be written as

$$V_2(s) = \frac{A}{s} + \frac{B}{s+1-j} + \frac{C}{s+1+j}$$

It can be seen from the partial fraction expansion above that the component of the response due to the input must be $\frac{A}{s}$, so evaluate coefficient A.

$$A = s\, V_2(s)\Big|_{s=0} = \frac{10(1)}{2} = 5$$

The response due to the source, $v_s(t)$ is found by taking the inverse Laplace transform.

$$v_s(t) = L^{-1}\left\{\frac{5}{s}\right\} = 5\ u(t)$$

(b) Using the same method as in (a), first find $v_1(s) = L\{10t\ u(t)\} = \frac{10}{s^2}$. The output response becomes,

$$V_2(s) = \frac{10}{s^2}\left(\frac{(s+1)\ (s+1)}{s^2 + 2s + 2}\right)$$

$$V_2(s) = \frac{A}{s} + \frac{B}{s^2} + \frac{C}{s+1-j} + \frac{D}{s+1+j}$$

$$V_s(s) = \frac{A}{s} + \frac{B}{s^2}$$

$$A = s^2 V_2(s)\Big|_{s=0} = \frac{10(1)}{2} = 5$$

$$B = \frac{1}{(2-1)!}\ \frac{d}{ds}\ [s^2 V_2(s)]\Big|_{s=0} =$$

$$B = \frac{(20s+20)(s^2+2s+2)-(2s+2)(s^2+2s+1)10}{(s^2+2s+2)^2}\Bigg|_{s=0}$$

$$B = \frac{40-20}{2^2} = 5$$

$$V_s(s) = \frac{5}{s} + \frac{5}{s^2}$$

$$v_s(t) = L^{-1}\left\{\frac{5}{s} + \frac{5}{s^2}\right\} = (5+5t)\ u(t)$$

(c) Find the response due to source, $v_1(t) = (10 + 10t)u(t)$ by using superposition.

$v_s(t)$ due to 10 u(t) from part (a) +

$v_s(t)$ due to 10 ut (t) from part (b) = $v_s(t)$ due to (10 + 10t) u(t)

$$v_s(t) = (5 + 5 + 5t)\ u(t)$$

$$= (10 + 5t)\ u(t)$$

(d) Find $V_1(s) = L\{5e^{-3t}\} = \frac{5}{s+3}$, and the response

$$V_2(s) = \frac{5}{(s+3)}\left[\frac{(s+1)(s+1)}{s^2+2s+2}\right]$$

$$V_2(s) = \frac{A}{s+3} + \frac{B}{s+1-j} + \frac{C}{s+1+j}$$

$$V_s(s) = \frac{A}{s+3}$$

$$A = (s+3)\ V_2(s)\Big|_{s=-3} = \frac{5(-2)(-2)}{9-6+2} = \frac{20}{5} = 4$$

$$v_s(t) = L^{-1}\left\{\frac{4}{s+3}\right\} = 4e^{-3t}\ u(t)$$

(e) Using superposition yields

$$v_s(t) = (5 + 5t + 4e^{-3t})\ u(t)$$

• PROBLEM 12-30

Calculate i(t) for the network below if $V_1(s) = \frac{1}{s}$ and initial conditions are zero.

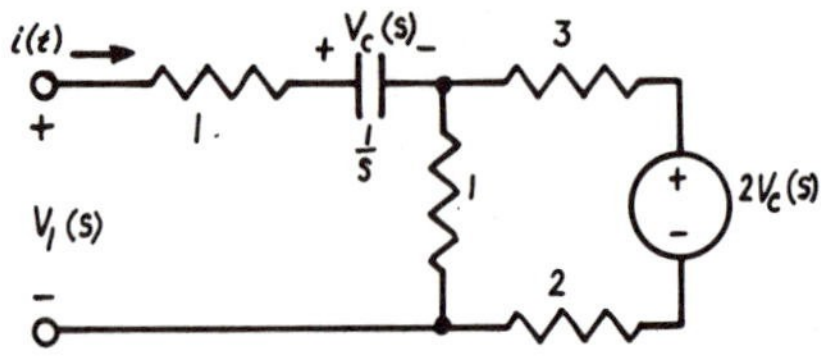

Solution: Call the voltage at the upper node (to the right of the capacitor) $V_A(s)$. The Laplace transform of the desired current is I(s). A current summation at the above node gives

$$I(s) - \frac{V_A(s)}{1} - \frac{[V_A(s) - 2\ V_C(s)]}{5} = 0. \qquad (1)$$

However, $V_A(s)$ can be represented by other quantities. Applying KVL around the left hand loop gives:

$$V_1(s) - 1\,I(s) - V_C(s) - V_A(s) = 0 \qquad (2)$$

Solving for $V_A(s)$ in (2) and substituting for $V_A(s)$ in (1) gives:

$$I(s) - V_1(s) + I(s) + V_C(s) - \frac{V_1(s)}{5} + \frac{I(s)}{5} + \frac{V_C(s)}{5}$$

$$+ \frac{2\,V_C(s)}{5} = 0$$

Collecting terms, we obtain

$$11\,I(s) + 8\,V_C(s) = 6\,V_1(s). \qquad (3)$$

But $V_C(s)$ is the capacitor voltage and $I(s)$ is its current. In the s domain, the element relation is $I(s) = V_C(s)/H(s)$ which, in this case, is $V(s) = I(s)/s$. $V_1(s)$ is given as 1/s.

Hence, equation (3) becomes

$$11\,I(s) + \frac{8\,I(s)}{s} = \frac{6}{s};$$

therefore, $I(s) = \frac{6}{8 + 11s}$, and, from the transform table,

$$i(t) = L^{-1}\left[\frac{\frac{6}{11}}{\frac{8}{11} + s}\right] = \frac{6}{11}\,e^{-\frac{8}{11}t}\,u(t).$$

INVERSE LAPLACE TRANSFORMS

• **PROBLEM** 12-31

Find the impulse response of the networks shown. The response to be found is shown as current and/or voltage on the figure.

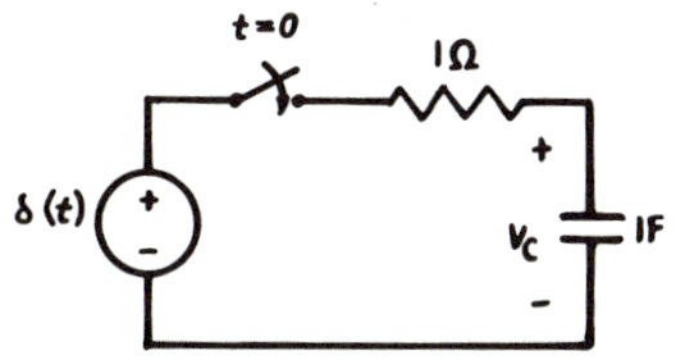

(a)

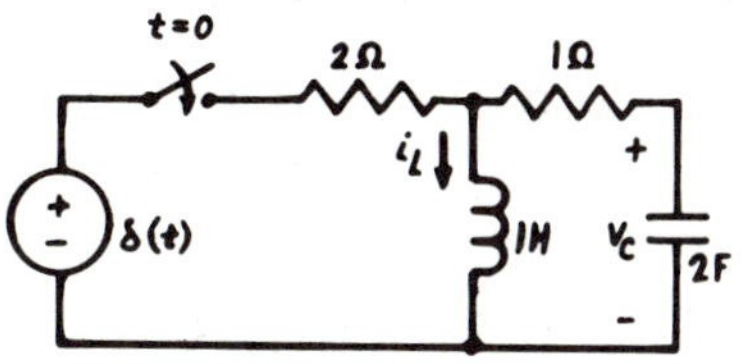

(b)

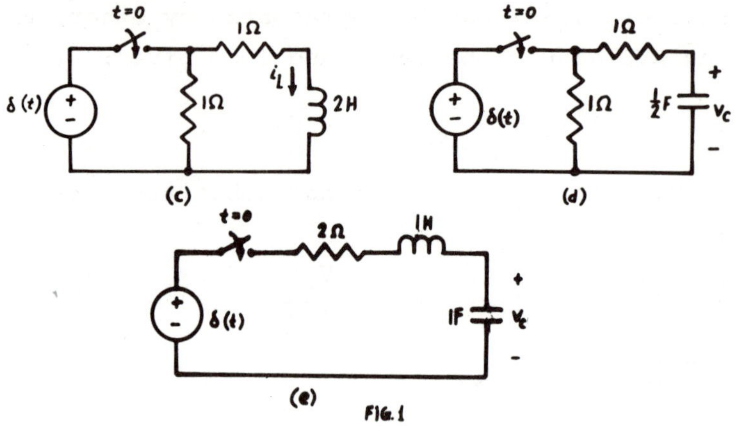

FIG. 1

Solution: (a) The governing equation is

$$R\left(C\frac{dv_C}{dt}\right) + v_C = \delta(t)$$ which, with R = 1 and C = 1, is

$$\frac{dV_C}{dt} + v_C(t) = \delta(t).$$

Taking the Laplace transform of both sides gives

$$s\,V(s) + V(s) = 1$$

so that $$V(s) = \frac{1}{s+1}.$$

From a transform table the inverse gives

$$v_C(t) = e^{-t}$$

(b) Designate the transform of $i_L(t)$ and $v_C(t)$ as I(s) and V(s). Applying KVL to the right hand loop of the circuit, we have

$$L\frac{di_L}{dt} - i_C R - v_C = 0.$$

Again, utilizing the element to express $i_C = C\frac{dv_C}{dt}$ and putting in the circuit values, we obtain

$$1\frac{di_L}{dt} = 2\frac{dv_C}{dt} + v_C$$

Taking the Laplace transform results in

$$s\,I(s) = 2s\,V(s) + V(s)$$

or, $$V(s) = \frac{s\ I(s)}{2s + 1}.$$

The current in the 2Ω resistor is $i_L + i_C = i_L + C\,\frac{dv_C}{dt}$. Writing the KVL equation for the left gives

$$1\,\frac{di_L}{dt} + 2\left(i_L + 2\,\frac{dv_C}{dt}\right) = \delta(t)$$

which is transformed to

$$s\ I(s) + 2\ I(s) + 4s\ V(s) = 1.$$

Now, put in the relation for V(s) found above in terms of I(s).

$$2\ I(s) + \frac{4s^2}{2s + 1}\ I(s) + s\ I(s) = 1$$

so that reorganizing gives

$$I(s) = \frac{2s + 1}{6s^2 + 5s + 2}.$$

The denominator has complex conjugate roots given by $s_1, s_2 = -\frac{5}{12} \pm \frac{\sqrt{23}}{12}$, and these are of the form

$$s_1 = a + j\omega$$

$$s_2 = a - j\omega$$

In general, it is shown that, for a transform function $F(s) = \frac{A(s)}{B(s)}$, it can be expanded by partial fractions to give

$$F(s) = \frac{K_1}{s - s_1} + \frac{K_2}{s - s_2} \ldots.$$

It is also shown that the qth constant is given by

$$K_1 = \left.\frac{A(s)}{\frac{d}{ds}\,B(s)}\right|_{s=s_q}.$$

It is also shown in transform calculus that if s_1 and s_2 are complex conjugates then so are K_1 and K_2 of the form $K_1 = \alpha + j\beta$ and $K_2 = K_1^* = \alpha - j\beta$.

For this case, then,

$$K_1 = \frac{2s + 1}{12s + 5}\Bigg|_{s=-\frac{5}{12} - j\frac{\sqrt{23}}{12}}$$

$$= \frac{-\frac{1}{6} - j\frac{23}{6}}{-j\,23} = +\frac{1}{6} + \frac{j}{6\sqrt{23}} \quad \text{and, thus,}$$

$$K_2 = \frac{1}{6} - \frac{j}{6\sqrt{23}}$$

$$\text{Further, } I(s) = \frac{K_1}{s - s_1} + \frac{K_2}{s - s_2} = \frac{2\alpha s - 2(\alpha a + \beta\omega)}{(s - a)^2 + \omega^2}$$

when a, ω, α and β are as above. This function has the inverse transform.

$$i(t) = 2\alpha e^{at} \cos \omega t - 2\beta e^{at} \sin \omega t.$$

Now substituting all the numerical values finally results in

$$i_L(t) = \frac{1}{3}\left[e^{-\frac{5}{12}t} \cos \frac{\sqrt{23}}{12}t - \frac{1}{\sqrt{23}} e^{-\frac{5}{12}t} \sin \frac{\sqrt{23}}{12}t\right]$$

(c) Around the outer loop, the equation is

$$1\, i_L + 2\frac{di_L}{dt} = \delta(t).$$

Taking the Laplace transform of both sides gives

$$I(s) + 2s\, I(s) = 1$$

$$I(s) = \frac{1}{1 + 2s} = \frac{\frac{1}{2}}{\frac{1}{2} + s}$$

From the transform table,

$$i_L(t) = \frac{1}{2} e^{-\frac{1}{2}t}$$

Note that the parallel branch containing only a 1Ω resistor, but no energy storage element, has nothing to do with the result. It is true that a current f(t)/1Ω exists in that branch at the impulse time (t = 0), but at all other times that branch is paralleled with a short circuit, the ideal source.

(d) The remarks of (c) pertain again.

The differential equation is:

$$R\left(C\,\frac{dv_C}{dt}\right) + v_C(t) = \delta(t)$$

Substituting the given values and Laplace transforming gives

$$\frac{s}{2}\,V(s) + V(s) = 1$$

$$V(s) = \frac{1}{1 + \frac{1}{2}s} = \frac{2}{2 + s}$$

The inverse transform results in

$$v_C(t) = 2e^{-2t}$$

(e) The differential equation for the series RLC circuit is in terms of current.

$$Ri + L\,\frac{di}{dt} + \frac{1}{C}\int i\,dt = \delta(t)$$

Here we wish to find the impulse response of the capacitor voltage $v_C(t)$. The current is $i = C\,dv/dt$ so that we can write.

$$RC\,\frac{dv_C}{dt} + L\,\frac{d}{dt}\left(C\,\frac{dv_C}{dt}\right) + v_C(t) = \delta(t)$$

$$LC\,\frac{d^2v_C}{dt^2} + RC\,\frac{dv_C}{dt} + v_C(t) = \delta(t)$$

Again, using Laplace transforms after substitution of the circuit values gives

$$s^2\,V(s) + 2s\,V(s) + V(s) = 1$$

$$V(s) = \frac{1}{s^2 + 2s + 1} = \frac{1}{(s + 1)^2}$$

From the table of transforms, then,

$$v_C(t) = te^{-t}$$

For the networks below find the response to driving functions u(t):

(a) e^{-t}; (b) pulse of height 1 and duration 1s.

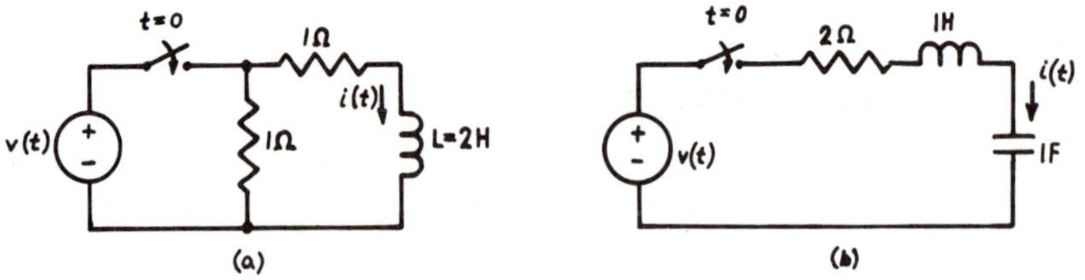

Solution: The currents can be found by formulating inpulse functions (or transfer functions) which remain the same for any input voltage. Then, each V(s) is multiplied by the transfer function, and the product is converted back to the time domain.

Circuit a

We can, by inspection, write down I(s).

$$I(s) = \frac{V(s)}{R + Ls} = \frac{V(s)}{1 + 2s}.$$

In terms of an impulse response H(s), we can write

$$I(s) = H(s)\ V(s)$$

where $$H(s) = \frac{1}{1 + 2s} = \frac{1}{2}\ \frac{1}{s + \frac{1}{2}}.$$

(a) $v(t) = e^{-t}$

There are two possibilities for proceding.

1. Convert v(t) to V(s), multiply if by H(s), and convert the product back to the time domain.

2. Recognize that H(s) V(s) is a convolution in the time domain and complete the solution by calculating the convolution integral.

We will use the latter approach, that is,

$$i(t) = \int_0^t h(t - \tau)\ v(\tau)\ d\tau.$$

In the time domain, the impulse response is

$$h(t) = \frac{1}{2} e^{-\frac{t}{2}}.$$

Hence, the integral becomes

$$i(t) = \int_0^t \frac{1}{2} e^{-\frac{1}{2}(t-\tau)} e^{-\tau} d\tau$$

$$= \frac{1}{2} e^{-\frac{t}{2}} \int_0^t e^{-\frac{\tau}{2}} d\tau$$

$$= \frac{1}{2} e^{-\frac{t}{2}} \left[-2 \left(e^{-\frac{t}{2}} - 1 \right) \right]$$

$$= e^{-\frac{t}{2}} - e^{-t} \text{ A}; \quad t \geq 0$$

which is the desired result.

(b) $v(t)$ = a pulse of height 1 and duration from $t = 0$ to $t = 1s$.

The first step in the solution is to formulate $V(s)$. $v(t)$ can be conveniently expressed in terms of unit step functions. Thus,

$$v(t) = u(t) - u(t-1).$$

The Laplace transform is

$$V(s) = \frac{1}{s} - \frac{e^{-s}}{s}$$

which can be determined from a table or by a straight forward evaluation using the transform integral. The current expression becomes

$$I(s) = H(s)\, V(s) = \frac{H(s)}{s} - \frac{H(s)}{s} e^{-s}.$$

For conveneience, let $F(s) = \frac{H(s)}{s}$ in the second term. Then, in the time domain,

$$i(t) = \int_0^t h(\tau)\, d\tau - f(t-1)\, u(t-1)$$

where $f(t)$ is the inverse transform of $F(s)$.

By partial fraction expansion,

$$F(s) = \frac{H(s)}{s} = \frac{1}{2} \frac{1}{s\left(s + \frac{1}{2}\right)} = \frac{1}{s} - \frac{1}{s + \frac{1}{2}}.$$

Then, $\quad f(t) = 1 - e^{-\frac{1}{2}t}.$

The integral is

$$\int_0^t h(\tau)\ d\tau = \frac{1}{2}\int_0^t e^{-\frac{\tau}{2}}\ d\tau = 1 - e^{-\frac{t}{2}};\ \text{for } t \geq 0$$

or, $\left(1 - e^{-\frac{t}{2}}\right) u(t).$

Putting the terms back together, we have, for the solution,

$$i(t) = \left(1 - e^{-\frac{t}{2}}\right) u(t) - \left[1 - e^{-\frac{1}{2}(t-1)}\right] u(t-1)\ A.$$

Circuit b

By inspection, the impulse response is

$$H(s) = \frac{1}{2 + s + \frac{1}{s}} = \frac{s}{s^2 + 2s + 1} = \frac{s}{(s+1)^2}.$$

By a partial fraction expansion,

$$H(s) = \frac{1}{s+1} - \frac{1}{(s+1)^2}.$$

In the time domain,

$$h(t) = e^{-t} - te^{-t}$$

The current is given by

$$i(t) = \int_0^t h(t-\tau)\ v(\tau)\ d\tau$$

(a) $v(t) = e^{-t}$

Substituting into the convolution integral, we obtain

$$i(t) = \int_0^t \left[e^{-(t-\tau)} - (t-\tau)e^{-(t-\tau)}\right] e^{-\tau}\ d\tau$$

$$= e^{-t}\int_0^t 1\ d\tau - te^{-t}\int_0^t 1\ d\tau$$

$$+ e^{-t}\int_0^t \tau\ d\tau$$

$$= te^{-t} - t^2e^{-t} + \frac{1}{2} t^2e^{-t}$$

$$= te^{-t} - \frac{1}{2} t^2e^{-t} \text{ A}$$

which is the desired result.

(b) $v(t) = u(t) - u(t-1)$

We will use the formulation developed in circuit a,

$$i(t) = \int_0^t h(\tau)\, d\tau - f(t-1)\, u(t-1)$$

The integral of h(t) is

$$\int_0^t h(\tau)\, d\tau = \int_0^t \left[e^{-\tau} - \tau e^{-\tau}\right] d\tau$$

$$= te^{-t}; \text{ for } t > 0$$

or,
$$= te^{-t}\, u(t)$$

The next step is to calculate f(t). By definition,

$$F(s) = \frac{H(s)}{s} = \frac{1}{(s + 1)^2}.$$

Then, $f(t) = te^{-t}.$

The complete result is

$$i(t) = te^{-t}\, u(t) - (t-1)e^{-(t-1)}\, u(t-1) \text{ A.}$$

• PROBLEM 12-33

The circuit shown has initial conditions $v_C(t_0) = 1V$, $i_L(t_0) = 0$, where $t_0 = 1$ sec. Calculate $i_L(t)$ using Laplace transforms.

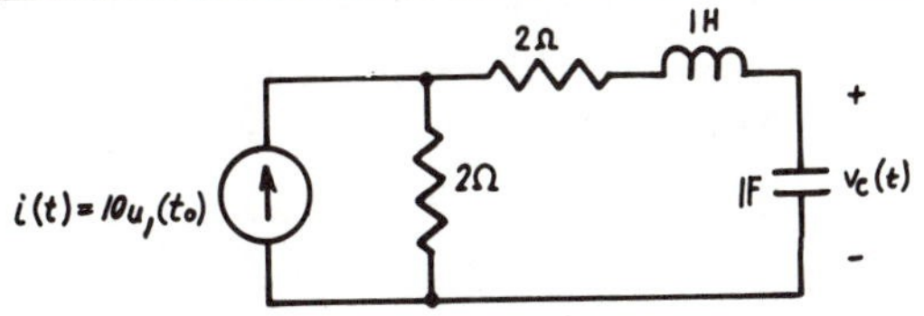

FIG. 1

Solution: The procedure in this problem will be to write two equations using the loop current method and then solve them for the current in the right hand loop. This equation will be Laplace transformed and solved for the current and then converted back to the time domain. The ramp function starts at $t_0 = 1$ sec and the initial conditions are given for $t_0 = 1$ sec. Hence we can solve the problem as if all processes started at $t = 0$ and in the final solution replace each t by t - 1.

The circuit is

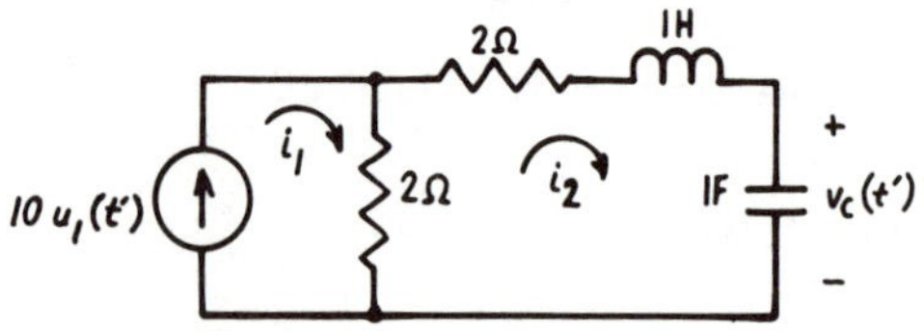

FIG. 2

where t' represents t - 1.

The equation for i_2 is

$$2i_2 + L\frac{di_2}{dt'} + \frac{1}{C}\int i_2 dt' + 2(i_2 - i_1) = 0. \qquad (1)$$

For loop 1,

$$i_1 = 10\,u_1(t') \qquad (2)$$

Substituting Eq. (1) into Eq. (2) and setting L = 1H and C = 1F yields

$$2i_2 + \frac{di_2}{dt'} + \int i_2 dt' + 2[i_2 - 10\,u_1(t')] = 0. \qquad (3)$$

The transform of the ramp function is

$$\int_0^\infty u_1(t')\, e^{-st'}\, dt' = \int_0^\infty t'\, u(t')\, e^{-st'}\, dt'$$

$$= \frac{1}{s^2}$$

where u(t') is the unit step function.

Next transform Eq. (3) and substitute the initial conditions. Thus

$$2I_2(s) + sI_2(s) - 0 + \frac{I_2(s)}{s} + \frac{1}{s} + 2\left[I_2(s) - \frac{10}{s^2}\right] = 0.$$

Combining coefficients of $I_2(s)$ and simplifying gives

$$\left(4 + s + \frac{1}{s}\right) I_2(s) = \frac{10}{s^2} - 1.$$

Solving for $I_2(s)$ results in

$$I_2(s) = \frac{10 - s}{s\,(s^2 + 4s + 1)}.$$

The quadratic term in the denominator is factored for a partial fraction expansion. This gives

$$I_2(s) = \frac{10 - s}{s\,(s + 2 + \sqrt{3})(s + 2 - \sqrt{3})}.$$

In the partial fraction expansion set

$$I_2(s) = \frac{10 - s}{s\,(s + 2 + \sqrt{3})(s + 2 - \sqrt{3})} = \frac{A}{s} + \frac{B}{s + 2 + \sqrt{3}}$$

$$+ \frac{C}{s + 2 - \sqrt{3}} . \qquad (4)$$

The solution for A, B, and C follows:

$$A = sI_2(s)\Big|_{s=0} = \frac{10 - s}{(s + 2 + \sqrt{3})(s + 2 - \sqrt{3})}\Bigg|_{s=0} = 10$$

$$B = (s + 2 + \sqrt{3})\ I_2(s)\Big|_{s=-2-\sqrt{3}} = \frac{10 - s}{s\ (s + 2 - \sqrt{3})}\Bigg|_{s=-2-\sqrt{3}}$$

$$= 1.062$$

$$C = (s + 2 - \sqrt{3})\ I_2(s)\Big|_{s=-2+\sqrt{3}} = \frac{10 - s}{s\ (s + 2 + \sqrt{3})}\Bigg|_{s=-2+\sqrt{3}}$$

$$= - 11.06$$

Substituting the coefficients into Eq. (4) gives

$$I_2(s)\ \frac{10}{s} + \frac{1.062}{s + 2 + \sqrt{3}} - \frac{11.06}{s + 2 - \sqrt{3}}.$$

Now transform back to the time domain. Using Laplace transform tables yields

$$i_2(t') = 10\ u(t') + 1.062e^{-(2 + \sqrt{3})t'} - 11.06e^{-(2 - \sqrt{3})t'}$$

or $\quad i_2(t') = 10\ u(t') + 1.062e^{-3.732t'} - 11.06e^{-0.268t'}.$

We obatin the final result by replacing t' by t - 1. Thus

$$i_2(t) = 10\ u(t-1) + 1.062e^{-3.732(t-1)} - 11.06e^{-0.268(t-1)} \text{A}.$$

• **PROBLEM 12-34**

Using Laplace transform techniques, find the current in resistor R_2 in Fig. 1. Assume that the initial current in the inductor, $i_L(0^-) = 1A$.

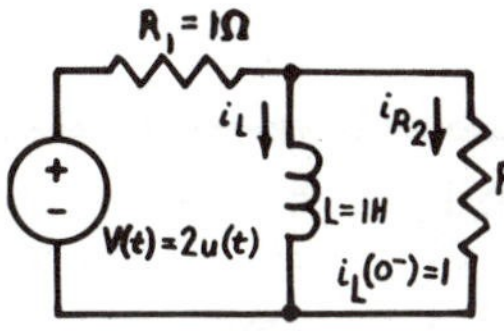

FiG. 1

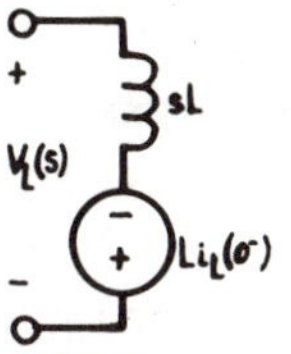

FiG.2

Solution: Note that when an initial condition is given for an inductor, transform the equation

$$v_L(t) = L\frac{di_L}{dt}$$

to
$$V_L(s) = L\,(sI_L(s) - i_L(0^-))$$

$$V_L(s) = sLI_L(s) - Li_L(0^-)$$

A schematic representation of the transformed inductor with initial condition is shown in Fig. 2.

Applying this concept to the circuit in Fig. 1 gives

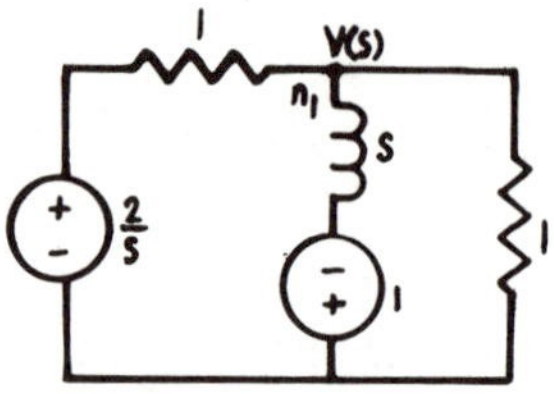

FIG. 3

Writing a KCL equation for node n_1

$$\frac{V(s) - \frac{2}{s}}{1} + \frac{V(s) + 1}{s} + \frac{V(s)}{1} = 0$$

$$V(s)\left[2 + \frac{1}{s}\right] = \frac{2}{s} - \frac{1}{s} = \frac{1}{s}$$

$$V(s) = \frac{\frac{1}{s}}{2 + \frac{1}{s}}$$

Multiply both numerator and denominator by $\frac{s}{2}$ in order to obtain the form $\frac{A}{s + B} \overset{L^{-1}}{\Rightarrow} Ae^{-Bt}$

$$V(s) = \frac{\frac{1}{2}}{s + \frac{1}{2}} \overset{L^{-1}}{\Rightarrow} \frac{1}{2}e^{-\frac{1}{2}t}$$

since
$$i_{R_2} = \frac{v(t)}{R_2} = \frac{v(t)}{1} = \frac{1}{2}e^{-\frac{1}{2}t}\ \text{A}.$$

• PROBLEM 12-35

A 1V voltage source is connected at $t = -\infty$ to a series RL circuit consisting $L_1 = 2H$ and $R_1 = 2\Omega$. At $t = 0$ another series RL circuit consisting of $L_2 = 3H$ and $R_2 = 2\Omega$ is switched in parallel with R_1. Find the current in L_2 for $t > 0$.

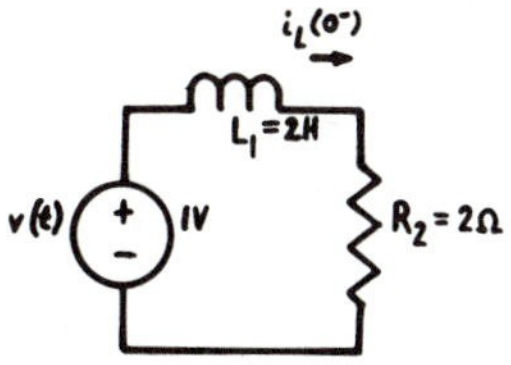

FIG. 1

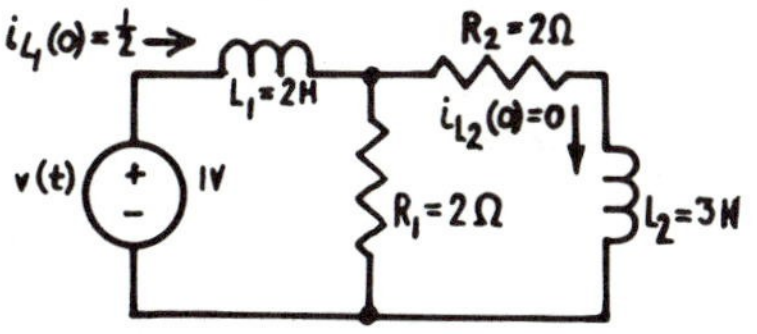

FIG. 2

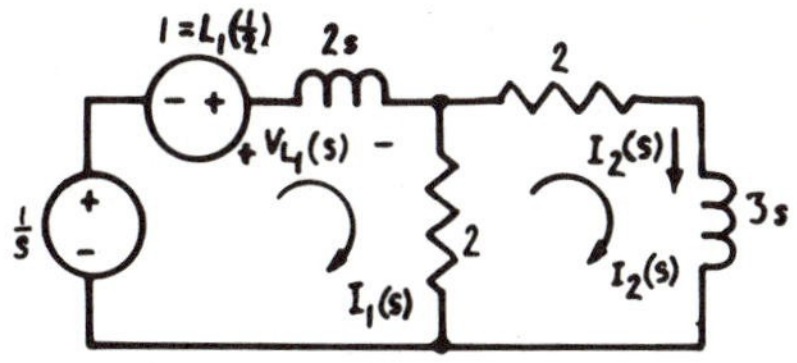

FIG. 3

<u>Solution</u>: At $t < 0$ we have the circuit shown in Fig. 1.

Since all transients have passed at $t = 0^-$ and the inductor is a short-circuit, the initial current through inductor L_1 is $\frac{v(t)}{R_2} = \frac{1}{2} = 0.5$ A.

At $t = 0$ the circuit is expanded to that which is shown in Fig. 2.

Since inductor L_2 was not connected to the circuit at $t = 0^-$, the initial current through it is zero.

Now draw the transformed network in Fig. 3.

Writing KVL equation for the two loops yields

$$\frac{1}{s} + 1 = I_1(s)\ (2s + 2) - I_2(s)\ (2)$$

$$0 = I_1(s)\ (-2) + I_2(s)\ (4 + 3s)$$

Using determinants to solve for $I_2(s)$

$$I_2(s) = \frac{\begin{vmatrix} 2s + 2 & \frac{1}{s} + 1 \\ -2 & 0 \end{vmatrix}}{\begin{vmatrix} 2s + 2 & -2 \\ -2 & 4 + 3s \end{vmatrix}} = \frac{-(-2)\left(\frac{1}{s} + 1\right)}{(2s + 2)\ (4 + 3s) - (-2)\ (-2)}$$

$$I_2(s) = \frac{\frac{2}{s} + 2}{6s^2 + 8s + 6s + 8 - 4} = \frac{\frac{2}{s} + 2}{6s^2 + 14s + 4}$$

multiplying numerator and denominator by $\frac{s}{2}$ gives,

$$I_2(s) = \frac{s + 1}{s\ (3s^2 + 7s + 2)}$$

Normalize the denominator polynomial, and multiply the numerator by $\frac{1}{3}$.

$$I_2(s) = \frac{\frac{1}{3}(s+1)}{s\left(\frac{3}{3}s^2 + \frac{7}{3}s + \frac{2}{3}\right)}$$

Use the quadratic formula to find the roots of the quadratic in the denominator.

$$r_1 r_2 = \frac{-b \pm \sqrt{b^2 - 4aC}}{2a}$$

$$r_1 r_2 = \frac{-\frac{7}{3} \pm \sqrt{\frac{49}{9} - \frac{8}{3}}}{2}$$

$$r_1 r_2 = \frac{-\frac{7}{3} \pm \sqrt{\frac{25}{9}}}{2} = \frac{-7}{6} \pm \frac{5}{6}$$

$$r_1 = -\frac{1}{3}$$

$$r_2 = -2$$

Factoring the expression for $I_2(s)$ gives

$$I_2(s) = \frac{\frac{1}{3}(s+1)}{s\left(s+\frac{1}{3}\right)(s+2)}$$

Using partial fraction expansion, write

$$I_2(s) = \frac{A}{s} + \frac{B}{s+\frac{1}{3}} + \frac{C}{s+2}$$

where $$A = S(I_2(s))\Big|_{s=0} = \frac{\frac{1}{3}(s+1)}{\left(s+\frac{1}{3}\right)(s+2)}\Bigg|_{s=0} = \frac{\frac{1}{3}}{\frac{1}{3}(2)}$$

$$A = \frac{1}{2}$$

$$B = \left(s+\frac{1}{3}\right)(I_2(s))\Big|_{s=-\frac{1}{3}} = \frac{\frac{1}{3}(s+1)}{s(s+2)}\Bigg|_{s=-\frac{1}{3}}$$

$$B = \frac{\left(\frac{1}{3}\right)\left(\frac{2}{3}\right)}{-\frac{1}{3}\left(\frac{5}{3}\right)} = -\frac{2}{5}$$

$$C = (s+2)\,(I_2(s))\Big|_{s=-2} = \frac{\frac{1}{3}(s+1)}{s\left(s+\frac{1}{3}\right)}\Bigg|_{s=-2}$$

$$C = \frac{\frac{1}{3}(-1)}{(-2)\left(-\frac{5}{3}\right)} = -\frac{1}{10}$$

giving, $$I_2(s) = \frac{\frac{1}{2}}{s} + \frac{-\frac{2}{5}}{s+\frac{1}{3}} + \frac{-\frac{1}{10}}{s+2}.$$

Since, $\frac{A}{s+B} \overset{L^{-1}}{\Rightarrow} Ae^{-Bt}$, we can write

$$I_2(s) \overset{L^{-1}}{\Rightarrow} i_2(t) = \left(\frac{1}{2} - \frac{2}{5}e^{-\frac{1}{3}t} - \frac{1}{10}e^{-2t}\right) u(t).$$

• PROBLEM 12-36

A periodic waveform shown in Fig. 1 is applied to the RLC network shown in Fig. 2. Find the voltage $v_C(t)$ by using Laplace transform techniques.

Solution: The Laplace transform of the circuit in Fig. 2 is shown in Fig. 3.

The voltage across the inductor V_2 is found by use of the voltage division concept, hence, the capacitor voltage is

$$V_C = \frac{\frac{1}{s}}{\left(1+\frac{1}{s}\right)} V_L = \frac{\frac{1}{s}}{\left(1+\frac{1}{s}\right)} \left[\frac{\dfrac{s\left(1+\frac{1}{s}\right)}{s+1+\frac{1}{s}}}{1+\dfrac{s\left(1+\frac{1}{s}\right)}{s+1+\frac{1}{s}}}\right] V$$

where V is the source voltage shown in Fig. 1. The transfer function $H(s) = \frac{V_C(s)}{V(s)}$ is

$$\frac{\frac{1}{s}\left[\dfrac{s\left(1+\frac{1}{s}\right)}{s+1+\frac{1}{s}}\right]}{\left(\frac{s+1}{s}\right)\left\{1+\left[\dfrac{s\left(1+\frac{1}{s}\right)}{s+1+\frac{1}{s}}\right]\right\}} = \frac{\frac{1}{2}s}{s^2+s+\frac{1}{2}}$$

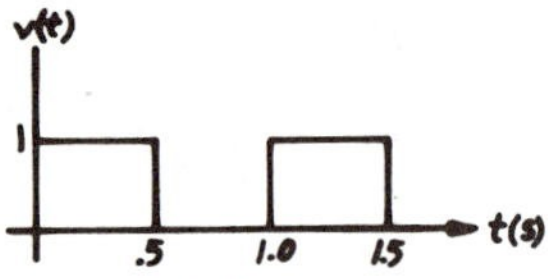

FIG.1

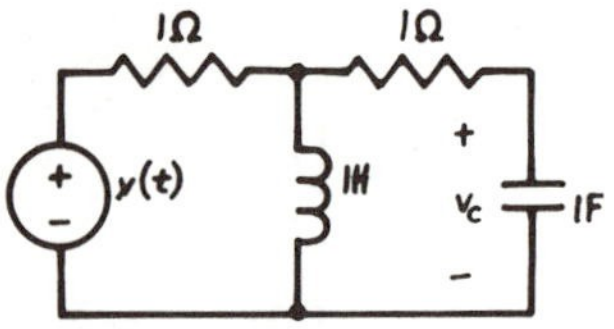

FIG.2

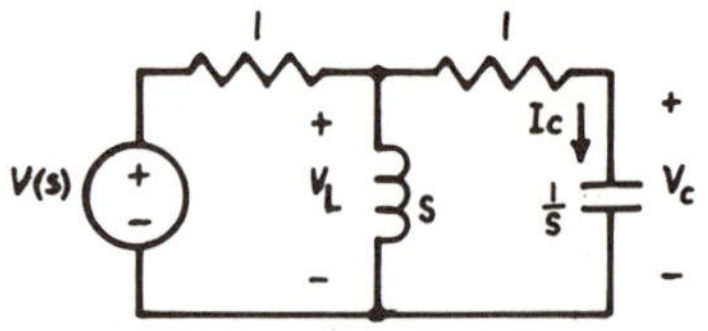

FIG.3

The periodic voltage waveform shown in Fig. 1 can be transformed by use of Laplace transforms. Hence,

$$L\{v(t)\} = V(s) = \frac{F(s)}{\left(1 - e^{-sT}\right)}$$

where T is the period, and F(s) is the Laplace transform of V(s) over one period. Here, using the time-shift property of the Laplace transform, $F(s) = \frac{1}{s} - \frac{1}{s}e^{-s/2}$. Since T = 1 sec., then

$$V(s) = \frac{1 - e^{-\frac{s}{2}}}{s(1 - e^{-s})}$$

The solution $v_C(t)$ can be found by taking the inverse Laplace transform of H(s) V(s). Hence,

$$H(s)\ V(s) = V_C(s) = \frac{\left(1 - e^{-\frac{s}{2}}\right)}{\cancel{s}\ (1 - e^{-s})} \quad \frac{\frac{1}{2}\cancel{s}}{\left(s^2 + s + \frac{1}{2}\right)}$$

$$V_C(s) = \frac{\frac{1}{2}\left(1 - e^{-\frac{s}{2}}\right)}{\left(s + \frac{1}{2} - \frac{j}{2}\right)\left(s + \frac{1}{2} + \frac{j}{2}\right)\ (1 - e^{-s})}$$

Using partial fraction expansion yields

$$V_C(s) = \frac{A}{\left(s + \frac{1}{2} - \frac{j}{2}\right)} + \frac{A^*}{\left(s + \frac{1}{2} + \frac{j}{2}\right)} + \frac{V_s(s)}{(1 - e^{-s})} \qquad (1)$$

where $\frac{A}{\left(s + \frac{1}{2} - \frac{j}{2}\right)} + \frac{A^*}{\left(s + \frac{1}{2} + \frac{j}{2}\right)}$ is the transient response

$V_t(s)$, and $\frac{V_s(s)}{(1 - e^{-s})}$ is the periodic or steady-state response $V_p(s)$. Multiplying Eq. (1) by it's characteristic polynomial $\left(s^2 + s + \frac{1}{2}\right)(1 - e^{-s})$ gives

$$\frac{1}{2}\left(1 - e^{-\frac{s}{2}}\right) = A\left(s + \frac{1}{2} + \frac{j}{2}\right)(1 - e^{-s})$$

$$+ A^*\left(s + \frac{1}{2} - \frac{j}{2}\right)(1 - e^{-s})$$

$$+ [V_s(s)]\left(s^2 + s + \frac{1}{2}\right) \qquad (2)$$

If we let $s = -\frac{1}{2} + \frac{j}{2}$, then

$$\frac{1}{2}\left(1 - e^{\frac{1}{4}} e^{-\frac{j}{4}}\right) = A_j\left(1 - e^{\frac{1}{2}} e^{-\frac{j}{2}}\right)$$

$$A = \frac{\frac{1}{2}\left(1 - e^{\frac{1}{4}} e^{-\frac{j}{4}}\right)}{j\left(1 - e^{\frac{1}{2}} e^{-\frac{j}{2}}\right)}.$$

A can be evaluated by substituting

$e^{-\frac{1}{4}} = \cos\frac{1}{4} - j\sin\frac{1}{4}$. Hence, $A = 0.031 - j\,0.218$ and the conjugate $A^* = 0.031 + j\,0.218$.

Now, solving for $V_s(s)$ in Eq. (2) we find the steady state response over one period. Thus,

$$V_s(s) = \frac{\frac{1}{2}\left(1 - e^{-\frac{s}{2}}\right) - A\left(s + \frac{1}{2} + \frac{j}{2}\right)\left(1 - e^{-s}\right)}{s^2 + s + \frac{1}{2}}$$

$$\frac{- A^*\left(s + \frac{1}{2} - \frac{j}{2}\right)\left(1 - e^{-s}\right)}{s^2 + s + \frac{1}{2}}$$

$$V_s(s) = \frac{\frac{1}{2}\left(1 - e^{-\frac{s}{2}}\right)}{s^2 + s + \frac{1}{2}} - \frac{A\,(1 - e^{-s})}{s + \frac{1}{2} - \frac{j}{2}} - \frac{A^*\,(1 - e^{-s})}{s + \frac{1}{2} + \frac{j}{2}}$$

$$V_s(s) = \frac{\frac{1}{2j}\left(1 - e^{-\frac{s}{2}}\right)}{s + \frac{1}{2} - \frac{j}{2}} - \frac{\frac{1}{2j}\left(1 - e^{-\frac{s}{2}}\right)}{s + \frac{1}{2} + \frac{j}{2}} - \frac{A(1 - e^{-s})}{s + \frac{1}{2} - \frac{j}{2}}$$

$$- \frac{A^*(1 - e^{-s})}{s + \frac{1}{2} + \frac{j}{2}}$$

Taking the inverse Laplace transform of $V_s(s)$ yields

$$v_s(t) = \left[\frac{1}{2j} e^{-\frac{1}{2}t} e^{\frac{j}{2}t} - \frac{1}{2j} e^{-\frac{1}{2}t} e^{-\frac{j}{2}t} - 0.31e^{-\frac{1}{2}t} e^{\frac{j}{2}t}\right.$$

$$+ j\, 0.218e^{-\frac{1}{2}t} e^{\frac{j}{2}t} - 0.31e^{-\frac{1}{2}t} e^{-\frac{j}{2}t}$$

$$\left. - j\, 0.218\, e^{-\frac{1}{2}t} e^{-\frac{j}{2}t}\right] u(t)$$

$$+ \left[-\frac{1}{2j} e^{-\frac{1}{2}\left(t-\frac{1}{2}\right)} e^{\frac{j}{2}\left(t-\frac{1}{2}\right)} + \frac{1}{2j} e^{-\frac{1}{2}\left(t-\frac{1}{2}\right)} e^{-\frac{j}{2}\left(t-\frac{1}{2}\right)}\right] u\left(t-\frac{1}{2}\right)$$

$$+ \left[0.031e^{-\frac{1}{2}(t-1)} e^{\frac{j}{2}(t-1)} - j\, 0.218e^{-\frac{1}{2}(t-1)} e^{\frac{j}{2}(t-1)}\right.$$

$$\left. + 0.31e^{-\frac{1}{2}(t-1)} e^{-\frac{j}{2}(t-1)} + j\, 0.218e^{-\frac{1}{2}(t-1)} e^{-\frac{j}{2}(t-1)}\right]$$

$u(t-1)$

Making use of the identities,

$$\frac{e^{jt} - e^{-jt}}{2j} = \sin t$$

$$\frac{e^{jt} + e^{-jt}}{2} = \cos t \quad \text{allows us to write,}$$

$$v_s(t) = \left[0.564e^{-\frac{1}{2}t} \sin \tfrac{1}{2}t - 0.062e^{-\frac{1}{2}t} \cos \tfrac{1}{2}t\right] u(t)$$

$$- \left[e^{-\frac{1}{2}\left(t-\frac{1}{2}\right)} \sin \frac{1}{2}\left(t-\frac{1}{2}\right) \right] u\left(t-\frac{1}{2}\right) + \left[0.436e^{-\frac{1}{2}(t-1)} \sin \frac{1}{2}(t-1) \right.$$

$$\left. + 0.062e^{-\frac{1}{2}(t-1)} \cos \frac{1}{2}(t-1) \right] u(t-1).$$

The inverse Laplace transform of the transient response

$$V_t(s) = \frac{A}{s + \frac{1}{2} - \frac{j}{2}} + \frac{A^*}{s + \frac{1}{2} + \frac{j}{2}}$$

is found similarly. Hence,

$$v_t(t) = \left[0.436e^{-\frac{1}{2}t} \sin \frac{1}{2}t + 0.062e^{-\frac{1}{2}t} \cos \frac{1}{2}t \right] u(t) \text{ V}$$

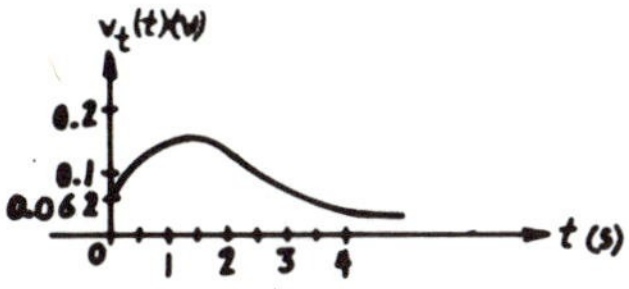

FIG. 4

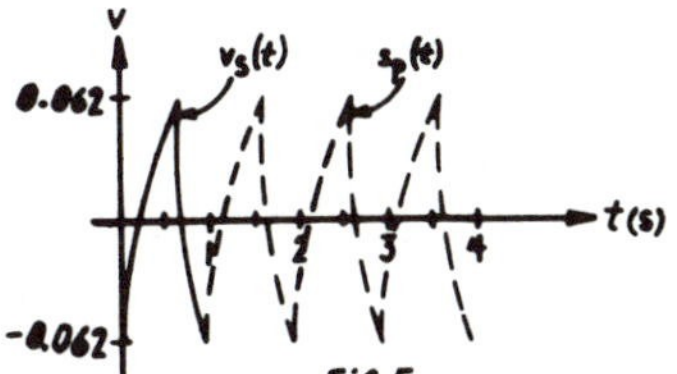

FIG. 5

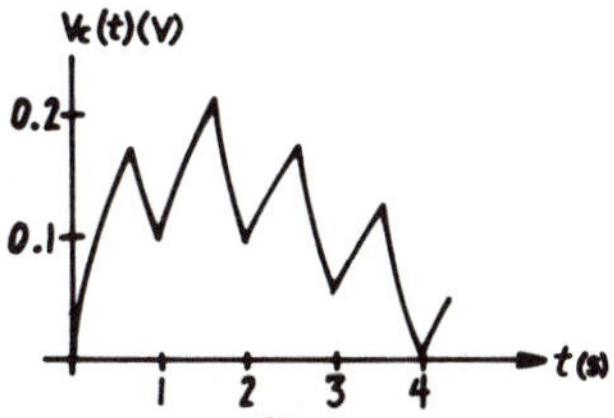

FIG. 6

Fig. 4 shows the transient response of the circuit. Fig. 5 shows the periodic response, and Fig. 6 shows the total response.

• PROBLEM 12-37

Use Laplace transform techniques to solve for $i_1(t)$ in the circuit of Fig. 1. Assume the initial conditions $v_C(0) = 1$, $i_1(0) = 0$ and $i_2(0) = 0$. Sketch $i_1(t)$.

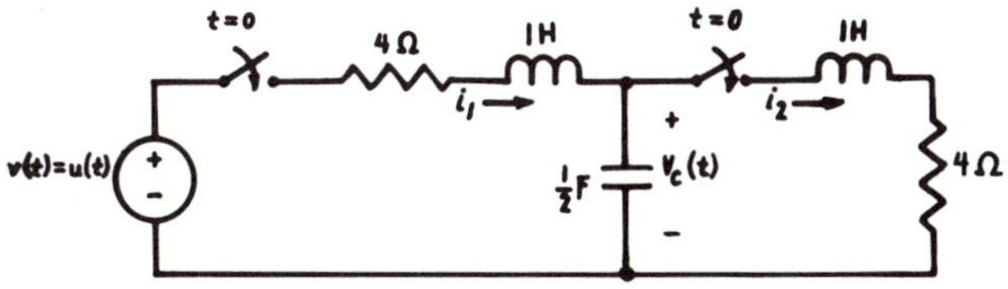

FIG. 1

Solution: First, transform the circuit to the frequency domain by use of Laplace transforms. Fig. 2 shows the transformed circuit.

To account for the initial condition $v_C(0) = 1$, note that

$$v_C(t) = v_C(0) + \frac{1}{C}\int_0^t i_C(t)\,dt$$

can be Laplace transformed to

$$V_C(s) = \frac{V_C(0)}{s} + \frac{1}{sC} I_C(s).$$

This in turn, can be represented by a capacitor and initial condition generator shown in Fig. 3.

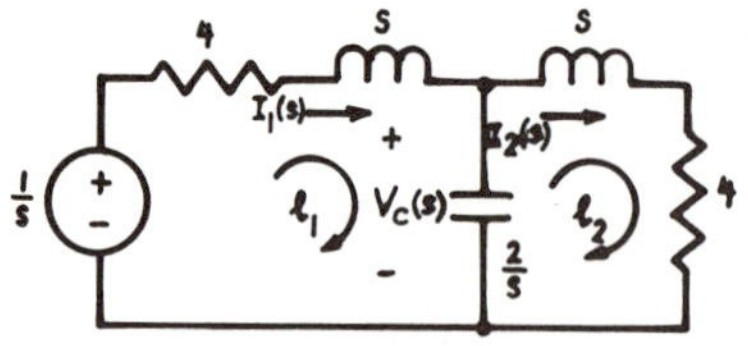

Fig. 2

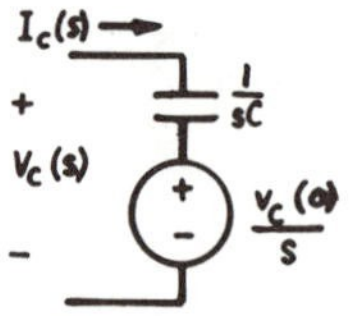

Fig. 3

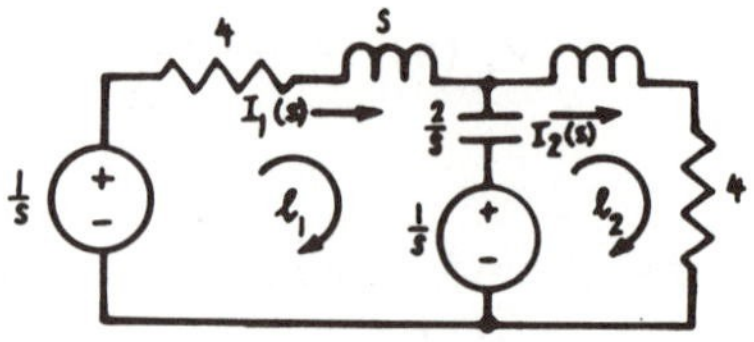

Fig. 4

Inserting this into the circuit of Fig. 2 yields a completely transformed circuit shown in Fig. 4.

Writing the two loop equation around the loops shown yields

$$\ell_1:\quad 0 = I_1(s)\left[4 + s + \frac{2}{s}\right] - I_2(s)\left[\frac{2}{s}\right]$$

In the loop equation, the sources in Fig. 4 $\frac{1}{s}$ and $\frac{1}{s}$ for loop 1 cancel. Therefore they do not appear.

$$\ell_2:\quad \frac{1}{s} = -I_1(s)\left[\frac{2}{s}\right] + I_2(s)\left[\frac{2}{s} + s + 4\right]$$

Writing the above equations in matrix form enables us to use Cramer's rule to solve for $I_1(s)$. Thus we can write

$$I_1(s) = \frac{\begin{vmatrix} 0 & -\frac{2}{s} \\ \frac{1}{s} & \frac{2}{s} + s + 4 \end{vmatrix}}{\begin{vmatrix} \frac{2}{s} + s + 4 & -\frac{2}{s} \\ -\frac{2}{s} & \frac{2}{s} + s + 4 \end{vmatrix}}$$

$$I_1(s) = \frac{\frac{2}{s^2}}{\left(\frac{2}{s} + s + 4\right)^2 - \frac{4}{s^2}}.$$

Manipulate $I_1(s)$ in order to write it in the rational polynomial form

$$F(s) = \frac{s^n + k_1 s^{n-1} + k_2 s^{n-2} + \ldots + k_n}{s^j + m_1 s^{j-1} + m_2 s^{j-2} + \ldots + m_j}$$

where $n < j$. This allows $I_1(s)$ to be expanded by partial fraction and inverse Laplace transformed to obtain $i_1(t)$. Hence,

$$I_1(s) = \frac{\frac{2}{s^2}}{\left(\frac{s^2 + 4s + 2}{s}\right)^2 - \frac{4}{s^2}}$$

$$I_1(s) = \frac{\frac{2}{s^2}}{\frac{(s^2 + 4s + 2)^2 - 4}{s^2}}$$

$$I_1(s) = \frac{\frac{2}{\not{s}^2}}{\frac{s^4 + 8s^3 + 20s^2 + 16s}{\not{s}^2}}$$

$$I_1(s) = \frac{2}{s\,(s^3 + 8s^2 + 20s + 16)}$$

In order to expand by partial fractions, it is necessary to factor $(s^3 + 8s^2 + 20s + 16)$. Remember, the cubic polynomial must have at least one real root.

In fact, it can be demonstrated that the cubic polynomial can be factored to

$$(s + 4)\ (s + 2)^2.$$

Hence, $$I_1(s) = \frac{2}{s\ (s + 4)\ (s + 2)^2}.$$

This can be expanded by partial fractions into the form,

$$I_1(s) = \frac{A}{s} + \frac{B}{s + 4} + \frac{C}{(s + 2)^2} + \frac{D}{s + 2}$$

where $$A = sI_1(s)\Big|_{s=0} = \frac{2}{(s + 4)(s + 2)^2}\Big|_{s=0} = \frac{1}{8}$$

$$B = (s + 4)\ I_1(s)\Big|_{s=-4} = \frac{2}{s(s+2)^2}\Big|_{s=-4} = -\frac{1}{8}$$

$$C = (s + 2)^2\ I_1(s)\Big|_{s=-2} = \frac{2}{s(s+4)}\Big|_{s=-2} = -\frac{1}{2}$$

$$D = \frac{1}{(2-1)!}\ \frac{d}{ds}\left[(s+2)^2\ I_1(s)\right]\Big|_{s=-2}$$

$$D = \frac{-2\ (2s+4)}{s^4 + 8s^3 + 16s^2}\Big|_{s=-2} = 0$$

Hence, $$I_1(s) = \frac{\frac{1}{8}}{s} - \frac{\frac{1}{8}}{s+4} - \frac{\frac{1}{2}}{(s+2)^2}$$

is inverse Laplace transformed, term-by-term, to obtain

$$i_1(t) = \frac{1}{8} - \frac{1}{8}\,e^{-4t} - \frac{1}{2}\,te^{-2t}.$$

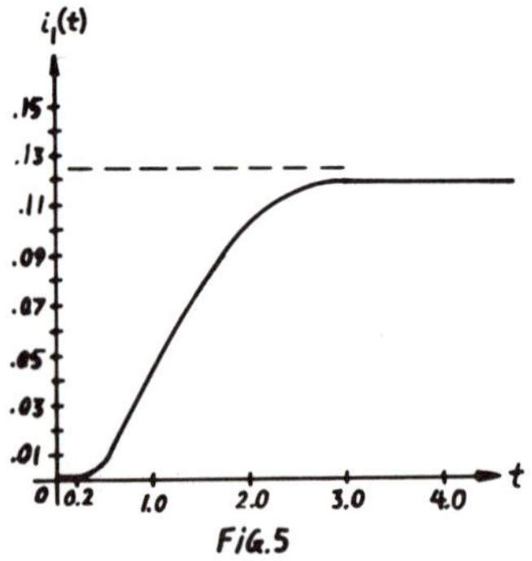

FIG.5

A sketch of $i_1(t)$ is shown in Fig. 5.

• PROBLEM 12-38

In the circuit of Fig. 1, $i_1(0^-) = 4A$, $i_2(0^-) = 2A$, and $v_{ab}(0^-) = 10V$. The source is given by $v(t) = 10 \sin 4t$. Solve for $i_2(t)$ for $t \geq 0^+$, using mesh analysis.

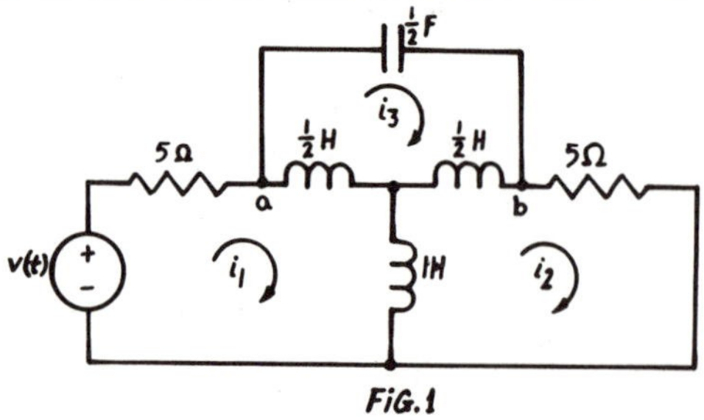

FIG.1

Solution: By transformation of the initial - condition sources, illustrated in Figs. 2 and 3 for the capacitor and inductors, we can find the Laplace transform of the circuit in Fig. 1.

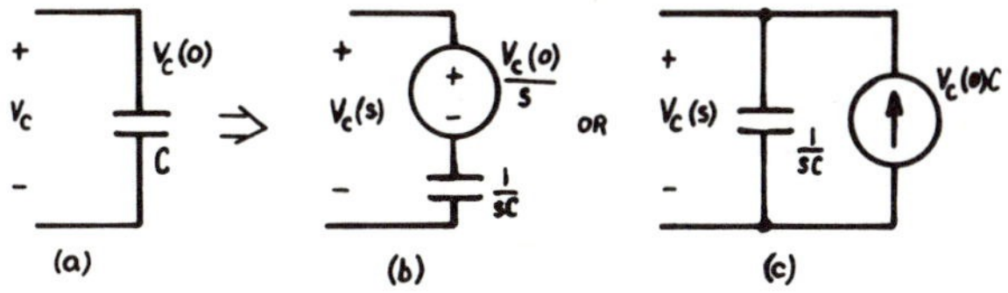

FiG.2

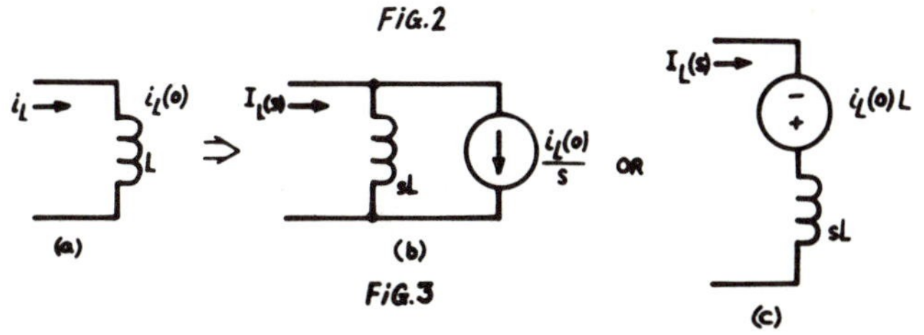

FiG.3

The complete transformed circuit is shown in Fig. 4. Note that the initial condition source transformations used for the capacitor and inductors were those illustrated in Figs. 2(a) and 3(c).

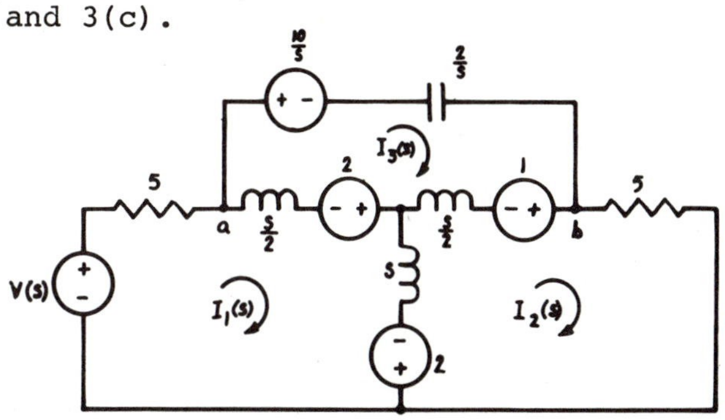

FiG.4

The entire circuit was transformed to include only passive elements and voltage sources allowing the solution to be obtained from mesh analysis.

Using KVL around each loop we can write the three equations:

$$\ell_1:\quad I_1(s)\left[5+\frac{3}{2}s\right] - I_2[s] - I_3\left[\frac{s}{2}\right] = V(s) + 4$$

$$\ell_2:\quad -I_1(s)\,[s] + I_2\left[5+\frac{3}{2}s\right] - I_3\left[\frac{s}{2}\right] = -1$$

$$\ell_3:\quad -I_1(s)\left[\frac{s}{2}\right] - I_2\left[\frac{s}{2}\right] + I_3\left[s+\frac{2}{s}\right] = -3-\frac{10}{s}$$

Writing the loop equations in matrix form yields,

$$\begin{bmatrix} V(s)+4 \\ -1 \\ -3-\frac{10}{s} \end{bmatrix} = \begin{bmatrix} s+\frac{3}{2}s & -s & -\frac{s}{2} \\ -s & 5+\frac{3}{2}s & -\frac{s}{2} \\ -\frac{s}{2} & -\frac{s}{2} & s+\frac{2}{s} \end{bmatrix} \begin{bmatrix} I_1(s) \\ I_2(s) \\ I_3(s) \end{bmatrix}$$

for which the solution of $I_2(s)$ can be written directly in determinate form as

$$I_2(s) = \frac{\begin{vmatrix} 5 + \frac{3}{2}s & V(s) + 4 & -\frac{s}{2} \\ -s & -1 & -\frac{s}{2} \\ -\frac{s}{2} & -3 - \frac{10}{s} & s + \frac{2}{s} \end{vmatrix}}{\begin{vmatrix} 5 + \frac{3}{2}s & -s & -\frac{s}{2} \\ -s & 5 + \frac{3}{2}s & -\frac{s}{2} \\ -\frac{s}{2} & -\frac{s}{2} & s + \frac{2}{s} \end{vmatrix}}$$

Evaluating the determinants

$$I_2(s) = \frac{-(V(s) + 4)\left[(-s)\left(s + \frac{2}{s}\right) - \frac{s^2}{4}\right] - \left[\left(5 + \frac{3}{2}s\right)\left(s + \frac{2}{s}\right) - \frac{s^2}{4}\right]}{-(-s)\left[(-s)\left(s + \frac{2}{s}\right) - \frac{s^2}{4}\right] + \left(5 + \frac{3}{2}s\right)\left[\left(5 + \frac{3}{2}s\right)\left(s + \frac{2}{s}\right) - \frac{s^2}{4}\right]}$$

$$\frac{-\left(-3 - \frac{10}{s}\right)\left[\left(5 + \frac{3}{2}s\right)\left(-\frac{s}{2}\right) - \frac{s^2}{2}\right]}{-\left(-\frac{s}{2}\right)\left[\left(5 + \frac{3}{2}s\right)\left(-\frac{s}{2}\right) - \frac{s^2}{2}\right]}$$

$$I_2(s) = \frac{-V(s) - 4\left[-\frac{5}{4}s^2 - 2\right] - \left[\frac{5}{4}s^2 + 5s + 3 + \frac{10}{s}\right]}{s\left[-\frac{5}{4}s^2 - 2\right] + \left(5 + \frac{3}{2}s\right)\left[\frac{5}{4}s^2 + 5s + 3 + \frac{10}{s}\right]}$$

$$\frac{3 + \frac{10}{s}\left[-\frac{5}{4}s^2 - \frac{5}{2}s\right]}{\frac{s}{2}\left[-\frac{5}{4}s^2 - \frac{5}{2}s\right]}$$

$$I_2(s) = \frac{-25s - 20 - \frac{10}{s} + V(s)\left(\frac{5}{4}s^2 + 2\right)}{\frac{25}{2}s^2 + \frac{55}{2}s + 30 + \frac{50}{s}}$$

and substituting

$$V(s) = L\{v(t)\} = L\{10 \sin 4t\} = \frac{40}{(s^2 + 16)}$$

yields,

$$I_2(s) = \frac{-\,25s - 20 - \dfrac{10}{s} + \dfrac{50s^2 + 80}{s^2 + 16}}{\dfrac{25}{2}s^2 + \dfrac{55}{2}s + 30 + \dfrac{50}{s}}$$

$$I_2(s) = \frac{-\,25s^2 - 20s - 10 + \dfrac{50s^3 + 80s}{s^2 + 16}}{\dfrac{25}{2}s^3 + \dfrac{55}{2}s^2 + 30s + 50}$$

$$I_2(s) = \frac{2(25s^2 + 20s + 10)\ (-\,s^2 - 16) + 100s^3 + 160s}{(s^2 + 16)\ (25s^3 + 55s^2 + 60s + 100)}$$

$$I_2(s) = \frac{-\,50s^4 + 60s^3 - 820s^2 - 480s - 160}{(s^2 + 16)\ (25s^3 + 55s^2 + 60s + 100)}$$

$$I_2(s) = \frac{-\,2s^4 + 2.4s^3 - 32.8s^2 - 19.2s - 6.4}{(s^2 + 16)\ (s^3 + 2.2s^2 + 2.4s + 4)}$$

$$I_2(s) = \frac{-\,2s^4 + 2.4s^3 - 32.8s^2 - 19.2s - 6.4}{(s^2 + 16)^2\ (s + 2)\ (s^2 + 0.2s + 2)}$$

Using partial fraction expansion we obtain

$$I_2(s) = \frac{As + B}{s^2 + 16} + \frac{C}{s + 2} + \frac{Ds + E}{(s + 0.1)^2 + (1.41)^2}$$

By evaluating the above coefficients we can find the inverse transform. Hence,

$$i_2(t) = A\cos 4t + B\sin 4t + Ce^{-2t}$$

$$De^{-0.1t}\cos 1.41t + \left(\frac{E - D(0.1)}{1.41}\right) \cdot$$

$$e^{-0.1t}\sin 1.41t.$$

For simultaneous linear equations can be otained for solving the coefficients A through E. Written in matrix form they are,

$$\begin{bmatrix} 1 & 0 & 1 & 1 & 0 \\ 2.2 & 1 & 0.2 & 2 & 1 \\ 2.4 & 2.2 & 18 & 16 & 2 \\ 4 & 2.4 & 3.2 & 32 & 16 \\ 0 & 4 & 32 & 0 & 32 \end{bmatrix} \begin{bmatrix} A \\ B \\ C \\ D \\ E \end{bmatrix} \doteq \begin{bmatrix} -\,2 \\ 2.4 \\ -\,32.8 \\ -\,19.2 \\ -\,6.4 \end{bmatrix}$$

From which we find

$A = 0.48$, $B = 3.14$, $C = -1.34$, $D = -1.14$ and $E = 0.75$.

Hence,

$$i_2(t) = 0.48 \cos 4t + 3.14 \sin 4t - 1.34e^{-2t}$$

$$- 1.14e^{-0.1t} \cos 1.41t + 0.61e^{-0.1t} \sin 1.41t \text{ A.}$$

• PROBLEM 12-39

> Calculate $v_1(t)$ and $v_2(t)$ in the circuit shown in Fig. 1, assuming zero initial conditions and $R_1 = R_2 = R_3 = 1\Omega$, $C = 2F$, and $L = 1H$. Assume $v_a(t) = v_b(t) = u_0(t)$.

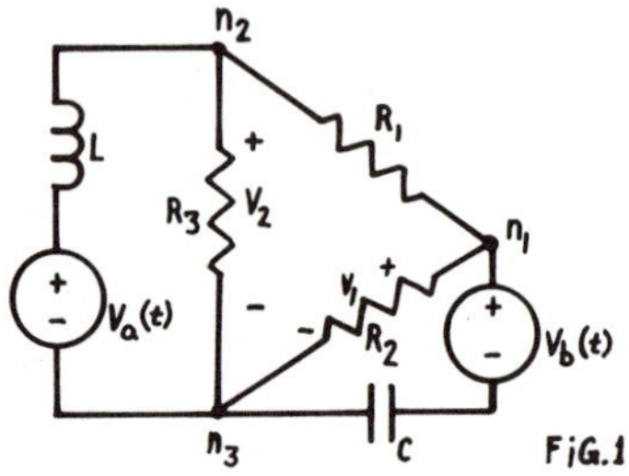

FiG.1

Solution: In order to perform nodal analysis, convert the voltage sources to current sources as illustrated in Fig. 2.

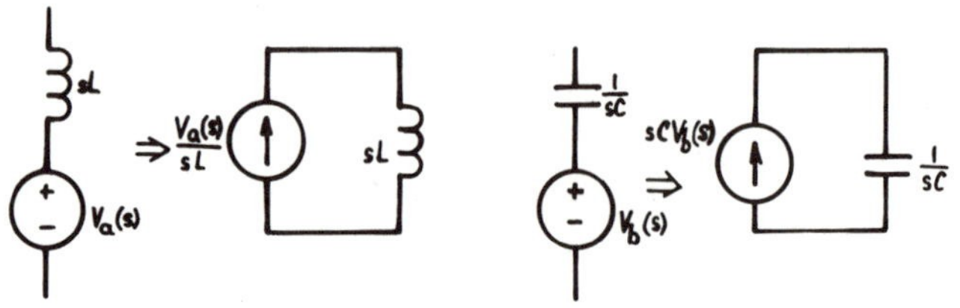

FiG.2

The equivalent transform network, with sources converted, is drawn in Fig. 3.

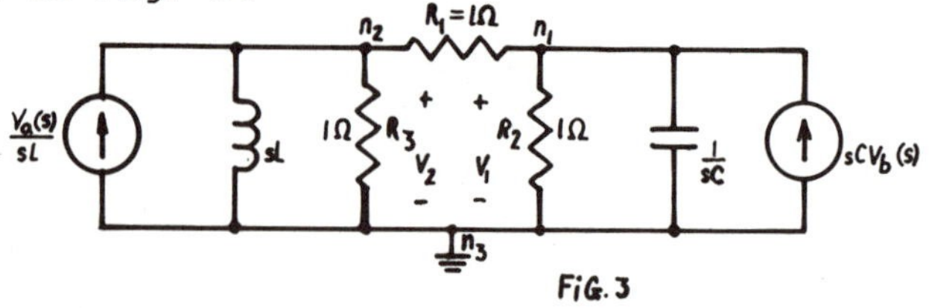

FiG.3

Writing KCL equations for nodes n_1 and n_2

$$n_1: \quad \frac{V_1 - V_2}{1} + \frac{V_1}{1} + \frac{V_1}{\frac{1}{2s}} = 2sV_b(s).$$

$$n_2: \quad \frac{V_2 - V_1}{1} + \frac{V_2}{1} + \frac{V_2}{s} = \frac{V_a(s)}{s}$$

combining terms yields

$$V_1(2 + 2s) + V_2\ (-1) = 2sV_b(s) \tag{1}$$

$$V_1\ (-1) + V_2\left(2 + \frac{1}{s}\right) = \frac{V_a(s)}{s} \tag{2}$$

$V_a = V_B = \frac{1}{s}$; we can write Eqs. (1) and (2) in matrix form, giving

$$\begin{bmatrix} 2 + 2s & -1 \\ -1 & s + \frac{1}{s} \end{bmatrix} \begin{bmatrix} V_1(s) \\ V_2(s) \end{bmatrix} = \begin{bmatrix} 2 \\ \frac{1}{s^2} \end{bmatrix}$$

solving for V_1 and V_2

$$V_1 = \frac{\begin{vmatrix} 2 & -1 \\ \frac{1}{s^2} & 2 + \frac{1}{s} \end{vmatrix}}{\begin{vmatrix} 2 + 2s & -1 \\ -1 & 2 + \frac{1}{s} \end{vmatrix}} = \frac{2\left(2 + \frac{1}{s}\right) - (-1)\left(\frac{1}{s^2}\right)}{(2 + 2s)\left(2 + \frac{1}{s}\right) - 1}$$

$$V_1(s) = \frac{4 + \frac{2}{s} + \frac{1}{s^2}}{4 + 4s + \frac{2}{s} + 2 - 1} = \frac{4s^2 + 2s + 1}{4s^3 + 5s^2 + 2s}$$

$$V_2 = \frac{\begin{vmatrix} 2 + 2s & 2 \\ -1 & \frac{1}{s^2} \end{vmatrix}}{\begin{vmatrix} 2 + 2s & -1 \\ -1 & 2 + \frac{1}{s} \end{vmatrix}} = \frac{\left[(2 + 2s)\frac{1}{s^2} - (-1)(2)\right]s^2}{4s^3 + 5s^2 + 2s}$$

$$V_2(s) = \frac{s^2\left[\frac{2}{s^2} + \frac{2}{s} + 2\right]}{4s^3 + 5s^2 + 2s} = \frac{2s^2 + 2s + 2}{4s^3 + 5s^2 + 2s}$$

In the expressions for $V_1(s)$ and $V_2(s)$, divide the numerator and the denominator by 4 and factor by solving each of the quadratic expressions:

$$V_1(s) = \frac{s^2 + \frac{1}{2}s + \frac{1}{4}}{s\left(s^2 + \frac{5}{4}s + \frac{1}{2}\right)}$$

$$= \frac{s^2 + \frac{1}{2}s + \frac{1}{4}}{s\left(s + \frac{5}{8} - j\frac{\sqrt{7}}{8}\right)\left(s + \frac{5}{8} + j\frac{\sqrt{7}}{8}\right)}$$

$$V_2(s) = \frac{\frac{1}{2}s^2 + \frac{1}{2}s + \frac{1}{2}}{s\left(s^2 + \frac{5}{4}s + \frac{1}{2}\right)}$$

$$V_2(s) = \frac{\frac{1}{2}s^2 + \frac{1}{2}s + \frac{1}{2}}{s\left(s + \frac{5}{8} - j\frac{\sqrt{7}}{8}\right)\left(s + \frac{5}{8} + j\frac{\sqrt{7}}{8}\right)}$$

Using partial fraction expansion yields

$$V_1(s) = \frac{A}{s} + \frac{B}{\left(s + \frac{5}{8} - j\frac{\sqrt{7}}{8}\right)} + \frac{C}{\left(s + \frac{5}{8} + j\frac{\sqrt{7}}{8}\right)} \quad (3)$$

$$V_2(s) = \frac{D}{s} + \frac{E}{\left(s + \frac{5}{8} - j\frac{\sqrt{7}}{8}\right)} + \frac{F}{\left(s + \frac{5}{8} + j\frac{\sqrt{7}}{8}\right)} \quad (4)$$

where,

$$A = s(V_1(s))\Big|_{s=0} = \frac{\frac{1}{4}}{\frac{1}{2}} = \frac{1}{2}$$

$$B = \left(s + \frac{5}{8} - j\frac{\sqrt{7}}{8}\right) V_1(s)\Big|_{s = -\frac{5}{8} + j\frac{\sqrt{7}}{8}}$$

$$B = \frac{\left(-\frac{5}{8} + j\frac{\sqrt{7}}{8}\right)^2 + \frac{1}{2}\left(-\frac{5}{8} + j\frac{\sqrt{7}}{8}\right) + \frac{1}{4}}{\left(-\frac{5}{8} + j\frac{\sqrt{7}}{8}\right)\left(j\frac{2\sqrt{7}}{8}\right)}$$

$$B = \frac{\frac{18}{64} - j\frac{10\sqrt{7}}{64} - \frac{20}{64} + j\frac{4\sqrt{7}}{64} + \frac{16}{64}}{-\frac{14}{64} - j\frac{10\sqrt{7}}{64}}$$

$$B = \frac{\frac{14}{64} - j\frac{6\sqrt{7}}{64}}{-\frac{14}{64} - j\frac{10\sqrt{7}}{64}}$$; multiply through by 64 gives

$$B = \frac{14 - j\,6\sqrt{7}}{-14 - j\,10\sqrt{7}}; \text{ converting to polar form:}$$

$$B = \frac{8\sqrt{7}\;\underline{/\tan^{-1} - \frac{3}{\sqrt{7}}}}{\sqrt{128}\sqrt{7}\;\underline{/\left[\tan^{-1}\frac{5}{\sqrt{7}}\right] + 180°}} = \frac{8\sqrt{7}\;\underline{/\tan^{-1} - \frac{3}{\sqrt{7}}}}{8\sqrt{2}\sqrt{7}\;\underline{/\left[\tan^{-1}\frac{5}{\sqrt{7}}\right] + 180°}}$$

$$B = \frac{1}{\sqrt{2}}\;\underline{/\tan^{-1} - \frac{3}{\sqrt{7}} - \tan^{-1}\frac{5}{\sqrt{7}} - 180°}$$

$$B = \frac{1}{\sqrt{2}}\;\underline{/\theta} = \frac{1}{\sqrt{2}}\cos\theta + j\,\frac{1}{\sqrt{2}}\sin\theta$$

$$\theta = \tan^{-1} - \frac{3}{\sqrt{7}} - \tan^{-1}\frac{5}{\sqrt{7}} - 180°$$

we can add 360° to θ without changing the angle

$$\theta = \tan^{-1} - \frac{3}{\sqrt{7}} - \tan^{-1}\frac{5}{\sqrt{7}} - 180° + 360°$$

$$\theta = \tan^{-1} - \frac{3}{\sqrt{7}} - \tan^{-1}\left(\frac{5}{\sqrt{7}}\right) + 180° \qquad (5)$$

Using an inverse trigonometric identity obtains;

$$\tan^{-1}\left(-\frac{3}{\sqrt{7}}\right) = -90° - \tan^{-1}\left(-\frac{\sqrt{7}}{3}\right)$$

$$\text{and} \quad -\tan^{-1}\frac{5}{\sqrt{7}} = -\left(90° - \tan^{-1}\left(\frac{\sqrt{7}}{5}\right)\right)$$

Substituting back into Eq. (5) gives

$$\theta = -90° - \tan^{-1}\left(-\frac{\sqrt{7}}{3}\right) - 90° + \tan^{-1}\left(\frac{\sqrt{7}}{5}\right) + 180°$$

$$\theta = \tan^{-1}\left(\frac{\sqrt{7}}{5}\right) - \tan^{-1}\left(-\frac{\sqrt{7}}{3}\right)$$

Allowing $\tan^{-1}\left(\frac{\sqrt{7}}{5}\right) = \alpha$

and $\quad \tan^{-1}\left(-\frac{\sqrt{7}}{3}\right) = \beta,$

then, $\theta = \alpha - \beta$

and $\beta = \frac{1}{\sqrt{2}} \cos(\alpha - \beta) + j \frac{1}{\sqrt{2}} \sin(\alpha - \beta).$

From trigonometry:

if $\alpha = \tan^{-1} x$ then $\sin \alpha = \frac{x}{\sqrt{1 + x^2}}$;

$$\cos \alpha = \frac{1}{\sqrt{1 + x^2}}$$

and the angle difference relations are

$$\sin(\alpha - \beta) = \sin \alpha \cos \beta - \cos \alpha \sin \beta$$

$$\cos(\alpha - \beta) = \cos \alpha \cos \beta + \sin \alpha \sin \beta$$

Now, substitute the identities for inverse trignometric functions into the angle difference relations.

$$\sin(\alpha - \beta) = \frac{x}{\sqrt{1 + x^2}} \frac{1}{\sqrt{1 + y^2}} - \frac{1}{\sqrt{1 + x^2}} \frac{y}{\sqrt{1 + y^2}}$$

simplifying gives

$$\sin(\alpha - \beta) = \frac{x - y}{\sqrt{x^2y^2 + x^2 + y^2 + 1}}$$

where $x = \tan \alpha = \frac{\sqrt{7}}{5}$

$y = \tan \beta = -\frac{\sqrt{7}}{3}$

$$\sin(\alpha - \beta) = \frac{\frac{\sqrt{7}}{5} + \frac{\sqrt{7}}{3}}{\sqrt{\frac{7^2}{25.9} + \frac{7}{25} + \frac{7}{9} + 1}}$$

$$\sin(\alpha - \beta) = \frac{\frac{3\sqrt{7} + 5\sqrt{7}}{15}}{\sqrt{\frac{7^2 + 9.7 + 25.7 + 25.9}{25.9}}} = \frac{\frac{3\sqrt{7} + 5\sqrt{7}}{15}}{\frac{\sqrt{512}}{\sqrt{5^2 \cdot 3^2}}}$$

$$\sin(\alpha - \beta) = \frac{3\sqrt{7} + 5\sqrt{7}}{\sqrt{512}} = \frac{3\sqrt{7} + 5\sqrt{7}}{\sqrt{64}\cdot\sqrt{8}} = \frac{8\sqrt{7}}{8\sqrt{8}}$$

$$\sin(\alpha - \beta) = \frac{\sqrt{7}}{\sqrt{8}}$$

$$\cos(\alpha - \beta) = \frac{1}{\sqrt{1 + x^2}}\,\frac{1}{\sqrt{1 + y^2}} + \frac{x}{\sqrt{1 + x^2}}\,\frac{y}{\sqrt{1 + y^2}}$$

simplifying,

$$\cos(\alpha - \beta) = \frac{1 + xy}{\sqrt{x^2y^2 + x^2 + y^2 + 1}}$$

$$\cos(\alpha - \beta) = \frac{1 - \frac{\sqrt{7^2}}{15}}{\frac{\sqrt{512}}{15}} = \frac{1 - \frac{7}{15}}{\frac{\sqrt{512}}{15}}$$

$$\cos(\alpha - \beta) = \frac{\frac{8}{15}}{\frac{\sqrt{512}}{15}} = \frac{8}{\sqrt{512}} = \frac{8}{8\sqrt{8}} = \frac{1}{\sqrt{8}}$$

Now, find B in cartesian form exactly:

$$B = \frac{1}{\sqrt{2}} \cos(\alpha - \beta) + j\,\frac{1}{\sqrt{2}} \sin(\alpha - \beta)$$

$$B = \frac{1}{\sqrt{2}}\,\frac{1}{\sqrt{8}} + j\,\frac{1}{\sqrt{2}}\,\frac{\sqrt{7}}{\sqrt{8}}$$

$$B = \frac{1}{\sqrt{16}} + j\,\frac{\sqrt{7}}{\sqrt{16}}$$

$$B = \frac{1}{4} + j\,\frac{\sqrt{7}}{4}$$

Using partial fraction methods, find the remaining values for C, D, E and F using the methods used for A and B above

$$C = \left(s + \frac{5}{8} + j\,\frac{\sqrt{7}}{8}\right) V_1(s)\Bigg|_{s= -\frac{5}{8} - j\frac{\sqrt{7}}{8}}$$

$$C = \frac{\left(-\frac{5}{8} - j\,\frac{\sqrt{7}}{2}\right)^2 + \frac{1}{2}\left(-\frac{5}{8} - j\,\frac{\sqrt{7}}{8}\right) + \frac{1}{4}}{\left(-\frac{5}{8} - j\,\frac{\sqrt{7}}{8}\right)\left(-j\,\frac{2\sqrt{7}}{8}\right)}$$

$$C = \frac{\frac{18}{64} + j\,\frac{10\sqrt{7}}{64} - \frac{20}{64} - j\,\frac{4\sqrt{7}}{64} + \frac{16}{64}}{-\frac{14}{64} + j\,\frac{10\sqrt{7}}{64}}$$

$$C = \frac{\frac{14}{64} - \frac{6\sqrt{7}}{64}}{-\frac{14}{64} + j\,\frac{10\sqrt{7}}{64}} = \frac{14 + j\,6\sqrt{7}}{-14 + j\,10\sqrt{7}}$$

$$C = \frac{8\sqrt{7}\;\angle \tan^{-1}\frac{3}{\sqrt{7}}}{\sqrt{128}\sqrt{7}\;\angle\left[\tan^{-1} - \frac{5}{\sqrt{7}}\right] - 180°}$$

$$C = \frac{1}{\sqrt{2}}\angle\; \tan^{-1}\frac{3}{\sqrt{7}} - \left[\tan^{-1} - \frac{5}{\sqrt{7}}\right] + 180°$$

$$C = \frac{1}{4} - j\,\frac{\sqrt{7}}{4}$$

$$D = s\,(V_2(s))\Big|_{s=0} = \frac{\frac{1}{2}}{\frac{1}{2}} = 1$$

$$E = \left(s + \frac{5}{8} - j\,\frac{\sqrt{7}}{8}\right)\Big|_{s=\,-\frac{5}{8} + j\,\frac{\sqrt{7}}{8}}$$

$$E = \frac{\frac{1}{2}\left(-\frac{5}{8} + j\,\frac{\sqrt{7}}{8}\right)^2 + \frac{1}{2}\left(-\frac{5}{8} + j\,\frac{\sqrt{7}}{8}\right) + \frac{1}{2}}{\left(-\frac{5}{8} + \frac{\sqrt{7}}{8}\right)\left(j\,\frac{2\sqrt{7}}{8}\right)}$$

$$E = \frac{\frac{9}{16} - j\,\frac{5\sqrt{7}}{64} - \frac{20}{64} + j\,\frac{4\sqrt{7}}{64} + \frac{32}{64}}{-\frac{14}{64} - j\,\frac{10\sqrt{7}}{64}}$$

$$E = \frac{\frac{21}{64} - j\,\frac{\sqrt{7}}{64}}{-\frac{14}{64} - j\,\frac{10\sqrt{7}}{64}} = \frac{21 - j\,\sqrt{7}}{-14 - j\,10\sqrt{7}}$$

$$E = \frac{8\sqrt{7}\;\angle \tan^{-1} - \frac{1}{3\sqrt{7}}}{\sqrt{128}\sqrt{7}\;\angle\left(\tan^{-1}\frac{5}{\sqrt{7}}\right) + 180°}$$

$$E = \frac{1}{\sqrt{2}} \underline{\Big/ \tan^{-1} - \frac{1}{3\sqrt{7}} - \left(\tan^{-1} \frac{5}{\sqrt{7}}\right) - 180°}$$

$$E = -\frac{1}{4} + j\frac{\sqrt{7}}{4}$$

$$F = \left(s + \frac{5}{8} + j\frac{\sqrt{7}}{8}\right) V_2(s) \Bigg|_{s = -\frac{5}{8} - j\frac{\sqrt{7}}{8}}$$

$$F = \frac{\frac{1}{2}\left(-\frac{5}{8} - j\frac{\sqrt{7}}{8}\right)^2 + \frac{1}{2}\left(-\frac{5}{8} - j\frac{\sqrt{7}}{8}\right) + \frac{1}{2}}{-\frac{14}{64} + j\frac{10\sqrt{7}}{64}}$$

$$F = \frac{\frac{9}{64} + \frac{5\sqrt{7}}{64} - \frac{20}{64} - \frac{4\sqrt{7}}{64} + \frac{32}{64}}{-\frac{14}{64} + j\frac{10\sqrt{7}}{64}}$$

$$F = \frac{\frac{21}{64} + j\frac{\sqrt{7}}{64}}{-\frac{14}{64} + j\frac{10\sqrt{7}}{64}} = \frac{21 + j\sqrt{7}}{-14 + j10\sqrt{7}}$$

$$F = \frac{8\sqrt{7} \Big/ \tan^{-1} \frac{1}{3\sqrt{7}}}{\sqrt{128}\sqrt{7} \underline{\Big/ \left[\tan^{-1} - \frac{5}{\sqrt{7}}\right] - 180°}}$$

$$F = \frac{1}{\sqrt{2}} \underline{\Big/ \tan^{-1} \frac{1}{3\sqrt{7}} - \left[\tan^{-1} - \frac{5}{\sqrt{7}}\right] + 180°}$$

$$F = -\frac{1}{4} - j\frac{\sqrt{7}}{4}$$

Substituting A, B, C, D, E, and F into Eqs. (3) and (4) gives

$$V_1(s) = \frac{\frac{1}{2}}{s} + \frac{\frac{1}{4} + j\frac{\sqrt{7}}{4}}{\left(s + \frac{5}{8} - j\frac{\sqrt{7}}{8}\right)} + \frac{\frac{1}{4} - j\frac{\sqrt{7}}{4}}{\left(s + \frac{5}{8} + j\frac{\sqrt{7}}{8}\right)}$$

$$V_2(s) = \frac{1}{s} + \frac{-\frac{1}{4} + j\frac{\sqrt{7}}{4}}{\left(s + \frac{5}{8} - j\frac{\sqrt{7}}{8}\right)} + \frac{-\frac{1}{4} - j\frac{\sqrt{7}}{4}}{\left(s + \frac{5}{8} + j\frac{\sqrt{7}}{8}\right)}$$

Taking the inverse Laplace transform of the two equations above yields,

$$V_1(t) = \frac{1}{2} + \left(\frac{1}{4} + j\,\frac{\sqrt{7}}{4}\right) e^{t\left(-\frac{5}{8} + j\,\frac{\sqrt{7}}{8}\right)} + \left(\frac{1}{4} - j\,\frac{\sqrt{7}}{4}\right) e^{t\left(-\frac{5}{8} - j\,\frac{\sqrt{7}}{8}\right)}$$

$$V_2(t) = 1 + \left(-\frac{1}{4} + j\,\frac{\sqrt{7}}{4}\right) e^{t\left(-\frac{5}{8} + j\,\frac{\sqrt{7}}{8}\right)} + \left(-\frac{1}{4} - j\,\frac{\sqrt{7}}{4}\right) e^{t\left(-\frac{5}{8} - j\,\frac{\sqrt{7}}{8}\right)}$$

Multiplying out gives:

$$V_1(t) = \frac{1}{2} + \frac{1}{4}\left(e^{-\frac{5}{8}t}\, e^{j\frac{\sqrt{7}}{8}t}\right) + j\,\frac{\sqrt{7}}{4}\left(e^{-\frac{5}{8}t}\, e^{j\frac{\sqrt{7}}{8}t}\right) + \frac{1}{4}\left(e^{-\frac{5}{8}t}\, e^{-j\frac{\sqrt{7}}{8}t}\right) - j\,\frac{\sqrt{7}}{4}\left(e^{-\frac{5}{8}t}\, e^{-j\frac{\sqrt{7}}{8}t}\right) \qquad (6)$$

$$V_2(t) = 1 - \frac{1}{4}\left(e^{-\frac{5}{8}t}\, e^{j\frac{\sqrt{7}}{8}t}\right) + j\,\frac{\sqrt{7}}{4}\left(e^{-\frac{5}{8}t}\, e^{j\frac{\sqrt{7}}{8}t}\right) - \frac{1}{4}\left(e^{-\frac{5}{8}t}\, e^{-j\frac{\sqrt{7}}{8}t}\right) - j\,\frac{\sqrt{7}}{4}\left(e^{-\frac{5}{8}t}\, e^{-j\frac{\sqrt{7}}{8}t}\right). \qquad (7)$$

It is known that:

$$\sin \alpha t = \frac{e^{\alpha t} - e^{-\alpha t}}{2j}$$

$$\cos \alpha t = \frac{e^{\alpha t} + e^{-\alpha t}}{2}$$

Notice that Eqs. (6) and (7) can be written in this form:

$$V_1(t) = \frac{1}{2} + \frac{1}{2}e^{-\frac{5}{8}t}\left(\frac{e^{j\frac{\sqrt{7}}{8}t} + e^{-j\frac{\sqrt{7}}{8}t}}{2}\right) - \frac{\sqrt{7}}{2}e^{-\frac{5}{8}t}\left(\frac{e^{j\frac{\sqrt{7}}{8}t} - e^{-j\frac{\sqrt{7}}{8}}}{2j}\right)$$

$$V_2(t) = 1 - \frac{1}{2}e^{-\frac{5}{8}t}\left(\frac{e^{j\frac{\sqrt{7}}{8}t} + e^{-j\frac{\sqrt{7}}{8}t}}{2}\right)$$

$$-\frac{\sqrt{7}}{2}e^{-\frac{5}{8}t}\left(\frac{e^{j\frac{\sqrt{7}}{8}t} - e^{-j\frac{\sqrt{7}}{8}t}}{2j}\right)$$

Finally,

$$V_1(t) = \frac{1}{2}\left[1 + e^{-\frac{5}{8}t}\left(\cos\frac{\sqrt{7}}{8}t - \sqrt{7}\sin\frac{\sqrt{7}}{8}t\right)\right]u(t)$$

$$V_2(t) = \left[1 - \frac{1}{2}e^{-\frac{5}{8}t}\left(\cos\frac{\sqrt{7}}{8}t + \sqrt{7}\sin\frac{\sqrt{7}}{8}t\right)\right]u(t).$$

• PROBLEM 12-40

The circuit shown in Fig. 1 is in the steady state at $t = 0^-$ with the switch in position 1. At $t = 0$, the switch is thrown to position 2. Use Laplace transforms to formulate an expression for $v_C(s)$ for any source $v_x(s)$.

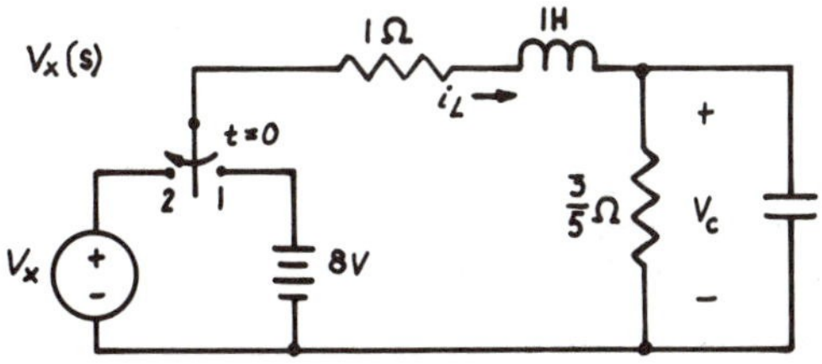

Fig. 1

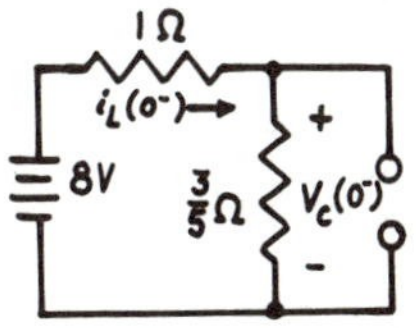

Fig. 2

Solution: Use Kirchhoff's Laws to find $V_C(s)$ in the tranformed circuit for $t > 0$. But first, find the two initial conditions, $i_L(0^-)$, and $v_C(0^-)$ for the circuit at $t = 0^-$ shown in Fig. 2.

Note that

$$i_L(0^-) = \frac{8}{1 + \frac{3}{5}} = 5 \text{ A}$$

$$v_C(0^-) = \left(\frac{8}{1 + \frac{3}{5}}\right)\frac{3}{5} = (5)\,\frac{3}{5} = 3 \text{ V},$$

and since, $i_L(0^-) = i_L(0^+)$ and $v_C(0^-) = v_C(0^+)$, we can add these inital conditions to our transformed circuit as independent current and voltage sources in parallel and series with the inductor and capacitor respectively.

Fig. 3 shows the transformed circuit

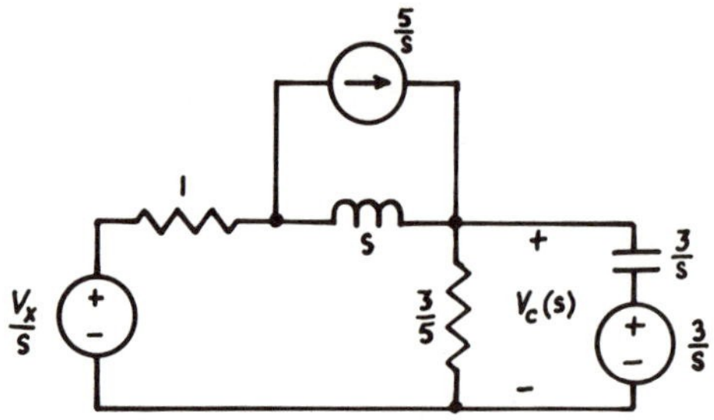

FIG.3

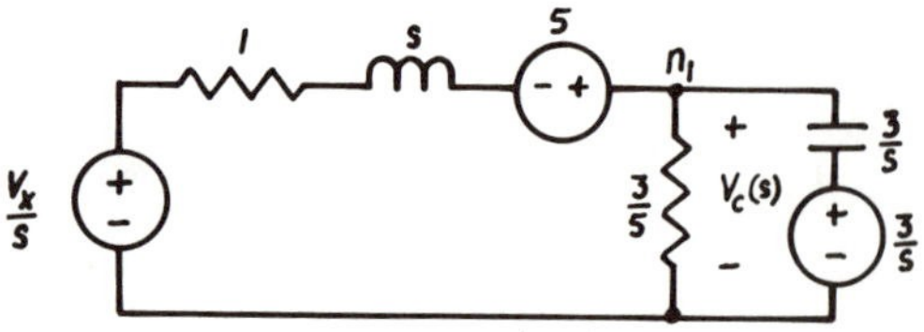

FIG.4

Making use of source transformation theory, transform the current source to a voltage source to give a circuit which is easier to work with.

The transformed current source is now a voltage source that is equivalent to a voltage spike in the time-domain equal to 5 $\delta(t)$.

Applying KCL to node n_1 in Fig. 4 gives the equation

$$\frac{\left[\frac{V_x}{s} + 5\right] - V_C}{1 + s} = \frac{V_C}{\frac{3}{5}} + \frac{V_C - \frac{3}{s}}{\frac{3}{s}}$$

Simplifying gives:

$$\frac{V_x + 5s}{s(s + 1)} - \frac{V_C}{(s + 1)} = \frac{5V_C}{3} + \frac{V_C s}{3} - 1.$$

Collecting terms gives:

$$V_C\left(\frac{5}{3} + \frac{s}{3} + \frac{1}{s + 1}\right) = \frac{V_x + 5s}{s(s + 1)} + 1$$

$$V_C\left(\frac{s^2 + 6s + 8}{3(s + 1)}\right) = \frac{s^2 + 6s + V_x}{s(s + 1)}$$

$$V_C(s) = \frac{3(s^2 + 6s + V_x)}{s(s^2 + 6s + 8)} = \frac{3s + 18 + 3\frac{V_x}{s}}{s^2 + 6s + 8}.$$

Since $V_x(s) = \frac{V_x}{s}$, substitute any source $V_x(s)$ into the expression

$$V_C(s) = \frac{3s + 18 + 3V_x(s)}{s^2 + 6s + 8}$$

to find $V_C(s)$.

Use Laplace transform techniques to find v(t) in the circuit of Fig. 1. At t = 0, switch (a) closes, and switch (b) opens.

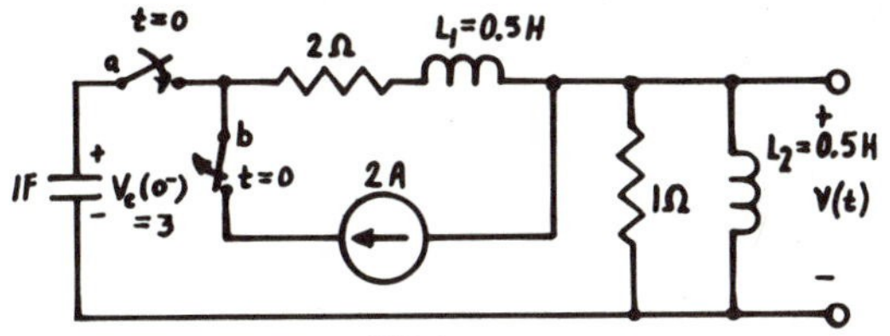

FIG.1

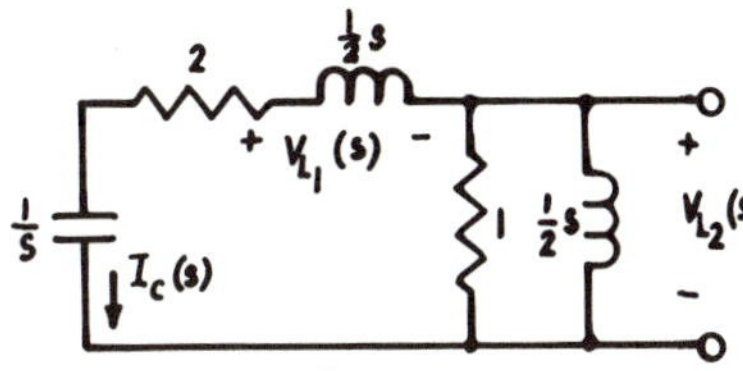

FIG.2

Solution: The information given about the circuit in Fig. 1 enables us to find the initial conditions for the capacitor and inductors.

The capacitor initial condition is given as $v_C(0^-) = 3V$. The current through inductor L_1 is $i_{L_1}(0^-) = 2A$. No current flows through L_2 at $t = 0^-$.

Now transform the circuit in Fig. 1 remembering that $V_{L_1}(s) = sL_1 I_{L_1}(s) - L_1 i_{L_1}(0^-)$, $V_{L_2}(s) = SL_2 I_{L_2}(s)$ and $I_C(s) = sCV_C(s) - Cv_C(0^-)$.

Fig. 2 shows the transformed circuit, since $V_{L_1}(s) = SL_1 I_2(s) - L_1 i_L(0^-)$ we can represent the inductor as shown in Fig. 3:

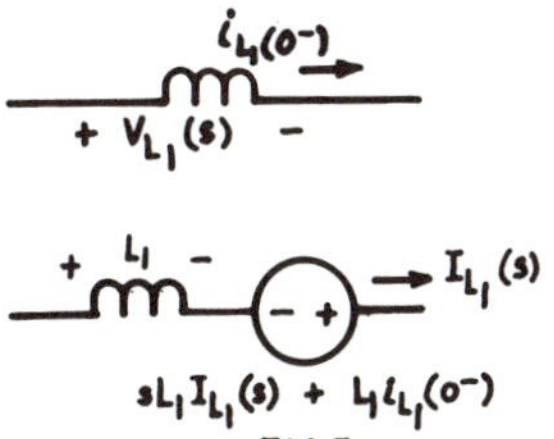

FIG.3

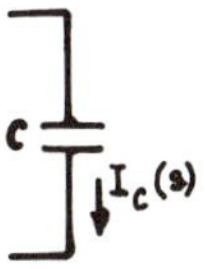

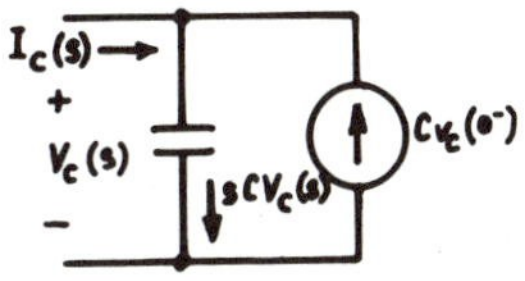

FIG.4

and the capacitor can be represented as:

By using source transformation theory, transform the parallel susceptance and current source in Fig. 4 to a series reactance and voltage source, as shown in Fig. 5.

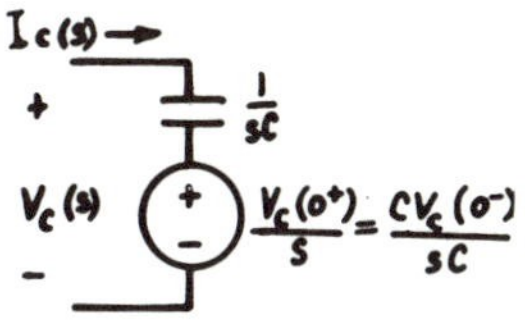

FIG.5

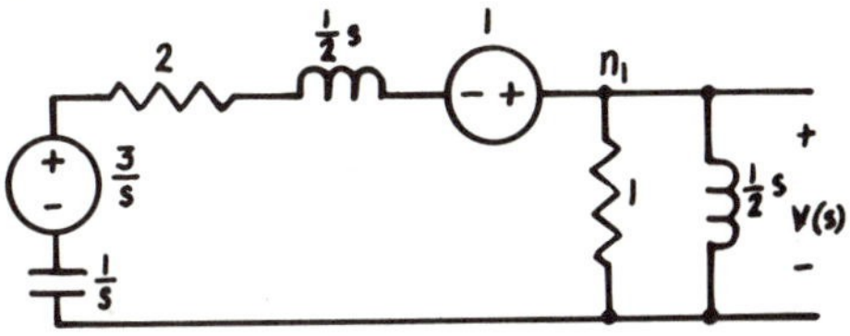

FIG.6

Using this to transform the circuit in Fig. 2;

Writing a KCL equation at node n_1 in Fig. 6 yields

$$\frac{V(s)}{\frac{s}{2}} + \frac{V(s)}{1} + \frac{V(s) - 1 - \frac{3}{s}}{\frac{s}{2} + 2 + \frac{1}{s}} = 0$$

collecting terms gives

$$V(s)\left[\frac{1}{\frac{s}{2}} + 1 + \frac{1}{\frac{s}{2} + 2 + \frac{1}{s}}\right] = \frac{1 + \frac{3}{s}}{\frac{s}{2} + 2 + \frac{1}{2}}$$

$$V(s) = \frac{\dfrac{1 + \frac{3}{s}}{\frac{s}{2} + 2 + \frac{1}{s}}}{\frac{2}{s} + 1 + \dfrac{1}{\frac{s}{2} + 2 + \frac{1}{s}}}$$

simplifying gives

$$V(s) = \frac{\dfrac{1 + \frac{3}{s}}{\frac{s}{2} + 2 + \frac{1}{s}}}{\dfrac{2\left(\frac{s}{2} + 2 + \frac{1}{s}\right) + s\left(\frac{s}{2} + 2 + \frac{1}{s}\right) + s}{s\left(\frac{s}{2} + 2 + \frac{1}{s}\right)}}$$

$$V(s) = \frac{s\left(1 + \frac{3}{s}\right)}{s + 4 + \frac{2}{s} + \frac{s^2}{2} + 2s + 1 + s} = \frac{\left(1 + \frac{3}{s}\right)s}{\frac{s^2}{2} + 4s + 5 + \frac{2}{s}}$$

multiplying numerator and denominator by 2s gives

$$V(s) = \frac{2s^2 + 6s}{s^3 + 8s^2 + 10s + 4} = \frac{2s\,(s + 3)}{s^3 + 8s^2 + 10s + 4}$$

In order to find the inverse Laplace transform of V(s), factor the denominator, $\Delta(s) = s^3 + 8s^2 + 10s + 4$. Remember that the roots of the polynomial D(s) must all be negative or imaginary since $D(0) = 4$ and $D(\infty) = +\infty$ give no sign change.

By trial and error, we arrive at the root 6.571.

By dividing:

$$\frac{s^3 + 8s^2 + 10s + 4}{(s + 6.571)} = \text{quadratic}$$

$$\begin{array}{r|l} & s^2 + 1.429s + 0.609 \\ \hline s + 6.571 & s^3 + 8.000s^2 + 10.000s + 4.000 \\ & \underline{s^3 + 6.571s^2} \\ & \quad 1.429s^2 + 10.000s \\ & \quad \underline{1.429s^2 + 9.391s} \\ & \qquad 0.609s + 4.000 \\ & \qquad \underline{0.609s + 4.000} \\ & \qquad\qquad 0 \end{array}$$

$D(s) = (s + 6.571)\ (s^2 + 1.429s + 0.609)$

factoring the quadratic yields

$$r_1 r_2 = \frac{-b \pm \sqrt{b^2 - 4ac}}{2a}$$

$$r_1 r_2 = \frac{-1.429 \pm \sqrt{2.042 - 2.436}}{2}$$

$$r_1 r_2 = \frac{-1.429 \pm \sqrt{-0.394}}{2} = \frac{-1.429}{2} \pm j\,\frac{0.628}{2}$$

$$r_1 = -0.715 + j\,0.314$$

$$r_2 = -0.715 - j\,0.314$$

$$V(s) = \frac{2s\ (s + 3)}{(s + 6.571)(s + 0.715 - j\,0.314)(s + 0.715 + j\,0.314)}$$

Using partial fraction expansion

$$V(s) = \frac{A}{(s + 6.571)} + \frac{B}{(s + 0.715 - j\,0.314)} + \frac{C}{(s + 0.715 + j\,0.314)}$$

where;

$$A = (s + 6.571)\ V(s) \Big|_{s= -6.571}$$

$$= \frac{46.93}{(-5.856 - j\,0.314)(-5.856 + j\,0.314)}$$

$$A = \frac{46.93}{34.39} = 1.365$$

$$B = (s + 0.715 - j\,0.314)\ V(s) \Big|_{s= -0.715 + j\,0.314}$$

$$B = \frac{(-1.430 + j\,0.628)(2.285 + j\,0.314)}{(5.856 + j\,0.314)(j\,0.628)}$$

in polar form

$$B = \frac{(1.562\underline{/156.29°})\ (2.306\underline{/7.82°})}{(5.864\underline{/3.07°})\ (0.628\underline{/90°})} = \frac{3.602\underline{/164.11°}}{3.683\underline{/93.07°}}$$

$$B = 0.978\underline{/71.04} = 0.318 + j\,0.925$$

to obtain C, use the fact that B* = C, therefore

$$C = +\,0.318 - j\,0.925$$

$$V(s) = \frac{1.365}{(s + 6.571)} + \frac{0.318 + j\,0.925}{(s + 0.715 - j\,0.314)} + \frac{0.318 - j\,0.925}{(s + 0.715 + j\,0.314)}$$

Using the inverse Laplace transform yields

$$v(t) = 1.365e^{-6.571t} + (+0.318 + j0.925)e^{(-0.715 + j0.314)t}$$

$$+(+\,0.318 - j0.925)e^{(-0.715 - j0.314)t}$$

expanding

$$v(t) = 1.365e^{-6571t} + 0.318e^{(-0.715 + j0.314)t}$$

$$+ j0.925e^{(-0.715 + j0.314)t} + 0.318e^{(-0.715 - j0.314)t}$$

$$- j0.925e^{(-0.715 - j0.314)t}$$

$$v(t) = 1.365e^{-6.571t} + 0.318e^{-0.715t}\ e^{j0.314t}$$

$$+ j0.925e^{-0.715t}\ e^{j0.314t} + 0.318e^{-0.715t}\ e^{-j0.314t}$$

$$- j0.925e^{-0.715t}\ e^{-j0.314t}$$

collecting terms,

$$v(t) = 1.365e^{-6.571t} + 0.318e^{-0.715t}\left(e^{j0.314t} + e^{-j0.314t}\right)$$

$$j0.925e^{-0.715t}\left(e^{j0.314t} - e^{-j0.314t}\right).$$

Multiplying the second term by $\frac{2}{2}$ and the last terms by $\frac{2j}{2j}$ gives

$$v(t) = 1.365e^{-6.571t} + 0.636e^{-0.715t}\left(\frac{e^{j0.314t} + e^{-j0.314t}}{2}\right)$$

$$+ (2j)\ j0.925e^{-0.715t}\left(\frac{e^{j0.314t} - e^{-j0.314t}}{2j}\right).$$

Note that since,

$$\sin \omega t = \frac{e^{j\omega t} - e^{-j\omega t}}{2j} \quad \text{and}$$

$$\cos \omega t = \frac{e^{j\omega t} + e^{-j\omega t}}{2}$$

then,

$$v(t) = 1.365e^{-6.571t} + 0.636e^{-0.715t} \cos 0.314t$$

$$- 1.85e^{-0.715t} \sin 0.314t.$$

• PROBLEM 12-42

In the circuit of Fig. 1, $v_C(0^+) = 4$ V, and the source voltage is plotted in Fig. 2. Use Laplace transform techniques to find $i_x(t)$ for $t \geq 0^+$.

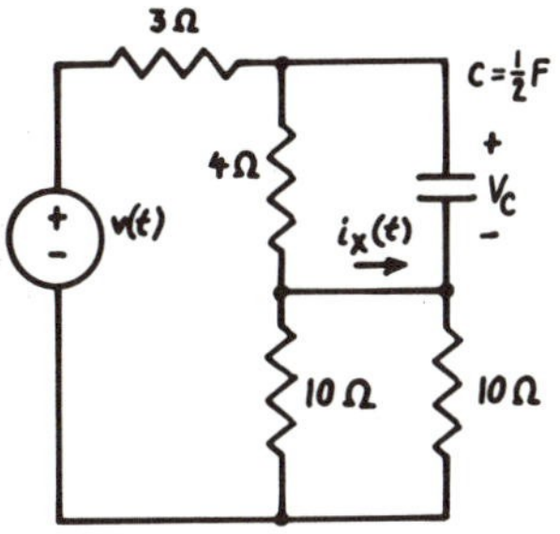

FiG.1

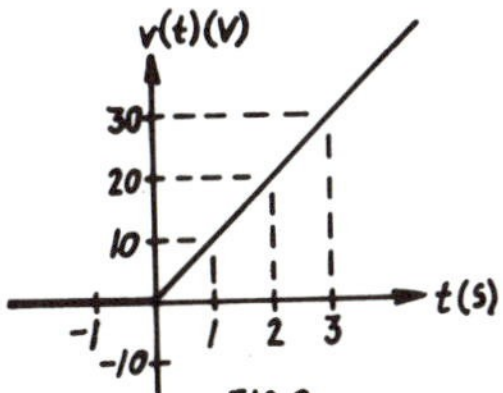

FiG.2

Solution: In Fig. 2 the ramp v(t) can be represented by using unit function notation as 10t u(t).

Transform the circuit in Fig. 2 to the frequency domain noting that

$$V_C(s) = \frac{v_C(0^+)}{s} + \frac{1}{sC} I_C(s).$$

Fig. 3 shows how the capacitor is transformed. Fig. 4 shows the complete transformed circuit.

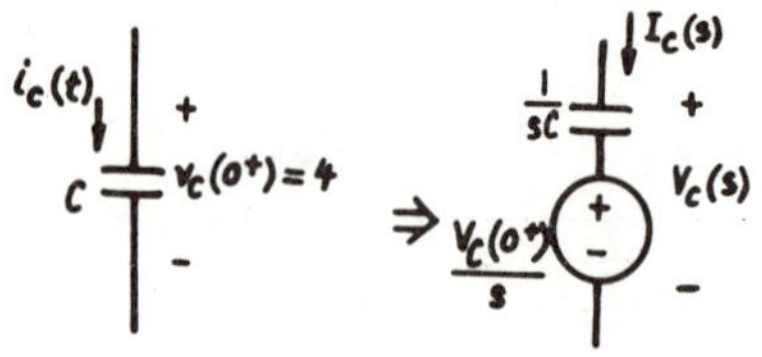

FiG.3

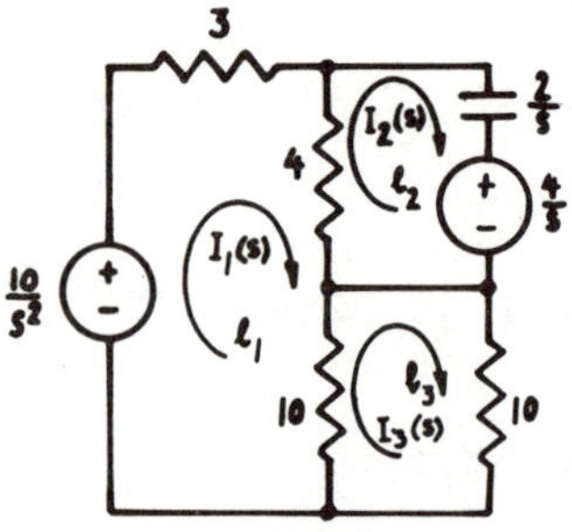

FiG.4

Note that the ramp function 10t u(t) transforms to $\frac{10}{s^2}$, also $I_x(t) = I_3(t) - I_2(t)$.

Using the three loops indicated in Fig. 4, write three KVL equations.

$$\ell_1: \quad \frac{10}{s^2} = 17I_1 - 4I_2 - 10I_3$$

$$\ell_2: \quad -\frac{4}{s} = -4I_1 + \left(4 + \frac{2}{s}\right) I_2$$

$$\ell_3: \quad 0 = -10I_1 \qquad + 20I_3$$

Writing these equations in matrix form yields

$$\begin{bmatrix} 17 & -4 & -10 \\ -4 & 4 + \frac{2}{s} & 0 \\ -10 & 0 & 20 \end{bmatrix} \begin{bmatrix} I_1(s) \\ I_2(s) \\ I_3(s) \end{bmatrix} = \begin{bmatrix} \frac{10}{s^2} \\ -\frac{4}{s} \\ 0 \end{bmatrix}$$

solving for I_3 using determinants yields

$$I_3(s) = \frac{\begin{vmatrix} 17 & -4 & \frac{10}{s^2} \\ -4 & 4 + \frac{2}{s} & -\frac{4}{s} \\ -10 & 0 & 0 \end{vmatrix}}{\begin{vmatrix} 17 & -4 & -10 \\ -4 & 4 + \frac{2}{s} & 0 \\ -10 & 0 & 20 \end{vmatrix}}$$

$$I_3(s) = \frac{17[0] + 4[0] - 10\left[\frac{16}{s} - \frac{10}{s^2}\left(4 + \frac{2}{s}\right)\right]}{17\left[80 + \frac{40}{s}\right] + 4\,[-80] - 10\left[10\left(4 + \frac{2}{s}\right)\right]}$$

$$I_3(s) = \frac{-\frac{160}{s} + \frac{400}{s^2} + \frac{200}{s^3}}{640 + \frac{480}{s}}$$

dividing through by s^3 yields

$$I_3(s) = \frac{-160s^2 + 400s + 200}{640s^3 + 480s^2}$$

solving for I_2 gives

$$I_2(s) = \frac{\begin{vmatrix} 17 & \frac{10}{s^2} & -10 \\ -4 & -\frac{4}{s} & 0 \\ -10 & 0 & 20 \end{vmatrix}}{640 + \frac{480}{s}}$$

$$I_2(s) = \frac{17\left[-\frac{80}{s}\right] + 4\left[\frac{200}{s^2}\right] - 10\left[-\frac{40}{s}\right]}{640 + \frac{480}{s}}$$

$$I_2(s) = \frac{-\frac{960}{s} + \frac{800}{s^2}}{640 + \frac{480}{s}}$$

dividing through by s^3 yields

$$I_2(s) = \frac{-960s^2 + 800s}{640s^3 + 480s^2}$$

Since $I_x(s) = I_3(s) - I_2(s)$

$$I_x(s) = \frac{(-160s^2 + 400s + 200) - (-960s^2 + 800s)}{640s^3 + 480s^2}$$

$$I_x(s) = \frac{800s^2 - 400s + 200}{640s^3 + 480s^2}$$

dividing through by $\frac{1}{640}$ gives

$$I_x(s) = \frac{\frac{5}{4}s^2 - \frac{10}{16}s + \frac{5}{16}}{s^3 + \frac{3}{4}s^2}$$

$$I_x(s) = \frac{\frac{5}{4}s^2 - \frac{10}{16}s + \frac{5}{16}}{s^2\left(s + \frac{3}{4}\right)}$$

Using partial fraction expansion:

$$I_x(s) = \frac{A}{s^2} + \frac{B}{s} + \frac{C}{s + \frac{3}{4}} = \frac{\frac{5}{4}s^2 - \frac{10}{16}s + \frac{5}{16}}{s^2\left(s + \frac{3}{4}\right)}$$

where,

$$A = s^2\,(I_x(s))\Big|_{s=0} = \frac{\frac{5}{16}}{\frac{3}{4}} = \frac{5}{12}$$

$$B = \frac{1}{(2-1)!}\frac{d}{ds}\left[s^2\,(I_x(s))\right]\Bigg|_{s=0}$$

$$B = (1)\frac{d}{ds}\left[\frac{\frac{5}{4}s^2 + \frac{10}{16}s + \frac{5}{16}}{s + \frac{3}{4}}\right]\Bigg|_{s=0}$$

$$B = \frac{\left(-\frac{10}{4}s + \frac{10}{16}\right)\left(s + \frac{3}{4}\right) - \left(\frac{5}{4}s^2 + \frac{10}{16}s + \frac{5}{16}\right)(1)}{\left(s + \frac{3}{4}\right)^2}\Bigg|_{s=0}$$

$$B = \frac{\left(-\frac{10}{16}\right)\left(\frac{3}{4}\right) - \frac{5}{16}}{\left(\frac{3}{4}\right)^2} = \frac{-\frac{30}{64} - \frac{5}{16}}{\frac{9}{16}}$$

$$B = \frac{-\frac{7.5}{16} - \frac{5}{16}}{\frac{9}{16}} = -\frac{12.5}{16}\cdot\frac{16}{9} = -\frac{25}{18}$$

$$C = \left(s + \frac{3}{4}\right) I_x(t)\Big|_{s=-\frac{3}{4}}$$

$$C = \frac{\frac{5}{4}\left(-\frac{3}{4}\right)^2 - \frac{10}{16}\left(-\frac{3}{4}\right) + \frac{5}{16}}{\left(-\frac{3}{4}\right)^2} = \frac{\frac{45}{64} + \frac{30}{64} + \frac{20}{64}}{\frac{36}{64}}$$

$$C = \frac{95}{64} \cdot \frac{64}{36} = \frac{95}{36}$$

Substituting the values found for A, B, and C into our partial fraction gives,

$$I_x(s) = \frac{\frac{5}{12}}{s^2} - \frac{\frac{25}{18}}{s} + \frac{\frac{95}{36}}{s + \frac{3}{4}}$$

Transforming back to the time-domain gives the desired result.

$$i_x(t) = \frac{5}{12}t - \frac{25}{18} + \frac{95}{36} e^{-\frac{3}{4}t}; \quad t \geq 0^+$$

• PROBLEM 12-43

Find $i_L(t)$ by solving a differential equation for $i_L(t)$.

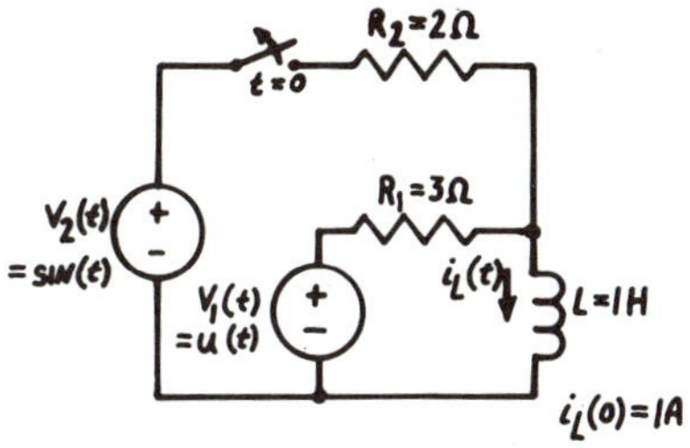

Solution: The differential equation can be found by writing a nodal equation since the voltage across the inductance is $v_L = L \frac{di_L}{dt}$.

At $t = 0^-$,

$$i_L + \frac{L\frac{di_L}{dt} - v_1(t)}{R_1} + \frac{L\frac{di_L}{dt} - v_2(t)}{R_2} = 0$$

$L = 1h$, $R_1 = 3\Omega$, $v_1(t) = u(t)$, $R_2 = 2\Omega$, $v_2(t) = \sin(t)$

Hence, $$i_L + \frac{1}{3}\frac{di_L}{dt} - \frac{u(t)}{3} + \frac{1}{2}\frac{di_L}{dt} - \frac{\sin(t)}{2} = 0$$

Collecting terms and rearranging yields

$$\frac{di_L}{dt} + \frac{6}{5} i_L = \frac{2}{5} u(t) + \frac{3}{5} \sin t$$

However $u(t) = 1$ for $t \geq 0$ and $i_L(0) = 1$; therefore,

$$\frac{di_L}{dt} + \frac{6}{5}\, i_L = \frac{2}{5} + \frac{3}{5}\sin t, \ t \geq 0;\ i_L(0) = 1$$

To find $i_L(t)$ by solving the differential equation, take the Laplace transform and expand in partial fractions. Then find the inverse Laplace transform from the table of Laplace transform pairs.

The Laplace transform is

$$sI_L(s) - i_L(0) + \frac{6}{5}\, I_L(s) = \frac{2}{5}\,\frac{1}{s} + \frac{3}{5}\,\frac{1}{s^2 + 1}.$$

Solve for $I_L(s)$ with $i_L(0) = 1$

$$I_L(s) = \frac{2}{5}\,\frac{1}{s\left(s + \frac{6}{5}\right)} + \frac{1}{s + \frac{6}{5}} + \frac{3}{5}\,\frac{1}{\left(s + \frac{6}{5}\right)(s^2 + 1)}$$

From the expansion theorem, $I_L(s)$ is

$$I_L(s) = \frac{1}{3}\,\frac{1}{s} - \frac{1}{3}\,\frac{1}{s + \frac{6}{5}} + \frac{1}{s + \frac{6}{5}} + \frac{15}{61}\,\frac{1}{s + \frac{6}{5}} + \frac{-\frac{15}{122} - j\,\frac{18}{122}}{s + j} + \frac{-\frac{15}{122} + j\,\frac{18}{122}}{s - j}.$$

Collecting terms and combining and rearranging the last two terms, yields

$$I_L(s) = \frac{1}{3}\,\frac{1}{s} + \frac{167}{183}\,\frac{1}{s + \frac{6}{5}} + \frac{18}{61}\,\frac{1}{s^2 + 1} - \frac{15}{61}\,\frac{s}{s^2 + 1}.$$

Taking the inverse Laplace and rearranging,

(b) $$i_L(t) = \frac{167}{183}\, e^{-\left(\frac{6}{5}\right)t} + \frac{1}{3}\, u(t) + \frac{18}{61}\sin t - \frac{15}{61}\cos t,\ A,$$

$t \geq 0.$

• **PROBLEM** 12-44

Find the current i(t) for all t in the circuit shown in Figure 1.

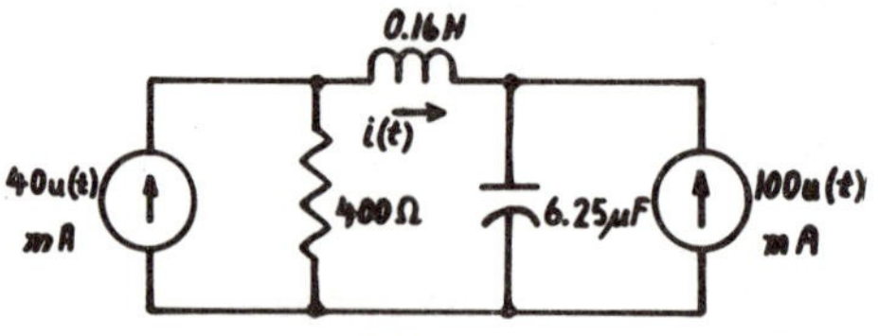

FiG.1

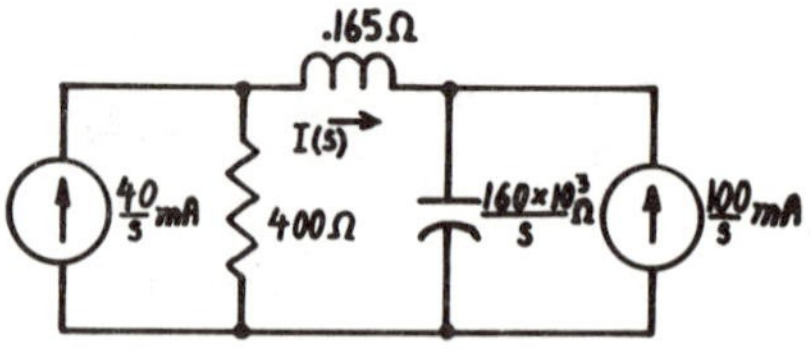

FIG.2

Solution: Since we are asked to calculate the response i(t) over a restricted time interval (t > 0), the solution will involve transient analysis. The most useful tools for such a solution are complex impedance and the Laplace transform. The transformed equivalent circuit is shown in Figure 2. Note that the initial current in the inductor is zero, i(0) = 0, since both current sources have zero value for t < 0. Due to the fact that the desired quantity is the current in a series element, we will apply Thevenin's Theorem to both current sources. The corresponding equivalent circuit is shown in Figure 3.

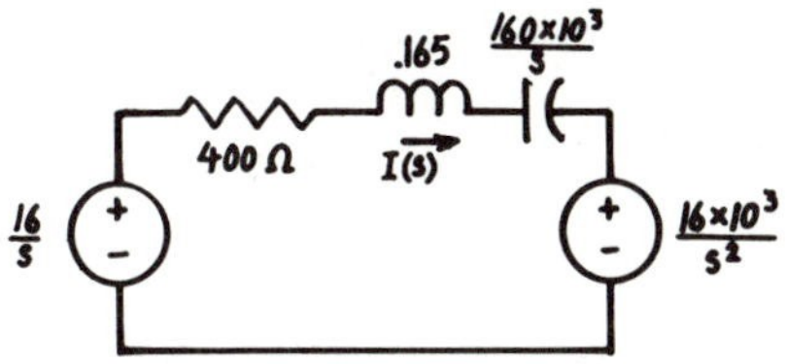

FiG.3

Now directly solve for I(s).

$$I(s) = \frac{\frac{16}{s} - \frac{16 \times 10^3}{s^2}}{400 + .16s + \frac{160 \times 10^3}{s}} \tag{1}$$

Simplifying, gives

$$I(s) = \frac{100\,(s - 1000)}{s\,(s^2 + 2500s + 10^6)} \tag{2}$$

or, factoring the denominator,

$$I(s) = \frac{100\,(s - 1000)}{s\,(s + 500)\,(s + 2000)} \tag{3}$$

Now expand into partial fractions.

$$I(s) = \frac{A}{s} + \frac{B}{s + 500} + \frac{C}{s + 2000}$$

Thus, $A = sI(s)\Big|_{s=0} = -\ 0.1$

$$B = (s + 500)\ I(s)\Big|_{s=-500} = 0.2$$

$$C = (s + 2000)\ I(s)\Big|_{s=-2000} = -\ 0.1$$

so $I(s) = -\ \frac{0.1}{s} + \frac{0.2}{s + 500} - \frac{0.1}{s + 2000}$

or, in the time domain,

$$i(t) = -0.1 + 0.2e^{-500t} - 0.1e^{-2000t}\ \text{A},$$

$$t > 0.$$

• PROBLEM 12-45

Find the mesh equations for the network of Fig. 1 and write them in matrix form.

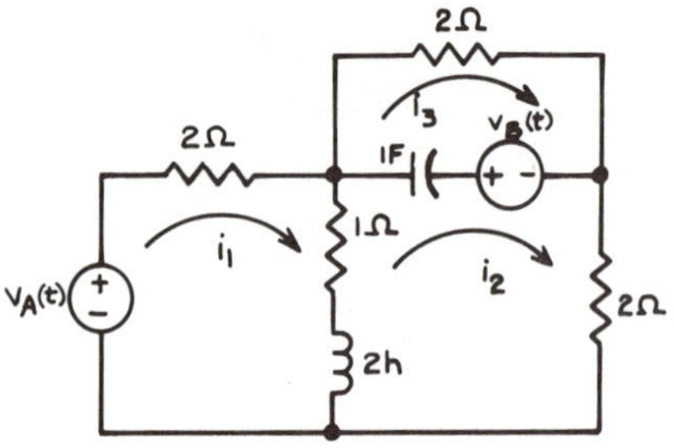

Fig. 1

Solution: Writing an equation for each loop, sum the voltages across each of the branches making up the loop. The voltage across each branch is found by multiplying the branch current by the impedance of the branch.

$$V_A = 2i_1 + (i_1 - i_2)\,1 + 2s$$

$$-V_B - 2i_2 + (i_2 - i_1)\,1 + 2s + (i_2 - i_3)\,\frac{1}{s}$$

$$V_B = 2i_3 + (i_3 - i_2)\,\frac{1}{s}$$

Re-arranging the terms

$$V_A = i_1[2+1+2s] + i_2[-1-2s] + i_3[o] \qquad (1)$$

$$-V_B = i_1[-1-2s] + i_2[2+1+2s+\tfrac{1}{s}] + i_3[-\tfrac{1}{s}] \qquad (2)$$

$$V_B = i_1[o] \qquad + i_2[-\tfrac{1}{s}] \qquad + i_3[2+\tfrac{1}{s}] \qquad (3)$$

Simplifying each of the bracketed branch impedances, write the equations above in matrix form

$$\begin{bmatrix} 2s+3 & -(2s+1) & o \\ -(2s+1) & 2s+\frac{1}{s}+3 & -\frac{1}{s} \\ o & -\frac{1}{s} & \frac{1}{s}+2 \end{bmatrix} \begin{bmatrix} i_1 \\ i_2 \\ i_3 \end{bmatrix} = \begin{bmatrix} V_A \\ -V_B \\ V_B \end{bmatrix}$$

Notice that the branch impedance in Eq.(1) for i_2 is equal to that of i_1 in Eq.(2). The branch impedance for i_2 in Eq.(3) is equal to that of i_3 in Eq.(2). Also, two zero impedances appear in our equations. The only unique branch impedances are for i_1-i_3 in equations (1)-(3) respectively. Note that except for these unique branch impedances (Z_{11}, Z_{22}, Z_{33} etc.), all the other impedances are either negative or zero. By realizing that all loop equations written from planar networks have these properties, write a general set of loop equations that can be applied to all planar networks.

$$
\begin{array}{ccccccc}
Z_{11}(s)i_1 + & Z_{12}(s)i_2 + & Z_{13}(s)i_3 & \cdots & Z_{1n}(s)i_n & = & V_1 \\
Z_{21}(s)i_1 + & Z_{22}(s)i_2 + & Z_{23}(s)i_3 & \cdots & Z_{2n}(s)i_n & = & V_2 \\
Z_{31}(s)i_1 + & Z_{32}(s)i_2 + & Z_{33}(s)i_3 & \cdots & Z_{3n}(s)i_n & = & V_3 \\
\vdots & \vdots & \vdots & & \vdots & & \vdots \\
Z_{n1}(s)i_1 + & Z_{n2}(s)i_2 + & Z_{n3}(s)i_3 & \cdots & Z_{nn}(s)i_n & = & V_n
\end{array}
$$

Where Z_{11}, Z_{22}, Z_{33} . . . Z_{nn} are unique branch impedances and all impedances with the same two loop subscripts are equal to the negative of the branch impedance that is common to the two loops. Therefore $Z_{12} = Z_{21}$, $Z_{32} = Z_{23}$, $Z_{83} = Z_{38}$, etc. The impedances of common branches are negative when the currents through them are in opposite directions.

• PROBLEM 12-46

Find the node equations for the circuit of Fig. 1, and write them in matrix form.

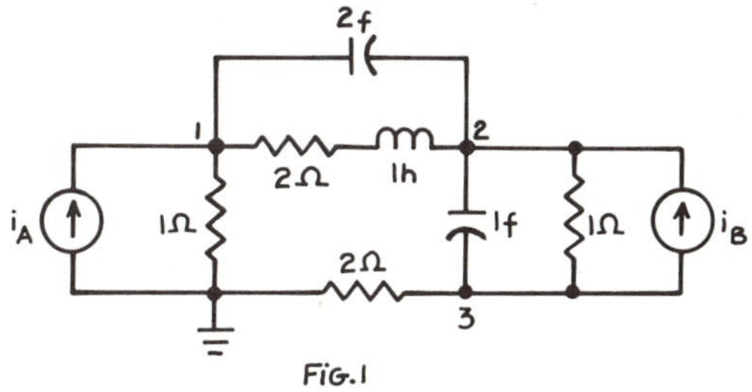

FiG.1

Solution: Writing an equation for each node, by summing the currents in each branch. The current in each branch is found by multiplying the voltage across the branch by the admittance of the branch.

$$i_A = \frac{V_1}{1} + (V_1 - V_2)\frac{1}{s+2} + (V_1 - V_2)\,2s$$

$$i_B = \frac{(V_2-V_3)}{1} + (V_2-V_3)s + (V_2-V_1)\frac{1}{s+2} + (V_2-V_1)2s$$

$$-i_B = \frac{V_3}{2} + \frac{(V_3-V_1)}{1} + (V_3-V_2)s$$

Re-arranging terms, obtains

$$i_A = V_1[1+\frac{1}{s+2}+2s] + V_2[-\frac{1}{s+2}-2s] + V_3[o] \qquad (1)$$

$$i_B = V_1[-\frac{1}{s+2}-2s] + V_2[1+s+\frac{1}{s+2}+2s] + V_3[-1-s] \qquad (2)$$

$$-i_B = V_1[o] \qquad + V_2[-1-s] \qquad + V_3[\frac{1}{2}+1+s] \qquad (3)$$

Simplify each of the bracketed node admittances, and write the equations above in matrix form.

$$\begin{bmatrix} \frac{2s^2+5s+3}{s+2} & -\frac{2s^2+4s+1}{s+2} & o \\ \frac{-2s^2+4s+1}{s+2} & \frac{3s^2+7s+3}{s+2} & -(s+1) \\ o & -(s+1) & \frac{2s+3}{2} \end{bmatrix} \begin{bmatrix} V_1 \\ V_2 \\ V_3 \end{bmatrix} = \begin{bmatrix} i_A \\ i_B \\ -i_B \end{bmatrix}$$

Notice that the node admittance in Eq.(1) for V_2 is equal to that of V_1 in Eq.(2). The node admittance for V_2 in Eq.(3) is equal to that of V_3 in Eq.(2). Also, two zero admittances appear in our equations. The only unique node admittances are for V_1-V_3 in equation (1)-(3) respectively. Note that except for these unique node admittances (Y_{11}, Y_{22}, Y_{33} etc.), all the other admittances are either negative or zero, **because** reference currents are opposite one another. By realizing that all node equations written from planar networks have these properties, write a general set of node equations that can be applied to all planar networks.

$$Y_{11}(s)V_1 + Y_{12}(s)V_2 + Y_{13}(s)V_3 \;.\;.\;.\; Y_{1n}(s)V_n = i_1$$

$$Y_{21}(s)V_1 + Y_{22}(s)V_2 + Y_{23}(s)V_3 \ldots Y_{2n}(s)V_n = i_2$$

$$Y_{31}(s)V_1 + Y_{32}(s)V_2 + Y_{33}(s)V_3 \ldots Y_{3n}(s)V_n = i_3$$

$$\vdots$$

$$Y_{n1}(s)V_1 + Y_{n2}(s)V_2 + Y_{n3}(s)V_3 \ldots Y_{nn}(s)V_n = i_n$$

Where $Y_{11}, Y_{22}, Y_{33}, \ldots, Y_{nn}$ are unique node admittances (also called self-conductances). All other admittances with the same two node subscripts are equal to the negative of the summed admittances of branches connecting those two nodes. These are also called mutual conductances. Therefore $Y_{12} = Y_{21}$, $Y_{45} = Y_{54}$, $Y_{97} = Y_{79}$ etc. The admittances between two nodes will be negative as long as the reference directions for the current at each node are opposite one another.

• PROBLEM 12-47

Find the transfer function $\frac{V_2(s)}{V_1(s)}$ for the circuit in Fig. 1.

Solution: Write the loop equations for the loops shown in Fig. 1.

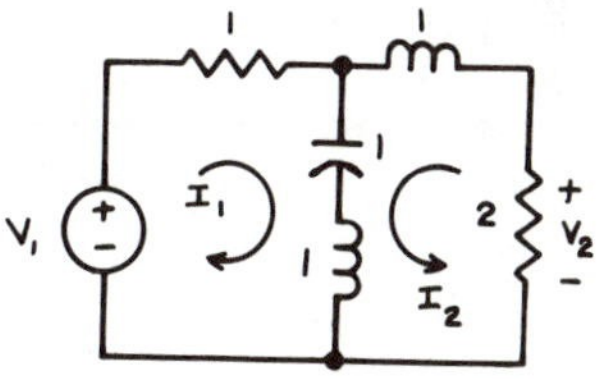

FIG.1

$$V_1 = I_1\left(1 + \frac{1}{s} + s\right) + I_2\left(\frac{1}{s} + s\right)$$

$$0 = I_1\left(\frac{1}{s} + s\right) + I_2\left(2 + 2s + \frac{1}{s}\right)$$

Using **determinants** solve for I_2, since solving -or I_2 will give us V_1 times some impedance Z. Multiply I_2 by the impedance across V_2 to give us V_2. $I_2 = V_1(Z)$; $V_2 = -2V_1(Z)$; $\frac{V_2}{V_1} = -2(Z)$.

$$I_2 = \frac{\begin{vmatrix} 1+\frac{1}{s}+s & V_1 \\ \frac{1}{s}+s & 0 \end{vmatrix}}{\begin{vmatrix} 1+\frac{1}{s}+s & \frac{1}{s}+s \\ \frac{1}{s}+s & 2+2s+\frac{1}{s} \end{vmatrix}} = \frac{-V_1\left(\frac{1}{s}+s\right)}{\left(1+\frac{1}{s}+s\right)\left(2+2s+\frac{1}{s}\right) - \left(\frac{1}{s}+s\right)^2}$$

$$I_2 = \frac{-V_1\left(\frac{1}{s}+s\right)}{\left|\frac{(s^2+s+1)}{s}\right|\left|\frac{2s^2+2s+1}{s}\right| - \left|\frac{(s^2+1)^2}{s^2}\right|} = \frac{-V_1\left(\frac{s^2+1}{s}\right)}{\frac{s^3+4s^2+3s+3}{s}}$$

Simplifying gives

$$I_2 = \frac{-V_1(s^2+1)}{s^3 + 4s^2 + 3s + 3} \qquad (1)$$

Since $V_2 = 2(-I_2)$ substituting into Eq.(1) gives

$$V_2 = \frac{V_1 2(s^2+1)}{s^3 + 4s^2 + 3s + 3}$$

$$\frac{V_2}{V_1}(s) = \frac{2(s^2+1)}{s^3 + 4s^2 + 3s + 3}$$

• PROBLEM 12-48

Using source transformations and impedance combinations, convert the circuit in Fig. 1 to a single loop circuit to find the transfer impedance which relates I(s) and $V_o(s)$.

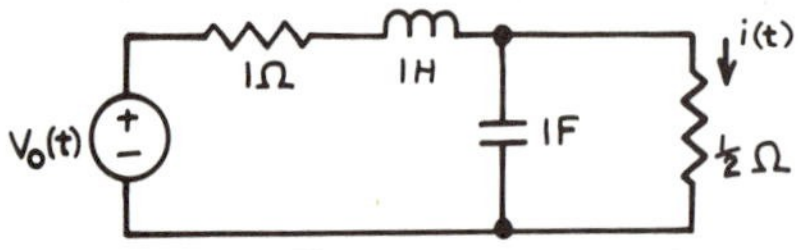

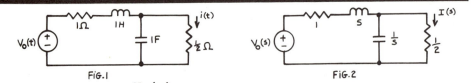

FiG.1

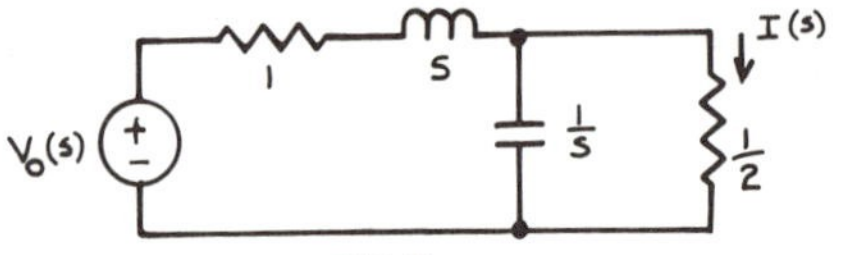

FiG.2

Solution: Find $\frac{V_o(s)}{I(s)}$. Transform the circuit of Fig. 1 to obtain that which is shown in Fig. 2. Transform the practical voltage source to the left of the capacitor into a practical current source, $I_o(s) = \frac{1}{s+1} V_o(s)$, as shown in Fig. 3. Combine impedances, $[s+1||\frac{1}{s}] = \frac{(s+1)}{s(s+1)+1}$, as shown in Fig. 4. Another source transformation where, $V_s(s) = I_o(s)Z_1$ is the new voltage source gives the single loop circuit shown in Fig. 5.

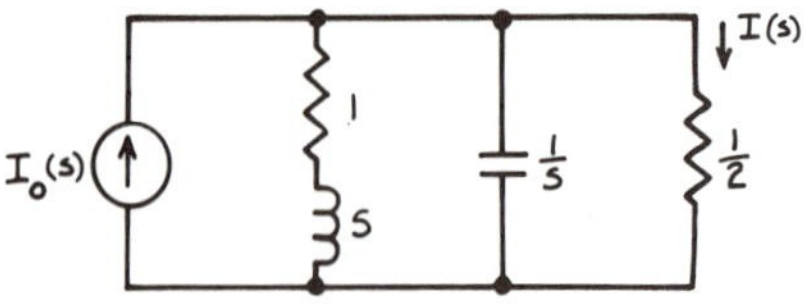

FiG.3

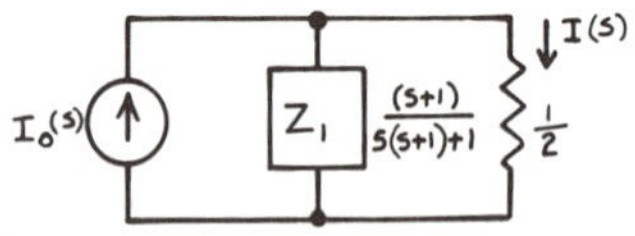

FiG.4

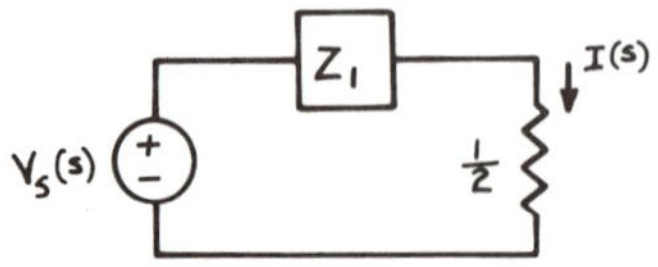

FiG.5

Using Kirchoffs Voltage Law, write

$$V_s(s) = I(s)(Z_1 + \frac{1}{2}).$$

With $V_s(s) = I_o(s)Z_1$ and $Z_1 = \frac{(s+1)}{s(s+1)+1}$,

$$I_o\left[\frac{(s+1)}{s(s+1)+1}\right] = I(s)\left[\frac{(s+1)}{s(s+1)+1} + \frac{1}{2}\right].$$

Since $I_o = \frac{1}{s+1} V_o(s)$, the equation becomes

$$V_o\left[\frac{1}{s(s+1)+1}\right] = I(s)\left[\frac{(s+1)}{s(s+1)+1} + \frac{1}{2}\right]$$

and therefore

$$\frac{V_o(s)}{I(s)} = \frac{\frac{(s+1)}{s(s+1)+1} + \frac{1}{2}}{\frac{1}{s(s+1)+1}}$$

Simplifying obtains

$$\frac{V_o(s)}{I(s)} = \frac{s^2 + 3s + 3}{2}$$ the desired transfer impedance.

• PROBLEM 12-49

Using superposition and Laplace transform techniques, calculate the voltage $v_c(t)$ in Fig. 1.

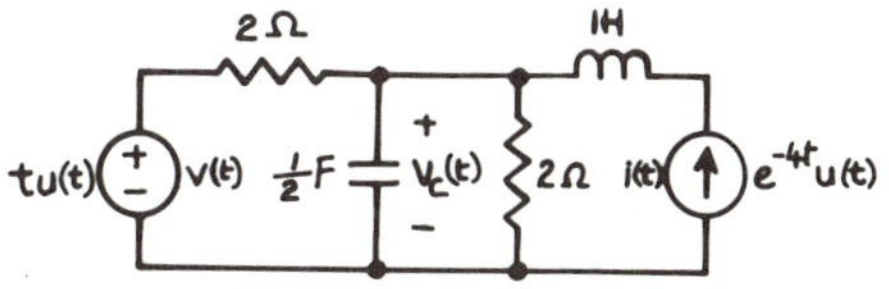

FIG. 1

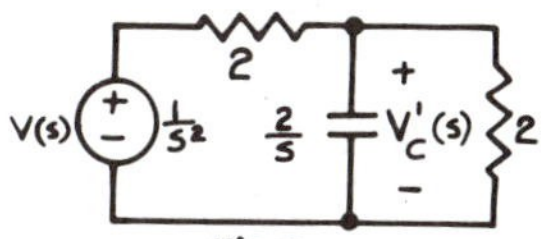

FIG. 2

Solution: First find the response due to the source v(t) by setting i(t) equal to zero. The equivalent transformed circuit for this case is shown in Fig. 2.

Applying voltage division, the transform of voltage $v_c'(t)$ is found

$$V_c'(s) = \left[\frac{2||\frac{2}{s}}{2||\frac{2}{s} + 2}\right]\frac{1}{s^2} = \left[\frac{\frac{2}{s+1}}{\frac{2}{s+1} + 2}\right]\frac{1}{s^2}$$

$$V_c'(s) = \left[\frac{\frac{2}{s+1}}{\frac{2+2(s+1)}{s+1}}\right]\frac{1}{s^2} = \left(\frac{2}{2s+4}\right)\frac{1}{s^2} = \left(\frac{1}{s+2}\right)\frac{1}{s^2}$$

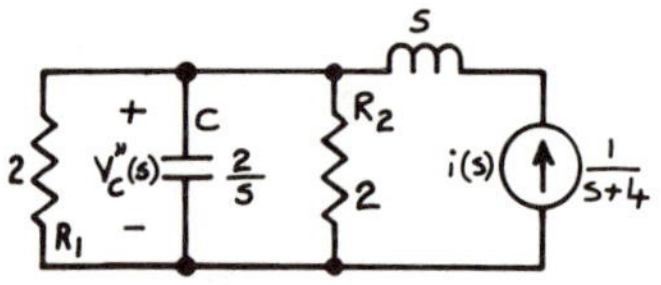

FIG.3

Next, set v(t) equal to zero and calculate the response to i(t). The equivalent transformed circuit is shown in Fig. 3. The current i(s) in Fig. 3 flows through the parallel combination of R_1, C_1 and R_2. Hence, the transform of voltage $v_c'(t)$ is

$$V_c''(s) = (2||2||\tfrac{2}{s})\frac{1}{s+4} = (1||\tfrac{2}{s})\frac{1}{s+4}$$

$$V_c''(s) = \left|\frac{\frac{2}{s}}{1+\frac{2}{s}}\right| \frac{1}{s+4} = \left(\frac{2}{s+2}\right)\left(\frac{1}{s+4}\right)$$

Using superposition,

$$V_c(s) = V_c'(s) + V_c''(s)$$

$$V_c(s) = \left(\frac{1}{s+2}\right)\left(\frac{1}{s^2}\right) + \left(\frac{2}{s+2}\right)\left(\frac{1}{s+4}\right)$$

Expand the two fractions by using partial fraction expansion.

$$\left(\frac{1}{s+2}\right)\frac{1}{s^2} = \frac{A}{s+2} + \frac{B}{s^2} + \frac{C}{s}$$

where $A = \left| s+2\left(\frac{1}{s+2}\right)\frac{1}{s^2}\right|_{s=-2} = \frac{1}{4}$

$$B = \left| s^2\left(\frac{1}{s+2}\right)\frac{1}{s^2}\right|_{s=0} = \frac{1}{2}$$

$$C = \left|\frac{1}{(2-1)!}\frac{d}{ds}\left[s^2\left(\frac{1}{s+2}\right)\left(\frac{1}{s^2}\right)\right]\right|_{s=0} = \left|-\frac{1}{(s+2)^2}\right|_{s=0}$$

$$C = -\frac{1}{4}$$

$$\left(\frac{2}{s+2}\right)\left(\frac{1}{s+4}\right) = \frac{D}{s+2} + \frac{E}{s+4}$$

where $D = \left| s+2 \left(\frac{2}{s+2}\right)\left(\frac{1}{s+4}\right) \right|_{s=-2} = 1$

$E = \left| s+4 \left(\frac{2}{s+2}\right)\left(\frac{1}{s+4}\right) \right|_{s=-4} = -1$

yielding

$$V_c(s) = \frac{\frac{1}{4}}{s+2} + \frac{\frac{1}{2}}{s^2} + \frac{-\frac{1}{4}}{s} + \frac{1}{s+2} + \frac{-1}{s+4}$$

transforming to the time domain yields

$$v_c(t) = \left[\frac{1}{4}e^{-2t} + \frac{1}{2}t - \frac{1}{4} + e^{-2t} - e^{-4t}\right] u(t)$$

$$v_c(t) = \left[\frac{5}{4}e^{-2t} - e^{-4t} + \frac{1}{2}t - \frac{1}{4}\right] u(t)$$

• PROBLEM 12-50

Using superposition and Laplace transform techniques, calculate the voltage $v_2(t)$ in Fig. 1.

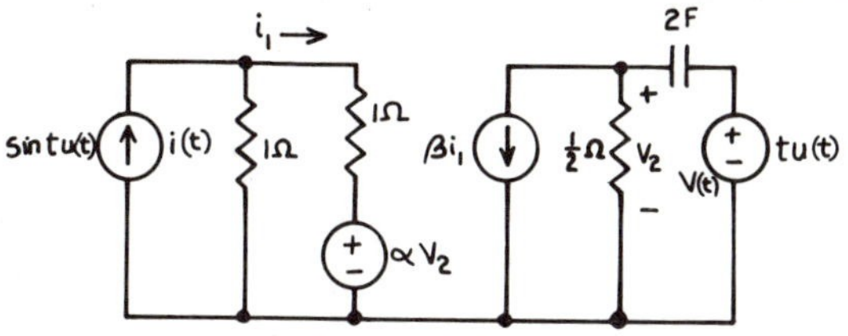

$\alpha = 1$, $\beta = 2$ FIG. 1

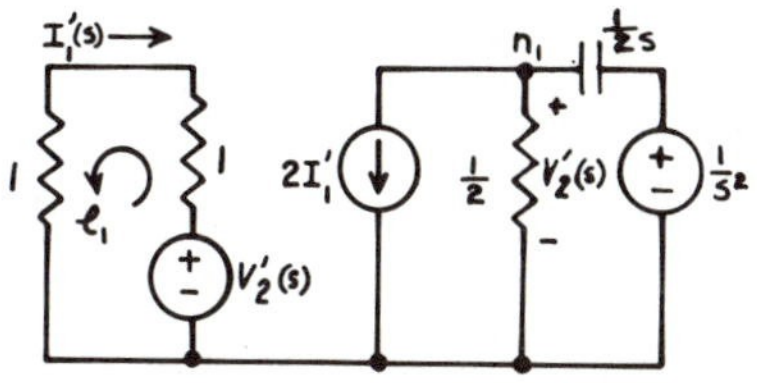

FIG. 2

Solution: Setting the source i(t) equal to zero results in the transformed network shown in Fig. 2.

Writing a KVL equation for loop ℓ_1 and a KCL equation for node n_1 in Fig. 2 yields

$$\ell_1: \quad V_2'(s) = -I_1'(s)\ [1 + 1]$$

$$n_1: \quad -2I_1'(s) = 2V_2'(s) + 2s[V_2'(s) - \frac{1}{s^2}]$$

Substituting for I_1 from the first equation into the second yields

$$-2\left[\frac{-V_2'(s)}{2}\right] = V_2'(s)[2 + 2s] - \frac{2}{s}$$

$$V_2'(s) - V_2(s)[2 + 2s] = -\frac{2}{s}$$

$$V_2'(s)[-1 - 2s] = -\frac{2}{s}$$

$$V_2'(s)[2s + 1] = \frac{2}{s}$$

$$V_2'(s) = \frac{2}{s(2s+1)} = \frac{1}{s(s+\frac{1}{2})}$$

Setting the source v(t) equal to zero results in the transformed network shown in Fig. 3.

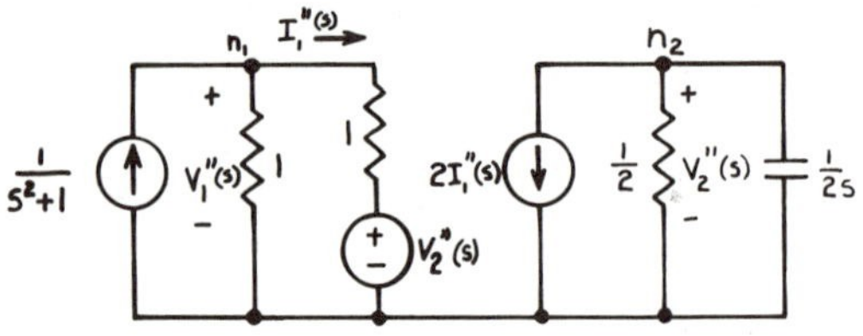

FIG.3

Writing a KCL equation for node n_1 allows us to find $I_1''(s)$ in terms of $V_2''(s)$. Finally we obtain $V_2''(s)$ by substituting for $I_1''(s)$ in the KCL equation for node n_2 in Fig. 3. The final result is found by superposition

$$V_2(s) = V_2'(s) + V_2''(s)$$

$$n_1: \quad \frac{1}{s^2+1} = V_1'' + V_1'' - V_2'' \; ; \quad I_1'' = V_1'' - V_2''$$

$$n_2: \quad -2I_1'' = 2V_2'' + 2sV_2''$$

from node n_1 obtain

$$I_1'' = \frac{1}{2(s^2+1)} - \frac{V_2}{2}$$

Substituting into the equation for node n_2 yields

$$V_2''(s) = -\frac{1}{(s^2+1)}\,\frac{1}{(2s+1)} = -\frac{1}{2}\;\frac{1}{(s^2+1)}\;\frac{1}{(s+\frac{1}{2})}$$

By superposition

$$V_2(s) = V_2'(s) + V_2''(s)$$

$$V_2(s) = \frac{1}{s(s+\frac{1}{2})} + \frac{-\frac{1}{2}}{(s^2+1)(s+\frac{1}{2})}$$

Expand each of the two fractions into partial fractions:

$$\frac{1}{s(s+\frac{1}{2})} = \frac{A}{s} + \frac{B}{s+\frac{1}{2}}$$

where $A = s \left.\frac{1}{s(s+\frac{1}{2})}\right|_{s=0} = 2$

$$B = (s+\tfrac{1}{2}) \left.\frac{1}{s(s+\frac{1}{2})}\right|_{s=-\frac{1}{2}} = -2$$

$$\frac{-\frac{1}{2}}{(s^2+1)(s+\frac{1}{2})} = \frac{-\frac{1}{2}}{(s+j)(s-j)(s+\frac{1}{2})} = \frac{C}{s+j} + \frac{C^*}{s-j} + \frac{D}{s+\frac{1}{2}}$$

where $C = s+j \left.\frac{-\frac{1}{2}}{(s^2+1)(s+\frac{1}{2})}\right|_{s=-j} = 0.2 - 0.1j$

$$C^* = .2 + .1j$$

$$D = s+\tfrac{1}{2} \left.\frac{-\frac{1}{2}}{(s^2+1)(s+\frac{1}{2})}\right|_{s=-\frac{1}{2}} = -.4$$

$$V_2(s) = \frac{2}{s} + \frac{-2}{s+\frac{1}{2}} + \frac{-0.4}{s+\frac{1}{2}} + \frac{0.2-j0.1}{s+j} + \frac{0.2+j0.1}{s-j}$$

transforming back to the time domain yields

$$v_2(t) = 2 - 2e^{-\frac{1}{2}t} - 0.4e^{-\frac{1}{2}t} + (0.2-j0.1)e^{-jt} + (0.2+j0.1)e^{jt}$$

$$v_2(t) = 2 - 2.4e^{-\frac{1}{2}t} + .2(e^{j}+e^{-j}) + .1j(e^{j}-e^{-j})$$

$$v_2(t) = 2 - 2.4e^{-\frac{1}{2}t} + .4(\frac{e^{j}+e^{-j}}{2}) - .2(\frac{e^{j}-e^{-j}}{2j})$$

Since $\cos t = \frac{e^{j}+e^{-j}}{2}$ and $\sin t = \frac{e^{j}-e^{-j}}{2j}$ gives

$$v_2(t) = 2 - 2.4e^{-\frac{1}{2}t} + .4 \cos t - .2 \sin t$$

Using phasors to represent the sinusoidals, where

$A \cos(t+\emptyset) - R_e\{Ae^{j(t+\emptyset)}\} \Rightarrow Ae^{j\emptyset} \Rightarrow A\underline{/\emptyset}$, giving,

$$v_2(t) = 2 - 2.4e^{-\frac{1}{2}t} + 0.4\underline{/0°} - -0.2\underline{/-90°}$$

In rectangular form:

$$v_2(t) = 2 - 2.4e^{-\frac{1}{2}t} + 0.4 + j0.2$$

$$v_2(t) = 2 - 2.4e^{-\frac{1}{2}t} + \sqrt{0.2^2 + 0.4^2}\ \cos\ (t+\tan^{-1}\frac{1}{2})$$

$$v_2(t) = [2 - 2.4e^{-\frac{1}{2}t} + \frac{1}{\sqrt{5}}\ \cos\ (t+26.56°)]u(t)$$

$$v_2(t) = [2 - 2.4e^{-\frac{1}{2}t} - \frac{1}{\sqrt{5}}\ \sin\ (t-63.44°)]u(t)$$

• PROBLEM 12-51

Calculate the driving-point current i(t) for the network in Fig. 1 by converting the initial conditions into equivalent sources and applying the superposition theorem.

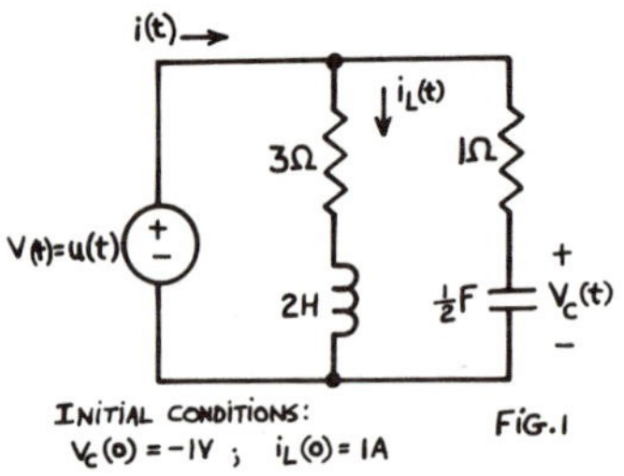

FIG.1

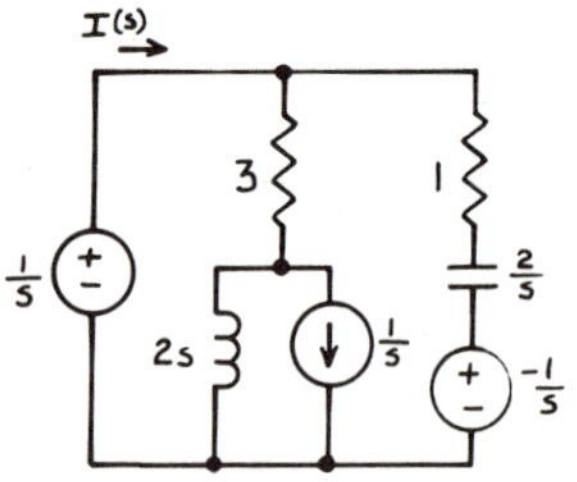

FIG.2

Solution: Since one can write

$$v_c(t) = v_c(o) + \frac{1}{C}\int_o^t i_c(\tau)d\tau$$

for the capacitor and initial voltage, we can express it by use of Laplace transforms as

$$V_c(s) = \frac{v_c(o)}{s} + \frac{1}{sC}\ I_c(s)$$

a capacitor impedance $\frac{1}{sC}$ in series with an initial condition generator $\frac{v_c(o)}{s}$.

For the inductor

$$i_L(t) = i_L(o) + \frac{1}{L}\int_o^t V_L(\tau)d\tau$$

is $I_L(s) = \frac{i_c(o)}{s} + \frac{1}{sL} V_L(s)$

an inductor impedance sL in parallel with an initial condition generator $\frac{i_L(o)}{s}$.

The initial conditions are shown converted in equivalent sources in Fig. 2. Using superposition, calculate then sum the responses I'(s), I"(s), I'''(s) when each source is "on" obtaining the desired response I(s).

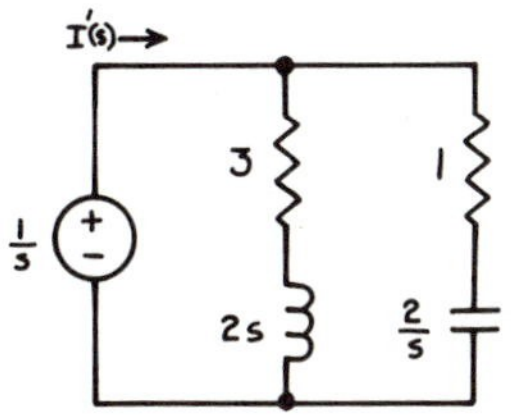

FiG.3

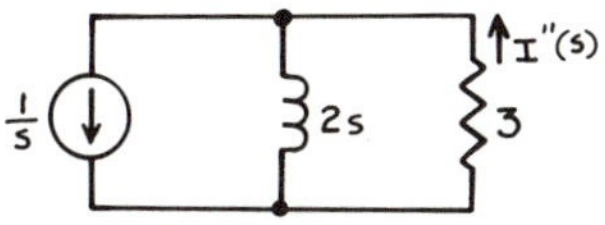

FiG.4

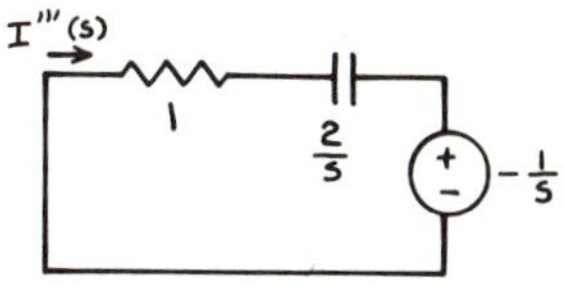

FiG.5

Figs. 3-5 show the equivalent circuits used in finding I'(s), I"(s) and I'''(s) when only the input voltage source, inductor initial condition generator, and capacitor initial condition generator is "on" respectively.

Writing the node equation for the network in Fig. 3 gives

$$I'(s) = \frac{\frac{1}{s}}{3 + 2s} + \frac{\frac{1}{s}}{1 + \frac{2}{s}} .$$

Using the current division concept in Fig. 4 yields

$$I''(s) = - \frac{\frac{1}{s}(2s)}{2s + 3}$$

Writing a loop equation for the network in Fig. 5 and solving for I'''(s) gives

$$I'''(s) = \frac{\frac{1}{s}}{(1 + \frac{2}{s})}$$

Since I(s) = I'(s) + I"(s) + I'''(s) we write

$$I(s) = \frac{\frac{1}{s}}{3+2s} + \frac{\frac{1}{s}}{1+\frac{2}{s}} - \frac{2}{2s+3} + \frac{\frac{1}{s}}{1+\frac{2}{s}}$$

$$I(s) = \frac{\frac{1}{s}-2}{2s+3} + \frac{\frac{2}{s}}{1+\frac{2}{s}}$$

$$I(s) = \frac{1-2s}{s(2s+3)} + \frac{2}{s+2} = \frac{\frac{1}{2}-s}{s(s+\frac{3}{2})} + \frac{2}{s+2}$$

Using partial fraction expansion gives

$$I(s) = \frac{A}{s} + \frac{B}{s+\frac{3}{2}} + \frac{2}{s+2}$$

$$A = \left.\frac{\frac{1}{2}-s}{s+\frac{3}{2}}\right|_{s=0} = \frac{1}{3}\,, \quad B + \left.\frac{\frac{1}{2}-s}{s}\right|_{s=-\frac{3}{2}} = \frac{2}{3}$$

Hence $I(s) = \dfrac{\frac{1}{3}}{s} + \dfrac{\frac{2}{3}}{s+\frac{3}{2}} + \dfrac{2}{s+2}$

is $i(t) = \frac{1}{3} + \frac{2}{3}e^{-\frac{3}{2}t} + 2e^{-2t}$ in the time domain.

• PROBLEM 12-52

Find the Laplace transform of the output of an ideal full-wave rectifier with input $V_m \sin \omega t u(t)$.

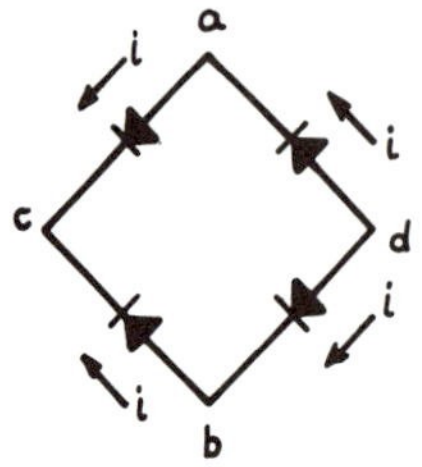

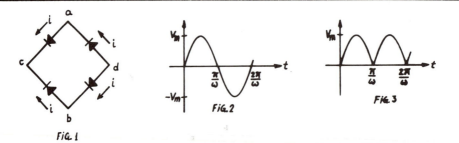

FiG. 1

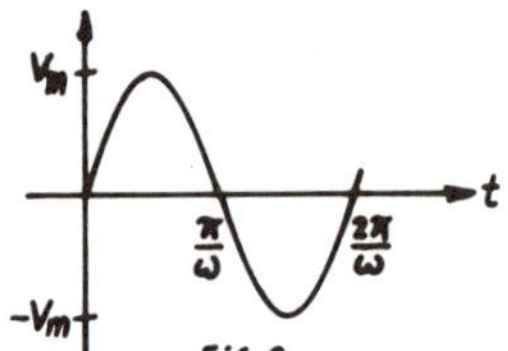

FiG. 2

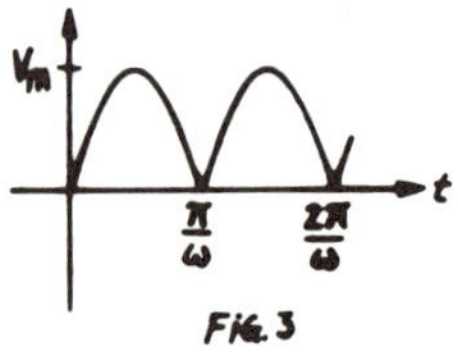

FiG. 3

Solution: An ideal full-wave rectifier is the electronic device shown in Fig. 1, made-up of four ideal diodes. A diode is a two-terminal device which allows current to flow through it in one direction only.

If we apply $V_m \sin \omega t u(t)$ to terminals a-b then the full-wave rectified output will appear at terminals c-d. Fig. 2 shows the input $V_m \sin \omega t u(t)$ and Fig. 3 shows the full-wave rectified output.

To find the Laplace transform of the periodic function in Fig. 3, find the transform for one period and divide by $1 - e^{-as}$, where a is the period. Stated mathematically:

$$f(t + a) = f(t) \text{ then}$$

$$F(s) = \frac{\int_0^a e^{-st} f(t)\,dt}{1 + e^{-as}}, \qquad (1)$$

specifying further,

$$G(s) = \int_0^a e^{-st} f(t)\,dt = L\left\{ f(t)\,[u(t) - u(t - a)]\right\} \qquad (2)$$

Eq. (2) tells us to find the Laplace transform for a single pulse (as shown in Fig. 4) for the wave in Fig. 3.

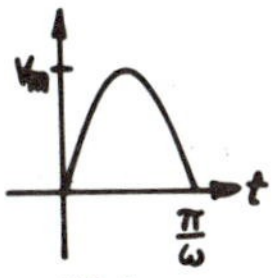

Fig.4

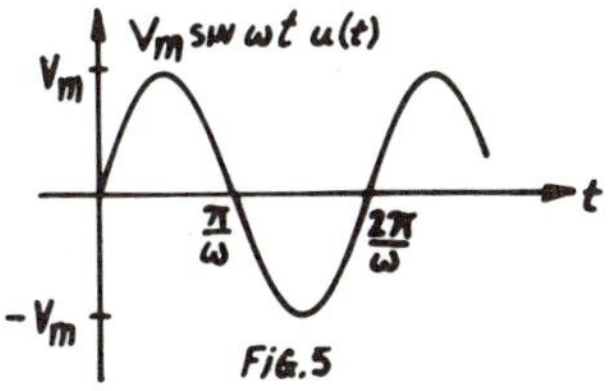

Fig.5

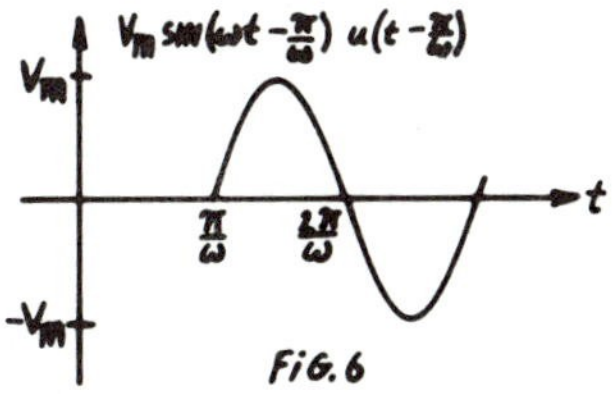

Fig.6

The pulse in Fig. 4 can be obtained by adding the two waves shown in Fig. (5-6).

$$\text{Since } L\left\{\frac{C}{a}\sin at\right\} = \frac{C}{s^2 + \omega^2}, \text{ then}$$

$$L\left\{V_m \sin \omega t u(t)\right\} = \frac{V_m \omega}{s^2 + \omega^2} \text{ and } L\left\{V_m \sin\left(\omega t - \frac{\pi}{\omega}\right) u\left(t - \frac{\pi}{\omega}\right)\right\}$$

$$= e^{-s\frac{\pi}{\omega}} \frac{V_m \omega}{s^2 + \omega^2} \text{ giving}$$

$$G(s) = \frac{V_m \omega}{s^2 + \omega^2}\left[1 + e^{-s\frac{\pi}{\omega}}\right], \text{ for the single pulse and,}$$

$$F(s) = \frac{G(s)}{1 + e^{-s\frac{\pi}{\omega}}} = \frac{V_m \omega}{s^2 + \omega^2}\left[\frac{1 + e^{-s\frac{\pi}{\omega}}}{1 - e^{-s\frac{\pi}{\omega}}}\right]$$

for the full-wave rectified output.

THÉVENIN'S AND NORTON'S EQUIVALENT CIRCUITS

Find the Thevenin equivalent of the circuit in Fig. 1 and use it to find the complete response of the following networks connected across terminals a-b: (a) 1Ω resistor (b) series combination of 1Ω resistor and 1H inductor (inductor initially de-energized). The capacitor is de-energized just before the networks are connected to terminals a-b. Use Laplace transform techniques.

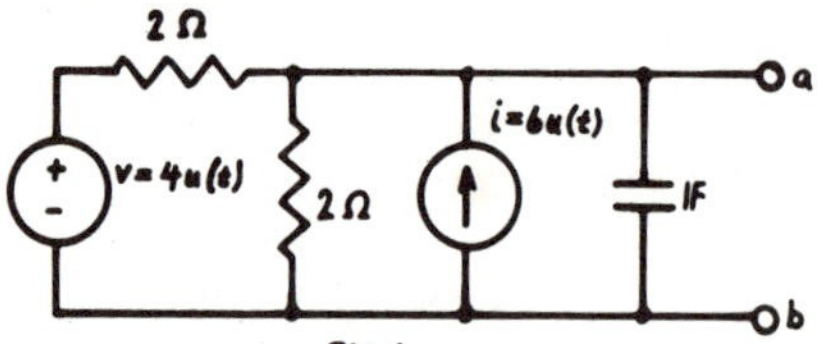

FIG.1

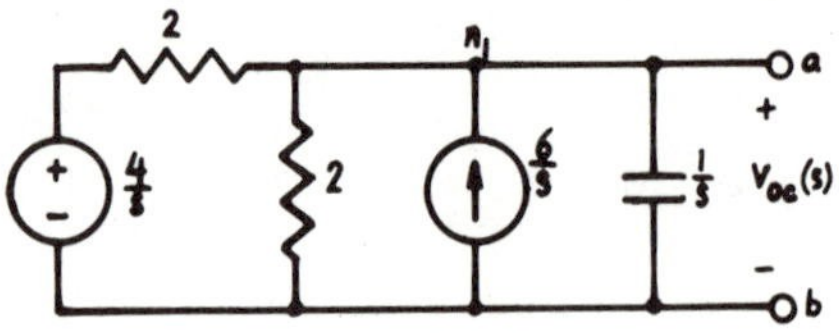

FIG.2

Solution: To find the Thevenin equivalent of any circuit, first, find the open circuit voltage V_{0C} and then, short-circuit all independent voltage sources and open-circuit all independent current sources to find the equivalent impedance Zth across the output terminals a-b. Thevenin's theorem states that the network can be replaced by an independent voltage source V_{0C} in series with the equivalent impedance Zth.

To find V_{0C} in the circuit of Fig. 1, transform the circuit to the frequency domain. Note that, since we are told in the question that the capacitor is de-energized just before t = 0, do not include any initial voltage generators.

Write a KCL equation for node n_1 in the circuit of Fig. 2.

$$\frac{6}{s} = \frac{V_{0C} - \frac{4}{s}}{2} + \frac{V_{0C}}{2} + \frac{V_{0C}}{\frac{1}{s}}$$

$$\frac{6}{s} = V_{0C}\left(\frac{1}{2} + \frac{1}{2} + s\right) - \frac{2}{s}$$

solving for V_{0C},

$$V_{0C} = \frac{\frac{6}{s} + \frac{2}{s}}{\frac{1}{2} + \frac{1}{2} + s} = \frac{\frac{8}{s}}{s + 1} = \frac{8}{s(s + 1)}$$

Fig. 3 shows the circuit in Fig. 2 with all independent sources "dead".

$$Zth = (2||2)||\frac{1}{s} = 1||\frac{1}{s} = \frac{1}{s + 1}$$

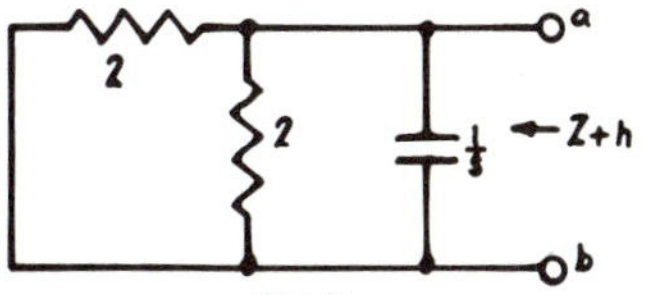

Fig. 3

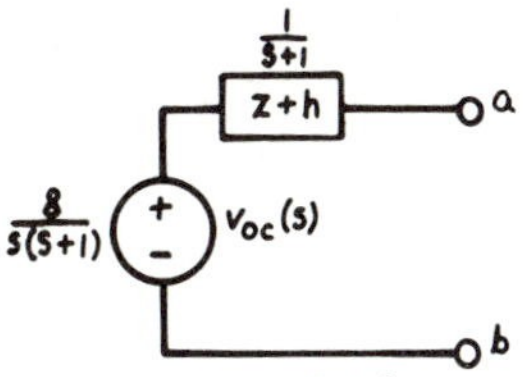

Fig. 4

According to Thevenin's theorem the circuit in Fig. 2 can be replaced by the circuit in Fig. 4 above.

(a) Using the circuit in Fig. 4, find the voltage response across a 1Ω resistor connected across terminals a-b.

From Fig. 5, using the concept of voltage division yields:

$$V_{ab}(s) = V_{0C}(s)\left[\frac{1}{Zth + 1}\right] \quad \text{where } Zth = \frac{1}{s + 1} \text{ and}$$

$$V_{0C}(s) = \frac{8}{s(s + 1)}$$

$$V_{ab}(s) = \frac{8}{s(s + 1)}\left[\frac{1}{\frac{1}{s + 1} + 1}\right] = \frac{8}{s(s + 1)}\,\frac{s + 1}{s + 2}$$

$$V_{ab}(s) = \frac{8}{s(s + 2)} = \frac{A}{s} + \frac{B}{s + 2}$$

evaluating the constants A & B yields

$$A = s\left(V_{ab}(s)\right)\Big|_{s=0} = 4$$

$$B = (s + 2)\left(V_{ab}(s)\right)\Big|_{s=-2} = -4$$

Therefore, $V_{ab}(s) = \frac{4}{s} - \frac{4}{s + 2}$

$$\text{or} \quad v_{ab}(t) = L^{-1}\left\{\frac{4}{s} - \frac{4}{s + 2}\right\} = (4 - 4e^{-2t})\,u(t).$$

(b) A series combination of 1Ω resistance and a 1H inductor is shown connected to the Thevenin circuit of Fig. 4.

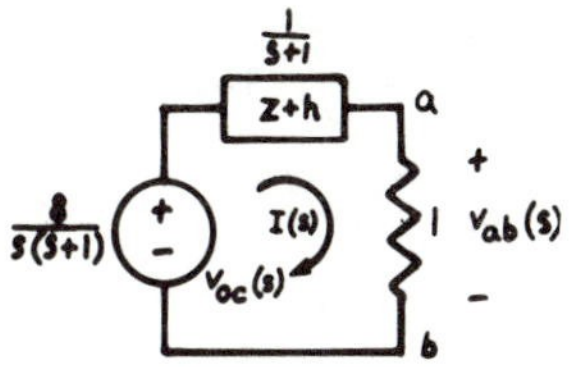

Fig. 5

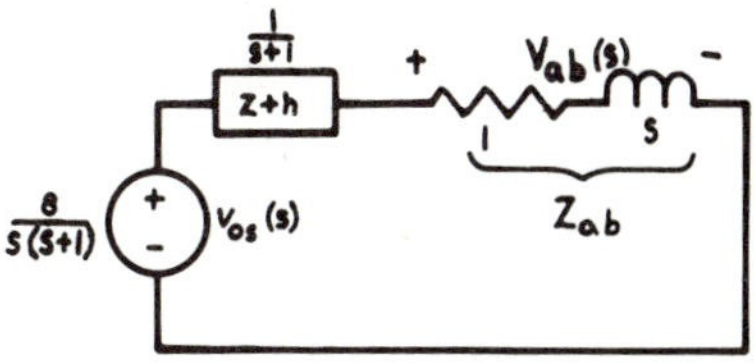

Fig. 6

Using voltage division for circuit in Fig. 6:

$$V_{ab}(s) = V_{OC}(s)\left[\frac{Z_{ab}}{Z_{th} + Z_{ab}}\right] \quad \text{where } Z_{ab} = (1 + s)$$

$$V_{ab}(s) = \frac{8}{s(s + 1)}\left[\frac{s + 1}{\frac{1}{s + 1} + s + 1}\right]$$

simplifying the expression for $V_{ab}(s)$ gives

$$V_{ab}(s) = \frac{8\,(s + 1)}{s(s^2 + 2s + 2)}$$

factoring: $V_{ab}(s) = \dfrac{8\,(s + 1)}{s(s + 1 - j)(s + 1 + j)}$

Using partial fraction expansion yields

$$V_{ab}(s) = \frac{A}{s} + \frac{B}{s + 1 - j} + \frac{B^*}{s + 1 + j}$$

where $A = (s)\; v_{ab}(s)\Big|_{s=0} = 4$

$$B = (s + 1 - j)\; v_{ab}(s)\Big|_{s=-1+j}$$

$$B = \frac{8(j)}{(-1 + j)(2j)} = \frac{8\underline{/90°}}{(\sqrt{2}\underline{/135°})(2\underline{/90°})}$$

$$B = 2\sqrt{2}\underline{/-135°} = -2 - j2$$

$$B^* = -2 + j2$$

$$V_{ab}(s) = \frac{4}{s} + \frac{-2 - j2}{s + 1 - j} + \frac{-2 + j2}{s + 1 + j}$$

in the time domain

$$v_{ab}(t) = 4 + (-2 - j2)e^{(-1+j)t} + (-2 + j2)e^{(-1-j)t}$$

expanding gives:

$$v_{ab}(t) = 4 - 2e^{-t}e^{jt} - j2e^{-t}e^{jt} - 2e^{-t}e^{-jt}$$

$$+ j2e^{-t} e^{-jt}$$

Noting that four terms can be combined into the form

$$\cos t = \frac{e^{jt} + e^{-jt}}{2}; \ \sin t = \frac{e^{jt} - e^{-jt}}{2j}$$

which results in

$$v_{ab} = 4 - 4e^{-t}\left(\frac{e^{jt} + e^{-jt}}{2}\right) + 4e^{-t}\left(\frac{e^{jt} - e^{-jt}}{2j}\right)$$

finally the following is obtained

$$v_{ab}(t) = 4 - 4e^{-t} \cos t + 4e^{-t} \sin t$$

The results can be simplified further by using trigonometric identities.

It is possible to change $4e^{-t} \sin t$ to $4e^{-t} \cos (t-90°)$ and use the function difference relation

$$\cos \alpha - \cos \beta = - 2 \sin \frac{1}{2} (\alpha + \beta) \sin \frac{1}{2} (\alpha - \beta)$$

where $\alpha = t - 90°$; $\beta = t$

$$4e^{-t} (\cos (t-90°) - \cos t)$$

$$= 4e^{-t} \left(- 2 \sin \frac{1}{2} (2t-90°) \sin \frac{1}{2} (- 90°)\right)$$

$$= 4e^{-t} (- 2 \sin (t-45°) \sin (- 45°))$$

$$\sin (- 45°) = \frac{-1}{\sqrt{2}}$$

so that $= 4e^{-t} (\sqrt{2} \sin (t-45°))$ changing

$\sin (t-45°)$ to $\cos (t-45°-90°) = \cos (t-135°)$ gives

$$v_{ab}(t) = 4 \left[1 + \sqrt{2}\, e^{-t} \cos (t-135°)\right]$$

• PROBLEM 12-54

Develop Norton's equivalent network and Thevenin's equivalent network for the network shown. For each case find $I_L(s)$.

Solution: (a) Thevenin's Theorem applies to networks with two output terminals. To apply Thevenin's Theorem to this problem, we disconnect the inductor at points A and B. This leaves a circuit with two output terminals. The Thevenin Equivalent of the remaining portion of the circuit may now be constructed.

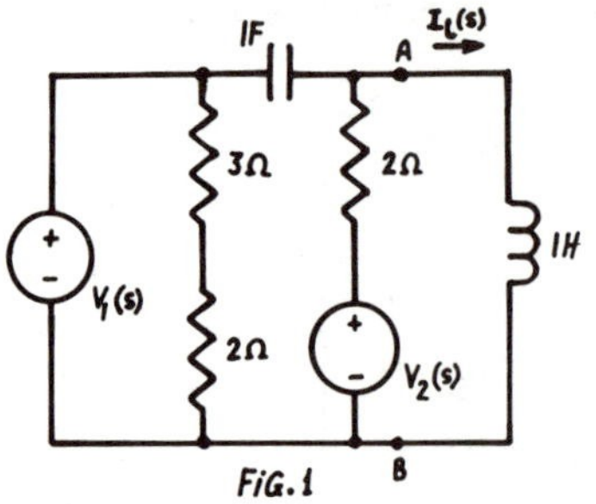

FiG. 1

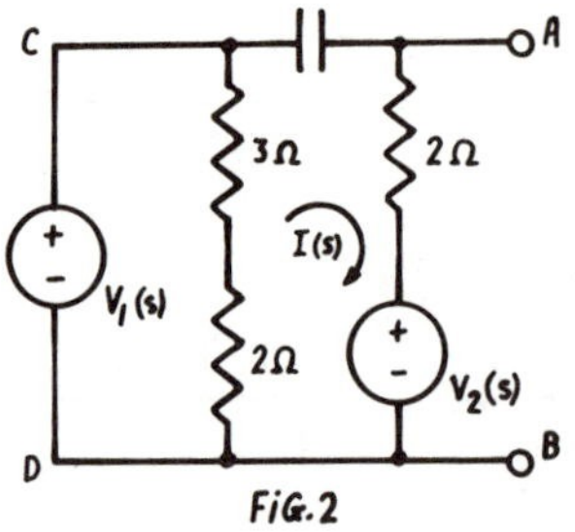

FiG. 2

The Thevenin Equivalent potential is the open-circuit potential, i.e., the potential between points A and B with the inductance removed. This is the potential across the series combination of $V_2(s)$ and the 2Ω resistor. That is,

$$V_{th}(s) = V_2(s) - (2\Omega)\ [I(s)] \qquad (1)$$

where I(s) is shown in Figure 2.

The current I(s) is found using Kirchhoff's voltage sum rule about the loop ACDB:

$$V_2(s) - (2\Omega)\ I(s) - \left(\frac{1}{sC}\right) I(s) - V_1(s) = 0. \qquad (2)$$

Solving for I(s) and substituting C = 1F,

$$I(s) = s\ [V_2(s) - V_1(s)]/\ (2s + 1),$$

and substituting into (1) gives

$$V_{th}(s) = V_2(s) - 2s\ [V_2(s) - V_1(s)]/(2s + 1).$$

Putting the two terms over a common denominator,

$$V_{th}(s) = \frac{(2s + 1)\ V_2(s) - 2sV_2(s) + 2s\ V_1(s)}{(2s + 1)}$$

results in the Thevenin Equivalent voltage,

$$V_{th}(s) = \frac{2s\ V_1(s) + V_2(s)}{(2s + 1)} \qquad (3)$$

The Thevenin Equivalent impedance is found by assuming that the internal impedance of the voltage sources in the network is zero and calculating the resistance between terminals A and B. Setting the voltage sources to zero alters the circuit as shown in Figure 3a.

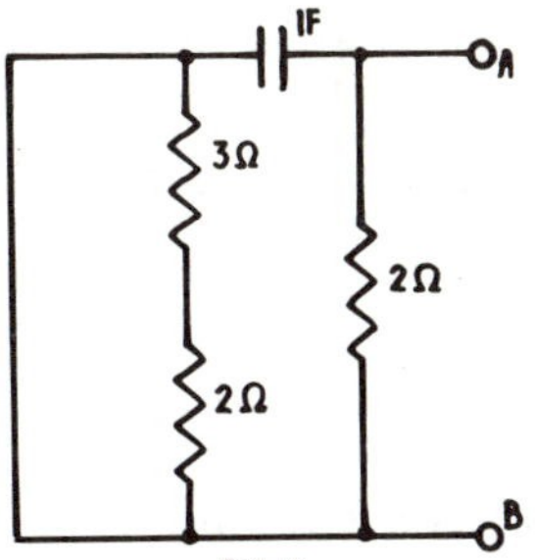

FiG.3a

From this, it can be seen that the 3Ω and 2Ω resistors do not contribute to the Thevenin impedance, and the circuit becomes

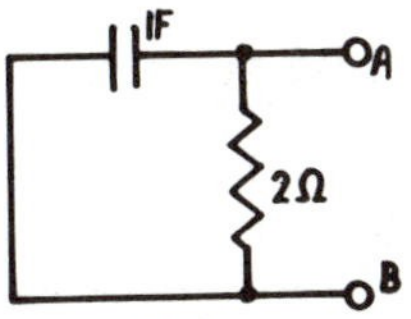

FiG.3b

Therefore, the Thevenin impedance is the parallel combination of the 2Ω resistor and 1F capacitor,

$$Z_{th}(s) = \frac{(1/s)\ (2\Omega)}{(1/s) + (2\Omega)}.$$

Multiplying both the numerator and denominator by s gives

$$Z_{th}(s) = \frac{2}{2s + 1}. \tag{4}$$

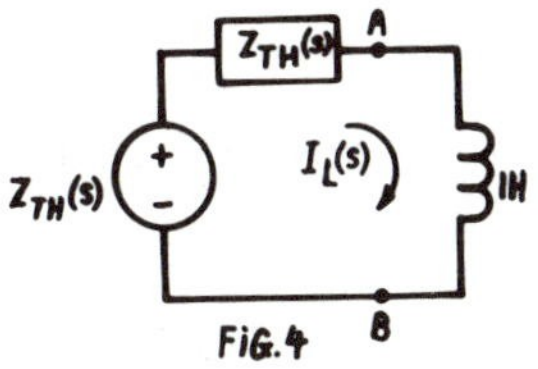

FiG.4

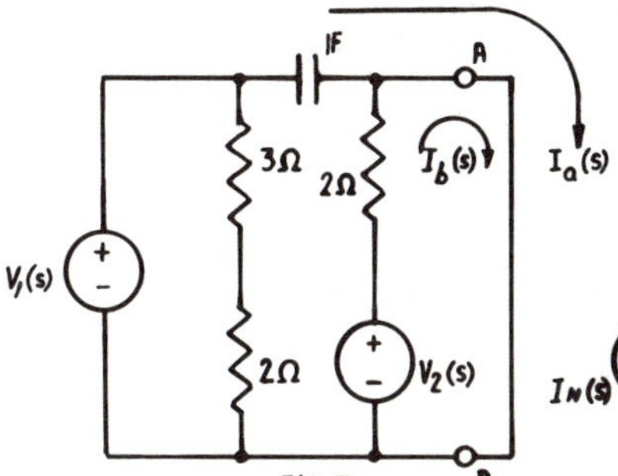

FiG.5

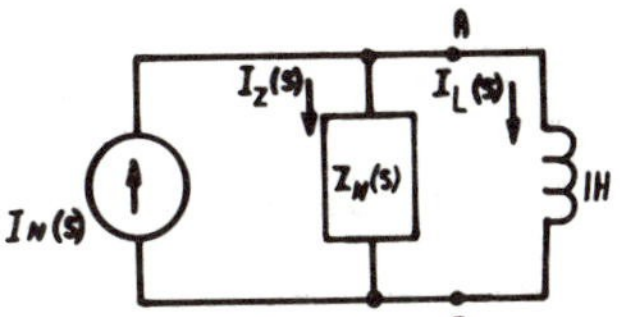

FiG.6

The Thevenin Equivalent network is shown in Figure 4 where $V_{th}(s)$ and $Z_{th}(s)$ have the values calculated above.

The current flowing in the inductor $I_L(s)$ may now be found, using the Thevenin Equivalent. From Figure 4 and Ohm's Law,

$$I_L(s) = V_{th}(s)/Z_T(s) \tag{5}$$

where $Z_T(s)$ is the total impedance in the circuit:

$$Z_T(s) = Z_{th}(s) + Z_L(s), \tag{6}$$

$$Z_T(s) = \frac{2}{2s + 1} + sL.$$

Substituting L = 1H and putting both terms under a common denominator,

$$Z_T(s) = \frac{2s^2 + s + 2}{2s + 1}. \quad (7)$$

Therefore, substituting (3) and (7) into (5) gives

$$I_L(s) = \frac{[2s\ V_1(s) + V_2(s)]/(2s + 1)}{[2s^2 + s + 2]/(2s + 1)}. \quad (8)$$

Multiplying both numerator and denominator by (2s + 1) gives $I_L(s)$:

$$I_L(s) = \frac{2s\ V_1(s) + V_2(s)}{2s^2 + s + 2}.$$

(b) The circuit is analyzed with Norton's Theorem in the same manner, except that the Norton Equivalent network is constructed. The Norton equivalent current source is found by shorting the output terminals, A and B, in Figure 1 and determining the short circuit current, $I_N(s)$. The short circuit current,

$$I_N(s) = I_a(s) + I_b(s), \quad (9)$$

is illustrated in Figure 5.

Using Kirchhoff's voltage sum rule about the I_a loop gives

$$V_1(s) - \left(\frac{1}{s}\right) I_a(s) = 0 \quad (10)$$

and about the I_b loop gives

$$V_2(s) - (2\Omega)\ I_b(s) = 0. \quad (11)$$

From (10),

$$I_a(s) = s\ V_1(s) \quad (12)$$

and from (11),

$$I_b(s) = V_2(s)/2. \quad (13)$$

Combining (9), (12), and (13) gives the Thevenin Equivalent current source,

$$I_N(s) = s\ V_1(s) + V_2(s)/2. \quad (14)$$

The Norton Equivalent impedance R_N is found by the same method as the Thevenin Equivalent impedance. That is,

$$Z_N(s) = Z_{th}(s). \quad (15)$$

Therefore, from (4) and (15),

$$Z_N(s) = \frac{2}{2s + 1}. \quad (16)$$

The Norton Equivalent network is shown in Figure 6 where $I_N(s)$ and $Z_N(s)$ have the values calculated above.

When the inductor is connected across points A and B, the potentials across $Z_N(s)$ and the inductor are equal. That is,

$$V_Z(s) = V_L(s) \quad (17)$$

where $V_Z(s)$ is the voltage drop across $Z_N(s)$ and $V_L(s)$ is the voltage drop across the inductor. From Ohm's Law,

$$V_Z(s) = I_Z(s) \cdot Z_N(s), \quad (18)$$

$$V_L(s) = I_L(s) \cdot Z_L(s). \quad (19)$$

Substituting $Z_L(s) = sL = s(1H)$ and combining (17), (18), and (19) gives

$$I_Z(s) \cdot Z_N(s) = sI_L(s). \quad (20)$$

Now, from Kirchhoff's current sum rule,

$$I_N(s) = I_Z(s) + I_L(s). \quad (21)$$

Solving (21) for $I_Z(s)$ and substituting into (20) gives

$$[I_N(s) - I_L(s)]\ Z_N(s) = sI_L(s).$$

Solving for $I_L(s)$,

$$I_L(s)\ [s + Z_N(s)] = I_N(s) \cdot Z_N(s)$$

$$I_L(s) = I_N(s) \cdot Z_N(s)/[s + Z_N(s)]. \quad (22)$$

Substituting (14) and (16) into (22) gives

$$I_L(s) = \frac{[s\ V_1(s) + V_2(s)/2][2/(2s+1)]}{s + \frac{2}{(2s+1)}}$$

which, with minor algebraic manipulation, becomes

$$I_L(s) = \frac{[2s\ V_1(s) + V_2(s)]/(2s+1)}{\frac{s(2s+1)}{(2s+1)} + \frac{2}{(2s+1)}}.$$

Collecting terms in the denominator,

$$I_L(s) = \frac{[2s\ V_1(s) + V_2(s)]/(2s+1)}{(2s^2 + s + 2)/(2s+1)}.$$

Multiplying both the numerator and denominator by $(2s + 1)$ gives $I_L(s)$:

$$I_L(s) = \frac{2s\ V_1(s) + V_2(s)}{2s^2 + s + 1}.$$

• PROBLEM 12-55

Use Thévenin's equivalent network to calculate the voltage $V_3(s)$ in Fig. 1. Assume zero initial conditions.

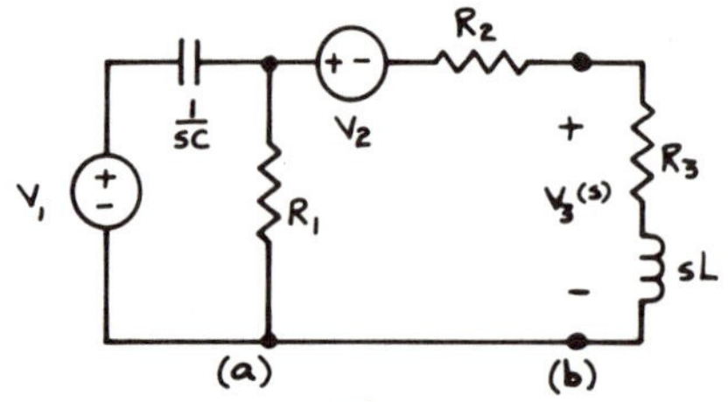

FIG.1

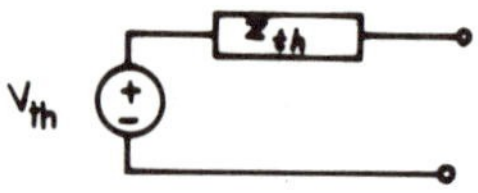

FIG.2

Solution: The circuit in Fig. 1(a) can be replaced by its Thévenin equivalent shown in Fig. 2.

Where Z_{th} is the impedance looking into the circuit and V_{th} is the voltage at the circuit's terminals. Hence,

$$V_{th} = \frac{sR_1C\ V_1}{sR_1C + 1} - V_2$$

and $$Z_{th} = R_2 + \frac{R_1}{sR_1C + 1}$$

Then the solution is,

$$V_3 = \frac{R_3 + sL}{R_3 + sL + Z_{th}} V_{th}$$

$$V_3 = \frac{R_3 + sL}{R_3 + sL + R_2 + \frac{R_1}{sR_1C+1}} \left[\frac{sR_1CV_1}{sR_1C + 1} - V_2\right] .$$

• PROBLEM 12-56

Use Norton's equivalent network to calculate the current $I_3(s)$ in Fig. 1. Assume zero initial conditions.

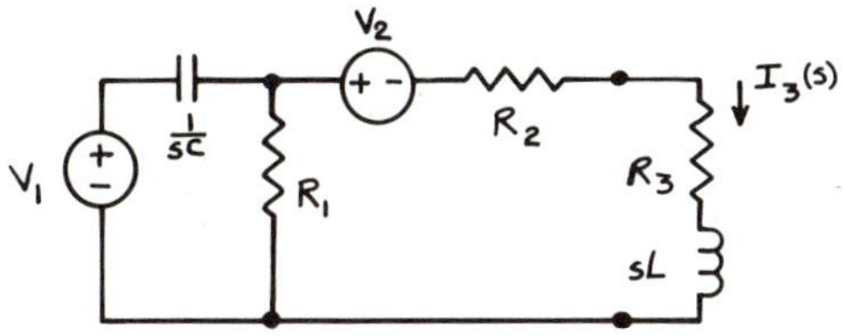

FiG. 1

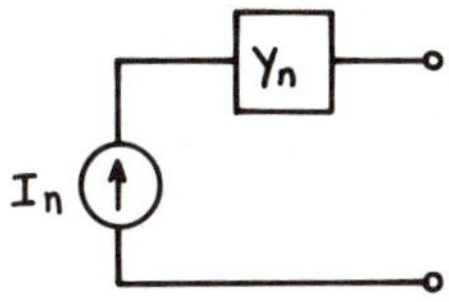

FiG. 2

Solution: The circuit in Fig. 1(a) can be replaced by its Norton equivalent shown in Fig. 2.

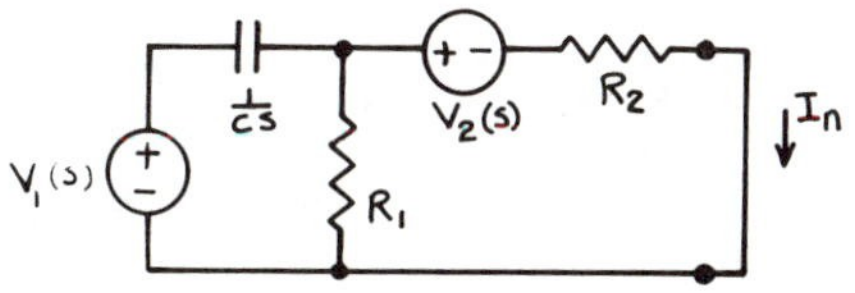

FiG. 3

Where Y_n is the admittance looking into the circuit and I_n is the current at the circuits terminals. The network of Fig. 3 is used to find I_n. The component of I_n due to the V_1 source is

$$I_{nV_1} = \frac{V_1(s)}{R_2} \left[\frac{R_1||R_2}{R_1||R_2 + \frac{1}{cS}}\right]$$

$$= \frac{V_1(s)}{R_2} \left[\frac{\frac{R_1R_2}{R_1+R_2}}{\frac{R_1R_2}{R_1+R_2} + \frac{1}{cS}}\right]$$

$$= V_1(s) \left[\frac{sR_1C}{sR_1R_2C + R_1 + R_2}\right]$$

The component of I_n due to the V_2 source is

$$I_{nV_2} = \frac{-V_2(s)}{R_2 + R_1 || \frac{1}{sC}} = -V_2(s) \frac{(sR_1C + 1)}{sCR_1R_2 + R_2 + R_1} .$$

Hence,

$$I_n(s) = I_{nV_2} + I_{nV_1}$$

$$I_n(s) = \frac{sR_1C\ V_1(s)}{R_1 + sR_1R_2C + R_2} - \frac{(sR_1C+1)V_2(s)}{sCR_1R_2 + R_2 + R_1}$$

and $$Y_n(s) = \frac{sR_1C + 1}{sR_1R_2C + R_1 + R_2}$$

Then the solution is

$$I_3(s) = I_n(s) \frac{\frac{1}{R_3+sL}}{\frac{1}{R_3+sL} + Y_n}$$

$$I_3 = \frac{sR_1CV_1 - (sR_1C+1)V_2}{(sR_1C+1)(R_3+sL)(sR_1R_2C+R_1+R_2) + (sR_1R_2C+R_1+R_1)^2}$$

• PROBLEM 12-57

Calculate Thévenin and Nortons equivalent networks for the circuit below.

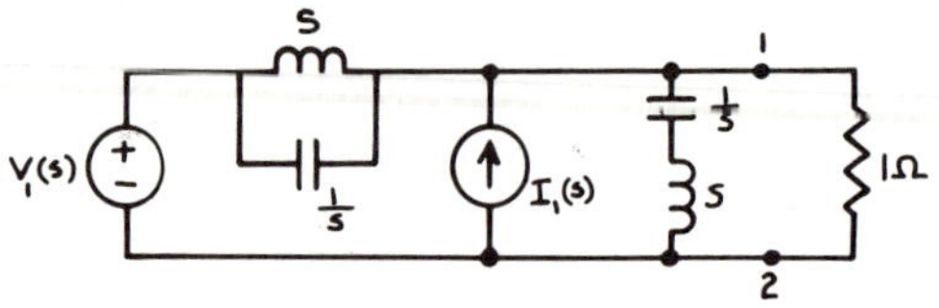

Solution: In order to obtain Norton's equivalent internal impedance $Z_e(s)$, the voltage source $V_1(s)$ and the current source $I_1(s)$ must be turned off. That is the voltage source terminals must be short circuited and the current source terminals must be open circuited. Under these conditions, the internal impedance of the network, which is at the left side of terminals 1 and 2, is the parallel impedance of an LC series circuit and LC parallel circuit. Therefore the total impedance is

$$\frac{1}{Z_e(s)} = \frac{1}{s} + \frac{1}{\frac{1}{s}} + \frac{1}{\frac{1}{s} + s} = \frac{1}{s} + s + \frac{s}{s^2 + 1}$$

$$= \frac{s^2 + 1 + s^2(s^2+1) + s^2}{s(s^2+1)} = \frac{s^4 + 3s^2 + 1}{s(s^2+1)} \qquad (1)$$

$$Z_e(s) = \frac{s(s^2+1)}{s^4 + 3s^3 + 1} \qquad (2)$$

This is the Norton equivalent internal impedance.

In order to obtain Thévenin's equivalent open circuit voltage, the principle of superposition will be employed. First, the load resistor 1Ω will be removed from the circuit to open terminals 1 and 2. Then, short circuit the voltage source $V_1(s)$. The voltage across terminals 1 and 2, due only to the current source $I_1(s)$, is then

$$V_{cs} = I_1(s)Z_e(s) \qquad (3)$$

where $Z_e(s)$ is the Norton equivalent impedance given by Eq.(2).

Now, remove the short circuit across the voltage source $V_1(s)$ turning $V_1(s)$ back on. Then turn off the current source $I_1(s)$ by replacing it by an open circuit. Then the voltage across the terminal 1 and 2, due only to the voltage source $V_1(s)$, is obtained by the application of the voltage division principle between the parallel LC circuit impedance and the series LC circuit impedance. The voltage across the series LC circuit is therefore

$$V_{vs} = V_1(s) \cdot \frac{\frac{1}{s} + s}{\frac{1}{\frac{1}{s} + \frac{1}{\frac{1}{s}}} + \frac{1}{s} + s} = V_1(s) \frac{\frac{1}{s} + s}{\frac{s}{s^2+1} + \frac{1}{s} + s}$$

$$= V_1(s) \frac{\frac{s^2 + 1}{s}}{\frac{s^2 + s^2 + 1 + s^2(s^2+1)}{s(s^2+1)}}$$

$$= V_1(s) \frac{s^2+1}{s} \cdot \frac{s(s^2+1)}{s^4 + 3s^2 + 1} \qquad (4)$$

Combining Eqs.(2) and (4)

$$V_{vs} = V_1(s) \frac{s^2+1}{s} Z_e(s) \qquad (5)$$

By the principle of superposition, when both the voltage source $V_1(s)$ and the current source $I_1(s)$ are turned on as shown in the figure given, the open circuit voltage across the terminals 1 and 2 without the load resistance of 1 ohm, or the Thévenin equivalent open circuit voltage is

$$V_{oc} = V_{cs} + V_{vs} = I_1(s)Z_e(s) + V_1(s)\frac{s^2+1}{s} Z_e(s) \quad (6)$$

where $Z_e(s)$ is given by Eq.(2).

STATE EQUATIONS

• PROBLEM 12-58

In the circuit of Fig. 1; (a) write the state equations. (b) Obtain from the state equations the transfer impedance relating V_3 to i_s; (c) the driving point impedance at the source terminals.

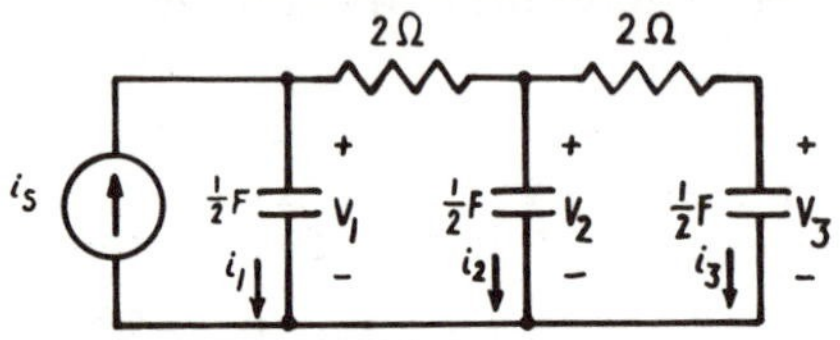

Fig. 1

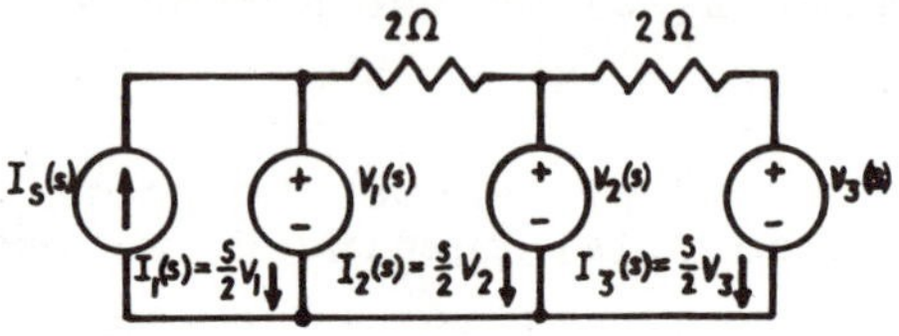

Fig. 2

Solution: (a) Using the transformed circuit of Fig. 2, where all the capacitors are replaced by voltage sources, use superposition to find the responses $I_1(s)$, $I_2(s)$, and $I_3(s)$ which give the state equations.

The equations obtained from Fig. 2 are,

$$sV_1 = -V_1 + V_2 + 2I_s$$

$$sV_2 = V_1 - 2V_2 + V_3$$

$$sV_3 = V_2 - V_3.$$

Write them in matrix form as

$$\begin{bmatrix} sV_1 \\ sV_2 \\ sV_3 \end{bmatrix} = \begin{bmatrix} -1 & 1 & 0 \\ 1 & -2 & 1 \\ 0 & 1 & -1 \end{bmatrix} \begin{bmatrix} V_1 \\ V_2 \\ V_3 \end{bmatrix} + \begin{bmatrix} 2I_s \\ 0 \\ 0 \end{bmatrix} \quad (1)$$

or

$$\begin{bmatrix} s+1 & -1 & 0 \\ -1 & s+2 & -1 \\ 0 & -1 & s+1 \end{bmatrix} \begin{bmatrix} V_1 \\ V_2 \\ V_3 \end{bmatrix} = \begin{bmatrix} 2I_s \\ 0 \\ 0 \end{bmatrix} \quad (2)$$

(b) Find the transfer impedance relating V_3 to I_s by solving for V_3 in matrix (2).

$$V_3(s) = \frac{\begin{vmatrix} s+1 & -1 & 2I_s \\ -1 & s+2 & 0 \\ 0 & -1 & 0 \end{vmatrix}}{\begin{vmatrix} s+1 & -1 & 0 \\ -1 & s+2 & -1 \\ 0 & -1 & s+1 \end{vmatrix}}$$

$$= \frac{2I_s}{(s+1)((s+2)(s+1)-1) + (-1)(s+1)}$$

$$V_3(s) = \frac{2I_s}{s^3 + 4s^2 + 3s}$$

$$Z(s) = \frac{V_3(s)}{I_s(s)} = \frac{2}{s^3 + 4s^2 + 3s}$$

(c) The driving point impedance is the impedance looking into the circuit from the current source.

The driving point impedance can be expressed as $\frac{V_1(s)}{I_s(s)}$. Thus, solve matrix (2) for V_1

$$V_1(s) = \frac{\begin{vmatrix} 2I_s & -1 & 0 \\ 0 & s+2 & -1 \\ 0 & -1 & s+1 \end{vmatrix}}{s^3 + 4s^2 + 3s} = \frac{2I_s[(s+2)(s+1) - 1]}{s^3 + 4s^2 + 3s}$$

$$V_1(s) = \frac{2s^2 + 6s + 2}{s^3 + 4s^2 + 3s} I_s$$

$$Z_d(s) = \frac{2s^2 + 6s + 2}{s^3 + 4s^2 + 3s}$$

For the circuit of Fig. 1 find the state equations using the state variables x_1, x_2, and x_3 indicated.

Show that the state equations yield the characteristic equation for the circuit D(s), and the network equations for $I_1(s)$, $I_2(s)$, and V(s).

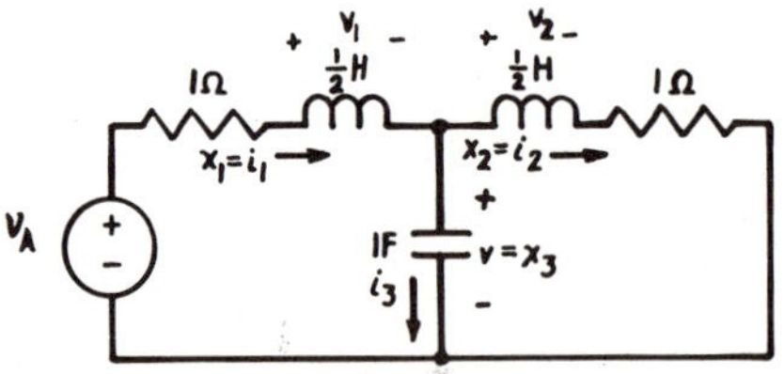

FiG.1

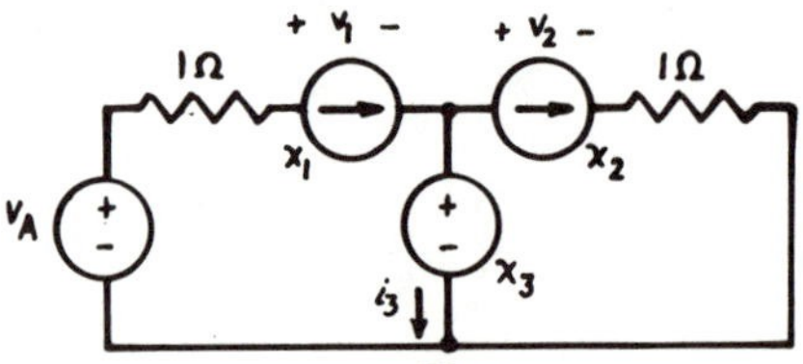

FiG.2

Solution: Write the state equations by applying the theory of superposition. Replace each energy-storing element (capacitors and inductors) by an independent source equal to it's state variable, as shown in Fig. 2.

Express $V_1(s)$, $V_2(s)$ and $I_3(s)$ in terms of state variables and impedance

$$V_1(s) = sL_1x_1 \quad V_2(s) = sL_2x_2 \qquad I_3(s) = sCx_3.$$

Find V_1, V_2, and I_3 in terms of the state variables and circuit parameters by allowing only one source to be "on" at a time. The sum of these expressions for V_1, V_2 and I_3 give the state equations. The table below gives the expressions obtained.

	V_A "on"	x_1 "on"	x_2 "on"	x_3 "on"
$V_1 = \frac{s}{2} x_1$	V_A	$-x_1$	0	$-x_3$
$V_2 = \frac{s}{2} x_2$	0	0	$-x_2$	x_3
$I_3 = s x_3$	0	x_1	$-x_2$	0

The equations obtained from above become

$$\frac{s}{2} x_1 = -x_1 - x_3 + V_A$$

$$\frac{s}{2} x_2 = -x_2 + x_3$$

$$sx_3 = x_1 - x_2$$

Writing the matrix in the form $\dot{\bar{x}} = \bar{A}\bar{x} + \bar{b}$ gives

$$\begin{bmatrix} sx_1 \\ sx_2 \\ sx_3 \end{bmatrix} = \begin{bmatrix} -2 & 0 & -2 \\ 0 & -2 & 2 \\ 1 & -1 & 0 \end{bmatrix} \begin{bmatrix} x_1 \\ x_2 \\ x_3 \end{bmatrix} + \begin{bmatrix} 2V_A \\ 0 \\ 0 \end{bmatrix}.$$

By manipulating the matrices as follows

$$s\bar{x} = \bar{A}\bar{x} + \bar{b}$$

$$s\bar{x} - \bar{A}\bar{x} = \bar{b}$$

$$(\overline{s - A})\bar{x} = \bar{b} \quad \text{we obtain}$$

$$\bar{D}x = b$$

The determinate of the matrix $\bar{D}$ is the characteristic equation for the circuit where

$$x_1 = \frac{\begin{vmatrix} b & x_2 & x_3 \\ b & x_2 & x_3 \\ b & x_2 & x_3 \end{vmatrix}}{|\bar{D}|}$$

$$x_2 = \frac{\begin{vmatrix} x_1 & b & x_3 \\ x_1 & b & x_3 \\ x_1 & b & x_3 \end{vmatrix}}{|\bar{D}|} \quad \text{and}$$

$$x_3 = \frac{\begin{vmatrix} x_1 & x_2 & b \\ x_1 & x_2 & b \\ x_1 & x_2 & b \end{vmatrix}}{|\bar{D}|}$$

By manipulating the matrices as indicated

$$\begin{aligned} x_1(s + 2) \qquad\qquad\qquad & + 2x_3 = 2V_A \\ x_2(s + 2) & - 2x_3 = 0 \\ -x_1 \qquad + x_2 \qquad\qquad & + x_3 s = 0 \end{aligned}$$

yields
$$\begin{bmatrix} s+2 & 0 & 2 \\ 0 & s+2 & -2 \\ -1 & 1 & s \end{bmatrix} \begin{bmatrix} x_1 \\ x_2 \\ x_3 \end{bmatrix} = \begin{bmatrix} 2V_A \\ 0 \\ 0 \end{bmatrix}$$

The characteristic equation is

$$|\bar{D}| = \begin{vmatrix} s+2 & 0 & 2 \\ 0 & s+2 & -2 \\ -1 & 1 & s \end{vmatrix} =$$

$$(s + 2)((s + 2)s + 2) - 1 (- 2 (s + 2))$$

$$= (s + 2)(s^2 + 2s + 2) + 2s + 4$$

$$= s^3 + 4s^2 + 8s + 8.$$

The state variables in terms of the other circuit parameters are

$$x_1 = I_1(s) = \frac{\begin{vmatrix} 2V_A & 0 & 2 \\ 0 & s+2 & -2 \\ 0 & 1 & s \end{vmatrix}}{s^3 + 4s^2 + 8s + 8} = \frac{2V_A[(s + 2)s + 2]}{s^3 + 4s^2 + 8s + 8}$$

$$= \frac{V_A(2s^2 + 4s + 4)}{s^3 + 4s^2 + 8s + 8}$$

$$x_2 = I_2(s) = \frac{\begin{vmatrix} s+2 & 2V_A & 2 \\ 0 & 0 & -2 \\ -1 & 0 & s \end{vmatrix}}{s^3 + 4s^2 + 8s + 8} = \frac{-1\,(2V_A\,(-2))}{s^3 + 4s^2 + 8s + 8}$$

$$= \frac{4V_A}{s^3 + 4s^2 + 8s + 8}$$

$$x_3 = V(s) = \frac{\begin{vmatrix} s+2 & 0 & 2V_A \\ 0 & s+2 & 0 \\ -1 & 1 & 0 \end{vmatrix}}{s^3 + 4s^2 + 8s + 8} = \frac{-1\,(-2V_A\,(s + 2))}{s^3 + 4s^2 + 8s + 8}$$

$$= \frac{V_A\,(2s + 4)}{s^3 + 4s^2 + 8s + 8}$$

This example illustrates the advantages of state representation.

CHAPTER 13

FREQUENCY DOMAIN ANALYSIS

POLES & ZEROS

• PROBLEM 13-1

Obtain the pole-zero plot for

$$H(s) = \frac{s(s^2 + \sqrt{2}s + 1)}{s^2 + 3s + 2}$$

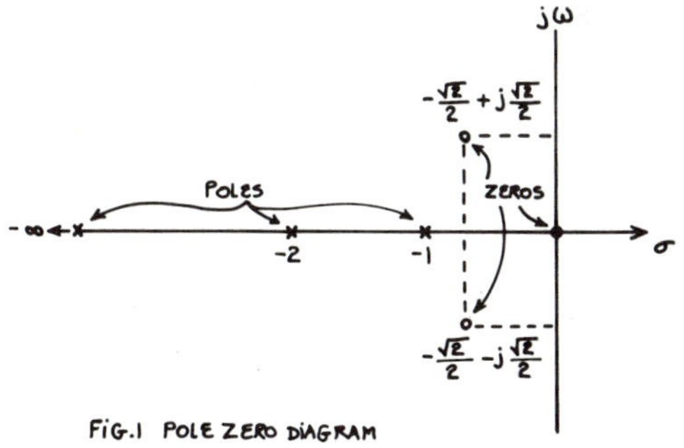

FIG.1 POLE ZERO DIAGRAM

Solution: The denominator and numerator polynomials must be factored into the form

$$H(s) = \frac{(s \pm a_1)(s \pm a_2)(s \pm a_3) \dots}{(s \pm b_1)(s \pm b_2)(s \pm b_3) \dots}.$$

The poles are indicated in the denominator as the roots of that polynomial. The zeros are indicated in the numerator as the roots of that polynomial.

Factoring the denominator gives

$$(s^2 + 3s + 2) = (s + 1)(s + 2),$$

poles = -1, -2 and ∞.

Factoring the numerator yields

$$s(s^2 + \sqrt{2}s + 1) = (s + 0)(s + (-z_1))(s + (-z_2))$$

$$z_1, z_2 = \frac{-b \pm \sqrt{b^2 - 4ac}}{2a} = \frac{-\sqrt{2} \pm \sqrt{2 - 4}}{2}$$

$$z_1, z_2 = \frac{-\sqrt{2}}{2} \pm J\frac{\sqrt{-2}}{2} = \frac{-\sqrt{2}}{2} \pm J\frac{\sqrt{2}}{2},$$

which gives $(s + 0)(s + \frac{\sqrt{2}}{2} - j\frac{\sqrt{2}}{2})(s + \frac{\sqrt{2}}{2} + j\frac{\sqrt{2}}{2})$,

zeroes $= 0, \frac{-\sqrt{2}}{2} + j\frac{\sqrt{2}}{2}, \frac{-\sqrt{2}}{2} - j\frac{\sqrt{2}}{2}$.

The pole-zero diagram is shown in fig. 1.

• PROBLEM 13-2

When a source voltage $v(t) = 5e^{-2t}$ is impressed on a network, the complete response v_2 is given by $v_2(t) = 3te^{-2t} + 2e^{-3t}\sin 6t$. Locate in the s-plane the poles of the voltage gain function relating v_2 to v.

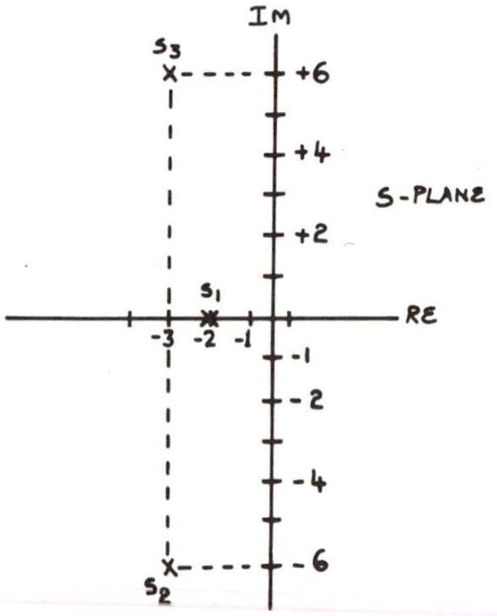

FIG. 1

Solution: The voltage gain function $H(s) = \frac{V_2(s)}{V_1(s)}$ is

$$\frac{L\{3te^{-2t} + 2e^{-3t}\sin 6t\}}{L\{5e^{-2t}\}}.$$

Hence,

$$H(s) = \frac{\frac{3}{(s+2)^2} + \frac{2}{(s+3-j6)(s+3+j6)}}{\frac{5}{(s+2)}}$$

can be written as

$$\frac{3(s + 3 + j6)(s + 3 - j6) + 2(s + 2)^2}{5(s + 2)(s + 3 + j6)(s + 3 - j6)} .$$

The poles of the function are found in the denominator:

$$s_1 = -2$$

$$s_2 = -3 - j6$$

$$s_3 = -3 + j6.$$

Each pole is located in the s-plane as shown in Fig. 1.

• PROBLEM 13-3

> The voltage ratio $\frac{\vec{V}_2}{\vec{V}_1}$ for the circuit shown in Fig. 1 has a pole at -100 + j700. If R = 500Ω, find L and C.

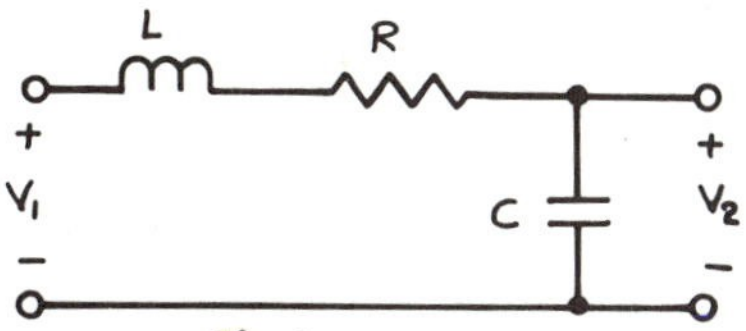

FIG. 1

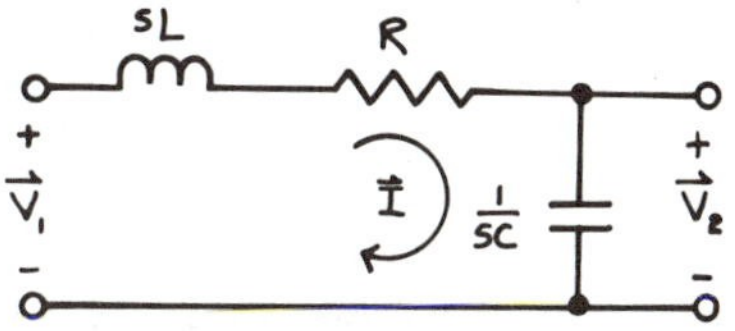

FIG. 2

Solution: Since complex poles are always found in conjugate pairs another pole can be found at -100 - j700. We can find the values of L and C by solving for them from the equation for $\frac{\vec{V}_2}{\vec{V}_1}$. Hence, from Fig. 2,

$$\vec{V}_1 = \vec{I}(sL + R + \frac{1}{sC})$$

$$\vec{V}_2 = \vec{I}(\frac{1}{sC}) ,$$

$$\frac{\vec{V}_2}{\vec{V}_1} = \frac{\frac{1}{sC}}{sL + R + \frac{1}{sC}} = \frac{\frac{1}{LC}}{s^2 + s\frac{R}{L} + \frac{1}{LC}} .$$

$s^2 + s\frac{R}{L} + \frac{1}{LC}$ can be factored into the form (s - p)(s - p*) where p and p* are the poles -100 + j700 and -100 -j700.

Using the quadratic formula

$$p, p^* = \frac{-\frac{R}{L} \pm \sqrt{(\frac{R}{L})^2 - \frac{4}{LC}}}{2} = -\frac{R}{2L} \pm \sqrt{(\frac{R}{2L})^2 - \frac{4}{LC}} .$$

Since R = 500Ω and Real part, $-\frac{R}{2L}$, is -100 we can solve for L.

Hence,

$$-\frac{500}{2L} = -100$$

$$L = \frac{5}{2}\ \text{H.}$$

Now we can solve for C as follows:

$$\left|\left(\frac{R}{2L}\right)^2 - \frac{1}{LC}\right| = 700^2$$

$$\left|\left(\frac{500}{5}\right)^2 - \frac{1}{\frac{5}{2}C}\right| = 700^2$$

$$\left|10^4 - \frac{2}{5C}\right. = 700^2$$

$$\left|-\frac{2}{5C}\right| = 700^2 - 10^4$$

$$C = \left|\frac{-2}{5(700^2 - 10^4)}\right|$$

$$C = 0.83\ \mu\text{F}$$

• PROBLEM 13-4

An impedance consists of a 2.5 mH inductor in series with the parallel combination of a 5-Ω resistor and a 50-μF capacitor. Draw a pole-zero configuration for Z(s).

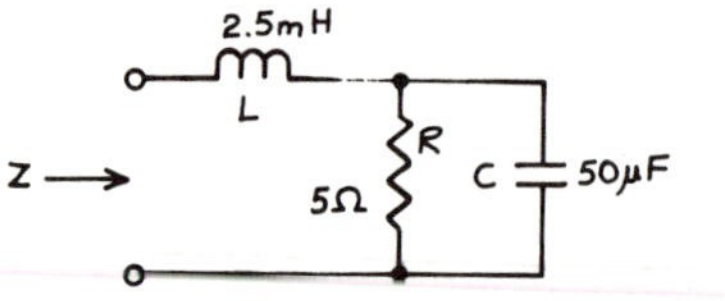

Fig.1

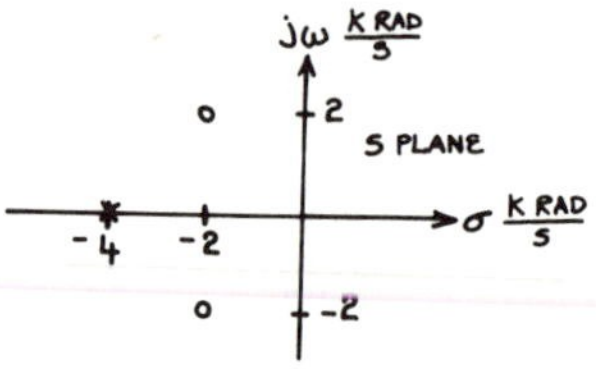

Fig.2

Solution: The circuit is drawn in fig. 1.

Write an expression for Z(s):

$$Z(s) = Z_L + Z_R \| Z_C$$

$$Z(s) = sL + \frac{\frac{R}{SC}}{R + \frac{1}{SC}}$$

$$Z(s) = \frac{sL(sRC + 1) + R}{sRC + 1}$$

$$Z(s) = \frac{s^2 LRC + sL + R}{sRC + 1}$$

$$Z(s) = L\left[\frac{s^2 + s\frac{1}{RC} + \frac{1}{LC}}{s + \frac{1}{RC}}\right].$$

Substituting the values for R, C, and L gives

$$Z(s) = 2.5 \times 10^{-3}\,\frac{s^2 + 4 \times 10^3 s + 8 \times 10^6}{s + 4 \times 10^3}.$$

Factoring the numerator as follows,

$$Z(s) = 2.5 \times 10^{-3}\,\frac{(s+2x10^3-j2x10^3)(s+2x10^3+j2x10^3)}{(s + 4x10^3)}$$

yields

zeros: $-2 + j2 \quad K\frac{rad}{s}$

$-2 - j2 \quad K\frac{rad}{s}$

pole: $-4 \quad K\frac{rad}{s}$.

The pole-zero configuration for Z(s) is shown in fig. 2.

FREQUENCY RESPONSE

• PROBLEM 13-5

In a network, a current source I(s) produces a response V(s) such that

$$V(s) = \frac{s}{s^2 + s + 1}\, I(s).$$

Use the graphical method to determine the component of the response due to: (a) a sinusoidal source i(t) = 10 cos (t + 30°)A; (b) $i(t) = 10e^{-t}$ cos t A.

Solution: The procedure for the graphical solution is to plot the poles and zeros of the transfer function on the complex s plane and evaluate it for the desired frequency and then multiply this response by the input current function.

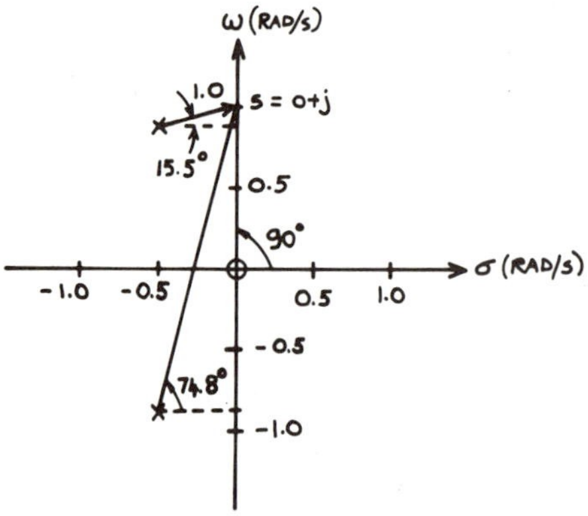

FIG.1 COMPLEX S PLANE FOR PART a

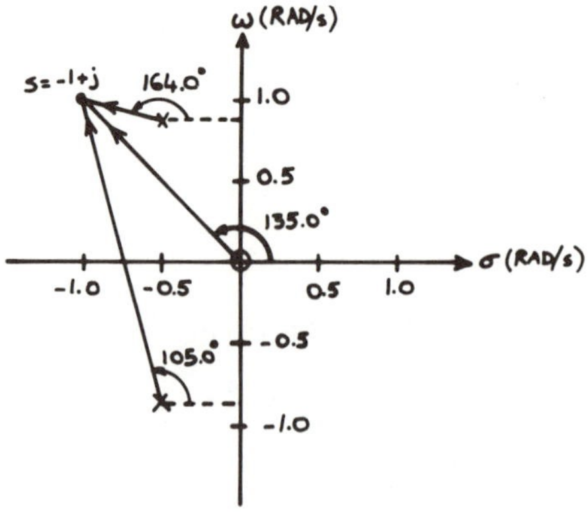

FIG.2 COMPLEX S PLANE FOR PART b

The transfer function F(s) is

$$F(s) = \frac{s}{s^2 + s + 1} = \frac{s}{(s+0.5+j0.866)(s+0.5-j0.866)} .$$

A plot of the poles and zero is shown in Fig. 1.

(a) The current source is $i(t) = 10 \cos(t + 30°)$A with an angular frequency of 1 rad/s. The response of the transfer function to $\omega = 1$ rad/s is found by graphically evaluating F(s) at $s = j1$.

$$F(j1) = \frac{1 \underline{/90°}}{(0.525 \underline{/15.5°})(1.93 \underline{/74.8°})}$$

$$= 0.987 \underline{/-0.3°} = 0.987e^{-j0.3°}$$

The complex voltage is found by multiplying F(j1) by the complex current, $10e^{j(t+30°)}$, giving

$$9.87e^{j(t+29.7°)} .$$

v(t) is then determined by taking the real part:

$$v(t) = 9.87 \cos (t + 29.7°) \text{ V.}$$

(b) The exponentially decaying current, $i(t) = 10e^{-t} \cos t$ A, will cause the voltage to decay also. In complex notation,

$$i(t) = 10e^{-t}e^{jt} = 10e^{(-1+j)t} \text{ A.}$$

In this situation F(s) is evaluated at $s = -1+j$ as shown in Fig. 2.

$$F(-1+j) = \frac{1.43 \underline{/135°}}{(0.525 \underline{/164°})(1.93 \underline{/105})}$$

$$= 1.41 \underline{/-134°} = \text{Re}\, 1.41e^{-j135°}$$

Multiplying this by the original complex vector gives

$$14.1e^{(-1+j)t}e^{-j135°} = 14.1e^{-t}e^{j(t-134°)}.$$

The final result for v(t) is found by taking the real part:

$$v(t) = 14.1e^{-t} \cos (t-134°) \text{ V}.$$

• PROBLEM 13-6

Find $\frac{|I_2|}{|I_s|}$ in the circuit shown in fig. 1 if ω = : (a) 200; (b) 2000; (c) $20x10^3$ rad/s.

Solution: We can write an expression for $\frac{I_2}{I_s}$ by inspection if we recognize that the circuit is a current divider. Hence, the resistor and inductor can be combined to form an impedance:

$$Z = \frac{sRL}{sL + R},$$

then

$$I_2 = \frac{ZI_s}{Z + \frac{1}{sC}}$$

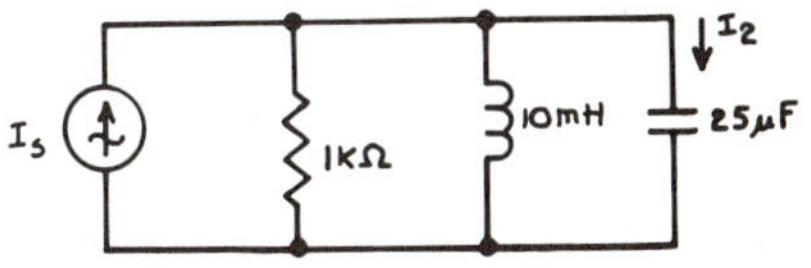

FIG. 1

and

$$\frac{I_2}{I_s} = \frac{Z}{Z + \frac{1}{sC}}$$

$$\frac{I_2}{I_s} = \frac{\frac{sRL}{sL + R}}{\frac{sRL}{sL + R} + \frac{1}{sC}} = \frac{s^2}{s^2 + s\frac{1}{RC} + \frac{1}{LC}}.$$

Substituting R = 1KΩ, L = 10mH and C = 25 mF:

$$\frac{I_2}{I_s} = \frac{s^2}{s^2 + 40s + 4x10^6}.$$

Factoring the denominator we obtain the complex poles:

$$\frac{s^2}{(s + 20 - j2x10^3)(s + 20 + j2x10^3)}$$

poles: at $s = -20 + j2x10^3$

and $s = -20 - j2x10^3$

zeros: two zeros at $s = 0$.

The pole-zero diagram for $\frac{I_2}{I_s}$ is shown in fig. 2.

The magnitude of $\frac{I_2}{I_s}$, at a specific frequency ω, can be found quickly from the pole-zero diagram by finding the product of the distances for the zeros to the desired frequency ω on the $j\omega$-axis, and dividing by the product of the distances from the poles to the frequency ω on the $j\omega$-axis.

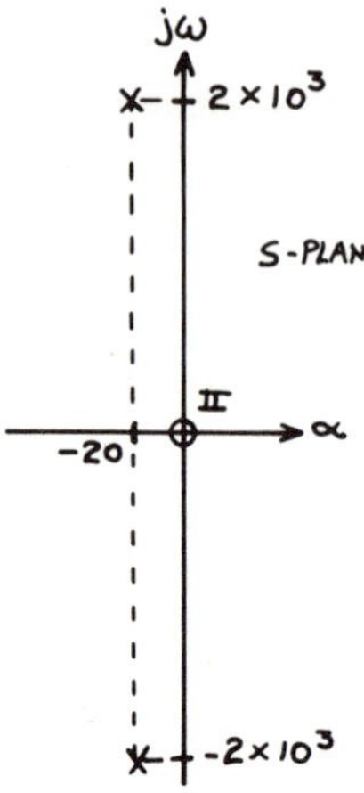

FIG. 2

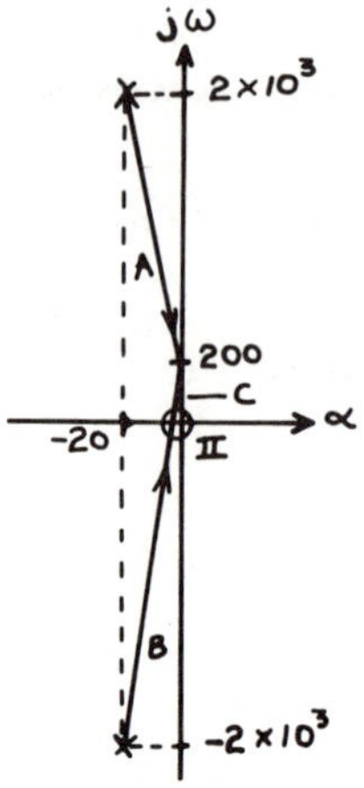

FIG. 3

(a) To find $\frac{I_2}{I_s}$ at $\omega = 200 \frac{\text{rad}}{\text{s}}$ we find the distances shown in Fig. 3.

$$\left|\frac{I_2}{I_s}\right| = \frac{C^2}{AB} = \frac{200^2}{(1.8\times10^3)(2.2\times10^3)} = 0.0101$$

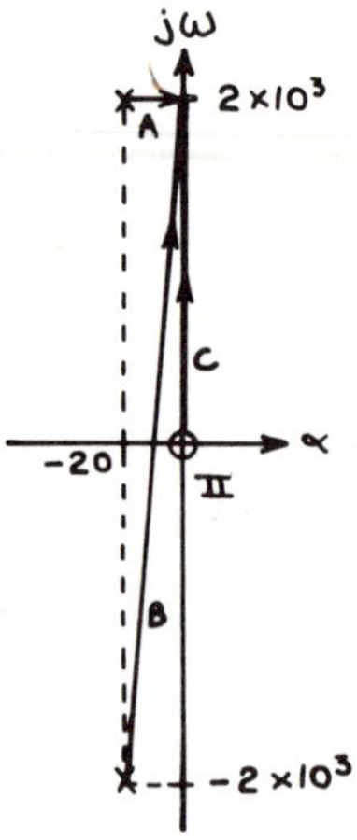

FIG. 4

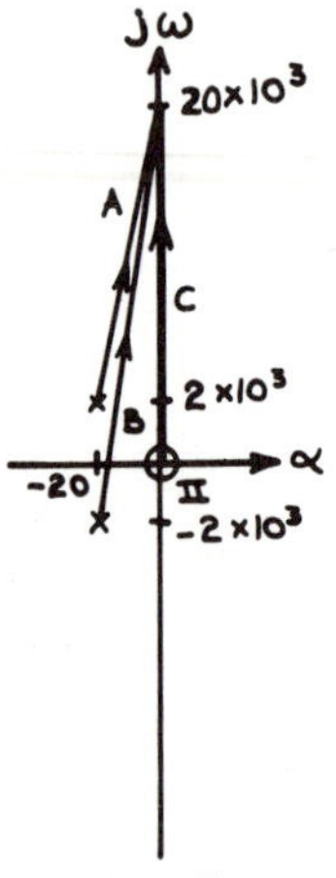

FIG. 5

(b) To find $\left|\frac{I_2}{I_s}\right|$ at $\omega = 2\text{x}10^3$ rad/s find the distances A, B, and C in fig. 4

$$\frac{|I_2|}{|I_s|} = \frac{C^2}{AB} = \frac{2000^2}{20(4\times10^3)} = 50$$

(c) To find $\frac{|I_2|}{|I_s|}$ at $\omega = 20\times10^3$ rad/s find the distances A, B, and C in fig. 5

$$\frac{|I_2|}{|I_s|} = \frac{C^2}{AB} = \frac{(20\times10^3)^2}{(18\times10^3)(22\times10^3)} = 1.01$$

• PROBLEM 13-7

For a certain response-source relationship the transform network function is $H(s) = \frac{s+1}{(s+2)(s^2+2s+10)}$. Calculate graphically the component of the response due to each of the source functions: (a) $20e^{-3t}$; (b) $20 \cos 3t$; (c) $20 \cos(3t - 30°)$; (d) $20e^{-3t} \cos t$.

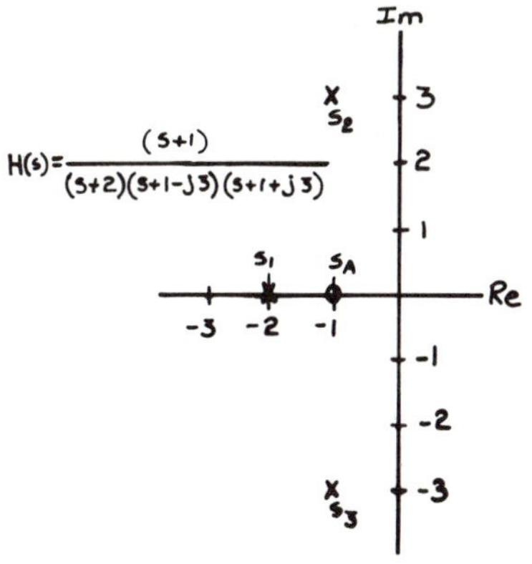

FIG.1

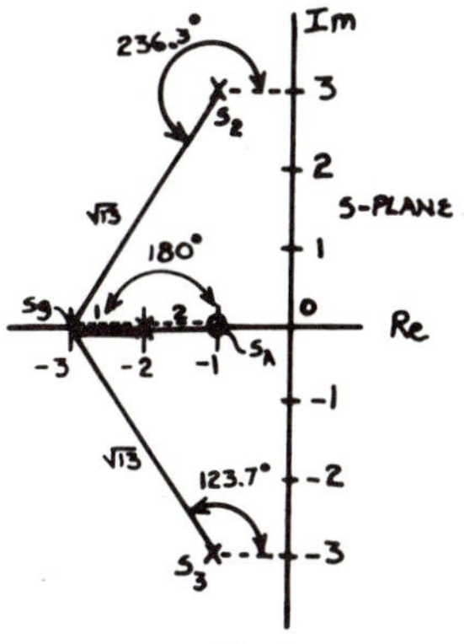

FIG.2

Solution: If a certain source function $Ae^{s_g t}$ is impressed on a network, then if s_g is not a pole of the network function, the response component due to the source is

$$y_s(t) = \frac{N(s_g)}{D(s_g)} Ae^{s_g t} = H(s_g)Ae^{s_g t}$$

or $$y_s(t) = H \frac{(s_g - s_A)(s_g - s_B)\ . \ . \ . \ (s_g - s_N)}{(s_g - s_1)(s_g - s_2)\ . \ . \ . \ (s_g - s_M)} Ae^{s_g t}.$$

Hence, the evaluation of the complete response depends on the evaluation of products of the form $\prod_k (s_g - s_k)$. Since each term $(s_g - s_k)$ may be a complex number, it is convenient to carry out the multiplication and division in polar form.

(a) The network function H(s) has a zero at $s_A = -1$ and three poles, $s_1 = -2$, $s_2 = -1 + j3$, $s_3 = -1 - j3$, as shown in fig. 1.

The source function $20e^{-3t}$ can be represented by a pole at $s_g = -3$. Hence, the response due to the source is:

$$y_s(t) = \frac{|s_g - s_A| \angle\theta_{g-A}}{|s_g - s_1||s_g - s_2||s_g - s_3| \angle\theta_{g-1} \angle\theta_{g-2} \angle\theta_{g-3}} 20e^{-3t} .$$

The above calculation is illustrated graphically in fig. 2.

$$y_s(t) = \frac{2 \angle 180°}{1 \cdot \sqrt{13} \cdot \sqrt{13} \angle 180° \angle 236.3° \angle 123.7°} 20e^{-3t}$$

$$y_s(t) = \frac{2 \angle 180°}{13 \angle 540°} 20e^{-3t} = \frac{2}{13} 20e^{-3t}$$

$$y_s(t) = 3.08e^{-3t}$$

(b) The source function $20\cos 3t$ can be represented by a pole at $s_g = j3$. The procedure for finding the response due to this source is illustrated below in fig. 3.

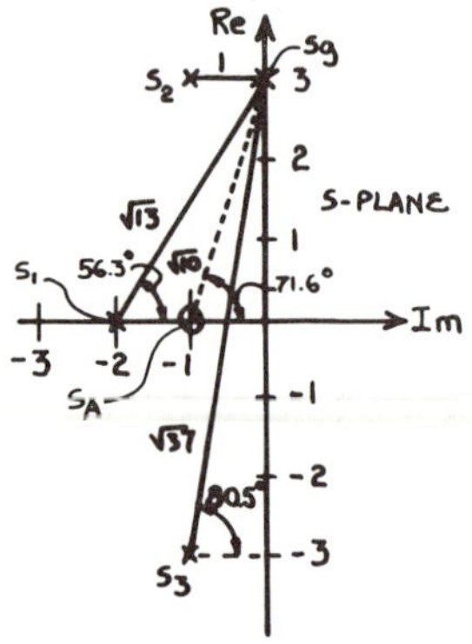

FiG.3

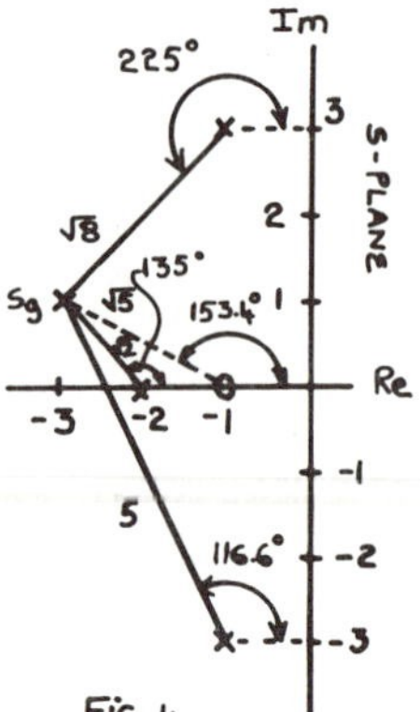

FiG.4 , 3-7

$$y_s(t) = \frac{\sqrt{10} \angle 71.6°}{\sqrt{13} \cdot 1 \cdot \sqrt{37} \angle 56.3° \angle 0° \angle 80.5°} 20 \cos 3t$$

$$y_s(t) = \frac{\sqrt{10}}{\sqrt{13}\ \sqrt{37}} 20 \cos [3t + 71.6° - (56.3°+80.5°)]$$

$$y_s(t) = 2.88 \cos (3t - 65.2°)$$

(c) To find the response due to the source 20 cos (3t - 30°), we use the same procedure as in (b):

$$y_s(t) = \frac{\sqrt{10}}{\sqrt{13}\ \sqrt{37}}\ 20\cos\left[3t-30°+71.6°-(56.3°+80.5°)\right]$$

$$y_s(t) = 2.88\cos(3t - 95.2°).$$

(d) The source $20e^{-3t}\cos t$ can be represented by a pole at $s_g = -3 + j$. The procedure for finding the response due to the source is shown in fig. 4.

$$y_s(t) = \frac{\sqrt{5}\ \underline{/153.4°}}{\sqrt{2}\cdot\sqrt{8}\cdot 5\ \underline{/135°}\ \underline{/225°}\ \underline{/116.6°}}\ 20e^{-3t}\cos t$$

$$y_s(t) = \frac{\sqrt{5}}{\sqrt{2}\cdot\sqrt{8}\cdot 5}\ 20e^{-3t}\cos\left[t+153.4°-(135°+225°+116.6°)\right]$$

$$y_s(t) = 2.24e^{-3t}\cos(t + 36.8°)$$

• PROBLEM 13-8

Make an s-plane plot of the poles and zeros of the transfer admittance, $Y = \frac{I_1}{V_s}$, for the circuit shown in fig. 1. Using this plot, sketch $|Y|$ versus ω; indicate values at $\omega = 0$ and $\omega = \infty$.

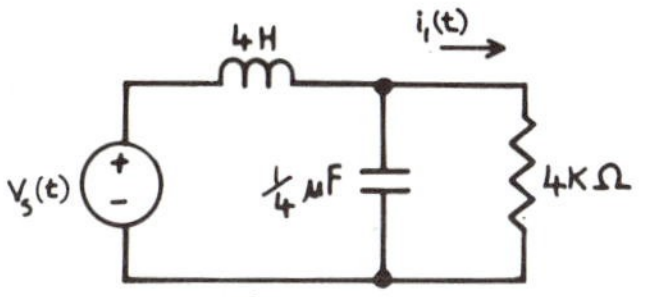

Fig. 1

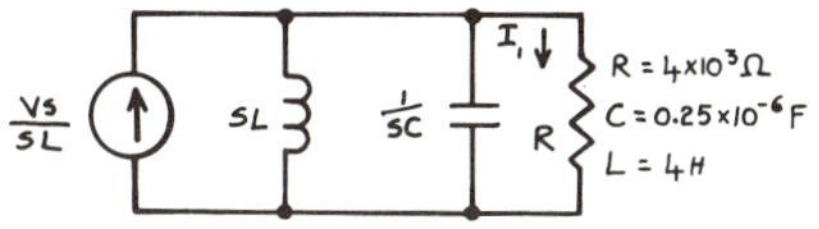

Fig. 2

Solution: We can write the expression for the transfer admittance $Y = \frac{I_1}{V_s}$ by transforming the circuit in fig. 1 into the frequency domain and then transforming the voltage source into a current source and parallel impedance as shown in fig. 2.

The current I_1 is found by use of current division. Hence,

$$Y = \frac{I_1}{V_1} = \frac{\dfrac{\frac{L}{C}}{sL + \frac{1}{sC}} \cdot \dfrac{1}{sL}}{\dfrac{\frac{L}{C}}{sL + \frac{1}{sC}} + R} = \frac{\frac{1}{RLC}}{s^2 + s\frac{1}{RC} + \frac{1}{LC}} .$$

Substituting the values for R, L and C gives

$$\frac{I_1}{V_1} = \frac{250}{s^2 + 1000s + 1x10^6} .$$

Factoring yields:

$$\frac{I_1}{V_1} = \frac{250}{(s + 500 - j866)(s + 500 + j866)} .$$

We find that there are two poles,

$$s_1 = -500 + j866$$

and $s_2 = -555 - j866.$

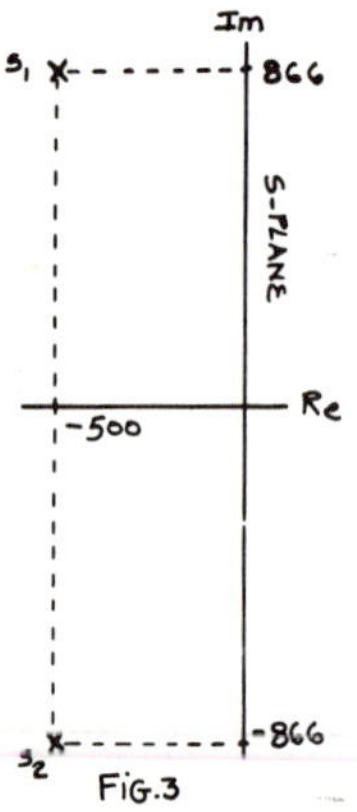

FiG.3

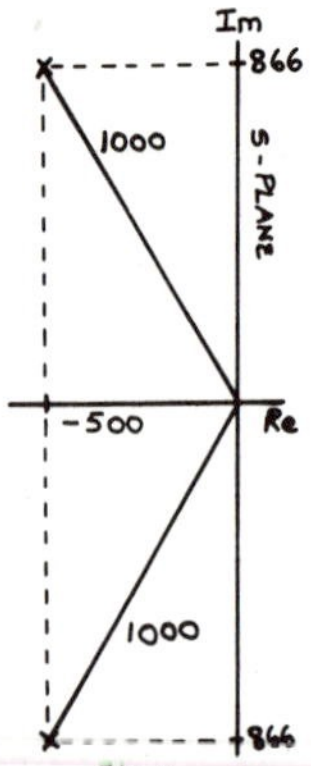

FiG.4

We can plot $|Y|$ vs. ω by obtaining information graphically from fig. 3. For example, if we wish to know $|Y|$ for $\omega = 0$ we find the product of the distances from each pole to $\omega = 0$ (the origin) and divide into 250. Hence, from fig. 4, at $\omega = 0$:

$$|Y| = \frac{250}{|1000||1000|} = 0.25 \text{ m mho} .$$

The value for $|Y|$ at $\omega = 866$ is (from fig. 5)

$$|Y(866)| = \frac{250}{(500)(1800)} = 0.28 \text{ m mho} .$$

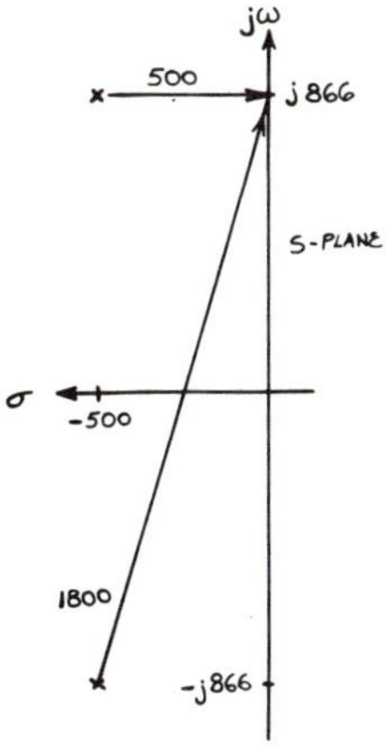

Fig.5

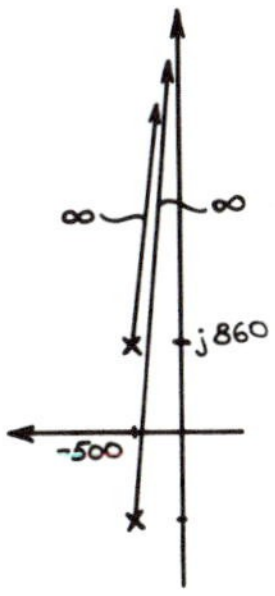

Fig.6

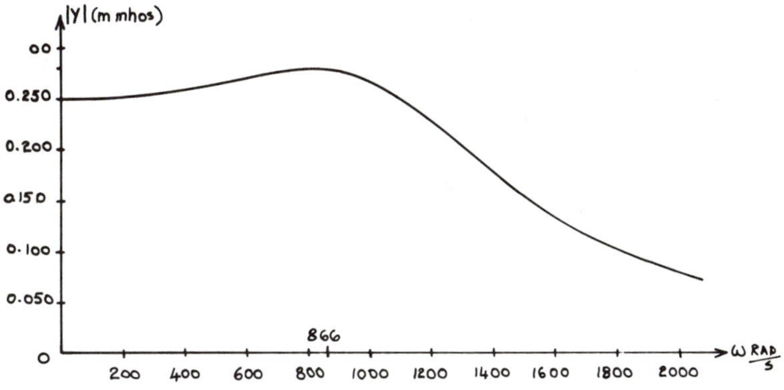

Fig.7

The value of $|Y|$ at $\omega = \infty$ is (from fig. 6)

$$|Y(\infty)| = \frac{250}{\infty^2} = 0.$$

The sketch of $|Y|$ vs ω is shown in fig. 7.

• PROBLEM 13-9

In a certain low-loss network the response $v_2(t)$ is related to the source $v_1(t)$ by the equation

$$v_2(t) = \frac{1}{3}\,\frac{p(p^2 + 2p + 200)}{(p^2 + p + 50)(p^2 + 3p + 500)}\,v_1(t).$$

Use the approximate pole-zero-diagram method to sketch the amplitude response curve $|H(j\omega)| = h(\omega)$.

Solution: In this problem $|H(j\omega)| = \dfrac{v_2(t)}{v_1(t)}$ and

$$H(s) = \frac{s(s^2 + 2s + 200)}{3(s^2 + s + 50)(s^2 + 3s + 500)}.$$

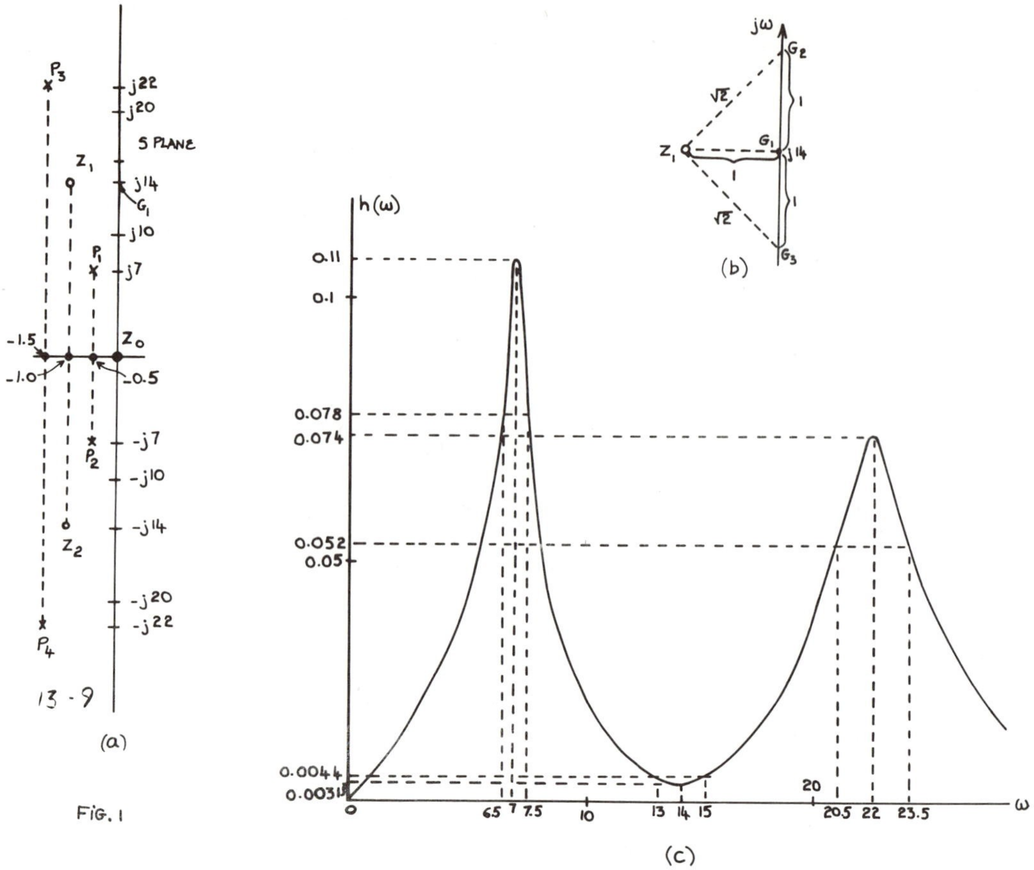

FIG. 1

The zeros of H(s) are the roots of the numerator:

$$s_1 = 0; \quad s_{2,3} = \frac{-2 \pm \sqrt{(2)^2 - 4(200)}}{2} \cong -1 \pm j14.$$

The poles of H(s) are the roots of the denominator:

$$s_{12} = \frac{-1 \pm \sqrt{1 - 4(50)}}{2} \cong -1 \pm j7$$

$$s_{34} = \frac{-3 \pm \sqrt{9 - 4(500)}}{2} \cong -1.5 \pm j22.$$

The pole-zero diagram is shown in fig. 1(a).

The function $h(\omega)$ will peak when ω is equal to the imaginary part of a pole. Hence, $h(\omega)$ will peak at $\omega \cong 7$ and $\omega \cong 22$. Similarly $h(\omega)$ will be small when ω is equal to the imaginary part of a zero. Hence $h(\omega)$ will be small when $\omega = 0$ and $\omega = 14$.

$h(\omega)$ at any ω (say for example ω_1) is the product of the distances from each zero to ω_1 divided by the product of the distances from each pole to ω_1. The minimum values of $h(\omega)$ will be calculated first. Note that:

$$h(14) = \frac{(Z_0G_1)(Z_1G_1)(Z_2G_1)}{3(P_1G_1)(P_2G_1)(P_3G_1)(P_4G_1)}$$

where Z_0G_1 is the distance from Z_0 to the point (0,j14), Z_1G_1 is the distance from Z_1 to (0,j14), etc. h(14) can be approximated as

$$h(14) \cong \frac{(14) \times (1) \times (28)}{3(7) \times (21) \times (8) \times (36)} = 0.0031.$$

A minimum also occurs at $\omega = 0$:

$$h(0) = \frac{(0) \times (14) \times (14)}{(7) \times (7) \times (22) \times (22)} = 0.$$

A maximum occurs at $\omega \cong 7$ and $\omega \cong 22$:

$$h(7) = \frac{(Z_0G_2)(Z_1G_2)(Z_2G_2)}{3(P_1G_2)(P_2G_2)(P_3G_2)(P_4G_2)}$$

$$\cong \frac{(7) \times (7) \times (21)}{3(0.5) \times (14) \times (1.5) \times (29)}$$

$$\cong 0.11$$

$$h(22) = \frac{(Z_0G_3)(Z_1G_3)(Z_2G_3)}{3(P_1G_3)(P_2G_3)(P_3G_3)(P_4G_3)}$$

$$\cong \frac{(22) \times (8) \times (36)}{3(15) \times (29) \times (1.5) \times (44)}$$

$$\cong 0.074$$

Now that all the maximum and minimum values are obtained, more values must be obtained to sketch $h(\omega)$. In the neighborhood of $\omega = 14$, the distance from Z_1 to (0,j14) controls the magnitude of $h(\omega)$. The distances from all the other poles and zeros can be assumed constant.

From fig. 1b it is noted that

$$h(15) \cong h(13) \cong h(14) \times \sqrt{2} = 0.0044.$$

Similar conditions hold in the neighborhood of $\omega = 7$ and $\omega = 22$:

$$h(6.5) \cong h(7.5) = \frac{h(7)}{\sqrt{2}} = 0.078$$

$$h(20.5) \cong h(23.5) = \frac{h(22)}{\sqrt{2}} = 0.052.$$

For each ω we have the following values:

ω	0	6.5	7	7.5	13	14
$h(\omega)$	0	0.078	0.11	0.078	0.0044	0.0031

ω	15	20.5	22	23.5
$h(\omega)$	0.0044	0.052	0.074	0.052

The sketch of the response curve $h(\omega)$ is shown in fig. 1c.

• PROBLEM 13-10

Sketch a curve of $|Y_{in}|$ versus ω for the network shown in fig. 1 if K = (a) 0; (b) 100.

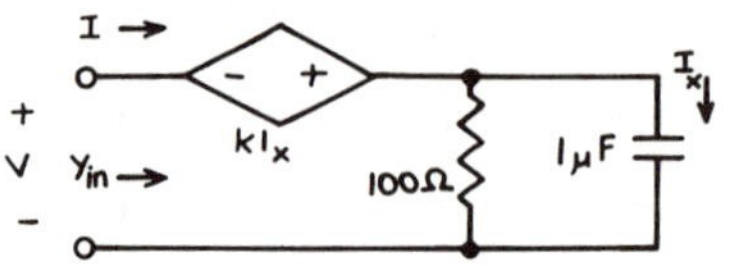

Fig. 1

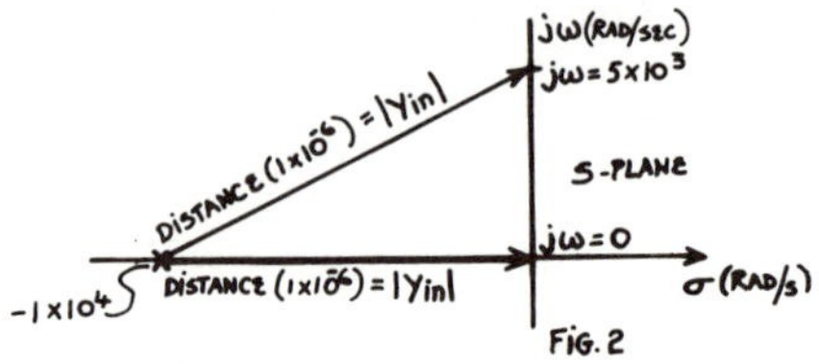

Fig. 2

Solution: The input admittance of the network in fig. 1 is defined as $Y_{in} = \frac{I}{V}$, which can be found by eliminating I_x from the loop equations

$$V + KI_x = RI - RI_x \tag{1}$$

$$0 = RI_x + I_x \frac{1}{sC} - RI. \tag{2}$$

Solving for I_x in equation (2) yields

$$I_x = I \frac{R}{R + \frac{1}{sC}}$$

Substituting for I_x in equation (1) gives

$$V = I \left[R - \frac{R^2}{R + \frac{1}{sC}} - \frac{KR}{R + \frac{1}{sC}}\right].$$

Thus,

$$Y_{in} = \frac{I}{V} = \frac{1}{R + \frac{R^2}{R + \frac{1}{sC}} - \frac{KR}{R + \frac{1}{sC}}}.$$

Simplifying the above expression for Y_{in}:

$$Y_{in} = \frac{sRC + 1}{sKRC + R}.$$

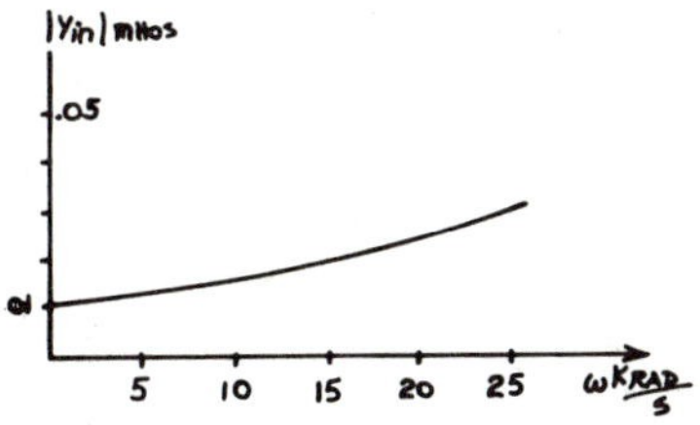

FiG.3

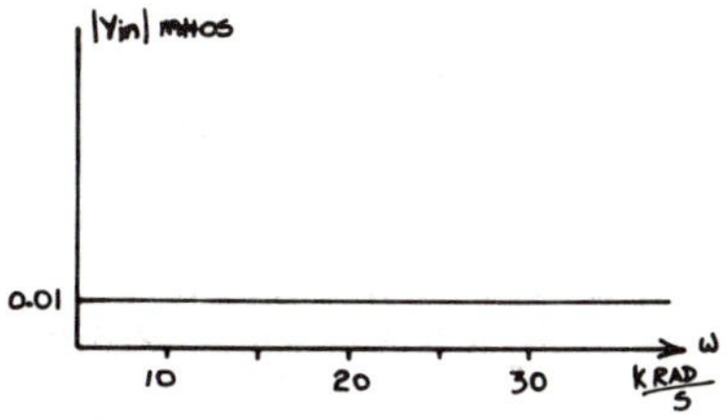

FiG.4

(a) For K = 0, R = 100Ω and C = 1×10^{-6}F.

$$Y_{in} = 1 \times 10^{-6}(s + 1 \times 10^4)$$

The single pole of $s = -1 \times 10^4$ shown in Fig. 2 allows us to calculate the admittance $|Y_{in}|$ by multiplying the constant 1×10^{-6} times the distance from the pole to the specific frequency on the jω axis.

Fig. 3 shows a sketch of $|Y_{in}|$ vs. ω.

(b) When K = 100,

$$Y_{in} = \frac{1}{100} \frac{(s + 1 \times 10^4)}{(s + 1 \times 10^4)} = \frac{1}{100} \text{ mho}$$

hence $|Y_{in}|$ is a constant for all ω as shown in Fig. 4.

• PROBLEM 13-11

Sketch the frequency response of v_{ab} for the circuit in fig. 1. The source is an ideal current source. Use and justify approximate methods (assume R = 0).

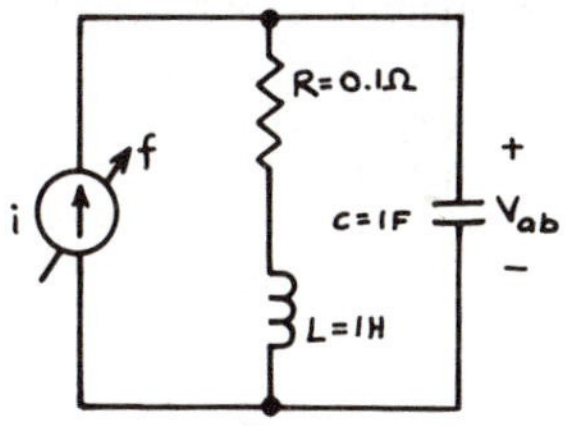

Fig.1

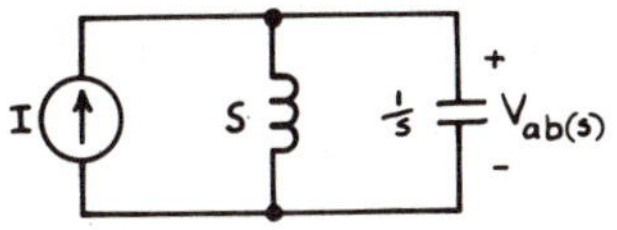

FiG.2

Solution: Using approximate methods we set R = 0 and obtain the transformed LC circuit of fig. 2.

Since the voltage $V_{ab}(s)$ is proportional to Z(s) we can sketch |Z| vs. ω as follows:

$$Z(s) = s \,||\, \frac{1}{s} = \frac{1}{s + \frac{1}{s}} = \frac{1}{\frac{s^2 + 1}{s}} = \frac{s}{s^2 + 1} \ .$$

Substituting jω for s (pure sinusoid) we obtain:

$$Z(j\omega) = \frac{j\omega}{(j\omega)^2 + 1} = \frac{j\omega}{1 - \omega^2} .$$

Hence

$$|Z(j\omega)| = Z(\omega) = \frac{\omega}{|1 - \omega^2|} ,$$

where the poles are obtained by finding the roots of the denominator polynomial and the zeros are obtained by finding the roots of the numerator polynomial.

For the circuit of fig. 1 we would expect a similar pole-zero diagram, except that it would have poles and zero in the left side of the plane. We know that this must be true since this is the region that gives stable (monotonically decreasing) response for the RLC circuit. (Monotonically increasing natural response is impossible when some resistance exists in a circuit with passive elements.)

Plotting $Z(\omega)$ for values of ω gives the sketch in fig. 3.

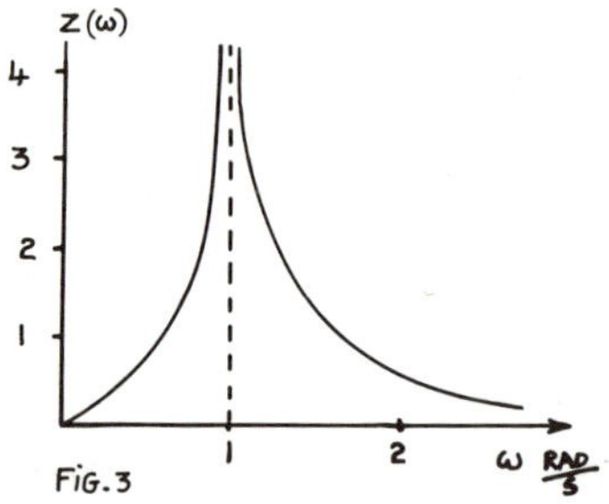

FIG.3

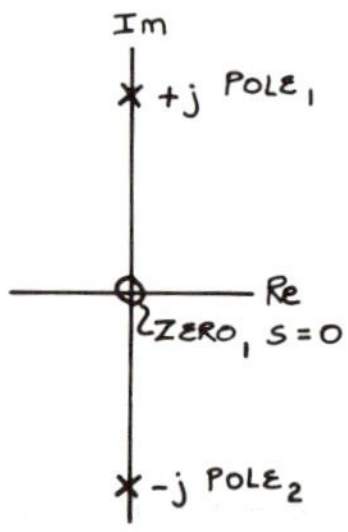

FIG.4

A pole-zero diagram for the function $Z(s) = \frac{s}{s^2 + 1}$ is shown in fig. 4.

The frequency response would have a resonant frequency at $\omega = 1$ rad/s and some finite value for the impedance at that point. For $\omega > 1$ rad/s the impedance would decrease until it is zero at $\omega = \infty$ as in the LC circuit.

To obtain an accurate plot of $Z(\omega)$ we find $Z(s) = (0.1 + s) \| 1/s$:

$$Z(s) = \frac{\frac{0.1 + s}{s}}{0.1 + s + \frac{1}{s}} = \frac{0.1 + s}{s^2 + 0.1s + 1} .$$

There is a zero = -0.1 and two poles = $-0.05 \pm j0.999$.

Fig. 5 shows the pole-zero diagram for the RLC circuit,

$$Z(j\omega) = \frac{j\omega + 0.1}{(j\omega)^2 + 0.1j\omega + 1}$$

$$Z(j\omega) = Z(\omega) = \frac{\sqrt{\omega^2 + 0.01}}{\sqrt{(1 - \omega^2)^2 + 0.01\omega^2}} .$$

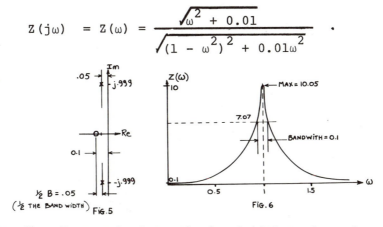

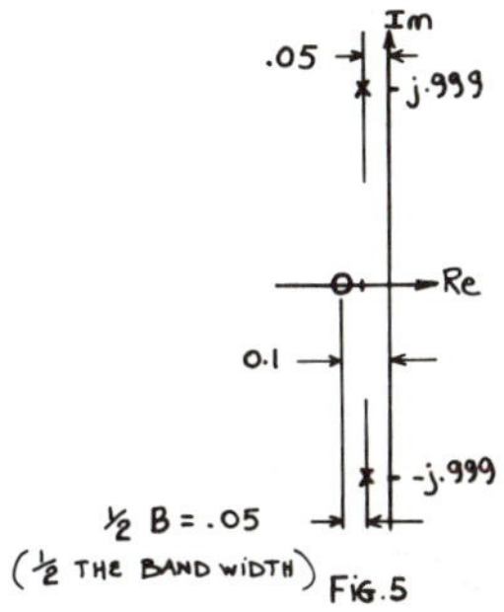

Fig. 5

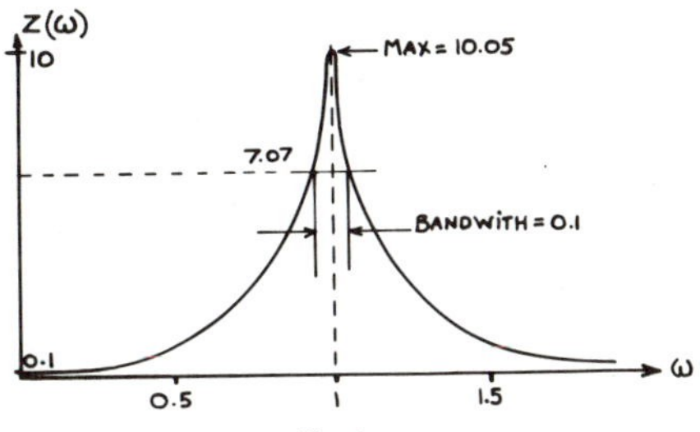

Fig. 6

In fig. 5 we calculate the bandwidth to be twice the real distance between the poles and the zero; β = 0.1. In fig. 6 we plot Z(ω) and measure the bandwidth graphically to confirm our results.

• PROBLEM 13-12

Use Laplace transform techniques to find $H(s) = \frac{V_O(s)}{V(s)}$ in the circuit shown in Fig. 1. Write an equation that would give values of $|H(j\omega)|$ when applying a sinusoidal driving function at a frequency, ω.

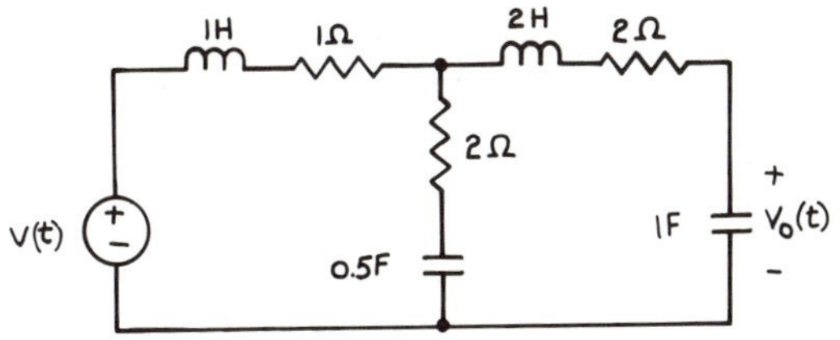

Fig. 1

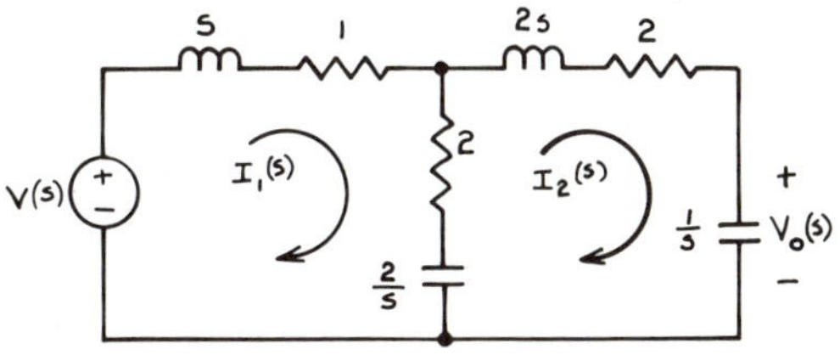

Fig. 2

Solution: of Fig. 1 Using Laplace transforms, transform the circuit

$$v(t) \Rightarrow V(s)$$

$$v_O(t) \Rightarrow V_O(s)$$

$$1H \Rightarrow sL = s$$

$$0.5F \Rightarrow \frac{1}{sC} = \frac{2}{s}$$

$$1F \Rightarrow \frac{1}{sC} = \frac{1}{s}$$

Hence, the circuit of fig. 2 is obtained.

Using KVL around the two loops in fig. 2 gives,

$$V(s) = I_1(s + 1 + 2 + \frac{2}{s}) - I_2(2 + \frac{2}{s})$$

$$0 = -I_1(2 + \frac{2}{s}) + I_2(\frac{2}{s} + 2 + 2s + 2 + \frac{1}{s})$$

Hence,

$$I_2 = \frac{\begin{vmatrix} 3+\frac{2}{s}+s & V(s) \\ -(2+\frac{2}{s}) & 0 \end{vmatrix}}{\begin{vmatrix} 3+\frac{2}{s}+s & -(2+\frac{2}{s}) \\ -(2+\frac{2}{s}) & 4+\frac{3}{s}+2s \end{vmatrix}} = \frac{V(s)(2+\frac{2}{s})}{(3+\frac{2}{s}+s)(4+\frac{3}{s}+2s) - (2+\frac{2}{s})^2}$$

$$I_2 = \frac{sV(s)(2s + 2)}{(2s^4 + 10s^3 + 15s^2 + 9s + 2)}$$

Since $V_0(s) = \frac{I_2}{s}$

$$sV_0(s) = \frac{sV(s)(2s + 2)}{(2s^4 + 10s^3 + 15s^2 + 9s + 2)}$$

Hence

$$H(s) = \frac{V_0(s)}{V(s)} = \frac{2s + 2}{(2s^4 + 10s^3 + 15s^2 + 9s + 2)}$$

When the driving source is sinusoidal, $s = j\omega$, then,

$$H(j\omega) = \frac{2j\omega + 2}{(2\omega^4 - 10j\omega^3 - 15\omega^2 + 9j\omega + 2)},$$

$$H(j\omega) = \frac{2 + j(2\omega)}{2\omega^4 - 15\omega^2 + 2 + j(9\omega - 10\omega^3)}$$

and, $$|H(j\omega)| = \frac{\sqrt{2^2 + (2\omega)^2}}{\sqrt{(2\omega^4 - 15\omega^2 + 2)^2 + (9\omega - 10\omega^3)^2}}$$

• PROBLEM 13-13

In the circuit of fig. 1, $v(t) = V_m \cos\omega t$ V and, in the steady state, $v_{ab}(t) = (V_{ab})_m \cos(\omega t - \Theta)$ V. (a) Find the transform network function that relates v_{ab} to v. (b) Plot $(V_{ab})m/V_m$ as a function of ω. (c) Plot Θ as a function of ω.

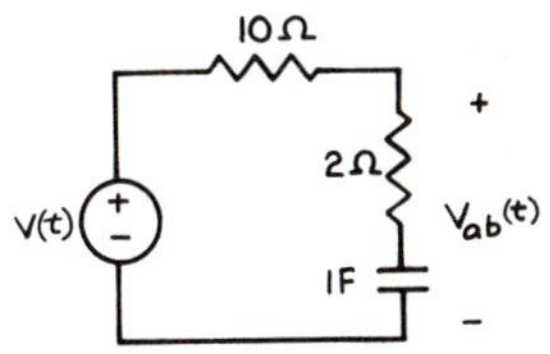

FiG.1

<u>Solution</u>: (a) The transfer function $H(s) = \frac{V_{ab}(s)}{V(s)}$ is found by the concept of voltage division,

$$V_{ab}(s) = \frac{\frac{1}{s} + 2}{\frac{1}{s} + 12} V(s)$$

$$H(s) = \frac{V_{ab}(s)}{V(s)} = \frac{\frac{1}{s}+2}{\frac{1}{s}+12} = \frac{2s+1}{12s+1} = \frac{1}{6}\frac{(s+\frac{1}{2})}{(s+\frac{1}{12})}.$$

(b) By substituting $j\omega$ for s we can plot the transfer function $|H(j\omega)|$ as a function of ω.

$$H(j\omega) = \frac{(V_{ab})_m}{V_m} = \frac{1}{6}\frac{j\omega+\frac{1}{2}}{j\omega+\frac{1}{12}} = \frac{2j\omega+1}{12j\omega+1}$$

$$|H(j\omega)| = \frac{\sqrt{1 + 4\omega^2}}{\sqrt{1 + 144\omega^2}}$$

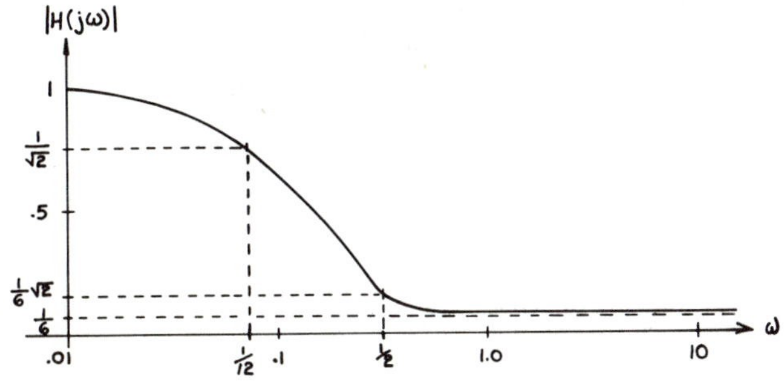

FiG.2

(c) The plot of the phase angle, Θ, of the transfer function, $H(j\omega)$, is in the form

$$H(j\omega) = \frac{A(j\omega)}{B(j\omega)}$$

where

$$\Theta = \tan^{-1}\frac{Im[A(j\omega)]}{Re[A(j\omega)]} - \tan^{-1}\frac{Im[B(j\omega)]}{Re[B(j\omega)]} .$$

Hence, for

$$H(j\omega) = \frac{2j\omega + 1}{12j\omega + 1}$$ we obtain

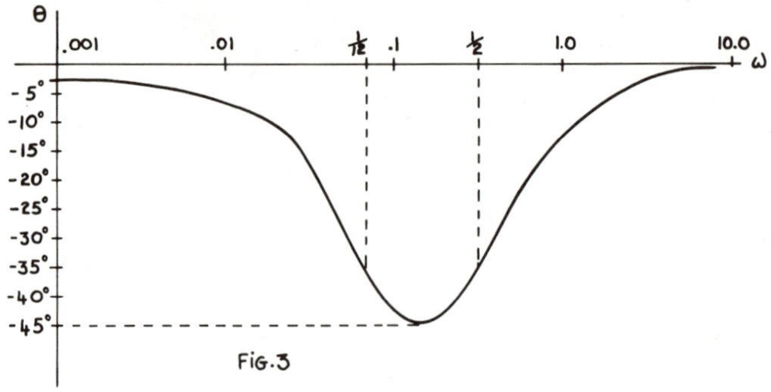

FIG.3

$$\Theta = \tan^{-1}\frac{2\omega}{1} - \tan^{-1}\frac{12\omega}{1} \quad .$$

Fig. 3 is the plot of Θ versus ω.

• PROBLEM 13-14

Plot the magnitude and phase of

$$H(s) = \frac{s^2 - 2s + 1}{s^3 + s^2 + s + 1}$$

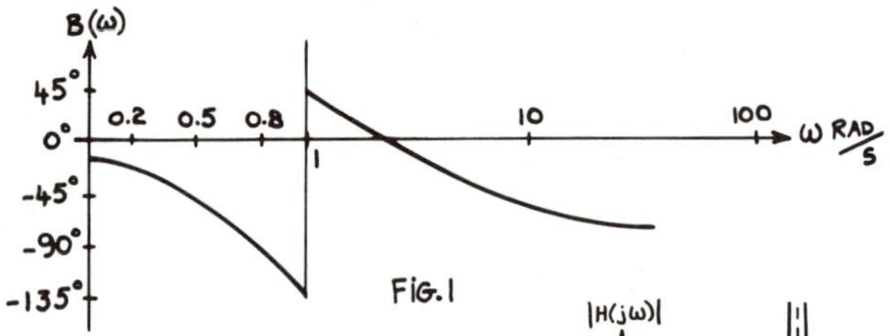

FIG.1

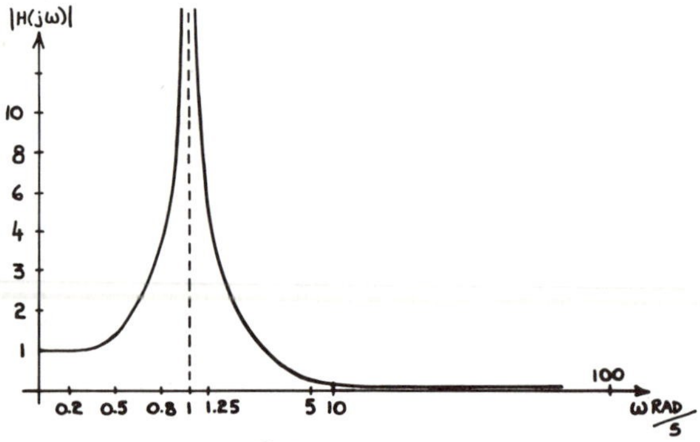

FIG.2

Solution: Substituting $j\omega$ for s in the equation above yields:

$$H(j\omega) = \frac{(1 - \omega^2) + j\omega(-2)}{(1 - \omega^2) + j\omega(1 - \omega^2)} \quad .$$

Thus, the magnitude is

$$|H(j\omega)| = \sqrt{\frac{(1 - \omega^2)^2 + 4\omega^2}{(1 - \omega^2)^2 + \omega^2(1 - \omega^2)^2}}$$

$$|H(j\omega)| = \sqrt{\frac{(1-\omega^2)^2 + 4\omega^2}{(1+\omega^2)(1-\omega^2)^2}}$$

$$|H(j\omega)| = \sqrt{\frac{\omega^4 + 2\omega^2 + 1}{(1+\omega^2)(1-\omega^2)^2}}$$

$$|H(j\omega)| = \frac{\sqrt{1+\omega^2}}{1-\omega^2}$$

The phase function $\beta(\omega)$ can be obtained from

$$H(j\omega) = \frac{(1-\omega^2) + j\omega(-2)}{(1-\omega^2) + j\omega(1-\omega^2)} .$$

Hence,
$$\beta(\omega) = \tan^{-1}\frac{-2\omega}{1-\omega^2} - \tan^{-1}\frac{\omega(1-\omega^2)}{(1-\omega^2)}$$

$$\beta(\omega) = \tan^{-1}\frac{-2\omega}{1-\omega^2} - \tan^{-1}\omega .$$

It is convenient to tabulate the calculations of magnitude and phase, as shown below.

ω	$\lvert H(j\omega)\rvert$	$\beta(\omega)$
0	1	0°
0.1	1.01	-17.3°
0.2	1.06	-34.0°
0.5	1.49	-79.6°
0.8	3.57	-116.0°
0.99	70.7	-134.1
1.0	∞	
1.01	70.7	44.1°
1.25	2.85	26.0°
2.00	0.75	-10.3°
5.00	0.21	-56°

The phase and magnitude of $H(\omega)$ appear in fig. 1 and 2, respectively.

• PROBLEM 13-15

The circuit of Fig. 1 shows an equivalent circuit for a transformer. Determine the poles and zeros and sketch the magnitude versus ω curve for the ratio:

(a) $\frac{I_2}{I_1}$; (b) $\frac{V_o}{V_1}$.

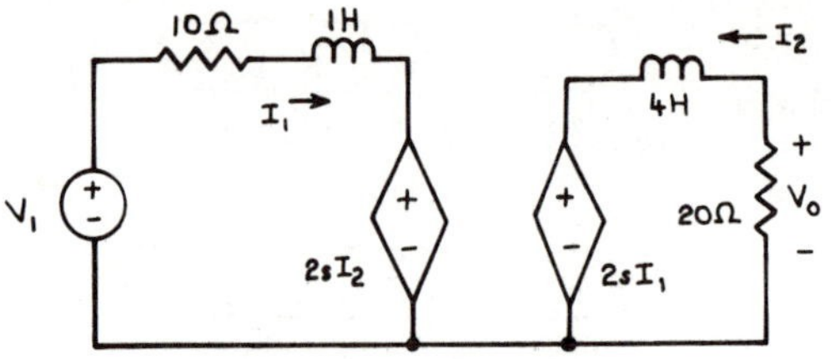

Fig.1

Solution: Using KVL we write the two loop equations:

$$\ell_1: \quad V_1 = I_1 10 + I_1 s1 + 2sI_2 \qquad (1)$$

$$\ell_2: \quad 0 = I_2 20 + I_2 s4 + 2sI_1 \,. \qquad (2)$$

(a) We can write $\frac{I_2}{I_1}$ immediately from Eq.(2):

$$\frac{I_2}{I_1} = \frac{-2s}{20 + 4s} = \frac{-\frac{1}{2}s}{s + 5} \,.$$

The zero at s = 0 indicates that $\left|\frac{I_2}{I_1}\right| = 0$ at $\omega = 0$ rad/s. This equation also tells us that as ω approaches ∞ $\left|\frac{I_2}{I_1}\right|$ approaches $\frac{1}{2}$ and at $\omega = 5$ rad/s (pole s = -5) $\left|\frac{I_2}{I_1}\right| = \frac{1}{2\sqrt{2}} = 0.35$. From the above information we plot $\left|\frac{I_2}{I_1}\right|$ in fig. 2.

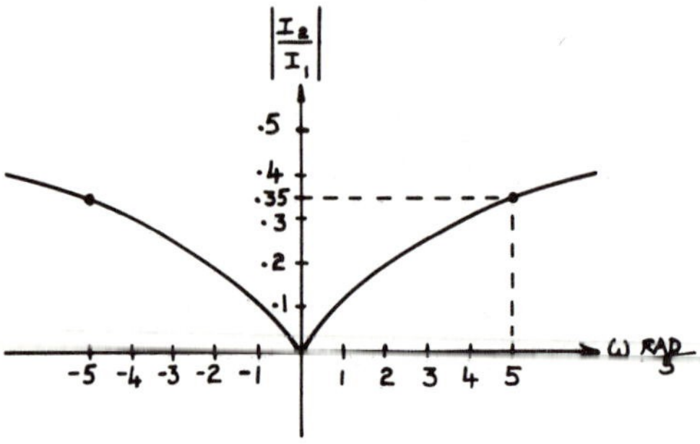

Fig. 2

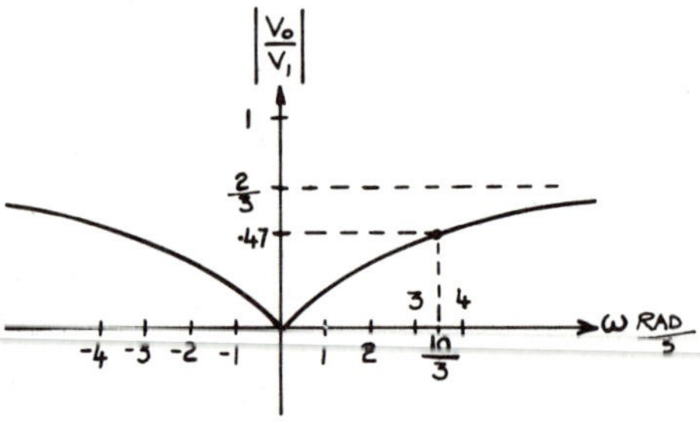

Fig. 3

(b) By substituting $I_2 = \frac{-2s}{20+4s} I_1$ from part (a) into Eq.(1) and solving for V_1 we obtain

$$V_1 = I_1 \left(10 + s - \frac{4s^2}{20+4s}\right).$$

Since $\qquad V_O = -I_2 20 = I_1 \left(\frac{2s}{20+4s} \cdot 20\right),$

then $$\frac{V_O}{V_1} = \frac{\dfrac{40s}{20 + 4s}}{10 + s - \dfrac{4s^2}{40 + 4s}}$$

$$\frac{V_O}{V_1} = \frac{40s}{60s + 200} = \frac{\frac{2}{3}s}{s + \frac{10}{3}} .$$

The resulting relation is in the same form as that for $\frac{I_2}{I_1}$ in part (a) with a zero at $s = 0$ and a pole at $s = -\frac{10}{3}$. And again,

at $\omega = 0$ rad/s; $\left|\frac{V_O}{V_1}\right| = 0$

at $\omega = \frac{10}{3}$ rad/s; $\left|\frac{V_O}{V_1}\right| = \frac{2}{3\sqrt{2}} = 0.47$

as ω approaches ∞; $\left|\frac{V_O}{V_1}\right|$ approaches $\frac{2}{3}$. The plot for $\left|\frac{V_O}{V_1}\right|$ is shown in Fig. 3.

• PROBLEM 13-16

Sketch curves of the magnitude and phase angle of V_1 as functions of ω for the circuit in fig. 1. Identify the half-power frequency.

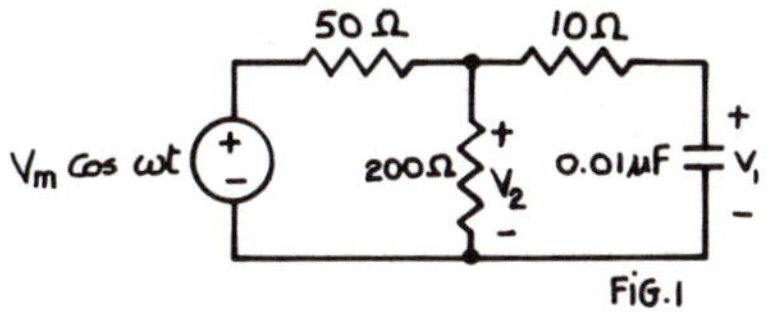

FiG.1

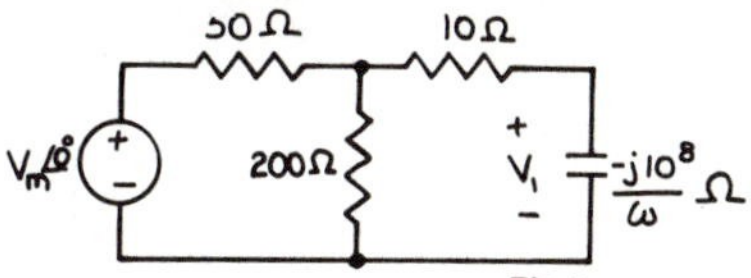

FiG.2

Solution: A phasor equivalent circuit, with ω left as a parameter, is shown in fig. 2. A Thevenin equivalent for the voltage source and 50 and 200 Ω resistors, followed by combining the Thevenin resistance and the 10Ω resistance simplifies the circuit to be analyzed, as shown in fig. 3. Then by the voltage divider rule,

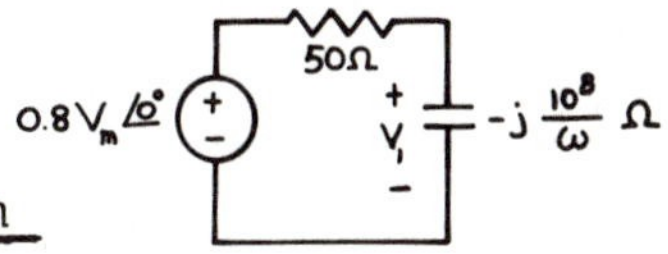

FiG.3

$$V_1 = \frac{-j\,\dfrac{10^8}{\omega}}{50 - j\dfrac{10^8}{\omega}}\,(0.8V_m) = \frac{1.6 \times 10^6 V_m}{2 \times 10^6 + j\omega}$$

from which

$$|V_1| = \frac{1.6 \times 10^6 V_m}{\sqrt{4 \times 10^{12} + \omega^2}}, \text{ Angle } (V_1) = \tan^{-1}\left(\frac{-\omega}{2 \times 10^6}\right).$$

The half-power frequency is that value of ω that reduces $|V_1|$ to $\frac{1}{\sqrt{2}}$ of its peak value, and is seen to be 2×10^6 rad/s.

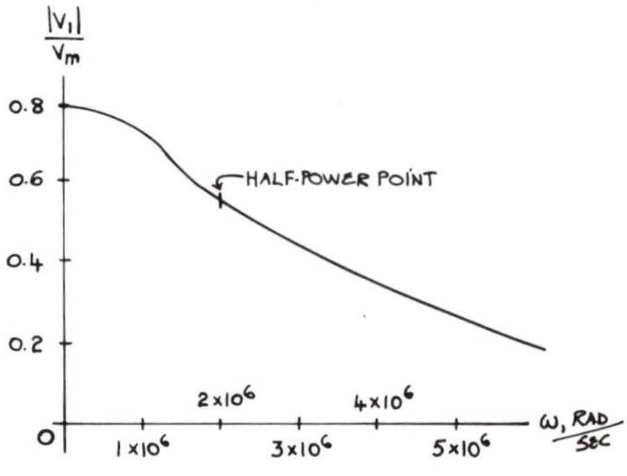

FIG. 4

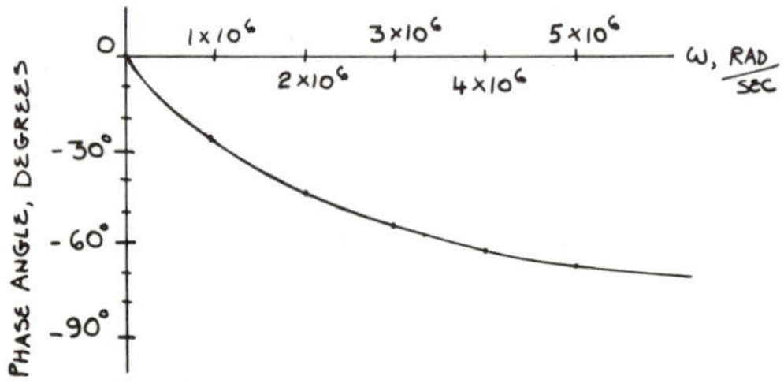

FIG. 5

Sketches of $|V_1|$ and Angle (V_1) are shown in figs. 4 and 5, respectively.

• PROBLEM 13-17

The drawing in fig. 1 shows a simple low-pass filter. (a) Calculate the steady-state response $v_R(t)$. (b) Determine the steady-state transfer function of the network, $H(j\omega)$, and sketch the amplitude and phase response.

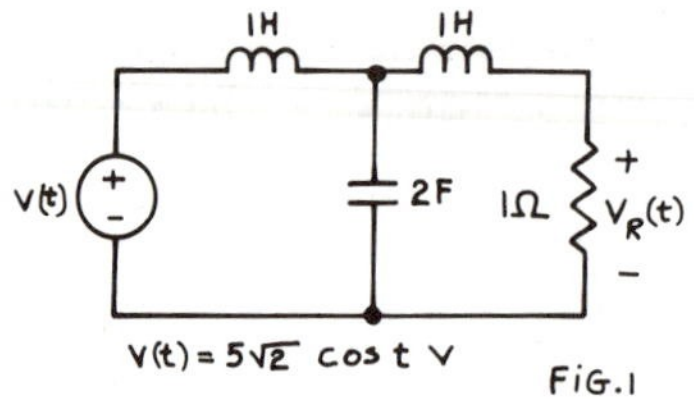

FIG. 1

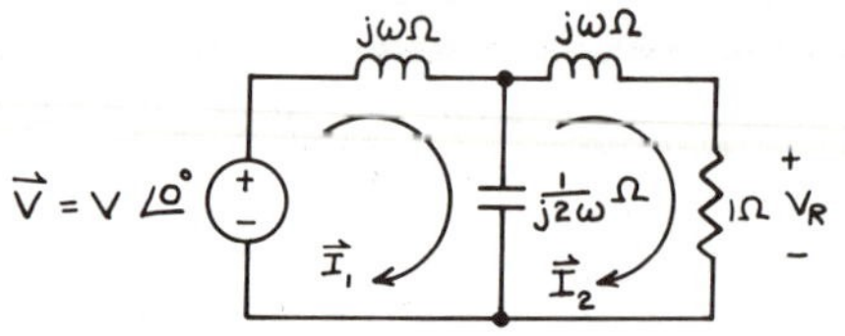

FIG. 2

Solution: The circuit of fig. 1 is operating in the ac steady state. Once the transfer function is determined, it can be used to calculate the steady state response. Hence we will first derive $H(j\omega)$. The steady state equivalent circuit is shown in Figure 2. For derivation of $H(j\omega)$, note that one must assume an input of the complex phasor form shown. We must derive the phasor $\vec{V}_R$.

Now, since this circuit has two loops, we can use the loop currents indicated in Figure 2. The resulting loop equations are:

$$(j\omega + \frac{1}{j2\omega})\vec{I}_1 - \frac{1}{j2\omega}\vec{I}_2 = \vec{V} \tag{1}$$

$$-\frac{1}{j2\omega}\vec{I}_1 + (1 + j\omega + \frac{1}{j2\omega})\vec{I}_2 = 0. \tag{2}$$

Solving (2) for $\vec{I}_1$, we get

$$\vec{I}_1 = (2j\omega + 2(j\omega)^2 + 1)\vec{I}_2. \tag{3}$$

Now inserting this value of $\vec{I}_1$ in (1) we obtain:

$$[[j\omega + \frac{1}{j2\omega}][2j\omega + 2(j\omega)^2 + 1] - \frac{1}{j2\omega}]\vec{I}_2 - \vec{V} \tag{4}$$

or
$$[2(j\omega)^3 + 2(j\omega)^2 + 2(j\omega) + 1]\vec{I}_2 + \vec{V}. \tag{5}$$

But $\vec{V}_R = \vec{I}_2$, so

$$H(j\omega) = \frac{\vec{V}_R}{\vec{V}} = \frac{\frac{1}{2}}{(j\omega)^3 + (j\omega)^2 + j\omega + \frac{1}{2}}, \tag{6}$$

which is the solution to part (b).

In order to compute the steady state response to $v(t) = 5\sqrt{2}\cos t$ V, simply let $\vec{V} = 5\sqrt{2}\,\underline{/0°}$ V in fig. 2 and note that $\omega = 1$ rad/s, hence

$$\vec{V}_R = \frac{\frac{1}{2}}{j^3 + j^2 + j + \frac{1}{2}} \times 5\sqrt{2}\,\underline{/0°} \tag{7}$$

or
$$\vec{V}_R = -5\sqrt{2}\,\underline{/0°}\text{ V}. \tag{8}$$

Hence $v_R(t) = \text{Re}\{V_R e^{jt}\} = -5\sqrt{2}\cos t$ V. (9)

The amplitude and phase response are obtained from $H(j\omega) - A(j\omega)e^{j\beta(j\omega)}$, where $A(j\omega)$ is the amplitude and $\beta(j\omega)$ is the phase. Therefore,

$$H(j\omega) = \frac{\frac{1}{2}}{(\frac{1}{2} - \omega^2) + j(\omega - \omega^3)} \tag{10}$$

or

$$H(j\omega) = \frac{\frac{1}{2}}{\sqrt{(\frac{1}{2} - \omega^2)^2 + \omega^2(1 - \omega^2)^2}\ \underline{/\tan^{-1}\frac{\omega(1-\omega^2)}{\frac{1}{2} - \omega^2}}}. \tag{11}$$

so

$$A(j\omega) = \frac{1}{\sqrt{(\frac{1}{2} - \omega^2)^2 + \omega^2(1 - \omega^2)^2}} \tag{12}$$

and

$$\beta(j\omega) = -\tan^{-1}\left[\frac{\omega - (1 - \omega^2)}{\frac{1}{2} - \omega^2}\right] . \tag{13}$$

The frequency response plots are shown in fig. 3.

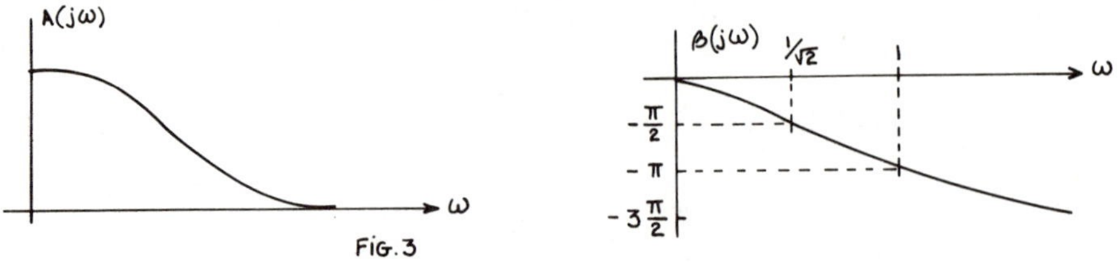

Fig. 3

Note that in sketching $\beta(j\omega)$ the inverse tangent function has several branches. Do not interpret (13) as the principal value.

• PROBLEM 13-18

Plot the phase response for the network of fig. 1.

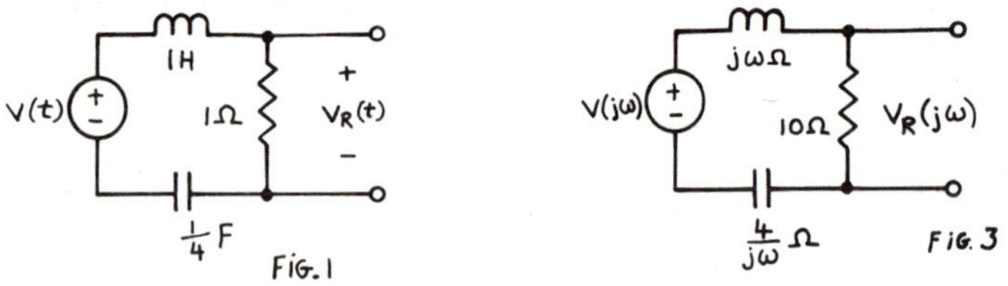

Fig. 1 Fig. 3

Solution: The steady-state equivalent of the circuit in fig. 1 is shown in fig. 3.

The network transfer function $H(j\omega) = \frac{V_R(j\omega)}{V(j\omega)}$ is

$$\frac{1}{j\omega + 1 + \frac{4}{j\omega}} = \frac{j\omega}{(4 - \omega^2) + j\omega} .$$

The phase response is

$$\Theta(\omega) = \tan^{-1}\left[\frac{\text{Im}[H(j\omega)]}{\text{Re}[H(j\omega)]}\right] \tag{1}$$

but since $H(j\omega)$ is in the form $\frac{B(j\omega)}{A(j\omega)}$, then

$$\Theta(\omega) = \tan^{-1}\left[\frac{\text{Im}[B(j\omega)]}{\text{Re}[B(j\omega)]}\right] - \tan^{-1}\left[\frac{\text{Im}[A(j\omega)]}{\text{Re}[A(j\omega)]}\right] .$$

Hence,

$$\Theta(\omega) = \tan^{-1}\left[\frac{\omega}{0}\right] - \tan^{-1}\left[\frac{\omega}{4-\omega^2}\right]$$

$$\Theta(\omega) = \tan^{-1}[\infty] - \tan^{-1}\left[\frac{\omega}{4-\omega^2}\right]$$

$$\Theta(\omega) = \frac{\pi}{2} - \tan^{-1}\left[\frac{\omega}{4-\omega^2}\right] . \quad (2)$$

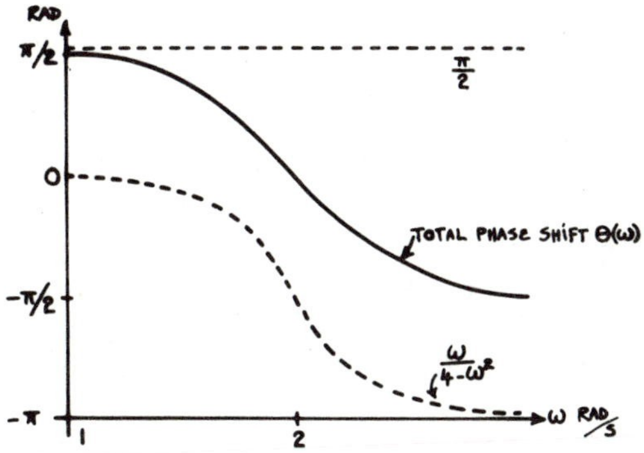

FIG. 2

Fig. 2 shows a plot of the phase response.

• PROBLEM 13-19

Plot the Bode diagrams for the network function

$$H(s) = \frac{4(s + 2)}{s^2 + 4s + 3} .$$

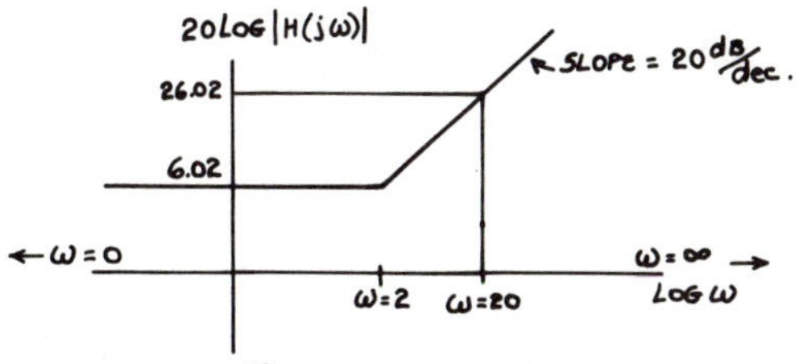

FIG. 1

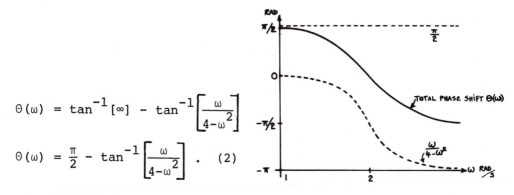

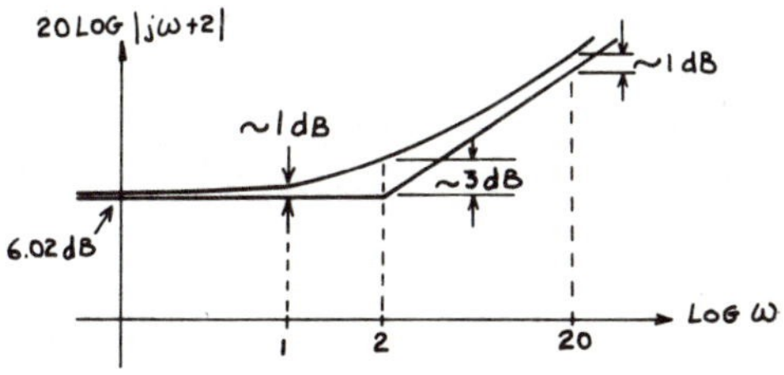

FIG. 2

Solution: We factor the denominator of H(s) and obtain

$$H(s) = \frac{4(s + 2)}{(s + 1)(s + 3)}$$

H(s) has poles at s = -1, s = -3, and a zero at s = -2. We take s = jω, since we are concerned with the sinusoidal frequency response and we obtain:

$$H(j\omega) = \frac{4(j\omega+2)}{(j\omega+1)(j\omega+3)} = \frac{8 + j4\omega}{(-\omega^2+3) + j4\omega}$$

Taking the absolute value yields:

$$|H(j\omega)| = H(\omega) = \frac{\sqrt{8^2 + 4^2\omega^2}}{\sqrt{(-\omega^2+3)^2 + 4^2\omega^2}} .$$

In the magnitude plot for our Bode diagram we must represent the magnitude in decibels. Thus

$$20 \log H(\omega) = 20 \left[\frac{1}{2} \log (64 + 16\omega^2)\right.$$

$$- \frac{1}{2} \log \left((-\omega^2+3)^2 + 16\omega^2\right)]$$

$$H(\omega)_{dB} = 10 \log(64+16\omega^2) - 10 \log((-\omega^2+3)^2 + 16\omega^2).$$

This equation gives us the exact values of the magnitude of $H(j\omega)$ in dB. By writing $20 \log|H(j\omega)|$ in another form we can draw an approximate Bode diagram for the magnitude of $H(j\omega)$. We take $H(j\omega)$ and take 20 times the log:

$$H(j\omega) = \frac{4(j\omega + 2)}{(j\omega + 1)(j\omega + 3)}$$

$$20 \log H(j\omega) = 20 \log 4 + 20 \log(j\omega+2)$$

$$- 20 \log(j\omega+1) - 20 \log(j\omega+3).$$

Thus,

$$20 \log|H(j\omega)| = 20 \log 4 + 20 \log|j\omega+2|$$

$$- 20 \log|j\omega+1| - 20 \log|j\omega+3|. \quad (1)$$

We note that $20 \log 4 = 12.04$dB is represented by a horizontal line on our Bode diagram. Looking at $20 \log|j\omega+2| = 20 \log\left[\sqrt{4+\omega^2}\right]$, we note that at $\omega = 0$, $20 \log\left[\sqrt{4+\omega^2}\right] = 20 \log 2 = 6.02$dB, and for values of $0 < \omega < 2$, $20 \log |H(j\omega)|$ does not change in magnitude as much as for values of $2 < \omega < \infty$. If we consider $20 \log|H(j\omega)|$ to be linear in these two regions then we note that when $\omega = 0$ the slope of the line in the region $0 < \omega < 2$ is

$$\frac{20 \log|H(j\omega)|}{\log \omega} = \frac{20 \log|H(j\omega)|}{\log 0} = \frac{20 \log|H(j\omega)|}{\infty} = 0.$$ Hence,

the magnitude is constant in this region. When $\omega = \infty$ we note that the slope for the region $2 < \omega < \infty$ is 20. Fig. 1 shows the Bode diagram for the magnitude of $H(j\omega) = (j\omega+2)$.

If we superimpose the exact plot of the magnitude of $(j\omega+2)$ over the approximation in fig. 1 we obtain the information shown in fig. 2.

Using the above technique for each term of Eq.(1) we obtain the plots shown in fig. 3.

To plot the angle of $H(\omega)$ note that

$$H(\omega) = 4 \frac{\sqrt{\omega^2+4} \;\underline{/\tan^{-1} \frac{\omega}{2}}}{\sqrt{\omega^2+1} \;\underline{/\tan^{-1} \omega} \;\sqrt{\omega^2+9} \;\underline{/\tan^{-1} \frac{\omega}{3}}}.$$

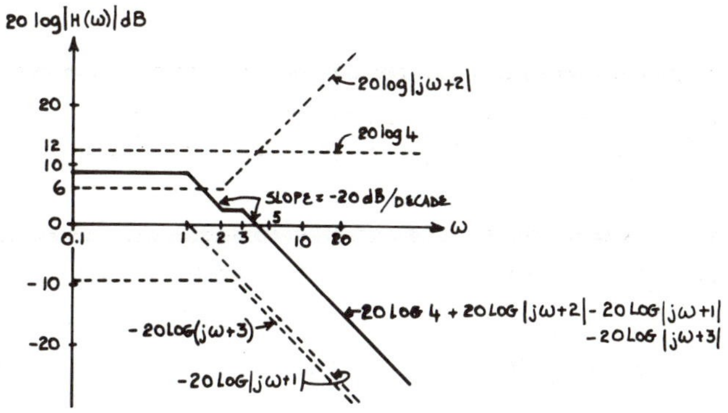

Fig.3

Therefore the angle of H(ω) is:

$$\measuredangle H(\omega) = \tan^{-1}\frac{\omega}{2} - \tan^{-1}\omega - \tan^{-1}\frac{\omega}{3}\,.$$

In this form we can make a plot of each tangent-function and then sum each function to get $\measuredangle H(\omega)$.

As ω approaches zero $\tan^{-1}\frac{\omega}{3}$, $\tan^{-1}\omega$, and $\tan^{-1}\frac{\omega}{2}$ approach zero. As ω approaches infinity $\tan^{-1}\frac{\omega}{2}$, $\tan^{-1}\omega$ and $\tan^{-1}\frac{\omega}{3}$ approach 90 degrees. Fig. 4 shows a table of values for $\tan^{-1}\frac{\omega}{2}$, $\tan^{-1}\omega$ and $\tan^{-1}\frac{\omega}{3}$ for a given value of ω.

ω (rad/s)	$\tan^{-1}\frac{\omega}{2}$	$-\tan^{-1}\omega$	$-\tan^{-1}\frac{\omega}{3}$	arg H(ω)
0.1	0^+	0^{-}°	0^{-}°	0^{-}°
0.5	14.0°	-26.6°	-9.5°	-22.0°
1.0	26.5°	-45.0°	-18.4°	-36.9°
1.5	36.9°	-56.3°	-26.6°	-46.0°
3.0	56.3°	-71.6°	-45.0°	-60.3°
4.0	63.4°	-76.0°	-53.1°	-65.7°
6.0	71.6°	-80.5°	-63.4°	-72.4°
∞	90.0°	-90.0°	-90.0°	-90.0°

Fig. 4

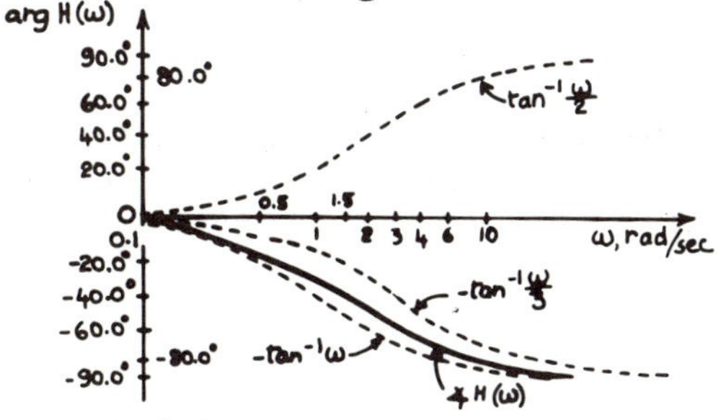

Fig.5

The plot of each $\tan^{-1}$ function, and the resulting angle of H(ω) are shown in Fig. 5.

Plot the sinusoidal-steady state amplitude response for the network shown in fig. 1. Let v(t) be the input and $v_R(t)$ be the response.

Solution: The steady state equivalent circuit is shown in fig. 2.

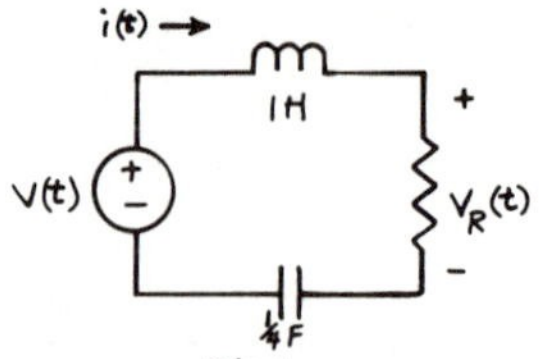

FIG. 1

Using KVL around the loop, we obtain

$$\vec{V} = \vec{I}\ [j\omega + 1 - \frac{j4}{\omega}].$$

Since $\vec{V}_R = \vec{I}$ then,

$$\vec{V} = \vec{V}_R\ [j\omega + 1 - \frac{j4}{\omega}]$$

$$H(j\omega) = \frac{\vec{V}_R}{\vec{V}} = \frac{1}{j\omega + 1 - \frac{j4}{\omega}}$$

$$H(j\omega) = \frac{j\omega}{(4-\omega^2) + j\omega}.$$

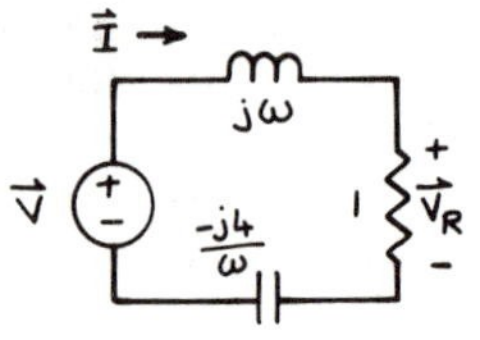

FIG. 2

Hence, the magnitude of the response of the network is

$$|H(j\omega)| = \frac{\omega}{\sqrt{(4-\omega^2)^2 + \omega^2}}.$$

In order to plot the amplitude response as a function frequency, it is conventional to plot 20 times the logarithm of the magnitude, rather than the magnitude itself. This, of course, permits a wide variation of the function $|H(j\omega)|$ to be plotted on a compact graph.

Define the units of the function

$$20 \log_{10}|H(j\omega)| = 10 \log_{10}|H(j\omega)|^2$$

as decibels abbreviated as dB.

In general, the amplitude response of any function $H(j\omega)$ can be plotted by noting that there are only three types of terms that appear in the transfer function $H(j\omega)$. These terms are

1) s^n

2) $(s + a)^n$

3) $s^2 + 2\zeta\omega_n s + \omega_n^2$.

If $s = j\omega$ then the three terms are:

1) $(j\omega)^n$ (1)

2) $(j\omega + a)^n$ (2)

3) $(\omega_n^2 - \omega^2) + j2\zeta\omega_n\omega.$ (3)

Since the magnitude function can be written

$$|H(j\omega)| = \left|\frac{B(j\omega)}{A(j\omega)}\right| ,$$

then

$$20 \log|H(j\omega)| = 20 \log|B(j\omega)| - 20 \log |A(j\omega)|.$$

It has been assumed that

$$B(j\omega) = B_1(j\omega)B_2(j\omega) \ . \ . \ . \ B_m(j\omega)$$

and $$A(j\omega) = A_1(j\omega)A_2(j\omega) \ . \ . \ . \ A_n(j\omega)$$

where each $B_i(j\omega)$ and $A_i(j\omega)$ is given by one of the terms (1), - (3).

Thus, a plot of the function $20 \log|H(j\omega)|$ consists of the addition and subtraction of plots corresponding to the magnitude of terms (1), - (3).

$$20 \log|H(j\omega)| = 10 \log \frac{\omega^2}{(4-\omega^2)+j\omega}$$

$$20 \log|H(j\omega)| = 20 \log|\omega| - 10 \log|4-\omega^2+j\omega|$$

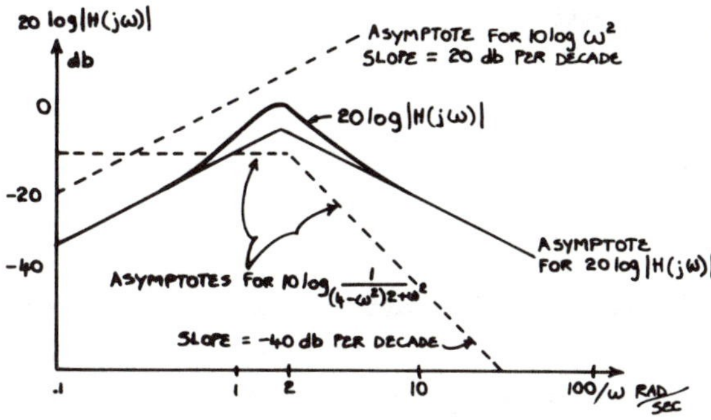

Fig. 3

The plot of this function is shown in fig. 3.

RESONANCE

• PROBLEM 13-21

The Impedances $\vec{Z}_1 = 20 + j10\Omega$ and $\vec{Z}_2 = 10 - j30\Omega$ are connected in parallel as shown in fig. 1. Find the value of Z_3 which will produce resonance at the terminal a,b.

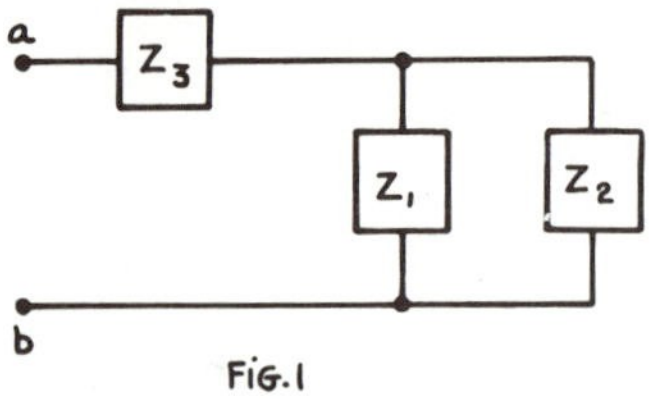

FIG. 1

Solution: The parallel combination of $\vec{Z}_1$ and $\vec{Z}_2$ represents an impedance of

$$\frac{\vec{Z}_1\vec{Z}_2}{\vec{Z}_1+\vec{Z}_2} = \frac{(20+j10)(10-j30)}{(20+j10)+(10-j30)} = 10\left[\frac{(2+j1)(1-j3)}{3-j2}\right]$$
$$= 10\left[\frac{5-j5}{3-j2}\right]. \qquad (1)$$

Next connect this impedance to rectangular form by multiplying the numerator and denominator by the complex conjugate of the denominator:

$$\frac{\vec{Z}_1\vec{Z}_2}{\vec{Z}_1+\vec{Z}_2} = 50\left[\frac{1-j1}{3-j2}\right]\left[\frac{3+j2}{3+j2}\right] = 50\left[\frac{5-j1}{13}\right] = \frac{250}{13} - j\frac{50}{13}. \qquad (2)$$

The total impedance

$$\vec{Z}_{total} = \vec{Z}_3 + \frac{\vec{Z}_1\vec{Z}_2}{\vec{Z}_1+\vec{Z}_2}, \qquad (3)$$

will thus be in resonance (i.e., will be purely resistive) for $\vec{Z}_3$ equal to the conjugate of the reactive part of $\frac{\vec{Z}_1\vec{Z}_2}{\vec{Z}_1+\vec{Z}_2}$. Therefore

$$Z_3 = +j\frac{50}{13} \text{ or } +j3.847\Omega.$$

• PROBLEM 13-22

Determine the resonant frequency of the circuit in fig. 1.

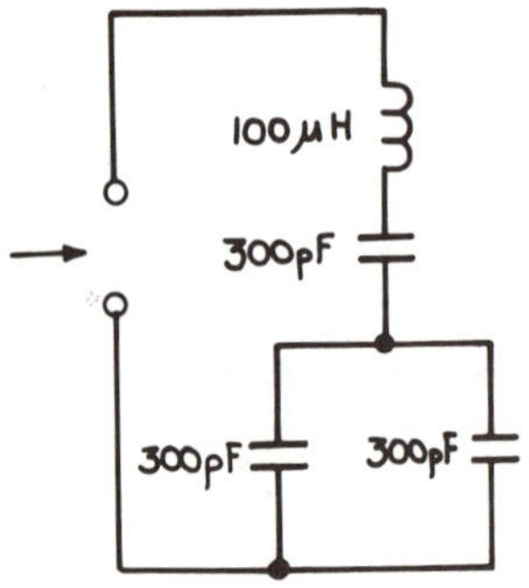

FIG. 1

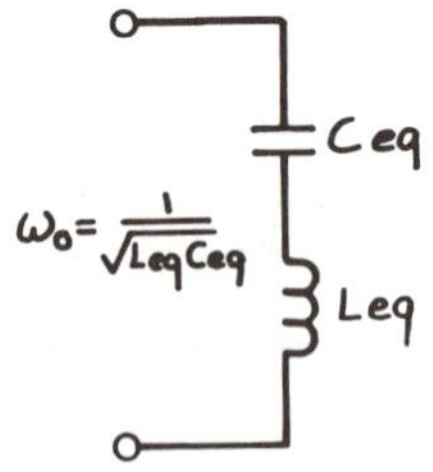

FIG. 2

Solution: The resonant frequency of the LC circuit in fig. 1 is found by combining inductors and capacitors to obtain the circuit shown in fig. 2.

The two parallel 300 pF capacitors in fig. 1 combine to firm a 600 pF capacitor. The resulting 600 pF and 300 pF capacitors in series combine to form a 200 pF capacitor. Hence, the resonant frequency is now given by the formula $\frac{1}{\sqrt{LC}}$ where L = 100 μH and C = 200 pF. Hence,

$$\omega_o = \frac{1}{\sqrt{100x10^{-6} \cdot 200x10^{-2}}} = 7.071x10^6 \text{ rad/s}$$

$$f_o = \frac{\omega_o}{2\pi} = 1.125x10^6 \text{ Hz} = 1.125 \text{ MHz}.$$

• PROBLEM 13-23

Determine the resonance frequency of the circuit in fig. 1.

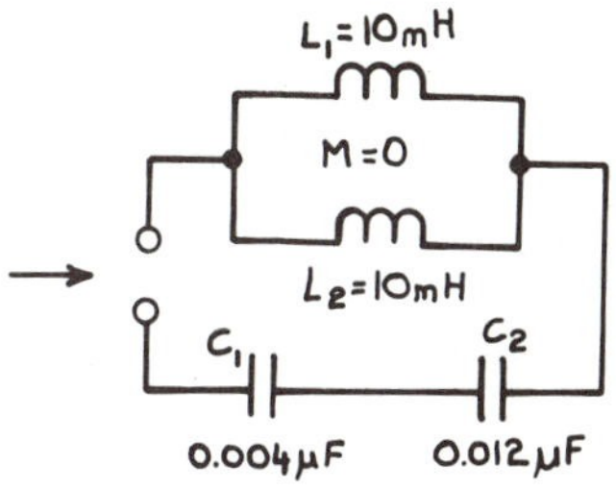

FiG.1

Solution: The resonance frequency for an inductor and capacitor in series is that frequency at which the inductive reactance just equals the capacitive reactance. First reduce the two parallel inductors into a single equivalent inductor:

$$L_{eq} = \frac{L_1L_2}{L_1+L_2} = \frac{(10\text{mH})(10\text{mH})}{20\text{mH}} = 5 \text{ mH}.$$

Similarly, reduce the two series capacitors to a single equivalent capacitor:

$$C_{eq} = \frac{C_1C_2}{C_1+C_2} = \frac{(.004\mu F)(.012\mu F)}{.016\mu F} = 0.003 \ \mu F.$$

Then for equal inductive and capacitive reactances $\omega_r L = 1/(\omega_r C)$, so the resonant frequency is

$$\omega_r = 1/\sqrt{LC}$$

$$= 1/\sqrt{(5x10^{-3})(3x10^{-9})}$$

$$= 0.2582 \times 10^6 \text{ rad/s}$$

or equivalently, since freq. (Hz) equals freq. (rad/s) divided by 2π,

$$f_r = \frac{0.2582 \times 10^6}{2\pi} = 41.1 \text{ kHz.}$$

• PROBLEM 13-24

Determine the resonance frequency of the circuit in fig. 1.

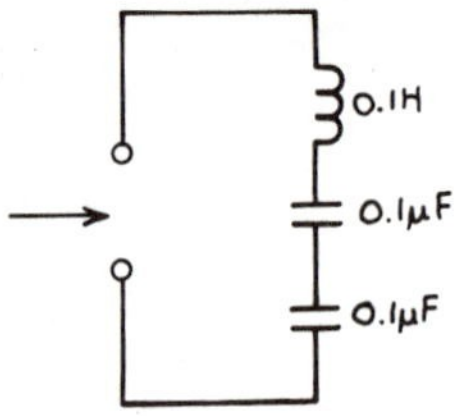

FIG.1

Solution: The two 0.1μF capacitors can be combined into an equivalent capacitance of

$$C = \frac{1}{\frac{1}{0.1\mu F} + \frac{1}{0.1\mu F}} = 0.05 \ \mu F.$$

Series resonance will occur at a frequency where the terminal impedance Z is zero:

$$Z = j\omega L - j\left(\frac{1}{\omega C}\right) = 0\Omega$$

Thus $\omega L = \frac{1}{\omega C}$

or $\omega = \frac{1}{\sqrt{LC}} = \frac{1}{\sqrt{(0.1H)(5 \times 10^{-8} F)}} = 14.14 \text{ krad/s}$

and $f = \frac{\omega}{2\pi} = 2.25 \text{ KHz.}$

• PROBLEM 13-25

(a) Express the resonant frequency and Q_o of the network shown in Fig. 1 as functions of R, L, C, and k. (b) Repeat if the positions of R and C are interchanged.

FIG.1

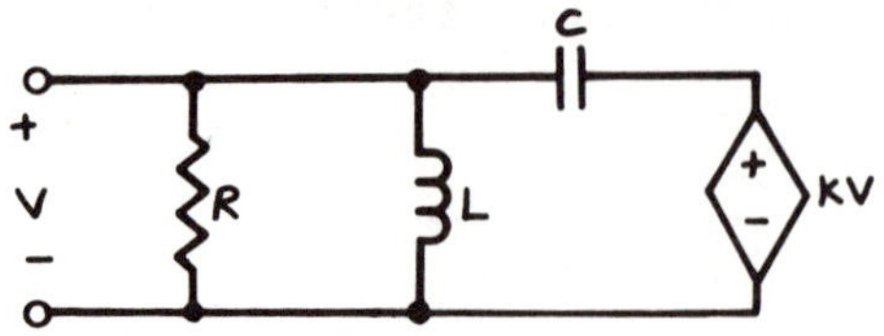

Solution: If the circuit between the terminals can be reduced to an equivalent resistance, inductance, and capacitance, in parallel with each other, then ω_o and Q_o can be written by well known relationships.

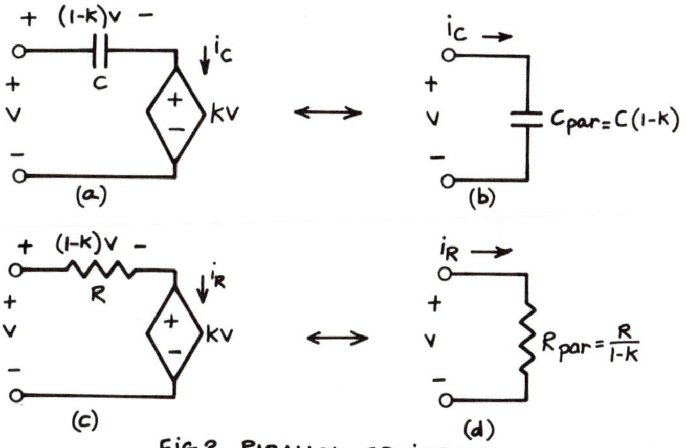

FIG. 2 PARALLEL EQUIVALENCE

(a) Since R and L are already in parallel, we will concentrate on the series combination of C and the dependent source (see Fig. 2(a)).

The current is

$$I_C = (1-k)V[j\omega C] = V[j\omega C(1-k)]. \quad (1)$$

An equivalent parallel capacitance is defined as

$$C_{par} = C(1-k).$$

We note that if this capacitance is placed across the terminals (see Fig. 2(b)), the same amount of current will flow. Thus the series circuit of C and the dependent source can be replaced by simply C_{par}, as far as the terminal relationship between V and I_C is concerned.

The original circuit is now shown to be equivalent to a parallel combination of R, L, and C_{par} between the terminals. From this we know that

$$\omega_0 = \frac{1}{\sqrt{LC_{par}}} = \frac{1}{\sqrt{LC(1-k)}}$$

and $$Q_o = \frac{R}{\omega_o L} = R\sqrt{\frac{C(1-k)}{L}}.$$

(b) An approach similar to that in (a) is taken. The current through R (see Fig. 2(c)) is

$$I_R = (1-k)V/R = \frac{V}{R/(1-k)} \quad (2)$$

From equation (2) the equivalent parallel resistance (see

Fig. 2(d)) is seen to be

$$R_{par} = R/(1-k). \tag{6}$$

The original circuit is now equivalent to a parallel combination of R_{par}, L and C. Hence

$$\omega_o = \frac{1}{\sqrt{LC}} \tag{7}$$

and

$$Q_o = \frac{R_{par}}{\omega_o L} = \frac{R}{(1-k)}\sqrt{\frac{C}{L}} \tag{8}$$

Note that k must be less than unity, or else negative circuit elements appear in the parallel equivalents of parts (a) and (b).

• PROBLEM 13-26

Between terminals a and b is a capacitance C in series with the parallel combination of a 2000-Ω resistor and a 2-mH coil. Find C so that the susceptive part of the input admittance is zero at 20KHz.

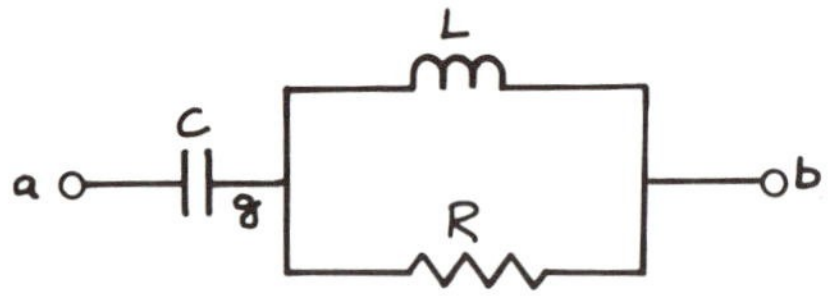

FiG.1

Solution: For this problem first find the complex impedance of the whole circuit (from a to b in Fig. 1) and then solve for the numerical value of capacitance which makes the imaginary part of that impedance equal to zero. Making the input impedance have a zero reactive part is the same as making the input admittance have a zero susceptive part, since both the impedance and admittance will have a zero angle under this circumstance. The impedance approach is preferred because the unknown quantity, C, is in series with the rest of the circuit.

First, find the impedance of the parallel branches between points b and g in Fig. 1 by obtaining the reciprocal of the admittance:

$$\vec{Z}_{bg} = \frac{1}{\vec{Y}_{bg}} = \frac{1}{\frac{1}{R} + \frac{1}{j\omega L}} = \frac{1}{\frac{1}{2000} + \frac{1}{j2\pi(20\text{x}10^{3})2\text{x}10^{-3}}} \tag{1}$$

$$\vec{Z}_{bg} = \frac{1}{.50\text{x}10^{-3} - j4.0\text{x}10^{-3}} = \frac{1}{4.03\text{x}10^{-3}\ \underline{/-82.9°}} \tag{2}$$

$$\vec{Z}_{bg} = 248 \underline{/82.9} = 30.8 + j246\Omega \ . \quad (3)$$

Next add this to the reactance of the capacitor to obtain the total impedance between a and b:

$$\vec{Z}_{ab} = \vec{Z}_{bg} = \vec{Z}_{ga} = (30.8+j246)-j\frac{1}{\omega C} = 30.8+j(246-\frac{1}{\omega C}) \ . \quad (4)$$

Now all we have to do is to solve equation (4) for the value of C which makes the imaginary part equal to zero:

$$246 - \frac{1}{\omega C} = 0 \quad (5)$$

$$C = \frac{1}{2\pi(20\text{x}10^3)246} = 3.24\text{x}10^{-8} \ . \quad (6)$$

Therefore, C = .032 μF. (7)

As a check, note that now

$$\vec{Z}_{ab} = 30.8 + jo = 30.8 \underline{/0°} \ \Omega \quad (8)$$

and $$\vec{Y}_{ab} = \frac{1}{\vec{Z}_{ab}} = \frac{1}{30.8\underline{/0°}} = .0325\underline{/0°} = .0325+j0 \text{ mho} \quad (9)$$

which fulfills the demands of the problem, since it has a zero susceptive part.

• PROBLEM 13-27

Find L and C so that the input impedance of the circuit shown in fig. 1 is 2R at $\omega = \omega_o$.

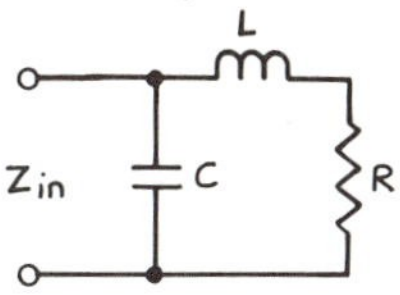

FIG.1

Solution: Since the impedance is to be resistive at the specified frequency ω_o, the imaginary part of the impedance at ω_o must be zero, and the real part must be the specified resistance, 2R. We must, therefore, calculate the impedance at ω_o. Due to the presence of the parallel capacitor, calculate the admittance $Y(j\omega_o)$ and then require that $Y(j\omega_o) = \frac{1}{2R}$:

$$Y(j\omega_o) = j\omega_o C + \frac{1}{R+j\omega_o L} \quad (1)$$

Rationalizing the second term by multiplying by $\frac{R-j\omega_o L}{R-j\omega_o L}$, we get

$$Y(j\omega_o) = j\omega_o C + \frac{R-j\omega_o L}{R^2+\omega_o^2 L^2} \tag{2}$$

or $$Y(j\omega_o) = \frac{R}{R^2+\omega_o^2 L^2} + j\omega_o\left[C - \frac{L}{R^2+\omega_o^2 L^2}\right] \tag{3}$$

Now applying the second of the two conditions alluded to above, we get

$$\frac{1}{2R} = \frac{R}{R^2 + \omega_o^2 L^2}, \tag{4}$$

from which we derive the fact that

$$L = \frac{R}{\omega_o}. \tag{5}$$

If we use the above value of L and require that the imaginary part be zero, we get

$$C = \frac{\frac{R}{\omega_o}}{R^2 + \omega_o^2 (\frac{R}{\omega_o})^2} \tag{6}$$

or $$C = \frac{1}{2R\omega_o}. \tag{7}$$

Equations (5) and (7) are the desired values of L and C.

• PROBLEM 13-28

How much capacitance is required to resonate a coil of 0.001 mH at a frequency of 2MHz?

Solution: Resonance requires that $X_L = -X_C$. The angular frequency is $\omega = 2\pi f = 2\pi(2\text{x}10^6) = 1.257\text{x}10^7$ rad/s. Thus $X_L = \omega L = (1.257\text{x}10^7)(10^{-6}) = 12.57\Omega$. For the capacitor, $-X_C = \frac{1}{\omega C}$, so that $C = \frac{1}{\omega(-X_C)} = \frac{1}{\omega X_L} = \frac{1}{(1.257\text{x}10^7)(12.57)}$ $= 6.33\text{x}10^{-9} = 6{,}330\text{pF}$. Note that if the coil and capacitor are in series, the combination will have zero reactance at the resonant frequency; if the coil and capacitor are placed in parallel, the combination will have infinite reactance at the resonant frequency.

What inductance is required to resonate at 10 MHz with a capacitance of 100 pF?

Solution: The resonate frequency of an LC circuit is given as:

$$\omega_o = \frac{1}{\sqrt{LC}}$$

$$f_o = \frac{1}{2\pi\sqrt{LC}}$$. We are given f_o and C and must find L.

First we solve for L and write a generalized expression:

$$f_o 2\pi = \frac{1}{\sqrt{LC}}$$

$$\frac{1}{f_o 2\pi} = \sqrt{LC}$$

$$\frac{1}{(f_o 2\pi)^2} = LC$$

$$\frac{1}{C(f_o 2\pi)^2} = L$$

$$\frac{1}{C f_o^2 4\pi^2} = L .$$

Substituting $10\text{x}10^6$Hz for f_o and $100\text{x}10^{-12}$F for C we obtain:

$$L = \frac{1}{(100\text{x}10^{-12})(10\text{x}10^6)^2 4\pi^2}$$

$$L = \frac{1}{(100\text{x}10^{-12})(100\text{x}10^{12})4\pi^2}$$

$$L = \frac{1}{(10000)4\pi^2} = \frac{1}{(1\text{x}10^4)4\pi^2}$$

$$L = \frac{1}{4\pi^2}(1\text{x}10^{-4}) = 0.0253(1\text{x}10^{-4})$$

$$L = (2.53\text{x}10^{-2})(1\text{x}10^{-4}) = 2.53\text{x}10^{-6} = 2.53\ \mu\text{H}.$$

Find the impedance at ω = 1 M rad/s, of: (a) a 2-kΩ resistor in parallel with a 1-mH inductor; (b) the series combination of a 0.001 μF capacitor and the network of (a) above; (c) the parallel combination of a 0.002 μF capacitor and the network of (b) above.

Solution: (a) Parallel impedance of $\vec{Z}_1$ and $\vec{Z}_2$ is

$$\vec{Z}_a = \frac{\vec{Z}_1 \vec{Z}_2}{\vec{Z}_1 + \vec{Z}_2} \tag{1}$$

In this case $\vec{Z}_1$ is a resistor and $\vec{Z}_2$ is an inductor. If the resistance is R Ohms and the inductance is L Henrys, then

$$\vec{Z}_1 = R\Omega \tag{2}$$

$$\vec{Z}_2 = j\omega L\Omega \tag{3}$$

Substituting Eqs. (2) and (3) into Eq. (1) yields:

$$\vec{Z}_a = \frac{j\omega LR}{R+j\omega L} = \frac{j\omega LR(R-j\omega L)}{(R+j\omega L)(R-j\omega L)}$$

$$= \frac{\omega^2 L^2 R + j\omega LR^2}{R^2+\omega^2 L^2} = \frac{\omega LR}{R^2+\omega^2 L^2}(\omega L + jR)$$

$$\vec{Z}_a \equiv Z_a \underline{/\Theta a}\ \Omega \tag{4}$$

where

$$Z_a = \frac{\omega LR}{R^2+\omega^2 L^2}\sqrt{R^2+\omega^2 L^2} = \frac{\omega LR}{\sqrt{R^2+\omega^2 L^2}} \tag{5}$$

$$\Theta_a = \tan^{-1}\frac{R}{\omega L}\,. \tag{6}$$

In this case ω = 1 M rad/s = 10^6 rad/s

$$R = 2k\Omega = 2000\Omega \tag{7}$$

$$L = 1mH = 10^{-3}\ H$$

$$Z_a = \frac{10^6 x 10^{-3} x 2000}{\sqrt{2^2 x 10^6 + 10^{12} x 10^{-6}}} = \frac{2x10^6}{\sqrt{5x10^6}} = 0.894x10^3 \tag{8}$$

$$= 0.984\ k\Omega.$$

$$\Theta_a = \tan^{-1}\frac{2000}{10^6 x 10^{-3}} = \tan^{-1}2 = 63.4° \quad (9)$$

or $$\vec{Z}_a = 0.894\underline{/63.4°}\ k\Omega \quad (10)$$

(b) Impedance of a capacitance C_b Farads is

$$\vec{Z}_3 = \frac{1}{j\omega C_b} = -j\frac{1}{\omega C_b}\ \Omega \quad (11)$$

Series impedance with this and $\vec{Z}_a$ is

$$\vec{Z}_b = \vec{Z}_3 + \vec{Z}_a$$

$$= -j\frac{1}{\omega C_b} + Z_a \angle \Theta_a$$

$$= -j\frac{1}{\omega C_b} + Z_a \cos \Theta_a + jZ_a \sin \Theta_a$$

$$= Z_a \cos \Theta_a + j \{Z_a \sin \Theta_a - \frac{1}{\omega C}\}$$

$$= \sqrt{(Z_a\cos\Theta_a)^2 + (Z_a\sin\Theta_a - \frac{1}{\omega C_b})^2} \angle \tan^{-1}\frac{Z_a\sin\Theta_a - \frac{1}{\omega C}}{Z_a\cos\Theta_a}$$

$$= \sqrt{(0.894\cos 63.4°)^2 + (0.894\sin 63.4° - \frac{1x10^{-3}}{10^6 x 0.001x10^{-6}})^2}\ k\Omega$$

$$\angle \tan^{-1}\frac{0.894\sin 63.4° - \frac{1x10^{-3}}{10^6 x 0.001x10^{-6}}}{0.894\cos 63.4°}$$

$$= \sqrt{(0.4)^2 + (-0.2)^2} \angle \tan^{-1}\frac{-0.2}{0.4}$$

$$= 0.447\ \underline{/-26.6°}\ \ k\ \Omega. \quad (12)$$

It should be noted that the impedance was calculated in the unit of $k\Omega(10^{+3}\Omega)$ instead of Ω. Therefore a factor of 10^{-3} is needed. Otherwise, the term's unit is only Ω and not $k\Omega$.

(c) If 0.002 μF capacitor is added in parallel to $\vec{Z}_b$, the parallel impedance is

$$\vec{Z}_C = \frac{\vec{Z}_b \frac{1}{j\omega C_C}}{\vec{Z}_b + \frac{1}{j\omega C_C}} \tag{13}$$

Substituting the value of $C_C = 0.002x10^{-6}F$ into this equation yields:

$$\vec{Z}_C = \frac{Z_b \angle\Theta_b \cdot \frac{1}{\omega C_C} \angle -90°}{Z_b\cos\Theta_b + j(Z_b\sin\Theta_b - \frac{1}{\omega C_C})}$$

$$= \frac{\frac{Z_b}{\omega C_C} \angle \Theta_b - 90°}{\sqrt{(Z_b\cos\Theta_b)^2 + (Z_b\sin\Theta_b - \frac{1}{\omega C_C})^2} \angle \tan^{-1} \frac{Z_b\sin\Theta_b - \frac{1}{\omega C_C}}{Z_b\cos\Theta_b}}$$

$$= \frac{\frac{Z_b}{\omega C_C}}{\sqrt{(Z_b\cos\Theta_b)^2 + (Z_b\sin\Theta_b - \frac{1}{\omega C_C})^2}} \angle \left(\Theta_b - 90° - \tan^{-1} \frac{Z_b\sin\Theta_b - \frac{1}{\omega C_C}}{Z_b\cos\Theta_b}\right)$$

$$= \frac{\frac{0.447x10^{-3}}{10^6 x0.002x10^{-6}}}{\sqrt{(0.447\cos(-26.6°))^2 + (0.447\sin(-26.6°) - \frac{10^{-3}}{10^6 x0.002x10^{-6}})^2}}$$

$$\angle \left(-26.6° - 90° - \tan^{-1} \frac{0.447\sin(-26.6°) - \frac{10^3}{10^6 x0.002x10^{-6}}}{0.447\cos(-26.6°)}\right)$$

$$= \frac{0.2235}{\sqrt{0.4^2 + (-0.7)^2}} \angle(-116.6° - (-60.3°))$$

$$= \frac{0.2235}{0.8062} \angle -56.3°$$

$$Z_C = 0.277 \angle -56.3° \text{ k } \Omega$$

• **PROBLEM 13-31**

The admittance and impedance of the network shown in Fig. 1 are equal at every frequency. Find R and C.

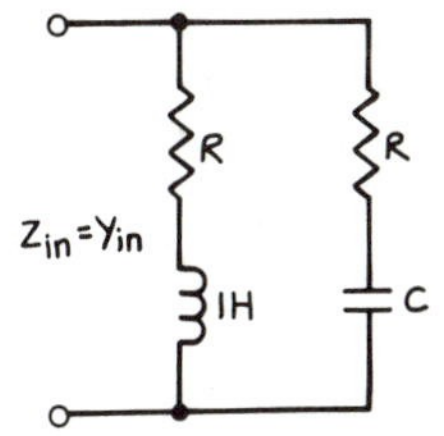

FIG.1

Solution: This problem nicely illustrates the relationship of the poles and zeros of Z(s) and Y(s). Since Z(s) = Y(s) for all frequencies, and since the poles and zeros of Z(s) are the zeros and poles, respectively, of Y(s), the poles and zeros of Y(s) (or Z(s)) must be identical. Thus, Y(s) is a constant. The question is, what values of R and C result in constant Y(s)?

$$Y(s) = \frac{1}{R + s} + \frac{1}{R + \frac{1}{sC}} \tag{1}$$

Or, after rationalizing,

$$Y(s) = \frac{1}{R}\,\frac{s^2 + 2Rs + \frac{1}{C}}{s^2 + (R + \frac{1}{RC})s + \frac{1}{C}}\,. \tag{2}$$

Evidently, if $2R = R + \frac{1}{RC}$, $Y(s) = \frac{1}{R}$, a constant. Thus, we require that $R^2 = \frac{1}{C}$.

However, because Y(s) = Z(s), it is necessary that $\frac{1}{R} = R$, or R = 1. Thus, R = 1Ω and C = 1F.

• PROBLEM 13-32

Given a 10-Ω resistor in series with a 10-μF capacitor, determine the two-element parallel equivalent if ω =: (a) 200; (b) 1000; (c) 5000 rad/s.

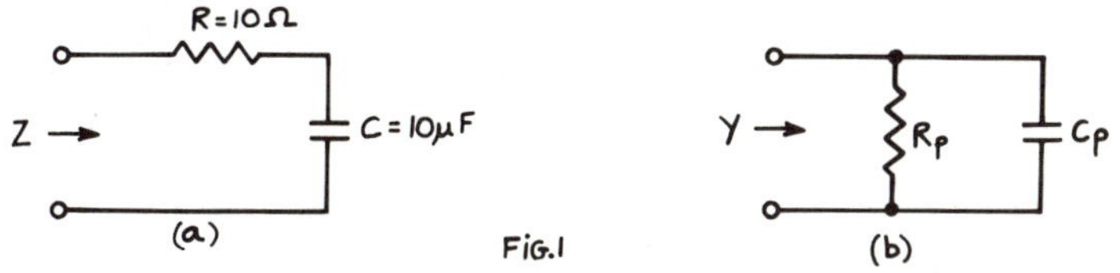

FIG.1

Solution: In this problem we are speaking of equivalence with respect to the terminals. Therefore, both the series and the parallel network must have the same impedance Z (or the same admittance Y). Now

$$Z = R + \frac{1}{j\omega C} = \frac{1 + j\omega RC}{j\omega C} \quad \text{and}$$

$$Y = \frac{1}{Z} = \frac{j\omega C}{1 + j\omega RC} \tag{1}$$

It is necessary to express Y in terms of a real and an imaginary part, corresponding to a conductance and susceptance, respectively, in the parallel equivalent circuit. Thus the numerator and denominator of equation (1) are multiplied by the complex conjugate of the denominator:

$$Y = \frac{j\omega C(1-j\omega RC)}{(1+j\omega RC)(1-j\omega RC)} = \frac{\omega^2 RC^2 + j\omega C}{1+\omega^2 R^2 C^2}$$

$$\text{or} \qquad Y = G+jB = \frac{\omega^2(10)(10^{-5})^2}{1+\omega^2(10)^2(10^{-5})^2} + j\omega\left[\frac{10^{-5}}{1+\omega^2(10)^2(10^{-5})^2}\right]. \qquad (2)$$

Now $R_p = \frac{1}{G}$ (see fig. 1(b)) and B is positive in equation (2), so the parallel reactance must be a capacitor with value $C_p = B/\omega$. The numerical values for R_p and C_p are given in the table below for the three given frequencies:

ω (rad/s)	$R_p = \dfrac{1+10^{-8}\omega^2}{10^{-9}\omega^2}$	$C_p = \dfrac{10^{-5}}{1+10^{-8}\omega^2}$
200	25 kΩ	10 μF
1000	1.01 kΩ	9.9 μF
5000	50 Ω	8 μF

• PROBLEM 13-33

In order to indicate that active elements may be used to increase the Q of resonant circuits, find Q_o for the circuit shown in fig. 1 by an inspection of the input admittance presented to the independent source if k =: (a) 0; (b) $4\text{x}10^{-5}$; (c) $-2\text{x}10^{-4}$.

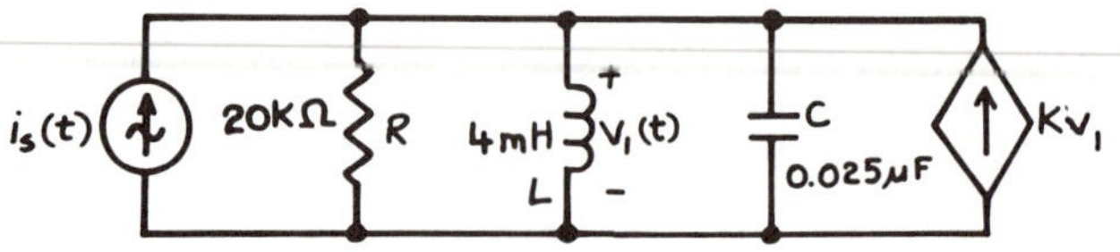

FiG.1

Solution: The admittance looking into the circuit from the independent source is

$$\frac{I_s}{V_1} = Y = \frac{1}{R} + \frac{1}{sL} + sC - k$$

$$Y = \frac{s^2 + s\left[\frac{1}{RC} - \frac{k}{C}\right] + \frac{1}{LC}}{\frac{1}{C}s}.$$

From this equation it is found that Q_o is the value of ω_o

divided by the bracketed value $(\frac{1}{RC} - \frac{k}{C})$.

(a) When k = 0, the admittance of the parallel RLC circuit is

$$Y = \frac{s^2 + s\left[\frac{1}{RC}\right] + \frac{1}{LC}}{\frac{1}{C}s} .$$

The Q is

$$Q_o = \omega_o RC = \frac{1}{\sqrt{LC}} RC = 50.$$

(b) When $k = 4x10^{-5}$

$$Q_o = \omega_o \left[\frac{1}{\frac{1}{RC} - \frac{k}{C}}\right] = 250$$

(c) When $k = -2x10^{-4}$,

$$Q_o = \omega_o \left[\frac{1}{\frac{1}{RC} - \frac{k}{C}}\right] = 10.$$

• **PROBLEM 13-34**

For the circuit shown in fig. 1, determine: the resonant frequency; the Q_o; the bandwidth; and sketch $|I|$ versus ω. What would the Q_o of the circuit be if the 20Ω resistor were short-circuited? What would it be if the 200 kΩ resistor were open-circuited?

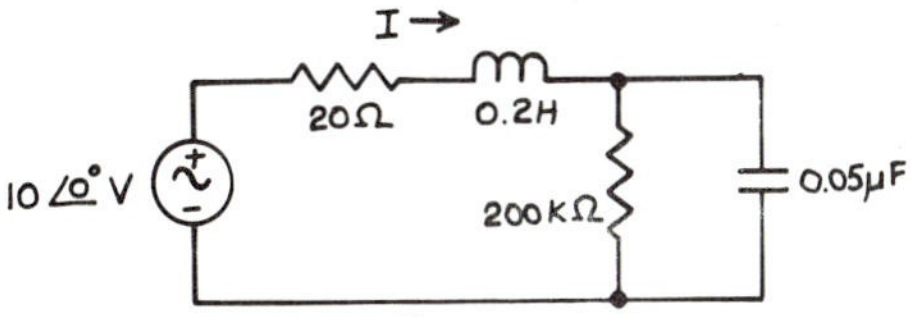

FIG.1

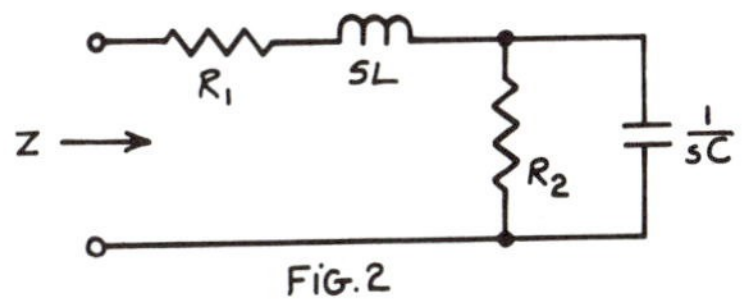

FIG.2

Solution: Since the circuit is an RLC circuit containing one inductor and one capacitor but two resistors we cannot use the formula for the resonant frequency in a series-parallel RLC circuit, $\frac{1}{\sqrt{LC}}$. Instead, we must write an expression for the impedance of the circuit in fig. 1 and compare it to the impedance of a series RLC circuit. Hence, from fig. 2

$$Z = R_1 + sL + \frac{\frac{R_2}{sC}}{R_2 + \frac{1}{sC}}$$

can be written

$$Z = \frac{s^2 + s\left(\frac{R_1}{L} + \frac{1}{R_2C}\right) + \frac{R_1 + R_2}{R_2LC}}{\frac{1}{L}(s + \frac{1}{R_2C})} \qquad (1)$$

Note that the impedance for a series RLC circuit in fig. 3 is of similar form to Eq. (1) above:

$$Z_s = \frac{s^2 + s\frac{R}{L} + \frac{1}{LC}}{\frac{1}{L}s} \qquad (2)$$

The resonant frequency of the series RLC circuit is $\frac{1}{\sqrt{LC}}$.

Since $R_1 << R_2$ in the circuit of fig. 1 the resonant frequency is the square root of the last term in the numerator of Eq. (1). Hence,

$$\omega_o = \sqrt{\frac{R_1 + R_2}{R_2LC}} \cong \frac{1}{\sqrt{LC}} = 10{,}000 \text{ rad/s}.$$

The Q_o of the series RLC circuit is $\frac{1}{\sqrt{LC}}\ \frac{L}{R} = \frac{\omega_o}{\frac{R}{L}}$. Note that the pole at $-\frac{1}{R_2C}$ in the impedance expression for the circuit in fig. 1 is -100 rad/s which is far enough away from its resonant frequency so as not to interfere with that portion of the response. Hence, the Q_o for the circuit is $\omega_o\left[\frac{1}{\frac{R_1}{L} + \frac{1}{R_2C}}\right] = 50$. The RLC circuit in fig. 1 will behave similarly to an equivalent series RLC circuit with $R_{eq} = 40\Omega$, $L_{eq} = .2H$ and $C_{eq} = 0.05\mu F$. The bandwidth of an RLC circuit is defined as $\frac{\omega_o}{Q_o} = \frac{10{,}000}{50} = 200$ rad/s.

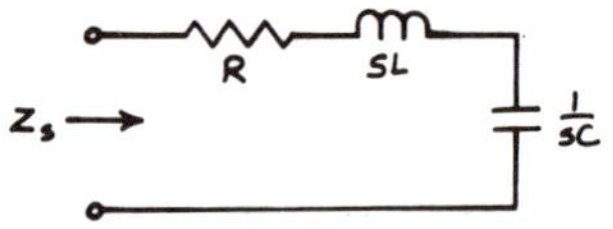

FiG.3

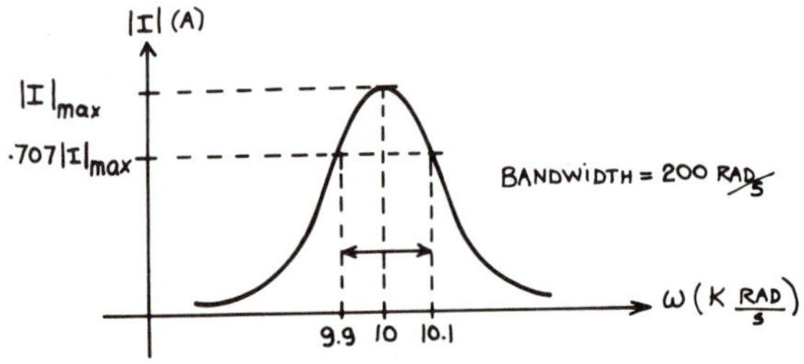

FiG.4

Fig. 4 shows a plot of $|I|$ vs. ω, drawn from the information found above.

The value for $|I|$ at the resonant frequency $|I|_{max}$ is found by themethods of sinusoidal steady-state analysis:

$$\frac{\vec{V}}{\vec{Z}} = \vec{I} = \frac{10\ \underline{/0°}}{(20 + j\omega 0.2) + \dfrac{\dfrac{200\text{x}10^3}{j\omega 0.05\text{x}10^{-6}}}{200\text{x}10^3 + \dfrac{1}{j\omega 0.05\text{x}10^{-6}}}}$$

Letting $\omega = 10{,}000$, yields:

$$\vec{I}_{max} = \frac{10\ \underline{/0°}}{(20 + j2000) + \dfrac{-j(200\text{x}10^3)(2\text{x}10^3)}{200\text{x}10^3 - j2\text{x}10^3}}$$

$$\vec{I}_{max} = \frac{10\ \underline{/0°}}{\dfrac{(20 + j2\text{x}10^3)(200\text{x}10^3 - j2\text{x}10^3) - j4\text{x}10^8}{200\text{x}10^3 - j2\text{x}10^3}}$$

$$|\vec{I}|_{max} = \frac{1}{4}\ \text{A}\ .$$

When the 20Ω resistor is short-circuited,

$$Z = \frac{s^2 + s\dfrac{1}{R_2C} + \dfrac{1}{LC}}{\dfrac{1}{L}\left(s + \dfrac{1}{R_2C}\right)}\ ,$$

hence, $$Q_o = \frac{\omega_o}{\dfrac{1}{R_2C}} = \frac{1}{\sqrt{LC}} R_2C = 100.$$

When the 200 kΩ resistor is open-circuited,

$$Z = \frac{s^2 + s\dfrac{R_1}{L} + \dfrac{1}{LC}}{\dfrac{1}{L}s}\ ,$$

the impedance of a series RLC circuit. Hence,

$$Q_o = \frac{\omega_o}{\dfrac{R_1}{L}} = 100.$$

• **PROBLEM 13-35**

For the circuit shown in fig. 1 perform the following:
(a) Find the resonant frequency of the circuit.

(b) Find the quality Factor Q_o.

(c) With $i_R(t)$ as the desired circuit response, find and sketch the phase and amplitude transfer characteristics of the network.

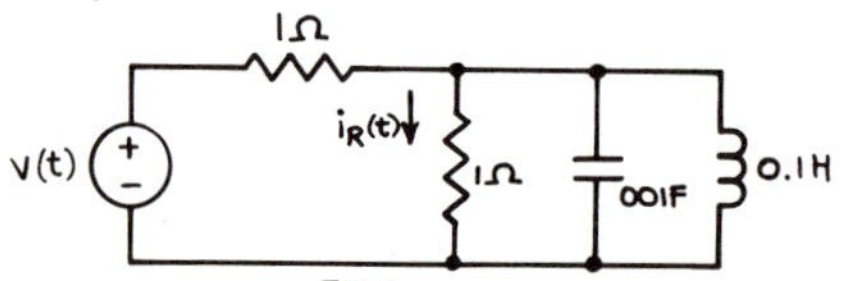

Fig.1

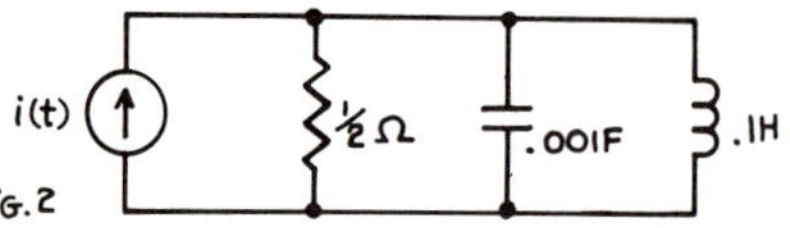

Fig.2

Solution: (a) The resonant frequency of the circuit is defined as

$$\omega_o = \frac{1}{\sqrt{LC}}$$

$$\omega_o = \frac{1}{\sqrt{(.001)(0.1)}} = \frac{1}{.01} = 100 \text{ rad/s.}$$

(b) The Q_o of the circuit can be found by noting that the Q_o of a parallel RLC circuit is $\omega_o R_p C_p$. Convert the circuit of fig. 1 to a parallel RLC circuit shown in fig. 2 by making a source transformation and combining resistors. Hence

$$Q_o = \omega_o R_p C_p = 100(\tfrac{1}{2})(.001)$$

$$Q_o = (50)(.001) = 0.05.$$

(c) In fig. 1 we are interested in sketching the phase and amplitude of $i_R(t)$ as we vary ω of the voltage source. Since we are interested in the sinusoidal steady state solution of $i_R(t)$ we transform the circuit of fig. 1 to its sinusoidal steady-state equivalent, shown in fig. 3, by finding $\frac{\vec{I}_R}{\vec{V}_R} = H(j\omega)$, the transfer function for i_R in the circuit of fig. 3.

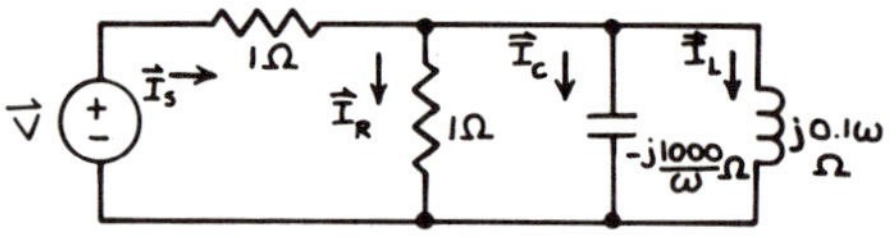

Fig.3

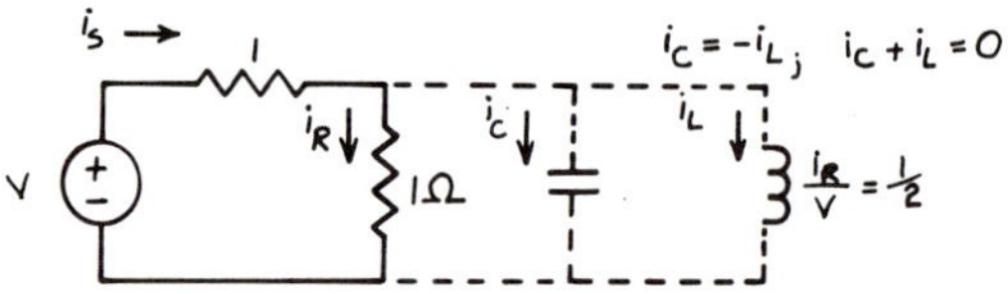

Fig.4

Using KCL we can write

$$\vec{I}_s = \vec{I}_R + \vec{I}_C + \vec{I}_L \tag{1}$$

where
$$I_C = \frac{\vec{V} - \vec{I}_s}{-j\,\frac{1000}{\omega}} = \frac{j(\vec{V} - I_s)\omega}{1000} \tag{2}$$

$$I_L = \frac{\vec{V} - \vec{I}_s}{j0.1\omega} = \frac{-j(\vec{V} - \vec{I}_s)}{0.1\omega} \quad (3)$$

$$I_R = \frac{\vec{V} - \vec{I}_s}{R} = \vec{V} - \vec{I}_s. \quad (4)$$

Hence $$I_s = \vec{V} - \vec{I}_s + \frac{j(\vec{V} - \vec{I}_s)\omega}{1000} - \frac{j(\vec{V} - \vec{I}_s)}{0.1\omega} \quad (5)$$

$$\vec{I}_s = \frac{\vec{V}(1 + \frac{j\omega}{1000} - \frac{j}{0.1\omega})}{(2 + \frac{j\omega}{1000} - \frac{j}{0.1\omega})} . \quad (6)$$

By substituting Eq. (6) and (4) into Eq. (5) we obtain:

$$\frac{\vec{V}(1 + \frac{j\omega}{1000} - \frac{j}{0.1\omega})}{(2 + \frac{j\omega}{1000} - \frac{j}{0.1\omega})} = \vec{I}_R \; (1 + \frac{j\omega}{1000} - \frac{j}{0.1\omega})$$

$$\frac{\vec{I}_R}{\vec{V}} = H(j\omega) = \frac{1}{2 + \frac{j\omega}{1000} - \frac{j}{0.1\omega}}$$

$$H(j\omega) = \frac{j0.1}{(1 - 0.0001\omega^2) + j0.2\omega} .$$

We can write the magnitude $|H(j\omega)|$ as

$$|H(j\omega)| = \frac{0.1\omega}{\sqrt{(1 - 0.0001\omega^2) + (0.2\omega)^2}} .$$

We note that at the resonant frequency $\omega_o = 100$ rad/s $|H(j\omega)| = 0.5$. It can be seen that this is equivalent to the circuit shown in fig. 4, since $\vec{Z}_C = -j10 = -\vec{Z}_L = j10\Omega$.

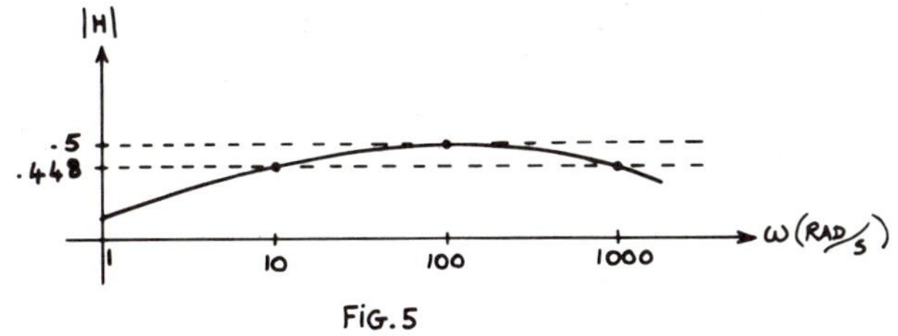

FIG. 5

Fig. 5 shows a plot of magnitude vs. ω for $\vec{I}_R$.

CHAPTER 14

FOURIER ANALYSIS

FOURIER TECHNIQUES

• PROBLEM 14-1

Represent the triangular waveform shown in Fig. 1 as a Fourier series.

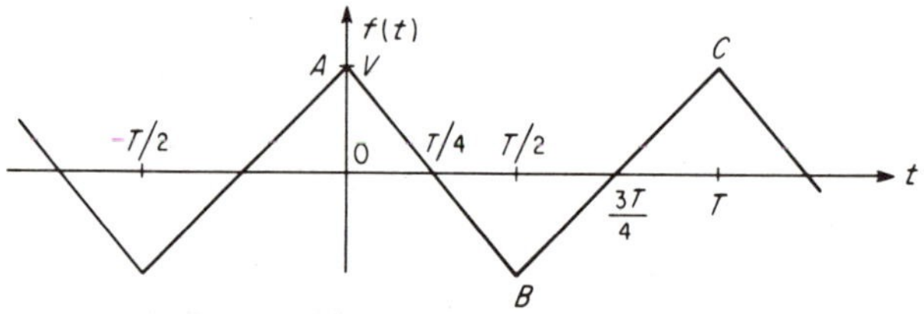

Fig. 1

Solution: We choose the origin to make the function even; therefore, $b_n = 0$. Also the function has half-wave symmetry; therefore, $a_o = 0$. So we can write the Fourier series

$$f(t) = \sum_{n=1}^{\infty} a_n \cos n\omega t; \quad \omega = \frac{2\pi}{T}$$

where $a_n = \frac{2}{T}\int_{o}^{T} f(t) \cos n\omega t \, dt.$

We can choose the periodic interval $[t_o, t_o+T]$ rather than $[0,T]$. In this case it is convenient to choose $[-\frac{T}{2}, \frac{T}{2}]$ because of the form of f(t). Hence we can write the alternate form of a_n as

$$a_n = \frac{2}{T}\int_{-\frac{T}{2}}^{\frac{T}{2}} f(t) \cos n\omega t \, dt.$$

Since both f(t) and cos nωt are even functions, the integral must have the same value between $-\frac{T}{2}$ to 0 and between 0 and

$\frac{T}{2}$. We then have

$$a_n = \frac{4}{T}\int_0^{\frac{T}{2}} f(t) \cos n\omega t \, dt.$$

Hence,

$$f(t) = -\frac{4V}{T}\left(t - \frac{T}{4}\right) \qquad \text{where}$$

$-\frac{4V}{T}$ is the slope of the line for the interval $0 \le t \le \frac{T}{2}$. Therefore,

$$a_n = -\frac{16V}{T^2}\int_0^{\frac{T}{2}} \left(t - \frac{T}{4}\right) \cos n\omega t \, dt$$

$$a_n = -\frac{16V}{T^2}\int_0^{\frac{T}{2}} t \cos n\omega t \, dt + \frac{4V}{T}\int_0^{\frac{T}{2}} \cos n\omega t \, dt$$

$$a_n = -\frac{16V}{T^2}\left[\frac{1}{n^2\omega^2}\cos n\omega t + \frac{t}{n\omega}\sin n\omega t\right]_0^{\frac{T}{2}}$$

$$+ \frac{4V}{T}\left[\frac{\sin n\omega t}{n\omega}\right]_0^{\frac{T}{2}} \quad ; \quad \omega = \frac{2\pi}{T}$$

$$a_n = -\frac{16V}{T^2}\left[\frac{T^2}{n^2 4\pi^2}\{\cos n\pi - 1\} + \frac{T^2}{n4\pi}\{\sin n\pi\}\right]$$

$$+ \frac{2V}{\pi n}\{\sin n\pi\}.$$

Hence, $\sin n\pi = 0$ for all n and $\cos n\pi = 1$ for n = even, and -1 for n = odd, which yields

$$f(t) = \sum_{n \text{ odd}}^{\infty} \frac{8V}{n^2\pi^2} \cos n\omega t$$

$$f(t) = \frac{8V}{\pi^2}\left[\cos \omega t + \frac{1}{9}\cos 3\omega t + \frac{1}{25}\cos 5\omega t + \ldots\right].$$

• **PROBLEM 14-2**

Write the Fourier series for the three voltage waveforms in Fig. 1.

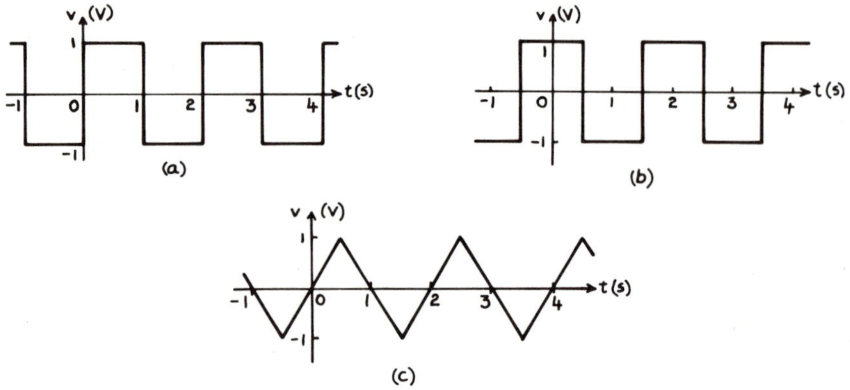

Fig. 1

Solution: In general, the Fourier coefficients are given by

$$a_n = \frac{2}{T}\int_0^T f(t)\cos n\omega_o t\,dt \tag{1}$$

$$b_n = \frac{2}{T}\int_0^T f(t)\sin n\omega_o t\,dt \tag{2}$$

$$a_o = \frac{1}{T}\int_0^T f(t)\,dt \tag{3}$$

(a) This waveform is an odd function, which means that $a_o = 0$ and $a_n = 0$. The period, T, is 2 and the waveform over one period given by $v(t) = \begin{cases} 1; & 0 < t < T/2 \\ -1; & T/2 < 0 < T \end{cases}$.

Also, since the function is odd, $b_n = \frac{4}{T}\int_0^{T/2} f(t)\sin n\omega_o t\,dt$.

Then $b_n = \frac{4}{2}\int_0^1 \sin n\omega_o\, t\,dt = 2\left[\frac{1}{n\omega_o}(-\cos n\omega_o t)\Big|_0^1\right]$

Recall that $\omega_o = \frac{2\pi}{T} = \frac{2\pi}{2} = \pi$

$$b_n = \frac{2}{n\pi}(-\cos n\pi + \cos 0)$$

$$b_n = \frac{2}{n\pi}(1 - \cos n\pi)$$

$$\cos n\pi = \begin{cases} 1 & n \text{ even} \\ -1 & n \text{ odd} \end{cases}.$$

$$\text{Then} \quad b_n = \begin{cases} \frac{2}{n\pi}(0), & n \text{ even} \\ \frac{2}{n\pi}(2), & n \text{ odd} \end{cases}$$

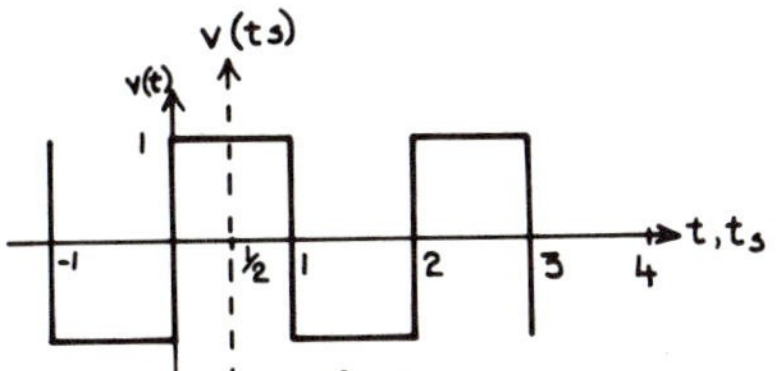

FiG. 2

$$b_n = \begin{cases} 0, & n \text{ even} \\ \frac{4}{n\pi}, & n \text{ odd.} \end{cases}$$

So the Fourier series is

$$v(t) = \frac{4}{\pi}(\sin\pi t + \frac{1}{3}\sin 3\pi t + \frac{1}{5}\sin 5\pi t + \ldots.).$$

(b) For this problem, we can make use of the results of the previous problem and use a very nice trick that eliminates the integration. (The integration in this problem is easy, but this method can be of great use when the integrations involved are more tedious.)

Use the waveform of part (a) and shift the v-axis so that it intersects the t-axis at $t = \frac{1}{2}$ s. (See Fig. 2.)

Then $\quad t = t_s + \frac{1}{2}$. We know from part (a) that

$$v(t) = \frac{4}{\pi}(\sin \pi t + \frac{1}{3}\sin 3\pi t + \frac{1}{5}\sin 5\pi t + \ldots.).$$

The required waveform for part (b) is $v(t_s)$. Therefore

$$v(t_s) = \frac{4}{\pi}(\sin\pi(t_s+\frac{1}{2}) + \frac{1}{3}\sin 3\pi(t_s+\frac{1}{2}) + \frac{1}{5}\sin 5\pi(t_s+\frac{1}{2}) + \ldots.)$$

$$= \frac{4}{\pi}(\sin(\pi t_s+\frac{\pi}{2}) + \frac{1}{3}\sin(3\pi t_s+\frac{3\pi}{2}) + \frac{1}{5}\sin(5\pi t_s+\frac{5\pi}{2}) + \ldots.).$$

Dropping the subscript s for simplicity, we obtain

$$v(t) = \frac{4}{\pi}(\cos\pi t - \frac{1}{3}\cos 3\pi t + \frac{1}{5}\cos 5\pi t + \ldots.).$$

(c) Again, this waveform is odd so $a_n = a_o = 0$ and

$$b_n = \frac{4}{T}\int_0^{\frac{T}{2}} f(t)\sin n\omega_o t\, dt = \frac{4}{T}\int_{-\frac{T}{4}}^{\frac{T}{4}} f(t)\sin n\omega_o t\, dt$$

$$f(t) = v(t) = 2t \quad \text{for} \quad -\frac{T}{4} < t < \frac{T}{4}$$

$$b_n = \frac{4}{2}\int_{-\frac{1}{2}}^{\frac{1}{2}} 2t \sin n\omega_o t \, dt = 4\int_{-\frac{1}{2}}^{\frac{1}{2}} t \sin n\omega_o t \, dt.$$

Integrating by parts: $u = t \quad du = dt$

$$dv = \sin n\omega_o t \, dt \qquad v = -\frac{1}{n\omega_o} \cos n\omega_o t.$$

Then

$$b_n = 4\left[\frac{-t}{n\omega_o} \cos n\omega_o t\Big|_{-\frac{1}{2}}^{\frac{1}{2}} + \frac{1}{n\omega_o}\int_{-\frac{1}{2}}^{\frac{1}{2}} \cos n\omega_o t \, dt\right];$$

$$\omega_o = \frac{2\pi}{T} = \pi$$

$$b_n = 4\left[\frac{-t}{n\pi} \left(\cos\left(\frac{n\pi}{2}\right) - \cos\left(\frac{-n\pi}{2}\right)\right) + \frac{1}{n^2\omega_o^2}(\sin n\omega_o t)\Big|_{-\frac{1}{2}}^{\frac{1}{2}}\right.$$

Recall that $\cos\theta = \cos(-\theta)$

$$b_n = 4\left[\frac{1}{n^2\pi^2} \left(\sin \frac{n\pi}{2} - \sin \left(\frac{-n\pi}{2}\right)\right]\right.$$

$$\sin(-\theta) = -\sin\theta .$$

So, $\quad b_n = \dfrac{8}{n^2\pi^2} \sin \dfrac{n\pi}{2}$

$$\sin \frac{n\pi}{2} = \begin{cases} 0, \ n \text{ even} \\ 1, \ n=1, 5, 9..... \\ -1, \ n=3, 7, 11..... \end{cases}$$

So,

$$b_n = \begin{cases} 0, \ n \text{ even} \\ \dfrac{8}{n^2\pi^2}, \ n=1, 5, 9..... \\ \dfrac{-8}{n^2\pi^2}, \ n=3, 7, 11..... \end{cases}$$

and the Fourier series is

$$v(t) = \frac{8}{\pi^2} \left(\sin \pi t - \frac{1}{9} \sin 3\pi t + \frac{1}{25} \sin 5\pi t - \ \ldots.\right).$$

Determine the coefficient c_k in the complex Fourier series for the **waveforms** shown.

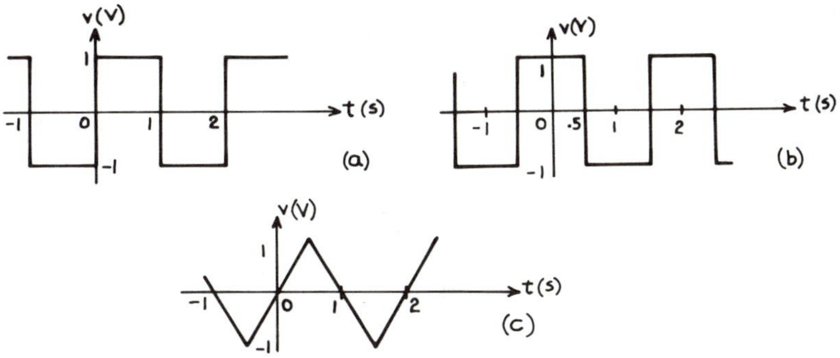

Solution: The general expression for the coefficient is

$$c_k = \frac{1}{\tau}\int_{\tau}^{\tau_1+\tau} v(t)e^{-jk\omega_o t}dt$$

where τ is the period, $\omega_o=2\pi/\tau$ is the fundamental angular frequency and τ_1 can be chosen to facilitate evaluation of c_k.

(a) $\tau=2s$ and $\omega_o=\pi$ rad/s. τ_1 is chosen as -1s, so that v(t) changes values only once over the interval of integration. Thus

$$c_k = \frac{1}{2}\int_{-1}^{1} v(t)e^{-jk\pi t}dt$$

$$= \frac{1}{2}\left[-\int_{-1}^{0} e^{-jk\pi t}dt + \int e^{-jk\pi t}dt\right].$$

Evaluation of the integrals gives

$$c_k = \frac{1}{2}\left[\frac{1}{jk\pi}\, e^{-jk\pi t}\Big|_{-1}^{0} - \frac{1}{jk\pi}\, e^{-jk\pi t}\Big|_{0}^{1}\right]$$

$$= \frac{1}{2jk\pi}\ [(1 - e^{jk\pi}) + (1 - e^{-jk\pi})].$$

This expression is rearranged to collect the exponential terms:

$$c_k = \frac{1}{jk\pi}\left[1 - \left(\frac{e^{jk\pi} + e^{-jk\pi}}{2}\right)\right].$$

The exponential form of cos kπ is recognized, so that

$$c_k = -j\frac{1}{k\pi}(1 - \cos k\pi).$$

(b) This is the same waveform as part (a), but advanced by $\frac{1}{2}$ s. The time τ, is thus chosen as $-\frac{1}{2}$ s, and

$$c_k = \frac{1}{2}\left[\int_{-\frac{1}{2}}^{\frac{1}{2}} e^{-jk\pi t}dt - \int_{\frac{1}{2}}^{\frac{3}{2}} e^{-jk\pi t}dt\right]$$

$$c_k = \frac{-1}{2jk\pi}\left[e^{-jk\pi t}\Big|_{-\frac{1}{2}}^{\frac{1}{2}} - e^{-jk\pi t}\Big|_{\frac{1}{2}}^{\frac{3}{2}}\right]$$

$$= \frac{-1}{2jk\pi}\left[2e^{-jk\frac{\pi}{2}} - e^{jk\frac{\pi}{2}} - e^{-j\frac{3k\pi}{2}}\right].$$

To simplify this, we first note that

$$e^{-j\frac{3k\pi}{2}} \quad e^{j\frac{k\pi}{2}} = e^{-j2k\pi} = e^{j\frac{k\pi}{2}} \qquad k = 0, \pm 1, \pm 2, \ldots$$

Therefore

$$c_k = \frac{-1}{jk\pi}\left[e^{-jk\frac{\pi}{2}} - e^{j\frac{k\pi}{2}}\right] = \frac{2}{k\pi}\left[\frac{e^{j\frac{k\pi}{2}} - e^{-j\frac{k\pi}{2}}}{2j}\right].$$

The exponential form of sin $(\frac{k\pi}{2})$ is recognized, so that

$$c_k = \frac{2}{k\pi}\sin\left(\frac{k\pi}{2}\right).$$

(c) The period τ, and hence ω_o, are unchanged from the values of the first two parts. In considering a choice of τ_1, it is noted that v(t) = 2t for $-\frac{1}{2} \leq t \leq \frac{1}{2}$ and v(t) = 2 - 2t for $\frac{1}{2} \leq t \leq \frac{3}{2}$. Thus $\tau_1 = -\frac{1}{2}$ is a reasonable choice, and

$$c_k = \frac{1}{2}\left[\int_{-\frac{1}{2}}^{\frac{1}{2}} 2te^{-jk\pi t}dt + \int_{\frac{1}{2}}^{\frac{3}{2}} (2-2t)e^{-jk\pi t}dt\right].$$

This can be evaluated making use of the integral

$$\int xe^{ax}dx = \frac{e^{ax}}{a^2}(ax-1).$$

However, we will approach the evaluation of c_k in another way. Let $v_b(t)$ be the waveform of part (b) and $v_c(t)$, that of part (c). Now if $v_b(t)$ were doubled in amplitude, it would be exactly the time derivative of $v_c(t)$. That is,

$$2v_b(t) = \frac{d}{dt} v_c(t).$$

In this equation, $v_b(t)$ and $v_c(t)$ are replaced by their complex Fourier series representation, with the known values of c_k (for the $v_b(t)$ series) inserted:

$$2\sum_{k=-\infty}^{\infty}\left(\frac{2}{k\pi}\right)\sin\left(\frac{k\pi}{2}\right)e^{jk\pi t} = \frac{d}{dt}\sum_{k=-\infty}^{\infty} c_k e^{jk\pi t}.$$

The differentiation can be carried inside the summation to give

$$\sum_{k=-\infty}^{\infty}\left(\frac{4}{k\pi}\right)\sin\left(\frac{k\pi}{2}\right)e^{jk\pi t} = \sum_{k=-\infty}^{\infty} jk\pi c_k e^{jk\pi t}.$$

The coefficients of $e^{jk\pi t}$ are equated term for term:

$$\frac{4}{k\pi}\sin\left(\frac{k\pi}{2}\right) = jk\pi c_k$$

or $$c_k = -j\frac{4}{k^2\pi^2}\sin\left(\frac{k\pi}{2}\right).$$

• PROBLEM 14-4

Find the Fourier series representation of a periodic function with period 2π, such that

$$f(t) = t^2 \qquad -\pi \leq t \leq \pi.$$

Also compute the Fourier coefficients.

Solution: The Fourier series representation for a periodic function f(t) whose period is τ is

$$f(t) \sim a_o + \sum_{n=1}^{\infty} [a_n \cos(\frac{2\pi nt}{\tau}) + b_n \sin(\frac{2\pi nt}{\tau})] \tag{1}$$

where
$$a_o = \frac{1}{\tau}\int_{t_o}^{t_o+\tau} f(t)dt \tag{2}$$

$$a_n = \frac{2}{\tau}\int_{t_o}^{t_o+\tau} f(t)\cos(\frac{2\pi nt}{\tau})dt \tag{3}$$

$$b_n = \frac{2}{\tau}\int_{t_o}^{t_o+\tau} f(t)\sin(\frac{2\pi nt}{\tau})dt. \tag{4}$$

Using Eq.(2) with $\tau = 2\pi$, $f(t) = t^2$ yields

$$a_o = \frac{1}{2\pi}\int_{-\pi}^{\pi} t^2 dt. \tag{5}$$

Evaluate the integral:

$$a_o = \frac{1}{2\pi} \left.\frac{t^3}{3}\right|_{-\pi}^{\pi} = \frac{\pi^2}{3} . \tag{6}$$

Using Eq.(3) with $\tau = 2\pi$, $f(t) = t^2$ yields:

$$a_n = \frac{2}{2\pi}\int_{-\pi}^{\pi} t^2 \cos(nt)dt. \tag{7}$$

The expression can be simplified slightly by taking advantage of symmetry. The integral of an even function over a symmetric interval is twice the integral of the function over the upper half of the interval. Therefore,

$$a_n = \frac{2}{\pi}\int_{0}^{\pi} t^2 \cos(nt)dt. \tag{8}$$

The integral can be evaluated by a double application of integration by parts, summarized as follows:

$$\frac{2}{\pi}\int_0^{\pi} t^2\cos(nt)dt = \frac{2t^2\sin(nt)}{\pi n}\Bigg|_0^{\pi} - \frac{4}{\pi n}\int_0^{\pi} t\ \sin(nt)dt$$

$$= 0 - \frac{4}{\pi n}\left[\frac{-t\ \cos\ (nt)}{n}\Bigg|_0^{\pi} + \frac{1}{n}\int_0^{\pi}\cos(nt)dt\right]$$

$$= -\frac{4}{\pi n}\left(\frac{-\pi\cos(n\pi)}{n} + 0\right)$$

$$= \frac{4\ \cos\ (n\pi)}{n^2}\ . \tag{9}$$

The expression $\cos(n\pi)$ takes the value +1 if n is even, -1 if n is odd; therefore, we can write $\cos(n\pi) = (-1)^n$. Eq. (9) is then

$$a_n = (-1)^n \frac{4}{n^2}\ . \tag{10}$$

The expression for b_n involves integrating an odd function over a symmetric interval; so by symmetry $b_n = 0$.

$$b_n = \frac{2}{2\pi}\int_{-\pi}^{\pi} t^2\ \sin(nt)dt = 0 \tag{11}$$

The Fourier series is thus

$$f(t) \sim \frac{\pi^2}{3} + \sum_{n=1}^{\infty}\left[(-1)^n \frac{4}{n^2}\cos(nt)\right].$$

Note: We use the symbol ∿, "corresponds to," rather than =, "equals," because at discontinuities in f(t) the series converges to the mean of $f(t^-)$ and $f(t^+)$. Thus if f(t) is continuous, the Fourier series is equal to f(t). If f(t) has discontinuities, the series produces the same value at all values of t except those at which discontinuities occur, where it produces the average of $f(t^-)$ and $f(t^+)$.

• PROBLEM 14-5

Find the Fourier series of the half-wave rectified sinusoid as shown in Fig. 1.

Fig. 1

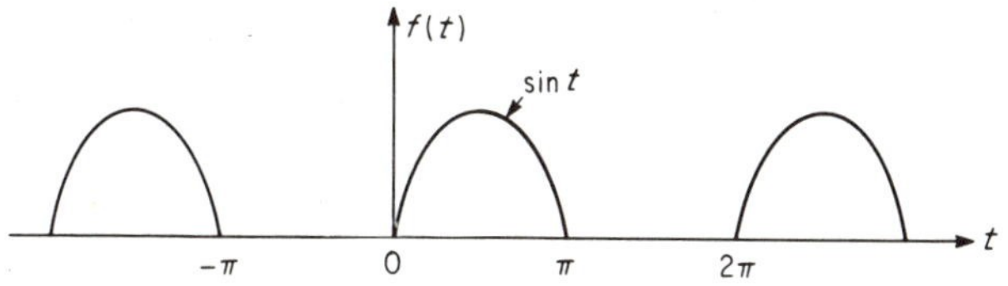

Solution: The function f(t) can be represented over one period as

$$f(t) = \begin{cases} \sin t & 0 \le t < \pi \\ 0 & \pi \le t < 2\pi \end{cases}$$

Since there is no half-wave symmetry we can expect a d.c. value. Hence,

$$a_o = \frac{1}{\tau}\int_0^{\tau} f(t)dt = \frac{1}{2\pi}\int_0^{\pi} \sin t\ dt + \frac{1}{2\pi}\int_{\pi}^{2\pi} 0\ dt$$

$$a_o = \frac{1}{2\pi}\,[-\cos t]_o^{\pi} = \frac{1}{2\pi}\,[1 + 1] = \frac{1}{\pi}\ .$$

However we cannot say that the function has odd or even symmetry. Therefore, we must determine a_n and b_n.

The Fourier representation of a periodic function is

$$a_o + \sum_{n=1}^{\infty} (a_n \cos n\omega_o t + b_n \sin n\omega_o t)$$

where

$$a_n = \frac{2}{\tau}\int_0^{\tau} f(t) \cos n\omega_o t\ dt$$

and

$$b_n = \frac{2}{\tau}\int_0^{\tau} f(t) \sin n\omega_o t\ dt.$$

First we find

$$b_n = \frac{2}{2\pi}\int_0^{\pi} \sin(t) \sin(nt)\ dt; \qquad \omega_o = 1$$

by the trigonometric function-product relationship

$$\sin t \sin nt = \frac{1}{2}\cos(t-nt) - \frac{1}{2}\cos\ (t+nt).$$

We obtain

$$b_n = \frac{1}{\pi}\int_0^{\pi} [\frac{1}{2}\cos t(1-n) - \frac{1}{2}\cos t\ (1+n)]\ dt$$

$$b_n = \frac{1}{\pi}\,[\frac{1}{2(1-n)}\sin t(1-n) - \frac{1}{2(1+n)}\sin t(1+n)]_o^{\pi}$$

$$b_n = \frac{1}{\pi}\left[\frac{1}{2(1-n)}\sin(\pi-n\pi) - \frac{1}{2(1+n)}\sin(\pi+n\pi) - 0 + 0\right]$$

$$b_n = \frac{1}{2\pi}\left(\frac{1}{1-n}\sin(\pi-n\pi) - \frac{1}{1+n}\sin(\pi+n\pi)\right).$$

When we substitute any positive value for n into the above expression we obtain $b_n = 0$; for $n = 1$ we obtain the undetermined form $\frac{0}{0}$ for the first term. Using L'Hospital's rule we can evaluate $\frac{1}{1-n}\sin(\pi-n\pi)$ for $n = 1$ as follows:

Let $$g(n) = \frac{g_1(n)}{g_2(n)} = \frac{\sin(\pi-n\pi)}{1-n}$$

where $g_1(n) = \sin(\pi-n\pi)$ and $g_2(n) = 1-n$.

By L'Hospital's rule we can evaluate

$$G(n) = \frac{g_1'(n)}{g_2'(n)} = \frac{-\pi\cos(\pi-n\pi)}{-1}.$$

If $G(1)$ exists then $g(1) = G(1)$. Hence.

$$g(1) = \pi\cos(\pi-\pi) = \pi \quad \text{and}$$

$$b_1 = \frac{1}{2\pi}(\pi) = \frac{1}{2}.$$

The only sine component in the Fourier series is $\frac{1}{2}\sin t$. We proceed to find a_n:

$$a_n = \frac{1}{\pi}\int_0^{\pi}\sin(t)\cos(nt)\,dt$$

$$a_n = \frac{1}{\pi}\int_0^{\pi}\left[\frac{1}{2}\sin t(1+n) + \frac{1}{2}\sin t(1-n)\right]dt$$

$$a_n = -\frac{1}{2\pi}\left[\frac{1}{1+n}\cos t(1+n) + \frac{1}{1-n}\cos t(1-n)\right]_0^{\pi}$$

$$a_n = -\frac{1}{2\pi}\left[\frac{1}{1+n}\cos(\pi+n\pi) + \frac{1}{1-n}\cos(\pi-n\pi) - \left(\frac{1}{1+n} + \frac{1}{1-n}\right)\right].$$

When $n = 1$ we obtain

$$a_1 = -\frac{1}{2\pi}\left[\frac{1}{2} + \frac{1}{0} - \left(\frac{1}{2} + \frac{1}{0}\right)\right]$$

where the two indeterminate terms are both equal to zero by L'Hospital's rule, giving $a_1 = 0$.

When n = 2 we obtain

$$a_2 = -\frac{1}{2\pi}\left[-\frac{1}{3} + 1 - \frac{1}{3} + 1\right] = -\frac{2}{3\pi}.$$

When n = 3 we again find

$$a_3 = -\frac{1}{2\pi}\left[\frac{1}{4} - \frac{1}{3} - \frac{1}{4} + \frac{1}{3}\right] = 0.$$

Continuing in this way we find that for all odd harmonics (n = 1,3,5,...) a_n is zero and all even harmonics (n = 2,4,6...)

$$a_n = -\frac{1}{2\pi}\left[-\frac{1}{1+n} - \frac{1}{1-n} - \frac{1}{1+n} - \frac{1}{1-n}\right]$$

$$a_n = -\frac{1}{2\pi}\left[-\frac{2}{1+n} - \frac{2}{1-n}\right] = \frac{1}{2\pi}\left[\frac{2}{1+n} + \frac{2}{1-n}\right]$$

$$a_n = \frac{1}{2\pi}\left[\frac{2(1-n)+2(1+n)}{n^2-1}\right] = \frac{2}{\pi}\,\frac{1}{1-n^2}.$$

Hence, we can express the half-wave rectified sine wave as

$$f(t) = \frac{1}{\pi} + \frac{1}{2}\sin t + \frac{2}{\pi}\sum_{n=2,4,6...}\frac{1}{1-n^2}\cos nt.$$

• PROBLEM 14-6

Find the Fourier series for the sawtooth waveform of Fig. 1.

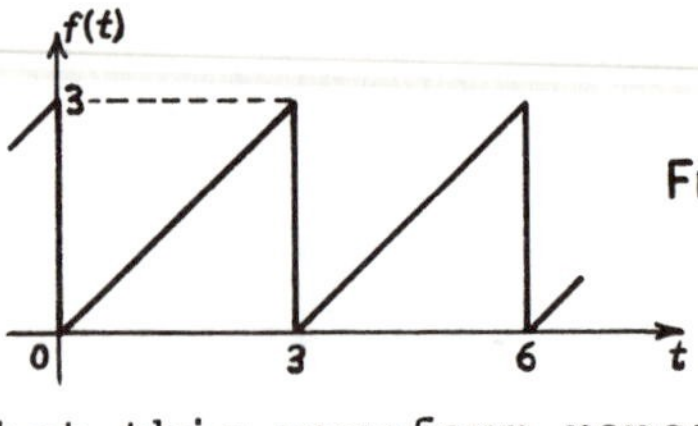

Fig.1

Solution: We note that this waveform repeats every 3 seconds. Thus the period τ = 3s and since $\omega_o = \frac{2\pi}{\tau}$, $\omega_o = \frac{2\pi}{3}$.

A periodic function can be represented as a Fourier series in the form

$$f(t) = a_o + \sum_{n=1}^{\infty}(a_n \cos n\omega_o t + b_n \sin n\omega_o t) \qquad (1)$$

where a_o is the average value of the function f(t) and is defined as

$$a_o = \frac{1}{\tau}\int_o^{\tau} f(t)dt. \qquad (2)$$

f(t) represents one period of the entire periodic function, thus f(t) = t; 0 < t < 3.

Substituting these values yields:

$$a_o = \frac{1}{3}\int_o^{3} t\ dt$$

$$a_o = \frac{1}{3}\left[\frac{t^2}{2}\right]_o^3$$

$$a_o = \frac{1}{3}\left[\frac{9}{2} - 0\right] = \frac{3}{2}.$$

The coefficients b_n and a_n are defined as

$$b_n = \frac{2}{\tau}\int_o^{\tau} f(t)\ \sin n\ \omega_o t\ dt \qquad (3)$$

$$a_n = \frac{2}{\tau}\int_o^{\tau} f(t)\ \cos n\ \omega_o t\ dt. \qquad (4)$$

Hence

$$b_n = \frac{2}{3}\int_o^{3} t\ \sin n\ \left(\frac{2\pi}{3}\right)t\ dt$$

$$b_n = \frac{2}{3}\left[\frac{1}{\left(\frac{n2\pi}{3}\right)^2}\sin n\ \left(\frac{2\pi}{3}\right)t - \frac{t}{\frac{n2\pi}{3}}\cos n\ \left(\frac{2\pi}{3}\right)t\right]_o^3$$

$$b_n = \frac{2}{3}\left[\frac{1}{\left(\frac{n2\pi}{3}\right)^2}\sin n2\pi - \frac{9}{n2\pi}\cos n2\pi\right]$$

$$b_n = \frac{3}{n^2 2\pi^2}\sin n\ 2\pi - \frac{3}{n\pi}\cos n\ 2\pi.$$

The sine term is zero for all n since any multiple of 2π in the sine term is zero.

Hence

$$b_n = -\left(\frac{3}{\pi}\frac{1}{n}\right) \qquad n = 1,2,3....$$

since the cos term is 1 for any multiple of 2π.

$$a_n = \frac{2}{3}\int_o^3 t \cos n\omega_o t \, dt$$

$$a_n = \frac{2}{3}\left[\frac{1}{\left(\frac{n2\pi}{3}\right)^2} \cos \frac{n2\pi}{3} t + \frac{t}{\frac{n2\pi}{3}} \sin \frac{n2\pi}{3} t\right]_o^3$$

$$a_n = \frac{2}{3}\left[\frac{9}{n^2 4\pi^2} \cos n2\pi - \frac{9}{n^2 4\pi^2} \cos 0\right]$$

But $\cos n2\pi = \cos 0$ for all n; therefore, $a_n = 0$ for all n.

The Fourier representation of this waveform is written

$$f(t) = \frac{3}{2} - \frac{3}{\pi}\sum_{n=1}^{\infty} \frac{1}{n} \sin n \frac{2\pi}{3} t$$

$$f(t) = \frac{3}{2} - \frac{3}{\pi}\left(\sin \frac{2\pi}{3} t + \frac{1}{2} \sin \frac{4\pi}{3} t + \frac{1}{3} \sin \frac{6\pi}{3} + ...\right).$$

• PROBLEM 14-7

Find the Fourier series for the sawtooth waveform shown in Fig. 1.

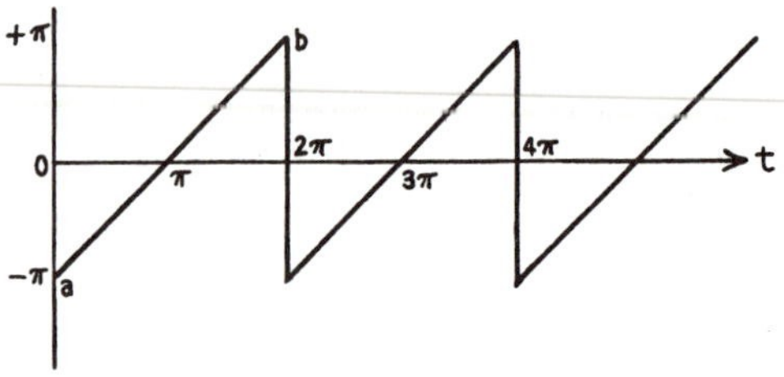

FIG. 1

Solution: We note that the sawtooth waveform has odd-symmetry and its average value is zero; thus $a_o = 0$, $a_n = 0$, and

$$b_n = \frac{2}{\tau}\int_o^{\tau} f(t) \sin n\omega_o t \, dt.$$

Hence, we can express the sawtooth waveform as the Fourier series

$$\sum_{n=1}^{\infty} b_n \sin n\omega_o t.$$

Note that the waveform is a straight-line variation, ranging from $-\pi$ to $+\pi$ over one complete cycle; thus $\tau = 2\pi$ and $\omega_o = \frac{2\pi}{\tau} = 1$.

The straight-line variation that represents one period of the waveform can be written as

$$f(t) = t - \pi \quad ; \quad 0 < t < 2\pi.$$

Therefore

$$b_n = \frac{2}{\tau}\int_0^{\tau} f(t) \sin n\omega_o t \, dt = \frac{1}{\pi}\int_0^{2\pi} (t-\pi) \sin nt \, dt$$

$$b_n = \frac{1}{\pi}\int_0^{2\pi} t \sin nt \, dt - \int_0^{2\pi} \sin nt \, dt$$

$$b_n = \left[\frac{1}{n^2\pi} \sin nt - \frac{t}{n\pi} \cos nt + \frac{1}{n} \cos nt\right]_0^{2\pi}$$

$$b_n = -\frac{2}{n} \cos (2\pi n) + \frac{1}{n} \cos (2\pi n) - \frac{1}{n}$$

$$b_n = -\frac{2}{n} \quad \text{for } n = 1,2,3,4,\ldots$$

Hence, the Fourier series which represents the sawtooth wave is

$$\sum_{n-1}^{\infty} -\frac{2}{n} \sin nt = -2(\sin t + \frac{1}{2} \sin 25 + \frac{1}{3} \sin 3t + \ldots)$$

• PROBLEM 14-8

First obtain the Fourier series of the waveform below. Then find the sum of the first four terms of the series at $t = 2$.

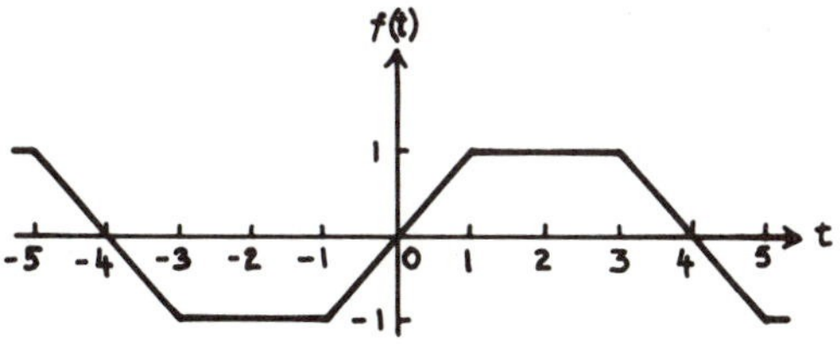

Solution: The general Fourier series expansion is

$$f(t) = a_o + \sum_{n=1}^{\infty} (a_n \cos(n\omega_o t) + b_n \sin(n\omega_o t))$$

where $\omega_o = 2\pi/\tau = 2\pi/8 = \pi/4$ rad/s.

Note that the waveform has odd symmetry (i.e. $f(t) = -f(-t)$; therefore a_n for all n, including a_o, are zero. The waveform also has half-wave symmetry (i.e. $f(t) = -f(t \pm \tau/2)$, so only the fundamental and odd harmonics are present. The b coefficients are then given by

$$b_n = \frac{4}{\tau}\int_0^{\tau/2} f(t) \sin(n\omega_o t)dt = \frac{1}{2}\int_0^4 f(t) \sin\left(\frac{n\pi t}{4}\right)dt \quad \text{for odd } n$$

$$b_n = 0 \text{ for even } n$$

To perform this integration, f(t) is expressed by three different functions over the range $0 \le t \le 4$:

$$f(t) = t \quad 0 \le t \le 1$$

$$f(t) = 1 \quad 1 \le t \le 3$$

$$f(t) = 4-t \quad 3 \le t \le 4.$$

The integral is broken up into three corresponding parts:

$$b_n = \frac{1}{2}\int_0^1 t \sin\left(\frac{n\pi t}{4}\right)dt + \frac{1}{2}\int_1^3 \sin\left(\frac{n\pi t}{4}\right)dt$$

$$+ \frac{1}{2}\int_3^4 (4-t)\sin\left(\frac{n\pi t}{4}\right)dt \text{ for n odd.}$$

Evaluation of the first and last integrals is facilitated by making use of the following relation, from integral tables:

$$\int x \sin x\, dx = \sin x - x \cos x$$

where x would be made equal to $n\pi t/4$.

The first of the three integrals becomes

$$\frac{1}{2}\int_0^1 t \sin\left(\frac{n\pi t}{4}\right)dt = \frac{4^2}{2n^2\pi^2}\int_0^1 \left(\frac{n\pi t}{4}\right)\sin\left(\frac{n\pi t}{4}\right)\left(\frac{n\pi dt}{4}\right)$$

$$= \frac{8}{n^2\pi^2}\left[\sin\left(\frac{n\pi t}{4}\right) - \left(\frac{n\pi t}{4}\right)\cos\left(\frac{n\pi t}{4}\right)\right]_{t=0}^{1}$$

$$= \frac{8}{n^2\pi^2}\left[\sin(\frac{n\pi}{4}) - (\frac{n\pi}{4})\cos(\frac{n\pi}{4})\right].$$

The second integral is

$$\frac{1}{2}\int_1^3 \sin(\frac{n\pi t}{4})\,dt = \frac{2}{n\pi}\int_1^3 \sin(\frac{n\pi t}{4})\,(\frac{n\pi dt}{4})$$

$$= -\frac{2}{n\pi}\cos(\frac{n\pi t}{4})\Bigg|_1^3 = \frac{2}{n\pi}\,[\cos(\frac{n\pi}{4}) - \cos(\frac{3n\pi}{4})].$$

The third integral is

$$\frac{1}{2}\int_3^4 (4-t)\sin(\frac{n\pi t}{4})\,dt = 2\int_3^4 \sin(\frac{n\pi t}{4})\,dt - \frac{1}{2}\int_3^4 t\,\sin(\frac{n\pi t}{4})\,dt$$

$$= \frac{-8}{n\pi}\cos(\frac{n\pi t}{4})\Bigg|_3^4 - \frac{8}{n^2\pi^2}\left[\sin(\frac{n\pi t}{4}) - (\frac{n\pi t}{4})\cos(\frac{n\pi t}{4})\right]_3^4$$

$$= \frac{8}{n\pi}\,[\cos(\frac{3n\pi}{4}) - \cos(n\pi)] + \frac{8}{n^2\pi^2}\,[n\pi\cos(n\pi)$$

$$+ \sin(\frac{3n\pi}{4}) - \frac{3n\pi}{4}\cos(\frac{3n\pi}{4})]$$

$$= \frac{2}{n\pi}\cos(\frac{3n\pi}{4}) + \frac{8}{n^2\pi^2}\sin(\frac{3n\pi}{4}).$$

The results of these three integrations are collected to obtain

$$b_n = \frac{8}{n^2\pi^2}\,[\sin(\frac{n\pi}{4}) + \sin(\frac{3n\pi}{4})] \qquad n \text{ odd}.$$

When evaluated for n = 1, 3, 5 and 7, the coefficients are

$$b_1 = \frac{8}{\pi^2}\,[\frac{\sqrt{2}}{2} + \frac{\sqrt{2}}{2}] = \frac{8\sqrt{2}}{\pi^2}$$

$$b_3 = \frac{8}{9\pi^2}\,[\frac{\sqrt{2}}{2} + \frac{\sqrt{2}}{2}] = \frac{8\sqrt{2}}{9\pi^2}$$

$$b_5 = \frac{8}{25\pi^2}\,[-\frac{\sqrt{2}}{2} - \frac{\sqrt{2}}{2}] = -\frac{8\sqrt{2}}{25\pi^2}$$

$$b_7 = \frac{8}{49\pi^2}\left[-\frac{\sqrt{2}}{2} - \frac{\sqrt{2}}{2}\right] = \frac{8\sqrt{2}}{49\pi^2}$$

The first four terms of the Fourier series are therefore

$$f(t) \simeq \frac{8\sqrt{2}}{\pi^2}\left[\sin(\omega_o t) + \frac{1}{9}\sin(3\omega_o t) - \frac{1}{25}\sin(5\omega_o t) - \frac{1}{49}\sin(5\omega_o t)\right]$$

$$\simeq \frac{8\sqrt{2}}{\pi^2}\left[\sin\left(\frac{\pi t}{4}\right) + \frac{1}{9}\left(\frac{3\pi t}{4}\right) - \frac{1}{25}\sin\left(\frac{5\pi t}{4}\right) - \frac{1}{49}\sin\left(\frac{7\pi t}{4}\right)\right].$$

The value of f(t) at t = 2 sec is

$$f(t=2) \simeq \frac{8\sqrt{2}}{\pi^2}\left[1 - \frac{1}{9} - \frac{1}{25} + \frac{1}{49}\right] = 0.9965.$$

• **PROBLEM 14-9**

Find the exponential Fourier series of the periodic waveform below.

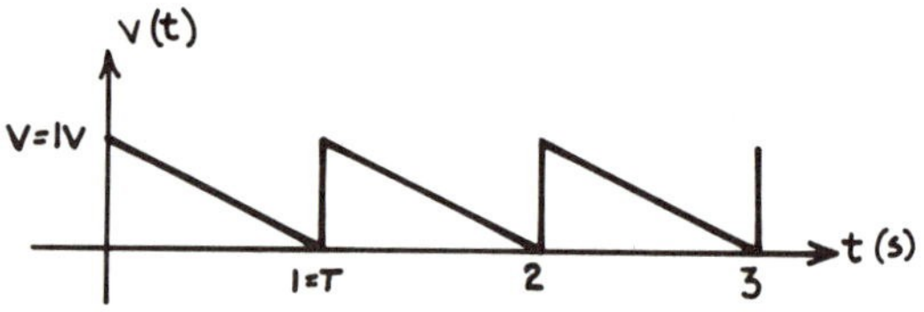

Solution: The Fourier series may be expressed in exponential form as follows:

$$v(t) = c_o + \sum_{n=-\infty}^{\infty} c_n e^{jn\omega t} \tag{1}$$

where c_o is the average value of v(t), ω is the fundamental frequency in radians per second, and c_n is the coefficient of the nth harmonic. The values for c_n are determined from

$$c_n = \frac{1}{\tau}\int_0^{\tau} v(t)\, e^{-jn\omega t}\, dt \quad (n=0,\pm1,\pm2,\pm3,\ldots). \tag{2}$$

First, evaluate the average value of the waveform. By inspection, this can be seen to be 1/2. A more formal approach would be

$$c_o = \frac{1}{\tau}\int_0^{\tau} v(t)dt \tag{3}$$

where the graph shows

$$v(t) = 1 - t \qquad (t=0 \text{ to } t=\tau). \tag{4}$$

Therefore,

$$c_o = \frac{1}{\tau}\int_o^{\tau}(1-t)\,dt = \left[\frac{1}{\tau}\,t - \frac{t^2}{2}\right]_o^{\tau} \tag{5}$$

and evaluating yields

$$c_o = \frac{1}{\tau}\left[\tau - \frac{\tau^2}{2}\right] = 1 - \frac{\tau}{2} = \frac{1}{2} \tag{6}$$

since $\tau = 1$ second.

Similarly, evaluate equation (2):

$$c_n = \frac{1}{\tau}\int_o^{\tau}(1-t)e^{-jn\omega t}dt, \tag{7}$$

$$c_n = \frac{1}{\tau}\int_o^{\tau}e^{-jn\omega t}dt - \frac{1}{\tau}\int_o^{\tau}te^{-jn\omega t}dt \tag{8}$$

by using integral tables, where we find

$$\int e^{ax}dx = \frac{e^{ax}}{a} \qquad \text{and} \tag{9}$$

$$\int xe^{ax}dx = \frac{xe^{ax}}{a} - \frac{e^{ax}}{a^2}\,. \tag{10}$$

Following the integral tables, integrate

$$c_n = \frac{1}{\tau}\left[\frac{e^{-jn\omega t}}{(-jn\omega)} - \frac{te^{-jn\omega t}}{(-jn\omega)} + \frac{e^{-jn\omega t}}{(-jn\omega)^2}\right]_o^{\tau} \tag{11}$$

and evaluate:

$$c_n = \frac{1}{\tau}\left[\frac{e^{-jn\omega\tau}-e^o}{(-jn\omega)} - \frac{\tau e^{-jn\omega\tau}-0}{(-jn\omega)} + \frac{e^{-jn\omega\tau}-e^o}{(-jn\omega)^2}\right]. \tag{12}$$

We should seek ways to simplify equation (12) before proceeding further. One way is recalling that

$$\omega = \frac{2\pi}{\tau} \qquad \text{rad/s.} \tag{13}$$

Since $\tau = 1$ in this instance, from the graph.

$$\omega = 2\pi \qquad \text{rad/s.} \tag{14}$$

Substituting equation (14) into equation (12),

$$c_n = \left[\frac{e^{-j2\pi n}-1}{(-j2\pi n)} - \frac{e^{-j2\pi n}}{(-j2\pi n)} + \frac{e^{-j2\pi n}-1}{(-j2\pi n)^2}\right] . \tag{15}$$

A further simplification may be made by evaluating the exponential quantity with the use of Euler's equation,

$$e^{jx} = \cos x + j \sin x, \tag{16}$$

$$e^{-j2\pi n} = \cos(-2\pi n) + j \sin(-2\pi n) = +1 \tag{17}$$

since the cosine or sine of $\pm 2\pi$ or any whole-number multiple thereof is +1 or zero, respectively. Now, substituting equation (17) into equation (15),

$$c_n = \left[\frac{1-1}{(-j2\pi n)} - \frac{1}{(-j2\pi n)} + \frac{1-1}{(-j2\pi n)^2}\right] \quad \text{and} \tag{18}$$

$$c_n = \frac{1}{j2\pi n} . \tag{19}$$

A further step in simplification can be made by recalling that

$$\frac{1}{j} = -j = 0 - j = e^{-2\frac{\pi}{2}} \tag{20}$$

which can be proven with equation (16). Substituting equation (20) into equation (19),

$$c_n = \frac{1}{2\pi n}\, e^{-j\frac{\pi}{2}} \tag{21}$$

Finally, substitute equations (6), (14) and (21) into equation (1):

$$v(t) = \frac{1}{2} + \sum_{n=-\infty}^{\infty} \left(\frac{1}{2\pi n}\right) e^{-j\frac{\pi}{2}}\, e^{j2\pi nt} . \tag{22}$$

This can be rewritten as

$$v(t) = \frac{1}{2} + \sum_{n=-\infty}^{\infty} \left(\frac{1}{2\pi n}\right) e^{j(2\pi nt - \frac{\pi}{2})} \quad \text{volts} \tag{23}$$

since $e^a e^b = e^{a+b}$. (24)

Determine the Fourier series and sketch the line frequency spectrum for each waveform shown in Fig. 1.

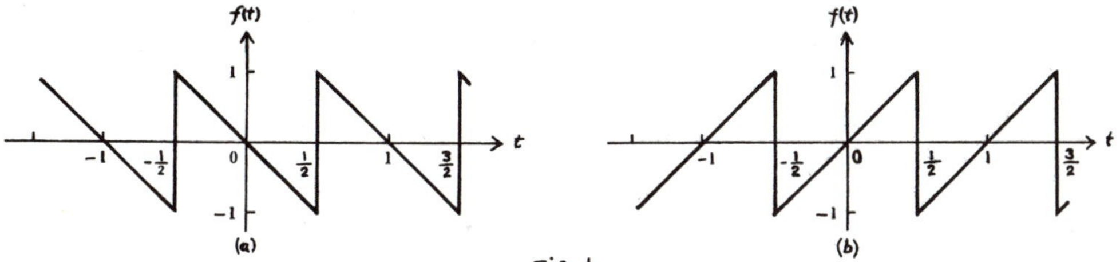

FIG. 1

Solution: Using the trigonometric form of the Fourier series, the function of f(t) can be written as $f(t) = a_o + \sum_{n=1}^{\infty} (a_n \cos n\omega_o t + b_n \sin \omega_o t)$, where the coefficients can be evaluated as follows:

$$a_o = \frac{1}{\tau}\int_0^{\tau} f(t)\, dt$$

$$a_n = \frac{2}{\tau}\int_0^{\tau} f(t) \cos n\omega_o t\, dt, \; n = 1,2,3...$$

$$b_n = \frac{2}{\tau}\int_0^{\tau} f(t) \sin n\omega_o t\, dt, \; n = 1,2,3...$$

Note that the function satisfies the Dirichlet conditions and can be expanded in a Fourier series. The function is an odd function since $f(-t) = -f(t)$ and is periodic with $\tau = 1$ and $\omega_1 = 2\pi$.

Therefore, for Fig.1(a), $f(t) = -2t$, $-\frac{\tau}{2} = -\frac{1}{2} \leq t \leq \frac{\tau}{2} = \frac{1}{2}$.

By symmetry, $a_o = a_n = 0$. Hence

$$b_n = 2\int_{-\frac{1}{2}}^{\frac{1}{2}} (-2t) \sin 2\pi nt\, dt.$$

Now compute the integral: from a table of integrals, the form is

$$\int x(\sin ax)dx = \frac{1}{a^2} \sin ax - \frac{x}{9} \cos ax.$$

Let x = t, a = 2πn.

Then, $$-4\int_{-\frac{1}{2}}^{\frac{1}{2}} t\sin 2\pi nt\,dt = -4\left[\frac{t}{(2\pi n)^2}\sin 2\pi nt\Big|_{-\frac{1}{2}}^{\frac{1}{2}} - \frac{t}{2\pi n}\cos 2\pi nt\Big|_{-\frac{1}{2}}^{\frac{1}{2}}\right].$$

But,

$$t\sin 2\pi nt\Big|_{-\frac{1}{2}}^{\frac{1}{2}} = 0 \text{ for all } n$$

$$t\cos 2\pi nt\Big|_{-\frac{1}{2}}^{\frac{1}{2}} = \begin{cases} -1, & n=1,3,5\ldots \\ 1, & n=2,4,6\ldots \end{cases}$$

Therefore,

$$b_n = \begin{cases} -\dfrac{2}{\pi n}, & n=1,3,5\ldots \\ \dfrac{2}{\pi n}, & n=2,4,6\ldots \end{cases}.$$

Hence

$$f(t) = \frac{2}{\pi}\left(-\sin 2\pi t + \frac{1}{2}\sin 4\pi t - \frac{1}{3}\sin 6\pi t + \ldots\right). \qquad \text{(a)}$$

For the line spectrum, $f_o = \frac{\omega_o}{2\pi} = 1$ and the harmonic amplitude versus the harmonic frequencies will be as shown in Table 1 or as sketched in Figure 2.

Harmonic Amplitude	Frequency
$\frac{2}{\pi} = 0.64$	1
$\frac{1}{\pi} = 0.32$	2
$\frac{2}{3\pi} = 0.21$	3
$\frac{1}{2\pi} = 0.16$	4

Table 1.

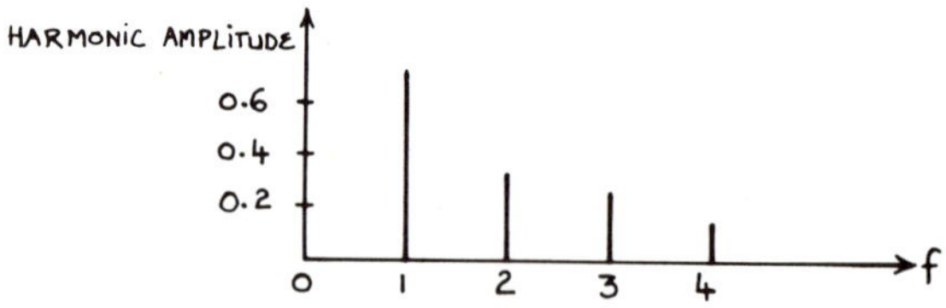

FIG. 2

For Figure 1(b)

$$f(t) = 2t.$$

Therefore the procedure will be the same except for the sign, and

$$b_n = \begin{cases} \frac{2}{\pi n}, & n=1,3,5\ldots \\ -\frac{2}{\pi n}, & n=2,4,6\ldots \end{cases}$$

The line spectrum will be the same as Fig.2.

• PROBLEM 14-11

A waveform having half-wave symmetry is given as shown in Fig. 1. Determine the first four coefficients of Fourier series, taking the 30° steps shown. Then draw the frequency spectrum.

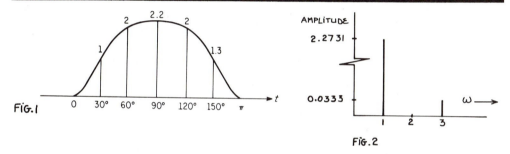

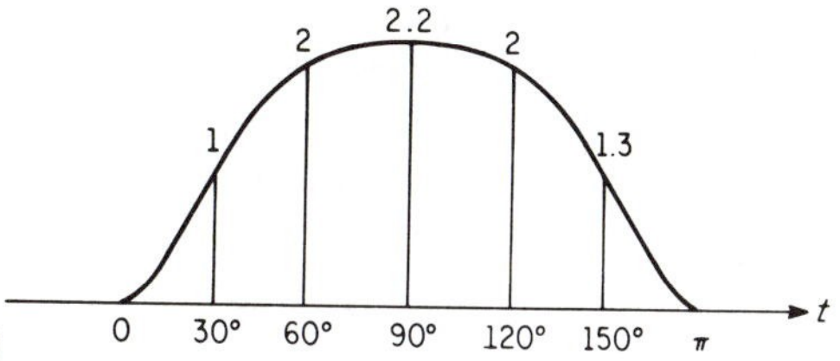

FIG.1

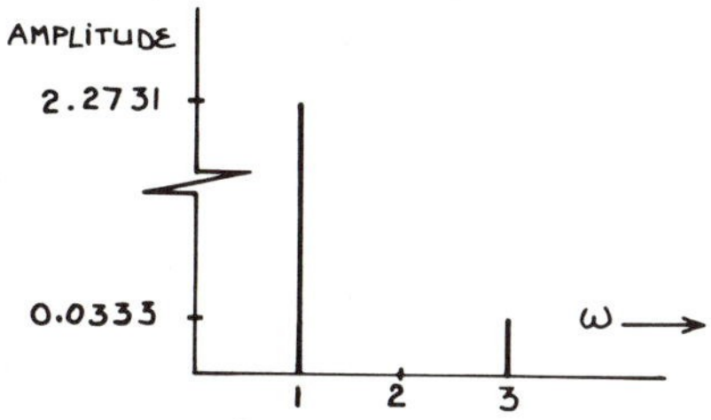

FIG.2

Solution: The waveform has a period τ of 2π sec., making the fundamental frequency equal to $\omega_o = 2\pi/\tau = 1$ rad/s. Since half-wave symmetry exists, even harmonics and the dc offset terms are zero. The first four terms are therefore

$$y(t) \approx b \sin t + a_1 \cos t + b_3 \sin 3t + a_3 \cos 3t. \quad (1)$$

The coefficients in equation (1) could be calculated by the following series:

$$b_n = \frac{2}{k} \sum_{m=1}^{k} y_m \sin\left(\frac{n\ m\ 360°}{k}\right)$$

$$a_n = \frac{2}{k} \sum_{m=1}^{k} y_m \cos\left(\frac{n\ m\ 360°}{k}\right) \qquad (2)$$

where k is the number of intervals into which the whole period has been divided and y_m is the ordinate of the waveform at the m-th interval. However, we can take advantage of the fact that half-wave symmetry exists and simplify equation (2) to

$$b_n = \frac{2}{k} \sum_{m=1}^{k} y_m \sin \left(\frac{n\ m\ 180°}{k}\right) \qquad (3a)$$

$$a_n = \frac{2}{k} \sum_{m=1}^{k} y_m \cos \left(\frac{n\ m\ 180°}{k}\right) \qquad (3b)$$

where k is now the number of intervals in a half period of the waveform.

For the waveform given, k = 6 and

$y_1 = 1$ $y_4 = 2$

$y_2 = 2$ $y_5 = 1.3$

$y_3 = 2.2$ $y_6 = 0$

Use of these values in equations (3a) and (3b) gives

$$b_1 = \frac{2}{6} [1 \sin 30° + 2 \sin 60° + 2.2 \sin 90° + 2 \sin 120° + 1.3 \sin 150°] = 2.2714$$

$$a_1 = \frac{2}{6} [1 \cos 30° + 2 \cos 60° + 2.2 \cos 90° + 2 \cos 120° + 1.3 \cos 150°] = -0.0866$$

$$b_3 = \frac{2}{6} [1 \sin 90° + 2 \sin 180° + 2.2 \sin 270° + 2 \sin 360° + 1.3 \sin 450°] = 0.0333$$

$$a_3 = \frac{2}{6} [1 \cos 90° + 2 \cos 180° + 2.2 \cos 270° + 2 \cos 360° + 1.3 \cos 450°] = 0.$$

Since the value of a_3 was zero, we have only three non-zero terms of y(t) through the third harmonic. We would not be justified, however, in attempting to calculate b_5 and a_5 by equation (3). The reason for this is that we are not given a sufficient number of waveform ordinates or samples to correctly calculate the coefficients of higher ordered harmonics. For example, an attempt to calculate b_5 from equation (3) would yield a value of -1.0536. There is obviously not this amount of 5th harmonic "ripple" on the waveform.

When the fundamental and 3rd harmonic coefficients are substituted into equation (1), the following is obtained:

$$y(t) \simeq 2.2714 \sin t - 0.0886 \cos t + 0.0333 \sin 3t.$$

A comparison of y(t) with the waveform, at the sampling points, is given in the table below:

t	y(t)	waveform
0	-0.0866	0
$\pi/6$	1.0923	1.0
$\pi/3$	1.9228	2.0
$\pi/2$	2.2381	2.2
$2\pi/3$	2.0114	2.0
$5\pi/6$	1.2457	1.3
π	0.0886	0

The fundamental and harmonic amplitudes for the frequency spectrum are obtained from

$$c_n = \sqrt{a_n^{\,2} + b_n^{\,2}} \;.$$

Hence,

$$c_1 = \sqrt{(-0.0886)^2 + (2.2714)^2} = 2.2731$$

$$c_3 = b_3 = 0.0333.$$

The spectrum is displayed in Figure 2.

• PROBLEM 14-12

Modify the Fourier series representation of the waveforms below if the origin is shifted to 0'.

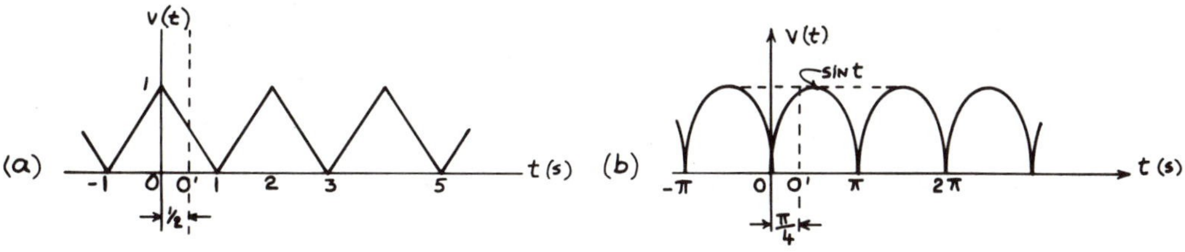

Solution: For the triangular waveform, we see it is an even function, hence, $b_n \equiv 0$. It has an average value of $\frac{1}{2}$, therefore, $a_o = \frac{1}{2}$. The period $\tau = 2$ and $\omega = \pi$, the fundamental frequency of the series expansion. The remaining Fourier coefficients are thus found by

$$a_n = \int_{-1}^{o} (1+t)\cos n\pi t \, dt + \int_{o}^{1} (1-t)\cos n\pi t \, dt$$

Expanding to four integrals, we have:

1: $\int_{-1}^{0} \cos n\pi t \, dt$ which is found to be zero.

2: $\int_{-1}^{0} t \cos n\pi t \, dt = \left[\frac{1}{n^2\pi^2} \cos n\pi t + \frac{1}{n\pi} t \sin n\pi t\right]_{-1}^{0}$

3: $\int_{0}^{1} \cos n\pi t \, dt = 0$

4: $\int_{0}^{1} t \cos n\pi t \, dt = \left[\frac{1}{n^2\pi^2} \cos n\pi t + \frac{1}{n\pi} t \sin n\pi t\right]_{0}^{1}$

2 and 4 each yield $\frac{2}{n^2\pi^2}$ if n is odd, and 0 if n is even; therefore, $a_n = \frac{4}{n^2\pi^2}$ and the representation is

$$f(t) = \frac{1}{2} + \frac{4}{\pi^2} \cos \pi t + \frac{4}{9\pi^2} \cos 3\pi t + \frac{4}{25\pi^2} \cos 5\pi t$$

Now, if the origin is shifted to 0' as indicated, the time functions are shifted a quarter period. The maximum value of f(t) will now occur at $t = -\frac{1}{2}$; thus, each term in the series must be shifted a like amount. The new argument in time is $t' = t + \frac{1}{2}$ and in ωt is $\omega t = \frac{\pi}{2}$. The general term $\frac{4}{n^2\pi^2} \cos n\pi t$ (n=odd) is

$$\frac{4}{n^2\pi^2} \cos\left(n\pi t + \frac{\pi}{2}\right) = -\frac{4}{n^2\pi^2} \sin n\pi t \quad \text{where n=odd.}$$

A new representation would then be

$$f(t) = \frac{1}{2} + \sum_{n \text{ odd}}^{\infty} \frac{4}{n^2\pi^2} (-1)^n \sin n\pi t.$$

It is easy to see that at the new t = 0 all terms $\sin n\pi t$ are 0 and $f(t=0') = \frac{1}{2}$. This is as it should be.

(b) For the sinusoidal rectified form,

$a_o = \frac{2}{\pi}$ (the average value);

$b_n = 0$ (since it is an even function).

The a_n are found from

$$a_n = \frac{2}{\pi} \int_0^{\pi} \sin t \cos 2nt \, dt$$

since the ω of the fundamental of the harmonic series is twice the unrectified save ω = 2. Using trigonometric identities, this integral is equivalent to

$$a_n = \frac{2}{\pi} \int_0^{\pi} \frac{1}{2} \sin (1+2n)t + \frac{1}{2} \sin (1-2n)t \, dt$$

$$= \frac{1}{\pi} \left[- \frac{1}{1+2n} \cos(1+2n)t - \frac{1}{1-2n} \cos(1-2n)t \right] \Bigg|_0^{\pi}$$

$$= \frac{1}{\pi} \left[- \frac{1}{1+2n} \cos(2n+1)\pi - \frac{1}{1-2n} \cos\ (1-2n)\pi \right]$$

$$- \frac{1}{\pi} \left[- \frac{1}{1+2n} \cos 0 - \frac{1}{1-2n} \cos 0 \right]$$

$$a_n = \frac{1}{\pi} \left[\frac{4}{1-4n^2} \right]$$

The expression for f(t) is, thus,

$$f(t) = \frac{2}{\pi} [1 + \sum_{n=1}^{\infty} \frac{2}{1-4n^2} \cos 2nt] \qquad \text{or since}$$

$$\frac{1}{1-4n^2} = - \frac{1}{4n^2-1}$$

$$= \frac{2}{\pi} [1 - \frac{2}{3} \cos 2t - \frac{2}{15} \cos 4t - \frac{2}{32} \cos 6t....]$$

As before, the series can be modified to accommodate the shift in zero by adding the appropriate shift to each term. Thus, the sum of terms at each instant in time is the correct value.

$$f(t) = \frac{2}{\pi} [1 - \frac{2}{3} \cos(2t + \frac{\pi}{4}) - \frac{2}{15} \cos(4t + \frac{\pi}{4}) - ...]$$

• **PROBLEM 14-13**

(a) Find the Fourier series for the current waveform of Fig. 1. (b) If this waveform is shifted 4 ms to the right, determine the Fourier series.

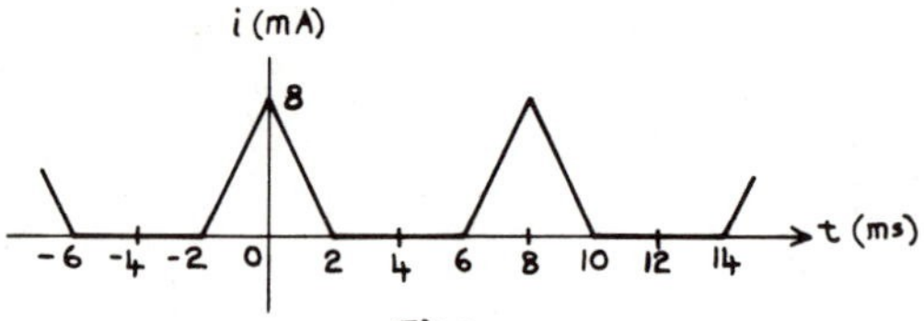

FIG. 1

Solution: (a) Since i(t) is an even function, its Fourier expansion consists of only a dc and cosine terms. Then

$$\frac{i(t)}{8\text{mA}} = \frac{a_o}{2} + \sum_{n=1}^{\infty} a_n \cos(n\omega_o t); \qquad \omega_o = \frac{2\pi}{\tau}$$

where

$$a_n = \frac{2}{\tau}\int_{-\frac{\tau}{2}}^{\frac{\tau}{2}} i(t)\cos n\omega_o t\,dt = \frac{4}{\tau}\int_{0}^{\frac{\tau}{2}} i(t)\cos n\omega_o t\,dt$$

for n = 0,1,2,....

Thus,

$$a_n = \frac{4}{\tau}\int_{0}^{\frac{\tau}{4}} \left(1 - \frac{4t}{\tau}\right)\cos n\omega_o t\,dt$$

Letting $u = \omega_o t$, we get $du = \omega_o dt$ and (noting that $\omega_o t = 2\pi$),

$$a_n = \frac{2}{\pi}\int_{0}^{\frac{\pi}{2}} \left(1 - \frac{2u}{\pi}\right)\cos nu\,du$$

If n=0, $a_n = a_o = \frac{2}{\pi}\int_{0}^{\frac{\pi}{2}} \left(1-\frac{2u}{\pi}\right)du = \frac{2}{\pi}\left(\frac{\pi}{2}-\frac{\left(\frac{\pi}{2}\right)^2}{\pi}\right) = 1-\frac{1}{2} = \frac{1}{2}$.

This could, of course, be derived by inspection of triangle areas. For $n \geq 1$,

$$a_n = \frac{2}{\pi}\int_{0}^{\frac{\pi}{2}} \cos nu\,du - \frac{4}{\pi^2}\int_{0}^{\frac{\pi}{2}} u\cos nu\,du$$

let

$$\alpha = u \qquad d\beta = \cos nu\,du$$

$$d\alpha = du \qquad \beta = \frac{1}{n}\sin u$$

$$a_n = \frac{2}{n\pi}\sin nu\Big|_{0}^{\frac{\pi}{2}} - \frac{4}{\pi^2}\left[\frac{1}{n}u\sin nu\Big|_{0}^{\frac{\pi}{2}} - \frac{1}{n}\int_{0}^{\frac{\pi}{2}}\sin nu du\right]$$

$$= \frac{2}{n\pi} \sin(n\frac{\pi}{2}) - \frac{2}{n\pi} \sin(n\frac{\pi}{2}) + \frac{4}{n\pi^2} \quad (-\frac{1}{n}\cos nu)\Big|_0^{\frac{\pi}{2}}$$

$$= -(\frac{2}{n\pi})^2 [\cos(n\frac{\pi}{2})-1] = (\frac{2}{n\pi})^2 [1-\cos(n\frac{\pi}{2})].$$

Now since $\omega_o = \frac{2\pi}{\tau} = \frac{2\pi}{8x10^{-3}} = \frac{\pi}{4} \times 1000 = 250\pi$, we get

$$i(t) = 8[\frac{1}{4} + \sum_{n=1}^{\infty} (\frac{2}{n\pi})^2 [1-\cos(n\frac{\pi}{2})]\cos(250n\pi t)]$$

$$= 2 + \frac{32}{\pi^2} [\cos 250\pi t + \frac{2}{4} \cos(500\pi t) + \ldots] \text{ mA.}$$

(b) One can derive this result by noting that a shift of 4 ms is a shift of $\frac{\tau}{2}$, and by then investigating

$$\cos n\omega_o(t-\frac{\tau}{2}) = \cos n\omega_o t \cos n\frac{\omega_o\tau}{2} + \sin n\omega_o t \sin n\frac{\omega_o\tau}{2}$$

$$= \cos(n\pi)\cos(n\omega_o t) + \sin n\pi \sin n\omega_o t$$

$$= (-1)^n \cos n\omega_o t$$

Thus,

$$i(t) = 8[\frac{1}{4} + \sum_{n=1}^{\infty} (-1)^n(\frac{2}{n\pi})^2 [1-\cos \frac{n\pi}{2}]\cos(250n\pi t)]$$

$$= 2 + \frac{32}{\pi^2} [-\cos(250\pi t) + \frac{2}{4} \cos(500n\pi t) - \frac{1}{9} \cos(750\pi t) + \ldots].$$

• PROBLEM 14-14

Find the Fourier series for the function f(t) shown in Fig. 1 by the translation of axis method.

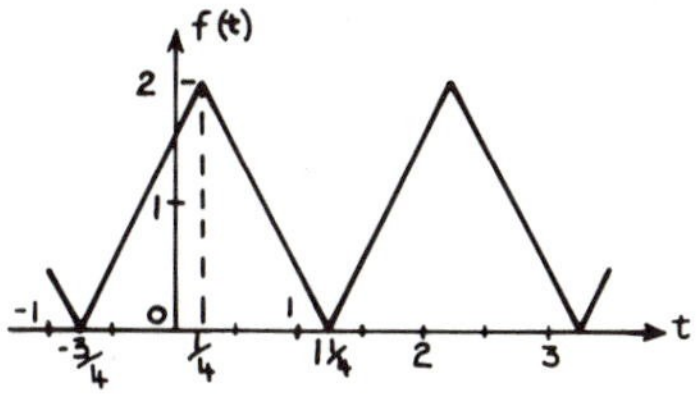

Fig. 1

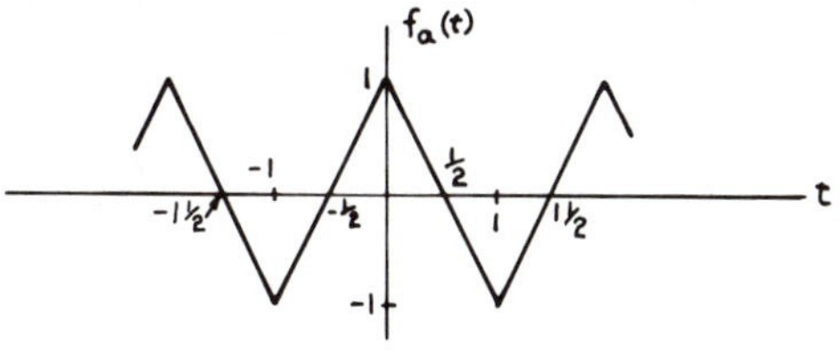

Fig. 2

Solution: We can find the Fourier series for a triangle wave that has even-symmetry and zero average value more readily than we can for the triangle wave in Fig. 1.

By translating the axis in Fig. 1 we obtain the triangle wave shown in Fig. 2.

We note that $f(t) = 1 + f_a(t - \frac{1}{4})$; thus by finding the Fourier series of $f_a(t)$ and then translating it we have the Fourier series of f(t).

Since $f_a(t)$ has even-half-wave symmetry, $a_o = 0$ and $b_n = 0$. Thus,

$$f_a(t) = \sum_{n=1}^{\infty} a_n \cos n\omega_o t$$

where $a_n = \frac{2}{\tau}\int_o^{\tau} g(t) \cos n\omega_o t \, dt$, $\tau = 2$, and $\omega_o = \frac{2\pi}{\tau} = \pi$.

The function g(t) defines $f_a(t)$ over one period; thus,

$$g(t) = \begin{cases} 1 - 2t & 0 < t < 1 \\ 1 + 2t & -1 < t < 0 \end{cases} .$$

Therefore

$$a_n = \int_{-1}^{1} g(t) \cos n\pi t \, dt$$

$$a_n = \int_{-1}^{o} (1+2t) \cos n\pi t \, dt + \int_o^1 (1-2t) \cos n\pi t \, dt$$

$$a_n = \int_{-1}^{o} \cos n\pi t \, dt + 2\int_{-1}^{o} t \cos n\pi t \, dt + \int_o^1 \cos n\pi t \, dt$$
$$- 2\int_o^1 t \cos n\pi t \, dt$$

$$a_n = \left[\frac{1}{n\pi} \sin n\pi t + \frac{2}{n^2\pi^2} \cos n\pi t + \frac{2t}{n\pi} \sin n\pi t\right]_{-1}^{o}$$
$$+ \left[\frac{1}{n\pi} \sin n\pi t - \frac{2}{n^2\pi^2} \cos n\pi t - \frac{2t}{n\pi} \sin n\pi t\right]_o^1$$

$$a_n = \frac{2}{n^2\pi^2} - \frac{2}{n^2\pi^2} \cos(-n\pi) - \frac{2}{n^2\pi^2} \cos(n\pi) + \frac{2}{n^2\pi^2}$$

$$a_n = \frac{2}{n^2\pi^2}[2 - \cos(-n\pi) - \cos(n\pi)]$$

$$a_n = \begin{cases} \frac{8}{n^2\pi^2} & \text{for } n = 1,3,5... \\ 0 & \text{for } n = 2,4,6... \end{cases}$$

Hence,

$$f_a(t) = \sum_{n=1,3,5...}^{\infty} \frac{8}{n^2\pi^2} \cos n\pi t$$

and,

$$f(t) = 1 + f_a(t-\tfrac{1}{4}) = 1 + \sum_{n=1,3,5...}^{\infty} \frac{8}{n^2\pi^2} \cos[n\pi(t-\tfrac{1}{4})].$$

• PROBLEM 14-15

Convert the Fourier series

$$f(t) = \sum_{k=-\infty}^{+\infty} \frac{1}{1+jk} e^{jkt}$$

to trigonometric form.

Solution: We know that the Fourier series is

$$f(t) = a_o + \sum_{n=1}^{\infty} (a_n \cos n\omega_o t + b_n \sin n\omega_o t)$$

$$= \sum_{n=-\infty}^{+\infty} c_n e^{jn\omega_o t} \qquad (1)$$

where $$c_n = \frac{1}{2}(a_n - jb_n) \qquad (2)$$

$$c_o = a_o. \qquad (3)$$

The expression on the left in Eq.(1) is known as the trigonometric form of the Fourier series. We must convert

$$f(t) = \sum_{k=-\infty}^{+\infty} \frac{1}{1+jk} e^{jkt}$$ to trigonometric form.

Making use of Eq.(2) we have

$$\frac{1}{1+jk} = \frac{1}{2} a_n - \frac{1}{2} jb_n ,$$

which can be written in polar form as

$$\frac{1}{\sqrt{1+k^2}\,\angle\tan^{-1}(k)} = \frac{1}{2}a_n - \frac{1}{2}jb_n$$

$$\frac{1}{\sqrt{1+k^2}}\,\angle -\tan^{-1}(k) = \frac{1}{\sqrt{1+k^2}}\,\angle\tan^{-1}(-k) = \frac{1}{2}a_n - \frac{1}{2}jb_n$$

$$= \frac{1}{\sqrt{1+k^2}}\left[\cos(\tan^{-1}(-k)) + j\sin(\tan^{-1}(-k))\right]$$

$$\cos(\tan^{-1}(-k)) = \frac{1}{\sqrt{1+k^2}}\;;\quad \sin(\tan^{-1}(-k)) = \frac{-k}{\sqrt{1+k^2}}$$

giving

$$\frac{1}{2}a_n - \frac{1}{2}jb_n = \frac{1}{1+k^2} - j\frac{k}{1+k^2}$$

$$a_n = \frac{2}{1+k^2}\;;\qquad a_o = c_o = \frac{1}{1} = 1;\qquad b_n = \frac{2k}{1+k^2}\,.$$

Substituting into Eq.(1) yields

$$f(t) = 1 + 2\sum_{k=1}^{\infty}\frac{\cos kt + k\sin kt}{1+k^2}\,.$$

• **PROBLEM 14-16**

A pulsed communication signal consists of 250-k Hz energy every 1.25 ms that lasts 0.2 ms. Determine the frequency spectrum of this signal, noting the atmospheric pollution it produces.

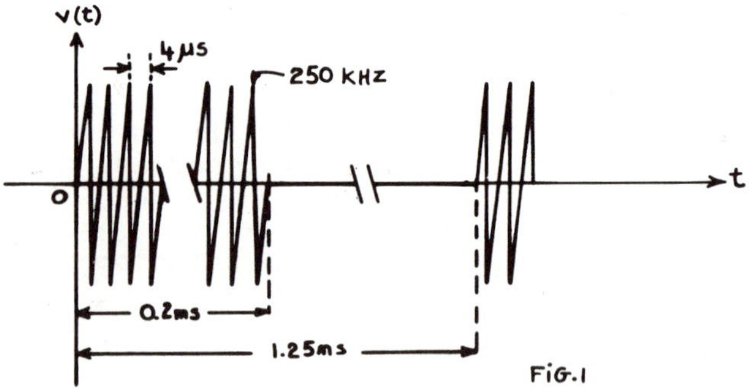

FIG.1

Solution: A sketch of the pulsed communication signal is shown in Fig. 1.

$$v(t) = \begin{cases} \sin(250\times10^3\ 2\pi\ t); & 0 \le t \le 0.2\text{ ms} \\ 0 & ;\ 0.2\text{ ms} < t \le 1.25\text{ ms} \end{cases} \qquad (1)$$

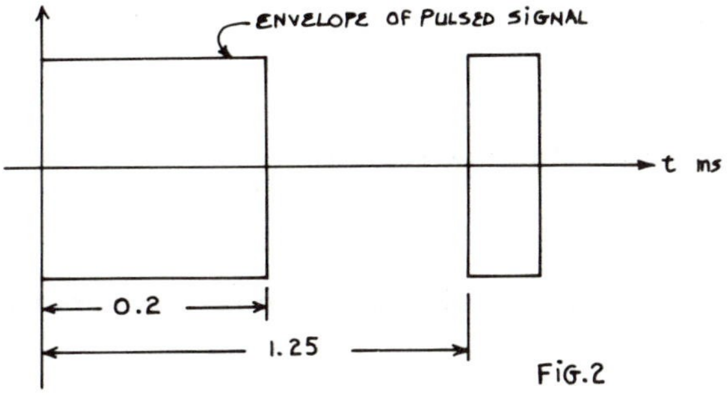

FIG.2

Eq.(1) above is an expression for one cycle of the pulsed signal. If we view the envelope of the pulsed signal shown in Fig. 2, we have a square wave with odd-symmetry. Hence, $a_o = 0$ and $a_n = 0$ in the Fourier representation of the signal

$$a_o + \sum_{n=1}^{\infty} (a_n \cos n\omega_o t + b_n \sin n\omega_o t).$$

Hence,

$$b_n = \frac{2}{\tau} \int_0^{\tau} v(t) \sin n\omega_o t\ dt$$

$$b_n = \frac{2}{1.25\times10^{-3}} \int_0^{0.2\times10^{-3}} \sin(250\times10^3 2\pi t) \sin(n \frac{2\pi}{1.25\times10^{-3}} t) dt$$

$$b_n = \frac{2}{1.25\times10^{-3}} \left[\frac{\sin(250\times10^3 2\pi - \frac{n2\pi}{1.25\times10^{-3}}) 0.2\times10^{-3}}{2(250\times10^3 2\pi - \frac{n2\pi}{1.25\times10^{-3}})} - \frac{\sin(250\times10^3 2\pi + \frac{n2\pi}{1.25\times10^{-3}}) 0.2\times10^{-3}}{2(250\times10^3 2\pi + \frac{n2\pi}{1.25\times10^{-3}})} \right]$$

$$b_n = \frac{\sin(50(2\pi) - n2\pi(0.16))}{625\pi - n2\pi} - \frac{\sin(50(2\pi) + n2\pi(0.16))}{625\pi + n2\pi}$$

$$b_n = - \frac{[(625\pi - n2\pi) + (625\pi + n2\pi)] \sin(n2\pi(.16)]}{(625\pi)^2 - (n2\pi)^2}$$

$$b_n = -\frac{1250\pi \sin(n2\pi(.16))}{390625\pi^2 - n^2 4\pi^2}$$

$$b_n = -\frac{1250 \sin(n2\pi(.16))}{390625\pi - n^2 4\pi} \quad .$$

• PROBLEM 14-17

Determine the Fourier transform for: (a) 10 cos 5t; (b) 10 sin 5t; (c) $10 \cos(5t + \frac{\pi}{3})$.

Solution: (a) Note that the Fourier transform of the unit impulse $\delta(t-t_o)$ is

$$F(\delta(t-t_o)\} = \int_{-\infty}^{\infty} e^{-j\omega t}\delta(t-t_o)dt$$

$$= \cos \omega t_o - j \sin \omega t_o = e^{-j\omega t_o}$$

which means that

$$\delta(t-t_o) <=> e^{-j\omega t_o}$$

is a Fourier transform pair.

Also, the unit impulse in the frequency domain $\delta(\omega-\omega_o)$ produces the transform pair

$$\frac{1}{2\pi} e^{j\omega_o t} <=> \delta(\omega-\omega_o)$$

Hence,

$$10 \cos 5t = 5e^{j5t} + 5e^{-j5t} <=> 10\pi\ \delta(\omega-5) + 10\pi\ \delta(\omega+5).$$

(b)

$$10 \sin 5t = -5je^{j5t} + 5je^{-j5t} <=> -10\pi j\ \delta(\omega-5) + 10\pi j\ \delta(\omega+5).$$

(c)

$$10 \cos(5t+\frac{\pi}{3}) = 5e^{j5t} e^{j\frac{\pi}{3}} + 5e^{-j5t}e^{-j\frac{\pi}{3}}$$

$$<=> 10\pi e^{j\frac{\pi}{3}}\ \delta(\omega-5) + 10\pi e^{-j\frac{\pi}{3}}\ \delta(\omega+5).$$

Determine the Fourier transform of the waveform shown.

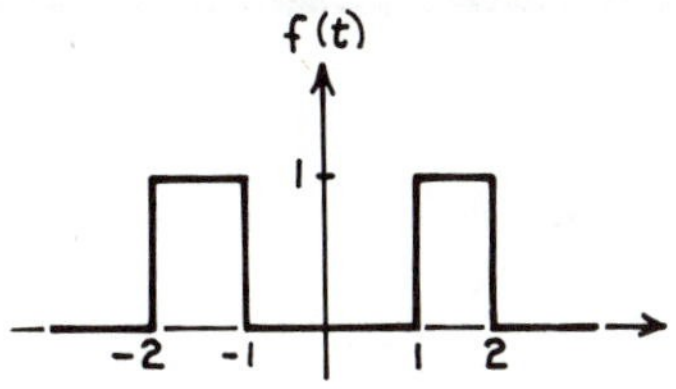

Solution: The definition of $F(j\omega)$ which in the Fourier transform of $f(t)$ is given by

$$F(j\omega) \equiv \int_{-\infty}^{\infty} f(t)e^{-j\omega t}dt. \tag{1}$$

In the given function, $f(t) = 1$ in ranges of $t = -2$ to -1 and $+1$ to $+2$. In the rest of t, $f(t) = 0$. Therefore

$$F(j\omega) \equiv \int_{-\infty}^{-2} 0\cdot e^{-j\omega t}dt + \int_{-2}^{-1} 1\cdot e^{-j\omega t}\,dy + \int_{-1}^{+1} 0\cdot e^{-j\omega t}dt$$

$$+ \int_{+1}^{+2} 1\cdot e^{-j\omega t}dt + \int_{+2}^{+\infty} 0\cdot e^{-j\omega t}dt$$

$$= 0 - \frac{1}{j\omega}\,(e^{+j\omega} - e^{+2j\omega}) + 0 - \frac{1}{j\omega}\,(e^{-2j\omega} - e^{-j\omega}) + 0$$

$$= -\frac{1}{j\omega}\{(e^{+j\omega} - e^{-j\omega}) - (-e^{2j\omega} - e^{-j\omega})\} \tag{2}$$

Using Euler's identity

$$e^{j\theta} = \cos\theta + j\sin\theta, \tag{3}$$

yields:

$$F(j\omega) = \frac{j}{\omega}\{(\cos\omega + j\sin\omega - \cos(-\omega) - j\sin(-\omega))$$

$$- (\cos 2\omega + j\sin 2\omega - \cos(-2\omega) - j\sin(-2\omega))\}$$

$$= \frac{j}{\omega}\,(2j\sin\omega - 2j\sin 2\omega)$$

$$= \frac{2}{\omega}\,(-\sin\omega + \sin 2\omega)$$

$$= \frac{2}{\omega}\,(\sin 2\omega - \sin\omega). \tag{4}$$

Find the Fourier transforms for the waveforms illustrated in Fig. 1.

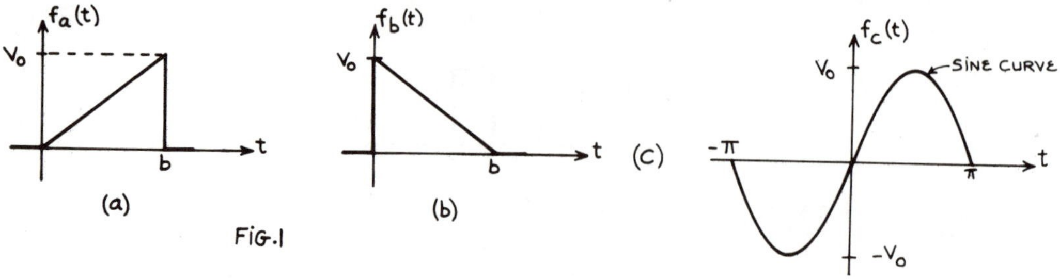

FIG.1

Solution: The Fourier transform of a function is found by applying

$$F(j\omega) = \int_{-\infty}^{\infty} f(t)e^{-j\omega t}dt.$$

(a) For the function in Fig. 1(a),

$$f_a(t) = \frac{V_o}{b} t; \qquad 0 \leq t \leq b.$$

Hence,

$$F_a(j\omega) = \int_0^b \frac{V_o}{b} t\ e^{-j\omega t}dt.$$

Making the substitution $u = \frac{V_o}{b} t$ and $dv = e^{-j\omega t}dt$ we can write

$$\int_0^b \frac{V_o}{b} + e^{-j\omega t}dt = \int_0^b udv = \left[uv - \int vdu\right]_0^b$$

where $v = -\frac{1}{j\omega} e^{-j\omega t}$ and $du = \frac{V_o}{b} dt.$

Hence

$$F_a(j\omega) = \left[\frac{V_o}{b}\left(\frac{t}{-j\omega} e^{-j\omega t} + \int \frac{1}{j\omega} e^{-j\omega t}dt\right)\right]_0^b$$

$$F_a(j\omega) = \left[\frac{V_o}{b}\left(\frac{t}{-j\omega} e^{-j\omega t} - \frac{1}{(j\omega)^2} e^{-j\omega t}\right)\right]_0^b$$

$$F_a(j\omega) = \frac{V_o}{b}\left[\frac{b}{-j\omega} e^{-j\omega b} + \frac{1}{\omega^2} e^{-j\omega b} - \frac{1}{\omega^2}\right].$$

(b) For the function in Fig. 1(b),

$$f_b(t) = -\frac{V_o}{b}(t-b); \quad 0 \le t \le b.$$

Hence,

$$F_b(j\omega) = -\frac{V_o}{b}\int_0^b (t-b)e^{-j\omega t}dt$$

$$F_b(j\omega) = -\frac{V_o}{b}\int_0^b te^{-j\omega t} + V_o\int_0^b e^{-j\omega t}dt$$

$$F_b(j\omega) = -F_a(j\omega) + V_o\int_0^b e^{-j\omega t}dt$$

$$F_b(j\omega) = -F_a(j\omega) + V_o\left[\frac{1}{-j\omega} e^{-j\omega t}\right]_0^b$$

$$F_b(j\omega) = -F_a(j\omega) + V_o\left[\frac{1}{-j\omega} e^{-j\omega b} - \frac{1}{j\omega}\right]$$

$$F_b(j\omega) = +\frac{V_o}{j\omega} e^{-j\omega b} - \frac{V_o}{b\omega^2} e^{-j\omega b} + \frac{V_o}{b\omega^2} - \frac{V_o}{j\omega} e^{-j\omega b} - \frac{V_o}{j\omega}$$

$$F_b(j\omega) = V_o\left[-\frac{1}{b\omega^2} e^{-j\omega b} + \frac{1}{b\omega^2} - \frac{1}{j\omega}\right].$$

(c) For the functions in Fig. 1(c),

$$f_c(t) = \sin t; \quad -\pi \le t \le \pi.$$

Hence,

$$F_c(j\omega) = V_o\int_{-\pi}^{\pi}\sin t\, e^{-j\omega t}dt$$

$$F_c(j\omega) = V_o\int_{-\pi}^{\pi}\left(\frac{e^{jt}-e^{-jt}}{2j}\right) e^{-j\omega t}dt$$

$$F_c(j\omega) = V_o \int_{-\pi}^{\pi} \frac{e^{jt(1-\omega)} - e^{-jt(1+\omega)}}{2j}\, dt$$

$$F_c(j\omega) = V_o \left[\frac{e^{jt(1-\omega)}}{2j(j(1-\omega))} + \frac{e^{-jt(1+\omega)}}{2j(j(1+\omega))} \right]_{-\pi}^{\pi}$$

$$F_c(j\omega) = V_o \left| - \frac{e^{jt(1-\omega)}}{2(1-\omega)} - \frac{e^{-jt(1+\omega)}}{2(1+\omega)} \right|_{-\pi}^{\pi}$$

$$F_c(j\omega) = V_o \ -\left(\frac{e^{j\pi(1-\omega)}}{2(1-\omega)} - \frac{e^{-j\pi(1+\omega)}}{2(1+\omega)} + \frac{e^{-j\pi(1-\omega)}}{2(1-\omega)} + \frac{e^{-j\pi(1+\omega)}}{2(1+\omega)} \right)$$

$$F_c(j\omega) = \frac{V_o j}{(1-\omega)} \left[\frac{-e^{j\pi(1-\omega)} + e^{-j\pi(1-\omega)}}{2j} \right] + \frac{V_o j}{(1+\omega)} \left[\frac{e^{j\pi(1+\omega)} - e^{-j\pi(1+\omega)}}{2j} \right] .$$

$$F_c(j\omega) = -\frac{jV_o}{1-\omega} \sin \pi(1-\omega) + \frac{jV_o}{1+\omega} \sin \pi(1+\omega)$$

can be simplified by combining the two terms

$$F_c = \frac{-jV_o(1+\omega)\sin\pi(1-\omega) + jV_o(1-\omega)\sin\pi(1+\omega)}{(1-\omega)(1+\omega)}$$

$$= \frac{-jV_o(1+\omega)\sin\pi\omega - jV_o(1-\omega)\sin\pi\omega}{1-\omega^2}$$

$$F_c(j\omega) = \frac{-j2V_o \sin\pi\omega}{1-\omega^2} .$$

APPLICATIONS TO CIRCUIT THEORY

• **PROBLEM** 14-20

Given the current pulse, $i(t) = te^{-bt}u(t)$: (a) find the total 1Ω energy associated with this waveform; (b) what fraction of this energy is present in the frequency band from -b to b rad/s?

Solution: The total 1-Ω energy associated with either a current or voltage waveform can be found by use of Parseval's theorem,

$$W_{1\Omega} = \frac{1}{2\pi}\int_{-\infty}^{\infty} |F(j\omega)|^2 dw$$

where F(jω) is the Fourier transform of the current or voltage waveform.

Hence, the Fourier transform of the current is

$$I(j\omega) = \int_{0}^{\infty} t\ e^{-bt}\ e^{-j\omega t} dt$$

$$I(j\omega) = \int_{0}^{\infty} t\ e^{-t(j\omega+b)} dt$$

$$I(j\omega) = \left[\frac{-te^{-j(j\omega+b)}}{(j\omega+b)} - \int \frac{e^{-t(j\omega+b)}}{-(j\omega+b)}\ dt\right]_{0}^{\infty}$$

$$I(j\omega) = \left[-\frac{te^{-t(j\omega+b)}}{j\omega+b} - \frac{e^{-t(j\omega+b)}}{(j\omega+b)^2}\right]_{0}^{\infty}$$

$$I(j\omega) = \frac{1}{(j\omega+b)^2}$$

$$|I(j\omega)|^2 = \frac{1}{(b^2-\omega^2)^2+4\omega^2b^2} = \frac{1}{(b^2+\omega^2)^2}\ .$$

The total energy associated with the current is

$$W = \frac{1}{2\pi}\int_{-\infty}^{\infty} |I(j\omega)|^2 d\omega.$$

Since

$$W = \frac{1}{\pi}\int_{0}^{\infty} |I(j\omega)|^2 d\omega$$

then

$$W = \frac{1}{\pi}\int_{0}^{\infty} \frac{1}{(b^2+\omega^2)^2}\ d\omega.$$

If we make the trigonometric substitution,

$\omega = b \tan\theta$, then

$$(b^2+\omega^2)^2 = (b^2\tan^2\theta + b^2)^2 = (b^2\sec^2\theta)^2 .$$

Also, $d\omega = b\sec^2\theta d\theta$. Hence

$$W = \frac{1}{\pi}\int \frac{b\sec^2\theta}{b^4\sec^4\theta}\, d\theta$$

$$W = \frac{1}{\pi}\int \frac{1}{b^3\sec^2\theta}\, d\theta$$

$$W = \frac{1}{\pi b^3}\int \cos^2\theta\, d\theta$$

$$W_\tau = \frac{1}{\pi b^3}\left(\frac{1}{2}\theta + \frac{1}{4}\sin 2\theta\right).$$

Since $\omega = b\tan\theta$

and $\theta = \arc\tan\frac{\omega}{b}$

then $$W_\tau = \left[\frac{1}{\pi b^3}\left(\frac{1}{2}\tan^{-1}\frac{\omega}{b} + \frac{1}{4}\sin 2(\tan^{-1}\frac{\omega}{b})\right)\right]_0^\infty .$$

Since $\tan^{-1}\infty = \frac{\pi}{2}$

then $$W_\tau = \frac{1}{\pi b^3}\left(\frac{1}{2}\,\frac{\pi}{2}\right)$$

$$W_\tau = \frac{1}{4b^3} .$$

(b) To find the energy present in the frequency band $-b < f < b$ we use Parseval's theorem and integrate:

$$W_b = \frac{1}{2\pi}\int_{-b}^{b} |I(j\omega)|^2\, d\omega$$

$$W_b = \frac{1}{\pi}\int_{0}^{b} |I(j\omega)|^2\, d\omega$$

$$W_b = \left[\frac{1}{\pi b^3}\left(\frac{1}{2}\tan^{-1}\frac{\omega}{b} + \frac{1}{4}\sin 2(\tan^{-1}\frac{\omega}{b})\right)\right]_0^b$$

$$W_b = \frac{1}{\pi b^3}\left[\left(\frac{1}{2}\tan^{-1}1 + \frac{1}{4}\sin 2(\tan^{-1}1)\right) - \left(\frac{1}{2}\tan^{-1}0 + \frac{1}{4}\sin 2(\tan^{-1}0)\right)\right]$$

$$W_b = \frac{1}{\pi b^3}\left[\frac{\pi}{8} + \frac{1}{4} - 0 - 0\right]$$

$$W_b = \left[\frac{\pi+2}{8}\right]\frac{1}{\pi b^3} .$$

The fraction of energy present is

$$\frac{W_o}{W_\tau} = \frac{\left[\frac{\pi+2}{8}\right]\frac{1}{\pi b^3}}{\frac{1}{4b^3}} = \frac{\pi+2}{2\pi} = 0.818.$$

• PROBLEM 14-21

(a) If $a > 0$ show that the Fourier transform of $f(t) = e^{-at}(\cos\omega_d t)u(t)$ is $(a+j\omega)/[(a+j\omega)^2+\omega_d^2]$. Find the total 1-Ω energy associated with $e^{-t}(\cos t)u(t)$ by using: (b) time-domain integration; (c) frequency-domain integration.

Hint: $\int_0^\infty [(x^2+1)]/(x^4+4)dx = \frac{3\pi}{8}$

Solution: (a) Using the relation

$$F(j\omega) = \int_{-\infty}^{\infty} f(t)e^{-j\omega t}dt,$$

we can find the Fourier transform of $f(t) = e^{-at}(\cos\omega_d t)$ $u(t)$. Hence,

$$F(j\omega) = \int_0^\infty e^{-at}(\cos\omega_d t)e^{-j\omega t}dt$$

$$F(j\omega) = \int_0^\infty e^{-at}\left(\frac{e^{j\omega_d t}+e^{-j\omega_d t}}{2}\right) e^{-j\omega t}dt$$

$$F(j\omega) = \int_0^\infty \left[\frac{e^{(j\omega_d-j\omega-a)t}}{2} + \frac{e^{(-j\omega_d-j\omega-a)t}}{2}\right]dt$$

$$F(j\omega) = \left[\frac{e^{(j\omega_d - j\omega - a)}}{2(j\omega_d - j\omega - a)} + \frac{e^{(-j\omega_d - j\omega - a)t}}{2(-j\omega_d - j\omega - a)}\right]_o^{\infty}$$

$$F(j\omega) = \frac{-1}{2(j\omega_d - j\omega - a)} - \frac{1}{2(-j\omega_d - j\omega - a)}$$

$$F(j\omega) = \frac{a + j\omega}{(a + j\omega)^2 + \omega_d^2}$$

(b) In the time domain we find the total energy by integrating $[f(t)]^2$ as follows:

$$W = \int_{-\infty}^{\infty} [f(t)]^2 dt$$

$$W = \int_{o}^{\infty} [e^{-t}(\cos\ t)]^2 dt$$

$$W = \int_{o}^{\infty} e^{-2t}(\tfrac{1}{2} + \tfrac{1}{2}\cos\ 2t)dt$$

$$W = \int_{o}^{\infty} [\frac{e^{-2t}}{2} + \frac{e^{-2t}}{2}(\frac{e^{+j2t}}{2} + \frac{e^{-j2t}}{2})]dt$$

$$W = \left[-\frac{e^{-2t}}{4} + \frac{e^{-2t}e^{j2t}}{4(-2+j2)} - \frac{e^{-2t}e^{-j2t}}{4(2+j2)}\right]_o^{\infty}$$

$$W = \frac{8-(-2-j2)+(2-j2)}{32} = \frac{12}{32} = \frac{3}{8}\ J.$$

(c) In the frequency domain the total energy is found by integrating

$$W = \frac{1}{2\pi}\int_{-\infty}^{\infty} |F(j\omega)|^2 d\omega$$

where $F(j\omega)$ is the Fourier transform of the function $f(t)$. Hence, we found

$$F(j\omega) = \frac{a + j\omega}{(a + j\omega)^2 + \omega_d^2}$$ if $a = 1$ and $\omega_d = 1$. Then,

$$F(j\omega) = \frac{1+j\omega}{(1+j\omega)^2+1}$$

and $$|F(j\omega)|^2 = \frac{1+\omega^2}{(2-\omega^2)^2+4\omega^2} = \frac{1+\omega^2}{4+\omega^4}.$$

The energy is

$$W = \frac{1}{2\pi}\int_{-\infty}^{\infty} \frac{1+\omega^2}{4+\omega^4}\,d\omega = \frac{1}{\pi}\int_{0}^{\infty} \frac{1+\omega^2}{4+\omega^4}\,d\omega$$

since $\displaystyle\int_0^{\infty} \frac{x^2+1}{x^4+4}\,dx = \frac{3\pi}{8}$.

Then, $W = \frac{3}{8}$ J.

• PROBLEM 14-22

If $Y(j\omega) = F(j\omega)H(j\omega)$, find $y(t)$ for $F(j\omega) = H(j\omega) = \frac{1}{1+\omega^2}$.

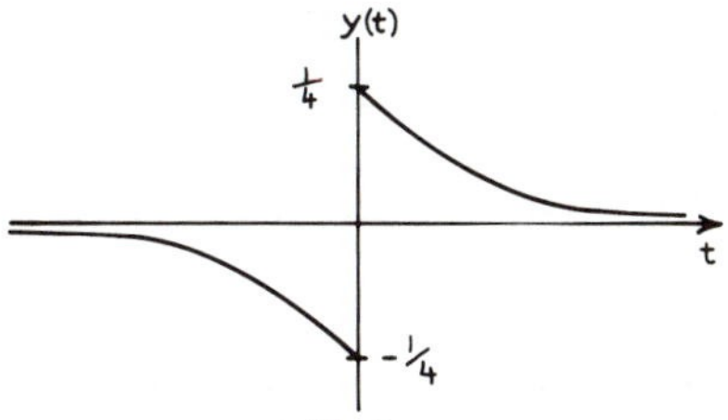

Fig.1

Solution: The response $Y(j\omega)$ is the product of the system function, $H(j\omega)$, and the driving function, $F(j\omega)$. Since

$F(j\omega) = H(j\omega) = \frac{1}{1+\omega^2}$, then $Y(j\omega) = \frac{1}{(1+\omega^2)^2}$.

Hence,

$$y(t) = F^{-1}\left\{\frac{1}{(1+\omega^2)^2}\right\}, \text{ the inverse Fourier transform.}$$

Since $1+\omega^2$ can be expanded into the form

$$-(j\omega+1)(j\omega-1)$$

then

$$y(t) = F^{-1}\left\{\frac{1}{(j\omega+1)^2(j\omega-1)^2}\right\}.$$

Partial fraction expansion yields

$$y(t) = F^{-1}\left\{\frac{\frac{1}{4}}{(j\omega+1)^2} + \frac{\frac{1}{4}}{(j\omega+1)} + \frac{\frac{1}{4}}{(j\omega-1)^2} - \frac{\frac{1}{4}}{(j\omega-1)}\right\}.$$

Noting that $F^{-1}\left\{\frac{\frac{1}{4}}{(j\omega+1)^2}\right\} = \frac{1}{4}\, te^{-t}u(t)$ and

$F^{-1}\left\{\frac{\frac{1}{4}}{j\omega+1}\right\} = \frac{1}{4}\, e^{-t}u(t)$, we obtain

$$y(t) = \frac{1}{4}\, te^{-t}u(t) + \frac{1}{4}\, e^{-t}u(t) + \frac{1}{4}\, te^{t}u(t) - \frac{1}{4}\, e^{t}u(t)$$

$$y(t) = \frac{1}{4}\, e^{-t}(t+1)u(t) + \frac{1}{4}\, e^{t}(t-1)u(t),$$

if we let $x(t) = \frac{1}{4}\, e^{-t}(t+1)u(t)$ and

$$z(t) = \frac{1}{4}\, e^{t}(t-1)u(t).$$

Hence,

$$y(t) = -z(-t) + z(t)$$

$$y(t) = x(t) - x(-t)$$

and $\quad y(t) = -z(-t) - x(-t) = x(t) + z(t).$

$y(t) = -y(-t)$ is an odd function (see Fig. 1).

• PROBLEM 14-23

Find the system function $H(j\omega) = \frac{V_i(j\omega)}{I_i(j\omega)}$, and impulse response of a one-port network if the input voltage, $v_i(t) = 100 \cos \omega_o t$ V, produces the input current, $I_i(j\omega) = 100\pi[\delta(\omega+\omega_o) + \delta(\omega-\omega_o)](1-j/2\omega)$.

Solution: The Fourier transform of $v_i(t) = 100 \cos \omega_o t$ is

$$V_i(j\omega) = 100\pi[\,\delta(\omega+\omega_o) + \delta(\omega-\omega_o)].$$

Hence, the system function $H(j\omega)$ is

$$\frac{V_i(j\omega)}{I_i(j\omega)} = \frac{1}{1 - \frac{j}{2\omega}} = \frac{j2\omega}{1+j2\omega}\,.$$

The impulse response is found by taking the inverse Fourier transform of $H(j\omega)$. Hence,

$$h(t) = F^{-1}\left\{\frac{j2\omega}{1 + j2\omega}\right\} = F^{-1}\left\{1 - \frac{1}{1+j2\omega}\right\}.$$

Using the transform pairs yields

$\delta(t) <=> 1.$ Since $\delta(t-t_o) <=> e^{-j\omega t_o}$

and $e^{-at}u(t) <=> \dfrac{1}{a+j\omega}$, we obtain

$$h(t) = \delta(t) - \frac{1}{2}e^{-\frac{1}{2}t}.$$

• PROBLEM 14-24.

> Use Fourier transform methods to find the time-domain response of a network having a system function $j2\omega/(1+2j\omega)$, if the input is: (a) a unit impulse; (b) a unit-step function; (c) cos t.

Solution: The time-domain response for a particular input v(t) can be obtained by finding the product of the system function $H(j\omega)$ and the Fourier transform of the input. The inverse Fourier transform of the resulting function is the time-domain response.

(a) For a unit impulse $\delta(t)$, the Fourier transform pair, $\delta(t) <=> 1$, shows that the response must be the inverse Fourier transform of the system function. Hence,

$$f(t) = F^{-1}\left\{\frac{j2\omega}{1+2j\omega}\right\} = F^{-1}\left\{\frac{j\omega}{j\omega+\frac{1}{2}}\right\}$$

$$f(t) = F^{-1}\left\{1 - \frac{\frac{1}{2}}{j\omega+\frac{1}{2}}\right\} = \delta(t) - 0.5e^{-\frac{1}{2}t}.$$

(b) For a unit step, u(t), the Fourier transform pair $u(t) <=> \pi\delta(\omega) + \dfrac{1}{j\omega}$ allows us to find the response

$$f(t) = F^{-1}\left\{\frac{j2\omega}{1+2j\omega}\left(\pi\delta(\omega) + \frac{1}{j\omega}\right)\right\}$$

$$f(t) = F^{-1}\left\{\frac{j2\pi\omega\delta(\omega) + 2}{1 + 2j\omega}\right\}$$

$$f(t) = F^{-1}\left\{\frac{j2\pi\omega\delta(\omega)}{1+2j\omega} + \frac{1}{\frac{1}{2} + j\omega}\right\}$$

$$= F^{-1}\left\{\frac{j\pi\omega\delta(\omega)}{\frac{1}{2} + j\omega} + \frac{1}{\frac{1}{2} + j\omega}\right\}$$

$$= F^{-1}\left\{\pi\delta(\omega) - \frac{\frac{\pi\delta(\omega)}{2}}{\frac{1}{2} + j\omega} + \frac{1}{\frac{1}{2} + j\omega}\right\}.$$

Making use of the sifting property of the unit impulse, we obtain

$$f(t) = \frac{1}{2} - \frac{1}{2} + e^{-\frac{1}{2}t}\, u(t)$$

$$f(t) = e^{-\frac{1}{2}t}\, u(t).$$

(c) For a sinusoidal input cos t, the Fourier transform pair $\cos t \Leftrightarrow \pi[\delta(\omega+1) + \delta(\omega-1)]$ allows us to find the response

$$f(t) = F^{-1}\left\{\frac{j2\omega}{1+2j\omega}\pi(\delta(\omega+1) + \delta(\omega-1))\right\}$$

$$f(t) = F^{-1}\left\{\frac{j2\pi\omega\ \delta(\omega+1)}{1 + 2j\omega} + \frac{j2\pi\omega\ \delta(\omega-1)}{1 + 2j\omega}\right\}$$

$$f(t) = F^{-1}\left\{\frac{j\pi\omega\ \delta(\omega+1)}{\frac{1}{2} + j\omega} + \frac{j\pi\omega\ \delta(\omega-1)}{\frac{1}{2} + j\omega}\right\}$$

Using the sifting property of the unit impulse, we obtain:

$$f(t) = F^{-1}\left\{-\frac{j\pi\ \delta(\omega+1)}{\frac{1}{2} - j} + \frac{j\pi\ \delta(\omega-1)}{\frac{1}{2} + j}\right\}$$

$$f(t) = F^{-1}\left\{-\frac{j\pi\ \delta(\omega+1)(\frac{1}{2}+j)}{\frac{1}{4} + 1} + \frac{j\pi\ \delta(\omega-1)(\frac{1}{2}-j)}{\frac{1}{4} + 1}\right\}$$

$$f(t) = F^{-1}\left\{\frac{\pi\ \delta(\omega+1)}{\frac{5}{4}} - \frac{j\frac{1}{2}\pi\ \delta(\omega+1)}{\frac{5}{4}} + \frac{\pi\ \delta(\omega-1)}{\frac{5}{4}} + \frac{j\frac{1}{2}\pi\ \delta(\omega+1)}{\frac{5}{4}}\right\}$$

$$f(t) = F^{-1}\left\{\frac{4}{5}\pi(\delta(\omega+1)+\delta(\omega-1)) - \frac{2}{5}\pi(j\ \delta(\omega+1)-j\ \delta(\omega-1))\right\}$$

$$f(t) = \frac{4}{5}\cos t - \frac{2}{5}\sin t.$$

Find $v_o(t)$ in the LC circuit shown in Fig. 1 if $v_i(t) = \delta(t)$. Use Fourier transform methods.

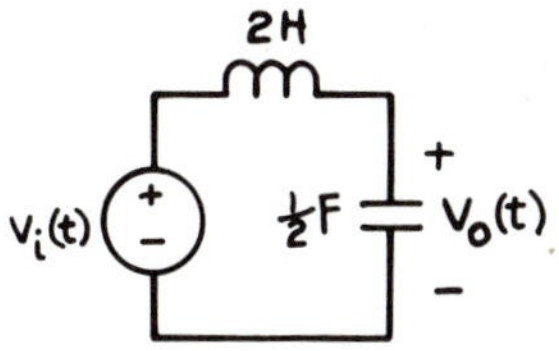

Fig. 1

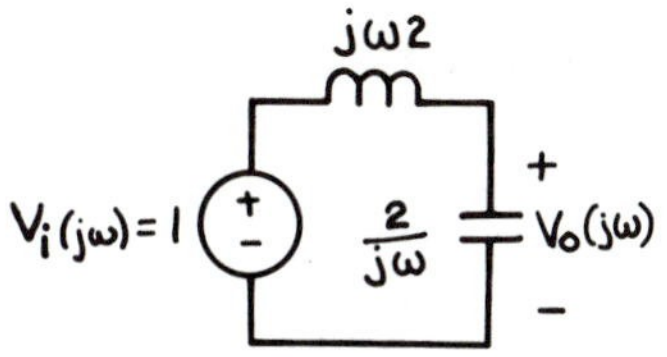

Fig. 2

Solution: The Fourier transform of each element in the circuit is found and shown in Fig. 2.

Hence, the output $v_o(t)$ is found by taking the inverse transform of $V_o(j\omega)$:

$$v_o(t) = F^{-1}\left[\frac{\frac{2}{j\omega}}{j\omega 2 + \frac{2}{j\omega}}\right]$$

$$v_o(t) = F^{-1}\left[\frac{1}{(j\omega)^2 + 1}\right] \qquad (1)$$

$$v_o(t) = F^{-1}\left[\frac{1}{(j\omega+j)(j\omega-j)}\right].$$

Expanding into partial fractions yields

$$v_o(t) = F^{-1}\left[\frac{\frac{1}{2j\omega}}{j\omega - j} + \frac{\frac{1}{2j\omega}}{j\omega + j}\right]$$

$$v_o(t) = F^{-1}\left[\frac{1}{2j\omega(j\omega-j)} + \frac{1}{2j\omega(j\omega+j)}\right].$$

Expand again into partial fractions to obtain

$$v_o(t) = F^{-1}\left[\frac{j}{2j\omega} + \frac{\frac{1}{2j}}{j\omega-j} - \frac{j}{2j\omega} - \frac{\frac{1}{2j}}{j\omega+j}\right]$$

$$v_o(t) = F^{-1}\left[\frac{1}{2j}\left(\frac{1}{j\omega-j} - \frac{1}{j\omega+j}\right)\right].$$

Since $\qquad e^{-at}u(t) \iff \frac{1}{j\omega+a}$

and $\qquad e^{at}u(t) \iff \frac{1}{j\omega-a}$

are Fourier transform pairs, then

$$v_o(t) = \frac{1}{2j}(e^{jt} - e^{-jt})u(t)$$

$$v_o(t) = \sin t\ u(t).$$

(Of course, we could have looked up Eq.(1) in a table and found that

$$\frac{a}{(j\omega)^2+a^2} \Leftrightarrow \sin at\ u(t)$$

are Fourier transform pairs.)

• PROBLEM 14-26

Find $i_o(t)$ in the circuit shown in Fig. 1 using Fourier transform methods if $v_i(t) =$ (a) $\delta(t)V$; (b) $u(t)V$; (c) $e^{-t}u(t)V$.

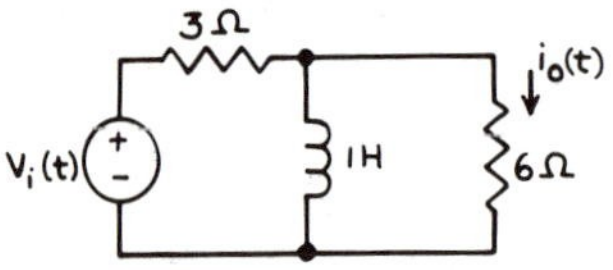

FiG.1

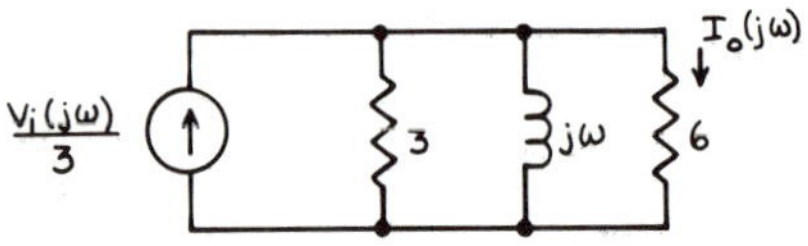

FiG.2

Solution: (a) The solution to an input $\delta(t)V$ is known as the impulse response. The input response of a network is its transfer function $H(j\omega)$, where

$$H(j\omega) = \frac{I_o(j\omega)}{V_i(j\omega)}\ .$$

Since $F\{\delta(t)\} = 1$,

$$H(j\omega) = I_o(j\omega) \quad \text{for } v_i(t) = \delta(t).$$

The transfer function of the circuit in Fig. 1 can be found by transforming the voltage source as shown in Fig. 2.

The output current is found using the concept of current division. Hence,

$$I_o(j\omega) = \frac{\frac{3j\omega}{3+j\omega}\frac{V_i}{3}}{\frac{3j\omega}{3+j\omega} + 6}$$

$$H(j\omega) = \frac{\frac{j\omega}{3+j\omega}}{\frac{3j\omega}{3+j\omega} + 6} = \frac{1}{9}\frac{j\omega}{j\omega+2}\ .$$

For $v_i(t) = \delta(t)$,

$$i_o(t) = F^{-1}\{\frac{1}{9}\frac{j\omega}{j\omega+2}\}$$

$$i_o(t) = F^{-1}\{\frac{1}{9}(\frac{-2}{j\omega+2} + 1)\}$$

$$i_o(t) = \frac{1}{9}\delta(t) - \frac{2}{9}e^{-2t}u(t),$$

since $F^{-1}\{\frac{1}{9}\} = \frac{1}{9}\delta(t)$ and $F^{-1}\{\frac{-\frac{2}{9}}{j\ \ +2}\} = -\frac{2}{9}e^{-2t}u(t)$.

(b) For a step input $v_i(t) = u(t)V$,

$$V_i(j\omega) = \pi\delta(\omega) + \frac{1}{j\omega}$$

and

$$I_o(j\omega) = H(j\omega)V_i(j\omega)$$

$$I_o(j\omega) = \frac{1}{9}(\frac{j\omega}{j\omega+2})(\pi\delta(\omega) + \frac{1}{j\omega}) .$$

Hence,

$$i_o(t) = F^{-1}\{\frac{1}{9}(\frac{j\omega}{j\omega+2})(\pi\delta(\omega) + \frac{1}{j\omega})\}$$

$$i_o(t) = F^{-1}\{\frac{1}{9}\frac{1}{j\omega+2} + \frac{1}{9}\frac{\pi\delta(\omega)\ j\omega}{j\omega+2}\} .$$

Using the sifting property of the unit impulse, we obtain

$$i_o(t) = F^{-1}\{\frac{1}{9}\frac{1}{j\omega+2} + 0\}$$

$$i_o(t) = \frac{1}{9}e^{-2t}u(t).$$

(c) For an exponential input $v_i(t) = e^{-t}u(t)$,

$$V_i(j\omega) = \frac{1}{j\omega+1} .$$

Hence,

$$I_o(j\omega) = H(j\omega)V_i(j\omega)$$

$$I_o(j\omega) = (\frac{1}{9}\frac{j\omega}{j\omega+2})(\frac{1}{j\omega+1})$$

$$I_o(j\omega) = \frac{A}{j\omega+2} + \frac{B}{j\omega+1} .$$

Partial fraction expansion yields

$$I_o(j\omega) = \frac{\frac{2}{9}}{j\omega+2} - \frac{\frac{1}{9}}{j\omega+1} .$$

Hence,

$$i_o(t) = F^{-1} \quad \frac{\frac{2}{9}}{j\omega+2} - \frac{\frac{1}{9}}{j\omega+1}$$

$$i_o(t) = (\tfrac{2}{9} e^{-2t} - \tfrac{1}{9} e^{-t})u(t).$$

• PROBLEM 14-27

A rectangular voltage pulse, $v(t) = 10k[u(t)-u(t-\frac{1}{k})]V$, is applied in series with a 100-μF capacitor and a 20-kΩ resistor. Find the capacitor voltage at t = 1 as follows: (a) if k = 1; (b) if k = 10; (c) if k = 100; (d) in the limit as $k \to \infty$.

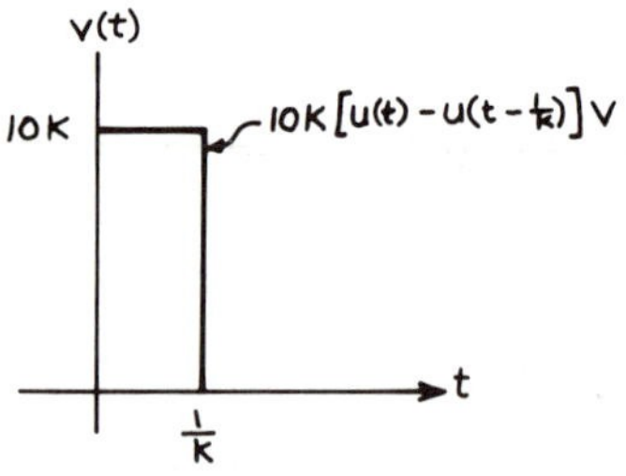

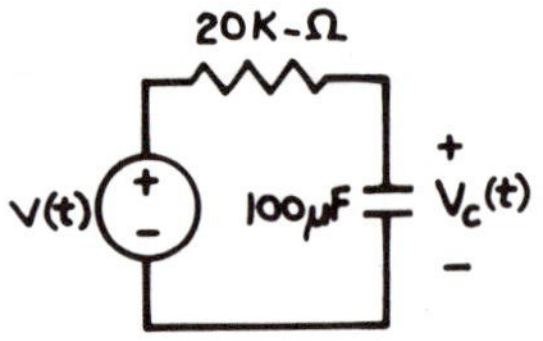

FiG.2

FiG.1

Solution: The pulse shown in Fig. 1 is applied to the circuit of Fig. 2.

For the capacitor voltage output shown the system function is

$$H(j\omega) = \frac{V_c(j\omega)}{V(j\omega)} = \frac{\frac{1}{(20x10^3)(100x10^{-6})}}{j\omega + \frac{1}{(20x10^3)(100x10^{-6})}}$$

$$H(j\omega) = \frac{\frac{1}{2}}{j\omega + \frac{1}{2}} .$$

The Fourier transform of the input pulse is

$$F\ \{10k[u(t) - u(t-\tfrac{1}{k})\} =$$

$$10k[\pi\ \delta(\omega) + \frac{1}{j\omega}] - 10k[\pi\ \delta(\omega) + \frac{1}{j\omega}]e^{-j\omega\frac{1}{k}}\ .$$

Hence, the output is

$$v_c(t) = F^{-1}\{\frac{\frac{1}{2}}{j\omega+\frac{1}{2}}\ [10k(\pi\delta(\omega) + \frac{1}{j\omega}) - 10k(\pi\delta(\omega) + \frac{1}{j\omega})e^{-j\omega\frac{1}{k}}]\}.$$

We can split the response into two parts:

$$v_{co}(t) = F^{-1}\{\frac{\frac{1}{2}}{j\omega+\frac{1}{2}}\ 10k(\pi\delta(\omega) + \frac{1}{j\omega})\},$$

and $$v_{cl}(t) = F^{-1}\{-\ \frac{\frac{1}{2}}{j\omega+\frac{1}{2}}\ 10k(\pi\delta(\omega) + \frac{1}{j\omega})e^{-j\omega\frac{1}{k}}\}.$$

Evaluating $v_{co}(t)$ yields

$$v_{co}(t) = F^{-1}\{\frac{\frac{1}{2}10k\pi\delta(\omega)}{j\omega+\frac{1}{2}} + \frac{\frac{1}{2}10k}{j\omega(j\omega+\frac{1}{2})}\}$$

$$v_{co}(t) = F^{-1}\{10k\pi\delta(\omega) + \frac{10k}{j\omega} - \frac{10k}{j\omega+\frac{1}{2}}\}$$

$$v_{co}(t) = 10k\ u(t) - 10ke^{-\frac{1}{2}t}u(t),\ \text{hence}$$

$$v_{cl}(t) = -[10k\ u(t-\frac{1}{k}) - 10k\ e^{-\frac{1}{2}(t-\frac{1}{2})}u(t-\frac{1}{k})].$$

(a) When k = 1,

$$v_c(t) = 10u(t) - 10e^{-\frac{1}{2}t}u(t) - 10u(t-1) + 10e^{-\frac{1}{2}(t-1)}u(t-1).$$

Hence,

$$v_c(1) = 10 - 10e^{-\frac{1}{2}} - 10 + 10 = 3.93\ \text{V}.$$

(b) When k = 10,

$$v_c(t) = 100u(t) - 100e^{-\frac{1}{2}t}u(t) - 100u(t-\frac{1}{10})$$
$$+ 100e^{-\frac{1}{2}(t-\frac{1}{10})}u(t-\frac{1}{10})$$

and

$$v_c(1) = 100 - 100e^{-\frac{1}{2}} - 100 + 100e^{-\frac{9}{20}} = 3.11\ \text{V}.$$

(c) When k = 100,

$$v_c(t) = 1000u(t) - 1000e^{-\frac{1}{2}t}u(t) - 1000u(t-\frac{1}{100}) + 1000e^{-\frac{1}{2}(t-\frac{1}{100})}u(t-\frac{1}{100})$$

and

$$v_c(1) = 1000 - 1000e^{-\frac{1}{2}t} - 1000 + 1000e^{-\frac{99}{200}} = 3.04 \text{ V.}$$

(d)

$$v_c(t) = 10k - 10ke^{-\frac{1}{2}t} - 10ke^{-\frac{1}{2}(t-\frac{1}{k})} - 10k$$

$$= 10k(-e^{-\frac{1}{2}t} + e^{-\frac{1}{2}(t-\frac{1}{k})})$$

$$= 10k\ e^{-\frac{1}{2}t}(e^{\frac{1}{2k}} - 1).$$

The Taylor series expansion for e^x is

$$e^x = 1 + x + \frac{x^2}{2} + \frac{x^3}{6} + \ldots$$

Substituting this expansion for $e^{\frac{1}{2}k}$ where $x = \frac{1}{2k}$

$$v_c(t) = 10k\ e^{-\frac{1}{2}t}(-1 + 1 + \frac{1}{2k} + \frac{1}{8k^2} + \frac{1}{24k^3} + \ldots)$$

neglecting higher terms as $k \to \infty$

$$v_c(t) = 10k\ e^{-\frac{1}{2}t}(\frac{1}{2k})$$

$$v_c(t) = 5e^{-\frac{1}{2}t} \qquad \text{Therefore}$$

$$v_c(1) = 5e^{-\frac{1}{2}t} = 3.033 \text{ V.}$$

• PROBLEM 14-28

The source in the circuit shown in Fig. 1 is $v_i(t) = 100[u(t) - u(t-0.001)]$V. Sketch curves as a function of ω of the amplitude of $V_i(j\omega)$, $H(j\omega) = \frac{V_o(j\omega)}{V_i(j\omega)}$, and $V_o(j\omega)$. In what way does the RC network affect the spectrum of the input signal?

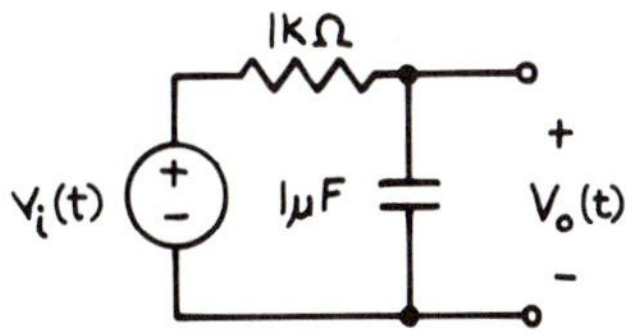

FiG.1

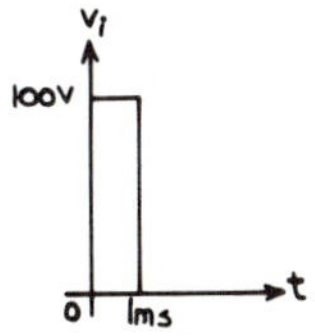

FIG.2

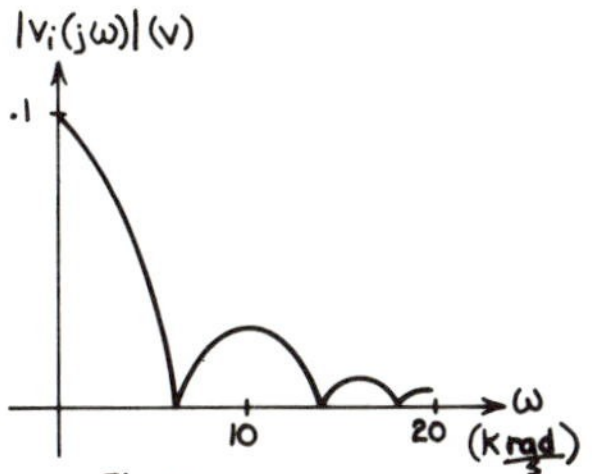

FiG.3

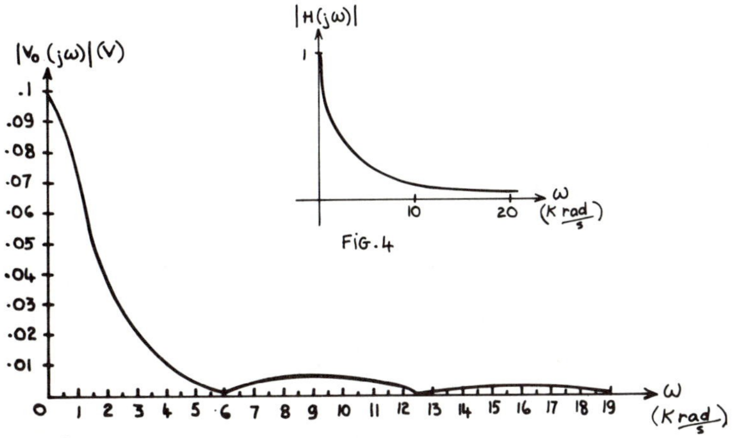

FiG.5

Solution: The source voltage, $v_i(t)$, is a single 1 ms pulse shown in Fig. 2. The Fourier transform of this input pulse is found by use of the Fourier integral

$$F(j\omega) = \int_{-\infty}^{\infty} f(t)\ e^{-j\omega t} dt.$$

Hence,

$$V_i(j\omega) = 100 \int_{0}^{.001} e^{-j\omega t}$$

$$V_i(j\omega) = 100\left[-\frac{1}{j\omega}\ e^{-j\omega(.001)} + \frac{1}{j\omega}\right]$$

$$V_i(j\omega) = \frac{100}{\omega}\ [\sin\ \omega(.001) + j(-1+\cos\ \omega(.001))],$$

and the magnitude is

$$|V_i(j\omega)| = \frac{100}{\omega}\sqrt{\sin^2\omega(.001)+\cos^2\omega(.001)-2\cos\omega(.001)+1}$$

$$|V_i(j\omega)| = \left|\frac{200}{\omega}\ \sin\ \frac{\omega(.001)}{2}\right| = \left| .1\ \frac{\sin\ \frac{\omega(.001)}{2}}{\frac{\omega(.001)}{2}}\right| .$$

Since the circuit is a voltage divider we can write

$$V_o(j\omega) = H(j\omega)V_i(j\omega)$$

$$V_o(j\omega) = \frac{\frac{1}{j\omega c} V_i(j\omega)}{\frac{1}{j\omega c} + R} .$$

Hence,

$$H(j\omega) = \frac{\frac{1}{j\omega c}}{\frac{1}{j\omega c} + R} = \frac{1000}{j\omega + 1000} ,$$

and $$|H(j\omega)| = \frac{1000}{\sqrt{\omega^2 + 1000^2}} .$$

The plots of the frequency spectrum of the input and transfer functions are shown in Figures 3 and 4.

The output frequency spectrum is the product of the two spectrums shown above. Hence,

$$|V_o(j\omega)| = \frac{1000}{\sqrt{\omega^2 + 1000^2}} \left(\left| .1 \frac{\sin \frac{\omega(.001)}{2}}{\frac{\omega(.001)}{2}} \right| \right) .$$

$V_o(j\omega)$ is plotted in Fig. 5.

Note that the circuit in Fig. 1 attenuates high frequency components of the input more than low frequency components; in fact, frequency components less than 1000 rad/s are not attenuated. The circuit in Fig. 1 is known as a low-pass filter.

• PROBLEM 14-29

The rectangular waveform of Fig. 1 is applied to an RC network as shown in Fig. 2. Find the steady-state value of the current i.

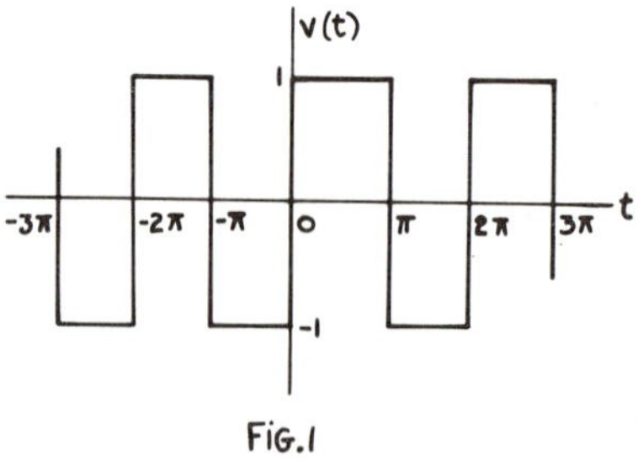

Fig.1

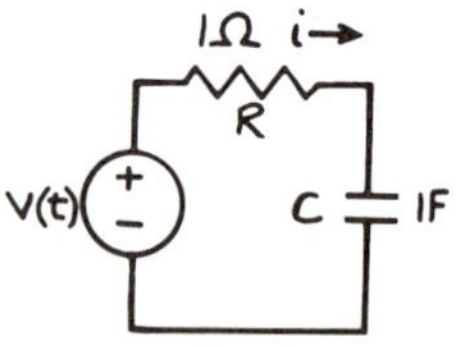

Fig. 2

Solution: Find the Fourier representation of the rectangular waveform of Fig. 1. Note that the d.c. value of the waveform, a_o, is zero, and that the waveform has odd-symmetry. Therefore, its Fourier series must be composed of sine terms only; hence $a_n = 0$. We find that

$$b_n = \frac{2}{\tau} \int_{-\frac{\tau}{2}}^{\frac{\tau}{2}} f(t) \sin \omega_o nt \, dt$$

where $\omega_o = 1 \quad \tau = 2\pi$, and

$$f(t) = \begin{cases} 1 & 0 < t < \pi \\ -1 & -\pi < t < 0 \end{cases} .$$

Thus,

$$b_n = \frac{1}{\pi} \int_{-\pi}^{\pi} f(t) \sin nt \, dt = \frac{1}{\pi} \int_{-\pi}^{0} -\sin nt \, dt + \frac{1}{\pi} \int_{0}^{\pi} \sin nt \, dt$$

$$b_n = \left[\frac{1}{n\pi} \cos nt\right]_{-\pi}^{0} + \left[-\frac{1}{n\pi} \cos nt\right]_{0}^{\pi}$$

$$b_n = \frac{1}{n\pi} [1-\cos(-n\pi)] + \frac{1}{n\pi} [1-\cos(n\pi)].$$

Because $\cos(\theta) = \cos(-\theta)$ we can combine terms to obtain:

$$b_n = \frac{2}{n\pi} [1-\cos(n\pi)]$$

$$b_n = \begin{cases} \frac{4}{n\pi} & n = 1,3,5\ldots \\ 0 & n = 2,4,6\ldots \end{cases}$$

Hence,

$$V(t) = \sum_{n=1,3,5\ldots}^{\infty} \frac{4}{n\pi} \sin nt \quad .$$

The impedance Z for a frequency ω is

$$Z = 1 + \frac{1}{j\omega} = \sqrt{1 + \frac{1}{\omega^2}} \underline{/\phi} \quad ,$$

where $\phi = \tan^{-1}(-\frac{1}{\omega})$.

If v(t) is sinusoidal the current would be simply

$$i = I_m \left[\frac{Ve^{j\omega t}}{Z}\right] = \frac{V}{\sqrt{1 + \frac{1}{\omega^2}}} \sin(\omega t - \phi).$$

However v(t) is the sum of sinusoids. Note that $V = \frac{4}{n\pi}$ and $\omega = n$. The current i_n, due to the n-th voltage term, is

$$i_n = \frac{4}{n\pi\sqrt{1 + \frac{1}{n^2}}} \sin(nt - \phi_n).$$

Where $\phi_n = \tan^{-1}(-\frac{1}{n})$,

$$i_n = \frac{4}{\pi\sqrt{1 + n^2}} \sin(nt - \phi_n).$$

The total steady-state current is

$$i = \sum_{n \text{ odd}} \frac{4}{\pi\sqrt{1 + n^2}} \sin(nt - \phi_n)$$

$$= \sum_{n \text{ odd}} I_n \sin(nt - \phi_n),$$

where $$I_n = \frac{4}{\pi\sqrt{1 + n^2}}$$

$$\phi_n = \tan^{-1}(-\frac{1}{n}).$$

• PROBLEM 14-30

> The half-wave rectified sinusoidal voltage of Fig. 1 is applied to the RL network in Fig. 2. Find the steady-state RMS value of the current.

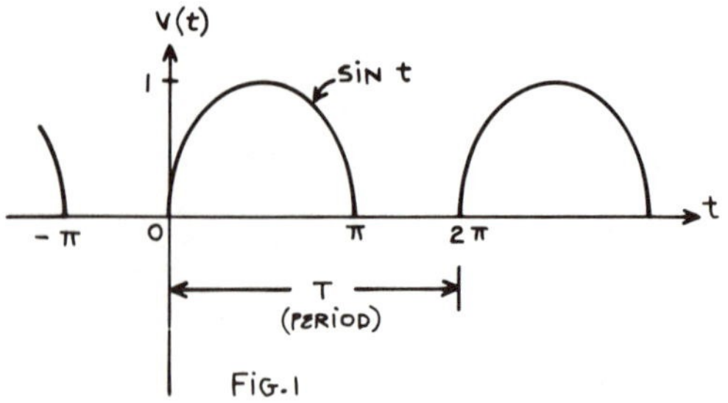

Fig. 1

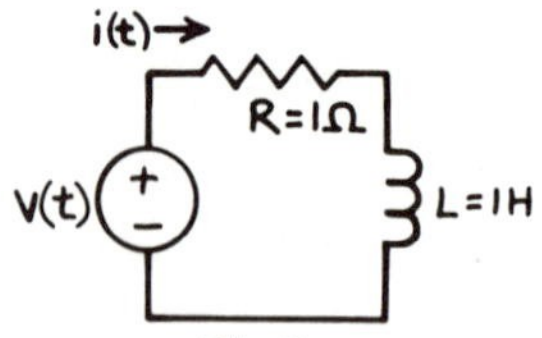

Fig. 2

Solution: First, we must find the Fourier series of the half-wave rectified sinusoid. Using the Fourier series in exponential form, we can express v(t) as

$$v(t) = \sum_{k=-\infty}^{\infty} c_k\, e^{jk\omega t} \qquad (1)$$

where

$$c_k = \frac{1}{\tau}\int_0^{\tau} v(t)e^{-jk\omega t}dt; \qquad (2)$$

$\tau = 2\pi$ from Fig. 1; and

$$v(t) = \begin{cases} \sin t & 0 \le t < \pi \\ 0 & \pi \le t < 2\pi \end{cases},$$

which represents the voltage being applied to the RL circuit over one period. Hence,

$$c_k = \frac{1}{2\pi}\int_0^{\pi}(\sin t)e^{-jk\omega t}dt + \frac{1}{2\pi}\int_{\pi}^{2\pi}(0)e^{-jk\omega t}$$

$$c_k = \frac{1}{2\pi}\int_0^{\pi}(\sin t)e^{-jk\omega t}dt$$

can be written

$$c_k = \frac{1}{2\pi}\int_0^{\pi}\left(\frac{e^{jt} - e^{-jt}}{2J}\right) e^{-jkt}dt$$

$$c_k = \frac{1}{2\pi}\int_0^{\pi}\left(\frac{e^{jt(1-k)}}{2j} - \frac{e^{-jt(1+k)}}{2j}\right) dt$$

$$c_k = \frac{1}{2\pi}\left[\frac{e^{jt(1-k)}}{2j(j-jk)} - \frac{e^{-jt(1+k)}}{2j(-j-jk)}\right]_0^{\pi}$$

$$c_k = \frac{1}{4\pi}\left[\frac{(1+k)(e^{j\pi(1-k)}-1) - (-1+k)(e^{-j\pi(1+k)}-1)}{-1+k^2}\right]$$

$$c_k = \frac{1}{4\pi(-1+k^2)}\left[e^{j\pi(1-k)} + e^{-j\pi(1+k)} + ke^{j\pi(1-k)} - ke^{-j\pi(1-k)} - 2\right]$$

$$c_k = \frac{1}{4\pi(-1+k^2)}\left[(e^{j\pi}+e^{-j\pi})e^{-jk\pi}+k(e^{j\pi}-e^{-j\pi})e^{-jk\pi}-2\right].$$

Multiply through by $\frac{2j}{2j}$ to obtain:

$$c_k = \frac{j}{2\pi(-1+k^2)}\left[\left(\frac{e^{j\pi}+e^{-j\pi}}{2}\right)\frac{e^{-jk\pi}}{j} + k\left(\frac{e^{j\pi}-e^{-j\pi}}{2j}\right)e^{-jk\pi} - \frac{1}{j}\right]$$

$$c_k = \frac{j}{2\pi(-1+k^2)}\left[(\cos\pi)\frac{e^{-jk\pi}}{j} + k(\sin\pi)e^{-jk\pi} - \frac{1}{j}\right].$$

$$c_k = \frac{j}{2\pi(-1+k^2)}\left[\frac{-e^{-jk\pi}}{j} - \frac{1}{j}\right] = \frac{-e^{-jk\pi}-1}{2\pi(-1+k^2)} = \frac{e^{-jk\pi}+1}{2\pi(1-k^2)} .$$

Hence,

$$v(t) = \sum_{k=-\infty}^{\infty} \frac{e^{-jk\pi}+1}{2\pi(1-k^2)} e^{jkt} .$$

When k = 0

$$v(t) = \frac{1}{\pi} \text{, called the D.C. value.}$$

When k = 1, by le Hopital's rule,

$$v(t) = \frac{-j\pi e^{-j\pi}}{-4\pi} e^{jt} = \frac{j}{4} e^{-j\pi}e^{jt} .$$

When k = -1,

$$v(t) = -\frac{j}{4} e^{j\pi}e^{-jt} .$$

Hence, the first harmonic is

$$v(t) = -\frac{j}{4} e^{j(\pi-t)} + \frac{j}{4} e^{-j(\pi-t)} ; \quad \text{for } k \pm 1$$

$$v(t) = \frac{1}{2}\left(\frac{e^{j(\pi-t)}-e^{-j(\pi-t)}}{2j}\right) = \frac{1}{2}\sin(\pi-t) ; \quad k \pm 1$$

$$= -\frac{1}{2}\sin(t-\pi)$$

$$= \frac{1}{2}\sin t \text{ V}$$

When k = 2,

$$v(t) = \frac{(e^{-j2\pi}+1)e^{j2t}}{-6\pi} .$$

When k = -2,

$$v(t) = \frac{(e^{j2\pi}+1)e^{-j2t}}{-6\pi} .$$

Hence, the second harmonic is

$$v(t) = \frac{1}{-6\pi}\left[(e^{-j2\pi}+1)e^{j2t} + (e^{j2\pi}+1)e^{-j2t}\right] \quad ; \quad k = \pm 2$$

$$v(t) = \frac{2}{-6\pi}\left[\frac{e^{j(2\pi-2t)}+e^{-j(2\pi-2t)}}{2} + \frac{e^{j2t}+e^{-j2t}}{2}\right]$$

$$v(t) = -\frac{1}{3\pi}\left[\cos(2\pi-2t) + \cos 2t\right]$$

$$v(t) = -\frac{2}{3\pi}\cos 2t.\ V$$

When $k = 3$,

$$v(t) = \frac{(e^{-j3\pi}+1)e^{j3t}}{-16\pi}\ .$$

When $k = -3$,

$$v(t) = \frac{(e^{j3\pi}+1)e^{-j3t}}{-16\pi}\ .$$

Hence, the third harmonic is

$$v(t) = \frac{(e^{=j3\pi}+1)e^{j3t}}{-16\pi} + \frac{(e^{j3\pi}+1)e^{-j3t}}{-16\pi} \quad ; \quad k \pm 3$$

$$v(t) = \frac{2}{-16\pi}\left[\frac{e^{j(3\pi-3t)}+e^{-j(3\pi-3t)}}{2} + \frac{e^{j3t}+e^{-j3t}}{2}\right]$$

$$v(t) = \frac{1}{-8\pi}\left[\cos(3\pi-3t) + \cos 3t\right]\ V.$$

Since $\cos\alpha = \cos -\alpha$

$$\cos(3\pi-3t) = \cos(3t-3\pi) = -\cos 3t,$$

then $v(t) = 0$ when $k = \pm 3$.

If we continue in this way we find that

$$v(t) = \frac{1}{\pi} + \frac{1}{2}\sin t + \frac{2}{\pi}\sum_{n=2,4,6...}\frac{1}{1-n^2}\cos nt.$$

Note that there are no odd harmonics (3t, 5t etc.).

The current in the circuit of Fig. 2 can be found from

$$i(t) = \frac{v(t)}{z(t)}\ .$$

Since we have expressed v(t) as a Fourier series, the current can be found by summing each sinusoidal term expressed as a phasor divided by the impedance $1+jn = \sqrt{1+n^2}\,\underline{/\tan^{-1}n}$ for each harmonic, n=2,4,6....

Hence,

$$i(t) = \frac{1}{\pi} + \frac{\frac{1}{2}\underline{/-90°}}{\sqrt{2}\,\underline{/45°}} + \frac{2}{\pi}\sum_{n=2,4,6...}\frac{\frac{1}{1-n^2}\,\underline{/0°}}{\sqrt{1+n^2}\,\underline{/\tan^{-1}n}}$$

$$i(t) = \frac{1}{\pi} + \frac{1}{2\sqrt{2}}\sin(t-135°)$$

$$+ \frac{2}{\pi}\sum_{n=2,4,6...}\frac{1}{1-n^2\sqrt{1+n^2}}\cos(nt - \tan^{-1}n).$$

We note that the current due to the D.C. term $\frac{1}{\pi}$ is the current through the 1-Ω resistor. Also, each sinusoidal term is displaced by -45°.

The Root Mean Square (RMS) of a periodic function is defined as

$$F_{rms} = \sqrt{\frac{1}{\tau}\int_0^{\tau}[f(t)]^2 dt}\ .$$

Using this definition we obtain

$$[I_{rms}]^2 = \frac{1}{2\pi}\int_0^{2\pi}\frac{1}{\pi^2}\,dt + \frac{1}{16\pi}\int_0^{2\pi}\sin^2(t-135°)dt$$

$$+ \frac{1}{2\pi}\int_0^{2\pi}\{\frac{2}{\pi}\sum_{n=2,4,6...}\frac{1}{(1-n^2)\sqrt{1+n^2}}\cos(nt - \tan^{-1}n)\}^2 dt,$$

where

$$\frac{1}{2\pi}\int_0^{2\pi}\frac{1}{\pi^2}\,dt = \frac{1}{2\pi}\left[\frac{t}{\pi^2}\right]_0^{2\pi} = \frac{1}{\pi^2}$$

and $\frac{1}{16\pi}\int_0^{2\pi}\sin^2(t-135°)dt$

$$= \frac{1}{16\pi}\int_0^{2\pi}\{\cos(-135°)\sin t - \sin(-135°)\cos t\}^2 dt$$

$$= \frac{1}{16\pi}\int_0^{2\pi}\{\tfrac{1}{2}\sin^2 t - \sin t\cos t + \tfrac{1}{2}\cos^2 t\}dt$$

$$= \frac{1}{16\pi}\int_0^{2\pi}\{\tfrac{1}{4}(1-\cos 2t) - \tfrac{1}{2}\sin 2t + \tfrac{1}{4}(1+\cos 2t)\}dt$$

$$= \frac{1}{16\pi}\left[\tfrac{1}{4}t - \tfrac{1}{8}\sin 2t + \tfrac{1}{4}\cos 2t + \tfrac{1}{4}t + \tfrac{1}{8}\sin 2t\right]_0^{2\pi}$$

$$= \frac{1}{16\pi}\left[\frac{\pi}{2} + \frac{1}{4} + \frac{\pi}{2} - \frac{1}{4}\right] = \frac{1}{16}\ .$$

Also, since

$$\{\sum_{n=2,4,6\ldots} k_n\cos(nt - \tan^{-1}n)\}^2 =$$

$$\{k_2\cos(2t-\tan^{-1}2) + k_4\cos(4t-\tan^{-1}4) + \ldots + k_n\cos(nt-\tan^{-1}n)\}^2$$

$$= \sum_{\substack{n\text{ even}\\ n\neq m\\ m,n>0}}\ \sum_{m\text{ even}} k_n k_m\cos(nt-\tan^{-1}n)\cos(mt-\tan^{-1}m)$$

$$+ \sum_{n\text{ even}} k_n^2\cos^2(nt-\tan^{-1}n),\ \text{where}$$

$$k_n = \frac{1}{1-n^2\sqrt{1+n^2}}\ ,$$

$$[I_{rms}]^2 = \frac{2}{\pi^3}\int_0^{2\pi}\sum_{\substack{n\text{ even}\\ n\neq m;\ n,m>0}}\ \sum_{m\text{ even}} k_n k_m\cos(nt-\tan^{-1}n)\cos(mt-\tan^{-1}m)dt$$

$$+ \frac{2}{\pi^3}\int_0^{2\pi}\sum_{\substack{n\text{ even}\\ n>0}} k_n^2\cos^2(nt-\tan^{-1}n)dt + \frac{1}{16} + \frac{1}{\pi^2}$$

By orthogonality:

$$\cos(n-\tan^{-1}n)\cos(mt-\tan^{-1}m) = 0;\ n \neq m$$

Hence,

$$[I_{rms}]^2 = \frac{1}{\pi^2} + \frac{1}{16} + \frac{2}{\pi^3}\int_0^{2\pi}\sum_{\substack{n\text{ even}\\ n>0}} k_n^2\cos^2(nt-\tan^{-1}n)dt$$

where

$$\frac{2}{\pi^3} \sum_{\substack{n \text{ even} \\ n>0}} k_n^{\,2} \int_0^{2\pi} \cos^2(nt-\tan^{-1}n)\,dt =$$

$$\frac{2}{\pi^3} \sum_{n \text{ even}} k_n^{\,2} \int_0^{2\pi} \{\cos(nt)\cos(\tan^{-1}n)+\sin(nt)\sin(\tan^{-1}n)\}^2 dt.$$

If $\tan^{-1}n = \alpha$, then $\cos\alpha = \frac{1}{\sqrt{1+n^2}}$ and $\sin\alpha = \frac{n}{\sqrt{1+n^2}}$.

Thus

$$\frac{2}{\pi^3} \sum_{\substack{n \text{ even} \\ n>0}} k_n^{\,2} \int_0^{2\pi} \{\frac{1}{\sqrt{1+n^2}} \cos nt + \frac{n}{\sqrt{1+n^2}} \sin nt\}^2 \, dt$$

$$= \frac{2}{\pi^3} \sum_{\substack{n \text{ even} \\ n>0}} k_n^{\,2} \int_0^{2\pi} \{\frac{1}{1+n^2} \cos^2(nt) + \frac{2n}{1+n^2} \cos(nt)\sin(nt) + \frac{n^2}{1+n^2} \sin^2(nt)\}dt$$

$$= \frac{2}{\pi^3} \sum_{\substack{n \text{ even} \\ n>0}} k_n^{\,2} \int_0^{2\pi} \frac{1}{2(1+n^2)} (1+\cos 2nt) + \frac{n}{1+n^2} \sin(2nt) + \frac{n^2}{2(1+n^2)} (1-\cos 2nt)\,dt$$

$$= \frac{2}{\pi^3} \sum_{\substack{n \text{ even} \\ n>0}} k_n^{\,2} \left[\frac{t}{2(1+n^2)} + \frac{1}{4n(1+n^2)} \sin 2nt - \frac{1}{2(1+n^2)} \cos(2nt) + \frac{n^2 t}{2(1+n^2)} - \frac{n}{4(1+n^2)} \sin(2nt)\right]_0^{2\pi}$$

$$= \frac{2}{\pi^3} \sum_{\substack{n \text{ even} \\ n>0}} k_n^{\,2} \left[\frac{\pi}{1+n^2} - \frac{1}{2(1+n^2)} + \frac{n^2\pi}{1+n^2} + \frac{1}{2(1+n^2)}\right]$$

$$= \frac{2}{\pi^3} \sum_{\substack{n \text{ even} \\ n>0}} k_n^{\,2} \left[\frac{\pi(1+n^2)}{(1+n^2)}\right]$$

$$= \frac{2}{\pi^2} \sum_{\substack{n \text{ even} \\ n>0}} k_n^2 .$$

Hence,

$$I_{rms} = \sqrt{\frac{1}{\pi^2} + \frac{1}{16} + \frac{2}{\pi^2} \sum_{\substack{n \text{ even} \\ n>0}} k_n^2}$$

$$I_{rms} = \sqrt{\frac{1}{\pi^2} + \frac{1}{16} + 2 \sum_{\substack{n \text{ even} \\ n>0}} \left[\frac{1}{\pi(1-n^2)\sqrt{1+n^2}}\right]^2} .$$

• PROBLEM 14-31

A low-pass filter is shown in Fig. 1. The input is given in Fig. 2. Find the value C such that the peak value of the largest a-c component of $v_o(t)$ is 1/25 of the d-c component.

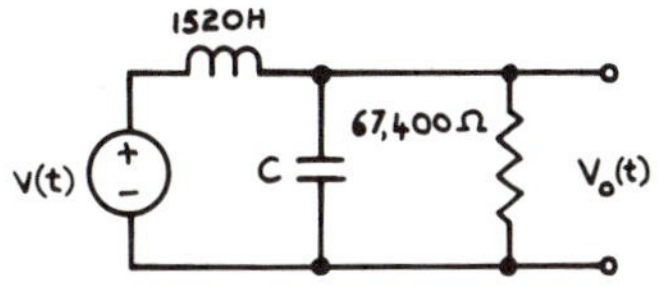

Fig.1

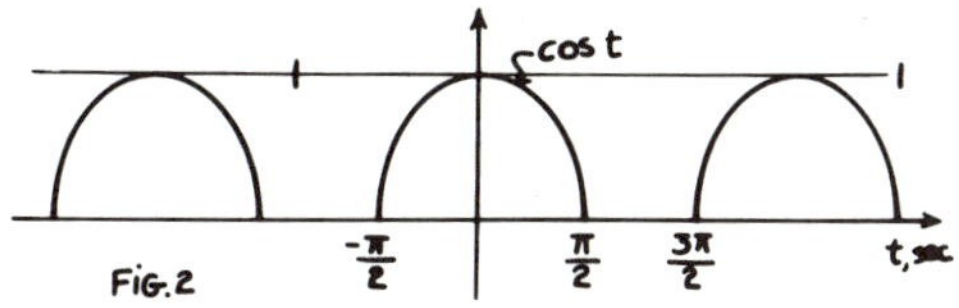

Fig.2

Solution: The first step in the solution is to determine the largest a-c component and the d-c component of the half-wave rectified cosine input voltage. Next, the portions of the output voltage due to each of these input components will be found, and then their ratio will be set equal to 1/25. The resulting equation will be solved for the capacitance, C.

The components of the input voltage will be found by Fourier analysis. Since the input is an even function, the Fourier series will consist of a d-c component and cosine terms only. Thus we have

$$v(t) = \frac{a_o}{\tau} + \frac{2}{\tau} \sum_{n=1}^{\infty} a_n \cos n\omega_o t \tag{1}$$

where

$$a_o = \int_{-\frac{\tau}{2}}^{\frac{\tau}{2}} v(t)\,dt \tag{2}$$

$$a_n = \int_{-\frac{\tau}{2}}^{\frac{\tau}{2}} v(t) \cos n\omega_o t \, dt \tag{3}$$

τ = period of v(t) which is 2π sec

$\omega_o = 2\pi/\tau = 1$ rad/sec.

Substituting into Eq.(2),

$$a_o = \int_{-\pi}^{-\frac{\pi}{2}} (0)dt + \int_{-\frac{\pi}{2}}^{\frac{\pi}{2}} \cos t\, dt + \int_{\frac{\pi}{2}}^{\pi} (0)dt$$

$$= \sin t \Big|_{-\frac{\pi}{2}}^{\frac{\pi}{2}} = 2.$$

Note that the only nonzero contribution of v(t) to the integral is in the region $-\pi/2 \le t \le \pi/2$. Substituting v(t) into Eq.(3) yields

$$a_n = \int_{-\frac{\pi}{2}}^{\frac{\pi}{2}} \cos t \cos n(1)t\, dt.$$

To integrate, apply the identity

$$\cos A \cos B = \frac{1}{2} [\cos (A+B) + \cos (A-B)].$$

Then,

$$a_n = \frac{1}{2} \int_{-\frac{\pi}{2}}^{\frac{\pi}{2}} [\cos(1+n)t + \cos(1-n)t]dt.$$

Integrating yields

$$a_n = \frac{1}{2} \left[\frac{\sin(1+n)t}{1+n} + \frac{\sin(1-n)t}{1-n}\right] \Bigg|_{-\frac{\pi}{2}}^{\frac{\pi}{2}} .$$

Evaluating the expression at the limits yields:

$$a_n = \frac{1}{2}\left[\frac{\sin(1+n)\frac{\pi}{2} + \sin(1+n)\frac{\pi}{2}}{1 + n} + \frac{\sin(1-n)\frac{\pi}{2} + \sin(1-n)\frac{\pi}{2}}{1 - n}\right] . \tag{4}$$

Applying the identity

$$\sin(A+B) = \sin A \cos B + \cos B \sin A$$

and collecting terms we obtain

$$a_n = \frac{1}{2}\left[\frac{2 \cos n\frac{\pi}{2}}{1+n} + \frac{2 \cos n\frac{\pi}{2}}{1-n}\right] .$$

Putting the terms over a common denominator gives

$$a_n = \frac{2 \cos n\frac{\pi}{2}}{1-n^2} . \qquad (5)$$

Determine several of the coefficients.

$$a_1 = \lim_{n\to1} \frac{2 \cos n\frac{\pi}{2}}{1-n^2}$$

The expression is of the indeterminate form 0/0 so L'Hospital's rule must be applied:

$$a_1 = \lim_{n\to1} \frac{\frac{d}{dn}\, 2 \cos n\frac{\pi}{2}}{\frac{d}{dn}\,(1-n^2)} = \lim_{n\to1} \frac{-2\frac{n\pi}{2} \sin n\frac{\pi}{2}}{-2n} = \frac{\pi}{2} .$$

This result could be in error since the second term in Eq. (4) is 0/0 for $n = 1$ and algebraic manipulations were performed on this equation. However, an independent evaluation of Eq.(3) with m initially set equal to one confirms the result that $a_1 = \pi/2$.

Continue the evaluation of coefficients using Eq.(5):

$$a_2 = \frac{2 \cos 2\frac{\pi}{2}}{1-2^2} = \frac{2}{3}$$

$$a_3 = \frac{2 \cos 3\frac{\pi}{2}}{1-2^3} = 0$$

$$a_4 = \frac{2 \cos 4\frac{\pi}{2}}{1-2^4} = - \frac{2}{15} .$$

As n increases, the denominator increases rapidly and the numerator cannot exceed 2, so no coefficient will be larger than $a_1 = \pi/2$. Since the given circuit is a low-pass filter, the $n = 1$ term will remain the largest a-c component in the output.

If we substitute the coefficient into Eq.(1), the components of interest are

d-c component: $\frac{a_o}{\tau} = \frac{2}{2\pi} = \frac{1}{\pi}$ volts

a-c component: $\frac{2}{\tau}\, a_1 \cos (1)(1)t$

$= \frac{2}{2\pi}\, \frac{\pi}{2} \cos t = \frac{1}{2}$ volts peak.

The next step is to find the contributions of each

of these components to the output voltage. First, consider the d-c output component.

The d-c equivalent circuit has the inductor short circuited and the capacitor open circuited. Hence the d-c output is the same as the d-c input.

$$(v_o)_{dc} = \frac{1}{\pi} \text{ volts}$$

Now examine the largest a-c output component.

The output can be determined from a voltage divider approach. Let $Z_{||}$ be the parallel impedance of the resistor and capacitor. Representing the largest a-c component by $(v_o)_{ac}$, we have

$$(v_o)_{ac} = v_{in} \frac{Z_{||}}{j\omega_o 1520 + Z_{||}} . \tag{6}$$

Calculating $Z_{||}$ yields:

$$Z_{||} = \frac{67400 \frac{1}{j\omega_o C}}{67400 + \frac{1}{j\omega_o C}} = \frac{67400}{1 + j\omega_o C67400} . \tag{7}$$

Using $\omega_o = 1$ and $v_{in} = \frac{1}{2}$, which correspond to the a_1 component, and substituting Eq. (7) into Eq. (6) we obtain

$$(v_o)_{ac} = \frac{1}{2} \frac{\frac{67400}{1+jC67400}}{j1520 + \frac{67400}{1 + jC67400}} .$$

Multiplying by $1 + jC67400$,

$$(v_o)_{ac} = \frac{33700}{67400 + j1520(1+jC67400)} . \tag{8}$$

Applying the constraint given in the problem we must have

$$|(v_o)_{ac}| = \frac{1}{25} (v_o)_{dc} . \tag{9}$$

Substituting Eq. (8), which is the peak a-c value, and the d-c output of $\frac{1}{\pi}$ into Eq. (9), we have

$$\left|\frac{33700}{67400 + j1520(1+jC67400)}\right| = \frac{1}{25} \left(\frac{1}{\pi}\right). \tag{10}$$

Rearranging yields

$$|67400 + j1520(1+jC67400)| = 25\pi(33700) = 2.647 \times 10^6$$

Collect real terms on the left hand side,

$$|67400 - (1520)(67400)C + j1520| = 2.647 \times 10^6.$$

Then determine the magnitude:

$$[(67400 - 1.024 \times 10^8 C)^2 + (1520)^2]^{\frac{1}{2}} = 2.647 \times 10^6.$$

Squaring both sides and rearranging yields

$$(67400 - 1.024 \times 10^8 C)^2 = 7.007 \times 10^{12}.$$

The 1520^2 term is negligible. Taking the square root gives

$$\pm (67400 - 1.024 \times 10^8 C) = \pm 2.647 \times 10^6.$$

The two possible positive values for C are

$$C = \frac{2.580 \times 10^6}{1.024 \times 10^8} = 0.0252 \text{ F}$$

and

$$C = \frac{2.714 \times 10^6}{1.024 \times 10^8} = 0.0265 \text{ F}.$$

The correct result can be determined by substituting each possible solution into Eq. (10) and checking for the result that equals $\frac{1}{25\pi} = 0.0127$. By doing this we find that the only solution is

$$C = 0.0265 \text{ F}.$$

• PROBLEM 14-32

The waveform shown in Fig. 1 is applied to the circuit in Fig. 2. Find: (a) rms value of the current I; (b) the current in each branch; (c) power dissipated in each branch; (d) total power dissipated; (e) the equation of the resultant current.

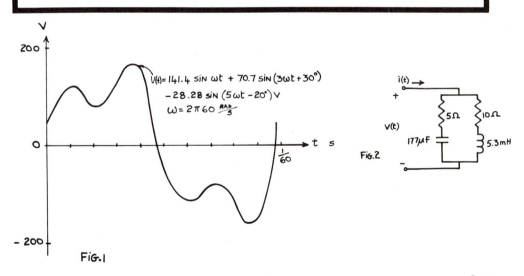

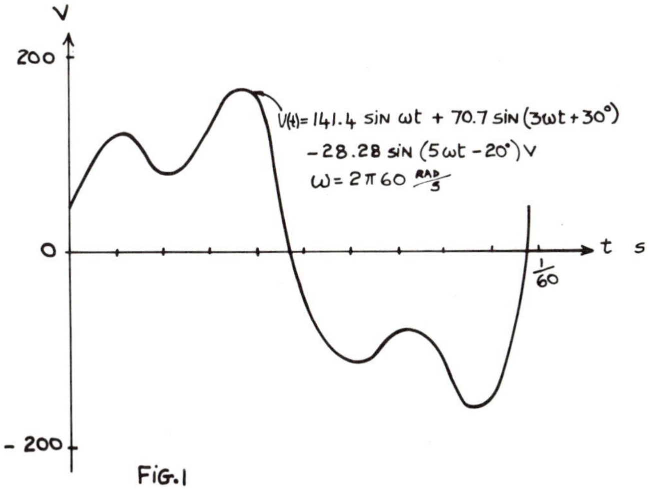

FIG. 1

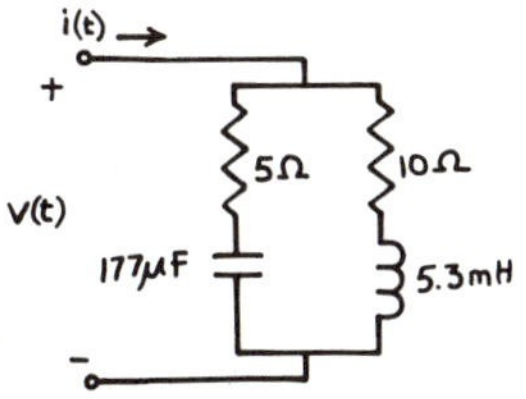

FIG. 2

Solution: Since the waveform of Fig. 1 is composed of three sinusoidal waves, the fundamental 141.4 sin ωt, the third harmonic 70.7 sin$(3\omega t+30°)$, and the fifth harmonic -28.28 sin$(5\omega t-20°)$, we can use superposition to find the current and power by adding the effect of each harmonic on the circuit.

Fig. 3 shows the fundamental wave driving the circuit and Fig. 4 shows the same circuit in phasor form.

With $V_1 = 100\ V_{rms}$, the currents are:

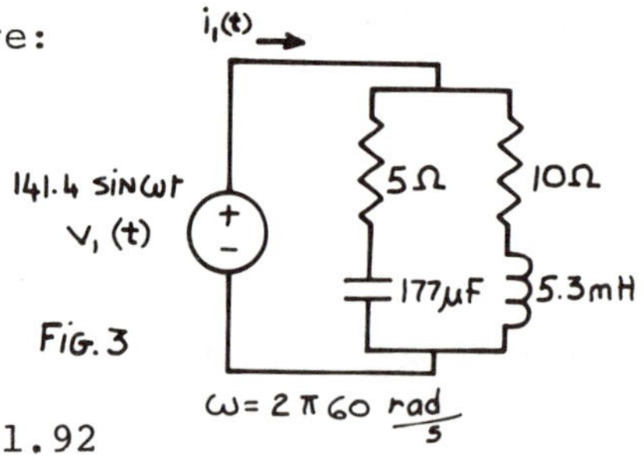

FiG.3

$$\vec{I}_{a1} = \frac{V_1}{\vec{Z}_{a1}} = \frac{100}{5-j15} = 2 + j6$$

$$|I_{a1}| = 6.32\ A$$

$$\vec{I}_{b1} = \frac{V_1}{\vec{Z}_{b1}} = \frac{100}{10+j2} = 9.62 - j1.92$$

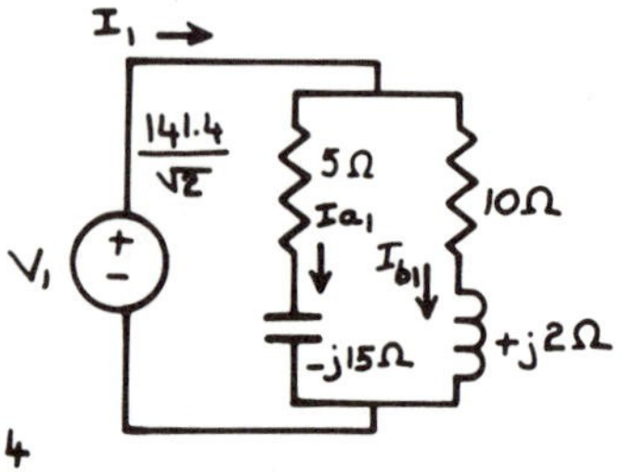

Fig.4

$$|I_{b1}| = 9.81\ A$$

$$\vec{I}_1 = \vec{I}_{a1} + \vec{I}_{b1} = 11.62 + j4.08$$

$$|I_1| = 12.32\ A$$

The powers are, by $P = ei + e'i'$ (where $V = e+je'$, $I = i+ji'$):

$$P_{a1} = 200 \qquad P_{b1} = 962 \qquad P_1 = 1162.$$

$$P_{b1} = P_{b1} + jQ_{b1} = V^*_{b1} = 100\underline{/-90°}(-1.92+j9.62) = 962 + j192$$

We make the same calculations for the third harmonic as we did for the Fundamental, except $\omega_{3rd\ har} = 3(2\pi 60)$. Take V_3 along the reference axis for the third harmonic.

The circuit used for the third harmonic is shown in Fig. 5.

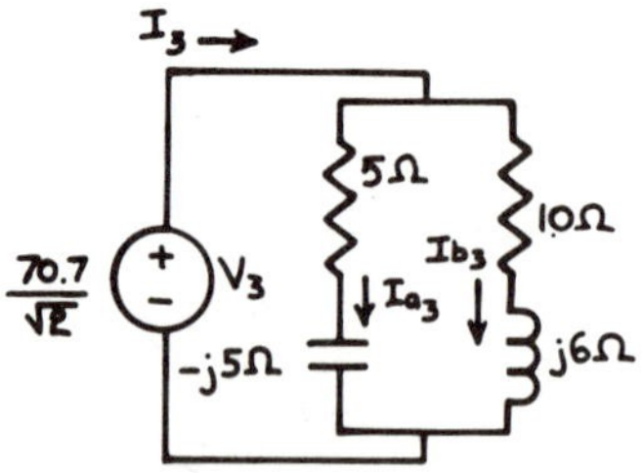

FiG.5

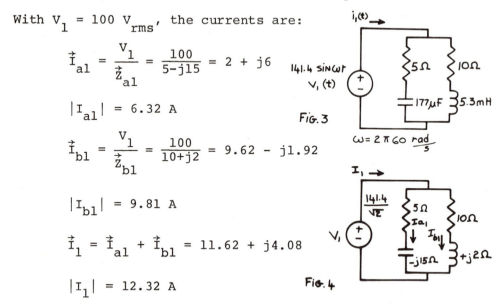

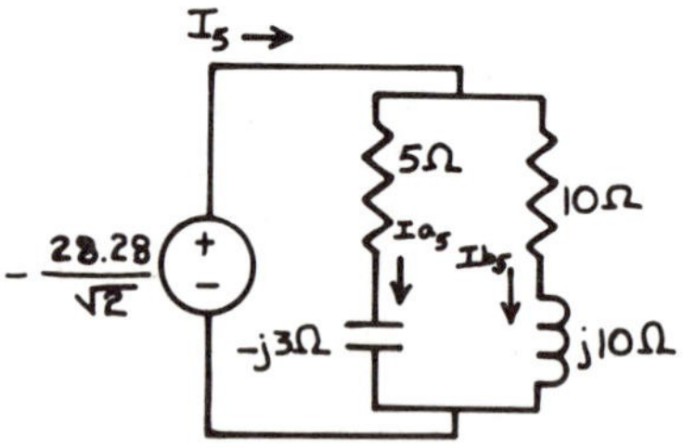

FiG.6

With $V_3 = 50\ V_{rms}$ the currents are:

$$I_{a3} = \frac{V_3}{Z_{a3}} = \frac{50}{5-j5} = 5 + j5$$

$$|I_{a3}| = 7.07 \text{ A}$$

$$I_{b3} = \frac{V_3}{Z_{b3}} = \frac{50}{10+j6} = 3.68 - j2.21$$

$$|I_{b3}| = 4.29 \text{ A}$$

$$I_3 = I_{a3} + I_{b3} = 8.68 + j2.79$$

$$|I_3| = 9.11 \text{ A}$$

The powers are

$$P_{a3} = 250, \quad P_{b3} = 184, \quad P_3 = 434.$$

The circuit used for the fifth harmonic is shown in Fig. 6. Take V_5 along the reference axis for the fifth harmonic. With $V_5 = 20\ V_{rms}$ the currents are

$$I_{a5} = \frac{20}{5-j3} = 2.97 + j1.76$$

$$|I_{a5}| = 3.43 \text{ A}$$

$$I_{b5} = \frac{20}{10+j10} = 1 - j1$$

$$|I_{b5}| = 1.41 \text{ A}$$

$$I_5 = I_{a5} + I_{b5} = 3.94 + j.76$$

$$|I_5| = 4.01 \text{ A}.$$

The powers are

$$P_{a5} = 58.8, \quad P_{b5} = 20, \quad P_5 = 78.8.$$

The total currents are

$$I_{aT} = \sqrt{I_{a1}^2 + I_{a3}^2 + I_{a5}^2} = \sqrt{6.32^2 + 7.07^2 + 3.43^2}$$

$$= 10.08$$

$$I_{bT} = \sqrt{9.81^2 + 4.29^2 + 1.41^2} = 10.80$$

$$I_T = \sqrt{12.32^2 + 9.11^2 + 4.01^2} = 15.84$$

The total powers are

$$P_{aT} = P_{a1} + P_{a3} + P_{a5} = 200 + 250 + 58.8 = 508.8$$

$$P_{bT} = 962 + 184 + 20 = 1166$$

$$P_T = 1162 + 434 + 78.8 = 1674.8.$$

To find the current equation we first find how much each current leads its respective voltage component:

$$\phi_1 = \tan^{-1} \frac{4.08}{11.62} = 19.35°$$

$$\phi_3 = \tan^{-1} \frac{2.79}{8.68} = 17.82°$$

$$\phi_5 = \tan^{-1} \frac{.76}{3.94} = 10.92°.$$

Therefore

$$i_1 = \sqrt{2}\ |I_1|\ \sin \omega t + \phi_1 = \sqrt{2}\ 12.32 \sin \omega t + 19.35°$$

$$i_3 = \sqrt{2}\ |I_3|\ \sin 3\omega t + 30° + \phi_3$$

$$= \sqrt{2}\,(9.11) \sin 3\omega t + 30° + 17.82°$$

$$i_5 = \sqrt{2}\ |I_5|\ \sin 5\omega t - 20° + \phi_5$$

$$= \sqrt{2}\ 4.01 \sin 5\omega t - 20° + 10.92°.$$

And $\quad i = i_1 + i_3 + i_5$

$$= 17.42 \sin \omega t + 19.35° + 12.88 \sin 3\omega t + 47.82°$$

$$+ 5.67 \sin 5\omega t - 9.08°.$$

CHAPTER 15

DISCRETE SYSTEMS AND Z-TRANSFORMS

DISCRETE ELEMENTS & EQUATIONS

• **PROBLEM** 15-1

Find the output w[n] of the discrete non-recursive system in fig. 1 to a step sequence input

$$x[n] = u[n]$$

shown in fig. 2, given $b_0 = b_1 = 1$.

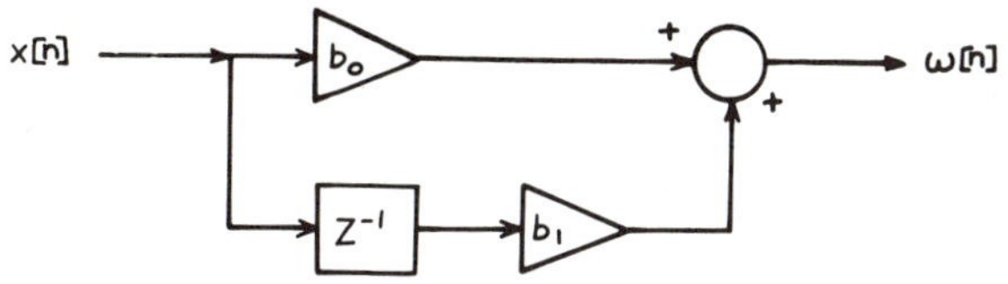

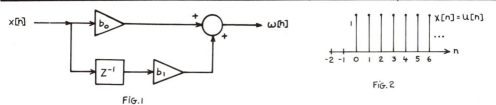

FIG. 1

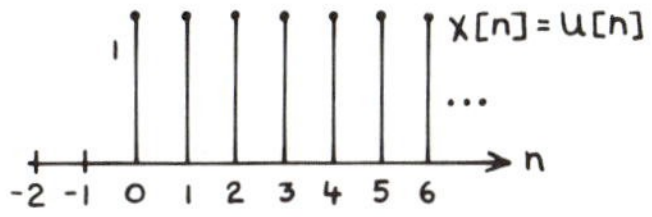

FIG. 2

Solution: Fig. 3 shows all the signals in the system in terms of the input.

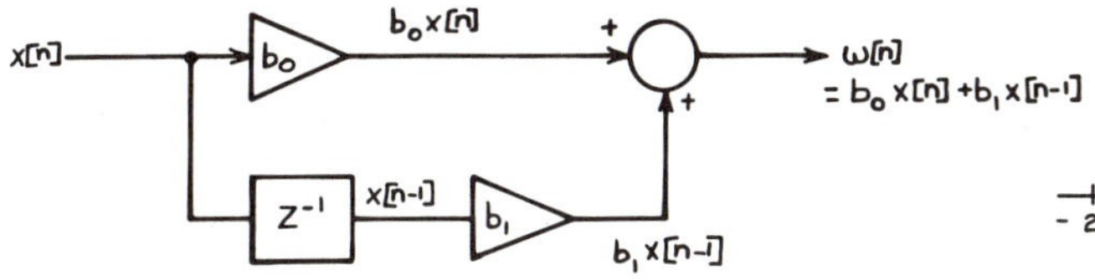

FIG. 3

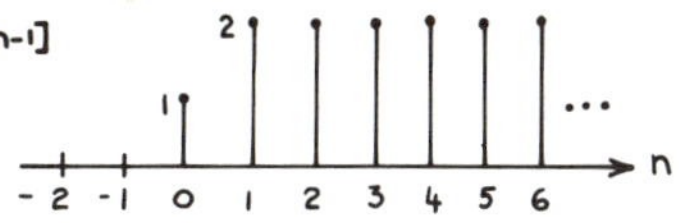

FIG. 4

The response, w[n], to a step sequence can be found by evaluating w[n] for each n.

For n = 0 x[0] = 1 and x[0-1] = 0. Since

$$w[0] = b_0 x[a] + b_1 x[0-1],$$

then $w[0] = b_0 = 1$.

For n = 1 x[1] = 1 and x[1-1] = x[0] = 1. Since $w[1] = b_0 x[1] + b_1 x[0]$ then $w[1] = b_0 + b_1 = 2$.

Noting that w[n] = 2 for all n ≥ 1 and w[0] = 1 we can plot the output w[n] shown in fig. 4.

This sequence can be written as

$2u[n] - \delta[n]$

where $2u[n]$ is a step sequence defined as

$$2u[n] \begin{cases} 2 & n \geq 0 \\ 0 & n < 0 \end{cases}$$

and $\delta[n]$ is the delta sequence defined as $\delta[n] = \begin{cases} 1 & n = 0 \\ 0 & n \neq 0. \end{cases}$

The delta sequence should not be confused with the time dependent delta function $\delta(t)$.

• PROBLEM 15-2

The block diagram of the discrete non-recursive system shown in fig. 1 consists of a discrete source, $x[n]$; two multipliers; and one delay element. Express the signals at points i, j, and k in terms of the input $x[n]$. Also express the output $w[n]$ in terms of the input $x[n]$.

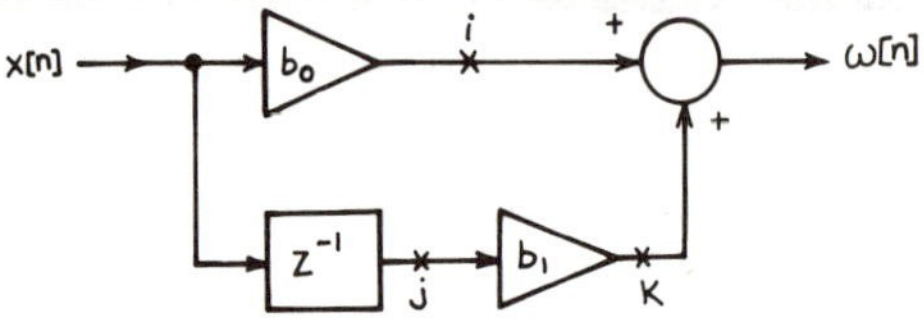

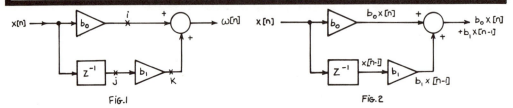

Fig.1

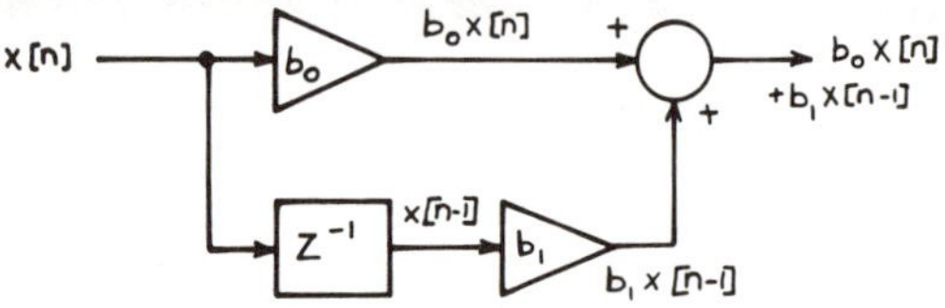

Fig.2

Solution: Starting at the input, $x[n]$ is multiplied by b_0 and is presented to the summer as $b_0x[n]$ in the top branch. In the bottom branch, $x[n]$ is delayed by one time unit, multiplied by b_1, and then presented to the summer as $b_1x[n-1]$. Hence, $w[n]$ can be written as $b_0x[n] + b_1x[n-1]$. Fig. 2 shows all the signals present in the system in terms of $x[n]$.

• PROBLEM 15-3

Find the output $y[n]$ of the discrete recursive system in fig. 1 given the input function $w[n]= 2u[n] -\delta[n]$, the multiplication factor $a_1 = -2$, and the initial condition $y[-1] = 0$.

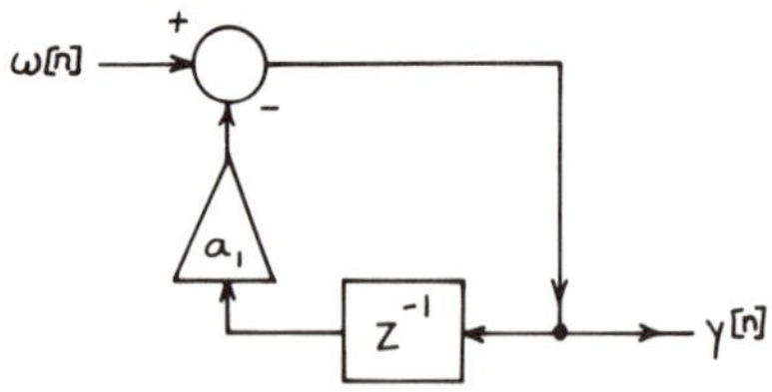

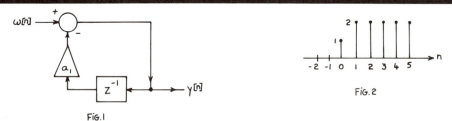

Fig.1

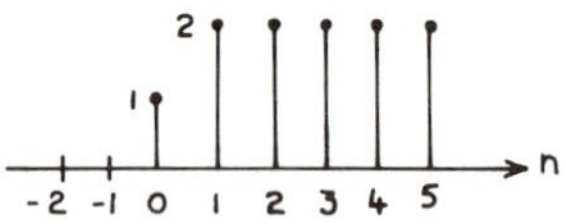

Fig.2

Solution: Starting at the output, $y[n]$ is the input to the delay element; hence, the output of the delay element must be $y[n-1]$.

After amplification by $a_1 = -2$, $-2y[n-1]$ is subtracted from w[n]. The recursion equation for the system is

$$y[n] = w[n]+ 2y[n-1],$$

where $w[n] = 2u[n] -\delta[n]$ and $y[-1] = 0$. The input, w[n], is shown in fig. 2 below.

The output can be found recursively by substituting the values of w[n] and the initial condition y[-1] for n = 0,1,2, 3, ...

For n = 0,

$$y[0] = w[0]+ 2y[-1] = 1.$$

For n = 1,

$$y[1]= 2 + 2 = 4.$$

For n = 2

$$y[2]= 2 + 8 = 12.$$

Hence,

$$y[n] = 1,4,12, \ldots .$$

• PROBLEM 15-4

Write a recursion equation for the discrete system shown in fig. 1.

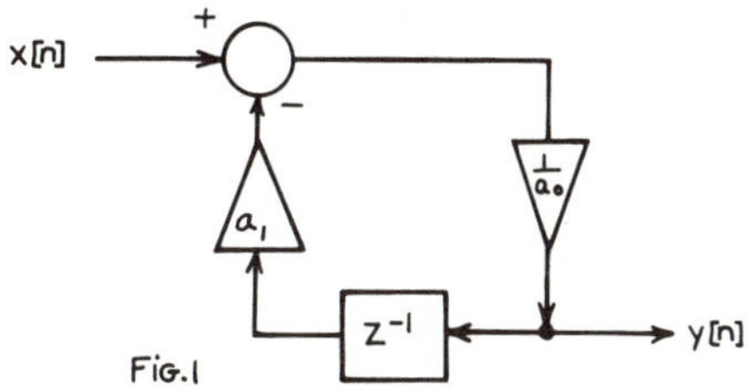

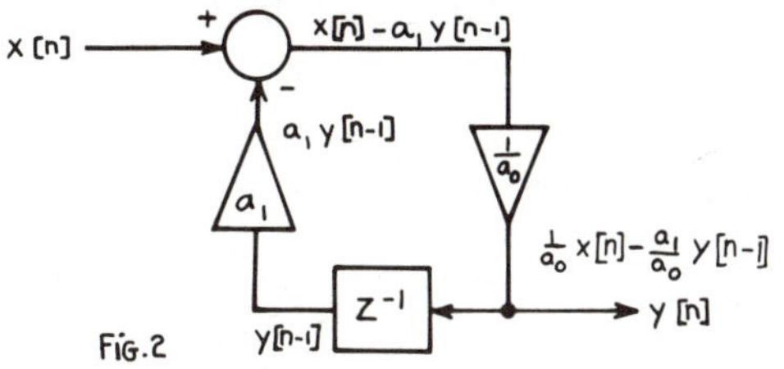

Solution: Starting at the output, note that the input to the delay element is y[n]. Hence, the output of the delay unit must be y[n-1]. y[n-1] is then multiplied by a_1 and subtracted from x[n] at the input. The signal present at the input to the second multiplier $\left(\frac{1}{a_0}\right)$ is $x[n] - a_1 y[n-1]$.

After the multiplier we have the output

$$y[n] = \frac{1}{a_0}x[n] - \frac{a_1}{a_0}y[n-1]$$

which can be written

$$x[n] = a_0 y[n] + a_1 y[n-1].$$

Fig. 2 shows all the signals in the system.

Two discrete systems shown in fig. 1 and 2 are cascaded (i.e., connected in series). Write a single recursion equation for the resulting system and solve recursively for y[n] using $b_0 = b_1 = 1$, $a_1 = 2$, x[n] = u[n] and y[-1] = 0.

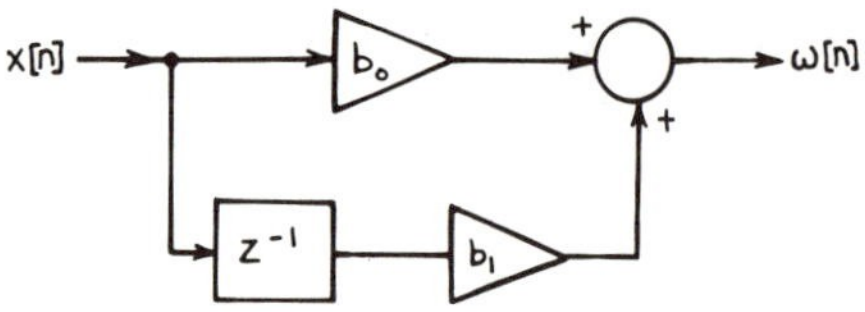

FIG.1

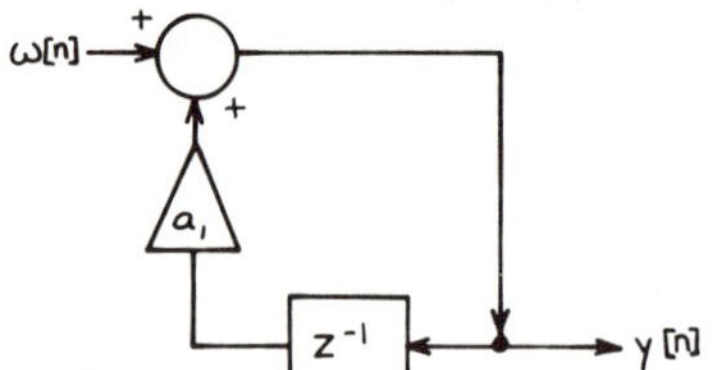

FIG.2

Solution: Connecting the two systems in cascade as shown in fig. 3 and substituting $b_0 = b_1 = 1$, and $a_1 = 2$ we can write a single recursion equation for the entire simplified system shown in fig. 4.

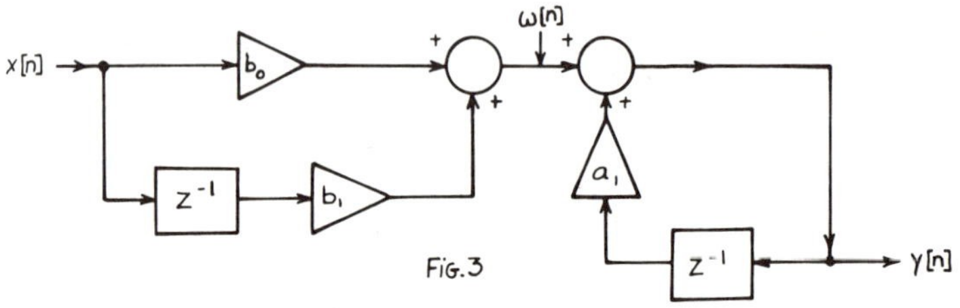

FIG.3

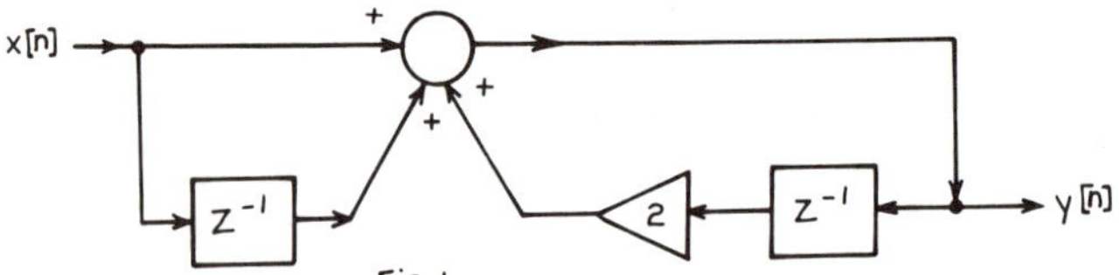

FIG.4

Starting at the output, note that the input to the delay element on the right is y[n]; hence, the output of that delay unit must be y[n-1]. The output of the multiplier is 2y[n-1]. The output of the delay element on the left is x[n-1]. Hence, the signals present at the summer are x[n], x[n-1], and 2y[n-1] producing the output

$$y[n]= x[n]+ x[n-1] + 2y[n-1], \qquad (1)$$

which can be written

$$y[n]- 2y[n-1]= x[n]+ x[n-1]. \qquad (2)$$

Eq. (1) can be solved recursively for y[n]by setting n = 0, 1,2, ...

We are given x[n] = u[n] and y[-1]= 0. Hence, for n = 0

$$y[0] = x[0] + x[-1] + 2y[-1].$$

Since x[0] = 1, x[-1] = 0, and y[-1] = 0 then y[0] = 1. For n = 1,

$$y[1] = x[1] + x[0] + 2y[0].$$

Since we know that $x[n] = 1$ for all $n \geq 0$, and we previously found the value of $y[0]$, then $y[1] = 1 + 1 + 2 = 4$.

For $n = 2$,

$$y[2] = 2 + 2y[1] = 10.$$

For $n = 3$

$$y[3] = 2 + 2y[2] = 22.$$

For $n = 4$

$$y[4] = 2 + 2y[3] = 46.$$

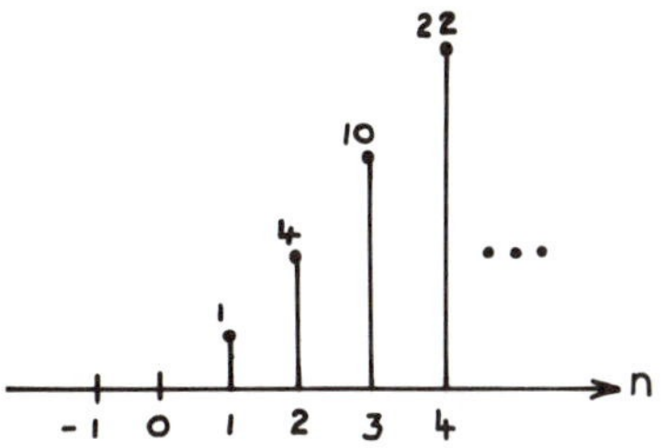

FIG. 5

The sequence 1, 4, 10, 22, 46 ... can be written $-2 + 3 \cdot 2^n$ for $n \geq 0$. The solution is

$$y[n] = -2u[n] + 3 \cdot 2^n u[n],$$

shown in fig. 5.

• PROBLEM 15-6

Show that the recursion equation for the discrete system shown in fig. 1 is

$$y[n] + a_1 y[n-1] = b_0 x[n] + b_1 x[n-1] \quad \text{for } n \geq 1.$$

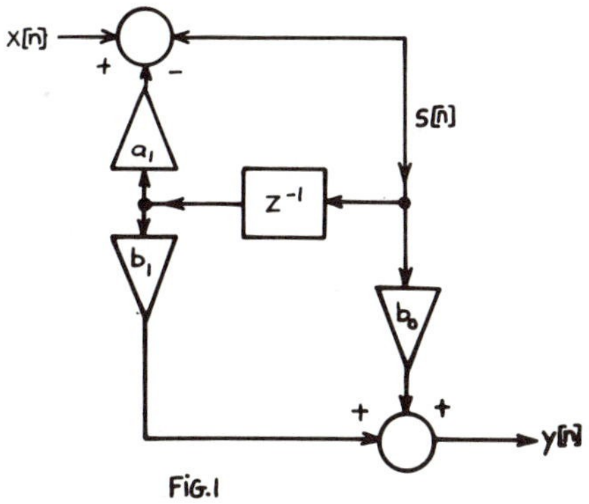

FIG. 1

Solution: We note that the output of the delay element must be $s[n-1]$. $s[n-1]$ is routed to the two amplifiers a_1 and b_1 and then to the two summers which yield

$$x[n] - a_1 s[n-1] = s[n] \; ; \quad n \geq 0 \tag{1}$$

$$b_0 s[n] + b_1 s[n-1] = y[n] \; ; \; n \geq 0. \tag{2}$$

Write Eq. (1) as

$$x[n] = s[n] + a_1 s[n-1] \quad ; \quad n \geq 0. \tag{3}$$

Changing n to n-1 in Eqs. (2) and (3) yields

$$y[n-1] = b_0 s[n-1] + b_1 s[n-2] \; ; \quad n-1 \geq 0 \tag{4}$$

$$x[n-1] = s[n-1] + a_1 s[n-2] \quad ; \quad n-1 \geq 0. \qquad (5)$$

Multiplying Eq.(4) by a_1 and adding the result to Eq. (2) yields:

$$y[n]+ a_1 y[n-1] = b_0 s[n]+ b_1 s[n-1] + a_1 b_0 s[n-1] + a_1 b_1 s[n-2]$$

$$= b_0\{s[n] + a_1 s[n-1]\} + b_1\{s[n-1] + a_1 s[n-2]\}. \quad (6)$$

Substituting Eqs. (3) and (5) into equation (6) gives the desired result:

$$y[n]+ a_1 y[n-1] = b_0 x[n]+ b_1 x[n-1] \quad ; \; n-1 \geq 0$$

$$; \; n \geq 1.$$

• **PROBLEM** 15-7

Write a recursion equation for the discrete system shown in fig. 1. Solve for y[n] recursively for $a_1 = -\frac{5}{6}$, $a_2 = \frac{1}{6}$, $x[n] = 2u[n]$, $y[-1] = 1$, and $y[-2] = -7$.

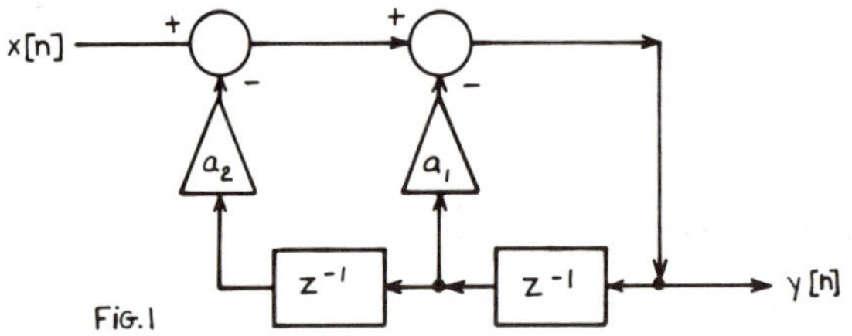

FIG.1

Solution: Starting at the output, notice that y[n] passes through two delay elements. Fig. 2 shows the signals present in the system.

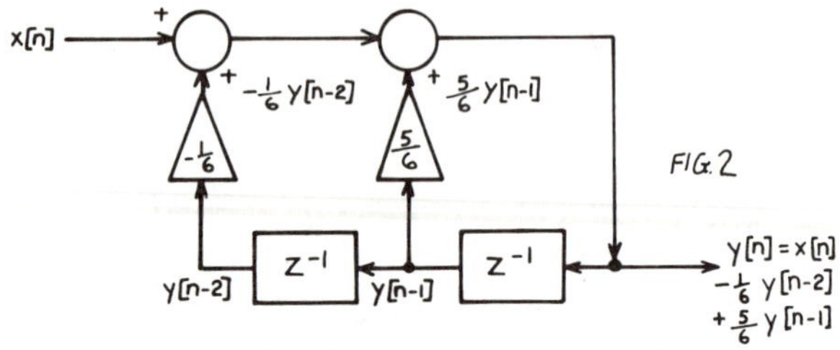

FIG.2

The recursion equation

$$x[n] = y[n]- \frac{5}{6}y[n-1] +\frac{1}{6}y[n-2]$$

can be solved recursively for y[n] given the input $x[n] = 2u[n]$ and the initial conditions $y[-1] = 1$, and $y[-2] = -7$.

For n = 0

$$y[0] = x[0]- \frac{1}{6}y[-2] + \frac{5}{6}y[-1]$$

$$y[0] = 2 - \frac{1}{6}(-7) + \frac{5}{6} =4.$$

For n = 1

$$y[1] = 2 - \frac{1}{6} + \frac{5}{6}(4) = \frac{31}{6}.$$

For n = 2

$$y[2] = 2 - \frac{1}{6}(4) + \frac{5}{6}\cdot\frac{31}{6} = \frac{203}{36}.$$

Hence,

$$y[n] = 4,\ \frac{31}{6},\ \frac{203}{36}\ \ldots \qquad ; n \geq 0.$$

• **PROBLEM** 15-8

Find a recursion equation satisfying y[n] where y[n] is the sum of the squares of all integers from 0 to n.

Solution: The sequence y[n], where y[n] is the sum of the squares of all integers from 0 to n, can be denoted:

$$y[n] = \sum_{k=0}^{n} k^2 = 0 + 1^2 + 2^2 + \ldots + (n-1)^2 + n^2.$$

We can also write

$$y[n-1] = \sum_{k=0}^{n-1} k^2 = 0 + 1^2 + 2^2 + \ldots + (n-1)^2.$$

Hence, y[n] satisfies the recursion equation

$$y[n-1] = y[n] - n^2 \quad ; \quad n \geq 1$$

where y[0] = 0.

• **PROBLEM** 15-9

Find recursively the sequence y[n] such that y[-1] = 4 and $y[n] - \frac{1}{2}y[n-1] = 6$ for $n \geq 0$.

Solution: The first order recursion equation

$$y[n] - \frac{1}{2}y[n-1] = 6$$

can be solved for the sequence y[n] by substituting n=0,1,2, 3, ...

For n = 0,

$y[0] = \frac{1}{2}y[-1] + 6.$ Since y[-1] = 4, then

$$y[0] = \frac{1}{2}(4) + 6 = 8.$$

For n = 1

$$y[1] = \frac{1}{2}y[0] + 6. \quad \text{Since } y[0] = 8, \text{ then}$$

$$y[1] = \frac{1}{2}(8) + 6 = 4 + 6 = 10.$$

Also,

$$y[2]= \frac{1}{2}y[1] + 6 = 5 + 6 = 11.$$

Hence,

$$y[n] = 8, 10, 11 \ldots . \qquad ; n \geq 0.$$

• PROBLEM 15-10

Find recursively the sequence y[n] such that y[1] = 3, y[0] = 2, and y[n] -3y[n-1] + 2y[n-2] = 0; n ≥ 2.

Solution: Setting n = 2, 3, 4, ... in the above equation we obtain:

$$y[2] = 3y[1]- 2y[0] = (3\cdot3) - (2\cdot2) = 5$$

$$y[3] = 3y[2]- 2y[1] = (3\cdot5) - (2\cdot3) = 9$$

$$y[4] = 3y[3]- 2y[2] = (3\cdot9) - (2\cdot5) = 17.$$

Hence, y[n] = 5, 9, 17, ... ; n ≥ 2.

• PROBLEM 15-11

Find a mathematical expression for the sequence y[n], such that y[-1] = 6 and $y[n] - \frac{1}{2}y[n-1] = 0$ for n ≥ 0.

Solution: The first order homogeneous recursion equation,

$$y[n]- \frac{1}{2}y[n-1] = 0$$

can be solved in a manner similar to that used in solving ordinary differential equations.

It can be shown that if z is properly chosen, then $y[n] = z^n$ is a solution to all homogeneous recursion equations in the form,

$$y[n] + a_1y[n-1] + \ldots + a_my[n-m] = 0. \qquad (1)$$

Since, $z^n + a_1z^{n-1} + \ldots + a_mz^{n-m} = z^n(1 + a_1z^{-1} + \ldots + a_mz^{-m}) = 0$, then the solution of the characteristic equation

$$1 + a_1z^{-1} + \ldots + a_mz^{-m} = 0 \qquad (2)$$

provides the roots $z_1, z_2, \ldots, z_m$.

Therefore, the m sequences

$z_1^n, z_2^n, \ldots, z_m^n$ are solutions to Eq. (1). By superposition the sum

$$y[n] = c_1 z_1^n + c_2 z_2^n + \ldots + c_m z_m^n \qquad (3)$$

is also a solution to Eq. (1) for arbitrary constants $c_1, \ldots, c_m$.

The characteristic equation for the first order equation

$$y[n] - \frac{1}{2} y[n-1] = 0$$

is $\quad 1 - \frac{1}{2} z^{-1} = 0.$

Hence
$$1 = \frac{1}{2} z^{-1},$$

thus $\quad z = \frac{1}{2}.$

The solution is in the form of equation (3). Therefore, $y[n] = cz^n = c\left(\frac{1}{2}\right)^n.$

The constant c can be found by applying the initial condition $y[-1] = 6$.

Hence, $y[-1] = c\left(\frac{1}{2}\right)^{-1} = 6$

$$y[-1] = 2c = 6$$

$$c = 3.$$

The solution is

$$y[n] = 3\left(\frac{1}{2}\right)^n.$$

• PROBLEM 15-12

Find a mathematical expression for the sequence y[n] such $y[n] - 7y[n-1] + 10y[n-2] = 0$, $y[-1] = 16$, $y[-2] = 5$.

Solution: In this case, the characteristic equation for this second order homogeneous recursion equation is

$$1 - 7z^{-1} + 10z^{-2} = 0.$$

We can solve for z by use of the quadratic formula,

$$z_1, z_2 = \frac{-b \pm \sqrt{b^2 - 4ac}}{2a}$$

where $a = 1$, $b = -7$, and $c = 10$.

Hence,

$$z_1, z_2 = \frac{7 \pm \sqrt{49 - 40}}{2} = \frac{7}{2} \pm \frac{3}{2}$$

$$z_1 = 2$$

$$z_2 = 5.$$

The solution must be in the form

$$y[n] = c_1 z_1^n + c_2 z_2^n = c_1 2^n + c_2 5^n. \qquad (1)$$

By applying the initial conditions $y[-1] = 16$ and $y[-2] = 5$ we can find the value of the constants c_1, and c_2.

For $n = -1$, the solution [Eq(1)] becomes

$$y[-1] = \frac{c_1}{2} + \frac{c_2}{5} = 16, \qquad (2)$$

and for $n = -2$

$$y[-2] = \frac{c_1}{4} + \frac{c_2}{25} = 5. \qquad (3)$$

Hence, using determinants,

$$c_1 = \frac{\begin{vmatrix} 16 & \frac{1}{5} \\ 5 & \frac{1}{25} \end{vmatrix}}{\begin{vmatrix} \frac{1}{2} & \frac{1}{5} \\ \frac{1}{4} & \frac{1}{25} \end{vmatrix}} = \frac{\frac{16}{25} - 1}{\frac{1}{50} - \frac{1}{20}} = \frac{-\frac{9}{25}}{-\frac{3}{100}} = 12$$

and $$c_2 = \frac{\begin{vmatrix} \frac{1}{2} & 16 \\ \frac{1}{4} & 5 \end{vmatrix}}{\frac{-3}{100}} = \frac{\frac{5}{2} - 4}{\frac{-3}{100}} = \frac{-\frac{3}{2}}{-\frac{3}{100}} = 50.$$

After substituting the values of c_1 and c_2 into Eq.(1) the solution is found to be:

$$y[n] = 12(2)^n + 50(5)^n.$$

• PROBLEM 15-13

Find a sequence $y[n]$ such that $y[n] - 6y[n-1] + 9y[n-2] = 0$, $y[-1] = 1$, $y[-2] = 0$.

Solution: The characteristic equation obtained from the second order recursion equation is

$$1 - 6z^{-1} + 9z^{-2} = 0.$$

Solving for z we find that the two roots are equal,

$z_1 = z_2 = 3.$

Hence, the solution $y[n] = c_1 z_1^n + c_2 z_2^n$ no longer holds true.

It can be shown that

$$y[n] = nz_1^n$$

is a solution to the equation,

$$y[n] + by[n-1] + cy[n-2] \qquad (1)$$

where $b^2 = 4c$ and $z_1 = z_2 = -\frac{b}{2}$.

Substituting $y[n] = nz_1^n$ into Eq(1) gives

$$nz_1^n + b(n-1)z_1^{n-1} + c(n-2)z_1^{n-2} = 0,$$

which can be written

$$(1 + bz_1^{-1} + cz_1^{-2})nz_1^n - (bz_1^{-1} + 2cz_1^{-2})z_1^n = 0.$$

Since $1 + bz_1^{-1} + cz_1^{-2} = 0$ has the solutions $z_1 = z_2 = -\frac{b}{2}$ we can substitute into $bz_1^{-1} + 2cz_1^{-2} = 0$

to obtain

$$-2 + \frac{c8}{b^2} = 0.$$

Therefore $b^2 = 4c$. This proves that $y[n] = nz_1^n$ is indeed a solution, as is $y[n] = z_1^n$. Hence, z_1^n and nz_1^n are linearly independent and the general solution is:

$$y[n] = c_1 z_1^n + c_2 n z_1^n.$$

In our problem, since $z_1 = 3$ we have

$$y[n] = c_1 3^n + c_2 n 3^n$$

where

$$y[-1] = \frac{c_1}{3} - \frac{c_2}{3} = 1$$

and $$y[-2] = \frac{c_1}{3^2} - 2\frac{c_2}{3^2} = 0$$

can be solved simultaneously for the constants c_1 and c_2. Hence,

$$c_1 = \frac{\begin{vmatrix} 1 & -\frac{1}{3} \\ 0 & -\frac{2}{9} \end{vmatrix}}{\begin{vmatrix} \frac{1}{3} & -\frac{1}{3} \\ \frac{1}{9} & -\frac{2}{9} \end{vmatrix}} = \frac{-\frac{2}{9}}{-\frac{2}{27} + \frac{1}{27}} = 6$$

$$\text{and} \quad c_2 = \frac{\begin{vmatrix} \frac{1}{3} & 1 \\ \frac{1}{9} & 0 \end{vmatrix}}{-\frac{1}{27}} = \frac{-\frac{1}{9}}{-\frac{1}{27}} = 3.$$

The solution is

$$y[n] = (6 + 3n)3^n \quad ; \quad n \geq 0.$$

STEADY STATE & HOMOGENEOUS SOLUTIONS

• PROBLEM 15-14

> Use the sum of the steady-state solution and the homogeneous solution to find a sequence y[n] such that $y[n] - \frac{1}{2}y[n-1] = 3$ and $y[-1] = 8$.

Solution: The solution to a first order homogeneous recursion equation is in the form ca^n where the characteristic equation is $1 - az^{-1} = 0$.

Hence, for

$$y[n] - \frac{1}{2}y[n-1] = 0$$

the characteristic equation is

$$1 - \frac{1}{2}z^{-1} = 0$$

and the solution is

$$y_h[n] = c\left(\frac{1}{2}\right)^n .$$

The steady-state solution to a first order recursion equation, which has the form

$$y[n] - a_1 y[n-1] = E$$

where E is a constant, is

$$y_{ss}[n] = EH(1) = \frac{E}{1 - a_1} .$$

H(z) is defined as the inverse of the characteristic equation N(z); hence,

$$H(z) = \frac{1}{1 - az^{-1}} \text{ and } H(1) = \frac{1}{1 - a} .$$

For the recursion equation

$$y[n] - \frac{1}{2}y[n-1] = 3,$$

the steady-state solution is

$$y_{ss}[n] = \frac{3}{1 - \frac{1}{2}} = 6.$$

Hence, the solution is the sum of the homogeneous and steady-state solutions:

$$y[n] = y_{ss}[n] + y_h[n] = 6 + c\left(\frac{1}{2}\right)^n.$$

To find the constant c, we note that

$$y[-1] = 8 \text{ and}$$

$$y[-1] = 6 + 2c = 8$$

$$2c = 2$$

$$c = 1.$$

The solution is

$$y[n] = 6 + \left(\frac{1}{2}\right)^n.$$

• PROBLEM 15-15

Use the sum of the steady-state and homogeneous solutions to find a sequence y[n] such that $y[n] - \frac{5}{6}y[n-1] + \frac{1}{6}y[n-2] = 2$, $y[-1] = 1$, and $y[-2] = -7$.

Solution: The characteristic equation,

$$1 - \frac{5}{6}z^{-1} + \frac{1}{6}z^{-2} = 0.$$

is solved for z_1 and z_2 as follows

$$z_1, z_2 = \frac{\frac{5}{6} \pm \sqrt{\frac{25}{36} - \frac{4}{6}}}{2} = \frac{\frac{5}{6} \pm \frac{1}{6}}{2}$$

$$z_1 = \frac{1}{2}$$

$$z_2 = \frac{1}{3}$$

giving the homogeneous solution

$$y_h[n] = c_1\left(\frac{1}{2}\right)^n + c_2\left(\frac{1}{3}\right)^n.$$

The steady-state solution is

$$y_{ss}[n] = EH(1) = \frac{2}{1 - \frac{5}{6} + \frac{1}{6}} = 6.$$

Hence,

$$y[n] = 6 + c_1\left(\frac{1}{2}\right)^n + c_2\left(\frac{1}{3}\right)^n.$$

c_1 and c_2 are solved from the initial conditions:

$$y[-1] = 1 = 6 + 2c_1 + 3c_2$$

$$y[-2] = -7 = 6 + 4c_1 + 9c_2$$

$$c_1 = c_2 = -1.$$

The solution is

$$y[n] = 6 - \left(\frac{1}{2}\right)^n - \left(\frac{1}{3}\right)^n.$$

• PROBLEM 15-16

Find y[n] such that $y[n] - \frac{5}{6}y[n-1] + \frac{1}{6}y[n-2] = 3^n$ and $y[-1] = y[-2] = 0$, using the sum of the steady-state and homogeneous solutions.

Solution: The homogeneous solution is found from the characteristic equation

$$1 - \frac{5}{6}z^{-1} + \frac{1}{6}z^{-2} = 0.$$

Solving for z_1 and z_2 in the quadratic above yields:

$$y_h[n] = c_1 z_1{}^n + c_2 z_2{}^n = c_1\left(\frac{1}{2}\right)^n + c_2\left(\frac{1}{3}\right)^n.$$

The steady-state solution to a geometric sequence input is

$y_{ss}[n] = H(r)r^n$ where r^n is the input and $H(r)$ is

$$\frac{1}{1 - \frac{5}{6}r^{-1} + \frac{1}{6}r^{-2}}$$

Hence,

$$y_{ss}[n] = H(3)3^n = \frac{3^n}{1 - \frac{5}{6}\left(\frac{1}{2}\right) + \frac{1}{6}\left(\frac{1}{9}\right)} = \frac{27}{20}3^n.$$

The solution is the sum

$$y[n] + y_{ss}[n] + y_h[n] = \frac{27}{20}3^n + c_1\left(\frac{1}{2}\right)^n + c_2\left(\frac{1}{3}\right)^n.$$

Applying the initial conditions

$$y[-1] = 0 = \frac{9}{20} + 2c_1 + 3c_2$$

$$y[-2] = 0 = \frac{3}{20} + 4c_1 + 9c_2$$

and solving for c_1 and c_2 gives the solution

$$y[n] = \frac{27}{20}3^n - \frac{3}{5}\left(\frac{1}{2}\right)^n + \frac{1}{4}\left(\frac{1}{3}\right)^n.$$

• PROBLEM 15-17

> Find the steady-state solution of the equation $y[n] + \frac{1}{\sqrt{3}} y[n-1] = \cos \frac{\pi}{6} n$.

Solution: The input $x[n] = \cos\frac{\pi}{6}n$ can be written

$$\frac{e^{j\frac{\pi}{6}n} + e^{-j\frac{\pi}{6}n}}{2} = \frac{r^{+an} + r^{-an}}{2}.$$

Using superposition,

$$y_{ss}[n] = \frac{1}{2} H(r^a)\, r^{an} + \frac{1}{2}H(r^{-a})\, r^{-an}$$

$$y_{ss}[n] = \frac{1}{2}H(e^{j\frac{\pi}{6}})\, e^{j\frac{\pi}{6}n} + \frac{1}{2}H(e^{-j\frac{\pi}{6}})\, e^{-j\frac{\pi}{6}n}$$

where

$$H(e^{j\frac{\pi}{6}}) = \frac{1}{1 + \frac{1}{\sqrt{3}} e^{-j\frac{\pi}{6}}}$$

and $$H(e^{-j\frac{\pi}{6}}) = \frac{1}{1 + \frac{1}{\sqrt{3}} e^{j\frac{\pi}{6}}}.$$

$H(e^{j\beta})$ can be written in a phasor form as

$$A(\beta)e^{j\phi(\beta)}.$$

Also, $H(e^{-j\beta})$ can be written in the form

$$A(\beta)e^{-j\phi(\beta)}.$$

Since the coefficients of the recursion equation are real

$$H(e^{-j\beta}) = H^*(e^{j\beta}).$$

This allows us to write $y_{ss}[n] = A(\beta) \cos (\beta n + \phi(\beta))$.

Hence,

$$H(e^{j\frac{\pi}{6}}) = \frac{1}{1 + \frac{1}{\sqrt{3}}e^{-j\frac{\pi}{6}}} = \frac{1}{1 + \frac{1}{\sqrt{3}} \cos \frac{\pi}{6} - j\frac{1}{\sqrt{3}} \sin \frac{\pi}{6}}$$

$$= \frac{1}{\sqrt{\left(1+\frac{1}{\sqrt{3}}\cos\frac{\pi}{6}\right)^2 + \left(-\frac{1}{\sqrt{3}}\sin\frac{\pi}{6}\right)^2}} e^{j\tan^{-1}\frac{\frac{1}{\sqrt{3}}\sin\frac{\pi}{6}}{1+\frac{1}{\sqrt{3}}\cos\frac{\pi}{6}}}$$

$$= \sqrt{\frac{3}{7}}\, e^{j\tan^{-1}\frac{1}{3\sqrt{3}}}$$

and $H(e^{-j\frac{\pi}{6}}) = \sqrt{\frac{3}{7}}\, e^{-j\tan^{-1}\frac{1}{3\sqrt{3}}}$.

The steady-state solution is

$$y_{ss}[n] = \sqrt{\frac{3}{7}} \cos\left(\frac{\pi}{6}n + \tan^{-1}\frac{1}{3\sqrt{3}}\right).$$

• PROBLEM 15-18

Find y[n] such that

$$y[n] - 4y[n-1] + 3y[n-2] = 8$$

when $y[-1] = y[-2] = 12$.

Solution: The homogeneous solution is found by solving the characteristic equation

$$1 - 4z^{-1} + 3z^{-2} = 0.$$

We find that $z_1 = 1$, and $z_2 = 3$; hence

$$y_h[n] = c_1 + c_2 3^n.$$

In this recursion equation we notice that the particular solution is not a constant r becaues the Hurwitz polynomial $H(z) = \frac{1}{1 - 4z^{-1} + 3z^{-2}}$ is $\frac{1}{0}$ for H(1).

It can be shown that the particular solution is a linearly increasing sequence, nr. Substituting nr for y[n] in the recursion equation we obtain

$$rn - 4r(n-1) + 3r(n-2) = 8$$

$$rn - 4rn + 4r + 3rn - 6r = 8$$

$$(4 - 4)rn - 2r = 8$$

$$r = -4.$$

Hence $-4n = y_p[n]$.

The total solution is the sum of the homogeneous and particular solutions

$$y[n] = c_1 + c_2 3^n - 4n.$$

To find constants c_1 and c_2 we apply the initial conditions to obtain

$$y[-1] = c_1 + \frac{c_2}{3} + 4 = 12$$

$$c_1 + \frac{c_2}{3} = 8$$

$$y[-2] = c_1 + \frac{c_2}{9} + 8 = 12$$

$$c_1 + \frac{c_2}{9} = 4,$$

two equations and two unknowns. Solving the above equations yields

$$c_1 = 2 \text{ and } c_2 = 18.$$

Hence, the solution is

$$y[n] = 2 + 18(3)^n - 4n.$$

• PROBLEM 15-19

Find y[n] for the input x[n], shown in fig. 1, by first finding the delta response h[n] of the system.

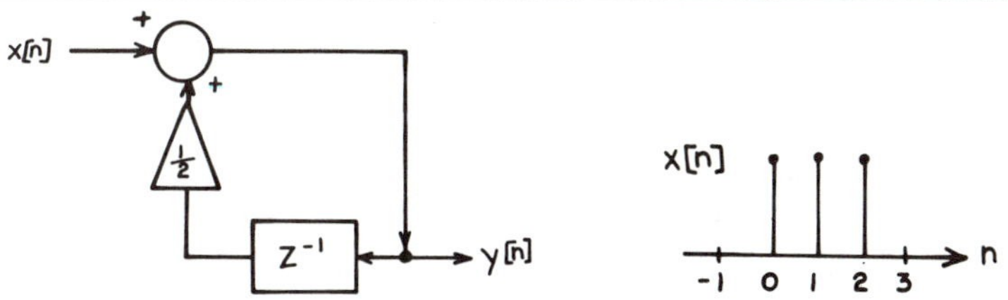

FIG. 1

Solution: The output y[n] of the system in fig. 1 satisfies the equation

$$y[n] - \frac{1}{2}y[n-1] = x[n].$$

The delta response, h[n], can be found by solving the homogeneous equation

$$1 - \frac{1}{2}z^{-1} = 0$$

$$z = \frac{1}{2}$$

and $h[n] = c\left(\frac{1}{2}\right)^n$ where $h[0] = c = 1$.

Hence, $h[n] = \left(\frac{1}{2}\right)^n u[n]$.

For the input to be

$$x[n] = \delta[n] + \delta[n-1] + \delta[n-2] = \begin{cases} 1, & 0 \le n \le 2 \\ 0, & \text{otherwise} \end{cases}$$

the output must be

$$y[n] = h[n] + h[n-1] + h[n-2] .$$

Since $y[n] = 0$ for $n < 0$ and $h[n] = 0$ for $n < 0$ then

$$y[0] = h[0] = \left(\frac{1}{2}\right)^0 = 1,$$

$$y[1] = h[1] + h[0] = \left(\frac{1}{2}\right)^1 + \left(\frac{1}{2}\right)^0 = 1.5, \text{ and}$$

$$y[n] = \left(\frac{1}{2}\right)^n + \left(\frac{1}{2}\right)^{n-1} + \left(\frac{1}{2}\right)^{n-2} = 7\left(\frac{1}{2}\right)^n \text{for } n \geq 2.$$

• PROBLEM 15-20

Find the delta response h[n] of the equation

$$y[n] - \frac{5}{6}y[n-1] + \frac{1}{6}y[n-2] = x[n], \quad \text{where } h[-1] = 0.$$

Solution: To find the delta response, note that $x[n] = \delta[n]$ and

$$\delta[n] = \begin{cases} 1 & n = 0 \\ 0 & n \neq 0 . \end{cases}$$

Since the delta response is defined as $y[n] = h[n]$, we can write

$$h[n] - \frac{5}{6}h[n-1] + \frac{1}{6}h[n-2] = \delta[n]. \tag{1}$$

The solution to Eq. (1) for $n \geq 1$ is the solution to the homogeneous equation

$$h[n] - \frac{5}{6}h[n-1] + \frac{1}{6}h[n-2] = 0 \tag{2}$$

because $\delta[n] = 0$ for $n \geq 1$.

Hence, $h[n] = c_1 z_1{}^n + c_2 z_2{}^n$ where z_1 and z_2 are found by solving the characteristic equation

$$1 - \frac{5}{6}z^{-1} + \frac{1}{6}z^{-2} = 0. \tag{3}$$

The solution of h[n] is

$$c_1\left(\frac{1}{2}\right)^n + c_2\left(\frac{1}{3}\right)^n. \tag{4}$$

Now apply the initial condition to equation 4:

$$h[-1] = 2c_1 + 3c_2 = 0. \tag{5}$$

From Eq. (1),

$$h[0] = 1,$$

hence

$$h[0] = c_1 + c_2 = 1. \tag{6}$$

To find c_1 and c_2 combine equations (5) and (6).

$$c_1 = 3 \qquad c_2 = -2$$

Hence, the delta response is

$$h[n] = 3\left(\frac{1}{2}\right)^n - 2\left(\frac{1}{3}\right)^n \quad ; \quad n \geq 0,$$

which can be written as

$$h[n] = 3\left(\frac{1}{2}\right)^n u[n] - 2\left(\frac{1}{3}\right)^n u[n].$$

• PROBLEM 15-21

Use the formula for discrete convolution,

$$y[n] = \sum_{k=0}^{n} x[n-k]h[k],$$

to find the solution of the equation

$$y[n] - \frac{1}{2}y[n-1] = u[n] - u[n-3].$$

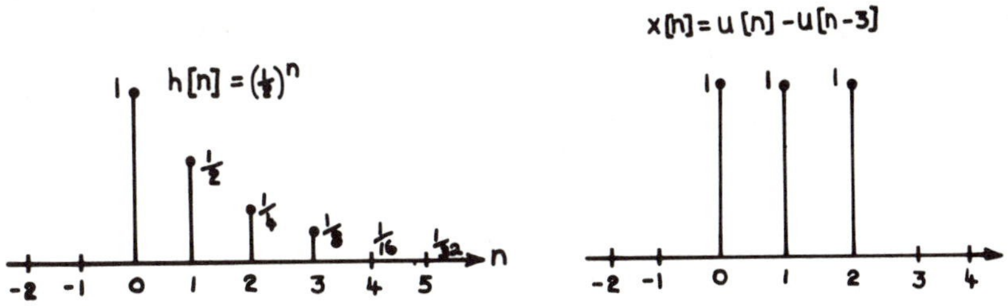

FIG. 1

Solution: To find the solution $y[n]$ we convolve the impulse $h[n]$, with the input $u[n] - u[n-3]$. We find that the impulse response $h[n]$ for $n \geq 1$ is the solution to

$$h[n] - \frac{1}{2}h[n-1] = \delta[n] = 0.$$

Hence,

$$h[n] = c\left(\frac{1}{2}\right)^n$$

$$h[0] = \delta[0] + \frac{1}{2}h[-1] = 1 + 0 = 1$$

$$h[n] = \left(\frac{1}{2}\right)^n \; ; \; n \geq 0.$$

We must convolve the two functions shown in fig. 1.

We can write the convolution formula as

$$y[n] = \sum_{k=0}^{n} h[n-k]x[k].$$

For $n \le 2$, $x[k] = 1$ for $0 \le k \le n$, hence

$$y[n] = \sum_{k=0}^{n} h[n-k] = \sum_{k=0}^{n} \left(\frac{1}{2}\right)^{n-k}.$$

We can show that

$$\sum_{k=0}^{n} \left(\frac{1}{2}\right)^{n-k} = \sum_{k=0}^{n} \left(\frac{1}{2}\right)^{k}$$ since we notice

$$\left(\frac{1}{2}\right)^{n-0} + \left(\frac{1}{2}\right)^{n-1} + \cdots + \left(\frac{1}{2}\right)^{n-(n-1)} + \left(\frac{1}{2}\right)^{0}$$

$$= \left(\frac{1}{2}\right)^{0} + \left(\frac{1}{2}\right)^{1} + \cdots + \left(\frac{1}{2}\right)^{n-1} + \left(\frac{1}{2}\right)^{n},$$

a one-to-one correspondence between the terms of the two series. We obtain

$$\sum_{k=0}^{n} \left(\frac{1}{2}\right)^{k} = -\left(\frac{1}{2}\right)^{n} + \sum_{k=0}^{\infty} \left(\frac{1}{2}\right)^{k}$$ where

$\sum_{k=0}^{\infty} \left(\frac{1}{2}\right)^{k}$ is a geometric progression which converges to $\frac{1}{1-\frac{1}{2}} = 2.$

Hence, $y[n] = \sum_{k=0}^{n} \left(\frac{1}{2}\right)^{n-k} = 2 - \left(\frac{1}{2}\right)^{n}$, for $n \le 2$.

For $n > 2$, $x[k] = 1$ for $0 \le k \le 2$ and $x[k] = 0$ for $k > 2$; hence,

$$y[n] = \sum_{k=0}^{2} h[n-k] = \sum_{k=0}^{2} \left(\frac{1}{2}\right)^{n-k}$$

$$= \left(\frac{1}{2}\right)^{n} + \left(\frac{1}{2}\right)^{n-1} + \left(\frac{1}{2}\right)^{n-2}$$

$$= \left(\frac{1}{2}\right)^{n} + 2\left(\frac{1}{2}\right)^{n} + 4\left(\frac{1}{2}\right)^{n} = 7\left(\frac{1}{2}\right)^{n}.$$

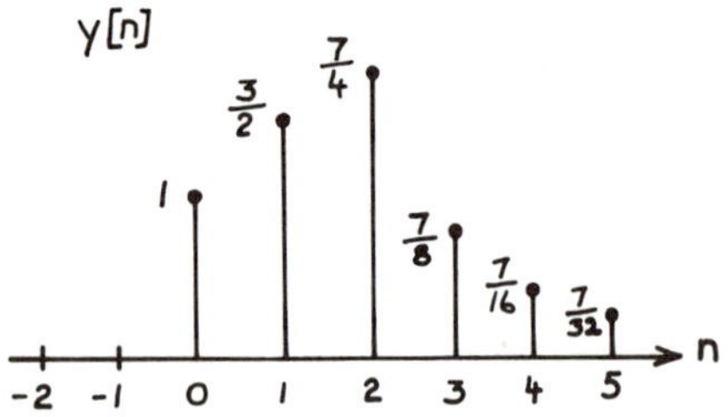

FIG. 2

Fig. 2 shows the output signal y[n].

Find the solution y[n] of the equation $y[n] - \frac{1}{2}y[n-1] = x[n]$ when $x[n] = u[n] - u[n-3]$ by finding the step response, r[n], first.

Solution: The step response of the system

$$y[n] + a_1 y[n-1] + \ldots + a_m r[n-m] = x[n] \tag{1}$$

can be found by letting x[n] = u[n], then y[n] = r[n]. Hence, it can be shown that

$$r[n] + a_1 r[n-1] + \ldots + a_m r[n-m] = u[n]. \tag{2}$$

With the initial conditions,

$$r[-1] = r[-2] = \ldots = r[-m] = 0 \tag{3}$$

we know that the particular solution of Eq. (2) for $n \geq 0$ is the constant sequence

$$H(1) = \frac{1}{1 + a_1 + \ldots + a_m} .$$

Using superposition,

$$r[n] = H(1) + c_1 z_1^n + \ldots + c_m z_m^n . \quad ; n \geq 0.$$

To find the constants, we insert $c_1, c_2, \ldots, c_m$ into Eq(3) to obtain the set of equations,

$$\frac{c_1}{z_1} + \frac{c_2}{z_2} + \ldots + \frac{c_m}{z_m} = -H(1)$$

$$\vdots \qquad \vdots \qquad \vdots$$

$$\frac{c_1}{z_1^m} + \frac{c_2}{z_2^m} + \ldots + \frac{c_m}{z_m^m} = -H(1).$$

Consider the equation in the example

$y[n] - \frac{1}{2}y[n-1] = x[n]$. The step response is r[n] = $H(1) + c\left(\frac{1}{2}\right)^n$, where $H(1) = \frac{1}{1 - \frac{1}{2}} = 2$: Hence, $r[n] = 2 + c\left(\frac{1}{2}\right)^n$.

To find c, we apply the initial condition $r[-1] = 2 + 2c = 0$. This yields c = -1 and the step response

$$r[n] = \left[2 - \left(\frac{1}{2}\right)^n\right] u[n].$$

If $x[n] = u[n] - u[n-3]$, then

$$y[n] = r[n] - r[n-3].$$

Thus, for $0 \leq n \leq 2$

$$y[n] = r[n] = 2 - \left(\frac{1}{2}\right)^n.$$

For $n > 2$

$$y[n] = r[n] - r[n-3] = 2 - \left(\frac{1}{2}\right)^n - 2 + \left(\frac{1}{2}\right)^{n-3}$$

$$= 2 - \left(\frac{1}{2}\right)^n - 2 + 8\left(\frac{1}{2}\right)^n = 7\left(\frac{1}{2}\right)^n .$$

DIGITAL SOLUTION OF ANALOG SYSTEMS

• PROBLEM 15-23

Find a digital system that will simulate the RL circuit shown in fig. 1, where v(t) is the desired response. Use the approximation

$$\frac{dv(t)}{dt} \cong \frac{v(t) - v(t - T)}{T}$$

in the differential equation describing the RL circuit.

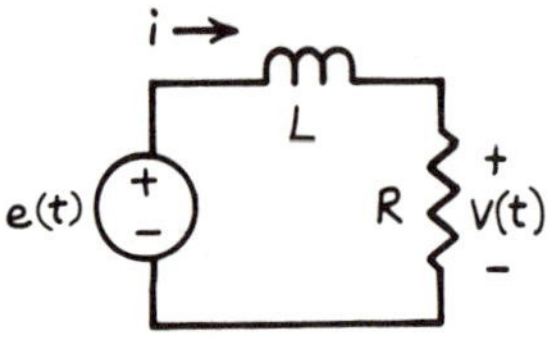

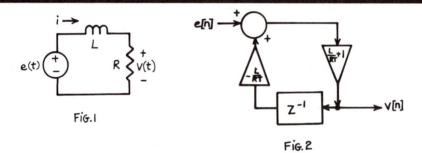

Fig.1

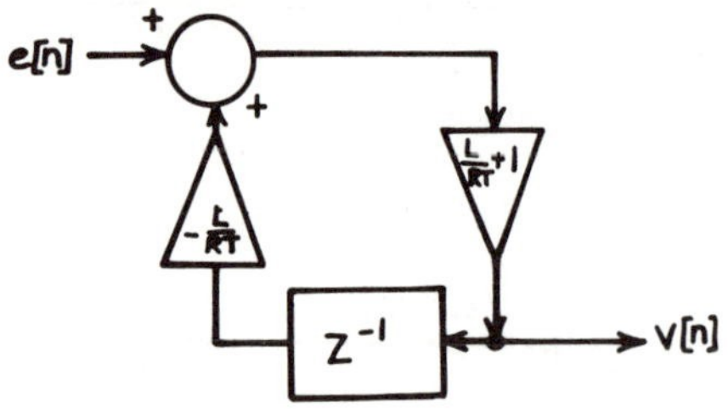

Fig.2

Solution: We obtain the differential equation by writing

$$e(t) = L\frac{di}{dt} + v(t).$$

Since $v(t) = Ri$ then $\frac{1}{R}\frac{d\ v(t)}{dt} = \frac{di}{dt}$ and we obtain,

$$e(t) = \frac{L}{R}\frac{dv(t)}{dt} + v(t). \tag{1}$$

Substituting $\frac{dv(t)}{dt} \cong \frac{v(t) - v(t - T)}{T}$

into Eq.(1) yields:

$$e(t) \cong \frac{L}{R}\left[\frac{v(t) - v(t - T)}{T}\right] + v(t)$$

$$e(t) \cong \left[\frac{L}{RT} + 1\right] v(t) - \frac{L}{RT} v(t - T). \tag{2}$$

We can write a discrete system recursion equation for the continuous system described by Eq.(2) by sampling the forcing function e(t) every T seconds to obtain e(nT). When this forcing function is placed in the circuit the response is also v(nT). Sampling every T seconds produces discrete signals where $e[n] = e(nT)$ and $v[n] = v(nT)$. Hence, Eq.(2) becomes the discrete recursion equation,

$$e[n] = \left[\frac{L}{RT} + 1\right] v[n] - \frac{L}{RT} v[n-1]$$

defining the discrete system of fig. 2.

• PROBLEM 15-24

Use the approximation

$$\frac{d\,y(t)}{dt} \cong \frac{y(t) - y(t - T)}{T}$$

to solve numerically the differential equation

$$\frac{dy(t)}{dt} + ay(t) = x(t) \; ; \qquad y(0) = 0$$

where (a) $x(t) = u(t)$; (b) $x(t) = \cos \omega_0 t$.

Solution: Substituting the approximation for $\frac{dy(t)}{dt}$ into the the differential equation above gives

$$\frac{y(t) - y(t-T)}{T} + ay(t) = x(t)$$

$$\left[\frac{1}{T} + a\right] y(t) - \frac{1}{T} y(t - T) = x(t).$$

Making the substitution $t = nT$ yields the recursion equation

$$\left[\frac{1}{T} + a\right] y[n] - \frac{1}{T} y[n-1] = x[n]. \tag{1}$$

The homogeneous solution is found by solving the characteristic equation

$$\left(\frac{1}{T} + a\right) - \frac{1}{T} z^{-1} = 0$$

$$\frac{1}{T} z^{-1} = \frac{1}{T} + a$$

$$z = \frac{\frac{1}{T}}{\frac{1}{T} + a} = \frac{1}{1 + aT}.$$

Hence $y_h[n] = c\left(\frac{1}{1 + aT}\right)^n.$

(a) For $x[n] = u[n]$,

$$y[n] = H(1) + c\left(\frac{1}{1 + aT}\right)^n$$

$$y[n] = \frac{1}{\frac{1}{T} + a - \frac{1}{T}} + c\left(\frac{1}{1 + aT}\right)^n = \frac{1}{a} + c\left(\frac{1}{1 + aT}\right)^n.$$

Since $y[0] = 0 = \frac{1}{a} + c$, $c = -\frac{1}{a}$.

The solution of the recursion equation to a step sequence input is

$$y[n] = \frac{1}{a}\left[1 - \left(\frac{1}{1 + aT}\right)^n\right] u[n].$$

The solution of the differential equation to a step function input is

$$y(t) = \frac{1}{a}(1 - e^{-at})\ u(t).$$

We note that e^{-anT} can be approximated by $\left(\frac{1}{1 + aT}\right)^n$ when $aT \ll 1$.

(b) For $x(t) = \cos \omega_0 t$, we again make the substitution $t = nT$ into equation (1) obtaining

$$\left[\frac{1}{T} + a\right] y[n] - \frac{1}{T} y[n-1] = \cos \omega_0 nT.$$

The steady state solution $y_{ss}[n]$ can be obtained by finding $H(e^{j\omega_0 T})$ and writing it in phasor form, $A(\omega_0 T)\ e^{j\phi(\omega_0 T)}$, then $y_{ss}[n] = A \cos (\omega_0 Tn + \phi(\omega_0 T))$.

Hence,

$$H(z) = \frac{1}{\left(\frac{1}{T} + a\right) - \frac{1}{T} z^{-1}}$$

$$H(e^{j\omega_0 T}) = \frac{1}{\left(\frac{1}{T} + a\right) - \frac{1}{T} e^{-j\omega_0 T}}$$

$$H(e^{j\omega_0 T}) = \frac{1}{\left(\frac{1}{T} + a\right) - \frac{1}{T}[\cos \omega_0 T - j \sin \omega_0 T]}.$$

To convert to polar form we find the magnitude to be

$$A = \frac{1}{\sqrt{\left(\frac{1}{T} + a - \frac{1}{T} \cos \omega_0 T\right)^2 + \left(\frac{1}{T} \sin \omega_0 T\right)^2}}$$

$$A = \frac{T}{\sqrt{(1 + aT)^2 + 1 - 2(1+aT)\cos \omega_0 T}}$$

and the angle,

$$\phi = \tan^{-1}\left[-\frac{\frac{1}{T} \sin \omega_0 T}{\frac{1}{T} + a - \frac{1}{T} \cos \omega_0 T}\right]$$

$$\phi = \tan^{-1}\left[-\frac{\sin \omega_0 T}{1 + aT - \cos \omega_0 T}\right].$$

The steady state solution becomes

$$y_{ss}[n] = \frac{T\cos(n\,\omega_0 T + \phi)}{\sqrt{(1+aT)^2 + 1 - 2(1+aT)\cos\omega_0 T}}$$

where $\phi = \tan^{-1}\left[-\frac{\sin\ \omega_0 T}{1 + aT - \cos\omega_0 T}\right]$.

Comparing $y_{ss}[n]$ with the continuous steady state solution,

$$y_{ss}(t) = \frac{1}{\sqrt{\omega_0^{\ 2} + a^2}}\cos(\omega_0 t + \phi_c)$$

where $\phi_c = \tan^{-1}\left[-\frac{\omega_0}{a}\right]$.

We note that when $aT \ll 1$ and $\omega_0 T \ll 1$ then $y_{ss}[n] \cong y_{ss}(nT)$.

Z-TRANSFORM DEFINITIONS & PROPERTIES

• PROBLEM 15-25

Show that the unilateral z - transform

$$X(z) = \sum_{n=0}^{\infty} x[n]z^{-n}$$

can be obtained from the definition of the Laplace transform

$$X(s) = \int_0^{\infty} x(t)e^{-st}\,dt.$$

Hint: Take any analog signal x(t) and sample it every T seconds (multiply by an impulse train $a(t) = \delta(t) + \delta(t-T) + \delta(t+2T) + \ldots + \delta(t+nT)$) and find the Laplace transform, $L\{x(t)a(t)\}$.

Solution: If we take any analog (causal signal) function and sample it every T seconds, as shown in fig. 1, we obtain a function

$$x_n(t) = x(t)a(t) = x_0\delta(t) + x_1\delta(t-T) + x_2\delta(t-2T) + \ldots$$

$$x_n\delta(t-nT) = \sum_{n=0}^{\infty} x_n\,\delta(t-nT),$$

where $x_0, x_1, x_2, \ldots, x_n$ are samples taken at times 0, T, 2T, ..., nT respectively. Hence, we can call the sequence $x_0, x_1, x_2, \ldots, x_n$ the discrete signal x[n].

Taking the Laplace transform of $X_n(t)$ yields:

$$X(s) = \int_0^{\infty}\sum_{n=0}^{\infty} x_n\,\delta(t-nT)e^{-st}\,dt$$

$$X(s) = \sum_{n=0}^{\infty} x_n \int_0^{\infty}\delta(t-nT)e^{-st}\,dt.$$

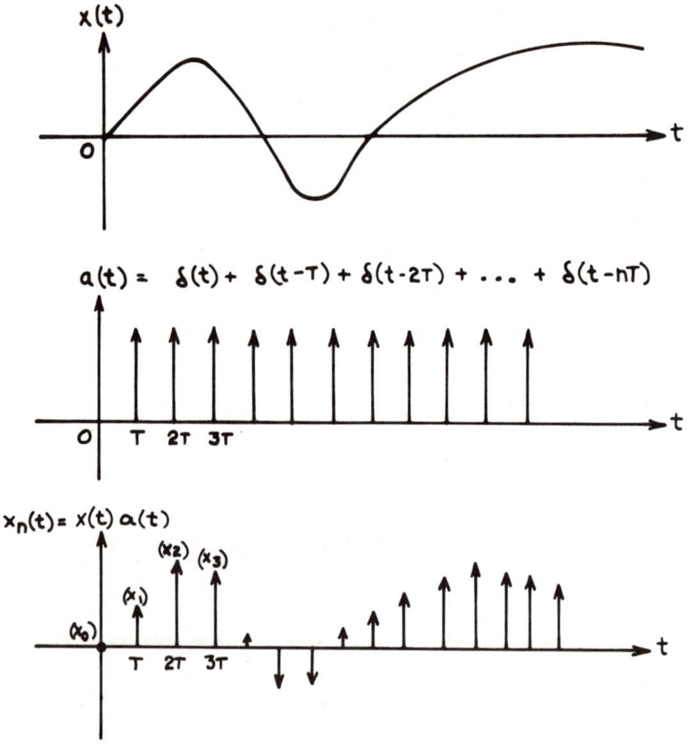

FiG.1

Using the time shift property of the Laplace transform stated as,

$$\int_0^\infty f(t-a)e^{-st}\,dt = e^{-sa}\int_0^\infty f(t)e^{-st}\,dt$$

yields

$$X(s) = \sum_{n=0}^{\infty} x_n e^{-snT} \int_0^\infty \delta(t)e^{-st}\,dt.$$

Since $\int_0^\infty \delta(t)e^{-st}\,dt = 1$

then, $X(s) = \sum_{n=0}^{\infty} x_n e^{-snT}$.

By replacing e^{-st} with z^{-1} and x_n with $x[n]$ we obtain the unilateral z - transform

$$X(z) = \sum_{n=0}^{\infty} x[n]z^{-n}$$

• **PROBLEM** 15-26

Using the definition of the bilateral z transform

$$F(z) = \sum_{n=-\infty}^{\infty} f[n]z^{-n},$$

find the z - transform of

$$f[n] = 3\delta[n+2] + 4\delta[n-1] + 2\delta[n-3].$$

Solution: Since

$$\delta[n-k] = \begin{cases} 1 & n = k \\ 0 & n \neq k \end{cases}$$

then,

$$F(z) = \sum_{n=-\infty}^{\infty} 3\delta[n+2]z^{-n} + \sum_{n=-\infty}^{\infty} 4\delta[n-1]z^{-n} + \sum_{n=-\infty}^{\infty} 2\delta[n-3]z^{-n}$$

$$F(z) = 3z^2 + 4z^{-1} + 2z^{-3}.$$

• **PROBLEM** 15-27

Using the definition of the unilateral Z transform,

$$F(z) = \sum_{n=0}^{\infty} f[n]z^{-n}.$$

find the z - transform of the step-sequence

$$f[n] = u[n].$$

Solution: Since,

$$u[n] = \begin{cases} 1 & n \geq 0 \\ 0 & n < 0 \end{cases}$$

then

$$F(z) = \sum_{n=0}^{\infty} u[n]z^{-n} = \sum_{n=0}^{\infty} z^{-n}.$$

The geometric progression,

$$\sum_{n=0}^{\infty} r^n, \text{ converges to } \frac{1}{1-r}$$

when $r < 1$, hence,

$$F(z) = \sum_{n=0}^{\infty} z^{-n} = \frac{1}{1 - z^{-1}} = \frac{z}{z-1}.$$

Therefore,

$$u[n] \leftrightarrow \frac{z}{z-1}.$$

(a) Show that the z - transform of

$$f[n] = x^n u[n] \text{ is}$$

$$\frac{z}{z - x}.$$

(b) Find the z - transform of

$$f[n] = 3 \cdot 2^n u[n] + 5 \cdot 4^n u[n].$$

Solution: (a) Since

$$x^n u[n] = \begin{cases} x^n & n \geq 0 \\ 0 & n < 0 \end{cases}$$

then $F(z) = \sum_{n=0}^{\infty} x^n z^{-n} = \sum_{n=0}^{\infty} \left(\frac{x}{z}\right)^n$ converges

to $\dfrac{1}{1 - \frac{x}{z}} = \dfrac{z}{z - x}$.

Hence,

$$x^n u[n] \leftrightarrow \frac{z}{z - x}.$$

(b) The z-transform of $3 \cdot 2^n u[n] + 5 \cdot 4^n u[n]$ can be found using the procedure in part (a),

$$F(z) = 3 \sum_{n=0}^{\infty} 2^n z^{-n} + 5 \sum_{n=0}^{\infty} 4^n z^{-n}$$

$$F(z) = \frac{3}{1 - \frac{2}{z}} + \frac{5}{1 - \frac{4}{z}}$$

$$F(z) = \frac{3z}{z - 2} + \frac{5z}{z - 4} = \frac{8z^2 - 22z}{z^2 - 6z + 8}.$$

• PROBLEM 15-29

(a) Show that the z-transform of

$$f[n] = nx^{n-1} u[n] \text{ is}$$

$$\frac{z}{(z-w)^2}.$$

(b) Find the z-transform of

$$F[n] = (3 + 5n) u[n].$$

Solution: (a) We can immediately write

$$F(z) = \sum_{n=0}^{\infty} nx^{n-1}z^{-n} = z^{-1}\sum_{n=0}^{\infty} nx^{n-1}z^{-(n-1)}.$$

Then, $F(z) = z^{-1}\sum_{n=1}^{\infty} n\left(\frac{x}{z}\right)^{n-1}$ since $F[0] = 0$.

We can show that

$$\sum_{n=1}^{\infty} nar^{n-1} = a + 2ar + \ldots + nar^{n-1} = \frac{a}{(1-r)^2}.$$

Hence,

$$F(z) = z^{-1}\sum_{n=1}^{\infty} n\left(\frac{x}{z}\right)^{n-1} = \frac{z^{-1}}{\left[1 - \left(\frac{y}{z}\right)\right]^2}$$

$$F(z) = \frac{z^{-1}}{\left(1 - \frac{x}{z}\right)^2} = \frac{z}{(z-x)^2}$$

Hence,

$$nx^{n-1}u[n] \leftrightarrow \frac{z}{(z-x)^2}.$$

(b) Using the above transform and $u[n] \leftrightarrow \frac{z}{z-1}$ we obtain

$$F(z) = \sum_{n=1}^{\infty} (3 + 5n)u[n]z^{-n}$$

$$F(z) = \frac{3z}{z-1} + \frac{5z}{(z-1)^2} = \frac{3z^2 + 2z}{(z-1)^2}.$$

• PROBLEM 15-30

Use partial fraction expansion to find the inverse z - transform of

$$F(z) = \frac{2z^2 - 2z}{(z-3)(z-5)^2}.$$

Solution: Using the transform pairs,

$$u[n] \leftrightarrow \frac{z}{z-1}$$

$$a^n u[n] \leftrightarrow \frac{z}{z-a}$$

$$na^{n-1}u[n] \leftrightarrow \frac{z}{(z-a)^2}$$

$$\frac{n(n-1)}{2}a^{n-2}u[n] \leftrightarrow \frac{z}{(z-a)^3}$$

$$\vdots$$

$$\frac{n(n-1)\ldots(n-m+1)}{m!}a^{n-m}u[n] \leftrightarrow \frac{z}{(z-a)^{m+1}},$$

we can find the inverse z-transform of the rational function F(z) by expanding it using partial fractions.

Hence, F(z) can be expanded into the form

$$\frac{Az}{(z-3)} + \frac{Bz}{(z-5)^2} + \frac{Cz}{(z-5)}.$$

We find the constants as follows:

$$F(z) = \frac{2z^2 - 2z}{(z-3)(z-5)^2} = \frac{Az}{(z-3)} + \frac{Bz}{(z-5)^2} + \frac{Cz}{(z-5)}.$$

Multiply both sides of the equation by $(z-3)(z-5)^2$:

$$2z^2 - 2z = Az(z-5)^2 + Bz(z-3) + Cz(z-3)(z-5)$$

$$2z^2 - 2z = A(z^3 - 10z^2 + 25z) + B(z^2 - 3z) + C(z^3 - 8z^2 + 15z)$$

$$2z^2 - 2z = z^3(A + C) + z^2(-10A + B - 8C) + z(25A-3B+15C).$$

From the last equation, we compare coefficients and write three equations:

$$\begin{aligned} A + C &= 0 \\ -10A+B-8C &= 2 \\ 25A-3B+15C &= -2. \end{aligned}$$

Solving simultaneously we obtain A = 1, B = 4, and C = -1.

Hence,

$$F(z) = \frac{z}{z-3} + \frac{4z}{(z-5)^2} - \frac{z}{(z-5)}.$$

Obtain the inverse z-transform by use of the transform pairs above. Hence

$$f[n] = 3^n u[n] + 4n5^{n-1} u[n] - 5^n u[n]$$

can be written

$$f[n] = 3^n u[n] + \left(\frac{4}{5}n - 1\right) 5^n u[n].$$

• PROBLEM 15-31

Use partial fraction expansion to find the inverse z-transform of

$$F(z) = \frac{30z^2 - 12z}{6z^2 - 5z + 1}.$$

Solution: The function F(z) is expanded into:

$$F(z) = d_0 + \frac{d_1 z}{z-z_1} + \frac{d_2 z}{z-z_2}, \qquad (1)$$

where z_1 and z_2 are the roots of the denominator of F(z) and

$$d_0 = F(0)$$

$$d_1 = \left.\frac{F(z)(z-z_1)}{z}\right|_{z=z_1}$$

$$d_2 = \left.\frac{F(z)}{z}(z-z_2)\right|_{z=z_2}$$

Once equation(1) is calculated, the inverse z-transform is easily found because of the known transform pair:

$$w^n u[n] \leftrightarrow \frac{z}{z-w} .$$

First expand F(z):

$$F(z) = \frac{30z^2 - 12z}{6z^2 - 5z + 1} = \frac{5z^2 - 2z}{z^2 - \frac{5}{6}z + \frac{1}{6}} = \frac{5z^2 - 2z}{(z-z_1)(z-z_2)} .$$

The roots of the denominator are:

$$z_1, z_2 = \frac{5}{12} \pm \sqrt{\frac{25}{144} - \frac{1}{6}} = \frac{1}{2}, \frac{1}{3} .$$

Hence

$$d_0 = F(0) = 0$$

$$d_1 = \left.\frac{5z^2 - 2z}{z\left(z - \frac{1}{3}\right)}\right|_{z=\frac{1}{2}} = 3$$

$$d_2 = \left.\frac{5z^2 - 2z}{z\left(z - \frac{1}{2}\right)}\right|_{z=\frac{1}{3}} = 2$$

From equation(1)

$$F(z) = \frac{3z}{z - \frac{1}{2}} + \frac{2z}{z - \frac{1}{3}}$$

It is known that

$$w^n u[n] = \frac{z}{z-w} ,$$

so

$$\frac{3z}{z - \frac{1}{2}} \Leftrightarrow 3\left(\frac{1}{2}\right)^n u[n]$$

$$\frac{2z}{z - \frac{1}{3}} \Leftrightarrow 2\left(\frac{1}{3}\right)^n u[n].$$

Therefore

$$f(n) = \left[3\left(\frac{1}{2}\right)^n + 2\left(\frac{1}{3}\right)^n\right]u[n].$$

• **PROBLEM** 15-32

Determine the inverse z-transform of the function:

$$F(z) = \frac{z(z+1)}{(z-1)\left(z^2 - z + \frac{1}{4}\right)}.$$

Solution: Use partial fractions expansion. The roots of the denominator are

$$z_1 = 1$$

$$z_2,\ z_3 = \frac{1}{2} \pm \sqrt{\left(\frac{1}{2}\right)^2 - \frac{1}{4}} = \frac{1}{2}, \frac{1}{2}.$$

F(z) is rewritten as

$$F(z) = \frac{z(z+1)}{(z-1)\left(z-\frac{1}{2}\right)^2} = \frac{Az}{z-1} + \frac{Bz}{\left(z-\frac{1}{2}\right)^2} + \frac{Cz}{z-\frac{1}{2}} + D.$$

The values of A, B and D are:

$$A = \left.\frac{F(z)(z-1)}{z}\right|_{z=1} = \left.\frac{z(z+1)}{z\left(z-\frac{1}{2}\right)^2}\right|_{z=1} = 8$$

$$B = \left.\frac{F(z)\left(z-\frac{1}{2}\right)^2}{z}\right|_{z=\frac{1}{2}} = \left.\frac{z(z+1)}{z(z-1)}\right|_{z=\frac{1}{2}} = -3$$

$$D = F(z)\Big|_{z=0} = 0.$$

Since F(z) has multiple poles, the value of C cannot be found in the above manner. With the information that has already been calculated, however, the value of C can be found. It is now known that

$$F(z) = \frac{8z}{z-1} - \frac{3z}{\left(z-\frac{1}{2}\right)^2} + \frac{Cz}{z-\frac{1}{2}} = \frac{z(z+1)}{(z-1)\left(z-\frac{1}{2}\right)^2}.$$

The value of C can be found by setting z, in this equation, equal to any arbitrary number that is not a pole. Hence, with z = 2.

$$F(2) = \frac{2(2+1)}{(2-1)\left(2-\frac{1}{2}\right)^2} = \frac{8(2)}{2-1} - \frac{3(2)}{\left(2-\frac{1}{2}\right)^2} + \frac{c(2)}{2-\frac{1}{2}}$$

$$\frac{24}{9} = 16 - \frac{24}{9} + \frac{4c}{3}$$

$$c = -8.$$

So the expanded function is

$$F(z) = \frac{8z}{z-1} - \frac{3z}{\left(z - \frac{1}{2}\right)^2} - \frac{8z}{z - \frac{1}{2}}$$

It is known that:

$$\frac{z}{z-1} \quad <=> \quad u[n]$$

$$\frac{z}{(z-x)^2} <=> nx^{n-1}u[n]$$

$$\frac{z}{z-x} <=> x^n [u]n .$$

Hence

$$f(n) = \{8 - n3\left(\frac{1}{2}\right)^{n-1} - 8\left(\frac{1}{2}\right)^n\} u[n]$$

or

$$f(n) = \{8 - n6\left(\frac{1}{2}\right)^n - 8\left(\frac{1}{2}\right)^n\}u[n].$$

• **PROBLEM** 15-33

Determine the inverse z-transform of

$$F(z) = \frac{(z+5)}{\left(z - \frac{1}{2}\right)\left(z^2 - \frac{1}{2}z + \frac{1}{16}\right)}.$$

Solution: F(z) is simplified into sums of smaller equations by using the partial fraction expansion method. Once simplified, the inverse z-transform can be easily found. First find the roots of the denominator of F(z):

$$z_1 = \frac{1}{2} \qquad z_2, z_3 = \frac{1}{4} \pm \sqrt{\left(\frac{1}{4}\right)^2 - \frac{1}{16}} = \frac{1}{4}, \frac{1}{4} .$$

Now, F(z) is rewritten as

$$F(z) = \frac{z + 5}{\left(z-\frac{1}{2}\right)\left(z- \frac{1}{4}\right)^2} = A + \frac{Bz}{z - \frac{1}{2}} + \frac{Cz}{\left(z - \frac{1}{4}\right)^2} + \frac{Dz}{z - \frac{1}{4}} .$$

Now find the constants A, B, and C:

$$A = F(z)\Big|_{z = 0} = -160$$

$$B = \frac{F(z)}{z}\left(z-\frac{1}{2}\right)\Bigg|_{z=\frac{1}{2}} = \frac{z+5}{z\left(z-\frac{1}{4}\right)^2}\Bigg|_{z=\frac{1}{2}} = 176$$

$$C = \frac{F(z)}{z}\left(z-\frac{1}{4}\right)^2\Bigg|_{z=\frac{1}{4}} = \frac{z+5}{z\left(z-\frac{1}{2}\right)}\Bigg|_{z=\frac{1}{4}} = -84.$$

To find D calculate F(z) at any arbitrary value of z that is not a pole or zero of F(z). Pick 1 as the arbitrary value of z. Hence,

$$F(1) = \frac{1+5}{\left(1-\frac{1}{2}\right)\left(1-\frac{1}{4}\right)^2} = \frac{64}{3}.$$

Now calculate D using the formula

$$\frac{64}{3} = -160 + \frac{176(1)}{1-\frac{1}{2}} - \frac{84(1)}{\left(1-\frac{1}{4}\right)^2} + \frac{D(1)}{1-\frac{1}{4}} :$$

$$D = \frac{3}{4}\left[\frac{64}{3} + 160 - 352 + \frac{488}{3}\right]$$

$$D = -16.$$

The entire function F(z) is expanded to

$$F(z) = -160 + \frac{176z}{z-\frac{1}{2}} - \frac{84z}{\left(z-\frac{1}{4}\right)^2} - \frac{16z}{z-\frac{1}{4}}.$$

We use the transform pairs:

$$1 \leftrightarrow \delta(n),$$

$$\frac{z}{z-x} \leftrightarrow x^n u(n),$$

and $\frac{z}{(z-x)^2} \leftrightarrow nx^{n-1}u(n),$

to calculate F(n) from the expanded function F(z):

$$f(n) = -160\delta(n) + 176\left(\frac{1}{2}\right)^n u(n) - 84n\left(\frac{1}{4}\right)^{n-1}u(n) - 16\left(\frac{1}{4}\right)^n u(n)$$

or

$$f(n) = -160\delta(n) + 176\left(\frac{1}{2}\right)^n u(n) - 336n\left(\frac{1}{4}\right)^n u(n) - 16\left(\frac{1}{4}\right)^n u(n).$$

Z-TRANSFORM APPLICATIONS

• **PROBLEM 15-34**

Find y[n] such that

$$y[n] - \frac{1}{2}y[n-1] = u[n]$$

using z transforms.

Solution: We can transform the recursion equation by use of z-transforms. Hence,

$$Y(z)\ 1 - \frac{1}{2}z^{-1} = \frac{z}{z-1}$$

$$Y(z) = \frac{z}{z-1} \cdot \frac{1}{1-\frac{1}{2}z^{-1}}$$

$$Y(z) = \frac{z^2}{(z-1)\left(z-\frac{1}{2}\right)}.$$

Expanding by partial fractions yields:

$$Y(z) = \frac{z^2}{(z-1)\left(z-\frac{1}{2}\right)} = \frac{Az}{z-1} + \frac{Bz}{z-\frac{1}{2}}$$

$$z^2 = Az\left(z-\frac{1}{2}\right) + Bz(z-1)$$

$$z^2 = A\left(z^2 - \frac{1}{2}z\right) + B(z^2 - z)$$

$$z^2 = z^2(A + B) + z\left(-\frac{1}{2}A - B\right).$$

Thus

$$A + B = 1$$

$$\frac{1}{2}A + B = 0$$

and $A = 2,\ B = -1.$

Hence,

$$Y(z) = \frac{2z}{z-1} - \frac{z}{z-\frac{1}{2}}$$

can be inverse z-transformed to yield

$$y[n] = 2u[n] - \left(\frac{1}{2}\right)^n u[n]\ .$$

• PROBLEM 15-35

Find the system function $H(z) = \frac{Y(z)}{X(z)}$ and the delta response $h[n]$ of the system shown in fig. 1.

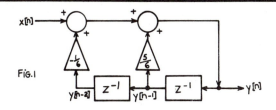

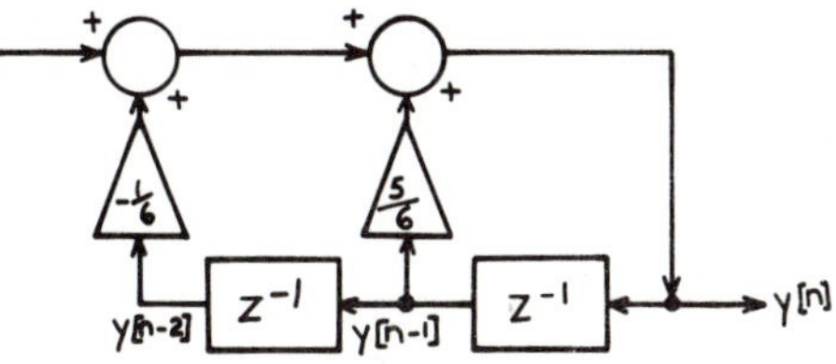

Fig.1

Solution: Note that the output, y[n], satisfies the recursion equation

$$y[n] - \frac{5}{6}y[n-1] + \frac{1}{6}y[n-2] = x[n]$$

which can be transformed into

$$Y(z)\left(1 - \frac{5}{6}z^{-1} + \frac{1}{6}z^{-2}\right) = X(z).$$

The system function is

$$H(z) = \frac{Y(z)}{X(z)} = \frac{1}{1 - \frac{5}{6}z^{-1} + \frac{1}{6}z^{-2}}$$

$$H(z) = \frac{z^2}{z^2 - \frac{5}{6}z + \frac{1}{6}} = \frac{z^2}{\left(z - \frac{1}{2}\right)\left(z - \frac{1}{3}\right)}.$$

Using partial fraction expansion,

$$H(z) = \frac{3z}{z - \frac{1}{2}} - \frac{2z}{z - \frac{1}{3}}.$$

The delta response h[n] can be found by obtaining the inverse z-transform of H(z). Hence,

$$h[n] = 3\left(\frac{1}{2}\right)^n u[n] - 2\left(\frac{1}{3}\right)^n u[n].$$

• PROBLEM 15-36

Find the output y[n] of the system shown in fig. 1 if x[n] = u[n].

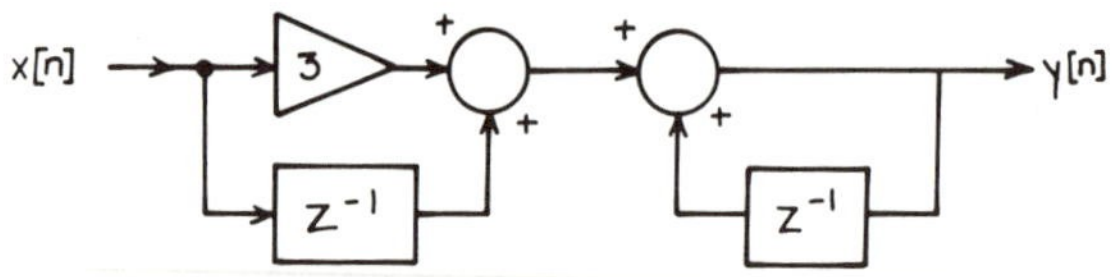

Fig.1

Solution: We can shown that the output y[n] satisfies the recursion equation

$$3x[n] + x[n-1] + y[n-1] = y[n].$$

Transforming the above equation yields

$$X(z)(3 + z^{-1}) = Y(z)(1 - z^{-1}),$$

and the system function

$$H(z) = \frac{Y(z)}{X(z)} = \frac{3 + z^{-1}}{1 - z^{-1}}.$$

Since x[n] = u[n], then $X(z) = \frac{z}{z-1}$

and $Y(z) = H(z)X(z) = \frac{3 + z^{-1}}{1 - z^{-1}} \cdot \frac{z}{z - 1}$

$$Y(z) = \frac{3z+1}{z - 2 + z^{-1}} = \frac{(3z + 1)z}{(z - 1)^2} .$$

Using partial fraction expansion yields

$$Y(z) = \frac{4z}{(z - 1)^2} + \frac{3z}{z - 1}$$

which transforms to

$$y[n] = (4n + 3)u[n].$$

• **PROBLEM** 15-37

Use z-transforms to find a sequence y[n] such that

$$y[n] - \frac{5}{6}y[n-1] + \frac{1}{6}y[n-2] = 3^n ; \quad n \geq 0$$

where $y[-1] = y[-2] = 0$.

Solution: Since the initial conditions are zero, we can assume that the above recursion equation holds for every n provided $y[n] = 0$ for $n<0$. Hence, we can replace the right side of the equation with $3^n u[n]$.

Transforming the entire equation gives

$$Y(z)(1 - \frac{5}{6}z^{-1} + \frac{1}{6}z^{-2}) = \frac{z}{z - 3} .$$

Solving for Y(z) yields

$$Y(z) = \frac{z^3}{(z-3)\left(z^2-\frac{5}{6}z+\frac{1}{6}\right)} = \frac{z^3}{(z-3)(z-\frac{1}{2})(z-\frac{1}{3})} .$$

Using partial fraction expansion,

$$Y(z) = \frac{\frac{27}{20}z}{z - 3} - \frac{\frac{3}{5}z}{z - \frac{1}{2}} + \frac{\frac{1}{4}z}{z - \frac{1}{3}} .$$

The inverse z-transform is

$$y[n] = \frac{27}{30}(3)^n - \frac{3}{5}\left(\frac{1}{2}\right)^n + \frac{1}{4}\left(\frac{1}{3}\right)^n , \text{ for } n \geq 0.$$

• **PROBLEM** 15-38

Use z transforms to find a sequence y[n] such that

$$y[n] - \frac{1}{2}y[n-1] = 3 ; \quad n \geq 0$$

where $y[-1] = 8$.

Solution: Transforming both sides of the recursion equation yields

$$Y(z)(1 - \frac{1}{2}z^{-1}) - \frac{1}{2}y[-1] = \frac{3z}{z - 1}$$

$$Y(z)(1 - \frac{1}{2}z^{-1}) - 4 = \frac{3z}{z - 1}$$

$$Y(z) = \frac{3z}{(z-1)(1 - \frac{1}{2}z^{-1})} + \frac{4}{(1 - \frac{1}{2}z^{-1})} .$$

Hence,

$$Y(z) = \frac{6z}{z - 1} + \frac{z}{z - \frac{1}{2}}$$

is obtained using partial fraction expansion.

Hence,

$$6\,u[n] + \left(\frac{1}{2}\right)^n u[n]$$

is the inverse transform to the above equation,

and $y[n] = 6 + \left(\frac{1}{2}\right)^n$; $n \geq 0$ is the solution.

• PROBLEM 15-39

Use z-transforms to solve the homogeneous equation

$$y[n] - 7y[n-1] + 10y[n-2] = 0$$

where $y[-1] = 16$, and $y[-2] = 5$.

Solution: The z-transform of the homogeneous equation is:

$$Y[z] - 7\{z^{-1}Y[z] + y[-1]\} + 10\{z^{-2}Y[z] + z^{-1}y[-1] + y[-2]\} = 0.$$

Combining terms yields:

$$Y[z] = \frac{7y[-1] - 10\{z^{-1}y[-1] + y[-2]\}}{1 - 7z^{-1} + 10z^{-2}}.$$

Substituting 16 for y[-1] and 5 and for y[-2] and combining terms yields:

$$Y[z] = \frac{62z^2 - 160z}{z^2 - 7z + 10} .$$

Expanding into partial fractions, we obtain

$$Y[z] = \frac{12z}{z-2} + \frac{50z}{z-5}$$

and $y[n] = \{12(2)^n + 50(5)^n\}u[n]$.

Use z-transforms to solve the equation

$$y[n] - 6y[n-1] + 9y[n-2] = 0 \; ; \quad n \geq 0$$

where $y[-1] = 1$, and $y[-2] = 0$.

Solution: The above recursion equation can be written

$$y[n] = 6y[n-1] - 9y[n-2].$$

Taking into account the initial conditions, the z transform of the equation is

$$Y(z) = 6\{z^{-1}Y(z) + y(-1)\} - 9\{z^{-2}Y(z) + z^{-1}Y(-1) + Y(-2)\}$$

Combining terms yields:

$$Y(z) = Y(z)\{6z^{-1} - 9z^{-2}\} + 6y(-1) - 9\{z^{-1}y(-1) + y(-2)\}.$$

Hence,

$$Y(z) = \frac{6y[-1] - 9\left(z^{-1}y[-1] + y[-2]\right)}{1 - 6z^{-1} + 9z^{-2}}$$

Substituting 1 and 0 for $y[-1]$ and $y[-2]$ yields

$$Y(z) = \frac{6 \quad - 9z^{-1}}{1 - 6z^{-1} + 9z^{-2}}$$

Hence,

$$Y(z) = \frac{6z^2 - 9z}{z^2 - 6z + 9} = \frac{6z^2 - 9z}{(z - 3)^2}.$$

Using partial fraction expansion we obtain

$$Y(z) = \frac{6z}{z - 3} + \frac{9z}{(z - 1)^2} .$$

The inverse transform gives the solution

$$y[n] = 6(3)^n + 9n(3)^{n-1} = (6 + 3n)3^n \; ; \; n \geq 0.$$

• **PROBLEM** 15-41

Use z-transforms to solve the equation

$$y[n] - 4y[n-1] + 3y[n-2] = 8 \; ; \quad n \geq 0$$

where $y[-1] = y[-2] = 12$.

Solution: The solution to the above non-homogeneous recursion equation can be written as the sum of the particular and homogeneous solutions.

Hence, the particular solution is

$$Y_p(z) = \frac{E(z)}{C(z)}$$ where $E(z)$ is the z-transform constant term $\frac{8z}{(z-1)}$ and $C(z)$ is the characteristic equation.

The particular solution

$$Y_p(z) = \frac{8z}{(1 - 4z^{-1} + 3z^{-2})(z - 1)}$$

can be expanded into partial functions:

$$Y_p(z) = \frac{18z}{(z-3)} - \frac{4z}{(z-1)^2} - \frac{10z}{(z-1)} .$$

The homogeneous solution is found by writing the recursion equation in its homogeneous form,

$$y[n] = 4y[n-1] - 3y[n-2]$$

and taking the z-transform of both sides of the equation. Hence,

$$Y(z) = [4y[-1] - 3(z^{-1}y[-1] + y[-2])] Y(z).$$

The homogeneous solution is

$$Y_h(z) = \frac{4y[-1] - 3(z^{-1}y[-1] + y[-2])}{1 - 4z^{-1} + 3z^{-2}} ,$$

with initial conditions

$$Y_h(z) = \frac{4(12) - 3(z^{-1}(12) + 12)}{1 - 4z^{-1} + 3z^{-2}}$$

$$Y_h(z) = \frac{12z^2 - 36z}{(z-3)(z-1)} = \frac{12z(z-3)}{(z-3)(z-1)}$$

$$Y_h(z) = \frac{12z}{(z-1)} .$$

The solution is

$$Y(z) = Y_p(z) + Y_h(z) = \frac{18z}{(z-3)} - \frac{4z}{(z-1)^2} + \frac{2z}{(z-1)}$$

Taking the inverse transformation

$$y(n) = 18(3)^n - 4n + 2$$

CHAPTER 16

TWO-PORT NETWORKS

TRANSFORMERS & MUTUAL INDUCTANCE

• **PROBLEM** 16-1

Find the loop equation for the network shown.

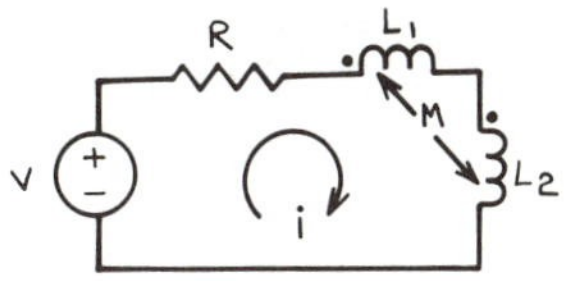

FiG.1

Solution: The current flowing into the dotted terminal of an inductor in a magnetically coupled circuit induces a voltage due to mutual inductance in the second inductor. The positive side is at the dotted terminal; hence, we write the loop equation,

$$v = Ri + (L_1 + L_2 + 2M) \frac{di}{dt} \quad .$$

Note that the voltage induced due to mutual inductance is present in both inductors and sums to $2M\frac{di}{dt}$.

• **PROBLEM** 16-2

Perform the mesh analysis on the network shown.

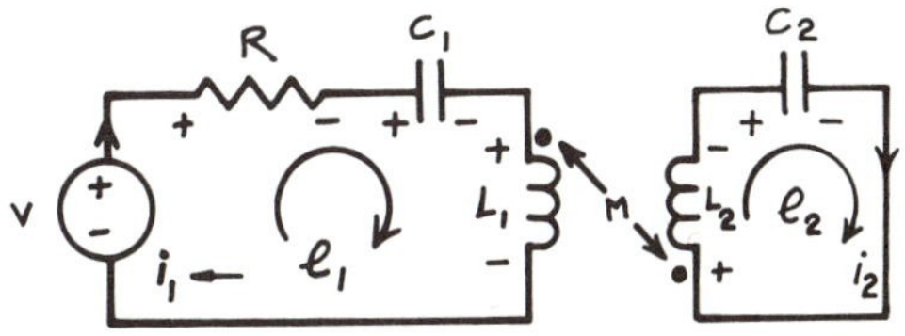

Solution: Pay careful attention to the signs of the mutual inductances. Applying KVL yields

$$l_1: \quad v = Ri_1 + \frac{1}{C_1} \int_o^t i_1(\tau)d\tau + v_{C_1}(o) + L_1 \frac{di_1}{dt} + M \frac{di_2}{dt}$$

$$l_2: \quad o = \frac{1}{C_2}\int_o^t i_2(\tau)d\tau + v_{C_2}(o) + L_2 \frac{di_2}{dt} + M \frac{di_1}{dt}$$

The sign on M is positive if both currents enter, or both leave, on a dotted terminal.

• PROBLEM 16-3

> Refer to the coupled coils shown in Fig. 1 and let L_1 = 4H, L_2 = 3H, and M = 2H. Find v_2 if (a) i_1 = 5 cos 6t A, i_2 = 0; (b) i_1 = 0, i_2 = 3 cos 6t A; (c) i_1 = 5 cos 6t, i_2 = 3 cos 6t A.

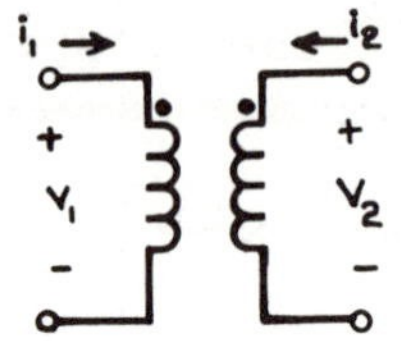

FIG.1

Solution: With the winding polarity dots as shown, the terminal voltages and currents are related as follows:

$$v_1 = L_1\frac{di_1}{dt} + M\frac{di_2}{dt} = 4\frac{di_1}{dt} + 2\frac{di_2}{dt} \tag{1}$$

$$v_2 = M\frac{di_1}{dt} + L_2\frac{di_2}{dt} = 2\frac{di_1}{dt} + 3\frac{di_2}{dt} \tag{2}$$

Since the currents are given for this problem and the secondary voltage v_2 is required, only eq. (2) is needed.

(a) i_1 = 5 cos 6t A, i_2 = 0

$$\frac{di_1}{dt} = -30 \sin 6t$$

Substituting this derivative into the equation for v_2 yields v_2 = -60 sin 6t V.

(b) i_1 = 0, i_2 = 3 cos 6t

$$\frac{di_2}{dt} = -18 \sin 6t$$

Thus v_2 = -54 sin 6t V.

(c) i_1 and i_2 can be considered as being supplied by current sources at the input and output terminals, respectively.

Since their values, in this part of the problem, are the sums of the values in the previous two parts, the principle of superposition can be used. That is, v_2 will have a value equal to the sum of its values in part (a) and (b).

$$v_2 = -60 \sin 6t - 54 \sin 6t = -114 \sin 6t \text{ V}.$$

• PROBLEM 16-4

Calculate the voltages v_1 and v_2.

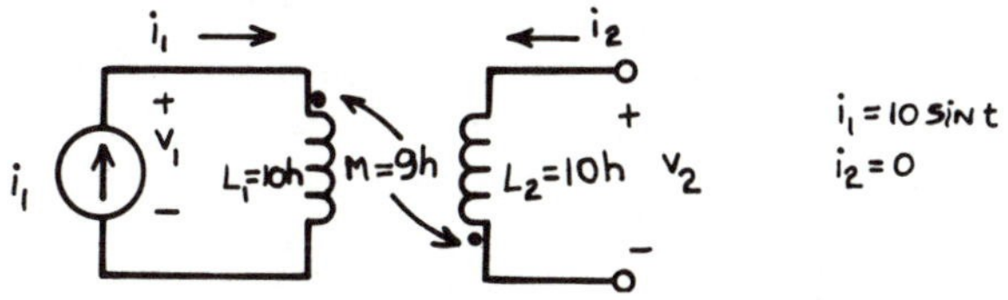

Solution: It is necessary to establish the proper sign for the self- and mutual-terms of the voltages.

Thus,

$$v_1 = L_1\frac{di_1}{dt} - M\frac{di_2}{dt} = 10\frac{d}{dt}(10 \sin t) - 9\frac{d}{dt}(o)$$

$$= 100 \text{ cost volts.}$$

The mutual term is negative since i_2 enters the undotted terminal and i_1 is positive at the undotted terminal; hence the mutual portion of v_1, $M\frac{di_2}{dt}$, must be negative.

$$v_2 = -L_2\frac{di_2}{dt} - M\frac{di_1}{dt}$$

$$= -10\frac{d}{dt}(o) - 9\frac{d}{dt}(10 \sin t)$$

$$= -90 \cos t.$$

The self-term is negative since the positive current i_2 enters the unmarked-terminal of coil L_2. The mutual portion of v_2, $M\frac{di_1}{dt}$, is negative since i_1 entering the dotted terminal of coil L_1 produces a positive polarity at the dotted terminal of coil L_2.

• PROBLEM 16-5

Assume an ideal transformer, find $a = \frac{V_2}{V_1} = \frac{I_1}{I_2}$ for the conditions shown.

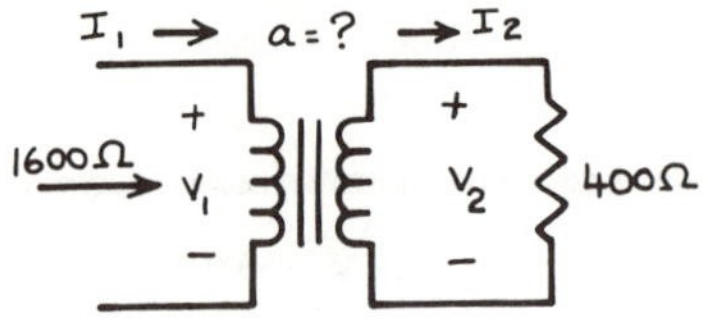

FIG. 1

Solution: If $a = \frac{V_2}{V_1} = \frac{I_1}{I_2}$ then $V_2 = aV_1 = I_2\ 400\Omega$ substituting $\frac{I_1}{a}$ for I_2, gives

$$aV_1 = \frac{I_1}{a}\ 400\Omega. \tag{1}$$

Knowing the input impedance $\frac{V_1}{I_1} = 1600\Omega$ and solving Eq. (1) for $\frac{V_1}{I_1}$ gives

$$1600\Omega = \frac{V_1}{I_1} = \frac{400\Omega}{a^2}$$

Hence,

$$a^2 = \frac{400}{1600} = \frac{1}{4}$$

$$a = \frac{1}{2}\ .$$

• PROBLEM 16-6

Find I_1, I_2 and V_2 in the circuit below. (Assume ideal transformer.)

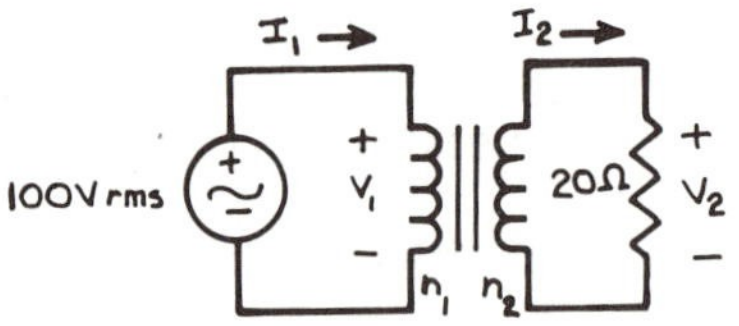

Solution: For an ideal transformer, $\frac{V_2}{V_1} = \frac{n_2}{n_1}$ and $\frac{I_2}{I_1} = \frac{n_1}{n_2}$. Thus for $n_1 = 10{,}000$ and $n_2 = 5000$,

$$\frac{V_2}{V_1} = 0.5, \quad \frac{I_2}{I_1} = 2.$$

Also, by Ohm's law, $I_2 = \frac{V_2}{R} = \frac{V_2}{20}$.

Thus for $V_1 = 100$ Vrms, $V_2 = 50$ Vrms

$I_2 = 2.5$ A rms

$I_1 = 1.25$ A rms

• PROBLEM 16-7

Find the turns ratio for the ideal transformer in Fig. 1 required to match the 200Ω source impedance to the 8Ω load.

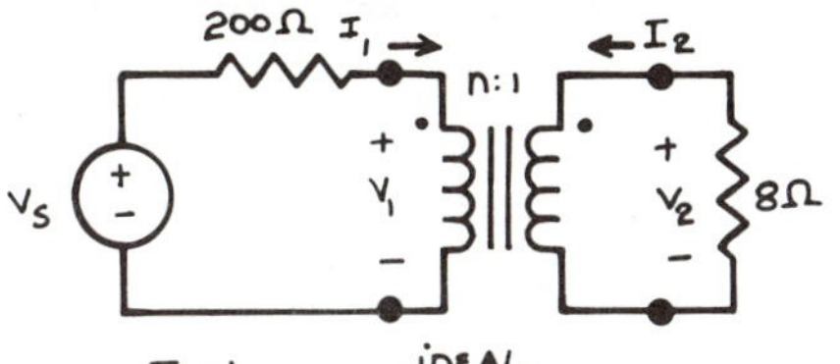

Fig. 1

Solution: Note that an ideal transformer transforms voltage and current in the following manner,

$$V_2 = \frac{1}{n} V_1$$

$$I_2 = nI_1$$

Since we are asked to match source impedance to the load impedance, $Z_p = \frac{V_1}{I_1} = 200\Omega$ looking into the primary as shown in Fig. 2 and $Z_s = \frac{V_2}{I_1} = 8\Omega$ looking into the secondary as shown in Fig. 3.

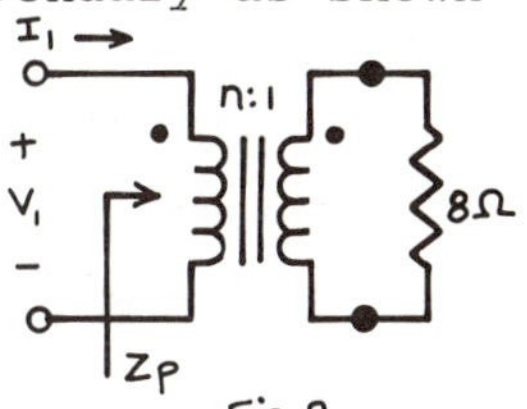

Fig. 2

Hence, solve for n

$$Z_s = \frac{V_2}{I_2} = 8 = \frac{\frac{1}{n}V_1}{nI_1} = \frac{1}{n^2}\frac{V_1}{I_1} = \frac{1}{n^2} Z_p$$

$$Z_s = 8 = \frac{1}{n^2}(200)$$

$$n^2 = \frac{200}{8}$$

$$n = \sqrt{\frac{200}{8}} = 5$$

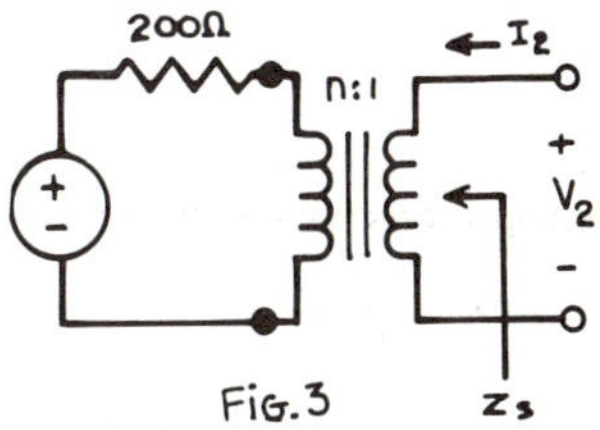

Fig. 3

to find that a turns ratio of 5 enables us to match the load impedance to the source.

• PROBLEM 16-8

Calculate the admittance $Y(s) = \frac{I_1(s)}{V_1(s)}$ for the network shown.

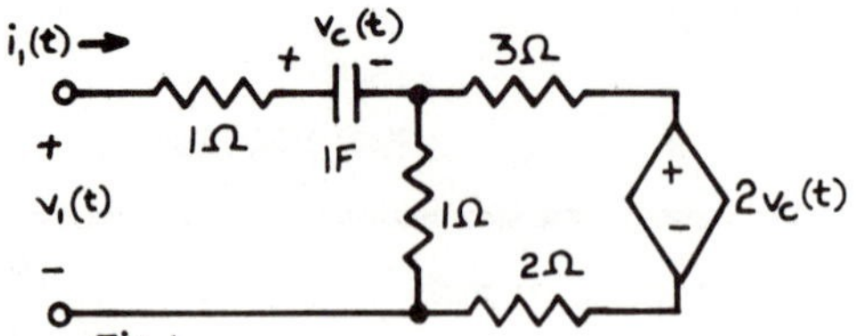

FIG.1

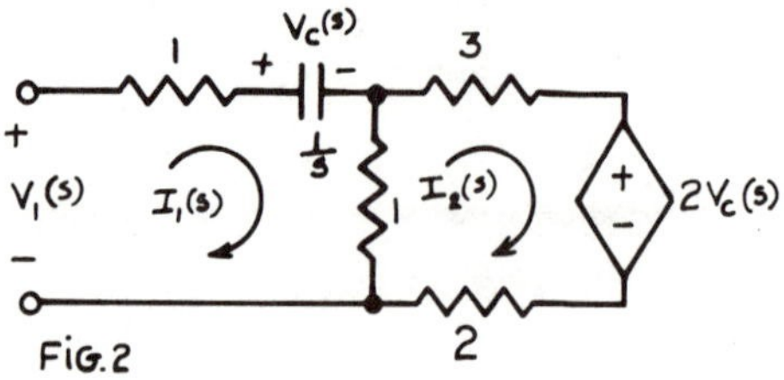

FIG.2

Solution: The circuit is transformed, as shown in Fig. 2. Writing the KVL loop equations,

$$\left(2 + \frac{1}{s}\right)I_1 - I_2 = V_1(s) \tag{1}$$

$$-I_1 + 6I_2 = -2V_c(s) \tag{2}$$

Since $V_c(s) = I_1\frac{1}{1s}$, substitute this expression for $V_c(s)$ in Eq. (2) and solve for I_2 in terms of I_1. Hence,

$$-I_1 + 6I_2 = -\frac{2}{s}I_1$$

$$I_2 = I_1\left[\frac{1}{6} - \frac{2}{6s}\right]$$

Substituting for I_2 in Eq.(1) yields

$$\left[2 + \frac{1}{s} - \frac{1}{6} + \frac{2}{6s}\right]I_1 = V_1(s)$$

the admittance is,

$$Y = \frac{I_1(s)}{V_1(s)} = \frac{1}{2 + \frac{1}{s} - \frac{1}{6} + \frac{2}{6s}}$$

$$Y = \frac{6s}{11s + 8}\ .$$

• PROBLEM 16-9

Find the input impedance Z(s) for the network shown in Fig. 1 as presented at terminal pair (a) a-b; (b) c-d; (c) a-c.

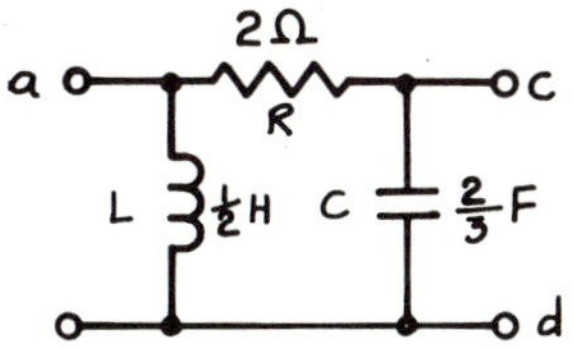

FIG. 1

Solution: The solution to this problem consists primarily of recognizing series and parallel combinations of elements.

a) $$Z_{a-b} = \frac{1}{2}s \,||\, (2+\frac{3}{2s}) = \frac{\frac{1}{2}s \times (2+\frac{3}{2s})}{\frac{1}{2}s+2+\frac{3}{2s}} = \frac{\frac{1}{2}s(4s+3)}{s^2+4s+3}$$

$$= \frac{s(4s+3)}{2(s^2+4s+3)}$$

b) $$Z_{c-d} = \frac{3}{2s} \,||\, (2+\frac{1}{2}s) = \frac{\frac{3}{2s} \times (2+\frac{1}{2}s)}{\frac{3}{2s}+2+\frac{1}{2}s} = \frac{3(s+4)}{6+8s+2s^2}$$

$$= \frac{3(s+4)}{2(s^2+4s+3)}$$

c) $$Z_{a-c} = 2 \,||\, (\frac{1}{2}s+\frac{3}{2s}) = \frac{2(\frac{1}{s}s+\frac{3}{2s})}{2+\frac{1}{2}s+\frac{3}{2s}} = \frac{2(s^2+3)}{s^2+4s+3}$$

• PROBLEM 16-10

In the circuit shown in Fig. 1 determine Z_{ab}, I_1, I_2.

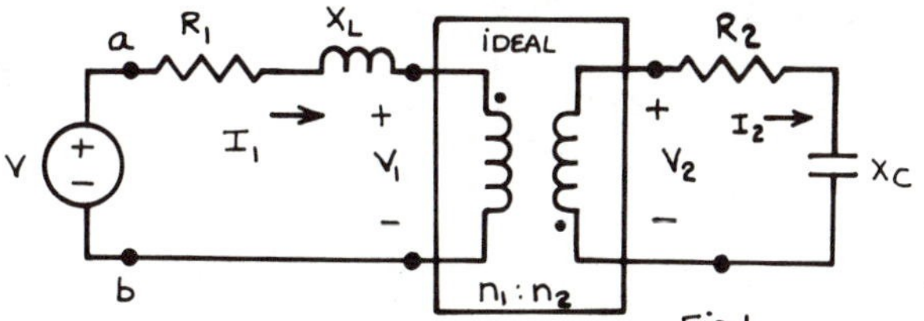

FIG. 1

Solution: The driving-point impedance at the input terminals of the ideal transformer Z_{ab}, is found by first reflecting the secondary impedance $R_2 + X_c$ to the primary side of the transformer.

Since

$$I_2 = \frac{n_1}{n_2} I_1$$

$$V_2 = \frac{n_2}{n_1} V_1$$

Therefore $$\frac{V_1}{I_1} = \frac{V_2 \frac{n_1}{n_2}}{I_2 \frac{n_2}{n_1}} = \frac{V_2}{I_1} \left(\frac{n_1}{n_2}\right)^2$$

The impedance seen at the primary side of the transformer is

$$\frac{V_2}{I_1} \left(\frac{n_1}{n_2}\right)^2 = (R_2 + jX_c) \left(\frac{n_1}{n_2}\right)^2$$

Hence, the total input impedance is

$$Z_{ab} = R_1 + R_2 \left(\frac{n_1}{n_2}\right)^2 + jX_L + jX_c \left(\frac{n_1}{n_2}\right)^2$$

and $$I_1 = \frac{V}{Z_{ab}} .$$

• PROBLEM 16-11

Find the impedance looking into the primary winding in the air-core transformer circuit of Fig. 1. Assume $L_s = 0$ and $f = 100k$ Hz.

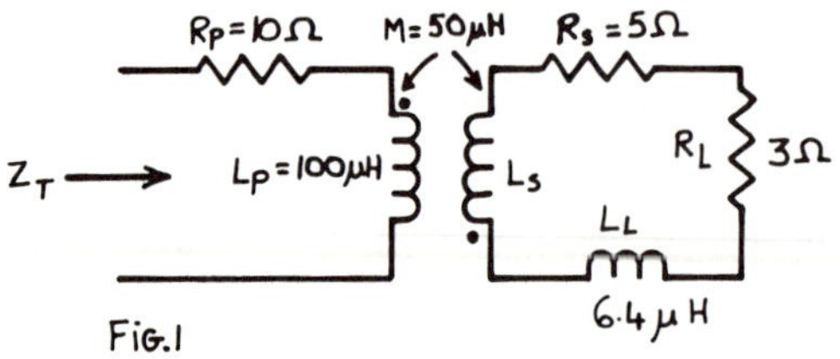

Fig.1

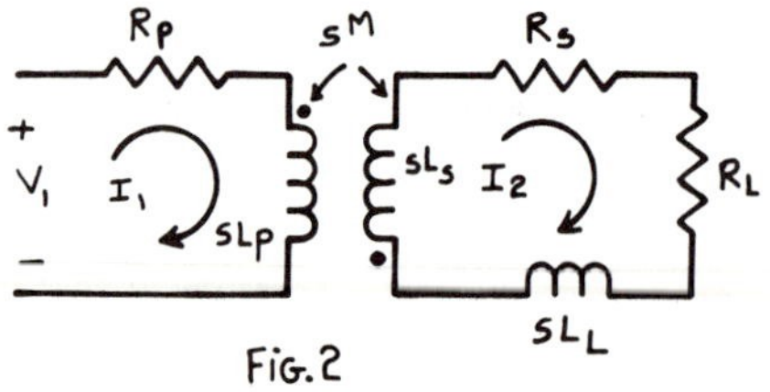

Fig.2

Solution: Writing the loop equations around the loops chosen in Fig. 2

$$V_1 = I_1(R_p + sL_p) + I_2 sM$$

$$0 = I_1 sM + I_2(R_s + sL_s + R_L + sL_L)$$

Substituting $s = j\omega = j2\pi(100 \times 10^3)$, $R_p = 10\Omega$, $R_s = 5\Omega$, $M = 50\mu H$, $L_p = 100\mu H$, $L_L = 6.4\mu H$, $R_L = 3\Omega$, and $L_s = 0$, gives

$$V_1 = I_1(10 + j62.8) + I_2(j31.4) \quad (1)$$

$$0 = I_1(j31.4) + I_2(8 + j4) \quad (2)$$

Solve for $Z_T = \frac{V_1}{I_1}$ as follows, from Eq.(2):

$$I_2 = \frac{-j31.4}{8 + j4} I_1$$

substituting into Eq.(1)

$$V_1 = I_1(10 + j62.8) - I_1 \frac{(j31.4)(j31.4)}{8 + j4}$$

Hence,

$$Z_T = \frac{V_1}{I_1} = (10 + j62.8) - \frac{(j31.4)(j31.4)}{8 + j4}$$

$$Z_T = \frac{V_1}{I_1} = 10 + j62.8 + \frac{986}{8 + j4}$$

$$Z_T = \frac{V_1}{I_1} = 10 + j62.8 + \frac{986(8 - j4)}{8^2 + 4^2}$$

$$Z_T = \frac{V}{I_1} = 10 + j62.8 + 98.6 - j49.3$$

$$Z_T = \frac{V}{I_1} = 108.6 + j13.5\Omega$$

• PROBLEM 16-12

Two inductors, $L_1 = 25L_2 = 1H$, are mutually coupled with a mutual impedance of $\frac{1}{5}\Omega$. An impedance $\vec{Z}_L$ is connected across L_2. Find the input impedance at the terminals of L_1 at $\omega = 1000$ rad/s if $\vec{Z}_L$ =: (a) 4; (b) j4; (c) -j4Ω.

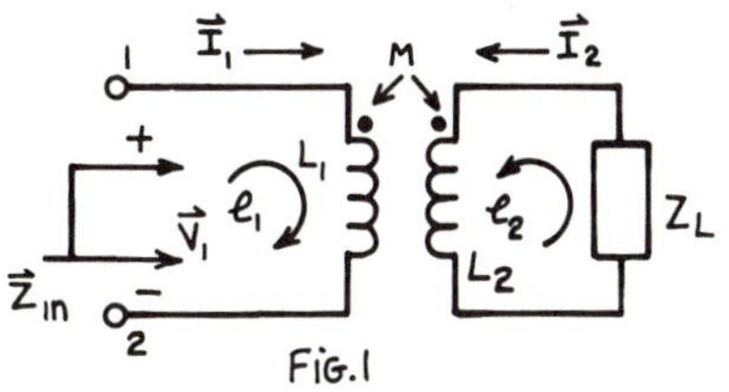

FIG.1

Solution: The circuit for this problem is shown in Fig. 1.

The input impedance which is the impedance looking into points 1 and 2, is the ratio of $\vec{V}_1$ to $\vec{I}_1$. To find this ratio use KVL around loops ℓ_1 and ℓ_2.

$$\vec{V}_1 = \vec{I}_1(j\omega L_1) + \vec{I}_2(j\omega M) = \vec{I}_1(j1000) + \vec{I}_2(j200) \quad (1)$$

$$0 = \vec{I}_1(j\omega M) + \vec{I}_2(Z_L + j\omega L_2) \quad (2)$$

Since L_1 = 1H and $L_1 = 25L_2$; then $L_2 = \frac{1}{25}$ H. Equation (2) is rewritten as:

$$0 = \vec{I}_1(j200) + \vec{I}_2(Z_L + j40) \quad (3)$$

or $$\vec{I}_2 = -\vec{I}_1\left(\frac{j200}{Z_L + j40}\right) \quad (4)$$

Substituting this into equation (1) yields:

$$\vec{V}_1 = \vec{I}_1\left(j1000 - \frac{(j200)(j200)}{Z_L + j40}\right)$$

Now Z_{in} can be found.

$$\vec{Z}_{in} = \frac{\vec{V}_1}{\vec{I}_1} = j1000 + \frac{200^2}{Z_L + j40} = \frac{Z_L(j1000)}{Z_L + j40} \quad (5)$$

a) For this part $Z_L = 4\Omega$. Inserting this value into equation (5) yields:

$$\vec{Z}_{in} = \frac{j4000}{4 + j40} = \frac{j1000}{1 + j10} = \frac{1000\angle 90°}{10.05\angle 84.3°} = 99.5\angle 5.71°\Omega$$

or $$\vec{Z}_{in} = 99.0 + j9.90\ \Omega$$

b) For this part $Z_L = j4\Omega$. Inserting this value into equation (5) yields:

$$\vec{Z}_{in} = \frac{(j4)j(1000)}{j4 + j40} = \frac{-4000}{j44} = \frac{4000\angle 180°}{44\angle 90°} = 90.9\angle 90°\ \Omega$$

or $$\vec{Z}_{in} = j90.9\ \Omega$$

c) For this part $Z_L = -j4\Omega$. Inserting this value into equation (5) yields:

$$Z_{in} = \frac{(-j4)(j1000)}{-j4 + j40} = \frac{4000}{j36} = \frac{4000\angle 0°}{36\angle 90°} = 111.1\angle -90° \ \Omega$$

or $\quad Z_{in} = -j111.1 \ \Omega$

• PROBLEM 16-13

> Find Z_{ab} for the network of Fig. 1.

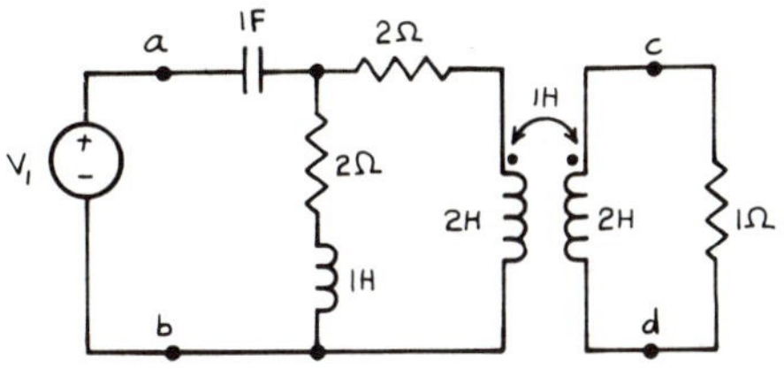

Fig. 1

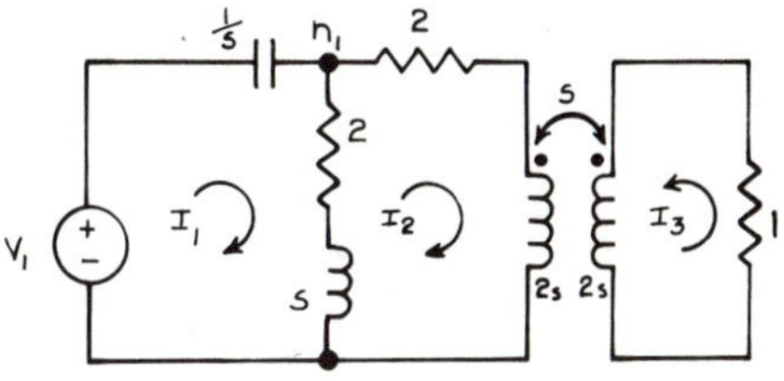

Fig. 2

Solution: Before we find Z_{ab}, assign currents in the loops of the circuit in Fig. 1. Fig. 2 shows the Laplace transformed circuit with currents assigned.

First find the equivalent impedance to the right of node n_1 in Fig. 2, by writing the following loop equations

$$V_2 = I_2(2s + 2) + I_3(s)$$

$$0 = I_2(s) \qquad + I_3(2s + 1)$$

Hence,

$$\frac{V_2}{I_2} = 2s + 2 + s\left(\frac{-s}{2s+1}\right)$$

$$\frac{V_2}{I_2} = \frac{2s(2s+1) + 2(2s+1) - s^2}{2s + 1} = \frac{3s^2 + 6s + 2}{2s + 1}$$

Now replace every thing to the right of node n_1 in Fig. 2 by the impedance $\frac{3s^2+4s+2}{2s+1}$, and proceed to find $\frac{V_1}{I_1}$ in Fig. 4.

Fig. 3

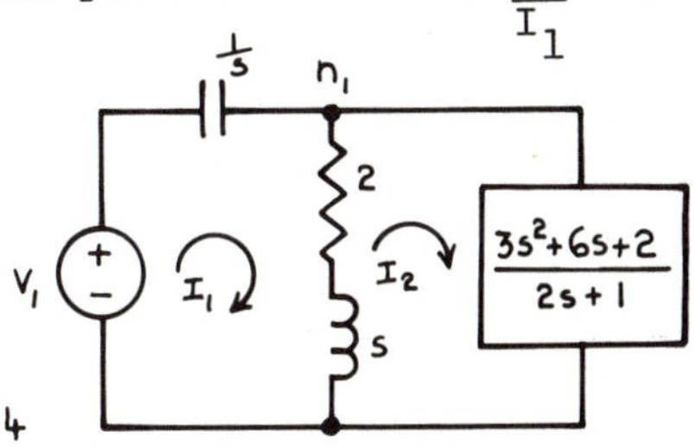

Fig. 4

Writing the loop equations yields

$$V_1 = I_1\left(\frac{s^2+2s+1}{s}\right) - I_2(s+2)$$

$$0 = I_1(s+2) + I_2\left(\frac{3s^2+6s+2}{2s+1} + s + 2\right)$$

Hence,

$$Z_{ab} = \frac{V_1}{I_1} = \frac{s^2+2s+1}{s} - (s+2)\frac{(s+2)}{\frac{3s^2+6s+2}{2s+1} + s + 2} .$$

$$Z_{ab} = \frac{V_1}{I_1} = \frac{s^2+2s+1}{s} - \frac{(s+2)^2}{\frac{3s^2+6s+2+s(2s+1)+2(2s+1)}{2s+1}}$$

$$Z_{ab} = \frac{V_1}{I_1} = \frac{(s^2+2s+1)(5s^2+11s+4) - s(s+2)^2(2s+1)}{5s^3 + 11s^2 + 4s}$$

$$Z_{ab} = \frac{V_1}{I_1} = \frac{3s^4+12s^3+19s^2+15s+4}{5s^3+11s^2+4s}$$

• **PROBLEM 16-14**

In the circuit shown in Fig. 1 find: (a) $\vec{I}_2$; (b) $\vec{I}_1$; (c) $\vec{I}_R$.

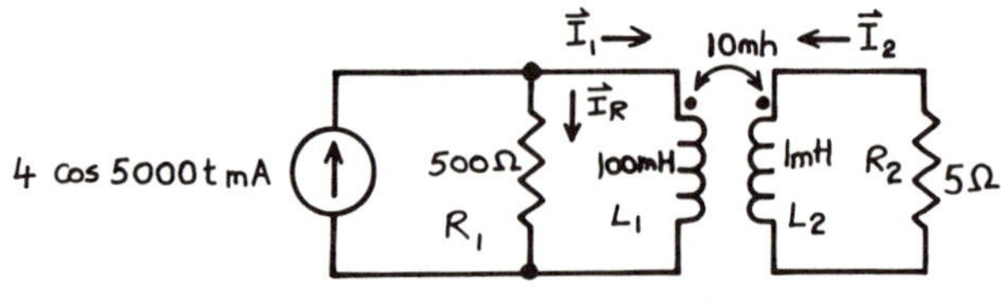

FIG. 1

Solution: This problem will be solved by simultaneous equations. The voltage across L_1, which will be called $\vec{V}_1$, is composed of two parts.

$$\vec{V}_1 = \vec{I}_1(j\omega(.1)) + I_2(j\omega(.01)) \tag{1}$$

In this equation .1H is the self-inductance of L_1 and .01H is the mutual inductance of L_1 and L_2. The voltage across L_2, which will be called $\vec{V}_2$, is also composed of two parts.

$$\vec{V}_2 = \vec{I}_2(j\omega(.001)) + \vec{I}_1(j\omega(.01))$$

In this equation .001H is the self-inductance of L_2 and,

again, .01H is the mutual inductance of L_1 and L_2. KVL around the loop containing L_2 and R_2 gives:

$$0 = \vec{I}_2(5 + j\omega(.001)) + \vec{I}_1(j\omega(.01)) \qquad (2)$$

Notice that the voltage across L_1 is also the voltage across R_1

$$\vec{V}_1 = \vec{I}_R 500 \qquad (3)$$

There are now 3 equations and 4 unknowns: variables $\vec{I}_1$, $\vec{I}_R$, $\vec{I}_2$, and $\vec{V}_1$. For the fourth equation:

$$\vec{I}_1 + \vec{I}_R = 4\underline{/0°}\text{ mA} = .004\underline{/0°}\text{ A} \qquad (4)$$

Therefore,

$$\vec{I}_R = .004 = \vec{I}_1 \qquad (5)$$

Putting equation (5) into (3) gives $\vec{V}_1$ as a function of $\vec{I}_1$.

$$\vec{V}_1 = 500(.004-\vec{I}_1) = 2-500\vec{I}_1 \qquad (6)$$

Substituting $\vec{V}_1$ in (1) by equation (6) gives

$$2-500\vec{I}_1 = \vec{I}_1(j\omega(.1)) + \vec{I}_2(j\omega(.01))$$

Solving for $\vec{I}_1$ gives:

$$\vec{I}_1 = \frac{2 - \vec{I}_2(j\omega(.01)}{500 + j\omega(.1)} \qquad (7)$$

Substituting $\vec{I}_1$ in (2) by (7) gives

$$0 = \vec{I}_2(5+j\omega(.001)) + \frac{[j\omega(.01)][2-\vec{I}_2(j\omega(.01))]}{500 + j\omega(.1)}$$

Hence,

$$\vec{I}_2 = \frac{-j\omega(.02)}{2500+j\omega} \qquad (8)$$

The current source is 4 cos ωt = 4 cos 5000 t; hence $\omega = 5000$.

$\vec{I}_2$ can now be solved from equation (8)

$$\vec{I}_2 = \frac{j(5000)(.02)}{2500 + j(5000)} = \frac{-j}{25 + j50}$$

$$\vec{I}_2 = \frac{1\angle -90°}{56\angle 63.43°} = 17.9\angle -153.4° \text{ mA}$$

b) To find $\vec{I}_1$, substitute the value of $\vec{I}_2$ into equation (2)

$$\vec{I}_2(5 + j5) = -\vec{I}_1(j50)$$

$$\vec{I}_1 = -\vec{I}_2 \frac{(5+j5)}{j50} = \frac{+j}{25+j50}\left[\frac{5+j5}{j50}\right]$$

$$\vec{I}_1 = \frac{-5+j5}{-2500+j1250} = \frac{7.05\angle 135°}{2795\angle 153.4°} = 2.53\angle -18.4° \text{ mA}$$

c) To find $\vec{I}_R$, substitute the value of $\vec{I}_1$ into equation (5)

$$\vec{I}_R = 4\angle 0° - 2.53\angle -18.4°$$

$$= 4 - [2.4 - j0.8]$$

$$= 1.6 + j0.8$$

$$\vec{I}_R = 1.79\angle 26.6° \text{ mA}$$

• PROBLEM 16-15

For the circuit shown solve for i_2.

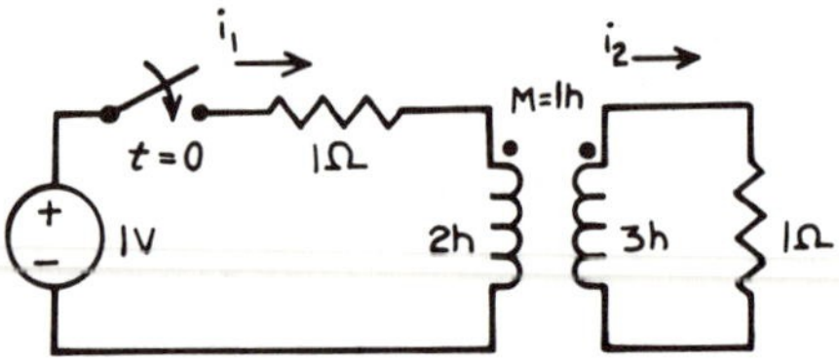

Solution: The first step of solving this type of problem is to set up the loop equations and then Laplace transform these equations.

For the primary circuit,

$$R_1I_1(s) + sL_1I_1(s) - sMI_2(s) = \frac{1}{s} \quad (1)$$

where $I_1(s) = L[i_1(t)]$

$$I_2(s) + L[i_2(t)] \quad (2)$$

$$\frac{1}{s} = L[u(t)]$$

Equation (1) is further modified as follows:

$$(R_1+sL_1)I_1(s) - sMI_2(s) = \frac{1}{s} \tag{3}$$

In this case

$$R_1 = 1\ (\Omega)$$

$$L_1 = 2\ (H) \tag{4}$$

$$M = 1\ (H)$$

$$(1+2s)I_1(s) - sI_2(s) = \frac{1}{s} \tag{5}$$

For the secondary circuit,

$$sL_2I_2 + R_2I_2(s) - sMI_1(s) = 0$$

$$-sMI_1(s) + (R_2+sL_2)I_2(s) = 0 \tag{6}$$

where $R_2 = 1\ (\Omega)$

$$L_2 = 3\ (H) \tag{7}$$

with Eqs. (4), (6) and (7),

$$-sI_1(s) + (1+3s)I_2(s) = 0 \tag{8}$$

$$I_1(s) = \frac{(3s+1)}{s} I_2(s) \tag{9}$$

Substituting Eq.(9) into Eq.(5),

$$\left\{\frac{(2s+1)(3s+1)}{s} - s\right\} I_2(s) = \frac{1}{s} \tag{10}$$

$$\{(2s+1)(3s+1) - s^2\} I_2(s) = 1$$

$$(5s^2+5s+1)I_2(s) = 1$$

$$I_2(s) = \frac{1}{5s^2+5s+1} = \frac{1}{5(s^2+5+\frac{1}{5})} = \frac{0.2}{(s^2+5+0.2)}$$

$$\equiv \frac{0.2}{(s+a)(s+b)} \tag{11}$$

Then $a + b = 1$ (12)

$ab = 0.2$ (13)

From Eq.(12) $b = 1-a$ (14)

Substituting Eq.(14) in Eq.(13),

$$a(1-a) = 0.2$$

$$-a^2+a = 0.2$$

$$-a^2+a-0.2 = 0$$

$$a = \frac{-1 \pm \sqrt{1-4x0.2}}{-2} = 0.5 \pm 0.2236$$

$$= 0.73 \quad (15)$$
or
$$0.27 \quad (16)$$

Substituting (15) into Eq.(14)

$$b = 0.27 \quad (17)$$

Substituting (16) into Eq.(14)

$$b = 0.73 \quad (18)$$

From Eqs.(11), (15) and (17)

$$I_2(s) = \frac{0.2}{(s+0.73)(s+0.27)} \quad (19)$$

Using partial fractions

$$\frac{0.2}{(s+0.73)(s+0.27)} = \frac{A}{s+0.73} + \frac{B}{s+0.27} \quad (20)$$

Multiplying through by (s+0.73)

$$\frac{0.2}{s+0.27} = A + \frac{B(s+0.73)}{s+0.27} \quad (20)$$

Let s = -0.73

$$\frac{0.2}{-0.46} = A, \quad A = -0.435 \quad (21)$$

Multiplying through by (s+0.27) in Eq.(20),

$$\frac{0.2}{s+0.73} = \frac{A(s+0.27)}{s+0.73} + B \quad (22)$$

Let s = -0.27

$$\frac{0.2}{0.46} = B \qquad B = 0.435 \quad (23)$$

Substituting Eqs.(21) and (23) into Eq.(20)

$$I_2(s) = \frac{-0.435}{s+0.73} + \frac{0.435}{s+0.27} \quad (24)$$

$$= 0.435 \left(\frac{1}{s+0.27} - \frac{1}{s+0.73}\right) \tag{25}$$

$$i_2(t) = L^{-1}[I_2(s)]$$

$$= L^{-1}\left[0.435 \left(\frac{1}{s+0.27} - \frac{1}{s+0.73}\right)\right] \tag{26}$$

$$= 0.435 \left[L^{-1}\left[\frac{1}{s+0.27}\right] - L^{-1}\left[\frac{1}{s+0.73}\right]\right] \tag{27}$$

Knowing that $L^{-1}\left[\frac{1}{s+a}\right] = e^{-at}u(t)$ (28)

$$i_2(t) = 0.435 \left[e^{-0.27t}u(t) - e^{-0.73t}u(t)\right]$$

$$= 0.435 \left[e^{-0.27t} - e^{-0.73t}\right] u(t) \tag{29}$$

It should be noted that all initial conditions are zero. If not, these conditions must be taken into consideration in the Laplace transform procedure.

• PROBLEM 16-16

A step-up transformer has a primary winding resistance of 0.05 Ω and a secondary winding resistance of 1 Ω. Primary and secondary leakage reactances are 0.01 and 2 Ω, respectively. Turns ratio is 10; shunt resistance referred to the primary is 100 Ω and shunt reactance is j200 Ω. The primary source voltage is 100 V and the load on the secondary is $50\underline{/25°}$ Ω. Determine the current drawn from the source and the impedance looking into the primary.

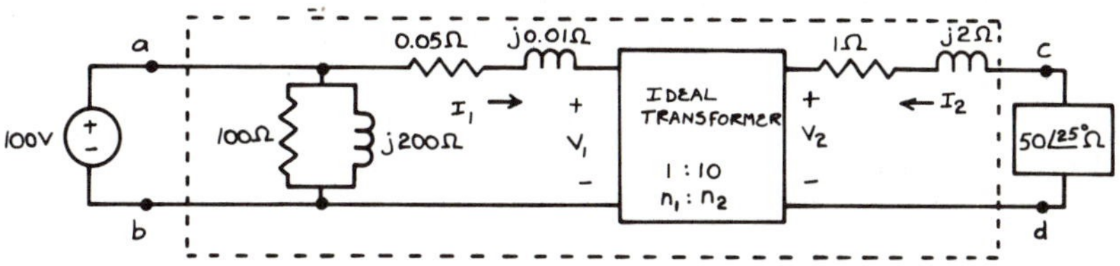

FIG.1

Solution: Using the information provided about the transformer, draw an equivalent circuit containing an ideal transformer, resistors, and reactances.

In an ideal transformer the primary and secondary voltages and currents are related in the following way:

$$V_2 = \frac{n_2}{n_1} V_1$$

$$I_2 = \frac{I_1}{\frac{n_2}{n_1}}$$

Hence, the impedance looking into the primary winding of an ideal transformer is

$$\frac{V_1}{I_1} \text{ which is } \frac{\frac{V_2}{\frac{n_2}{n_1}}}{\frac{n_2}{n_1} I_2} = \frac{n_1^2}{n_2^2} \frac{V_2}{I_2}$$

Applying this information to the equivalent circuit of Fig. 1 gives the circuit shown in Fig. 2.

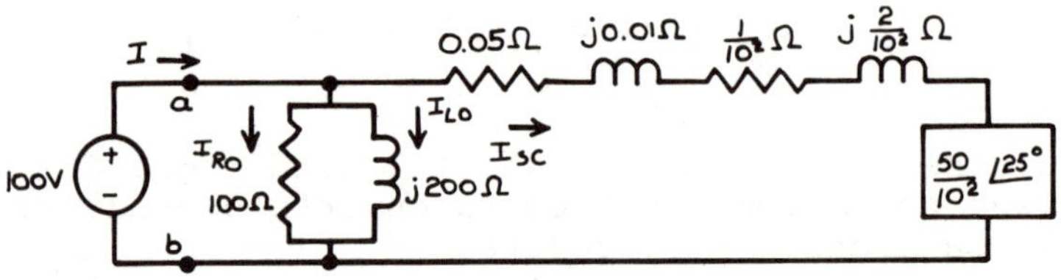

FIG. 2

The impedance looking into the primary is now the equivalent impedance to the right of terminals a, b. Converting the load impedance to rectangular form gives

$$\frac{R_L}{10^2} = \frac{Z_L}{10^2} \cos 25° = \frac{50}{10^2} (0.906) = \frac{45.32}{10^2} \ \Omega$$

$$\frac{X_L}{10^2} = \frac{Z_L}{10^2} \sin 25° = \frac{50}{10^2} (0.423) = \frac{21.13}{10^2} \ \Omega .$$

Summing the series combination of the transformed load, transformed secondary winding impedance, and primary winding impedance gives

$$\frac{45.32}{10^2} + j\frac{21.13}{10^2}$$

$$\frac{1}{10^2} + j\frac{2}{10^2}$$

$$\underline{0.05 \quad + j0.01}$$

$$0.513 + j0.241 \ \Omega$$

In polar form $0.513 + j0.241 = 0.567\underline{/25.16°}$.

The current flowing in the series portion of the circuit is

$$I_{sc} = \frac{100\angle 0^\circ}{0.567\angle 25.16^\circ} = 176\angle -25.16^\circ .$$

The currents flowing through the shunt resistance and shunt reactance are

$$I_{Ro} = \frac{100\angle 0^\circ}{100} = 1$$

$$I_{Lo} = \frac{100\angle 0^\circ}{200\angle 90^\circ} = 0.5\angle -90^\circ = -j0.5$$

The sum of the currents at terminal a gives the current drawn from the source

$$I = I_{sc} + I_{Ro} + I_{Lo} = 176\angle -25.16^\circ + 1 - j0.5$$

$$I = 159.3 - j74.83 + 1 - j0.5$$

$$I = 160.3 - j75.33 = 177\angle -25.17^\circ \text{ A}$$

The total impedance looking into terminals a, b is

$$\frac{100\angle 0^\circ}{177\angle -25.17^\circ} = 0.565\angle 25.17^\circ \ \Omega$$

NETWORK PARAMETERS

• PROBLEM 16-17

Given the resistive two-port network of Fig. 1, determine (a) y_{11}; (b) y_{12}; (c) y_{22}.

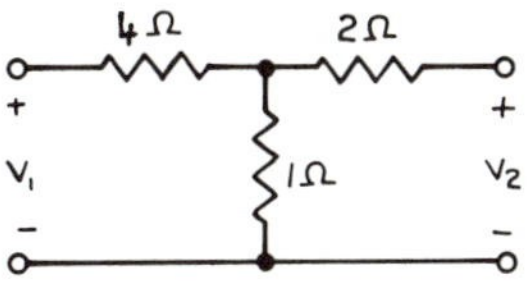

FiG.1

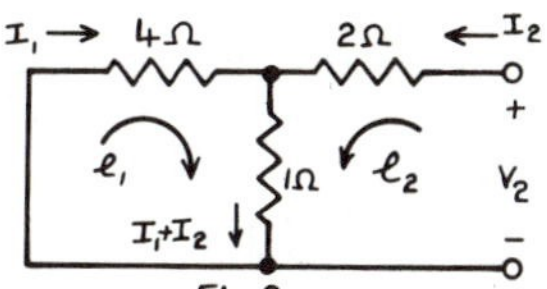

FiG. 2

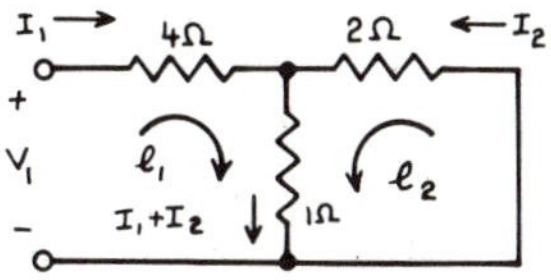

FiG.3

Solution: The Y-parameters are given by

$$I_1 = y_{11}V_1 + y_{12}V_2$$

$$I_2 = y_{21}V_1 + y_{22}V_2$$

Then $y_{11} = \left.\frac{I_1}{V_1}\right|_{V_2 = 0}$, $y_{12} = \left.\frac{I_1}{V_2}\right|_{V_1 = 0}$, and $y_{22} = \left.\frac{I_2}{V_2}\right|_{V_1 = 0}$.

First look at the network with $V_1 = 0$ (see Fig. 2), writing loop equations:

$$\ell_1: \quad 4I_1 + i_1 + I_2 = 0 \qquad (1)$$

$$\ell_2: \quad V_2 = 2I_2 + I_1 + I_2 \qquad (2)$$

from (1): $-5I_1 = I_2$

inserting this into (2) we have

$$V_2 = 2(-5I_1) + I_1 - 5I_1$$

$$V_2 = -14I_1$$

so $\quad y_{12} = \dfrac{I_1}{V_2} = -\dfrac{1}{14}$ mhos

Next, inserting $I_1 = -\dfrac{I_2}{5}$ into (2),

$$V_2 = 2I_2 - \frac{I_2}{5} + I_2$$

$$V_2 = \frac{14}{5} I_2$$

so $\quad y_{22} = \dfrac{I_2}{V_2} = \dfrac{5}{14}$ mho

Next, look at the network for $V_2 = 0$ (see Fig. 3). Again writing loop equations:

$$(1) \quad V_1 = 4I_1 + I_1 + I_2 \qquad (3)$$

$$(2) \quad 2I_2 + I_1 + I_2 = 0 \qquad (4)$$

From (4) we have

$$I_2 = \frac{-I_1}{3}$$

so from (3) $\quad V_1 = 4I_1 + I_1 - \dfrac{I_1}{3}$

$$V_1 = \frac{14}{3} I_1$$

and $\quad y_{11} = \dfrac{I_1}{V_1} = \dfrac{3}{14}$ mho

• PROBLEM 16-18

Find the hybrid parameters for the network shown in Fig. 1.

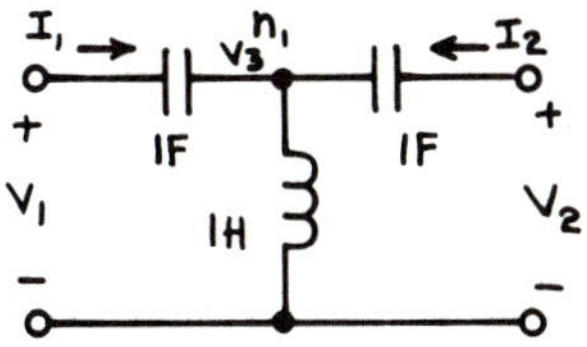

Fig.1

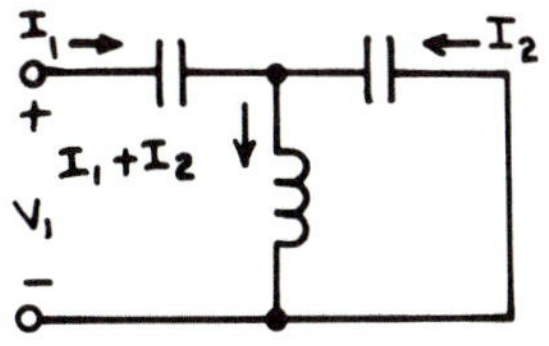

Fig.2

Solution: The hybrid parameters are defined as

$$h_{11}(s) = \left.\frac{V_1}{I_1}\right|_{V_2=0} \qquad h_{12}(s) = \left.\frac{V_1}{V_2}\right|_{I_1=0}$$

$$h_{21}(s) = \left.\frac{I_2}{I_1}\right|_{V_2=0} \qquad h_{22}(s) = \left.\frac{I_2}{V_2}\right|_{I_1=0}$$

Note that the parameter $\left.\frac{V_1}{I_1}\right|_{V_2=0}$ is nothing more than the input impedance when the output is short-circuited. Hence,

$$h_{11}(s) = \left.\frac{V_1}{I_1}\right|_{V_2=0} = \frac{1}{s} + [\tfrac{1}{s}||s]$$

$$h_{11}(s) = \frac{1}{s} + \frac{1}{\frac{1}{s} + s}$$

$$h_{11}(s) = \frac{1}{s} + \frac{s}{s^2 + 1}$$

$$h_{11}(s) = \frac{2s^2 + 1}{s(s^2+1)}$$

The parameter h_{12} is the inverse voltage gain when the input is open-circuited. The voltage at node n_1 equal to V_1 and the current through the inductor is I_2, thus

$$h_{12}(s) = \left.\frac{sI_2}{(s + \frac{1}{s})I_2}\right|_{I_1 = 0} = \frac{s}{s + \frac{1}{s}} = \frac{s^2}{s^2 + 1}$$

Note that the parameter $h_{21}(s)$ is the current gain when the output is short-circuited. Using the circuit shown in Fig. 2 below write the following two equations:

$$V_1 = I_1(\tfrac{1}{s}) + (I_1+I_2)(s) \qquad (1)$$

$$I_2 = \frac{V_1 - I_1(\frac{1}{s})}{\frac{1}{s}} = sV_1 - I_1 \qquad (2)$$

Solving Eqs. (1) and (2) for I_1 and I_2 in terms of V_1 gives

$$I_1 = V_1 s(1-s^2)$$

$$I_2 = V_1 s(s^2)$$

Hence,

$$h_{21}(s) = \left.\frac{I_2}{I_1}\right|_{V_2=0} = \frac{s^2}{1-s^2} = \frac{-s^2}{s^2+1}$$

The parameter h_{22} is the admittance looking into the output with the input open-circuited. Hence,

$$h_{22}(s) = \left.\frac{I_2}{V_2}\right|_{I_1=0} = \frac{1}{s+\frac{1}{s}} = \frac{s}{s^2+1} .$$

• PROBLEM 16-19

Find the h-parameters for the network shown.

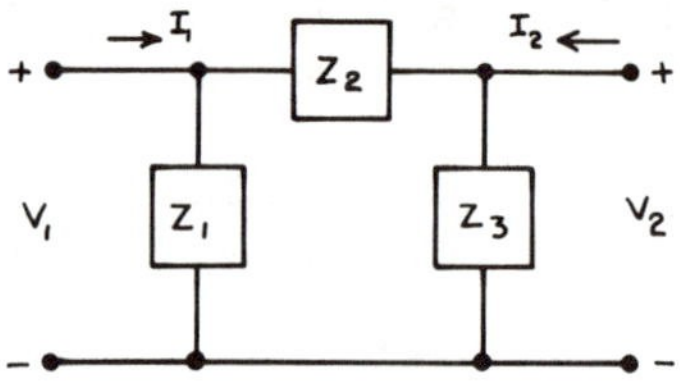

Solution: To solve this problem, it is important to know the definitions of h-parameters.

$$h_{11} \equiv \left.\frac{V_1}{I_1}\right|_{V_2=0} \quad \text{(self impedance at port-1)} \qquad (1)$$

By **short** circuiting Z_3, h_{11} is the parallel impedance of Z_1 and Z_2

$$h_{11} = \frac{Z_1 Z_2}{Z_1 + Z_2} \qquad (2)$$

$$h_{22} \equiv \left.\frac{I_2}{V_2}\right|_{I_1=0} \quad \text{(self admittance at port-2)} \qquad (3)$$

Now let $I_1 = 0$ (port 1 open). Thus h_{22} is the parallel admittance of Z_3 and series connection of Z_2 and Z_1

$$h_{22} = \frac{Z_1+(Z_2+Z_3)}{(Z_1+Z_2)Z_3} \tag{4}$$

$$h_{12} \equiv \left.\frac{V_1}{V_2}\right|_{I_1=0} \quad \text{(voltage transfer ratio from Port 2 to Port 1)} \tag{5}$$

Using the current division principle, the current in (Z_1+Z_2) with $I_1=0$ is,

$$I_2 = \frac{Z_3}{Z_1+Z_2+Z_3} \tag{6}$$

Then, the voltage across Z_1 is,

$$V_1 = V_2 - Z_2 \left(I_2 \frac{Z_3}{Z_1+Z_2+Z_3}\right) \tag{7}$$

$$I_2 = \frac{V_2}{\dfrac{Z_3(Z_1+Z_2)}{Z_1+Z_2+Z_3}} = V_2 \frac{Z_1+Z_2+Z_3}{Z_3(Z_1+Z_2)} \tag{8}$$

Substituting Eq. (8) into Eq. (7)

$$V_1 = V_2 - Z_2V_2 \frac{Z_1+Z_2+Z_3}{Z_3(Z_1+Z_2)} \; \frac{Z_3}{Z_1+Z_2+Z_3} = V_2 - \frac{Z_2}{Z_1+Z_2} V_2$$

$$= \left(1 - \frac{Z_2}{Z_1+Z_2}\right)V_2 = \frac{Z_1}{Z_1+Z_2} V_2 \tag{9}$$

So,
$$h_{12} \equiv \left.\frac{V_1}{V_2}\right|_{I_1=0} = \frac{Z_1}{Z_1+Z_2} \tag{10}$$

Now

$$h_{21} \equiv \left.\frac{I_2}{I_1}\right|_{V_2=0} \quad \text{(short circuit current transfer ratio from Port 1 to Port 2)} \tag{11}$$

Since $V_2=0$ or Port 2 is short circuited, I_2 flows in a direction opposite to that originally assigned. So, h_{21} is ordinarily negative. By the current division principle,

$$-I_2 = I_1 \frac{Z_1}{Z_1+Z_2} \tag{12}$$

Combining Eq. (12) with Eq. (11),

$$-h_{21} \equiv \left.\frac{I_2}{I_1}\right|_{V_2=0} = \frac{Z_1}{Z_1+Z_2} \tag{12}$$

h-parameters are also called hybrid parameters and originated from the following simultaneous equations.

$$\begin{aligned} V_1 &= h_{11}I_1 + h_{12}V_2 \\ I_2 &= h_{21}I_1 + h_{22}V_2 \end{aligned} \tag{13}$$

• PROBLEM 16-20

Find the four short-circuit admittance parameters for the circuit shown.

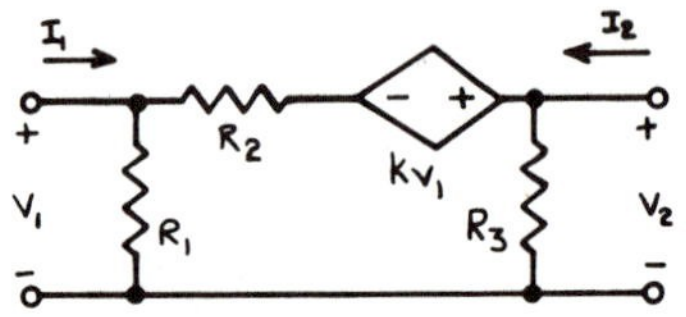

Solution: If the primary current I_1 and the secondary current I_2 are assigned as shown in the figure, the current equations involving the four short-circuited admittance parameters are given, by definition, as follows:

$$I_1 = y_{11}V_1 + y_{12}V_2 \tag{1}$$

$$I_2 = y_{21}V_1 + y_{22}V_2 \tag{2}$$

1. Admittance y_{11} is obtained by letting

$$V_2 = 0 \tag{3}$$

or by short circuiting the secondary terminal. In Eq. (1), under this condition

$$I_1 = y_{11}V_1$$

or
$$y_{11} = \frac{I_1}{V_1} \tag{4}$$

With R_3 short circuited in the given figure, the current flowing through R_1 is

$$\frac{V_1}{R_1} \tag{5}$$

The current flowing through R_2 is

$$\frac{V_1 + kV_1}{R_2} \tag{6}$$

Then, the summation of current (5) and current (6) must be equal to the input current I_1. So

$$I_1 = \frac{V_1}{R_1} + \frac{V_1 + kV_1}{R_2} = (\frac{1}{R_1} + \frac{1 + k}{R_2})V_1 \tag{7}$$

$$\frac{I_1}{V_1} = \frac{1}{R_1} + \frac{1 + k}{R_2} \tag{8}$$

Combining Eqs. (4) and (8),

$$y_{11} = \frac{1}{R_1} + \frac{1 + k}{R_2} \tag{9}$$

2. Admittance y_{21} is obtained by letting

$$V_2 = 0 \tag{10}$$

or by short circuiting the secondary terminal and removing R_3. Under this condition, in Eq. (2)

$$I_2 = y_{21}V_1 \tag{11}$$

So $$y_{21} = \frac{I_2}{V_1} \tag{12}$$

Under this condition I_2 is assigned in the opposite direction of current (6)

so $$I_2 = - \frac{V_1 + kV_1}{R_2} = - \frac{1 + k}{R_2} V_1$$

$$\frac{I_2}{V_1} = - \frac{1 + k}{R_2} \tag{13}$$

by combining (12) and (13),

$$y_{21} = - \frac{1 + k}{R_2} \tag{14}$$

3. Admittance y_{12} is obtained by letting

$$V_1 = 0 \tag{15}$$

in Eq. (1), or by short circuiting the primary terminals and removing R_1 in the figure shown. Under this condition

$$I_1 = -\frac{V_2 - kV_1}{R_2} \tag{16}$$

But due to (15), $V_1=0$ and

$$I_1 = -\frac{V_2}{R_2} \tag{17}$$

$$\frac{I_1}{V_2} = -\frac{1}{R_2} \tag{18}$$

In Eq.(1) with $V_1=0$, $y_{12} = \frac{I_1}{V_2}$ (19)

Combining Eqs.(18) and (19),

$$y_{12} = -\frac{1}{R_2}$$

4. Admittance y_{22} is obtained by letting

$$V_1 = 0 \tag{20}$$

in Eq.(2). Then $y_{22} = \left.\frac{I_2}{V_2}\right|_{V_1=0}$ (21)

or by short circuiting the primary terminal and removing R_1. Under this condition I_2 is the summation of the current flowing through R_2 which is $\frac{V_2 - kV_1}{R_2} = \frac{V_2}{R_2}$ due to $V_1=0$ and the current going through R_3 which is $\frac{V_2}{R_3}$.

$$I_2 = \frac{V_2}{R_2} + \frac{V_2}{R_3} = \left(\frac{1}{R_2} + \frac{1}{R_3}\right)V_2 \tag{22}$$

$$\frac{I_2}{V_2} = \frac{1}{R_2} + \frac{1}{R_3} \tag{23}$$

With combination of Eqs.(21) and (23),

$$y_{22} = \frac{1}{R_2} + \frac{1}{R_3} \tag{24}$$

• PROBLEM 16-21

Determine z_{11}, z_{22}, y_{11}, y_{22}, z_{12}, and y_{12} for the two-ports of Fig. 1a and b.

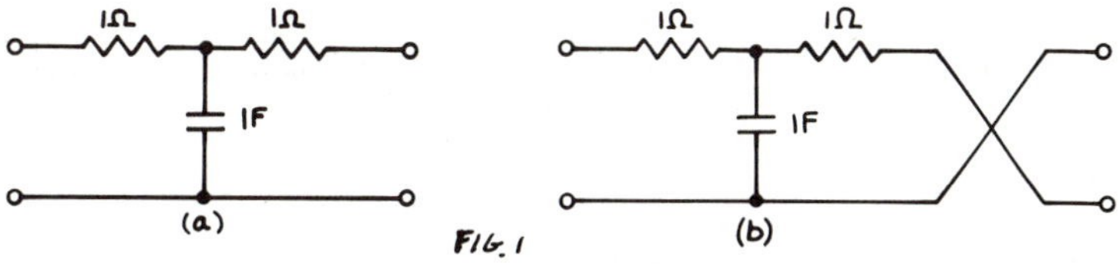

FIG. 1

Solution: For the network of Fig. 1(a), we have, by inspection,

$$z_{11} = z_{22} = 1 + \frac{1}{s} = \frac{s + 1}{s}$$

$$y_{11} = y_{22} = \frac{s + 1}{s + 1 + 1} = \frac{s + 1}{s + 2}$$

$z_{11} = z_{22}$ since the network is reciprocal, and the open-circuit impedance that is common to both ports is $\frac{1}{s}$, thus $z_{12} = z_{21} = \frac{1}{s}$. y_{12} is defined as $\left.\frac{I_1}{V_2}\right|_{V_1=0}$. Writing a node equation for the top node

$$I_1 = \frac{V_1 - I_1}{\frac{1}{s}} + \frac{V_1 - I_1 - V_2}{1}$$

take; $V_1 = 0$;

$$I_1 = -I_1(s+1) - V_2$$

$$\frac{I_1}{V_2}[1+s+1] = -1$$

$$\left.\frac{I_1}{V_2}\right|_{V_1=0} = -\frac{1}{s+2}\ .$$

For the network in Fig. 1(b) we note that

$$z_{11} = z_{22} = \frac{s+1}{s} \quad \text{remain the same}$$

also $y_{11} = y_{22} = \frac{s+1}{s+2}$ remain the same.

z_{12} however, is now $-\frac{1}{s}$ and y_{12} is $\frac{1}{s+2}$.

• PROBLEM 16-22

Find the hybrid parameters for the network of Fig. 1.

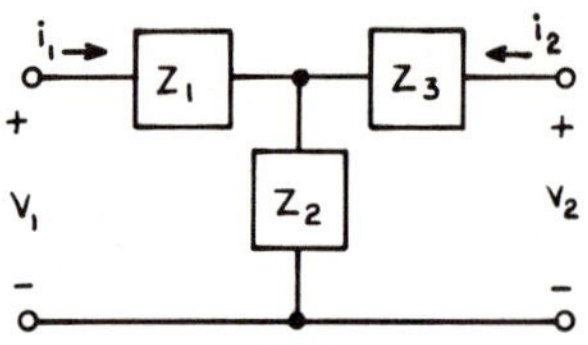

Fig. 1

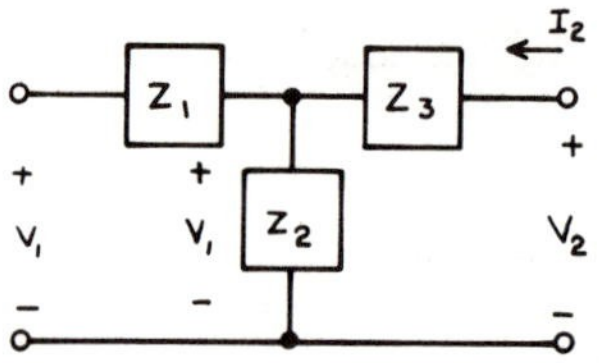

Fig. 3

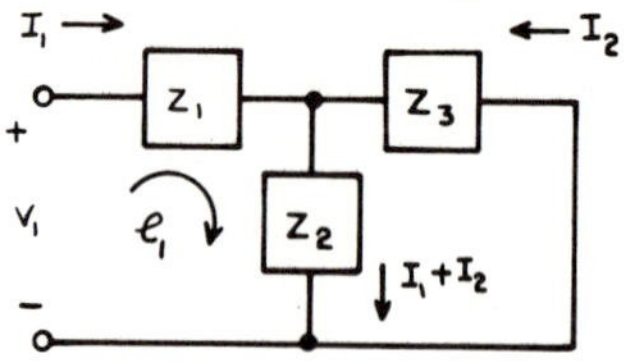

Fig. 2

<u>Solution</u>: The hybrid parameters are given by

$$V_1 = h_{11}I_1 + h_{12}V_2$$

$$I_2 = h_{21}I_1 + h_{22}V_2$$

Note that these can also be written in the following way:

$$h_{11} = \left.\frac{V_1}{I_1}\right|_{V_2=0} \qquad h_{21} = \left.\frac{I_2}{I_1}\right|_{V_2=0}$$

$$h_{12} = \left.\frac{V_1}{V_2}\right|_{I_1=0} \qquad h_{22} = \left.\frac{I_2}{V_2}\right|_{I_1=0}$$

Thus by looking at the network first with $V_2=0$ (output terminals shorted) and then with $I_1=0$ (input terminals open), the hybrid parameters are easily determined.

With $V_2 = 0$: see Fig. 2. Now, $h_{11} = \frac{V_1}{I_1}$ is simply the input impedance, which is Z_1 in series with $Z_2 \| Z_3$.

$$\text{Thus} \quad h_{11} = Z_1 = \frac{Z_2Z_3}{Z_2+Z_3}$$

Next, $h_{21} = \left.\frac{I_2}{I_1}\right|_{V_2=0.}$ Writing KVL in loop 1 of Fig. 2:

$$\ell_1 \quad V_1 = I_1Z_1 + I_1Z_2 + I_2Z_2 \tag{1}$$

To eliminate V_1, we notice that

$$h_{11} = \frac{V_1}{I_1} = Z_1 + \frac{Z_2Z_3}{Z_2+Z_3}$$

$$\text{so } V_1 = I_1 \left(Z_1 + \frac{Z_2Z_3}{Z_2+Z_3}\right)$$

Then (1) becomes

$$I_1(Z_1 + \frac{Z_2Z_3}{Z_3+Z_3} - Z_1 - Z_2) = I_2Z_2$$

so $$\left(\frac{I_2}{I_1} = \frac{Z_2Z_3}{Z_2+Z_3} - Z_2\right)/Z_2 = \frac{Z_3}{Z_2+Z_3} - 1 = \frac{Z_3-Z_2-Z_3}{Z_2+Z_3}$$

$$h_{21} = \frac{-Z_2}{Z_2+Z_3}$$

With $I_1 = 0$: see Fig. 3. First, $h_{12} = \left.\frac{V_1}{V_2}\right|_{I_1=0.}$ As indicated in Fig. 3, V_1 is now the voltage across Z_2 since $I_1 = 0$. This is a simple voltage divider and,

since $V_1 = \frac{Z_2}{Z_2+Z_3} V_2$,

$$h_{12} = \frac{Z_2}{Z_2+Z_3}$$

Finally, $h_{22} = \left.\frac{I_2}{V_2}\right|_{I_1=0.}$ This is simply the output admittance given by $h_{22} = \frac{1}{Z_3+Z_2}$.

• **PROBLEM 16-23**

Find the transmission parameters for the network in Fig. 1.

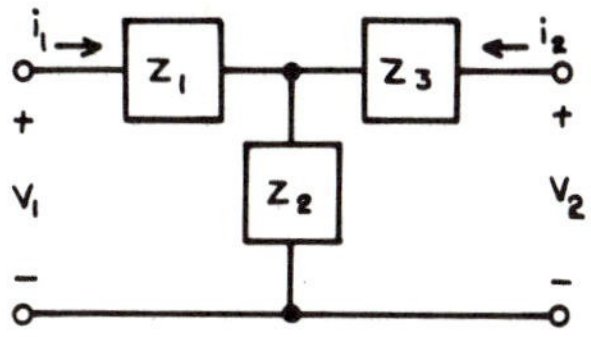

FIG.1

Solution: The transmission parameters are written as follows:

$$v_1 = Av_2 - Bi_2$$

$$i_1 = Cv_2 - Di_2$$

These parameters can be solved for a network by setting one of the independent variables, v_2 or i_2, equal to zero as follows:

$$A = \left.\frac{v_1}{v_2}\right|_{i_2=0} \tag{1}$$

$$B = \left.\frac{v_1}{-i_2}\right|_{v_2=0} \tag{2}$$

$$C = \left.\frac{i_1}{v_2}\right|_{i_2=0} \tag{3}$$

$$D = \left.\frac{i_1}{-i_2}\right|_{v_2=0} \tag{4}$$

From equation (1) with $i_2 = 0$

$$v_1 = i_1(Z_1+Z_2) \tag{5}$$

$$v_2 = i_1 Z_2 \tag{6}$$

Dividing equation (5) by equation (6)

$$A = \left.\frac{v_1}{v_2}\right|_{i_2=0} = \frac{i_1(Z_1+Z_2)}{i_1 Z_2} = \frac{Z_1+Z_2}{Z_2}$$

From equation (2) with $v_2 = 0$ (terminals shorted)

$$i_1 = \frac{v_1}{Z_1 + \frac{Z_2 Z_3}{Z_2+Z_3}} = \frac{Z_2+Z_3}{Z_1Z_2+Z_1Z_3+Z_2Z_3} \quad v_1 \tag{7}$$

By current division

$$-i_2 = \frac{Z_2}{Z_2+Z_3} \quad i_1 \tag{8}$$

Substituting equation (8) into equation (7)

$$-i_2\left(\frac{Z_2+Z_3}{Z_2}\right) = \frac{Z_2+Z_3}{Z_1Z_2+Z_1Z_3+Z_2Z_3} \quad v_1$$

and $$B = \left.\frac{v_1}{-i_2}\right|_{v_2=0} = \frac{Z_1Z_2+Z_1Z_3+Z_2Z_3}{Z_2} \text{ ohms}$$

From equation (3) with $i_2=0$

$$v_2 = i_1 Z_2$$

or $$C = \left.\frac{i_1}{v_2}\right|_{i_2=0} = \frac{1}{Z_2} \text{ mhos}$$

From equation (4) with $v_2 = 0$ (terminals shorted) using current division

$$-i_2 = \frac{Z_2}{Z_2+Z_3}\, i_1$$

and $$D = \left.\frac{i_1}{-i_2}\right|_{v_2=0} = \frac{Z_2+Z_3}{Z_3}$$

• PROBLEM 16-24

Find the transmission parameters of the network shown in Fig. 1.

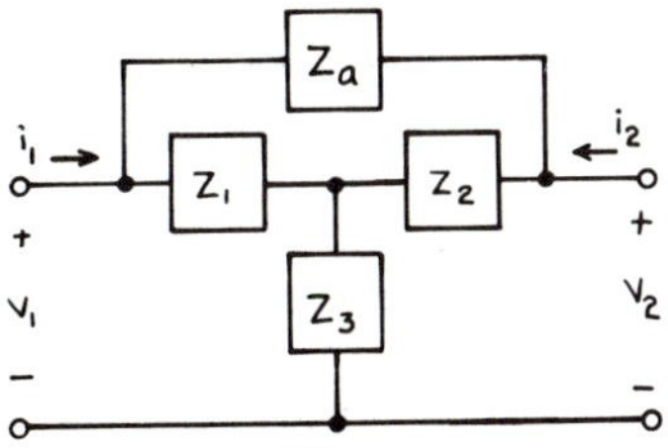

Fig. 1

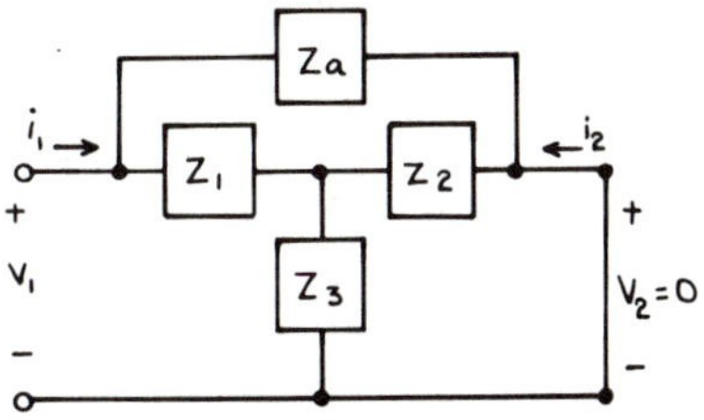

Fig. 2

Solution: The transmission parameters or ABCD parameters, are defined by

$$A = \left.\frac{v_1}{v_2}\right|_{i_2=0} \qquad B = -\left.\frac{v_1}{i_2}\right|_{v_2=0}$$

$$C = \left.\frac{i_1}{v_2}\right|_{i_2=0} \qquad D = -\left.\frac{i_1}{i_2}\right|_{v_2=0}$$

Note that we must either open circuit or short circuit port 2.

In order to calculate A or C, open circuit port 2. For the calculation of A, drive port 1 with a voltage source and calculate v_2. Hence,

$$v_2 = \frac{z_3}{z_3+z_1||(z_a+z_2)}\, v_1 + \frac{v_1}{z_3+z_1||(z_a+z_2)} \times \frac{z_1}{z_1+z_a+z_2}(z_2)$$

so $$\frac{v_2}{v_1} = \frac{z_3(z_1+z_a+z_2)}{z_3(z_1+z_a+z_2)+z_1(z_a+z_2)} + \frac{z_1z_2}{z_3(z_1+z_a+z_2)+z_1(z_a+z_2)}$$

Hence,

$$A = \frac{v_1}{v_2} = \frac{z_az_3+z_az_1+\Delta z}{z_az_3+\Delta z}$$

when $\Delta z = z_1z_2+z_2z_3+z_1z_3$

For calculation of C, drive port 1 with a current source and calculate v_2. Thus,

$$v_2 = z_3i_1 + \frac{z_1}{z_1+z_a+z_2} i_1z_2 = \frac{z_1z_3+z_az_3+z_2z_3+z_1z_2}{z_1+z_a+z_2} i_1$$

and $$C = \frac{z_1+z_a+z_2}{z_az_3+\Delta z}$$

In order to calculate B and D, short port 2. The resulting network is sketched in Fig. 2.

Driving port 1 with a voltage source, yields

$$i_2 = -\frac{v_1}{z_a} - \frac{v_1 \frac{z_3}{z_1+z_3}}{z_1||z_3+z_2} = -v_1 \left| \frac{1}{z_a} + \frac{z_3/z_1+z_3}{z_2 + \frac{z_1z_3}{z_1+z_3}} \right|$$

Therefore,

$$\frac{-i_2}{v_1} = \frac{1}{z_a} + \frac{z_3}{z_2(z_1+z_3)+z_1z_3} = \frac{(z_1+z_3)z_2+z_1z_3+z_az_3}{z_a[z_2(z_1+z_3)+z_1z_3]}$$

and $$B = \frac{z_a\Delta z}{z_az_3+\Delta z}$$

Finally, driving port 1 with a current source, gives

$$-i_2 = \frac{-i_2}{v_1} \times v_1 = \frac{z_az_3+\Delta z}{z_a\Delta z} \times \frac{z_a(z_1 + \frac{z_2z_3}{z_2+z_3})}{z_a+z_1 + \frac{z_2z_3}{z_2+z_3}} i_1$$

$$= \frac{(z_az_3+\Delta z)}{z_a(z_2+z_3)+z_1(z_2+z_3)+z_2z_3}$$

Thus, $$D = \frac{z_a(z_2+z_3)+\Delta z}{z_az_3+\Delta z}$$

Find the open-circuit impedance parameters for the network in Fig. 1.

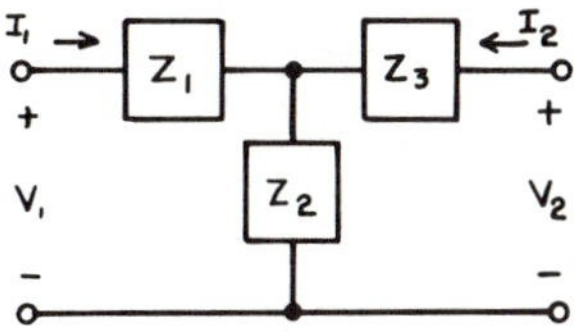

FIG. 1

Solution: The open-circuit impedance parameters (Z-parameters) are defined as

$$Z_{11} = \left.\frac{V_1}{I_1}\right|_{I_2=0}$$

the impedance looking into port 1 with port 2 open.

$$Z_{12} = \left.\frac{V_1}{I_2}\right|_{I_1=0}$$

$$Z_{21} = \left.\frac{V_2}{I_1}\right|_{I_2=0}$$

and

$$Z_{22} = \left.\frac{V_2}{I_2}\right|_{I_1=0}$$

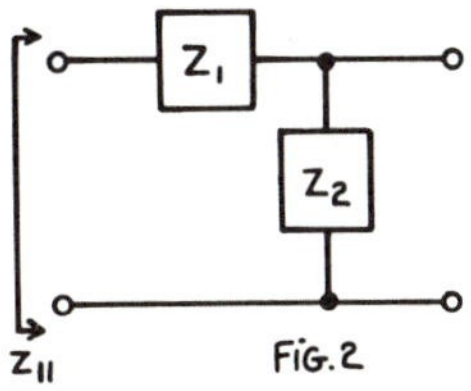

FIG. 2

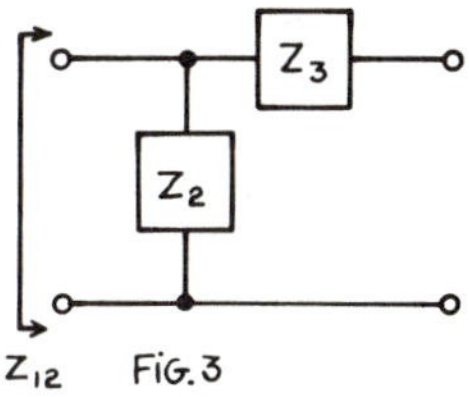

FIG. 3

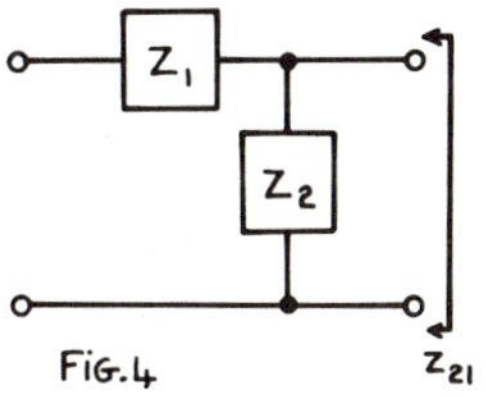

FIG. 4

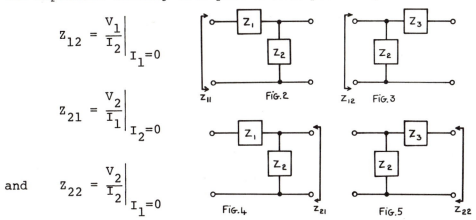

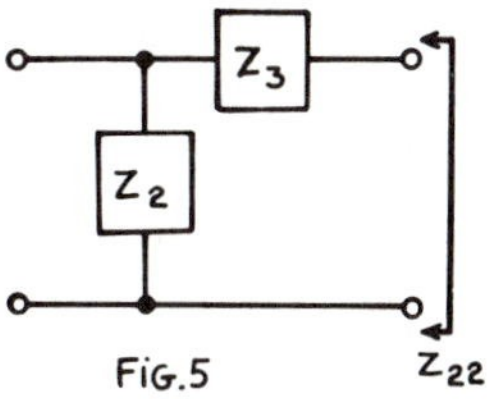

FIG. 5

the impedance looking into port 2 with port 1 open.

Fig. 2 shows the circuit for calculating Z_{11}.

$$Z_{11} = Z_1 + Z_2$$

Fig. 3 shows the circuit for calculating Z_{12}.

$$Z_{12} = Z_2$$

Fig. 4 shows the circuit for calculating Z_{21}.

$$Z_{21} = Z_2$$

Fig. 5 shows the circuit for calculating Z_{22}.

$$Z_{22} = Z_2 + Z_3$$

Find the open-circuit impedance parameters for the network shown.

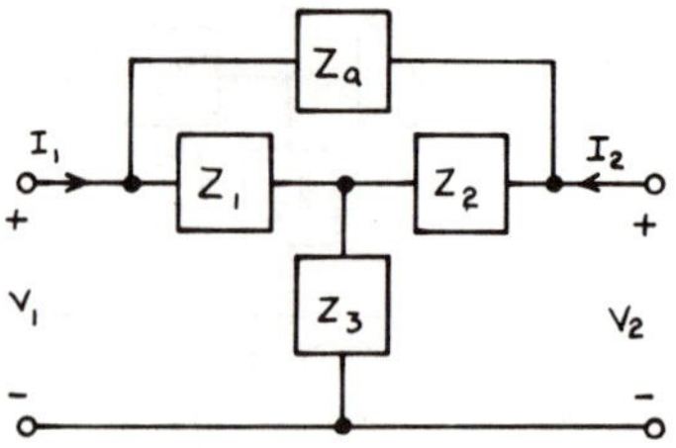

Solution: The open circuit impedance parameters are given by

$$V_1 = Z_{11}I_1 + Z_{12}I_2$$

$$V_2 = Z_{21}I_1 + Z_{22}I_2$$

which can also be written in the following form:

$$Z_{11} = \left.\frac{V_1}{I_1}\right|_{I_2=0} \qquad Z_{21} = \left.\frac{V_2}{I_1}\right|_{I_2=0}$$

$$Z_{12} = \left.\frac{V_1}{I_2}\right|_{I_1=0} \qquad Z_{22} = \left.\frac{V_2}{I_2}\right|_{I_1=0}$$

Thus examine the circuit first for $I_1 = 0$ and then for $I_2 = 0$ and determine the Z-parameters.

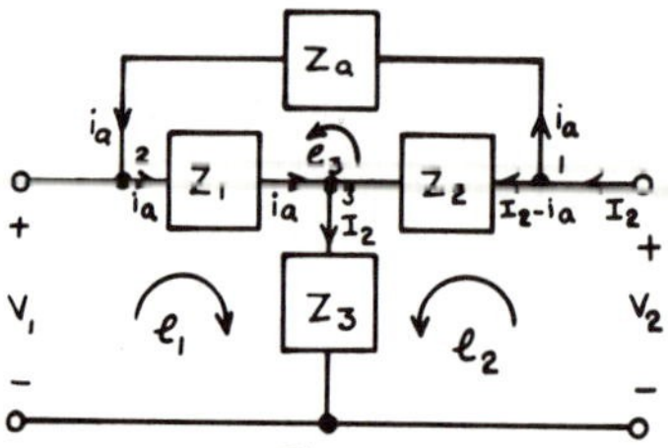

FIG. 1

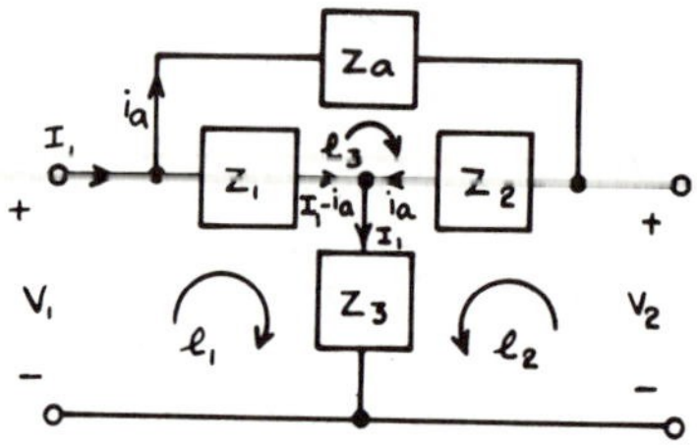

FIG. 2

(a) $I_1 = 0$: Refer to Fig. 1. The indicated currents are obtained by applying KCL at nodes 1, 2, and 3 and calling the current in Z_a i_a. First consider $Z_{12} = \frac{V_1}{I_2}$. Writing loop equations for the indicated loops:

$$\ell_1: \; V_1 = Z_1 i_a + Z_3 I_2 \qquad (1)$$

$$\ell_3: \quad 0 = i_a Z_a + i_a Z_1 - (I_2 - i_a) Z_2 \tag{2}$$

Simplifying Eq.(2) and gathering terms:

$$i_a(Z_a+Z_1+Z_2) = Z_2 I_2 \ , \ i_a = \left(\frac{Z_2}{Z_a+Z_1+Z_2}\right) I_2 \tag{3}$$

Now substituting this into Eq.(1):

$$V_1 = Z_1 \left(\frac{Z_2}{Z_a+Z_1+Z_2}\right) I_2 + Z_3 I_2$$

so $$Z_{12} = \left.\frac{V_1}{I_2}\right|_{I_1=0} = \frac{Z_1 Z_2}{Z_a+Z_1+Z_2} + Z_3$$

Next, $Z_{22} = \dfrac{V_2}{I_2}$. Writing KCL for ℓ_2:

$$\ell_2: \quad V_2 = Z_2 I_2 - Z_2 i_a + Z_3 I_2 \tag{4}$$

using the value of i_a given in Eq.(3), Eq.(4) becomes

$$V_2 = Z_2 I_2 - Z_2 \left(\frac{Z_2}{Z_a+Z_1+Z_2}\right) I_2 + Z_3 I_2$$

Then $$Z_{22} = \frac{V_2}{I_2} = Z_2 - \frac{Z_2^{\ 2}}{Z_a+Z_1+Z_2} + Z_3$$

(b) Next consider the circuit at $I_2 = 0$: Refer to Fig. 2, which was arrived at through the same reasoning as Fig. 1. We need $Z_{11} = \dfrac{V_1}{I_1}$.

$$\ell_1: \quad V_1 = Z_1 I_1 - Z_1 i_a + Z_3 I_1$$

$$\ell_3: \quad 0 = Z_a i_a + Z_2 i_a - Z_1 I_1 + Z_1 i_a$$

then $$i_a = \left(\frac{Z_1}{Z_a+Z_2+Z_1}\right) I_1 \tag{5}$$

and ℓ_1 $$V_1 = (Z_1+Z_3) I_1 - \left(\frac{Z_1^{\ 2}}{Z_a+Z_2+Z_1}\right) I_1$$

so $$Z_{11} = \left.\frac{V_1}{I_1}\right|_{I_2=0} = Z_1 + Z_3 - \frac{Z_1^{\ 2}}{Z_a+Z_2+Z_1}$$

Finally, $Z_{21} = \frac{V_2}{I_1}$

$$\ell_2: \quad V_2 = Z_2 i_a + Z_3 I_1 \tag{6}$$

and inserting Eq.(5) into Eq.(6) yields

$$V_2 = Z_2 \left| \frac{Z_1}{Z_a+Z_2+Z_1} \right| I_1 + Z_3 I_1$$

so $$Z_{21} = \frac{V_2}{I_1} = \frac{Z_1 Z_2}{Z_a+Z_2+Z_1} + Z_3$$

Note that the network is reciprocal since $Z_{21} = Z_{12}$.

• **PROBLEM 16-27**

Obtain the Y-parameters of the network below.

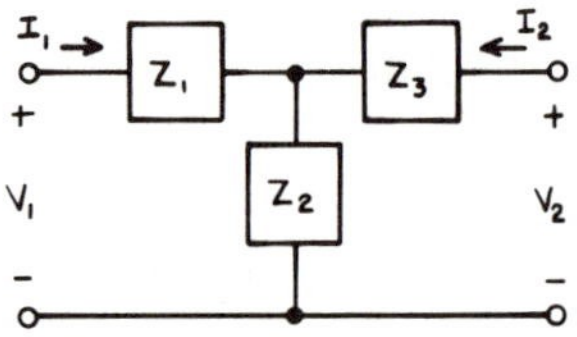

Solution: Since the Y (admittance) parameters are required, first write the equations for I_1 and I_2 in terms of Y's, V_1, and V_2. Then, by individually setting V_1 and V_2 equal to zero, solve for the Y's in terms of the Z's.

(a) The general equations for I_1 and I_2 can be written as functions of V_1 and V_2 as follows:

$$I_1 = Y_{11}V_1 + Y_{12}V_2 \tag{1}$$

$$I_2 = Y_{21}V_1 + Y_{22}V_2 \ . \tag{2}$$

The system of Y subscripts is chosen to identify each Y with the corresponding I and V. Thus, the first subscript is 1 where I_1 is involved and 2 where I_2 is involved. The second subscript is 1 where V_1 is involved and 2 where V_2 is involved. This notation is standardized.

(b) To solve for the various Y's, use the following scheme: note that

$$Y_{11} = I_1/V_1 \quad \text{for} \quad V_2 = 0 \tag{3}$$

in equation (1). Now, mathematically setting $V_2 = 0$ has the electrical counterpart of placing a short circuit across the two right-hand terminals. Placing this constraint on the network, redraw the circuit as shown in Fig. 1.

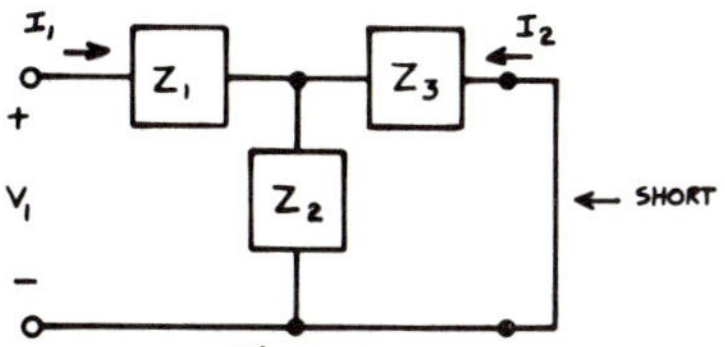

Fig. 1

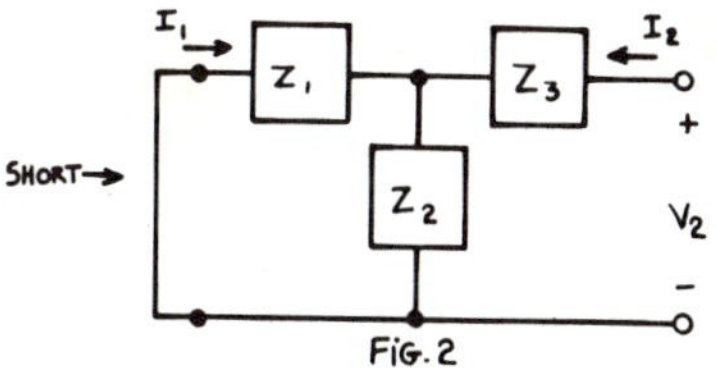

Fig. 2

(c) The next task is solving for the ratio of I_1/V_1 which we recognize as the reciprocal of the impedance looking into the left set of terminals. Drawing upon our previous experience with circuits of this kind, we can write by inspection:

$$1/Y_{11} = Z_1 + \frac{Z_2 Z_3}{Z_2+Z_3} \quad \text{ohms} \tag{4}$$

which becomes, with algebraic manipulation,

$$Y_{11} = \frac{Z_2+Z_3}{Z_1Z_2 + Z_2Z_3 + Z_1Z_3} \quad \text{mhos.} \tag{5}$$

(d) In a similar manner, choose from equation (2),

$$Y_{22} = I_2/V_2 \quad \text{for} \quad V_1 = 0 \tag{6}$$

and redraw the circuit with the <u>left</u> terminals shorted as seen in Figure 2.

Again, by similar analysis, we arrive at the solution

$$1/Y_{22} = Z_3 + \frac{Z_1 Z_2}{Z_1+Z_2} \quad \text{ohms} \tag{7}$$

which becomes

$$Y_{22} = \frac{Z_1+Z_2}{Z_1Z_2+Z_2Z_3+Z_1Z_3} \quad \text{mhos.} \tag{8}$$

(e) Following the same general line of reasoning, set $V_1 = 0$ in equation (1), thereby solving for

$$Y_{12} = I_1/V_2 \quad \text{for } V_1 = 0. \tag{9}$$

Since $V_1 = 0$, the circuit of Figure 2 still applies, but this time it is necessary to solve for I_1. Studying the network, we see that the current I_2 divides at the junction into two components, one of which is $(-I_1)$. This brings to mind the current division principle, which yields

$$-I_1 = I_2 \left(\frac{Z_2}{Z_1+Z_2}\right) . \tag{10}$$

Looking at our objective as seen in equation (9), recognize

that it can be obtained by dividing both sides of equation (10) by $(-V_2)$:

$$Y_{12} = I_1/V_2 = -\frac{I_2}{V_2}\left(\frac{Z_2}{Z_1+Z_2}\right). \tag{11}$$

Looking back over the equations already obtained, recognize the I_2/V_2 factor in equation (11) has appeared before in equation (6), and, therefore, equation (11) can be rewritten as

$$Y_{12} = -(Y_{22})\left(\frac{Z_2}{Z_1+Z_2}\right). \tag{12}$$

Now, substituting equation (8) into (12),

$$Y_{12} = -\frac{Z_2}{Z_1Z_2+Z_2Z_3+Z_1Z_3} \text{ mhos}. \tag{13}$$

(f) Using the same approach for Y_{21}, set $V_2 = 0$ in equation (2). This yields

$$Y_{21} = I_2/V_1 \quad \text{for } V_2 = 0. \tag{14}$$

Since $V_2 = 0$, the circuit of Figure 1 still applies, and we must solve for I_2 in this circuit. Again, it is observed that the current division principle can be applied, resulting in

$$-I_2 = I_1\left(\frac{Z_2}{Z_2+Z_3}\right). \tag{15}$$

Dividing both sides of equation (15) by $(-V_1)$ to obtain Y_{21},

$$Y_{21} = I_2/V_1 = -\frac{I_1}{V_1}\left(\frac{Z_2}{Z_2+Z_3}\right). \tag{16}$$

Now recognize the factor I_1/V_1 is Y_{11} and substitute equation (5) into equation (16):

$$Y_{21} = -\frac{Z_2+Z_3}{Z_1Z_2+Z_2Z_3+Z_1Z_3} \times \frac{Z_2}{Z_2+Z_3} \tag{17}$$

for a final result of

$$Y_{21} = -\frac{Z_2}{Z_1Z_2+Z_2Z_3+Z_1Z_3} \text{ mhos}. \tag{18}$$

(g) A quick glance over the four results indicates that $Y_{12} = Y_{21}$. This is worth noting for future reference.

Obtain the Y parameters from its Z-parameters.

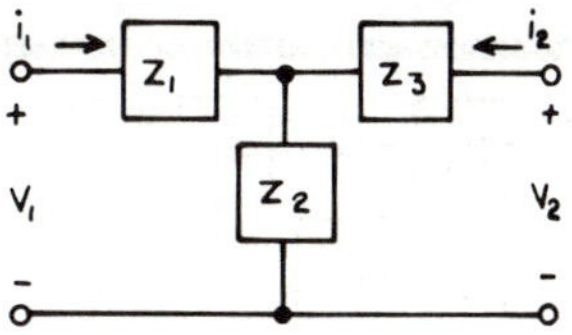

Solution: Use the theory of matrices to solve this problem.

Let $\bar{V}$ = voltage matrix

$\bar{I}$ = current matrix

$\bar{Z}$ = Z-parameter matrix

$\bar{Y}$ = Y-parameter matrix

Then $(\bar{V}) = (\bar{Z})(\bar{I})$

or $(\bar{I}) = (\bar{Z})^{-1}(\bar{V}) = (\bar{Y})(\bar{V})$

Hence $(\bar{Y}) = (\bar{Z})^{-1}$

From the theory of matrices, the inverse of a matrix $(\bar{A})$ is

$$(\bar{A})^{-1} = \frac{(\bar{A}_{jk})^T}{\det(\bar{A})}$$

Where $(\bar{A}_{jk})$ is the matrix of cofactors $\bar{A}_{jk}$ and $(\bar{A}_{jk})^T$ is its transpose. $\det(\bar{A})$ is the determinant of the $(\bar{A})$ matrix.

Consequently,

$$(\bar{Z}) = \begin{bmatrix} Z_1 + Z_2 & Z_2 \\ Z_2 & Z_2 + Z_3 \end{bmatrix}$$

$$\det(\bar{Z}) = \Delta\bar{Z} = (Z_1+Z_2)(Z_2+Z_3) - Z_2^2 = Z_1Z_2+Z_2Z_3+Z_1Z_3$$

$$(\bar{Z})^{-1} = \frac{\begin{bmatrix} Z_2+Z_3 & -Z_2 \\ -Z_2 & Z_1+Z_2 \end{bmatrix}}{\Delta Z} = (\bar{Y}) = \begin{bmatrix} Y_{11} & Y_{12} \\ Y_{21} & Y_{22} \end{bmatrix}$$

Comparing entries,

$$Y_{11} = \frac{Z_2+Z_3}{\Delta\bar{Z}}, \quad Y_{22} = \frac{Z_1+Z_2}{\Delta\bar{Z}}, \quad Y_{12} = Y_{21} = -\frac{Z_2}{\Delta\bar{Z}}$$

Note that this also shows that the network is reciprocal.

Use the Table for Conversion of Two-Port Parameters to find the ABCD parameters of the network below from its Z parameters.

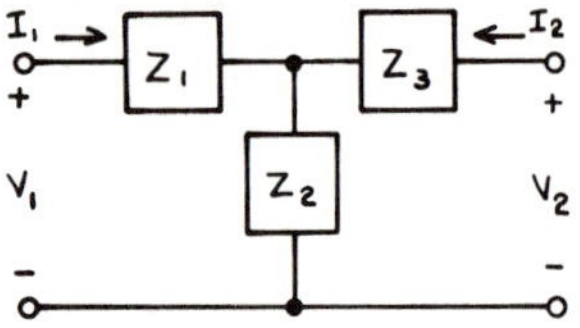

FIG. 1

TABLE CONVERSION CHART FOR TWO-PORT PARAMETERS.

	Z		Y		T		H		G	
Z	z_{11}	z_{12}	$\frac{y_{22}}{\Delta Y}$	$\frac{-y_{12}}{\Delta Y}$	$\frac{A}{C}$	$\frac{\Delta T}{C}$	$\frac{\Delta H}{h_{22}}$	$\frac{h_{12}}{h_{22}}$	$\frac{1}{g_{11}}$	$\frac{-g_{12}}{g_{11}}$
	z_{21}	z_{22}	$\frac{-y_{21}}{\Delta Y}$	$\frac{y_{11}}{\Delta Y}$	$\frac{1}{C}$	$\frac{D}{C}$	$\frac{-h_{21}}{h_{22}}$	$\frac{1}{h_{11}}$	$\frac{g_{21}}{g_{11}}$	$\frac{\Delta G}{g_{11}}$
Y	$\frac{z_{22}}{\Delta Z}$	$\frac{-z_{12}}{\Delta Z}$	y_{11}	y_{12}	$\frac{D}{B}$	$\frac{-\Delta T}{B}$	$\frac{1}{h_{11}}$	$\frac{-h_{12}}{h_{11}}$	$\frac{\Delta G}{g_{22}}$	$\frac{g_{12}}{g_{22}}$
	$\frac{-z_{21}}{\Delta Z}$	$\frac{z_{11}}{\Delta Z}$	y_{21}	y_{22}	$\frac{-1}{B}$	$\frac{A}{B}$	$\frac{h_{21}}{h_{11}}$	$\frac{\Delta H}{h_{11}}$	$\frac{-g_{21}}{g_{22}}$	$\frac{1}{g_{22}}$
T	$\frac{z_{11}}{z_{21}}$	$\frac{\Delta Z}{z_{21}}$	$\frac{-y_{22}}{y_{21}}$	$\frac{1}{-y_{21}}$	A	B	$\frac{-\Delta H}{h_{21}}$	$\frac{h_{11}}{h_{21}}$	$\frac{1}{g_{21}}$	$\frac{g_{22}}{g_{21}}$
	$\frac{1}{z_{21}}$	$\frac{z_{22}}{z_{21}}$	$\frac{-\Delta Y}{y_{21}}$	$\frac{-y_{11}}{y_{21}}$	C	D	$\frac{-h_{22}}{h_{21}}$	$\frac{-1}{h_{21}}$	$\frac{g_{11}}{g_{21}}$	$\frac{\Delta G}{g_{21}}$
H	$\frac{\Delta Z}{z_{22}}$	$\frac{z_{12}}{z_{22}}$	$\frac{1}{y_{11}}$	$\frac{-y_{12}}{y_{11}}$	$\frac{B}{D}$	$\frac{\Delta T}{D}$	h_{11}	h_{12}	$\frac{g_{22}}{\Delta G}$	$\frac{-g_{12}}{\Delta G}$
	$\frac{-z_{21}}{z_{22}}$	$\frac{1}{z_{22}}$	$\frac{y_{21}}{y_{11}}$	$\frac{\Delta Y}{y_{11}}$	$\frac{-1}{D}$	$\frac{C}{D}$	h_{21}	h_{22}	$\frac{-g_{21}}{\Delta G}$	$\frac{g_{11}}{\Delta G}$
G	$\frac{1}{z_{11}}$	$\frac{-z_{12}}{z_{11}}$	$\frac{\Delta Y}{y_{22}}$	$\frac{y_{12}}{y_{22}}$	$\frac{C}{A}$	$\frac{-\Delta T}{A}$	$\frac{h_{22}}{\Delta H}$	$\frac{-h_{12}}{\Delta H}$	g_{11}	g_{12}
	$\frac{z_{21}}{z_{11}}$	$\frac{\Delta Z}{z_{11}}$	$\frac{-y_{21}}{y_{22}}$	$\frac{1}{y_{22}}$	$\frac{1}{A}$	$\frac{B}{A}$	$\frac{-h_{21}}{\Delta H}$	$\frac{h_{11}}{\Delta H}$	g_{21}	g_{22}

Solution:

By inspection the Z-parameters of the network of Fig. 1 are

$$z_{11} = Z_1 + Z_2$$

$$z_{12} = z_{21} = Z_2$$

$$z_{22} = Z_3 + Z_2$$

From the Table, the ABCD parameters in terms of the Z-parameters are

$$A = \frac{z_{11}}{z_{21}}, \quad B = \frac{\Delta Z}{z_{21}}, \quad C = \frac{1}{z_{21}}, \quad D = \frac{z_{22}}{z_{21}}$$

where $\Delta Z = z_{11}z_{22} - z_{12}z_{21}$

Hence, the ABCD parameters of the network in Fig. 1 are,

$$A = \frac{Z_1+Z_2}{Z_2}, \quad B = \frac{(Z_1+Z_2)(Z_3+Z_2)-Z_2^{\,2}}{Z_2}$$

$$C = \frac{1}{Z_2}, \quad \text{and } D = \frac{Z_3+Z_2}{Z_2}$$

The admittance parameters of a bilateral two port are: $y_{11} = 0.25$ mho, $y_{12} = -0.05$ mho, $y_{22} = 0.1$ mho. Find the z-parameters.

Solution: Using the conversion chart for two-port parameters find that

$$z_{11} = \frac{y_{22}}{\Delta y} = \frac{y_{22}}{y_{11}y_{22}-y_{12}y_{21}}$$

$$z_{12} = z_{21} = \frac{-y_{12}}{\Delta y}$$

$$z_{22} = \frac{y_{11}}{\Delta y}$$

$$\Delta y = y_{11}y_{22} - y_{12}y_{21} = y_{11}y_{22} - y_{12}^{\,2}$$

$$\Delta y = (0.25)(0.10) - (-0.05)^2 = 0.0225$$

Hence

$$z_{11} = \frac{0.1}{.0225} = 4.44\ \Omega$$

$$z_{12} = z_{21} = \frac{0.05}{.0225} = 2.22\ \Omega$$

and $$z_{22} = \frac{0.25}{.0225} = 11.11\ \Omega$$

• PROBLEM 16-31

For a bilateral two port, $y_{11} = 0.5$ mho, $y_{12} = -0.4$ mho, and $y_{22} = 0.6$ mho. Find: (a) h_{11}; (b) h_{21}; (c) h_{22}.

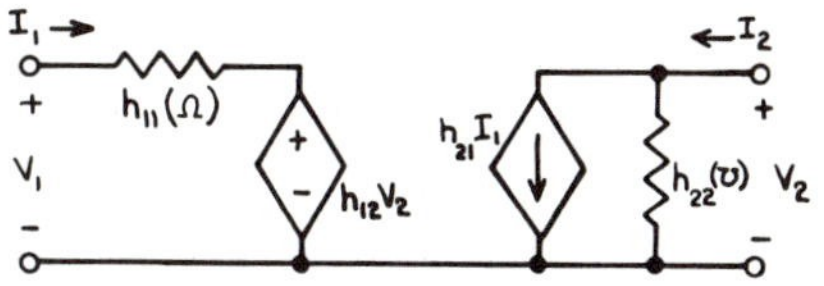

Solution: For a bilateral network, the admittance parameters y_{12} and y_{21} are equal. In terms of terminal voltages and currents, the y parameters are defined as

$$y_{11} = \left.\frac{I_1}{V_1}\right|_{V_2=0} = 0.5 \text{ mho} \qquad y_{12} = \left.\frac{I_1}{V_2}\right|_{V_1=0} = -0.4 \text{ mho}$$

$$y_{21} = \left.\frac{I_2}{V_1}\right|_{V_2=0} = -0.4 \text{ mho} \qquad y_{22} = \left.\frac{I_2}{V_2}\right|_{V_1=0} = 0.6 \text{ mho}$$

In the figure, is shown the hybrid (h-parameter) equivalent circuit. Proceed by applying the electrical conditions, stated above for the y parameters, to the hybrid equivalent circuit. This will give relationships between the y and h parameters.

y_{11} is the ratio of the current I_1 to an applied voltage V_1, with the output port shorted. Applying this condition to the hybrid circuit we obtain

$$h_{11} = \left.\frac{V_1}{I_1}\right|_{V_2=0} = \frac{1}{y_{11}} = 2\ \Omega$$

y_{21} is the ratio of the current I_2, to an applied voltage V_1, with the output port shorted. Under this condition, it is noted that $I_2 = h_{21}I_1$ and $I_1 = \frac{1}{h_{11}} V_1$. Thus $I_2 = \frac{h_{22}}{h_{11}} V_1 = 0.5\ h_{21}V_1$. Therefore

$$y_{21} = -0.4 \text{ mho} = \left.\frac{I_2}{V_1}\right|_{V_2=0} = 0.5\ h_{21}\ ,\ \text{or } h_{21} = -0.8.$$

y_{12} is the ratio of the current I_1, to the applied voltage V_2, with the input port shorted. From the figure, it is noted that $I_1 = \frac{h_{12}}{h_{11}} V_2 = -0.5\ h_{12}V_2$. Thus

$$y_{12} = -0.4 \text{ mho} = \left.\frac{I_1}{V_2}\right|_{V_1=0} = -0.5\ h_{12},\ \text{or } h_{12} = 0.8.$$

Finally, y_{22} is the ratio of the current I_2, to an applied voltage V_2, with the input port again shorted. This means that $I_2 = h_{22}V_2 + h_{21}I_1$. From above, for a shorted input, it was determined that $I_1 = -\frac{h_{12}}{h_{11}} V_2$. Therefore

$$I_2 = \left| h_{22} - \frac{h_{12}h_{21}}{h_{11}} \right| V_2$$

and $$y_{22} = \left.\frac{I_2}{V_2}\right|_{V_1=0} = 0.6 \text{ mho} = h_{22} - \frac{h_{12}h_{21}}{h_{11}} = h_{22} + 0.32$$

or $h_{22} = 0.6 - 0.32 = 0.28$ mho

• PROBLEM 16-32

Given the ABCD parameters

$$\begin{bmatrix} V_1 \\ I_1 \end{bmatrix} = \begin{bmatrix} A & B \\ C & D \end{bmatrix} \begin{bmatrix} V_2 \\ -I_2 \end{bmatrix} \qquad (1)$$

for a two port network. Find the G parameters,

$$\begin{bmatrix} I_1 \\ V_2 \end{bmatrix} = \begin{bmatrix} g_{11} & g_{12} \\ g_{21} & g_{22} \end{bmatrix} \begin{bmatrix} V_1 \\ I_2 \end{bmatrix} \qquad (2)$$

in terms of the ABCD parameters.

Solution: Since the G-parameters are defined as

$$g_{11} = \left.\frac{I_1}{V_1}\right|_{I_2=0} \qquad g_{12} = \left.\frac{I_1}{I_2}\right|_{V_1=0}$$

$$g_{21} = \left.\frac{V_2}{V_1}\right|_{I_2=0} \qquad g_{22} = \left.\frac{V_2}{I_2}\right|_{V_1=0}$$

and $V_1 = AV_2 - BI_2$, $I_1 = CV_2 - DI_2$ from (1).

We can write

$$g_{11} = \left.\frac{CV_2 - DI_2}{AV_2 - BI_2}\right|_{I_2=0} = \frac{C}{A}$$

If $V_1 = 0$ in the ABCD parameter than we obtain the equation

$$0 = AV_2 - BI_2 \ , \ BI_2 = AV_2$$

Hence,

$$g_{22} = \left.\frac{V_2}{I_2}\right|_{V_1=0} = \frac{B}{A}$$

If $I_2 = 0$ yields the equation

$$V_1 = AV_2$$

Hence,

$$g_{21} = \left.\frac{V_2}{V_1}\right|_{I_2=0} = \frac{1}{A}$$

When $V_1 = 0$ gives the two equations

$$AV_2 = BI_2 \quad (3)$$

$$I_1 = CV_2 - DI_2 \quad (4)$$

Solving Eq.(3) for V_2 and substituting in Eq.(4) gives

$$I_1 = I_2(C\frac{B}{A} - D)$$

Hence,

$$g_{12} = \left.\frac{I_1}{I_2}\right|_{V_1=0} = \frac{CB}{A} - D$$

$$g_{12} = \frac{CB - AD}{A}$$

Since $\begin{vmatrix} A & B \\ C & D \end{vmatrix} = AD - BC = \Delta T$ the determinate of the transmission matrix, we can write

$$y_{12} = \frac{-\Delta T}{A}$$

• **PROBLEM 16-33**

Consider the networks N and N' shown in Fig. 1(a) and (b). Find the Y parameters of their parallel combination.

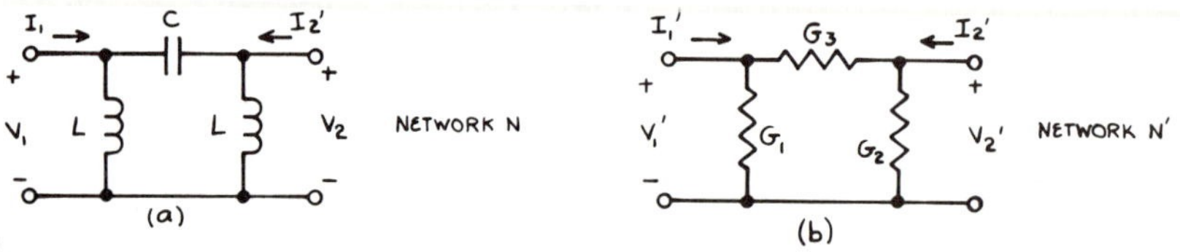

FIG.1

Solution: **Y parameters** are often referred to as the short circuit admittance parameters and are defined by

$$\begin{bmatrix} I_1(s) \\ I_2(s) \end{bmatrix} = \begin{bmatrix} y_{11}(s) & y_{12}(s) \\ y_{21}(s) & y_{22}(s) \end{bmatrix} \begin{bmatrix} V_1(s) \\ V_2(s) \end{bmatrix} .$$

To obtain $y_{11}(s)$ we apply a current source $I_1(s)$ at terminal #1, short-circuit the terminals #2 and measure the voltage

$V_1(s)$ across terminals #1.

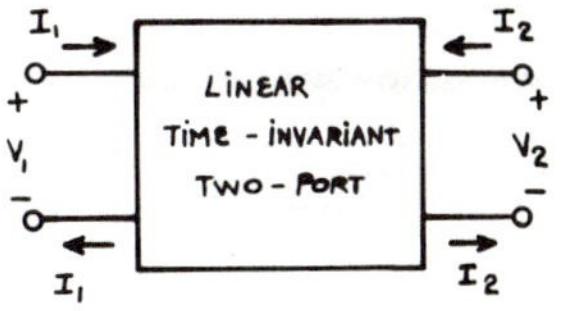

Then

$$y_{11}(s) = \left.\frac{I_1(s)}{V_1(s)}\right|_{V_2(s)=0}$$

$$y_{12}(s) = \left.\frac{I_1(s)}{V_2(s)}\right|_{V_1(s)=0}$$

$$y_{21}(s) = \left.\frac{I_2(s)}{V_1(s)}\right|_{V_2(s)=0}$$

$$y_{22}(s) = \left.\frac{I_2(s)}{V_2(s)}\right|_{V_1(s)=0}$$

From the equations and definitions we find the Y parameters for the network in part (a)

$$\bar{y} = \begin{bmatrix} \frac{1}{Ls} + Cs & -Cs \\ -Cs & \frac{1}{Ls} + Cs \end{bmatrix}$$

and similarly for part (b).

$$\bar{y}' = \begin{bmatrix} G_1 + G_3 & -G_3 \\ -G_3 & G_2 + G_3 \end{bmatrix}$$

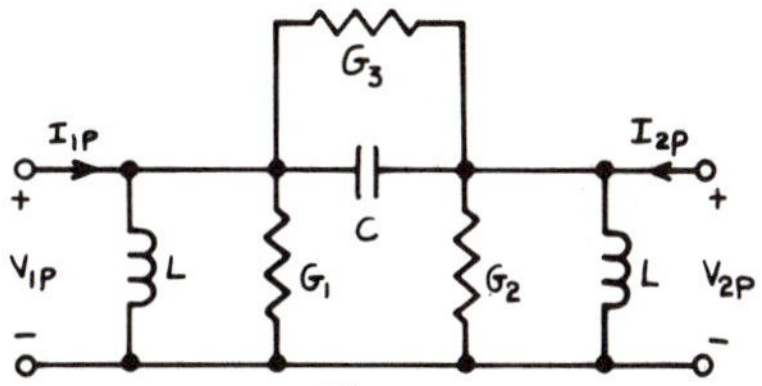

Fig.2

If the networks N and $\underline{N}'$ are connected in parallel, as shown in Fig. 2, the matrix $\bar{y}_p$ for the new network is the addition of two matrices previously found

$$\bar{y}_p = \bar{y} + \bar{y}'$$

that is

$$\begin{bmatrix} y_{11}^{p} & y_{12}^{p} \\ y_{21}^{p} & y_{22}^{p} \end{bmatrix} = \begin{bmatrix} y_{11}+y'_{11} & y_{12}+y'_{12} \\ y_{21}+y'_{21} & y_{22}+y'_{22} \end{bmatrix}$$

and the answer is:

$$\bar{y}^p = \begin{bmatrix} Cs + \frac{1}{Ls} + G_1 + G_3 & -(Cs + G_3) \\ -(Cs + G_3) & Cs + \frac{1}{Ls} + G_2 + G_3 \end{bmatrix}$$

Consider the simple networks N and N' shown in Fig. 1(a) and (b). Find the Z parameters of their series connection.

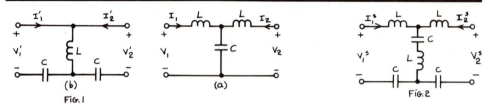

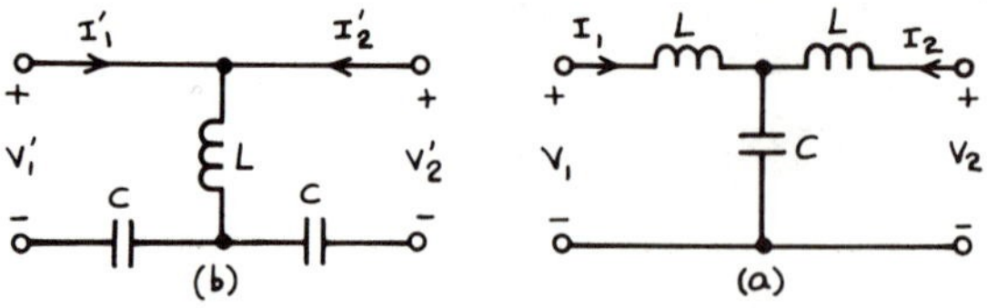

FIG. 1

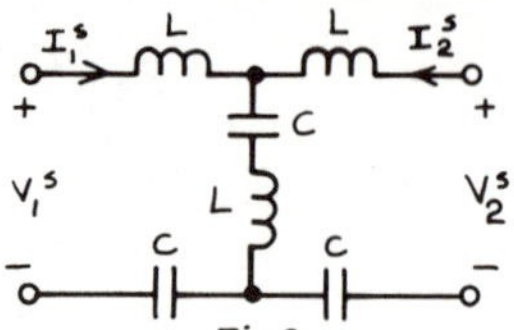

FIG. 2

Solution: For a basic two port network shown below we can write

$$\begin{bmatrix} V_1(s) \\ V_2(s) \end{bmatrix} = \begin{bmatrix} Z_{11}(s) & Z_{12}(s) \\ Z_{21}(s) & Z_{22}(s) \end{bmatrix} \begin{bmatrix} I_1(s) \\ I_2(s) \end{bmatrix}$$

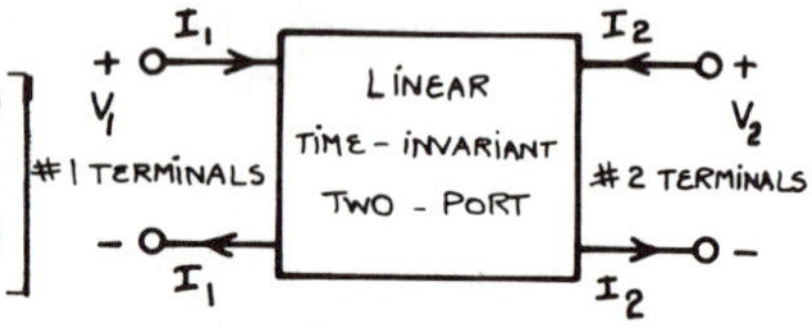

where elements of matrix $\overline{Z}$ are called open-circuit impedance functions (Z parameters)

$$Z_{11}(s) = \left.\frac{V_1(s)}{I_1(s)}\right|_{I_2(s)=0}$$

That is apply a voltage source $V_1(s)$ at #1 terminals, open-circuit the #2 terminals and measure the current in #1 terminals.

$$Z_{12} = \left.\frac{V_1(s)}{I_2(s)}\right|_{I_1(s)=0}$$

Apply a current source I_2 at #2 terminals, open-circuit the #1 terminals and measure the voltage across #1 terminals.

Similarly,

$$Z_{21}(s) = \left.\frac{V_2(s)}{I_1(s)}\right|_{I_2(s)=0}$$

$$Z_{22}(s) = \left.\frac{V_2(s)}{I_2(s)}\right|_{I_1(s)=0}$$

we find

$$\overline{Z}(s) = \begin{bmatrix} Z_{11}(s) = sL = \frac{1}{Cs} & Z_{12}(s) = \frac{1}{Cs} \\ Z_{21}(s) = \frac{1}{Cs} & Z_{22}(s) = sL + \frac{1}{Cs} \end{bmatrix}$$

for the network in part (a).

Similarly the following parameters can be found for part (b)

$$\overline{Z}'(s) = \begin{bmatrix} z_{11}(s) = aL + \frac{1}{Cs} & z_{12}(s) = sL \\ z_{21}(s) = sL & z_{22}(s) = sL + \frac{1}{Cs} \end{bmatrix}$$

when two networks are connected in series as shown in Fig. 2.

It is known that the matrix $\overline{Z}^s$ for the new network is the addition of two matrices previously found. That is

$$\overline{Z}^s = \overline{Z} + \overline{Z}'$$

or

$$\begin{bmatrix} z^s_{11} & z^s_{12} \\ z^2_{21} & z^s_{22} \end{bmatrix} = \begin{bmatrix} z_{11}+z'_{11} & z_{12}+z'_{12} \\ z_{21}+z'_{21} & z_{22}+z'_{22} \end{bmatrix}$$

which gives the result

$$\overline{Z}^s(s) = \begin{bmatrix} 2(sL + \frac{1}{Cs}) & sL + \frac{1}{Cs} \\ sL + \frac{1}{Cs} & 2(sL + \frac{1}{Cs}) \end{bmatrix}$$

• PROBLEM 16-35

Two two-port networks with ABCD parameters A, B, C, D and A', B', C', D' are connected in series as shown in Fig. 1. Find the ABCD parameters for the new two-port network.

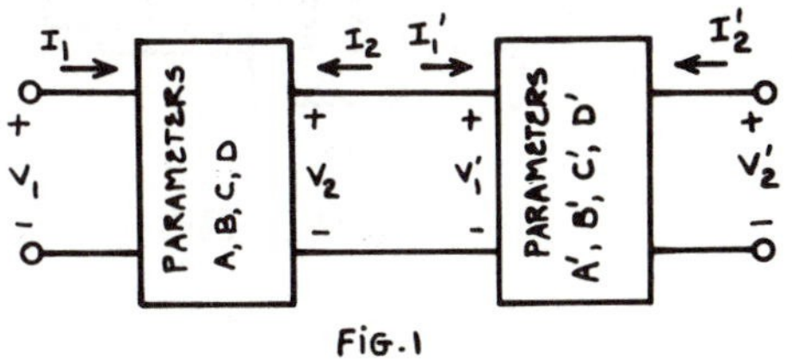

Fig.1

Solution: The ABCD parameters of each network are described by the equations in matrix form

$$\begin{bmatrix} V_1 \\ I_1 \end{bmatrix} = \begin{bmatrix} A & B \\ C & D \end{bmatrix} \begin{bmatrix} V_2 \\ -I_2 \end{bmatrix} \quad \text{and} \quad \begin{bmatrix} V'_1 \\ I'_1 \end{bmatrix} = \begin{bmatrix} A' & B' \\ C' & D' \end{bmatrix} \begin{bmatrix} V'_2 \\ -I'_2 \end{bmatrix}$$

where $\begin{bmatrix} A & B \\ C & D \end{bmatrix}$ and $\begin{bmatrix} A' & B' \\ C' & D' \end{bmatrix}$ are called transmission matrices.

Since $I_2 = I_1'$ and $V_2 = V_1'$ we can write

$$\begin{bmatrix} V_1' \\ I_1' \end{bmatrix} = \begin{bmatrix} A' & B' \\ C' & C' \end{bmatrix} \begin{bmatrix} V_2' \\ -I_2' \end{bmatrix} \qquad \text{as}$$

$$\begin{bmatrix} V_2 \\ -I_2 \end{bmatrix} = \begin{bmatrix} A' & B' \\ C' & D' \end{bmatrix} \begin{bmatrix} V_2' \\ -I_2' \end{bmatrix}$$

Hence,

$$\begin{bmatrix} V_1 \\ I_1 \end{bmatrix} = \begin{bmatrix} A & B \\ C & D \end{bmatrix} \begin{bmatrix} V_2 \\ -I_2 \end{bmatrix} = \begin{bmatrix} A & B \\ C & D \end{bmatrix} \begin{bmatrix} A' & B' \\ C' & D' \end{bmatrix} \begin{bmatrix} V_2' \\ -I_2 \end{bmatrix}$$

the new ABCD parameters are the product of each networks transmission matrix

$$\begin{bmatrix} A & B \\ C & D \end{bmatrix} \begin{bmatrix} A' & B' \\ C' & D' \end{bmatrix} = \begin{bmatrix} AA'+BC' & AB'+BD' \\ CA'+DC' & CB'+DD' \end{bmatrix}$$

• PROBLEM 16-36

Given a two-port network terminated in a load impedance Z_ℓ find the input impedance of the two-port network in terms of Z_ℓ and its Z-parameters.

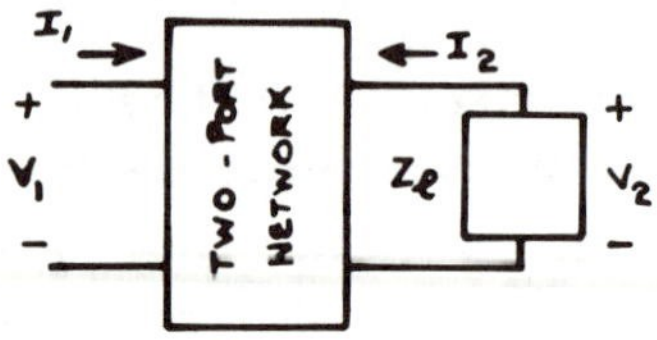

FIG. 1

Solution: The Z-parameters of a two-port network are described as

$$\begin{bmatrix} V_1 \\ V_2 \end{bmatrix} = \begin{bmatrix} Z_{11} & Z_{12} \\ Z_{21} & Z_{22} \end{bmatrix} \begin{bmatrix} I_1 \\ I_2 \end{bmatrix}$$

But since $V_2 = -Z_\ell I_2$ from Fig. 1 we can write the equation

$$\begin{bmatrix} V_1 \\ 0 \end{bmatrix} = \begin{bmatrix} Z_{11} & Z_{12} \\ Z_{21} & Z_{22}+Z_\ell \end{bmatrix} \begin{bmatrix} I_1 \\ I_2 \end{bmatrix}$$

Since the input impedance is $\frac{V_1}{I_1}$ we can write $V_1 = Z_{11}I_1 + Z_{12}I_2$ from the matrix above. Also from the matrix

$$\frac{V_1}{I_1} = Z_{11} + Z_{12}\frac{I_2}{I_1}$$

where $I_2 = \dfrac{\begin{vmatrix} Z_{11} & V_1 \\ Z_{21} & 0 \end{vmatrix}}{\begin{vmatrix} Z_{11} & Z_{12} \\ Z_{21} & Z_{22}+Z_\ell \end{vmatrix}}$ and

$$I_1 = \frac{\begin{vmatrix} V_1 & Z_{12} \\ 0 & Z_{22}+Z_\ell \end{vmatrix}}{\begin{vmatrix} Z_{11} & Z_{12} \\ Z_{21} & Z_{22}+Z_\ell \end{vmatrix}}$$

Hence,

$$\frac{V_1}{I_1} = Z_{12}\frac{\begin{vmatrix} Z_{11} & V_1 \\ Z_{21} & 0 \end{vmatrix}}{\begin{vmatrix} V_1 & Z_{12} \\ 0 & Z_{22}+Z_\ell \end{vmatrix}} + Z_{11}$$

$$\frac{V_1}{I_1} = Z_{12}\frac{(-V_1Z_{21})}{V_1(Z_{22}+Z_\ell)} + Z_{11} = -\frac{Z_{12}Z_{21}}{Z_{22}+Z_\ell} + Z_{11}$$

$$\frac{V_1}{I_1} = \frac{Z_{11}(Z_{22}+Z_\ell)-Z_{11}Z_{21}}{Z_{22}+Z_\ell} = \frac{Z_{11}Z_{22}-Z_{12}Z_{21}+Z_{11}Z_\ell}{Z_{22}+Z_\ell}$$

Since $Z_{11}Z_{22} - Z_{12}Z_{21}$ is the determinant of the original matrix,

$$\begin{vmatrix} Z_{11} & Z_{12} \\ Z_{22} & Z_{22} \end{vmatrix}$$

the input impedance can be written as

$$\frac{V_1}{I_1} = \frac{|Z| + Z_{11}Z_\ell}{Z_{22} + Z_\ell}$$

where $|Z| = Z_{11}Z_{22} - Z_{12}Z_{21}$

• PROBLEM 16-37

Find Z-parameters for the network shown.

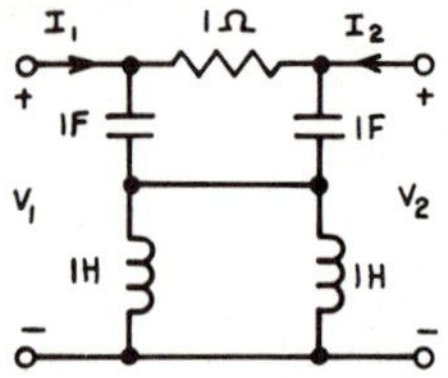

Solution: Z-parameters Z_{11} and Z_{22} are, by definition, the input impedance parameters of a given network with the output open circuited. By inspection of the given network, it is known that the circuit structure is symmetrical. Therefore:

$$Z_{11} = Z_{22} \tag{1}$$

With $I_2 = 0$ or Port 2 open circuited, Z_{11} consists of a parallel circuit containing 1F and series connection of 1 ohm and 1F in series with a parallel combination of two 1H inductances. Assigning an s parameter to 1 ohm and 1H respectively, and $\frac{1}{s}$ parameter to the 1F capacitor,

$$Z_{11} = \frac{1}{\frac{1}{\frac{1}{s}} + \frac{1}{s + \frac{1}{s}}} + \frac{1}{\frac{1}{s} + \frac{1}{s}} = \frac{1}{s + \frac{s}{s^2 + 1}} + \frac{1}{\frac{2}{s}}$$

$$= \frac{s^2 + 1}{s(s^2+1) + s} + \frac{s}{2} = \frac{s^2+1}{s(s^2+2)} + \frac{s}{2} = Z_{22} \tag{2}$$

The symmetry of the given network yields

$$Z_{12} = Z_{21} \tag{3}$$

The Z-parameter Z_{12} is the impedance parameter.

$Z_{12} = \left.\frac{V_1}{I_2}\right|_{I_1=0}$. The voltage V_1 is equal to V_2 minus the voltage drop across the 1 ohm resistor.

With $I_1 = 0$ or Port 1 open circuited, I_2 is divided into 1F capacitor and a series combination of a 1 ohm resistor and 1F capacitor. The current in the 1 ohm resistor is found by the principle of current division,

$$I_2 \quad \frac{\frac{1}{s}}{s + \frac{1}{s} + \frac{1}{s}} = I_2 \frac{1}{s^2 + 2} \tag{4}$$

The voltage drop across the 1 ohm resistor is then

$$sI_2 \frac{1}{s^2 + 2} \tag{5}$$

$$V_1 = V_2 - sI_2 \frac{1}{s^2 + 2} \tag{6}$$

$$Z_{12} \equiv \left.\frac{V_1}{I_2}\right|_{I_1=0} = \left.\frac{V_2}{I_2}\right|_{I_1=0} - \frac{s}{s^2 + 2} = Z_{22} - \frac{s}{s^2 + 2} \tag{7}$$

Substituting Eq.(2) in Eq.(7)

$$Z_{12} = \frac{s^2+1}{s(s^2+2)} + \frac{s}{2} - \frac{s}{s^2+2} = \frac{s^2+1-s^2}{s(s^2+2)} + \frac{s}{2}$$

$$= \frac{1}{s(s^2+2)} + \frac{s}{2} \tag{8}$$

• PROBLEM 16-38

Find the Z-parameters for the two-port network in Fig. 1, by noting that $Z_{12} = Z_{21}$ and $y_{12} = y_{21}$ for reciprocal two-ports.

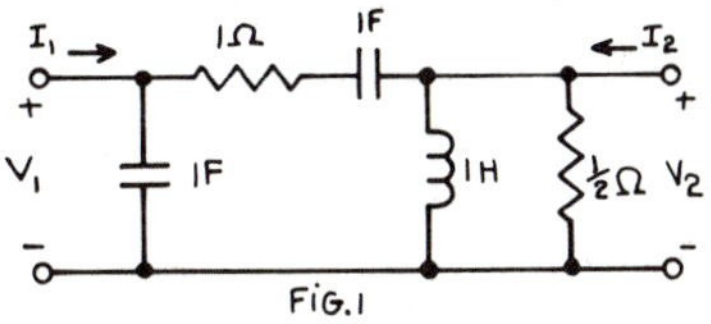

FiG.1

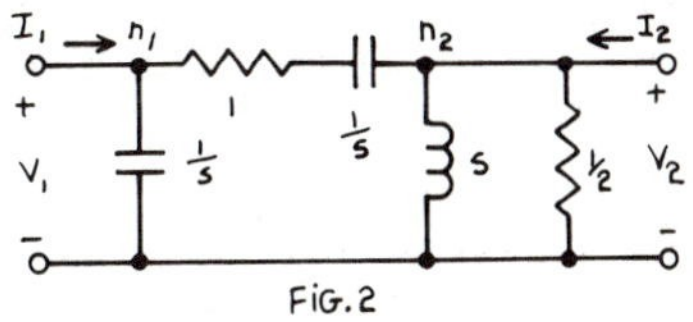

FiG.2

Solution: It is easier to determine the y-parameters than the Z-parameters here. Thus, by inspection,

$$y_{11} = \left.\frac{I_1}{V_1}\right|_{V_2=0} = s + \frac{s}{s+1} = \frac{s^2+2s}{s+1}$$

Since $y_{12} = y_{21}$ for this reciprocal two-port, $y_{12} = \left.\frac{I_1}{V_2}\right|_{V_1=0}$ can be found from the node equation for node n_1 in the circuit in Fig. 2.

Hence,

$$n_1: \quad I_1 = \frac{V_1}{\frac{1}{s}} + \frac{V_1 - V_2}{\frac{1}{s}+1}$$

$$y_{12} = \left.\frac{I_1}{V_2}\right|_{V_1=0} = -\frac{1}{\frac{1}{s}+1} = -\frac{s}{s+1}$$

and $y_{21} = y_{12} = -\frac{s}{s+1}$

also $y_{22} = \left.\frac{I_2}{V_2}\right|_{V_1=0} = \frac{1}{s} + 2 + \frac{s}{s+1} = \frac{3s^3+3s+1}{s(s+1)}$

The Z-parameters can be written in terms of the y-parameters as follows:

Since $I_1 = V_1y_{11} + V_2y_{22}$

$I_2 = V_1y_{21} + V_2y_{22}$

we can write,

$$z_{11} = \left.\frac{V_1}{I_1}\right|_{I_2=0} = \frac{V_1}{V_1y_{11}+V_2y_{12}} = \frac{V_1}{V_1y_{11}+(-V_1\frac{y_{21}}{y_{22}})y_{12}}$$

$$z_{11} = \frac{1}{y_{11} - \frac{y_{21}y_{12}}{y_{22}}} = \frac{y_{22}}{y_{11}y_{22} - y_{21}y_{12}} = \frac{y_{22}}{\Delta y}$$

Similarly,

$$z_{12} = \frac{-y_{12}}{\Delta y}, \; z_{21} - \frac{-y_{21}}{\Delta y}, \text{ and } z_{22} = \frac{y_{11}}{\Delta y}.$$

Hence,

$$z_{11} = \frac{3s^2+3s+1}{s(s+1)\left[\frac{s^2+2s}{(s+1)} \cdot \frac{3s^2+3s+1}{s(s+1)} - \frac{s^2}{(s+1)^2}\right]}$$

$$z_{11} = \frac{(3s^2+3s+1)(s+1)}{(s^2+2s)(3s^2+3s+1)-s^3} = \frac{(3s^2+3s+1)(s+1)}{s(3s^3+8s^2+7s+2)}$$

$$z_{11} = \frac{3s^2+3s+1}{s(3s^2+5s+2)}$$

$$z_{12} = z_{21} = \frac{\frac{s}{s+1}}{\frac{3s^3+8s^2+7s+2}{(s+1)^2}} = \frac{s}{3s^2+5s+2}$$

$$z_{22} = \frac{(s^2+2s)(s+1)}{3s^3+8s^2+7s+2} = \frac{s^2+2s}{3s^2+5s+2}$$

Using mesh analysis, in the circuit of Fig. 1, calculate the impedance matrix Z(s). Assume zero initial conditions.

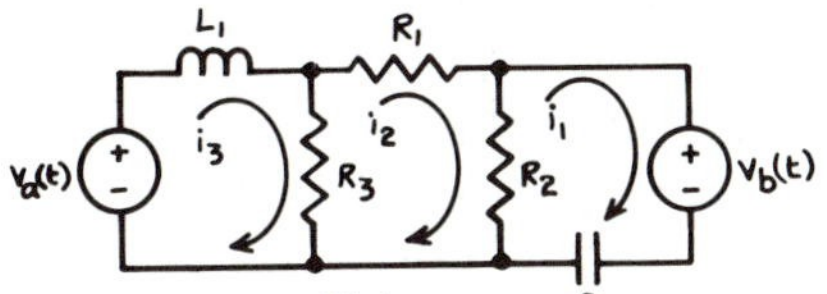

Fig.1

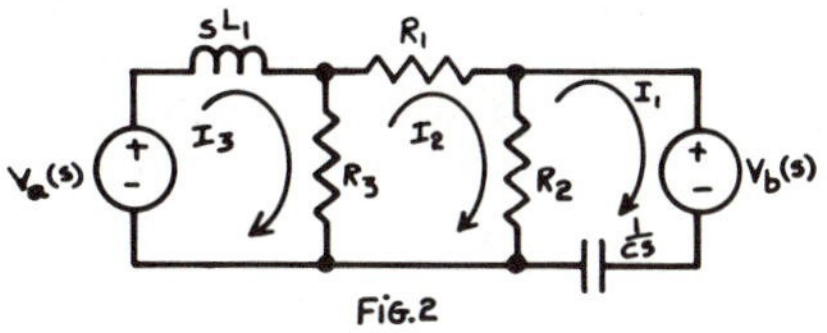

Fig.2

Solution: Write three KVL equations for the three loops in the transformed network shown in Fig. 2:

$$\ell_1: \quad -V_b(s) = I_1(R_2 + \frac{1}{Cs}) - I_2(R_2)$$

$$\ell_2: \quad 0 = I_1(R_2) + I_2(R_1+R_2+R_3) - I_3(R_3)$$

$$\ell_3: \quad V_a(s) = - I_2(R_3) + I_3(sL_1+R_3)$$

Writing the matrix directly from the three equations yields

$$Z(s) = \begin{bmatrix} R_2 + \frac{1}{CS} & -R_2 & 0 \\ -R_2 & R_1+R_2+R_3 & -R_3 \\ 0 & -R_3 & sL_1+R_3 \end{bmatrix}$$

For the circuit shown in Fig. 1, calculate the admittance matrix Y(s) by using nodal analysis. Assume zero initial conditions.

Solution: The general form of an admittance matrix of a circuit that has n + 1 nodes is

$$Y(s) = \begin{bmatrix} y_{11} & y_{12} & y_{13} & \cdots & y_{1n} \\ y_{21} & & & & \\ y_{31} & & & & \\ \vdots & & & & \\ y_{n1} & y_{n2} & y_{n3} & \cdots & y_{nn} \end{bmatrix} .$$

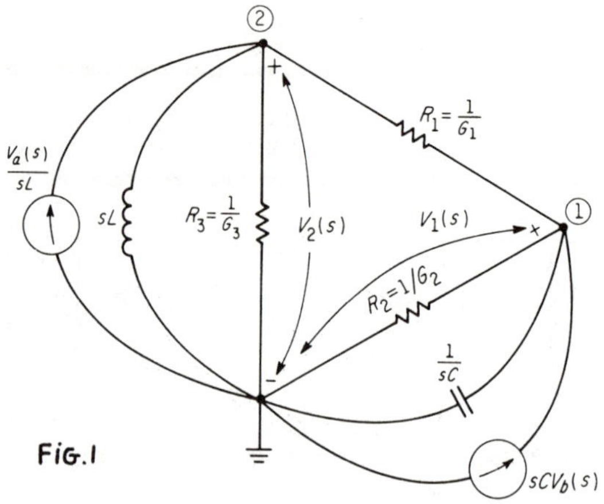

FiG.1

Each diagonal element (y_{kk} where k=1, 2, 3, . . ., n) is the sum of all of the admittances connected to node k. As an example y_{22} is the sum of all the admittances connected to node 2. Each non-diagonal element (y_{ij} where i = 1, 2, . . ., n and j = 1, 2 . . . n and i+j) is the negative of the admittance connected between nodes i and j. As an example y_{13} and y_{31} are the negative of the admittance connected between nodes 1 and 3.

With this information Y(s) for the circuit in Fig. 1 can be found. It is seen from Fig. 1 that there are three nodes. Two nodes lie on either side of R_1 and another node lies where R_2 and R_3 are connected to each other. There are three nodes; hence the matrix is 2 x 2.

$$Y(s) = \begin{bmatrix} y_{11} & y_{12} \\ y_{21} & y_{22} \end{bmatrix}$$

The diagonal elements y_{11} and y_{22} represent the sum of the admittances connected to nodes 1 and 2 respectively. Hence,

$$y_{11} = \frac{1}{R_1} + \frac{1}{R_2} + sC = G_1 + G_2 = sC$$

$$y_{22} = \frac{1}{R_1} + \frac{1}{R_3} + \frac{1}{sL} = G_1 + G_3 + \frac{1}{sL}$$

The non-diagonal elements y_{12} and y_{21} represent the negative of the admittance connected between nodes 1 and 2. Hence

$$y_{12} = y_{21} = -\frac{1}{R_1} = -G_1.$$

So the admittance matrix is

$$Y(s) = \begin{bmatrix} G_1 + G_2 + sC & -G_1 \\ -G_1 & G_1 + G_3 + \frac{1}{sL} \end{bmatrix} .$$

• PROBLEM 16-41

Calculate the short-circuit admittance parameters for the transformer circuit shown in Fig.1.

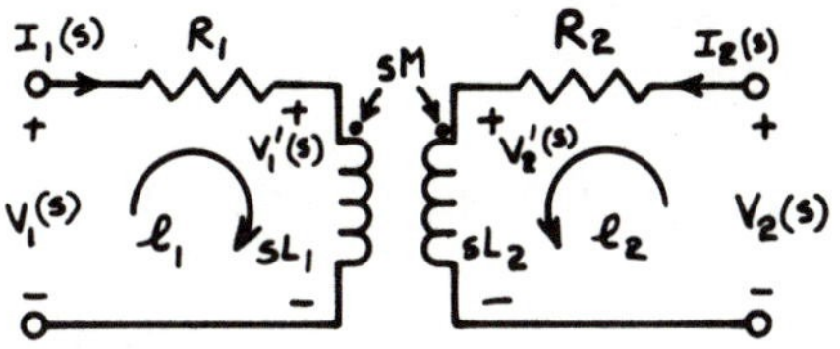

FiG.1

Solution: The short-circuit admittance or Y parameters are defined by

$$\begin{bmatrix} I_1(s) \\ I_2(s) \end{bmatrix} = \begin{bmatrix} y_{11}(s) & y_{12}(s) \\ y_{21}(s) & y_{22}(s) \end{bmatrix} \begin{bmatrix} V_1(s) \\ V_2(s) \end{bmatrix}$$

or $$I\overline{(s)} = \overline{Y}(s)\overline{V}(s)$$

The elements of the Y(s) matrix are

$$y_{11}(s) = \left.\frac{I_1(s)}{V_1(s)}\right|_{V_2(s)=0} \qquad y_{12}(s) = \left.\frac{I_1(s)}{V_2(s)}\right|_{V_1(s)=0}$$

$$y_{21}(s) = \left.\frac{I_2(s)}{V_1(s)}\right|_{V_2(s)=0} \qquad y_{22}(s) = \left.\frac{I_2(s)}{V_2(s)}\right|_{V_1(s)=0}$$

The terminal V-I relationships are written as

$$V_1(s) = (R_1+sL_1)I_1(s) + sMI_2(s) \tag{1}$$

and $$V_2(s) = sMI_1(s) + (R_2+sL_2)I_2(s) \tag{2}$$

Using (2) and setting $V_2(s) = 0$ yields

$$I_2 = \frac{-sM}{R_2+sL_2} I_1 \tag{3}$$

Substituting this result into (1) yields

$$V_1(s) = \frac{(R_1+sL_1)(R_2+sL_2) - s^2M^2}{R_2+sL_2} \quad I_1(s)$$

and therefore

$$y_{11}(s) = \left.\frac{I_1(s)}{V_1(s)}\right|_{V_2(s)=0} = \frac{R_2+sL_2}{(R_1+sL_1)(R_2+sL_2) - s^2M^2}$$

since (3) can be written as

$$I_1 = -\frac{(R_2+sL_2)}{sM} I_2 \tag{4}$$

This result, when substituted into (1), yields

$$y_{21} = \left.\frac{I_2(s)}{V_1(s)}\right|_{V_2(s)=0} = \frac{-sM}{(R_1+sL_1)(R_2+sL_2) - s^2M^2}$$

Similarly, using (1) and setting $V_1(s) = 0$ yields

$$I_1 = \frac{-sM}{R_1+sL_1} I_2 \tag{5}$$

Substituting this result into (2) yields

$$V_2(s) = \left(\frac{(R_1+sL_1)(R_2+sL_2) - s^2M^2}{R_1+sL_1}\right) I_2(s)$$

and therefore

$$y_{22}(s) = \left.\frac{I_2(s)}{V_2(s)}\right|_{V_1(s)=0} = \frac{R_1 + sL_1}{(R_1+sL_1)(R_2+sL_2) - s^2M^2}$$

Since (5) can be written as

$$I_2(s) = -\frac{R_1+sL_1}{sM} I_1 \tag{6}$$

substitution of (6) into (2) yields

$$V_2(s) = \frac{(R_2+sL_2)(R_1+sL_1) - s^2M^2}{-sM} \quad I_1$$

or

$$y_{12}(s) = \left.\frac{I_1(s)}{V_2(s)}\right|_{V_1(s)=0} = \frac{-sM}{(R_2+sL_2)(R_1+sL_1) - s^2M^2}$$

Since the denominator of all the admittance relations is the same, it can be designated as $\Delta = (R_1+sL_1)(R_2+sL_2) - s^2M^2$. The results, in matrix notation are:

$$Y(s) = \frac{1}{\Delta}\begin{bmatrix} R_2+sL_2 & -sM \\ -sM & R_1+sL_1 \end{bmatrix}$$

The Z parameters were found by applying a voltage source to the ports and open circuiting the appropriate port depending on the parameter for which the solution is sought. The dual procedure gives the short-circuit Y parameters. Note that the $\bar{Y}(s)$ matrix is the inverse of the $\bar{Z}(s)$ matrix.

• **PROBLEM** 16-42

Find the open-circuit impedance parameters for the network shown.

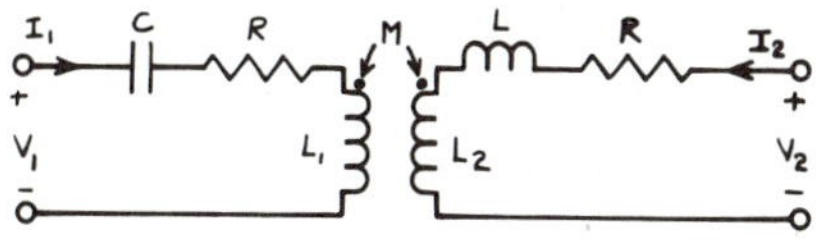

Solution: The impedance, or open circuit parameters, are calculated from $Z_{11} = \left.\frac{V_1}{I_1}\right|_{I_2=0}$, $Z_{12} = \left.\frac{V_1}{I_2}\right|_{I_1=0}$, $Z_{21} = \left.\frac{V_2}{I_1}\right|_{I_2=0}$, and $Z_{22} = \left.\frac{V_2}{I_2}\right|_{I_1=0}$. Hence,

$$Z_{11} = \frac{1}{sC} + R + sL_1 \quad (\text{since } I_2=0)$$

$$Z_{22} = R + sL + sL_2 = R + s(L+L_2) \quad (\text{since } I_1=0)$$

In order to calculate Z_{12}, apply a current source at port 2 and calculate the voltage V_1 with port 1 open.

Hence, $V_1 = MsI_2$, or

$$Z_{12} = sM .$$

• **PROBLEM** 16-43

Find the Z parameters for the two-port network shown in Fig. 1.

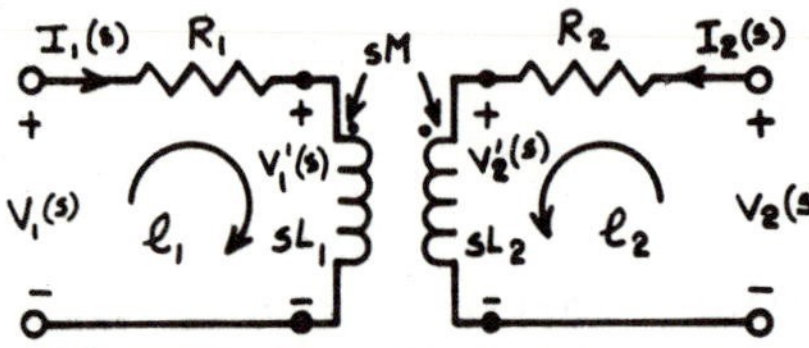

FIG.1 TRANSFORMER NETWORK

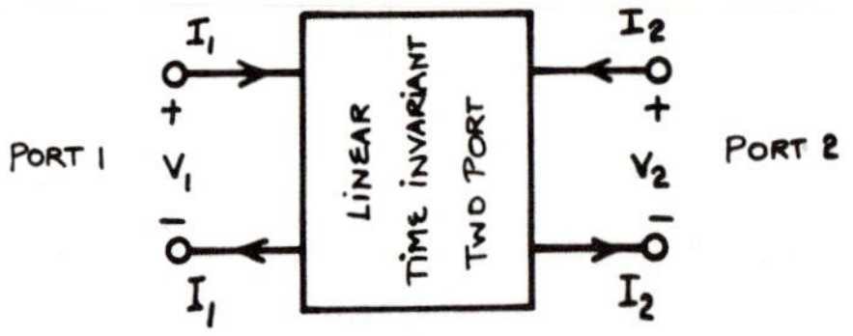

FIG. 2

Solution: Generally, a linear time invariant two-port network such as that depicted in Fig. 2, can be represented by its impedance or admittance matrix.

It can be seen that

$$\begin{bmatrix} V_1(s) \\ V_2(s) \end{bmatrix} = \begin{bmatrix} Z_{11}(s) & Z_{12}(s) \\ Z_{21}(s) & Z_{22}(s) \end{bmatrix} \begin{bmatrix} I_1(s) \\ I_2(s) \end{bmatrix}$$

or in short form

$$\overline{V(s)} = \overline{Z(s)}\ \overline{I(s)}$$

where the elements of Z(s) are the open-circuit impedance or Z parameters. With a voltage source $V_1(s)$ applied to port 1 and port 2 open circuited $(I_2(s)=0)$,

$$Z_{11}(s) = \left.\frac{V_1(s)}{I_1(s)}\right|_{I_2(s)=0}$$

With a current source $I_2(s)$ applied to port 2 and port 1 open circuited $(I_1(s)=0)$,

$$Z_{12}(s) = \left.\frac{V_1(s)}{I_2(s)}\right|_{I_1(s)=0}$$

Similarly we obtain

$$Z_{21}(s) = \left.\frac{V_2(s)}{I_1(s)}\right|_{I_2(s)=0}$$

and $$Z_{22}(s) = \left.\frac{V_2(s)}{I_2(s)}\right|_{I_1(s)=0}$$

For this example, start by writing the equations which characterize the operation of the transformer. Doing so yields the following:

$$V_1'(s) = sL_1 I_1(s) + sMI_2(s) \tag{1}$$

and $$V_2'(s) = sMI_1(s) + sL_2 I_2(s) \tag{2}$$

By writing a KVL equation around the primary loop ℓ_1 of Fig. 1 one obtains

$$V_1(s) = I_1(s)R_1 + V_1'(s)$$

or $\qquad V_1'(s) = V_1(s) - I_1(s)R_1 \qquad\qquad\qquad (3)$

Similarly, for loop ℓ_2 of Fig. 1

$$V_2(s) = I_2(s)R_2 + V_2'(s)$$

or $\qquad V_2'(s) = V_2(s) - I_2(s)R_2 \qquad\qquad\qquad (4)$

Substituting for $V_1'(s)$ and $V_2'(s)$ in Eqs. (1) and (2) yields

$$V_1(s) = (R_1 + sL_1)I_1(s) + sM\,I_2(s) \qquad\qquad (5)$$

and $\qquad V_2(s) = sM\,I_1(s) \qquad\qquad + (R_2 + sL_2)I_2(s) \qquad (6)$

Find the Z-parameter $Z_{11}(s)$ defined above as

$$\left. \frac{V_1(s)}{I_1(s)} \right|_{I_2(s)=0}$$

by solving for $I_1(s)$ in equation (5) above and substituting into the above definition. Hence,

$$Z_{11}(s) = \left.\frac{V_1(s)}{I_1(s)}\right|_{I_2(s)=0} = \left.\frac{V_1(s)}{\dfrac{V_1(s) - sMI_2(s)}{R_1 + sL_1}}\right|_{I_2(s)=0}$$

$$Z_{11}(s) = \frac{\dfrac{V_1(s)}{V_1(s)}}{R_1 + sL_1} = R_1 + sL_1$$

Similarly,

$$Z_{12}(s) = \left.\frac{V_1(s)}{I_2(s)}\right|_{I_1(s)=0} = sM$$

$$Z_{21}(s) = \left.\frac{V_2(s)}{I_1(s)}\right|_{I_2(s)=0} = sM$$

$$Z_{22}(s) = \left.\frac{V_2(s)}{I_2(s)}\right|_{I_1(s)=0} = R_2 + sL_2$$

From this, it is obvious that $Z_{12}(s) = Z_{21}(s) = sM$. Two-port networks which satisfy the condition

$$Z_{12}(s) = Z_{21}(s)$$

are generally called reciprocal two-ports.

• PROBLEM 16-44

Find the z-parameters of the network shown in Fig. 1.

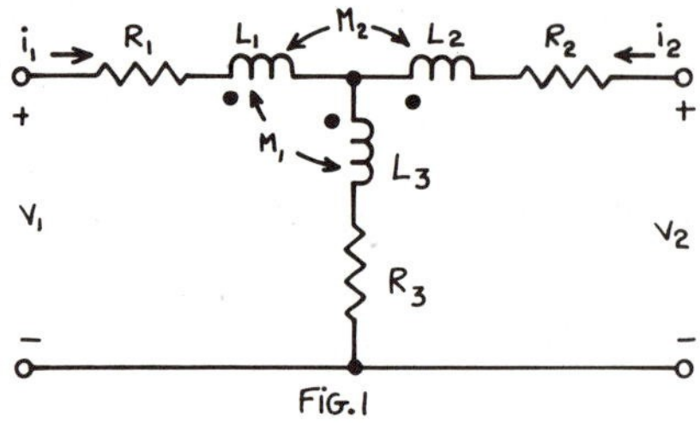

FiG. 1

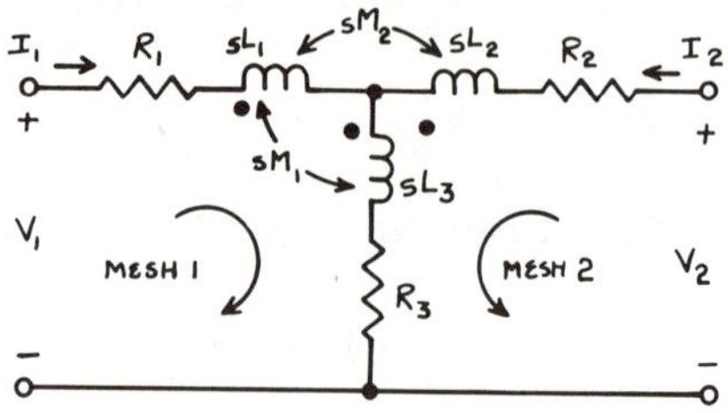

FiG. 2

Solution: Take the Laplace transform of the circuit in Fig. 1 to obtain the circuit shown in Fig. 2 below. The z-parameters are defined as:

$$z_{11} = \left.\frac{V_1}{I_1}\right|_{I_2=0} = \text{(sum of impedances in mesh 1 when } I_2=0,$$

added in the direction of mesh 1);

$$z_{12} = \left.\frac{V_1}{I_2}\right|_{I_1=0} = \text{(sum of impedances in mesh 1 when } I_1=0,$$

added in the direction of mesh 1);

$$z_{21} = \left.\frac{V_2}{I_1}\right|_{I_2=0} = \text{(sum of impedances in mesh 2 when } I_2=0,$$

added in the direction of mesh 2);

$$z_{22} = \left.\frac{V_2}{I_2}\right|_{I_1=0} = \text{(sum of impedances in mesh 2 when } I_1=0,$$

added in the direction of mesh 2).

In order to write z-parameters defined above, note that in a mutual inductor, when a current enters the dotted side of one coil, a voltage is produced which is sensed positively at the dotted side of the second coil.

The z_{11} parameter is written

$$z_{11}(s) = R_1 + R_3 + s(L_1 + L_3 + 2M_1),$$

the sum of the impedances around mesh 1. Note that M_2 has no effect, since we set $I_2=0$.

The z_{12} parameter is written

$$z_{12}(s) = R_3 + s(L_3 - M_2 + M_1).$$

Note that when $I_1=0$ the impedance around mesh 1 includes

both M_1 and M_2. Obtain the M_2 mutual inductance from the current flowing in L_2, which in turn induces a voltage across L_1 giving the sM_2 impedance when summing around mesh 1. The current I_2 flowing in L_3 induces a voltage across L_1 giving the sM_1 term for the impedance around mesh 1.

Note that when $I_2=0$ and we sum the impedances around mesh 2 we obtain z_{21} which is equal to z_{12} found above. Hence,

$$z_{21}(s) = R_3 + s(L_3 - M_2 + M_1)$$

The impedance around mesh 2 when $I_1=0$ is

$$z_{22}(s) = R_2 + R_3 + s(L_2 + L_3)$$

Note that since $I_1=0$ no voltages are induced in mesh 2 due to L_1. From the z-parameters the loop equations in matrix form are

$$\begin{bmatrix} V_1 \\ V_2 \end{bmatrix} = \begin{bmatrix} R_1+R_3+s(L_1+L_3+2M_1) & R_3+s(L_3-M_2+M_1) \\ R_3+s(L_3-M_2+M_1) & R_2+R_3+s(L_2+L_3) \end{bmatrix} \begin{bmatrix} I_1 \\ I_2 \end{bmatrix}$$

• PROBLEM 16-45

For the transistor T-equivalent shown, let $r_e = 20\ \Omega$, $r_b = 800\ \Omega$, $r_c = 500\ k\Omega$, and $\alpha = 0.98$. Find: (a) h_{12}; (b) h_{21}; (c) h_{22}.

Solution: The definitions of the h parameters will be applied to the circuit for their evaluation. These are

$$h_{12} = \left.\frac{V_1}{V_2}\right|_{I_1=0 \text{ (open)}}$$

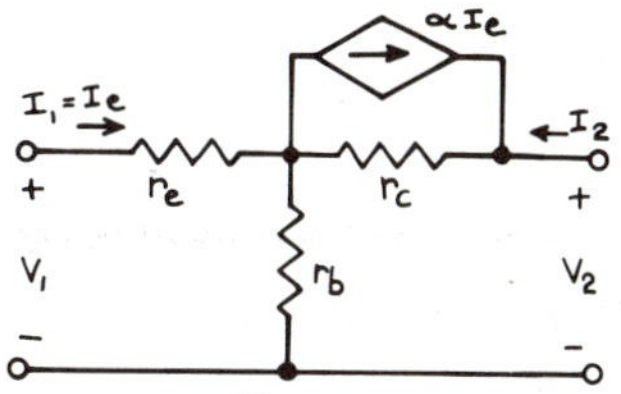

Fig.1

$$h_{22} = \left.\frac{I_2}{V_2}\right|_{I_1=0 \text{ (open)}}$$

$$h_{21} = \left.\frac{I_2}{I_1}\right|_{V_2=0 \text{ (short)}}$$

For evaluation of h_{12} and h_{22}, the input terminals are open; thereby making the dependent current source zero.

The ratio of V_1 to V_2 is obtained by simple voltage division

$$h_{12} = \left.\frac{V_1}{V_2}\right|_{I_1=0} = \frac{r_b}{r_b+r_c} = \frac{800}{800+5x10^5} \simeq 1.6x10^{-3}$$

The ratio of I_2 to V_2 is the output admittance, which is the reciprocal of the sum of the two resistances, r_b and r_c.

$$h_{22} = \left.\frac{I_2}{V_2}\right|_{I_1=0} = \frac{1}{r_b+r_c} = \frac{1}{800+5x10^5} \simeq 2x10^{-6} \text{ mho}$$

h_{21} is defined in terms of a shorted output. With this done, the currents out of the node labeled V_x are summed and equated to zero.

$$-I_1 + \frac{V_x}{r_b} + \frac{V_x}{r_c} + \alpha I_1 = 0$$

or
$$V_x = \frac{(1-\alpha)\ I_1}{\frac{1}{r_b} + \frac{1}{r_c}} = \frac{(1-\alpha) r_b r_c}{r_b + r_c} I_1$$

The short circuit current I_2 is then

$$I_2 = -\alpha I_1 - \frac{V_x}{r_c} = -\left[\alpha + \frac{(1-\alpha) r_b}{r_b+r_c}\right] I_1 = -\left[\frac{\alpha r_c + r_b}{r_b+r_c}\right] I_1$$

Thus

$$h_{21} = \left.\frac{I_2}{I_1}\right|_{V_2=0} = -\left|\frac{\alpha r_c + r_b}{r_b+r_c}\right| = -\left|\frac{(0.98)(5x10^5)+800}{800+5x10^5}\right|$$

$$\simeq -0.981$$

• **PROBLEM 16-46**

For the circuit shown, find (a) Z_{12}; (b) Z_{21}; (c) Z_{22}.

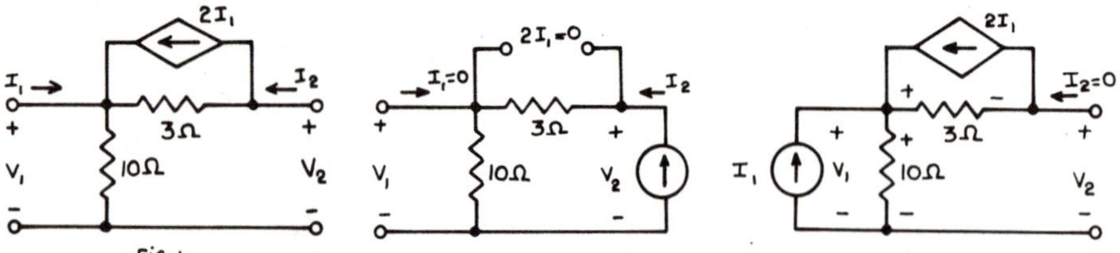

Fig.1 Fig.2 Fig.3

Solution: The two port impedance parameters are defined by

$$Z_{11} = \left.\frac{V_1}{I_1}\right|_{I_2=0}, \quad Z_{12} = \left.\frac{V_1}{I_2}\right|_{I_1=0}, \quad Z_{21} = \left.\frac{V_2}{K_1}\right|_{I_2=0}, \text{ and}$$

$$Z_{22} = \left.\frac{V_2}{I_2}\right|_{I_1=0}.$$

a) In order to calculate Z_{12}, change the general circuit of figure 1 to reflect the constraint that $I_1=0$. This means that port 1 must be open circuited, as shown in figure 2. The current source has been added to indicate that I_2 is considered to be the independent variable. Then $V_1 = 10I_2$, so

$$Z_{12} = 10\ \Omega.$$

b) To calculate Z_{21}, proceed in the reverse fashion: we open circuit port 2, and drive port 1 with a current source as in figure 3. The drop across the 3 Ω resistor has the polarity shown and is given by $3 \text{x} 2I_1 = 6I_1$. That across the 10 Ω resistor is given by $10I_1$ (note that the total current through the 10 Ω resistor is simply I_1 --- since $I_2=0$). Thus, $V_2 = -6I_1+10I_1 = 4I_1$, so

$$Z_{21} = 4\ \Omega.$$

c) One can calculate Z_{22} from the circuit of figure 2, but in this case we compute $\frac{V_2}{I_2}$. Now, since $I_1=0$, $V_2=13I_2$, so

$$Z_{22} = 13\ \Omega.$$

• PROBLEM 16-47

For the T-equivalent circuit of a transistor in the common-base arrangement, find: (a) Z_{11}; (b) Z_{12}; (c) Z_{21}.

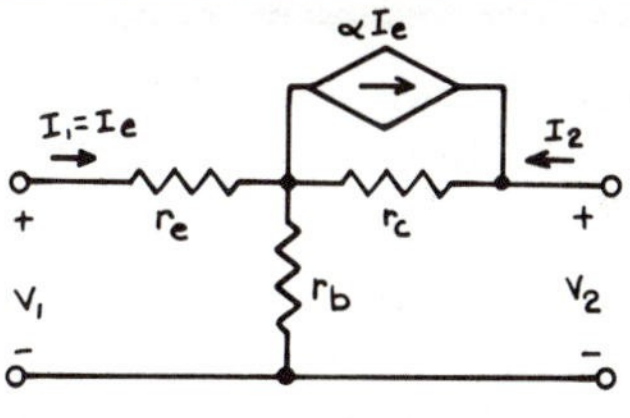

Solution: The Z-parameters for the two-port network are given as follows:

$$V_1 = Z_{11}I_1 + Z_{12}I_2$$

$$V_2 = Z_{21}I_1 + Z_{22}I_2$$

Solve for these parameters by setting one of the independent currents equal to zero as follows:

$$Z_{11} = \left.\frac{V_1}{I_1}\right|_{I_2=0} \quad (1)$$

$$Z_{12} = \left.\frac{V_1}{I_2}\right|_{I_1=0} \quad (2)$$

$$Z_{21} = \left.\frac{V_2}{I_1}\right|_{I_2=0} \quad (3)$$

From equation (1) with $I_2 = 0$

$$V_1 = I_1(r_e+r_b) \quad \text{volts}$$

or, $$Z_{11} = \left.\frac{V_1}{I_1}\right|_{I_2=0} = r_e+r_b \quad \text{ohms}$$

From equation (2) with $I_1 = I_e = 0$

$$V_1 = I_2r_b \quad \text{volts}$$

or $$Z_{12} = \left.\frac{V_1}{I_2}\right|_{I_1=0} = r_b \quad \text{ohms}$$

From equation (3) with $I_1 = I_e$ and $I_2 = 0$

$$V_1 = I_1r_b + \alpha I_er_c = I_1(r_b+\alpha r_c) \quad \text{volts}$$

or $$Z_{21} = \left.\frac{V_2}{I_1}\right|_{I_2=0} = r_b+\alpha r_c \quad \text{ohms}$$

• **PROBLEM** 16-48

Consider the circuit shown in Fig. 1. Assume that all elements marked R are 1Ω and that all elements marked X_c are -jΩ. Assume that the phasor V_{db} is known and that the expression V_{ab}, in terms of V_{db}, is desired.

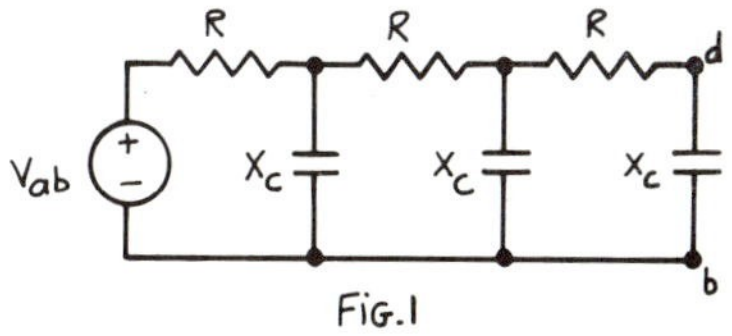

Fig.1

Solution: This type of problem is suitable for a matrix approach which will be used here. The ABCD matrix representation is the most convenient matrix for cascaded networks.

Consider one section of the entire circuit. The equations for the ABCD parameters are

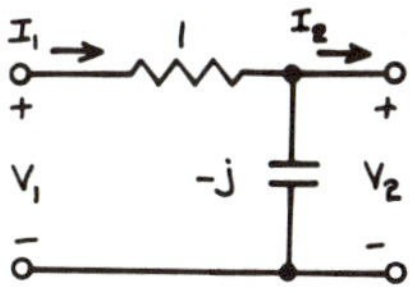

$$V_1 = AV_2 + BI_2$$

$$I_1 = CV_2 + DI_2$$

Next, solve for A, B, C, and D.

$$A = \left.\frac{V_1}{V_2}\right|_{I_2=0} = \frac{V_1}{V_1 \frac{-j}{1-j}} = j(1-j)$$

$$B = \left.\frac{V_1}{I_2}\right|_{V_2=0} = \frac{(1)I_1}{I_1} = 1$$

$$C = \left.\frac{I_1}{V_2}\right|_{I_2=0} = \frac{I_1}{V_1-(1)I_1} = \frac{I_1}{(1-j)I_1-I_1} = j$$

$$D = \left.\frac{I_1}{I_2}\right|_{V_2=0} = 1$$

The given circuit consists of three cascaded sections of the simple element. Multiplying three of the ABCD matrices gives the ABCD matrix for the entire circuit. Thus,

$$\begin{bmatrix} A & B \\ C & D \end{bmatrix} = \begin{bmatrix} j(1-j) & 1 \\ j & 1 \end{bmatrix}\begin{bmatrix} j(1-j) & 1 \\ j & 1 \end{bmatrix}\begin{bmatrix} j(1-j) & 1 \\ j & 1 \end{bmatrix}$$

$$= \begin{bmatrix} j(1-j) & 1 \\ j & 1 \end{bmatrix}\begin{bmatrix} 3j & j+2 \\ -1+2j & j+1 \end{bmatrix}$$

$$= \begin{bmatrix} 4-5j & 4j+2 \\ -4+2j & 3j \end{bmatrix}$$

The overall circuit equations for the entire three-section network are

$$V_1 = (4-5j)V_2 + (4j+2)I_2$$

$$I_1 = (-4+2j)V_2 + 3jI_2$$

where V_1 , I_1 and V_2 , I_2 are the input and output voltage and currents, respectively. In this problem $V_1 = V_{ab}$, $V_2 = V_{db}$, and $I_2 = 0$. Hence, using the first equation, we may write the solution

$$V_{ab} = (4-j5)V_{db} = 6.40\underline{/128.7°}.$$

• PROBLEM 16-49

A simplified linear equivalent circuit for a transistor is shown below. Two identical two ports are used in cascade (output of first connected directly to input of second). If the combination is terminated in $R_L = 10\ k\Omega$ and $h_{11} = 10^3\Omega$, $h_{21} = 10^2$, $h_{22} = 10^{-5}$ mho, find V_L/V_1.

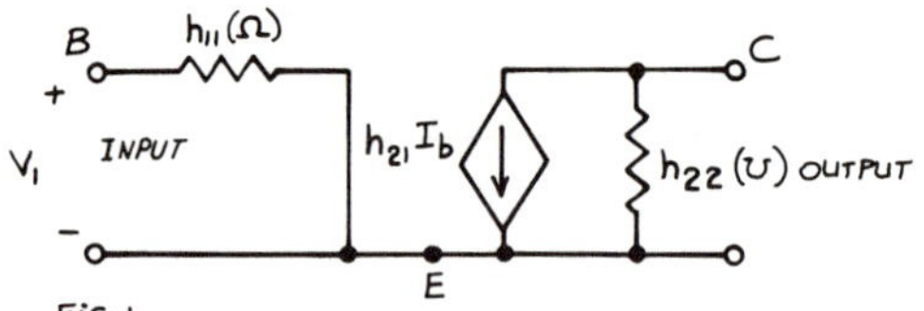

FIG. 1

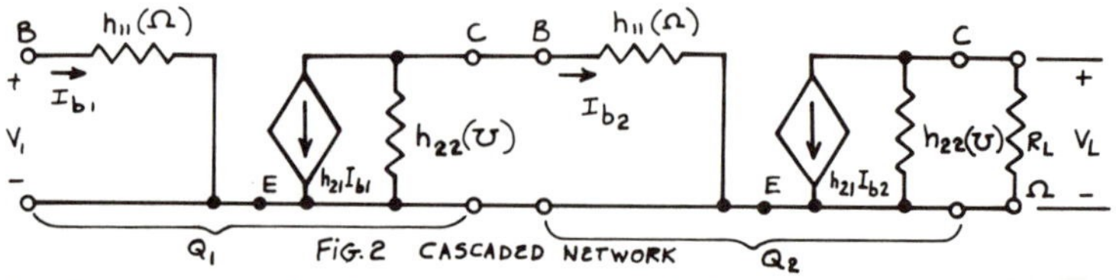

FIG. 2 CASCADED NETWORK

Solution: Sketch the network for the cascaded transistors using the model shown in Figure 1. See figure 2. V_1 and V_L are input and output voltages respectively with polarities as indicated. I_{b1} and I_{b2} are the base currents of the input and output transistors, Q_1 and Q_2, respectively with positive directions as shown. The steps in the solution are given below.

Find I_{b1} by Ohm's Law.

$$I_{b1} = \frac{V_1}{h_{11}} = \frac{V_1}{10^3} \text{ Amps.}$$

The dependent current source of Q_1 is $h_{21}I_{b1} = (10^2)\frac{V_1}{10^3} = \frac{V_1}{10}$ Amps. By current division,

$$I_{b2} = -(h_{21}I_{b1}) \frac{h_{22}^{-1}}{h_{22}^{-1}+h_{11}} = -\frac{V_1}{10} \frac{10^5}{10^5+10^3}$$

$$= -9.9\text{x}10^{-2}V_1 \text{ Amps.}$$

Note current is negative since it is opposite to the assumed positive direction.

The dependent current source of transistor Q_2 is $h_{21}I_{b2} = (10^2)(-9.9\text{x}10^{-2}V_1) = -9.9V_1$ Amps. Find R_L' , the equivalent resistance of $R_L || h_{22}^{-1}$

$$R_L' = \frac{h_{22}^{-1} \text{ x } R_L}{h_{22}^{-1} + R_L} = \frac{(10^5)(10^4)}{10^5 + 10^4} = 9{,}091\ \Omega$$

Finally,

$$-V_L = h_{21}I_{b2} \text{ x } R_L' = (-9.9V_1)(9{,}091) = -90{,}000V_1 \text{ volts}$$

Note that the negative sign for V_L is due to the assumed polarities of the voltages and the positive directions of the currents.

Finally,

$$\frac{V_L}{V_1} = 90{,}000$$

RECIPROCITY

• PROBLEM 16-50

Show that the network in the Figure is reciprocal.

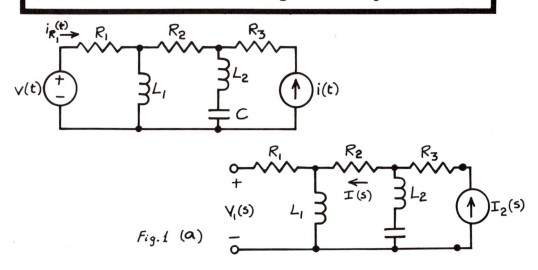

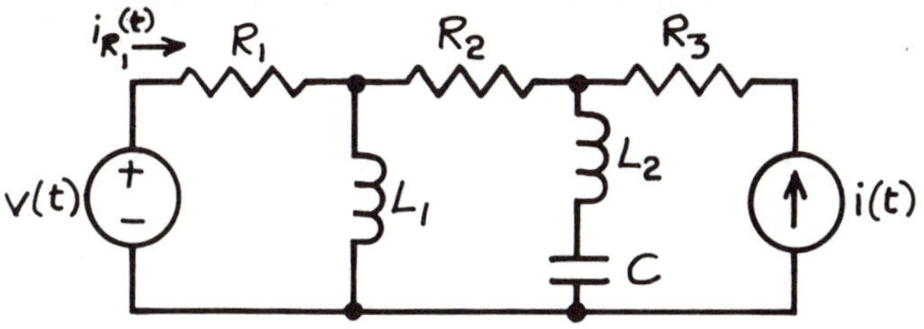

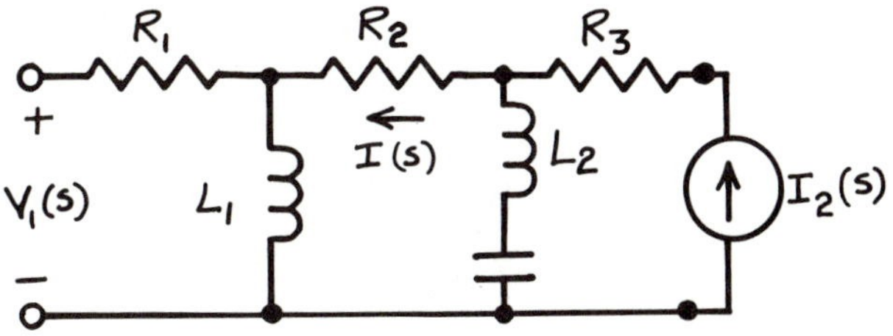

Fig. 1 (a)

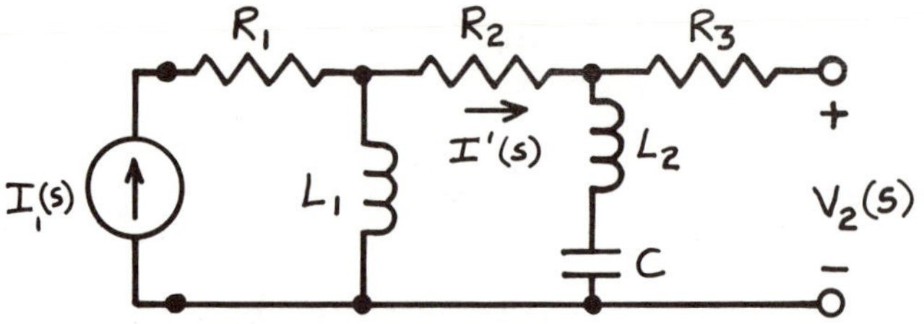

Fig. 1 (b)

Solution: For any linear passive circuit, with no initial conditions, reciprocity can be demonstrated many ways. Taking the given circuit as a two-port network, we will first apply a current $I_2(s)$ to the output port (see Fig. 1(a)). To make the results more general, the Laplace transform variable s will be used. The ratio of the open-circuit input voltage $V_1(s)$ to $I_2(s)$ is then obtained and designated $Z_{12}(s)$. Second we will apply a current source $I_1(s)$ to the input port (see Fig. 1(b)) and determine the ratio $Z_{21}(s)$, of the open-circuit voltage $V_2(s)$ to $I_1(s)$. Finally, if $Z_{21}(s) = Z_{12}(s)$, the network is shown to be reciprocal.

Referring to Fig. 1(a), I(s) is obtained by current division as

$$I(s) = I_2(s)\,\frac{sL_2 + \frac{1}{sC}}{s(L_1+L_2) + R_2 + \frac{1}{sC}}$$

$$= I_2(s)\,\frac{s^2L_1C + 1}{s^2C(L_1+L_2) + sR_2C + 1}$$

This is multiplied by sL_1 to obtain $V_1(s)$ and divided by $I_2(s)$ to obtain $Z_{12}(s)$

$$Z_{12}(s) = \frac{V_1(s)}{I_2(s)} = \frac{sL_1(s^2L_2C + 1)}{s^2C(L_1+L_2) + sR_2C + 1} \qquad (1)$$

Going now to Fig. 1(b), and again applying current division

$$I'(s) = I_1(s)\,\frac{sL_1}{s(L_1+L_2) + R_2 + \frac{1}{sC}}$$

$$= I_1(s)\,\frac{s^2L_1C}{s^2C(L_1+L_2) + sR_2C + 1}$$

This is multiplied by $(sL_2 = \frac{1}{sC})$ to obtain $V_2(s)$ and divided by $I_1(s)$ to obtain $Z_{21}(s)$

$$Z_{21}(s) = \frac{V_2(s)}{I_1(s)} = \frac{s^2L_1C(sL_2+\frac{1}{sC})}{s^2C(L_1+L_2) + sR_2C + 1}$$

$$= \frac{sL_1(s^2L_2C+1)}{s^2C(L_1+L_2) + sR_2C + 1} \qquad (2)$$

Finally, comparing equations (1) and (2), we see that $Z_{12}(s) = Z_{21}(s)$ and the network is reciprocal.

Alternately, the network could be shown reciprocal by applying voltage rather than current sources. The opposite port would be short-circuited, and the short-circuit current obtained. The voltage source is moved to the other port and the procedure repeated. This would lead to two transfer admittances, which are the ratio of the short-circuit current to the applied voltage. Equality of the two admittances shows reciprocity.

• PROBLEM 16-51

Consider the network shown in Fig. 1. Determine whether this network is reciprocal.

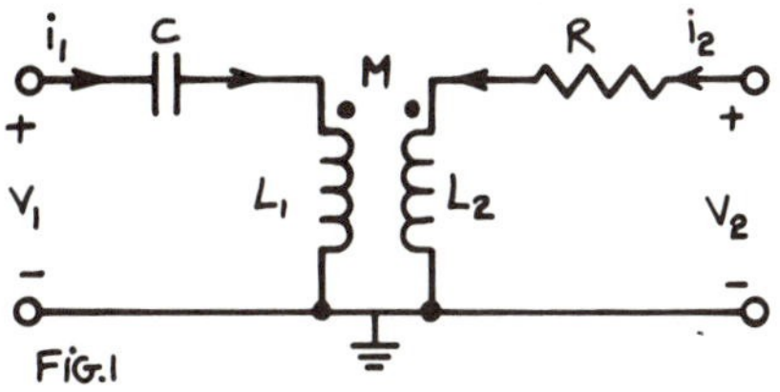

FiG.1

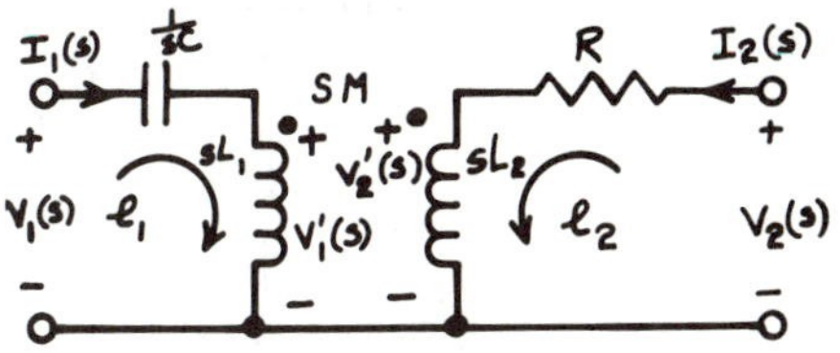

FiG.2

Solution: After transforming the network of Fig. 1 to that of Fig. 2 above, begin by writing the equations which characterize the network.

The equations which describe our network are:

$$V_1'(s) = sL_1I_1(s) + sMI_2(s) \qquad (1)$$

and

$$V_2'(s) = sMI_1(s) + sL_2I_2(s) \qquad (2)$$

Using KVL around loop ℓ_1 yields

$$V_1'(s) = V_1(s) - \frac{1}{sC}I_1(s) \qquad (3)$$

and

$$V_2'(s) = V_2(s) - RI_2(s) \qquad (4)$$

Substituting Eqs.(3) and (4) into (1) and (2) yields

$$V_1(s) = (\frac{1}{sC} + sL_1)I_1(s) + sMI_2(s) \qquad (5)$$

and $$V_2(s) = sMI_1(s) + (R + sL_2)I_2(s) \tag{6}$$

Since the impedance matrix is written as

$$Z(s) = \begin{bmatrix} Z_{11}(s) & Z_{12}(s) \\ Z_{21}(s) & Z_{22}(s) \end{bmatrix}$$

Write the Z(s) matrix for our network from Eqs.(5) and (6). This yields

$$Z(s) = \begin{bmatrix} (\frac{1}{sC} + sL_1) & sM \\ sM & R + sL_2 \end{bmatrix}$$

It can be seen from matrix Z(s) that $Z_{12}(s) = Z_{21}(s) = sM$. Hence the network considered is reciprocal.

• PROBLEM 16-52

Consider the small-signal equivalent circuit of the common-emitter transistor shown in Fig. 1. Show that this network is nonreciprocal.

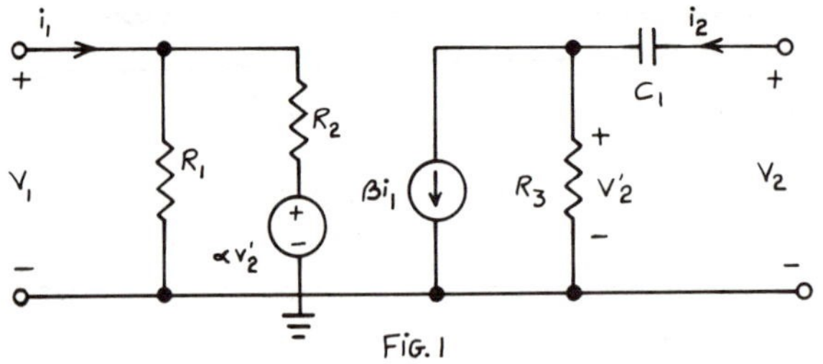

FIG. 1

Solution: Recall that the two-port Z parameters are defined as

$$Z_{11}(s) = \left.\frac{V_1(s)}{I_1(s)}\right|_{I_2(s)=0} \qquad Z_{12}(s) = \left.\frac{V_1(s)}{I_2(s)}\right|_{I_1(s)=0}$$

$$Z_{21}(s) = \left.\frac{V_2(s)}{I_1(s)}\right|_{I_2(s)=0} \qquad Z_{22}(s) = \left.\frac{V_2(s)}{I_2(s)}\right|_{I_1(s)=0}$$

where $Z_{12}(s) = Z_{21}(s)$ for reciprocal two-ports.

For our example, the network of Fig. 1 is transformed to that of Fig. 2 below. Proceed to obtain the relationships between terminal voltages and currents for this transformed network.

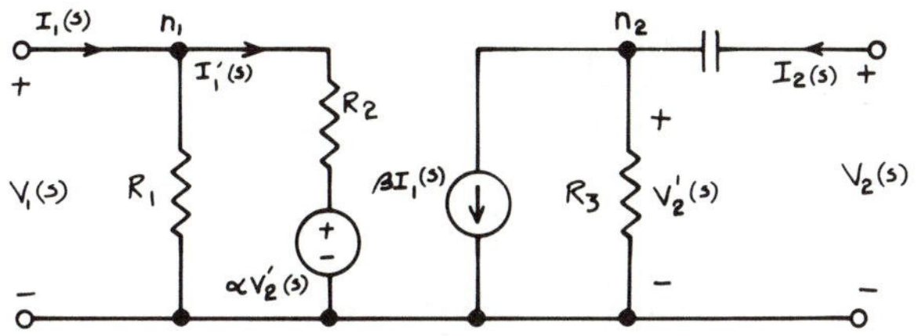

Fig. 2

Looking first at the left side of the network observe that

$$V_1(s) = I_1'(s)R_2 + \alpha V_2'(s) \tag{1}$$

Applying KCL at node n_1 yields

$$I_1(s) = I_1'(s) + \frac{V_1(s)}{R_1}$$

or $$I_1(s) = I_1'(s) + V_1(s)G_1 \tag{2}$$

where G_1 is defined as the conductance $\frac{1}{R_1}$.

Looking at the right side of the network

$$V_2(s) = \frac{I_2(s)}{sC_1} + V_2'(s) \tag{3}$$

and by KCL at node n_2

$$I_2(s) = \frac{V_2'(s)}{R_3} + \beta I_1'(s) \tag{4}$$

Equations (1) - (4) characterize the network and can be used to obtain the Z parameters for the network.

From Eqs. (2) and (3)

$$I_1'(s) = I_1(s) - V_1(s)G_1 \tag{5}$$

and $$V_2'(s) = V_2(s) - \frac{I_2(s)}{sC_1} \tag{6}$$

Substituting Eqs. (5) and (6) into Eqs. (1) and (4) and solving the resulting simultaneous equations for $V_1(s)$ and $V_2(s)$ yields

$$V_1(s) = \left(\frac{R_2-\alpha\beta R_3}{1+G_1R_2-\alpha\beta R_3G_1}\right)I_1(s) + \left(\frac{\alpha R_3}{1+G_1R_2-\alpha\beta R_3G_1}\right)I_2(s) \tag{7}$$

and

$$V_2(s) = \left(\frac{\beta R_3 G_1 (G_2 - \alpha\beta R_3)}{1+G_1R_2-\alpha\beta R_3 G_1} - \beta R_3\right) I_1(s) +$$

$$\left(\frac{\alpha\beta R_3^2 G_1}{1+G_1R_2-\alpha\beta R_3G_1} + R_3 + \frac{1}{sC_1}\right) I_2(s) \qquad (8)$$

With the Z parameters as defined previously, we see from Eqs. (7) and (8) that

$$Z_{11}(s) = \frac{R_2-\alpha\beta R_3}{1+G_1R_2-\alpha\beta R_3G_1} \qquad Z_{12}(s) = \frac{\alpha R_3}{1+G_1R_2-\alpha\beta R_3G_1}$$

$$Z_{21}(s) = \frac{\beta R_3G_1(G_2-\alpha\beta R_3)}{1+G_1R_2-\alpha\beta R_3G_1} - \beta R_3 \qquad Z_{22} = \frac{\alpha\beta R_3^2 G_1}{1+G_1R_2-\alpha\beta R_3G_1} + R_3 + \frac{1}{sC_1}$$

From these parameters it can be seen that $Z_{12} \neq Z_{21}$. Since this is indeed so for non-reciprocal networks, our network is therefore non-reciprocal.

• PROBLEM 16-53

For the network shown, verify the reciprocity property by: (a) calculating the load current $I_L(s)$ with a load connected across terminals 3,4 and a matched source connected across terminals 1,2; and (b) interchanging load and matched source position as well as recalculating the load current. Take $Z_L = 5\Omega$, $v(t) = tu(t)$.

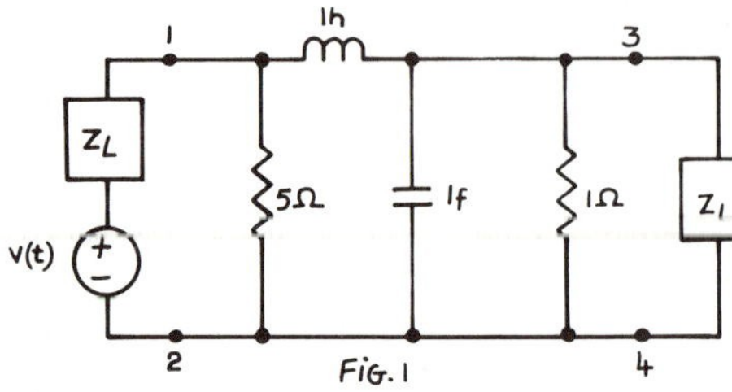

Fig. 1

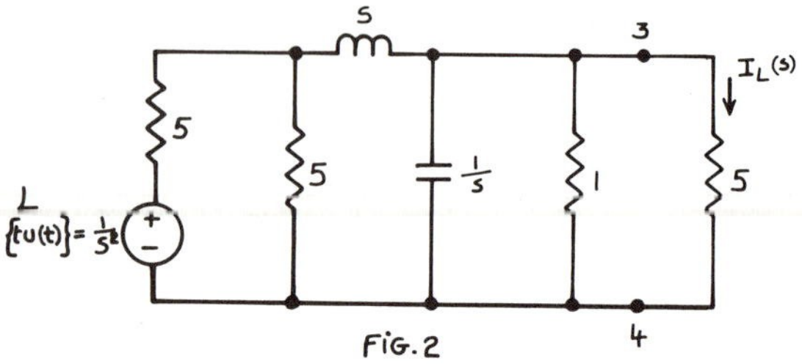

Fig. 2

Solution: (a) By use of Laplace transforms, transform the circuit of Fig. 1 to that shown in Fig. 2.

Since we are interested in finding only $I_2(s)$, find the Thévenin equivalent circuit which will replace the circuit to the left of terminals 3,4 in Fig. 2.

The open circuit voltage is found by forming a single loop circuit combining impedances, and making source transformations as shown in Figs. 3-5 below.

First, combine the impedances shown below in Fig. 3. Second, transform the voltage source and series impedance

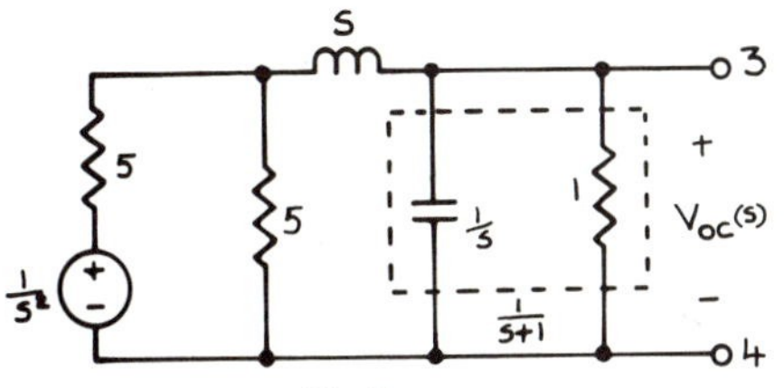

FiG.3

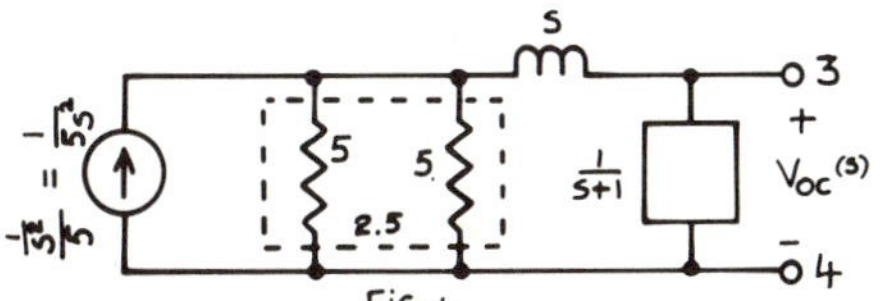

FiG.4

into a current source and parallel impedance as shown in Fig. 4. Third, combine the two parallel impedances shown in Fig. 4 and transform the resulting current source and parallel impedance into a voltage source and series impedance shown in Fig. 5.

The open circuit voltage in Fig. 5 can be found by using the voltage division concept. Doing so yields

$$V_{oc}(s) = \frac{\frac{1}{2s^2}\left(\frac{1}{s+1}\right)}{\frac{1}{s+1} + s + 2.5}$$

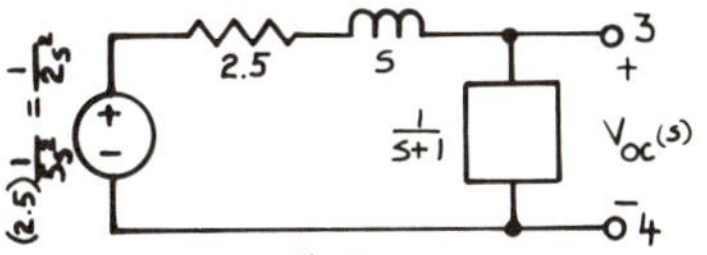

FiG.5

$$V_{oc}(s) = \frac{\frac{1}{2s^2(s+1)}}{\frac{s^2+3.5s+3.5}{s+1}}$$

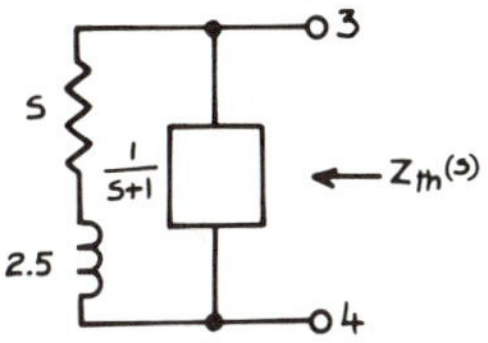

FiG.6

$$V_{oc}(s) = \frac{1}{2s^2(s^2+3.5s+3.5)}$$

The Thévenin equivalent impedance can be found by short circuiting the voltage source in Fig. 5 and calculating the impedance looking into terminals 3,4, as in Fig. 6. Hence,

$$Z_{th}(s) = \frac{\frac{s+2.5}{s+1}}{\frac{1}{s+1} + s + 2.5}$$

$$Z_{th}(s) = \frac{s+2.5}{s^2+3.5s+3.5}$$

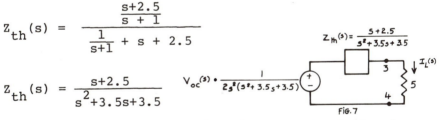

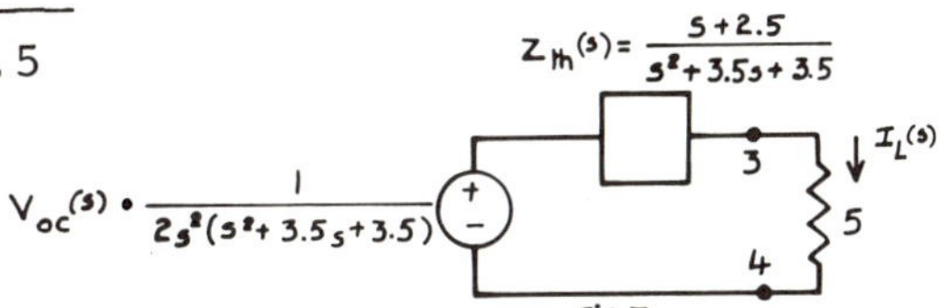

FiG.7

The Thévenin equivalent circuit for the network of Fig. 3 is shown in Fig. 7. Now, by attaching the load to the Thévenin circuit it is possible to find the load current $I_L(s)$ by the loop equation

$$\frac{1}{2s^2(s^2+3.5s+3.5)} = I_L(s)\left|\frac{s+2.5}{s^2+3.5s+3.5} + 5\right| .$$

Solving for $I_L(s)$,

$$I_L(s) = \frac{\dfrac{1}{2s^2(s^2+3.5s+3.5)}}{\dfrac{s+2.5}{s^2+3.5s+3.5} + 5}$$

$$I_L(s) = \frac{\dfrac{1}{2s^2(s^2+3.5s+3.5)}}{\dfrac{s+2.5+5(s^2+3.5s+3.5)}{s^2+3.5s+3.5}}$$

$$I_L(s) = \frac{1}{2s^2(5s^2+18.5s+20)}$$

gives,

$$I_L(s) = \frac{1}{s^2(10s^2+37s+40)} \quad .$$

(b) The reciprocity property to be verified is illustrated in Fig. 8 below. In the network of Fig. 1 we are asked to perform the test shown in Fig. 8. Move v(t) from one side of the network to the other and find the current in the load. If the circuit is reciprocal, the two currents must be equal.

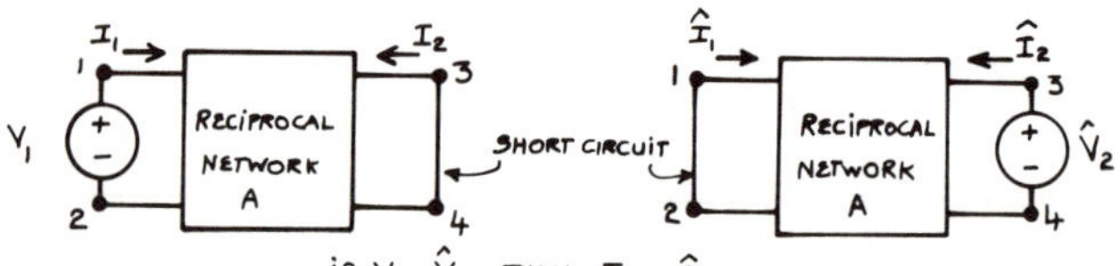

if $V_1 = \hat{V}_2$ THEN $I_2 = \hat{I}_1$

Fig. 8

Using the same method used in part (a), find the Thévenin equivalent circuit for the network in Fig. 9 to solve for $I_L(s)$.

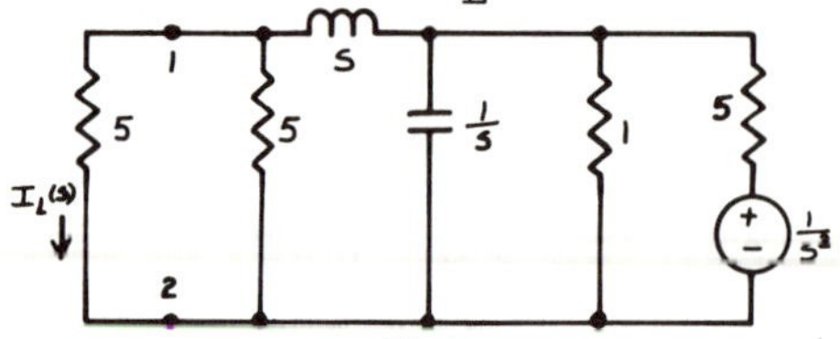

Fig. 9

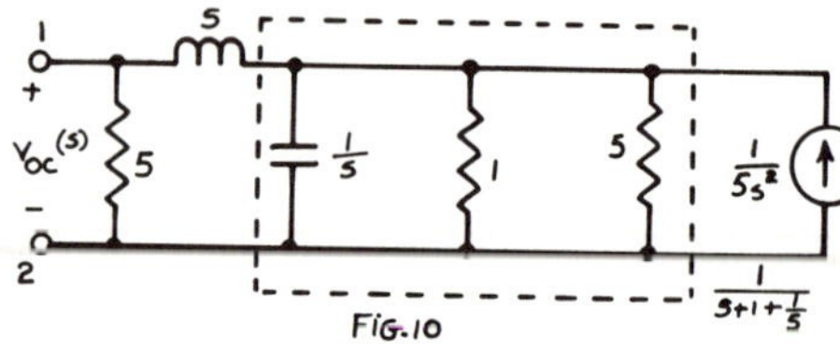

Fig. 10

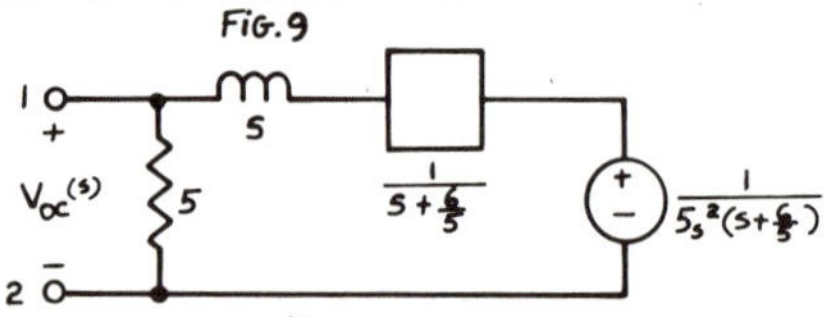

Fig. 11

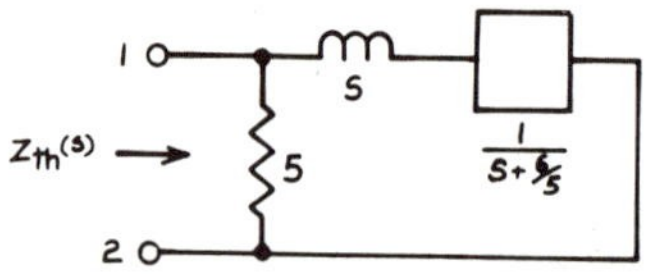

Fig. 12

Figs. 10-12 outline the steps taken in obtaining the Thévenin equivalent circuit.

$$V_{oc}(s) = \frac{\dfrac{1}{s^2(s+\frac{6}{5})}}{\dfrac{1}{s+\frac{6}{5}} + s + 5}$$

$$V_{oc}(s) = \frac{\frac{1}{s^2(s+\frac{6}{5})}}{\frac{s^2 + s\frac{31}{5} + 7}{s+\frac{6}{5}}} = \frac{1}{s^2(s^2 + s\frac{31}{5} + 7}$$

$$Z_{th}(s) = \frac{5\left[s + \frac{1}{s+\frac{6}{5}}\right]}{5 + s + \frac{1}{s+\frac{6}{5}}}$$

$$Z_{th}(s) = \frac{\frac{5s^2+6s+5}{s+\frac{6}{5}}}{\frac{s^2+s\frac{31}{5}+7}{s+\frac{6}{5}}} = \frac{5s^2 + 6s + 5}{s^2 + \frac{31}{5}s + 7}$$

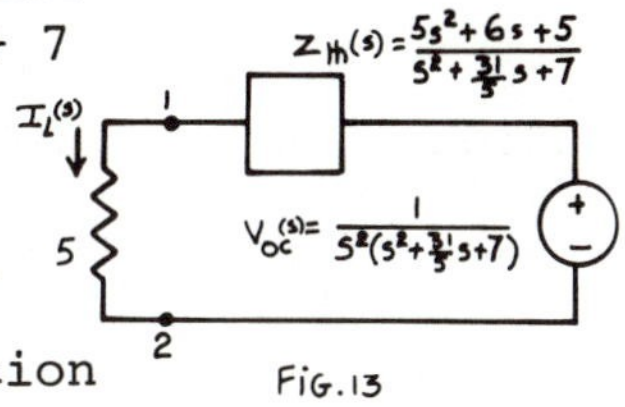

FiG.13

Fig. 13 shows the Thévenin circuit found.

$I_L(s)$ is found from the loop equation

$$\frac{1}{s^2(s^2 + \frac{31}{5} + 7} = I_L(s)\left[5 + \frac{5s^2 + 6s + 5}{s^2 + \frac{31}{5} + 7}\right]$$

Solving for $I_L(s)$

$$I_L(s) = \frac{\frac{1}{s^2(s^2 + \frac{31}{5} + 7)}}{5 + \frac{5s^2 + 6s + 5}{s^2 + \frac{31}{5} + 7}}$$

$$I_L(s) = \frac{\frac{1}{s^2(s^2 + \frac{31}{5} + 7)}}{\frac{5(s^2 + \frac{31}{5}s + 7) + 5s^2 + 6s + 5}{s^2 + \frac{31}{5}s + 7}}$$

$$I_L(s) = \frac{1}{s^2(10s^2 + 37s + 40)}$$

Since $I_L(s)$ found in part (a) equals $I_L(s)$ found above, the reciprocity property for this network has been verified.

π-T CONVERSION

• PROBLEM 16-54

Using π-T conversion, put the network shown in Fig. 1 into the form in Fig. 2 and thus solve for the current i(t).

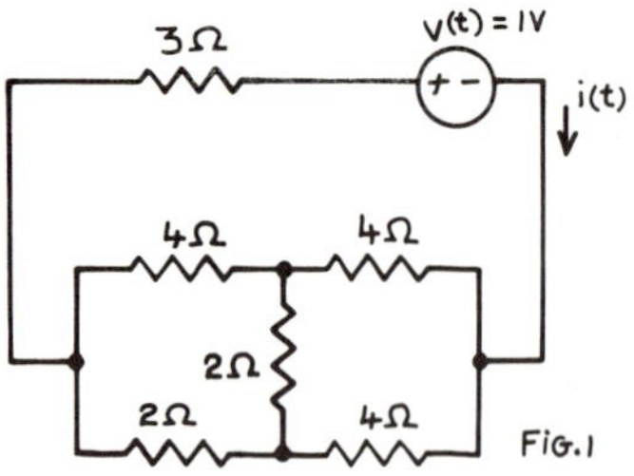

Fig.1

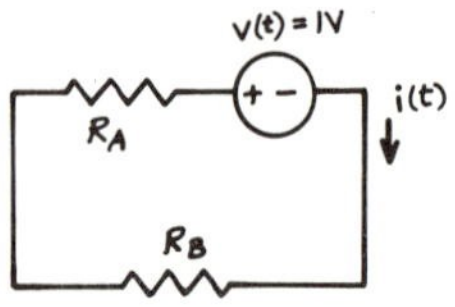

Fig.2

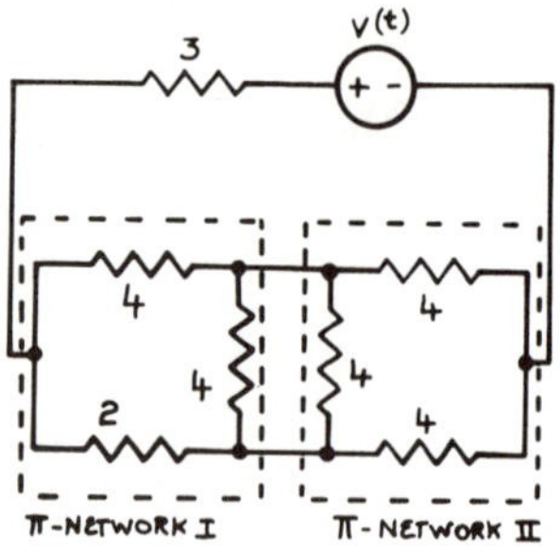

Fig.3

Solution: The circuit in Fig. 1 can be viewed as a resistor in series with two back-to-back π networks as shown in Fig. 3. Each π-network can be converted to a T-network shown in Fig. 5. The resulting circuit is shown in Fig. 6.

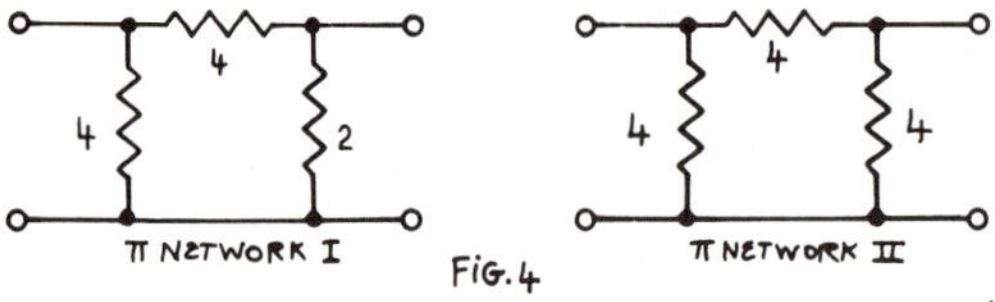

Fig.4

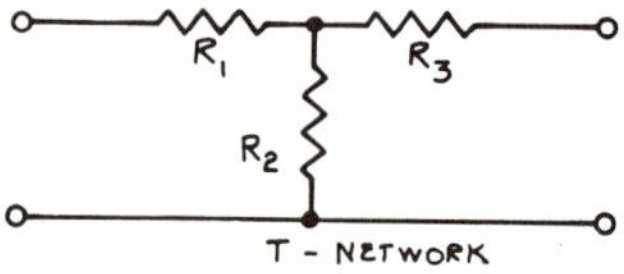

Fig.5

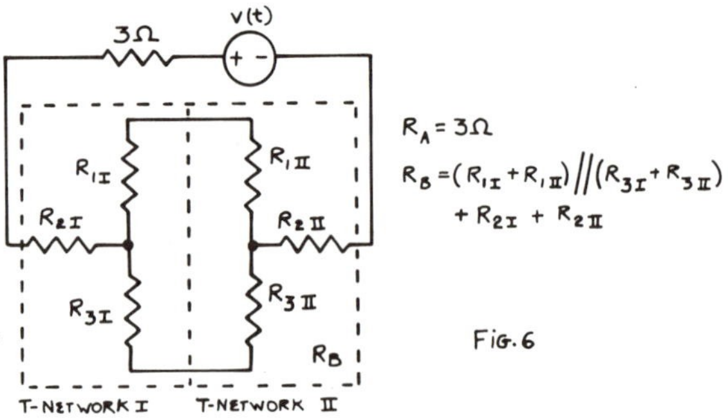

Fig.6

The formulas for converting from π-networks to T-networks are

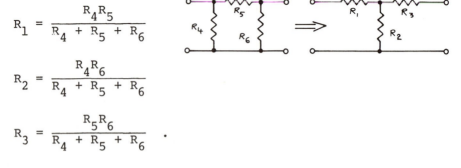

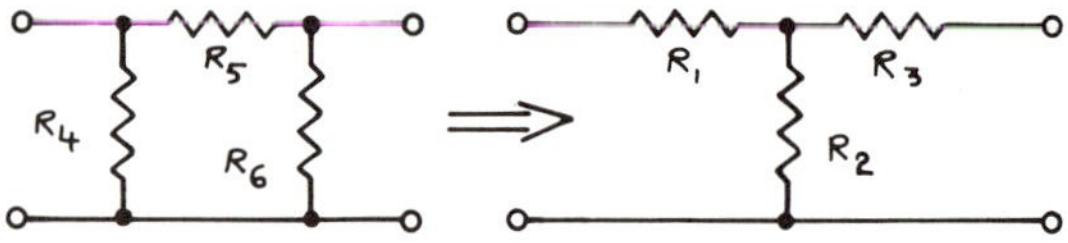

$$R_1 = \frac{R_4 R_5}{R_4 + R_5 + R_6}$$

$$R_2 = \frac{R_4 R_6}{R_4 + R_5 + R_6}$$

$$R_3 = \frac{R_5 R_6}{R_4 + R_5 + R_6} .$$

From these we find

$$R_1 I = 1.6\ \Omega \qquad R_1 II = \frac{4}{3}\ \Omega$$

$$R_2 I = 0.8\ \Omega \qquad R_2 II = \frac{4}{3}\ \Omega$$

$$R_3 I = 0.8\ \Omega \qquad R_3 II = \frac{4}{3}\ \Omega$$

Hence $R_B = (1.6 + \frac{4}{3}) || (0.8 + \frac{4}{3}) + 0.8 + \frac{4}{3}$

$$R_B = 3.37\ \Omega$$

and $i = \dfrac{-v(t)}{R_A + R_B} = \dfrac{1}{6.37} = -.157\ A$

• PROBLEM 16-55

Derive a T and equivalent π network for a transformer with self-inductance values L, mutual inductance M and resistance R as shown in Fig. 1.

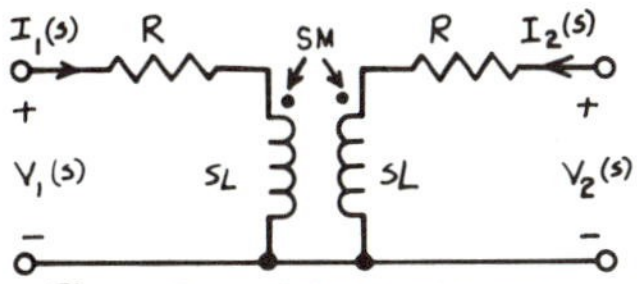

FIG. 1 TRANSFORMER NETWORK

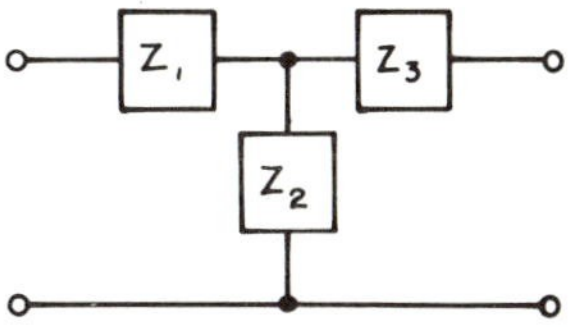

FIG. 2

Solution: We have seen previously, for this network

$$Z_{11} = R + sL$$

$$Z_{12} = Z_{21} = sM$$

$$Z_{22} = R + sL$$

Derive the T-network equivalent shown in Fig. 2, where

$$Z_{11} = Z_1 + Z_2$$

$$Z_{12} = Z_{21} = Z_2$$

$$Z_{22} = Z_2 + Z_3$$

Hence, $Z_{11} = Z_1 + Z_2 = R + sL$

$$Z_{12} = Z_{21} = Z_2 = sM$$

$$Z_{22} = Z_2 + Z_3 = R + sL$$

Solving for Z_1, Z_2, and Z_3 gives the elements of the T-network:

$Z_1 = R + sL - Z_2 = R + sL - sM$

$Z_1 = R + s(L-M)$

$Z_2 = sM$

$Z_3 = R + sL - Z_2 = R + sL - sM$

$Z_3 = R + s(L-M)$

The T-equivalent of the transformer is shown in Fig. 3.

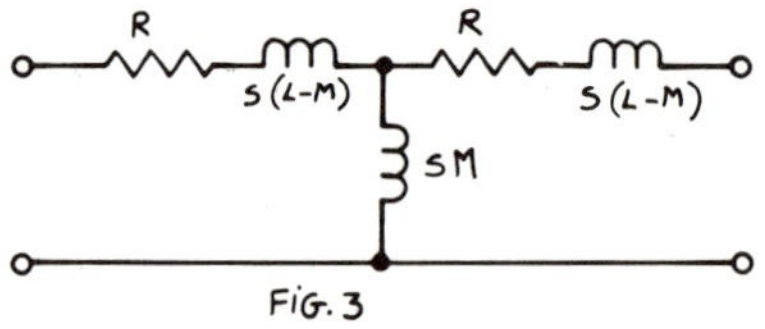

Fig. 3

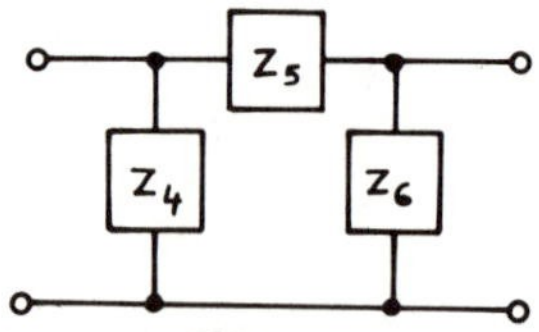

Fig. 4

Derive the equivalent π-network shown in Fig. 4, where

$$z_{11} = \frac{Z_4(Z_5 + Z_6)}{Z_4 + Z_5 + Z_6}$$

$$z_{21} = z_{12} = \frac{Z_4 Z_6}{Z_4 + Z_5 + Z_6}$$

$$z_{22} = \frac{Z_6(Z_4 + Z_5)}{Z_4 + Z_5 + Z_6}$$

Hence,

$$z_{11} = R + sL = \frac{Z_4(Z_5 + Z_6)}{Z_4 + Z_5 + Z_6}$$

$$z_{21} = z_{12} = sM = \frac{Z_4 Z_6}{Z_4 + Z_5 + Z_6} \qquad (1)$$

$$z_{22} = R + sL = \frac{Z_6(Z_4 + Z_5)}{Z_4 + Z_5 + Z_6}$$

Solve for Z_4, Z_5 and Z_6 as follows,

since, $R + sL = \dfrac{Z_4 Z_5}{Z_4 + Z_5 + Z_6} + \dfrac{Z_4 Z_6}{Z_4 + Z_5 + Z_6}$

$$R + sL = \frac{Z_4 Z_5}{Z_4 + Z_5 + Z_6} + z_{21}$$

then, $$R + sL = \frac{Z_4 Z_5}{Z_4 + Z_5 + Z_6} + sM \qquad (2)$$

also $$R + sL = \frac{Z_6 Z_4}{Z_4 + Z_5 + Z_6} + \frac{Z_6 Z_5}{Z_4 + Z_5 + Z_6}$$

$$R + sL = z_{21} + \frac{Z_6 Z_5}{Z_4 + Z_5 + Z_6}$$

$$R + sL = sM + \frac{Z_6 Z_5}{Z_4 + Z_5 + Z_6} \qquad (3)$$

From equations (2) and (3) it follows that

$$Z_4 Z_5 = Z_6 Z_5$$

$$Z_4 = Z_6$$

Taking Eq.(1) and solving for Z_5 yields

$$sM = \frac{Z_4^{\,2}}{2Z_4 + Z_5}$$

$$\frac{Z_4^{\,2} - 2Z_4 sM}{sM} = Z_5. \qquad (4)$$

Substitute Z_5 in either Eq.(2) or (3) to obtain

$$R + sL - sM = \frac{Z_4\left(\dfrac{Z_4^{\,2} - 2Z_4 sM}{sM}\right)}{2Z_4 + \dfrac{Z_4^{\,2} - 2Z_4 sM}{sM}}$$

$$R + sL - sM = \frac{Z_4\left(\dfrac{Z_4 - 2sM}{sM}\right)}{2 + \dfrac{Z_4 - 2sM}{sM}}$$

$$R + sL - sM = Z_4 - 2sM$$

$$R + s(L+M) = Z_4 = Z_6$$

Substituting Z_4 into equation (4) yields

$$Z_5 = \frac{[R + s(L+M)]^2 - 2[R + s(L+M)]sM}{sM}$$

$$Z_5 = \frac{R^2 + 2Rs(L+M) + s^2(L+M)^2 - 2RsM - 2s^2M(L+M)}{sM}$$

$$Z_5 = \frac{R^2}{sM} + \frac{2R(L+M)}{M} + \frac{s(L+M)^2}{M} - 2R - 2s(L+M)$$

$$Z_5 = 2R\,\frac{L+M}{M} - 1 + \frac{1}{s}\,\frac{R^2}{M} + s\left(\frac{(L+M)^2}{M} - 2(L+M)\right)$$

$$Z_5 = \frac{2RL}{M} + \frac{1}{s}\,\frac{R^2}{M} = s\left(\frac{L^2}{M} - M\right)$$

Fig. 5 shows the equivalent π-network for the transformer.

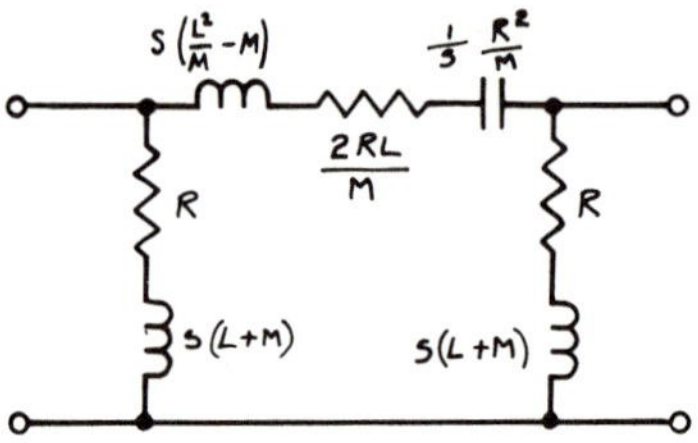

FIG. 5

CHAPTER 17

STATE EQUATIONS

DEFINITIONS & PROPERTIES

• **PROBLEM 17-1**

Write the following integrodifferential equation in the state form.

$$\frac{d^2v}{dt^2} + a\frac{dv}{dt} + bv + C\int_0^t v(t')\,dt' = \sin t$$

Solution: First differentiate the given expression:

$$\dddot{v} + a\ddot{v} + b\dot{v} + cv = \cos t \qquad (1)$$

There is no unique choice of state variables. One particular choice is

$$q_1 = v \qquad (2)$$

$$q_2 = \dot{v} = \dot{q}_1 \qquad (3)$$

$$q_3 = \ddot{v} = \dot{q}_2 \qquad (4)$$

Substituting these expressions into (1),

$$\dot{q}_3 + aq_3 + bq_2 + cq_1 = \cos t \qquad (5)$$

Solving for $\dot{q}_3$,

$$\dot{q}_3 = -cq_1 - bq_2 - aq_3 + \cos t \qquad (6)$$

Equations 3, 4 and 6 are in state format. Restating them in matrix form, in the standard state format $\dot{\bar{q}} = \bar{A}q + \bar{B}u$ where [q] is a state fector and [u] an input vector,

$$\begin{bmatrix} \dot{q}_1 \\ \dot{q}_2 \\ \dot{q}_3 \end{bmatrix} = \begin{bmatrix} 0 & 1 & 0 \\ 0 & 0 & 1 \\ -C & -b & -a \end{bmatrix} \begin{bmatrix} q_1 \\ q_2 \\ q_3 \end{bmatrix} + \begin{bmatrix} 0 \\ 0 \\ 1 \end{bmatrix} [\cos t]$$

• PROBLEM 17-2

Find the original differential equation for which the state representation is given by

$$\begin{bmatrix} \dot{x}_1 \\ \dot{x}_2 \\ \dot{x}_3 \end{bmatrix} = \begin{bmatrix} 0 & 1 & 0 \\ 0 & 0 & 1 \\ -1 & -2 & -3 \end{bmatrix} \begin{bmatrix} x_1 \\ x_2 \\ x_3 \end{bmatrix} + \begin{bmatrix} 0 \\ 0 \\ 1 \end{bmatrix} u(t)$$

$$y(t) = x_1(t)$$

Solution: Expand the matrix equation which will give three differential equations as follows:

$$\dot{x}_1 = x_2 \qquad (1)$$

$$\dot{x}_2 = x_3 \qquad (2)$$

$$\dot{x}_3 = - x_1 - 2x_2 - 3x_3 + u(t) \qquad (3)$$

From equation (2) and differentiating equation (1)

$$x_3 = \dot{x}_2 = \ddot{x}_1 \qquad (4)$$

Differentiating equation (4)

$$\dot{x}_3 = \dddot{x}_1 \qquad (5)$$

Also from equation (1)

$$x_2 = \dot{x}_1 \qquad (6)$$

Substituting equations (4), (5) and (6) into equation (3)

$$\dddot{x}_1 = - x_1 - 2\dot{x}_1 - 3\ddot{x}_1 + u(t) \qquad (7)$$

Rearranging equation (7) in descending order of the derivative

$$\dddot{x}_1 + 3\ddot{x}_1 + 2\dot{x}_1 + x_1 = u(t) \qquad (8)$$

Substitute $y(t) = x_1(t)$ in equation (8)

$$\dddot{y} + 3\ddot{y} + 2\dot{y} + y = u(t) \qquad (9)$$

Finally, $\dddot{y} = \frac{d^3y}{dt^3}$, $\ddot{y} = \frac{d^2y}{dt^2}$, $\dot{y} = \frac{dy}{dt}$

and equation (9) becomes

$$\frac{d^3y}{dt^3} + 3\frac{d^2y}{dt^2} + 2\frac{dy}{dt} + y = u(t)$$

• PROBLEM 17-3

Given the matrix,

$$\bar{A} = \begin{bmatrix} -1 & -1 \\ 2 & -4 \end{bmatrix}$$

Find, (a) $\bar{A}^{-1}$; (b) the **eigenvalues** of $\bar{A}$, λ_1 and λ_2.

(c) Show that $\bar{A}^{-1} = \alpha_0\bar{I} + \alpha_1\bar{A}$ where $\lambda_1^{-1} = \alpha_0 + \alpha_1\lambda_1$ and $\lambda_2^{-1} = \alpha_0 + \alpha_1\lambda_2$ are used to evaluate α_0 and α_1.

(d) It can be shown that any function of the above matrix $f(\bar{A})$ can be found from

$$F(\bar{A}) = \alpha_0\bar{I} + \alpha_1\bar{A}$$

provided α_0 and α_1 are found by solving

$$f(\lambda) = \alpha_0 + \alpha_1\lambda$$

for α_0 and α_1 with **eigenvalues** λ_1 and λ_2.

Using this information, evaluate $e^{\bar{A}}$ and $e^{\bar{A}t}$.

Solution: (a) Using the development method matrix, $\bar{A}$ can be reduced together with the identity matrix $\bar{I}$ using Gaussian Elimination to obtain $\bar{A}^{-1}$

Hence,
$$\left[\begin{array}{cc:cc} \overbrace{-1 \quad -1}^{\bar{A}} & & \overbrace{1 \quad 0}^{\bar{I}} & \\ 2 & -4 & 0 & 1 \end{array}\right]$$

can be reduced by multiplying the first row by 2 and adding to the second to obtain:

$$\left[\begin{array}{cc:cc} -1 & -1 & 1 & 0 \\ 0 & -6 & 2 & 1 \end{array}\right]$$

By dividing the second row by - 6 and adding it to the first obtains,

$$\left[\begin{array}{cc:cc} -1 & 0 & \frac{2}{3} & -\frac{1}{6} \\ 0 & 1 & -\frac{1}{3} & -\frac{1}{6} \end{array}\right]$$

Finally, multiply the first row by - 1 and obtain:

$$\left[\begin{array}{cc:cc} \underbrace{\begin{matrix} 1 & 0 \\ 0 & 1 \end{matrix}}_{\bar{I}} & \underbrace{\begin{matrix} -\frac{2}{3} & \frac{1}{6} \\ -\frac{1}{3} & -\frac{1}{6} \end{matrix}}_{\bar{A}^{-1}} \end{array}\right]$$

(b) Obtain the eigenvalues of a square matrix by solving

$$\det(\bar{A} - \bar{I}\lambda) = 0 \text{ for } \lambda_1, \lambda_2, \ldots, \lambda_n$$

Hence, $\det \begin{bmatrix} -1-\lambda & -1 \\ 2 & -4-\lambda \end{bmatrix} = 0$

$$0 = (-1-\lambda)(-4-\lambda) - (-1)(2)$$

$$0 = \lambda^2 + 5\lambda + 6$$

$$\lambda_1\lambda_2 = \frac{-5 \pm \sqrt{5^2 - (4)(6)}}{2}$$

$$\lambda_1\lambda_2 = -\frac{5}{2} \pm \frac{1}{2};\ \lambda_1 = -2 \text{ and } \lambda_2 = -3.$$

(c) The two equations,

$$\lambda_1^{-1} = \alpha_0 + \alpha_1\lambda_1 \text{ and } \lambda_2^{-1} = \alpha_0 + \alpha_1\lambda_2$$

can be written in matrix form

$$\begin{bmatrix} -\frac{1}{2} \\ -\frac{1}{3} \end{bmatrix} = \begin{bmatrix} 1 & -2 \\ 1 & -3 \end{bmatrix} \begin{bmatrix} \alpha_0 \\ \alpha_1 \end{bmatrix}$$

and solved by use of determinates for α_0 and α_1.

Hence,

$$\alpha_0 = \frac{\begin{vmatrix} -\frac{1}{2} & -2 \\ -\frac{1}{3} & -3 \end{vmatrix}}{\begin{vmatrix} 1 & -2 \\ 1 & -3 \end{vmatrix}} = \frac{\frac{3}{2} - \frac{2}{3}}{-3 + 2} = -\frac{5}{6}$$

$$\text{and} \quad \alpha_1 = \frac{\begin{vmatrix} 1 & -\frac{1}{2} \\ 1 & -\frac{1}{3} \end{vmatrix}}{-1} = \frac{-\frac{1}{3} + \frac{1}{2}}{-1} = -\frac{1}{6}$$

Now, show that

$$\bar{A}^{-1} = \alpha_0 \bar{I} + \alpha_1 \bar{A}$$

Hence,

$$\bar{A}^{-1} = -\frac{5}{6} \begin{bmatrix} 1 & 0 \\ 0 & 1 \end{bmatrix} - \frac{1}{6} \begin{bmatrix} -1 & -1 \\ 2 & -4 \end{bmatrix}$$

$$\bar{A}^{-1} = \begin{bmatrix} -\frac{5}{6} + \frac{1}{6} & \frac{1}{6} \\ -\frac{1}{3} & -\frac{5}{6} + \frac{4}{6} \end{bmatrix} = \begin{bmatrix} -\frac{2}{3} & \frac{1}{6} \\ -\frac{1}{3} & -\frac{1}{6} \end{bmatrix}.$$

This checks with $\bar{A}^{-1}$ computed in part (a).

(d) Since we are asked to find $e^{\bar{A}}$, solve

$$\begin{bmatrix} e^{-2} \\ e^{-3} \end{bmatrix} = \begin{bmatrix} 1 & -2 \\ 1 & -3 \end{bmatrix} \begin{bmatrix} \alpha_0 \\ \alpha_1 \end{bmatrix}$$

for α_0 and α_1.

Hence,

$$\alpha_0 = \frac{\begin{vmatrix} e^{-2} & -2 \\ e^{-3} & -3 \end{vmatrix}}{-1} = \frac{-3e^{-2} + 2e^{-3}}{-1} = 3e^{-2} - 2e^{-3}$$

and $$\alpha_1 = \frac{\begin{vmatrix} 1 & e^{-2} \\ 1 & e^{-3} \end{vmatrix}}{-1} = \frac{e^{-3} - e^{-2}}{-1} = e^{-2} - e^{-3}$$

Now, solve for

$$e^{\bar{A}} = \alpha_0 \bar{I} + \alpha_1 \bar{A}$$

$$e^{\bar{A}} = \begin{bmatrix} 3e^{-2} - 2e^{-3} & 0 \\ 0 & 3e^{-2} - 2e^{-3} \end{bmatrix}$$

$$+ (e^{-2} - e^{-3}) \begin{bmatrix} -1 & -1 \\ 2 & -4 \end{bmatrix}$$

$$e^{\bar{A}} = \begin{bmatrix} 2e^{-2} - e^{-3} & -e^{-2} + e^{-3} \\ 2e^{-2} - 2e^{-3} & -e^{-2} + 2e^{-3} \end{bmatrix} = \begin{bmatrix} 0.220 & -0.085 \\ 0.170 & -0.035 \end{bmatrix}$$

Find $e^{\bar{A}t}$ by similar computation to be

$$e^{\bar{A}t} = \begin{bmatrix} 2e^{-2t} - e^{-3t} & -e^{-2t} + e^{-3t} \\ 2e^{-2t} - 2e^{-3t} & -e^{-2t} + 2e^{-3t} \end{bmatrix}$$

• PROBLEM 17-4

Find the solution for the equation $\dot{\bar{X}} = \bar{\bar{A}}\bar{X} + \bar{b}u(t)$; $\bar{X}(0) = \bar{X}_0$ where

$$\bar{A} = \begin{bmatrix} 0 & -\omega \\ -\omega & 0 \end{bmatrix}, \quad \bar{b} = \begin{bmatrix} 0 \\ 1 \end{bmatrix}, \quad \bar{X}_0 = \begin{bmatrix} 1 \\ 1 \end{bmatrix}.$$

Solution: Solve the two simultaneous differential equations,

$$\frac{dX_1}{dt} = -\omega X_2 \qquad ; X_1(0) = X_2(0) = 1$$

$$\frac{dX_2}{dt} = -\omega X_1 + 1 \qquad ; \text{ for } t \geq 0$$

written in matrix form,

$$\dot{\bar{X}} = \underbrace{\begin{bmatrix} 0 & -\omega \\ -\omega & 0 \end{bmatrix}}_{\bar{A}} \bar{X} + \underbrace{\begin{bmatrix} 0 \\ 1 \end{bmatrix}}_{\bar{b}} u(t); \; \bar{X}_0 = \begin{bmatrix} 1 \\ 1 \end{bmatrix}$$

where,

$$\dot{\bar{X}} = \begin{bmatrix} \frac{dX_1}{dt} \\ \frac{dX_2}{dt} \end{bmatrix}, \; \bar{X} = \begin{bmatrix} X_1 \\ X_2 \end{bmatrix}, \text{ and } \bar{X}_0 = \begin{bmatrix} X_1(0) \\ X_2(0) \end{bmatrix}$$

The solution to these differential equations is given by

$$\bar{X}(t) = e^{\bar{A}t} \bar{X}_0 + \int_0^t e^{A(t-\tau)} \bar{b}u(\tau) \, d\tau \qquad (1)$$

hence, find, $e^{\bar{A}t}$.

The matrix $e^{\bar{A}t}$ can be found in three steps; first, compute the eigenvalues λ_i of matrix $\bar{A}$ from the equation

$$\det [\bar{A} - \lambda \bar{I}] = 0. \qquad (2)$$

Eq. (2) gives an i-th order equation, hence, i eigenvalues; second, if the eigenvalues are distinct, i.e., $\lambda_1 \neq \lambda_2 \neq \lambda_3 \neq \ldots \neq \lambda_i$; then the terms $\alpha_0(t)$, $\alpha_1(t)$, ..., $\alpha_{i-1}(t)$ can be found from the following i-linear algebraic equations:

$$\alpha_0 + \alpha_1\lambda_1 + \alpha_2\lambda_1^{\,2} + \ldots + \alpha_{i-1}\lambda_1^{\,i-1} = e^{\lambda_1 t}$$

$$\alpha_0 + \alpha_1\lambda_2 + \alpha_2\lambda_2^{\,2} + \ldots + \alpha_{i-1}\lambda_2^{\,i-1} = e^{\lambda_2 t}$$

$$\vdots \qquad\qquad \vdots \qquad\qquad \vdots$$

$$\alpha_0 + \alpha_1\lambda_i + \alpha_2\lambda_i^{\,2} + \ldots + \alpha_{i-1}\lambda_i^{\,i-1} = e^{\lambda_i t};$$

Third, the matrix $e^{\bar{A}t}$ is computed from the equation

$$e^{\bar{A}t} = \alpha_0(t)\bar{I} + \alpha_1(t)A + \alpha_2(t)A^2 + \dots + \alpha_{i-1}(t)A^{i-1}. \quad (3)$$

Taking the matrix $\bar{A}$,

$$\begin{bmatrix} 0 & -\omega \\ -\omega & 0 \end{bmatrix}$$

and applying Eq. (2) yields,

$$\det\left[\begin{bmatrix} 0 & -\omega \\ -\omega & 0 \end{bmatrix} - \lambda \begin{bmatrix} 1 & 0 \\ 0 & 1 \end{bmatrix}\right] = 0$$

$$\det\begin{bmatrix} -\lambda & -\omega \\ -\omega & -\lambda \end{bmatrix} = 0$$

$$\lambda^2 - \omega^2 = 0$$

$$\lambda^2 = \omega^2$$

the eigenvalues

$$\lambda_1 = \omega$$

$$\lambda_2 = -\omega.$$

Substituting into the i-linear algebraic equations.

$$\alpha_0 + \alpha_1\omega = e^{\omega t}$$

$$\alpha_0 - \alpha_1\omega = e^{-\omega t}$$

Solving for α_0 and α_1 as follows:

$$\alpha_0 = e^{-\omega t} + \alpha_1\omega$$

$$e^{-\omega t} + \alpha_1\omega + \alpha_1\omega = e^{\omega t}$$

$$e^{-\omega t} + 2\alpha_1\omega = e^{\omega t}$$

gives, $$\alpha_1 = \frac{e^{\omega t} - e^{-\omega t}}{2\omega}$$

and $$\alpha_0 = \frac{e^{\omega t} + e^{-\omega t}}{2}.$$

Hence, $$e^{\bar{A}t} = \frac{e^{\omega t} + e^{-\omega t}}{2}\begin{bmatrix} 1 & 0 \\ 0 & 1 \end{bmatrix} + \frac{e^{\omega t} - e^{-\omega t}}{2\omega}\begin{bmatrix} 0 & -\omega \\ -\omega & 0 \end{bmatrix}$$

$$e^{\bar{A}t} = \begin{bmatrix} \frac{e^{\omega t} + e^{-\omega t}}{2} & \frac{-e^{\omega t} + e^{-\omega t}}{2} \\ \frac{-e^{\omega t} + e^{-\omega t}}{2} & \frac{e^{\omega t} + e^{-\omega t}}{2} \end{bmatrix}$$

Finally solve Eq. (1) for $\bar{X}(t)$ as follows,

$$\bar{X}(t) = \begin{bmatrix} \frac{e^{\omega t} + e^{-\omega t}}{2} & \frac{-e^{\omega t} + e^{-\omega t}}{2} \\ \frac{-e^{\omega t} + e^{-\omega t}}{2} & \frac{e^{\omega t} + e^{-\omega t}}{2} \end{bmatrix}\begin{bmatrix} 1 \\ 1 \end{bmatrix}$$

$$+ \int_0^t \underbrace{\begin{bmatrix} \frac{e^{\omega(t-\tau)} + e^{-\omega(t-\tau)}}{2} & \frac{-e^{\omega(t-\tau)} + e^{-\omega(t-\tau)}}{2} \\ \frac{-e^{\omega(t-\tau)} + e^{\omega(t-\tau)}}{2} & \frac{e^{\omega(t-\tau)} + e^{-\omega(t-\tau)}}{2} \end{bmatrix}}_{e^{\bar{A}(t-\tau)}} \underbrace{\begin{bmatrix} 0 \\ 1 \end{bmatrix}}_{\bar{b}} d\tau$$

$$\bar{X}(t) = \begin{bmatrix} e^{-\omega t} \\ e^{-\omega t} \end{bmatrix} + \int_0^t \begin{bmatrix} \frac{-e^{\omega(t-\tau)} + e^{-\omega(t-\tau)}}{2} \\ \frac{e^{\omega(t-\tau)} + e^{-\omega(t-\tau)}}{2} \end{bmatrix} d\tau$$

$$\bar{X}(t) = \begin{bmatrix} e^{-\omega t} \\ e^{-\omega t} \end{bmatrix} + \begin{bmatrix} -\frac{e^{\omega t}}{2}\int_0^t e^{-\omega\tau} d\tau + \frac{e^{-\omega t}}{2}\int_0^t e^{\omega\tau} d\tau \\ \frac{e^{\omega t}}{2}\int_0^t e^{-\omega\tau} d\tau + \frac{e^{-\omega t}}{2}\int_0^t e^{\omega\tau} d\tau \end{bmatrix}$$

$$\bar{X}(t) = \begin{bmatrix} e^{-\omega t} \\ e^{-\omega t} \end{bmatrix} + \begin{bmatrix} -\frac{e^{\omega t}}{2} \left[-\frac{1}{\omega} e^{-\omega \tau} \right]_0^t + \frac{e^{-\omega t}}{2} \left[\frac{1}{\omega} e^{\omega \tau} \right]_0^t \\ \frac{e^{\omega t}}{2} \left[-\frac{1}{\omega} e^{-\omega \tau} \right]_0^t + \frac{e^{-\omega t}}{2} \left[\frac{1}{\omega} e^{\omega \tau} \right]_0^t \end{bmatrix}$$

$$\bar{X}(t) = \begin{bmatrix} e^{-\omega t} \\ e^{-\omega t} \end{bmatrix} + \begin{bmatrix} -\frac{e^{\omega t}}{2} \left(-\frac{1}{\omega} e^{-\omega t} + \frac{1}{\omega} \right) + \frac{e^{-\omega t}}{2} \left(\frac{1}{\omega} e^{\omega t} - \frac{1}{\omega} \right) \\ \frac{e^{\omega t}}{2} \left(-\frac{1}{\omega} e^{-\omega t} + \frac{1}{\omega} \right) + \frac{e^{-\omega t}}{2} \left(\frac{1}{\omega} e^{\omega t} - \frac{1}{\omega} \right) \end{bmatrix}$$

$$\bar{X}(t) = \begin{bmatrix} e^{-\omega t} \\ e^{-\omega t} \end{bmatrix} + \begin{bmatrix} \frac{1}{\omega} - \frac{e^{\omega t}}{2\omega} - \frac{e^{-\omega t}}{2\omega} \\ \frac{e^{\omega t}}{2\omega} - \frac{e^{-\omega t}}{2\omega} \end{bmatrix}$$

$$\bar{X}(t) = \begin{bmatrix} \frac{1}{\omega} + e^{-\omega t} - \frac{1}{\omega} \left(\frac{e^{\omega t} + e^{-\omega t}}{2} \right) \\ e^{-\omega t} + \frac{1}{\omega} \left(\frac{e^{\omega t} - e^{-\omega t}}{2} \right) \end{bmatrix},$$ which can be written

as $$\bar{X}(t) = \begin{bmatrix} \frac{1}{\omega} - \sinh \omega t + \left(\frac{\omega - 1}{\omega} \right) \cosh \omega t \\ \cosh \omega t + \left(\frac{1 - \omega}{\omega} \right) \sinh \omega t \end{bmatrix},$$

since, $e^{-\omega t} = \cosh \omega t - \sinh \omega t$,

$$\cosh \omega t = \frac{e^{\omega t} + e^{-\omega t}}{2} \quad \text{and}$$

$$\sinh \omega t = \frac{e^{\omega t} - e^{-\omega t}}{2}.$$

Find the solution for the equation $\dot{\bar{X}} = \bar{A}\bar{X} + \bar{b}u$; $\bar{X}(0) = X_0$ where

$$\bar{A} = \begin{bmatrix} 1 & 0 \\ -2 & 2 \end{bmatrix}, \quad \bar{b} = \begin{bmatrix} 1 \\ 2 \end{bmatrix}, \quad \bar{X}_0 = \begin{bmatrix} -1 \\ 0 \end{bmatrix},$$

$u(t) = \delta(t)$ (impulse function).

Solution: This problem will be solved by using matrix techniques. Exponential notation will be used in which

$$e^{\bar{A}} = \bar{I} - \bar{A}t + \frac{\bar{A}^2 t^2}{2!} - \ldots$$

where $\bar{A}$ is a matrix and $\bar{A}^2 = \bar{A}\bar{A}$.

The equation to be solved is

$$\dot{\bar{X}} = \bar{A}\bar{X} + \bar{b}u(t)$$

where the overbar indicates a vector and the dot means $\frac{d}{dt}$. The general solution for this equation is found by using an integration factor $e^{-\bar{A}t}$. Thus

$$e^{-\bar{A}t}\left(\frac{d\bar{X}}{dt} - \bar{A}\bar{X}\right) = \frac{d}{dt}\left(e^{-\bar{A}t}\bar{X}\right) = e^{-\bar{A}t}\bar{b}u(t).$$

Integration gives

$$e^{-At}\bar{X} = \bar{X}(0) + \int_0^t e^{-\bar{A}\tau}\bar{b}u(\tau)\,d\tau$$

or $$\bar{X} = e^{\bar{A}t}\bar{X}(0) + \int_0^t e^{\bar{A}(t-\tau)}\bar{b}u(\tau)\,d\tau$$

In our case $u(t) = \delta(t)$ so that solution simplifies to

$$\bar{X} = e^{\bar{A}t}\bar{X}(0) + e^{\bar{A}t}\bar{b} = e^{\bar{A}t}\left(\bar{X}(0) + \bar{b}\right).$$

The right hand factor is easily evaluated.

$$\bar{X}(0) - \bar{b} = \begin{bmatrix} -1 \\ 0 \end{bmatrix} + \begin{bmatrix} 1 \\ 2 \end{bmatrix} = \begin{bmatrix} 0 \\ 2 \end{bmatrix}.$$

$e^{\bar{A}t}$ will be evaluated using its definition,

$$e^{\bar{A}t} = \bar{I} - \bar{A}t + \frac{\bar{A}^2t^2}{2!} - \ldots$$

$$= \begin{bmatrix} 1 & 0 \\ 0 & 1 \end{bmatrix} - \begin{bmatrix} 1 & 0 \\ -2 & 2 \end{bmatrix} t + \begin{bmatrix} 1 & 0 \\ -2 & 2 \end{bmatrix} \begin{bmatrix} 1 & 0 \\ -2 & 2 \end{bmatrix} \frac{t^2}{2} - \ldots.$$

$$= \begin{bmatrix} 1 - t + \frac{t^2}{2} - \ldots & 0 \\ 2t - 6\,\frac{t^2}{2} + \ldots & 1 - 2t + 4\,\frac{t^2}{2!} - \ldots \end{bmatrix}$$

Multiplying by $(\bar{X}(0) - \bar{b})$ gives

$$\bar{X} = \begin{bmatrix} 0 \\ 1 - 2t + 2t^2 - \ldots \end{bmatrix}$$

$$= \begin{bmatrix} 0 \\ 2e^{2t} \end{bmatrix}.$$

APPLICATIONS TO RC CIRCUITS

• PROBLEM 17-6

Write the state equation for the given RC circuit.

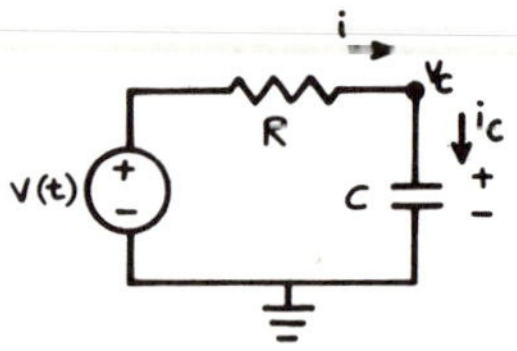

Solution: In the given circuit there is only one energy-storing element (the capacitor). Therefore define only one variable.

Choose the state variable to be the voltage across the capacitor (v_C)

Kirchhoff's Voltage Law around the loop gives

$$v(t) = Ri(t) + v_C(t). \quad (1)$$

Since $i(t) = i_C(t) = C \frac{dv_C}{dt}$ (2)

therefore substituting (2) into (1) yields

$$v(t) = RC \frac{dv_C}{dt} + v_C(t)$$

dividing both sides with RC and leaving $\frac{dv_C}{dt}$ alone

$$\frac{dv_C}{dt} = \frac{1}{RC} v(t) - \frac{1}{RC} v_C(t)$$

Denoting the state variable $v_C(t)$ by $x(t)$

$$\dot{x}(t) = \frac{1}{RC} (v(t) - x(t)).$$

• PROBLEM 17-7

Write the state equation for the network shown.

Solution: Show the currents, i_R and i_C, for R and C respectively on the diagram as indicated.

Then, $i = I$; $i_R = \frac{v_C}{R}$ and $i_C = C \frac{dv_C}{dt}$.

By Kirchhoff's Current Law,

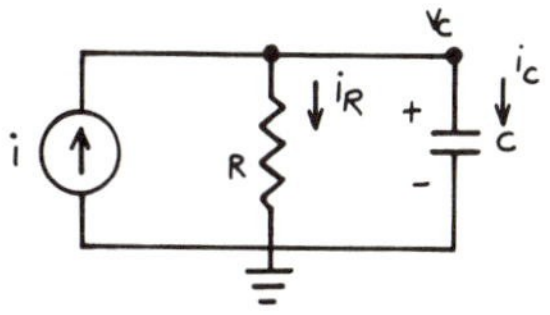

$$i = I = i_R + i_C = \frac{v_C}{R} + C \dot{v}_C$$

or $\dot{v}_C = -\frac{1}{RC} v_C + \frac{1}{C} I.$

APPLICATIONS TO RL CIRCUITS

• PROBLEM 17-8

In the circuit of Fig. 1 the two switches are thrown simultaneously at t = 0: after being in the position shown for a long time. Find the initial conditions $i_1(0^+)$ and $i_2(0^+)$ and write the state equations for i_1 and i_2 in the form $\dot{\bar{x}} = \bar{A}\bar{x} + \bar{b}\bar{u}$ and then solve for them using

$$\bar{x}(t) = e^{\bar{A}t}x(0) + \int_0^t e^{\bar{A}(t-\tau)}\bar{b}u(\tau)\, d\tau$$

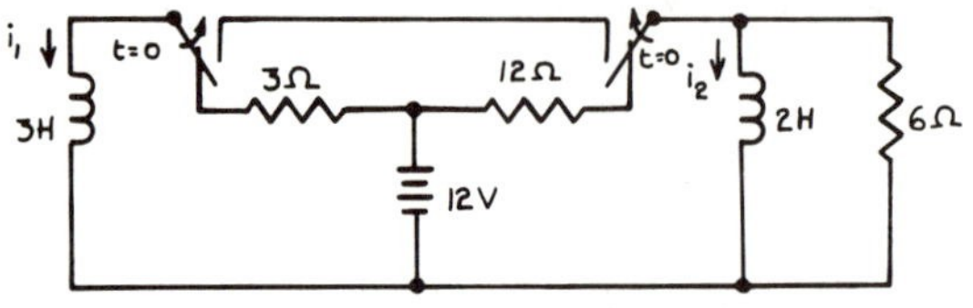

FiG.1

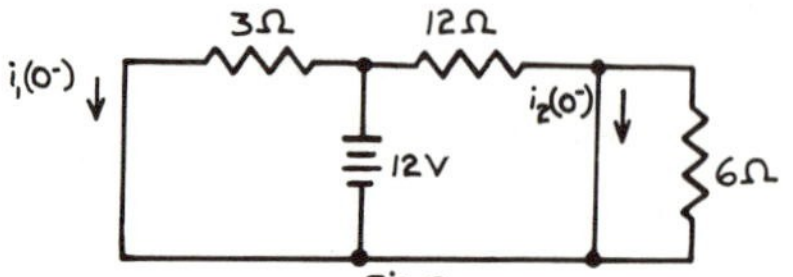

FiG.2

Solution: In order to find the initial conditions $i_1(0^+)$ and $i_2(0^+)$ we re-draw the circuit in Fig. 1 for $t = 0^-$ and find $i_1(0^-)$ and $i_2(0^-)$. Since i_1 and i_2 are both inductor currents, their values cannot change instantaneously for finite voltage values, thus $i_1(0^-) = i_1(0^+)$ and $i_2(0^-) = i_2(0^+)$.

At $t = 0^-$ the two inductors can be represented by short circuits as shown in Fig. 2.

Since the 12-V battery is in parallel with the 3Ω and 12Ω resistor (the 6Ω resistor is shorted by the 2H inductor) $i_1(0^-) = \frac{12V}{3\Omega} = 4A$ and $i_2(0^-) = \frac{12V}{12\Omega} = 1A$. Thus $i_1(0^+) = 4A$ and $i_2(0^+) = 1A$.

For t>0 the circuit becomes that shown in Fig. 3.

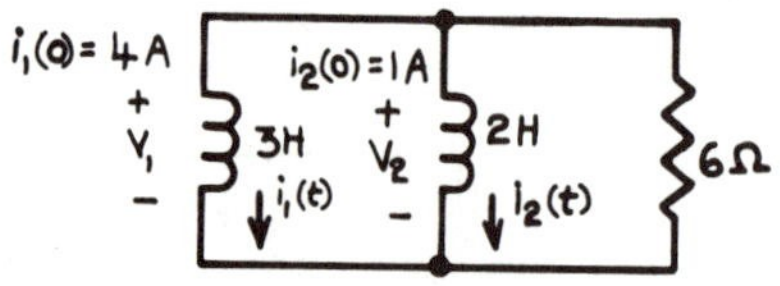

FiG.3

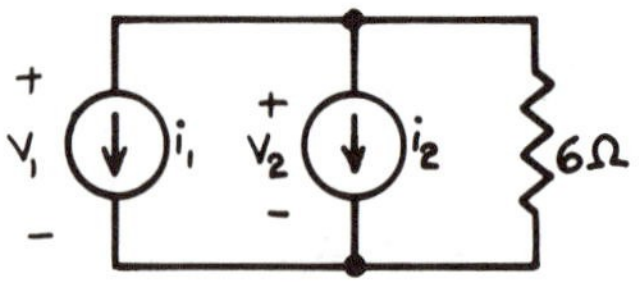

FiG.4

The two inductors cannot be combined to form a simple RL circuit since they each have different initial conditions and their parallel combination forms an additional inductive loop.

We can however use state variable techniques to solve for $i_1(t)$ and $i_2(t)$ for $t > 0$.

We define our state variables as v_1 and v_2 and replace the two inductors in the circuit of Fig. 3 by current sources i_1 and i_2 as shown in Fig. 4,

where $$v_1 = 3\frac{di_1}{dt} \text{ and } v_2 = 2\frac{di_2}{dt}.$$

Using KCL we can write two equations

$$i_1 + i_2 + \frac{v_1}{6} = 0$$

$$i_1 + i_2 + \frac{v_2}{6} = 0$$

substituting $3 \frac{di_1}{dt}$ and $2 \frac{di_2}{dt}$ for v_1 and v_2 we obtain our state equations

$$\frac{di_1}{dt} = - 2i_1 - 2i_2$$

$$\frac{di_2}{dt} = - 3i_1 - 3i_2$$

When we write the state equations in matrix form $\dot{\bar{x}} = \bar{A}\bar{x} + \bar{b}u$ we note that constant term matrix, $\bar{b}$ equals zero thus our equation for the solution x(t) becomes

$$\bar{x}(t) = e^{\bar{A}t}\bar{x}(0)$$

where $\bar{x}(0) = \begin{bmatrix} 4 \\ 1 \end{bmatrix}$ are our initial conditions.

All that is left for us to do in order to find the solution is to find $e^{\bar{A}t}$.

Solve the equation

$\det(\bar{A} - \lambda\bar{I}) = 0$ to find the **eigenvalues** of A,

thus, $\det(\bar{A} - \lambda\bar{I}) = \begin{vmatrix} -\lambda - 2 & -2 \\ -3 & -\lambda - 3 \end{vmatrix} = \lambda(\lambda + 5) = 0$

giving $\lambda_1 = 0$ and $\lambda_2 = -5$

Now find

$$e^{\bar{A}t} = \alpha_0\bar{I} + \alpha_1\bar{A}$$

where α_0 and α_1 can be found by solving the following two equations

$$\alpha_0 + \alpha_1\lambda_1 = e^{\lambda_1 t}$$
$$\alpha_0 + \alpha_1\lambda_2 = e^{\lambda_2 t}$$

Thus $$\alpha_0 = e^0 = 1$$

$$\alpha_0 - 5\alpha_1 = e^{-5t}$$

$$\alpha_1 = \frac{1}{5} - \frac{e^{-5t}}{5} \text{ and } \alpha_0 = 1$$

substituting into $e^{At} = \alpha_0 I + \alpha_1 A$ yields

$$e^{\bar{A}t} = 1\begin{bmatrix} 1 & 0 \\ 0 & 1 \end{bmatrix} + \frac{1}{5} - \frac{e^{-5t}}{5}\begin{bmatrix} -2 & -2 \\ -3 & -3 \end{bmatrix}$$

$$e^{\bar{A}t} = \begin{bmatrix} 1 & 0 \\ 0 & 1 \end{bmatrix} + \begin{bmatrix} -\frac{2}{5} + \frac{2e^{-5t}}{5} & -\frac{2}{5} + \frac{2}{5}e^{-5t} \\ -\frac{3}{5} + \frac{3}{5}e^{-5t} & -\frac{3}{5} + \frac{3}{5}e^{-5t} \end{bmatrix}$$

$$e^{\bar{A}t} = \begin{bmatrix} \frac{3}{5} + \frac{2}{5}e^{-5t} & -\frac{2}{5} + \frac{2}{5}e^{-5t} \\ -\frac{3}{5} + \frac{3}{5}e^{-5t} & \frac{2}{5} + \frac{2}{5}e^{-5t} \end{bmatrix}$$

Finally we obtain the solution

$$\bar{x}(t) = e^{\bar{A}t}\bar{x}(0) = \begin{bmatrix} \frac{3}{5} + \frac{2}{5}e^{-5t} & -\frac{2}{5} + \frac{2}{5}e^{-5t} \\ -\frac{3}{5} + \frac{3}{5}e^{-5t} & \frac{2}{5} + \frac{2}{5}e^{-5t} \end{bmatrix}\begin{bmatrix} 4 \\ 1 \end{bmatrix}$$

$$x(t) = \begin{bmatrix} 4\left(\frac{3}{5} + \frac{2}{5}e^{-5t}\right) - \frac{2}{5} + \frac{2}{5}e^{-5t} \\ 4\left(-\frac{3}{5} + \frac{3}{5}e^{-5t}\right) + \frac{2}{5} + \frac{2}{5}e^{-5t} \end{bmatrix}$$

$$x(t) = \begin{bmatrix} i_1 \\ i_2 \end{bmatrix} = \begin{bmatrix} 2 + 2e^{-5t} \\ -2 + 3e^{-5t} \end{bmatrix} ; \; t > 0$$

APPLICATIONS TO RLC CIRCUITS

• PROBLEM 17-9

Write the state equations for the given RLC circuit.

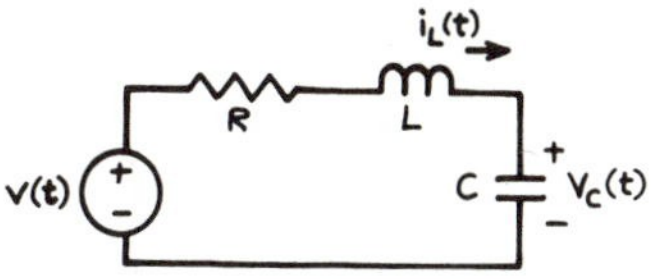

Solution: In the given circuit there are two energy-storing elements (the inductor and the capacitor). Therefore, solve for two variables.

The two state variables that must be chosen are: 1) The current through the inductor and 2) The voltage across the capacitor.

Kirchhoff's Voltage Law around the loop gives:

$$v(t) = Ri_L(t) + v_L(t) + v_C(t). \qquad (1)$$

It is known that $v_L(t) = L\dfrac{di_L(t)}{dt}$. (2)

Substituting equation (2) into equation (1) yields:

$$v(t) = Ri_L(t) + L\frac{di_L(t)}{dt} + v_C(t). \qquad (3)$$

Solving for $\dfrac{di_L(t)}{dt}$ yields:

$$\frac{di_L(t)}{dt} = -\frac{R}{L}\,i_L(t) - \frac{1}{L}\,v_C(t) + \frac{1}{L}\,v(t). \qquad (4)$$

It is known that

$$i_C(t) = C\frac{dv_C(t)}{dt}. \qquad (5)$$

Solving for $\dfrac{dv_C(t)}{dt}$

$$\frac{dv_C(t)}{dt} = \frac{1}{C}\,i_C(t) \qquad (6)$$

In this problem, however, $i_C(t) = i_L(t)$. Therefore, rewrite equation (6) as

$$\frac{dv_C(t)}{dt} = \frac{1}{C} i_L(t). \qquad (7)$$

Denoting the state variables $i_L(t)$ by $x_1(t)$ and $v_C(t)$ by $x_2(t)$ yields for equation (7)

$$\dot{x}_2(t) = \frac{1}{C} x_1(t). \qquad (8)$$

Repeating the process for equation (4) gives

$$\dot{x}_1(t) = -\frac{R}{L} x_1(t) - \frac{1}{L} x_2(t) + \frac{1}{L} v(t). \quad (9)$$

• PROBLEM 17-10

Choose the charges and fluxes to write the state equations for the network shown in Figure 1.

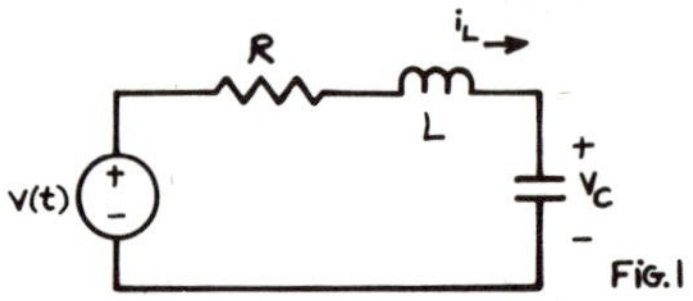

FIG. 1

Solution: Immediately observe that KVL around the single loop will give us the first equation needed.

$$v(t) = Ri_L + Li_L' + v_C \qquad (1)$$

Since this is a series circuit also notice that

$$i_L = Cv_C' \qquad (2)$$

Using $q_C = Cv_C(t)$, $\phi_L = Li_L(t)$ and their derivatives, substituting into (1) and (2) yields

$$\phi_L' = \left(-\frac{R}{L}\right) \phi_L - \left(\frac{1}{C}\right) q_C + v(t) \qquad (3)$$

and

$$q_C' = \frac{1}{L} \phi_L \qquad (4)$$

If we let the state variable $x_1 = \phi_L$ and $x_2 = q_C$, equations (3) and (4) become

$$x_1' = \left(-\frac{R}{L}\right) x_1 - \left(\frac{1}{C}\right) x_2 + v(t)$$

$$x_2' = \frac{1}{L} x_1$$

or, in matrix form:

$$\begin{bmatrix} x_1' \\ x_2' \end{bmatrix} = \begin{bmatrix} -\frac{R}{L} & -\frac{1}{C} \\ \frac{1}{L} & 0 \end{bmatrix} \begin{bmatrix} x_1 \\ x_2 \end{bmatrix} + \begin{bmatrix} 1 \\ 0 \end{bmatrix} v(t)$$

• PROBLEM 17-11

Choose the capacitor voltage and the inductor current to write the state equations for the network shown.

Solution: Writing the node voltage equation for v_C,

$$\frac{v_C - u(t)}{R} + C\frac{dv_C}{dt} + i_L = 0.$$

Dividing by C and rearranging gives

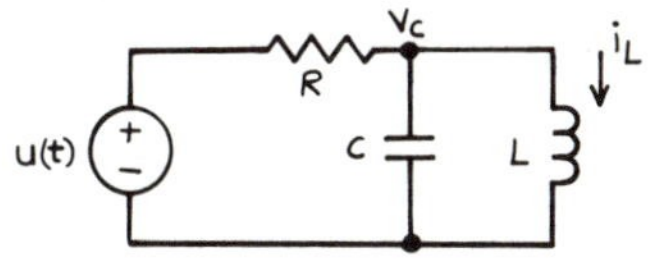

$$\frac{dv_C}{dt} + \frac{v_C}{RC} + \frac{i_L}{C} = \frac{u(t)}{RC}$$

or,
$$\frac{dv_C}{dt} = -\frac{i_L}{C} - \frac{v_C}{RC} + \frac{1}{RC}u(t).$$

But the inductor voltage and capacitor voltage are always equal since they are in parallel. Thus,

$$\frac{di_L}{dt} = \frac{1}{L}v_C.$$

If we select $x_1 = i_L$ and $x_2 = v_C$ as our state variables, these two equations look as follows:

$$\dot{x}_2 = -\frac{x_1}{C} - \frac{x_2}{RC} + \frac{1}{RC}u(t)$$

$$\dot{x}_1 = \frac{x_2}{L}.$$

These can be put easily into the matrix form:

$$\begin{bmatrix} \dot{x}_1 \\ \dot{x}_2 \end{bmatrix} = \begin{bmatrix} 0 & \frac{1}{L} \\ -\frac{1}{C} & -\frac{1}{RC} \end{bmatrix} \begin{bmatrix} x_1 \\ x_2 \end{bmatrix} + \begin{bmatrix} 0 \\ \frac{1}{RC} \end{bmatrix} u(t)$$

Choose the charges and fluxes to write the state equations for the network of Figure 1.

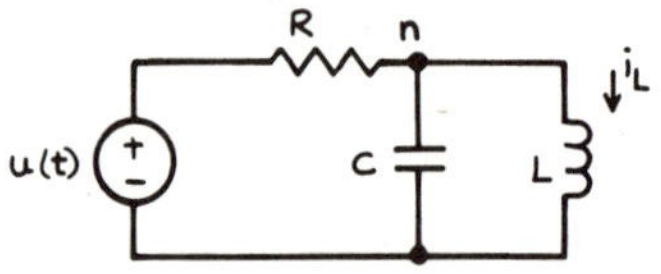

Fig.1

Solution: We know that $\phi_L = Li(t)$ and $q_C = Cv_C(t)$ where

ϕ_L = magnetic flux and

q_C = electric charge.

From this we get $\phi_L' = Li'(t)$ and $q_C' = Cv_C'(t)$. By KCL at node n we have

$$\frac{1 - v_C}{R} = Cv_C' + i_L \qquad (1)$$

Substituting the relations for charge and flux into (1) yields

$$q_C' = \left(-\frac{1}{RC}\right) q_C - \left(\frac{1}{L}\right) \phi_L - \frac{1}{R} \qquad (2)$$

Since the capacitor is in parallel with the inductor we also know that

$$Li_L' = v_C \qquad (3)$$

Again, using the charge and flux relations, eq. (3) becomes

$$\phi_L' = \frac{q_C}{C} \qquad (4)$$

Denoting the state variables as follows

$$x_1 = \phi_L$$

$$x_2 = q_C$$

equations (2) and (4) become the state equations (5) and (6).

$$x_2' = \left(-\frac{1}{RC}\right) x_2 - \left(\frac{1}{L}\right) x_1 + \frac{1}{R} \qquad (5)$$

$$x_1' = \left(\frac{1}{C}\right) x_2 \qquad (6)$$

The result, in matrix form, is therefore:

$$\begin{bmatrix} x_1' \\ x_2' \end{bmatrix} = \begin{bmatrix} 0 & \frac{1}{C} \\ -\frac{1}{L} & -\frac{1}{RC} \end{bmatrix} \begin{bmatrix} x_1 \\ x_2 \end{bmatrix} + \begin{bmatrix} 0 \\ \frac{1}{R} \end{bmatrix} u(t).$$

• **PROBLEM 17-13**

Choose v_C and i_L and write the state equations for the network shown in Figure 1.

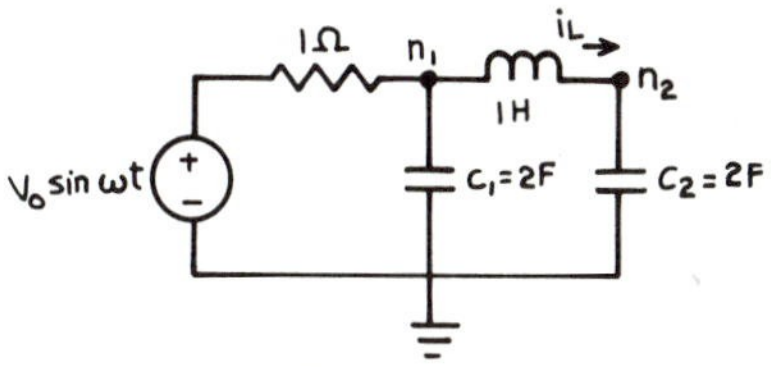

FIG. 1

Solution: An nth order differential equation is not generally suitable for computer solution nor is it easily solvable using conventional methods. It is therefore best to obtain a set of n first order simultaneous differential equations from the given nth order equation by introducing a new set of variables called state variables.

The state variables are usually chosen to represent parameters of energy-storing elements such as capacitors and inductors. Using the chosen state variables, write the matrix form of n first order equations from the relations obtained by KVL, KCL and the defining equations for the network in question.

For our network, choose v_C and i_L as the state variables and denote them as follows:

$$x_1 = i_L$$

$$x_2 = v_{C1}$$

$$x_3 = v_{C2}$$

Now proceed to write the equations which define our network. Since there are three parameters v_{C1}, v_{C2} and v_L, three equations are needed.

Applying KCL at node n_1 yields

$$\frac{v_{C1} - V_0 \sin \omega t}{R} + Cv_{C1}' + i_L = 0$$

or $$v_{C1}' = -\frac{v_{C1}}{RC} + \frac{V_0 \sin \omega t}{RC} - \frac{1}{C} i_L \quad (1)$$

Similarly, KCL at node n_2 gives

$$i_L = Cv_{C2}'$$

or $$v_{C2}' = \frac{1}{C} i_L \quad (2)$$

To obtain the equation for the third parameter i_L, notice that the inductor voltage can be written as

$$Li_L' = v_{C1} - v_{C2}$$

or $$i_L' = \frac{1}{L} v_{C1} - \frac{1}{L} v_{C2} \quad (3)$$

Using the chosen state variables and their first derivatives,

$$i_L' = x_1'$$

$$v_{C1}' = x_2'$$

$$v_{C2}' = x_3'$$

Obtain the state equations which follow by substituting into eqs. (1), (2) and (3).

$$x_1' = \frac{1}{L} x_2 - \frac{1}{L} x_3$$

$$x_2' = -\frac{1}{C} x_1 - \frac{1}{RC} x_2 + \frac{V_0}{RC} \sin \omega t$$

$$x_3' = \frac{1}{C} x_1$$

Hence we have three first order differential equations, which can be solved simultaneously for the state variables. These can then be used to obtain the original three unknowns. Substituting the given element values, write the state equations in matrix form as shown below.

$$\begin{bmatrix} x_1' \\ x_2' \\ x_3' \end{bmatrix} = \begin{bmatrix} 0 & 1 & -1 \\ \frac{1}{2} & -\frac{1}{2} & 0 \\ \frac{1}{2} & 0 & 0 \end{bmatrix} \begin{bmatrix} x_1 \\ x_2 \\ x_3 \end{bmatrix} + \begin{bmatrix} 0 \\ \frac{1}{2} \\ 0 \end{bmatrix} V_0 \sin \omega t$$

• **PROBLEM 17-14**

Write the state equations for the network using the charges on the capacitors and the flux in the inductor as state variables.

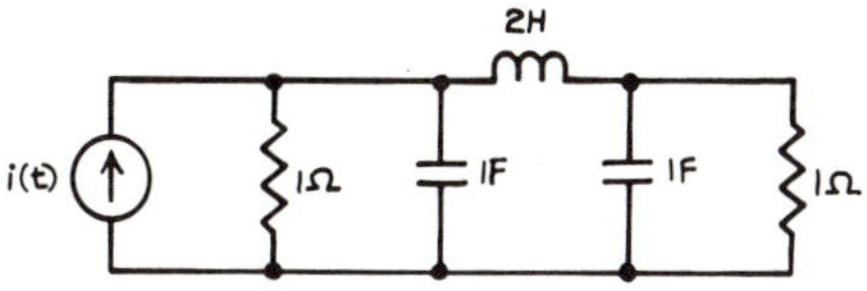

Solution: Write nodal equations at the two nodes, i.e., apply Kirchhoff's current law, expressing the currents (except for the current source) in terms of the capacitor charges and inductor flux, as appropriate.

(a) the left-hand node voltage can be expressed in terms of the capacitor charge,

$$v_1 = \frac{q_1}{C_1} = q_1 \tag{1}$$

since $C_1 = 1$ F; similarly, the right-hand node voltage will be

$$v_3 = \frac{q_3}{C_3} = q_3 \tag{2}$$

To express inductor current in terms of flux, recall that one definition of inductance is

$$L = \frac{\phi}{i}, \tag{3}$$

i.e., the number of flux linkages per ampere of current, which, by transposition, becomes

$$i_L = \frac{\phi_2}{L} = \frac{\phi_2}{2} . \tag{4}$$

Recall also that capacitor current is

$$i_C = \frac{dq}{dt}. \tag{5}$$

(b) Now apply Kirchhoff's current law at node 1:

i (current source) - i (resistor) - i (capacitor) - i (inductor) = 0 (6)

$$i - \frac{v_1}{R} - \frac{dq_1}{dt} - \frac{\phi_2}{2} = 0 \quad (7)$$

$$i - q_1 - \frac{dq_1}{dt} - \frac{\phi_2}{2} = 0. \quad (8)$$

(c) Next, turn attention to node 3:

i (inductor) - i (capacitor) - i (resistor) = 0 (9)

$$\frac{\phi_2}{2} - \frac{dq_3}{dt} - \frac{v_3}{R} = 0 \quad (10)$$

$$\frac{\phi_2}{2} - \frac{dq_3}{dt} - q_3 = 0 \quad (11)$$

(d) Finally, to obtain a third equation (since there are three unknowns), express inductor voltage in terms of the two node voltages:

$$v_L = v_1 - v_3 \quad (12)$$

$$\frac{d\phi_2}{dt} = q_1 - q_3, \quad (13)$$

recalling equations (1) and (2) and that inductor voltage is equal to the rate at which flux linkages are changing.

(e) The "state equations" are written by gathering equation (8), (11) and (13) together into a set of simultaneous equations with the first derivatives of the state variables on the left.

$$\frac{dq_1}{dt} = - q_1 - \frac{\phi_2}{2} + i \quad (14)$$

$$\frac{d\phi_2}{dt} = q_1 - q_3 \quad (15)$$

$$\frac{dq_3}{dt} = \frac{\phi_2}{2} - q_3. \quad (16)$$

Written in matrix form with the notation

$$\frac{dx}{dt} = \dot{x}, \tag{17}$$

$$\begin{bmatrix} \dot{q}_1 \\ \dot{\phi}_2 \\ \dot{q}_3 \end{bmatrix} = \begin{bmatrix} -1 & -\frac{1}{2} & 0 \\ 1 & 0 & -1 \\ 0 & \frac{1}{2} & -1 \end{bmatrix} \begin{bmatrix} q_1 \\ q_2 \\ q_3 \end{bmatrix} + \begin{bmatrix} 1 \\ 0 \\ 0 \end{bmatrix} i(t) \tag{18}$$

• **PROBLEM 17-15**

For the circuit shown in Fig. 1, write the state equations for state variables $v_1(t)$, $v_3(t)$, and $i_2(t)$ in matrix form, $\dot{\bar{x}} = \bar{A}\bar{x} + \bar{b}u(t)$

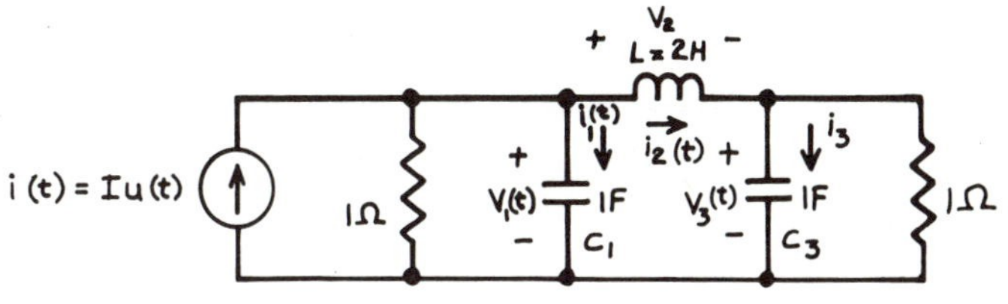

Fig. 1 $\bar{x}_0 = 0$

Solution: By replacing each energy storing element in Fig. 1 with a voltage or current source corresponding to its state variable and by using superposition, enough information from the circuit can be obtained to write the state equations.

Fig. 2, below, shows the energy storing elements replaced by corresponding voltage and current sources.

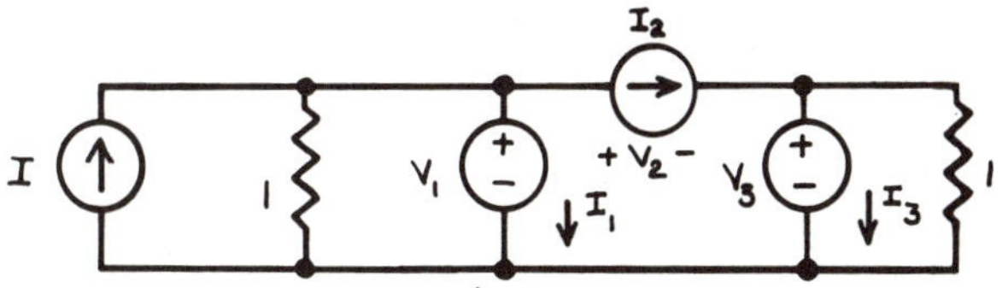

Fig. 2

Using superposition, solve for I_1, I_3 and V_2 for each of the four circuits shown in Fig. 3-6, and add the results to obtain

$$I_1 = -V_1 - I_2 + I$$

$$I_3 = -V_3 + I_2$$

$$V_2 = -V_1 - V_3$$

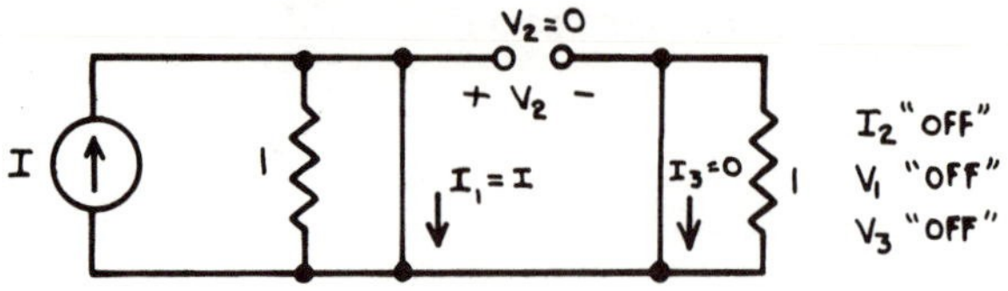

FiG. 3

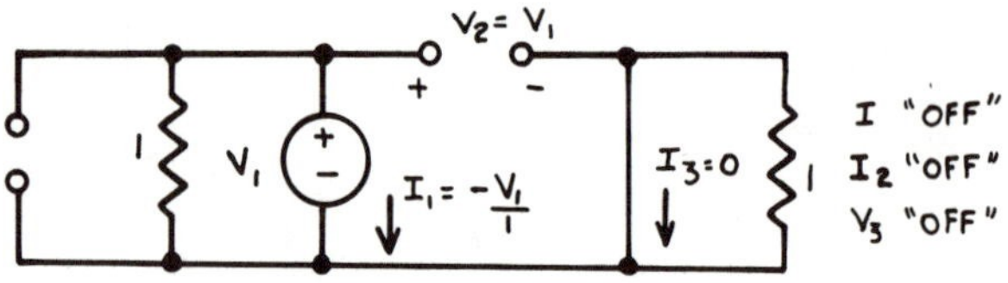

FiG. 4

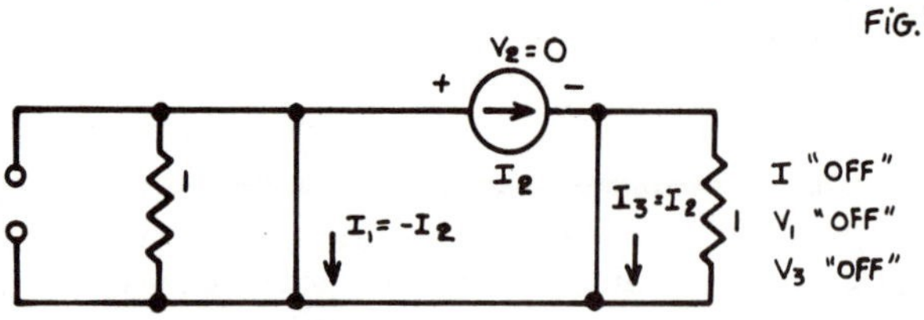

FiG. 5

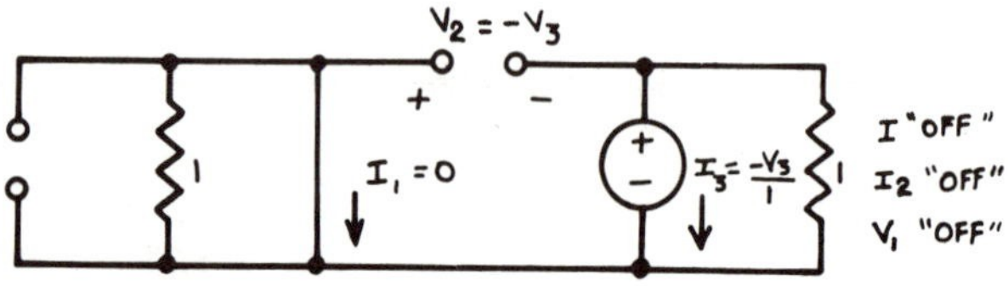

FiG. 6

Since $I_1 = C_1 \frac{dv_1}{dt}$, $I_3 = C_3 \frac{dv_3}{dt}$, and $V_2 = L \frac{dI_2}{dt}$, substitute into the above equations and write

$$C_1 \frac{dV_1}{dt} = -V_1 - I_2 + I$$

$$C_3 \frac{dV_3}{dt} = -V_3 + I_2$$

$$L \frac{dI_2}{dt} = V_1 - V_3$$

Dividing through by C_1, C_2 and L in each corresponding equation and writing them in matrix form yields the following:

$$\underbrace{\begin{bmatrix} \dot{V}_1 \\ \dot{V}_3 \\ \dot{I}_2 \end{bmatrix}}_{\dot{\bar{X}}(t)} = \underbrace{\begin{bmatrix} -1 & 0 & -1 \\ 0 & -1 & 1 \\ \frac{1}{2} & -\frac{1}{2} & 0 \end{bmatrix}}_{\bar{A}} \underbrace{\begin{bmatrix} V_1 \\ V_3 \\ I_2 \end{bmatrix}}_{X(t)} + \underbrace{\begin{bmatrix} I \\ 0 \\ 0 \end{bmatrix}}_{bu(t)} .$$

(a) Obtain the state equations for the circuit of Fig. 1.

(b) Find $v_2(0^+)$, $\frac{dv_2(0^+)}{dt}$, $\frac{d^2v_2(0^+)}{dt^2}$ if $i_1(0^+) = 2A$, $i_2(0^+) = -1A$ and $v_C(0^+) = 3V$.

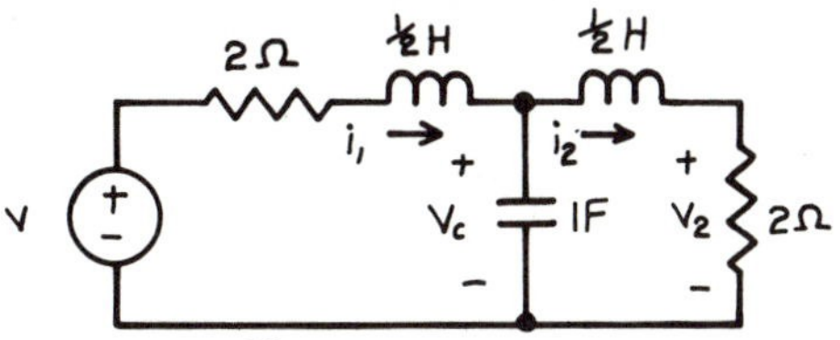

FIG. 1

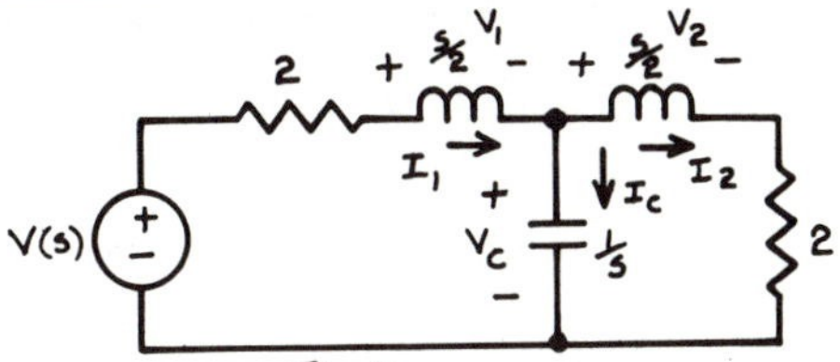

FIG. 2

Solution: (a) Begin by transforming the circuit in Fig. 1 to obtain the circuit shown in Fig. 2.

We can write,

$$V_1(s) = \frac{s}{2} I_1(s);\ I_1 = V_1(s)\frac{2}{s}$$

$$V_2(s) = \frac{s}{2} I_2(s);\ I_2 = V_2(s)\frac{2}{s}$$

$$I_C(s) = s\, V_C(s);\ V_C = I_C(s)\frac{1}{s}$$

and replace the inductors with current sources and capacitors with voltage sources as shown in Fig. 3.

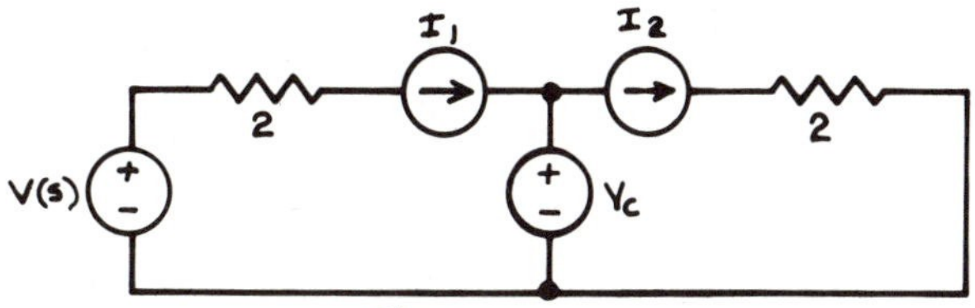

FIG. 3

By allowing only one source to be "on" at a time sum the responses by use of the superposition theorem. Figs. 4-7 show the responses obtained.

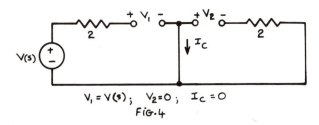

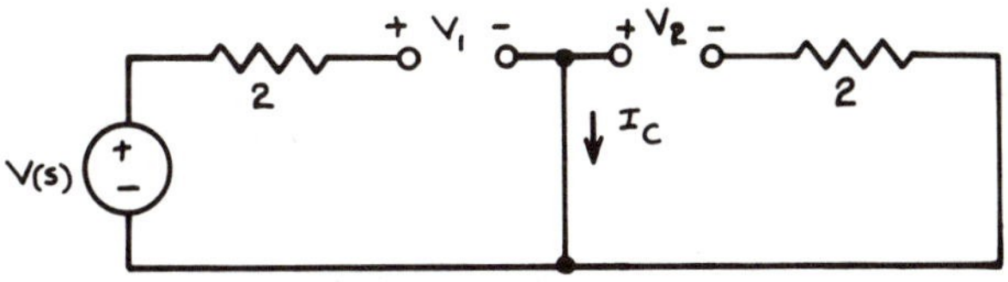

$V_1 = V(s)$; $V_2 = 0$; $I_C = 0$

FIG. 4

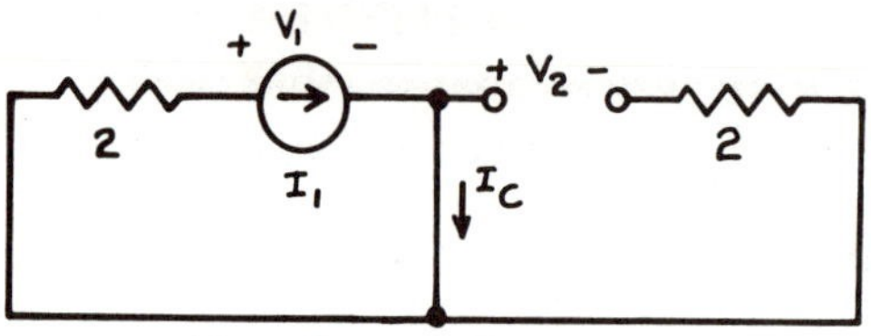

$V_1 = -2I_1$; $V_2 = 0$; $I_C = I_1$

FIG. 5

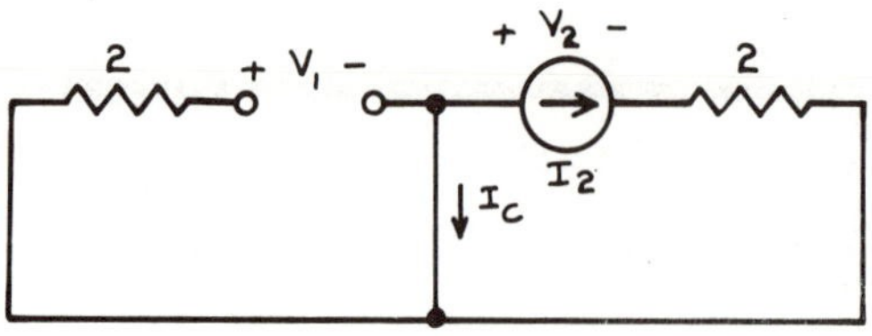

$V_1 = 0$; $V_2 = -2I_2$; $I_C = -I_2$

FIG. 6

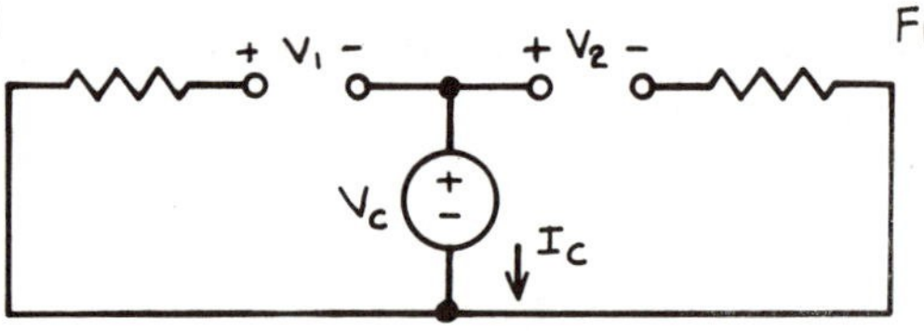

$V_1 = -V_C$; $V_2 = V_C$; $I_C = 0$

FIG. 7

Hence, by superposition,

$$V_1 = \frac{s}{2} I_1 = -2I_1 - V_C + V$$

$$V_2 = \frac{s}{2} I_2 = -2I_2 + V_C$$

$$I_C = \frac{1}{s} V_C = I_1 - I_2$$

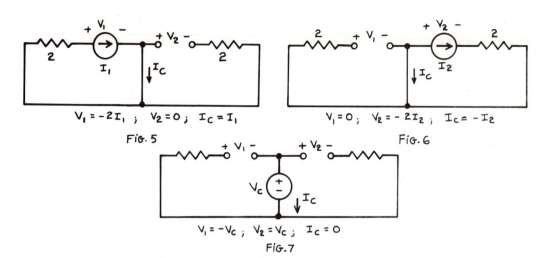

Since $sI_1 \Rightarrow \frac{di_1}{dt}$, then the state equations are,

$$\begin{bmatrix} \frac{di_1}{dt} \\ \frac{di_2}{dt} \\ \frac{dv_C}{dt} \end{bmatrix} = \begin{bmatrix} -4 & 0 & -2 \\ 0 & -4 & 2 \\ 1 & -1 & 0 \end{bmatrix} \begin{bmatrix} i_1 \\ i_2 \\ v_C \end{bmatrix} + \begin{bmatrix} 2V \\ 0 \\ 0 \end{bmatrix} \qquad (1)$$

(b) Since $v_2(0^+) = 2i_2(0^+)$, and $i_2(0^+)$ is given as $-1A$, then $v_2(0^+) = -2V$.

Also, since

$\frac{dv_2}{dt} = 2\frac{di_2}{dt}$ we have from the state equations,

$$\frac{di_2(0^+)}{dt} = -4i_2(0^+) + 2v_C(0^+) \qquad (2)$$

$$\frac{di_2(0^+)}{dt} = -4(-1) + 2(3) = 10 \text{ A/s}$$

then $$\frac{dv_2(0^+)}{dt} = 20 \text{ V/s}.$$

The second derivative of Eq. (2) yields

$$\frac{d^2 i_2(0^+)}{dt^2} = -4\,\frac{di_2(0^+)}{dt} + 2\,\frac{dv_C(0^+)}{dt}$$

The value of $\frac{di_2(0^+)}{dt}$ has already been found to be 10 A/s. Since,

$$\frac{dv_C(0^+)}{dt} = i_1(0^+) - i_2(0^+) = 3\text{V/s then,}$$

$$\frac{d^2 v_2(0^+)}{dt^2} = 2\left[\frac{d^2 v_C(0^+)}{dt^2}\right] = \left[-8\,\frac{di_2(0^+)}{dt} + 2\,\frac{dv_C(0^+)}{dt}\right] 2$$

$$= -68 \text{ V/s}^2.$$

• PROBLEM 17-17

Find the impulse response of the network in Fig. 1. Use the impulse response to obtain the step response. Assume zero initial conditions.

Solution: To find the impulse response, $v_C = h(t)$, make $v(t) = \delta(t)$ and write the differential equations describing the network in the form:

$$\frac{d\bar{x}}{dt} = \bar{A}\bar{x} + \bar{b}u(t).$$

Then, the impulse response is found by

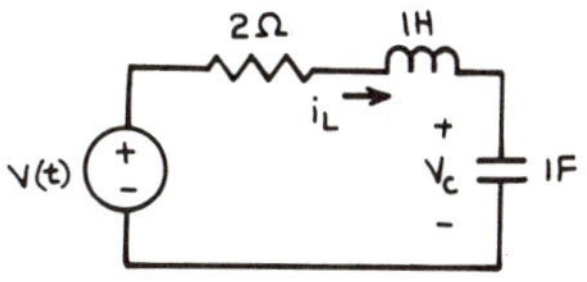

Fig. 1

$$h(t) = \bar{x}(t) = e^{\bar{A}t}\,\bar{b}u(t)$$

Using KVL around the single loop in Fig. 1 write,

$$v(t) = 2i_L + \frac{di_L}{dt} + v_C \qquad (1)$$

Also, $i_L = \frac{dv_C}{dt}$. Eq. (1) can be written as

$$\frac{di_L}{dt} = -2i_L - v_C + v(t), \qquad (2)$$

and the second equation

$$\frac{dv_C}{dt} = i_L \qquad (3)$$

In matrix form, equations (2) and (3) become

$$\begin{bmatrix} \frac{di_L}{dt} \\ \\ \frac{dv_C}{dt} \end{bmatrix} = \underbrace{\begin{bmatrix} -2 & -1 \\ \\ 1 & 0 \end{bmatrix}}_{\bar{A}} \begin{bmatrix} i_L \\ \\ v_C \end{bmatrix} + \underbrace{\begin{bmatrix} 1 \\ \\ 0 \end{bmatrix}}_{\bar{b}} v(t) \qquad (4)$$

$e^{\bar{A}t}$ is found by first computing the **eigenvalues** λ_1 and λ_2 in

$$\det [\bar{A} - \lambda\bar{I}] = 0$$

where $\bar{I}$ is the identity matrix

$$\begin{bmatrix} 1 & & & & \\ & 1 & & 0 & \\ & & 1 & & \\ & 0 & & \ddots & \\ & & & & 1 \end{bmatrix}$$

Hence, $\det \begin{bmatrix} -2-\lambda & -1 \\ \\ 1 & -\lambda \end{bmatrix} = (-2-\lambda)(-\lambda) + 1 = 0$

$$\lambda^2 + 2\lambda + 1 = 0$$

$$(\lambda + 1)(\lambda + 1) = 0$$

$\lambda_1 = -1,\ \lambda_2 = -1$ are the **eigenvalues**.

For multiple **eigenvalues** solve

$$\alpha_0 + \alpha_1\lambda_1 = e^{\lambda_1 t} \qquad (5)$$

$$0 + \alpha_1 = te^{\lambda_2 t} \qquad (6)$$

for α_0 and α_1 and then, solve for $e^{\bar{A}t} = \alpha_0\bar{I} + \alpha_1\bar{A}$. (7)

From Eq. (6-7), $\alpha_1 = te^{-t}$ and $\alpha_0 = e^{-t} + te^{-t}$.
Hence,

$$e^{At} = e^{-t} + te^{-t}\begin{bmatrix} 1 & 0 \\ 0 & 1 \end{bmatrix} + te^{-t}\begin{bmatrix} -2 & -1 \\ 1 & 0 \end{bmatrix}$$

$$e^{\bar{A}t} = \begin{bmatrix} e^{-t} - te^{-t} & -te^{-t} \\ te^{-t} & e^{-t} + te^{-t} \end{bmatrix}$$

The impulse response is

$$h(t) = \bar{x}(t) = e^{\bar{A}t}\,\bar{b}u(t) = \begin{bmatrix} e^{-t} - te^{-t} \\ te^{-t} \end{bmatrix},$$

thus, when the input is $\delta(t)$, the outputs are $i_L(t) = e^{-t} - te^{-t}$ and $v_C(t) = te^{-t}$. Since it is established that

$$h(t) = \frac{d}{dt}\ \{\text{step response}\}$$

then, step response $= \int_0^t h(\tau)\,d\tau$

$$\{v_C(t)\text{ step response}\} = \int_0^t \tau e^{-\tau}\,d\tau$$

$$\{v_C(t)\text{ step response}\} = \Big[e^{-\tau}(-\tau-1)\Big]_0^t = e^{-t}(-t-1) - 1(-1)$$

$$\{v_C(t)\text{ step response}\} = -te^{-t} - e^{-t} + 1$$

Also, $\{i_L(t)\text{ step response}\} = \int_0^t (e^{-\tau} - \tau e^{-\tau})\,d\tau$

$$\{i_L(t) \text{ step response}\} = \left[\cancel{e^{-t}} + te^{-t} + \cancel{e^{-t}} + \cancel{1} - \cancel{1}\right]_0^t$$

$$\{i_L(t) \text{ step response}\} = te^{-t}.$$

• PROBLEM 17-18

> Using the state variable method of describing network response, write the state equation $\dot{x} = \bar{A}\bar{x} + \bar{B}\bar{u}$ for the network shown and calculate the matrix $e^{\bar{A}t}$.

Solution: The state variables chosen are the inductor current and the capacitor voltage since they establish the stored energy at any instant of time. Differential equations are written for the state variables and converted to matrix notation. The matrix $e^{\bar{A}t}$ is calculated by using the inverse Laplace transformation:

$$e^{\bar{A}t} = L^{-1}\left[(s\bar{I} - \bar{A})^{-1}\right].$$

(a) Using the state variables i_L and v_C, write the Kirchhoff's current equation at the top node:

$$i_R + i_L + i_C = 0$$

$$\frac{v_C}{R} + i_L + C\frac{dv_C}{dt} = 0.$$

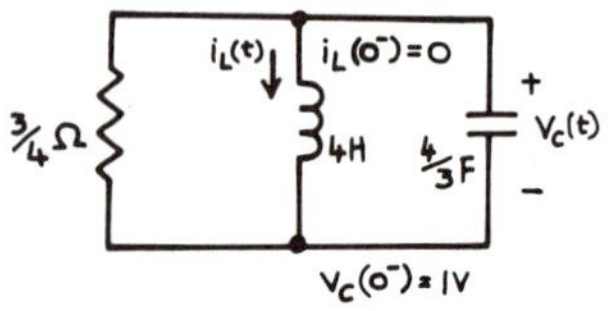

Now, substituting know values,

$$\frac{4}{3}v_C + i_L + \frac{4}{3}\frac{dv_C}{dt} = 0,$$

and, solving for $\frac{dv_C}{dt}$,

$$\frac{dv_C}{dt} = -v_C - \frac{3}{4}i_L$$

which is the first state equation.

To write the second equation, we have the inductor relationship

$$v_L = L\frac{di_L}{dt}$$

which becomes

$$v_L = 4 \frac{di_L}{dt}$$

and rearranges to

$$\frac{di_L}{dt} = \frac{1}{4} v_L = \frac{1}{4} v_C.$$

Now, convert the two differential equations to matrix form as follows:

$$\begin{bmatrix} \dot{i}_L \\ \dot{v}_C \end{bmatrix} = \begin{bmatrix} 0 & 0.25 \\ -0.75 & -1 \end{bmatrix} \begin{bmatrix} i_L \\ v_C \end{bmatrix}.$$

This conforms to the general state equation form:

$$\dot{\bar{x}} = \bar{A}\bar{x} + \bar{B}\bar{u}$$

where $\bar{B}$ is zero in this instance and

$$\bar{A} = \begin{bmatrix} 0 & 0.25 \\ -0.75 & -1 \end{bmatrix}.$$

(b) To calculate the value of $e^{\bar{A}t}$, use the relationship

$$e^{\bar{A}t} = L^{-1}[(s\bar{I} - \bar{A})^{-1}] \qquad (1)$$

which includes the symbol $\bar{I}$, called the "identity matrix"

$$\bar{I} = \begin{bmatrix} 1 & 0 \\ 0 & 1 \end{bmatrix}$$

and the complex frequency symbol s used in the Laplace transform method. To describe the meaning of equation (1) in words, we say that the value of the matrix exponential is the inverse Laplace transform of the inverse of the matrix $(s\bar{I} - \bar{A})$. Begin by evaluating this matrix according to the principles of matrix analysis:

$$(s\bar{I} - \bar{A}) = \begin{bmatrix} s & 0 \\ 0 & s \end{bmatrix} - \begin{bmatrix} 0 & \frac{1}{4} \\ -\frac{3}{4} & -1 \end{bmatrix} = \begin{bmatrix} s & -\frac{1}{4} \\ \frac{3}{4} & s+1 \end{bmatrix} . \quad (2)$$

Next, obtain the inverse of this matrix by using the relationship

$$(\bar{M})^{-1} = \frac{(\mathrm{Adj}\ \bar{M})}{(\mathrm{Det}\ \bar{M})} \quad (3)$$

where (Det $\bar{M}$) is the determinant of $\bar{M}$ and (Adj $\bar{M}$) is the "adjoint matrix of $\bar{M}$", defined as

$$(\mathrm{Adj}\ \bar{M}) = \begin{bmatrix} M_{11} & M_{21} \\ M_{12} & M_{22} \end{bmatrix},$$

where M_{12}, etc. are the co-factors of the elements of m_{12} etc. in:

$$\bar{M} = \begin{bmatrix} m_{11} & m_{12} \\ m_{21} & m_{22} \end{bmatrix}.$$

Evaluating these co-factors results in

$M_{11} = m_{22}$ and $M_{22} = m_{11}$, but

$M_{21} = -m_{12}$ and $M_{12} = -m_{21}$.

Applying these principles to our situation, gives

$$\mathrm{Adj}\ (s\bar{I} - \bar{A}) = \mathrm{Adj} \begin{bmatrix} s & -\frac{1}{4} \\ \frac{3}{4} & (s+1) \end{bmatrix} = \begin{bmatrix} s+1 & \frac{1}{4} \\ -\frac{3}{4} & s \end{bmatrix} \quad (4)$$

where we have used equation (2) for $(s\bar{I} - \bar{A})$.

Next, evaluate the determinant of $(s\bar{I} - \bar{A})$:

$$\text{Det}\,(s\bar{I} - \bar{A}) = \text{Det}\begin{bmatrix} s & -\frac{1}{4} \\ \frac{3}{4} & (s+1) \end{bmatrix} = s\,(s+1) + \frac{1}{4}\left(\frac{3}{4}\right)$$

which becomes

$$\text{Det}\,(s\bar{I} - \bar{A}) = s^2 + s + \frac{3}{16} = \left(s + \frac{1}{4}\right)\left(s + \frac{3}{4}\right) \tag{5}$$

by factoring.

Finally, substitute equations (4) and (5) into equation (3) to obtain the inverse of the matrix $(s\bar{I} - \bar{A})$:

$$(s\bar{I} - \bar{A})^{-1} = \frac{1}{\left(s + \frac{1}{4}\right)\left(s + \frac{3}{4}\right)}\begin{bmatrix} s+1 & \frac{1}{4} \\ -\frac{3}{4} & s \end{bmatrix}.$$

Now, carrying out the indicated multiplication,

$$(s\bar{I} - \bar{A})^{-1} = \begin{bmatrix} \dfrac{s+1}{\left(s + \frac{1}{4}\right)\left(s + \frac{3}{4}\right)} & \dfrac{1/4}{\left(s + \frac{1}{4}\right)\left(s + \frac{3}{4}\right)} \\ \dfrac{-\frac{3}{4}}{\left(s + \frac{1}{4}\right)\left(s + \frac{3}{4}\right)} & \dfrac{s}{\left(s + \frac{1}{4}\right)\left(s + \frac{3}{4}\right)} \end{bmatrix} \tag{6}$$

(c) Now, we are ready to take the inverse Laplace transform of equation (6). Since this requires taking each element of the matrix in turn, identify them as shown in equation (7):

$$(s\bar{I} - \bar{A})^{-1} = \begin{bmatrix} a_{11} & a_{12} \\ a_{21} & a_{22} \end{bmatrix}. \tag{7}$$

Starting with a_{11} and using partial fractions,

$$a_{11} = \frac{s+1}{\left(s + \frac{1}{4}\right)\left(s + \frac{3}{3}\right)} = \frac{A}{\left(s + \frac{1}{4}\right)} + \frac{B}{\left(s + \frac{3}{4}\right)},$$

where $A = \frac{s+1}{s+\frac{3}{4}} \left(\text{for } s = -\frac{1}{4}\right) = \frac{\frac{3}{4}}{\frac{2}{4}} = \frac{3}{2}$

and $B = \frac{s+1}{s+\frac{1}{4}} \left(\text{for } s = -\frac{3}{4}\right) = \frac{\frac{1}{4}}{-\frac{2}{4}} = -\frac{1}{2},$

making $a_{11} = \frac{1.5}{s+0.25} - \frac{0.5}{s+0.75}$.

Finally, taking the inverse Laplace transform,

$$L^{-1}(a_{11}) = 1.5\,e^{-.25t} - 0.5\,e^{-.75t}. \qquad (8)$$

Next, using partial fractions with a_{12},

$$a_{12} = \frac{\frac{1}{4}}{\left(s+\frac{1}{4}\right)\left(s+\frac{3}{4}\right)} = \frac{C}{\left(s+\frac{1}{4}\right)} + \frac{D}{\left(s+\frac{3}{4}\right)},$$

where $C = \frac{\frac{1}{4}}{s+\frac{3}{4}} \left(\text{for } s = -\frac{1}{4}\right) = \frac{\frac{1}{4}}{\frac{2}{4}} = \frac{1}{2}$

and $D = \frac{\frac{1}{4}}{s+\frac{1}{4}} \left(\text{for } s = -\frac{3}{4}\right) = \frac{\frac{1}{4}}{-\frac{2}{4}} = -\frac{1}{2},$

resulting in $a_{12} = \frac{0.5}{s+0.25} - \frac{0.5}{s+0.75}$.

Finally, taking the inverse Laplace transform,

$$L^{-1}(a_{12}) = 0.5e^{-.25t} - 0.5e^{-.75t}. \qquad (9)$$

The third element is

$$a_{21} = \frac{-\frac{3}{4}}{\left(s+\frac{1}{4}\right)\left(s+\frac{3}{4}\right)} = \frac{E}{s+\frac{1}{4}} + \frac{F}{s+\frac{3}{4}},$$

where $E = \frac{-\frac{3}{4}}{s+\frac{3}{4}} \left(\text{for } s = -\frac{1}{4}\right) = \frac{-\frac{3}{4}}{\frac{2}{4}} = -\frac{3}{2}$

and $$F = \frac{-\frac{3}{4}}{s + \frac{1}{4}} \left(\text{for } s = -\frac{3}{4}\right) = \frac{-\frac{3}{4}}{-\frac{2}{4}} = \frac{3}{2},$$

making $$a_{21} = \frac{-1.5}{s + 0.25} + \frac{1.5}{s + 0.75}.$$

Finally, taking the inverse Laplace transorm,

$$L^{-1}(a_{21}) = -1.5e^{-.25t} + 1.5e^{-.75t}. \quad (10)$$

The fourth and last element is

$$a_{22} = \frac{s}{\left(s + \frac{1}{4}\right)\left(s + \frac{3}{4}\right)} = \frac{G}{s + \frac{1}{4}} + \frac{H}{s + \frac{3}{4}},$$

where $$G = \frac{s}{s + \frac{3}{4}} \left(\text{for } s = -\frac{1}{4}\right) = \frac{-\frac{1}{4}}{\frac{2}{4}} = -\frac{1}{2}$$

and $$H = \frac{s}{s + \frac{1}{4}} \left(\text{for } s = -\frac{3}{4}\right) = \frac{-\frac{3}{4}}{-\frac{2}{4}} = \frac{3}{2},$$

resulting in $$a_{22} = \frac{-0.5}{s + 0.25} + \frac{1.5}{s + 0.75}.$$

Taking the inverse Laplace transorm of this term, we have

$$L^{-1}(a_{22}) = -0.5e^{-.25t} + 1.5e^{-.75t}. \quad (11)$$

(d) The final step in the solution is to substitute the values found for the inverse Laplace transforms from equations (8), (9), (10) and (11) into the matrix

$$e^{\bar{A}t} = L^{-1}\begin{bmatrix} a_{11} & a_{12} \\ a_{21} & a_{22} \end{bmatrix}$$

$$e^{\bar{A}t} = \begin{bmatrix} 1.5e^{-.25t} - 0.5e^{-.75t} & 0.5e^{-.25t} - 0.5e^{-.75t} \\ -1.5e^{-.25t} + 1.5e^{-.75t} & -0.5e^{-.25t} + 1.5e^{-.75t} \end{bmatrix}$$

In the circuit of Fig. 1 write the state equations in the form

$$\dot{\bar{x}} = \bar{A}\bar{x} + \bar{b}\bar{u} \quad \text{and}$$

then solve them using

$$x(t) = e^{\bar{A}t} x(0) + \int_0^t e^{A(t-\tau)} bu(\tau)\, d\tau$$

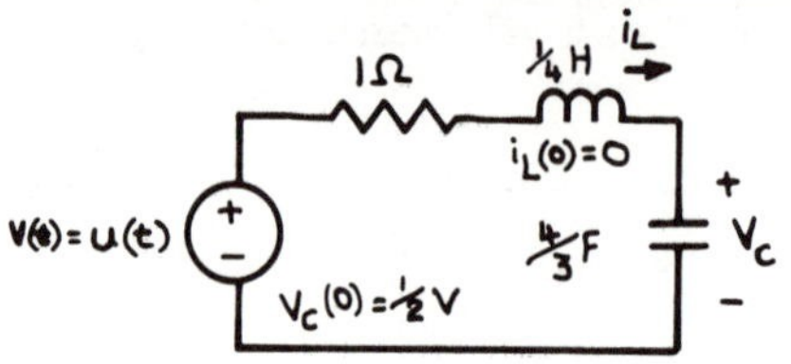

FIG. 1

Solution: In order to write the state equations make use of the relations

$$i_L = C\frac{dv_C}{dt} \qquad (1)$$

and

$$v_C = L\frac{di_L}{dt} \qquad (2)$$

Substituting Eq. (2) into the KVL equation,

$$v(t) = v_R + v_L + v_C, \qquad (3)$$

for v_L and $i_L R$ for v_R we obtain

$$v(t) = i_L R + L\frac{di_L}{dt} + v_C. \qquad (4)$$

Equation (4) along with Eq. (1) give the required state equations. Writing them in matrix form yields,

$$\underbrace{\begin{bmatrix} \frac{di_L}{dt} \\ \frac{dv_C}{dt} \end{bmatrix}}_{\dot{\bar{x}}} = \underbrace{\begin{bmatrix} -\frac{R}{L} & -\frac{1}{L} \\ \frac{1}{C} & 0 \end{bmatrix}}_{\bar{A}} \underbrace{\begin{bmatrix} i_L \\ v_C \end{bmatrix}}_{\bar{x}} + \underbrace{\begin{bmatrix} \frac{1}{L} \\ 0 \end{bmatrix}}_{\bar{b}} u(t)$$

Substituting for R, L, and C yields the matrix

$$\bar{A} = \begin{bmatrix} -4 & -4 \\ \frac{3}{4} & 0 \end{bmatrix} .$$

Using the formula

$$\det [\bar{A}\bar{I} - \lambda] = 0$$

solve for the eigenvalues λ_1 and λ_2 as follows

$$AI - \lambda = \begin{bmatrix} -4-\lambda & -4 \\ \frac{3}{4} & -\lambda \end{bmatrix}$$

$$\det [AI - \lambda] = -\lambda(-4-\lambda) + 3 = 0$$

$$\lambda^2 + 4\lambda + 3 = 0$$

$$(\lambda + 3)(\lambda + 1) = 0$$

hence, $\lambda_1 = -3$, and $\lambda_2 = -1$.

These distinct eigenvalues are then substituted into the set of linear equations below

$$\alpha_0 + \alpha_1\lambda_1 = e^{\lambda_1 t}$$

$$\alpha_0 + \alpha_1\lambda_2 = e^{\lambda_2 t}$$

and solved for α_0, and α_1 as follows

$$\alpha_0 - \alpha_1 3 = e^{-3t}$$

$$\alpha_0 - \alpha_1 = e^{-t}$$

hence $$\alpha_0 = \frac{3e^{-t} - e^{-3t}}{2}$$

$$\alpha_1 = \frac{e^{-t} - e^{-3t}}{2}$$

Now the matrix $e^{\bar{A}t}$ is found from

$$\alpha_0\bar{I} + \alpha_1\bar{A}.$$

Hence, $e^{\bar{A}t} = \frac{3e^{-t} - e^{-3t}}{2}\begin{bmatrix} 1 & 0 \\ 0 & 1 \end{bmatrix} + \frac{e^{-t} - e^{-3t}}{2}\begin{bmatrix} -4 & -4 \\ \frac{3}{4} & 0 \end{bmatrix}$

$$e^{\bar{A}t} = \begin{bmatrix} -\frac{1}{2}e^{-t} + \frac{3}{2}e^{-3t} & 2e^{-t} - 2e^{-3t} \\ \frac{3}{8}e^{-t} - \frac{3}{8}e^{-3t} & \frac{3}{2}e^{-t} - \frac{1}{2}e^{-3t} \end{bmatrix}$$

The solution $\bar{x} = \begin{bmatrix} i_L \\ v_C \end{bmatrix}$ is obtained by substituting into the equation

$$\bar{x}(t) = e^{\bar{A}t}\,\bar{x}(0) + \int_0^t e^{\bar{A}(t-\tau)}\bar{b}u(\tau)\,d\tau$$

hence, $\bar{x}(t) = \begin{bmatrix} -\frac{1}{2}e^{-t} + \frac{3}{2}e^{-3t} & -2e^{-t} + 2e^{-3t} \\ \frac{3}{8}e^{-t} - \frac{3}{8}e^{-3t} & \frac{3}{2}e^{-t} - \frac{1}{2}e^{-3t} \end{bmatrix}\begin{bmatrix} 0 \\ \frac{1}{2} \end{bmatrix}$

$$+ \int_0^t \begin{bmatrix} -\frac{1}{2}e^{-(t-\tau)} + \frac{3}{2}e^{-3(t-\tau)} & -2e^{-(t-\tau)} + 2e^{-3(t-\tau)} \\ \frac{3}{8}e^{-(t-\tau)} - \frac{3}{8}e^{-3(t-\tau)} & \frac{3}{2}e^{-(t-\tau)} - \frac{1}{2}e^{-3(t-\tau)} \end{bmatrix}\begin{bmatrix} 4 \\ 0 \end{bmatrix} d\tau$$

$$\bar{x}(t) = \begin{bmatrix} -e^{-t} + e^{-3t} \\ \frac{3}{4}e^{-t} - \frac{1}{4}e^{-3t} \end{bmatrix} + 4\int_0^t \begin{bmatrix} -\frac{1}{2}e^{-(t-\tau)} + \frac{3}{2}e^{-3(t-\tau)} \\ \frac{3}{8}e^{-(t-\tau)} - \frac{3}{8}e^{-3(t-\tau)} \end{bmatrix} d\tau$$

$$\bar{x}(t) = \begin{bmatrix} -e^{-t} + e^{-3t} \\ \frac{3}{4}e^{-t} - \frac{1}{4}e^{-3t} \end{bmatrix} + 4\begin{bmatrix} \frac{1}{2}e^{-(t-\tau)} - \frac{1}{2}e^{-3(t-\tau)} \\ -\frac{3}{8}e^{-(t-\tau)} + \frac{1}{8}e^{-3(t-\tau)} \end{bmatrix}\Bigg|_0^t$$

$$\bar{x}(t) = \begin{bmatrix} -e^{-t} + e^{-3t} \\ \frac{3}{4}e^{-t} - \frac{1}{4}e^{-3t} \end{bmatrix} + 4\begin{bmatrix} 0 \\ -\frac{1}{4} \end{bmatrix} - 4\begin{bmatrix} \frac{1}{2}e^{-t} - \frac{1}{2}e^{-3t} \\ -\frac{3}{8}e^{-t} + \frac{1}{8}e^{-3t} \end{bmatrix}$$

$$\bar{x}(t) = \begin{bmatrix} -e^{-t} + e^{-3t} \\ \frac{3}{4}e^{-t} - \frac{1}{4}e^{-3t} \end{bmatrix} + 4\begin{bmatrix} -\frac{1}{2}e^{-t} + \frac{1}{2}e^{-3t} \\ -\frac{1}{4} + \frac{3}{8}e^{-t} - \frac{1}{8}e^{-3t} \end{bmatrix}$$

$$\bar{x}(t) = \begin{bmatrix} -e^{-t} + e^{-3t} \\ \frac{3}{4}e^{-t} - \frac{1}{4}e^{-3t} \end{bmatrix} + \begin{bmatrix} -2e^{-t} + 2e^{-3t} \\ -1 + \frac{3}{2}e^{-t} - \frac{1}{2}e^{-3t} \end{bmatrix}$$

$$\bar{x}(t) = \begin{bmatrix} -3e^{-t} + 3e^{-3t} \\ -1 + \frac{9}{4}e^{-t} - \frac{3}{4}e^{-3t} \end{bmatrix}$$

• PROBLEM 17-20

Find v(t) for t>0 in the circuit of Fig. 1 if the given initial conditions are: (a) v(0) = 1, v'(0) = - 2, v"(0) = 6; (b) v(0) = 2, i(0) = 3/20, i'(0) = 1/60.

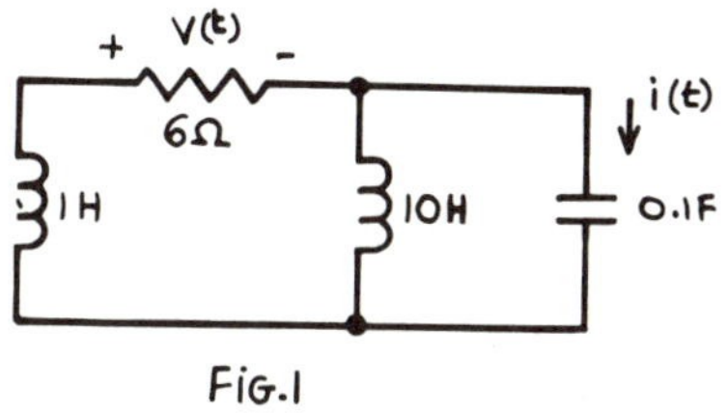

Fig.1

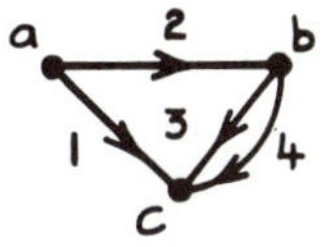

Fig.2

Solution: This method involves drawing a system graph and selecting a tree such that no current state variables are branches of the trees and all voltage state variables are in the tree. In this circuit we have three energy storing components. The two current state variables are through the inductors and the voltage state variable is across the capacitor.

Draw the system graph as shown in Figure 2.

Elements 1 - 4 are the 1H inductance, 6Ω resistance, 10H inductance and 0.1F capacitance, respectively. Branches of the tree are elements 1 and 2. Write the fundamental circuit equations.

$$v_1 = v_2 + v_4$$

$$v_3 = v_4$$

Write the fundamental cut set equations.

$$i_1 + i_2 = 0$$

$$i_1 + i_3 + i_4 = 0$$

Write the component equations.

$$\frac{di_1}{dt} = \frac{1}{L_1} v_1$$

$$\frac{di_3}{dt} = \frac{1}{L_3} v_3$$

$$\frac{dv_4}{dt} = \frac{1}{C_4} i_4$$

$$v_2 = i_2 R_2$$

using the cut set and circuit equations express the right-hand side of the three state equations in terms of the state variables.

The result is

$$\frac{di_1}{dt} = -\frac{R_2}{L_1} i_1 + 0\, i_3 + \frac{1}{L_1} v_4$$

$$\frac{di_3}{dt} = 0\, i_1 + 0\, i_3 + \frac{1}{L_3} v_4$$

$$\frac{dv_4}{dt} = -\frac{1}{C_4} i_1 - \frac{1}{C_4} i_3 + 0\, v_4$$

or in matrix form

$$\frac{d}{dt}\begin{bmatrix} i_1 \\ i_3 \\ v_4 \end{bmatrix} = \begin{bmatrix} -\frac{R_2}{L_1} & 0 & \frac{1}{L_3} \\ 0 & 0 & \frac{1}{L_3} \\ -\frac{1}{C_4} & -\frac{1}{C_4} & 0 \end{bmatrix} \begin{bmatrix} i_1 \\ i_3 \\ v_4 \end{bmatrix}$$

Before solving we need the initial conditions on the state variables given the initial conditions on $v = v_2$ and its first and second derivative. Proceed as follows,

$$v_2 = i_2R_2 = - i_1R_2$$

$$i_1(0) = - \frac{v_2(0)}{R_2} = - \frac{v(0)}{R_2} = - \frac{1}{6} \text{ A}$$

$$v_4 = v_1 - v_2 = L_1 \frac{di_1}{dt} - v = - \frac{L_1}{R_2} \dot{v} - v$$

$$v_4(0) = - \frac{1}{6} (- 2) - 1 = \frac{1}{3} - 1 = - \frac{2}{3} \text{ V}$$

$$i_3 = - i_1 - i_4 = - i_1 - C_4 \frac{dv_4}{dt}$$

$$= - i_1 - C_4 \left(- \frac{L_1}{R_2} \ddot{v} - \dot{v}\right)$$

$$i_3(0) = - i_1(0) + \frac{C_4L_1}{R_2} \ddot{v}(0) + C_4\dot{v}(0)$$

or,
$$i_3(0) = \frac{1}{6} + 0.1 - 0.2 = \frac{1}{6} - \frac{1}{10} = \frac{4}{60} \text{ A}$$

Substitute in the coefficient matrix $R_2 = 6\Omega$, $L_1 = 1H$, $L_3 = 10H$, $C = 0.1F$ and call this the matrix (A).

$$(A) = \begin{pmatrix} -6 & 0 & 1 \\ 0 & 0 & 0.1 \\ -10 & -10 & 0 \end{pmatrix}$$

The matrix equation can be written as $\frac{d\bar{x}}{dt} = \bar{A}\bar{x}$ where $\bar{x}$ is the state vector.

The solution is of the form

$$\bar{x}(t) = e^{At}\bar{x}(0)$$

First form $(su - \bar{A})$ where (u) is the identity matrix

$$(su - \bar{A}) = \begin{pmatrix} s+6 & 0 & -1 \\ 0 & s & -0.1 \\ 10 & 10 & s \end{pmatrix}$$

then $\bar{x}(t) = L^{-1}\{(su - \bar{A})^{-1}\bar{x}(0)\}$

First find the inverse of $(su - \bar{A})$ as

$$(su - \bar{A})^{-} = \frac{1}{(s + 1)(s + 2)(s + 3)} \begin{bmatrix} s^2 + 1 & -10 & s \\ -10 & s^2 + 6 + 10 & 0.1s + 0.6 \\ -10s & -10(s + 6) & s^2 + 6s \end{bmatrix}$$

Now $\bar{X}(s) = \begin{bmatrix} I_1(s) \\ I_3(s) \\ I_4(s) \end{bmatrix}$ and $\bar{x}(0) = \begin{bmatrix} -\frac{1}{6} \\ \frac{4}{60} \\ -\frac{2}{3} \end{bmatrix}$

We are interested only in $I_1(s)$ to find $V_2(s)$ and hence $v_2(t) = v(t)$ and so taking it out of the matrix equation

$$I_1(s) = \frac{-\frac{s^2 + 1}{6} - \frac{4}{6} - \frac{2}{3}s}{(s + 1)(s + 2)(s + 3)}$$

But $V(s) = -6I_1(s) = \dfrac{s^2 + 1 + 4 + 4s}{(s + 1)(s + 2)(s + 3)}$

$$= \frac{s^2 + 4s + 5}{(s + 1)(s + 2)(s + 3)}$$

By partial fraction expansion

$$V(s) = \frac{1}{s + 1} - \frac{1}{s + 2} + \frac{1}{s + 3}$$

$$v(t) = L^{-1}\{V(s)\} = e^{-t} - e^{-2t} + e^{-3t}\ V \qquad \text{(a)}$$

For part (b) find the initial conditions on the state variables as follows:

$$i_1 = -\frac{v_2}{R_2} = -\frac{v}{6}$$

$$i_1(0) = -\frac{2}{6} = -\frac{1}{3} \text{ A}$$

$$i_3 = -i_1 - i_4 = -i_1 - i$$

$$i_3(0) = -i_1(0) - i(0) = \frac{1}{3} - \frac{1}{20} = \frac{20}{60} - \frac{9}{60} = \frac{11}{60} \text{ A}$$

$$i_4 = -i_1 - i_3$$

$$\frac{di_4}{dt} = -\frac{di_1}{dt} - \frac{di_3}{dt} = -\frac{v_1}{L_1} - \frac{v_3}{L_3} = -\frac{(v_2 + v_4)}{L_1} - \frac{v_4}{L_3}$$

or
$$\frac{di_4}{dt} = -\frac{v}{L_1} - \left(\frac{1}{L_1} + \frac{1}{L_3}\right) v_4$$

$$v_4(0) = \frac{1}{\left(\frac{1}{L_1} + \frac{1}{L_3}\right)} \left(-\frac{v(0)}{L_1} - i'(0)\right) = \frac{10}{11}\left(-2 - \frac{1}{60}\right)$$

or $v_4(0) = -\left(\frac{10}{11}\right)\left(\frac{121}{60}\right) = -\frac{11}{6}$ V

$$\bar{x}(0) = \begin{bmatrix} -\frac{1}{3} \\ \frac{11}{60} \\ -\frac{11}{6} \end{bmatrix}$$

and $\bar{V}(s) = \dfrac{2\,(s^2 + 1) + 11 + 11s}{(s + 1)(s + 2)(s + 3)} = \dfrac{2s^2 + 11s + 13}{(s + 1)(s + 2)(s + 3)}$

and by partial fraction expansion

$$\bar{V}(s) = \frac{2}{s + 1} + \frac{1}{s + 2} - \frac{1}{s + 3}$$

Then by inverse Laplace transform

$$v(t) = L^{-1}\{\bar{V}(s)\} = 2e^{-t} + e^{-2t} - e^{-3t} \text{ V} \qquad \text{(b)}$$

• PROBLEM 17-21

For the circuit of Fig. 1, apply the method of state variables to solve for $i_L(t)$ and $v_C(t)$, where $v(t) = u(t)$.

Compare the results with solutions obtained by conventional methods for solving RLC circuits.

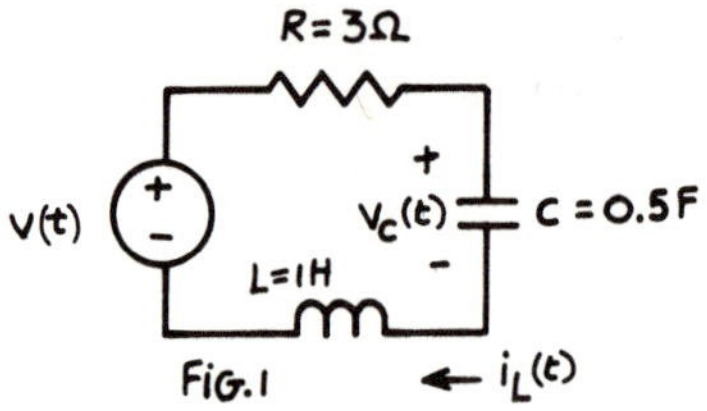

Fig. 1

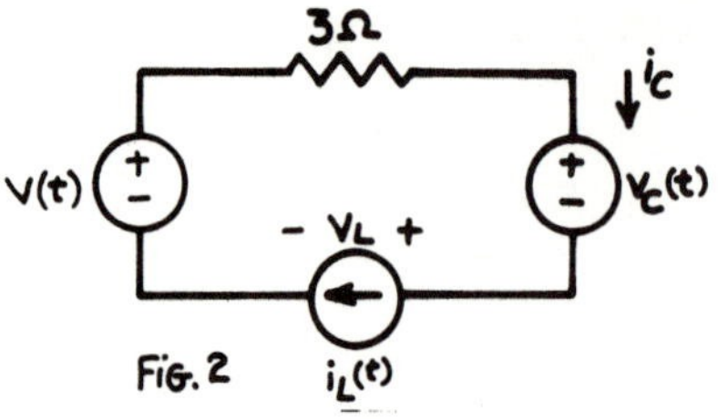

Fig. 2

Solution: Choose $i_L(t)$ and $v_C(t)$ as state variables and replace the inductor and capacitor with independent current and voltage sources respectively.

By using superposition, write the state equations directly. Figs. 3-5 show the circuit when each source is "on", one at a time.

Since i_C and v_L can be written as $C\,\frac{dv_C(t)}{dt}$ and $L\,\frac{di_L(t)}{dt}$ respectively, use superposition to write the state equations from the equations for Figs. 3-5.

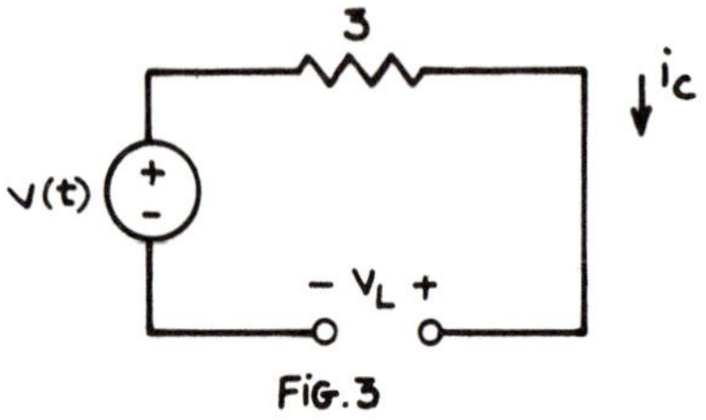

Fig. 3

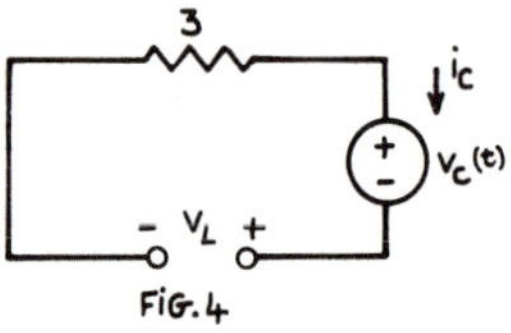

Fig. 4

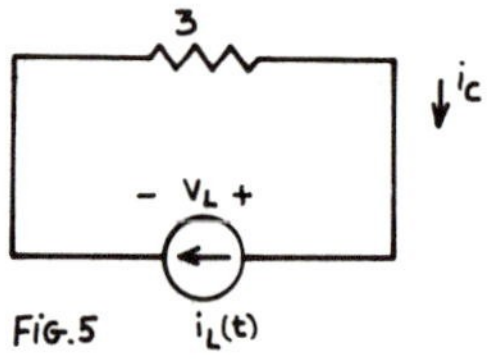

Fig. 5

for Fig. 3: $i_C = C\,\frac{dv_C}{dt} = 0$

$$V_L = L\,\frac{di_L}{dt} = v(t)$$

for Fig. 4: $i_C = C\,\frac{dv_C}{dt} = 0$

$$V_L = L\,\frac{di_L}{dt} = -\,v_C(t)$$

for Fig. 5: $i_C = C\,\frac{dv_C}{dt} = i_L$

$$V_L = L\,\frac{di_L}{dt} = -\,3i_L$$

Summing all the equations for i_C and V_L give the state equations.

$$C\,\frac{dv_C}{dt} = i_L$$

$$L\,\frac{di_L}{dt} = -\,v_C - 3i_L + v(t)$$

In matrix form, write

$$\frac{d}{dt}\begin{bmatrix} v_C \\ i_L \end{bmatrix} = \begin{bmatrix} 0 & \frac{1}{C} \\ -\frac{1}{L} & -\frac{3}{L} \end{bmatrix}\begin{bmatrix} v_C \\ i_L \end{bmatrix} + \begin{bmatrix} 0 \\ \frac{1}{L} \end{bmatrix} v(t)$$

Transforming by Laplace transforms gives

$$\begin{bmatrix} sV_C \\ sI_L \end{bmatrix} = \begin{bmatrix} 0 & \frac{1}{C} \\ -\frac{1}{L} & -\frac{3}{L} \end{bmatrix}\begin{bmatrix} V_C \\ I_L \end{bmatrix} + \begin{bmatrix} 0 \\ \frac{1}{L} \end{bmatrix} V(s).$$

This equation is in the form

$$s\bar{X} = \bar{A}\bar{X} + \bar{B}.$$

In order to solve for X, find

$$s\bar{X} - \bar{A}\bar{X} = \bar{B}$$

$(\bar{I}s - \bar{A})\bar{X} = \bar{B}$ where $\bar{I}$ is the identity matrix.

We define $e^{\bar{A}t} = L^{-1}\left[(\bar{I}s - \bar{A})^{-1}\right]$,

thus, $\bar{x}(t) = e^{\bar{A}t}\,\bar{x}(0) + \int_0^t e^{\bar{A}(t-\tau)}\,\bar{b}u(\tau)\,d\tau$

Since $\bar{x}(0) = 0$ for this example,

$$\bar{x}(t) = \int_0^t e^{\bar{A}(t-\tau)}\,\bar{b}u(\tau)\,d\tau.$$

First, find $\bar{I}s - \bar{A}$:

$$\bar{I}s - \bar{A} = s\begin{bmatrix} 1 & 0 \\ 0 & 1 \end{bmatrix} - \begin{bmatrix} 0 & \frac{1}{C} \\ -\frac{1}{L} & -\frac{3}{L} \end{bmatrix}$$

$$\bar{I}s - \bar{A} = \begin{bmatrix} s & \frac{1}{C} \\ -\frac{1}{L} & s + \frac{3}{L} \end{bmatrix}.$$

Substituting L = 1H and C = 0.5F obtains

$$\bar{I}s - \bar{A} = \begin{bmatrix} s & -2 \\ 1 & s + 3 \end{bmatrix}$$

Find the inverse of this square matrix by finding the determinant of the matrix. Then, by replacing each element of the matrix by its cofactor, find the adjoint and divide by the determinant.

$$[\bar{I}s - \bar{A}]^{-1} = \frac{1}{s^2 + 3s + 2} \begin{bmatrix} s + 3 & 2 \\ -1 & s \end{bmatrix}$$

$$[\bar{I}s - \bar{A}]^{-1} = \begin{bmatrix} \frac{s + 3}{(s + 1)(s + 2)} & \frac{2}{(s + 1)(s + 2)} \\ \frac{-1}{(s + 1)(s + 2)} & \frac{s}{(s + 1)(s + 2)} \end{bmatrix}$$

Finally,

$$e^{\bar{A}t} = L^{-1}\left([\bar{I}s - \bar{A}]^{-1}\right)$$

$$e^{\bar{A}t} = L^{-1} \begin{bmatrix} \frac{A}{s + 1} + \frac{B}{s + 2} & \frac{C}{s + 1} + \frac{D}{s + 2} \\ \frac{E}{s + 1} + \frac{F}{s + 2} & \frac{G}{s + 1} + \frac{H}{s + 2} \end{bmatrix}$$

where $A = \left[\frac{s + 3}{s + 2}\right]_{s=-1} = 2$

$$B = \left[\frac{s + 3}{s + 1}\right]_{s=-2} = -1$$

$$C = \left[\frac{2}{s+2}\right]_{s=-1} = 2$$

$$D = \left[\frac{2}{s+1}\right]_{s=-2} = -2$$

$$E = \left[\frac{-1}{s+2}\right]_{s=-1} = -1$$

$$F = \left[\frac{-1}{s+1}\right]_{s=-2} = 1$$

$$G = \left[\frac{s}{s+2}\right]_{s=-1} = -1$$

$$H = \left[\frac{s}{s+1}\right]_{s=-2} = 2$$

thus, $$e^{\bar{A}t} = \begin{bmatrix} 2e^{-t} - e^{-2t} & 2e^{-t} - 2e^{-2t} \\ -e^{-t} + e^{-2t} & -e^{-t} + 2e^{-2t} \end{bmatrix}.$$

Substituting into

$$\bar{x}(t) = \int_0^t e^{\bar{A}(t-\tau)} \bar{b}u(\tau)\, d\tau,$$

where $\bar{b}u(t) = \begin{bmatrix} 0 \\ 1 \end{bmatrix}$.

$$\bar{x}(t) = \int_0^t \begin{bmatrix} 2e^{-(t-\tau)} - e^{-2(t-\tau)} & 2e^{-(t-\tau)} - 2e^{-2(t-\tau)} \\ -e^{-(t-\tau)} + e^{-2(t-\tau)} & -e^{-(t-\tau)} + 2e^{-2(t-\tau)} \end{bmatrix} \begin{bmatrix} 0 \\ 1 \end{bmatrix} d\tau$$

$$\bar{x}(t) = \int_0^t \begin{bmatrix} 2e^{-(t-\tau)} - 2e^{-2(t-\tau)} \\ -e^{-(t-\tau)} + 2e^{-2(t-\tau)} \end{bmatrix} d\tau$$

$$\bar{x}(t) = \begin{bmatrix} v_C \\ i_L \end{bmatrix} = \begin{bmatrix} 2 - 1 \\ -1 + 1 \end{bmatrix} - \begin{bmatrix} 2e^{-t} - e^{-2t} \\ -e^{-t} + e^{-2t} \end{bmatrix}$$

$$v_C(t) = 1 - 2e^{-t} + e^{-2t} \text{ V}$$

$$i_L(t) = e^{-t} - e^{-2t} \text{ A}$$

Check the results by applying conventional methods. At $t < 0$, $v(t) = 0$.At $t = 0$ $v(t) = 1V$ thus $v_C(0) = 0$ and $i_L(0) = 0$.

Since the solution to $i_L(t)$ is in the form $Ae^{s_1 t} + Be^{s_2 t}$ where $s_1 s_2 = -\alpha \pm \sqrt{\alpha^2 - \omega_0^2}$ and $\alpha = \frac{R}{2L}$, $\omega_0 = \frac{1}{\sqrt{LC}}$.

Here $\alpha = \frac{3}{2}$ and $\omega_0 = \sqrt{2}$

$$s_1 = -\frac{3}{2} + \sqrt{\frac{9}{4} - \frac{8}{4}} = -1$$

$$s_2 + -\frac{3}{2} - \sqrt{\frac{9}{4} - \frac{8}{4}} = -2$$

$$i_L(t) = Ae^{-t} + Be^{-2t}$$

$$i_L(0) = 0 = A + B$$

$$A + B = 0 \qquad (1)$$

$$\frac{di_L}{dt} = -Ae^{-t} - 2Be^{-2t}$$

Since $L\frac{di_L}{dt} = V_L$ at $t = 0^+$ $V_L = 1V$,

$$L\frac{di_L(0^+)}{dt} = -A - 2B = 1$$

or $\quad -A - 2B = 1 \qquad (2)$

Solving equations (1) and (2) yields

$A = 1$ and $B = -1$ thus

$i_L(t) = e^{-t} - e^{-2t}$ A, the desired result.

Since $i_L(t) = i_C(t) = C \dfrac{dv_C(t)}{dt}$

find $v_C(t)$ by integrating $V_C(t) = v_C(0^+) + \frac{1}{C}\int_0^t i_L(\tau)\,d\tau$

$$v_C(t) = 0 + 2\int_0^t (e^{-\tau} - e^{-2\tau})\,d\tau$$

$$v_C(t) = 2\left[-e^{-\tau} + \frac{1}{2}e^{-2\tau}\right]_0^t$$

$$v_C(t) = 2\left[-e^{-t} + \frac{1}{2}e^{-2t} - \left(-1 + \frac{1}{2}\right)\right]$$

$$v_C(t) = 2\left[\frac{1}{2} - e^{-t} + \frac{1}{2}e^{-2t}\right]$$

or $\quad v_C(t) = 1 - 2e^{-t} + e^{-2t}$ V

APPLICATIONS TO NONLINEAR & TIME-VARYING CIRCUITS

• PROBLEM 17-22

Write the state equations for the time varying network shown in Fig. 1. Choose the charges and fluxes as the state variables and let $C(t) = 1 + 0.5 \sin 2t$ F and $L(t) = 2 + \sin 2t$ H.

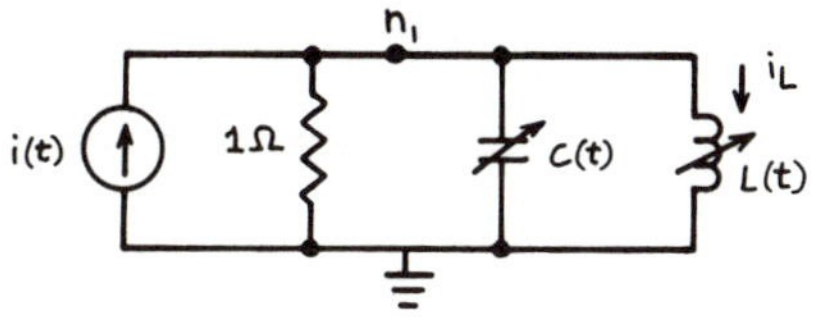

FIG.1

Solution: Recall that $q_C = C\,v_C(t)$ and $\phi_L = Li(t)$. These can be substituted into the equations describing our network.

By KCL at node n_1

$$i = \frac{v_C}{R} + i_C + i_L.$$

But, if $q_C = C\,v_C(t)$, $q_C' = i_C$ and $\phi_L = Li(t)$, $\phi_L' = v_L$

then we have

$$q_C' = -\frac{1}{RC(t)}\,q_C - \frac{1}{L(t)}\,\phi_L + i \tag{1}$$

Since this is a parallel circuit

$$v_C = v_L.$$

But because $\phi_L' = v_L$ and $q_C = C\,v_C$

$$\frac{1}{C(t)}\,q_C = \phi_L' \tag{2}$$

Choosing $x_1 = \phi_L$, $x_1' = \phi_L'$

and $x_2 = q_C$, $x_2' = q_C'$

rewrite equations (1) and (2) as

$$x_1' = \frac{1}{C(t)}\,x_2$$

and $$x_2' = -\frac{1}{RC(t)}\,x_2 - \frac{1}{L(t)}\,x_1 + i$$

In matrix form, our result is

$$\begin{bmatrix} x_1' \\ x_2' \end{bmatrix} = \begin{bmatrix} 0 & \frac{1}{C(t)} \\ -\frac{1}{L(t)} & -\frac{1}{RC(t)} \end{bmatrix} \begin{bmatrix} x_1 \\ x_2 \end{bmatrix} + \begin{bmatrix} 0 \\ 1 \end{bmatrix} i$$

• PROBLEM 17-23

Write the state equations, in matrix form, for the time-varying, second-order circuit shown in Fig. 1 below.

Note: choose charge q_C and flux ϕ_2 as the state variables.

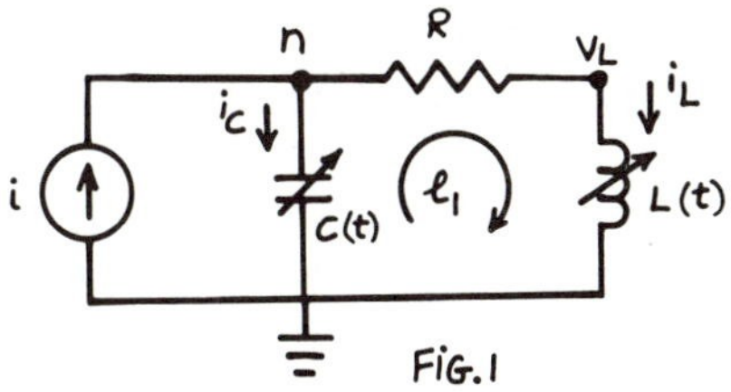

FIG.1

Solution: Begin by writing the KCL equation for node n.

$$i = i_C + i_L \quad (1)$$

By KVL around loop ℓ_1 we also have

$$\frac{v_C - v_L}{R} = i_L$$

which simplifies to

$$Li_L' = - R\, i_L + v_C \quad (2)$$

With $\phi_L = L\, i_L(t)$, $q_C = C\, v_C(t)$, $\phi_L' = v_L$ and $q_C' = i_C$, equations (1) and (2) become:

$$q_C' = - \frac{1}{L(t)} \phi_L + i \quad (3)$$

and
$$\phi_L' = - \frac{R}{L(t)} \phi_L + \frac{1}{C(t)} q_C \quad (4)$$

Since we are choosing ϕ_L and q_C as the state variables, we shall denote them as follows:

$$x_1 = \phi_L$$

$$x_2 = q_C$$

Substituting these into equations (3) and (4) gives us the state equations

$$x_1' = \left(- \frac{R}{L(t)}\right) x_1 + \frac{1}{C(t)} x_2$$

$$x_2' = \left(- \frac{1}{L(t)}\right) x_1 + i$$

or in matrix form

$$\begin{bmatrix} x_1 \\ x_2 \end{bmatrix} = \begin{bmatrix} -\frac{R}{L(t)} & \frac{1}{C(t)} \\ -\frac{1}{L(t)} & 0 \end{bmatrix} \begin{bmatrix} x_1 \\ x_2 \end{bmatrix} + \begin{bmatrix} 0 \\ 1 \end{bmatrix} i$$

• PROBLEM 17-24

> Write the state equations for the time-varying network shown. Choose the charges and the fluxes as the state variables and let $C(t) = 1 + 0.5 \sin 2t$ F and $L(t) = 2 + \sin 2t$ H

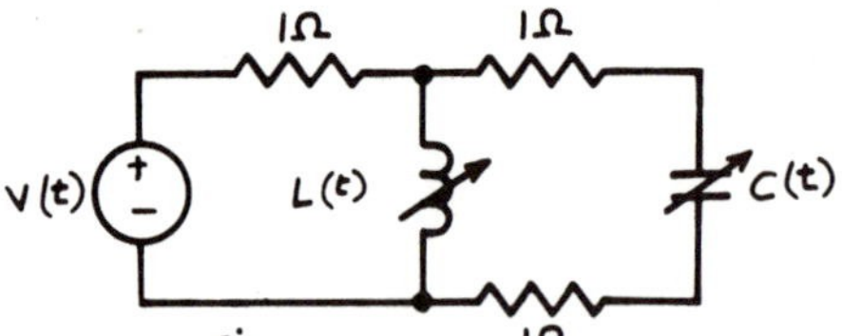

FiG.1

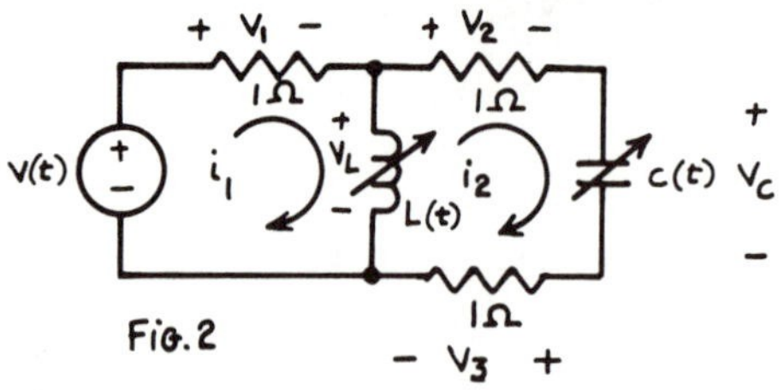

Fig.2

Solution: Assign the two loop currents shown in Fig. 2 and write the two independent KVL equations for them.

$$\ell_1: \quad v(t) = 1\, i_1 + v_L \tag{1}$$

$$\ell_2: \quad 0 = 1\, i_2 + 1\, i_2 - v_L + v_C \tag{2}$$

The inductor current is defined as

$$i_1 - i_2 = i_L = \frac{\phi}{L(t)} \tag{3}$$

and the capacitor voltage is defined as

$$v_C = \frac{q}{C(t)} \tag{4}$$

where ϕ is the flux in the inductor and q is the charge in the capacitor. Also, the inductor voltage is defined as

$$v_L = \frac{d\phi}{dt} \tag{5}$$

and the capacitor current as

$$i_2 = \frac{dq}{dt}. \tag{6}$$

Substituting (6) in (3) for i_2 and solving for i_1 gives

$$i_1 = \frac{\phi}{L(t)} + \frac{dq}{dt}$$

Now eliminate current terms, and write the loop equations (1) and (2) in terms of charge and flux. Hence,

$$\ell_1: \quad v(t) = \frac{\phi}{L(t)} + \frac{dq}{dt} + \frac{d\phi}{dt} \tag{7}$$

$$\ell_2: \quad 0 = 2\frac{dq}{dt} - \frac{d\phi}{dt} + \frac{q}{C(t)} \tag{8}$$

In order to write the state equations in the form $\dot{x} = Ax + v(t)$ solve Eq. (7) for $\frac{dq}{dt}$ and substitute into Eq. (8). Hence,

$$\frac{dq}{dt} = v(t) - \frac{\phi}{L(t)} - \frac{d\phi}{dt}$$

substituted in Eq. (8) gives

$$0 = 2\,v(t) - \frac{2\phi}{L(t)} - \frac{3d\phi}{dt} + \frac{q}{C(t)} \tag{9}$$

the first state equation.

Solving for $\frac{d\phi}{dt}$ in Eq. (8) and substituting into Eq. (7) gives the second state equation

$$v(t) = \frac{\phi}{L(t)} + \frac{3dq}{dt} + \frac{q}{C(t)} \tag{10}$$

Re-arranging the terms in equations (9) and (10) gives

$$\frac{dq}{dt} = -\frac{q}{3C(t)} - \frac{\phi}{3L(t)} + \frac{1}{3}v(t)$$

$$\frac{d\phi}{dt} = \frac{q}{3C(t)} - \frac{2\phi}{3L(t)} + \frac{2}{3}v(t).$$

In matrix form,

$$\begin{bmatrix} \dot{q} \\ \dot{\phi} \end{bmatrix} = \begin{bmatrix} -\frac{1}{3C(t)} & -\frac{1}{3L(t)} \\ \frac{1}{3C(t)} & -\frac{2}{3L(t)} \end{bmatrix} \begin{bmatrix} q \\ \phi \end{bmatrix} + \begin{bmatrix} \frac{1}{3} \\ \frac{2}{3} \end{bmatrix} v(t)$$

where $C(t) = 1 + 0.5 \sin 2t$ and $L(t) = 2 + \sin 2t$.

• PROBLEM 17-25

Consider the network shown in Fig. 1. Assume that the terminating resistors are linear and time invariant, the inductor is flux controlled and the capacitors are charge-controlled, i.e., we assume that

$$v_1 = f_1(q_1)$$

$$v_3 = f_3(q_3) \qquad (7.3.22)$$

and $$i_2 = f_2(\phi_2)$$

Write the state equations for Fig. 1.

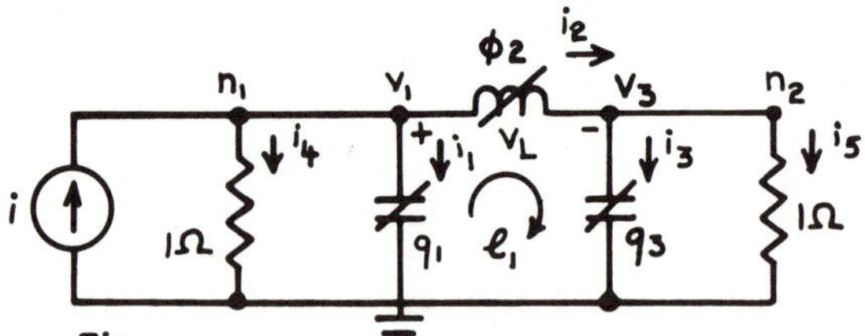

Fig.1

Solution: We must note that the values of inductance and capacitance in this network are functions of flux ϕ and charge q and are time invariant. As stated in the problem

$$v_1 = f_1(q_1)$$

$$v_3 = f_3(q_3)$$

and $$i_2 = f_2(\phi_2)$$

Write the independent node and loop equations:

$$\left.\begin{aligned} n_1: &\quad i = i_4 + i_1 + i_2 \\ n_2: &\quad i_2 = i_3 + i_5 \\ \ell_1 &\quad v_1 = v_2 + v_3 \end{aligned}\right\} \qquad (1)$$

Using the branch voltage-current relations, express all the variables appearing in (1) in terms of the state variables q_1, ϕ_2, q_3 and their derivatives:

$$i_1 = q_1' \qquad v_1 = f_1(q_1) \qquad i_4 = f_1(q_1)$$

$$i_2 = f_2(\phi_2) \qquad v_3 = f_3(q_3) \qquad i_5 = f_3(q_3)$$

$$i_3 = q_3' \qquad v_2 = \phi_2'$$

We now use the above relations and (1) to eliminate all variables that are not state variables. Doing so yields

$$i = f_1(q_1) + q_1' + f_2(\phi_2)$$

$$f_2(\phi_2) = q_3' + f_3(q_3)$$

$$f_1(q_1) = \phi_2' + f_3(q_3)$$

or by rearranging

$$q_1' = -f_1(q_1) - f_2(\phi_2) + i$$
$$\phi_2' = f_1(q_1) - f_3(q_3)$$
$$q_3' = f_2(\phi_2) - f_3(q_3)$$

Putting these results in matrix form yields

$$\begin{bmatrix} q_1' \\ \phi_2' \\ q_3' \end{bmatrix} = \begin{bmatrix} -1 & -1 & 0 \\ 1 & 0 & -1 \\ 0 & 1 & -1 \end{bmatrix} \begin{bmatrix} f_1(q_1) \\ f_2(\phi_2) \\ f_3(q_3) \end{bmatrix} + \begin{bmatrix} 1 \\ 0 \\ 0 \end{bmatrix} i$$

CHAPTER 18

TOPOLOGICAL ANALYSIS

DEFINITIONS

• **PROBLEM** 18-1

Consider the network shown.

(a) Draw a graph of this network and number the branches and nodes arbitrarily.

(b) Indicate the degree of each node.

(c) Draw as many trees of this graph as you can.

(d) What is the number of branches in each of the trees you found in (c).

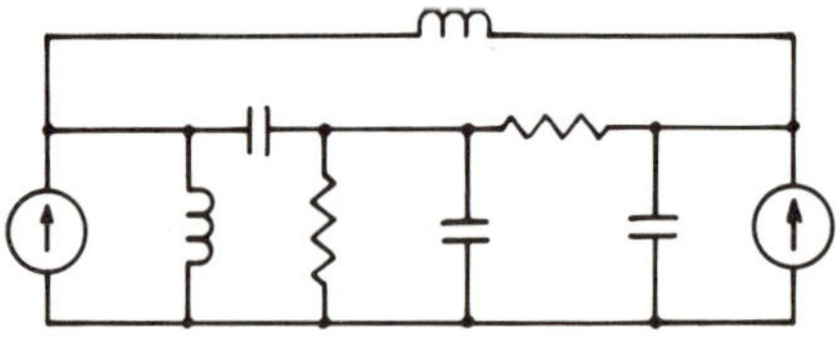

FiG.1

Solution: Notice that the circuit of fig. 1 has 4 nodes and 9 branches. Fig. 2 shows the circuit with the branches and nodes picked arbitrarily.

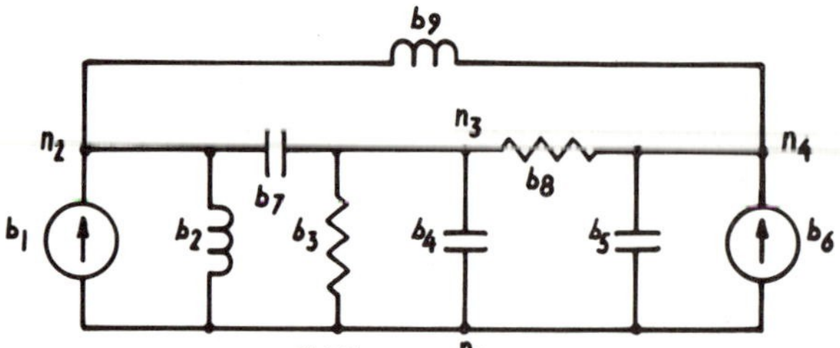

FiG.2

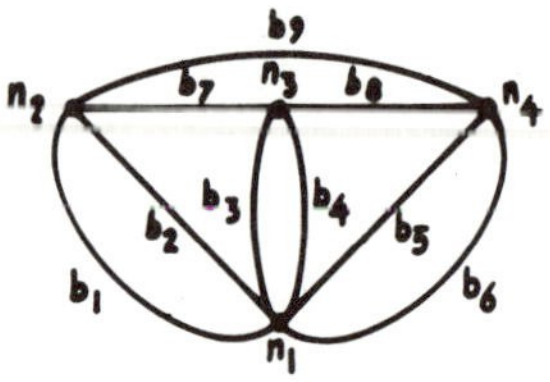

FiG.3

Fig. 3 shows a complete drawing of the graph for the circuit of fig. 2.

The degree of each node is the number of branches connected to it. In our case:

$n_1 = 6$ $n_3 = 4$

$n_2 = 4$ $n_4 = 4$

By definition, a tree contains all nodes and no loops. The following are some of the possible trees for our graph. Notice that the number of branches in each tree is n-1 or 3.

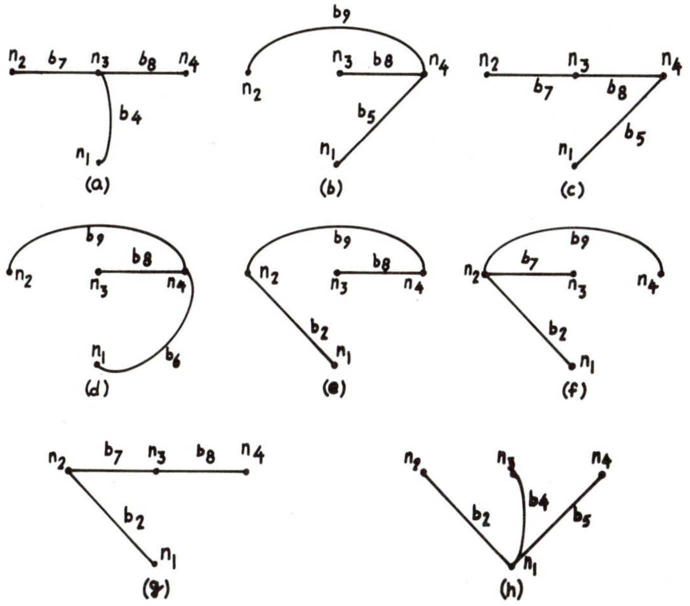

• PROBLEM 18-2

Find all the cutsets for an arbitrary tree of the network of fig. 1.

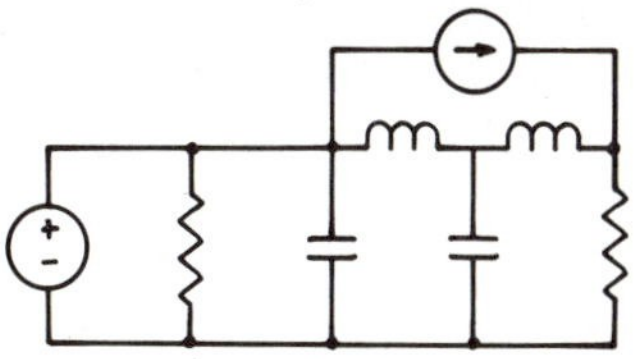

FiG. 1

Solution: By definition, a cutset of a connected graph G is a set of the minimum number of branches whose removal will separate the graph into two disjoint subgraphs.

Fig. 2 shows the network with specified nodes and branches.

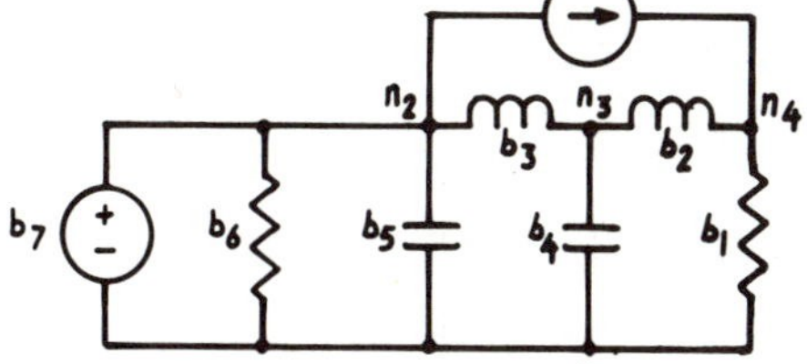

FiG. 2

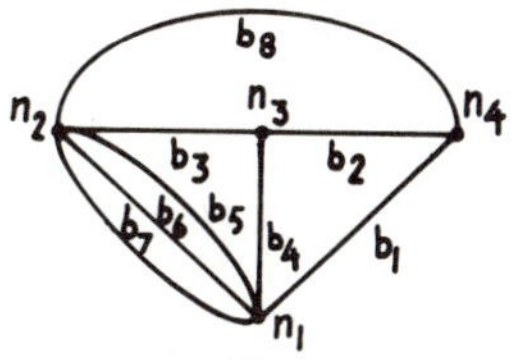

FiG. 3

For the network in fig. 2, the graph G is shown in fig. 3

By inspection observe that there are three possible cutsets for fig. 3.;

cutset #1 = {2, 4, 5, 6, 7, 8}

cutset #2 = {1, 2, 3, 5, 6, 7}

cutset #3 = {1, 4, 3, 8}

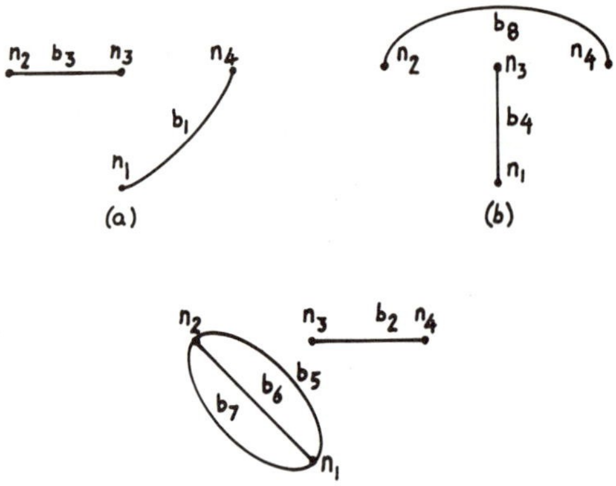

FIG. 4

The pictorial representation of these graphs after the cutsets have been removed can be seen in fig. 4.

THE INCIDENCE MATRIX

• PROBLEM 18-3

Develop the augmented incidence matrix for the network in fig. 1. Use the branches and nodes as labeled.

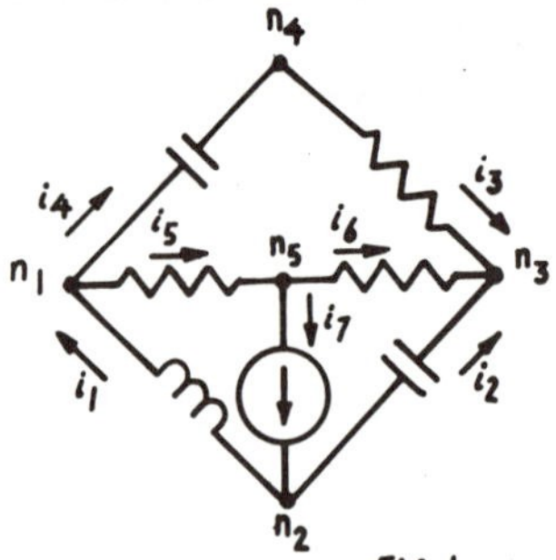

FIG. 1

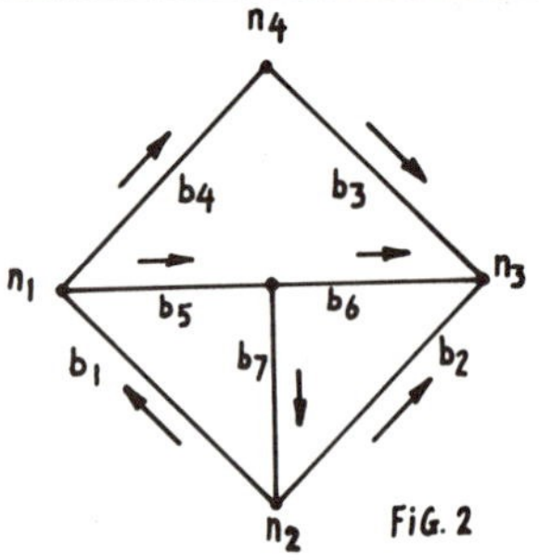

FIG. 2

Solution: Using the definition of the augmented incidence matrix, we find A_a is a 5 x 7 matrix since our circuit has 5 nodes and 7 branches. Begin by drawing the graph shown in fig. 2 with the current directions as indicated. The $\overline{A_a}$ matrix for fig. 2 is as follows:

$$\overline{A_a} = \begin{array}{c} \\ n_1 \\ n_2 \\ n_3 \\ n_4 \\ n_5 \end{array} \begin{array}{c} \begin{array}{ccccccc} b_1 & b_2 & b_3 & b_4 & b_5 & b_6 & b_7 \end{array} \\ \begin{bmatrix} -1 & 0 & 0 & 1 & 1 & 0 & 0 \\ 1 & 1 & 0 & 0 & 0 & 0 & -1 \\ 0 & -1 & -1 & 0 & 0 & -1 & 0 \\ 0 & 0 & 1 & -1 & 0 & 0 & 0 \\ 0 & 0 & 0 & 0 & -1 & 1 & 1 \end{bmatrix} \end{array}$$

For the network given in fig. 1, write down the augmented incidence matrix.

Solution: By definition an augmented incidence matrix, $\bar{A}_a$, is an N x B matrix of a directed graph where:

$a_{ij} = 1$ when branch b_j is incident to node n_i and the reference current, i_j, leaves the node.

$a_{ij} = -1$ when branch b_j is incident to node n_j and the reference current, i_j, enters the node.

$a_{ij} = 0$ when branch b_j is not incident to node n_i.

Note that N and B denote the N nodes and B branches of any planar network.

Using this definition the augmented incidence matrix looks like the following

$$A_a = \begin{array}{c} \\ n_1 \\ n_2 \\ n_3 \\ n_4 \\ \vdots \\ n_i \end{array} \begin{array}{c} b_1 \quad b_2 \quad b_3 \quad b_4 \quad \cdots \quad b_j \\ \left[\begin{array}{c} \\ \\ \\ \\ \\ \\ \end{array} \right] \end{array}$$

In our problem we have 4 nodes and 6 branches. Therefore our augmented incidence matrix is as follows:

$$A_a = \begin{array}{c} \\ n_1 \\ n_2 \\ n_3 \\ n_4 \end{array} \begin{array}{c} \begin{array}{cccccc} b_1 & b_2 & b_3 & b_4 & b_5 & b_6 \end{array} \\ \begin{bmatrix} 1 & 1 & 1 & 0 & 0 & 0 \\ -1 & 0 & 0 & 1 & 0 & -1 \\ 0 & -1 & 0 & -1 & 1 & 0 \\ 0 & 0 & -1 & 0 & -1 & 1 \end{bmatrix} \end{array}$$

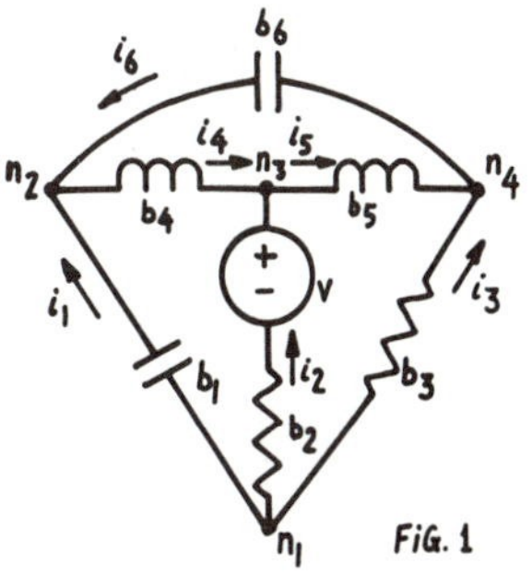

FIG. 1

For the network of fig. 1, write down the independent node equations in matrix form.

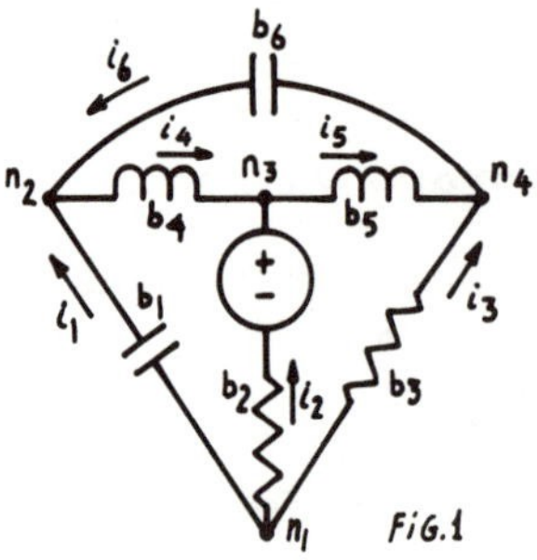

FiG.1

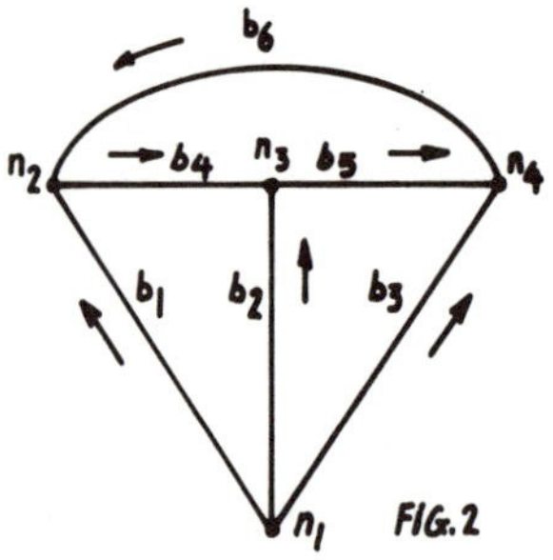

FIG.2

Solution: Start by drawing the graph of the network. Doing this gives the graph shown in fig. 2.

From this write the augmented incidence matrix A_a.

$$A_a = \begin{array}{c} \\ n_1 \\ n_2 \\ n_3 \\ n_4 \end{array} \begin{bmatrix} b_1 & b_2 & b_3 & b_4 & b_5 & b_6 \\ 1 & 1 & 1 & 0 & 0 & 0 \\ -1 & 0 & 0 & 1 & 0 & -1 \\ 0 & -1 & 0 & -1 & 1 & 0 \\ 0 & 0 & -1 & 0 & -1 & 1 \end{bmatrix}$$

Recall that the incidence matrix A is an (N-1)x B matrix formed by deleting the row corresponding to the reference node n_i from A_a. The node equations of a network are represented by the equation

$$Ai_b = 0$$

where this denotes a set of linearly independent KCL equations.

Using n_1 as the reference node yields the incidence matrix for the graph of fig. 2,

Hence,

$$A = \begin{bmatrix} -1 & 0 & 0 & 1 & 0 & -1 \\ 0 & -1 & 0 & -1 & 1 & 0 \\ 0 & 0 & -1 & 0 & -1 & 1 \end{bmatrix}$$

Because the network has N = 4 nodes and B = 6 branches, there exists N-1 or 3 independent node equations. Thus in matrix form,

$$\begin{bmatrix} -1 & 0 & 0 & 1 & 0 & -1 \\ 0 & -1 & 0 & -1 & 1 & 0 \\ 0 & 0 & -1 & 0 & -1 & 1 \end{bmatrix} \begin{bmatrix} i_1 \\ i_2 \\ i_3 \\ i_4 \\ i_5 \\ i_6 \end{bmatrix} = 0$$

which is the desired result.

• PROBLEM 18-6

For the network shown in fig. 1, take n_1 as the reference node and write down the incidence matrix.

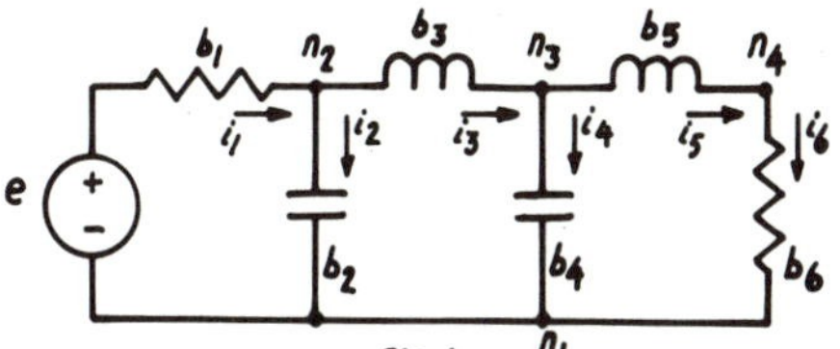

FiG. 1

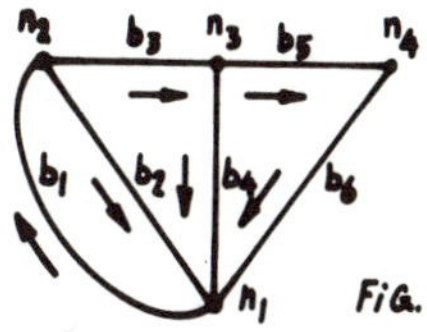

FiG. 2

Solution: Begin by drawing the graph for the circuit of fig. 1. Doing so yields the graph of fig. 2.

Notice that we have 4 nodes and 6 branches. This therefore means that A_a is a 4 x 6 matrix. $\bar{A}_a$, the augmented incidence matrix is shown in fig. 3 below.

$$\bar{A}_a = \begin{array}{c} \\ n_1 \\ n_2 \\ n_3 \\ n_4 \end{array} \begin{array}{c} \begin{matrix} b_1 & b_2 & b_3 & b_4 & b_5 & b_6 \end{matrix} \\ \begin{bmatrix} 1 & -1 & 0 & -1 & 0 & -1 \\ -1 & 1 & 1 & 0 & 0 & 0 \\ 0 & 0 & -1 & 1 & 1 & 0 \\ 0 & 0 & 0 & 0 & -1 & 1 \end{bmatrix} \end{array}$$

By definition, if the row a_i corresponding to the reference node n_i is deleted from $\bar{A}_a$, the result is the incidence matrix $\bar{A}$. Therefore, deleting the row corresponding to our reference node n_1 from $\bar{A}_a$ gives us $\bar{A}$, the incidence matrix.

$$A = \begin{bmatrix} -1 & 1 & 1 & 0 & 0 & 0 \\ 0 & 0 & -1 & 1 & 1 & 0 \\ 0 & 0 & 0 & 0 & -1 & 1 \end{bmatrix}$$

• **PROBLEM 18-7**

For the network in fig. 1, write the independent node equations in matrix form. Take n_1 as the reference node.

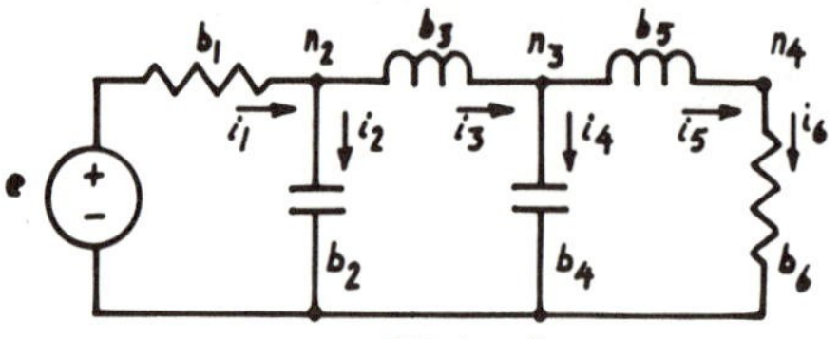

FiG. 1 n_1

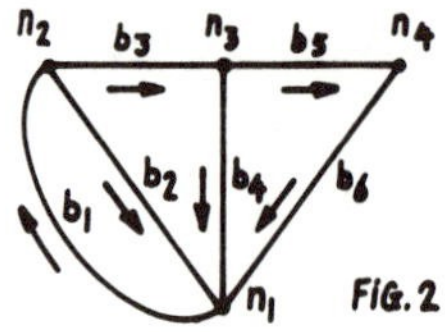

FIG. 2

Solution: By definition, the independent node or KCL equations are given by $\bar{A}\,\bar{i}_b = 0$ where $\bar{A}$ is the incidence matrix. From the graph shown in fig. 2, write the matrix $\bar{A}$ directly by excluding the row corresponding to the reference node n_1.

Recalling the designation criteria for the matrix elements a_{ij}, use the indicated branch directions to yield the matrix $\bar{A}$

$$\bar{A}=\begin{bmatrix} -1 & 1 & 1 & 0 & 0 & 0 \\ 0 & 0 & -1 & 1 & 1 & 0 \\ 0 & 0 & 0 & 0 & -1 & 1 \end{bmatrix}$$

As stated before the node equations are given by $\bar{A}\,\bar{i}_b = 0$, therefore $\bar{A}\,\bar{i}_b$ equals

$$\begin{bmatrix} -1 & 1 & 1 & 0 & 0 & 0 \\ 0 & 0 & -1 & 1 & 1 & 0 \\ 0 & 0 & 0 & 0 & -1 & 1 \end{bmatrix} \begin{bmatrix} i_1 \\ i_2 \\ i_3 \\ i_4 \\ i_5 \\ i_6 \end{bmatrix} = 0$$

• **PROBLEM 18-8**

The augmented incidence matrix of a graph G is given by

$$A_a=\begin{bmatrix} 1 & 0 & -1 & 1 & -1 & 0 \\ -1 & 1 & 0 & 0 & 1 & -1 \\ 0 & -1 & 1 & -1 & 0 & 1 \end{bmatrix}$$

Draw the graph G.

Solution: By definition the augmented incidence matrix has dimensions N x B where N = # of nodes and B = # of branches. In this matrix, N = 3 and B = 6. From this, begin drawing the graph.

Labeling the columns and rows of our matrix yields:

$$A_a = \begin{array}{c} \\ n_1 \\ n_2 \\ n_3 \end{array} \begin{array}{c} \begin{array}{cccccc} b_1 & b_2 & b_3 & b_4 & b_5 & b_6 \end{array} \\ \begin{bmatrix} 1 & 0 & -1 & 1 & -1 & 0 \\ -1 & 1 & 0 & 0 & 1 & -1 \\ 0 & -1 & 1 & -1 & 0 & 1 \end{bmatrix} \end{array} .$$

It is simpler to look at the branch columns first and note what nodes they connect. Be sure to remember that nodes are connected only when a +1 or -1 appears in the position a_{ij}, a_{ij} = 0 means that the nodes are not connected. For example, b_1 connects nodes n_1 and n_2 and since a_{11} = +1 the reference current flows from n_1 to n_2. Following this procedure for all branches in the matrix yields the graph shown in the figure.

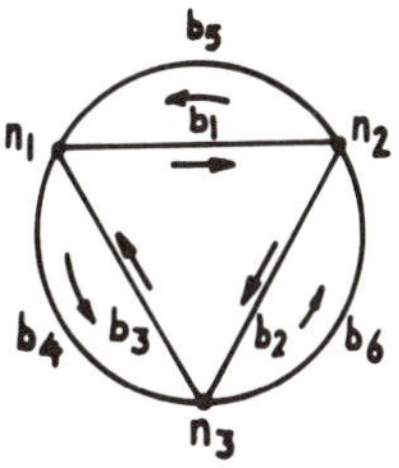

THE LOOP MATRIX

• **PROBLEM** 18-9

Find the augmented loop matrix of the network shown in fig. 1

Solution: The graph of the given network is shown in fig. 2

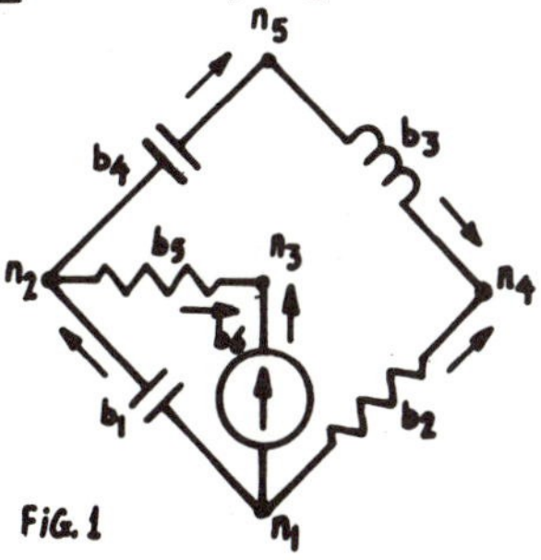

FiG. 1

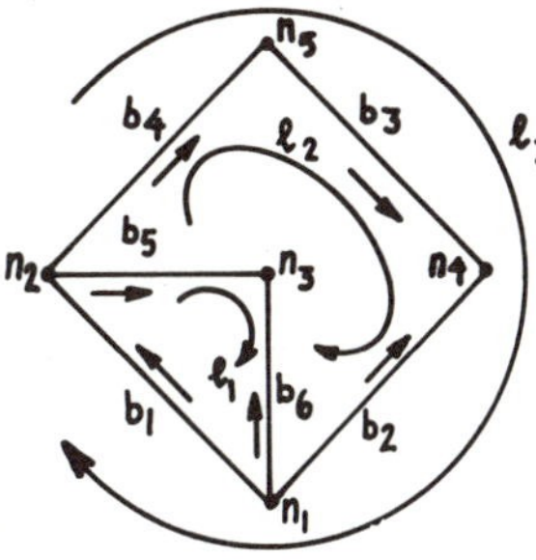

FiG. 2

The possible loops for this graph are also labeled in fig 2. Since the augmented loop matrix is an L x B matrix, where L is the number of loops and B is the number of branches, our matrix is 3 x 6. Applying the criteria for the matrix B_a discussed in earlier problems yields the B_a matrix:

$$B_a = \begin{matrix} & b_1 & b_2 & b_3 & b_4 & b_5 & b_6 \\ \ell_1 \\ \ell_2 \\ \ell_3 \end{matrix} \begin{bmatrix} 1 & 0 & 0 & 0 & 1 & -1 \\ 0 & -1 & 1 & 1 & -1 & 1 \\ 1 & -1 & 1 & 1 & 0 & 0 \end{bmatrix}.$$

• PROBLEM 18-10

For the network shown in fig. 1

(a) Specify all the loops

(b) Choose arbitrary orientations for the loops and write down the augmented loop matrix.

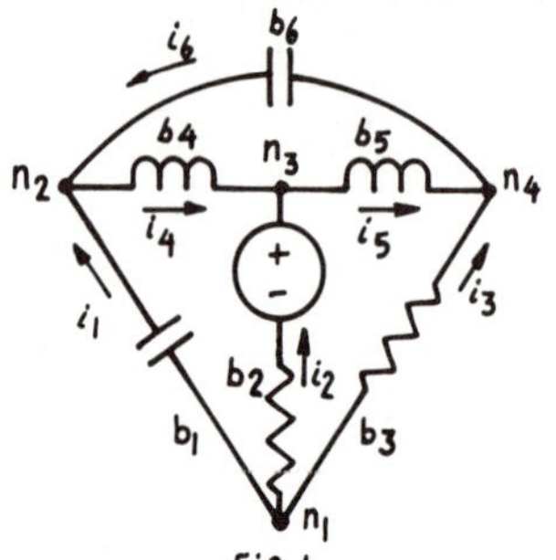

FIG. 1

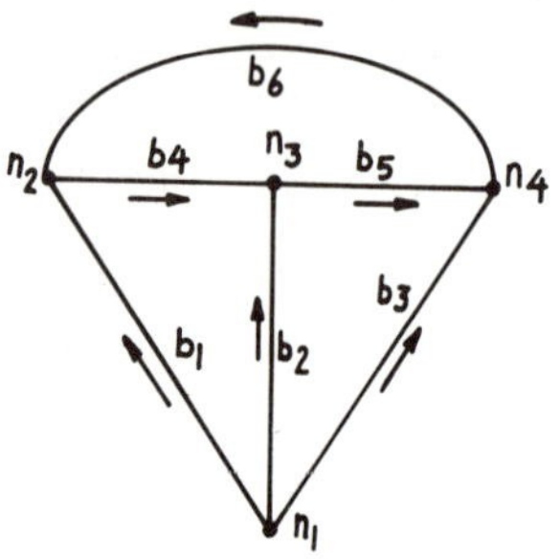

FIG. 2

Solution: After drawing the graph for the network shown in fig. 2, begin by specifying all the loops for the graph.

Refer to the diagrams in fig. 3 for the designated loops and their directions.

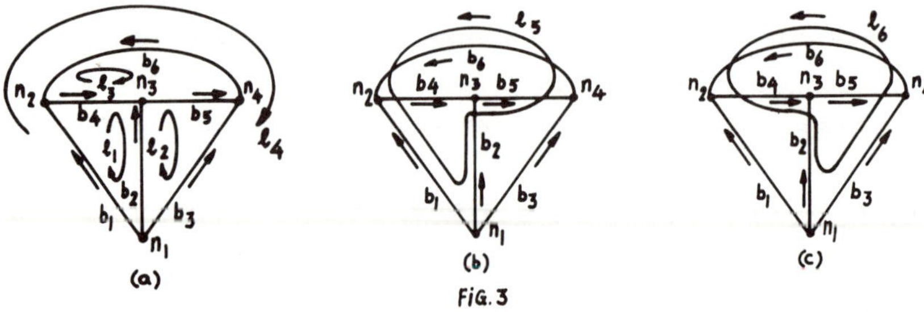

FIG. 3

By definition, an augmented loop matrix is an L x B matrix, where L = # of loops (here L = 6), $\overline{B}_a = (b_{ij})$ and

$$b_{ij} = \begin{cases} 1 & \text{when branch } b_j \text{ is in loop } \ell_i \text{ and is oriented in same direction} \\ -1 & \text{when branch } b_j \text{ is in loop } \ell_i \text{ and is oriented in opposite direction} \\ 0 & \text{when branch } b_j \text{ is not in loop } \ell_i \end{cases}$$

Looking at fig. 3 (a)-(c), begin writing the matrix $\overline{B}_a$.

Loop ℓ_1 contains branches b_1, b_4 and b_2. Observe that b_1 and b_4 are oriented in the same direction as ℓ_1 and b_2 in the opposite direction. Therefore positions b_{11} and b_{14} are +1 and b_{12} is -1 by the criteria described earlier. Using this procedure for all loops yields the matrix B_a shown below.

$$\bar{B}_a = \begin{array}{c} \\ l_1 \\ l_2 \\ l_3 \\ l_4 \\ l_5 \\ l_6 \end{array} \begin{array}{c} \begin{array}{cccccc} b_1 & b_2 & b_3 & b_4 & b_5 & b_6 \end{array} \\ \begin{bmatrix} 1 & -1 & 0 & 1 & 0 & 0 \\ 0 & 1 & -1 & 0 & 1 & 0 \\ 0 & 0 & 0 & -1 & -1 & -1 \\ 1 & 0 & -1 & 0 & 0 & -1 \\ -1 & 1 & 0 & 0 & 1 & 1 \\ 0 & -1 & 1 & 1 & 0 & 1 \end{bmatrix} \end{array}$$

• PROBLEM 18-11

For the graph G shown in fig. 1, choose a tree and develop the fundamental loop matrix.

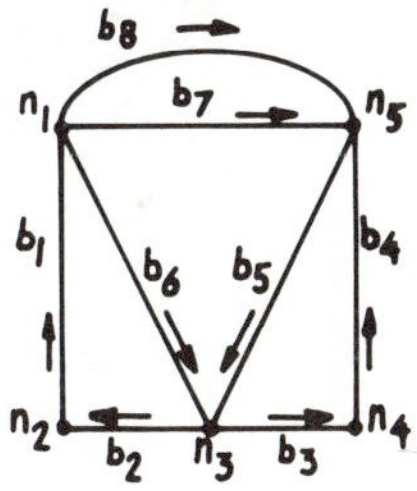

FIG. 1

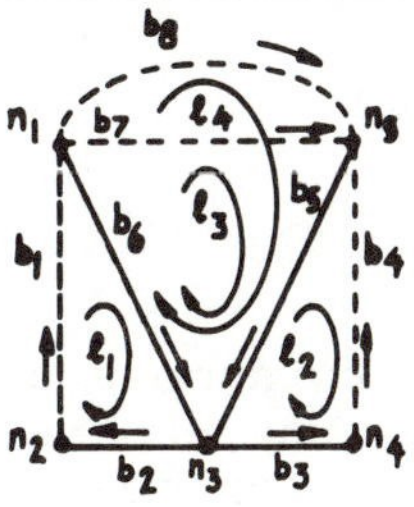

FIG. 2

Solution: Begin by choosing an arbitrary tree for the given graph. One such tree is shown in fig. 2 by the solid lines. The dashed lines indicated the cotree branches or chords.

Recalling that a fundamental loop cannot contain more than one chord, we see that there are only four loops possible. These are also shown in fig. 2. The fundamental loop matrix B_f is a [B-N] x B matrix where B is the number of branches and N + 1 is the number of nodes. Taking arbitrary directions assigned to each loop, the corresponding B_f matrix is

$$B_f = \begin{array}{c} \\ \ell_1 \\ \ell_2 \\ \ell_3 \\ \ell_4 \end{array} \begin{array}{c} \begin{array}{cccccccc} b_1 & b_2 & b_3 & b_4 & b_5 & b_6 & b_7 & b_8 \end{array} \\ \begin{bmatrix} 1 & 1 & 0 & 0 & 0 & 1 & 0 & 0 \\ 0 & 0 & 1 & 1 & 1 & 0 & 0 & 0 \\ 0 & 0 & 0 & 0 & 1 & -1 & 1 & 0 \\ 0 & 0 & 0 & 0 & 1 & -1 & 0 & 1 \end{bmatrix} \end{array}$$

For the network shown in fig. 1,

(a) Specify all the fundamental loops

(b) Choose arbitrary orientations for the loops and write the fundamental loop matrix.

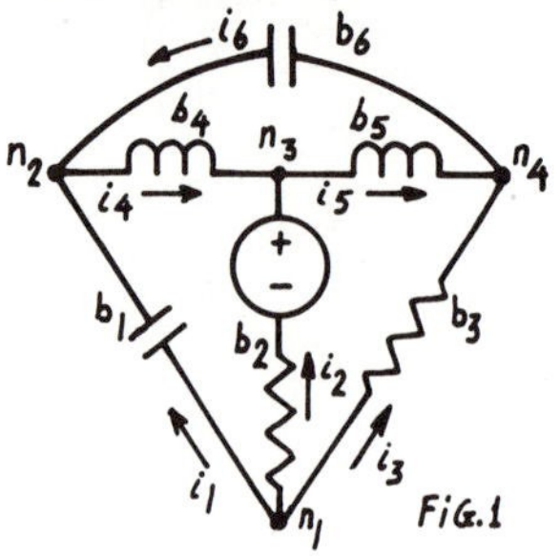

FiG.1

Solution: Begin by drawing the graph G corresponding to the network given. Doing so yields

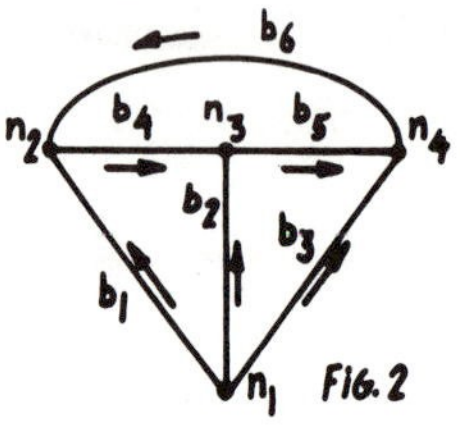

FIG.2

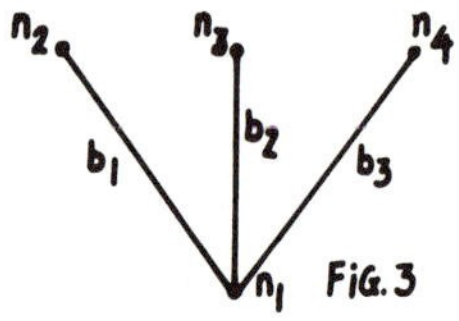

FiG.3

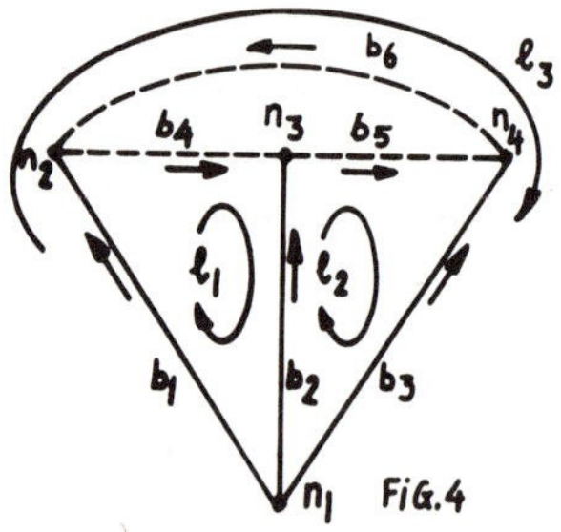

FiG.4

The next step is to choose a tree for the graph of fig 2. Fig. 3 shows the tree chosen. Note that this is not the only possible tree for our graph.

In order to proceed first define a "cotree" and a "chord."

A cotree - is the set of all branches not in tree T.
A chord - is any branch of a cotree.

Since our tree contains branches b_1, b_2, and b_3, the cotree contains b_4, b_5, and b_6. The graph of fig. 2 can be represented in a slightly different manner using solid lines for the tree branches and dashed lines for the cotree branches or chords. This is shown in fig. 4.

A fundamental loop cannot contain more than one chord. Recall that a chord is one branch of the cotree in fig. 4 drawn by dashed lines. Refering to fig. 4, the 3 possible fundamental loops $\ell_1 = (b_1, b_4, b_2)$, $\ell_2 = (b_2, b_5, b_3)$ and $\ell_3 = (b_1, b_6, b_3)$ were found based on the above criteria and given the arbitrary directions indicated.

To answer part b first define the fundamental loop matrix, B_f:

B_f is a B-(n-1) x B matrix where $\bar{B}_f = (b_{ij})$

and $b_{ij} = \begin{cases} 1 & \text{when branch } b_j \text{ is in the fundamental loop } \ell_i \text{ and is oriented in the same direction} \\ -1 & \text{when branch } b_j \text{ is in the fundamental loop } \ell_i \text{ and is oriented in the opposite direction} \\ 0 & \text{when branch } b_j \text{ is not in the fundamental loop } \ell_i \end{cases}$

$\bar{B}_f$ is represented in the following form:

$$\bar{B}_f = \begin{array}{c} \\ \ell_1 \\ \ell_2 \\ \ell_3 \\ \vdots \\ \ell_i \\ \vdots \\ \ell_{B-(N-1)} \end{array} \begin{array}{c} b_1 \quad b_2 \quad b_3 \quad \dots \; b_j \; \dots \; b_B \\ \left[\begin{array}{c} \\ \\ \\ \\ \\ \\ \\ \end{array} \right] \end{array}$$

Now proceed to develop the matrix B_f for our network. Using the definition given previously , the augmented loop matrix, B_a, for the graph in fig. 4 is

$$B_a = \begin{array}{c|cccccc} & b_1 & b_2 & b_3 & b_4 & b_5 & b_6 \\ \hline \ell_1 & 1 & -1 & 0 & 1 & 0 & 0 \\ \ell_2 & 0 & 1 & -1 & 0 & 1 & 0 \\ \ell_3 & 1 & 0 & -1 & 0 & 0 & -1 \end{array}$$

To give one final illustration of how this matrix was obtained we look at loop ℓ_1 in fig. 4. Notice that this loop contains branches b_1, b_2, and b_4. With ℓ_1 oriented in a clockwise direction, branch b_1 is found to be in the same direction; hence the entry in the ℓ_1, b_1 position is +1. Following this procedure for all loops until all have been covered and making entries in the appropriate loop, branch position on the matrix gives the final result.

For the network of Prob. 12, denote the voltage drop in branch b_j by v_j and write a set of linearly independent loop equations in matrix form.

Solution: Using the resulting fundamental loop matrix from Prob. 12, write the linearly independent loop equations in the form

$$\bar{B}_f \bar{v}_b = 0$$

For our network, this turns out to be

$$\begin{bmatrix} 1 & -1 & 0 & 1 & 0 & 0 \\ 0 & 1 & -1 & 0 & 1 & 0 \\ 1 & 0 & -1 & 0 & 0 & -1 \end{bmatrix} \begin{bmatrix} v_1 \\ v_2 \\ v_3 \\ v_4 \\ v_5 \\ v_6 \end{bmatrix} = 0$$

APPLICATIONS

• PROBLEM 18-14

Consider the network shown in Fig. 1. Let $C_1 = C_3 = \frac{1}{2}f$, $R_2 = 1\Omega$, $R_5 = 2\Omega$, $L_4 = 1$ h, $L_6 = 2$ h, $v_{C_1}(0) = 1$ volt, $v_{C_3}(0) = 0$, $i_{L_4}(0) = 1$ amp, and $i_{L_6}(0) = 0$. The problem is to find

$$\bar{I}_b = [i_1 i_2 i_3 i_4 i_5 i_6]^T$$

and $$\bar{v}_b = [v_1 v_2 v_3 v_4 v_5 v_6]^T$$

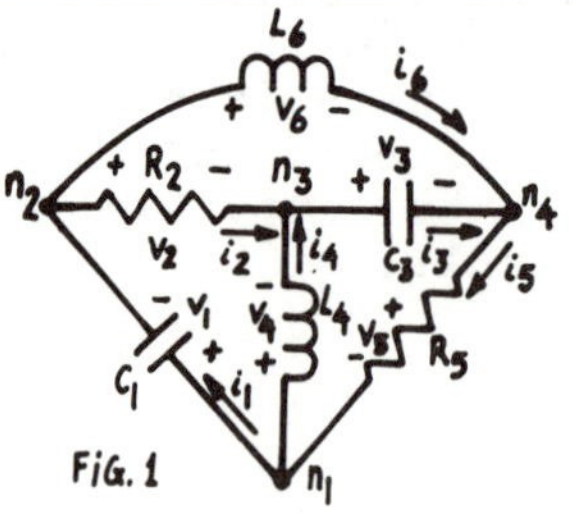

FiG. 1

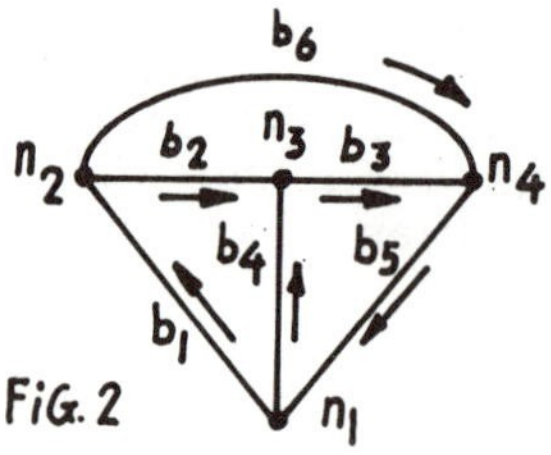

FiG. 2

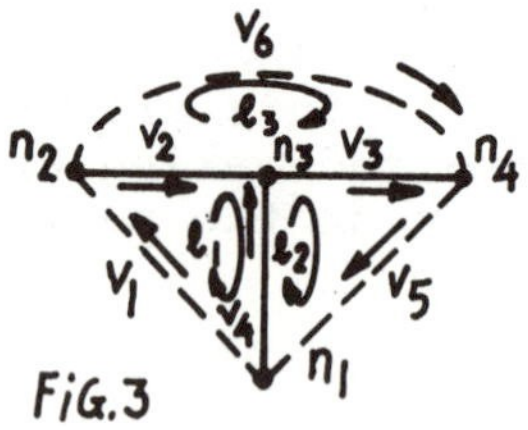

FiG. 3

Solution: Begin by first writing the KCL and KVL equations from the graph of fig. 2 with n_1 = reference node.

For KCL $\bar{A}\,\bar{i}_b = 0$ where $\bar{A}$ is the incidence matrix. Hence

$$\bar{A}\,\bar{i}_b = \underbrace{\begin{bmatrix} -1 & 1 & 0 & 0 & 0 & 1 \\ 0 & -1 & 1 & -1 & 0 & 0 \\ 0 & 0 & -1 & 0 & 1 & -1 \end{bmatrix}}_{\bar{A}} \underbrace{\begin{bmatrix} i_1 \\ i_2 \\ i_3 \\ i_4 \\ i_5 \\ i_6 \end{bmatrix}}_{\bar{i}_b} = 0 \qquad (1)$$

For KVL $\bar{B}_f\bar{v}_b = 0$ where $\bar{B}_f$ is the fundamental loop matrix. Choosing the tree and the loop directions indicated in fig. 3 proceed to write the KVL equations in matrix form.

Hence

$$\bar{B}_f\bar{v}_b = \underbrace{\begin{bmatrix} 1 & 1 & 0 & -1 & 0 & 0 \\ 0 & 0 & 1 & 1 & 1 & 0 \\ 0 & -1 & -1 & 0 & 0 & 1 \end{bmatrix}}_{\bar{B}_f} \underbrace{\begin{bmatrix} v_1 \\ v_2 \\ v_3 \\ v_4 \\ v_5 \\ v_6 \end{bmatrix}}_{\bar{v}_b} = 0 \qquad (2)$$

Thus far we have twelve unknowns and six equations given by (1) and (2). We therefore need six additional linearly independent equations in order to solve this problem. These equations are obtained from the branch voltage-current relations; the voltage across any branch is related to the current through it by the branch impedance. Using Laplace transforms to represent the branch voltage-current relations we have

$$\begin{aligned} &\text{branch 1} \quad V_1(s) = \frac{1}{sC_1} I_1(s) + \frac{1}{s} v_{C_1}(0) \\ &\text{branch 2} \quad V_2(s) = R_2 I_2(s) \\ &\text{branch 3} \quad V_3(s) = \frac{1}{sC_3} I_3(s) + \frac{1}{s} v_{C_3}(0) \\ &\text{branch 4} \quad V_4(s) = L_4 s I_4(s) - L_4 i_4(0) \\ &\text{branch 5} \quad V_5(s) = R_5 I_5(s) \\ &\text{branch 6} \quad V_6(s) = L_6 s I_6(s) - L_6 i_6(0) \end{aligned} \qquad (3)$$

Substituting the given initial conditions and the element values the equations in (3) become

$$
\begin{aligned}
V_1(s) &= \frac{2}{s} I_1(s) + \frac{1}{s} \\
V_2(s) &= I_2(s) \\
V_3(s) &= \frac{2}{s} I_3(s) \\
V_4(s) &= sI_4(s) - 1 \\
V_5(s) &= 2I_5(s) \\
V_6(s) &= 2sI_6(s)
\end{aligned} \tag{4}
$$

Since there are 6 equations for the impedance of the 6 branches represent them in matrix form. That is, the matrix $\overline{Z}_b(s)$ is a B x B matrix where B is the number of branches.

$$
\overline{Z}_b(s) = \begin{bmatrix} \frac{2}{s} & & & & & \\ & 1 & & & 0 & \\ & & \frac{2}{s} & & & \\ & & & s & & \\ & 0 & & & 2 & \\ & & & & & 2s \end{bmatrix} \tag{5}
$$

Also $\overline{v}_C(0)$, the capacitor initial voltage matrix, $\overline{i}_L(0)$, the inductor initial current matrix, and $\overline{M}$, the mutual inductance matrix, are

$$\overline{v}_C(0) = [1 \quad 0 \quad 0 \quad 0 \quad 0 \quad 0]^T \tag{6}$$

$$\overline{i}_L(0) = [0 \quad 0 \quad 0 \quad 1 \quad 0 \quad 0]^T \tag{7}$$

$$
\overline{M} = \begin{bmatrix} 0 & & & & & \\ & 0 & & & 0 & \\ & & 0 & & & \\ & & & 1 & & \\ & 0 & & & 0 & \\ & & & & & 2 \end{bmatrix} \tag{8}
$$

Taking the Laplace transform of both sides of Eqs. (1) and (2) yields

$$\overline{A}\, \overline{I}_b(s) = 0 \tag{9}$$

and $$\overline{B}_f \overline{V}_b(s) = 0. \tag{10}$$

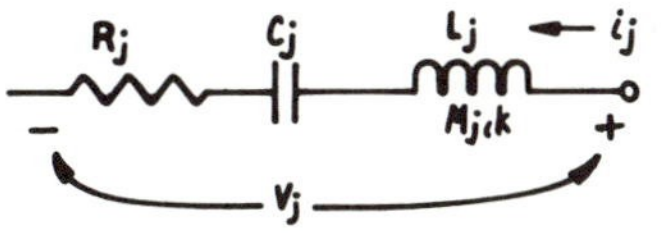

FIG.4

From fig. 4, the voltage for each branch can be seen as

$$V_j(s) = (R_j + \frac{1}{C_j s} + sL_j)\ I_j(s) + \frac{1}{s} v_{C_j}(0) + \sum_{\substack{k=1 \\ k\neq j}}^{B} M_{j,k}[sI_k(s) - i_k(0)] - L_j i_j(0)$$

Rearranging terms and using matrices 5, 6, 7, and 8

$$\overline{V}_b(s) = \overline{Z}_b \overline{I}_b(s) + \frac{1}{s} \overline{V}_c(0) - \overline{M}\ I_L(0) \qquad (11)$$

for any one branch (fig. 4) in the circuit. Substituting $\overline{v}_b$ from Eq. (11) into Eq (10) yields

$$\overline{B}_f \overline{Z}_b \overline{I}_b(s) = -\frac{1}{s} \overline{B}_f \overline{v}_c(0) + \overline{B}_f \overline{M} i_L(0).$$

This equation together with Eq (9) can be combined into an augment matrix

$$\begin{bmatrix} \overline{B}_f \overline{Z}_b \\ \hdashline \overline{A} \end{bmatrix} \overline{I}_b(s) = \begin{bmatrix} -\frac{1}{s} \overline{B}_f \overline{v}_c(0) + \overline{B}_f \overline{M} \overline{i}_L(0) \\ \hdashline 0 \end{bmatrix} \qquad (12)$$

Assuming that $\begin{bmatrix} \overline{B}_f \overline{Z}_b \\ \hdashline A \end{bmatrix}$ is nonsingular, solve for the branch currents, $\overline{I}_b(s)$, hence,

$$\overline{I}_b(s) = \begin{bmatrix} \overline{B}_f \overline{Z}_b \\ \hdashline \overline{A} \end{bmatrix}^{-1} \begin{bmatrix} -\frac{1}{s} \overline{B}_f \overline{v}_C(0) + \overline{B}_f \overline{M i}_L(0) \\ \hdashline 0 \end{bmatrix} \qquad (13)$$

The branch voltages can then be obtained from Eq (11).

To solve for $\overline{I}_b(s)$, find

$$\overline{B}_f \overline{Z}_b = \begin{bmatrix} 1 & 1 & 0 & -1 & 0 & 0 \\ 0 & 0 & 1 & 1 & 1 & 0 \\ 0 & -1 & -1 & 0 & 0 & 1 \end{bmatrix} \begin{bmatrix} \frac{2}{s} & 0 & 0 & 0 & 0 & 0 \\ 0 & 1 & 0 & 0 & 0 & 0 \\ 0 & 0 & \frac{2}{s} & 0 & 0 & 0 \\ 0 & 0 & 0 & s & 0 & 0 \\ 0 & 0 & 0 & 0 & 2 & 0 \\ 0 & 0 & 0 & 0 & 0 & 2s \end{bmatrix}$$

$$\bar{B}_f\bar{Z}_b = \begin{bmatrix} \frac{2}{s} & 1 & 0 & -s & 0 & 0 \\ 0 & 0 & \frac{2}{s} & s & 2 & 0 \\ 0 & -1 & -\frac{2}{s} & 0 & 0 & 2s \end{bmatrix}$$

The augment matrix $\begin{bmatrix} \bar{B}_f\bar{Z}_b \\ \hline \bar{A} \end{bmatrix}$ is,

$$\left[\begin{array}{cccccc} \frac{2}{s} & 1 & 0 & -s & 0 & 0 \\ 0 & 0 & \frac{2}{s} & s & 2 & 0 \\ 0 & -1 & -\frac{2}{s} & 0 & 0 & 2s \\ \hdashline -1 & 1 & 0 & 0 & 0 & 1 \\ 0 & -1 & 1 & -1 & 0 & 0 \\ 0 & 0 & -1 & 0 & 1 & -1 \end{array}\right]$$

also

$$-\frac{1}{s}\bar{B}_f\bar{v}_C(0) = -\frac{1}{s}\begin{bmatrix} 1 & 1 & 0 & -1 & 0 & 0 \\ 0 & 0 & 1 & 1 & 1 & 0 \\ 0 & -1 & -1 & 0 & 0 & 1 \end{bmatrix}\begin{bmatrix} 1 \\ 0 \\ 0 \\ 0 \\ 0 \\ 0 \end{bmatrix}$$

$$= -\frac{1}{s}\begin{bmatrix} 1 \\ 0 \\ 0 \end{bmatrix}$$

and $$\bar{B}_f\bar{M}_L\bar{i}(0) = \begin{bmatrix} 1 & 1 & 0 & -1 & 0 & 0 \\ 0 & 0 & 1 & 1 & 1 & 0 \\ 0 & -1 & -1 & 0 & 0 & 1 \end{bmatrix}\begin{bmatrix} 0 & & & & & 0 \\ & 0 & & & & \\ & & 0 & & & \\ & & & 1 & & \\ 0 & & & & 0 & \\ & & & & & 2 \end{bmatrix}\begin{bmatrix} 0 \\ 0 \\ 0 \\ 1 \\ 0 \\ 0 \end{bmatrix}$$

$$= \begin{bmatrix} 0 & 0 & 0 & -1 & 0 & 0 \\ 0 & 0 & 0 & 1 & 0 & 0 \\ 0 & 0 & 0 & 0 & 0 & 2 \end{bmatrix} \begin{bmatrix} 0 \\ 0 \\ 0 \\ 1 \\ 0 \\ 0 \end{bmatrix}$$

$$= \begin{bmatrix} -1 \\ 1 \\ 0 \end{bmatrix}$$

hence,

$$-\frac{1}{s}\bar{B}_f\bar{v}_C(0) + \bar{B}_f\bar{M}\ \bar{i}_L = -\frac{1}{s}\begin{bmatrix} 1 \\ 0 \\ 0 \end{bmatrix} + \begin{bmatrix} -1 \\ 1 \\ 0 \end{bmatrix}$$

$$= \begin{bmatrix} -1 - \frac{1}{s} \\ 1 \\ 0 \end{bmatrix}$$

The augmented matrix $\begin{bmatrix} -\frac{1}{s}\ \bar{B}_f\bar{v}_C(0) + \bar{B}_f\bar{M}\ \bar{i}_L\ (0) \\ \hline 0 \end{bmatrix}$

is $\begin{bmatrix} -1 - \frac{1}{s} \\ 1 \\ 0 \\ \hline 0 \\ 0 \\ 0 \end{bmatrix}$

Finally the branch currents can be found by solving

$$\bar{I}_b(s) = \begin{bmatrix} \frac{2}{s} & 1 & 0 & -s & 0 & 0 \\ 0 & 0 & \frac{2}{s} & s & 2 & 0 \\ 0 & -1 & -\frac{2}{s} & 0 & 0 & 2s \\ -1 & 1 & 0 & 0 & 0 & 1 \\ 0 & -1 & 1 & -1 & 0 & 0 \\ 0 & 0 & -1 & 0 & 1 & -1 \end{bmatrix}^{-1} \begin{bmatrix} -(1 + \frac{1}{s}) \\ 1 \\ 0 \\ 0 \\ 0 \\ 0 \end{bmatrix}$$

Consider the network shown in Fig. 1. Let the resistors be 1 ohm, the capacitor 1 farad and the inductor 1 henry. The problem is to find

$$\bar{I}_b = [i_1 \; i_2 \; i_3 \; i_4 \; i_5]^T$$

and $$\bar{V}_b = [v_1 \; v_2 \; v_3 \; v_4 \; v_5]^T$$

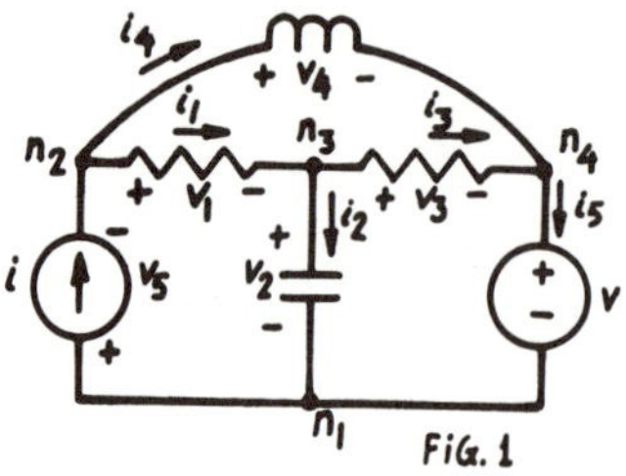

FiG. 1

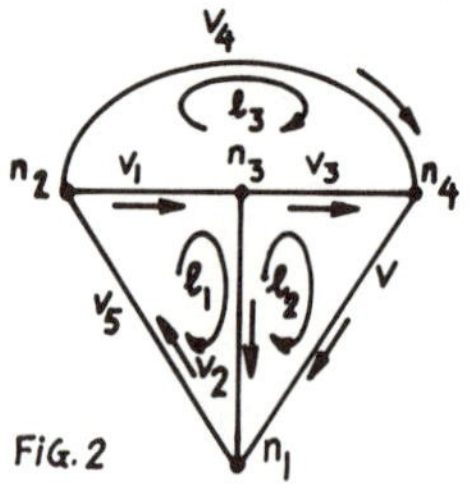

FiG. 2

Solution: Take n_1 as the reference node and write the three independent KCL node equations:

$$n_2: \qquad i_1 + i_4 = i$$

$$n_3: \quad -i_1 + i_2 + i_3 = 0$$

$$n_4: \quad -i_3 - i_4 + i_5 = 0$$

Writing the KCL equations in matrix form $\bar{A}\,\bar{I}_b = \bar{I}$ where $\bar{A}$ is the incidence matrix, yields,

$$\bar{A}\,\bar{I}_b = \underbrace{\begin{bmatrix} 1 & 0 & 0 & 1 & 0 \\ -1 & 1 & 1 & 0 & 0 \\ 0 & 0 & -1 & -1 & 1 \end{bmatrix}}_{\bar{A}} \underbrace{\begin{bmatrix} i_1 \\ i_2 \\ i_3 \\ i_4 \\ i_5 \end{bmatrix}}_{\bar{I}_b} = \underbrace{\begin{bmatrix} i \\ 0 \\ 0 \end{bmatrix}}_{\bar{I}} \qquad (1)$$

Writing the KVL equations for each fundamental loop shown in fig. 2 gives

$$\ell_1: \quad v_1 + v_2 + v_5 = 0$$

$$\ell_2: \qquad - v_2 + v_3 = -v$$

$$\ell_3: \; -v_1 - v_3 + v_4 = 0$$

Now write the equations in the matrix form $\bar{B}_f \bar{V}_b = \bar{V}$ where $\bar{B}_f$ is the fundamental loop matrix, hence

$$\bar{B}_f\bar{v}_b = \underbrace{\begin{bmatrix} 1 & 1 & 0 & 0 & 1 \\ 0 & -1 & 1 & 0 & 0 \\ -1 & 0 & -1 & 1 & 0 \end{bmatrix}}_{\bar{B}_f} \underbrace{\begin{bmatrix} v_1 \\ v_2 \\ v_3 \\ v_4 \\ v_5 \end{bmatrix}}_{\bar{v}_b} = \underbrace{\begin{bmatrix} 0 \\ -v \\ 0 \end{bmatrix}}_{\bar{V}} \qquad (2)$$

Find the branch impedance matrix $\bar{Z}_b(s)$ by using the four branch voltage-current relations,

$V_1(s) = I_1(s),\ V_2(s) = \frac{1}{s}I_2(s),\ V_3(s) = I_3(s),\ V_4(s) = sI_4(s),$

This yields

$$\begin{bmatrix} V_1 \\ V_2 \\ V_3 \\ V_4 \end{bmatrix} = \underbrace{\begin{bmatrix} 1 & 0 & 0 & 0 \\ 0 & \frac{1}{s} & 0 & 0 \\ 0 & 0 & 1 & 0 \\ 0 & 0 & 0 & s \end{bmatrix}}_{\bar{Z}_b(s)} \begin{bmatrix} I_1 \\ I_2 \\ I_3 \\ I_4 \end{bmatrix}. \qquad (3)$$

Now write Eq (2) in the form,

$$\begin{bmatrix} 1 & 1 & 0 & 0 \\ 0 & -1 & 1 & 0 \\ -1 & 0 & -1 & 1 \end{bmatrix} \begin{bmatrix} v_1 \\ v_2 \\ v_3 \\ v_4 \end{bmatrix} + \begin{bmatrix} 1 \\ 0 \\ 0 \end{bmatrix} v_5 = \begin{bmatrix} 0 \\ -V \\ 0 \end{bmatrix}. \qquad (4)$$

Now take the Laplace transform of Eq (4) and substitute Eq (3) to obtain

$$\begin{bmatrix} 1 & 1 & 0 & 0 \\ 0 & -1 & 1 & 0 \\ -1 & 0 & -1 & 1 \end{bmatrix} \begin{bmatrix} 1 & 0 & 0 & 0 \\ 0 & \frac{1}{s} & 0 & 0 \\ 0 & 0 & 1 & 0 \\ 0 & 0 & 0 & s \end{bmatrix} \begin{bmatrix} I_1 \\ I_2 \\ I_3 \\ I_4 \end{bmatrix} + \begin{bmatrix} 1 \\ 0 \\ 0 \end{bmatrix} V_5 = \begin{bmatrix} 0 \\ -V \\ 0 \end{bmatrix}$$

hence

$$\begin{bmatrix} 1 & \frac{1}{s} & 0 & 0 \\ 0 & -\frac{1}{s} & 1 & 0 \\ -1 & 0 & -1 & s \end{bmatrix} \cdot \begin{bmatrix} I_1 \\ I_2 \\ I_3 \\ I_4 \end{bmatrix} + \begin{bmatrix} 1 \\ 0 \\ 0 \end{bmatrix} V_5 = \begin{bmatrix} 0 \\ -V \\ 0 \end{bmatrix}$$

or,

$$\begin{bmatrix} 1 & \frac{1}{s} & 0 & 0 & 0 & 1 \\ 0 & -\frac{1}{s} & 1 & 0 & 0 & 0 \\ -1 & 0 & -1 & s & 0 & 0 \end{bmatrix} \begin{bmatrix} I_1 \\ I_2 \\ I_3 \\ I_4 \\ V_5 \end{bmatrix} = \begin{bmatrix} 0 \\ -V \\ 0 \end{bmatrix} \qquad (5)$$

Equation (1) can also be Laplace transformed and modified to read,

$$\begin{bmatrix} 1 & 0 & 0 & 1 & 0 & 0 \\ -1 & 1 & 1 & 0 & 0 & 0 \\ 0 & 0 & -1 & -1 & 1 & 0 \end{bmatrix} \begin{bmatrix} I_1 \\ I_2 \\ I_3 \\ I_4 \\ V_5 \end{bmatrix} = \begin{bmatrix} I \\ 0 \\ 0 \end{bmatrix} \qquad (6)$$

Combining Eqs. (5) and (6) gives

$$\begin{bmatrix} 1 & \frac{1}{s} & 0 & 0 & 0 & 1 \\ 0 & -\frac{1}{s} & 1 & 0 & 0 & 0 \\ -1 & 0 & -1 & s & 0 & 0 \\ 1 & 0 & 0 & 1 & 0 & 0 \\ -1 & 1 & 1 & 0 & 0 & 0 \\ 0 & 0 & -1 & -1 & 1 & 0 \end{bmatrix} \begin{bmatrix} I_1 \\ I_2 \\ I_3 \\ I_4 \\ I_5 \\ V_5 \end{bmatrix} = \begin{bmatrix} 0 \\ -V \\ 0 \\ I \\ 0 \\ 0 \end{bmatrix}.$$

Now solve,

$$\begin{bmatrix} I_1 \\ I_2 \\ I_3 \\ I_4 \\ I_5 \\ V_5 \end{bmatrix} \begin{bmatrix} 1 & \frac{1}{s} & 0 & 0 & 0 & 1 \\ 0 & -\frac{1}{s} & 1 & 0 & 0 & 0 \\ -1 & 0 & -1 & s & 0 & 0 \\ 1 & 0 & 0 & 1 & 0 & 0 \\ -1 & 1 & 1 & 0 & 0 & 0 \\ 0 & 0 & -1 & -1 & 1 & 0 \end{bmatrix}^{-1} \begin{bmatrix} 0 \\ -V \\ 0 \\ I \\ 0 \\ 0 \end{bmatrix}$$

and take the inverse Laplace transform to obtain

$[i_1, i_2, i_3, i_4, i_5, v_5]$

The four remaining unknowns, $[v_1, v_2, v_3, v_4]$ can be calculated from Eq (2).

CHAPTER 19

NUMERICAL METHODS

TRIAL & ERROR PROCEDURE

• PROBLEM 19-1

A nonlinear resistance with a voltage-current relationship given by

$$V = I/\sqrt{2 + I^2}$$

is connected to the network shown below. Find the current, I, flowing in this resistance.

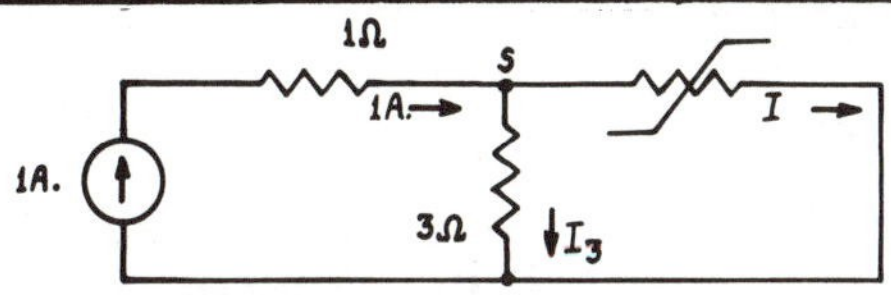

Solution: The 3Ω resistor and the nonlinear resistor are in parallel; therefore, the same potential drop is found across each resistor.

$$V_{3\Omega} = V \quad (1)$$

Using Ohm's Law for the 3Ω resistor, and the voltage-current relationship given in the problem for the nonlinear resistance results in the expression

$$3I_3 = I/\sqrt{2 + I^2} \quad (2)$$

A current of 1 A is supplied by the constant current source. This 1 A current flows through the 1Ω resistor, then branches at point s. From KCL

$$I + I_3 = 1. \quad (3)$$

Substituting (3) into (2), yields

$$3(1 - I) = I/\sqrt{2 + I^2}. \quad (4)$$

Dividing each side by (1 - I) gives

$$I/(1 - I)\sqrt{2 + I^2} = 3. \quad (5)$$

This equation can be algebraically manipulated to produce a quartic equation in I, i.e., an equation containing I^4, I^3, etc. which can then be solved by conventional (algebraic) methods. With use of hand calculators, however, it is often practical to find the solution to equations such as (5) numerically.

To solve (5) numerically, proceed as follows. Define

$$f(I) \equiv I/(1 - I)\sqrt{2 + I^2}. \qquad (6)$$

Now, (5) becomes

$$f(I) = 3,$$

and the solution to this is equation is the value of I required for this relationship to hold. To find the value of I that satisfies (5), we make a table of f(I) versus I:

I	f(I)
0	0
0.2	0.18
0.4	0.45
0.6	0.98
0.8	2.46

It is obvious from this table that the value of I where f(I) = 3 lies between 0.8 and 1.0. Therefore, we construct a second table with smaller increments in I for this region:

I	f(I)
0.81	2.62
0.83	2.98
0.85	3.43

The solution to (5) is $I \simeq 0.83$ and can be found more accurately by a third trial:

I	f(I)
0.830	2.98
0.831	3.00
0.832	3.02

The current flowing in the nonlinear resistance is I = 0.831 A (accurate to three significant figures).

A network with two nonlinear resistors is shown. Find the current i by trial-and-error numerical procedure.

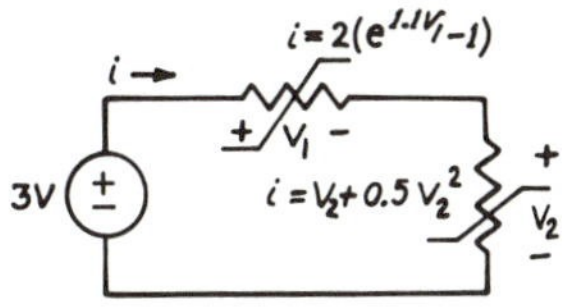

Solution: Two equations can be found by equating the currents through the two series resistors and by setting the sum of the voltage drops across the resistors equal to the source voltage.

Equating currents we get

$$2(e^{1.1v_1} - 1) = v_2 + 0.5v_2^2. \tag{1}$$

Summing voltage drops gives

$$3 = v_1 + v_2. \tag{2}$$

Solve for v_1 from Eq. (2) and substitute this into Eq. (1).

$$2[e^{1.1(3-v_2)} - 1] = v_2 + 0.5v_2^2.$$

Rearrange to get

$$e^{1.1(3-v_2)} = 1 + 0.5v_2 + 0.25v_2^2.$$

By trial and error substitute values of v_2 until the left hand side (LHS) equals the right hand side (RHS). Several attempts are shown below.

v_2	LHS	RHS
0	27.11	1
1	9.03	1.75
2	3.00417	3.00000
.		
.		
.		
2.000865	3.00131	3.00130

Thus to five significant digits

$$v_2 = 2.0009 \text{ V}$$

and substituting this into Eq. (2) gives

$$v_1 = 0.9991 \text{ V}.$$

The current i is:

$$i = v_2 + 0.5v_2^2 = 2.0009 + 0.5(2.0009)^2 = 4.0027 \text{ A}$$

or $i = 2(e^{1.1v_1} - 1) = 2(e^{(1.1)(0.9991)} - 1) = 4.0024$ A.

In both cases the current i is $\simeq$ 4 A.

NEWTON'S METHOD

• PROBLEM 19-3

Use numerical analysis to find the roots of the polynomial:

$$s^4 + 3s^3 + 2s^2 + 2s + 1 = 0.$$

Solution: The function f(s) will be designated as:

$$f(s) = s^4 + 3s^3 + 2s^2 + 2s + 1.$$

Since all the coefficients of f(s) are of the same sign and real, two conclusions may be drawn:

1. Complex roots occur in complex conjugate pairs.
2. Real roots, if any, are negative.

Since there must be 0, 2 or 4 real roots, we will first investigate the presence of these, since they are generally the easiest to find, if they exist. Newton's method will be used. That is,

$$s_{i+1} = s_i - \frac{f(s_i)}{f'(s_i)}$$

where s_i is the previous approximation (or guess) of the root, s_{i+1} is the new approximation and $f'(s_i) = 4s_i^3 + 9s_i^3 + 4s_i + 2$ is the derivative of f(s) evaluated at $s = s_i$. First make a guess of s_0 as equal to -1, and the following table is calculated:

i	s_i	$f(s_i)$	$f'(s_i)$	$s_{i+1} = s_i - \frac{f(s_i)}{f'(s_i)}$
0	-1	-1	3	-2/3
1	-2/3	-0.136	2.148	-0.603
2	-0.603	-0.0043	1.983	-0.6008
3	-0.6008	0.000018		

Since $s_3 = -0.6008$ makes $f(s_3)$ very close to zero, it will be taken as a root of $f(s) = 0$. We know there must be at least one more real root. To simplify matters we will take the known factor out of $f(s)$ by direct division.

$$
\begin{array}{r|l}
 & s^3 + 2.3992s^2 + 0.5586s + 1.6644 \\
\hline
s + 0.6008 & s^4 + 3s^3 + 2s^2 + 2s + 1 \\
 & \underline{s^4 + .6008s^3} \\
 & 2.3992s^3 + 2s^2 \\
 & \underline{2.3992s^3 + 1.4414s^2} \\
 & \qquad 0.5586s^2 + 2s \\
 & \qquad \underline{0.5586s^2 + 0.3356s} \\
 & \qquad\qquad 1.6644s + 1 \\
 & \qquad\qquad \underline{1.6644s + 0.99997} \\
 & \qquad\qquad\qquad 0.00003
\end{array}
$$

The remainder is, of course, negligible. A new polynomial $f_1(s)$ is obtained, where

$$f_1(s) = s^3 + 2.3992s^2 + 0.5586s + 1.6644 = 0$$

and has the remaining three roots of $f(s) = 0$. The derivative is $f_1'(s) = 3s^2 + 4.7984s + 0.5586$. A first guess of a root of $f_1(s) = 0$ will be made at $s_0 = -2$, and a similar table is calculated.

i	s_i	$f_1(s_i)$	$f_1'(s_i)$	$s_{i+1} = s_i - \dfrac{f_1(s_i)}{f_1'(s_i)}$
0	-2	2.144	2.9618	-2.724
1	-2.724	-2.267	9.748	-2.491
2	-2.491	-0.297	7.22	-2.45
3	-2.45	-0.0091	6.81	-2.4486
4	-2.4486	0.0004		

Thus $s_4 = -2.4486$ is another root of $f(s) = 0$. Again, by direct division

$$
\begin{array}{r|l}
 & s^2 - 0.0494s + 0.6796 \\
\hline
s + 2.4486 & s^3 + 2.3992s^2 + 0.5586s + 1.6644 \\
 & \underline{s^3 + 2.4486s^2} \\
 & \quad - 0.0494s^2 + 0.5586s \\
 & \quad \underline{- 0.0494s^2 - 0.1210s} \\
 & \qquad\qquad 0.6796s + 1.6645 \\
 & \qquad\qquad \underline{0.6796s + 1.6641} \\
 & \qquad\qquad\qquad 0.0004
\end{array}
$$

Finally, the roots of $s^2 - 0.0494s + 0.6796 = 0$ are obtained by the quadratic equation.

$$s = \frac{0.0494 \pm \sqrt{(0.0494)^2 - 4(1)(0.6796)}}{2} = 0.0247 \pm j0.824$$

• PROBLEM 19-4

Obtain the partial fraction expansion of

$$D(s) = \frac{s^3 + 2s^2 + 3s + 7}{s^4 + 5s^3 + 11s^2 + 12s + 4.5}.$$

Use Newton's Method to find the roots of the quartic equation in the denominator. Find d(t).

Note: Use of a calculator is suggested for solution of this problem.

<u>Solution:</u> The first step in finding the roots of the polynomial $F(s) = s^4 + 5s^3 + 11s^2 + 12s + 4.5$, by Newton's Method, is to start with a guess, s_0, then calculate

$h_0 = -\frac{F(s_0)}{F'(s_0)}$, where $F'(s) = \frac{d}{ds}F(s)$. This gives a new value, $s_1 = h_0 + s_0$. Again we iterate and find $h_1 = -\frac{F(s_1)}{F'(s_1)}$ which gives $s_2 = h_1 + s_1$. Depending on the initial guess, the root, s_n can be converged upon, after several iterations with an accuracy to at least 3 decimal places., The initial guess will converge, as long as the sign of $h_0F'(s)$ is the same as the sign of $F''(s_0)$.

If starting with the guess $s_0 = 0$,

$$h_0 = -\frac{F(0)}{F'(0)} = -\frac{4.5}{12} = -0.3750.$$

Hence, $s_1 = s_0 + h_0 = 0 - 0.3750 = -0.3750$, thus

$$h_1 = -\frac{F(-0.375)}{F'(-0.375)} = -\frac{1.3030}{5.6484} = -0.2307$$

and $s_2 = s_1 + h_1 = -0.3750 - 0.2307 = -0.6057$

$$h_2 = -\frac{F(-0.6057)}{F'(-0.6057)} = -\frac{0.2907}{3.2888} = -0.0884.$$

Hence, $s_3 = s_2 + h_2 = -0.6057 - 0.0884 = -0.6941$

and $h_3 = -\frac{F(-0.6941)}{F'(-0.6941)} = -\frac{0.0304}{2.6188} = -0.0116$

thus, $s_4 = s_3 + h_3 = -0.6941 - 0.0116 = -0.7057$

$$h_4 = -\frac{F(-0.7057)}{F'(-0.7057)} = -\frac{0.0005}{2.5390} = -0.0002.$$

Hence, $s_5 = s_4 + h_4 = -0.7057 - 0.0002 = -0.7059$. Since $F(s_4)$ was already very close to the root, we can safely choose -0.706 with accuracy to at least 3 decimal places.

Since we have found one real root to the quartic equation, at least one other real root must exist. Hence, there are either two real roots and a pair of complex roots or four real roots to the quartic equation. Divide the quartic polynomial by s + 0.706 to obtain a cubic equation G(s).

$$\begin{array}{r|l}
 & s^3 + 4.294s^2 + 7.968s + 6.374 \\
s + 0.706 & s^4 + 5s^3 + 11s^2 + 12s + 4.5 \\
 & \underline{s^4 + 0.706s^3} \\
 & \quad 4.294s^3 + 11s^2 \\
 & \quad \underline{4.294s^3 + 3.032s^2} \\
 & \qquad 7.968s^2 + 12s \\
 & \qquad \underline{7.468s^2 + 5.626s} \\
 & \qquad\quad 6.374s + 4.500 \\
 & \qquad\quad \underline{6.374s + 4.500}
\end{array}$$

Now, use Newton's Method to find one of the roots of the cubic equation, $G(s) = s^3 + 4.294s^2 + 7.968s + 6.374 = 0$. Then using this root, again divide the polynomial to obtain a quadratic which can be solved by the quadratic formula.

Starting the iteration with $s_0 = 0$, yields

$$h_0 = -\frac{G(0)}{G'(0)} = -\frac{6.374}{7.968} = -0.8000,$$

thus $s_1 = s_0 + h_0 = 0 - 0.8000 = -0.8000$.

Hence, $h_1 = -\frac{G(-0.8000)}{G'(-0.8000)} = -\frac{2.2358}{3.0176} = -0.7409$

thus, $s_2 = s_1 + h_1 = -0.8000 - 0.7409 = -1.5409$.

Hence, $h_2 = -\frac{G(-1.5409)}{G'(-1.5409)} = -\frac{0.6330}{1.8579} = -0.3407$

thus, $s_3 = s_2 + h_2 = -1.5409 - 0.3407 = -1.8816$.

Therefore, $h_3 = -\frac{G(-1.8816)}{G'(-1.8816)} = -\frac{-.0777}{2.4301} = +0.0320$

thus $s_4 = s_3 + h_3 = -1.8816 + 0.0320 = -1.8496.$

Then, $h_4 = -\dfrac{G(-1.8496)}{G'(-1.8496)} = -\dfrac{-0.0013}{2.3467} = +0.0006$

giving $s_5 = s_4 + h_4 = -1.8496 + 0.0006 = -1.8490,$

since, $G(-1.8490) = 0.0001$ one can safely assume that the root -1.849 is accurate to 3 decimal places.

Dividing the cubic by $s + 1.849$ gives,

$$\begin{array}{r l}
 & s^2 + 2.445s + 3.447 \\
s + 1.849 \,\big)\!\!\!\! & s^3 + 4.294s^2 + 7.968s + 6.374 \\
 & \underline{s^3 + 1.849s^2} \\
 & 2.445s^2 + 7.968s \\
 & \underline{2.445s^2 + 4.521s} \\
 & 3.447s + 6.374 \\
 & \underline{3.447s + 6.374}
\end{array}$$

Solving for the roots of the quadratic $s^2 + bs + c$ by the formula:

$$\frac{-b \pm \sqrt{b^2 - 4c}}{2}$$

obtains the two remaining complex roots

$$s = \frac{-2.445 + \sqrt{2.445^2 - 13.788}}{2} = -1.223 + j1.397$$

$$s^* = -1.223 - j1.397.$$

Hence,

$$s^4 + 5s^3 + 11s^2 + 12s + 4.5 = (s+0.706)(s+1.849)(s+1.223-j1.397)(s+1.223+j1.397)$$

Partial fraction expansion yields

$$D(s) = \frac{A}{s+0.706} + \frac{B}{s+1.849} + \frac{C_s + D}{s^2+2.445s+3.447}.$$

Multiplying both sides by $s^4 + 5s^3 + 11s^2 + 12s + 4.5$ gives

$$s^3 + 2s^2 + 3s + 7 = A(s^2 + 2.445s + 3.447)(s + 1.849)$$

$$+ B(s^2 + 2.445s + 3.447)(s + 0.706)$$

$$+ (Cs + D)(s + 0.706)(s + 1.849).$$

Multiplying out gives,

$$s^3 + 2s^2 + 3s + 7 = A(s^3 + 4.294s^2 + 7.968s + 6.374)$$

$$+ B(s^3 + 3.151s^2 + 5.173s + 2.434)$$

$$+ s^3C + s^2(2.555C + D) + s(2.555D + 1.305C)$$

$$+ 1.305D.$$

Factoring gives,,

$$s^3 + 2s^2 + 3s + 7 = s^3(A + B + C)$$

$$+ s^2(4.294A + 3.151B + 2.555C + D)$$

$$+ s(7.968A + 5.173B + 2.555D + 1.305C)$$

$$+ s^0(6.374A + 2.434B + 1.305D),$$

from which the following 4 equations with four unknowns are obtained:

$$1 = A + B + C$$

$$2 = 4.294A + 3.151B + 2.555C + D$$

$$3 = 7.968A + 5.173B + 1.305C + 2.555D$$

$$7 = 6.374A + 2.434B + 1.305D$$

Form an augmented matrix which can be solved using Gaussian elimination.

$$\left[\begin{array}{cccc|c} 1 & 1 & 1 & 0 & 1 \\ 4.294 & 3.151 & 2.555 & 1 & 2 \\ 7.968 & 5.173 & 1.305 & 2.555 & 3 \\ 6.374 & 2.434 & 0 & 1.305 & 7 \end{array}\right]$$

Obtain the lower triangular form

$$\left[\begin{array}{cccc|c} 1 & 1 & 1 & 0 & 1 \\ 0 & -1.143 & -1.739 & 1 & -2.294 \\ 0 & 0 & -2.411 & 0.110 & 0.642 \\ 0 & 0 & 0 & -2.159 & 8.433 \end{array}\right]$$

Hence, $D = \dfrac{8.433}{-2.159} = -3.906$

$$C = \frac{0.642 - (-3.906)(0.110)}{-2.411} = -0.444$$

$$B = \frac{-2.294 - (-3.906) - (-0.444)(-1.734)}{-1.143} = -0.735$$

and $\quad A = 1 + 0.444 + 0.735 = 2.179$

gives

$$D(s) = \frac{2.179}{s + 0.706} - \frac{0.735}{s + 1.849} - \frac{0.444(s - 3.906)}{s^2 + 2.445s + 3.447} .$$

The expression, $\dfrac{0.444(s - 3.906)}{s^2 + 2.445s + 3.447}$ can be written

$\dfrac{0.444(s - 3.906)}{(s + 1.223)^2 + (1.397)^2}$, and transformed to the time domain

by use of inverse Laplace transforms. The following inverse Laplace transform can be found in most Laplace transform tables

$$Ce^{-at} \cos bt + \frac{D - Ca}{b} e^{-at} \sin bt \Leftrightarrow \frac{Cs + d}{(s + a)^2 + b^2}.$$

We obtain the inverse Laplace transform of D(s),

$$\begin{aligned} d(t) = 2.179e^{-0.706t} &- 0.735e^{-1.849t} \\ &- 0.444e^{-1.223t} \cos 1.397t \\ &- 2.407e^{-1.223t} \sin 1.397t. \end{aligned}$$

SIMPSON'S RULE

• **PROBLEM** 19-5

(a) Find the energy delivered to a linear time-invariant capacitor (C = 1 F), from $t_0 = 0$ to $t_1 = 5$ s, if the voltage across the capacitor is $v(t) = t$.

(b) Use Simpson's Rule of Integration to find the energy delivered to the capacitor in part (a) from $t_0 = 0$ to $t_1 = 2$ s, if the voltage across the capacitor is $v(t) = t\,e^{-t^2}$.

Solution: (a) Since energy is defined as

$$E(t_0,t) = \int_{t_0}^{t} p(t)dt \quad \text{where} \quad p(t) = v(t)i(t)$$

gives $v(t) = t$ and $i(t) = C\frac{dv(t)}{dt} = C = 1$ A. Hence,

$$E(0,5) = \int_0^5 t\ dt = \frac{t^2}{2}\Big|_0^5 = \frac{25}{2} = 12.5\ \text{J}.$$

The energy could have been found from

$$E(t_0,t) = \int_{t_0}^{t} p(t)dt = C\int_{t_0}^{t} v\frac{dv}{dt}dt = c\int_{v(t_0)}^{v(t)} v\ dv$$

$$= \frac{1}{2}C[\{v(t)\}^2 - \{v(t_0)\}^2]$$

Thus,

$$E(0,5) = \frac{1}{2}(1)[(5)^2 - 0] = \frac{25}{2} = 12.5\ \text{J}.$$

(b) We are asked to numerically integrate for the energy, thus

$$E(t_o,t) = \int_{t_0}^{t} p(t)dt = C\int_0^2 t\ e^{-t^2}\ \frac{d}{dt}[t\ e^{-t^2}]dt$$

where $\frac{d}{dt}\ t[e^{-t^2}] = e^{-t^2} - 2t^2\ e^{-t^2}$ gives

$$E(0,2) = \int_0^2 e^{-2t^2}[t - 2t^3]dt.$$

One can find the approximate value of this integration by using Simpson's Rule which states:

$$\int_{x_0=a}^{x=b} y(x)dx \cong \frac{h}{3}(y_0 + 4y_1 + 2y_2 + 4y_3 + 2y_4 + \ldots + 2y_{n-2} + 4y_{n-1} + y_n)$$

where $h = \frac{b - a}{n}$ and $y_0 = y(a)$, $y_1 = y(a + h)$,

$y_2 = y(a + 2h)\ \ldots\ y_n = y(a + nh) = y(b)$. It is necessary to choose n to be an even integer. The larger n is, the more accurate the results will be.

Choosing $n = 20$, $h = \frac{2 - 0}{20} = 0.1$ gives

$$E(0,2) = \int_0^2 e^{-2t^2}[t - 2t^3]dt = \int_0^2 y(t)dt$$

$$\simeq \frac{0.1}{3}(y(0) + 4y(0.1) + 2y(0.2) + 4y(0.3) + \ldots + 4y(1.9) + y(2)).$$

The following table gives the values obtained.

n	t	y(t)		
0	0	0	x1 =	0
1	0.1	0.0961	x4 =	0.3842
2	0.2	0.1699	x2 =	0.3397
3	0.3	0.2055	x4 =	0.8219
4	0.4	0.1975	x2 =	0.3950
5	0.5	0.1516	x4 =	0.6065
6	0.6	0.0818	x2 =	0.1635
7	0.7	0.0053	x4 =	0.0210
8	0.8	-0.0623	x2 =	-0.1246
9	0.9	-0.1104	x4 =	-0.4417
10	1.0	-0.1353	x2 =	-0.2707
11	1.1	-0.1389	x4 =	-0.5556
12	1.2	-0.1266	x2 =	-0.2533
13	1.3	-0.1053	x4 =	-0.4214
14	1.4	-0.0811	x2 =	-0.1622
15	1.5	-0.0583	x4 =	-0.2333
16	1.6	-0.0394	x2 =	-0.0788
17	1.7	-0.0251	x4 =	-0.1004
18	1.8	-0.0151	x2 =	-0.0303
19	1.9	-0.0086	x4 =	-0.0346
20	2.0	-0.0047	x1 =	-0.0047
			SUM:	0.0202

$$E(0,2) \quad \frac{0.1}{3}(0.0202) = 0.000673 \text{ J}$$

To check the results of this example E(0,2) can be calculated exactly;

$$E(0,2) = \frac{1}{2}C\{(v(2))^2 - (v(0))^2\}$$

$$E(0,2) = \frac{1}{2}\{(2\ e^{-4})^2 - 0\}$$

$$E(0,2) = 2\ e^{-8} = 0.0006709 \text{ J.}$$

Note that the error for n = 20 is

$$\frac{0.000673 - 0.0006709}{0.0006709} \times 100 = 0.3\%$$

If n = 16 had been chosen, the result from Simpson's Rule would have given E(0,2)   0.000706 J, an error of 5.2%! Other numerical methods of integration yield accurate results in fewer iterations and are advantageous when integrating "fast" varying functions which require greater accuracy; also where computer time or memory allocation is critical.

Simpson's Rule, however, is a simple and reasonably accurate method which can be used to write programs for digital computers or programmable calculators. The flow chart shown below is for Simpson's Rule of Integration.

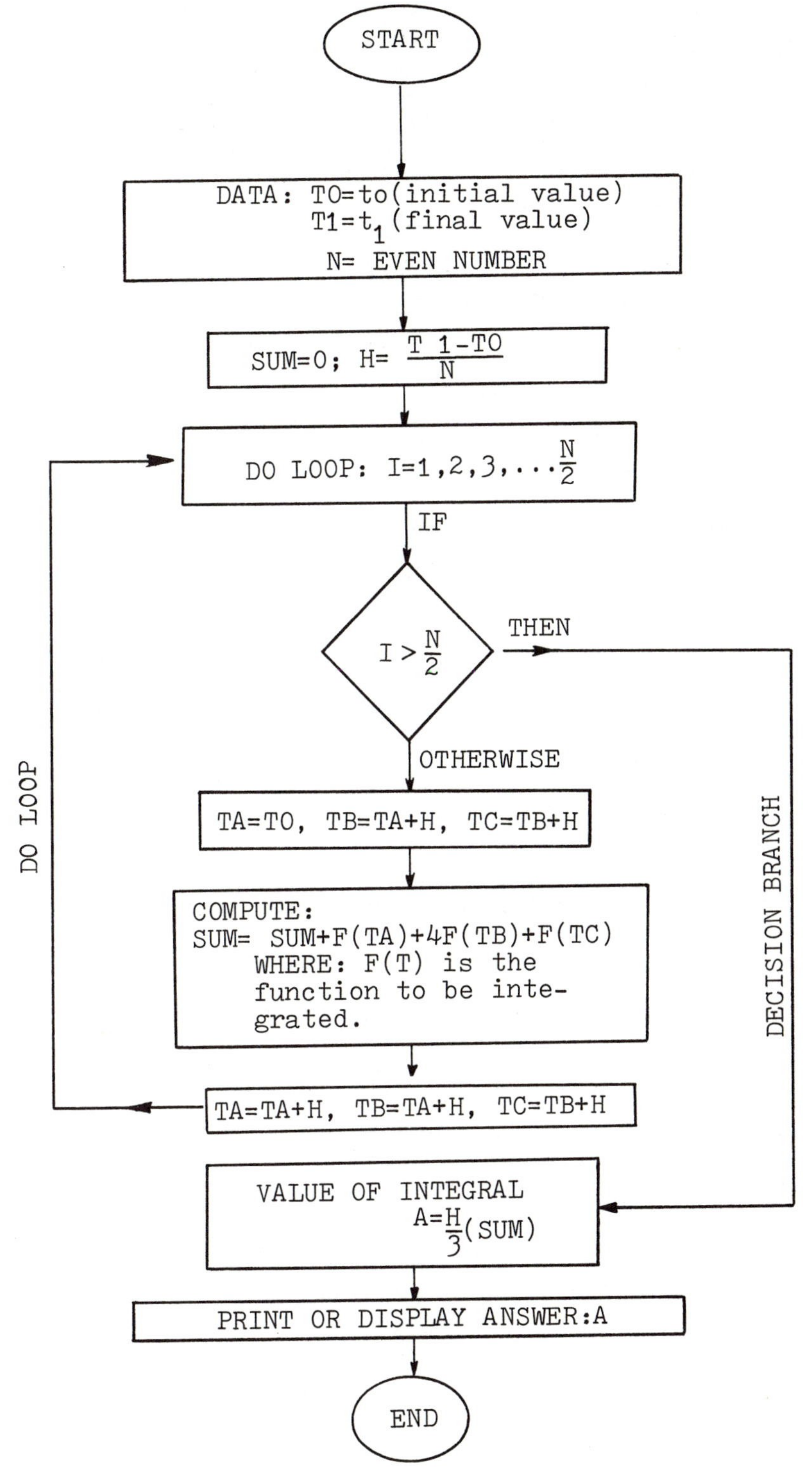
START
DATA: TO=to(initial value)
T1=t1(final value)
N= EVEN NUMBER
SUM=0; H= T 1-TO / N
DO LOOP: I=1,2,3,...N/2
IF
I > N/2
THEN
OTHERWISE
TA=TO, TB=TA+H, TC=TB+H
COMPUTE:
SUM= SUM+F(TA)+4F(TB)+F(TC)
WHERE: F(T) is the function to be integrated.
TA=TA+H, TB=TA+H, TC=TB+H
DO LOOP
DECISION BRANCH
VALUE OF INTEGRAL
A=H/3(SUM)
PRINT OR DISPLAY ANSWER:A
END

The voltage across a certain branch of a network can be approximated by the waveform shown in Fig. 1. Find, (a) the average value of this voltage; (b) the rms value of the voltage. Use Simpson's rule to evaluate the integrals. (A calculator is suggested for the solution of this problem).

Solution: (a) We note that the waveform is periodic with period, 0.5s, thus the average value may be found by integrating over the period and then dividing by the period.

Hence

$$V_{avg} = \frac{1}{0.5}\int_0^{0.5} e^{-4t^2}\,dt$$

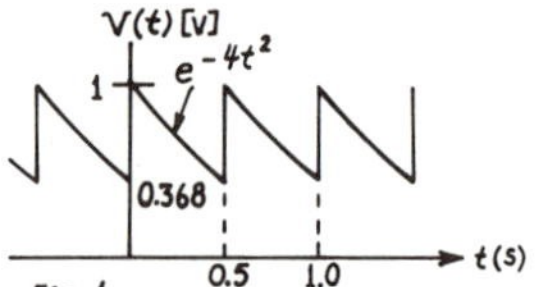

Fig.1

Since the function e^{-4t^2} is not easily evaluated, we use Simpson's rule which states

$$\int_a^b f(t)dt = \frac{h}{3}[f(a) + 4f(a+h) + 2f(a+2h) + 4f(a+3h) + 2f(a+4h) + \ldots + f(b)]$$

if $f(t) = y$

$$\int_a^b y\,dt = \frac{h}{3}[y_0 + 4y_1 + 2y_2 + 4y_3 + \ldots + y_n]$$

where $h = \frac{b-a}{n}$ where n is even. The more values between the limits of integration taken (the larger we make n), the less the error will be.

In our example, $f(t) = e^{-4t^2}$, $a = 0$, $b = 0.5$ if $n = 10$ is chosen the values listed below are obtained

t	f(t)		
0.00	1.0000	y_0	= 1.0000
0.05	0.9900	$4y_1$	= 3.9600
0.10	0.9608	$2y_2$	= 1.9216
0.15	0.9139	$4y_3$	= 3.6556
0.20	0.8521	$2y_4$	= 1.7042
0.25	0.7788	$4y_5$	= 3.1152
0.30	0.6977	$2y_6$	= 1.3954
0.35	0.6126	$4y_7$	= 2.4504
0.40	0.5273	$2y_8$	= 1.0546
0.45	0.4449	$4y_9$	= 1.7796
0.50	0.3679	y_{10}	= 0.3679

Summing the right column in the table above yields 22.4045. Hence

$$V_{avg} = \frac{1}{0.5} \; \frac{b-a}{n} \; \frac{1}{3}(22.4045)$$

$$V_{avg} = \frac{1}{0.5} \cdot \frac{0.5}{10} \cdot \frac{1}{3}(22.4045)$$

$$V_{avg} = \frac{1}{30}(22.4045) = 0.7468 \text{ V}$$

If we were to choose n = 5 then V_{avg} = 0.717 V and a substantial error would result. The answer would have been accurate to only one decimal place. If we were to choose n = 20 then the result is V_{avg} = 0.746828; not much more accurate than our result for n = 10.

(b) The rms value of a periodic waveform is defined as

$$\sqrt{\frac{1}{T}\int_0^T [f(t)]^2 dt}; \quad T = \text{period}$$

In our example,

$$V_{rms} = \sqrt{\frac{1}{0.5}\int_0^{0.5} e^{-8t^2}\, dt}$$

Using Simpson's rule evaluate the integral

$$\int_0^{0.5} e^{-8t^2}\, dt$$

thus $f(t) = e^{-8t^2}$ a = 0, b = 0.5

use n = 10.

The table below gives the values used in evaluating f(t) by Simpson's rule.

t	f(t)		
0.00	1.0000	y_0	= 1.0000
0.05	0.9802	$4y_1$	= 3.9208
0.10	0.9231	$2y_2$	= 1.8462
0.15	0.8353	$4y_3$	= 3.3412
0.20	0.7261	$2y_4$	= 1.4522
0.25	0.6065	$4y_5$	= 2.4260
0.30	0.4868	$2y_6$	= 0.9736
0.35	0.3753	$4y_7$	= 1.5012
0.40	0.2780	$2y_8$	= 0.5560
0.45	0.1979	$4y_9$	= 0.7416
0.50	0.1353	y_{10}	= 0.1353
			17.9441

Hence,

$$V_{rms} = \sqrt{\frac{1}{30}(17.9441)} = \sqrt{0.5981}$$

$$V_{rms} = 0.7734 \text{ V.}$$

• PROBLEM 19-7

Use numerical methods to find average and rms values of waveforms (a) and (b).

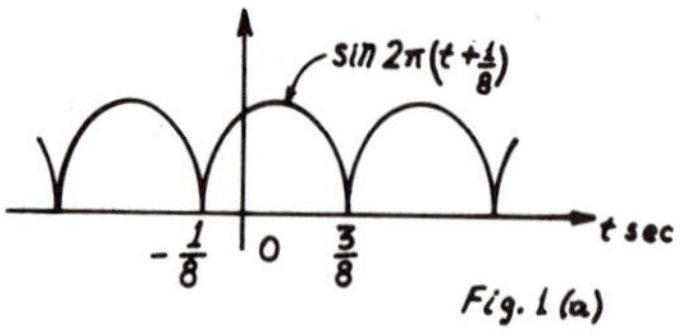

Fig. 1 (a)

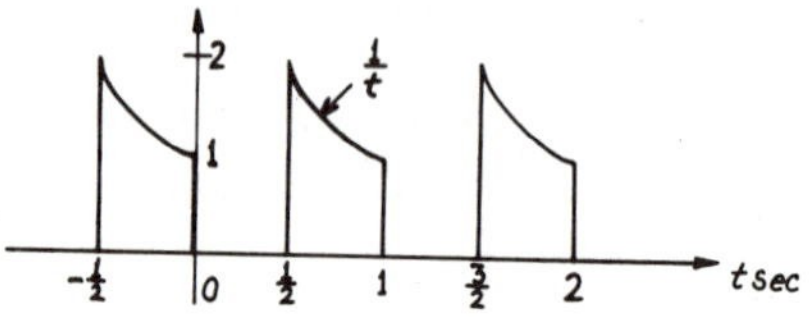

Fig. 1 (b)

Solution: (a) By observation, the given function,

$$f(t) = \sin 2\pi(t + \tfrac{1}{8}), \qquad (1)$$

has a period

$$T = \frac{4}{8} \quad \text{or} \quad 0.5 \text{ S.} \qquad (2)$$

The average value is then

$$\text{avg} = \frac{1}{T}\int_0^T f(t)dt \qquad (3)$$

To obtain the average value numerically

$$\text{avg} = \frac{1}{T}\sum_{n=0}^{10-1} f(t_n)\Delta t \qquad (4)$$

$$= \frac{1}{T}\sum_{n=0}^{10-1} \sin 2\pi(t_n + \tfrac{1}{8})\Delta t \qquad (5)$$

$$= \frac{\Delta t}{T}\sum_{n=0}^{10-1} \sin 2\pi(t_n + \tfrac{1}{8}) \qquad (6)$$

The numerical computation is listed as follows. $n = 10 - 1$ was chosen arbitrarily. The larger the value, the more exact the result. Let

$$\Delta t = \frac{T}{10} = 0.05 \text{ sec} \qquad (7)$$

n	t_n	$t_n + \frac{1}{8}$	$2\ (t_n + \frac{1}{8})$	$\sin 2\ (t_n + \frac{1}{8})$
0	0	0.125	0.785	0.7068
1	0.5	0.175	1.100	0.8910
2	0.1	0.225	1.414	0.9877
3	0.15	0.275	1.727	0.9878
4	0.2	0.325	2.042	0.8910
5	0.25	0.375	2.356	0.7072
6	0.3	0.425	2.670	0.4543
7	0.35	0.475	2.985	0.1559
8	0.4	0.525	3.299	0.1567
9	0.45	0.575	3.613	0.4541
10	0.5	0.625	3.927	0.7071

$$\sum_{n=0}^{10-1} f(t_n) = 6.3925 \qquad (8)$$

$$\text{avg} = \frac{\Delta t}{T} \sum_{n=0}^{10-1} f(t_n) = \frac{0.05}{0.5} \quad 6.3925 = 0.6393 \simeq 0.639 \qquad (9)$$

The value 0.639 is obtained by analytical methods. The root mean square value (rms) is given by

$$\text{rms} = \sqrt{\frac{1}{T} \int_0^T f(t)^2 dt} \qquad (10)$$

By numerical methods

$$\text{rms} = \sqrt{\frac{1}{T} \sum_{n=0}^{10-1} f(t_n)^2 \Delta t}$$

$$= \sqrt{\frac{\Delta t}{T} \sum_{n=0}^{10-1} f(t_n)^2} \qquad (11)$$

The table of $f(t_n)^2 = \sin^2 2\ (t_n + \frac{1}{8})$ is shown in the next table. Using these values,

n	$f(t_n)^2$
0	0.4996
1	0.5316
2	0.9756
3	0.9757
4	0.7939
5	0.5001
6	0.2064
7	0.0243
8	0.0246
9	0.2062
10	0.4999

$$\text{rms} = \sqrt{\frac{0.05}{0.5} \times \Sigma f(t_n)^2} = \sqrt{0.1 \times 4.738} = 0.6883 \tag{12}$$

If Eq. (10) is evaluated by the analytical method,

$$\text{rms} = 0.707. \tag{13}$$

In this problem, the period T was divided by 10. If the number of n divisions is larger, the result obtained by the numerical method approaches the result obtained by the analytical method.

(c) Here, the given function is, from the inspection of Fig. 1(c),

$$\left.\begin{aligned} f(t) &= 0 \quad \text{for} \quad 0 < t < \tfrac{1}{2} \\ f(t) &= \tfrac{1}{t} \quad \text{for} \quad 0 < t < 1 \end{aligned}\right\} \tag{14}$$

The period is T = 1 s. Let us divide the whole period into to uniformly spaced samples. The average value is,

$$\text{avg} = \frac{\Delta t}{T} \sum_{n=0}^{20-1} f(t_n) = \frac{0.05}{1}\left[\sum_{n=0}^{9} f(t_n) + \sum_{n=10}^{20-1} f(t_n)\right] \tag{15}$$

as seen from Fig. 1(c), $f(t_n) = 0$ for n = 0 to 9, so

$$\sum_{n=0}^{9} f(t_n) = 0 \tag{16}$$

$$\text{ave} = 0.05 \sum_{n=10}^{20-1} f(t_n) \tag{17}$$

The value of $f(t_n)$ is calculated and tabulated in the table below.

n	t_n(s)	$f(t_n)$
10	0.50	2.0000
11	0.55	1.8182
12	0.60	1.6667
13	0.65	1.5385
14	0.70	1.4286
15	0.75	1.3333
16	0.80	1.25
17	0.85	1.1765
18	0.90	1.1111
19	0.95	1.0526
20	1.00	1.0000

$$\sum_{n=10}^{20-1} f(t_n) = 14.3755 \tag{18}$$

$$\text{avg} = 0.05 \times 14.3755 = 0.7188 \tag{19}$$

By analytical methods

$$\text{avg} = 0.693 \tag{20}$$

The rms will be

$$\text{rms} = \sqrt{\frac{\Delta t}{T} \sum_{n=10}^{20-1} f^2(t_n)} \tag{21}$$

A table of $f^2(t_n)$ is shown below. From the table,

$$\sum_{n=10}^{20-1} f(t_n)^2 = 18.79$$

$$\text{rms} = \sqrt{\frac{0.05}{1} \quad 18.79} = \sqrt{0.9335} = 0.9662 \tag{22}$$

n	$f(t_n)^2$
10	4.0000
11	3.3059
12	2.7779
13	2.3669
14	2.0409
15	1.7777
16	1.5625
17	1.3842
18	1.2345
19	1.1079
20	1.0000

If this problem were done analytically, we would obtain

$$\text{rms} = 1.000. \tag{23}$$

The difference between the analytical approach and numerical method is mainly due to "truncation error" and the limited number of "samplings."

• PROBLEM 19-8

The i-v relation of a pn-junction diode is given by

$$i(t) = 0.1(e^{0.2v(t)} - 1).$$

Let $v(t) = \sin 3t.$

(a) Find the instantaneous power.
(b) Using numerical analysis, compute the energy dissipated in this diode from $t_0 = 0$ to $t_1 = 10$ s.

Solution: (a) The instantaneous power is

$$p(t) = v(t)i(t)$$

$$= 0.1 \sin 3t(e^{0.2 \sin 3t} - 1) \text{ W}$$

(b) The energy dissipated from $t = 0$ to $t_1 = 10$ s is given by the integral of the power over times

$$E = \int_0^{10} p(t)dt.$$

Use Simpson's composite formula to numerically evaluate the integral.

$$\int_0^{0+2nh} p(t)dt = \frac{h}{3}[p(0) + 4p(h) + 2p(2h)$$

$$+ 4p(3) + \ldots + p(2nh)].$$

Let $h = \frac{1}{2}$. There will be 20 intervals. Hence,

$$\int_0^{10} p(t)dt = \frac{1}{6}[p(0) + 4p(0.5) + 2p(1) + 4p(1.5)$$

$$+ 2p(2) + \ldots + p(10)].$$

Evaluating the instantaneous power at the points indicated and substituting into the above formula yields

$$E[0,10] = 0.1019 \text{ J}.$$

• PROBLEM 19-9

(a) Set up an integration that can be numerically solved for the energy required to charge a nonlinear capacitor from $v(t_0) = 1$ V to $v(t) = 3$V if the q-v relationship for the capacitor is given as

$$q = v + \tanh v.$$

(b) Use Simpson's Rule to solve the resulting integral. Note: part (b) requires the use of a calculator.

Solution: (a) Since the capacitor is nonlinear, $\frac{q}{v} \neq C$. However, one can find $C(v) = \frac{dq}{dv}$ by differentiating, $q = v + \tanh v$. Hence,

$$\frac{dq}{dv} = 1 + \operatorname{sech}^2 v$$

or

$$C(v) = 1 + \operatorname{sech}^2 v.$$

Using the integral defining the energy stored one can write

$$E = \int_{t_0}^{t} vi\ dt = \int_{t_0}^{t} C(v)v \frac{dv}{dt}\, dt$$

since $i = C(v)\frac{dv}{dt} = \frac{dq}{dv}\frac{dv}{dt}$

$$\int_{t_0}^{t} \frac{dq}{dv}\frac{dv}{dt} v\, dt = \int_{q(t_0)}^{q(t)} v\, dq$$

and since $dq = C(v)dv = (1 + \text{sech}^2 v)dv$ is found, we obtain

$$E = \int_{v(t_0)}^{v(t)} v(1 + \text{sech}^2 v)dv$$

which can be integrated over the limits

$$E(t_0,t) = \int_1^3 v(1 + \text{sech}^2 v)dv.$$

(b) Simpson's Rule allows one to integrate, numerically, the above expression

$$\int_{x_0=a}^{x=b} y(x)dx$$

$$\frac{1}{3}h(y_0 + 4y_1 + 2y_2 + 4y_3 + 2y_4 + \ldots + 2y_{n-1} + 4y_{n-1} + y_n)$$

where $h = \dfrac{b - a}{n}$ and $y_0 = y(a)$, $y_1 = y(a + h)$,

$y_2 = y(a + 2h) \ldots y_n = y(a + nh) = y(b)$; n = even.

If we choose n = 8 values shown in the following table are obtained:

n	V	$y = v(1 + \sec^2 v)$	
0	1.00	$y_0 = 1.4200$	$y_0 = 1.4200$
1	1.25	$y_1 = 1.6005$	$4y_1 = 6.4020$
2	1.50	$y_2 = 1.7711$	$2y_2 = 3.5421$
3	1.75	$y_3 = 1.9492$	$4y_3 = 7.7968$
4	2.00	$y_4 = 2.1413$	$2y_4 = 4.2826$
5	2.25	$y_5 = 2.3478$	$4y_5 = 9.3912$
6	2.50	$y_6 = 2.5665$	$2y_6 = 5.1330$
7	2.75	$y_7 = 2.7946$	$4y_7 = 11.1784$
8	3.00	$y_8 = 3.0296$	$y_8 = 3.0296$

SUM: 52.1757

Hence,

$$\int_1^3 v(1 + \text{sech}^2\, v)dv \cong \frac{h}{3}(52.1757) = \frac{0.25}{3}(52.1757) = 4.3480.$$

In this example, numerical integration offers a decided advantage, because one cannot easily evaluate the integral exactly.

RUNGE-KUTTA METHOD

• PROBLEM 19-10

The network of Fig. 1 consists of a linear 1 H inductor and a nonlinear resistance element, whose branch relation is

$$v = -i + i^3.$$

The v-i characteristic for the nonlinear resistor is sketched in Fig. 2. Solve for i(t); t > 0 numerically using the fourth order Runge-Kutta method for

$$i(0) = 2\ \text{A}.$$

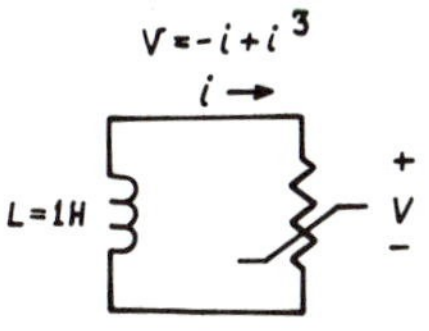

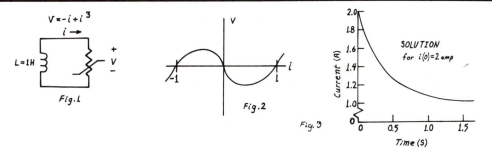

Fig.1

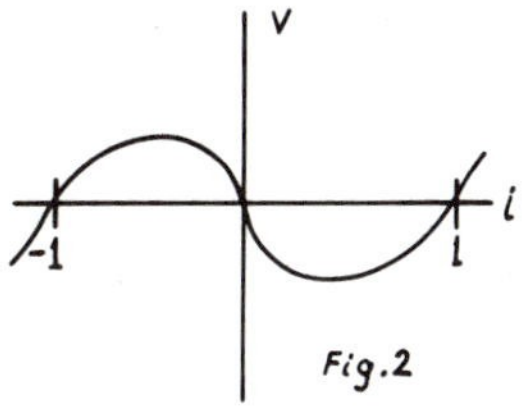

Fig.2

Fig. 3

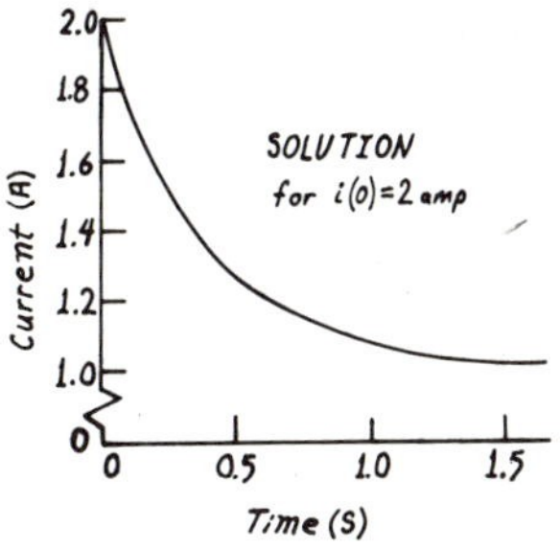

<u>Solution:</u> Using the KVL around the single loop in Fig. 1 yields

$$0 = L\frac{di}{dt} + v$$

since L = 1 $\qquad 0 = \frac{di}{dt} - i + i^3$

hence

$$\frac{di}{dt} = i - i^3.$$

A first order differential equation of the form,

$$\frac{dx}{dt} = f(x(t),\ u(t)),\ x(t_0) = x_0$$

where u is the input and x_0 is the initial state at time

t_0, can be solved for x(t) using the Runge-Kutta method as follows.

$$x(t_0 + h) = x_0 + \frac{1}{6}(K_1 + 2K_2 + 2K_3 + K_4)$$

where $K_1 = hf(x_0, u_0)$

$$K_2 = hf(x_0 + \frac{1}{2}K_1, u(t_0 + \frac{1}{2}h))$$

$$K_3 = hf(x_0 + \frac{1}{2}K_2, u(t_0 + \frac{1}{2}h))$$

$$K_4 = hf(x_0 + K_3, u(t_0 + h)).$$

h is the step-size. For the next iteration use the same formulas except substitute the newly calculated $x(t_0 + h)$ for x_0 and $t_0 + h$ for t_0 to obtain $x(t_0 + 2h)$ and so on. For i(0) = 2 A

$$x_0 = 2, \quad t_0 = 0 \quad \text{and} \quad u(t) = 0$$

hence, for h = .05 s

$$i(0.05) = 2 + \frac{1}{6}(K_1 + 2K_2 + 2K_3 + K_4)$$

where $K_1 = 0.05(2 - 2^3) = -0.300$

$$K_2 = 0.05(1.85 - 1.85^3) = -0.224$$

$$K_3 = 0.05(1.888 - 1.888^3) = -0.242$$

$$K_4 = 0.05(1.758 - 1.758^3) = -0.184$$

$$i(0.05) = 2 + \frac{1}{6}(-0.300 - 0.448 - 0.484 - 0.184)$$

$$i(0.05) = 2 - 0.236$$

$$i(0.05) = 1.764 \text{ A.}$$

Now i(0.10) can be calculated and so on. Fig. 3 shows the solution curve obtained.

• PROBLEM 19-11

Consider the network shown in Fig. 1. Develop a differential equation for v(t) at node n_1. The capacitor is voltage controlled with q-v relation

$$q = f(v) = v + \frac{1}{3}v^3.$$

Find the response v(t) by solving the differential euqation obtained above, using the Runge-Kutta method, for R = 4Ω, i(t) = u(t) and v(0) = 0. What is v(∞) for the above response?

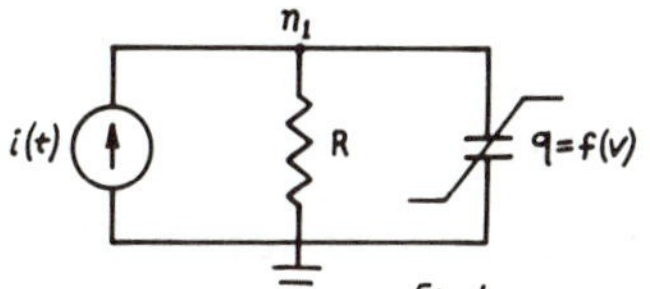

Fig.1

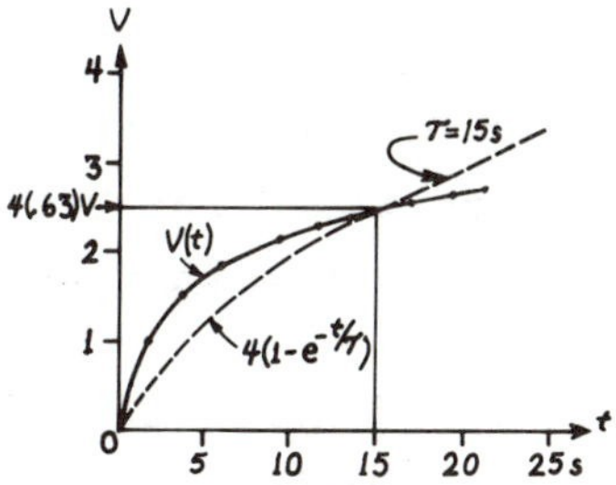

Fig.2

Solution: The KVL and KCL equations are independent of the nature of the elements. They hold for any network whether its elements are linear, nonlinear, or time-varying. Applying KCL at node n_1 yields

$$i(t) = i_1(t) + i_2(t)$$

where $$i_2(t) = \frac{v}{R}$$

and $$i_1(t) = \frac{d}{dt}q = \frac{d}{dv}f(v)\frac{dv}{dt}$$

$$i_1(t) = (1 + v^2)\frac{dv}{dt}\ .$$

Hence $$i(t) = (1 + v^2)\frac{dv}{dt} + \frac{1}{R}v.$$

This is a nonlinear differential equation that can be solved numerically for the voltage v.

To determine the response v(t), substitute R = 4Ω, i(t) = u(t)A and v(0) = 0. Hence,

$$u(t) = (1 + v^2)\frac{dv}{dt} + \frac{1}{4}v$$

$$\frac{dv}{dt} = \frac{1 - \frac{1}{4}v}{1 + v^2}$$

Using the fourth-order Runge-Kutta method we can find v(t) using the following formula;

$$K_1 = hf(v(t_0),\ u(t_0))$$

where h is the time step size, (choose an appropriate h.)

u(t) is the input. In our example, u(t) is constant, and is removed from the formulas to obtain,

$$K_2 = hf(v(t_0) + \frac{1}{2}K_1)$$

$$K_3 = hf(v(t_0) + \frac{1}{2}K_2)$$

$$K_4 = hf(v(t_0) + K_3)$$

where

$$v(t_0 + h) = v(t_0) + \frac{1}{6}(K_1 + 2K_2 + 2K_3 + K_4).$$

Using the value found for $V(t_0 + h)$, again, go through the method outlined above to find $v(t_0 + 2h)$, $v(t_0 + 3h)$, $v(t_0 + 4h)$ etc.

It can be expected that $v(\infty)$ is 4 V since the capacitor will eventually become an open circuit. Since the change varies by a cubic relationship with respect to voltage, we expect a relatively long charging time. With h = 2 seconds, the table below shows the calculated values of voltage v(t) using the Runge-Kutta method.

t(s)	v(t)(v)	K_1	K_2	K_3	K_4
0	0	2.000	0.750	1.589	0.341
2	1.170	0.597	0.401	0.457	0.325
4	1.610	0.333	0.268	0.279	0.231
6	1.886	0.232	0.199	0.204	0.178
8	2.084	0.178	0.159	0.161	0.144
10	2.249	0.146	0.131	0.133	0.121
12	2.382	0.121	0.112	0.123	0.104
14	2.498	0.104	0.097	0.097	0.091
16	2.595	0.091	0.085	0.086	0.081
18	2.681	0.081	0.076	0.076	0.072
20	2.756				

Fig. 2 is a plot of v(t) from the above table. In order to show that v(t) is not simply an exponential function, plot an exponential with a 15 sec. time constant whose final value is 4 V. Although not enough data points have been taken to show graphically that $v(\infty) = 4$ V. It can be shown that

$$\frac{dv}{dt} = \frac{1 - \frac{1}{4}v}{1 + v^2}$$

becomes zero when v = 4 V. Hence, if we stipulate $v(0) = 0$, v(t) can never rise above 4 V. This mathematical argument corresponds to our intuitive electrical argument.

OTHER METHODS

• PROBLEM 19-12

In the circuit of Fig. 1, let $R = 1\Omega$ and for the nonlinear resistor, let $i = e^{1.1v} - 1$. Find the current in the circuit if $V_1 = 1$ V.

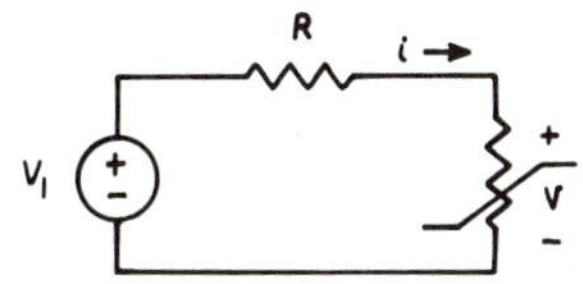

Fig. 1

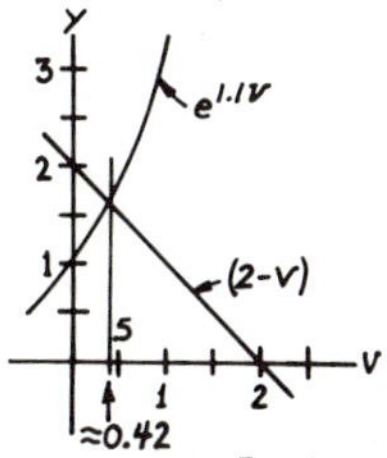

Fig. 2

Solution: Writing a KVL equation for the circuit in Fig. 1 gives,

$$1 = (1)i + V, \qquad (1)$$

substituting $e^{1.1v}-1$ for i yields

$$e^{1.1v} = 2 - v. \qquad (2)$$

From Eq. (2), note that v must be in the range 0 > v > 2. A graphic solution for v is shown in Fig. 2. Substituting this value into Eq. (1) gives

$$i = 1 - v = 1 - 0.42 = 0.58 \text{ A}.$$

• PROBLEM 19-13

Find i_x, the current through a new experimental, nonlinear, two-terminal device shown in the circuit of Fig. 1. The following experimental data are given for the new device labeled D1. The experimental set-up used to obtain the data is shown in Fig. 2.

V	i
-0.10 V	- 0.1 mA
-0.05	- 0.1 mA
-0.01	- 0.1 mA
0	0
0.01	10.0 mA
0.05	50.3 mA
0.10	.101 A
0.50	.533 A
0.70	.764 A
1.00	1.13 A
1.50	1.79 A
2.00	2.52 A
5.00	8.25 A

Assume i_v and v_a are negligable

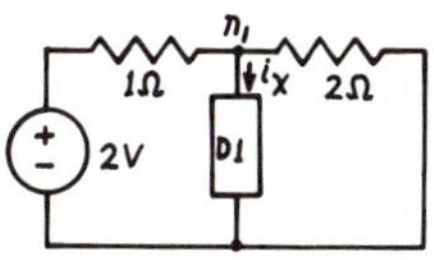

Fig. 1

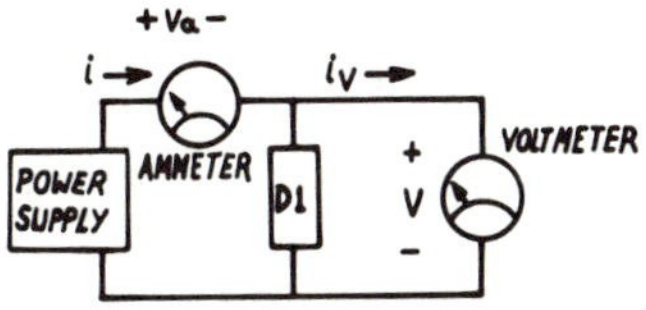

Fig. 2

Solution: Examination of the data shows that the device is "off" for V < 0 and "active" for V > 0. For small voltages

(.01 - .5 V), the device behaves much like a linear 1Ω resistor. The current increases rapidly for higher voltages which leads us to assume that the equation for the v-i characteristic of the device is a quadratic for $V > 0$.

Using least square approximation for a parabola, we can numerically find the v-i characteristic equation for the device D1.

Write the v-i characteristic equation in the general form,

$$i = av^2 + bv + c.$$

According to least square approximation, the three following equations can be written

$$\Sigma i = a\Sigma v^2 + b\Sigma v + nc$$

$$\Sigma vi = a\Sigma v^3 + b\Sigma v^2 + c\Sigma v$$

$$\Sigma v^2 i = a\Sigma v^4 + b\Sigma v^3 + c\Sigma v^2$$

where n is the number of data points, Σi is the sum of all the currents in the data, and Σv^2 is the sum of all the squares of the voltages in the data, and so on. Using determinants, solve for the coefficients a, b, and c.

$$\Delta = \begin{vmatrix} \Sigma v^2 & \Sigma v & n \\ \Sigma v^3 & \Sigma v^2 & \Sigma v \\ \Sigma v^4 & \Sigma v^3 & \Sigma v^2 \end{vmatrix}$$

$$a = \frac{\begin{vmatrix} \Sigma i & \Sigma v & n \\ \Sigma vi & \Sigma v^2 & \Sigma v \\ \Sigma v^2 i & \Sigma v^3 & \Sigma v^2 \end{vmatrix}}{\Delta}$$

$$b = \frac{\begin{vmatrix} \Sigma v^2 & \Sigma i & n \\ \Sigma v^3 & \Sigma vi & \Sigma v \\ \Sigma v^4 & \Sigma v^2 i & \Sigma v^2 \end{vmatrix}}{\Delta}$$

$$c = \frac{\begin{vmatrix} \Sigma v^2 & \Sigma v & i \\ \Sigma v^3 & \Sigma v^2 & \Sigma vi \\ \Sigma v^4 & \Sigma v^3 & \Sigma v^2 i \end{vmatrix}}{\Delta}$$

The data yields

$\Sigma i = 15.15$, $\Sigma vi = 50.92$, $\Sigma v^2 i = 222$, $\Sigma v = 10.86$

$\Sigma v^2 = 33$, $\Sigma v^3 = 137.8$, $\Sigma v^4 = 645.8$, and $n = 10$.

So that,

$$\Delta = \begin{vmatrix} 33 & 10.86 & 10 \\ 137.8 & 33 & 10.86 \\ 645.8 & 137.8 & 33 \end{vmatrix}$$

$$\Delta = 33(33^2 - 137.8[10.86]) - 137.8(10.86[33] - 1378)$$

$$+ 645.8(10.86^2 - 330)$$

$$\Delta = 33(-407.51) - 137.8(-1019.6) + 645.8(-212.1)$$

$$\Delta = -9920$$

$$a = \frac{\begin{vmatrix} 15.15 & 10.86 & 10 \\ 50.92 & 33 & 10.86 \\ 222 & 137.8 & 33 \end{vmatrix}}{\Delta}$$

$$a = \frac{15.15(-402.51) - 50.92(-1019.6) + 222(-212.1)}{-9920} = \frac{-1340}{-9920}$$

$$a \cong 0.13$$

$$b = \frac{\begin{vmatrix} 33 & 15.15 & 10 \\ 137.8 & 50.92 & 10.86 \\ 645.8 & 222 & 33 \end{vmatrix}}{\Delta}$$

$$b = \frac{33[50.92(33)-222(10.86)]-137.8[15.15(33)-2220]+645.8[15.15(10.86)-509.2]}{-9920}$$

$$b = \frac{33(-730.6) - 137.8(-1720) + 645.8(-344.7)}{-9920} = \frac{-9780}{-9920}$$

$$b = 0.98 \cong 1.0$$

$$c = \frac{\begin{vmatrix} 33 & 10.86 & 15.15 \\ 137.8 & 33 & 50.92 \\ 645.8 & 137.8 & 222 \end{vmatrix}}{\Delta}$$

$$c = \frac{33[33(222)-137.8(50.92)] - 137.8[10.86(222)-15.15(137.8)] + 645.8[10.86(50.92)-33(15.15)]}{-9920}$$

$$c = \frac{33(309.22) - 137.8(323.25) + 645.8(53.04)}{-9920} = \frac{-\ 86}{-9920}$$

$$c = 0.008 \cong 0.$$

The resulting values for a, b, and c gives the v-i characteristic equation for the device

$$i = 0.13v^2 + v; \qquad v > 0 \tag{1}$$

Now find i_x by writing a KCL equation for the node n_1 in the circuit of Fig. 1. This yields,

$$\frac{v - 2}{1} + i + \frac{v}{2} = 0.$$

Now $i = 0.13v^2 + v$. Substituting, yields

$$\frac{3v}{2} - 2 + v + 0.13v^2 = 0$$

$$0.13v^2 + 2.5v - 2 = 0$$

using the quadratic formula to solve for v, gives

$$v = \frac{-2.5 \pm \sqrt{6.25 + 1.04}}{0.26} = \frac{-2.5 \pm 2.7}{0.26}$$

$$v = \frac{0.2}{0.26}\ ; \quad -\frac{5.2}{0.26}$$

$$v = \frac{10}{13}\ , \ -20\ ;$$

since the characteristic equation holds for $v > 0$, $v = \frac{10}{13}V$

Substituting back into our characteristic equation, (1)

$$i_x = 0.13\left(\frac{100}{169}\right) + \frac{10}{13}$$

$$i_x = \frac{13}{169} + \frac{10}{13} = \frac{13 + 130}{169} = \frac{143}{169}$$

$$i_x = \frac{11}{13}\ A$$

is obtained.

• **PROBLEM 19-14**

Three nonlinear coupled inductive elements have the branch relations

$$i_{L_1} = \frac{\phi_2\phi_3}{1 + (\phi_1\phi_2\phi_3)^2}, \quad i_{L_2} = \frac{\phi_1\phi_3}{1 + (\phi_1\phi_2\phi_3)^2},$$

$$i_{L_3} = \frac{\phi_1\phi_2}{1 + (\phi_1\phi_2\phi_3)^2}.$$

Show that the magnetic energy is a state function yielding

$$E_m(\overline{\Phi}_2) = \tan^{-1} \phi_1\phi_2\phi_3$$

for each of the coupled inductors.

Solution: An example of a state function is the energy of a simple linear inductor defined as

$$E_m = \int_0^{\phi_L} i_L(\phi)d\phi = \tfrac{1}{2}L\phi_L^2.$$

Note that the energy depends only on the limits of integration and is independent of the path of integration. In network theory the above condition is fulfilled if the network or element is reciprocal. For a nonlinear network a symmetric incremental parameter matrix means the network is reciprocal.

For our coupled inductors, the incremental inductance parameters can be found from

$$\overline{L} = \frac{\partial \overline{\Phi}_L}{\partial \overline{I}_L} = \begin{bmatrix} L_{11} & L_{12} & \cdots & L_{1n} \\ L_{21} & L_{22} & & \\ \vdots & & & \\ L_{n1} & \cdots & \cdots & L_{nn} \end{bmatrix}$$

Since we are given the branch currents as a function of the flux linkages we can find the inverse incremental inductance matrix

$$\overline{\Gamma} = \frac{\partial \overline{I}_L}{\partial \overline{\Phi}_L} = \begin{bmatrix} \Gamma_{11} & \Gamma_{12} & \cdots & \Gamma_{1n} \\ \Gamma_{21} & \Gamma_{22} & & \\ \vdots & & \ddots & \\ \Gamma_{n1} & \cdots & \cdots & \Gamma_{nn} \end{bmatrix} \tag{1}$$

to see if the network is reciprocal. Equation (1) shows that it is possible to write the branch currents in the form

$$\begin{bmatrix} i_{L_1} \\ i_{L_2} \\ i_{L_3} \end{bmatrix} = \begin{bmatrix} \Gamma_{11} & \Gamma_{12} & \Gamma_{13} \\ \Gamma_{21} & \Gamma_{22} & \Gamma_{23} \\ \Gamma_{31} & \Gamma_{32} & \Gamma_{33} \end{bmatrix} \begin{bmatrix} \phi_1 \\ \phi_2 \\ \phi_3 \end{bmatrix}$$

or the total derivatives

$$i_{L_1} = \frac{\partial i_{L_1}}{\partial \phi_1}\phi_1 + \frac{\partial i_{L_1}}{\partial \phi_2}\phi_2 + \frac{\partial i_{L_1}}{\partial \phi_3}\phi_3$$

$$i_{L_2} = \frac{\partial i_{L_2}}{\partial \phi_1}\phi_1 + \frac{\partial i_{L_2}}{\partial \phi_2}\phi_2 + \frac{\partial i_{L_2}}{\partial \phi_3}\phi_3$$

$$i_{L_3} = \frac{\partial i_{L_3}}{\partial \phi_1}\phi_2 + \frac{\partial i_{L_3}}{\partial \phi_3}\phi_3 + \frac{\partial i_{L_3}}{\partial \phi_3}\phi_3 .$$

The magnetic energy is defined as

$$E_m = \int_0^{\phi_n} \sum_{n=1}^{3} i_{L_n} d\phi_n$$

and is a state function if $\overline{\Gamma}$ is symmetric. Since $\overline{\Gamma}$ is symmetric we find

$$E_m(\phi_1) = \int_0^{\phi_1} i_{L_1} d\phi_1 .$$

$$E_m(\phi_1) = \int_0^{\phi_1} \frac{\phi_2\phi_3}{1 + (\phi_1\phi_2\phi_3)^2} d\phi_1$$

can be written

$$E_m(\phi_1) = \int_0^1 \frac{\frac{1}{\phi_2\phi_3}}{\frac{1}{(\phi_2\phi_3)^2} + \phi_1^2} d\phi_1 .$$

Let $a = \frac{1}{\phi_2\phi_3}$

$$E_m(\phi_1) = a \int_0^{\phi_1} \frac{d\phi_1}{a^2 + \phi_1^2}$$

Substituting a tan x for ϕ_1 gives

$$d\phi_1 = a \sec^2 x\, dx$$

$$a^2 + \phi_1^2 = a^2 + a^2 \tan^2 x = a^2 \sec^2 x$$

Hence,

$$E_m(x) = a \int \frac{a \sec^2 x}{a^2 \sec^2 x} dx$$

$$E_m(x) = \int dx = x$$

$$E_m(\phi_1) = \tan^{-1} \frac{\phi_1}{a} = \tan^{-1} \phi_1\phi_2\phi_3$$

We find that $E_m(\phi_1) = E_m(\phi_2) = E_m(\phi_3)$ hence the state functions are

$$E_m(\bar{\phi}_L) = \tan^{-1} \phi_1\phi_2\phi_3.$$

Note we can obtain the branch relations again by differentiating the state functions.

• PROBLEM 19-15

Develop a numerical method for evaluating the Fourier coefficients of the periodic waveform f(t) in Fig. 1.
The values measured for f(t) are f(0) = 0, f(0.05) = 0.25, f(0.1) = 0.3, f(0.15) = 0.5, f(0.2) = f(0.25) = f(0.3) = 1.0, f(0.35) = 0.55, f(0.4) = 0.3, f(0.45) = 0, f(0.5) = -0.25, f(0.55) = f(0.6) = -0.3, f(0.65) = -0.2, f(0.7) = -0.1, f(0.75) = f(0.8) = f(0.85) = f(0.9) = f(0.95) = 0.

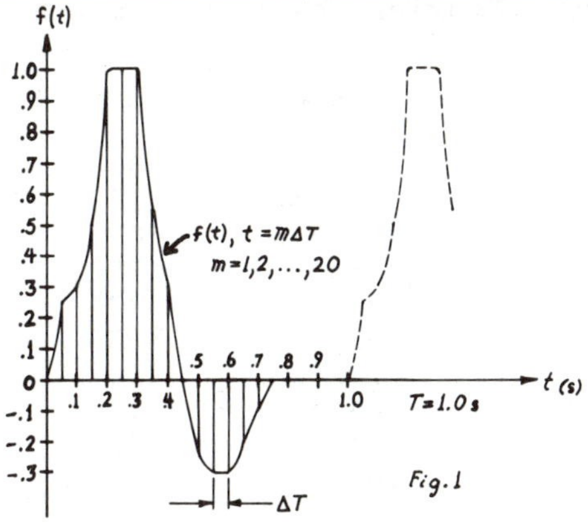

Fig. 1

Solution: Express the waveform in Fig. 1 as a Fourier series,

$$f(t) = a_0 + \sum_{n=1}^{\infty} (a_n \cos n\omega t + b_n \sin n\omega t) \quad (1)$$

where,

$$a_0 = \frac{1}{T}\int_0^T f(t)dt \quad (2)$$

$$a_n = \frac{2}{T}\int_0^T f(t)\cos n\omega t\, dt \quad (3)$$

$$b_n = \frac{2}{T}\int_0^T f(t)\sin n\omega t\, dt. \quad (4)$$

Since f(t) is not known as a mathematical function, we cannot evaluate the above integrals analytically. However, we can divide f(t) into small pulses of width ΔT as shown in Fig. 1. (The more pulses used, the better the approximation). The integrations indicated in Eqs. (2) - (4) can be replaced by summations

$$a_0 = \frac{1}{T}\sum_{m=1}^{K} f(m\Delta T)\Delta T \quad (5)$$

$$a_n = \frac{2}{T}\sum_{m=1}^{K} f(m\Delta T)\cos(n\frac{2}{T}m\Delta T)\Delta T \quad (6)$$

$$b_n = \frac{2}{T}\sum_{m=1}^{K} f(m\Delta T)\sin(n\frac{2}{T}m\Delta T)\Delta T. \quad (7)$$

Equations (5) - (7) must be evaluated for each (harmonic) value of n. The period T of the waveform is broken into (K-1) parts such that $(K-1)\Delta T = T$. Then using the values found for $f(m\Delta T)$, m = 1 to K in Fig. 1 and setting a limit on n, say 20, a computer program can be developed for evaluating the Fourier coefficients from the flow chart.

Some of the Fourier coefficients computed are

N	a_n	b_n
1	.07092	.50533
2	-.26274	-.11750
3	.02532	-.04067
4	.03654	.01790
5	-.00074	.03335
6	-.01567	.01030
7	-.02281	.02407
8	-.04237	-.01562
9	.01419	-.01915
10	.01419	.01915

INDEX

Numbers on this page refer to PROBLEM NUMBERS, not page numbers

Admittance, input, 16-8
Admittance matrix, 16-40
Admittance parameters, 16-17, 16-20, 16-21, 16-27, 16-28, 16-31, 16-33, 16-41
Analog systems, digital solutions of, 15-23, 15-24
Augmented incidence matrix, 18-3 to 18-8, 18-14, 18-15
Augmented loop matrix, 18-9 18-10
Average value, 8-1, 8-5, 8-9, 19-6, 19-7

Block diagram, 15-1 to 15-3
Bode plot, 13-19, 13-20
Branches, 1-18, 1-32

Capacitor, equivalent, 2-44, 2-45
Capacitor, non-linear, 4-13, 4-14, 4-23
Capacitor, voltage-current relationships, 4-9 to 4-20, 4-24
Charge, 1-25, 4-15, 4-20
Chords, 18-11, 18-12
Circuit reduction and voltage/current division, 1-10 to 1-17
Complete response, 6-16, 6-17, 6-19 to 6-23, 6-31, 7-3, 7-10, 7-12, 7-22, 12-9, 12-10, 12-27
Conductance, mutual, 3-14
Conductance, self, 3-14
Conductances, combination of, 1-11
Convolution, 11-20 to 11-22, 12-4, 12-18, 12-29, 15-21
Cotree, 18-11, 18-12
Coupled coils, 16-3, 16-4
Current division, 1-15, 1-16
Current, temperature dependent, 1-5
Cutsets, 18-2

Damping, 7-4 to 7-6, 7-8, 7-11, 7-19
Delta response, 6-24, 11-39, 12-25, 12-29, 12-31, 14-23, 15-1, 15-19, 15-20, 17-17
Dependent source, 1-26, 1-27, 2-23
Differential equations, 12-43, 17-2, 17-4
Digital solutions of analog systems, 15-23, 15-24
Discrete elements and equations, 15-1 to 15-13
Dual circuits, 5-18

Effective value, 8-1 to 8-9, 8-42, 19-6, 19-7
Eigenvalue, 3-9 to 3-11, 17-3
Energy, 1-7 to 1-9, 1-39, 1-41, 4-19, 4-21, 4-23 to 4-25, 4-27, 9-73, 14-20, 14-21
Energy, charge, and power in inductors and capacitors, 4-21 to 4-32
Energy in resistive circuits, 1-8, 1-9
Equivalent circuits, 9-7 to 9-11
Expansion by partial fractions, 11-23 to 11-27

Fourier applications to circuit theory, 14-20 to 14-32

Numbers on this page refer to PROBLEM NUMBERS, not page numbers

Fourier series, 14-1 to 14-16
Fourier series, complex, 14-3
Fourier series, exponential, 14-9, 14-30
Fourier series translation of axis, 14-12 to 14-14
Fourier techniques, 14-1 to 14-19
Fourier transform, 14-17 to 14-29, 14-31
Frequency response, 13-5 to 13-20
Frequency spectrum, 14-10, 14-11, 14-16, 14-28
Full wave rectifier, 12-52
Fundamental loop matrix, 18-11 to 18-13, 18-15

Gain parameter, 16-32

Half power frequency, 13-16
Homogeneity, 2-47
Homogeneous solution, 15-14 to 15-16, 15-18, 15-20, 15-41
Hybrid parameter, 16-18, 16-19, 16-22, 16-31, 16-45, 16-49

Impedance-admittance calculation, 9-1 to 9-6
Impedance, driving point, 16-10
Impedance, equivalent, 12-22
Impedance, input, 8-29, 8-32, 12-9, 16-9, 16-11 to 16-13, 16-16, 16-36
Impedance matching, 8-27
Impedance matrix, 16-39
Impedance parameters, 16-21, 16-25, 16-26, 16-28 to 16-30, 16-34, 16-36 to 16-38, 16-42 to 16-47
Impulse function, 6-3, 6-5
Impulse response, 6-24, 11-39, 12-25, 12-29, 12-31, 14-23, 15-1, 15-19, 15-20, 17-17
Incidence matrix, 18-3 to 18-8
Inductance, mutual, 16-1, 16-2, 16-12, 16-14
Inductor, equivalent, 2-45
Inductor, non-linear, 4-4
Inductors, voltage-current relationships, 4-1 to 4-10, 4-22
Initial conditions, 6-12, 6-14, 6-18, 6-25, 6-28, 7-15 to 7-17, 7-21, 12-33, 12-41, 17-8, 17-20
Integrodifferential equations, 2-1, 2-2, 2-19, 17-1
Inverse Laplace transforms, 11-28 to 11-39

KCL with matrices, 3-13, 3-14
KVL with matrices, 3-15 to 3-25
Kirchhoff's current law, 1-32 to 1-44, 2-1 to 2-3, 2-10, 2-32 to 2-36
Kirchhoff's voltage law, 1-18 to 1-31, 1-44, 2-19 to 2-21, 2-23, 2-24, 2-32

Laplace transform, 11-7 to 11-12, 11-14, 11-19
Laplace transform applications, mixed circuits, 12-22 to 12-52
Laplace transform applications, parallel circuits, 12-19 to 12-21
Laplace transform applications, series circuits, 12-1 to 12-18
Laplace transform applications, state equations, 12-58 to 12-61
Laplace transform applications, Thévenin's and Norton's equivalent, 12-53 to 12-57
Laplace transform frequency shift theorem, 11-38
Laplace transform initial value theorem, 11-16, 11-17
Laplace transform properties, 11-13 to 11-19
Laplace transform, time

Numbers on this page refer to PROBLEM NUMBERS, not page numbers

shifting theorem, 11-7, 11-11, 11-12, 12-2, 12-15
Laplace transforms, inverse, 11-28 to 11-39
Laplace transforms of parallel circuits, 12-19 to 12-52
Laplace transforms of series circuits, 12-1 to 12-18
Laplace transforms of time functions, 11-7 to 11-12
Least square approximation, 19-13
Loop matrix, 18-9 to 18-13
Low pass filter, 14-31

Matrix inverse, 17-3
Matrix math, 3-1 to 3-12
Mesh analysis, 2-19 to 2-37, 3-16, 3-17, 3-20, 12-38, 12-45, 16-2, 16-39

Network parameter, 16-17 to 16-49
Newton's method, 19-3, 19-4
Nodal analysis, 2-1 to 2-18, 2-30, 2-31, 2-34 to 2-36, 3-13, 3-14, 12-26, 12-46, 16-40
Nodes, 1-18, 1-19, 1-32, 1-37,
Norton's equivalent, 2-39 to 2-41, 9-52, 12-54, 12-56, 12-57
Norton's equivalent, Thévenin's equivalent, and superposition, 2-37 to 2-49
Numerical methods, 19-12 to 19-15

Ohm's Law, 1-1, 1-3

Parallel RLC circuits, 7-6 to 7-11
Parallel RLC circuits, combination series and, 7-11 to 7-26
Parseval's theorem, 14-20, 14-21
Partial fraction expansion, 11-23 to 11-38, 12-2, 12-3, 12-7, 12-8, 12-14, 12-23, 12-28, 12-29, 12-32, 12-35, 12-37, 12-39, 12-42, 12-53, 15-31 to 15-34, 19-4
Particular solutions, 15-18, 15-41
Phase angle, 6-26, 8-37, 9-13
Phasors, 8-10 to 8-26
Phasor diagram, 8-11, 8-22
Poles and zeroes, 13-1 to 13-4
Pole-zero plot, 13-1, 13-2, 13-4 to 13-11
Power, 1-2, 1-37, 1-40, 2-8, 2-23, 3-19, 4-22, 4-24, 4-26, 8-27 to 8-42, 9-64 to 9-75, 10-4, 10-11, 10-13, 10-15 to 10-17, 14-32
Power absorbed by an unknown element, 1-24, 1-38
Power factor, 8-37, 8-38, 9-68, 9-72, 9-75, 10-13, 10-17
π-T conversion, 16-54, 16-55

Quality factor, 13-33 to 13-35
Quartic equations, 19-1, 19-4

RC circuits, 5-1 to 5-18, 6-6 to 6-19, 6-23 to 6-31
RC circuits, forced response of, 6-20 to 6-31
RC circuits, natural response of, 5-8 to 5-18
Reciprococity, 16-50 to 16-53
Recursion equation, 15-4 to 15-13
Resistors, combination of, 1-10, 1-16, 1-28 to 1-31
Resistors, equivalent, 1-10, 1-12 to 1-14, 1-35, 2-44, 2-45
Resistors, non-linear, 1-4, 1-6, 19-1, 19-2, 19-10, 19-12

Resonance, 13-21 to 13-35
Response curve, 7-9
RL circuits, forced response of, 6-6 to 6-19
RL circuits, natural response of, 5-1 to 5-7
RLC circuits, series-parallel combinations, 7-11 to 7-26
RMS values, 8-1 to 8-9
Root mean square (RMS) value, 8-1 to 8-9, 8-42, 19-6, 19-7
Runge-Kutta method, 19-10, 19-11

Sawtooth waveform, 8-1 to 8-3, 11-12, 14-6, 14-7, 14-9, 14-10
Series RLC circuits, 7-1 to 7-5
Simple time functions, 11-1 to 11-6
Simpson's rule, 19-5 to 19-9
Single loop circuits, 1-21, 1-23, 1-28, 1-30, 1-31, 1-34
Source transformations, 2-30, 3-15
Square waves, 14-2, 14-3, 14-18
State equations, 12-58 to 12-61, 17-1, 17-6 to 17-25
State equations, applications to non-linear and time varying circuits, 17-22 to 17-25
State equations, applications to RC circuits, 17-6, 17-7
State equations, applications to RL circuits, 17-8
State equations, applications to RLC circuits, 17-9 to 17-21
State equations, definitions and properties, 17-1 to 17-5
Steady state and homogeneous solutions of discrete systems, 15-14 to 15-22
Steady state power, 9-64 to 9-75
Steady state response, 9-12 to 9-63, 12-40, 14-29, 14-30, 15-14 to 15-17
Step response, 15-22, 17-17
Step sequence, 15-1
Superposition, 2-46 to 2-49, 6-19, 12-49 to 12-51, 12-58
Symmetric incremental parameter matrix, 19-14
System function, 12-24, 12-47, 14-22 to 14-28, 14-30, 14-31

Thevenin's equivalent, 2-38 to 2-40, 2-42, 2-43, 9-8 to 9-11, 9-46, 9-60, 12-27, 12-44, 12-53 to 12-55, 12-57
Three phase circuits, combinations of Y and Δ connected, 10-15 to 10-18
Three phase circuits, Δ connected, 10-11 to 10-17
Three phase circuits, Δ-Y conversions, 10-12
Three phase circuits, Y connected, 10-1 to 10-10, 10-12, 10-15 to 10-17
Time varying networks, 17-22 to 17-25
Topological analysis, applications, 18-14, 18-15
Topological analysis, definitions of, 18-1, 18-2
Transfer function, 12-24, 12-47, 14-22 to 14-28, 14-30, 14-31
Transfer function plot, 13-8 to 13-11, 13-13 to 13-18, 13-35
Transformers and mutual inductance, 16-1 to 16-16
Transformers, ideal, 16-5 to 16-7, 16-10, 16-11, 16-13, 16-15, 16-41 to 16-43
Transformers, step up, 16-16
Transistor T equivalent, 16-45, 16-47
Transmission parameter, 16-23, 16-24, 16-29, 16-32, 16-35, 16-48
Trees, 18-1, 18-2
Trial and error procedure, 19-1, 19-2

Numbers on this page refer to PROBLEM NUMBERS, not page numbers

Triangular waves, 14-1 to 14-3, 14-12, 14-14
Turns ratio, 16-7
Two port network parameters, 16-17 to 16-49

Unit ramp function, 11-1
Unit step function, 6-1 to 6-5, 11-2 to 11-6

Voltage, current, and power relationships for resistors, 1-1 to 1-7
Voltage-current relationships for capacitors, 4-11 to 4-20
Voltage-current relationships for inductors, 4-1 to 4-10
Voltage division, 1-16

Z-transform, 15-25 to 15-41
Z-transform applications, 15-34 to 15-41
Z-transform definitions and properties, 15-25 to 15-33
Z-transform, bilateral, 15-26
Zero state and zero input response, 6-10